MW01361126

Jane's

High-Speed Marine Transportation 2008-2009

Edited by Stephen J Phillips

Forty-first Edition

Bookmark jhmt.janes.com today!

The title is also available on online/CD-rom (New Gen), JDS (Jane's Data Services), EIS (Electronic Information Service), and Intel centres. Online gives the capability of real-time editing, permitting frequent updating. We trust our readers will use these facilities to keep abreast of the latest changes as and when they occur.

Updates online: Any update to the content of this product will appear online as it occurs.

Jane's High-Speed Marine Transportation online site gives you details of the additional information that is unique to online subscribers and the many benefits of upgrading to an online subscription. Don't delay, visit jhmt.janes.com today and view the list of the latest updates to this online service.

ISBN 978 0 7106 2846 6

Printed and bound in Great Britain by Biddles Ltd, Kings Lynn

Front cover image: *Natchan Rera* - the first of two 112 m Incat wave-piercing catamarans built for the Libera Corporation for operation by Higashinihon Ferry of Japan. This is the largest fast ferry built by Incat. Powered by four MAN 20V 28/33D diesel engines each rated at 9000 kW and each driving a Wartsila LJX 1500 SRi water-jet via a ZF 60000 NR2H gearbox, these vessels have an operational speed of between 40 and 45 knots depending on the deadweight carried. Their maximum deadweight includes 900 passengers and 355 cars, although the vessels are arranged to carry 35 trucks with a reduced number of cars. (INCAT) 1120333

Contents

Jane's High-Speed Marine Transportation website: jhmt.janes.com

How To Use: Jane's High-Speed Marine Transportation

Content
The purpose of *Jane's High-Speed Marine Transportation* (*JHMT*) is to provide a comprehensive collection of information covering the design, build and operation of high-speed marine craft, worldwide.

Structure
Content in *JHMT* covers all organisations involved in the industry and is divided into four main sections, as follows:

Builders
This section is further divided into the following sections:

Air cushion vehicles: Largely commercial craft capable of exceeding 25 knots with a proportion of their weight supported by an air cushion.

Hydrofoils: Largely commercial craft capable of exceeding 25 knots with a proportion of their weight supported by hydrofoils.

High-speed multihull vessels: Mostly commercial multi-hulled craft mainly used as fast ferries, capable of speeds in excess of 25 knots.

SWATH vessels: Small- waterplane-area-twin-hull vessels.

High-speed monohull craft: Commercial craft such as fast ferries and crew supply boats that can exceed 25 knots.

Wing-in-ground-effect craft: Vessels that use the aerodynamic wing-in-ground-effect to support their weight, and have to date been built mainly for research and development purposes.

Fast interceptors: Military, coast guard, police, customs and workboat vessels up to 25 m in length that can exceed 30 knots. There is a degree of overlap between some large fast interceptors and the smaller patrol boats.

High-speed patrol craft: Military, coast guard, police or customs vessels, that are between 20 and 60 m in length and have a maximum speed of 25 knots or greater.

Civil operators of high-speed craft
Includes ferry operators, offshore support companies and operators of cruise and leisure craft, as well as coast guards, customs and police units.

Principal equipment and components for high-speed craft
This section is further divided into Engines, Transmissions, Air propellers, Marine propellers, Water-jet units, Ride control systems, Air cushion skirt systems, Marine escape systems, Construction materials and Lightweight seating.

Services
This section is further divided into Marine craft regulatory authorities, Consultants and designers and Societies involved with high-speed craft.

Under each of these subsections, entries are arranged by country and thereafter by organisation name.

A comprehensive Bibliography is also included which lists all the relevant books, technical papers and presentations that have been made public over the last six years.

Any particular entry may be located in the hardcopy version of *JHMT* by using on of the three indices covering Organisations, Craft type or Craft name.

Record Structure
The structure of records varies with section. However all entries will include contact information, key personnel listings and company background. This may be followed by details of the vessels or products that the company operates and specifications of its craft (including dimensions, speed, range, passenger capacity etc.).

Images
Photographs are provided for vessels or equipment, where possible. Line drawings and graphics are also provided in some cases. Images are annotated with a seven digit number which identifies them in Jane's image database.

Other Information
The *Jane's* Transport library is designed to provide the necessary information for you to explore your specific areas of interest, giving the fullest possible technical details and contact information. However, improvements can always be made and I welcome any constructive comments and feedback for this particular *Jane's* title. Please do not hesitate in contacting me with your views:

Stephen Phillips, CEng FRINA FIMarEST
Jane's Information Group
Sentinel House
163 Brighton Road
Coulsdon, Surrey
CR5 2YH, United Kingdom
Tel: (+44 208) 700 37 00
Fax: (+44 208) 700 39 00
e-mail: info@janes.com
Web: www.janes.com

Quality Policy

Jane's Information Group is the world's leading unclassified information integrator for military, government and commercial organisations worldwide. To maintain this position, the Company will strive to meet and exceed customers' expectations in the design, production and fulfilment of goods and services.

Information published by Jane's is renowned for its accuracy, authority and impartiality, and the Company is committed to seeking ongoing improvement in both products and processes.

Jane's will at all times endeavour to respond directly to market demands and will also ensure that customer satisfaction is measured and employees are encouraged to question and suggest improvements to working practices.

Jane's will continue to invest in its people through training and development to meet the Investor in People standards and changing customer requirements.

Jane's

EDITORIAL AND ADMINISTRATION

Chief Content Officer: Ian Kay, e-mail: ian.kay@janes.com

Group Publishing Director: Sean Howe, e-mail: sean.howe@janes.com

Publisher: Sara Morgan, e-mail: sara.morgan@janes.com

Compiler/Editor: Welcomes information and comments from users who should send material to:
Information Collection
Jane's Information Group, Sentinel House, 163 Brighton Road, Coulsdon, Surrey CR5 2YH, UK
Tel: (+44 20) 87 00 37 00 Fax: (+44 20) 87 00 39 00
e-mail: yearbook@janes.com

SALES OFFICES

Europe and Africa
Jane's Information Group, Sentinel House, 163 Brighton Road, Coulsdon, Surrey CR5 2YH, UK
Tel: (+44 20) 87 00 37 50 Fax: (+44 20) 87 00 37 51
e-mail: customer.servicesuk@janes.com

North/Central/South America
Jane's Information Group, 110 N Royal Street, Suite 200, Alexandria, Virginia 22314, US
Tel: (+1 703) 683 21 34 Fax: (+1 703) 836 02 97
Tel: (+1 800) 824 07 68 Fax: (+1 800) 836 02 97
e-mail: customer.servicesus@janes.com

Asia
Jane's Information Group, 78 Shenton Way, #10-02, Singapore 079120, Singapore
Tel: (+65) 63 25 08 66 Fax: (+65) 62 26 11 85
e-mail: asiapacific@janes.com

Oceania
Jane's Information Group, PO Box 3502, Rozelle Delivery Centre, New South Wales 2039 Australia
Tel: (+61 2) 85 87 79 00 Fax: (+61 2) 85 87 79 01
e-mail: oceania@janes.com

Middle East
Jane's Information Group, PO Box 502138, Dubai, United Arab Emirates
Tel: (+971 4) 390 23 36 Fax: (+971 4) 390 88 48
e-mail: mideast@janes.com

Japan
Jane's Information Group, Level 8 Pacific Century Place Marunouchi, 1-11-1 Marunouchi, Chiyoda-ku, Tokyo 100-6208, Japan
Tel: (+81 3) 68 60 82 00 Fax: (+81 3) 68 60 82 01
e-mail: japan@janes.com

ADVERTISEMENT SALES OFFICES

(Head Office)
Jane's Information Group
Sentinel House, 163 Brighton Road,
Coulsdon, Surrey CR5 2YH, UK
Tel: (+44 20) 87 00 37 00 Fax: (+44 20) 87 00 38 59/37 44
e-mail: defadsales@janes.com

Janine Boxall, Global Advertising Sales Director,
Tel: (+44 20) 87 00 38 52 Fax: (+44 20) 87 00 38 59/37 44
e-mail: janine.boxall@janes.com

Richard West, Senior Key Accounts Manager
Tel: (+44 1892) 72 55 80 Fax: (+44 1892) 72 55 81
e-mail: richard.west@janes.com

Nicky Eakins, Advertising Sales Manager
Tel: (+44 20) 87 00 38 53 Fax: (+44 20) 87 00 38 59/37 44
e-mail: nicky.eakins@janes.com

Carly Litchfield, Advertising Sales Executive
Tel: (+44 20) 87 00 39 63 Fax: (+44 20) 87 00 37 44
e-mail: carly.litchfield@janes.com

(US/Canada office)
Jane's Information Group
110 N Royal Street, Suite 200,
Alexandria, Virginia 22314, US
Tel: (+1 703) 683 37 00 Fax: (+1 703) 836 55 37
e-mail: defadsales@janes.com

US and Canada
Robert Silverstein, US Advertising Sales Director
Tel: (+1 703) 236 24 10 Fax: (+1 703) 836 55 37
e-mail: robert.silverstein@janes.com

Sean Fitzgerald, Southeast Region Advertising Sales Manager
Tel: (+1 703) 836 24 46 Fax: (+1 703) 836 55 37
e-mail: sean.fitzgerald@janes.com

Linda Hewish, Northeast Region Advertising Sales Manager
Tel: (+1 703) 836 24 13 Fax: (+1 703) 836 55 37
e-mail: linda.hewish@janes.com

Janet Murphy, Central Region Advertising Sales Manager
Tel: (+1 703) 836 31 39 Fax: (+1 703) 836 55 37
e-mail: janet.murphy@janes.com

Richard L Ayer
127 Avenida del Mar, Suite 2A, San Clemente, California 92672, US
Tel: (+1 949) 366 84 55 Fax: (+1 949) 366 92 89
e-mail: ayercomm@earthlink.com

Rest of the World
Australia: *Richard West* (UK Head Office)

Benelux: *Nicky Eakins* (UK Head Office)

Eastern Europe (excl. Poland): MCW Media & Consulting Wehrstedt
Dr Uwe H Wehrstedt
Hagenbreite 9, D-06463 Ermsleben, Germany
Tel: (+49 03) 47 43/620 90 Fax: (+49 03) 47 43/620 91
e-mail: info@Wehrstedt.org

France: Patrice Février
BP 418, 35 avenue MacMahon,
F-75824 Paris Cedex 17, France
Tel: (+33 1) 45 72 33 11 Fax: (+33 1) 45 72 17 95
e-mail: patrice.fevrier@wanadoo.fr

Germany and Austria: *MCW Media & Consulting Wehrstedt*
(see Eastern Europe)

Greece: *Nicky Eakins* (UK Head Office)

Hong Kong: *Carly Litchfield* (UK Head Office)

India: *Carly Litchfield* (UK Head Office)

Israel: *Oreet International Media*
15 Kinneret Street, IL-51201 Bene Berak, Israel
Tel: (+972 3) 570 65 27 Fax: (+972 3) 570 65 27
e-mail: admin@oreet-marcom.com
Defence: Liat Heiblum
e-mail: liat_h@oreet-marcom.com

Italy and Switzerland: *Ediconsult Internazionale Srl*
Piazza Fontane Marose 3, I-16123 Genoa, Italy
Tel: (+39 010) 58 36 84 Fax: (+39 010) 56 65 78
e-mail: genova@ediconsult.com

Japan: *Carly Litchfield* (UK Head Office)

Middle East: *Carly Litchfield* (UK Head Office)

Pakistan: *Carly Litchfield* (UK Head Office)

Poland: *Carly Litchfield* (UK Head Office)

Russia: Anatoly Tomashevich
1/3, appt 108, Zhivopisnaya Str, Moscow, 123103, Russia
Tel/Fax: (+7 495) 942 04 65
e-mail: to-anatoly@tochka.ru

Scandinavia: *Falsten Partnership*
23, Walsingham Road, Hove, East Sussex BN41 2XA, UK
Tel: (+44 1273) 77 10 20 Fax: (+ 44 1273) 77 00 70
e-mail: sales@falsten.com

Singapore: *Richard West* (UK Head Office)

South Africa: *Richard West* (UK Head Office)

Spain: Macarena Fernandez
VIA Exclusivas S.L., Virato, 69 - Sotano C, E-28010, Madrid, Spain
(+34 91) 448 76 22 Fax: (+34 91) 446 02 14
e-mail: macarena@viaexclusivas.com

Turkey: *Richard West* (UK Head Office)

ADVERTISING COPY
Kate Gibbs (UK Head Office)
Tel: (+44 20) 87 00 37 42 Fax: (+44 20) 87 00 38 59/37 44
e-mail: kate.gibbs@janes.com

For North America, South America and Caribbean only:
Tel: (+1 703) 683 37 00 Fax: (+1 703) 836 55 37
e-mail: us.ads@janes.com

Jane's Libraries

To assist your information gathering and to save you money, Jane's has grouped some related subject matter together to form 'ready-made' libraries, which you can access in whichever way suits you best – online, on CD-ROM, via Jane's EIS or through Jane's Data Service.

The entire contents of each library can be cross-searched, to ensure you find every reference to the subjects you are looking for. All Jane's libraries are updated according to the delivery service you choose and can stand alone or be networked throughout your organisation.

www.janes.com

Jane's Defence Equipment Library

Aero-Engines
Air-Launched Weapons
Aircraft Upgrades
All the World's Aircraft
Ammunition Handbook
Armour and Artillery
Armour and Artillery Upgrades
Avionics
C4I Systems
Electro-Optic Systems
Explosive Ordnance Disposal
Fighting Ships
Infantry Weapons
Land-Based Air Defence
Military Communications
Military Vehicles and Logistics
Mines and Mine Clearance
Naval Weapon Systems
Nuclear, Biological and Chemical Defence
Radar and Electronic Warfare Systems
Strategic Weapon Systems
Underwater Warfare Systems
Unmanned Aerial Vehicles and Targets

Jane's Defence Magazines Library

Defence Industry
Defence Weekly
Foreign Report
Intelligence Digest
Intelligence Review
International Defence Review
Islamic Affairs Analyst
Missiles and Rockets
Navy International
Terrorism and Security Monitor

Jane's Market Intelligence Library

Aircraft Component Manufacturers
All the World's Aircraft
Defence Industry
Defence Weekly
Electronic Mission Aircraft
Fighting Ships
Helicopter Markets and Systems
International ABC Aerospace Directory
International Defence Directory
Marine Propulsion
Naval Construction and Retrofit Markets
Police and Homeland Security Equipment
Simulation and Training Systems
Space Systems and Industry
Underwater Security Systems and Technology
World Armies
World Defence Industry

Jane's Security Library

Amphibious and Special Forces
Chemical-Biological Defense Guidebook
Facility Security
Fighting Ships
Intelligence Digest
Intelligence Review
Intelligence Watch Report
Islamic Affairs Analyst
Police and Homeland Security Equipment
Police Review
Terrorism and Security Monitor
Terrorism Watch Report
World Air Forces
World Armies
World Insurgency and Terrorism

Jane's Sentinel Library

Central Africa
Central America and the Caribbean
Central Europe and the Baltic States
China and Northeast Asia
Eastern Mediterranean
North Africa
North America
Oceania
Russia and the CIS
South America
South Asia
Southeast Asia
Southern Africa
The Balkans
The Gulf States
West Africa
Western Europe

Jane's Transport Library

Aero-Engines
Air Traffic Control
Aircraft Component Manufacturers
Aircraft Upgrades
Airport Review
Airports and Handling Agents –
  Central and Latin America (inc. the Caribbean)
  Europe
  Far East, Asia and Australasia
  Middle East and Africa
  United States and Canada
Airports, Equipment and Services
All the World's Aircraft
Avionics
High-Speed Marine Transportation
Marine Propulsion
Merchant Ships
Naval Construction and Retrofit Markets
Simulation and Training Systems
Transport Finance
Urban Transport Systems
World Airlines
World Railways

Executive Overview: High-Speed Marine Transportation

Welcome to the 41st edition of *Jane's High-Speed Marine Transportation* which continues to provide an unparalleled view of the market for high speed craft around the world. It is difficult to think of a year which has seen such a depth and breadth of activity in this industry sector, covering both commercial and military interests, from the very largest to the very smallest vessels. In particular the US Navy has launched new fast Littoral Combat Ships (the 115 m monohull *USS Freedom* and plans to launch in 2008 the 127 m trimaran *USS Independence*) and both the Austal and Incat shipyards (the main producers of commercial fast ferries) have launched their largest passenger/vehicle catamarans, the 107 m *Hawaii Superferry* and the 112 m wave-piercing catamaran *Natchan Rera* respectively. These enormous craft are all based on the experience gained in the construction of many previous similar fast ferries and are truly evolutionary – satisfyingly so, in an industry that has in the past been criticised, possibly with some justification, as being far too revolutionary (resulting in the true cost of innovation being paid for dearly by the operators of the early craft). The technology is now largely mature and future innovation will be associated more with the clever implementation, rather than the raw development, of high speed craft technology.

As in the past, I have divided my executive summary into two main areas dealing with the commercial and military sectors separately.

The Fast Ferry Industry

This year has seen the highest level of activity since the mid 1990s (when the construction of new craft was at its peak) – 2007 saw a total of 75 fast ferries delivered around the world, which compares with 45 in 2006 and 40 in 2005. The world fleet of fast ferries is larger than it has ever been with 2130 craft in total having been constructed to date, and with less than 200 known to have been scrapped. This gives a fleet of over 1,900 craft, although of course not all of these would be operational at any one time.

The top five nations producing these craft over the last 5 years include Australia (with 36 per cent), USA (with 24 per cent), Norway (with 15 per cent) and Italy and Singapore (both with 12 per cent). In terms of tonnage produced, the percentages for both Australia and the USA are even higher, due to the sizes of vessels constructed. Whilst costs of production are increasing slowly, it has been notable that the valuation of these craft is now largely quoted in Euros rather that the US dollar, due to the significant weakening of the latter over the past two years.

In terms of the operation of fast ferries, the top five countries are USA and China (both with 13 per cent), Italy (with 12 per cent), Norway with 9 per cent and Australia with 6 per cent. The rate of build of new craft for all these countries has accelerated over the last few years and it is the first time that Japan has not been in this list for at least 15 years.

Possibly the most notable delivery this year has been the Incat 112 m wave piercing catamaran, the *Natchan Rera*, which is the first of two such craft built for the Libera Corporation for operation by Higashinihon Ferry of Japan. The vessel is a direct development of the steadily increasing range of wave-piercing catamarans built by Incat, starting with the nine 74 m craft built between 1990 and 1993, three 78 m craft in 1995, three 81 m craft in 1996, four 86 m craft in 1996/1997, four 91 m craft in 1997/1998, six 96 m craft in 1998/2000, and six 98 m craft in 2000/2006. Each development has resulted in an improved fuel efficiency per tonne of payload and the 112 m craft can carry up to 1500 passengers and over 350 cars (a similar capacity to the giant Stena HSS 1500 vessels) with a fuel consumption of less than 8 tonnes per hour at full power – considerably less than the Stena vessels.

The Japanese operator Higashinihon Ferry has ordered two vessels with an option for a third vessel, all to operate between the Islands of Hokkaido and Honshu (from the ports of Hakodate and Aomori respectively), reducing the current conventional ferry voyage time of nearly 4 hours to 1 hour 45 minutes. It is expected that similar craft will be ordered from other operators in the region.

At the smaller end of the market, construction technology is also continuing to advance, with the delivery of three 27 m carbon-fibre built vessels and an order for a fifth similar 24.5 m vessel for Norwegian operator Stavangerske. With a combination of very light structural weight and efficient propeller-based propulsion systems, these vessels offer transport efficiency for this size of vessel. Other operators in the region are also known to be considering the benefits of this light-weight technology which offers potentially worthwhile operational cost savings.

Austal Ships of Australia (and USA) have maintained their dominant position in the fast ferry industry and continue to offer vessels of all sizes. The level of work has prompted their acquisition of the North West Bay Ships shipyard in Tasmania in order to satisfy the demand for aluminium fabrication.

Their most notable contract this year was an order for 10 identical 48 m catamaran craft for operation between Hong Kong and Macau's Cotai Strip, by the Venetian Marketing Services 'Cotai Waterjets'. Additionally, two 88 m catamarans were built for the Turkish operator IDO during the year, both with a capacity for 1200 passengers and 225 cars and an operational speed of 37 knots. As noted previously, Austal USA has also launched the first of two 107 m catamarans for Hawaii Superferry, which started full time operation in December 2007, after a delay of a several months due to pressure from environmental groups.

The most recent delivery from the yard is the 65 m catamaran *Shinas*, the first of two craft ordered by the Sultanate of Oman for operation on a 180 mile voyage between Muscat and the tip of the Musandam Peninsular in Oman. The vessels have a capacity for 200 passengers and 56 cars, including an aft deck strengthened for the carriage of trucks. Powered by four MTU 20V 1163 TB73L diesels driving Kamewa 90 SII water-jets, the first vessel reached a maximum speed of 56 knots, and a service speed of 52 knots on trials.

The most recent proposal from the yard is a 34 m trimaran for operation by Stagecoach in the UK between Torquay and Brixham. This craft would carry 150 passengers at an operational speed of about 30 knots, completing

Natchan Rera - the first of two 112 m Incat wave-piercing catamarans built for the Libera Corporation for operation by Higashinihon Ferry of Japan. This is the largest fast ferry built by Incat. Powered by four MAN 20V 28/33D diesel engines each rated at 9000 kW and each driving a Wartsila LJX 1500 SRi water-jet via a ZF 60000 NR2H gearbox, these vessels have an operational speed of between 40 and 45 knots depending on the deadweight carried. Their maximum deadweight includes 900 passengers and 355 cars, although the vessels are arranged to carry 35 trucks with a reduced number of cars. 1120333

Alakai is the first of two 107 m Austal catamarans being built for Hawaii Superferry. This vessel was introduced into service on 23 August but was subsequently withdrawn due to pressure from environmental protestors. It is understood that the vessel returned to service on 14 December for operation between Oahu and Manui in Hawaii and that its continued operation is the subject of an environmental impact study. The vessel is arranged for 866 passengers and 282 cars (or 28 trucks and a reduced number of cars) and is powered by four MTU 20V 8000 M70 engines each rated at 8200 kW and each driving a Kamewa 125 SII waterjet via a ZF 53800 NR2H gearbox, giving a service speed of approximately 35 knots. 1120348

Cotai Jet, the first of a series of 10 sister-ships ordered by Venetian Marketing Services from Austal Ships at a total cost of approximately USD 150 million. The 48 m catamarans are arranged to carry 411 passengers and a cargo of 12 tonnes at an operational speed of approximately 42 knots. They are designed to service the Cotai Strip, a new hotel and casino development at Taipa in Macau, and run from the Hong Kong Macau ferry terminal in Central Hong Kong. It is thought that this operation of fast catamaran ferries is likely to expand in both size and the number of routes operated, probably to include the Hong Kong airport and main China ferry terminal, although it is understood that this might be the subject of some negotiation with respect to the route licence distribution. 1120347

Silni (meaning Mighty in Croatian), is one of a new Axe-Bow design of fast crew boats supplied by Damen Shipyard of the Netherlands. Known as a Fast Crew Supplier (FCS 3507) the vessel is 35 m long and 7 m wide. The shipyard have delivered six of these craft ranging from 33 to 50 m in length. The Axe-Bow shape of the bow is designed to improve the seakeeping performance without compromising the powering characteristics of the vessel. Silni was handed over to her owner, Brodospas of Split, Croatia, earlier in the year and services various production platforms in the Adriatic. 1120344

the voyage in about 20 minutes. Competition from land transportation is significant and so it will be interesting to follow the progress of this proposal.

Fast Ferry Safety

As well as being one of the most active years in terms of fast ferry world operations, it has, unfortunately, also been one of the worst years from a safety perspective. On 5 January 2007, the 37 m fast passenger ferry *Dawn Fraser* hit a small vessel in Sydney Harbour resulting in one fatality and on 15 January 2007 the Italian monohull fast ferry *Segesta Jet* collided with a large merchant ship resulting in four fatalities and 86 injuries. Both incidents occurred in difficult visual conditions and the latter incident resulted in the fast ferry being declared a total loss. On 3 February 2007 in thick fog, a 74 m wave-piercing catamaran *Sea Express 1* was involved in a serious collision with a bulk carrier on the River Mersey in the UK, which resulted in the ferry flooding to a significant extent, although, thankfully only one minor injury occurred. On 10 February 2007 a hydrofoil carrying 25 people hit a cargo ship in Vietnam, killing three passengers and injuring a further five – again this incident occurred in thick fog. On 28 March 2007, the *Pam Burridge*, another Sydney Harbour passenger ferry, collided with a private motor yacht on the harbour in the dark, killing four people and injuring a further eight. On 12 April a South Korean Jetfoil *Kobee V*, when operating at high speed, hit a submerged object resulting in a rapid deceleration which caused one fatality and 99 minor injuries. Another hydrofoil, the *Giorgione*, this time in Italy, hit rocks and sank on the 9th August after a mechanical failure, with the loss of one life and 12 injuries. A new fast catamaran passenger ferry, the *Salih Reis-4*, collided with a cargo ship in Turkey on 13 August 2007, possibly as a result of a steering failure, injuring at least 30 passengers. The last serious incident in the year involved a fast crew boat (converted from a fast passenger ferry) which caught fire, burnt and sank with the loss of one life.

This catalogue of incidents, combined with a number of less serious ones, represents the worst year on record for personal injury and loss of life for the fast ferry industry. Whilst taking the last twenty years of incident records, the industry can still demonstrate a level of safety which is better than most other transport sectors throughout the world, it is expected that future regulation of high speed craft will include more onerous risk control measures for operation in poor visibility, as a result of these and other incidents.

Fast Naval Craft

The increased level of activity in the commercial sector is also reflected in the military world – particularly within the US where the build of the Littoral Combat Ship (LCS) prototype craft and the ongoing activity within the Joint High Speed Vessel (JHSV) programme are generating a high level of interest across the industry.

Whilst the LCS prototype programme has, during this year, been reduced in scope from four to two craft, due reportedly to cost over-runs, the results of the trials of both remaining craft are awaited with significant interest. The designs of these vessels are of a very different nature to any existing military craft, and indeed are rare within the commercial sector, but they have the potential to fill a range of 'capability gaps' within the US military arena.

The requirements of a similar, but separate, high speed vessel acquisition program in the US, the JHSV program, provide an interesting picture of such a 'capability gap'. This program has been running for over 6 years and significant operational experience with fast catamaran vessels has been gained through the lease of a number of commercial vessels including the Incat 96 m *Joint Venture*, the Austal 101 m *Westpac Express* and two Incat 98 m vessels, the *Spearhead* and *Swift*. It is interesting to note that at least one of these high speed vessels has been deployed in support of every major US contingency since the first charter in 2001. The programme has now reached the stage where a number of purpose-designed vessels are required. The specified requirements include:

- The vessels to be acquired under JHSV are to be non-combatant but are to support a range of US military services – the Navy, Army and the Marines, although via a Navy-led acquisition programme.
- The vessels will be based on a commercial, non-developmental platform, built to commercial standards and classified by the American Bureau of Shipping, without invoking normal naval design standards. Each platform is understood to cost between USD 130 million to 150 million, this being a critical consideration for the programme requirements.
- The performance requirements of the craft include a payload capacity of at least 600 tonnes, an average speed of 35 knots in Sea State 3 with a range of 1,200 nautical miles. An additional capability to transit nearly 5,000 nautical miles at 25 knots without payload is also required. Operation is expected in Sea State 5 with survival in Sea State 7.
- The vessels are to be able to gain access to austere ports and thus have been specified with a length of less than 140 m, a beam able to navigate the Panama canal (32 m) and a draft of less than 4.5 m. In order to load/off-load their payload in such ports the vessels are to have a rotating stern ramp capable of supporting an M1A2 tank. The ramp should be able to be deployed rapidly (in less than 5 minutes) and to work with a variety of quay heights and even onto lighters.
- Other capabilities include the ability to accommodate two H-60 helicopters for day/night landing with full service although limited refuelling, using a single landing spot, and the ability to launch and recover 12 m boats is also required. A mixed crew of about 40 personnel is expected, with seating for 312 passengers and permanent berthing for 104 passengers.
- The space requirements for the vessel include a total usable area of 1800 m_2 including an aft mission deck of at least 1000 m $_2$ with an overhead clearance of nearly 5 m. This is to allow the handling of M1A2 tanks, MTVR trucks, expeditionary fighting vehicles and the deployment of standard 20 ft containers.
- Whilst not a combatant, the vessel will be fitted with a range of self defence weapons.

It is understood that the vessels are to be procured under a two phase process – Phase 1 is to involve preliminary design contracts lasting about 6 months. Solicitation for this phase was started in September/October 2007 with the anticipated start date being within the first quarter of 2008. Phase 2 will revolve around the down selection to one contractor for the build of a lead ship with the option for up to seven follow-on ships – this phase is anticipated to start in late 2008 or early 2009.

It is not only the large navies that are taking advantage of high speed craft – many of the smaller navies are considering the acquisition of small fast vessels (such as fast patrol craft and fast interceptors), in some cases

The Henriksen HS 1365 2VD interceptor, a typical example of a new breed of fast interceptor craft, this one having been built for the Norwegian Home Guard. At 14 m in length and powered by two 480 kW Volvo Penta diesel engines, each driving a Kamewa K32 water-jet, this Gyda Class vessel has a crew of four and additional space for 10 fully equipped soldiers. 1120321

Chase 1 is a new interceptor craft built by Eraco Boat Builders in South Africa. Eraco was set up in 1997 to build sailing yachts but has subsequently diversified into the production of powered interceptors and patrol craft. Chase 1, a 14 m vessel, is constructed of aluminium and powered by a 1200 kW inboard engine to give speeds of up to 65 knots. This is a good example of the type of technology that would have been restricted to specialist yards only a decade ago but is now available to a much wider spectrum of builders. 1120305

resulting in the build of hundreds of craft for individual countries. It is also worthy of note that significant interest is also being shown in fast unmanned craft, another area that is set to grow significantly over the next decade.

The Future of the High-Speed Marine Craft Industry

Over the past 15 years as Editor of Jane's High-Speed Marine Transportation, I have seen the industry grow substantially in size – with varying annual rates of growth, but with a solid confidence that at the end of the next year, the world fleet would be larger than ever before. Looking back over the past year and forward to the year to come, it is difficult to see any substantial reason for not having that same level of confidence.

The popularity of fast ferries is closely related to the rate of tourist activity and, to a lesser extent, local maritime trading conditions and these continue to expand. Fast commercial workboats and fast recreational craft (including fast mega-yachts) are continuing to increase in popularity (the latter due to the rise in personal wealth, particularly in Russia). Most significantly, the flexibility offered by high speed is clearly teasing out significant new operational capabilities and requirements within the military sector and this is expected to continue to be a growth area for at least the next decade.

Whilst the threat of increasing fuel costs and competition from the economies of scale of large modern, more conventional, vessels are real, they are yet to dampen the demand for the perceived and proven benefits that modern high-speed marine craft can offer.

Finally, and with no apology for my enthusiasm for this industry, I am always grateful for your comments regarding the content and presentation of Jane's High-Speed Marine Transportation, whether positive or negative, and encourage you to contact me should you have any specific ideas relating to this database or, more generally, to this fascinating industry sector.

Stephen Phillips
January 2008

Acknowledgements

The accurate updating of *Jane's High Speed Marine Transportation* would not be possible without the assistance of all the organisations featured in this database. The gathering together and forwarding of information, drawings and photographs to the editor for inclusion is, itself, time consuming and to do this each year - and sometimes twice a year - is possibly past the call of duty. However, we are extremely grateful for those that do this.

My thanks also go to the editorial staff at *Jane's* who manage to make sense of my revisions and updates and continue to present the hardcopy and online information in such an attractive way. In particular I would like to thank Hannah Leech, Charles Hollosi and Jacqui Beard at Jane's, and Richard Daltry at Seaspeed for their hard work and patience in dealing with me and this database.

Stephen Phillips, CEng, FRINA, FIMarEST
Jane's Information Group
Sentinel House
163 Brighton Road
Coulsdon
Surrey CR5 2YH
Fax number: (+44 20) 87 00 39 59
email: yearbook@janes.com

Stephen Phillips, CEng, FRINA, FIMarEST

The Editor, Stephen Phillips, is an experienced naval architect specialising in the survey, valuation and performance assessment of large high-speed craft. He has been a Director of Seaspeed Marine Consulting Limited since 1990 and is responsible for all its technical and market research activities.

Mr Phillips has 30 years of experience within the marine industry covering both the commercial and military sectors. Over that time he has been closely involved in the research, design, build and testing of many fast commercial ferries, military patrol craft, crewboats and luxury yachts. He has published a number of technical papers covering the design, construction, operation and safety of high-speed craft and regularly chairs technical conferences and seminars on the subject.

He is a graduate of Southampton University and has previously worked for Vickers Shipbuilding and Engineering Group in Barrow-in-Furness, Cockatoo Dockyard and International Catamaran Designs in Sydney, Australia and VT Shipbuilding in Southampton and Portsmouth in the UK.

Preface: Jane's High-Speed Marine Transportation

Jane's High-Speed Marine Transportation aims to provide an informed, balanced and comprehensive data source for all those organisations with an interest in high-speed marine transport worldwide.

If your company or product is not included, but, you feel that it should be then please do contact me: I will be delighted to include all relevant information. There is no charge for editorial entries in this book.

While every effort is made to standardise and rationalise the information presented in this book, it is nonetheless not always possible to ensure that all the information is based on a common standard. For example, the operational speeds quoted for each craft may be for slightly different displacement definitions. However, each year we spend a considerable amount of time rationalising the data and hope that in the future this will lead to still further improvements in data quality.

What constitutes high-speed in the marine world has been the subject of much debate. It is clearly a relative term and for the purposes of this book, need not be defined quantitatively. However, other than Interceptor craft, vessels with speeds of less than 20 kt and with a length of less than 20 m are generally only included if they represent a development of particular interest.

In the International Code of Safety for High Speed Craft, a high-speed craft is one capable of a maximum speed equal to or exceeding $V=3.7(Displ)^{0.1667}$, where Displ is the maximum displacement in cubic metres and V is the speed in metres per second. Thus for a 100 tonne craft this speed is 15 kt, for a 1,000 tonne craft it is 23 kt and for a 10,000 tonne craft it is 35 kt. These values of speed relative to the size of the vessel have been set low to make the Code available to as wide a range of craft as possible. This Code of Safety provides an equivalance to the International SOLAS requirements by nature of a trade-off between design requirements (allowing for light weight construction) and restricted operating conditions.

Sea - General		Wind				Sea								
Sea State	Description	(Beaufort) Wind force	Description	Range	Wind Velocity	Wave height			Significant Range Periods	Periods of Maximum Energy of Spectra Tmax=Tc	Average of Period Tz	Average Wave Length Lw	Minimum Fetch	Minimum Duration
–	–	–	–	–	–	Average	Significant	Average of One-Tenth Highest	-	-	-	-	-	-
–	Sea like a mirror	U	Calm	1 kt	-	-	-	-	-	-	-	-	-	-
0	Ripples with the appearance of scales are formed, but without foam crests	1	Light airs	1-3 kt	2 kt	0.04 ft	0.01 ft 0.01 ft	0.09 ft	1.2 s	0.75	0.5	10 in	5 n miles	18 min
1	Small wavelets; short but pronounced crests have a glossy appearance, but do not break	2	Light breeze	4-6 kt	5 kt	0.3 ft	0.5 ft	0.6 ft	0.4-2.8 s	1.9	1.3	6.7 ft	8 n miles	39 min
–	Large wavelets; crests begin to break. Foam of glossy appearance. Perhaps scattered with horses	3	Gentle breeze	7-10 kt	8.5 kt 10 kt	0.8 ft 1.1 ft	1.3 ft 1.8 ft	1.6 ft 2.3 ft	0.8-5.0 s 1.0-6.0 s	3.2 3.2	2.3 2.7	20 ft 27 ft	9.8 n miles 10 n miles	1.7 h 2.4 h
2 3	Small waves, becoming larger; fairly frequent white horses	4	Moderate breeze	11-16 kt	12 kt 13.5 kt 14 kt 16 kt	1.6 ft 2.1 ft 2.3 ft 2.9 ft	2.6 ft 3.3 ft 3.6 ft 4.7 ft	3.3 ft 4.2 ft 4.6 ft 6.0 ft	1.0-7.0 s 1.4-7.6 s 1.5-7.8 s 2.0-8.8 s	4.5 5.1 5.3 6.0	3.2 3.6 3.8 4.3	40 ft 52 ft 59 ft 71 ft	18 n miles 24 n miles 28 n miles 40 n miles	3.8 h 4.8 h 5.2 h 6.6 h
4	Moderate waves, taking a more pronounced long form; many white horses are formed (chance of some spray)	5	Fresh breeze	17-21 kt	18 kt 19 kt 20 kt	3.7 ft 4.1 ft 4.6 kt	5.9 ft 6.6 ft 7.3 ft	7.5 ft 8.4 ft 9.3 ft	2.5-10.0 s 2.8-10.6 s 3.0-11.1 s	6.8 7.2 7.5	4.8 5.1 5.4	90 ft 99 ft 111 ft	55 n miles 65 n miles 75 n miles	8.3 h 9.2 h 10 h
5 6	Large waves begin to form; white crests are more extensive everywhere (probably some spray)	6	Strong	22-27 kt	22 kt 24 kt 24.5 kt 26 kt	5.5 ft 6.6 ft 6.8 ft 7.7 ft	8.8 ft 10.5 ft 10.9 ft 12.3 ft	11.2 ft 13.3 ft 13.8 ft 15.6 ft	3.4-12.2 s 3.7-13.5 s 3.8-13.6 s 4.0-14.5 s	8.3 9.0 9.2 9.8	5.9 6.4 6.6 7.0	134 ft 160 ft 164 ft 188 ft	100 n miles 130 n miles 140 n miles 180 n miles	12 h 14 h 15 h 17 h
7	Sea heaps up, and white foam from breaking waves begins to be blown in streaks along the direction of the wind (Spindrift begins to be seen)	7	Moderate gale	28-33 kt	28 kt 30 kt 30.5 kt 32 kt	8.9 ft 10.3 ft 10.6 ft 11.6 ft	14.3 ft 16.4 ft 16.9 ft 18.6 ft	18.2 ft 20.8 ft 21.5 ft 23.6 ft	4.5-15.5 s 4.7-16.7 s 4.8-17.0 s 5.0-17.5 s	10.6 11.3 11.5 12.1	7.5 8.0 8.2 8.6	212 ft 250 ft 258 ft 285 ft	230 n miles 280 n miles 290 n miles 340 n miles	20 h 23 h 24 h 27 h
7	Moderate high waves of greater length; edges of crests break into spindrift. The foam is blown in well-marked streaks along the direction of the wind. Spray affects visibility	8	Fresh gale	34-40 kt	34 kt 36 kt 37 kt 38 kt 40 kt	13.1 ft 14.8 ft 15.6 ft 16.4 ft 18.2 ft	21.0 ft 23.6 ft 24.9 ft 26.3 ft 29.1 ft	26.7 ft 30.0 ft 31.6 ft 33.4 ft 37.0 ft	5.5-18.5 s 5.8-19.7s 6-20.5 s 6.2-20.8 s 6.5-21.7 s	12.8 13.6 13.9 14.3 15.1	9.1 9.6 9.9 10.2 10.7	322 ft 363 ft 376 ft 392 ft 444 ft	420 n miles 500 n miles 530 n miles 600 n miles 710 n miles	30 h 34 h 37 h 38 h 42 h
8 9	High waves. Dense streaks of foamalong the direction of the wind. Sea begins to roll. Visibility affected. Very high waves with long over-hanging crests. The resulting foam is in great patches and is blown in dense white streaks along the direction of the wind. On the whole, the surface of the sea takes on a white appearance. The rolling of the sea becomes heavy and shocklike. Visibility is affected	9 10	Strong gale Whole[1] gale	41-37 kt 48-55 kt	42 kt 44 kt 46 kt 40 kt 50 kt 51.5 kt 52 kt 54 kt	20.1 ft 22.0 ft 24.1 ft 26.2 ft 28.4 ft 30.2 ft 30.8 ft 33.2 ft	32.1 ft 35.2 ft 38.5 ft 41.9 ft 45.5 ft 48.3 ft 49.2 ft 53.1 ft	40.8 ft 44.7 ft 48.9 ft 53.2 ft 57.8 ft 61.3 ft 62.5 ft 67.4 ft	7-23 s 7-24.2 s 7-25 s 7.5-26 s 7.5-27 s 8-28.2 s 8-28.2 s 8-29.5 s	15.8 16.6 17.3 18.1 18.8 19.4 19.6 20.4	11.3 11.8 12.3 12.9 13.4 13.8 13.9 14.5	492 ft 534 ft 590 ft 650 ft 700 ft 736 ft 750 ft 810 ft	830 n miles 960 n miles 1,110 n miles 1,250 n miles 1,420 n miles 1,560 n miles 1,610 n miles 1,800 n miles	47 h 52 h 57 h 63 h 69 h 73 h 75 h 81 h
–	Exceptionally high waves. Sea completely covered with long white patches of foam lying in direction of wind. Everywhere edges of wave crests are blown into froth. Visibility affected	11	Storm[1]	56-63 kt	56 kt 59.5 kt	35.7 ft 40.3 ft	57.1 ft 64.4 ft	72.5 ft 81.8 ft	8.5-31 s 10-32 s	21.1 22.4	15 15.9	910 ft 985 ft	2,100 n miles 2,500 n miles	88 h 101 h
–	Air filled with foam and spray. Sea white with driving spray. Visibility very seriously affected	12	Hurricane[1]	64-71 kt	<64 kt	<46.6 ft	74.5 ft	94.6 ft	10-35 s	24.1	17.2	-	-	-

For hurricane winds (and often whole gale and storm winds) required durations and reports are barely attained. Seas are therefore not fully arisen

DEFINITIONS OF SYMBOLS AND UNITS

Four of the seven base units of the SI (Système International d'Unités) which are used in this book are:

length	metre	m
mass	kilogram	kg
time	second	s
electric current	ampere	A
Decimal unit	Quantity	Formula
Pa (Pascal)	pressure or stress	N/m^2
N (Newton)	force	kg.m/s^2
W (Watt)	power	J/s
Hz(Hertz)	frequency	1/s (1 Hertz = 1 cycle per second in previous British practice)
V (Volt)	electric potential difference	W/A

Other units	Quantity
dB (decibel)	sound pressure level, re 0.0002 microbar
dBA (decibel)	sound level, A-weighted, re 0.0002 microbar

CONVERSIONS

Length

1 km	= 0.6214 statute mile = 0.540 nautical mile
1 m	= 3.281 ft
1 cm	= 0.3937 in
1 mm	= 0.0394 in

Area

1 ha (hectare = 10000 m^2)	= 2.471 acres
1 m^2	= 10.764 ft^2

Volume

1 m^3	= 35.315 ft^3
1 litre	= 0.220 Imperial gallon = 0.264 US gallon

Velocity

1 km/h	= 0.621 statute mile/h = 0.540 kt
1 m/s	= 3.281 ft/s

Acceleration

1 m/s^2	= 3.281 ft/s^2

Mass

1 t	(tonne)	= 1000 kg = 0.9842 long ton = 2204.62 lb = 1.1023 short tons
1 kg	(kilogram)	= 2.205 lb
1 g	(gram)	= 0.002205 lb

Force

1 MN	(meganewton)	= 100.36 long ton force
1 kgf	–	= 2.205 lbf
1 kp	(kilopond)	= 2.205 lbf
1 N	(newton)	= 0.2248 lbf (The Newton is that force which, applied to a mass of 1 kilogram, gives it an acceleration of 1 m/s^2.)

Moment of force (torque)

1 Nm	= 0.7376 lbf.ft

Pressure, stress

1 atm	(standard atmosphere)	= 14.696 lbf/in^2
1 bar	(10^5 pascal)	= 14.504 lbf/in^2
1 kPa	(kN/m^2)	= 20.885 lbf/ft^2
1 Pa	(N/m^2)	= 0.020885 lbf/ft^2

POWER

1 metric horsepower (ch, ps) = 0.7355 kW = 1.014 horsepower (550 ft lb/s)
1 kW = 1.341 horsepower (1 horsepower = 550 ft lb/s) = 1.360 metric horsepower

NAUTICAL MILE

The International Nautical Mile is equivalent to the average length of a minute of latitude and corresponds to a latitude of 45° and a distance of 1,852 m = 6,076.12 ft.

FUEL CONSUMPTION

Specific fuel consumption, 1.0 g/kWh = 0.001644 lb/hph (hp = 550 ft lb/s)
1.0 litre/h = 0.220 Imperial gallon/h = 0.264 US gallon/h

Jane's Users' Charter

This publication is brought to you by Jane's Information Group, a global company with more than 100 years of innovation and an unrivalled reputation for impartiality, accuracy and authority.

Our collection and output of information and images is not dictated by any political or commercial affiliation. Our reportage is undertaken without fear of, or favour from, any government, alliance, state or corporation.

We publish information that is collected overtly from unclassified sources, although much could be regarded as extremely sensitive or not publicly accessible.

Our validation and analysis aims to eradicate misinformation or disinformation as well as factual errors; our objective is always to produce the most accurate and authoritative data.

In the event of any significant inaccuracies, we undertake to draw these to the readers' attention to preserve the highly valued relationship of trust and credibility with our customers worldwide.

If you believe that these policies have been breached by this title, you are invited to contact the editor.

A copy of Jane's Information Group's Code of Conduct for its editorial teams is available from the publisher.

FREE ENTRY/CONTENT IN THIS PUBLICATION

Having your products and services represented in our titles means that they are being seen by the professionals who matter - both by those involved in the procurement and by those working for the companies that are likely to affect your business. We therefore feel that it is very much in the interest of your organisation, as well as Jane's, to ensure your data is current and accurate.

- **Don't forget** - You may be missing out on business if your entry in a Jane's product is incorrect because you have not supplied the latest information to us.
- **Ask yourself** - Can you afford not to be represented in Jane's printed and electronic products? And if you are listed, can you afford for your information to be out of date?
- **And most importantly** - The best part of all is that your entries in Jane's products are TOTALLY FREE OF CHARGE.

Please provide (using a photocopy of this form) the information on the following categories where appropriate:

1. Organisation name: ______
2. Division name: ______
3. Location address: ______
4. Mailing address if different: ______
5. Telephone (please include switchboard and main departmental contact numbers, for example Public Relations, Sales, and so on): ______
6. Facsimile: ______
7. E-mail: ______
8. Web sites: ______
9. Contact name and job title: ______
10. A brief description of your organisation's activities, products and services: ______
11. Jane's publications in which you would like to be included: ______

Please send this information to:
Jacqui Beard, Information Collection, Jane's Information Group
Sentinel House, 163 Brighton Road, Coulsdon, Surrey CR5 2YH, UK
Tel: (+44 20) 87 00 38 08
Fax: (+44 20) 87 00 39 59
e-mail: yearbook@janes.com

Copyright enquiries:
e-mail: copyright@janes.com

Please tick this box if you do not wish your organisation's staff to be included in Jane's mailing lists ☐

JHMT

BUILDERS

Air cushion vehicles
Hydrofoils
High-speed multihull vessels
Small-Waterplane-Area Twin-Hull (SWATH) vessels
High-speed monohull craft
Wing-in-ground-effect craft
Fast interceptors
High-speed patrol craft

AIR CUSHION VEHICLES

Company listing by country

Australia
Airlift Hovercraft

China
Dagu Shipyard
Dong Feng Shipyard
Huangpu Shipyard
Hudong-Zhonghua Shipyard
JianghuiShipyard
Qindao Beihai Shipbuilding Heavy Industries Ltd
Qiuxin Shipyard

Finland
Aker Finnyards Inc

France
DCN International (DCN)

Japan
Mitsubishi Heavy Industries Ltd
Mitsui Engineering & Shipbuilding Company Ltd

Korea, South
Hanjin Heavy Industries
Samsung Heavy Industries Co Ltd
Semo Company Ltd

Netherlands
Schelde Naval Shipbuilding (Royal Schelde)

Norway
Brødrene Aa
Ulstein Technology
Umoe Mandal

Russian Federation and Associated States (CIS)
Almaz Central Marine Design Bureau
Almaz Shipbuilding Company
Krasnoye Sormovo Shipyard
Krylov Shipbuilding Research Institute
Neptune Hovercraft
Pormornic Astrakhan Shipyard
Sosnovka Shipyard
Vympel Ship Design Company

Singapore
SCAT Technologies Pte Ltd
Singapore Technologies Marine Ltd

Ukraine
Feodosia Shipbuilding Association (MORYE)

United Kingdom
ABS Hovercraft Ltd
Aluminium Shipbuilders Ltd
Griffon Hovercraft Ltd
Hoverwork Ltd
Ingles Hovercraft Ltd
Slingsby Advanced Composites Ltd

United States
Hover Shuttle LLC
Kvichak Marine Industries
Northrop Grumman Ship Systems
Textron Marine & Land Systems

Australia

Airlift Hovercraft

24B August Lane, Alberton, 4207 Queensland, Australia

Tel: (+61) 7 3804 6636
Fax: (+61) 7 3804 6646
e-mail: sales@airlifthovercraft.com
Web: www.airlifthovercraft.com

Ross McLeod, *Principal*

Established in 1979, Airlift designs, builds and operates small to medium sized hovercraft. Designs are available for hovercraft from 5.8 m (600 kg, or six passengers and one crew) up to 14.5 m (3,500 kg, or 42 passengers plus two crew).

The company opened a new manufacturing facility in 2006, giving a wider scope for aluminium and glass-fibre fabrication as well as larger design offices.

Airlift 12 m Pioneer Hovercraft 1195584

China

Dagu Shipyard

Zeng No 1, Yangzhabei, Xigu, Tang-gu, Tianjin 300452, China

Tel: (+86 22) 25 89 56 67
Fax: (+86 22) 25 31 12 54

Young Tze-Wen, *Director*

The shipyard builds various types of hovercraft (amphibious and sidewall) to the designs of the Marine Design and Research Institute of China, and designs and builds small and medium-sized steel vessels.

Type 7203 fast ferry 0506560

TYPE 7203

Specifications

Length overall	22.2 m
Beam	6.9 m
Draught, hullborne	2.1 m
Draught, on-cushion	1.2 m
Weight, max	35 t
Passengers	81–100
Propulsive power	2 × 335 kW
Maximum speed	30 kt
Operational speed	26 kt
Range	180 n miles

Derived from Types 713 and 717 (built in the 1970s), Type 7203 is a high-speed passenger ferry for use on coastal and sheltered waters. Alternative applications include coastguard patrol and port/harbour firefighting duties.

Built at the Dagu Shipyard, Tianjin, the prototype was launched in September 1982 and underwent trials on the Hai river and in Tang-gu in late 1982. The vessel has a loop and segment type skirt and a multifan lift system with centrifugal fans at the bow, midships and stern allowing optimum power consumption for operation both in calm water and in waves. In calm conditions the bow fan is disengaged, saving over 50 per cent of the 188 kW of lift power installed. The craft is steered by twin rudders and differential control of the propellers. The seating is arranged in a three row configuration with two 0.8 m aisles.

Propulsion: The lift system is powered by a single 12150C high-speed diesel rated at 226 kW at 1,500 rpm. The lift engine directly drives the amidships and stern fans. The bow fan is driven via a hydraulic pump and motor. Total lift power is about 188 kW. Propulsive power is supplied by two 12150CZ water-cooled, turbocharged, high-speed marine diesels, each rated at 335 kW at 1,450 rpm. Each drives a three-bladed propeller via a V-type transmission.

Jinxiang

Jinxiang, a Type 7203, is a joint project of the Marine Design and Research Institute of China and the Dagu Shipyard, Tianjin. In 1983 it successfully completed a 128 km maiden voyage along the Yangtze, from Shanghai to Vantong, in under 3 hours.

Vessel type	Vessel name	Yard No.	Length (m)	Speed (kt)	Seats	Vehicles	Originally delivered to	Date of build
7203	*Jinxiang*	–	22.2	26.0	81	none	–	September 1982
722-11	*Dagu Class*	–	34.1	47.0	120	2 tanks	Chinese Navy	1979

Dong Feng Shipyard

A subsidiary of Chongquing Dongfeng Shipbuilding Corporation
Dongfeng village, Tang Guojianto, Chongquing Jianghei District, 400026, People's Republic of China

Tel: (+86 23) 6778 1007
Fax: (+86 23) 6778 2332
e-mail: office@cscdfship.com
Web: www.cscdfship.com

The company builds air cushion vehicles to the designs of the Marine Design and Research Institute of China.

TYPE 7210

Specifications

Length overall	9.9 m
Beam	3.4 m
Weight, max	4.7 t
Payload	0.8 t
Max speed	24 kt
Range	135 n miles

Structure: Built in medium-strength seawater-resistant aluminium alloy of riveted construction.

Propulsion: The craft is powered by air-cooled marine diesels. Lift is provided by a Deutz BF6L912 diesel engine, driving a centrifugal aluminium fan via a gearbox. Thrust is supplied by another diesel, a Deutz BF6L913 driving via a transmission shaft and elastic coupling, a 1.8 m, five-blade, ducted air propeller built in GRP.

Two fully amphibious utility hovercraft Type 7210, designed by MARIC were completed in May 1985. The hull structure is of a riveted construction with 0.5 m high bag and finger skirt system.

TYPE 717 II and 717 III

This craft is a water-jet-propelled, rigid sidewall air cushion vehicle, designed by MARIC as a high-speed inland water passenger ferry for use on shallow water. It is a development of the 717 design. The Type 717 *Chungqing* and the Type 717 II

Ming Jiang passenger ferry hovercraft were both completed in October 1984. They were delivered to Chongqing Ferry Boat Company as high-speed passenger craft operating on the rapids of the Yangtze river. A further Type 717 III was delivered in September 1989.

Type 717 II

Specifications

Length overall	20.4 m
Beam	4.5 m
Weight, max	21.2 t
Passengers	54–60
Propulsive power	2 × 224 kW
Max speed	24 kt
Range	220 n miles

TYPE 717 III

Specifications

Length overall	21.4 m
Beam	4.5 m
Weight, max	23 t
Passengers	70
Propulsive power	2 × 224 kW
Max speed	23 kt
Range	135 n miles

Structure: The sidewalls are built in GRP, but other parts of the hull and superstructure are built in riveted, high-strength aluminium alloy.

Propulsion: Integrated system powered by two 12V 150C marine diesels rated at 224 kW continuous. Two engines are mounted aft and each drives a 600 mm diameter centrifugal fan, Type 4-72 for lift and, via an elastic coupling, universal joint and transmission shaft, a mixed flow water-jet pump. The Type 717 III is powered by two Cummins high-speed diesel NTA-855-M engines 298 kW (maximum) each.

TYPE 717 IIIC

Yu Xiang

This SES vessel was delivered to Chongqing Ferry Boat Company in September 1989.

The transverse structure was made stronger than the previous Type 717 III so as to resist large stern waves.

Specifications

Length overall	21.4 m
Beam	4.5 m
Draught, hullborne	1.0 m
Draught, on-cushion	0.7 m
Weight, max	23.5 t
Crew	4
Max speed	22.7 kt
Range	220 n miles

Propulsion: Two Cummins NTA-855-M322 diesel engines manufactured by Sichuan Chongqing Automobile Motor Factory.

TYPE 7215

This craft was designed by MARIC and was completed in March 1993. The craft is operated on the upper reaches of the Yangtze River in Shi Chua Province from Chongqing to Fuling and Wan Xian.

Specifications

Length, overall	28.0 m
Beam overall	6.8 m
Height overall	6.4 m
Draught, hullborne	1.7 m
Draught, on-cushion	1.4 m
Crew	5

MARIC-designed Type 7210 built by Dong Feng Shipyard 0506561

MARIC Type 7226 0506947

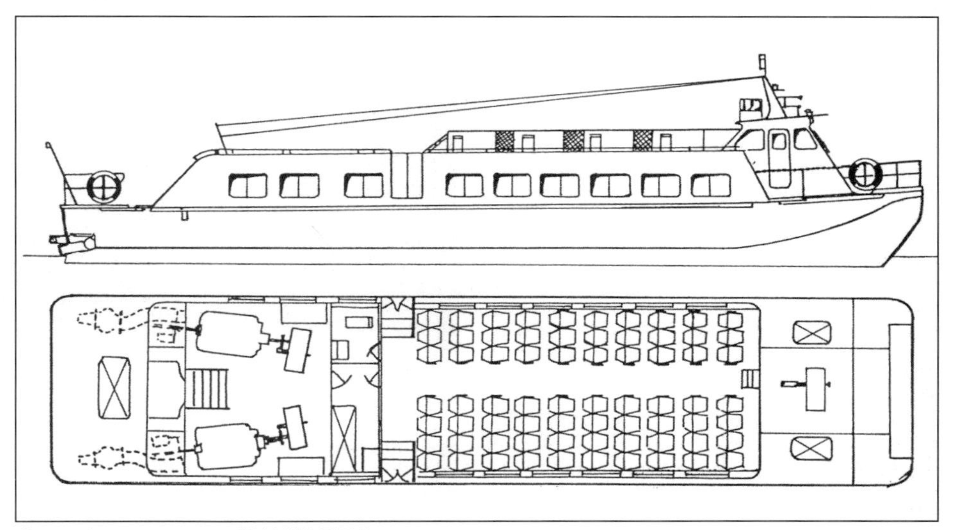

General arrangement of MARIC 717 II 0506562

MARIC Type 717 IIIC on its delivery trip 0506563

Vessel type	Vessel name	Yard No	Length (m)	Speed (kt)	Seats	Vehicles	Originally delivered to	Date of build
717III	*Chong Qing*	–	21.4	–	70	none	Chongqing Shipping Companies	October 1984
717II	*Ming Jiang*	–	20.4	–	60	none	Chongqing Shipping Companies	October 1984
7210	–	–	9.9	–	–	none	–	May 1985
7210	–	–	9.9	–	–	none	–	May 1985
717IIIC	*Yu Xiang*	–	21.4	–	70	none	Chongqing Shipping Companies	September 1989
717III	*Jin Ling Jiang*	–	21.4	–	70	none	Chongqing Shipping Companies	September 1989
7224	–	–	12.4	24.3	15	none	–	December 1992
7224	–	–	12.4	24.3	15	none	–	December 1992
7224	–	–	12.4	24.3	15	none	–	December 1992
7215	–	–	28.0	–	135	none	–	March 1993
7218	–	–	21.3	35.0	70	none	–	1994
7226	–	–	17.9	28.0	40	none	–	1995

Passengers	126–135
Max speed	27 kt
Range	243 n miles

Propulsion: Engines: Two Cummins KTA19-M diesels, 274 kW each at 2,034 rpm (made in China under license).
Thrust device: two directly driven marine propellers.
Lift engine: Cummins NT14-M diesel, 179 kW at 2,000 rpm (made in China under license).
Fans: two directly driven centrifugal fans, 0.85 m diameter.

MARIC Type 7215 0506948

TYPE 7218

This craft is a derivative of Type 716 II and was completed in 1994. This fully amphibious, all-aluminium vessel is operated along the coastal line of Qin Dao City as a passenger ferry.

Specifications

Length overall	21.3 m
Beam	8.8 m
Height overall	5.8 m
Weight, max	30 t
Crew	3
Passengers	70
Operational speed	35 kt
Range	140 n miles
Operational limitation	Beaufort 6, Sea State 3
Obstacle clearance	0.7 m

Classification: ZC (Sheltered Sea Area)
Propulsion: Three sets of Deutz air-cooled BF12L513C diesels, 386 kW at 2,300 rpm, one for lift and two for propulsion.

MARIC Type 7218 completed in 1994 0506949

TYPE 7224

This craft was designed by MARIC. Three were completed by December 1992, one for Zhenzhou City as a touring boat operating on the Yellow River and two for personnel transportation at the Lieu River oilfield.

Specifications

Length overall	12.4 m
Beam	4.5 m
Passengers	15
Operational speed	24.3 kt

Structure: Hull material: medium-strength aluminium alloy.
Hull construction: riveted.
Propulsion: Engines: two Deutz BF6L913C air-cooled diesel (lift and propulsion, mechanically integrated arrangement).

MARIC Type 7224 0506825

TYPE 7226

This fully amphibious all-aluminium vessel was delivered in 1995 and operates on the coastal line of Dalian City as a passenger and tourist ferry.

Specifications

Length overall	17.9 m
Beam	7.0 m
Height overall	4.6 m
Displacement, max	16.5 t
Crew	3
Passengers	40
Operational speed	28 kt
Range	120 n miles
Operational limitation	Beaufort 5, Sea State 2
Obstacle clearance	0.4 m

Classification: ZC (Sheltered Sea Area)
Propulsion: Two sets of Deutz air-cooled BF12L413F diesels, 282 kW at 2,500 rpm, both driving a lift fan and air ducted propeller.

Huangpu Shipyard

Changzhou Town, Huangpu District, Guangzhou 510336, People's Republic of China

Tel: (+86 20) 220 17 29
Fax: (+86 20) 220 13 87

Gao Feng, *Director*

One of three Huangpu shipyards in Guangdong, this yard builds lightweight high-speed craft.

The company builds air cushion vehicles to the designs of the Marine Design and Research Institute of China.

MARIC Type 7211 SES 0506564

TYPE 7211

A contract for designing and building a new 162-passenger SES ferry was signed on 15 April 1990 in Guangzhou. Ordered by China Merchants Development Company Ltd in Hong Kong, this Type 7211 passenger ferry was designed by MARIC and delivered in November 1992 to operate between Shekou and Hong Kong, a route operated by two Hovermarine International HM218s.

Vessel type	Vessel name	Yard No	Length (m)	Speed (kt)	Seats	Vehicles	Originally delivered to	Date of build
7211	*Yin Bin 4*	–	30.0	–	162	none	China Merchants Development Company Ltd	November 1992

Specifications

Length overall	30.0 m
Beam	7.6 m
Draught, hullborne	1.8 m
Draught, on-cushion	1.5 m
Passengers	162–171
Max speed	30 kt
Range	180 n miles

Classification: Designed and built to ZC rules.
Structure: Has a welded aluminium main structure and a riveted aluminium superstructure.
Propulsion: Powered by two MWM 12V TBD 234 diesel engines, each coupled to a propeller, and one MWM 6V TBD 234 diesel driving the lift fans.

Hudong-Zhonghua Shipbuilding (Group) Co Ltd

2851 Pudong Dadao, Shanghai 200129, China

Tel: (+86 21) 58 71 32 22
Fax: (+86 21) 58 71 26 03
e-mail: bmd@hz-shipgroup.com
Web: www.hudong.com.cn

Gu Baolong, *Chairman*

Hudong-Zhonghua Shipbuilding is one of the largest enterprises within the China Shipbuilding Group Corporation (CSSC), with a workforce of over 14,000 people. When the two companies merged, both the shipyards at Hudong and the one at Zhonghua were retained. While the core business of the shipyard is the design and construction of conventional steel ships, the company has built lightweight high-speed craft.

The MARIC 719 II steel hull SES 0506566

TYPE 716 II/III

Designed by MARIC, the amphibious hovercraft Type 716 II was completed at Hudong Shipyard in 1985. The craft underwent evaluation by the China Air Cushion Technology Development Corporation (CACTEC) and now operates in offshore areas, shallow water and marshes where it is used to transport people (up to 32 passengers without cargo) and equipment (up to 4 tonnes of cargo without passengers). This design has recently been updated to the Type 716 III.

Specifications

Length overall	18.4 m
Beam	7.7 m
Weight, max	19.4 t
Payload	4 t (no passengers)
Passengers	32 (no cargo)
Propulsive power	319 kW
Max speed	39 kt
Range	120 n miles

Structure: Riveted skin and stringer structure employing high-strength aluminium alloy sheet.
Propulsion: One Deutz BF12L413FC air-cooled marine diesel, 319 kW at 2,300 rpm, via a gearbox and transmission shaft drives a 2 m diameter centrifugal fan. Two identical engines, via transmission shafts, drive directly two four-blade 2.3 m diameter ducted air propellers.

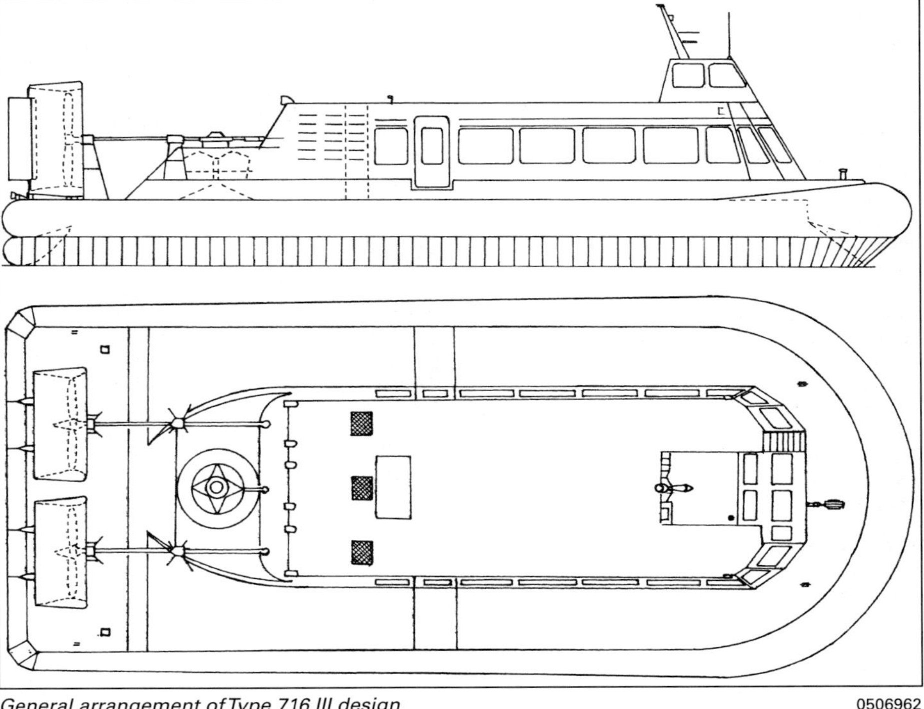

General arrangement of Type 716 III design 0506962

TYPE 719 II

In 1998, Zhonghua built a MARIC-designed 719 II SES *Hong Xiang* for operation between the Shanghai Municipality and Chong Ming Island. This 719 II sidewall hovercraft was the second such craft to be built with a steel hull. The vessel can carry 257 passengers and meets the requirements for ships operating in the Yangtze River Class A area.

The control of engines is by remote-control systems produced by HDW-Elektronik of Germany with manual back-up. The lift engine is directly connected to three double-intake centrifugal lift fans via a clutch system and there is no speed reduction device between them. The propulsion engines drive the propellers through a Type WVS 642 gearbox. The engine room also contains an auxiliary engine and an electric generator.

There are two double intake centrifugal fans in the fan room, arranged along the centreline of the ship.

Specifications

Length overall	40.0 m
Beam	8.3 m
Draught, hullborne	2.5 m
Draught, on-cushion	1.9 m
Displacement, max	123.5 t
Payload	22 t
Passengers	257
Max speed	27.5 kt
Operational speed	24 kt
Range	200 n miles
Operational limitation	–
Beaufort 7	wave height 1.5 m

Vessel type	Vessel name	Yard No	Length (m)	Speed (kt)	Seats	Vehicles	Originally delivered to	Date of build
716II	–	–	18.4	–	32	none	–	1985
719II	*Hong-Xiang*	–	40.0	24.0	257	none	–	1988

Jianghui Shipyard

Gangwan 2nd Road, Shekou Industrial Zone, Shen Zhen City, People's Republic of China

Tel: (+86 755) 26 82 91 58
Fax: (+86 755) 26 69 53 59
e-mail: sales@integritytrawlers.com
Web: www.jh-boat.com

In 1995, the company built a GRP air cushion vehicle for operation in Hong Kong.

200 passenger speed ferry

Designed by the Marine Design and Research Institute of China, this FRP sidewall hovercraft is operated in the Hong Kong–Shen Zhen river areas as a passenger vessel.

Specifications

Length overall	32.2 m
Length waterline	30.5 m
Beam	8.2 m
Height overall	6.9 m
Draught, on-cushion	1.8 m
Crew	5
Passengers	225
Max speed	28 kt
Range	200 n miles

Vessel type	Vessel name	Yard No	Length (m)	Speed (kt)	Seats	Vehicles	Originally delivered to	Date of build
200 Passenger Ferry	–	–	30.50	28	225	none	–	October 1995
160 Passenger Ferry	–	–	27.20	28	160	none	–	–
100 Passenger Ferry	–	–	18.35	30	100	none	–	–
46 Passenger Ferry	–	–	12.86	30	46	none	–	–

Structure: The hull and superstructure are constructed of GRP.

Propulsion: Two sets of MWM TBD 234 V16 diesels, 755 kW at 2,200 rpm each driving a fixed-pitch water propeller via a Reintjes WVS 430 gearbox. Single MWM TBD 234 V8 diesel, 380 kW at 2,100 rpm driving two fans for lift air.

160 passenger speed ferry

Specifications

Length overall	27.2 m
Beam	7.5 m
Draught, on-cushion	1.2 m
Crew	5
Passengers	160
Max speed	28 kt
Range	200 n miles

Propulsion: Two DDEC12V-92TA diesels, 605 kW at 2,200 rpm each driving a Hamilton 422 waterjet via a ZF BW 250 gearbox.

100 passenger speed ferry

Specifications

Length overall	18.35 m
Beam	6.1 m
Draught, on-cushion	0.65 m
Passengers	100
Max speed	30 kt
Range	250 n miles

Propulsion: Two GM12V92TA diesels at 1,900 rpm each driving a BM195V gearbox.

46 passenger speed ferry

Specifications

Length overall	12.86 m
Beam	3.75 m
Passengers	46
Max speed	30 kt
Range	400 n miles

Qindao Beihai Shipbuilding Heavy Industry Co Ltd

369 Lijiong East Road, Qingdao Economic and Technical Development Zone, 266520, China

Tel: (+86) 53 67 56 187
Fax: (+86) 53 67 56 199
e-mail: bhbsns@public.qd.sd.cn
Web: www.bhshipyard.com.cn

This shipyard builds air cushion vehicles to the designs of the Marine Design and Research Institute of China. Beihai Shipyard employs over 2,960 staff and works to ISO 9000 Quality Control Systems.

A major new construction facility, occupying 2.5 million m², is currently being built in Haixiwan Bay, opposite Qingdao. The ship repair division is expected to be operational in 2005 and the shipbuilding sector by 2007.

Type 7217

A Type 7217 craft was completed at Bei Hai Shipyard in December 1993 and operates as a passenger ferry at Qiao-Zhou Gulf, near Qing Dao City. A second vessel was completed in October 1995.

MARIC Type 7217 sidewall hovercraft 0506946

Specifications

Length overall	44.0 m
Length waterline	42.1 m
Beam	8.3 m
Height	7.0 m
Weight, max	145 t
Passengers	260
Max speed	25 kt
Operational speed	23.5 kt
Range	250 n miles
Operational limitation	Sheltered water service

Structure: The main hull is constructed of steel with an aluminium superstructure.

Propulsion: There are two sets of MWM TBD 234 V16 diesels rated at 755 kW at 2,200 rpm each for propulsion, and a single MWM TBD 234 V16 rated at 755 kW at 2,150 rpm for lift power.

Vessel type	Vessel name	Yard No	Length (m)	Speed (kt)	Seats	Vehicles	Originally delivered to	Date of build
7217	–	–	44.0	23.5	–	none	–	December 1993
7217	–	–	44.0	23.5	–	none	–	October 1995

Qiuxin Shipyard

132 Jichang Road, Nanshi District, Shanghai 200011, People's Republic of China

Tel: (+86 21) 63 78 32 28
Fax: (+86 21) 63 77 21 00
e-mail: business@qiuxin.com.cn

Shanghai Qiuxin Shipyard is a member of the China State Shipbuilding Corporation and builds a range of specialist ships up to about 115 m in length, including lightweight high-speed craft.

In 1996, the company built a GRP 32 m air cushion vehicle.

32 m SES

This diesel driven, composite hull SES is designed for operation in the relatively sheltered waters between Hong Kong and mainland China.

32 m SES at speed 0081002

Specifications

Length overall	32.15 m
Beam	8.2 m
Draught	2.2 m
Vehicles	225
Max speed	32 kt
Operational limitation	30 kt
Range	200 n miles

Propulsion: Two MWM TBD 234V16 diesel engines each rated at 755 kW and each driving a fixed-pitch propeller. Lift fans are powered by one MWM TBD 234V8 diesel rated at 405 kW.

Vessel type	Vessel name	Yard No	Length (m)	Speed (kt)	Seats	Vehicles	Originally delivered to	Date of build
32 m SES	–	–	32.2	30.0	225	none	–	–

Finland

Aker Yards, Finland

PO Box 666, FI-20101, Turku, Finland

Tel: (+358 10)67 00
Fax: (+358 10) 670 67 00
e-mail: finland@akeryards.com
Web: www.akeryards.com

Yrjö Julin, *Managing Director*
Juha Heikinheimo, *Sales Director*

In January 2005, Finland's three largest shipyards combined to form one company with ferries being built at the Helsinki shipyard. The company also has a division specialising in Arctic technology.

Aker Yards Finland is part of the international shipbuilding group Aker Yards, Europe's largest and one of the world's four largest shipbuilders.

30 m hovercraft combat vessel

The 30 m hovercraft delivered to the Finnish Navy in June 2002 is a prototype for a future series of fast combat vessels.

Also, another prototype for the new Squadron 2000 was built by Aker Finnyards – the missile boat *Hamina,* delivered in August 1998, with the second vessel, *Tornio,* delivered in May 2003 and the third, *Hanko,* in June 2005. The future squadron is planned to be four ships of missile boat type. FAC *Hamina* is a follow up to four Rauma class missile boats, which were built in Rauma during 1990–92. The yard's long cooperation with the navy started in the 1950s.

Aker Finnyards fast 30 m hovercraft Tuuli*, built for the Finnish Navy* 1027198

The hovercraft contract involves hovercraft technology transfer to Finland, from the US, in connection with the offset arrangements. The fundamental air cushion vehicle technology tested in the US was used to help the designers in Finland to fulfil the requirements of the Finnish Navy for year-round operation in harsh coastal conditions.

The price of the vessel was over FIM70 million (EUR11,773,155). It was constructed from welded panels of thin marine aluminium sheets and extrusions connected with lightweight composite constructions.

The length of the vessel is 27.4 m and beam 15.4 m, the height of the skirt being over 2 m. The propulsion and hovering mechanism consists of four gas turbines, two driving, with belts, the two CP air propellers in the thrusters, and two driving the lift fans and the manoeuvring thrusters in the bow. The 4,500 kW engines give the vessel a maximum speed of 50 kt.

The wheel house and the cabins as well as the operation, service and machinery spaces are situated in the middle of the ship. The spaces for weaponry and fans are towards the sides. The main armament consists of missiles, mines or torpedoes. Anti-aircraft weaponry is also to be installed.

Specifications

Length overall	27.4 m
Beam	15.4 m
Weight, max	95 t
Crew	10
Propulsive power	6,000 kW
Max speed	50 kt

The special features of the vessel are good mobility, independence from waterways and fixed port equipment, year-round operation and, owing to the advanced technology, a crew of only 10.

France

DCN

2 rue Sextius Michel, F-75732 Paris Cedex 15, France

Tel: (+33 1) 40 59 50 00
Fax: (+33 1) 40 59 56 48
e-mail: info@dcn.fr
Web: www.dcn.fr

J M Poimboeuf, *Chairman and CEO*

DCN International consists of five main shipyards in Toulon, Cherbourg, Lorient and Brest in France and Papeete in Tahiti. The company builds a wide range of naval vessels including corvettes, frigates submarines and aircraft carriers.

Whilst DCN no longer manufactures air cushion vehicles, the following details are held for interest.

AGNES 200 (in Brighton to Dieppe route, Summer 1992) entering Dieppe harbour (Roger Fayolle, STCAN) 0506578

AGNES 200 (ex-NES 200)

The AGNES 200 has been developed within the framework of an interministry programme, in which the following are participating: the Ministry of Defence, the Ministry of Research and the Ministry of Industry. The vessel was delivered at the end of 1990.

Involved in conducting the programme is Direction des Recherches, Etudes et Techniques (DRET).

Industrial concerns involved are DCN, in respect of the project design, and ACH and CMN shipyards, in construction.

AGNES 200 is designed in the form of a basic air cushion platform for both civil and military versions and with an aft deck area of 208 m² is able to support a 4 t helicopter.

The major components (propulsion system, lift fans and so on), all use existing technology and have therefore required only a limited degree of development. The evaluation prototype, named AGNES 200, has been built at Constructions Mécaniques de Normandie (CMN) and was launched on 2 July 1990. Maximum speed achieved during French Navy trials was 45 kt. The vessel is also fitted with a Codod propulsion system allowing the lift engines to drive the waterjets. The maximum speed in this mode is 15 kt.

During low-speed transits it is possible to lift the bow and stern seals to the wet-deck and operate as a catamaran. Supplied by Zodiac Espace, the bow seal consists of a double loop and six segments, the stern seal is a more conventional triple loop design.

Vessel type	Vessel name	Yard No	Length (m)	Speed (kt)	Seats	Vehicles	Originally delivered to	Date of build
MOLNES	*MOLENES*	–	12.1	–	–	none	DRET	1980
AGNES 200	–	–	51.0	–	192	none	French Navy	2 July 1990

After 18 months with the French Navy the vessel entered passenger service on a 68 n mile route between Dieppe and Brighton Marina, operated by Advanced Channel Express. The main modifications to prepare the vessel for its ferry role were the refitting of the existing bow saloon and the installation of a passenger saloon on the helicopter deck. Configured for 78 passengers in the bow saloon and 93 in the aft saloon, there was also a lounge on the upper deck with 12 first class seats for guests of the operator.

While on the English Channel the vessel had a crew of 12 comprising captain, mate, chief engineer, three sailors and six cabin crew.

Specifications

Length overall	51.0 m
Beam overall	13.0 m
Draught, hullborne	2.3 m
Draught, on-cushion	1.0 m
Displacement, max	250 t
Propulsive power	2 × 2,983 kW
Lift power	2 × 746 kW
Max speed	over 40 kt
Range	750 n miles

Propulsion: Main propulsion: two 2,983 kW MTU 16V 538 TB 93 diesels driving two Kamewa 71 S II water-jet units.

Lift: Two 746 kW MTU 8V 396 TB 83 diesels each driving one NEU centrifugal fan via a Renk two-stage reduction gearbox, 180,000 m³/h at 600 kg/m². Off-cushion, power transferable to the two Kamewa water-jet units.

MOLENES (Modèle Libre Expérimental de Navire à Effet de Surface)

In 1980, DCN built under contract from DRET a 5 tonne craft, which had its trials in 1981 in the Toulon area. MOLENES is a dynamic manned model capable of proving the NES (Navire à Effet de Surface) concept of a high length-to-beam ratio, and confirmed results obtained in the experimental tank. It has provided data on seaworthiness, performance and manoeuvrability, as well as acceleration levels and their effects on the structure, equipment and fittings.

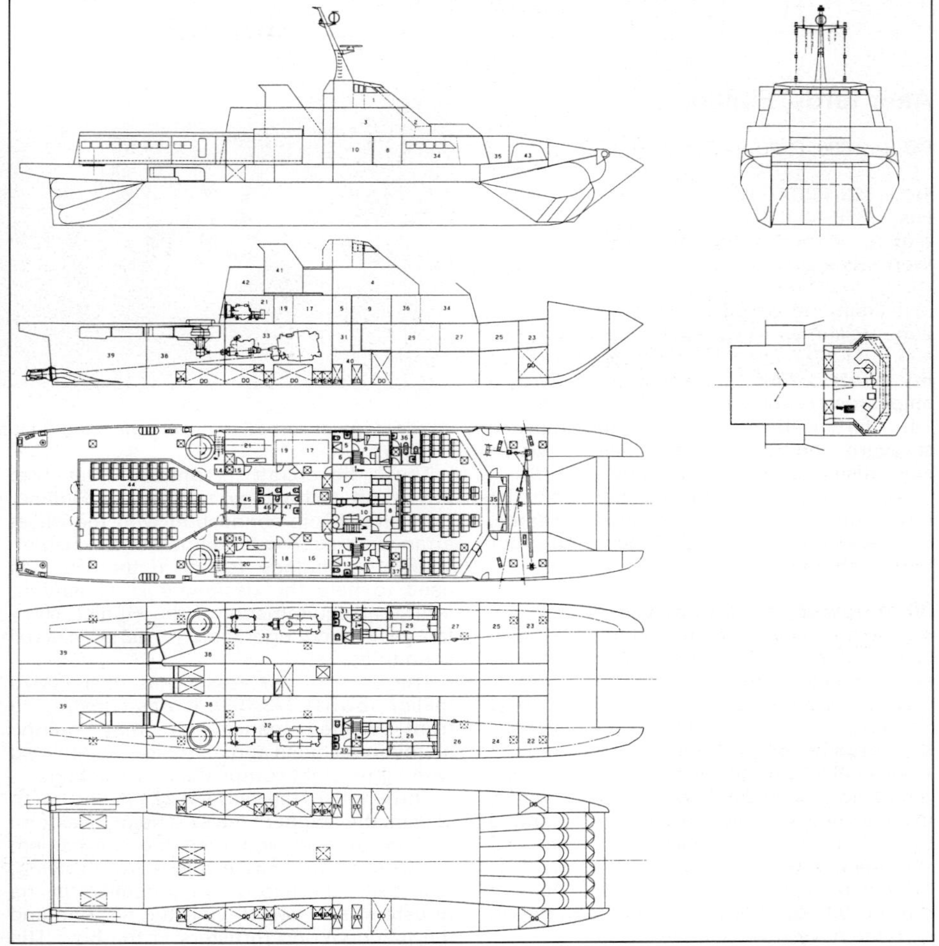

AGNES 200 in passenger ferry configuration 0506579

Specifications

Length overall	12.1 m
Beam	3.4 m
Displacement, max	5.5 t

Propulsion: Propulsion is provided by two 55 hp water-jet units. The air cushion is effected by two centrifugal fans on vertical axes, giving a pressure of 1,800 Pa (37.6 lb/ft²) and a flow of 4 m³/s (141 ft³/s).

Structure: The timber side keels are joined together by a tubular pyramid structure of light aluminium alloy 7020.

Japan

Mitsubishi Heavy Industries Ltd

16-5 Konan 2-chome, Minato-ku, Tokyo 108-8215, Japan

Tel: (+813) 67 16 31 11
Fax: (+813) 67 16 58 00
e-mail: ueda@ship.hq.mhi.co.jp
Web: www.mhi.co.jp

Naoki Ueda, *Manager*

Mitsubishi Heavy Industries has three main shipyards in Japan, Kobe, Nagasaki and Shimonoseki, with an additional dockyard at Yokohama. The company employs a total of over 10,000 people in all four facilities.

Mitsubishi Heavy Industries was a joint partner in the technological research association of Techno-Superliner. The company developed its own Surface Effect Ship (SES) technology with the construction of an 18.5 m SES in 1989.

The research and development project on SES at the technical Research and Development Institute (TRDI) of Japan Defence Agency started in 1976. From 1981, the project was carried out in collabration with Mitsubishi Heavy Industries Ltd (MHI) and Mitsui Engineering and Shipbuilding Company Ltd.

18.5 m SURFACE EFFECT SHIP

Méguro 2

In August 1989, the 18.5 m SES test craft *Méguro* was built at MHI Shimonoseki Shipyard and delivered to theTRDI, in order to verify the technological results in the research and development process.The technical trials completed successfully in March 1991.

Méguro 2 is a conversion of *Méguro* and was delivered to theTRDI in September 1994. The hull length was increased to 25 m and the existing prime movers, IHI IM-100, replaced by the Allison 501-KF gas-turbine engines. Following this conversion the craft achieved speeds of over 60 kt in September 1994 and further tests are currently under way.

In August 1989, the 18.5 m SES test craft *Méguro* was built at MHI Shimonoseki Shipyard and delivered to theTRDI, in order to verify the technological results in the research and development process. The technical trials completed successfully in March 1991.

Specifications

Length overall	25.0 m
Beam	7.5 m
Draught, hullborne	1.3 m
Draught, on-cushion	0.4 m
Max speed	60+ kt

Propulsion: The vessel is propelled by two Allison 501-KK gas turbine engines, each rated at 3,185 kW and each driving a water-jet via a reduction gearbox. Lift is supplied by six centrifugal fans, driven by two 1M100-1H gas turbines each rated at 772 kW.

74 m SURFACE EFFECT CRAFT

Kibo

Kibo is the converted Hisho which was the large-scale model of the Techno-Superliner, and was delivered to Shiznoka prefecture in 1997. Passenger accommodation and ro-ro facilities were added for high-speed ferry services.

Vessel type	Vessel name	Yard No	Length (m)	Speed (kt)	Seats	Vehicles	Originally delivered to	Date of build
18.5 m Surface Effect Ship	*Meguro 2* (ex-*Meguro*)	–	25.0	–	–	none	Japanese Defence Agency	August 1989
74 m Surface Effect Craft	*Kibo* (ex-*Hisho*)	–	74.0	–	260	30	Technological Research Association of Techno-Superliner	June 1994

Méguro 2 *during trails* 0506961

Kibo *(ex-*Hisho*)* 0024741

Specifications

Length overall	74.0 m
Beam	18.6 m
Draught	3.5 m
Passengers	260
Vehicles	30 cars
Propulsive Power	24,000 kW
Maximum speed	45.2 kt

Propulsion: The vessel is propelled by two gas turbines, each driving a water-jet. The lift system is powered by four 1,500 kW diesels driving eight centrifugal lift fans.

Mitsui Engineering & Shipbuilding Company Ltd

6-4 Tsukiji 5-chome, Chuo-ku, Tokyo 104-8439, Japan

Tel: (+81 3) 35 44 31 47
Fax: (+81 3) 35 44 30 50
e-mail: prdept@mes.co.jp
Web: www.mes.co.jp/

Superliner Dept
Tel: (+81 3) 35 44 34 62
Fax: (+81 3) 35 44 30 31
e-mail: tsl@mes.co.jp

Tamiyoshi Iwasaki, *Managing Director and General Manager, Ship and Ocean Project Headquarters*
Kiyoshi Hara, *Deputy Director and Deputy General Manager, Ship & Ocean Project Headquarters*
Mitsuru Kawamura, *General Manager, Governmental Ship and High-Speed Ship Sales Departments*
Kimio Uchida, *Manager, Governmental Ship and High-Speed Ship Sales Departments*

Mitsui Engineering & Shipbuilding Company Ltd has two main shipyards and a dockyard in Japan, at Chiba, Tamano and Yura respectively. Both the Chiba and Tamano shipyards have built lightweight high-speed craft.

Mitsui's Hovercraft Department was formed in May 1964, following the signing of a licensing agreement in 1963 with Hovercraft Development Ltd and Vickers Ltd, whose ACV interests were later merged with those of British Hovercraft Corporation. The company has been developing Mitsui hovercraft independently after terminating the licensing agreement in March 1986. The Mitsui ACVAS and MV-PP10 represent such developments. The company has built two MV-PP1s, 19 MV-PP5s, four MV-PP15s, two MV-PP05s, three MV-PP10s and two ACVAS.

Kibo *converted from the TSL-A70* Hisho 0007621

ACVAS (ACV with Aft Skegs)

A new form of hovercraft was developed by the Mitsui company in the early 1980s, with trials of the 10 m prototype *Eaglet* started in early 1986. The craft employs water-jet propulsion and is therefore not amphibious. The inlets for the two water-jet units are positioned in the underside of the two skegs which extend either side of the craft for approximately one-third of overall craft length. Over this length immersed areas of the skegs seal the air cushion; for the remainder of the cushion periphery a conventional loop and segment type skirt is employed. Unlike the sidewall hovercraft or surface effect ships, the craft is almost totally supported by its air cushion at cruising speed. The craft is provided with a Mitsui motion control system exerting control over cushion air pressure variation and is fitted with fin stabilisers. The skirt/skeg combination is also found to give good ride comfort over waves through its soft response.

The Mitsui ACVAS concept is aimed principally towards applications in shallow rivers, lakes and other smooth waters and for ultra-fast ferries for operation in coastal and inland sea routes.

Eaglet

Specifications

Length overall	10.9 m
Beam	5.1 m
Propulsive power	2 × 75 kW
Lift power	16 kW
Maximum speed	27 kt

Structure: GRP.
Propulsion: Main engines: two Nissan HA 120 (petrol).
Lift engine: two Robin petrol.
Water-jet: two Hamilton 771 units.

Sumidagawa

In August 1988, Mitsui received an order from the Tokyo Metropolitan Government Bureau of Construction for a river

Vessel type	Vessel name	Yard No	Length (m)	Speed (kt)	Seats	Vehicles	Originally delivered to	Date of build
11 m ACVAS	*Eaglet*	–	10.9	–	–	none	–	1986
20 m ACVAS	*Sumidagawa*	–	19.9	28.5	80	none	Tokyo Metropolitan Bureau of Construction	March 1989
MV-PP10	*Dream Aquamarine* (ex-*Dream No 1*)	–	23.1	45.0	125	none	Oita Hoverferry Company Ltd	March 1990
MV-PP10	*Dream Emerald* (ex-*Dream No 2*)	–	23.1	45.0	125	none	Oita Hoverferry Company Ltd	March 1991
MV-PP10	*Dream Ruby* (ex-*Dream No 3*)	–	23.1	45.0	125	none	Oita Hoverferry Company Ltd	October 1995
MV-PP10	*Dream Sapphire*	–	23.1	45.0	125	none	Oita Hoverferry Company Ltd	April 2002
Techno Superliner	*Super Liner Ogasawara*	1560	140.0	39.0	740	none	Techno-Seaways Inc	2005

observation craft that can be used for observation and inspection of rivers in the Metropolis. The craft is fitted out in a manner suitable for international conferences to be held on board. This ACVAS vessel, named *Sumidagawa*, was delivered in March 1989.

Specifications

Length overall	19.9 m
Beam	7.9 m
Draught, hullborne	1.5 m
Draught, on-cushion	0.5 m
Passengers	52–80
Propulsive power	2 × 346 kW
Lift power	2 × 272 kW
Max speed	30.4 kt
Operational speed	28.5 kt

Structure: The vessel structure is made of anti-corrosive aluminium. Upper structure is of plate construction consisting of welded and riveted antiflexure material.

A flexible skirt made of rubberised nylon cloth is provided around the vessel excepting the skegs at the aft. The front part of the skirt system consists of a bag and fingers and the rear part has lobe seal construction. Air is supplied to the lobe seal from the bag at the front through a duct. The forward skirt is designed to minimise water spray for ease of navigation in rivers.

Propulsion: Two high-speed diesel engines are used for propulsion. The power is transmitted to two water-jet units through reduction gears. One high-speed diesel engine drives two lift fans, DC generator and cooler compressors through clutch and flexible joint.

MV-PP10

The most recent hovercraft developed by Mitsui is the MV-PP10, which can accommodate 84 to 100 passengers and has a maximum speed of 50 kt. This hovercraft embodies the technological progress of the hovercraft developed by MES over the years and has been designed for low cost, easy maintenance and a comfortable ride.

The main hull of the craft is constructed of marine grade aluminium alloy as other MES high-speed craft. The MV-PP10 is fitted with a 1.2 m high flexible bag-and-finger skirt of improved durability and is powered by air-cooled, high-speed diesel engines. The vessel is steered by twin air rudders and air-bleed thruster ports, one forward and two aft.

The MV-PP10 can also be fitted out for such purposes as surveying and rescue operations.

Four MV-PP10 type craft, *Dream Aquamarine* (ex-*Dream No 1*), *Dream Emerald* (ex-*Dream No 2*), *Dream Ruby* (ex-*Dream No 3*) and *Dream Sapphire* were delivered in March 1990, March 1991, October 1995 and April 2002 respectively to the Oita Hoverferry Company Ltd.

Specifications

Length overall	23.1 m
Beam	11.0 m
Weight, max	40.0 t
Payload	9.0 t
Crew	3
Passengers	125
Propulsive power	2 × 441 kW
Maximum speed	50 kt
Operational speed	45 kt

Structure: The main hull of the craft is built in weldable anti-corrosive aluminium alloy and the superstructure is of riveted construction.

Propulsion: The MV-PP10 is powered by four Deutz BF12L513CP air-cooled turbocharged diesels, each rated at 441 kW (600 hp) at 2,300 rpm maximum and 383 kW (520 hp) continuous. On each side of the craft, one engine drives two double-entry mixed flow fans for lift and two engines aft drive two ducted variable-pitch propellers.

Manned test craft model TSL-A12 0506586

The TSL-A70 Hisho 0506866

MV-PP5 Mk II 0506582

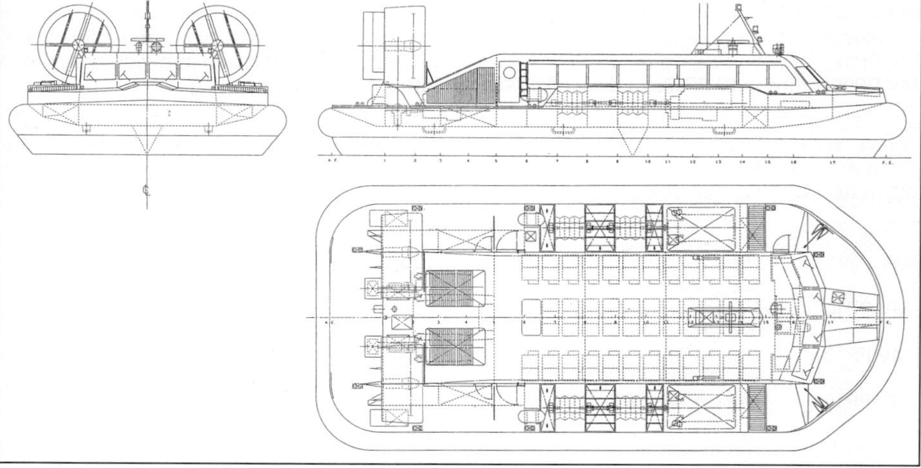
MV-PP10 general arrangement 0506585

MV-PP10 Dream No 3 0098430

Techno-Superliner Ogasawara 1176292

Techno-Superliner

The R&D programme for a very high-speed vessel named *Techno-Superliner* was initiated in Japan in 1989 involving seven leading Japanese shipbuilders. A target of the programme was to accomplish the technological foundation for developing and building a high-speed ocean-going cargo vessel with a speed of 50 kt, a payload of 1,000 tonnes and a range of 500 n miles. In addition, seaworthiness adequate to meet schedules in rough seas was desired.

The first phase of the programme ran from 1989 to 1996, throughout which the technological foundation for Techno-Superliner was established through analysis, experiments and tests. The last year of the programme consolidated these technologies by scale model tests in actual seas. The second phase of the programme ran between 1996 and 2002 when further operational experience was gained from the prototypes and the full size vessel was planned.

The third phase was programmed to run between 2002 and 2005 in which the full-scale vessel would be designed, built and operated. However, plans were cancelled because a large enough subsidy was not available from the Tokyo Metropolitan Government.

Kibo (ex-TSL-A70 Hisho)

The large scale sea model named TSL-A70 *Hisho* was constructed separately at two major shipyards. The fore part was constructed at Mitsui Tamano Shipyard, and the aft at Mitsubishi Nagasaki. These two blocks were connected at Nagasaki and the vessel was launched in June 1994. A maximum speed of 54.4 kt was achieved at the official speed trial. The TSL-A70 *Hisho* completed sea transportation trials in 1995. In 1997 the vessel was converted into a passenger ferry, *Kibo* at Mitsubishi's Nagasaki shipyard, to carry 260 passengers and 30 vehicles.

Specifications

Length overall	70.0 m (*Hisho*) 74.0 m (*Kibo*)
Beam	18.6 m
Draught, hullborne	3.5 m
Draught, on-cushion	1.1 m
Propulsive power	2 × 12,000 kW
Lift power	4 × 1,500 kW
Maximum speed	54.4 kt (*Hisho*) 45.2 kt (*Kibo*)
Range	500 n miles

Structure: The hull material is aluminium.

Propulsion: There are two propulsion engines, driving a water-jet each.

The lift engines are four 1,500 kW diesels driving eight centrifugal lift fans.

The skirt system is: full finger (bow), lobe (stern).

The Mitsui Sumidagawa *in operation with the Tokyo Metropolitan Government Bureau of Construction* 0506583

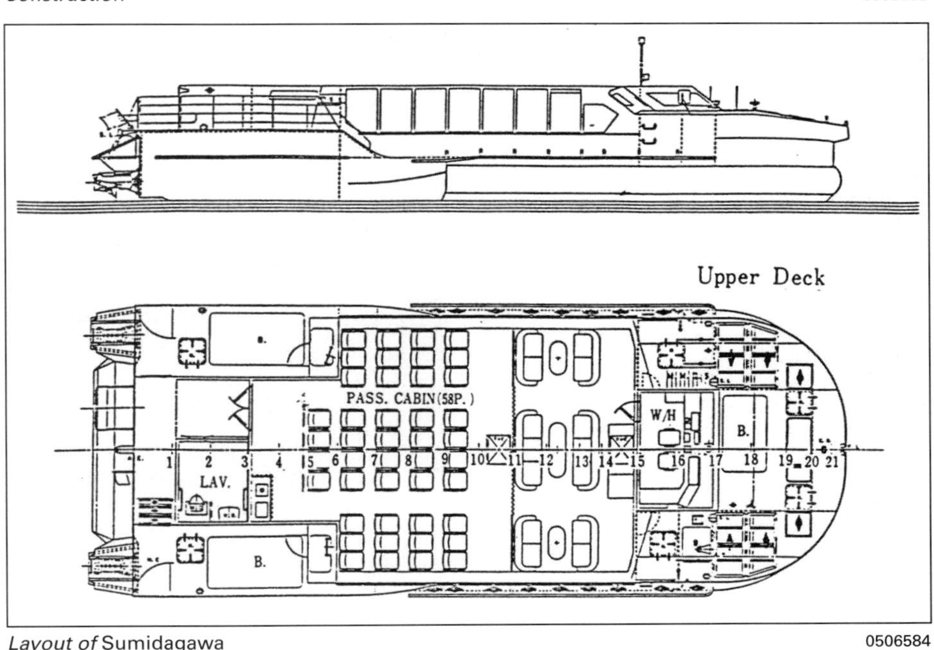

Layout of Sumidagawa 0506584

TSL-A140

In October 2005, Mitsui completed sea trials of the super high-speed passenger/cargo TSL (Techno Super Liner). Trials confirmed that all performance and specification requirements had been successfully met. The vessel, the longest aluminium craft in the world, achieved a maximum speed of 42.8 kt in wave heights of 2 m. This vessel had been ordered in 2003 by Techno-Seaways for operation by Ogasawara Kaiun Co Ltd in Japan. Delivery was made in 2005 for operation on the Tokyo to Ogasawara route, reducing the current crossing time of 26 hours to 17 hours. However, Techno-Seaways have announced that due to financial constraints they will be unable to operate this vessel.

Specifications

Length overall	140.0 m
Beam	29.8 m
Payload	210 t
Passengers	740
Propulsive power	2 × 25,180 kW
Maximum speed	42.8 kt
Range	1,200 n miles

Structure: The hull material is aluminium.

Propulsion: The propulsion engines are two 25 MW gas turbines, each driving a Rolls-Royce 23552 water-jet.

The lift engines are four high-speed diesel engines.

The skirt system is understood to be: full finger (bow), lobe (stern).

Korea, South

Hanjin Heavy Industries Shipbuilding Division

Head Office
29, 5-Ga, Bongnae-dong, Youngdo-ku, Pusan 606-796, South Korea

Special Ship Division
Tel: (+82 2) 20 06 71 14
Fax: (+82 2) 20 06 71 15
e-mail: kimjys@hanjinsc.com
Web: www.hanjinsc.com

J W Lee, *Special Support Sales Team*
Joo Young Kim, *Special Support Sales Team*
Hyeong Do Ki, *Special Support Sales Team*

KTMI 26 m SES 0506587

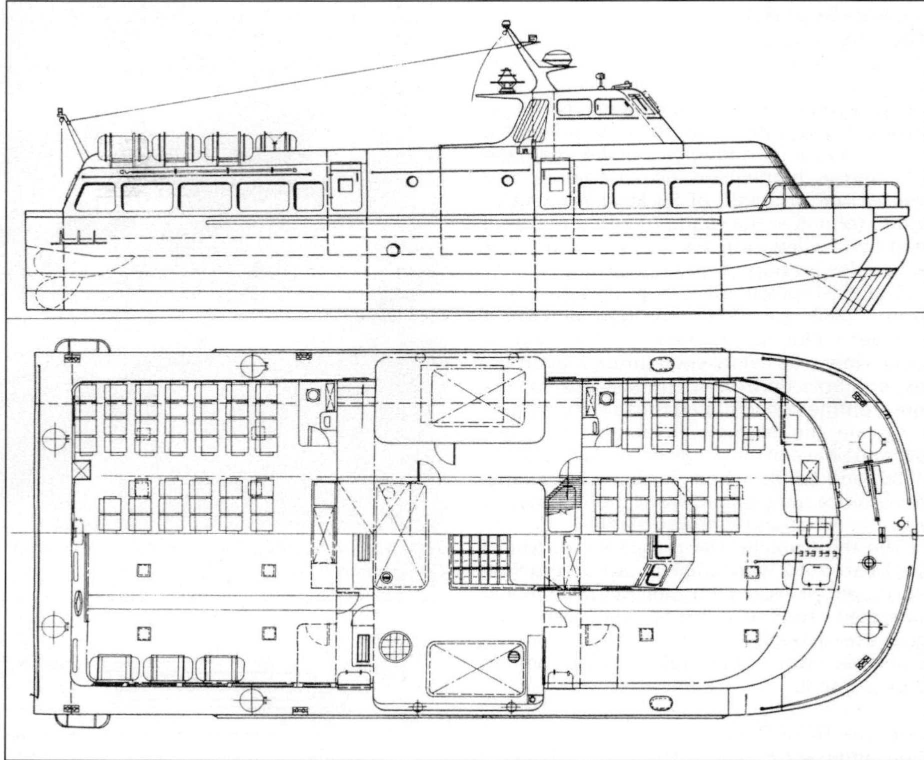

KTMI 26 m SES 0506588

In 1999 Hanjin merged with Korea Tacoma Marine Industries who had begun an ACV development programme in 1977 and designed and constructed a test surface effect ship in 1978.

Hanjin have consolidated the range offered to five ACV and SES designs aimed at military and commercial customers.

12 m hovercraft

Specifications

Length overall	12.7 m
Beam	7.0 m
Height, on-cushion	4.7 m
Weight, min	8.1 t
Passengers	5
Propulsive power	319 kW
Max speed	55 kt
Range	200 n miles

Propulsion: Power for the lift system is provided by a single 242 kW marine diesel.

14 m hovercraft

Specifications

Length overall	13.4 m
Beam	7.4 m
Height, on-cushion	3.9 m
Weight, min	8.5 t
Payload	2.0 t
Passengers	10
Propulsive power	2 × 203 kW
Max speed	50 kt
Range	150 n miles

Propulsion: Power for the lift system is provided by a single 203 kW marine diesel engine.

18 m SES

The 18 m design has buoyant catamaran-type sidewalls almost identical in shape to those of the smaller craft. The vessel is steered by twin water rudders and differential propeller thrust at low speeds.

Specifications

Length overall	18.1 m
Beam	7.4 m
Draught, on-cushion	1.3 m
Weight, min	27 t
Passengers	max 90
Propulsive power	2 × 380 kW
Max speed	35 kt
Range	150 n miles

Propulsion: Power for the lift system is provided by a single 194 kW marine diesel engine.

Turt IV Mk 2 0506589

26 m assualt hovercraft

Specifications

Length overall	26.5 m
Beam	13.8 m
Height, on-cushion	7.8 m
Weight, min	58 t
Payload	21 t
Passengers	100
Crew	8
Propulsive power	2 × 1,865 kW
Max speed	55 kt
Range	300 n miles

Propulsion: Power for the propulsion system is provided by twin gas turbines of 1,865 kW. Power for the lift system is provided by twin 1,865 kW gas turbines.

27 m SES

Specifications

Length overall	27.5 m
Beam	10.2 m
Height, on-cushion	2.7 m
Weight, min	58 t

Vessel type	Vessel name	Yard No	Length (m)	Speed (kt)	Seats	Vehicles	Originally delivered to	Date of build
12 m ACV	–	–	12.7	55	5	none	–	–
14 m ACV	–	–	13.4	50	10	none	–	–
18 m SES	–	–	18.1	35	90	none	–	–
26.5 m Assault Hovercraft	–	–	26.5	55	100	none	–	–
27 m SES	–	–	27.5	31	200	none	–	–

Passengers	200
Propulsive power	2 × 753 kW
Max speed	31 kt
Range	250 n miles

Propulsion: Power for the propulsion system is provided by twin diesels of 753 kW each. Power for the lift system is provided by a single 425 kW marine diesel.

Craft previously built by Korea Tacoma Marine Industries include:

Turt III

18 m SES

The 18 m design has buoyant catamaran-type sidewalls almost identical in shape to those of the smaller craft. The vessel is steered by twin water rudders and differential propeller thrust at low speeds.

Specifications

Length overall	18.1 m
Beam	9.0 m
Draught, hullborne	1.7 m
Draught, on-cushion	1.1 m
Weight, max	36 t
Passengers	max 90
Propulsive power	2 × 485 kW, 596 kW or 970 kW
Max speed	35, 40 or 50 kt

Propulsion: Power for the propulsion system is provided by twin diesels of 485, 596 or 970 kW, each driving a water propeller via a reversing gearbox and an inclined shaft. Power for the lift system is provided by a single marine diesel in the 298 to 373 kW range.

Structure: Main structure built in welded marine aluminium alloy. Segmented skirt at the bow and stern.

26 m SES

This 26 m craft has a single bag multisegment bow skirt and a multibag stern skirt. The stern skirt is inflated by two secondary 0.61 m diameter fans powered by an hydraulic system. The craft is normally steered by two water rudders and differential propeller thrust at low speeds.

Specifications

Length overall	25.7 m
Beam	10.2 m
Draught, hullborne	2.5 m
Draught, on-cushion	1.5 m
Displacement, max	65 t
Payload	16.5 t
Crew	10
Passengers	158
Propulsive power	2 × 755 kW
Lift power	380 kW
Max speed	35 kt
Range	250 n miles

Structure: The main structure is built in welded marine aluminium alloy and the superstructure is constructed in riveted marine aluminium alloy.

Propulsion: The lift system is powered by a single General Motors Detroit Diesel Allison 12V-71TI. This engine directly drives a dual 1.075 m diameter lift fan, to provide cushion air to the plenum chamber and, via toothed belts and a hydraulic system, two secondary 0.61 m diameter fans for stern skirt inflation. Propulsive power is supplied by two MTU 8V 396 TB 83 diesels, each driving a water propeller via a reversing gearbox and an inclined shaft.

Specifications

Length overall	12.7 m
Beam	7.0 m
Weight, min	8.1 t
Payload	1.5 t
Propulsive power	315 kW
Lift power	240 kW
Max speed	55 kt
Operational speed	40 kt
Range	200 n miles
Operational limitation	wave height 2 m

Structure: The main hull is built in welded aluminium alloy 5086-H116 (plate) and 6061-T6 (extrusion), and the deckhouse is riveted.

Propulsion: The lift system comprises two single 1.075 m fans driven by a Deutz air-cooled diesel engine. Thrust is supplied by a Deutz air-cooled diesel engine driving a 2.75 m ducted propeller.

KTMI 28 m SES (Modified 26 m SES) 0506590

Turt IV Mk 1 0506591

KTMI 17 m SES Que-Ryong II 0506592

Vessel type	Vessel name	Yard No	Length (m)	Speed (kt)	Seats	Vehicles	Originally delivered to	Date of build
18 m SES	*Sun Star*	–	18.1	–	60	none	Kumsan Hungup	–
26 m SES	*Duridoongsil* (ex-*Young-Kwang I*)	–	25.7	–	158	none	Semo Marine Co Ltd	–
28 m SES (mod)	*Dudoongsil* (ex-*Young-Kwang II*)	–	28.0	–	200	none	Semo Marine Co Ltd	–
18 m SES	*Air Ferry*	–	18.1	–	90	none	Sea-Kyung Ferry Co	–
18 m SES	*Cosmos*	–	18.1	–	60	none	Geo-Je Ferry Co	–
TURT IV Mk 2	–	–	15.3	35.0	–	none	–	–
TURT IV Mk 2	–	–	15.3	35.0	–	none	–	–
TURT IV Mk 2	–	–	15.3	35.0	–	none	–	–
18 m SES	*Phinex*	–	18.1	–	60	none	Geo-Je Development Co	–
26 m SES	*Soonpoong* (ex-*Tacoma III*)	–	25.7	–	158	none	Semo Marine Co Ltd	–
Turt IV	–	–	12.7	40.0	–	none	Korean Government	December 1984
TURT IV Mk 1	*Jun-Nam 540*	–	13.3	40.0	–	none	The Ministry of Health and Social Affairs	April 1988
17 m SES	*Que-Ryong II*	–	17.3	–	72	none	Kang Won Hungup	September 1988
TURT V	*Prototype*	–	26.5	50.0	–	none	–	November 1989
40 m SES	*DESIGN*	–	40.0	–	336	none	–	–

Turt IV Mk 1

As a sea ambulance version, Turt IV Mk 1 is intended for quick transportation of patients from islands to land. The distinctive features of the craft, compared with Turt IV, are the twin propulsion units for increased manoeuvrability and the reduced lift power due to the developed skirt. The skirt is an open loop and segment type with anti-bouncing ties built into the loop. Turt IV Mk 1 was constructed in 1987 and was delivered in April 1988.

Specifications

Length overall	13.3 m
Beam	7.4 m
Weight, min	11.0 t
Payload	2.0 t
Propulsive power	2 × 203 kW
Lift power	203 kW
Max speed	50 kt
Operational speed	40 kt
Range	150 n miles
Operational limitation	wave height 2 m

Structure: The main hull is built in welded aluminium alloy 5086-H116 (plate) and 6061-T6 (extrusion), and the deckhouse is riveted.

Propulsion: The lift system consists of two single 1.07 m diameter fans driven by a Deutz air-cooled diesel engine rated at 203 kW at 2,300 rpm. Thrust is supplied by two 205 kW Deutz air-cooled diesel engines at 2,300 rpm driving two 2.0 m diameter ducted propellers.

Turt IV Mk 2

This craft, of which three have been built, is a stretched version of the Turt IV Mk 1. Lift and propulsion systems are identical to those of the Turt IV Mk 1, but the hull has been lengthened by 1.75 m and propulsion engines have been upgraded from 203 kW (Deutz) to 254 kW to improve the performance in rough seas.

Specifications

Length overall	15.3 m
Beam	7.6 m
Weight, min	15.0 t
Payload	1.3 t
Propulsive power	2 × 254 kW
Lift power	203 kW
Max speed	40 kt
Operational speed	35 kt
Range	150 n miles
Operational limitation	wave height 2.0 m

Turt V

The prototype of the Turt V was constructed at the end of 1989.

Turt V is a multipurpose amphibious vehicle capable of operating in open sea and confined shallow waterways including swamps, scrub land, sandbanks and mudflats.

Specifications

Length overall	26.5 m
Beam	13.8 m
Weight, max	95.0 t
Payload	30.0 t
Propulsive power	2 × 1,864 kW
Lift power	2 × 1,211 kW
Max speed	60 kt
Operational speed	50 kt
Range	300 n miles

Structure: The main hull and the superstructure are built in welded aluminium alloy 5086-H116 (plate) and 6061-t6 (extrusion).

The skirt is an open loop and segment type with anti-bouncing webs built into the loop.

Propulsion: The lift system consists of six double inlet 1.07 m diameter fans driven by two gas turbines rated at 1,211 kW each. Thrust is supplied by two 1,864 kW gas turbines driving two 3.6 m diameter ducted controllable-pitch air propellers.

17 m SES

The first 17 m craft *Que-Ryong II* is a high-speed waterbus which was delivered and entered service on So-Yang man-made lake near Seoul in September 1988. The 17 m SES is designed as a very reliable, cost-effective, high-speed waterbus with good

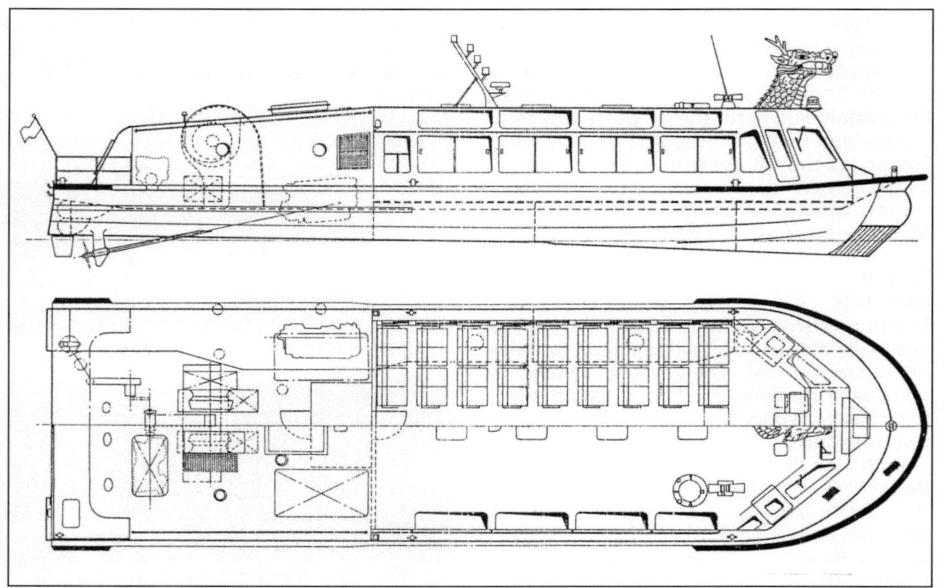

General arrangement of 40 m SES 0506593

Artist's impression of 40 m SES 0506867

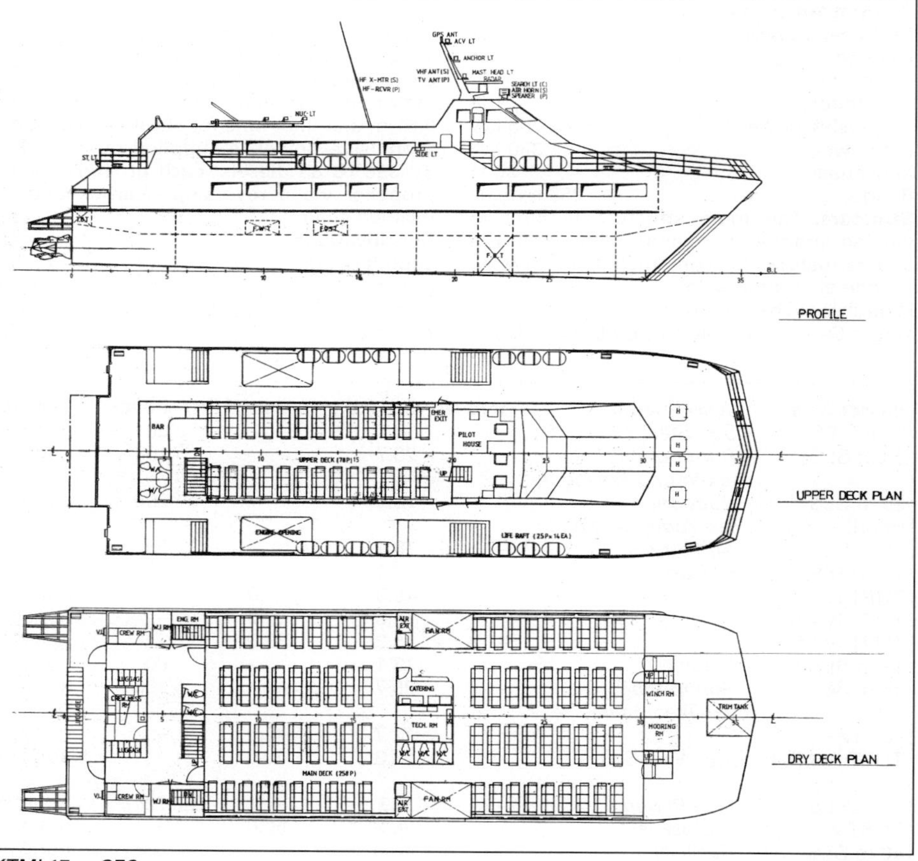

KTMI 17 m SES 0506826

transportation efficiency for operating on inland waters such as lakes and rivers. Good ride quietness and comfortable passenger space enhance this craft.

Specifications

Length overall	17.3 m
Beam	5.3 m
Crew	3
Passengers	72
Propulsive power	2 × 217 kW
Lift power	127 kW
Weight, max	21.5 t
Max speed	30.0 kt

Structure: The hull is built in welded marine grade aluminium alloy and the deckhouse is constructed in riveted marine grade aluminium alloy.

The bow and stern skirt consists of single-bag and multisegments which are attached to the bag and connected to the underside of the hull by straps.

Propulsion: Power for the lift fan is supplied by a single VOLVO PENTA TAMD 41A marine diesel engine. Propulsion engines are two VOLVO PENTA TAMD 71A marine diesels.

40 m SES (Design)

KTMI's 40 m SES was designed as a high-performance passenger craft.

Hull material is marine aluminium alloy rather than GRP for higher reliability and easier maintenance. Noise and vibration levels have been minimised and a ride control system is adopted on this craft.

Specifications

Length overall	40.0 m
Beam	12.0 m
Draught, hullborne	2.5 m
Draught, on-cushion	0.9 m
Passengers	336
Propulsive power	2 × 2,000 kW
Lift power	2 × 410 kW
Max speed	50 kt
Range	300 n miles

Samsung Heavy Industries Co Ltd

Headquarters:
17th Floor, Samsung Yeoksam Building, 647-9 Yeoksam-Dong, Gangnam-Gu, Seoul 135-080, South Korea

Tel: (+82 2) 34 58 60 00; 73 12
Fax: (+82 2) 6458 7200
e-mail: inseob.nam@samsung.com
Web: www.shi.samsung.co.kr

Shipyard:
530 Jangpyeong-ri, Sinhyeon-up, Geoje, Kyeongsangnam-do 656-717, South Korea

Tel: (+82 55) 630 31 14
Fax: (+82 558) 630 53 65
e-mail: inseob.nam@samsung.com

Jing-Wan Kim, *President*
Donald Mathias, *Senior Director, Passenger Ship Marketing*
Young Ryeol Joo, *Director, Passenger Ship Design*
Mun-kun Ha, *Senior Manager, Shipbuilding & Plant R&D Centre*

Dong Yang Gold 0506868

Samsung Heavy Industries Co Ltd was established in 1974 and operates three very large production facilities in South Korea.

The Koje shipyard builds various types of merchant ships and offshore structures, large-scale processing facilities and industrial machinery.

Since the shipbuilding business commenced in 1977, SHI has successfully built various types of vessels such as VLCCs, tankers, product carriers, full container vessels and bulk carriers.

In the area of high-speed craft, SHI has been performing research and development since 1991. In addition to conventional catamarans and monohull vessels, SHI has also developed and designed an SES-type high-speed passenger car ferry.

The SES project was started in 1992 and its first passenger vessel, *Dong Yang Gold*, was delivered to a domestic owner in May 1994.

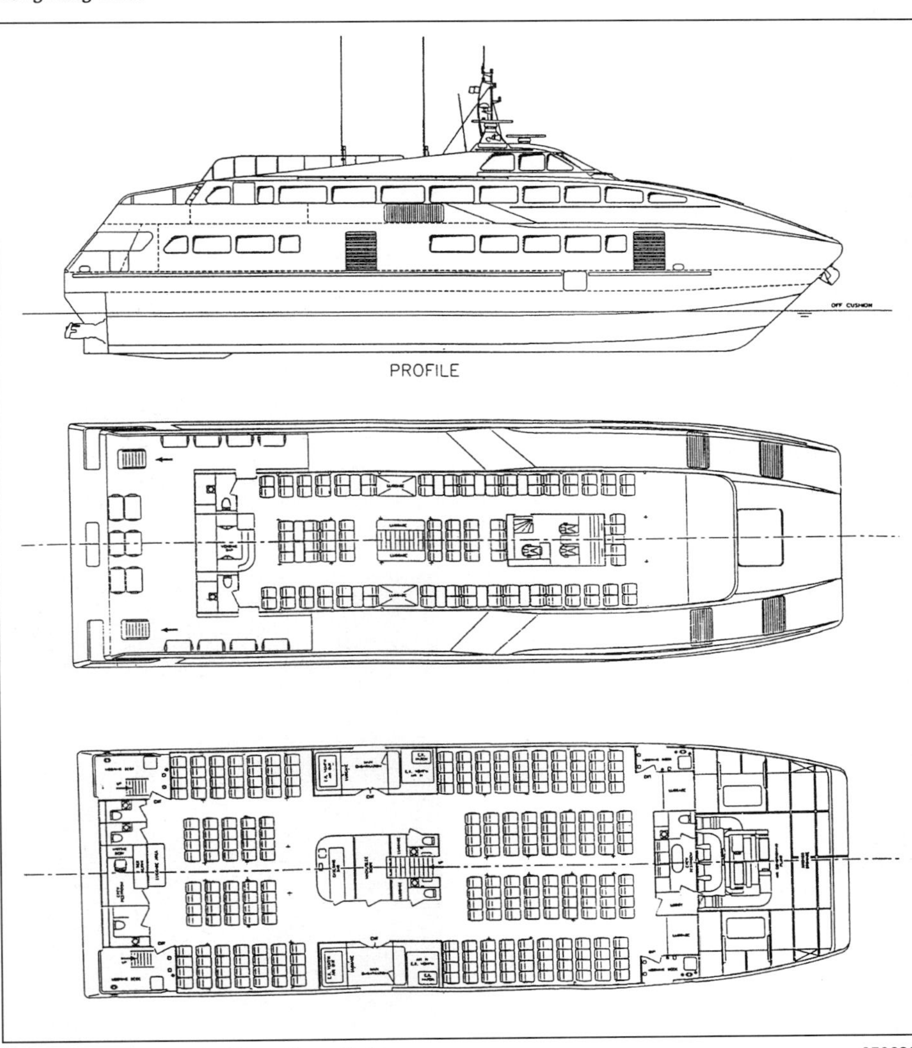

General arrangement of Dong Yang Gold 0506869

37 m SES

Dong Yang Gold

Ordered by Dong Yang Express Co Ltd, the 37 m SES vessel *Dong Yang Gold* was delivered in May 1994 and entered service on a route off the west coast of the country linking Mokpo with the islands of Heuksan-do and Hong-Do. In order to maximise passenger comfort, a ride control system is

Vessel type	Vessel name	Yard No	Length (m)	Speed (kt)	Seats	Vehicles	Originally delivered to	Date of build
37 m SES	*Dong Yang Gold*	–	36.5	45.0	352	none	Dong Yang Express Co Ltd	May 1994
80 m SES passenger car ferry	DESIGN	–	79.8	45.0	500	75	–	–

fitted to the craft, venting the main cushion overpressure.

Specifications

Length overall	36.5 m
Beam	12.2 m
Draught, hullborne	2.2 m
Draught, on-cushion	0.8 m
Crew	7
Passengers	352
Fuel capacity	7.2 t
Water capacity	0.8 t
Propulsive power	2 × 2,000 kW
Lift power	2 × 405 kW
Max speed	50 kt
Operational speed	45 kt
Range	250 n miles

Classification: Korea Register of Shipping.

Structure: FRP single skin in main hull, FRP sandwich in deck and superstructure.

Propulsion: Main engines: two MTU 16V 396 TE74L diesel engines, each driving a Kamewa 63SII water-jet.

Lift engine: two MTU 8V 183TE72 diesels.

80 m SES passenger car ferry (DESIGN)

SHI's 80 m SES was developed as a high-speed passenger car ferry in January 1994. A decision was made to select an SES, primarily because of the proven savings in power and cost, and its high development potential. The vessel is designed to carry a total of 500 passengers and 75 cars, and is capable of a fully loaded maximum speed at 100 per cent MCR in calm water of 50 kt. In order to maximise passenger comfort, a ride control system is fitted to the craft, venting the main cushion overpressure.

Specifications

Length overall	79.8 m
Beam	20.2 m
Draught, hullborne	2.8 m
Draught, on-cushion	1.0 m
Crew	16
Passengers	500
Vehicles	75
Propulsive power	2 × 13,500 kW
Lift power	3 × 1,700 kW
Max speed	50 kt
Operational speed	45 kt
Range	600 n miles
Operational limitation	Sea State 3

Classification: DnV +1A1 HSLC Passenger car ferry R2.

Structure: The hull is constructed in aluminium alloy.

Propulsion: Main engines: two gas turbines, each driving a water-jet.

Lift engine: three diesels, driving 3 × 2 × (1.3 m DWDIs).

Semo Company Ltd

A subsidiary of Semo Kimchi
4th Floor Dong Kwang Building, 798-1 Yeok sam-1-Dong, Kangnam-ku, Seoul 135-081, South Korea

Tel: (+82 2) 3458 8175
Fax: (+82 2) 569 1977
e-mail: semoship@semokimchi.com

Jae-Joong Yoon, *Managing Director*
J C Lee, *Marketing Manager*

Semo is known for its ferry operating division which includes, in a fleet of 30 ferries, two hydrofoils, nine SESs, one catamaran and two fast monohulls.

In 1992, Semo Company Ltd delivered the 36.4 m SES *Democracy* from its Kosungkun Shipyard, which then began service in its fleet operating between Inchon and Baknyung Island. Journey time for the route is approximately 3.5 hours.

The Semo Company yard had delivered a further three 40 m SESs (Democracy II, III and V) by the end of 1994.

36.4 m SES

Democracy

Specifications

Length overall	36.4 m
Beam	11.3 m
Draught, hullborne	2.2 m
Draught, on-cushion	0.7 m
Crew	10
Passengers	340
Fuel capacity	14,000 litres
Water capacity	2,000 litres
Propulsive power	2 × 1,680 kW
Lift power	2 × 373 kW
Max speed	50 kt
Operational speed	45 kt

Propulsion: Main engines: two MWM TBD 604B V16 diesels, 1,680 kW each at 1,800 rpm.

Lift engines: two DDC GM8V-92TA diesels, 373 kW each at 2,100 rpm.

40 m SES

Democracy II, III and V

Specifications

Length overall	40.0 m
Length waterline	33.8 m
Beam	11.6 m
Draught, hullborne	1.9 m
Draught, on-cushion	0.6 m
Crew	10
Passengers	390

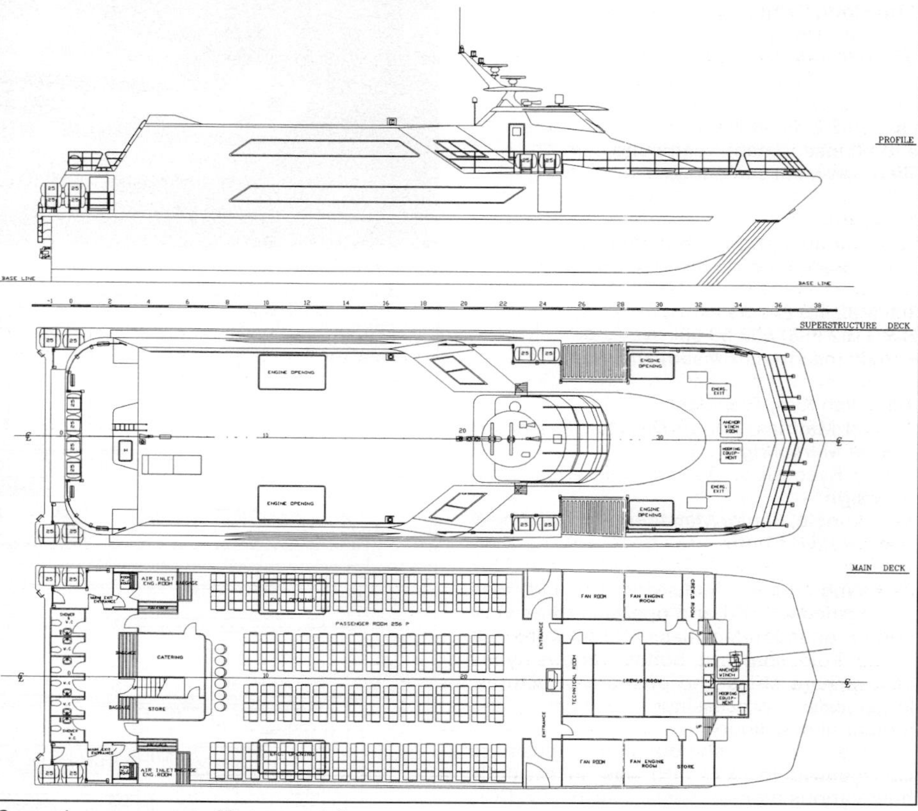

General arrangement of Democracy II 0506870

Democracy V *built by Semo Company Ltd* 0002636

Vessel type	Vessel name	Yard No	Length (m)	Speed (kt)	Seats	Vehicles	Originally delivered to	Date of build
36.4 m SES	*Democracy*	–	36.4	40.0	356	none	Semo Marine Co Ltd	December 1992
40 m SES	*Democracy II*	–	40.0	45.0	350	none	Semo Marine Co Ltd	April 1994
40 m SES	*Democracy III*	–	40.0	45.0	400	none	Semo Marine Co Ltd	August 1994
40 m SES	*Democracy V*	–	40.0	45.0	400	none	Semo Marine Co Ltd	November 1994

Propulsive power	1,970 kW
Lift power	529 kW
Max speed	50 kt
Operational speed	45 kt

Propulsion: Main engines: two MTU 16V 396 TE 74L rated at 1,970 kW each at 1,920 rpm. Lift engine: two DDC 12V-92TA rated at 529 kW each at 2,100 rpm.

45 m SES (Design)

Building on the experience of the 40 m craft, the following design has been developed.

Specifications

Length overall	45.0 m
Beam	11.6 m
Crew	10
Passengers	450
Fuel capacity	12,300 litres
Water capacity	2,000 litres
Max speed	50 kt
Operational speed	45 kt

Propulsion: Two MTU 16V 396 TE 74L rated at 1,970 kW each driving a Kamewa 71 S2 water-jet.

Netherlands

Schelde Naval Shipbuilding (Royal Schelde)

PO Box 555, NL-4380 AN Vlissingen, Netherlands

Tel: (+31 118) 48 2118
Fax: (+31 118) 48 50 20
e-mail: info@schelde.com
Web: www.schelde.com

W Laros, *Managing Director*
M Pater, *Marketing and Sales*

Royal Schelde has developed a range of high-speed vessels in the past, starting with 26 m SES craft. Subsequent to this vessel, Schelde Shipbuilding also developed a range of catamarans and monohulls, drawing on its experience of building naval frigates.

In view of current market developments, the company has committed itself to the marketing of large high-speed semi-displacement vessels such as the Cat 70, 90 and 105 catamarans and the 90 and 128 m monohull designs.

The 26 m demonstrator SES vessel, the Seaswift 23, was built in 1990 and operated for a period between Southampton and the Isle of Wight in the UK.

Royal Schelde Seaswift 23 0506963

Seaswift 23

The Seaswift 23 is a prototype/demonstrator vessel, used for demonstrating and testing the design of Royal Schelde's Seaswift series. It has also been used for demonstration purposes for prospective clients as well as having formed the basis of the further designs. In August 1991, the vessel entered service with Cowes Express, operating a passenger ferry service between Southampton and the Isle of Wight in the UK for a period of 12 months.

Specifications

Length overall	24.3 m
Length waterline	20.0 m
Beam	7.7 m
Beam overall	approx 7.9 m
Draught, hullborne	1.5 m
Draught, on-cushion	0.8 m
Crew	3
Passengers	132
Fuel capacity	3.6 m^3
Water capacity	0.5 m^3
Propulsive power	1,040 kW
Max speed	31 kt
Operational speed	26 kt

Classification: DnV +1A1 R30 Light Craft, ECO.

Structure: Welded marine grade aluminium alloy.

Propulsion: Three centrifugal lift fans, resiliently mounted, the forward two being driven directly by the lift system diesel engine and the aft one driven hydraulically by the same engine. Propulsion is provided by two MWM TBD 234 12V diesels, 520 kW each at 2,200 rpm MCR, driving Kamewa 40S water-jet units.

Vessel type	Vessel name	Yard No	Length (m)	Speed (kt)	Seats	Vehicles	Originally delivered to	Date of build
Seaswift 23	*Seaswift 23*	–	24.3	26.0	132	none	–	–

Norway

Brødrene Aa

6829 Hyen, Norway

Tel: (+47 57) 86 87 00
Fax: (+47 57) 86 99 14
e-mail: braa@braa.no
Web: www.braa.no

Tor Øyvin, *Sales Manager*

Brødrene Aa was formed as a boat building company in 1947 and pioneered the use of composite sandwich materials in 1975, resulting in the construction of a number of SES craft in the 1980s. Other composite applications included train nose cones, tunnel linings and modules for the offshore industry.

In 2002 the company restarted composite boat building using carbon fibre materials.

CIRR 120P

Thirteen of these composite SES craft were built between 1988 and 1991.

Vessel type	Vessel name	Yard No	Length (m)	Speed (kt)	Seats	Vehicles	Originally delivered to	Date of build
CIRR 105P	*Norcat*	170	–	–	264	–	TFDS	1984
CIRR 115P	*Ekwata*	184	–	–	290	–	Cabo Cat	1986
CIRR 120P	*Ekwata 2*	190	35.3	50.0	330	–	Norla Marine	1988
CIRR 60P	*Harpoon*	198	18.0	52.0	85	–	Brødrene Aa	1988
CIRR 120P	*San Pawl*	199	35.3	50.0	330	–	MTD MDC	1988
CIRR 120P	*Santa Maria*	200	35.3	50.0	330	–	MTD MDC	1989
CIRR 120P	*Sant Agata*	201	35.3	50.0	330	–	KS Carignor	1989
CIRR 120P	*San Pietro*	202	35.3	50.0	330	–	MORF	1989
CIRR 120P	*Sab Frangisk*	210	35.3	50.0	330	–	Virtu Ferries	1989
CIRR 120P	*Supercat One*	211	35.3	50.0	330	–	Supercats Marine	1990
CIRR 120P	*La Vikinga*	212	35.3	50.0	330	–	KS Venzunor	1990
CIRR 120P	*Perestroika*	213	35.3	50.0	368	–	Semo Marine	1990
CIRR 120P	*Nissho*	218	35.3	50.0	320	–	Yasuda Ocean Line	1990
CIRR 120P	Fjordkongen	219	35.3	50.0	320	–	TFDS	1990
CIRR 120P	–	204	35.3	50.0	349	–	Semo Marine	1991
CIRR 120P	–	226	35.3	50.0	330	–	–	1991

Specifications

Length overall	35.3 m
Beam	11.5 m
Draught, off cushion	2.1 m
Draught, on cushion	0.7 m
Passengers	330
Fuel capacity	60,000 litres
Water capacity	500 litres
Max speed	50 kt

Classification: DNV +1A1-R90-LC, EO, F-LC.

Structure: FRP sandwich construction.

Propulsion: Two MWM TBD604B 16V diesel engines each rated at 1,704 kW at 1,800 rpm, each driving a KaMeWa 56 water-jet via a ZF BW755D gearbox.

Ulstein Technology

Division of Rolls Royce Marine Systems
PO Box 158, N-6065 Ulsteinvik, Norway

Tel: (+47) 70 01 40 00
Fax: (+47) 70 01 40 02
Web: www.ulsteingroup.com
Web: www.rolls-royce.com

UlsteinTechnology is a division of Rolls Royce Marine Systems and no longer builds high-speed craft. However the design rights to the craft built to date are understood to be owned by the company. In the past, the organisation has delivered 18 air cushion catamarans in FRP sandwich construction. The UT 904 Air Cushion Catamaran (ACC) was the latest design, two having been launched in 1992.

A licence/technology transfer agreement was signed with International Shipyards of Australia in 1993 for the production of a series of UT 928 craft.

Ulstein UT 904 Ocean Flower 0506595

Ulstein UT 904 Santa Eleonora 0506596

UT 904

The UT 904 is designated an Air Cushion Catamaran by Ulstein International, a design in the category of surface effect ships and air cushion vehicles in general.

Construction of the first UT 904 started in September 1990. The vessel has accommodation for 320 passengers.

The lift fan engines power double-sided stainless steel centrifugal fans and in addition drive (hydraulically) two centrifugal booster fans which provide pressurising for the aft seal bag.

A hydraulically operated Ride Control System (RCS) is fitted, employing louvre valves, pressure sensors and a microprocessor control unit. This system is designed to reduce vertical accelerations.

Specifications

Length overall	39.0 m
Beam	12.0 m
Draught, hullborne	2.6 m
Draught, on-cushion	1.0 m
Passengers	320
Fuel capacity	13,800 litres

Vessel type	Vessel name	Yard No	Length (m)	Speed (kt)	Seats	Vehicles	Originally delivered to	Date of build
CIRR 105P	*Victoria Express* (ex-*Fjordkongen* ex-*Norcat*)	170	–	–	264	None	Troms Fylkes D/S	1984
CIRR 115P	*Santa Lucia* (ex-*Ekwata*)	184	–	–	290	None	Maritima Turistica del Mar de Cortes	1986
CIRR 120P	*Express La Pez* (ex-*San Pawl*)	199	–	–	315	None	Maritima Turistica del Mar de Cortes	1988
CIRR 120P	*Wight Queen* (ex-*Virgin Butterfly*)	190	–	–	280	None	–	1988
CIRR 60P	*Harpoon*	198	–	–	85	None	(Test vessel)	1988
CIRR 120P	*San Fragisk*	210	–	–	330	None	NIS	1989
CIRR 120P	*San Pietro*	202	–	–	330	None	Finnmark Fylkesrederi og Ruteselskap	1989
CIRR 120P	*Santa Maria*	200	–	–	330	None	Tidewater	1989
CIRR 120P	*Wight King* (ex-*Sant Agata*)	201	–	–	280	None	–	1989
CIRR 120P	*Nissho*	218	–	–	368	None	Yasuda Sangyo Kisen Company Ltd	1990
CIRR 120P	*Fjordkongen II* (ex-*Fjordkongen*)	219	–	–	320	None	Troms Fylkes D/S	1990
CIRR 120P	*Catamaran II* (ex-*Gloden Olympics*)	211	–	–	330	None	Piraiki Naftiliaki SA	1990
CIRR 120P	*La Vikinga*	212	–	–	330	None	KS Pantheon	1990
CIRR 120P	*Perestroika*	213	–	–	330	None	Semo Marine Craft	1990
CIRR 120P	*Sea Flower*	204	–	–	320	None	Dae-A Kwaesok Ferry	1991
CIRR 120P	*Catamaran I*	226	–	–	349	None	Piraiki Naftiliaki SA	1991
UT 904	*Ocean Flower*	208	39.0	46.0	360	None	Dae-A Ferry Company	1991
UT 904	*Santa Elorana*	208	39.0	46.0	342	None	Misano Alta Velocita	1992

Water capacity	2,000 litres
Propulsive power	2 × 2,000 kW
Lift power	3 × 380 kW
Max speed	50 kt
Operational speed	46 kt

Classification: DnV +1A1 R90 Light Craft EO.
Additional class:
SF-LC, stability and subdivision
F-LC, fire protection.
Regulations: National Flag Authorities.

Structure: The hulls are built in FRP/PVC foam sandwich or aluminium.
Propulsion: Two water-jet units.
The two lift fans are double-sided, stainless steel.

Umoe Mandal AS

Gismerøya, PO Box 902, N-4509 Mandal, Norway

Tel: (+47) 38 27 92 00
Fax: (+47) 38 26 03 88
e-mail: mandal@umoe.no
Web: www.mandal.umoe.no

Peter Klemsdal, *Managing Director*
Ole Ihme, *Contracts Manager*
Are Soreng, *Purchasing Manager*
Peter Reed-Larson, *Marketing Director*

Umoe Mandal specialises in the design and construction of advanced marine craft built from FRP composite materials. The company has built a number of lightweight high-speed craft including patrol craft and minehunters and minesweepers.

The company has developed specialist expertise in EMC/EMI, stealth technology, titanium welding, ship signatures, naval architecture and system integration.

Between 1993 and 1997 the company delivered nine SES mine countermeasure vessels.

Umoe Mandal 12 m manned test craft 0506822

SES Mine countermeasures vessel Oksøy and Alta class

A series consisting of four minehunters and five minesweepers has been produced at Umoe Mandal A/S for the Royal Norwegian Navy. The minehunters have the designation Oksøy class and the minesweepers Alta class. The design of the vessels is based on the Surface Effect Ship (SES) principle. The special characteristics of the SES concept are utilised to obtain advantages both with respect to operational capability and economy. Delivered between 1993 and 1997 these craft replace the eight United States-built 1950s MSC-60 coastal minesweepers/hunters.

Umoe Mandal moved into new production facilities in 1991, the yard is located at Gismerøya Island near Mandal. The production hall has been built for the particular purpose of building high-speed military and civil vessels in composite materials and to the highest quality standards.

Based on the existing design of the MCMVs, Umoe Mandal has further developed and modified the vessel design in order to meet

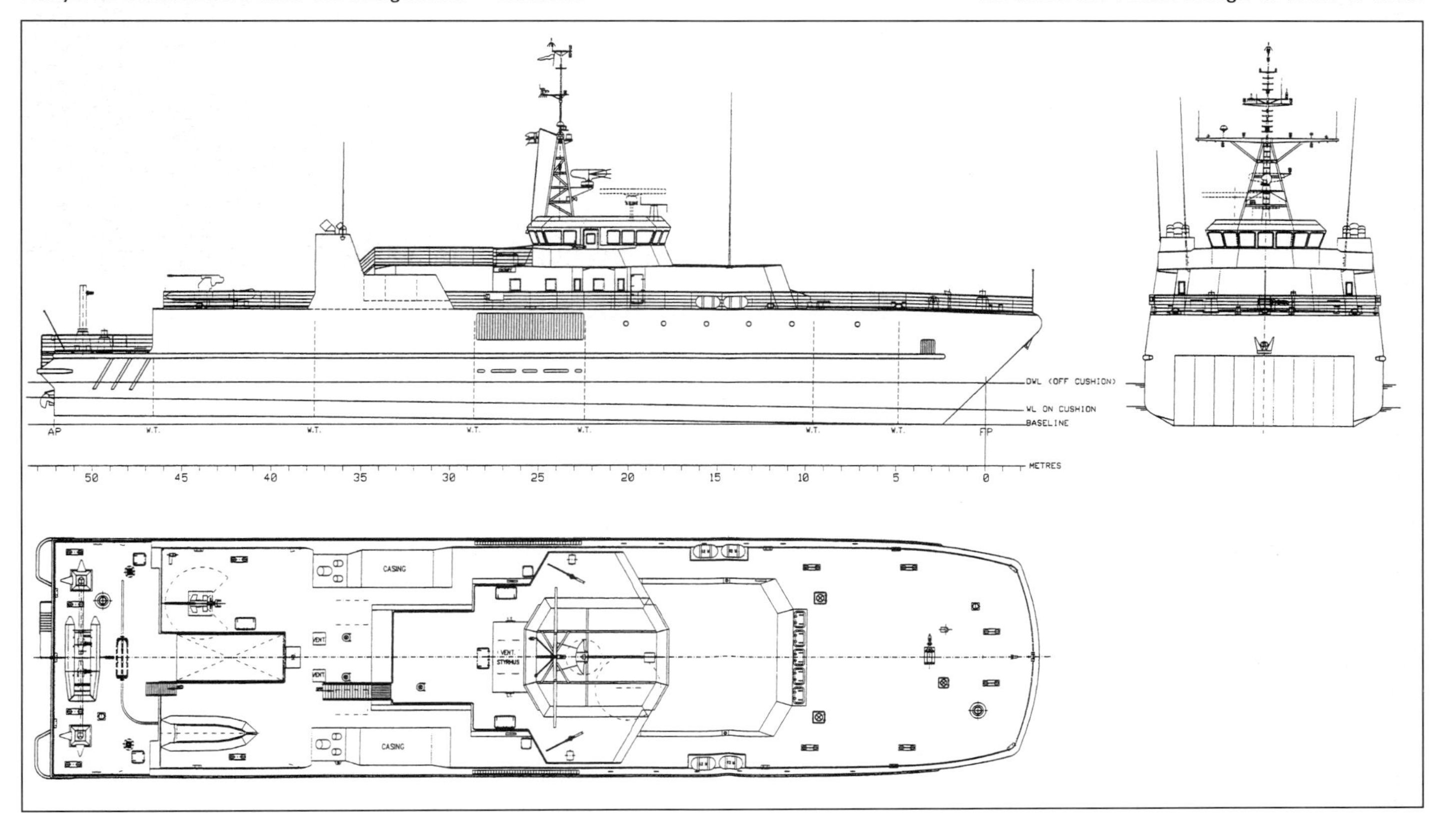

General arrangement of Umoe Mandal mine countermeasures SES 0506594

international demands. The first vessel, Oksøy, undertook sea trials in 1993 and was commissioned in August 1994. The last of the nine-vessel order was delivered in July 1997.

Specifications

Length overall	55.2 m
Beam	13.3 m
Draught, hullborne	2.15 m
Draught, on-cushion	0.9 m
Weight, max	370 t
Crew	29 (+12 extra capacity)
Water capacity	5,000 litres
Propulsive power	2 × 1,400 kW
Lift power	2 × 700 kW
Operational speed	20+ kt (transit)
–	5+ kt (minehunting)
–	12+ kt (minesweeping)
Range	1,400 n miles

Propulsion: The propulsion engines are two MTU 12V 396 TE 84, driving two Kværner Eureka water-jets.

Engines, lift: two MTU 8V 396TE 54, 700 kW each.

Oksøy 0506955

12 m manned SES test craft

On 20 January 1993, Umoe Mandal launched a one-third scale prototype of a new air cushion catamaran high-speed patrol vessel. The first of nine mine countermeasure vessels was launched two months later. The vessel is based on the air cushion catamaran principle (SES) and built of composite materials.

The favourable properties of air cushion catamarans are further enhanced in this vessel by a new single-skin hull construction of fibre-reinforced polymer, and by a novel hull geometry, developed by Umoe Mandal and its project partners.

To study materials, design and performance, Umoe Mandal has gone to the rather unusual step of building a 12 m, 40 kt manned model.

The Umoe Mandal new form of hull construction, gives a reduction of 25 per cent in the hull structural weight over conventional GRP hull structures and even more when compared to aluminium hulls.

The 12 m model is part of an international research project called 'Advanced FRP Composite Hull Structures for High-Speed Craft'. The study programme includes sea-going performance, speed and acceleration in calm and rough seas, dynamic stability in high waves, reactions to wave-induced slamming loads on the hull materials and improvement in air cushion design.

The other project participants are Du Pont, Conoco, Det Norske Veritas, Veritas Research, the Marine Consulting Group, Devold AMT, Jotun Polymer, the Norwegian Defence Research Establishment, the Norwegian Industry Fund and the Royal Norwegian Research Council.

Umoe Mandal has patented a new production method for series building under controlled conditions. The production hall allows indoor building of large air cushion catamarans side by side.

Vessel type	Vessel name	Yard No	Length (m)	Speed (kt)	Seats	Vehicles	Originally delivered to	Date of build
Prototype	Prototype	–	12.0	–	crew	none	–	January 1993
SES MCMV	*Oksoy*	–	55.2	–	crew	none	Norwegian Navy	March 1994
SES MCMV	*Karmoy*	–	55.2	–	crew	none	Norwegian Navy	October 1994
SES MCMV	*Maloy*	–	55.2	–	crew	none	Norwegian Navy	March 1995
SES MCMV	*Hinnoy*	–	55.2	–	crew	none	Norwegian Navy	September 1995
SES MCMV	*Alta*	–	55.2	–	crew	none	Norwegian Navy	January 1996
SES MCMV	*Otra*	–	55.2	–	crew	none	Norwegian Navy	November 1996
SES MCMV	*Rauma*	–	55.2	–	crew	none	Norwegian Navy	December 1996
SES MCMV	*Orkla**	–	55.2	–	crew	none	Norwegian Navy	April 1997
SES MCMV	*Glomma*	–	55.2	–	crew	none	Norwegian Navy	July 1997

*Destroyed by fire 19 November 2002

Russian Federation

Almaz Central Marine Design Bureau

50 Varshavskaya.str, St Petersburg 196128, Russian Federation

Tel: (+7 812) 389 55 02
Fax: (+7 812) 373 48 37
e-mail: office@almaz.kb.sp.ru
Web: www.almaz.info

Alexander V Shliakhtenko, *President*
Alexander M, Lazarev, *Chief Engineer*
Anatoly A, Gorodianko, *Marketing Director*

The Almaz Central Marine Design bureau was established in 1940 and specialises in the design and construction of high-speed commercial and military craft.

The company has constructed hovercraft, SESs, hydrofoils and SWATH vessels.

CHILIM HOVERCRAFT

This small multipurpose hovercraft was designed in 1999 to operate year round in coastal waters for fast speed patrol in territorial waters, search and interception of border trespassers, landing and evacuation in hard to reach areas, search and rescue operations and passenger transportation. The first vessel was built in 2000.

Specifications

Length overall	12.0 m
Beam	5.9 m
Hull beam	4.85 m
Weight, max	9.5 t
Crew	2
Max speed	43 kt
Operational speed	38 kt
Range	180 n miles
Operational limit	1.0 m seas

Classification: Russian Maritime Register of Shipping.

Propulsion: Powered by two Deutz BF82 L1513 diesel engines each rated at 235 kW.

YAMAL-150 AMPHIBIOUS CARGO CRAFT (DESIGN)

The Yamal-150 is designed for transportation from vessels on roadstead to stores ashore, transportation of cargo and passenger to inshore and shelf plate installations.

Specifications

Length overall	55.4 m
Beam	25.6 m
Weight, max	450 t
Crew	8
Maximum speed	50 kt
Operational speed	35 kt
Range	550 n miles
Operational limitation	1.0 m seas

Classification: Russian River Registration.

Vessel type	Vessel name	Yard No	Length (m)	Speed (kt)	Seats	Vehicles	Originally delivered to	Date of build
Chilim	–	–	12.0	43.0	12	none	–	2000
RSES-500	Design	–	93.0	50.0	700	170	–	–
Yamal-150	Design	–	55.4	50.0	–	Up to 150 t of any payload	–	–

Propulsion: Four gas turbine sets each comprising a gas turbine engine (7,353 kW) and a transmission system.

RSES-500 (DESIGN)

The company has also developed an extensive range of SES craft designs from 200 to 2,500 t displacement. These are known as the RSES series (Russian Surface Effect Ship). Brief details of the RSES-500 have been provided which can be configured for passenger or cargo transportation:

Specifications

Length overall	93.0 m
Beam	23.5 m
Draught, hullborne	4.1 m
Draught, on-cushion	1.8 m
Passengers	450–700
Vehicles	170 cars
Max speed	55 kt
Operational speed	50 kt
Range	600–1,000 n miles
Operational limitation	Sea State 6

Propulsion: The vessel is powered by four gas turbines with a total output power of 60,000 kW, driving water-jet propulsors. The lift power is derived from four marine diesels with a total output power of 10,000 kW.

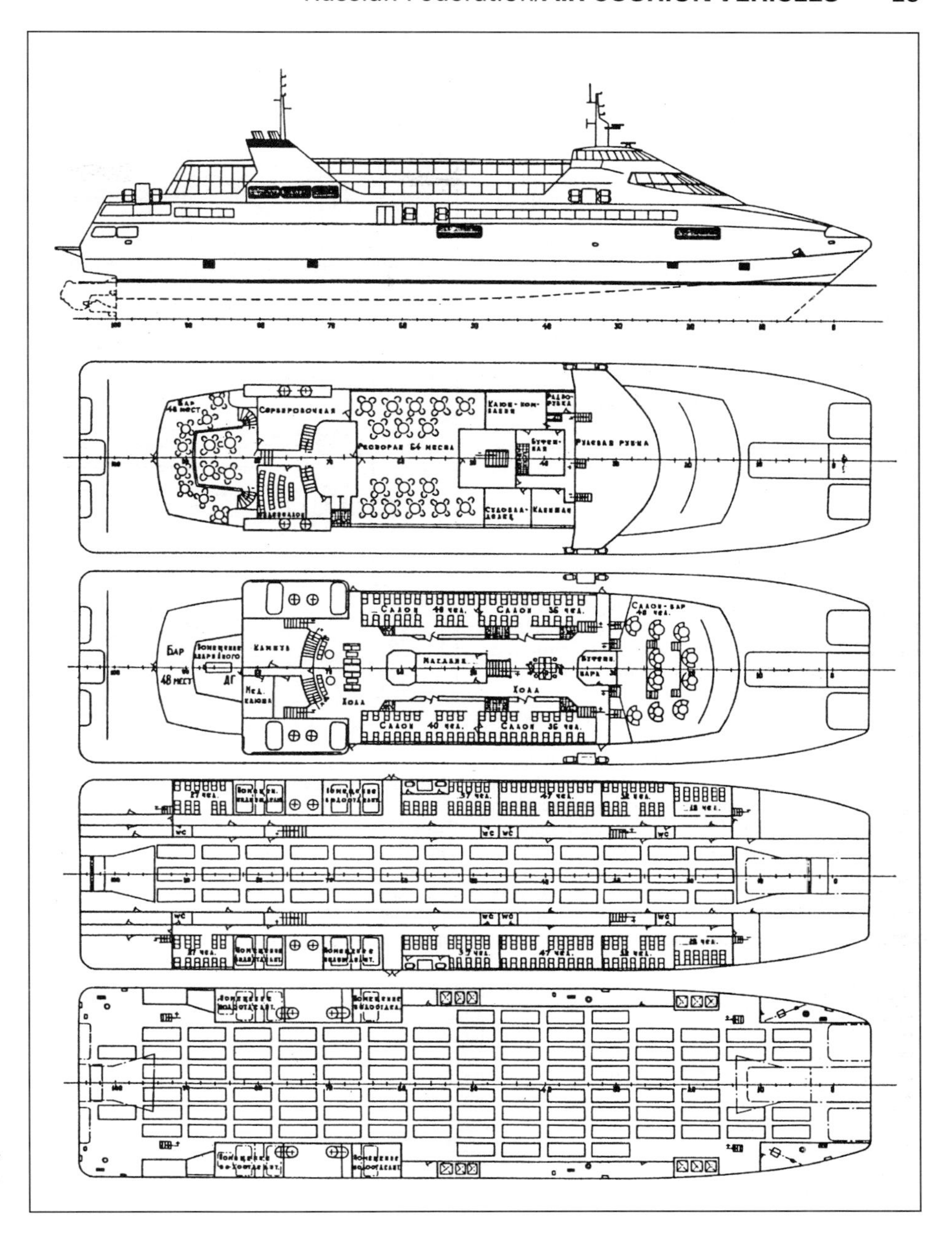

General arrangement of RSES-500
0506951

Almaz Shipbuilding Company

26 Petrovsky PR, St Petersburg 197110, Russian Federation

Tel: (+7 812) 235 48 20
Fax: (+7 812) 350 11 64
e-mail: market@almaz.spb.ru
Web: www.almaz.spb.ru

Leonid Grabovets, *Director General*
Valeri Demchenko, *Technical Director*
Alexei Ponomarev, *Production Director*
Sergei Galichenko, *Marketing Director*

Almaz Shipbuilding Company was established in 1901 although has been trading under the current private ownership since 1993. The shipyard has four main enclosed building areas, the largest having a length of 100 m with a width of 15 m. Building in steel and aluminium, the company specialises in lightweight fast patrol craft and passenger ferries.

Lynx pilot hovercraft

The Company handed over a new 13.5 m Lynx hovercraft to the Petersburg Pilot Association in January 1999. The craft was designed by Central Design Bureau Neptune for the transport of pilots across ice and open water areas adjacent to the port of St Petersburg.

The Almaz Lynx on trials
0045944

Specifications

Length overall	14.2 m
Beam	5.3 m
Weight, min	8.0 t
Payload	3,200 kg
Passengers	17
Fuel capacity	1,040 litres
Max speed	35 kt
Operational speed	30 kt
Obstacle clearance	0.7 m
Operational limitation	1.2 m wave height

Classification: Maritime Shipping Register of Russia, KM+KM1 CBIIA.

Structure: Main hull riveted aluminium, superstructure GRP.

Propulsion: Three VM Motori 100 kW diesel engines, two driving independent variable-pitch air propellers and one driving two lift fans. The two independent engines can also drive the lift fans in emergency conditions.

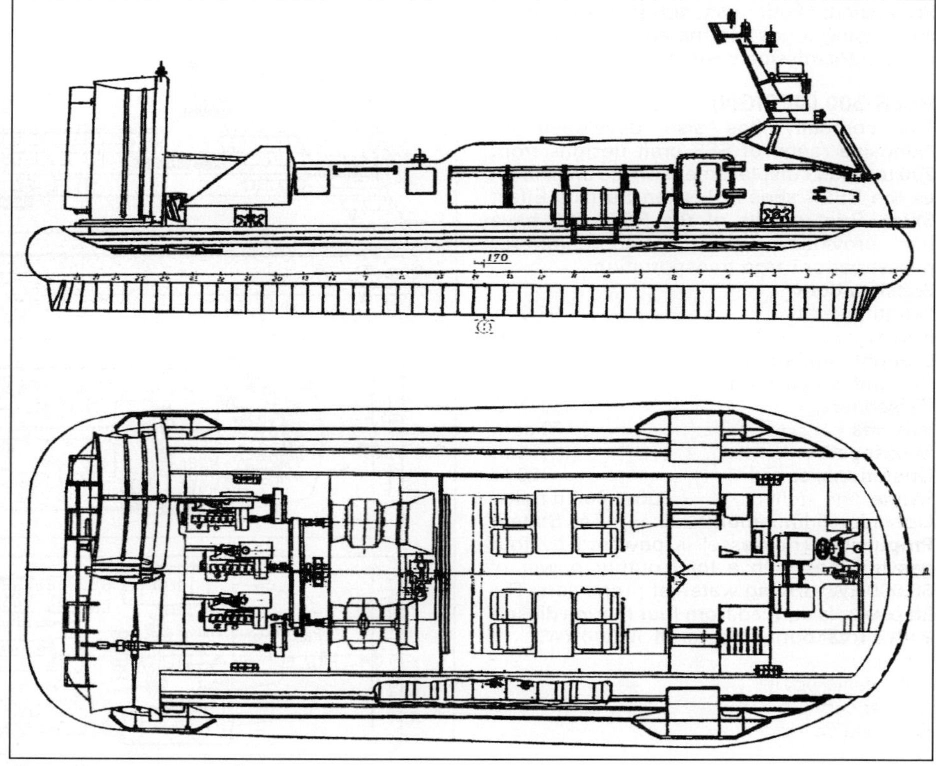

General arrangement of Lynx 0121768

Pomornik (ZUBR)

The Pormornik, also known as the Zubr or Project 1232.2 was designed by the Almaz Central Design Bureau in St Petersburg, with five having been built for the Russian Navy by Almaz Shipbuilding Company, the last delivered in 1994. Another was constructed at Feodosia in the Ukraine.

The Greek Navy ordered four Pormornik hovercraft in 2000, in the form of fast attack vessels. Three were existing Russian vessels, refurbished for their new role. The fourth is a new craft, and construction commenced in January 2003. The first refurbished vessel, *Kefallonia*, was delivered in December 2000, the second, *Zakintos*, was delivered in September 2001 and the third, *Kerkyra*, was delivered in 2005. In 2006 China placed an order for six Zubr hovercraft.

Specifications

Length overall	57.6 m
Beam	25.6 m
Displacement	550 t
Payload	130 t
Operational speed	55 kt
Maximum speed	60 kt
Range	300 n miles

Propulsion: Five Type NK-12MV gas turbines each rated at 8,826 kW (three for propulsion, two for lift).

Vessel type	Vessel name	Yard No	Length (m)	Speed (kt)	Seats	Vehicles	Originally delivered to	Date of build
Almaz Lynx	*Lynx*	–	14.2	30.0	17	none	Petersburg Pilot Association	1999
Pormornic	*Kefallonia*	–	57.6	55.0	–	–	Greek Navy	2000
Pormornic	*Zakintos*	–	57.6	55.0	–	–	Greek Navy	2001
Pormornic	*LCAC ZUBR L180*	–	57.6	55.0	–	–	Greek Navy	(2004)
15063	*Irbis*	–	17.5	32.0	32	none	–	–
14661	*Rys*	–	14.1	35.0	–	–	St Petersburg Pilots Association	1999
Murena	–	-	32.0	55.0	180	2	–	–
Pormornic	*Kerkyra*	–	57.6	55	–	–	Greek Navy	2005

Krasnoye Sormovo Shipyard

Part of the MNP Group
Barrikad Street 1, Nizhni Novgorod 603603, Russian Federation

Tel: (+7 8312) 25 31 53
Fax: (+7 8312) 25 13 41
e-mail: info@krsomovo.nnov.ru
Web: www.mnpglobal.com

N Zharkov, *General Director*
I Muchnik, *Business Director*

It is understood that this shipyard is still involved in high-speed craft production although details of the number of craft produced are unknown.

Zarnitsa 0506567

Zarnitsa

Evolved from Gorkovchanin, the Zarnitsa is a 48- to 50-seat water-jet-propelled rigid sidewall ferry designed to operate on shallow rivers, some less than 0.7 m (2 ft 3 in) deep. The vessel embarks and disembarks passengers from the bow in the centre of the wheelhouse structure. Series production is established and large numbers have been delivered.

The prototype was put into trial service on the Vyatka river, in the Kirov region, in the Summer of 1972, and the first production models began operating on shallow, secondary rivers later in the year. During 1973–74, Zarnitsas entered service on tributaries of the Kama, Lena and Volga. More than 100 are employed on almost all the river navigation lines of the Russian Federation as well as on the rivers of Ukraine, Moldova, Belarus and Kazakhstan.

Specifications

Length overall	22.3 m
Beam	3.9 m
Weight, min	9 t
Weight, max	15 t
Operational speed	19 kt

Structure, Propulsion, Control: Arrangements similar to those of the Gorkovchanin.

Vessel type	Vessel name	Yard No	Length (m)	Speed (kt)	Seats	Vehicles	Originally delivered to	Date of build
Zarnitsa	–	–	22.3	19.0	–	none	–	Series production

Krylov Shipbuilding Research Institute

Moskovskoye Shosse, 44, 196158 St Petersburg, Russian Federation

Tel: (+7 812) 727 96 47
Fax: (+7 812) 727 95 94
e-mail: krylov@krylov.spb.ru
Web: www.ksri.ru

V M Pashin, *Director*
A P Pylaev, *Deputy Director for Marketing*

The organisation is active in the design and build of air cushion vehicle prototype craft.

Sibir

The shipyard developed and built the cargo-carrying ACV *Sibir* in 1991 with ACV *Typhoon* as a prototype. The craft is designed for year-round fast delivery of tracked or wheeled vehicles and containerised cargo on rivers, lakes and reservoirs.

Specifications

Length overall	25.2 m
Beam	11.7 m
Weight, max	55 t
Payload	25 t
Propulsive power	1,874 kW
Max speed	43 kt
Obstacle clearance	1.3 m

Sibir in operation 0506877

Classification: Class P of the River Register of Russia.

Hull: The hull is built in 1561 weldable marine aluminium alloy with port and starboard superstructures. Each side structure contains an integrated axial turbine fan installation. The bow ramp provides roll-on, roll-off loading and unloading.

The skirt is a bag and finger tapered type, 1.2 m deep at the bow and 1.4 m at the stern. Longitudinal and transverse flexible keels divide the cushion into four compartments for stability.

Propulsion: Aerodynamically integrated system, powered by twin Ivchenko AI-24 gas turbines, maximum 1,874 kW and continuous 1,310 kW each. Each turbine drives a 2.5 m axial fan mounted on the turbine shaft. This pumps the air downwards into the cushion and backwards through the duct for propulsion.

Control: Rudders fitted in the duct provide directional control. The air flow is deflected for braking by horizontal vanes, hydraulically controlled from the wheelhouse.

Outfit: All controls are in a raised port wheelhouse.

Vessel type	Vessel name	Yard No	Length (m)	Speed (kt)	Seats	Vehicles	Originally delivered to	Date of build
Sibir	*Typhoon*	–	25.2	–	–	–	–	–
Sibir	*Sibir*	–	25.2	–	–	–	–	1991

Neptune Hovercraft

Proektiruemyi 4062, dom 6, Moscow, Russian Federation

Tel: (+7 495) 677 75 30
Fax: (+7 495) 677 39 49
e-mail: neptune@hovercraft.ru
Web: www.hovercraft.ru

V S Sokolov, *Director General*
A S Koudriavtsev, *Chief Designer*

The CDB (Central Design Bureau) Neptune Corp originally specialised in designing various types of small displacement motor boats and ships but recently hovercraft design and production have become its main option. Having its own shipyard near Moscow Neptune Hovercraft is co-operating with other shipyards for serial production of hovercraft. Over 200 hovercraft of various types have been built from Neptune's designs over the last 15 years.

Puma twin petrol engine hovercraft designed by the Neptune Central Design Bureau, Moscow 0506574

Gepard (Cheetah)

A multirole five-seater introduced in 1981, Gepard has been designed to provide convenient, reliable and inexpensive transport in more remote areas of the RFAS. Specialist professions in those areas were questioned about their transport needs before finalising the design. Early in 1983 a Gepard successfully underwent trials at Andreyevskoye Lake, near the centre of the Tyumen Oblast. About 60 Gepards had entered service by the end of 1990.

The craft is designed to operate in an ambient air temperature range from –40 to +40°C.

The following service life trials have been carried out:

Moscow to Lake Seleger and back: about 1,000 km
Along small rivers to the city of Vyshnii Volochek and back: about 1,000 km
Moscow by the Volga-Baltic route to St Petersburg: about 2,000 km
Moscow to Volgograd on the lower Volga River: about 3,500 km.

The five-seat, 42 kt Gepard is powered by a single 115 hp ZMZ-53 petrol engine 0506573

Nearly 150 units have been built up to now and operate in various areas in the European and Asian parts of Russia. Its remarkable duties include anti-pollution patrol in the city of Moscow and auxiliary service with a hydrological expedition on the Yamal Peninsula in the Arctic. Examples of the Gepard series are also operated out of the Russian Federation by ship owners in the US, Mexico and Canada.

It is understood that in 2002, new versions of this vessel with more powerful engines were introduced, with power up to 110 kW.

Specifications

Length overall	6.6 m
Beam	3.3 m
Weight, max	1.9 t
Crew	5
Propulsive power	86 kW
Max speed	32 kt
Operational speed	24 kt
Operational limitation	–40 to +40°C

Structure: Corrosion-resistant light-alloy hull. Moulded pigmented glass fibre cabin superstructure.

Propulsion: Integrated system, powered by a ZMZ-53 lorry engine (a widely used engine), aft of the cabin. Output is transmitted to a centrifugal lift fan (0.97 m diameter with GRP blades) and a duct-mounted, multibladed 0.95 m diameter glass fibre propeller for thrust. The blade leading edges are protected with stainless steel sheaths. Power transmission is by toothed-belt drives and a clutch is provided between the engine and propeller and fan drives.

Puma

Design and building of this craft was completed in less than a year. The first three variants underwent comprehensive trials in 1988. As an ambulance the Puma is equipped with an operating table and related medical apparatus including oxygen bottles, making it possible to provide urgent medical aid on board, including simple operations. The passenger variant is fitted with 16 aircraft-type seats, while the passenger/cargo variant has 10 folding seats.

Pumas are engaged in passenger transportation services in Siberia, on the Caspian Sea and in the far east of the Russian Federation. One craft is to be delivered to Mexico. The total number of craft built since 1985 is 11.

Specifications

Length overall	12.2 m
Beam	4.5 m
Weight, min	3.6 t
Weight, max	5 t
Payload	1.45 t
Passengers	16
Propulsive power	2 × 90 kW
Max speed	35 kt
Operational speed	22 kt
Obstacle clearance	0.6 m

Propulsion: Two ZMZ-53 90 kW petrol engines driving two centrifugal fans and two reversible-pitch propellers with toothed-belt drives.

Irbis

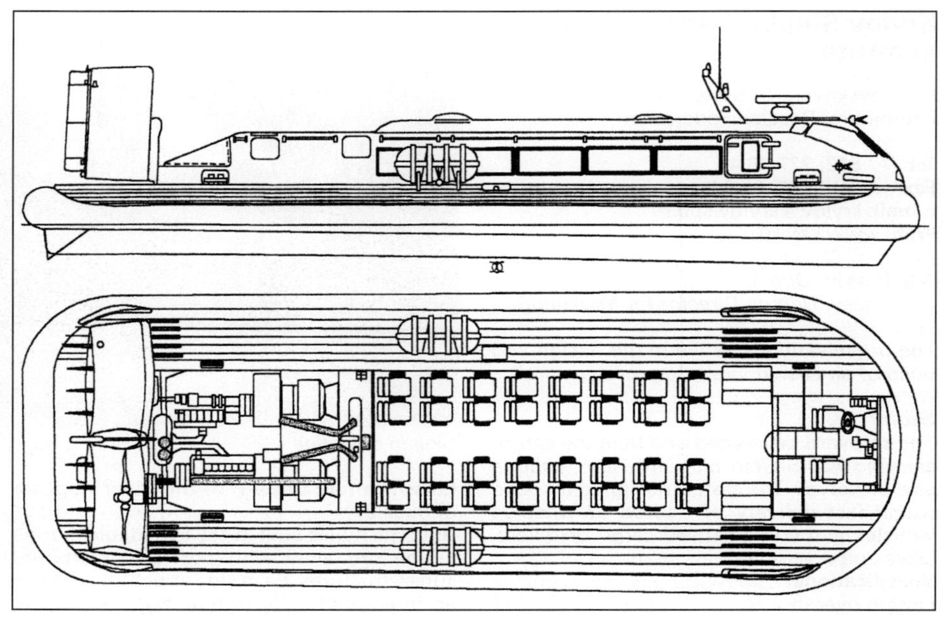

General arrangement of Irbis 0121769

Irbis 0506575

A twin-engined amphibious hovercraft designed to carry up to 32 passengers. Intended applications for this type include the transport of geologists and oil industry workers and their associated equipment, cargo shipment up to 2.5 tonnes and support for geophysical exploration. The craft is designed to operate in a wide range of operating conditions: in ambient temperatures from –40°C with maximum wind speeds of 29 kt, up to 30 miles from base; in significant wave heights up to 0.7 m when cushionborne, across deep or shallow water; on snow and compact ice; and on broken and sludge ice. The craft has folding sidebodies for ease of transport.

From their first appearance in 1989, these craft participated in many trials and commercial services in various areas of the Russian Federation, including: inland waterway routes Moscow–St Petersburg–Astrakhan; sea routes in the Finnish bay of the Baltic sea and on the Caspian Sea; as well as a scheduled passenger service on the North Dvina River; and on the Amur River as a ferry across the Russian–Chinese border.

One Irbis of four in operation is based in Olcott, New York, US and, after completing the demonstration programme on Lake Ontario, is passing a certification procedure to be assessed for operation in this area.

Classification: The craft is classed KM*1 SVPA Register of RFAS.

Propulsion: The engines are two Deutz BF6L913C air-cooled diesels, driving two four-blade, variable-pitch (forward and reverse thrust) ducted propellers.

Specifications

Length overall	16.5 m
Beam	6.2 m
Draught	0.3 m
Weight, min	7.4 t, passenger version
–	6.5 t, freight version
Weight, max	10.7 t
Passengers	32
Fuel capacity	0.6 t
Propulsive power	2 × 141 kW
Max speed	30 kt
Operational speed	24.3 kt
Range	250 n miles
Operational limitation	Sea State 3, wind speed 8 kt
Obstacle clearance	0.4 m

Argo

Designed for year-round transportation of six-nine passengers or 0.7 t of cargo on rivers and lakes. The hull is fabricated from corrosion resistant aluminium alloy with a GRP superstructure.

Specifications

Length overall	8.1 m
Beam	3.95 m
Crew	1
Passengers	9
Maximum speed	43 kt

Propulsion: Two VAZ 2112 petrol engines or two Steyr M14TCA diesel engines.

ALLIGATOR

The Alligator is designed for year-round transportation of 6–9 passengers or 0.7 t

Vessel type	Vessel name	Yard No	Length (m)	Speed (kt)	Seats	Vehicles	Originally delivered to	Date of build
Gepard	–	–	6.6	24.0	–	none	–	1981–series production
Puma	–	–	12.2	22.0	16	none	–	1988–series production
Irbis	–	–	16.5	24.3	32	none	–	1989
Irbis	–	–	16.5	24.3	32	none	–	1989
Irbis	–	–	16.5	24.3	32	none	–	1989
Argo	–	–	8.1	43.0	9	none	–	2004
Alligator	–	–	8.5	49.0	10	none	–	2002

of cargo on rivers and lakes. The hull is fabricated from corrosion resistant aluminium alloy with a GRP superstructure.

Specifications

Length overall	8.5 m
Beam	3.65 m
Crew	1
Passengers	9
Maximum speed	49 kt
Operational speed	45

Propulsion: Two VAZ 2112 petrol engines

Pormornic Astrakhan Shipyard

Part of the OMZ Group
20 Lenin Street, Astrakhan 414014, Russian Federation

Tel: (+7 8512) 22 02 29
Fax: (+7 8512) 22 55 20
e-mail: korabel@astranet.ru
Web: www.mnpglobal.com

Otherwise known as the Karl Marx shipyard, this company has in the past produced a number of air cushion vehicles. It is understood that the shipyard is still involved in high-speed craft production although details of the numbers produced are unknown. In 2002 the company was acquired by United Heavy Machinery (OMZ).

Luch 0506568

LUCH-1

A sidewall ACV for river use was completed at the Astrakhan Shipyard and underwent State trials in September 1983. Production of the ACV Luch-1 began at the Moscow Shipyard. Designed to replace the 12 year old Zarnitsa, the Luch, Design No 14351, carries two crew and up to 66 passengers, 15 standing. Of all-welded aluminium construction, it has a maximum operating weight of 22.6 tonnes and a top speed of 24 kt. The shallow on-cushion draught of just over 0.5 m allows it to run bow-on to flat sloping river banks so passengers can embark and disembark.

New design features permit operation on large rivers such as the Don, Kama, Oka and Volga, where higher waves are encountered.

Compared with the Zarnitsa, Luch has a more powerful diesel, the 382 kW 3KD12H-520 and uses a lighter form of construction. The overall dimensions of the craft permit rail transport to distant rivers and canals.

During trials, before the commissioning of the first of the class, it was demonstrated that its operating and technical performance is significantly superior to that of the Zarnitsa. The craft is built to the requirements of the R class of the RFAS River Craft Registry and has been produced in quantity.

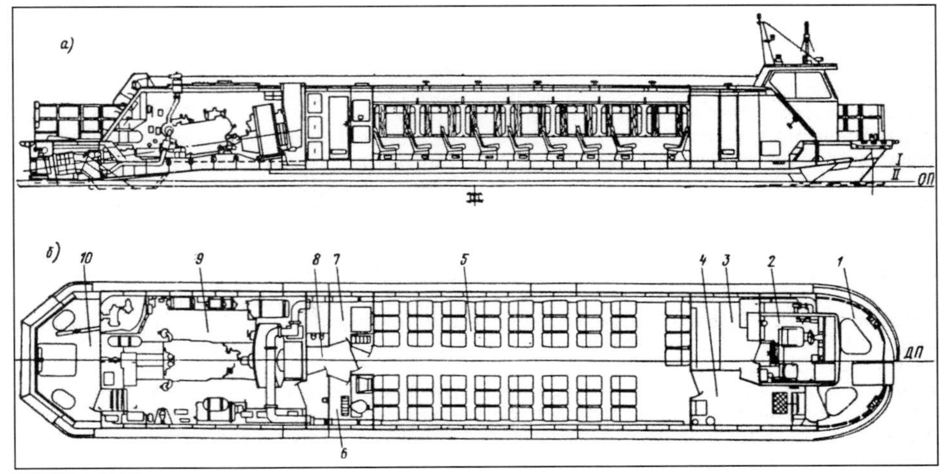

Inboard profile and plan of Luch: (a) longitudinal section; (b) deck plan; (I) waterline in displacement mode; (II) Waterline when cushionborne; (1) fore deck; (2) wheelhouse; (3) duty compartment; (4) vestibule; (5) Passenger saloon; (6) Toilet; (7) Storeroom; (8) Vestibule; (9) Engine room; (10) Aft deck 0506569

Specifications

Length overall	22.8 m
Beam	3.9 m
Draught, hullborne	0.7 m
Draught, on-cushion	0.5 m
Weight, min	15.4 t
Weight, max	22.6 t
Crew	2
Passengers	66
Propulsive power	382 kW
Max speed	23.8 kt
Range	162 n miles

Structure: The buoyancy structure is built in welded marine grade aluminium alloy. The superstructure is built in D16 alloy. Superstructure and bulkheads are riveted together. Pressed panels are employed throughout the hull to improve the external appearance and reduce the volume of assembly work necessary during construction.

The skirt is a fingered-type at bow, bag-type aft. Bow segments attached to an easily replaceable module.

Propulsion: The engine is a Type 81H12A diesel with a maximum power of 380 kW or a Type 3KD12H-520, 382 kW. Normal operating power 346 kW.

The engine is mounted aft and drives, via a transmission shaft, a centrifugal fan for lift and a water-jet rotor for propulsion. The water-jet rotor and its bearing can be replaced while the craft is in displacement condition without having to lift it out of the water.

Electrical system: One 1.2 kW 24 V DC engine-operated generator and batteries. During prolonged night stops power can be supplied by a shore-based 220 V, 50 Hz electrical supply.

Control: Twin rudder vanes in the water-jet stream control craft heading. Thrust reversal is achieved by water flow deflectors.

Vessel type	Vessel name	Yard No	Length (m)	Speed (kt)	Seats	Vehicles	Originally delivered to	Date of build
LUCH-1	*Luch-1*	-	22.8	-	66	none	-	Series production

Sosnovka Shipyard

Sosnovka Vyatsko-Polyansky, Kirovskaya Oblast 612980, Russian Federation

It is not known whether this shipyard continues to produce lightweight high-speed craft. The most recent craft for which details are available are the Rassvet, Plamya, Raduga-2 and Bargusin.

RASSVET (DAWN)

A water-jet-propelled sidewall passenger ferry, Rassvet is designed for local sea routes of limited water depth. It is an offshore counterpart to the Orion sidewall ACV. The Rassvet serves resort routes in the Crimea, on the Caspian and Baltic Seas as well as on large lakes and reservoirs. Like the Orion and Zarnitsa, its two predecessors, it can run bow-on to flat, sloping beaches to embark and disembark passengers. Landing on a beach is facilitated by an articulated gangway with a hydraulic drive.

Rassvet's features include shallow draught, good manoeuvrability and a relatively simple construction.

The water-jet reversing/steering system is specially protected to enable the craft to moor alongside existing berths built originally for small conventional displacement ferries.

Rassvet is designed to carry 80 passengers during daylight hours on coastal routes in conditions up to force 4. It complies with RFAS Registration classification KM*II Passenger ACV Class.

Specifications

Length overall	26.7 m
Beam	7.1 m
Draught, hullborne	1.27 m
Draught, on-cushion	0.8 m
Weight, max	47.5 t
Passengers	80
Propulsive power	2 × 383 kW

Vessel type	Vessel name	Yard No	Length (m)	Speed (kt)	Seats	Vehicles	Originally delivered to	Date of build
Raduga-2	–	–	–	–	–	–	–	–
Plamya	–	–	26.1	–	–	–	–	–
Rassvet	Series production	–	26.7	23.0	80	–	–	Series production
Orion	Series production	–	25.8	28.6	80	–	–	1973–series production
Bargusin	Series production	–	32.4	27.0	130	–	–	1989–series production began in 1991

Lift power	110 kW
Max speed	29 kt
Operational speed	23 kt
Range	190 n miles

Structure: The hull and superstructure are built in aluminium-magnesium alloy. The hull is of all-welded construction in AlMg-61 and the decks, superstructure, pilot-house and partitions are in AlMg-5 alloy

Propulsion: Power for the water-jet system is provided by two 3D12N-520 lightweight (3.54 kg/kW) irreversible, high-speed four-cycle V-type marine diesels each with a gas-turbine supercharger and a rated power of 383 kW at 1,500 rpm. Each powers a two-stage water-jet impeller. Water inlet scoops are arranged in the sidewalls and the pump ports, each comprising two rotors and two straightening devices, are installed in the sidewalls behind the transoms. Cushion air is generated by a single 110 kW PD6S-150A diesel driving an NTs6 centrifugal fan.

ORION-01

Design of the Orion, a rigid sidewall ACV with seats for 80 passengers, was approved in Moscow in Autumn 1970. The prototype, built in Leningrad, began trials in October 1973 and arrived at its port of registry, Kalinin, in late 1974, bearing the serial number 01.

The craft is used for passenger ferry services along shallow rivers, tributaries and reservoirs and can land and take on passengers bow-on from any flat sloping bank. It is faster than the Zarnitsa and its comfort and performance are less affected by choppy conditions. The cruising speed of the vessel, which is propelled by water-jets, is 28 kt. It belongs to the R class of the Soviet River Register.

Series production of this vessel is being undertaken at the Sosnovskaya Shipbuilding Yard in Kirovskaya Oblast.

Specifications

Length overall	25.8 m
Beam	6.5 m
Draught, hullborne	0.8 m
Draught, on-cushion	0.5 m
Weight, min	20.7 t
Weight, max	34.7 t
Crew	3
Passengers	80
Propulsive power	2 × 452 kW
Max speed	32 kt
Operational speed	28.6 kt
Range	216 n miles
Operational limitation	wave height 1.2 m

Structure: The hull is similar in overall appearance to Zarya and Zarnitsa types. All-welded structure in aluminium-magnesium alloy. Lateral framing throughout the hull with the exception of the bow and stern decks, where longitudinal frames have been fitted. The superstructure and wheelhouse are of welded and riveted duralumin construction on longitudinal framing.

Propulsion: Integrated system powered by two 3D12N-520 diesels mounted in an engine room aft. Each engine drives a Type Ts 39-13 centrifugal fan for lift via a cardan shaft, a semi-submerged single-stage water-jet rotor for propulsion.

Bargusin

0506576

Plamya

0506577

PLAMYA (FLAME) ACV RIVER FIRETENDER

The Central Design Bureau at Nizhni Novgorod has developed a river-going fire tender. The craft is based on the hull of the Orion sidewall-type passenger ferry, but the passenger cabin superstructure has been replaced by an open deck forward to accommodate a tracked or wheeled firefighting vehicle and its crew. Plamya is designed for the fast delivery of an off-road firefighting vehicle and its crew to points on lakes, reservoirs and major rivers near forest fires. The craft can be beached bow-on on the river bank and the firefighting vehicle or bulldozer offloaded across the bow ramp.

Specifications

Length overall	26.1 m
Beam	6.5 m
Draught, hullborne	0.9 m
Draught, on-cushion	0.7 m
Weight, max	34.5 t
Payload	7.3 t
Crew	3
Propulsive power	2 × 452 kW
Max speed	27 kt

Classification: Plamya meets the requirements of the Register of Shipping of the RFAS and is constructed to 'R' Class in the RSFSR Inland Waterways Register. It is in service on the River Kama.

Structure: All-welded structure in AlMg-61 aluminium-magnesium alloy. Frames, plates, partitions and roof of superstructure in D16 alloy.

The bow skirt is a triple-row, fully segmented type; stern, bag and finger type.

Propulsion: Integrated system powered by two 3D12N-520 marine diesels. Each engine drives a Type Ts 39-13 centrifugal fan for lift and, via a cardan shaft, a semi-submerged single-stage water-jet rotor for propulsion.

RADUGA-2

Little is known about this amphibious ACV, which was first reported in the Soviet press late Spring 1981. Designed at the Krasnoye Sormovo ship and ACV building facility at Nizhni Novgorod, Raduga-2, in common with the Gepard and Klest, is powered by an automotive engine, the Chayka.

BARGUSIN

The first production sidewall hovercraft (SES), Bargusin, was built in 1989 under the design developed by CKB 'Vympel'. The owner of the vessel is the Vostochno-Sibirskaya River Steamship Line.

The craft has a good seaworthiness and is intended for passenger ferry services along lakes, reservoirs and rivers. Having small draught it can land and take on passengers bow-on from any sloping bank.

The initial run of the Bargusin was organised in 1989 along the European waterways from Sosnovka to Belomorsk and then after transportation via motor-ship Kola by the North Sea Line from Tiksi to Yakutsk along the river Lena.

Bargusin's successful sea and strength trials took place at Baikal lake in Autumn 1990. Series production of this vessel started in 1991.

Specifications

Length overall	32.4 m
Beam	6.4 m
Draught, hullborne	1.4 m

Draught, on-cushion	0.8 m
Weight, min	54.5 t
Weight, max	70.8 t
Passengers	130
Propulsive power	2 × 590 kW
Lift power	220 kW
Operational speed	27 kt
Range	325 n miles
Operational limitation	wave height 1.2 m

Classification: M standards of the River Register of the Russian Soviet Federal Republic.

Structure: The hull and superstructure are built in AlMg-61 aluminium-magnesium alloy and are of all-welded construction.

Propulsion: Main engine: two m 401A-1 (12 CHSN 18/20) diesels, driving propellers via standard reverse-coupling.

A special centrifugal fan designed and manufactured by NPO Vint generates the air cushion. A 7D 12A (12 CHN 15/18) diesel is used for driving the fan. Pressure and air supply, necessary for the craft's stable motion both in calm water and in waves, are provided at 160 kW and 1,350 to 1,400 fan rpm. The five-blade propellers are within the draught of the sidewalls.

Vympel Ship Design Company

Nartov Str 6, Building 12, 603104 Nizhni Novgorod, Russian Federation

Tel: (+7 831) 233 41 49
Fax: (+7 831) 230 20 96
e-mail: info@vympel.ru
Web: www.vympel.ru

Viacheslav V Shatalov, *General Director*
Vladimir K Zoroastrov, *Chief Designer*

This Design Bureau has been responsible for the design and development of the following sidewall hovercraft: Zarnitsa, Orion, Chayka, Plamya, Luch, Luch-2, Altair, Olkhon and Bargusin. These craft have also been built by the Sosnovskaya, Astrakhan and Moscow shipyards.

The company has designed a 17.5 m ice-breaking hovercraft, Toros-1, with a displacement of 307 tonnes.

Luch-2 passenger sidewall ses

This vessel design won recognition at the Eureka 1994 exhibition for outstanding design. The water-jet-propelled craft is used for passenger transportation within the Russian river system.

Specifications

Length overall	23.3 m
Beam	3.9 m
Draught, hullborne	0.7 m
Draught, on-cushion	0.6 m
Weight, min	17.6 t
Passengers	57
Propulsive power	382 kW
Operational speed	22 kt
Range	173 n miles

Altair II passenger SES (DESIGN)

This fast passenger-carrying SES is a development of the Altair I version described in previous editions, for sea, lake and estuarial applications.

Specifications

Length overall	31.8 m
Beam	6.0 m
Draught, hullborne	1.4 m
Draught, on-cushion	0.8 m
Weight, min	60 t
Passengers	130
Propulsive power	2,700 kW
Lift power	240 kW
Operational speed	37 kt
Range	340 n miles

Olkhon Passenger SES (Design)

This is a double-deck, twin-screw vessel with passenger state cabins in the fore and middle portions of the ship. The craft has a bow platform so passengers can embark and disembark at an unequipped bank.

Specifications

Length overall	32.4 m
Beam	6.4 m
Draught, hullborne	1.5 m
Draught, on-cushion	0.8 m
Weight, min	65.2 t
Weight, max	75.8 t
Passengers	130
Propulsive power	1,470 kW
Lift power	220 kW
Operational speed	27 kt
Range	540 n miles

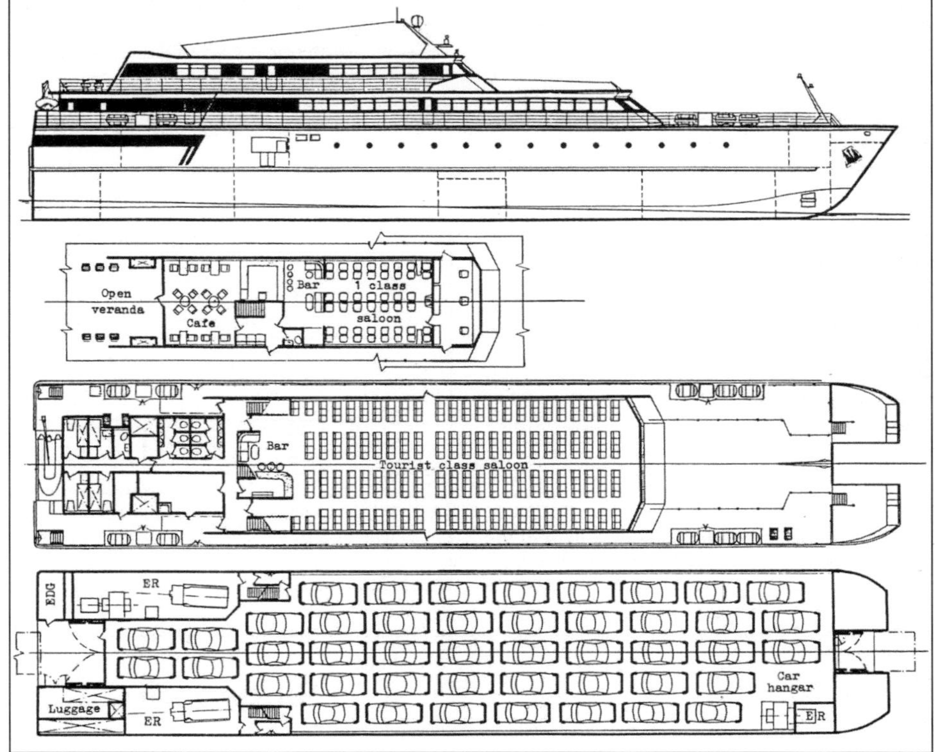

General arrangement of the Tourist SES (Design) 0506954

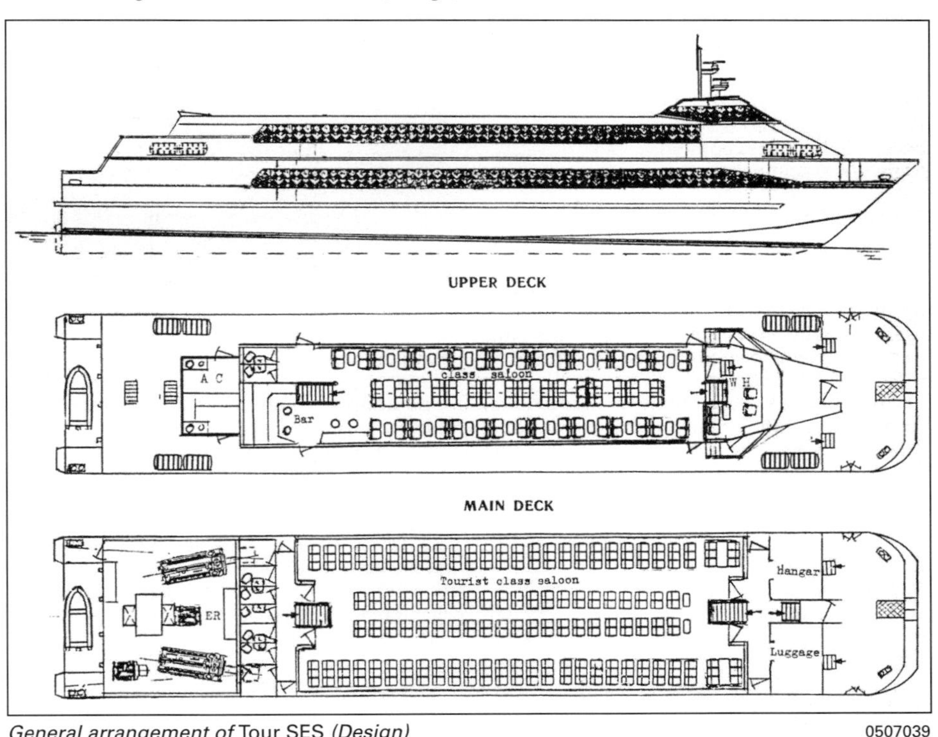

General arrangement of Tour SES *(Design)* 0507039

Vessel type	Vessel name	Yard No	Length (m)	Speed (kt)	Seats	Vehicles	Originally delivered to	Date of build
Luch-2 Passenger Sidewall SES	–	–	23.3	22.0	57	none	–	1994
Toros-1	–	–	17.5	–	–	–	–	–

Tourist car/passenger SES (DESIGN)

This vessel has been designed to meet the IMO HSC Code of Safety for operation in sea, river estuaries and coastal areas and is the largest SES design proposed by the Vympel Bureau.

Specifications

Length overall	64.0 m
Beam	12.0 m
Draught, hullborne	2.8 m
Draught, on-cushion	1.7 m
Crew	12
Passengers	370
Vehicles	45
Operational speed	40 kt
Range	600 n miles
Operational limitation	Beaufort 5

Luch-2 at speed 0506953

TOUR PASSENGER SES (DESIGN)

This vessel has been designed to carry a large number of passengers at speeds of over 30 kt on estuarial waters, coastal areas and lakes.

Specifications

Length overall	44.0 m
Beam	8.0 m
Draught, hullborne	2.0 m
Draught, on-cushion	1.2 m
Payload	50,000 kg
Passengers	360
Operational speed	30 kt
Range	500 n miles

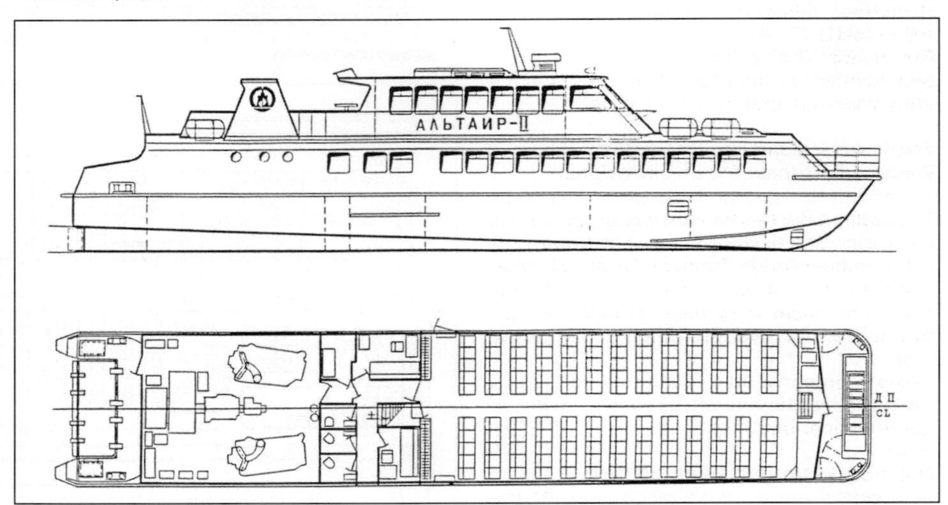

General arrangement of the Altair II SES (Design) 0507038

Singapore

Scat Technologies Pte Ltd

No 4 Jalan Samulum, Singapore 629121

Tel: (+65) 63 85 97 82
Fax: (+65) 63 85 97 81
e-mail: sales@scat.com.sg
Web: www.scat.com.sg

Captain T Chia, *President*

SCAT Technologies Pte Ltd has its headquarters in Singapore and is owned by the CTW Pte Ltd Group of companies. A marketing division in the US, SCAT USA Inc, has also been formed to sell hovercraft in the US and Europe.

The company has built a number of small hovercraft aimed at the recreational market but has recently started production of larger commercial craft including fire fighting, ambulance and passenger hovercraft.

SF3000 Falcon

Specifications

Length overall	15.0 m
Beam	4.8 m
Crew	2
Passengers	32
Range	300 n miles
Operational limitation	1.25 m wave height
Obstacle clearance	0.6 m

SF2000 0507041

Classification: ABS air cushion craft.

Structure: All-welded aluminium hull with GRP or aluminium superstructure.

Propulsion: The vessel is powered by two Deutz air-cooled diesels Type BF8L413F, each rated at 192 kW at 2,300 rpm and each driving a ducted variable-pitch propeller. Lift is supplied by two double air inlet centrifugal fans.

SF2000 Eagle

A 32 kt, 12 m craft has been delivered to a Singaporean company for a Hover Taxi service between Singapore and Batam Island, with a passenger capacity of 20 and a baggage capacity of 300 kg.

Specifications

Length overall	11.6 m
Crew	2
Passengers	20
Operational speed	32 kt
Range	300 n miles
Operational limitation	1.5 m wave height
Obstacle clearance	0.6 m

Vessel type	Vessel name	Yard No	Length (m)	Speed (kt)	Seats	Vehicles	Originally delivered to	Date of build
SF2000 Eagle	–	–	11.60	32.0	20	none	–	–
SF3000 Falcon	–	–	15.00	–	32	none	–	–
Fire Tender Hovercraft	*DESIGN*	–	11.30	28.0	2	none	–	–
Rescue XT Fire & Rescue	–	–	3.36	30.0	2	none	–	–
Sovereign 600 Ambulance	–	–	5.60	30.0	5	none	–	–
Sovereign 500	–	–	5.6	30.0	5	none	–	1989
Commuter 700 Shuttle	–	–	7.13	30.0	6	none	–	–

Classification: ABS air cushion craft.
Structure: All-welded aluminium hull with GRP superstructure.
Propulsion: The vessel is powered by two Deutz air-cooled diesels Type BF6L913C, each rated at 131 kW at 2,500 rpm and each driving a ducted variable-pitch propeller. Lift is supplied by two double air inlet centrifugal fans also powered by the main engines.

Rescue XT fire & rescue hovercraft

Specifications

Length overall	3.36 m
Payload	310 kg
Crew	1
Passengers	2
Operational speed	30 kt

Structure: Injection foam-filled one piece ABS plastic.
Propulsion: The craft is powered by a Rotax 582 two-cycle, cylinder oil injected engine with a 1.2 m diameter propeller.

Fire Tender Hovercraft (DESIGN)

Designated the SCAT 1060 this 11 m hovercraft is intended as an amphibious fire and rescue tender with pollution control capability. The craft's lift fan is powered via a hydrostatic drive from the main engines and two fire pumps are driven mechanically from the main engines, coupled to two fire monitors, one on either side of the craft.

Specifications

Length overall	11.3 m
Length waterline	10.6 m
Beam	6.0 m
Hull beam	4.62 m
Payload	2,500 kg
Crew	6
Passengers	2
Operational speed	28 kt

Structure: GRP and foam sandwich hull and superstructure.
Propulsion: The craft is powered by two General Motors 6.5L V8 turbo diesels each rated at 153 kW at 3,200 rpm.

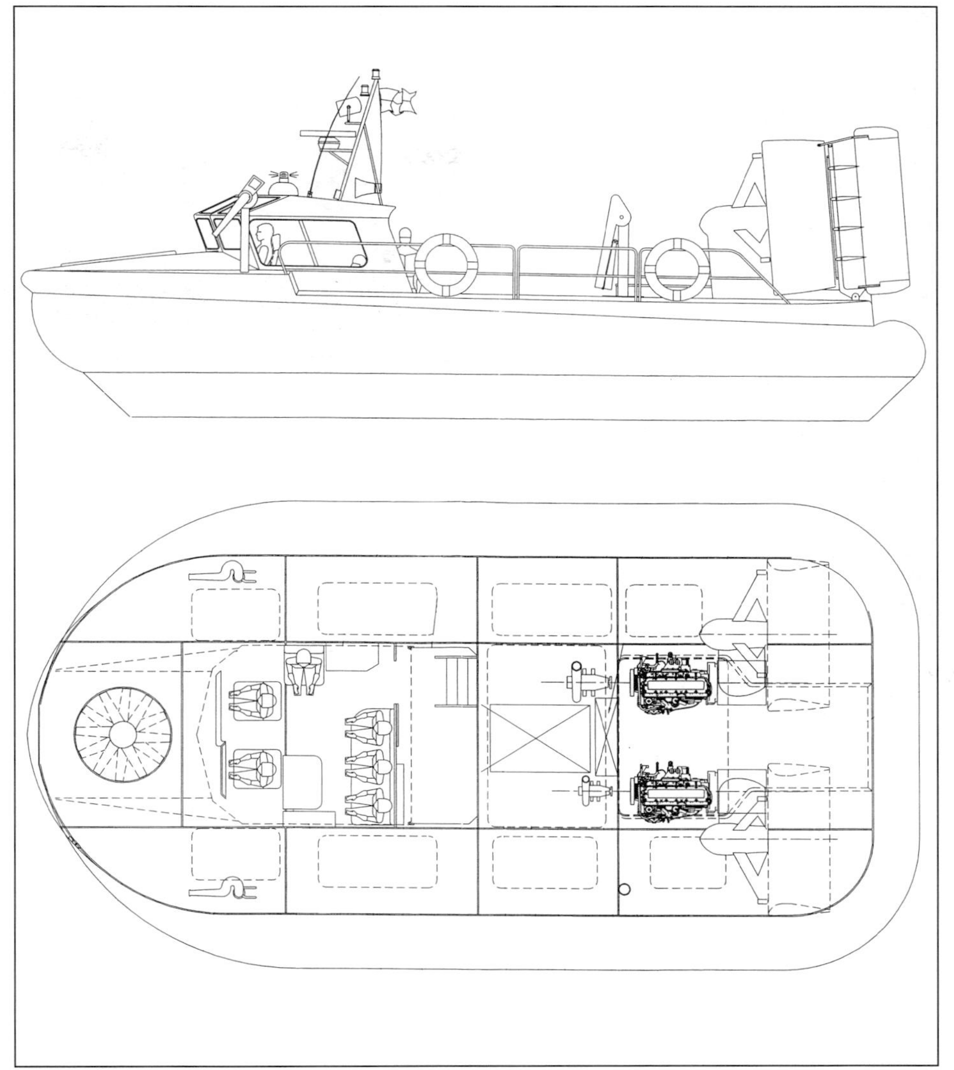

Fire Tender Hovercraft (Design) 0024740

Singapore Technologies Marine Ltd

7 Benoi Road, Singapore 629882

Tel: (+65) 861 22 44
Fax: (+65) 861 30 28
email: mktg.marine.stengg.com
Web: www.stengg.com

See Leong Teck, *President*
Han Kew Kwang, *Chief Operating Officer*
Larry Loon, *President, Naval Business Group*
Yew Shyan Teh, *Senior Vice-President (Tuas Yard)*
Twoon Kok Yang, *Vice-President (Benoi Yard)*
Tan Ching Eng, *Senior Vice-President (Design)*

Singapore Technologies Marine, the marine arm of Singapore Technologies Engineering, was established in 1968 as a specialist shipyard. Current areas of expertise include the building of specialised commercial vessels, military engineering equipment fabrication and the reconstruction and modernisation of old vessels. STSE is located in a 30 acre site at the mouth of the Benoi Basin.

In 1987, the company built an Air Vehicle Tiger 40 hovercraft under license from Air Vehicles of the UK.

Air Vehicles Tiger 40 Hovercraft built by Singapore Technologies 0506597

Tiger 40

The Air Vehicles Tiger 40 was built by Singapore Technologies Shipbuilding & Engineering Ltd for a leasing company in Singapore, SAL Leasing, which owns the craft. In 1987 this craft was leased to the Singapore Navy. Tiger 40 craft are marketed in Asia by Singapore Technologies Shipbuilding and Engineering Ltd.

Vessel type	Vessel name	Yard No	Length (m)	Speed (kt)	Seats	Vehicles	Originally delivered to	Date of build
Tiger 40	–	–	–	–	–	–	SAL leasing	–

Ukraine

Feodosia Shipbuilding Association (Morye)

Desantnikov Str 1, 98176 Feodosia, Crimea, Ukraine

Tel: (+380 6562) 325 56
Fax: (+380 6562) 323 73
e-mail: fsa@morye.kafa.crimea.ua
Web: www.morye.crimea.ua

Vitaly Krivenko, *Chairman*
Boris Kozlov, *Technical Director*
Grigory Klebanov, *Head of Marketing and Sales*

Pormornik U420 at speed 0077807

POMORNIK (ZUBR)

The Pormornik, also known as the Zubr or Project 1232.2 was designed by the Almaz Central Design Bureau in St Petersburg, with five having been built for the Russian Navy by Almaz Shipbuilding Company, the last delivered in 1994. Another was constructed at Feodosia in the Ukraine.

The Greek Navy ordered four Pormornik hovercraft in 2000, in the form of fast attack vessels. Two vessels were purchased from Feodosia and two from Almaz Shipbuilding Corporation.

Specifications

Length overall	57.6 m
Beam	25.6 m
Displacement	550 t
Payload	130 t
Operational speed	55 kt
Maximum speed	60 kt
Range	300 n miles

Propulsion: Five Type NK-12MV gas turbines each rated at 8,826 kW (three for propulsion, two for lift).

United Kingdom

ABS Hovercraft Ltd

The Mill Yard, Nursling Street, Southampton, Hampshire SO16 0AJ, United Kingdom

Tel: (+44 2380) 73 33 00
Fax: (+44 207) 504 80 08
e-mail: info@abs-hovercraft.com
Web: www.abs-hovercraft.com

J Cutts, *Technical Manager*

ABS Hovercraft Limited designed the world's next-generation hovercraft with an international team of designers taking full advantage of advanced composite FRP materials and technology. The award winning ABS M-10 and its variants are fully amphibious and designed for military applications.

In 2004 ABS were successful in a consortium bid to supply hovercraft and catamarans to Mumbai in India.

ABS M-10 Hydrographic Hovercraft BEASAC IV 0084061

ABS M-10 MULTIVARIANT HOVERCRAFT

All ABS M-10 hovercraft have a maximum operational speed of 50 kt (although speeds of 60 kt have been achieved), a cushion height of 1.0 m and a disposable load carrying ability of 10.0 tonnes. The vessel has performed well in rough weather conditions, operating on cushion, in gale force winds and swells of more than 4 m. The robust design of the ABS M-10 was proven during extensive sea evaluation trials by a number of military forces in the North Sea and Baltic and it is the only hovercraft to have successfully circumnavigated the Gulf of Bothnia under Winter conditions. Advantages over other hovercraft include its hull and superstructure construction comprising advanced FRP composites that do not suffer the fatigue or corrosion of aluminium alternatives. Hull durability and life are superior and through life maintenance costs are lower.

ABS M-10 Hovercraft 0007622

The standard M-10 can be operated by experienced boat crews after 40 hours of training. Powered by twin standard MTU diesel engines its 10 tonne disposable payload capability and the 2.5 m wide bow ramp allows rapid deployment of troops and vehicles well inland of the waterline.

Based around a standard hull, engine and transmission package, all ABS M-10 Hovercraft variants are available for any application with current designs suitable for Interception and Assault, Logistic Supply, Special Forces,

Vessel type	Vessel name	Yard No	Length (m)	Speed (kt)	Seats	Vehicles	Originally delivered to	Date of build
ABS M-10 Multivariant Hovercraft	–	–	20.6	50.0 (max)	As required	1 × 4WD	–	Series production

ABS M-10-29 Logistic Supply Hovercraft for the Sri Lanka Navy 0538377

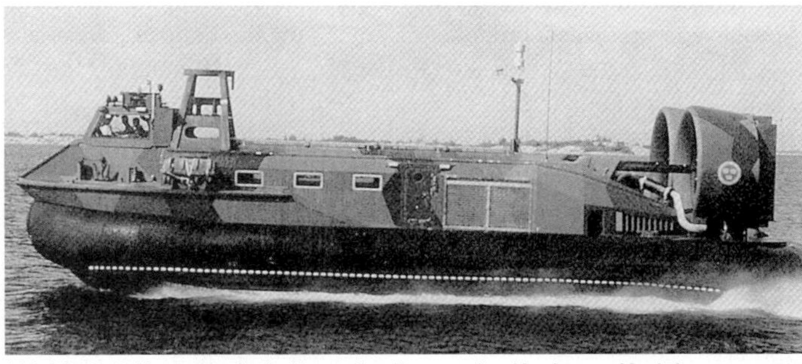

KA M-10-X Hovercraft, built under license by Kockums Karlskronavarvet, Sweden 0081001

Search and Rescue, Pollution Control, Disaster Relief, High-Speed Hydrographic and Coastguard applications. In addition, passenger and tourist versions are available.

ABS M-10 hovercraft are currently in use with the Swedish Coastal Artillery, the Sri Lanka Navy and in commercial operations in Belgium.

Specifications

Length overall	20.6 m
Beam	8.8 m
Payload	10 t
Crew	2
Fuel capacity	4,000 litres
Propulsive power	1,000 kW
Maximum speed	50 kt
Range	600 n miles
Obstacle clearance	1 m

Classification: Lloyd's +100 A1 SSC ACV Group 3.

Structure: Single-skin FRP hull, FRP foam sandwich deck and structure.

Aluminium Shipbuilders Ltd

Fishbourne Quay, Ashlake Copse Road, Fishbourne, Isle of Wight, PO33 4EY, United Kingdom

Tel: (+44 1983) 88 22 00
Fax: (+44 1983) 88 47 20

James Davies, *Managing Director*

Aluminium Shipbuilders Ltd was formed in 1983 by John Davies. Initially the company undertook a design and build contract for small specialist craft.

Aluminium Shipbuilders is known for its building of catamarans to the designs of International Catamaran Designs Pty Ltd, Australia.

The company has engaged in considerable hovercraft fabrication work. In particular, contracts have been received from Griffon Hovercraft Ltd and Westland Aerospace. By the end of 2001 ASL had completed over 50 hulls covering the current range of Griffon Hovercraft. For Westland Aerospace, ASL was involved in the total fabrication, fit out and trials of a BHC AP1-88 hovercraft for operation by The Northern Shipping Company in the White Sea area of Russia. The company's most recent delivery was a Hoverwork BMT 130 which was delivered in 2006.

Griffon Hovercraft Ltd

Head Office and Boatyard: 8 Hazel Road, Woolston, Southampton, Hampshire SO19 7GA, United Kingdom

Tel: (+44 23) 80 40 35 47
Fax: (+44 23) 80 40 67 47
e-mail: sales@griffonhovercraft.com
Web: www.griffonhovercraft.com

J H Gifford, *Chairman and Managing Director*
M I Gibson, *Sales Manager*

Founded in 1976, Griffon Hovercraft has concentrated on the design, development, manufacture and operation of small- to medium-sized amphibious hovercraft.

Since 1982, all Griffon hovercraft have been powered by diesel engines, to give the necessary manageability and reliability required for the harsh environments in which the hovercraft operate, ranging from ice floes in the Arctic to isolated jungle rivers in the Tropics.

Construction of the first diesel-engined hovercraft, the Griffon 1000 TD began in June 1982.

The company now has the largest range of manufactured hovercraft available today, with craft ranging from payloads of 375 kg up to 10 t. The company has concentrated on producing economical, easily maintainable craft for the workboat, survey, crash rescue, navy, army, paramilitary and ferry markets.

In 1997, to celebrate its 21st anniversary, Griffon Hovercraft Limited extended the present range of craft to include new Mk 2 versions of the 2000, 3000 and 4000 series craft. All craft in the Griffon range have commonality of design, systems and spare parts and are designed, wherever possible, to comply with the internationally recognised requirements of the IMO, the UK MCA and Lloyd's Register of Shipping.

Griffon Hovercraft recently licensed Kvichak Marine of Seattle, US to manufacture its range of amphibious hovercraft. The first order for a 20 seat Griffon 2000 TD was received from Crowley Marine Inc in 2003. Two further vessels have been ordered by the Polish Border Guard, as well as one by a commercial operator in Greenland and four have been delivered to the Pakistan Navy.

1000 TD for WAPDA, Pakistan 0506872

Griffon 2000 TDX with the Belgian Army recovering a drone 0506957

380 TD

The 380 TD (formerly the 375 TD) is the smallest craft in the Griffon range and is built on rigid inflatable boat lines. Propulsion is provided by a 63 kW VW water-cooled diesel engine and lift air by a Lombardini 12LD 14.9 kW air-cooled engine. This craft has a payload of 380 kg or five people. A hover-on/hover-off road trailer allows one-man deployment of the craft.

Specifications

Length overall	6.80 m
Beam	3.76 m
Payload	380 kg
Passengers, including driver	5
Speed	30 kt

Vessel type	Vessel name	Yard No	Length (m)	Speed (kt)	Seats	Vehicles	Originally delivered to	Date of build
1000TD	–	001	8.8	27.0	10	none	Normandeau, Canada	1984
1000TD	–	003	8.8	27.0	10	none	Geographical Surveys Inc, US	1984
1000TD	–	004	8.8	27.0	10	none	Geographical Surveys Inc, US	1984
1000TD	–	002	8.8	27.0	10	none	Geographical Surveys Inc, US	1984
1000TD	–	023	8.8	27.0	10	none	WAPDA, Pakistan	1984
1500TD	–	006	10.2	27.0	16	none	Chiriqui Hovercraft	1985
2500TD	*Rain Dance*	001	–	–	–	none	Hover Systems Inc, US	1985
1500TD	–	001	10.2	27.0	16	none	Clements	1987
1500TD	–	003	10.2	27.0	16	none	Griffon Hovercraft	1987
1000TD	–	008	8.8	27.0	10	none	WAPDA, Pakistan	1988
2000TD	–	011	10.6	27.0	22	none	Texaco	1988
2000TDX	–	010	10.6	35.0	22	none	HoverEire	1989
1000TD	–	012	8.8	27.0	10	none	Royal Thai Navy	1990
1000TD	–	014	8.8	27.0	10	none	Royal Thai Navy	1990
3000TD	–	017	15.0	35.0	30	none	Thai General Transport Ltd	1991
2000TDX	–	018	10.6	35.0	22	none	Sea Bus, Tunisia	1991
3000TD	–	016	15.0	35.0	30	none	Thai General Transport Ltd	1991
2000TDX	–	019	10.6	35.0	22	none	Sea Bus, Tunisia	1991
4000TD	*Shri Bajarangtasbapa*	–	17.9	35.0	51	none	Triton, India	1992
4000TD	*Shri Saibaba*	–	17.9	35.0	51	none	Triton, India	1992
2000TDX	–	022	10.6	35.0	22	none	Swedish Coast Guard	1992
1000TD	–	015	8.8	27.0	10	none	Royal Thai Navy	1993
2000TDX (M)	C24	029	10.6	35.0	22	none	British Royal Marines	1993
2000TDX (M)	C22	027	10.6	35.0	22	none	British Royal Marines	1993
2000TDX (M)	C21	026	10.6	35.0	22	none	British Royal Marines	1993
2000TDX	–	025	10.6	35.0	22	none	Swedish Coast Guard	1993
2000TDX (M)	C23	028	10.6	35.0	22	none	British Royal Marines	1993
2000TDX	–	024	10.6	35.0	22	none	Swedish Coast Guard	1993
375TD	–	030	6.6	–	4	none	Vigili del Fuoco, Italy	1994
2000TDX	–	034	10.6	35.0	22	none	Infaero, Brazil	1994
2000TDX	–	033	10.6	35.0	22	none	Infaero, Brazil	1994
2000TDX	–	031	10.6	35.0	22	none	Finland Frontier Guard	1994
2000TDX	–	032	10.6	35.0	22	none	Finland Frontier Guard	1994
2000TDX	–	035	10.6	35.0	22	none	Landsea Logistics, Nigeria	1995
375TD	–	036	6.6	–	4	none	NAU, Brazil	1995
2000TDX	–	039	10.6	35.0	22	none	Finland Frontier Guard	1995
2000TDX	–	038	10.6	35.0	22	none	Landsea Logistics, Nigeria	1995
2000TDX	–	037	10.6	35.0	22	none	Belgian Army	1995
2000TDX	–	040	10.6	35.0	22	none	Auckland International Airport	1996
375TD	–	044	6.6	35.0	4	none	Warsaw Police	1997
450TD	–	041	7.6	30.0	5	none	Landsea Logistics, Nigeria	1997
2000TDX Mk II	–	042	11.9	35.0	26	none	Flying Tiger Engineering, Burma	1997
450TD	–	043	7.6	30.0	5	none	Landsea Logistics, Nigeria	1998
375TD	–	045	6.6	–	4	none	Galicia State Government, Spain	1998
2000TDX Mk II	–	049	11.9	35.0	26	none	Klapeda State Seaport Authority, Lithuania	1998
2000TDX Mk II	–	053	11.9	35.0	26	none	Border Guard, Lithuania	1999
2000TDX Mk II	–	051	11.9	35.0	26	none	Border Guard, Estonia	1999
375TD	–	052	6.6	–	4	none	Hong Kong Department of Agriculture and Fisheries	1999
375TD	–	050	6.6	25.0	4	none	Dundee City Airport, UK	1999
3000TDX Mk II	–	048	15.4	37.0	30	none	Shell Petroleum Development Corporation, Nigeria	1999
3000TDX Mk II	–	047	15.4	37.0	30	none	Shell Petroleum Development Corporation, Nigeria	1999
3000TDX Mk II	–	046	15.4	37.0	30	none	Shell Petroleum Development Corporation, Nigeria	1999
8000TD(M)	–	054	19.9	42.0	84	none	Indian Coast Guard	2000
8000TD(M)	–	055	19.9	42.0	84	none	Indian Coast Guard	2000
8000TD(M)	–	056	19.9	42.0	84	none	Indian Coast Guard	2000
8000TD(M)	–	057	19.9	42.0	84	none	Indian Coast Guard	2000
450TD	–	060	7.6	30.0	5	none	Royal National Lifeboat Institute, UK	2000
8000TD(M)	–	058	19.9	42.0	84	none	Indian Coast Guard	2001
8000TD(M)	–	059	19.9	42.0	84	none	Indian Coast Guard	2001
8000TD(M)	–	061	19.9	42.0	80	none	Saudi Arabian Border Guard	2001
375TD	–	066	6.6	25.0	5	none	Shannon Airport, Ireland	2001
8000TD(K)	–	067	20.0	42.0	80	none	Korean National Maritime Police Agency	2001
8000TD(M)	–	062	19.9	42.0	80	none	Saudi Arabian Border Guard	2002
8000TD(M)	–	063	19.9	42.0	80	none	Saudi Arabian Border Guard	2002
470TD	–	069	8.04	30.0	5–7	none	Royal National Lifeboat Institute, UK	2002
470TD	–	070	8.04	30.0	5–7	none	Royal National Lifeboat Institute, UK	2002
8000TD(M)	–	064	19.9	42.0	80	none	Saudi Arabian Border Guard	2003
8000TD(M)	–	065	19.9	42.0	80	none	Saudi Arabian Border Guard	2003
8000TD	–	068	19.9	42.0	50	none	Changi Airport, Singapore	2003
2000TD	–	071	11.9	35.0	26	none	Crowley Marine for BP Alaska	2003
470TD	–	072	8.04	30.0	5–7	none	Royal National Lifeboat Institute, UK	2003
470TD	–	073	8.04	30.0	5–7	none	R.Hodson, UK	2003
2000TD	–	074	11.9	35.0	26	none	Navy, Pakistan	2004
2000TD	–	075	11.9	35.0	26	none	Navy, Pakistan	2004
2000TD	–	076	11.9	35.0	26	none	Navy, Pakistan	2004
2000TD	–	077	11.9	35.0	26	none	Navy, Pakistan	2004
8000	–	078	20.0	42.0	80	none	Korean National Maritime Police	2004
380TD	–	079	6.6	30.0	5	none	A.Hayward, Senegal	2004
470TD	–	080	8.0	30.0	6	none	Royal National Lifeboat Institute, UK	2005

Vessel type	Vessel name	Yard No	Length (m)	Speed (kt)	Seats	Vehicles	Originally delivered to	Date of build
470TD	–	081	8.0	30.0	6	none	Royal National Lifeboat Institute, UK	2005
380TD	–	082	6.8	30.0	5	none	Crimson Wings	2007
2000TD	–	083	12.7	35	25	none	KNI, Greenland	2005
2000TD	–	084	12.7	35	25	none	Polish Border Guard	2006
2000TD	–	085	12.7	35	25	none	Polish Border Guard	2006
8100TD	–	086	22.5	45	98	Humvee	Swedish Amphibious Battalion	2007
8100TD	–	087	22.5	45	98	Humvee	Swedish Amphibious Battalion	2007
8100TD	–	088	22.5	45	98	Humvee	Swedish Amphibious Battalion	2007
1000TD	–	089	9.0	27	11	none	Estonian Border Guard	2007
470TD	–	090	8.0	30	6	none	Korean Coast Guard	2007
470TD	–	091	8.0	30	6	none	Korean Coast Guard	2007
2000TD	–	092	12.7	35	25	none	Dr J K Hall, UK	2007

450TD

The 450 TD was designed and built in 1997. Powered by two 56 kW VW diesel engines, this craft was designed to have a 'get home' capability on one engine in case of machinery failure. A removable cabin roof and PVC side covers give weather protection to the passengers and cargo. Carrying a full payload of six people or 450 kg, the 450 TD can achieve a speed of 30 kt. Two of these craft are operated in Nigeria and one by the Royal National Lifeboat Institute in the UK.

Specifications

Length overall	7.60 m
Beam	3.40 m
Payload	450 kg
Passengers, including driver	6
Speed	30 kt

Griffon 2000TD Mk 1 (left) and the 2000TDX Mk II (right) 0007623

470TD

Three of these craft were delivered to the RNLI (UK) during 2002 and 2003 with a fourth vessel on order. The 470TD is a twin engined craft with a maximum speed in excess of 30 kt. Capable of seating six to eight or carrying a payload of 470 kg, and with a hovering height of 2.41 m and an obstacle clearance of 0.5 m, this vessel was designed specifically for the RNLI.

Specifications

Length overall	8.04 m
Beam	3.92 m
Payload	470 kg
Passengers, including driver	5–7
Speed	30 kt

1000TD

The 1000 TD carries a payload of 825 to 1,000 kg or 10 to 12 passengers and is the smallest Griffon hovercraft to be powered by an industrial diesel engine. The craft is available with a variety of superstructures. Folding side decks enable it to be loaded onto its purpose-built trailer and towed behind a small truck. It can also be fitted into a 40 ft container.

Specifications

Length overall	8.80 m
Beam	3.93 m
Payload	825–1,000 kg
Passengers	10
Cruise speed	27 kt

Propulsion: The 1000 TD is powered by a Deutz BF6L 913C, 141 kW air-cooled diesel engine.

1500TD

This variant is 1.8 m longer than the 1000TD, but with identical machinery. Capacity is increased to 1.5 tonnes or 16 persons. Folding side decks allow transportation within a 40 ft container.

These craft have been sold for tourism, survey work and pilot duties and have now been replaced by the Griffon 2000TD.

Details of the 1500TD are as for the 1000TD with the following exceptions:

Specifications

Length overall	10.2 m
Weight, min	2.3 t
Payload	1.5 t
Max speed	33 kt
Operational speed	27 kt

Griffon 4000TD 0506827

2000TD

The Griffon 2000 TD has the same engine and systems as the 1500TD, although with a larger propeller, a larger fan and an improved skirt design. These modifications enable the 2000TD to carry a 33 per cent larger payload but still retain the same performance. The cabin is also slightly enlarged to carry 20 to 24 passengers.

The engine, transmission, duct and pylon are mounted on a welded aluminium alloy subframe attached to the hull via resilient mounts.

Cooling air for the engine passes through a Knitmesh filter to remove spray and is drawn into the front of the engine by the engine cooling fan before passing over the cylinders and being drawn out from the rear of the engine bay by the propeller.

Sidebodies made from aluminium or composite materials fold upward for transport and are locked into the running position by use of struts either side. Forward and aft ballast tanks are fabricated from aluminium.

The skirts are a tapered HDL open loop type with similar segments at bow and sides.

Cones are fitted at the stern. Segments are fitted to the loop with stainless steel bolts and inner ends are attached using plastic shackles. All skirt maintenance can be done without lifting the craft.

Triple rudders in the GRP duct provide directional control. Elevators within the duct provide a degree of fore and aft trim augmented by a fuel ballast system. The craft is fitted with a skirt shift system operated by a small electrohydraulic ram which provides responsive control on roll trim, offsets crosswind effects and banks the craft into the turn.

Craft number 011 operated for 7½ years on the River Thames in London's Docklands as a shuttle bus for Texaco and Reuters and achieved an average reliability throughout this time of 99.5 per cent. This craft is now Griffon Hovercraft Ltd's demonstrator model.

Specifications

Length overall	10.6 m
Beam	4.5 m
Weight, min	2.6 t
Payload	2 t
Crew	2
Passengers	22
Propulsive power	140 kW
Max speed	31 kt
Operational speed	27 kt

Classification: IMO, the UK MCA and Lloyd's Register classification.

Propulsion: Integrated lift and propulsion system powered by a single Deutz BF6L913C air-cooled six-cylinder, in-line, turbocharged and intercooled diesel rated at 190 hp (140 kW) at 2,500 rpm. A 0.91 m diameter centrifugal lift fan is driven from the front of the crankshaft via an HTD toothed belt. Power for the 1.8 m diameter four-blade Robert Trillo Ltd designed ducted propeller is transmitted from the back of the engine via an automotive clutch to another HTD toothed belt transmission running inside the propeller support pylon.

Structure: Main hull is of welded riveted and bonded marine grade aluminium.

2000 TDX

This craft is virtually the same as the 2000 TD but is more powerful with a 239 kW or 265 kW V8 Deutz BF8L 513 engine giving it a superior into-wave and into-wind performance. A 2000 TDX was ordered by the Swedish Coast Guard in 1991 and, having operated the craft for nine months, an order for a further two was signed in January 1993.

In 1994–95 the Frontier Guard of Finland took delivery of three Griffon 2000 TDX craft and two further similar craft were delivered to Brazil as crash rescue craft at Rio de Janiero airport. In 1995 a number of Griffon 2000 TDXs were delivered to the Belgian Army, Elf Petroleum in Nigeria and to Nicaragua as a mobile medical clinic. In 1996 a Griffon 2000 TDX with a 2 m propeller was delivered to Auckland Airport, New Zealand, for an aircraft crash rescue role.

The UK Ministry of Defence ordered four Griffon 2000 TDX(M) hovercraft. Designed to carry a crew of two, driver and commander, plus 16 fully equipped Marine Commandos, each craft is equipped with a 7.62 mm GPMG (General Purpose Machine Gun), HF and VHF radios, radar, GPS and a variety of specialised equipment. Capable of speeds of over 35 kt, these craft are for 539 Assault Squadron (an integral part of Three Commando Brigade) and are used worldwide for high-speed amphibious assault, logistics support and other Commando-type applications. Each craft carries a payload of 2 t which can be quickly converted from the all-troop to the all-cargo configuration, or to a variable combination of both.

Four Griffon 2000 TDXs in service with the Royal Marines 0506873

Griffon 8000 TD(M) 0098036

Griffon 3000TD approaching beach 1326859

Griffon 380 TD 1326860

Details for the 2000 TDX are very similar to the 2000TD except for the propulsion and the following:

Specifications

Max speed	40 kt
Operational speed	35 kt

2000 TDX Mk II

The first Mk II version of the 2000 TDX was produced in 1997 and went into service as a passenger ferry in Burma. Further craft of this type are operating in Estonia and Lithuania.

The craft was designed to comply with the IMO code for high-speed craft and was understood to be the first such craft built to these standards. The significant differences between this design and the Mk I are: buoyancy tank redesign, a 2 m propeller, a raised 1.83 m wheelhouse, a longer, wider and higher cabin, a more powerful 265 kW engine, a deeper 0.73 m skirt and 19 forward-facing seats or 26 inward-facing seats.

Specifications

Length overall	11.9 m
Beam	4.76 m
Payload	2.2 t
Passengers	20–26
Cruise speed	35 kt

Propulsion: The 2000TDX Mk II is powered by a Deutz BF8L 513LC 265 kW diesel engine.

2500 TD

This craft has now been superseded by the Griffon 3000 TD. One craft of this type was built by Griffon Hovercraft Ltd and the other was built by a licensee in the US. Both are in operation in North America.

3000 TD

The Griffon 3000 TD design superseded the original 2500 TD and the 2500 TDX. Seating up to 37 passengers, the 3000 TD uses two Deutz BF8L 513 engines and 1.8 m propellers.

Similar controls to the single-engine craft are fitted except no elevators are used.

Three rudders mounted in each duct are operated by a steering yoke via stainless steel cables to provide directional control. Fore and aft trim is achieved by an electrically operated water ballast transfer system. Twin engine throttles are mounted near to the driver's left hand whilst clutch pedals, operated by the driver's feet, are provided for the propellers. By using only one propeller at low speed, together with the rudder and skirt shift, the craft can turn in its own length. Control systems are powerful enough to allow the craft to be operated and controlled on only one engine, although at a much reduced speed.

Two of these craft were built and the 3000 TD has now been superseded by the 3000TD Mk II and the 3000TDX Mk II.

Specifications

Length overall	15.0 m
Beam	7.0 m
Weight, min	8.2 t
Payload	3 t
Passengers	30–36
Propulsive power	2 × 239 kW
Max speed	40 kt
Operational speed	35 kt

Classification: IMO, the UK MCA and Lloyd's Register.

Structure: The main hull is of welded marine grade aluminium.

Propulsion: The 3000 TD uses two identical machinery units from the 2000 TDX, fitted next to each other.

3000 TD Mk II

The 3000TD Mk II replaces the 3000TD and is designed to meet the new IMO high-speed craft regulations. The main specification is similar but it now uses two 265 kW Deutz engines and 2 m propellers.

Griffon 470TD for RNLI 1326861

The superstructure can be designed to suit the customer's specification. The standard passenger cabin is constructed in light alloy with GRP mouldings at the front and back. Great emphasis has been placed on giving the passengers clear uninterrupted views through the large windows and narrow pillars are featured. A portion of the roof can be transparent if required and air conditioning and heating systems can be installed. Aircraft-type seating is arranged in up to six rows of six seats, three either side of a central aisle. Luggage racks, toilets and a galley may also be provided to suit customers' specific requirements.

Specifications

Length overall	15.35 m
Beam	7.34 m
Weight, min	8.2 t
Payload	3 t
Passengers	30–36
Propulsive power	2 × 265 kW
Max speed	40 kt
Operational speed	35 kt

Propulsion: The 3000 TD Mk II uses two identical machinery units to the 2000 TDX Mk II, fitted next to each other.

3000 TDX Mk II

The new Griffon 3000 TDX Mk II, like the other Mk II variants, has been designed to comply with the IMO High Speed Craft Code. It is similar in design to the 3000 TD Mk II but uses larger 386 kW engines and 2.2 m propellers, giving the craft improved into-wind and into-wave performance. It is steered via hydraulics from a joystick mounted in the cockpit. All other control methods are the same as the 3000TD Mk II. Three Griffon 3000TDX Mk II's are operated by Shell in Nigeria.

Specifications

Propulsive power	2 × 386 kW diesel engines
Operational speed	37 kt
Maximum speed	45 kt

4000 TD

The 4000 TD is a stretched version of the 3000 TD (and the proven 2500 TD). It can carry up to 60 passengers at a cruising speed of 35 kt. There are two of these craft in India operating along the coast at Bombay and within Bombay harbour.

Specifications

Length overall	17.9 m
Beam	7.3 m
Weight, max	15.9 t
Payload	4.0 t
Passengers	54–60
Propulsive power	2 × 298 kW
Max speed	38 kt
Operational speed	35 kt

Classification: IMO, the UK MSA and Lloyd's Register.

Structure: Exactly the same cross-section and construction method as the 3000TD but 2.85 m longer.

Propulsion: The same twin engine system as the 3000TD but Deutz engines BF10L 513V10 (298 kW each) are fitted.

Transmission details are similar to the smaller craft.

4000 TDX Mk II

The 4000TDX Mk II is a stretched version of the 4000TD carrying 60 to 80 passengers or 6 t of payload. It has two of the larger BF12L 513C, 386 kW engines installed and is 2 m longer than the 4000TD.

8000 TD(M)

The 8000TD(M) is a slightly larger and faster version of the 4000TDX Mk II, powered by two MTU water-cooled diesel engines, two 2.6 m propellers and four lift fans. Six of these craft are in operation with the Indian Coast Guard, five with the Saudi Arabian Border Guard, and two with the Korean National Maritime Police Agency. Commercial versions of this craft, the 8000TD, are also being sold. As with all sizes of Griffon hovercraft, the military, paramilitary and commercial versions of the craft are based on exactly the same hull and machinery, but the superstructure and fit-out are changed to reflect each customer's preferred specification.

Specifications

Length overall	19.85 m
Beam	8.70 m
Weight, max	27 t
Payload	8–10 t
Passengers	72–80
Propulsive power	2 × 493 kW
Max speed	50 kt
Operational speed	42 kt

Classification: IMO, the UK MSA and Lloyd's Register.

Propulsion: 2 × MTU 12 V 183 TB32 V12 water-cooled diesels.

All other systems similar to Griffon's smaller craft.

8100 TD

This is a stretched version of the Griffon 8000 TD. The 8100 TD is a multirole craft which carries a 12 t payload, 50 per cent more than the 8000TD, and is designed to take a tracked vehicle, general cargo, a container or troops. The 8100TD is powered by Iveco water-cooled diesel engines and is capable of speeds in excess of 40 kt. Griffon has recently been awarded a warrant to build three vessels for the Swedish Amphibious Battalion.

Specifications

Length overall	21.22 m
Beam	8.81 m
Displacement	31 t
Payload	8.1 t
Operational speed	40 kt

Propulsion: 2 × Iveco Vector 20C8 630 kW diesel engines driving 4 × Air Vehicles centrifugal fans and 2 × Hoffman CP air propellers.

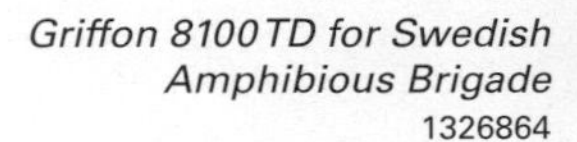

Griffon 8100TD for Swedish Amphibious Brigade
1326864

Hoverwork Ltd

Quay Road, Ryde, Isle of Wight, PO33 2HB, United Kingdom

Tel: (+44 1983) 56 30 51
Fax: (+44 1983) 81 28 59
e-mail: info@hoverwork.co.uk
Web: www.hoverwork.co.uk

C D J Bland, *Chairman*
R K Box, *Chief Executive*
P White, *Finance Director and Company Secretary*
R H Barton, *Engineering Director*
B A Jehan, *Operations Manager*

The roots of the hovercraft business extend back to the world's first hovercraft, the SR. N1, which was built by Saunders-Roe Limited in 1959, just prior to the company being taken over by Westland.

British Hovercraft Corporation (BHC) was formed in 1966, combining the hovercraft interests of Westland and Vickers and also involving NRDC. As a result of rationalisation, the Westland Aerospace organisation was set up in 1985. Westland Group was acquired by GKN plc in 1994.

From 2000, both the marketing and construction of the company's hovercraft were placed in the hands of Hoverwork Ltd under licence. Hoverwork Ltd now manages the hovercraft business of GKN Aerospace Services (formerly British Hovercraft Corporation) under a licence agreement. Hoverwork continues to develop new hovercraft designs.

Apart from SR.N1, all the early hovercraft were gas turbine powered. Of these, the SR.N4 and SR.N6 were in service for over 30 years.

In 1983 there was a change from gas turbine power and aircraft type construction to diesel engines and welded marine type structure with the production of the AP1-88. The latest development of this type, the Dash 400, has a payload of up to 20 t and a top speed of 50 kt in flat calm conditions. The first two craft of this type entered service with the Canadian Coast Guard in 1998.

Hoverwork Ltd is a wholly owned subsidiary of Hovertravel Ltd which provides a high-speed year-round hovercraft service between the Isle of Wight and Portsmouth. Founded in 1966, Hoverwork was established to provide hovercraft sales, charter services and consultancy on a worldwide basis. The company has trained over 60 hovercraft captains and has engaged in numerous hovercraft charter operations in diverse locations and environments. Charter duties

AP1-88/100 Liv Viking *after her refit for service with the Canadian Coast Guard* 0580422

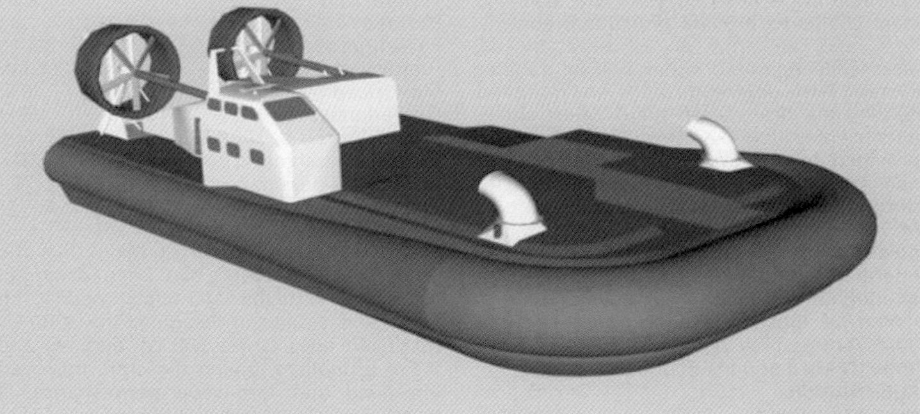

Artist's impression of BHT150 cargo version 0526358

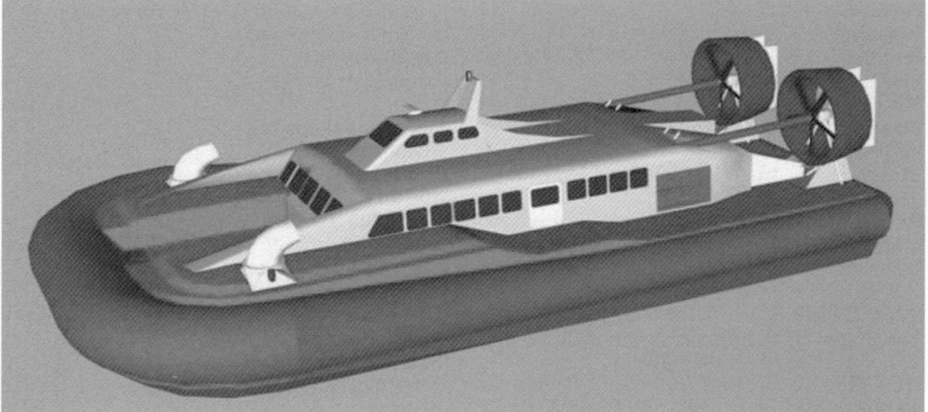

Artist's impression of the BHT150 passenger carrying version 0134520

Vessel type	Vessel name	Yard No	Length (m)	Speed (kt)	Seats	Vehicles	Originally delivered to	Date of build
SR.N6	Series production (57)	–	14.8	35.0	38	none	–	Series production from 1960
SR.N4 Mk 3	*The Princess Anne*	004	56.4	50.0	424	60	Hoverspeed Ltd	1969/1978 (mod)
SR.N4 Mk 3	*The Princess Margaret*	001	56.4	50.0	424	60	Hoverspeed Ltd	1968/1979 (mod)
SR.N4 Mk 2	*Sure*	3	39.7	–	282	37	Hoverspeed Ltd	1969/1974 (mod)
SR.N4 Mk 2	*Swift*	2	39.7	–	282	37	Hoverspeed Ltd	1969/1974 (mod)
BH 7	Series production (7)	–	23.9	–	–	none	–	Series production from 1972
SR.N4 Mk 2	*Sir Christopher*	5	39.7	–	282	37	Hoverspeed Ltd	1972/1974 (mod)
SR.N4 Mk 2	*The Prince of Wales*	6	39.7	–	282	37	Hoverspeed Ltd	1977
AP1-88/80	*Tenacity*	–	22.0	–	80	none	Hovertravel Ltd	1983
AP1-88/80	*Resolution*	–	22.0	–	80	none	Hovertravel Ltd	1983
AP1-88/100	*Freja Viking* (ex-*Svalan)*	–	24.2	–	84	none	A/S DSO	1984
AP1-88/100	*Liv Viking*	–	24.2	–	84	none	A/S DSO	1984
AP1-88/100	*Idun Viking* (ex-*Expo Spirit)*	–	24.2	–	84	none	A/S DSO	1985
AP1-88/80	*Perseverance*	–	22.0	–	80	none	Hovertravel Ltd	1985
AP1-88/200	*Waban-Aki*	–	24.2	–	–	none	Canadian Coast Guard	September 1987
AP1-88/100	*Double-O-Seven*	–	24.2	–	98	none	Hovertravel Ltd	1989
AP1-88/100	*Freedom 90*	–	24.2	–	98	none	Hovertravel Ltd	1990
AP1-88/100	*Siverko*	–	24.2	–	84	none	Northern Shipping, RFAS	1991
Dash 400	*Sipu Muin*	–	28.5	–	–	none	Canadian Coast Guard	April 1998
Dash 400	*Sijay*	–	28.5	–	–	none	Canadian Coast Guard	1999
BHT130	*Solent Express*	–	28.0	45.0	130	none	Hovertravel Ltd	2007
AP1-88 Dash 400	–	–	58.5	–	–	none	Canadian Coast Guard	(2008)
BHT130	Suna-X	–	30.7	45.0	49	1	Aleutions East Borough, Alaska	2007

include passenger services, mineral surveys, medical evaluation duties, feature film production and logistics support.

During 1998 the company provided captains in support of the Canadian Coast Guard AP1-88/400 hovercraft during manufacturer and customer's acceptance trials. During July 1999, Hoverwork Ltd operated a hovercraft service between Carnoustie and St Andrews in support of the British Open Golf Championship.

A recent project saw extensive refit and modification work to *Liv Viking* before it entered service with the Canadian Coast Guard. Recently Hoverwork have started to build a third AP1-88/400 hovercraft for the Canadian Coast Guard, for delivery in 2008.

Hoverwork Ltd has access to all hovercraft owned and operated by Hovertravel Ltd.

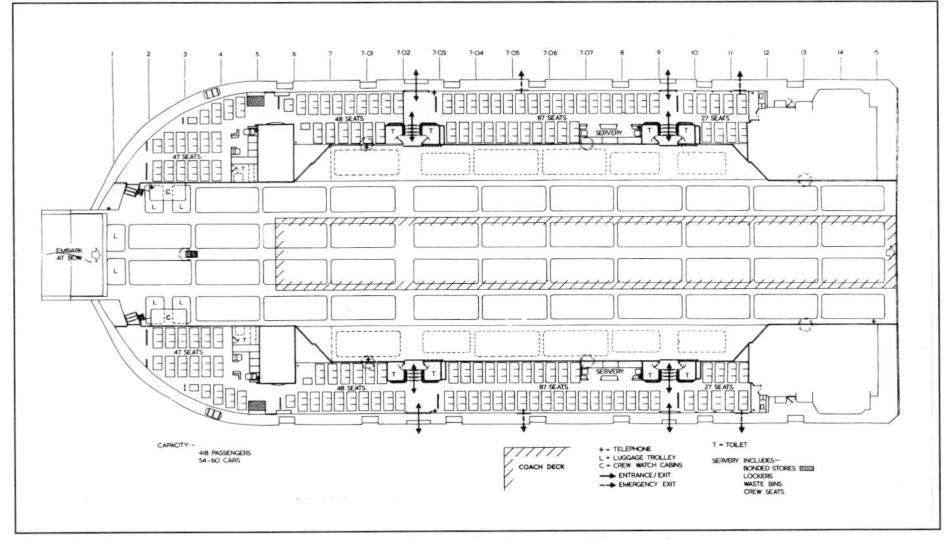

Layout of vehicle deck and passenger cabins on SR.N4 Mk 3 (Super 4) 0506601

SR.N4 Mk 3

This type has a payload of 424 passengers and 60 vehicles, a laden weight of 325 t and a top speed in excess of 65 kt. Both SR.N4 and Mk 3 hovercraft were finally withdrawn from service on 1 October 2000 after over 30 years of service.

Specifications

Length overall	56.4 m
Beam	23.2 m
Weight, max	325 t
Payload	112 t
Passengers	424
Vehicles	60
Fuel capacity	23,500 litres
Propulsive power	4 × 2,834 kW
Maximum speed	65 kt
Operational speed	50 kt
Operational limitation	
Beaufort 6	wave height 2.5 m

Propulsion: Motive power is supplied by four Rolls-Royce Marine Proteus Type 15M/529 free-turbine turboshaft engines, located in pairs at the rear of the craft on either side of the vehicle deck. Each engine is rated at 2,834 kW and is connected to one of four identical propeller/fan units, two forward and two aft. The propellers are four-bladed, controllable-pitch types. The lift fans are 12-bladed centrifugal types, 3.5 m in diameter. Maximum fuel tankage, 28.45 t; normal fuel allowing for ballast transfer is 18.29 t.

Northern Shipping AP1-88 Siverko 0506602

AP1-88

Major advances in hovercraft technology enabled a 10 t payload craft, with a performance equal to that of the well proven SR.N6. Built in welded aluminium alloy, AP1-88 is powered by four turbocharged diesels. This craft has low crew and maintenance requirements, footprint pressure and noise levels.

The AP1-88/100 in civil passenger configurations can seat up to 101 passengers. The first two AP1-88s *Tenacity* and *Resolution* began operating with Hovertravel between Ryde and Southsea in 1983. A similar craft, *Perseverance,* was built in 1985. These three craft are built to the 2.4 m shorter/80 configuration. Standard craft able to take up to 101 passengers have been in production since 1984.

The basic hull is formed by a buoyancy tank made almost entirely of very wide aluminium alloy extrusions. One extrusion is used for the I-beams, forming the transverse frames, and a second for the integrally stiffened planking used for the bottom and deck. The remainder of the rigid structure is built from smaller welded extrusions and plating, with the exception of the roof, made from riveted light gauge corrugated panels. The propeller ducts are a composite structure of light alloy and Kevlar-reinforced plastic. Marine alloys are used, including N8 plate and HE30 extrusions. In general, plate thicknesses are 2 or 3 mm except for the light-gauge roof plating. The structure is welded throughout to eliminate mechanical fastenings that can be sources of corrosion.

The skirt is a low-pressure ratio tapered skirt based on that of the Super 4. Mean cushion depth is 1.37 m.

Directional control is provided by two sets of triple aerodynamic rudder vanes mounted on the rear of the propeller ducts, differential propeller thrust and by swivelling bow thrusters. In the straight-aft position, the bow thrusters contribute to forward thrust. Trim is controlled by fuel ballast transfer.

Superstructure is divided into four main components: a large central accommodation area forward of the propulsion machinery bay, two sidebodies containing the lift system machinery and a control cabin mounted on top of the main cabin. In addition to the full-cabin and half well-deck versions, full well-deck variants and an open top forward cabin version is available. The commercial full-cabin version seats a maximum of 101 passengers.

In early 2004, Hoverwork delivered a refitted API-88/100S to the Canadian Coast Guard. The vessel, originally named *Liv Viking,* had been delivered to Danish Company DSO in the mid 1980s. The refit included MTU 12V 183 TB32 propulsion engines, variable-pitch propellers, a comprehensive communications package and high-power searchlights.

Specifications
(Standard AP1-88/200)

Length overall	24.4 m
Beam	11.0 m
Weight, min	29.5 t
Weight, max	40.8 t
Payload	11.3 t
Passengers	101
Fuel capacity	1,800 litres
Propulsive power	2 × 336 kW
Lift power	2 × 336 kW
Maximum speed	45 kt
Operational limitation	Sig wave height 1.5 m windspeed 30 kt

Classification: BHSR, IMO, and DnV.

Structure: Welded aluminium.

Propulsion: The AP1-88/100 craft is powered by four Deutz BF12L513FC 12-cylinder air-cooled diesels. The two 2.74 m diameter four-blade Hoffmann ducted propellers are each driven by one of the diesels via a toothed belt. On standard craft the propellers are of fixed-pitch type but ground adjustable through ±5°. The belt-drive reduction ratio is 1:0.6.

The Canadian Coast Guard Waban Aki *on oil clean-up duties* 0506956

Dash 400 in service with the Canadian Coast Guard 0024738

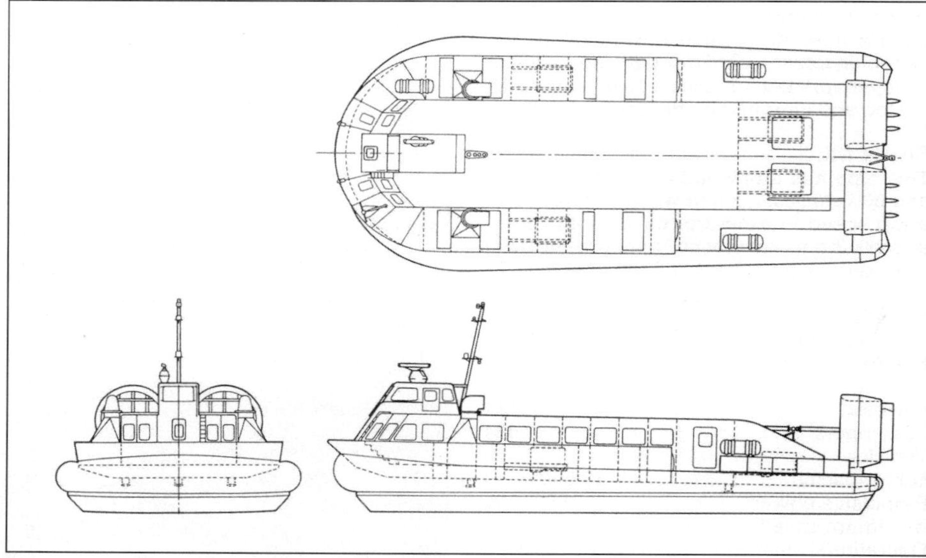

General arrangement of AP1-88/100 hovercraft 0506603

Two of the engines, housed in the side box structures, power the lift and bow thruster systems. On each side of the craft one engine drives three 0.84 m diameter double-entry centrifugal fans, two supply air to the cushion via the skirt system and the third supplies air to the rotatable bow thruster.

AP1-88/200

Waban-Aki

A half well-deck variant of the AP1-88, designated AP1-88/200, is in service with the Canadian Coast Guard undertaking search and rescue, navaid maintenance ice-breaking and oil-spill clean-up tasks on the St Lawrence River and its tributaries.

The craft is of a half well-deck configuration with accommodation for up to 12 crew members or technicians and up to 12 t of cargo. This AP1-88 is equipped with a hydraulic crane, a capstan and winch to facilitate the conduct of a variety of specialised coast guard tasks, and entered service with the Canadian Coast Guard in 1987.

Specifications

Length overall	24.5 m
Beam	11.2 m
Weight, max	47.1 t
Payload	12.5 t
Fuel capacity	5,912 litres
Propulsive power	4 × 441 kW
Maximum speed	50 kt

Propulsion: Four Deutz BF12L513CP air-cooled turbocharged diesels, 441 kW each, at 2,300 rpm. Two Hoffmann 2.75 m diameter controllable-pitch propellers Type HOV-254P2DFR/D275, ducted. Four centrifugal lift fans 0.885 m diameter. Two bow thrusters, centrifugal fans 0.840 m diameter.

AP1-88/300

This type is an open-top freighter developed from the standard AP1-88/100 design and sharing the same external dimensions and mechanical installation. The AP1-88/300 was produced under licence by NQEA Australia: this licence has now expired.

Two craft have been built to this design, being used for various duties in the mining and oil and gas industries.

DASH 400

The Dash 400 is an enlarged version of the AP1-88/200, with a longer well-deck and a payload capacity twice that of the original design. It is thus able to handle much larger buoys and other navigational aids. Two of these craft were built by Hike Metal Products of Wheatley, Ontario and are now in service with the Canadian Coast Guard. The craft are all aluminium with a well-deck length of 13.2 m and a ramp width of 3.8 m. Alternative configurations are available within the standard craft dimensions, for example offering a larger well-deck with a corresponding reduction in cabin length. An auxiliary hydraulic system powers a capstan, winch and removable deck crane.

Specifications

Length overall	28.5 m
Beam	12.0 m
Weight, min	43 t
Weight, max	70 t
Deadweight	27 t
Propulsive power	2,818 kW
Maximum speed	60 kt

Structure: Aluminium.

Propulsion: The vessels are powered by four Caterpillar 3412 TTA water-cooled diesels. Two engines for propulsion and two to drive the lift fans and bow thruster fans. Propellers are two Hoffmann 2.75 m diameter controllable-pitch, ducted-type HOV-254P2DFR/D275s.

BHT130, 150 and 200

The BHT130, 150 and 200 designs have been developed by Hoverwork Ltd to replace and improve on the existing AP1-88/100 currently employed by Hovertravel, the parent company, and to address the increasing interest received from many locations around the world for a hovercraft of this size and capability. The new diesel powered, commercial size amphibious hovercraft will be capable of carrying up to 200 passengers or up to 30 t cargo.

The design is available in a number of versions including: Passenger Accommodation, Open Deck Freighter, Passenger/Freight Mix, Ro-ro Logistics Support and Defence. The Defence version can be adapted for multirole operations (amphibious logistic support, amphibious assault, coastal patrol, anti-piracy and mine countermeasures).

SR.N4 Mk 3 (Super 4), The Princess Anne 0103311

BHT130

Solent Express

Built for Hovertravel, part of the Hoverwork group, in 2007, *Solent Express* was the first hovercraft to be designed in-house at Hoverwork. The vessel, which operates between Portsmouth and the Isle of Wight in the UK, has two MTU diesel engines providing power to two pairs of fans governing the lift system, and a further two diesel systems providing propulsion via Hoffman five-bladed air propellers.

Specifications

Length overall	29.3 m
Beam	15.0 m
Weight, max	75.0 t
Payload	20 t
Fuel capacity	3,600 l
Passengers	131
Operational speed	45 kt

Propulsion: Lift provided by two MTU 12V 2000-R1237K37 diesel engines each rated at 675 kW at 1800 rpm driving one pair of centrifugal fans (for bow thrusters and engine cooling) and two pairs of centrifugal fans supplying cushion air. Propulsion is provided by two MTU 16V 2000 diesels rated at 899 kW at 1800 rpm driving Hoffman 5-bladed, 3.5 m diameter variable pitch air propellers.

BHT130

Suna-X

This is the well-deck cargo carrying version of the BHT 130 configured to carry 49 passengers, one vehicle and freight. The area of operation for the hovercraft, between Lenard Harbour and Cold Bay in the Alaskan Aleution Islands, is difficult to reach by plane or conventional vessels in winter conditions and the hovercraft will also provide an emergency ambulance service.

Specifications

Length overall	30.7 m
Beam	15.0 m
Weight, max	75.0 t
Payload	20 t
Fuel capacity	3,600 l
Passengers	49
Vehicles	1
Operational speed	45 kt

Propulsion: Lift provided by two MTU 12V 2000-R1237K37 diesel engines each rated at 675 kW at 1800 rpm driving one pair of centrifugal fans (for bow thrusters and engine cooling) and two pairs of centrifugal fans supplying cushion air. Propulsion is provided by two MTU 16V 2000 diesels rated at 899 kW at 1800 rpm driving Hoffman 5-bladed, 3.5 m diameter variable pitch air propellers.

Ingles Hovercraft Ltd

101A High Street, Gosport, Hampshire PO12 1DS, United Kingdom

Tel: (+44 23) 92 51 05 93
Fax: (+44 23) 92 50 20 32
e-mail: hoveraid@internet.com
Web: www.hoveraid.co.uk

N A H Pool, *Director*
J W Wilson, *Director*
N J Smith, *Design Consultant*
David Wiltshire, *Administrator*

Ingles Hovercraft Ltd manufactures the River Rover hovercraft under licence to The HoverAid Trust, a registered charity, established to design appropriate technology hovercraft primarily for use in developing countries. Sales are made to missions and aid agencies on a non-profit basis. The River Rover design incorporates the 'Elevon' system of control.

The River Rover series has been developed over the last two decades to create a vehicle capable of reliably coping with the most arduous of river and delta conditions. The Elevon control system was first proven on a British joint services expedition to conquer the rapids of the Nepalese Kali Gandaki in 1978. Similar expeditions to the Amazonian headwaters in Peru, to the snowbound source of the Yangtze and on the Tibetan plateau have shown the River Rover to be unparalleled in terms of manoeuvrability.

The first River Rover Mk 4 was completed in April 1992. Several craft are at present in service with medical aid programmes. The River Rover Mk 5 was completed in 1995.

Vessel type	Vessel name	Yard No	Length (m)	Speed (kt)	Seats	Vehicles	Originally delivered to	Date of build
River Rover Mk 4	–	–	6.8	24.0	8	none	–	Several built
River Rover Mk 5	–	–	–	–	–	none	–	1995

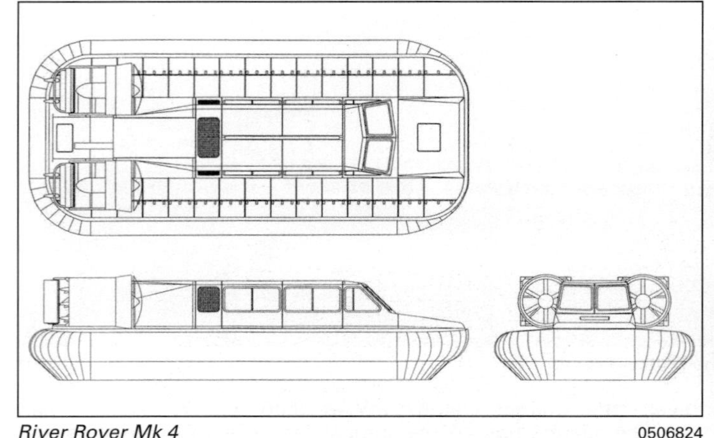

River Rover Mk 4 0506824

River Rover Mk 4 craft operating in Zambia, Central Africa 0044046

RIVER ROVER Mk 4

The River Rover Mk 4 is a natural development of the River Rover series. It is available with either a Volkswagen or Land Rover turbocharged and intercooled engine, providing greater economy, reliability and global parts availability with either installation. To fulfil its intended operational parameters, the River Rover Mk 4 has been designed for simple maintenance and repair. Removable side decks are built from two sets of identical deck panels, and the outer set can be lowered for containerisation or road transportation. There is sufficient closed cell protective foam contained in the vessel to prevent sinking, even when ruptured.

The River Rover Mk 4 is fitted with a unique twin cascade Elevon system which allows precise positional control. Pitch and roll are actively adjusted by vertical thrust vectoring, this effect is enhanced by the wide separation of the thrust fans. When combined with the conventional rudders a high degree of lateral control is achieved, allowing, for example, operation in crosswinds without side-slip. High-speed power turns can be performed without tail breakaway, an undesirable characteristic of many conventional hovercraft; this allows River Rover to negotiate intricate river systems well above hump speed.

River Rover Mk 5, powered by a Land Rover Gemini 3 diesel engine 0044047

Specifications

Length overall	6.8 m
Beam	2.8 m
Weight, min	1.0 t
Weight, max	1.6 t
Crew	1
Passengers	8
Fuel capacity	180 litres
Propulsive power	97 kW
Maximum speed	32 kt
Operational speed	24 kt
Range	190 n miles

Structure: The hull consists of a bolted aluminium spaceframe with repeated glass fibre box structures attached.

Propulsion: The operator has the choice of either a Volkswagen or Land Rover 2.4 litre turbocharged engine, developing 97 kW at 4,000 rpm. Two thrust fans and an axial flow pressure fan for lift, are driven directly from the engine via a clutch and a torsionally resilient coupling.

Slingsby Advanced Composites Ltd

Kirkbymoorside, Yorkshire YO62 6EZ, United Kingdom

Tel: (+44 1751) 43 24 74
Fax: (+44 1751) 43 11 73
e-mail: sales@slingsby.co.uk
Web: www.slingsby.co.uk

Steven Boyd, *Sales Director*

Since its formation in the 1930s as Slingsby Sailplanes, building gliders, Slingsby Advanced Composites Ltd (formerly Slingsby Aviation Ltd) has grown in size and greatly expanded its capabilities. The development into composite materials design and manufacturing, has led into building aerostructures, UAV airframes, detrainment doors, marine structures and hovercraft.

In 1984 the hovercraft development commenced and this has resulted in the SAH 2200 amphibious hovercraft, of which some 30 have been sold around the world, both for military and commercial applications.

SAH 2200

The SAH 2200 is a single diesel-engined hovercraft manufactured in a composite of glass and aramid fibre, epoxy resin and PVC foam. This combination produces an exceptionally strong, lightweight structure with excellent fatigue life and resistance to corrosion from saltwater. These characteristics give the hovercraft a very good payload and performance for the engine horsepower used, making it easy and inexpensive to maintain. With a maximum payload of 2,200 kg the craft is certificated to carry up to 22 passengers and can hold sufficient fuel for more than 12 hours of operations.

Current operations include coastal patrol work in areas ranging from Scandinavia, where winter use is over ice and snow, to the Persian Gulf, as a safety patrol craft on firing ranges, around coastal airports and as a training craft for crews on large military hovercraft.

Commercial and passenger-carrying operations have been established in the UK and other parts of Europe, Southeast Asia, the Caribbean and the Indian subcontinent. In these operations the craft has been used for transporting people and equipment, carrying out survey work and acting as a safety craft around shallow water drilling operations.

Operational areas have included large tidal ranges with associated sand and mud flats, inland waters with large areas of shallow, weed infested water, mangrove swamp coastal waters and major river complexes.

The skirt is of a loop and segment type.

The structure is strengthened with aramid fibre in high-load areas, and heavy-duty landing skids are provided. Four marine bollards on the deck form lifting rig attachments and guides for the integral jacking system. Side decks are rigid and enable large, bulky items to be carried outside the main load space, and these decks fold to allow transport of the craft by road vehicles or transport within a 40 ft flat

rack container. Packed dimensions with duct removed:
Length: 12.0 m
Height: 2.0 m
Beam: 2.4 m
Weight: 3.0 t

Rudders mounted in the propeller slipstream provide directional control; similarly mounted elevators provide fore and aft trim. Fuel ballast and roll control systems are incorporated to counteract adverse loading and to improve craft performance in high wind and sea states. A controllable-pitch propeller allows variable thrust both in forward and reverse giving enhanced control particularly over land and in difficult downwind sea conditions.

The seats for commander and passenger/navigator are located forwards in a self-contained wheelhouse. The load space is flexible in layout and two quick-release composite canopies allow conversion to three versions:

Passenger Version: Bench seating running fore and aft providing a total of 22 passenger seats. Access is port and starboard through gull wing doors mounted in canopies.

Supply Boat Version: Covered accommodation for eight passengers with open load space with a capacity for up to 1,400 kg of cargo.

Logistic Support Version: With canopies and seats removed, integral cargo lashing rails are provided to enable a disposable load of up to 2,200 kg to be carried.

In addition to the above, purpose-designed pods to fit the load space are manufactured to carry specialist equipment.

A high-roof craft is also available, manufactured to suit the US market. The executive version is outfitted to seat 10 passengers only, in armchair and settee comfort.

Specifications

Length overall	10.6 m
Beam	4.2 m
Payload	2.2 t
Passengers	22
Fuel capacity	510 litres
Propulsion power	142 kW
Max speed	40 kt
Range	500 n miles
Obstacle clearance	0.5 m

SAH 2200 operating with the Finnish Frontier Forces 0024739

High-roof executive SAH 2200 craft (in foreground) 0002637

Structure: Heavy-duty composite plastics structure.

Propulsion: Integrated propulsion and lift system. The standard SAH 2200 is powered by a single Deutz BF6L 913 series turbocharged air-cooled diesel, driving a centrifugal lift fan and a Hoffmann variable-pitch ducted propeller. Other engine variants such as the Volvo D7A TA Series and Cummins 6CTA turbocharged water-cooled diesels are also available.

United States

Hover Shuttle LLC

328 Ripon Road, Berlin, Wisconsin 54923, United States

Tel: (+1 920) 361 94 83
Fax: (+1 920) 361 94 83
e-mail: info@hover-shuttle.com
Web: www.hover-shuttle.com

Raymond R Jordison, *Vice-President*

Hover Shuttle have designed and built a demonstration version of their Hover-Rescue Model 1500 hovercraft.The company has also designed a Model 4500 hovercraft capable of carrying 42 passengers. Other designs include the Model-2500 and the Model-15000 which will be able to carry 160 passengers.

Specifications

Length overall	8.1 m
Beam	3.65 m
Height, on cushion	2.15 m
Passengers	12
Max wave height	1.2 m
Range	400 miles

Kvichak Marine Industries

469 NW Bowdoin Place, Seattle, Washington 98017, United States

Tel: (+1 206) 545 84 85
Fax: (+1 206) 545 35 04
e-mail: sales@kvichak.com
Web: www.kvichak.com

In 2003, Kvichak Marine was licensed to manufacture a range of amphibious hovercrafts designed by Griffon Hovercraft of the UK.

HOVERWORK BHT 130

Kvichack have built a Hoverwork BHT 130 under license for a remote service in Alaska. The vessel was fitted out to carry 47 passengers, two wheelchairs and up to 20 tonnes of cargo. The craft has a forward well deck and ramp, allowing it to carry an ambulance.

Specifications

Length overall	29.0 m
Payload	20 t
Passengers	47
Vehicles	1 ambulance

Griffon 2000 TD

Arctic Hawk

This vessel was built for Crowley Marine Services to operate on the North Slope of Alaska for British Petroleum. *Arctic Hawk* is used for carrying personnel, goods and equipment to Northstar Island, a man-made offshore production island in Prudhoe Bay. The hull is fabricated from aluminium and the skirt is segmented.The vessel is approved by the US Coast Guard for Subchapter T passenger Service.

Specifications

Length overall	10.6 m
Beam	4.5 m
Weight, min	2.6 t
Payload	2 t
Crew	2
Passengers	22
Propulsive power	268 kW
Maximum speed	35 kt
Operational speed	27 kt

Propulsion: The vessel is powered by a Deutz BF8L513LC engine rated at 268 kW at 2,100 rpm, driving a controllable pitch reversing propeller with wide chord blades.

Northrop Grumman Ship Systems

Avondale Operations
PO Box 50280, New Orleans, Louisiana 70150-0280, United States

Tel: (+1 504) 436 21 21
Fax: (+1 504) 436 52 00
e-mail:info@northropgrummanshipsystems.com
Web: www.ss.northropgrumman.com

Thomas M Kitchen, *President*
Ron McAlear, *Vice-President, Operations*
L J Derbes Jr, *Vice-President, Financial Controller*
Sara Price, *Director, Communication*
Brian Smith, *Director, Business Development*

Avondale Operations is a division of Northrop Grumman Ship Systems.

The Shipyard Systems Division is a diversified shipyard, having constructed a variety of large barges including self-unloaders; large crude carriers; complex chemical parcel carriers; large, high-speed container ships; offshore drill rigs. Also constructed are drill ships, a full range of Naval ships including destroyers, frigates, high-endurance Coast Guard cutters, fleet oilers, landing ship docks, an Oceanographic Research Vessel and others. Avondale's Boat Division builds ferries, gaming vessels, tugs, towboats, tug-supply boats, large fishing boats and other workboats.

In June 1994, Avondale entered into a co-operative agreement with Astilleros Españoles SA of Spain for the purpose of exchanging commercial ship designs, market analyses and ship production technology. This agreement is a significant step toward's Avondale's goal of acquiring commercial shipbuilding contracts.

LCAC 36, 1 of 15 LCAC craft delivered by Avondale 1000012

LCAC (LANDING CRAFT, AIR CUSHION)

From October 1988 to June 1993, Avondale Gulfport Marine, Inc delivered 15 LCAC vehicles to the US Navy.

Specifications

Length overall	26.8 m
Beam	14.3 m
Draught, hullborne	0.9 m
Weight, max	153.5 t
Payload	54.3 t
Propulsive power	4 × 3,955 kW
Maximum speed	50 kt

Propulsion: Four TF40B gas turbines are installed, driving two Dowty Rotol 3.582 m diameter four-blade variable, reversible-pitch propellers; and four double-entry centrifugal-type lift fans of 1.6 m diameter.

BHC AP1-88

In 1990 Avondale was appointed the US licensee for AP1-88 amphibious hovercraft.

Air Ride 109 400-passenger ferry 0506605

Vessel type	Vessel name	Yard No	Length (m)	Speed (kt)	Seats	Vehicles	Originally delivered to	Date of build
LCAC	LCAC-15	–	26.8	–	–	–	US Navy	–
LCAC	LCAC-21	–	26.8	–	–	–	US Navy	–
LCAC	LCAC-17	–	26.8	–	–	–	US Navy	–
LCAC	LCAC-18	–	26.8	–	–	–	US Navy	–
LCAC	LCAC-16	–	26.8	–	–	–	US Navy	–
LCAC	LCAC-20	–	26.8	–	–	–	US Navy	–
LCAC	LCAC-22	–	26.8	–	–	–	US Navy	–
LCAC	LCAC-23	–	26.8	–	–	–	US Navy	–
LCAC	LCAC-34	–	26.8	–	–	–	US Navy	–
LCAC	LCAC-35	–	26.8	–	–	–	US Navy	–
LCAC	LCAC-49	–	26.8	–	–	–	US Navy	–
LCAC	LCAC-50	–	26.8	–	–	–	US Navy	–
LCAC	LCAC-51	–	26.8	–	–	–	US Navy	–
LCAC	LCAC-19	–	26.8	–	–	–	US Navy	–
LCAC	LCAC-36	–	26.8	–	–	–	US Navy	–
Air Ride 109	–	–	33.2	44.0	360	none	Tri-State Marine Transport Ltd	1998
Air Ride 109	–	–	33.2	44.0	360	none	Tri-State Marine Transport Ltd	1998

AIR RIDE 109

In late 1988 Avondale Boat Division announced the building under licence from Air Ride Craft Inc of Air Ride Craft 109 (400-passenger) Surface Effect Ships (SES) for Tri-State Marine Transport Inc to operate between New York-JF Kennedy Airport and lower Manhattan. The Air Ride Craft concept developed by Don Burg represents a unique form of SES, in that a flexible seal or skirt is only used at the bow to contain the air cushion. The Avondale craft are built in marine grade aluminium alloy and form part of the diversified range of ship and boat building, conversion, repair, foundry and propeller work undertaken by the company. Ride control systems are fitted as standard.

Two Air Ride 109 SESs have been delivered by Avondale Boat Division.

Specifications

Length	33.2 m
Beam	10.4 m
Draught, on-cushion	0.9 m
Displacement	140 t (½ load)
Passengers	360–400
Fuel capacity	9,463 litres
Water capacity	1,514 litres
Propulsive power	2 × 1,603 kW
Lift power	336 kW
Operational speed	44 kt

Classification: USCG + ABS.

Structure: Hull: aluminium

Propulsion: The main engines are two Deutz MWM TBD 604B V16, driving Kamewa water-jet units. The lift engine is a Deutz MWM TBD 234 V12.

Textron Marine & Land Systems

Division of Textron Inc
19401 Chef Menteur Highway, New Orleans, Louisiana 70129, United States

Tel: (+1 504) 245 66 00
Fax: (+1 504) 254 80 00
Web: www.textronmarineandland.com

Richard J Millman, *President*
J O Smith, *Executive Chairman, Operations*
Mike Gelpi, *Senior Director Program Management*
Karen Deogracias, *Vice-President, Finance and Administration*
BC Moise, *Vice-President, Marine*
Karla Cooper, *Manager, Human Resources*
Gary Nelson, *Director, Business Development*
Scott White, *Director Supply Management*
David Whitaker, *Manager, Public Relations*

Textron Marine & Land Systems (TM&LS), a division of Textron Inc, began in 1969 as the New Orleans operations of Bell Aerospace Textron, which had been pursuing air cushion vehicle development programmes since 1958. It was renamed Textron Marine Systems in 1986 and acquired its manufacturing facility in 1988. In April 1994, TMS operations were joined with those of Cadillac Gage Textron under the name of Textron Marine & Land Systems. Cadillac Gage products include combat vehicles; weapon stations; and control, stabilisation and suspension systems for military vehicles.

The Textron Marine Systems Shipyard Operations, formerly Bell Halter Inc, was acquired in 1988 when TMS purchased it from Halter Marine Inc. Prior to that, the shipyard was operated as a joint venture of Bell Aerospace Textron and Halter Marine.

The TM&LS facility is located on a 10.1 ha (25 acre) site in eastern New Orleans, Louisiana, and has direct access to the Intra-coastal Waterway, the Mississippi River and the Gulf of Mexico. Products manufactured at the facility are in service with the US Army, US Navy, US Coast Guard and the commercial sector and have accumulated many thousands of hours of service.

TM&LS maintains a 5,202 m² storage facility in Slidell, Louisiana (32 km from the manufacturing facility).

Craft built by TM&LS range in size from the 7.32 m Utility Air Cushion Vehicle (UACV 1200) to the 26.8 m Landing Craft, Air Cushion (LCAC) currently in production for the US Navy SLEP (Ship Life Extension Program), to the 48.7 m SES-200 which has undergone modification for the US Navy.

TM&LS had delivered a total of 76 LCACs to the US Navy by September 1998, and redelivered three SLEP LCACs.

Of the 17 LCACs deployed to the Persian Gulf during the Gulf War, all were operational 100 per cent of the time.

LCACs provided relief to the outer islands of Bangladesh during the devastating cyclone of 1991. A total of 34 sorties was carried out in 14 days delivering 900 tonnes of desperately needed relief cargo. In 1992, LCACs also

LCAC-39 undergoing trials near New Orleans, Louisiana 0506606

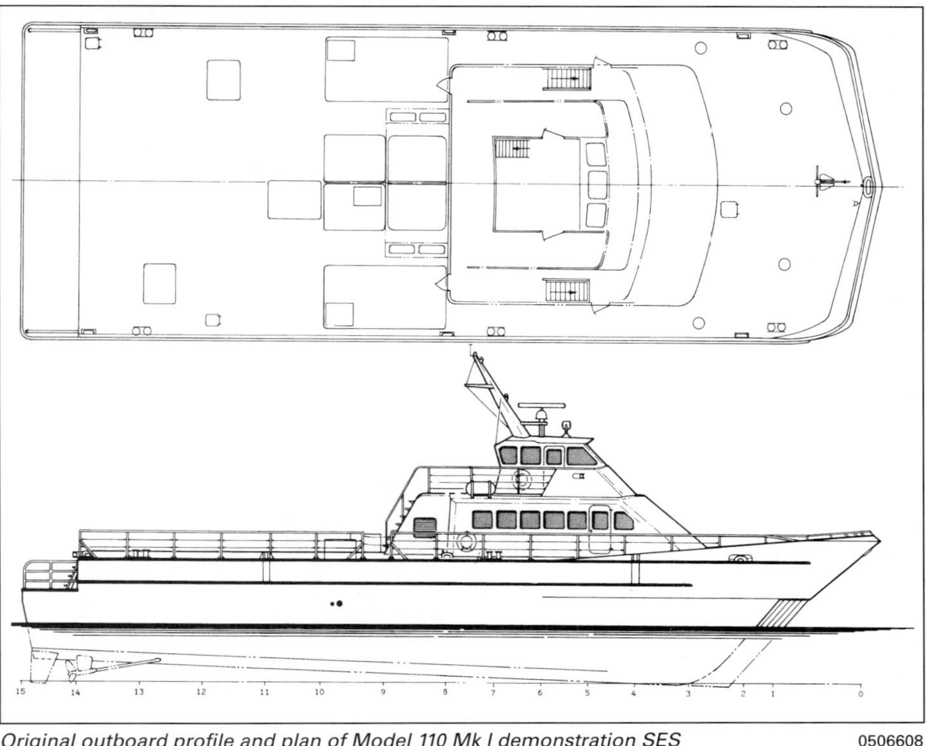

Original outboard profile and plan of Model 110 Mk I demonstration SES 0506608

LACV-30 craft at speed 0043242

played a vital role during Operation Restore Hope in Somalia.

LCAC Landing craft, air cushion

On 5 June 1981, TM&LS signed a USD40 million contract with the US Navy for the detail design and long-lead materials for an amphibious assault landing craft, designated by the navy as the LCAC (Landing Craft, Air Cushion). The LCAC is the production version of the JEFF(B) craft. The contract contained two options for the later construction of six craft. In February 1982, the US Navy exercised the first option for the production of three lead craft and, in October 1982, ordered a further three craft.

In March 1984, TM&LS was awarded a contract to build six LCACs. By December 1986, a further contract had been awarded to build two more LCACs bringing the total to 14. On 1 July 1987, the US Navy awarded a USD187 million contract to TM&LS to construct another 10 LCACs. This marked the start of full production for the LCAC programme.

On 15 December 1988, a second full-production contract for 12 LCACs was awarded to TM&LS, along with central procurement activities for long-lead time materials. This was valued at USD228 million. An option to this contract was exercised on 22 December 1989, for an additional nine LCACs. On 24 April 1991, another full-production contract, valued at USD139 million, for 12 LCACs was awarded to TM&LS with additional central procurement activities for long-lead time materials. A USD181 million option to this contract, for 12 LCACs, was exercised 22 May 1992.

Since March 1992, TM&LS has won all seven contracts in support of the LCAC mid-life overhaul and refurbishment programme aimed at upgrading the US Navy LCAC fleet and maintaining it in operational readiness. LCACs -1, -2, -3, -4, -6 and -14 have been overhauled at TM&LS West Coast Operations, and LCAC-8, -10 and -11 have been overhauled on the east coast.

In January 1993, TM&LS was awarded a USD117 million contract for the production of seven LCACs. This award made TM&LS the sole LCAC producer for the US government's FY92 acquisition.

In April 1996, the US Navy initiated a Service Life Extension Program (SLEP) for LCAC. TM&LS maintains its role as prime contractor for the LCAC, and is developing the necessary changes, upgrades and improvements. The objective of the SLEP is to increase the service life of LCAC from 20 to 30 years and reduce overall operating and maintenance costs. In March 2004 Textron were awarded a USD30 million contract to conduct SLEP on four LCAC craft. This contract took the total number of LCAC's having undergone SLEP to 14.

In April 1994, the Japan Defence Agency (JDA) awarded TM&LS a contract to build a large air cushion vehicle based on the US Navy LCAC. Valued at USD50 million, the contract includes delivery of engineering documentation, training manuals, reserve subsystems, spare parts and support and test equipment. The first ACV was delivered in October 1997. A second ACV was ordered by the JDA in November 1995 and was delivered in 1998.

The LCAC is a high-speed, ship-to-shore and over-the-beach amphibious landing craft, capable of carrying a 55 tonne payload. It can transport equipment, personnel and weapons systems (including the main battle tank) from ships located at increased standoff distances, through the surf zone and across the beach to hard landing points beyond the waterline. The craft is supported on a pressurised cushion of air and travels at much higher speeds than are presently possible with current conventional landing

Personnel transport module rigged MCAC in the well-deck of an LSD 0506958

Bow ramp of LCAC 0506607

MCAC firing an M-58 line charge on the beach 0506959

Vessel type	Vessel name	Yard No	Length (m)	Speed (kt)	Seats	Vehicles	Originally delivered to	Date of build
Model 212B (110 Mk II)	*Magaret Jill* (ex-*Speed Command*)	–	33.3	31.0	119	–	–	–
Model FR-7	–	–	16.0	–	50	–	Singapore Civil Aviation Authority	–
Model 212B (110 Mk II)	*Speed Tide* (ex-*Swift Command*)	–	33.3	31.0	119	–	–	–
552A US Coast Guard SES	*Shearwater*	–	33.5	–	–	–	US Coast Guard	October 1982
552A US Coast Guard SES	*Sea Hawk*	–	33.5	–	–	–	US Coast Guard	October 1982
552A US Coast Guard SES	*Petrel*	–	33.5	–	–	–	US Coast Guard	June 1983
LCAC	LCAC-1	–	26.8	–	–	–	US Navy	December 1984
LCAC	LCAC-2	–	26.8	–	–	–	US Navy	February 1986
LCAC	LCAC-3	–	26.8	–	–	–	US Navy	June 1986
LCAC	LCAC-4	–	26.8	–	–	–	US Navy	August 1986
LCAC	LCAC-5	–	26.8	–	–	–	US Navy	November 1986
LCAC	LCAC-6	–	26.8	–	–	–	US Navy	December 1986
LCAC	LCAC-7	–	26.8	–	–	–	US Navy	March 1987
LCAC	LCAC-8	–	26.8	–	–	–	US Navy	June 1987
LCAC	LCAC-9	–	26.8	–	–	–	US Navy	June 1987
LCAC	LCAC-10	–	26.8	–	–	–	US Navy	September 1987
LCAC	LCAC-12	–	26.8	–	–	–	US Navy	December 1987
LCAC	LCAC-11	–	26.8	–	–	–	US Navy	December 1987
LCAC	LCAC-13	–	26.8	–	–	–	US Navy	September 1988
LCAC	LCAC-14	–	26.8	–	–	–	US Navy	November 1988
LCAC	LCAC-24	–	26.8	–	–	–	US Navy	March 1990
LCAC	LCAC-26	–	26.8	–	–	–	US Navy	June 1990
LCAC	LCAC-25	–	26.8	–	–	–	US Navy	June 1990
LCAC	LCAC-27	–	26.8	–	–	–	US Navy	August 1990
LCAC	LCAC-28	–	26.8	–	–	–	US Navy	October 1990
LCAC	LCAC-30	–	26.8	–	–	–	US Navy	December 1990
LCAC	LCAC-29	–	26.8	–	–	–	US Navy	December 1990
SES-2000 conversion	–	–	48.7	–	–	–	US Navy	1978–1990 (final Conversion)
LCAC	LCAC-31	–	26.8	–	–	–	US Navy	February 1991
LCAC	LCAC-32	–	26.8	–	–	–	US Navy	May 1991
LCAC	LCAC-33	–	26.8	–	–	–	US Navy	June 1991
LCAC	LCAC-37	–	26.8	–	–	–	US Navy	July 1991
LCAC	LCAC-38	–	26.8	–	–	–	US Navy	September 1991
LCAC	LCAC-39	–	26.8	–	–	–	US Navy	September 1991
LCAC	LCAC-40	–	26.8	–	–	–	US Navy	November 1991
LCAC	LCAC-41	–	26.8	–	–	–	US Navy	November 1991
LCAC	LCAC-42	–	26.8	–	–	–	US Navy	December 1991
LCAC	LCAC-43	–	26.8	–	–	–	US Navy	February 1992
LCAC	LCAC-44	–	26.8	–	–	–	US Navy	February 1992
LCAC	LCAC-45	–	26.8	–	–	–	US Navy	March 1992
LCAC	LCAC-46	–	26.8	–	–	–	US Navy	May 1992
LCAC	LCAC-47	–	26.8	–	–	–	US Navy	June 1992
HM 221 Fire and Rescue Vessel	*John P Deveney*	–	21.2	20.0	–	–	Fire Department of the City of New York	July 1992
LCAC	LCAC-48	–	26.8	–	–	–	US Navy	July 1992
LCAC	LCAC-52	–	26.8	–	–	–	US Navy	September 1992
LCAC	LCAC-53	–	26.8	–	–	–	US Navy	October 1992
LCAC	LCAC-54	–	26.8	–	–	–	US Navy	October 1992
HM 221 Fire and Rescue Vessel	*Alfred E Ronaldson*	–	21.2	20.0	–	–	Fire Department of the City of New York	November 1992
LCAC	LCAC-55	–	26.8	–	–	–	US Navy	December 1992
LCAC	LCAC-56	–	26.8	–	–	–	US Navy	January 1993
LCAC	LCAC-57	–	26.8	–	–	–	US Navy	February 1993
LCAC	LCAC-58	–	26.8	–	–	–	US Navy	March 1993
LCAC	LCAC-59	–	26.8	–	–	–	US Navy	April 1993
LCAC	LCAC-60	–	26.8	–	–	–	US Navy	June 1993
LCAC	LCAC-61	–	26.8	–	–	–	US Navy	July 1993
LCAC	LCAC-62	–	26.8	–	–	–	US Navy	August 1993
LCAC	LCAC-63	–	26.8	–	–	–	US Navy	September 1993
LCAC	LCAC-64	–	26.8	–	–	–	US Navy	October 1993
LCAC	LCAC-65	–	26.8	–	–	–	US Navy	November 1993
LCAC	LCAC-66	–	26.8	–	–	–	US Navy	December 1993
LCAC	LCAC-67	–	26.8	–	–	–	US Navy	February 1994
LCAC	LCAC-68	–	26.8	–	–	–	US Navy	March 1994
LCAC	LCAC-69	–	26.8	–	–	–	US Navy	April 1994
LCAC	LCAC-70	–	26.8	–	–	–	US Navy	May 1994
LCAC	LCAC-71	–	26.8	–	–	–	US Navy	June 1994
LCAC	LCAC-72	–	26.8	–	–	–	US Navy	July 1994
LCAC	LCAC-73	–	26.8	–	–	–	US Navy	September 1994
LCAC	LCAC-74	–	26.8	–	–	–	US Navy	October 1994
LCAC	LCAC-75	–	26.8	–	–	–	US Navy	December 1994

Vessel type	Vessel name	Yard No	Length (m)	Speed (kt)	Seats	Vehicles	Originally delivered to	Date of build
Model C-7 (executive)	*Kasuari*	–	16.0	–	46	–	Freeport, Indonesia	1994
LCAC	LCAC-76	–	26.8	–	–	–	US Navy	February 1995
LCAC	LCAC-77	–	26.8	–	–	–	US Navy	March 1995
LCAC	LCAC-78	–	26.8	–	–	–	US Navy	May 1995
LCAC	LCAC-79	–	26.8	–	–	–	US Navy	July 1995
LCAC	LCAC-80	–	26.8	–	–	–	US Navy	August 1995
LCAC	LCAC-81	–	26.8	–	–	–	US Navy	October 1995
LCAC	LCAC-82	–	26.8	–	–	–	US Navy	December 1995
LCAC	LCAC-83	–	26.8	–	–	–	US Navy	February 1996
LCAC	LCAC-84	–	26.8	–	–	–	US Navy	April 1996
LCAC	LCAC-85	–	26.8	–	–	–	US Navy	July 1996
LCAC	LCAC-86	–	26.8	–	–	–	US Navy	September 1996
LCAC	LCAC-87	–	26.8	–	–	–	US Navy	November 1996
LCAC	LCAC-88	–	26.8	–	–	–	US Navy	February 1997
LCAC	LCAC-89	–	26.8	–	–	–	US Navy	April 1997
LCAC	LCAC-90	–	26.8	–	–	–	US Navy	September 1997
LCAC	LCAC-91	–	26.8	–	–	–	US Navy	December 2000
LCAC	–	–	26.8	–	–	–	Japan Defence Agency	October 1997
LCAC	–	–	26.8	–	–	–	Japan Defence Agency	December 1997
LCAC	–	–	26.8	–	–	–	Japan Defence Agency	October 2001
LCAC	–	–	26.8	–	–	–	Japan Defence Agency	January 2002
LCAC	–	–	26.8	–	–	–	Japan Defence Agency	October 2002
LCAC	–	–	26.8	–	–	–	Japan Defence Agency	January 2003

craft. Over-the-horizon assaults are made possible by the high-transit speeds of the LCAC.

The LCAC is capable of travelling over land, water, marshes and mud flats. Compared to conventional landing craft, the percentage of the world's shorelines suitable for landing is increased from 17 to 80 per cent. The LCACs will operate from well-deck-equipped amphibious ships. In the future LCACs may be deployed from shore-based facilities by navies having missions within the operating range of the craft.

By 2002 the majority of the LCAC fleet had undergone a Service Life Extension Programme (SLEP) which is understood to provide an improved performance envelope with reduced operating costs.

LCAC's multimission capability has led to the designation of Multipurpose Craft, Air Cushion (MCAC). Personnel transport modules can be fitted to the deck of the LCAC to carry up to 180 personnel or to convert the LCAC to a medical evacuation or mobile field hospital role. A mine countermeasures modular sweep deck, which can be installed in 12 hours or less, includes a winch and crane for sweep gear. For breaching missions, development programmes are under-way to incorporate Distributed Explosives Technology (DET) and MK-58 line charge systems. Targeted for deployment around 1999, these systems will follow the same MCAC philosophy of temporary installation for specific missions. Other potential roles for the MCAC include patrol, civil emergency, lighterage and coastal anti-submarine warfare.

By 2003 a total of 90 craft had been delivered.

Specifications

Length overall	26.8 m
Beam	14.3 m
Draught, hullborne	0.9 m
Weight, max	153.5 t
Payload	54.3 t
Propulsive power	4 × 3,955 kW
Maximum speed	50 kt

Propulsion: Four TF40B gas turbines are installed, driving two Dowty Rotol 3.582 m diameter four-blade variable reversible-pitch propellers; and four double-entry centrifugal-type lift fans of 1.6 m diameter.

LACV-30

In the early 1980s, TM&LS manufactured 26 LACV-30s for the US Army. These logistic support craft are capable of delivering 35 tonnes of payload to shore at speeds up to 40 kt, requiring neither piers nor dockage.

Model 212B, Margaret Jill

0506610

MCAC rigged with an AQS-14 minehunting sonar kit 0506960

After years of effective use, the US Army released these craft for civil use.

TM&LS has initiated design improvements and modifications for LACV-30 in order to reduce its airborne noise signature, improve manoeuvrability and increase cargo carrying capability. Design enhancements will include a new propeller and gearbox, a new planing stern seal and rudder as well as bow thruster changes.

LACV-30s can skim across snow and ice at high speeds without damaging the environment. The craft can operate in climates with high humidity, saltwater and extreme temperatures from –40°F to as hot as 120°F. Because of its high degree of operational flexibility, the upgraded LACV-30 is currently being studied by commercial enterprises and civil authorities for use in support of the oil industry, and as a supply vessel in the Arctic region. Other than standard logistics operations, LACV-30 has proven itself in operations such as aircraft recovery, icebreaking and missile deployment.

LACV-30 design upgrades include bow puff ports, noise reduced gearbox, propeller upgrade, double rudders and a planing stern seal.

Model 210A (110 Mk 1) demonstration SES

Launched in late 1978, the Model 210A demonstration boat has undergone extensive successful testing by both commercial operators and the US Coast Guard. The basic hull and machinery layout permits modification of the deckhouse and arrangement of the deck space for a number of alternative applications, from crewboat and 275-seat passenger ferry to fast patrol boat.

In September 1980, the US Navy purchased the Demonstration SES (110 Mk 1) to be used in a joint US Navy/Coast Guard programme. The US Coast Guard, designating the boat the USCG *Dorado* (WSES-1), conducted an operational evaluation of the craft for the first six months of the programme. The craft was modified to conform to USCG requirements for an operational evaluation vessel. On completion of its USCG evaluation a 15.24 m hull extension was added to the boat for the US Navy to assess the performance of a higher length-to-beam vessel. The craft has been designated the SES-200 by the US Navy (see later sub-entry for Model 730A and SES-200 conversion).

Model 522A US Coast Guard SES

In June 1981 the US Coast Guard awarded a contract for the purchase of three Model 522A (110 Mk 2) high-speed cutters, the first two of which were delivered in October 1982.

The craft, known as the 'Seabird' class, were designated *Sea Hawk* (WSES-2), *Shearwater* (WSES-3) and *Petrel* (WSES-4). *Petrel* was delivered in June 1983. These vessels, now decommissioned, were based at Key West, Florida, at a coastguard facility, where they were highly successful at intercepting drug runners operating in the Gulf of Mexico and the Caribbean Sea.

Specifications

Length overall	33.5 m
Beam	11.9 m
Draught, hullborne	2.5 m
Draught, on-cushion	1.7 m
Displacement, max	152 t
Crew	12
Fuel capacity	34,050 litres
Propulsive power	2 × 1,345 kW
Lift power	260 kW
Maximum speed	30 kt
Range	1,550 n miles

Structure: Primary structure is built in welded marine aluminium alloy 5086.

Propulsion: Motive power for the propulsion system is furnished by two DDA 16V-149TIB marine diesels. Each drives a 1.06 m diameter, 1.17 m pitch, three-blade, fixed-pitch NiBrAl propeller Gawn-Burrill Series type.

Sea Hawk *and* Shearwater, *the first two Model 522A cutters to be delivered to the US Coast Guard* 0506609

Textron Marine 48.78 m SES-200 before propulsion and hull structure change programme 0506611

Cushion lift is provided by two Detroit Diesel Allison 8V-92N marine diesels each driving a double-inlet centrifugal fan.

Model 212B (110 Mk II)

The Model 212B is a conversion of the Model 212A. The two craft converted are the *Margaret Jill* (ex-*Speed Command*) and *Speed Tide* (ex-*Swift Command*).

Specifications

Length overall	33.3 m
Beam	11.9 m
Draught, hullborne	2.5 m
Draught, on-cushion	1.9 m
Crew	6
Passengers	119
Fuel capacity	20,212 litres
Water capacity	3,181 litres
Propulsive power	2 × 1,230 kW
Lift power	2 × 260 kW
Operational speed	31 kt

Structure: 5086 aluminium.

Propulsion: Main engines: two Detroit Diesel Allison 16V-149TIB engines, diesel, turbocharged, intercooled, heat exchanger cooled, air start each driving a three bladed stainless propellers via a ZF BW 455 reduction gear. Input shafts – identical rotation, output shafts – opposite rotation 2:1 ratio. Lift engines: two GM 8V-92 engines, diesel, heat exchanger cooled, air start each driving a TMS 42 in welded aluminium centrifugal fan.

Model 730A (USN SES-200) high length-to-beam ratio test craft

The original 33.53 m SES demonstration boat is now owned by the US Navy. The boat was modified by adding a 15.24 m hull extension. The hull was cut amidships; the lift fans and engines remaining in the bow and the main engines remaining in the stern. Bow and stern sections were then moved apart and a 15.24 m plug section inserted between them. All major systems remained the same as they were in the original vessel, including the GM 16V 149TI engines.

In this configuration, the vessel had a 60 per cent greater disposable load than the demonstration boat, while its maximum speed was only reduced by 3 to 4 kt. At intermediate speeds, the power requirements were lower than for the shorter vessel, despite the greater displacement.

In 1986, the SES-200 successfully completed an eight month series of joint technical and operational trials. The trials, which were conducted in Canada, France, Germany, Norway, Spain, Sweden and the UK, provided each host nation opportunities for direct evaluation of the high length-to-beam SES in their own waters. Further details of the trials can be found in: *The United Kingdom Trials of the SES-200* by B J W Pingree, B J Russell and J B Wilcox. Paper given at the Fourth International Hovercraft Conference, 6–7 May 1987.

Specifications

Length overall	48.7 m
Beam	11.9 m
Draught, hullborne	2.8 m
Draught, on-cushion	1.7 m
Displacement, max	207 t
Fuel capacity	80,128 litres
Propulsive power	2,460 kW
Lift power	650 kW
Maximum speed	28 kt
Operational speed	22 kt
Range	3,850 n miles

Propulsion: The main engines are two DDC 16V 149TI, each driving 1.016 m diameter fixed-pitch propellers. The lift engines are four DDC 8V 92TI, each driving 1.067 m diameter Bell centrifugal fans.

SES-200 conversion

On 11 April 1990, Textron Marine & Land Systems announced the award of a USD1,744,858 contract by the US Army Corps of Engineers for the conversion, modification and upgrading of the propulsion system and hull structure of the 48.78 m SES-200.

The conventional propellers and existing 16V 149 TI diesel engines and gearboxes have been replaced by two MTU 16V 396 TB 94 diesel engines, using two ZF BW 755 gearboxes driving Kamewa 71 S62/6-SII water-jet systems. Following the conversion, which increases the propulsion power from 2,386 to 4,265 kW, the SES-200 achieved speeds in excess of 40 kt in calm water, had greater manoeuvrability, produced lower in-water noise emission and was able to operate in shallower waters.

To conduct the conversion/repowering work, the SES-200 was lifted from the water at the Textron Marine & Land Systems shipyard, transported 160 m overland and brought into position in a high bay construction building.

After successfully completing acceptance trials in the Gulf of Mexico, the SES-200 returned to its home port at the David Taylor Research Center (DTRC), located at the Naval Air Station, Patuxent River, Maryland on the Chesapeake Bay. The DTRC, the Navy's laboratory for advanced naval vehicle development, deploys the craft in evaluation programmes.

HM 221 Fire and rescue vessel

In 1992, TM&LS delivered two SES fire and rescue vessels to the City of New York. The Hovermarine International HM 221 hulls are equipped for firefighting by TM&LS. Highly automated to allow operation by a crew of two, the vessels will provide firefighting, search and rescue, harbour patrol and pollution monitoring capabilities.

Three fire pumps (port, starboard, forward) with capacity for 6,938 litres/min (1,833 US gallons/min) at 1,427 KPa (207 lb.ft/in^2). The port and starboard pumps are driven by the fire pump engine and the forward pump by the lift engine. The firefighting equipment is remotely controlled for rotation and elevation from the wheelhouse: one 20,818 litres/min (5,500 US gallons/min) master monitor mounted on the wheelhouse; two 9,629 litres/min (2,500 US gallons/min) monitors mounted on the foredeck; and two 9,629 litres/min (2,500 US gallons/min) monitors mounted on either side of the bow beneath the foredeck. All fire monitors, except the wheelhouse monitor, can be controlled from a straight stream to a 90° fog. There is a deck hydrant with connections for four 7.62 cm (3 in) and one 11.43 cm (4 ½ in) fire hoses.

The Fire Department of the City of New York accepted delivery of the *John P Devaney*, the first of two SES fire and rescue vessels, on 17 July 1992. TM&LS delivered the second, the *Alfred E Ronaldson*, on 24 November 1992. The contract was valued at USD6.5 million.

Specifications

Length overall	21.2 m
Beam	5.9 m
Draught, hullborne	1.8 m
Weight, max	36.5 t
Crew	2
Fuel capacity	2,460 litres
Water capacity	76 litres
Propulsive power	2 × 373 kW
Lift power	287 kW
Maximum speed	30 kt
Operational speed	20 kt
Range	60 n miles

SES-200 after propulsion upgrading in 1990 0506612

C-7 executive transporter 0506875

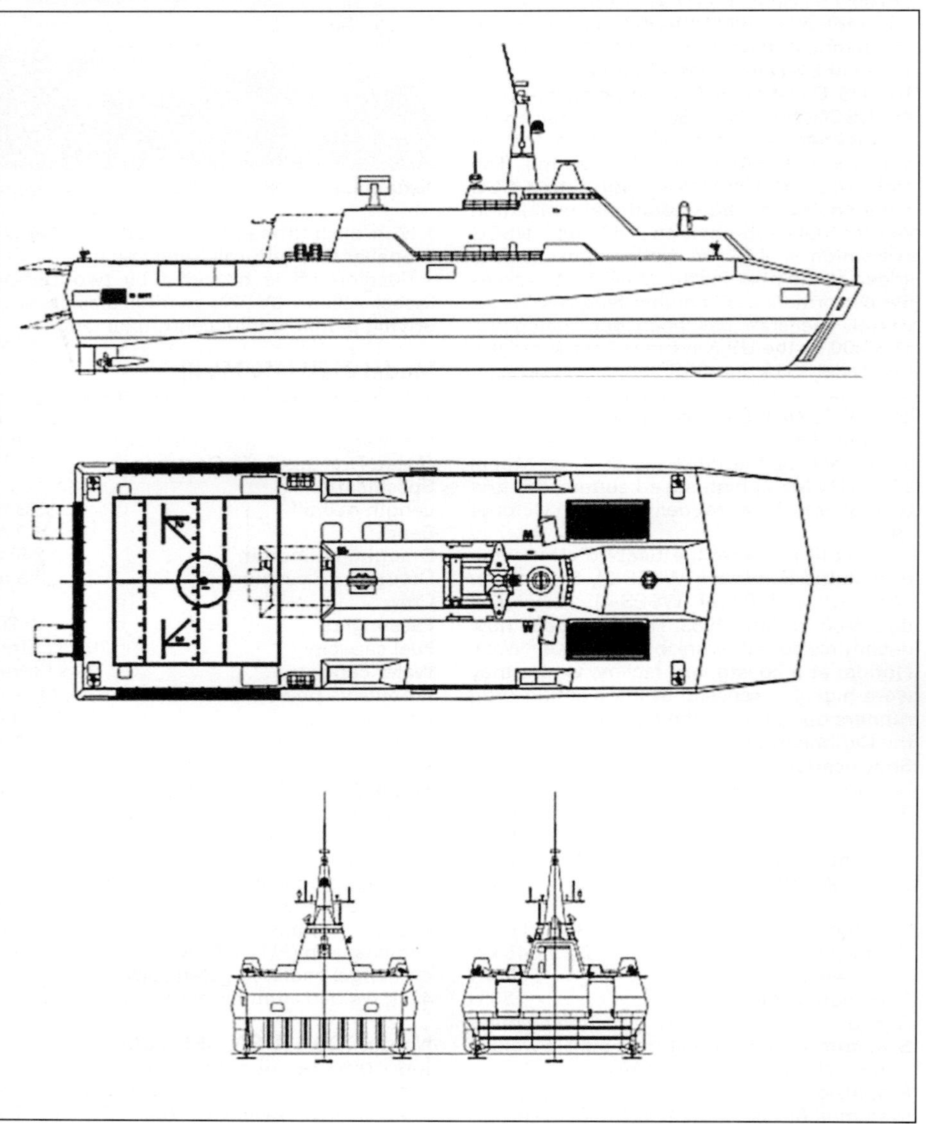

Hybrid Catamaran Air Cushion Ship 1120232

Structure: Shell mouldings, submouldings, frames, bulkheads and major attachments in GRP. Fendering is fitted to the gunwale.

Propulsion: The main engines are two Detroit Diesel Corporation 8V 92TA diesels driving 48.26 cm, three-blade propellers through direct-drive reversing gearboxes (ZF IRM 310 PL) and V-drive gearboxes (BPM VD/180; overdrive ratio 1.06:1).

Lift power is provided by a Detroit Diesel Corporation 6V 92TA which is also used as a pump engine. A second pump engine is provided by a Detroit Diesel Corporation 8V 92TA diesel.

Model C-7

The C-7 Hovercraft is designed for minimum-cost operation with high payloads and extreme environmental conditions. The 17.8 m craft combines a lightweight composite superstructure with a welded aluminium hull and is available in a number of configurations.

The concept chosen for the entire superstructure, propeller ducts and rudders is a Resin Transfer Moulding (RTM) process employing primarily an E-glass composition. The process uses resin injection under vacuum with a room temperature cure to consistently achieve a glass content of over 60 per cent. This compares to a conventional wet lay-up process that typically produces a 40 per cent glass content. Parts can be fabricated with virtually zero void content and an exceptional surface finish ready for paint.

The C-7 superstructure is composed of 12 basic sections, plus hatches, doors and bulkheads. The curved bow section, which is 139 × 247 in, is considered the largest production part ever fabricated using this process: it weighs 450 lb without the windows installed. To obtain maximum stiffness at minimum weight, carbon fibre composite is used in several transverse frames and in the centre ring section of the propeller shroud.

The basic C-7 hovercraft features a 40+ kt capacity, low noise levels, low operating costs and environmentally safe operations. It is available in four configurations: fire and rescue craft (FR-7), passenger ferry, executive transport and patrol craft (P-12).

In 1994, TM&LS delivered an executive C-7 transporter for use by Freeport Indonesia. The *Kasuari*, as it was named, features modern, quiet, first class accommodation, windows forward and on both sides, bar, kitchen and an interior design tailored to the customer.

C-7 fire and rescue craft (FR-7) 0506876

An FR-7 configured for firefighting and rescue was delivered to the Civil Aviation Authority of Singapore at Changi International airport. The FR-7 has a bow ramp, a fire monitor that delivers preselected concentrate strength foam at more than 1,500 litres/min at least 30 ft from the craft, and accommodates 50 passengers and 10 Stokes stretchers.

The P-12 patrol craft design, a stretched derivative of the C-7, features 45 kt speed and 6 tonne payload; crew berthing and galley; bow ramp for roll-on/off capability; carrying capacity for a light vehicle (HMMWV); and a 500 n mile range.

Specifications

Length overall	16.0 m
Beam	9.5 m
Weight, max	29.5 t
Payload	5.4 t
Crew	2
Passengers	46
Fuel capacity	3,785 litres
Water capacity	152 litres
Propulsive power	2 × 386 kW
Lift power	235 kW
Maximum speed	40 kt
Range	500 n miles

Classification: USCG Sub-chapter T Small Passenger Vessels (under 100 gross tons) and IMO Code of Safety for Dynamically Supported Craft.

Structure: Hull material: welded 5456 aluminium.

Superstructure material: advanced marine composite.

HCAC (design)

The Hybrid Catamaran Air Cushion Ship (HCAC) is designed to be fast and stealthy for operations in littoral (coastal) waters. The HCAC is a hybrid combining the catamaran and the surface effect ship. When operating as a catamaran the vessel has an efficient cruise speed of 18–20 kt, and when operating as a surface effect ship it can cruise at speeds up to 50 kt.

The hybrid hull provides a very shallow draft, allowing the vessel to be operated close to the shoreline. The vessel has a large flight deck, which can accommodate a number of aircraft types. The vessel also has a hanger capable of accommodating a MH-60 helicopter.

Specifications

Length overall	89.9 m
Beam	30 m
Displacement, max	1,640 t
Fuel capacity	55,000 litres
Max speed (SES mode)	55 kt
Max speed (Catamaran mode)	20 kt
Range at 50 kt	500 n miles
Range at 18 kt	5,250 nm

HYDROFOILS

Company listing by country

Belarus
Gomel Shipyard

Italy
Rodriquez Cantieri Navali SpA

Japan
Kawasaki Heavy Industries Ltd
Mitsubishi Heavy Industries Ltd
Sumitomo Heavy Industries Ltd
Universal Shipbuilding Corporation

Russian Federation and Associated States (CIS)
Volga Shipyard

Ukraine
Feodosia Shipbuilding Association (MORYE)

United States
Boeing
Navatek Ships Ltd

Belarus

Gomel Shipyard

Stolitca Group, 2-28 M-Tanka Street, Minsk, 220004 Belarus

Tel: (+375 17) 223 91 58
Fax: (+375 17) 223 91 58
e-mail: postmaster@capital.by
Web: www.capital.by/eng.shipyardengl.htm

Gomel Shipyard was being managed by the Stolitica Group in 2004, industry sources report.

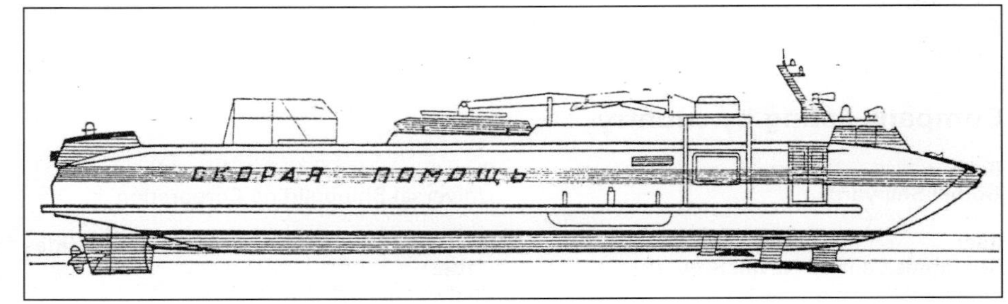

Malakhit river ambulance 0506819

BYELORUS

This craft was developed from the Raketa, via the Chaika, for fast passenger services on shallow winding rivers less than 1 m deep and too shallow for conventional vessels.

It was put into series production at the river shipyard at Gomel, Belarus in 1965. Byelorus was expected to be succeeded in service by the 53-seat Polesye.

Specifications

Length overall	18.6 m
Beam	4.6 m
Draught, hullborne	0.9 m
Draught, foilborne	0.3 m
Displacement, min	9.6 t
Displacement, max	14.5 t
Passengers	40
Propulsive power	708 kW
Operational speed	34 kt

Structure: Hull and superstructure are built in aluminium-magnesium alloy. The hull is of all-welded construction and the superstructure is riveted and welded.

The shallow draught submerged foil system consists of one bow foil and one rear foil.

Propulsion: Power is supplied by an M-50 F-3 or M-400 diesel, rated at 708 kW maximum and with a normal service output of 447 kW. The wheelhouse is fitted with an electrohydraulic remote-control system for the engine and fuel supply.

Outfit: Aircraft-type seats for 40 passengers. The prototype seated only 30.

Byelorus on Karakum Canal, Turkmenia 0506631

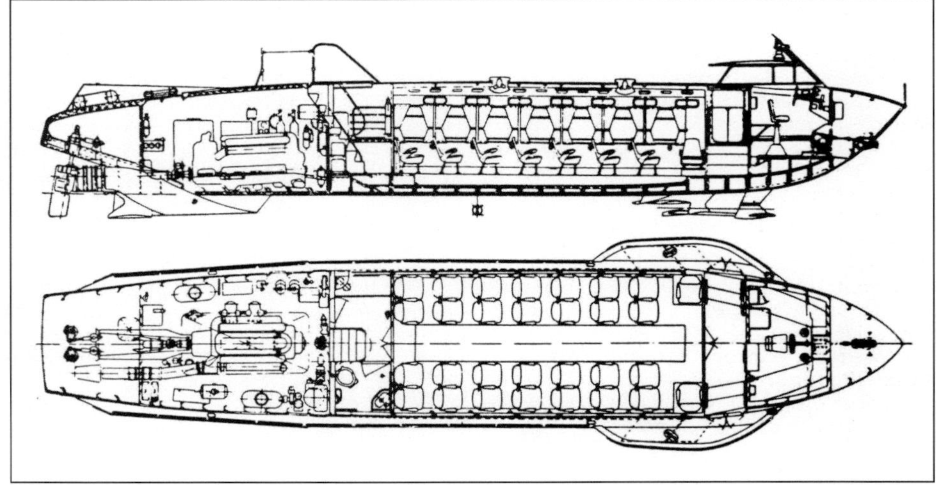

Profile and deck plan of Byelorus 0506632

POLESYE

This shallow-draught hydrofoil craft (27 built by early 1989) is intended for the high-speed transportation of passengers and tourists during daylight hours in the upper reaches of major rivers, river tributaries and freshwater reservoirs in regions with temperate climate. Two Polesyes entered service in 1992 with MAHART on the Budapest to Vienna route. The craft is classified *R on the RSFSR Register of River Shipping and is suitable for use in conditions with a wave height of 0.5 m when running on the hydrofoils and with a wave height of up to 1.2 m in the displacement mode.

The craft has capacity for a maximum of 53 passengers. The passengers are accommodated in a single lounge area in the mid-section of the vessel. In calm waters with wind conditions up to Beaufort 3, the vessel is capable of a speed of 35.1 kt. The vessel is capable of running on the hydrofoils on river channels with a radius of turn up to 100 to 150 m.

The time to accelerate from the stationary condition to becoming fully foilborne does not exceed 1.5 minutes. The distance from service speed to stop, with the propeller in reverse, is five to six boat lengths.

The hull is divided into five compartments by means of watertight bulkheads. The foils and side fenders are removable to facilitate overland transport.

The vessel is powered by a 12-cylinder 'V' diesel engine with a maximum capacity of 810 kW at 1,600 rpm. The engine is installed

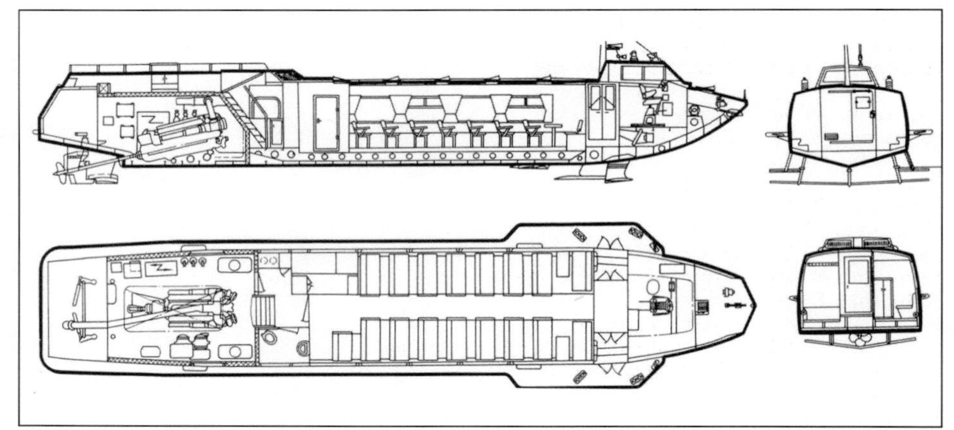

General arrangement of Polesye 0506633

Polesye 0506634

Vessel type	Vessel name	Yard No	Length (m)	Speed (kt)	Seats	Vehicles	Originally delivered to	Date of build
Byelorus	Series production	–	18.6	34.0	40	None	–	Series production
Polesys	Series production	–	21.3	35.0	53	None	–	Series production
Malakhit	DESIGN	–	22.0	35.0	–	None	–	–

at an angle of 12.5° to the horizontal and transmits power to the propeller via a direct-coupled reversing gear unit.

Specifications

Length overall	21.3 m
Beam	3.6 m
Foil width	5 m
Draught, hullborne	0.9 m
Draught, foilborne	0.4 m
Displacement, min	13.7 t
Displacement, max	20 t
Passengers	53
Propulsive power	810 kW
Operational speed	35 kt
Range	216 n miles
Operational limitation	wave height 0.5 m

MALAKHIT

A river ambulance hydrofoil designed for emergency medical calls for people living in remote regions and for passengers and crews of river-craft.

Malakhit is designed to take five patients. The project is based on the SPK Polesye series hydrofoil.

Specifications

Length	22.0 m
Beam	3.6 m
Draught, hullborne	0.9 m
Draught, foilborne	0.4 m
Crew	2
Operational speed	35 kt
Range	216 n miles

Italy

Rodriquez Cantieri Navali SpA

Via S Raineri 22, I-98122 Messina, Italy

Tel: (+39 090) 776 52 83
Fax: (+39 090) 776 52 93
e-mail: marketing@rodriquez.it
Web: www.rodriquez.it

Luciana La Noce, *President*
Gianclaudio Neri, *Managing Director*
Marco Pavoncelli, *Commercial Director*
Marco Ragazzini, *General Director*

Since 1956, when the shipyard started to build commercial fast ferries, Rodriquez has delivered over 270 high-speed craft and the company continues to be a world leader in the design, engineering and manufacture of such craft. The company has three shipyards in Italy: one in Messina in the south, one in Pietra Ligure in the north and the third is the Intermarine yard in Sarzana, close to La Spezia. Rodriquez also has a yard in Brazil for the South American market and has licence agreements in Spain and in the US.

In May 2004 RCN Finanziaria purchased a majority share of Rodriquez Cantieri Navali. The other shareholders are MRS Sviluppo and Ustica Lines.

Rodriquez Cantieri Navali SpA was the first company to produce hydrofoils in series and is the biggest hydrofoil builder outside the Russian Federation. In addition, the Aliscafi Shipping Company was established in Sicily to operate the world's first scheduled sea-going hydrofoil service in August 1956 between Sicily and the Italian mainland.

The service was operated by the first Rodriquez-built Supramar PT 20, *Freccia del Sole*. Cutting down the port-to-port time from Messina to Reggio di Calabria to one-quarter of that of conventional ferry boats, and completing 22 daily crossings, the craft soon proved its commercial viability. With a seating capacity of 72 passengers, the PT 20 carried between 800 and 900 passengers a day and conveyed a record number of some 31,000 passengers in a single month.

Eight main types of hydrofoil have now been produced: the Supramar PT 20 (44 craft) and PT 50 (31 craft), the RHS 70 (10 craft), RHS 110 (5 craft), RHS 140 (13 craft), RHS 150 (14 craft), RHS 160 (30 craft), RHS 200 (2 craft), MEC (1 craft) and Foilmaster (3 craft). Many of the early craft built are still operating.

In 2006 Rodriquez announced that it was in the process of designing and building two prototype 37 m, 280 passenger hydrofoil craft with fully submerged foils. The new craft, designed to achieve speeds of 50 knots, will maintain all the traditional advantages of hydrofoils but with increased efficiency and seakeeping performance. Both craft are being built under the supervision of RINA. The vessels feature an electronic control system to allow stabilisation and directional control of the fins including controlled trailing flaps. Performance will be optimised using a two-speed reduction gearbox with first gear designed for vessel take-off and second gear designed for high-speed performance. The fully submerged foil design is ideally suited to medium distance and coastal inshore routes.

Vessel type	Vessel name	Yard No	Length (m)	Speed (kt)	Seats	Vehicles	Originally delivered to	Date of build
Supramar PT 50	*Flecha Del Lago*	061	27.8	34.0	140	–	Naveca	1958
Supramar PT 50	*Flecha Mara*	060	27.8	34.0	140	–	Naveca	1958
Supramar PT 50	*Freccia di Sorrento*	062	27.8	34.0	125	–	Ministero del Mercantile Marine	1959
Supramar PT 50	*Frecchia di Messina*	059	27.8	34.0	125	–	–	1959
Supramar PT 50	*Freccia d'Oro*	063	27.8	34.0	130	–	–	1959
Supramar PT 50	*Freccia Atlantica*	064	27.8	34.0	125	–	Aliscafi SNAV SpA	1960
Supramar PT 50	*Freccia del Sud*	065	27.8	34.0	50	–	Aliscafi SNAV SpA	1960
Supramar PT 50	*Pisanello*	066	27.8	34.0	130	–	Siremar	1961
Supramar PT 50	*Queenfoil* (ex-*Sleipner*)	076	27.8	34.0	125	–	Transtour SA	1961
Supramar PT 50	*Freccia di Lipari*	077	27.8	34.0	125	–	Aliscafi SNAV SpA	1961
Supramar PT 50	*Flecha de Buenos Aires*	078	27.8	34.0	125	–	Alimar SA	1962
Supramar PT 50	*Flecha de Colonia* (ex-*Flecha de Montevideo*)	079	27.8	34.0	125	–	Alimar SA	1962
Supramar PT 50	*Flecha del Litoral*	080	27.8	34.0	125	–	–	1963
Supramar PT 50	*Freccia del Mediterraneo*	081	27.8	34.0	125	–	Aliscafi SNAV SpA	1963
Supramar PT 50	*Flying Albatross*	090	27.8	34.0	125	–	Hong Kong Macao Hydrofoil Company	1964
Supramar PT 50	*Freccia di Sicilia* (ex-*Condor 1*)	088	27.8	34.0	125	–	Aliscafi SNAV SpA	1964
Supramar PT 50	*Nibbio*	089	27.8	34.0	125	–	Adriatica di Navigazione SpA	1964
Supramar PT 50	*Svalan*	092	27.8	34.0	125	–	Tarnan Line, Limassol, Cyprus	1965
Supramar PT 50	*Flying Skimmer*	091	27.8	34.0	125	–	Hong Kong Macao Hydrofoil Company	1965
Supramar PT 50	*Fairlight*	120	27.8	34.0	140	–	–	1966
Supramar PT 50	*Tarnan*	118	27.8	34.0	120	–	Tarnan Line, Limassol, Cyprus	1966
Supramar PT 50	*Freccia delle Isole*	111	27.8	34.0	125	–	–	1966
Supramar PT 50	*Flying Condor*	093	27.8	34.0	125	–	Hong Kong Macao Hydrofoil Company	1966
Supramar PT 50	*Flying Flamingo*	121	27.8	34.0	125	–	ex-Hong Kong Macao Hydrofoil Company	1967

Vessel type	Vessel name	Yard No	Length (m)	Speed (kt)	Seats	Vehicles	Originally delivered to	Date of build
Supramar PT 50	*Star Capricorn*	122	27.8	34.0	117	–	Covemar	1967
Supramar PT 50	*Sun Arrow*	125	27.8	34.0	140	–	Aliscafi SNAV SpA	1968
Supramar PT 50	*Palm Beach*	131	27.8	34.0	140	–	State Transit Authority	1969
Supramar PT 50	*Long Reef*	124	27.8	34.0	140	–	NSW State Transit Authority	1969
Supramar PT 50	*Freccia Adriatica*	119	27.8	34.0	125	–	Aliscafi SNAV SpA	1969
Supramar PT 50	*Stilprins*	123	27.8	34.0	128	–	–	1970
RHS 110	*Flecha de Angra* (ex-*Flying Phoenix*)	148	25.6	37.0	140	–	–	1970
Supramar PT 50	*Dee Why*	132	27.8	34.0	140	–	State Transit Authority	1970
RHS 140	*Flying Dragon*	134	28.7	32.5	140	–	Hong Kong Macao Hydrofoil Company	1971
RHS 140	*Colonia del Sacramento* (ex-*Condor 3*)	133	28.7	32.5	140	–	Belt SA	1971
RHS 110	*Cacilhas*	147	25.6	37.0	110	–	–	1971
RHS 140	*Santa Maria del Buenos Aires* (ex-*Tyrving*)	153	28.7	32.5	116	–	Belt SA	1972
RHS 110	*Barca*	157	25.6	37.0	122	–	–	1972
RHS 140	*Flying Sandpiper*	180	28.7	32.5	125	–	Hong Kong Macao Hydrofoil Company	1972
RHS 140	*Curl-Curl*	155	28.7	32.5	140	–	NSW State Transit Authority	1972
RHS 140	*Farallon* (ex-*Loberan*)	154	28.7	32.5	111	–	Belt SA	1972
RHS 140	*Flying Egret*	152	28.7	32.5	125	–	Hong Kong Macao Hydrofoil Company	1972
RHS 70	*Shearwater 3*	150	25.6	32.4	67	–	Red Funnel Ferries	1972
RHS 140	*Rapido de Ibiza* (ex-*Viggen*)	161	28.7	32.5	120	–	Flebasa Lines	1973
RHS 140	*Flying Swift*	181	28.7	32.5	125	–	Hong Kong Macao Hydrofoil Company	1973
RHS 70	*Shearwater 4*	156	25.6	32.4	67	–	Red Funnel Ferries	1973
RHS 110	*Cerco*	159	25.6	37.0	111	–	–	1973
RHS 110	*Praia*	158	25.6	37.0	111	–	–	1973
RHS 70	*Freccia delle Camelie*	186	25.6	32.4	71	–	Navigazione Lago Maggiore	1974
RHS 160	*Princess Zoe* (ex-*Alijumbo Ustica* ex-*Lilau*)	181	30.9	32.0	160	–	Aliscafi SNAV SpA	1974
RHS 70	*Freccia del Benaco*	187	25.6	32.4	71	–	Navigazione Sul Lago di Garda	1974
RHS 140	*Condor 4*	184	28.7	32.5	136	–	Condor Ltd	1974
RHS 70	*Freccia delle Betulle*	185	25.6	32.4	71	–	Navigazione Lago di Como	1974
RHS 70	*Freccia delle Magnolie*	188	25.6	32.4	71	–	Navigazione Lago Maggiore	1975
RHS 140	*Flying Ibis*	182	28.7	32.5	125	–	Hong Kong Macao Hydrofoil Company Ltd	1975
RHS 160	*Diomedea* (ex-*Flying Phoenix*)	190	30.9	32.0	168	–	Adriatica di Navigazione SpA	1975
RHS 160	*Condor 5*	190	30.9	32.0	180	–	Condor Ltd	1976
RHS 70	*Freccia delle Gardenie*	189	25.6	32.4	71	–	Navigazione Lago di Como	1976
RHS 70	*Freccia dei Gerani*	196	25.6	32.4	71	–	Navigazione Sul Lago di Garda	1977
RHS 140	*Duccio* (ex-*Fabricia*)	193	28.7	32.5	140	–	Siremar	1977
RHS 140	*Albireo*	194	28.7	32.5	150	–	Caremar SpA	1977
RHS 160	*Algol*	190	30.9	32.0	180	–	Caremar	1978
RHS 160	*Linosa* (ex-*May W Craig* ex-*Alijumbo*)	190	30.9	32.0	161	–	Aliscafi SNAV SpA	1979
RHS 150SL	*Freccia delle Valli*	199	28.7	32.5	190	–	Navigazione Lago di Como	1979
RHS 160	*Alioth*	190	30.9	32.0	180	–	Caremar	1979
RHS 150	*Xel-Ha*	203	28.7	32.5	151	–	Secretaria de Turismo, Mexico	1980
RHS 150SL	*Freccia dei Giardini*	204	28.7	32.5	190	–	Navigazione Lago Maggiore	1980
RHS 70	*Shearwater 5*	197	25.6	32.4	67	–	Red Funnel Ferries	1980
RHS 150SL	*Freccia dei Gerani*	196	28.7	32.5	190	–	Navigazione Sul Lago del Garda	1980
RHS 160	*Botticelli*	190	30.9	32.0	184	–	Siremar	1980
RHS 160	*Donatello*	190	30.9	32.0	184	–	Siremar	1980
RHS 200	*Superjumbo Capri* (ex-*Superjumbo*)	092	35.8	35.0	254	–	Aliscafi SNAV SpA	1981
RHS 150SL	*Freccia delle Riviere*	206	28.7	32.5	190	–	Navigazione Sul Lago di Garda	1981
RHS 70	*Shearwater 6*	221	25.6	32.4	67	–	Red Funnel Ferries	1982
RHS 150SL	*Galileo Galilei*	208	28.7	32.5	190	–	Navigazione Sul Lago di Garda	1982
RHS 160	*Nicte-Ha*	190	30.9	32.0	160	–	Secretaria de Turismo, Mexico	1982
RHS 150SL	*Guglielmo Marconi*	207	28.7	32.5	190	–	Navigazione Lago di Como	1983
RHS 200	*Rapido de Mallorca* (ex-*San Cristobal* ex-*Stretto di Messina*)	209	35.8	35.0	254	–	Aliscafi SNAV SpA	1984
RHS 150SL	*Enrico Ferni*	220	28.7	32.5	190	–	Navigazione Lago Maggiore	1984
RHS 160F	*Sinai* (ex-*Manly*)	211	31.2	34.5	186	–	NSW State Transit Authority	1984
RHS 160F	*Condor 7*	217	31.2	34.5	200	–	Condor Ltd	1985
RHS 160F	*Fast Blu* (ex-*Sydney*)	216	31.2	34.5	238	–	NSW State Transit Authority	1985
RHS 160F	*Mega Dolphin XXXI* (ex-*Pez Volador* ex-*Aijumbo Eolie*)	218	31.2	34.5	220	–	Compania Naviera Mallorquina	1986
RHS 160F	*Alnilam*	227	31.2	34.5	210	–	Caremar	1986
RHS 160F	*Fabricia*	228	31.2	34.5	210	–	Toremar	1987

Vessel type	Vessel name	Yard No	Length (m)	Speed (kt)	Seats	Vehicles	Originally delivered to	Date of build
RHS 160F	*Aldebaran*	229	31.2	34.5	210	–	Caremar	1987
RHS 160F	*Alijumbo Stromboli* (ex-*Alijumbo Messian*)	M 001	31.2	34.5	210	–	Aliscafi SNAV SpA	1988
RHS 160F	*Masaccio*	230	31.2	34.5	210	–	Siremar	1988
RHS 150FL	*Goethe*	232	28.7	32.5	200	–	Navigazione Sul Lago di Garda	1988
RHS 160F	*Cris M* (ex-*Marrajo*)	M 003	31.2	34.5	204	–	Trasmediterranea SA	1989
RHS 160F	*Mategna*	231	31.2	34.5	210	–	Siremar	1989
RHS 160F	*Giorgione*	239	31.2	34.5	210	–	Siremar	1989
RHS 160F	*Monte Gargano*	240	31.2	34.5	212	–	Adriatica di Navigazione SpA	1989
RHS 160F	*Mega Dolphin XXXII* (ex-*Barracuda*)	M 002	31.2	34.5	201	–	Trasmediterranea SA	1989
RHS 150FL	*Voloire*	237	28.7	32.5	200	–	Navigazione Lago di Como	1989
RHS 160F	*Cittiships*	236	31.2	34.5	210	–	Aliscafi SNAV SpA	1990
RHS 160F	*Fiammetta M* (ex-*Tintorera*)	M 004	31.2	34.5	204	–	Trasmediterranea SA	1990
RHS 150FL	*Lord Byron*	238	28.7	32.5	200	–	Navigazione Lago Maggiore	1990
RHS 160F	*Alijumbo Zibibbo*	243	31.2	34.5	210	–	Aliscafi SNAV SpA	1991
RHS 160F	*Moretto 1* (ex-*Alijumbo Primo*)	244	31.2	34.5	210	–	Aliscafi SNAV SpA	1991
RHS 160F	*Alijumbo Eolie*	245	31.2	34.5	210	–	Aliscafi SNAV SpA	1991
RHS 160F	*Mega Dolphin XXX* (ex-*Alijumbo Messina*)	248	–	34.5	210	–	Aliscafi SNAV SpA	1992
MEC 1	*Mec Ustica*	242	25.0	–	146	–	Aliscafi SNAV SpA	1993
Foilmaster	*Tiziano*	246	31.4	38.0	219	–	Siremar	1994
Foilmaster	*Eduardo M*	262	31.4	38.0	219	–	Ustica Lines	1996
Foilmaster	*Adriana M*	274	31.4	38.0	219	–	Ustica Lines	1999
Foilmaster	*Natalie M*	275	31.4	38.0	219	–	Ustica Lines	2003
Foilmaster	*Ettore M*	276	31.4	38.0	219	–	Ustica Lines	2003
Foilmaster	*Antioco*	330	31.4	38.0	219	–	Siremar	2004
Foilmaster	*Calypso*	328	31.4	38.0	219	–	Siremar	2005
Foilmaster	*Eraclide*	327	31.4	38.0	219	–	Siremar	2005
Foilmaster	*Platone*	332	31.4	38.0	219	–	Siremar	2006
Foilmaster	*Atlantis*	331	31.4	38.0	219	–	Siremar	2006
Foilmaster	*Eschilio*	305	31.4	38.0	219	–	Siremar	2006
Foilmaster	*Mirella M*	344	31.4	38.0	219	–	Ustica Lines	2006
Foilmaster	–	345	31.4	38.0	219	–	Ustica Lines	2007
Fully Submerged Hydrofoil	–	325	37.5	48.0	260	–	–	(2008)
Fully Submerged Hydrofoil	–	326	37.5	48.0	260	–	–	(2008)

RHS 70

This is a 32 t coastal passenger ferry with seats for 71 passengers. Power is supplied by a single 1,066 kW MTU diesel and the cruising speed is 32.4 kt.

Specifications

Length overall	22.0 m
Foil width	7.4 m
Draught, hullborne	2.7 m
Draught, foilborne	1.2 m
Displacement, max	31.5 t
Payload	6 t
Passengers	71
Propulsive power	1,066 kW
Max speed	36.5 kt
Operational speed	32.4 kt

Structure: V-bottom hull of riveted light metal alloy construction. Watertight compartments are below the passenger decks and in other parts of the hull.

The foils are surface-piercing types in partly hollow welded steel.

Propulsion: A single MTU 12V 331 TC 82 diesel, developing 1,066 kW at 2,340 rpm, drives a three-blade bronze-aluminium propeller through a Zahnradfabrik W 800 H 20 gearbox.

Control: During operation, the angle of the bow foil can be adjusted within narrow limits from the steering position by means of a hydraulic ram operating on a foil support across the hull.

RHS 110

A 54 t hydrofoil ferry, the RHS 110 was originally designed to carry a maximum of 110 passengers at a cruising speed of 37 kt.

Specifications

Length overall	25.6 m
Beam	5.9 m
Foil width	9.2 m
Draught, hullborne	3.3 m
Draught, foilborne	1.2 m
Displacement, max	54 t
Passengers	110
Fuel capacity	3,600 litres
Propulsive power	2 × 1,006 kW
Max speed	40 kt
Operational speed	37 kt
Range	262 n miles

Foilmaster, Eduardo M, *as delivered to Ustica Lines* 0507046

Structure: V-bottom of high-tensile riveted light metal alloy construction, using Peraluman plate and anti-corrodal profiles. The upper deck plates are 3.5 mm thick Peraluman. Removable deck sections permit the lifting out and replacement of the main engines. The superstructure, which has a removable roof, is in 2 mm thick Peraluman plates, with L and C profile sections. Watertight compartments are below the passenger decks and other parts of the hull.

The foils are surface-piercing types, in partly hollow welded steel.

Propulsion: Power is supplied by two 12-cylinder supercharged MTU MB 12V 493 Ty 71 diesels, each with a maximum output of 1,006 kW at 1,500 rpm. Engine output is transferred to two three-blade bronze-aluminium propellers through Zahnradfabrik W 800 H20 gearboxes. Each propeller shaft is 90 mm in diameter and supported at three points by seawater-lubricated rubber bearings. Steel fuel tanks with a total capacity of 3,600 litres are aft of the engine room.

Control: Hydraulically operated flaps, attached to the trailing-edges of the bow and rear foils, are adjusted automatically by a Hamilton Standard stability augmentation system for the damping of heave, pitch and roll motions. The rear foil is rigidly attached to the transom, its incidence angle being determined during tests.

Auxiliary systems: Steering, variation of the foil flaps and the anchor windlass operation are all accomplished hydraulically from the wheelhouse. Plant comprises two Bosch pumps installed on the main engines that convey oil from a 60 litre tank under pressure to the control cylinders of the rudder, foil flaps and anchor windlass.

RHS 140

This 65 t hydrofoil passenger ferry seats up to 150 passengers and has a cruising speed of 32.5 kt.

Specifications

Length overall	28.7 m
Foil width	10.7 m
Draught, hullborne	3.5 m
Draught, foilborne	1.5 m
Displacement, max	65 t
Payload	12.5 t
Passengers	150
Propulsive power	2 × 1,007 kW
Max speed	36 kt
Operational speed	32.5 kt
Range	297 n miles

Structure: Riveted light metal alloy design framed on longitudinal and transverse formers.

The foils are surface-piercing V-foils of hollow welded steel construction.

Propulsion: Power is provided by two MTU 12V 493 Ty 71 12-cylinder supercharged engines, each developing 1,007 kW at 1,500 rpm. Engine output is transmitted to two, three-blade 700 mm diameter bronze propellers through Zahnradfabrik gearboxes.

Control: Lift of the bow foil can be modified by hydraulically operated trailing-edge flaps.

RHS 150

Combining features of both the RHS 140 and the RHS 160, the RHS 150 hydrofoil passenger ferry is powered by two 1,066 kW MTU supercharged four-stroke diesels. These give the craft a cruising speed of 32.5 kt and a cruising range of 130 n miles.

Specifications

Length overall	28.7 m
Foil width	11.0 m
Draught, hullborne	3.1 m
Draught, foilborne	1.4 m
Displacement, max	65.5 t
Passengers	150
Propulsive power	2 × 1,066 kW
Operational speed	32.5 kt
Range	130 n miles

Structure: Riveted light metal alloy design framed on longitudinal and transverse formers.

Surface-piercing W-foils of hollow welded steel construction.

Propulsion: Motive power is provided by two supercharged MTU MB 12V 331 TC 82 four-stroke diesels each developing 1,066 kW at 2,140 rpm continuous. Engine output is transmitted to two bronze propellers via two Zahnradfabrik BW 255L gearboxes.

Control: Lift of the bow foil can be modified by hydraulically operated trailing-edge flaps.

RHS 150SL

This variant has been designed for inland navigation, particularly on the Great Lakes in Northern Italy. Due to the less severe conditions on such waters, it has been possible to redesign the hull structure to allow for larger windows in the lower saloons and the superstructure, greatly increasing visibility for sightseeing. In addition, because the safety rules are less demanding than for open water routes, there is a saving in weight in the design allowing an increase in passenger numbers, so that 200 may be carried with lightweight seats fitted.

RHS 150F

This variant has a wider deck and the superstructure volume has been increased giving a more aesthetic shape as well as greater volume for passengers, giving greater comfort. Improvements have also been incorporated in this variant, increasing performance and reducing maintenance costs.

RHS 160

A 95 t passenger ferry with seats for up to 180 and a cruising speed of 32 kt.

In March and May 1986 two RHS 160s were used for oil spill clean-up trials.

RHS 150FL Voloire *delivered 1989 for service on Lake Como, Italy*  0506614

RHS 150F Salina *operated by Aliscafi SNAV SpA* 0506878

RHS 70 Shearwater 3, *sister vessel to three other RHS 70s delivered to Red Funnel Ferries* 0506613

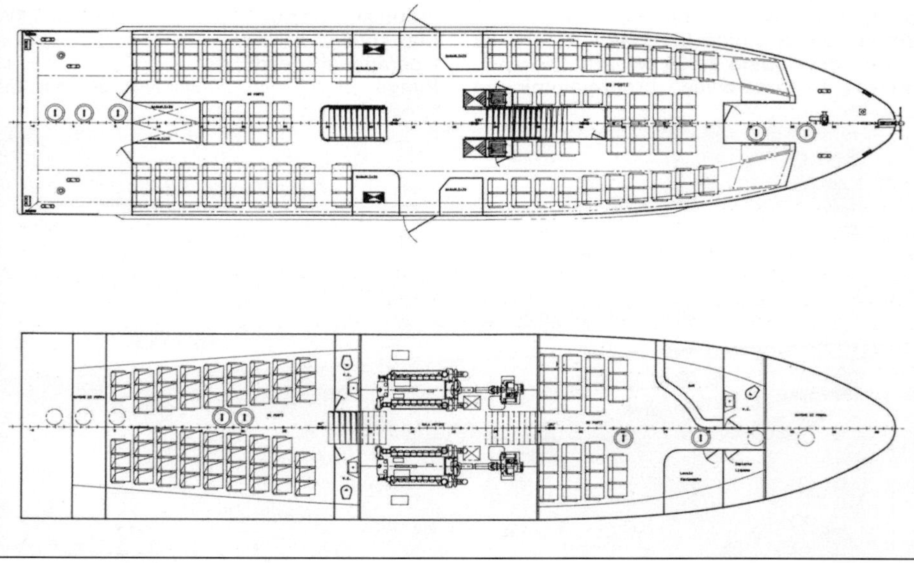

Foilmaster deck layouts 0506618

Specifications

Length overall	30.9 m
Beam	6.2 m
Foil width	12.6 m
Draught, hullborne	3.7 m
Draught, foilborne	1.4 m
Displacement, max	95 t
Payload	13.5 t
Passengers	180
Propulsive power	2 × 1,400 kW
Max speed	36 kt
Operational speed	32 kt
Range	260 n miles

Structure: Riveted light metal alloy longitudinal structure, welded in parts using inert gas. The hull shape of the RHS 160 is similar to the RHS 140 series. In the manufacture of the hull, plates of aluminium and magnesium alloy of 4.4 per cent are used, while angle bars are of a high-resistant aluminium, magnesium and silicon alloy.

The surface-piercing W-foils are of hollow welded steel construction.

Propulsion: Power is provided by two supercharged MTU MB 12V 652 TB 71 four-stroke diesel engines, each with a maximum output of 1,454 kW at 1,460 rpm under normal operating conditions. Engine starting is accomplished by compressed air starters. Engine output is transmitted to two three-blade bronze propellers through two Zahnradfabrik 900 HS 15 gearboxes.

Control: Craft in this series feature a bow rudder for improved manoeuvrability in congested waters. The bow rudder works simultaneously with the aft rudders. Hydraulically operated flaps, attached to the trailing-edges of the bow and rear foils, are adjusted automatically by a Hamilton Standard electronic stability augmentation system, for the damping of heave, pitch and roll motions in heavy seas.

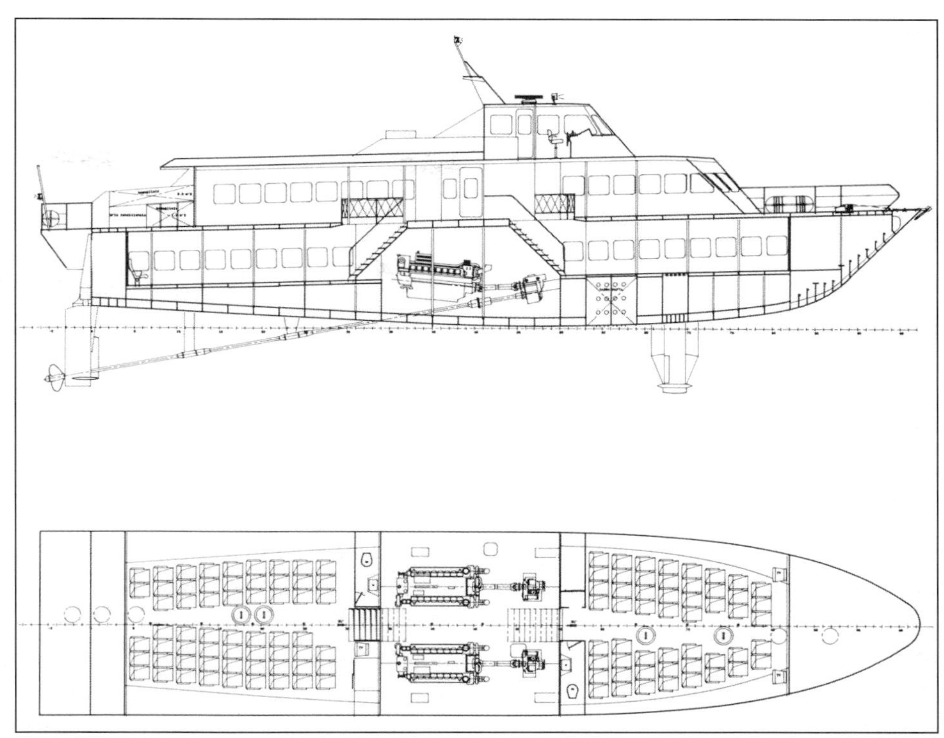

RHS 160F inboard profile and lower deck arrangement 0506879

RHS 160F Cittiships 0506880

RHS 160F Alijumbo Stromboli 0506881

RHS 160F

A further addition to the Rodriquez range is the RHS 160F, a 91.5 t passenger ferry with seats for up to 238 and a cruising speed of 34.5 kt.

Specifications

Length overall	31.2 m
Length waterline	26.6 m
Beam	6.7 m
Foil width	12.6 m
Draught, hullborne	3.8 m
Draught, foilborne	1.7 m
Displacement, max	91.5 t
Payload	17.8 t
Passengers	210–238
Propulsive power	2 × 1,400 kW
Max speed	38 kt
Operational speed	34.5 kt
Range	100 n miles

Structure: Riveted light metal alloy longitudinal structure, welded in parts using inert gas. The hull shape of the RHS 160F is similar to the RHS 140 series. In the manufacture of the hull, plates of aluminium alloy containing 4.4 per cent magnesium are used while angle bars are of a high-strength aluminium, magnesium and silicon alloy.

Surface-piercing W-foils of hollow welded steel construction.

Propulsion: Power is provided by two supercharged MTU 16V 396 TB 83 four-stroke diesel engines, each with a maximum output of 1,400 kW at 2,000 rpm under normal operating conditions. Engine starting is accomplished by compressed air starters. Engine output is transmitted to two three-blade bronze propellers through two Zahnradfabrik BW 7505 gearboxes, or alternatively, through two Reintjes WVS 1032U gearboxes as fitted to the RHS 160F craft supplied to the NSW State Transit Authority, Australia.

Control: Craft in this series feature a bow rudder for improved manoeuvrability in congested waters. The bow rudder works simultaneously with the aft rudders. Hydraulically operated flaps, attached to the trailing-edges of the bow and rear foils, are adjusted automatically by a Hamilton Standard electronic stability augmentation system, for the damping of heave, pitch and roll motions in heavy seas.

RHS 200

Powered by two supercharged MTU MB 16V 652 TB 71 four-stroke diesel engines, the 254-seat RHS 200 has a cruising speed of 35 kt.

Specifications

Length overall	35.8 m
Foil width	14.5 m
Draught, hullborne	4.6 m
Draught, foilborne	2.1 m
Displacement, max	130 t
Passengers	254
Propulsive power	2 × 1,938 kW
Max speed	37 kt
Operational speed	35 kt
Range	200 n miles

Structure: V-bottom hull of high-tensile riveted light metal alloy construction, employing Peraluman plates and anti-corrodal frames. The rake of the stem is in galvanised steel.

Surface-piercing W-foils of hollow welded steel construction.

Propulsion: Motive power is supplied by two supercharged MTU MB 16V 652 TB 71 four-stroke diesel engines, each with a maximum output of 1,938 kW at 1,460 rpm under normal operating conditions. Engine output is transferred to two super cavitating, controllable-pitch propellers.

Electrical system: Two generating sets: one 220 V, three-phase AC, for all consumer services, the second for charging 24 V battery sets and operating firefighting and hydraulic pumps. Power distribution panel in wheelhouse for navigation light circuits, cabin lighting, radar, RDF, gyrocompass and emergency circuits.

Control: Craft in this series feature a bow rudder for improved manoeuvrability in congested waters. This control operates simultaneously with the aft rudders. An advantage of the W configuration bow foil is its relatively shallow draught requirement in relation to the vessel's overall size. Hydraulically operated flaps are fitted to the trailing-edge of the bow foil to balance out longitudinal load shifting, assist take-off and adjust the flying height. The craft can also be equipped with the Hamilton Standard electronic stability augmentation system, which employs sensors and servo-mechanisms to position flaps automatically on the bow and stern foils for the damping of heave, pitch and roll motions in heavy seas.

Outfit: Seats for up to 400 passengers, according to the route served. In typical configuration there are three main passenger saloons and a bar. The standard seating arrangement allows for 116 in the main deck saloon, 58 in the aft lower saloon and 66 in the bow passenger saloon. Seating is normally four abreast in two lines with a central aisle. The bar, at the forward end of the wheelhouse belvedere superstructure, either has an eight-seat sofa or 19 seats.

The wheelhouse, which is raised to provide a 360° view, is reached from the main deck belvedere saloon by a short companionway. Controls and instrumentation are attached to a panel on the forward bulkhead extending the width of the wheelhouse. In the centre is the steering control and gyrocompass, on the starboard side are controls for the two engines, gearboxes and controllable-pitch propellers, and on the port side is the radar. Seats are provided for the captain, chief engineer and first mate. In the wheelhouse are a radiotelephone and a chart table.

Rodriquez MEC 1 Mec Ustica 0506615

Foilmaster Tiziano 0506882

Safety equipment: Fixed CO_2 firefighting self-contained automatic systems for power plant and fuel tank spaces, plus portable extinguishers for cabins and holds.

MEC 1 MAXIMUM EFFICIENCY CRAFT

Mec Ustica

Replacing the MEC 2 design previously reported, construction of MEC 1 started in July 1990. MEC 1 follows the same design principles and advances and is a new hydrofoil design developed in a joint effort with Marin (Netherlands), incorporating Rexroth hydrostatic power transmission (now continued by Hydromarine SA of Switzerland) as the Power Shaft Concept. The design embodies a new Rodriquez surface-piercing foil system and hull form (fully automatic welding construction), with the rear foil unit carrying the maximum possible weight. It has been shown to be desirable to have a very stiff (in response to waves) front foil with low damping and a very soft rear foil with high damping. It is then convenient to carry the maximum possible weight on the rear foil, the Canard lift distribution allowing a finer hull bow form to be used.

The hull form has been derived from the well known 65 Series.

In comparison with similar hydrofoils, it has a passenger capacity increase of about 25 per cent, a speed increase of nine per cent and displacement increases by less than five per cent. For equal passenger capacity, the installed power would be reduced by some 18 per cent with a consequent reduction in fuel consumption. Due to the aft location of the power plant a passenger cabin noise reduction of three to five dBA is expected.

Specifications

Length overall	25.0 m
Length waterline	20.8 m
Beam	6.7 m
Foil width	8.4 m
Draught, hullborne	2.8 m
Draught, foilborne	1.2 m
Displacement, max	55 t
Passengers	146
Operational speed	38 kt
Range	200 n miles

Propulsion: Engines: two Deutz MWM TBD 604B V8, driving hydrostatic power transmission.

Please see the Rexroth entry in the Transmission section for discussion of the basic concept and for details of the system as applied to an early Rodriquez Supramar PT 20 hydrofoil craft, *Aligrado*.

FOILMASTER

Eduardo M
Adriana M
Natalie M
Ettore M
Antioco
Tiziano
Calypso
Eraclide
Platone
Atlantis
Eschilio

The Foilmaster is an advanced variant of the RHS 160 surface-piercing hydrofoil. Improvements include resilient mounting of engines and gearboxes, exhaust gas silencers, the use of carbon fibre components in foil construction and more powerful engines. Tandem configuration foils are fitted with trailing-edge flaps. There are two sets of rudders provided, two flap rudders on the aft foil and a single flap rudder fitted to the fore foil. An air conditioning system is installed in all passenger saloons as well as in the wheelhouse.

In addition to the 11 craft delivered to date, an additional two vessels for Ustica Lines were ordered in March 2003 for delivery in 2006–07.

Specifications

Length overall	31.4 m
Length waterline	26.4 m
Beam	6.7 m
Foil width	13.3 m
Draught, hullborne	3.8 m
Draught, foilborne	1.6 m
Displacement, max	107 t
Crew	7
Passengers	219
Fuel capacity	2.5 t
Water capacity	0.6 t
Propulsive power	2 × 1,550 kW
Operational speed	38 kt
Range	150 n miles

Propulsion: Engines: two MTU 16V 396 TE 74.

37 m FULLY SUBMERGED HYDROFOIL

Two vessels are being constructed as part of Rodriquez's ongoing research into hydrofoil design. The vessels are expected to have a higher cruise speed than the Foilmaster hydrofoils, and to have superior seakeeping. The two vessels under construction will have alternative propulsion systems: one will have conventional shafts and propellers whilst the second will feature an azimuth propeller. Delivery is expected in early 2008.

Specifications

Length overall	35.5 m
Passengers	245
Operational speed	48 kt

Construction: Welded aluminium.

Propulsion: Two 2320 kW diesel engines driving either conventional shafts and propellers through a two speed gearbox or an azimuth propeller.

ALI SWATH (DESIGN)

The Ali Swath is a hybrid design combining a central SWATH hull and fully submerged foils. The first vessel will be a 64 m testcraft designed to carry 450 passengers and 60 cars at speeds up to 28 kt. The design is intended to reduce fuel consumption, wash and pollution whilst improving seakeeping. The vessel was developed jointly by Rodriquez and Registro Italiano Navale.

Japan

Kawasaki Heavy Industries Ltd

World Trade Centre Building, 4-1, Hamamatsu-cho 2-chome, Minato-ku, Tokyo 105-6116, Japan

Tel: (+81 3) 3435 21 11
Fax: (+81 3) 3436 30 37
Web: www.khi.co.jp

Matsamoto Tazaki, *President and CEO*

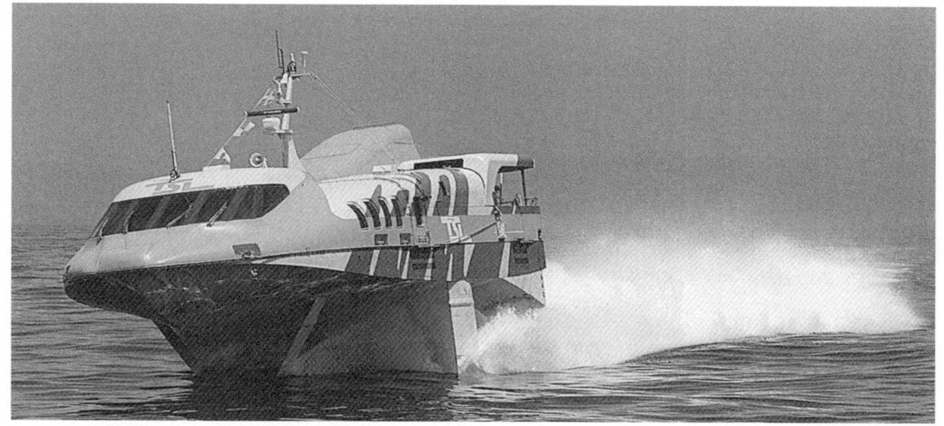

Prototype TSL-F trials craft Hayate

0506884

Kawasaki Heavy Industries has two main shipyards, one at Kobe and one at Sakaide. The Kobe yard specialises in naval and high-speed craft production.

In January 1987, Kawasaki Heavy Industries Ltd acquired a licence for the design, manufacture, marketing, maintenance and repair of Jetfoil 929-117 hydrofoil craft. By December 1998, 15 Kawasaki Jetfoils had been built. In 2002 the company decided to form a separate shipbuilding company called Kawasaki Shipbuilding Corporation.

JETFOIL 929-117

The majority of these craft have been built for fast passenger ferry services between mainland Japan and offshore islands. with two Jetfoils being exported to Spain in 1990/91.

Specifications

Length overall	27.4 m
Beam	9.1 m
Draught, hullborne	4.9 m
Max speed	45 kt

Propulsion: Two Allison 501-KF gas turbines, each rated at 2,795 kW at 13,120 rpm. Each is connected to a Kawasaki Powerjet 20 axial flow water-jet propulsor through a gearbox drive train.

Kawasaki Crystal Wing 0506883

TECHNO-SUPERLINER PROJECT

The Techno-Superliner Project in Japan involved seven leading shipbuilders, among

Vessel type	Vessel name	Yard No	Length (m)	Speed (kt)	Seats	Vehicles	Originally delivered to	Date of build
Kawasaki Jetfoil 929-117	*Tsubasa*	KJ01	27.4	–	266	–	Sado Kisen Kaisha	1989
Kawasaki Jetfoil 929-117	*Toppy*	KJ02	27.4	–	264	–	Kagoshima Shosen Co Ltd	1989
Kawasaki Jetfoil 929-117	*Beetle 3* (ex-*Pearl Wing*, ex-*Nagasaki*)	KJ03	27.4	–	230	–	Kyushu Railway Co	1989
Kawasaki Jetfoil 929-117	*Pegasus*	KJ04	27.4	–	265	–	Kyushu Shosen Co Ltd	1990
Kawasaki Jetfoil 929-117	*JB* (ex-*Sapphire Wing*, ex-*Beetle*)	KJ05	27.4	–	230	–	Kyushu Railway Co	1990
Kawasaki Jetfoil 929-117	*Princess Dacil*	KJ06	27.4	–	286	–	Trasmediterranea, SA	1990
Kawasaki Jetfoil 929-117	*Pegasus 2* (ex-*Unicorn*)	KJ07	27.4	–	233	–	Kyushu Shosen Kaisha	1990
Kawasaki Jetfoil 929-117	*Beetle II*	KJ08	27.4	–	232	–	Kyushu Railway Co	1991
Kawasaki Jetfoil 929-117	*Venus*	KJ09	27.4	–	263	–	Kyushu Yusen Co Ltd	1991
Kawasaki Jetfoil 929-117	*Suisei*	KJ10	27.4	–	260	–	Sado Kisen Kaisha	1991
Kawasaki Jetfoil 929-117	*Princess Teguise*	KJ11	27.4	–	286	–	Trasmediterranea, SA	1991
Kawasaki Jetfoil 929-117	*Toppy 2*	KJ12	27.4	–	244	–	Kagoshima Shosen Co Ltd	1992
Kawasaki Jetfoil 929-117	*Emerald Wing*	KJ15	27.4	–	230	–	Kaijo Access Co	1994
Kawasaki Jetfoil 929-117	*Crystal Wing*	KJ14	27.4	–	230	–	Kaijo Access Co	1994
Kawasaki Jetfoil 929-117	*Toppy 3*	KJ13	27.4	–	244	–	Kagoshima Shosen Co Ltd	1995

Vessel type	Vessel name	Yard No	Length (m)	Speed (kt)	Seats	Vehicles	Originally delivered to	Date of build
TSL-F Prototype	*Hayate*	–	17.1	–	–	–	–	July 1994

them Kawasaki Heavy Industries Ltd, NKK Corporation, Ishikawajima-Harima Heavy Industries Company Ltd, Sumitomo Heavy Industries Ltd and Hitachi Zosen Corporation. They jointly investigated the concept of a novel very high-speed ship with all the load being borne by a fully submerged hull and fully submerged foils, so called TSL-F.

The TSL-F superliner was designed to achieve the following performance targets:
Ship speed: 50 kt
Payload: approx 1,000 t
Endurance range: >500 n miles
Seaworthiness: regular service at Sea State 6.

For this purpose, the vessel was designed with a hull form that could reduce the influence of the waves, minimising motion and speed reduction in rough seas, giving at least a 98 per cent yearly operation rate on the expected route.

To avoid the influence of the sea surface, the main hull containing the cargo, machinery and navigation systems was supported at a height well above the waves. A lower hull for buoyancy and the foils for dynamic lift were deeply submerged, and the struts to connect the main hull, the lower hull and foils were placed vertically, as shown in the accompanying profile of the TSL-F concept. The planned TSL-F concept has been shown to have a high level of seaworthiness, with almost minimum speed reduction in relatively high sea states.

TSL-F PROTOTYPE CRAFT

Hayate

A large-scale model of a hydrofoil-type hybrid ship (TSL-F) named *Hayate* was completed in Kobe Works of Kawasaki Heavy Industries Ltd (KHI) at the beginning of July 1994.

Hayate, which is a one-sixth scale model of an R&D objective TSL-F ship, is composed of an upper hull, a fully submerged lower hull which bears buoyancy, fully submerged foils that generate dynamic lift and struts that connect the upper hull and the lower hull or the foils. The main propulsion system of *Hayate* consists of a water-jet propulsor, a reduction gear and a gas turbine.

Various at-sea tests of *Hayate* were conducted until March 1995 to acquire data unobtainable through laboratory tests and to evaluate and verify the overall performance of TSL-F as well as numerous elemental research findings.

Specifications

Length overall	17.1 m
Beam	6.2 m
Draught, hullborne	3.1 m
Draught, foilborne	1.6 m
Propulsive power	2,835 kW
Max speed	41 kt

Propulsion: One gas turbine, driving water-jet propulsor.

Mitsubishi Heavy Industries Ltd

16-5 Konan 2-chome, Minato-ku, Tokyo 108-8215, Japan

Tel: (+81 3) 6716 31 11
Fax: (+81 3) 6716 58 00
e-mail: ueda@ship.hq.mhi.co.jp
Web: www.mhi.co.jp

Naoki Ueda, *Manager*

Mitsubishi Heavy Industries has three main shipyards in Japan; Kobe, Nagasaki and Shimonoseki, with an additional dockyard at Yokohama. The company employs a total of over 10,000 people in all four facilities.

The company has considerable experience in building air cushion vehicles, hydrofoils, catamarans and monohull craft.

Super Shuttle 400 Rainbow 0506669

SUPER SHUTTLE 400

Rainbow

Rainbow II

Mitsubishi Super Shuttle 400 *Rainbow* was the world's first diesel-driven hydrofoil catamaran. It was delivered to Oki Shinko Ltd in March 1993 and entered service in April 1993. The main engines, water-jets, foils and computerised ride control system were newly developed and manufactured by Mitsubishi Heavy Industries Ltd for this project. A second vessel, Rainbow II, was delivered to the same operator in June 1998.

Specifications

Length overall	33.3 m
Beam overall	13.2 m
Foil width	12.8 m
Displacement	302 t (GRT)
Passengers	341
Max speed	45.4 kt

Propulsion: Four Mitsubishi S16R-MTK-S diesel engines, 2,125 kW each, driving two Mitsubishi MWJ-5000A water-jets.

Operator: Oki Kisen Ltd.

Vessel type	Vessel name	Yard No	Length (m)	Speed (kt)	Seats	Vehicles	Originally delivered to	Date of build
Super Shuttle 400	*Rainbow*	–	33.3	–	341	–	Oki Kisen KK	March 1993
Super Shuttle 400	*Rainbow II*	–	33.3	–	341	–	Oki Kisen KK	June 1998

Sumitomo Heavy Industries Ltd

4-7, Uraga-cho, Yokosuka, Kanagawa 239-0822, Japan

Tel: (+81 468) 46 20 01
Fax: (+81 468) 46 21 42
Web: www.shi.co.jp

Kenya Koseki, *General Manager*

Sumitomo Heavy Industries has two main shipyards in Japan, one in Uraga and one in Yokosuka. It is the Uraga shipyard that specialises in naval and high-speed craft.

The company received a licence from Fincantieri in Italy for the construction of Sparviero Class hydrofoils in the early 1990s.

PG Class

Three Sparviero Class hydrofoils (PG Class) were constructed and delivered for the Japanese Defence Agency.

Specifications

Length overall	21.8 m
Beam	7.0 m
Draught	1.4 m
Displacement	50 t
Crew	11
Max speed	46 kt

Propulsion: Two GE/IHI LM 500 gas turbines, driving a water-jet pump; and an Isuzu diesel 180PS for hullborne operations.

Vessel type	Vessel name	Yard No	Length (m)	Speed (kt)	Seats	Vehicles	Originally delivered to	Date of build
PG Class	*PG01*	821	21.8	–	–	–	Japanese Defence Agency	22 May 1993
PG Class	*PG02*	822	21.8	–	–	–	Japanese Defence Agency	22 May 1993
PG Class	*PG03*	823	21.8	–	–	–	Japanese Defence Agency	13 May 1995

Universal Shipbuilding Corporation

Head Office: Muza-Kawasaki Central Tower, 1310, Omiya-cho, Saiwai-ku, Kawasaki. Kanagawa 212-8554, Japan

Tel: (+81 44) 543 27 00
Fax: (+81 44) 543 27 10
Web: www.u-zosen.co.jp

Universal Shipbuilding Corporation, formerly Hitachi Zosen, has five main shipyards in Japan, the Ariake Works, Maizuru Works, Keihin Works, Innoshima Works and Tsu Works. While the Maizuru Works has built naval vessels, the Keihin Works specialises in commercial high-speed craft.

Hitachi Zosen was the Supramar licensee in Japan and, starting in 1961, built the Supramar PT 20, PT 32 and PT 50 hydrofoils. By 1981 some 43 hydrofoils had been built. The majority of these were built for fast passenger ferry services across the Japanese Inland Sea, cutting across deep bays which road vehicles might take 2 to 3 hours to drive round, and out to offshore islands. Other PT 20s and 50s have been exported to Australia, Hong Kong and South Korea for ferry services.

Specifications of the PT 32, PT 20 and PT 50 will be found under Supramar (Switzerland). The Hitachi Zosen craft are almost identical.

In 1974 the company completed the first PT 50 Mk II to be built at its Kawasaki yard. The vessel, *Hikari No 2,* is powered by two licence-built MTU MB 820Db diesels, carries 123 passengers plus a crew of seven and cruises at 33 kt. It was delivered to Setonaikai Kisen KK of Hiroshima in March 1975. Hitachi Zosen has constructed 25 PT 50s and 17 PT 20s.

In conjunction with Supramar, Hitachi Zosen developed a roll stabilisation system for the PT 50. The first to be equipped with this new system was completed in January 1983.

PTS 50 Mk II Housho 0506620

PT 20 Ryusei *operated by Ishizaki Kisen KK* 0506619

Roll-Stabilised Supramar PTS 50 Mk II

Housho, a PTS 50 Mk II, was delivered to Hankyu Kisen KK on 19 January 1983 and is operating on the Kobe–Naruto route.

The underside of the bow foil is fitted with two flapped fins to improve riding comfort. Operated by automatic sensors, the fins augment stability and provide side forces to dampen rolling and transverse motions.

Specifications

Length overall	27.6 m
Beam	5.8 m
Foil width	10.8 m
Draught, hullborne	3.5 m
Draught, foilborne	1.4 m
Displacement, max	62 t
Passengers	123
Propulsive power	2 × 1,029 kW
Max speed	38 kt

Vessel type	Vessel name	Yard No	Length (m)	Speed (kt)	Seats	Vehicles	Originally delivered to	Date of build
Hitatchi Supramar PT 20	*Hayate No 1*	–	20.8	34.0	–	–	Showa Kaiun Co Ltd	1962
Hitatchi Supramar PT 20	*Hibiki*	–	20.8	34.0	–	–	Setonaikai Kisen Company Ltd	1966
Hitatchi Supramar PT 50	*Ohtori*	–	27.8	34.0	113	–	–	1968
Hitatchi Supramar PT 20	*Hibiki No 3*	–	20.8	34.0	66	–	Setonaikai Kisen Company Ltd	1968
Hitatchi Supramar PT 50	*Kosei*	–	27.8	34.0	–	–	Ishizaki Kisen KK	1969
Hitatchi Supramar PT 20	*Shibuki No 2*	–	20.8	34.0	–	–	Boyo Kisen Co Ltd	1969
Hitatchi Supramar PT 50	*Ohtori No 2*	–	27.8	34.0	113	–	Setonaikai Kisen Company Ltd	1970
Hitatchi Supramar PT 20	*Myojo*	–	20.8	34.0	–	–	Ishizaki Kisen KK	1970
Hitatchi Supramar PT 50	*Zuihoh*	–	27.8	34.0	–	–	Hankyu Kisen KK	1972
Hitatchi Supramar PT 50	*Hoo*	–	27.8	34.0	–	–	Hankyu Kisen KK	1972
Hitatchi Supramar PT 50	*Condoru*	–	27.8	34.0	121	–	Setonaikai Kisen KK	1972
Hitatchi Supramar PT 20	*Kinsei*	–	20.8	34.0	–	–	Ishizaki Kisen KK	1966
Hitatchi Supramar PT 50	*Shibuki No 3*	–	27.8	34.0	–	–	Boyo Kisen Co Ltd	1972
Hitatchi Supramar PT 50	*Ohtori No 3*	–	27.8	34.0	–	–	Setonaikai Kisen Company Ltd	1972
Hitatchi Supramar PT 50	*Ohtori No 5*	–	27.8	34.0	–	–	Setonaikai Kisen Company Ltd	1973
Hitatchi Supramar PT 50	*Saisei*	–	27.8	34.0	–	–	Ishizaki Kisen KK	1974
Hitatchi Supramar PT 50	*Condor No 2*	–	27.8	34.0	121	–	Setonaikai Kisen Company Ltd	1974
Hitatchi Supramar PT 50	*Condoru No 3*	–	27.8	34.0	100	–	Setonaikai Kisen Company Ltd	1974
Hitatchi Supramar PT 50 Mk II	*Hikari No 2*	–	27.8	33.0	123	–	Setonaikai Kisen Company Ltd	1975
Hitatchi Supramar PT 50 Mk II	*Shunsei (ex-Kariyush I)*	–	27.8	33.0	123	–	Ishizaki Kisen KK	1976
Hitatchi Supramar PT 20	*Ryusei*	–	20.8	34.0	–	–	Ishizaki Kisen KK	1981
Roll stabilised Supramar PTS 50 Mk II	*Housho*	–	27.6	–	123	–	Hankyu Kisen KK	1983

Russian Federation

Volga Shipyard

Part of the Financial Industries Group high-speed ships

Head office
51 Svoboda St, GSP-565, 603600 Nizhny Novgorod, Russian Federation

Tel: (+7 8312) 73 08 23
Fax: (+7 8312) 73 13 72
e-mail: szvolga@mts.nn.ru
Web: www.volga-shipyard.com/

Moscow office
Office 9, 23 Sadovaya-Kudrinskaya Street, Moscow 123001, Russian Federation

Tel: (+7 095) 254 04 43
Fax: (+7 095) 254 99 20
e-mail: info@volga-shipyard.com
Web: www.volga-shipyard.com

Vitaliy A Surkov, *Director General*

Kolkhida 0506635

Kolkhida aft foil arrangement 0506639

Volga Shipyard is one of the oldest established shipyards in the Russian Federation and has constructed a wide range of medium and high-speed craft for the Russian Federation river fleet.

The yard has also constructed the world's widest range of passenger hydrofoils, many of which are equipped with the Alexeyev shallow draught submerged foil system. The late Dr Alexeyev started work at the end of 1945 on the design of his foil system and it had to be suitable for operation on smooth but open and shallow rivers and canals. He succeeded in making use of the immersion depth effect, or surface effect, for stabilising the foil immersion in calm waters by the use of small lift coefficients.

The system comprises two main horizontal lifting surfaces, one forward and one aft, with little or no dihedral, each carrying approximately half the weight of the vessel. A submerged foil loses lift gradually as it approaches the surface from a submergence of about one chord. This effect prevents the submerged foils from rising completely to the surface. Means therefore had to be provided to assist take-off and prevent the vessel from sinking back to the displacement condition. The answer lay in the provision of planing sub-foils of small aspect ratio in the vicinity of the forward struts, arranged so that when they are touching the water surface the main foils are submerged approximately to a depth of one chord.

The foils have good riding characteristics on inland and sheltered waters.

Passenger hydrofoils in production at yards on the Baltic and Black seas are the Zenit (designed to replace Meteor hydrofoils) and the Albatros and Kolkhida (Kometa replacements), with seats for 120 passengers. Smaller hydrofoils are also under development including the 50-seat Polesye at the river-craft shipyard at Gomel and Lastochka, which has been designed to supersede Voskhod.

Substantial numbers of hydrofoils have been exported, especially Kometas, Meteors, Raketas, Voskhods, Volgas and recently Kolkhidas which have been sold to Greek and Italian operators. Countries in which they are being or have been operated include Austria, Bulgaria, Cyprus, Czechoslovakia, Finland, France, Germany, Greece, Hungary, Iran, Italy, Morocco, Philippines, Poland, Romania, Spain, UK, US and Federal Republic of Yugoslavia.

LASTOCHKA M

Developed by the RE Alekseer Central Hydrofoil Design Bureau, the vessel is based on the Voskhod design.

Two vessels were constructed in 2004 (project SPK85), for a Chinese operator. The vessels were each equipped with two DPA-470PM diesel engines, producing a total of 1,980 kW. The engines were developed by the Zvezda Joint-stock Company for operation in tropical conditions.

RAKETA

The prototype Raketa was launched in 1957 and was the first multiseat passenger hydrofoil to employ the Alexeyev shallow draught submerged foil system. Several are still in service on the major rivers of the RFAS.

The prototype was in service and carried more than two million passengers. The distance travelled by the craft during the period was stated to be equal to '52 voyages around the equator'.

There was a substantial number of Raketas exported to Austria, Bulgaria, Czech Republic, Germany, Hungary, Poland, Romania, Slovakia and Federal Republic of Yugoslavia.

Production of the Raketa has now stopped.

The description that follows applies to the Raketa T, the standard export variant, powered by an M-401A diesel and with a cruising speed of about 58 km/h (32 kt).

Specifications

Length overall	27.0 m
Beam	5.0 m
Draught, hullborne	1.8 m
Draught, foilborne	1.1 m
Displacement, min	20.3 t
Displacement, max	27.1 t
Crew	4
Passengers	58
Fuel capacity	1,647 litres

Vessel type	Vessel name	Yard No	Length (m)	Speed (kt)	Seats	Vehicles	Originally delivered to	Date of build
Raketa-T	–	–	27.0	31.3	58	–	–	Series production
Meteor	–	–	34.6	35.1	116	–	–	Series production
Kometa-ME	–	–	35.1	31.0	120	–	–	Series production
Kometa-MT	–	–	35.1	31.0	102	–	–	Series production
Lastochka	–	–	29.0	–	64	–	–	Series production
Katran	–	–	34.5	35.0	140	–	American Virgin Islands	1995/1996
Kolkhida	–	–	34.5	34.0	155	–	–	–
Lastochka M	–	–	–	55.0	–	–	China	2004
Lastochka M	–	–	–	55.0	–	–	China	2004

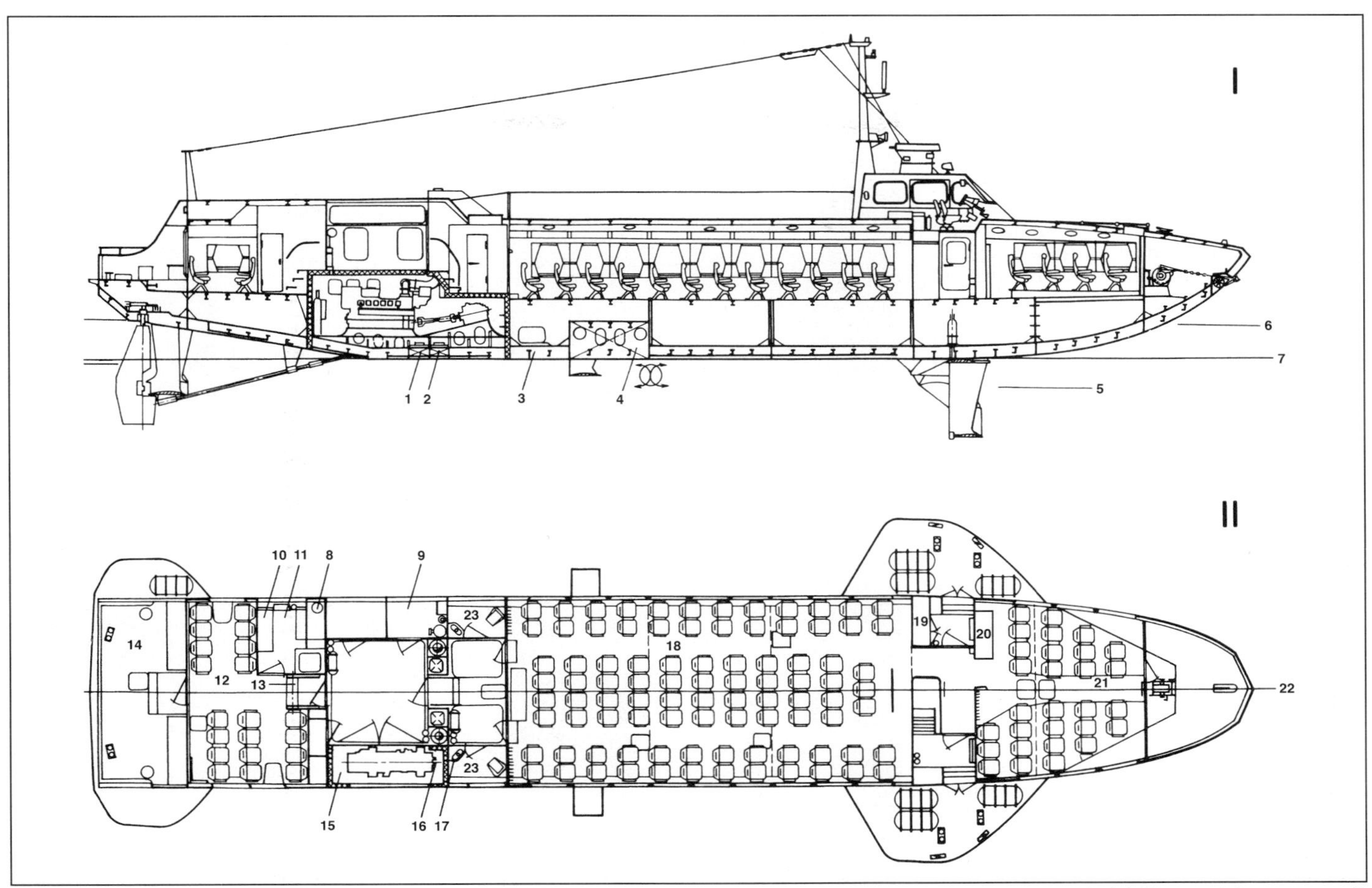

Kolkhida (area shown occupied by seats rows 12 and 13 is now a bar); Longitudinal section: (1) Waste oil collection tank; (2) Oil-containing water tank; (3) Sewage water tank; (4) Fuel tank; (5) Waterline when foilborne; (6) Waterline when hullborne; (7) Baseline, Main deck plan; (8) Hydraulic station; (9) Fuel and oil filling, waste water scavenging, firefighting station; (10) Conditioner; (11) Control post; (12) 20-seat passenger saloon; (13) VP; (14) Promenade platform; (15) Auxiliary unit room; (16) Gas exhaust trunk; (17) Air intake trunk; (18) 91-seat passenger saloon; (19) Aggregate room; (20) Conditioner; (21) 29-seat passenger saloon; (22) Central line; (23) Toilet 1304017

Propulsive power	671 kW
Operational speed	31.3 kt
Operational limitation	wave height 0.8 m

Propulsion: Power is supplied by a single M-401A water-cooled, supercharged 12-cylinder V-type diesel, with a normal service output of 671 kW. The engine drives, via a reverse gear and inclined stainless steel propeller shaft, a three-bladed cast bronze propeller.

METEOR

Dr Alexeyev's Meteor made its maiden voyage from Gorki to Moscow in 1960, bringing high performance and unprecedented comfort to river boat fleets, and setting the pattern for a family of later designs.

The craft was intended for use in daylight hours on local and medium-range routes up to 600 km in length. It meets the requirements of Class O, experimental type, on the Register of River Shipping in the RFAS.

Outside the RFAS, Meteors have been operated in Bulgaria, Hungary, Poland and Federal Republic of Yugoslavia.

Specifications

Length overall	34.6 m
Foil width	9.5 m
Draught, hullborne	2.4 m
Draught, foilborne	1.2 m
Displacement, min	36.4 t
Displacement, max	53.4 t
Crew	5
Passengers	116
Fuel capacity	3,000 litres
Propulsive power	2 × 820 kW
Operational speed	35.1 kt
Range	324 n miles
Operational limitation	Beaufort 3

Structure: Hull and superstructure are of riveted duralumin construction with welded steel members. The foils are in stainless steel and the subfoils in aluminium-magnesium alloy.

A Kometa of Kompas Line, Yugoslavia, arriving at Venice, June 1986 0506627

A Katran hydrofoil built by Volga Shipyard 0081003

Propulsion: Power is supplied by two M-401A 12-cylinder, four-stroke, supercharged, water-cooled diesels with reversing clutches. Each engine has a normal service output of 745 kW at 1,700 rpm and a maximum output of 820 kW at 1,800 rpm. Each engine drives its own inclined propeller shaft through a reverse clutch. The propeller shafts are in steel and the five-blade propellers are in brass. The drives are contrarotating.

KOMETA

Derived from the earlier Meteor, the Kometa was the first sea-going hydrofoil to be built in the Soviet Union. The prototype, seating 100 passengers, made its maiden voyage on the Black Sea in 1961, after which it was employed on various passenger routes on an experimental basis. Operating experience accumulated on these services led to the introduction of various modifications before the craft was put into series production.

Kometa operators outside the RFAS have included Kompas Line, Federal Republic of Yugoslavia; Alilauro SpA, Naples, Italy; and Transportes Touristiques Intercontinentaux, Morocco. Other vessels of this type have been supplied to Bulgaria, Cuba, Germany, Greece, Iran, Poland, Romania and Turkey. More than 60 have been exported.

Export orders have been mainly for the Kometa-ME, designed for service in countries with a moderate climate, which was introduced in 1968. Two distinguishing features of this model are the employment of new diesel engines with increased operating hours between overhauls and a completely revised surface-piercing foil system, with a trapeze bow foil instead of the former Alexeyev shallow draught submerged type.

A fully tropicalised and air conditioned version is now in production and designated Kometa-MT.

The present standard production Kometa-ME seats 116 to 120. Due to the additional weight of the Kometa-MT's air conditioning system and other refinements, the seating capacity is reduced in the interest of passenger comfort to 102.

Official designation of the Kometa in the RFAS is Hydrofoil Type 342. The craft meets the requirements of the Rules of the Register of Shipping of the RFAS and is constructed to Hydrofoil Class KM*211 Passenger Class under the Register's technical supervision.

The standard craft has proved to be exceptionally robust and has a good, all-round performance. On one charter, a Kometa-ME covered 2,867 n miles by sea and river in 127 hours. It can operate foilborne in waves up to 1.7 m and travel hullborne in waves up to 3.6 m.

One of the features of the more recent models is the relocation of the engine room aft to reduce the noise in the passenger saloons and the employment of a V-drive instead of the existing inclined shaft. The revised deck configuration allows more seats to be fitted. These modifications are also incorporated in the Kometa derivative, the Kolkhida, which is fitted with two 1,120 kW engines.

Kolkhida 0506636

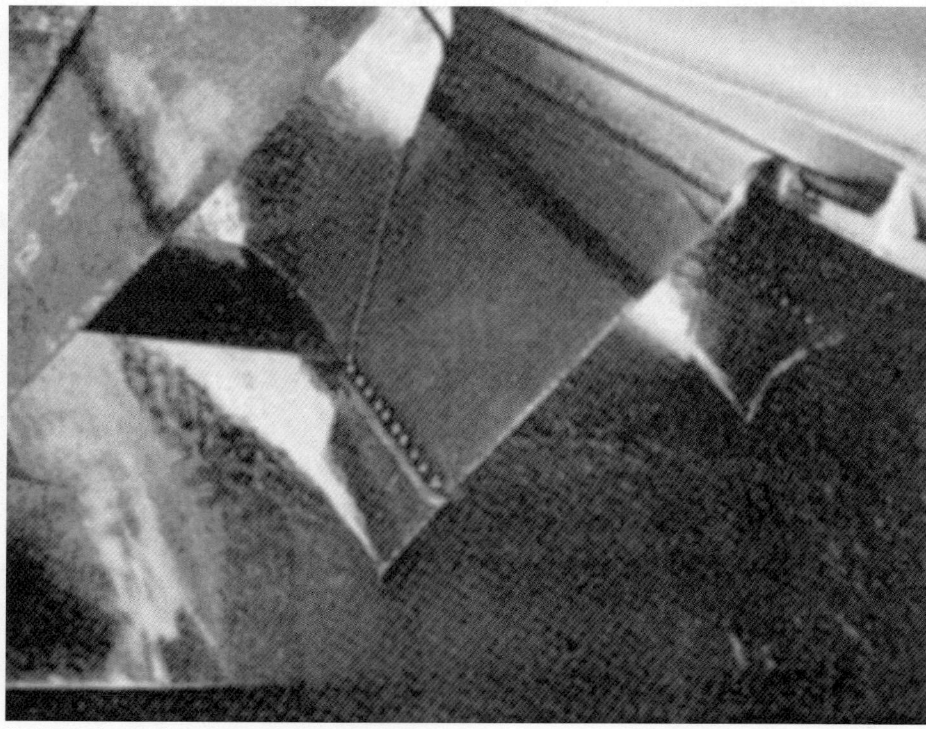

Kolkhida forward starboard foils and struts 0506637

Kometa-ME

Specifications

Length overall	35.1 m
Foil width	11.0 m
Draught, hullborne	3.6 m
Draught, foilborne	1.7 m
Displacement, min	44.5 t
Displacement, max	60 t
Crew	6
Passengers	102
Propulsive power	2 × 820 kW
Max speed	36 kt
Operational speed	31 kt
Range	200 n miles
Operational limitation	Sea State 4, Beaufort 5

Structure: The hull and superstructure are built in AlMg-61 and AlMg-6 alloys, and are of all-welded construction using contact and argon arc welding. The bow and stern foils are of hollow welded stainless steel construction. The midship and pitch stability foils and the upper components of the foil struts are in aluminium-magnesium alloy.

Propulsion: Power is supplied by two M-401A water-cooled, supercharged 12-cylinder V-type diesels, each with a normal service output of 745 kW at 1,550 rpm and a maximum output of 820 kW at 1,600 rpm. Each engine drives via a reverse gear, its own inclined steel shaft and the twin three blade brass propellers are contrarotating.

KOMETA-MT

Specifications

Length overall	35.1 m
Foil width	11.0 m
Draught, hullborne	3.6 m
Draught, foilborne	1.7 m
Displacement, min	45 t
Displacement, max	58.9 t
Crew	6
Passengers	102
Propulsive power	2 × 820 kW
Max speed	33 kt
Operational speed	31 kt
Range	130 n miles

LASTOCHKA

Successor to the 71-seat Voskhod, the first Lastochka was launched in 1986. The vessel is designed specifically for use over major rivers and reservoirs with wave heights that are unlikely to exceed 1.5 m. Loading conditions are optimised by means of a flap control on the bow foil arrangement.

Specifications

Length overall	29.0 m
Beam	4.4 m
Foil width	7.2 m
Draught, foilborne	2.5 m
Displacement, min	28 t
Displacement, max	37.3 t
Propulsive power	2 × 994 kW
Max speed	48 kt
Range	270 n miles

KATRAN

The Katran is a sea-going hydrofoil with the ability to operate under severe weather conditions in winds up to Force 5 and waves up to 2 m high foilborne or 3 m high hullborne.

Specifications

Length overall	34.5 m
Beam	10.3 m
Draught, hullborne	3.5 m
Draught, foilborne	1.9 m
Displacement, min	56 t
Displacement, max	72 t
Crew	5
Passengers	140
Operational Speed	35 kt
Range	200 n miles

Structure: Aluminium/Magnesium alloy.

Propulsion: The vessel is powered by two MTU 12V 396TE 74 diesel engines.

KOLKHIDA

Designed to replace the 20-year-old Kometa fast passenger ferry, Kolkhida is available in two versions: the Albatros, which operates on domestic services within the RFAS, and the Kolkhida, intended for export.

Kolkhida is faster than Kometa, seats more passengers, uses less fuel and can operate foilborne in higher sea states. Among the various design innovations are a new foil

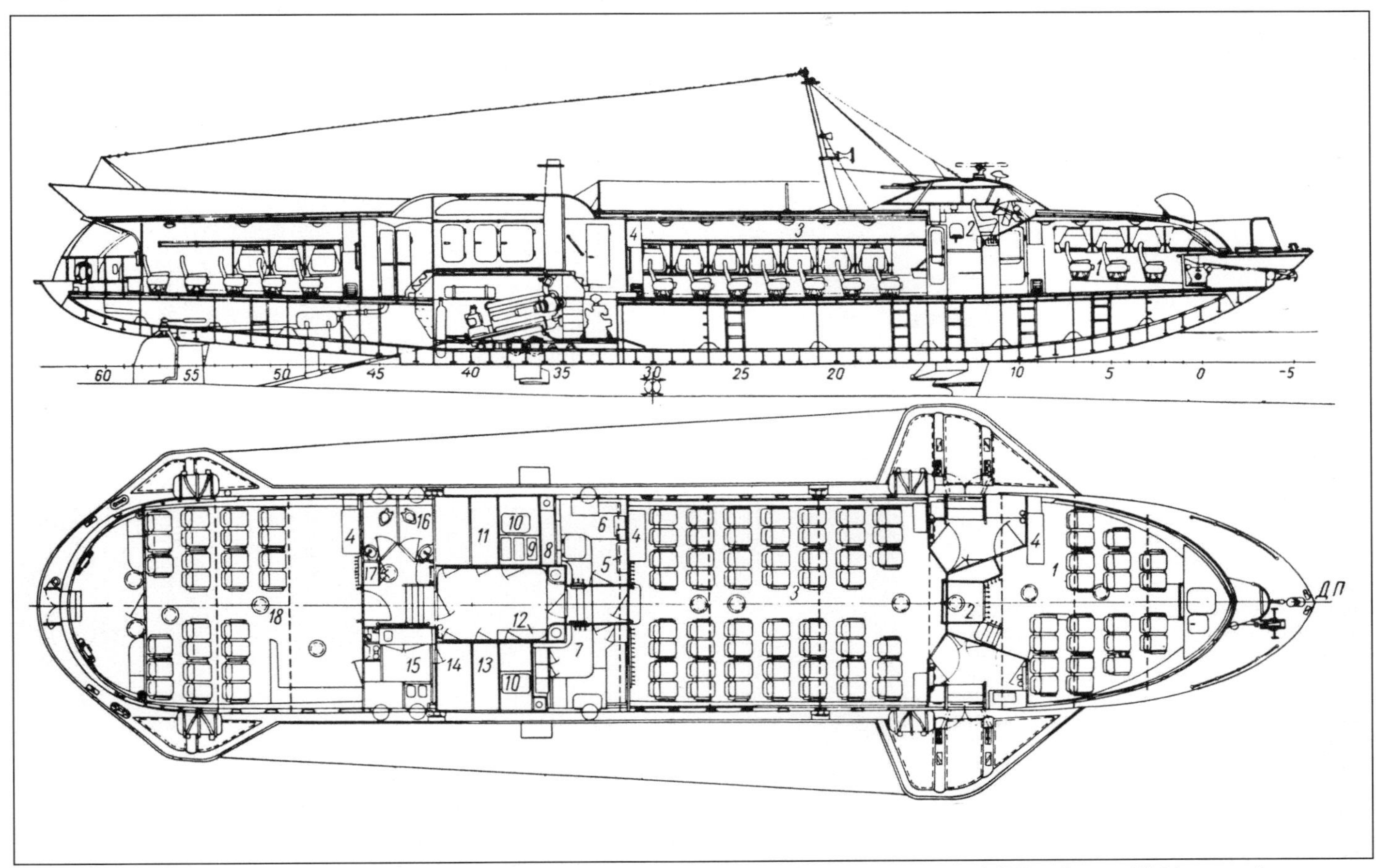

Internal arrangement of Kometa-MT, designed for tropical operation: (1) 22-seat forward passenger saloon; (2) Wheelhouse; (3) 54-seat main passenger saloon; (4) Luggage rack; (5) Engine room door; (6) Control position; (7) Duty cabin; (8) Liquid fire extinguisher bay; (9) Battery room; (10) Engine room; (11) Boiler room; (12) Installation point for portable radio; (13) Store; (14) Provision store; (15) Bar; (16) Toilet wash basin units; (17) Boatswain's store; (18) 26-seat aft passenger saloon 0506628

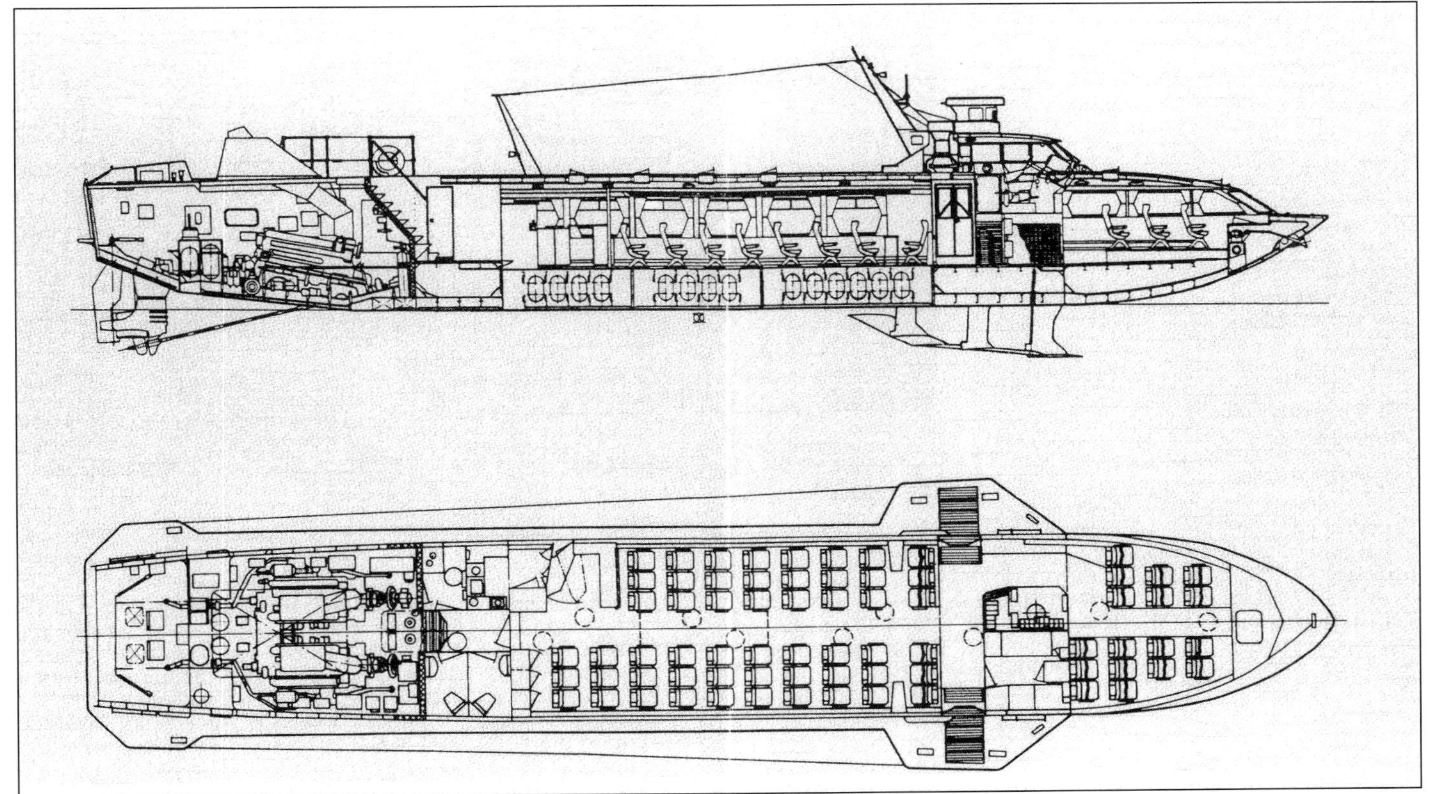

Layout of Lastochka 0506630

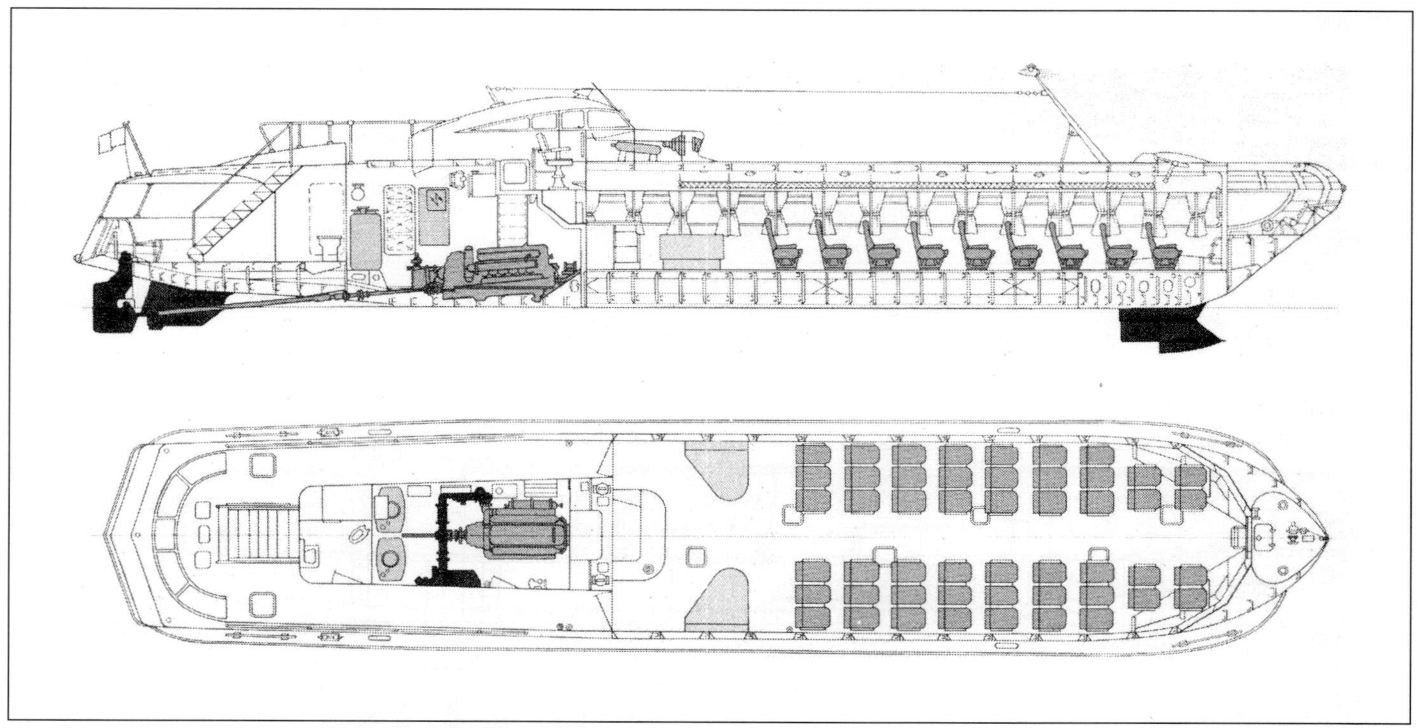

Inboard profile and plan view of standard 50-seat Raketa 0506624

system with automatic lift control, the use of new materials in the hull structure and a more rational cabin layout, permitting a substantial increase in seating capacity. The engine room is aft, as on the Voskhod, to reduce the noise level. Overall dimensions are almost identical to those of Kometa-M.

Trials of the Kolkhida prototype took place in the Baltic between March and June 1981 and the vessel has been in production since with sales being achieved in Greece, Italy and the Federal Republic of Yugoslavia.

Kolkhida is designed to operate under tropical and moderate climates up to 50 miles from a port of refuge in open seas and up to 100 miles from a port of refuge in inland seas and large lakes, with a permissible distance between two ports of refuge of not more than 200 miles.

Foilborne, the craft can operate in waves up to 2 m and winds up to Beaufort 5; hullborne it can operate in waves up to 3 m and winds up to Beaufort 6.

Specifications

Length overall	34.5 m
Beam	5.8 m
Foil width	10.3 m
Draught, hullborne	3.5 m
Draught, foilborne	1.9 m
Displacement, min	56 t
Displacement, max	72 t
Crew	6
Passengers	155
Propulsive power	2 × 1,050 kW
Operational speed	34 kt
Range	150 n miles
Operational limitation	Wave height 2 m, Beaufort 5

Classification: The craft meets the requirements of the Register of Shipping of the RFAS and is constructed to Hydrofoil Class KM* 2AS Passenger Class SPK under the Register's technical supervision. It complies fully with the IMO Code of Safety for Dynamically Supported Craft.

Structure: Double-chine, V-bottom type, with raked stern and streamlined superstructure. The hull and superstructure are built in aluminium-magnesium alloys. Framing is based on T and T-angle webframes. Frame spacing is 600 mm. Longitudinal framing of the sides, decks and hull bottom is based on stiffening ribs, keelson, stringers and deck girders. Below the main deck the hull is subdivided by watertight bulkheads into nine compartments. The craft will remain afloat with any two adjacent compartments flooded.

Lastochka 0506629

Bow foil and planing stabiliser foils of Raketa 0506622

The foil system, which is similar to that of Kometa, comprises a trapeze-type bow foil, an aft and amidships foil, close to the longitudinal centre of gravity to assist take-off. The foils are connected to the hull by struts and brackets. Bow and stern foil surfaces and the lower ends of the bow and stern foil struts are in steel alloy. The amidships foil, struts, upper sections of the bow and stern foil struts are in aluminium-magnesium alloy. A cast, balanced rudder in 40 mm thick aluminium-magnesium alloy is fitted. Total blade area is 2.75 m^2.

Propulsion: Power is supplied by two MTU 12V 396 TC 82 water-cooled, supercharged 12-cylinder V-type four-stroke marine diesels, each with a normal service output of 960 kW at 1,745 rpm and 1,050 kW at 1,800

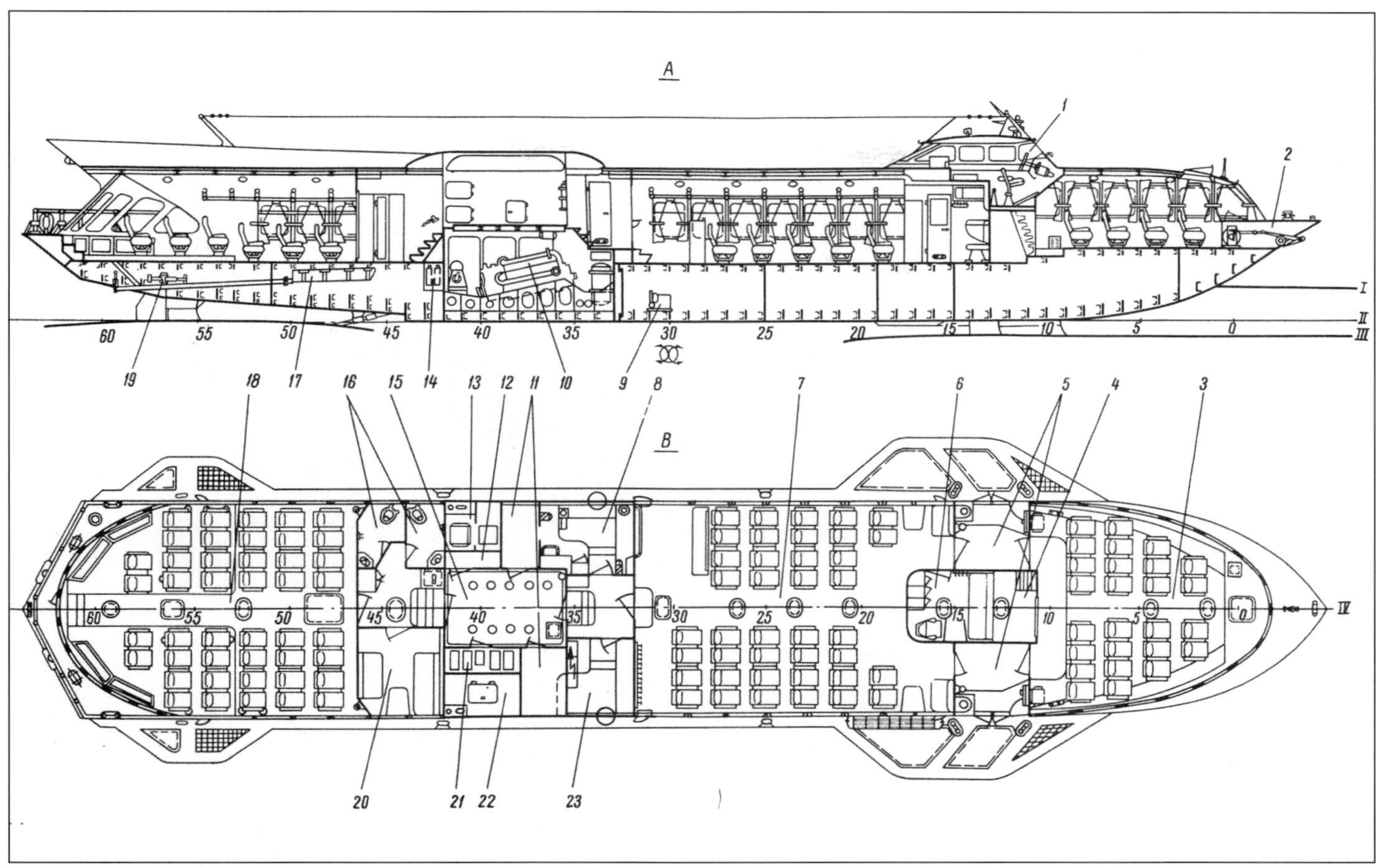

Meteor: (A) Inboard profile; (B) Main deck plan; (I) Waterline hullborne; (II) Hull base line; (III) Waterline foilborne; (IV) Longitudinal centreline; (1) Wheelhouse; (2) Anchor compartment; (3) Forward passenger saloon, 26 seats; (4) Luggage rack; (5) Embarkation companionway; (6) Crew duty room; (7) Midship passenger saloon, 42 seats; (8) Bar; (9) Refrigeration unit; (10) Engine room; (11) Pantry; (12) Boatswain's store; (13) Calorifier; (14) Firefighting equipment; (15) Promenade deck; (16) WCs; (17) Tank; (18) Aft passenger saloon, 44 seats; (19) Tiller gear; (20) Four-seat passenger cabin; (21) Storage batteries; (22) Hydraulic units; (23) Main switchboard 0506626

rpm maximum. Guaranteed service life of each engine before first major overhaul is 9,000 hours; maximum service life is 12 years. Output is transferred to twin 740 mm diameter contrarotating fixed-pitch propellers through reversible gearboxes that are remotely controlled from the wheelhouse. The propeller shafts are inclined at 14° and supported by rubber and metal bearings.

Electrical system: Engine-driven generators supply power while the craft is operating. The 4.5 kW generator included in the auxiliary unit supplies power when the craft is at rest. Acid storage batteries connected in series supply power during short stops.

Control: A sonic/electronic autopilot controls lift by operating trailing-edge flaps on the centre section of the bow foil and on the inner sections of the aft foil. The foil flaps are adjusted hydraulically to dampen heave, pitch, roll and yaw motions in heavy seas and provide co-ordinated turns.

Rudder movement is controlled hydraulically by any one of three systems: push-button, manual or via the autopilot.

DOLPHIN

This is a small sea-going hydrofoil craft marketed by the Volga Shipyard for limited passenger transportation on rivers and estuaries.

Specifications

Length overall	10.0 m
Beam	2.5 m
Draught, hullborne	1.1 m
Draught, foilborne	0.5 m
Displacement, load	3.1 t
Deadweight	1.0 t
Crew	1
Passengers	8
Maximum speed	35 kt
Range	200 n miles

Construction: All welded aluminium alloy.

Propulsion: Single Volvo Penta AD41/DP rated at 147 kW, driving a sterndrive.

Meteor 0506625

Raketa M 0506623

The Volga Dolphin hydrofoil at speed 0526359

Ukraine

Feodosia Shipbuilding Association (Morye)

Desantnikov Str 1, 98176 Feodosia, Crimea, Ukraine

Tel: (+380 6562) 325 56
Fax: (+380 6562) 323 73
e-mail: fsa@morye.kafa.crimea.ua
Web: www.morye.crimea.ua

Igor Grinuk, *Chairman*
Sergei Ostapenko, *Technical Director*

VOSKHOD-2

Designers of the Voskhod, which has been gradually replacing craft of the Raketa series, drew on engineering experience gained with the Raketa and also the more sophisticated Meteor and Kometa. Visually the Voskhod is more akin to a scaled-down Kometa with its engine room aft, replacing the rear passenger saloon.

Voskhod 14 was launched in June 1980 and delivered to the Amur Line for summer services along the Amur river.

The vessel is designed for high-speed passenger ferry services during daylight hours on rivers, reservoirs, lakes and sheltered waters. It meets the requirements of Soviet River Register Class O with the following wave restrictions (3 per cent safety margin): foilborne, 1.3 m; hullborne, 2 m.

Specifications

Length overall	27.6 m
Beam	6.2 m
Draught, hullborne	2.0 m
Draught, foilborne	1.1 m
Displacement, min	20 t
Displacement, max	28 t
Payload	5.9 t
Crew	4
Passengers	71
Fuel capacity	1,647 litres
Water capacity	138 litres
Propulsive power	810 kW
Max speed	37.8 kt
Operational speed	32.4 kt
Range	270 n miles
Operational limitation	wave height 1.3 m

Structure: The basic structure is largely in AlMg-61 aluminium-magnesium alloy, with extensive use made of arc and spot welding. The surface and lower parts of the foil struts and stabiliser are in Cr18Ni9Ti stainless steel, while the upper parts of the struts and stabiliser and also the amidships foil are in AlMg-61 plate alloy.

Propulsion: Power is supplied by a single M-401A four-stroke water-cooled, supercharged 12-cylinder V-type diesel, delivering 810 kW at 1,600 rpm maximum and 736 kW at 1,550 rpm cruising. The engine, which has a variable-speed governor and a reversing clutch, is sited aft with its shaft inclined at 9°. Output is transferred via a flexible coupling to a single six-blade variable-pitch propeller via an R-21 V-drive gearbox.

VOSKHOD-2M

This version is a sea-going hydrofoil for daylight operation within 25 miles of a port of refuge. The dimensions are the same as Voskhod-2.

Production of the sea-going Voskhod-2M started in 1993. In 2003 three craft were delivered to Connexxion-FFF of the Netherlands.

The gas-turbine powered 250-seat Cyclone at sea 0507043

Voskhod-2M 0506818

Voskhod-2 0077803

Olympia hydrofoil Jaanika *operating with Inreko* 0077805

Specifications

Length overall	27.6 m
Beam	6.8 m
Foil width	6.4 m
Draught	2.0 m
Displacement, min	24 t
Displacement, max	29 t
Passengers	65
Operational speed	33 kt
Range	247 n miles
Operational limitation	wave height 1.25 m, Beaufort 4

Propulsion: Engines: one MTU 12V 2000 M70 diesel.
Thrust device: fixed-pitch six-blade bronze propeller.

Vessel type	Vessel name	Yard No	Length (m)	Speed (kt)	Seats	Vehicles	Originally	Date of build
HYD	*Voskhod-2*	–	27.6	32.4	71	–	–	Series production
Cyclone	*Lilsa* (ex-*Cyclone*)	–	44.2	42.0	250	–	–	1986
Olympia	*Laura*	–	43.3	35.0	200	–	Linda Line Express	1993
Olympia	*Jaanika*	–	43.3	35.0	200	–	–	1994
HYD	*Voskhod-2M*	–	27.7	33.0	79	–	–	Series production
HYD	*Eurofoil*	–	27.7	33.0	79	–	–	–

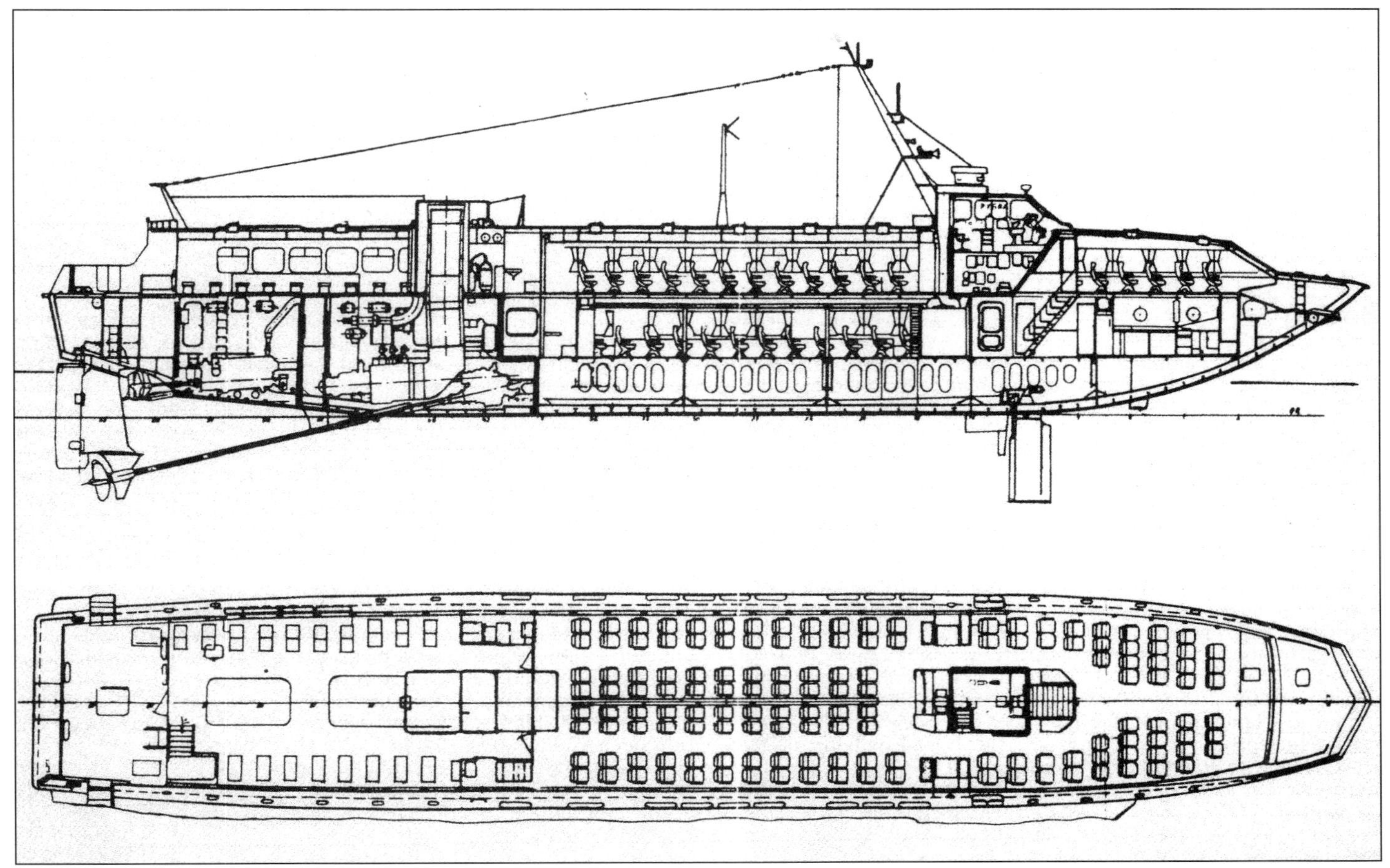

Layout of Cyclone 0506621

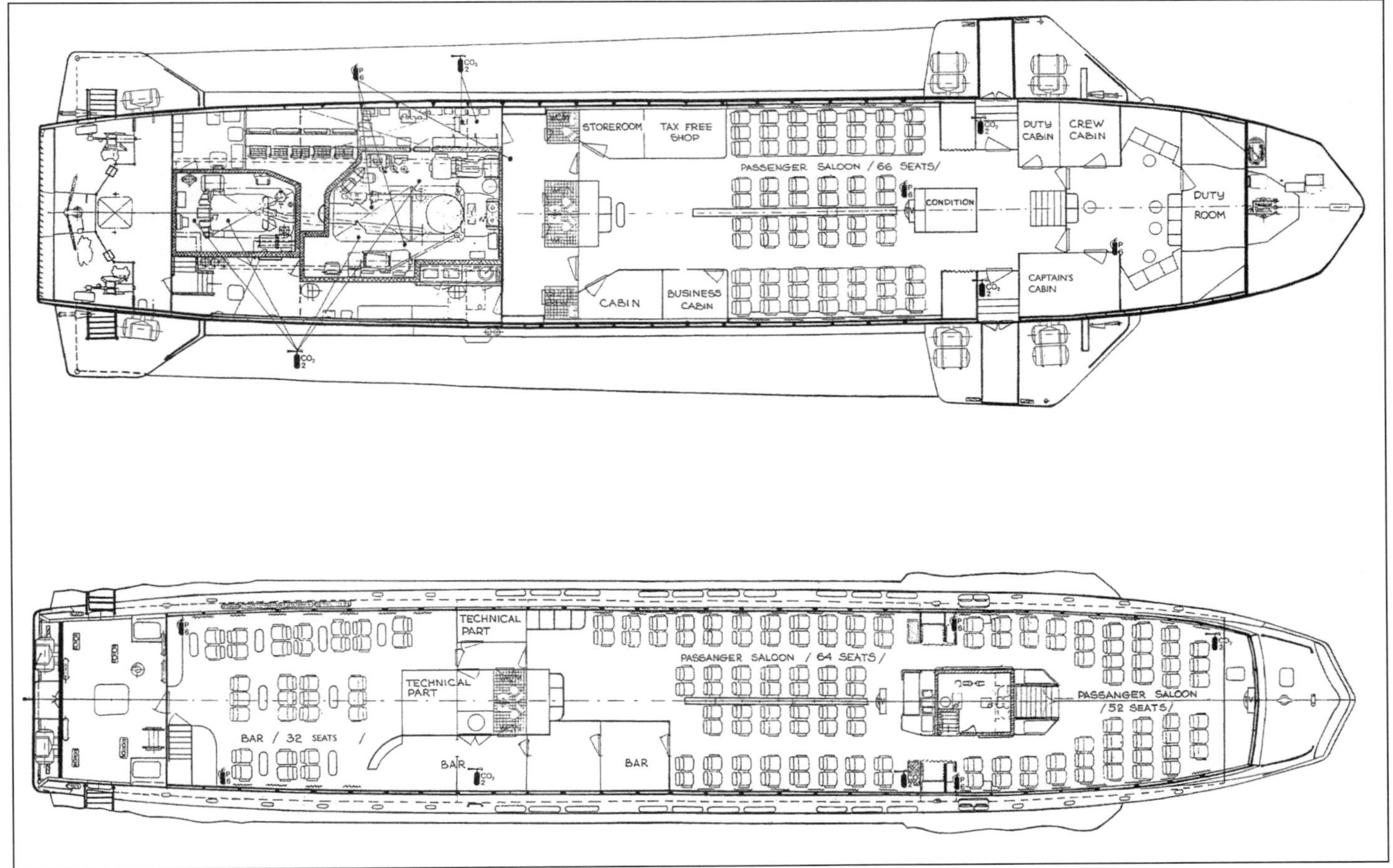

General arrangement of Lilsa 0507044

EUROFOIL

The Eurofoil has been specially designed for passenger shipping operations and sightseeing tours.

Specifications

Length overall	27.7 m
Beam	6.8 m
Draught	2.0 m
Displacement, min	23.8 t
Displacement, max	29.7 t
Passengers	79
Crew	2
Operational speed	33 kt
Range	247 n miles

Propulsion: Engines: one MTU 12V 2000 M70 diesel.

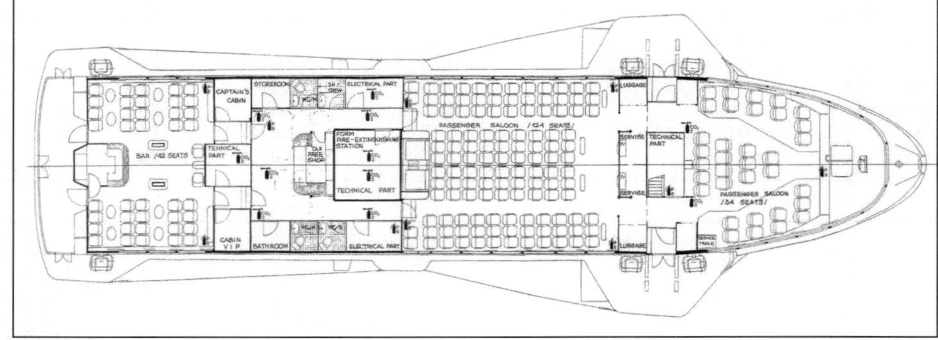

General arrangement of Laura 0507045

United States

Boeing

Although this company is no longer engaged in the marketing of commercial or military hydrofoil vessels, details of the Boeing Jetfoil are included here since these craft represented a most significant step in high-speed marine craft technology.

The Boeing Company has licensed Kawasaki Heavy Industries Ltd for the design, manufacture, marketing, maintenance and repair of Boeing Jetfoil 929-117 hydrofoil craft. Boeing's entry into the hydrofoil field was announced in June 1960, when the company was awarded a USD2 million contract for the construction of the US Navy's 120 ton PCH-1 *High Point*, a canard design which was the outcome of experiments with a similar arrangement in the US Navy test craft, *Sea Legs*.

As a result of Boeing's extensive hydrofoil programmes, two principal vessel types evolved, the 235 ton NATO/PHM patrol boat (six built, 1977 to 1982) and the Jetfoil type of which, by January 1990, 39 had been built or ordered. This included the Kawasaki vessels, almost entirely for ferry operations. Kawasaki has built 11 Jetfoil 929-117 craft.

JETFOIL 929-100

This is a 110 ton water-jet-propelled commercial hydrofoil for services in relatively rough waters. It employs a fully submerged, automatically controlled canard foil arrangement and is powered by two 2,767 kW Allison 501-K20A gas turbines. Normal foilborne cruising speed is 42 kt.

Typical interior arrangements include a commuter configuration with up to 350 seats and a tourist layout for 190 to 250 plus baggage.

Keel-laying of the first Jetfoil took place at the company's Renton, Washington, plant in January 1973 and the craft was launched in March 1974. After testing on Puget Sound and in the Pacific, the craft was delivered to Pacific Sea Transportation Ltd for inter-island services in Hawaii. High-speed foilborne tests began in Puget Sound in mid-July and it was reported that the vessel attained a speed of 48 kt during its runs.

During a rigorous testing programme to prove the boat's design and construction, Boeing No 001 operated for 470 hours, including 237 hours foilborne. The latter phase of testing was conducted in the rough waters of the straits of Juan de Fuca and the Pacific Ocean, where it encountered wave swells as high as 9.1 m, winds gusting up to 60 kt and wave chop averaging 1.8 m high.

The first operational Jetfoil service was successfully initiated in April 1975 by Far East Hydrofoil Company Ltd, Hong Kong, with Jetfoil 002, *Madeira*. Before this, the Jetfoil received its ABS classification, was certificated by the Hong Kong Marine Department and passed US Coast Guard certification trials. A US Coast Guard certificate was not completed as the craft would not be operating in US waters.

The first US service began in Hawaii in June 1975 and the tenth Jetfoil was launched in May 1977.

Specifications

Length overall	27.4 m
Beam	9.5 m
Draught, hullborne	5.0 m
Displacement	110 t
Passengers	190–350
Fuel capacity	15,140 litres
Propulsive power	2 × 2,767 kW
Max speed	50 kt
Operational speed	42 kt
Operational limitation	wave height 3.7 m

Vessel type	Vessel name	Yard No	Length (m)	Speed (kt)	Seats	Vehicles	Originally delivered to	Date of build
Jetfoil 929-100	*Flores* (ex-*Kalakoua*)	001	27.4	42.0	268	–	Turbojet, Hong Kong	1974
Jetfoil 929-100	*Madiera*	002	27.4	42.0	268	–	Turbojet, Hong Kong	1974
Jetfoil 929-100	*Corvo* (ex-*Kamahameha*)	003	27.4	42.0	268	–	Turbojet, Hong Kong	1975
Jetfoil 929-100	*Santa Maria*	005	27.4	42.0	268	–	Turbojet, Hong Kong	1975
Jetfoil 929-100	*Pico* (ex -*Kuhio*)	004	27.4	42.0	268	–	Turbojet, Hong Kong	1975
Jetfoil 929-100	*Sao Jorge* (ex-*Jet Caribe I*)	006	27.4	42.0	268	–	Turbojet, Hong Kong	1975
Jetfoil 929-100	*Urzela* (ex-*Flying Princess*)	007	27.4	42.0	268	–	Turbojet, Hong Kong	1976
Jetfoil 929-100	*Acores* (ex-*Jet Caribe II* ex-*Oriente*)	008	27.4	42.0	268	–	Turbojet, Hong Kong	1976
Jetfoil 929-100	*Guia* (ex-*Okesa*)	–	27.4	42.0	268	–	Turbojet, Hong Kong	1976
Jetfoil 929-100	*Ponta Delgarda* (ex-*Flying Princess II*)	010	27.4	42.0	268	–	Turbojet, Hong Kong	1977
Jetfoil 929-115	*Mikado*	011	27.4	–	260	–	Sado Kisen Kaisha	1978
Jetfoil 929-115	*Terceira* (ex-*Normandy Princess*)	–	27.4	–	268	–	–	1979
Jetfoil 929-115	*Funchal* (ex-*Jetferry One*)	–	27.4	–	268	–	–	1979
Jetfoil 929-115	*Lilau* (ex-*Speedy Princess* ex-*HMS Speedy*)	–	27.4	–	268	–	Royal Navy	1979
Jetfoil 929-115	*Ginga* (ex-*Cu na Mara*)	–	27.4	–	260	–	–	1979
Jetfoil 929-115	*Horta* (ex-*Jetferry Two*)	–	27.4	–	268	–	–	1980
Jetfoil 929-115	*Jet 7* (ex-*Spirit of Freindship* ex-*Aires* ex-*Montevideo Jet*)	–	27.4	–	268	–	Kato Kisen Company Ltd	1980
Jetfoil 929-115	*Calcilhas* (ex-*Princesa Guayarmina*)	–	27.4	–	268	–	–	1980
Jetfoil 929-115	*Alderney Blizzard* (ex-*Adler Blizzard* ex-*Princesse Clementine*)	–	27.4	–	268	–	Regie des Transports Maritimes	1981
Jetfoil 929-115	*Adler Wizard* (ex-*Prinses Stephanie*)	–	27.4	–	268	–	Regie des Transports Maritimes	1981
Jetfoil 929-115	*Taipa* (ex-*Princesa Guacimara*)	–	27.4	–	268	–	–	1981
Jetfoil 929-115	*Bima Sumudera I*	022	27.4	–	268	–	PT PAL Indonisia	1981
Jetfoil 929-115	*Jet 8* (ex-*Spirit of Discovery*)	–	27.4	–	268	–	Kato Kisen Company Ltd	1985
Jetfoil 929-120	–	–	27.4	–	–	–	PT PAL Indonisia	1986
Jetfoil 929-120	–	–	27.4	–	–	–	PT PAL Indonisia	1986

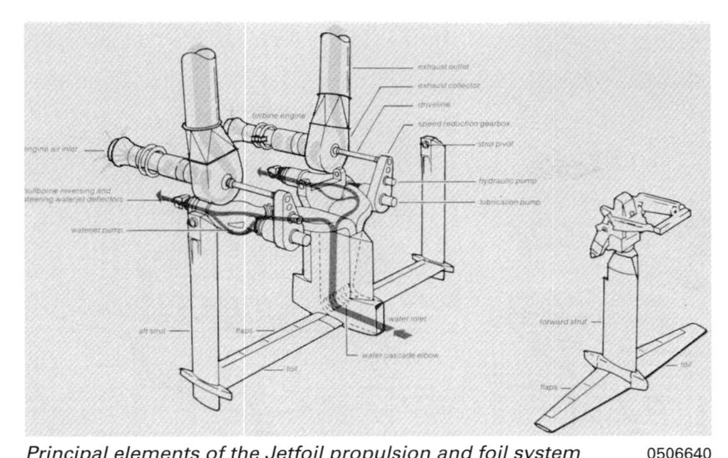

Principal elements of the Jetfoil propulsion and foil system 0506640

Boeing Jetfoil 929-115 Cu na Mara *renamed* Ginga *and in service with Sado Kisen Kaisha* 0506643

Structure: Hull and deckhouse in marine aluminium. Aircraft assembly techniques are used, including high-speed mechanised welding processes.

All structural components of the foil/strut system are in 15.5PH corrosion-resistant all-welded steel construction.

Propulsion: Power for the water-jet propulsion system is supplied by two Allison 501-K20A free-power gas turbines, each rated at 2,767 kW at 27°C at sea level. Each is connected to a Rocketdyne Powerjet 20 axial flow pump through a gearbox drive train. The system propels the craft in both foilborne and hullborne modes. When foilborne, water enters through the inlet at the forward lower end of the aft centre foil strut. At the top of the duct, the water is split into two paths and enters into each of the two axial flow pumps. It is then discharged at high pressure through nozzles in the hull bottom. The water path is the same during hullborne operations with the foils extended. When the foils are retracted, the water enters through a flush inlet located in the keel. Reversing and steering for hullborne operation only are accomplished by reverse-flow buckets located immediately aft of the water exit nozzles. A bow thruster is provided for positive steering control at low forward speeds.

A 15,140 litre integral fuel tank supplies the propulsion turbine and diesel engines. Coalescent-type water separating fuel filters and remote-controlled motor-operated fuel shut-off valves provide fire protection.

Electrical system: A 60 Hz, 440 V AC electrical system, supplied by two diesel-driven generators each rated at 62.5 kVA. Either is capable of supplying all vital electrical power. Capacity shore connection facilities of 90 kVA are provided and equipment can accept 50 Hz power. Transformer-rectifier units for battery charging provide 28 V DC from the AC system.

Control: The foil system is a fully submerged canard arrangement with a single inverted T strut/foil forward and a three-strut, full-span foil aft. The forward foil assembly is rotated hydraulically through 7° in either direction for steering. All foils have trailing-edge flaps for controlling pitch, roll and yaw and for take-off and landing. Foils and struts retract hydraulically above the waterline, the bow foil forward and the rear foil aft.

The craft is controlled by a three-axis automatic system while foilborne and during take-off and landing. The system senses the motion and position of the craft by gyros, accelerometers and height sensors, signals from which are combined in the control computer with manual commands from the helm. The resulting computer outputs provide control surface deflections through electrohydraulic servo actuators. Lift control is by full-span trailing-edge flaps on each foil. Forward and aft flaps operate differentially to provide pitch variation and height control. Aft flaps operate differentially to provide roll control for changes of direction.

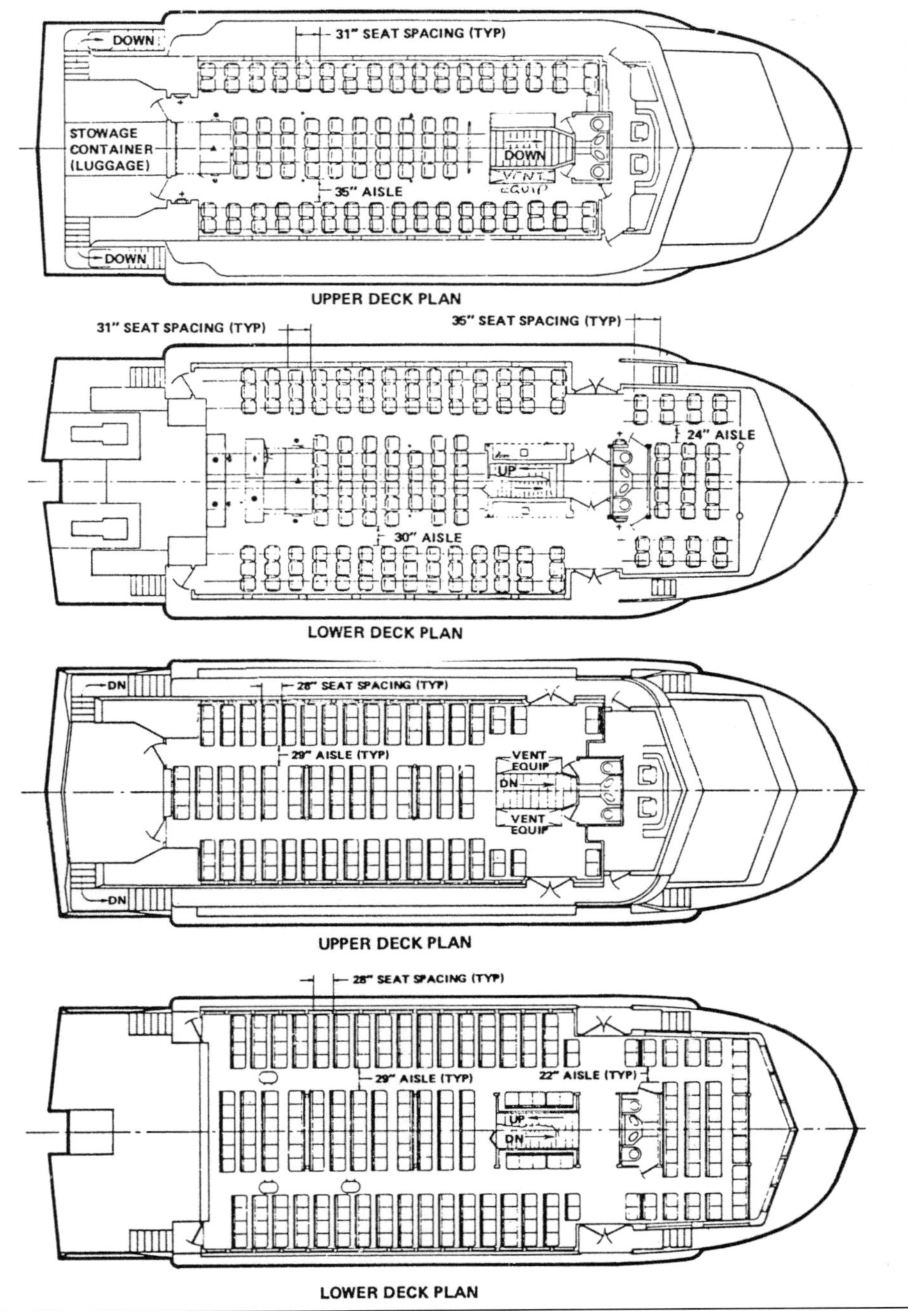

Jetfoil 929-100 interior arrangements 0506641

The vessel banks inwardly into all turns to ensure maximum passenger comfort. The ACS introduces the correct amount of bank and steering to co-ordinate the turn in full. Turn rates of up to 6°/s are attained within 1 second of providing a heading change command at the helm.

There are three basic controls required for foilborne operation: the throttle is employed to set the speed; the height command lever to set the required foil depth; and the helm to set the required heading. If a constant course is required, a 'heading hold' circuit accomplishes this automatically.

For take-off, the foil depth is set, the two throttles advanced and the hull clears the water in about 60 seconds. Acceleration continues until the craft automatically stabilises at the command depth and the speed dictated by the throttle setting. The

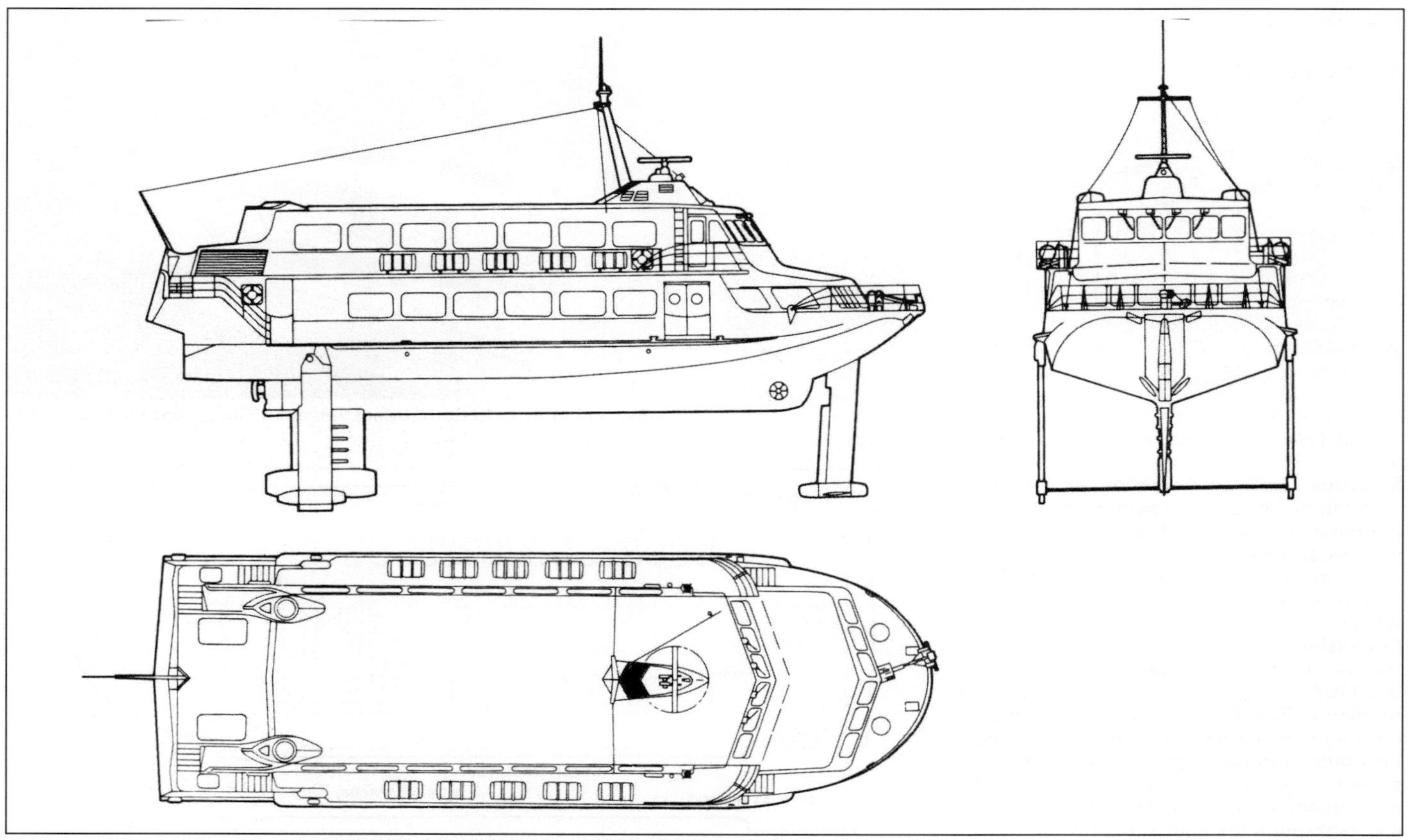

General arrangement of the Kawasaki Jetfoil Model 929-115 passenger ferry 0506644

throttle setting is reduced for landing, the craft settling as the speed drops. The speed normally diminishes from 45 kt (cruising speed) to 15 kt in approximately 30 seconds. In emergencies more rapid landings can be made by the use of the height command lever to provide hull contact within 2 seconds.

Quality of the ride in the craft is comparable with that of a Boeing 727 airliner. The vertical acceleration at the centre of gravity is very low and depends on sea state, for example, at 2 m significant wave height the vertical acceleration is only 0.05 *g* rms. Lateral acceleration is substantially less than vertical. Angles of pitch and roll are less than 1° rms. A structural fuse is provided limiting deceleration to less than 0.4 *g* longitudinally and 0.8 *g* vertically. In the event of the craft striking a major item of floating debris at full speed, the structural fuse, when actuated, allows the foil and strut to rotate backwards, preventing the system from sustaining significant damage.

Crew comprises a captain and first officer plus cabin attendants.

JETFOIL 929-115

The last of the Jetfoil 929-100 series was the 010 *Flying Princess II.* The first of the improved 929-115 series, Jetfoil 011 *Mikado*, was launched at Renton, Washington in June 1978, and is operated by Sado Kisen in the Sea of Japan.

The improved model Jetfoil has a lighter structure, allowing an increased payload and greater reliability, and is easier to maintain. Some of the modifications are listed below.

Specifications

Length overall	27.4 m
Beam	9.5 m
Draught, hullborne	5.2 m
Displacement, max	117 t
Max speed	43 kt

Structure: The bow structure design has been simplified to provide equivalent strength with increased payload and bulkhead, two have been revised for decreased stress levels. Based on a 7 minute evacuation time, in case of fire, the following fire protection provisions have been made:

The 1976-launched Acores Jetfoil 929-100 in service on the Hong Kong to Macao route 0506642

Propulsion: The propulsion system has been uprated to operate at 2,200 maximum intermittent pump rpm with an increase of three tons in maximum gross weight.

1. Glass fibre is used for thermal insulation where required throughout the passenger accommodation areas
2. Aluminium ceiling panels and air conditioner sleeves are employed throughout, together with aluminium doors and frames
3. One ½ in thick Marinite is employed in machinery spaces, with US Coast Guard-type felt added wherever required for insulation to comply with 30 minute fire test
4. External stiffeners on the foil struts have been eliminated and the bow foil has been changed from constant section to tapered planform for improved performance. Stress levels have been reduced for extended life.

JETFOIL 929-117

Since 1985 Boeing has not manufactured any model of the Jetfoil. The 929-117 model, an updated version, has been licensed for production outside the US, with Kawasaki Heavy Industries Ltd, Kobe, Japan and PT PAL, Indonesia. See Kawasaki entry for further details.

JETFOIL 929-320

***Lilau* (ex-*Speedy Princess*, ex-HMS *Speedy*)**

The Jetfoil 929-320 was delivered to the Royal Navy in June 1980 for use in fisheries patrol in the North Sea. Named HMS *Speedy,* the craft was a modified Model 929-115 commercial Jetfoil and was built on the commercial Jetfoil production line. The craft was decommissioned by the Royal Navy in April 1982 and eventually sold to the Far East Hydrofoil Company Ltd (FEH) in Autumn 1986. The craft has been converted to passenger configuration and operates on their Hong Kong to Macao route.

Navatek Ships Ltd

A subsidiary of Pacific Marine
Suite 1880, 841 Bishop Street, Honolulu, Hawaii 96813, United States

Tel: (+1 808) 531 70 01
Fax: (+1 808) 523 76 68
e-mail: schmicker@navatekltd.com
Web: www.navatekltd.com

Steven Loui, *President*
Gary Shimozono, *Vice President, R&D*
Todd Peltzer, *Director of Programs*
Michael Schmicker, *Vice President, Corporate Communications*

Navatek Ships Ltd researches, designs, tests and builds advanced marine hull forms for both commercial and military clients. These hull forms include Small Waterplane Area Craft (SWATH, SLICE) and HYbrid Small WAterplane Craft (HYSWAC, Midfoil, HDV 100 (Hybrid Deep-V)) employing lifting bodies. The company holds multiple patents on its proprietary hull forms.

Midfoil programme

Pacific Marine conducted independent research on different displacement foil configurations and developed a novel arrangement of a single large foil, positioned amidships, and a small stabilising forward foil on a small waterplane craft. The design became the Midfoil. Pacific Marine received a US patent on this invention in 1996.

The Midfoil is a hybrid of the low-waterplane and hydrofoil technologies. It uses the buoyancy of a thick submerged foil for support. The dynamic lift forces created by the hydrofoil-shaped underwater body, which are controlled by trailing edge flaps, eliminate the need for stabiliser fins. Since the vessel is a displacement ship there is no pronounced speed band as with a hydrofoil.

During 1994–96, Pacific Marine secured federal research funds form the US government's Center for Excellence in Research in Ocean Sciences (CEROS) programme to design, test and build a 65 ft manned model of the Midfoil. Construction of the manned model was completed in 1997. Sea trials began in early 1998.

The patented Midfoil concept provides high transportation efficiency (payload and passengers versus power required) and outstanding ride quality. In 2000, the vessel was refitted with a new underwater body to test the use of composite materials and the advantages of a more complex lifting body form. The Midfoil is in the process of ongoing sea trials to test the vessel for both speed, motion and to validate the use of Computational Fluid Dynamics (CFD) software used to design the vessel. The vessel has demonstrated the outstanding ride quality which is expected at all speeds and headings.

Specifications

Length overall	19.8 m
Beam	9.7 m
Draught	2.1 m
Displacement	52 t
Max speed	25 kt
Operational speed	20 kt
Range	500 n miles

Classification: USCG CFR Subchapter T (120 persons).
Construction: Aluminium superstructure, composite lower hull.
Propulsion: Two MTU 12V 183 TE 92 diesel engines each driving a fixed-pitch propeller via a ZF reverse reduction gearbox.

20 m Midfoil at sea

1176101

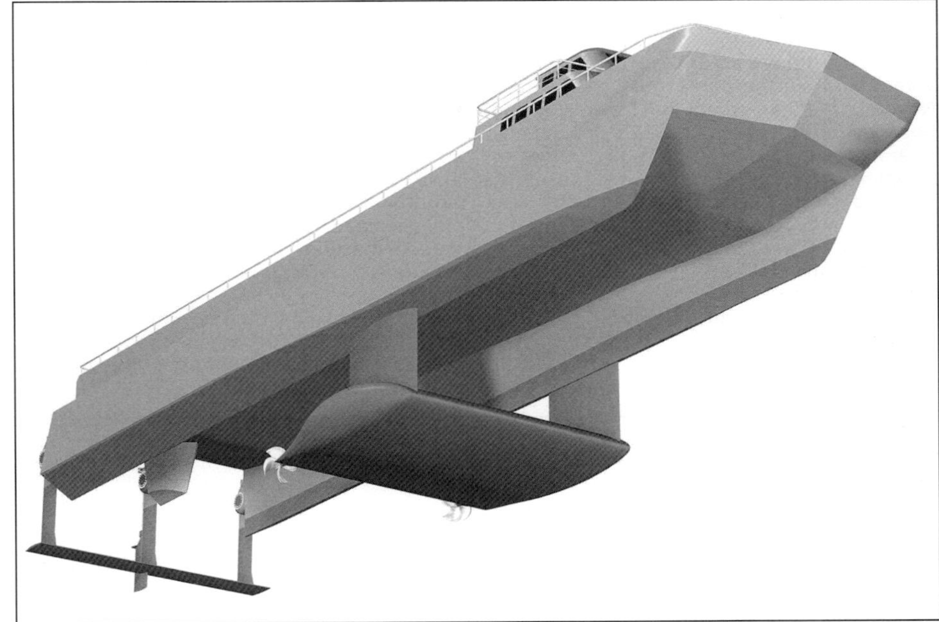

HYSWAC hull and lifting body arrangement 1111164

HYSWAC Sea Flyer during trials (Navatek) 1176093

HYSWAC lifting body programme

Following the success of the Midfoil small-scale lifting body technology demonstrator, Navatek moved to demonstrate its patented underwater lifting body technology on a large scale. Supported by funding from the US Navy's Office of Naval Research, Navatek began work in 2000, with US Navy technology demonstrator craft incorporating the company's proprietary, underwater lifting body technology. The 49 m, 30 kt-plus craft, called *Sea Flyer*, was launched in June 2003, and has a full load displacement of 340 litres.

The HYSWAC was designed to confirm the three large-scale benefits of underwater lifting bodies verified on an earlier, small-scale 65 ft, 50 litre Midfoil Navatek lifting body demonstrator craft, as well as through extensive computational fluid dynamics studies. These identified benefits include superior ride in all seas, all headings and all speeds (including zero/loiter to maximum speed); higher transport efficiency at all speeds and extended range/payload.

A former US Navy Surface Effect Ship (SES-200) provided the parent hull of the HYSWAC, reducing project costs. During the two year project, Navatek removed the existing SES air lift system and all related components, and installed a 170 ton Navatek underwater lifting body incorporating a new propulsion drive train (engines, gearboxes, shafts and propellers) within the lifting body. This allows the craft to be operated with variable immersion as speed increases with the parent hull fully out of the water at maximum speed. An aft crossfoil was also added for pitch stabilisation and control, along with a proprietary Navatek advanced ride control system.

In July 2004, the US Navy successfully completed sea trials on the 49 m, 320 ton technology demonstrator craft renamed *Sea Flyer*. Trials showed that even in rough seas the vessel could maintain a speed of 29 kt with exceptionally stable motion and good manoeuvrability.

The *Sea Flyer* subsequently departed Hawaii for mainland US where it is participating in Navy tests and demonstrations off the coast of California.

In addition to the multihull *Sea Flyer* lifting body demonstrator craft, Navatek has built a second, large scale, lifting body technology demonstrator craft, the *HDV 100*. This programme is designed to demonstrate the benefits of lifting body technology for monohull craft. The 30 m, 100 ton, 50 kt *HDV 100* employs an anti-slamming, deep-V monohull as the parent hull, mated to a Navatek blended-wing underwater lifting body. Launched in 2004, the *HDV 100* initially operated without a lifting body, allowing Navatek to conduct additional research on deep-V hull forms. The lifting body was installed in 2005 and sea trials are currently underway.

Specifications

Length overall	48.8 m
Displacement, full load	340 t
Max speed	30 kt
Operational speed	29 kt

HDV *100 on sea trials* (Navatek) 1176095

HDV 100 (Hybrid Deep-V)

In addition to the Navy's multihull Sea Flyer demonstrator, Navatek has built a second large-scale lifting body technology demonstrator craft, the HDV 100. Instead of a catamaran parent hull, the 100 ft, 100 ton, 50 kt HDV 100 employs an anti-slamming, deep-V monohull as the parent hull, mated to a Navatek blended-wing underwater lifting body. The concept was successfully tested on a 35 ft hybrid deep-V hull, modified with a Navatek lifting body, prior to beginning construction on the HDV 100.

The addition of an underwater lifting body, with a forward foil for heave and pitch control, improves the performance of the deep-V monohull. At zero/loiter speed, the added mass of the lifting body dampens motions, making the monohull more stable, allowing for safer, easier deployment and retrieval of autonomous unmanned vehicles, equipment packages and personnel. At high speeds, the lifting body provides enough lift to elevate the hull clear of the water and reduce hull drag. The lifting body lift-to-drag ratio is claimed to be higher than that of the hull and, as a result, less power is required to achieve speeds in excess of 40 kt. The underwater lifting body also offers a third benefit. The additional displacement from the underwater lifting body can increase the monohull's payload by 15 to 20 per cent, allowing it to carry more supplies, equipment, personnel, or fuel to increase its range of operations.

Launched in 2004, the HDV 100 initially operated without a lifting body allowing Navatek to conduct additional research on large deep-V hull forms. The lifting body was installed in Autumn 2005 and the sea trials are currently underway.

Specifications

Length overall	30.5
Displacement, full load	100 t
Max speed	50 kt

HIGH-SPEED MULTIHULL VESSELS

Company listing by country

Australia
Aluminium Boats Australia P/L
Aluminium Marine
Austal
Australus Catamarans
BSC Marine Group
Cougar Catamarans Australia Pty Ltd
Image Marine Pty Ltd
InCat
Lightning Boats
New Wave Catamarans
North West Bay Ships
NQEA Australia Pty Ltd
Richardson Devine Marine Constructions Pty Ltd
Sabre Catamarans Pty Ltd
SBF Shipbuilders
South Pacific Marine
Strategic Marine Pty Ltd

Brazil
Rodriquez Cantieri Navali do Brazil

Canada
Allied Shipbuilders Ltd
Chantier Naval Matane Inc
Sylte Shipyards
Washington Marine Group

Chile
Alwoplast SA
Asenav MR

Cuba
Damex Shipbuilding

Egypt
Mapso Marine

Finland
Aker Finnyards Inc

France
Construction Navale Bordeaux (CNB)
Iris Catamarans

Hong Kong
Afai Ships Ltd
Cheoy Lee Shipyards Ltd
Wang Tak Engineering & Shipbuilding Co Ltd

Italy
Cantiere Navale di Pesaro
Rodriquez Cantieri Navali SpA

Japan
IHI Company Ltd
Kawasaki Heavy Industries Ltd Ship Group
Mitsubishi Heavy Industries Ltd
Mitsui Engineering & Shipbuilding Company Ltd
Universal Shipbuilding Corporation
Yamaha Motor Company Ltd

Korea, South
Daewoo Shipbuilding and Marine Engineering Co Ltd
Hyundai Heavy Industries Company Ltd
Semo Company Ltd

Malaysia
Kay Marine
NGV Tech Sdn. BHD

Mauritius
Diogene Marine

Netherlands
Damen Shipyards
Luyt BV
Schelde Shipbuilding (Royal Schelde)

New Zealand
Fitzroy Yachts
Q-West

Norway
Båtservice Holding A/S
Båtutrustning Bømlo A/S
Brødrene Aa
Fjellstrand A/S
Holen Mek Verksted A/S
Lindstøls Skips- & Båtbyggeri A/S
Måløy Werft
Oma Båtbyggeri

Philippines
FBMA Marine Inc

Russian Federation
Almaz Marine Yard
Almaz Shipbuilding Company Ltd

Singapore
Damen Shipyards Singapore
Greenbay Marine Pte Ltd
Marinteknik Shipbuilders (S) Pte Ltd
Penguin Shipyard International Ltd
Precision Craft (88) Pte Ltd
Singapore Technologies Marine Ltd

South Africa
Farocean Marine

Spain
Navantia

Thailand
Asimar
Italthai Marine Ltd

Turkey
Pendik Shipyards

Ukraine
Feodosia Shipbuilding Association (MORYE)

United Kingdom
Alnmaritec
Aluminium Shipbuilders Ltd
FBM Babcock Marine Group
VT Shipbuilding

United States
All American Marine Inc
Allen Marine, Inc
Austal USA
Blount Boats Inc
Dakota Creek Industries
Derecktor Shipyards
Gambol Industries
Gladding-Hearn Shipbuilding
Gulf Craft
Island Boats Inc
Kitsap Catamarans
Knight and Carver Yacht Center
Kvichak Marine Industries
Marinette Marine Corporation
Midship Marine Inc
Nichols Brothers Boat Builders Inc
Royal Crown Yachts
VT Halter Marine Group Inc
Yank Marine.

Vietnam
Tam Bac Shipyard

Australia

Aluminium Boats Australia P/L

Brisbane Marine Industry Park, Hemmant, Queensland 4174, Australia

Tel: (+61 7) 33 93 69 09
Fax: (+61 7) 38 93 69 19
e-mail: roy@allyboats.com.au
Web: www.allyboats.com.au

Roy Whitewood, *Manager Director*

Aluminium Boats Australia was established in 1999 and has a facility in the Brisbane Marine Industry Park.

22 m WAVEPIERCING CATAMARAN

Tangalooma Express

Designed by Stuart Friezer Marine this vessel this vessel was built to Australian USL Code 1 Regulations and delivered to Tangalooma Wild Dolphin Resort in 2004.

Specifications

Length overall	22 m
Beam	7.5 m
Draught (static)	1.6 m
Crew	2
Passengers	118
Maximum speed	34 kt
Operational speed	29 kt

Propulsion: Two MAN D2842 LE401 diesel engines each rated at 5,88 kW at 2,100 rpm each driving a Henley four blade fixed pitch propeller via Twin Disk MG 5145 gearboxes.

28 m WAVEPIERCING CATAMARAN

Evolution

Designed by Stuart Friezer Marine and delivered to Haba Dive and Snorkel in August 2005, the vessel is equipped with an underwater observation platform.

Specifications

Length overall	28 m
Passengers	150
Operational speed	32 kt

Aluminium Marine

94 Beveridge Road, Thornlands, Queensland 4164, Australia

Tel: (+61 7) 32 07 06 00
Fax: (+61 7) 32 07 06 11
e-mail: aluminiummarine@bigpond.com.au

Steve Cordingley, *Managing Director*

Established in 1985, the company has built over 80 vessels including three fast ferries designed by Commercial Marine Consulting Services.

35 m catamaran, Cat Cacos 1324946

35 m CATAMARAN

Cat Cacos

Ordered by Inter Island Boats of the Seychelles, this vessel, built by Aluminium Marine in partnership with Wildcat Marine, was delivered in May 2007. The vessel features an MPI interceptor-based ride control system.

Specifications

Length overall	35.2 m
Length waterline	30.0 m
Beam	10.0 m
Draught	1.8 m
Passengers	350
Fuel capacity	5600 litres
Fresh water capacity	10,000 litres
Operational speed	35 kt

Propulsion: Two MTU 12V 4000 M70 diesel engines, each rated at 1,740 kW driving fixed pitch propellers via ZF 7500 gearboxes.

34 m CATAMARAN

Reef Voyager

Ordered by Voyager Hotels and Resorts, this 34 m catamaran was designed by InCat Crowther. The vessel is designed to operate on rough water routes and incorporates a raised wetdeck and hull strengthening.

Specifications

Length overall	34.25 m
Length waterline	28.8 m
Beam	9.5 m
Draught	1.7 m
Passengers	202
Operational speed	29 kt
Range	315 n miles

Propulsion: Two Caterpillar C32 diesel engines, each rated at 1,045 kW driving fixed pitch propellers.

23.5 m CATAMARAN

Arafura Pearl

The vessel, designed by Stephen & Gravler (Brisbane, Australia), was delivered to Australian operator Sea Cat Ferries & Charters in July 2004. The ferry operates on a route between Darwin and Nguiu (Bathurst Island).

Specifications

Length overall	23.5 m
Length waterline	22.2 m
Beam	7.4 m
Draught	1.2 m
Passengers	200
Fuel capacity	3,800 litres
Fresh water capacity	2,000 litres
Maximum speed	28 kt
Operational speed	25 kt

Propulsion: Two MTU 16V 183 TE72 engines, each producing 610 kW at 2,100 rpm and each driving a fixed pitch propeller via a ZF BW 665A gearbox.

24 m CATAMARAN

Marjorie B

Designed by Stephen & Gravler, this vessel is loosely based on *Arufur Pearl* and will be used on day trips to the Great Barrier Reef. The vessel uses surface drives for operation in shallow waters.

Specifications

Length overall	24 m
Length waterline	23.5 m
Draught	0.85 m
Passengers	75
Maximum speed	34 kt
Operational speed	30 kt

Propulsion: Two MTU 12V 183 TE92 diesels rated at 735 kW drive Seafury 360 surface drives via ZF BW 655H gearboxes.

17.5 m CATAMARAN

Matthew Flinders

Delivered to Inter Island Ferries, Melbourne, Australia in December 2005, this vessel is built to Australia's IC USC code.

Vessel type	Vessel name	Yard no	Length (m)	Speed (kt)	Seats	Vehicles	Originally delivered to	Date of build
CMCS 20 m	*Gren Garden Sangiang I*	–	20.0	27.0	100	none	Indonesian operator	1996
CMCS 25 m	*Reef Quest*	–	24.4	30.0	120	none	Australian operator	2001
CMCS 24 m	*Daydream Express*	–	24.0	30.0	150	none	Australian operator	2001
23.5 m Catamaran	*Arafura Pearl*	–	23.5	28.0	200	none	Sea Cat Ferries and Charters	2004
35 m Crowther Catamaran	*Five*	–	35.0	30.0	350	none	Fantasea Cruises	2005
17.5 m Catamaran	*Matthew Flinders*	–	17.65	25.0	98	none	Inter Island Ferries, Australia	2005
24 m Catamaran	*Cat Rose*	–	24.0	34.0	75	none	Inter Island Ferries, Australia	2006
35 m InCat Crowther Catamaran	*Cat Cacos*	–	35.2	35.0	352	none	Inter Island Boats, Seychelles	(2006)
CAT InCat Crowther 34 m	*Reef Voyager*	–	34.0	29.0	202	none	Voyager Resorts	(2008)

Specifications

Length overall	17.65 m
Length waterline	14.3 m
Beam	6.0 m
Draught	1.0 m
Passengers	98
Fuel capacity	2,400 litres
Fresh water capacity	800 litres
Operational speed	25 kt

Propulsion: Two Volvo Penta TAMD 74 CEDC diesel engines delivering 320 kW to fixed pitch propellers.

Austal

100 Clarence Beach Road, Henderson, Perth, Western Australia 6166, Australia

Tel: (+61 8) 94 10 11 11
Fax: (+61 8) 94 10 25 64
e-mail: marketing@austal.com
Web: www.austal.com

John Rothwell, *Executive Chairman*
Stephen Murdoch, *Chief Operating Officer*
Michael Atkinson, *Director, Financial Control*

Austal is a world leader in the design and construction of high-performance aluminium vessels, with over 160 vessels delivered throughout the world. The company's proven range of ferries and Auto Express vehicle-passenger catamaran ferries are established as market leaders and can be found operating in many parts of the world, including the Baltic, Irish Sea, English Channel, the Mediterranean, Spain, Venezuela, Asia and the Americas. Two of the largest Auto Express designs, 101 m, were deployed in Summer 2001. Austal's high-speed vessel platforms have a variety of military uses which the company is also developing. Austal's product range extends to cruise yachts, patrol and offshore craft and luxury yachts.

In 1999, Austal established its American joint venture yard in Mobile, Alabama.

In 2003, Austal received an order for the largest multihull vessel worldwide, a 127 m trimaran. *Benchijigua Express,* delivered in 2005, now operates in the Canary Islands. The vessel has a capacity for 1,291 passengers and 340 cars with a speed of 40.5 kt.

Auto Express 127

Benchijigua Express

This trimaran was the largest aluminium ship ever built at the time of its launch in 2004 and was delivered to Fred Olsen in the Canary Islands in mid-2005. The vessel was designed to significantly improve passenger comfort at a speed of approximately 40 kt. A development of this design, the *Flight 0 vessel,* was proposed to the US Navy for its Littoral Combat Ship (LCS) project and in October 2005 a contract was awarded for the first of two planned vessels for construction at Austal's United States shipyard.

Specifications

Length overall	126.7 m
Length waterline	116.9 m
Beam	30.4 m
Draught	4.2 m
Deadweight	1,000 t
Passengers	1,291
Vehicles	341 cars or 450 truck lane metres and 123 cars
Operational speed	40.5 kt

Classification: Germanischer Lloyd +100 A5, HSC-B OC3 High-Speed Passenger/Ro-Ro Type +MC, AUT.

Propulsion: Four MTU 20V 8000 M70 rated at 8,200 kW at 1,150 rpm each. Two engines drive a single Kamewa 180 BII water-jet via a Renk ASL 2X80 gearbox, and the other two engines each drive a Kamewa 125 SII water-jet via a Renk ASL 65 gearbox.

Auto Express 101

Two 101 m vessels, *Euroferrys Pacifica* and *Westpac Express* were built and delivered by

Vessel type	Vessel name	Yard No	Length (m)	Speed (kt)	Seats	Vehicles	Originally delivered to	Date of build
34 m Catamaran	*Danat Dubai* (ex-*Bali Hai I*)	003	33.6	20.0	300	none	Bali Hai Cruises, Indonisia	1990
38 m Catamaran	*Tong Zhou*	008	38.0	–	430	none	Nantong High-Speed passenger Ship Company, China	1990
40 m Catamaran	*Shun Shui*	018	40.1	31.4	354	none	Shun Gang Passenger Transportation Company, China	1991
40 m Catamaran	*Equator Triangle*	014	40.1	–	216	none	Singapore	1991
40 m Catamaran	*Hai Bin* (ex-*Shun De*)	028	40.1	31.4	354	none	Shun Gang Passenger Transportation Company, China	1992
40 m Catamaran	*Xin Duan Zhou*	038	40.1	32.0	338	none	Zhao Gang Steamer Navigation Company	1992
40 m Catamaran	*Hai Zhu*	288	40.1	31.4	338	none	Jiuzhou Port Administration Group	1992
40 m Catamaran	*Nan Gui*	048	40.1	32.0	338	none	Ping Gang Passenger Transportation Corporation	1992
Kai Ping Class	*Kai Ping*	068	39.9	32.0	318	none	China	1992
Kai Ping Class	*Cui Heng Hu*	078	39.9	32.0	354	none	China	1992
Kai Ping Class	*Lian Shan Hu*	088	39.9	32.0	354	none	China	1992
Kai Ping Class	*Nan Xing*	098	39.9	32.0	338	none	China	1993
Kai Ping Class	*Hai Chang*	099	39.9	32.0	338	none	China	1993
Kai Ping Class	*Hui Yang*	100	39.9	32.0	368	none	China	1993
Kai Ping Class	*Tai Shan*	101	39.9	32.0	354	none	Shenzhen Xunlong Shipping, China	1993
Kai Ping Class	*Gao Ming*	103	39.9	32.0	338	none	China	1993
Kai Ping Class	*Gang Zhou*	102	39.9	32.0	318	none	China	1993
Kai Ping Class	*Gui Feng*	109	39.9	32.0	318	none	China	1993
Kai Ping Class	*San Bu*	110	39.9	32.0	338	none	China	1993
40 m Gas Turbine Catamaran Class	*Yi Xian Hu*	108	40.0	39.0	354	none	China	1994
40 m Gas Turbine Catamaran Class	*Shun Jing*	105	40.0	39.0	354	none	China	1994
40 m Gas Turbine Catamaran Class	*Lian Gang Hu*	106	40.0	39.0	354	none	China	1994
36 m Catamaran	*Bali Hai II*	015	35.7	29.0	322	none	Bali Hai Cruises, Indonisia	1994
40 m Catamaran	*Free Flying*	168	40.1	32.0	338	none	Shanghai Free Ferry Company	1994
40 m Catamaran	*Xin He Shan*	111	40.1	32.0	338	none	China	1994
40 m Gas Turbine Catamaran Class	*Zhong Shan*	115	40.0	39.0	354	none	China	1994
43 m Catamaran	*Speeder*	039	43.0	42.0	331	none	Diamond Ferry, Japan	1995
40 m Catamaran	*Hai Yang*	118	40.1	32.0	338	none	Zhuhai Jinzhou Port Shipping Company	1995
40 m Gas Turbine Catamaran Class	*Shun De*	116	40.0	39.0	332	none	China	1995
Auto Express 79	*Tallink Auto Express* (ex-*Cat-Link III*)	037	78.7	34.2	600	163	DSB Rederi A/S, Denmark	1996

Vessel type	Vessel name	Yard No	Length (m)	Speed (kt)	Seats	Vehicles	Originally delivered to	Date of build
Auto Express 82	*Delphin*	046	82.3	38.0	600	175	TT-Line, Germany	1996
40 m Catamaran	*Zeng Cheng Yi Hao*	119	40.1	32.0	268	none	China	1996
Auto Express 82	*Croazia Jet* (ex-*Felix*)	052	82.3	38.0	700	160	Scandlines Danmark A/S	1996
40 m Catamaran	*Sinan Pasa*	056	40.1	34.5	450	none	Turkey	1996
40 m Catamaran	*Piyale Pasa*	057	40.1	34.5	450	none	Turkey	1996
Auto Express 82	*Boomerang*	053	82.3	38.0	700	175	Polferries	1997
40 m Catamaran	*Nan Hua*	121	40.1	34.5	318	none	China	1997
40 m Catamaran	*Tai Jian*	120	40.1	34.5	318	none	China	1997
Auto Express 60	*Turgut Reis 1*	054	59.9	33.0	450	94	Istanbul Deniz Otobusleri	1997
Auto Express 60	*Cezayirli Hasan Pasa 1*	055	59.9	33.0	450	94	Istanbul Deniz Otobusleri	1997
42 m Catamaran	*Zhao Qing*	122	42.1	34.0	352	none	Hong Kong	1997
Auto Express 82	*SuperStar Express*	060	82.3	38.0	900	175	Star Cruises, Malaysia	1997
Auto Express 48	*Jade Express*	061	47.6	39.0	329	10	L'Express Des Iles, Guadeloupe	1998
48 m Passenger Catamaran	*Flying Dolphin 2000*	064	47.6	–	516	none	Ceres Hydrofoil Joint Service	1998
Auto Express 86	*Adnan Menderes*	063	86.6	40.0	800	200	Istanbul Deniz Otobusleri	1998
40 m Catamaran	*Opal Express*	076	40.1	32.0	302	none	French Caribbean	1998
Auto Express 86	*Turgut Ozale*	071	86.6	40.0	800	200	Istanbul Deniz Otobusleri	1998
Auto Express 86	*Jonathan Swift*	094	86.6	40.0	800	200	Irish Ferries	1999
52 m Passenger Catamaran	*Cat No 1*	080	52.4	40.0	493	none	Reederei Norden-Frisia AG, Germany	1999
Auto Express 86	*Carmen Ernestima*	095	86.6	40.0	800	200	Conferry, Venuzuela	1999
44 m Cruise Catamaran	*Geka 2000*	062	44.0	30.0	500	none	Bounty Cruises, Bali	1999
42 m Catamaran	*Sleipner*	083	41.8	–	358	none	HSD, Norway	1999
42 m Catamaran	*Draupner*	083	41.8	–	358	none	HSD, Norway	1999
52 m Passenger Catamaran	*Betico*	079	52.4	34.0	366	none	Compagnie Maritime Des Iles, New Caledonia	1999
Auto Express 86	*Villum Clausen*	096	86.6	40.0	1,000	186	Bornholms Trafikken	2000
Auto Express 72	*Highspeed 2*	086	72.0	39.0	620	70	Minoan Flying Dolphins	2000
Auto Express 72	*Highspeed 3*	087	72.0	39.0	620	70	Minoan Flying Dolphins	2000
Auto Express 92	*Highspeed 4*	097	92.0	40.5	1,050	188	Minoan Flying Dolphins	2000
Auto Express 101	*Euroferrys Pacifica*	114	101.0	37.0	951	251	Euroferrys	2001
Auto Express 101	*Westpac Express*	130	101.0	37.0	951	251	US Marine Corps, Japan	2001
Auto Express 56	*Fares Al Salam*	239	56.0	38.0	430	–	El Salam Maritime	2002
47.5 m Catamaran	*New Ferry LXXXI*	144	47.5	40.0	416	none	New World First Ferry	2002
47.5 m Catamaran	*New Ferry LXXXII*	145	47.5	40.0	416	none	New World First Ferry	2002
47.5 m Catamaran	*New Ferry LXXXIII*	146	47.5	40.0	416	none	New World First Ferry	2002
41.5 m Catamaran	*Salten*	242	41.5	33.0	216	none	OVDS, Norway	2003
41.5 m Catamaran	*Steigtind*	243	41.5	33.0	216	none	OVDS, Norway	2003
Auto Express 86	*Lilia Concepcion*	140	86.6	40.0	868	243	Conferry Venezuela	2002
Auto Express 86	*Spirit of Ontario I*	251	86.6	47.0	774	238	Canadian American Transportation System	2004
Auto Express 66	*Bocayna Express*	196	66.2	31.0	450	69	Fred Olsen SA	2003
Auto Express 56	*Aremiti 5*	266	56.0	35.0	700	30	Aremiti Cruise	2004
47.5 m CAT	*New World LXXXV*	148	47.5	42.0	430	none	New World First Ferry	2004
47.5 m CAT	*New World LXXXVI*	150	47.5	42.0	430	none	New World First Ferry	2004
Auto Express 127 Trimaran	*Benchijigua Express*	260	126.7	40.5	1340	340	Lineas Fred Olsen	2005
Auto Express 85	*Highspeed 5*	285	85.0	39.0	810	154	Hellenic Seaways	2005
Auto Express 45	*Gold Express*	283	45.2	38.0	446	–	L'Express des Iles	2005
Auto Express 45	*Silver Express*	284	45.2	38.0	360	10	L'Express des Iles	2005
Auto Express 68	*Maria Dolores*	293	67.5	35.0	600	65	Virtu Ferries	2006
Auto Express 88	*Osman Gazi I*	294	87.8	36.0	1200	225	Istanbul Deniz Otobusleri	2007
Auto Express 88	*Orhan Gazi I-*	295	87.8	3.06	1200	225	Istanbul Deniz Otobusleri	2007
47.5 m Catamaran	*Cota Jet*	–	47.5	42.0	411	none	Venetian Marketing Services	2007
47.5 m Catamaran	–	–	47.5	42.0	411	none	Venetian Marketing Services	(2007)
47.5 m Catamaran	–	–	47.5	42.0	411	none	Venetian Marketing Services	(2007)
47.5 m Catamaran	–	–	47.5	42.0	411	none	Venetian Marketing Services	(2007)
47.5 m Catamaran	–	–	47.5	42.0	411	none	Venetian Marketing Services	(2007)
47.5 m Catamaran	–	–	47.5	42.0	411	none	Venetian Marketing Services	(2008)
47.5 m Catamaran	–	–	47.5	42.0	411	none	Venetian Marketing Services	(2008)
47.5 m Catamaran	–	–	47.5	42.0	411	none	Venetian Marketing Services	(2008)
47.5 m Catamaran	–	–	47.5	42.0	411	none	Venetian Marketing Services	(2008)
47.5 m Catamaran	–	–	47.5	42.0	411	none	Venetian Marketing Services	(2008)
47.5 m Catamaran	–	–	47.5	42.5	418	none	New World First Ferry	(2008)
47.5 m Catamaran	–	–	47.5	42.5	418	none	New World First Ferry	(2008)
Auto Express 69	–	263	65.0	50.0	203	56	Sultanate of Oman	(2009)
Auto Express 69	–	264	65.0	50.0	203	56	Sultanate of Oman	(2009)
Auto Express 88	–	340	88.0	–	1200	120	Sultanate of Oman	(2009)
Auto Express 88	–	341	88.0	–	1200	120	Sultanate of Oman	(2009)

Austal in 2001 and can carry 251 cars or 16 trucks and 96 cars.

Specifications

Length overall	101.0 m
Length waterline	88.7 m
Beam	26.7 m
Draught	4.2 m
Deadweight	750 t
Passengers	951
Vehicles	251
Operational speed	37 kt

Propulsion: Four caterpillar 3,618 diesel engines each rated at 7,200 kW at 1,050 rpm and each driving a Kamewa 125 SII water-jet via a Reintjes gearbox.

Auto Express 92

Highspeed 4

The 92 m Auto Express ferry *Highspeed 4* was delivered to Minoan Flying Dolphins of Greece in July 2000. It has a maximum deadweight of 470 tonnes and the capacity to carry 1,050 passengers (200 VIP, 196 Business, 654 Tourist) and 188 cars (or 150 cars and four coaches). *Highspeed 4* operates between Piraeus and the islands of Paros and Naxos. She features stern loading and unloading with scheduled turnaround times on the islands between 15 and 30 minutes. Four Caterpillar 3618 engines provide a service speed of 40.5 kt.

Specifications

Length overall	92.0 m
Length waterline	80.6 m
Beam	24.0 m
Draught	3.7 m
Deadweight	470 t
Passengers	1,050
Vehicles	188
Fuel capacity	160,000 litres
Operational speed	40.5 kt

Propulsion: The propulsion system comprises four Caterpillar 3618 diesel engines each rated at 7,200 kW and driving Kamewa 112 SII water-jet units via a Reintjes VLJ 6831 gearbox.

Auto Express 88
Osman Gazi I
Orhan Gazi I

Istanbul Metropolitan Municipality ordered two vessels which were delivered in 2007. The vessels are able to carry 1,200 passengers and 225 cars and operate at speeds in excess of 36 kt.

Specifications

Length overall	87.8 m
Length waterline	77.4 m
Beam	24.0 m
Draught	3.7 m
Deadweight	470 t
Passengers	1,200
Vehicles	225 cars
Crew	16
Max speed	36 kt

Classification: Germanischer Lloyd

Propulsion: Four MTU 20V 8000 M70R diesel engines delivering 7,200 kW to Lips 12E water-jets via Reintjes VLJ 6831 gearboxes.

Austal 47.5 m passenger catamaran operated by Venetian Marketing Services 1120347

40 m Tai Jian *and* Nan Hua 0007916

Auto Express 86
Adnan Menderes
Turgut Ozal
Jonathan Swift
Villum Clausen
Carmen Ernestina
Lilia Concepcion
Spirit of Ontario

The *Adnan Menderes* and *Turgut Ozal* were delivered in June and November 1998 respectively to Turkish operator Istanbul Deniz Otobüsleri. *Jonathan Swift* was delivered in April 1999 to Irish Ferries, and *Carmen Ernestina* was delivered to Conferry in Venezuela in June 1999; both these vessels were powered by Caterpillar 3618 engines.

The first gas turbine powered Auto Express class was introduced into Denmark in March 2000. This vessel, *Villum Clausen* is propelled by two GM LM2500 gas turbines with a capacity of 1,000 passengers and 186 cars and has a service speed of 41 kt. On her delivery voyage, *Villum Clausen* smashed the 24 hour distance record travelled by a commercial passenger ship. With an average speed of 44.29 kt, the vessel covered a distance of 1,063 n miles in a 24 hour period. A further 86 m vessel was ordered in September 2001 for Conferry, powered by Caterpillar 3618 engines. This vessel, *Lilia Concepcion,* was delivered to Conferry in June 2002. In 2004 a seventh vessel, *Spirit of Ontario,* was delivered to the Canadian American Transportation System for operation between Rochester, New York State and Toronto. Capable of a maximum speed of 47 knots, this vessel is powered by four MTU 8000 engines giving a combined output of 32,800 kW.

Specifications

Length overall	86.6 m
Length waterline	74.2 m
Beam	24.0 m
Draught	3.2 m
Deadweight	340 t
Passengers	800
Vehicles	200 cars or 10 coaches and 125 cars
Max speed	42 kt
Operational speed	40 kt

Classification: Germanischer Lloyd +100 A5 HSC-B OC3 High-Speed Passenger Craft/ Ro-Ro Type. (*Villum Clausen* is classified by DNV).

Propulsion: Four MTU 20V 1163 TB 73L each rated at 6,500 kW and each driving a Kamewa 112 S11 water-jet via a Reintjes VLJ4431 gearbox. *Jonathan Swift* and *Carmen Ernestina* are powered by four Caterpillar 3618 diesels with a rating of 7,200 kW each driving a Kamewa 112 S11 water-jet via a Reintjes VLJ 6831 gearbox. *Villum Clausen* is powered by two GE2500 gas turbines.

Auto Express 85
Highspeed 5

This vessel was ordered in 2004 by Hellas Flying Dolphins of Greece and delivered in July 2005. It features separate passenger and vehicle loading ramps at the stern.

Specifications

Length overall	85 m
Length waterline	78 m
Beam	21.2 m
Draught	2.9 m
Crew	35
Passengers	810
Vehicles	154 cars or 4 coaches and 131 cars
Operational speed	39 kt

Classification: Germanischer Lloyd.

Propulsion: Four Caterpillar 3618 diesel engines each driving a Rolls Royce Kamewa water-jet via a Reintjes gearbox.

Auto Express 82
Delphin
Felix
Boomerang
SuperStar Express

Four of these vessels have been constructed. The first *Delphin* for TT-Line of Germany was delivered in May 1996; the second, *Felix*, for the SweFerry/DSO partnership in October 1996; the third, *Boomerang,* for the Polish Baltic Shipping Company Polferries in May 1997; and the fourth, *SuperStar Express*, for Star Cruises of Malaysia in November 1997. *SuperStar Express* has been chartered by P&O European Ferries since 1998. *Boomerang* and *SuperStar Express* have an increased waterline length of 70.7 m and an increased seating capacity. The second vessel, *Felix,* has a bow door and a drive-through vehicle deck. This vessel has a slightly different capacity of 616 passengers and 156 cars. All others have a drive-around arrangement, entering and exiting at the stern.

Specifications

Length overall	82.3 m
Length waterline	69.0 m
Beam	23.0 m
Draught	2.7 m
Crew	24
Passengers	600 to 900
Vehicles	175 cars or 10 coaches and 70 cars
Fuel capacity	80,000 litres
Water capacity	4,000 litres
Max speed	40 kt
Operational speed	38 kt

Classification: Germanischer Lloyd + 100 A5 HSC-B OC3 High-Speed Passenger Craft/ Ro-Ro Type.

Propulsion: The vessels are powered by four MTU 20V 1163 TB 73 diesels each rated at 6,000 kW at 1,250 rpm and each driving a Kamewa 112 S11 water-jet via a Reintjes VLJ 4431 reduction gearbox. *SuperStar Express* was fitted with four MTU 20V 1163 TB 73L engines with an increased rating of 6,500 kW.

Auto Express 79
***Tallink Auto Express* (ex-*Cat-Link III*)**

Launched in April 1995, the Auto Express 79 was the first of Austal's Auto Express vehicle-passenger catamarans. Based on Austal's semi-swath hull form, featuring rounded bilge and bulbous bow, the design has since been followed by much larger designs with increased carrying capacities. *Cat-Link III* was delivered to DSB Rederi A/S in Denmark in April 1996. Since early 1999, the vessel has been redeployed to operate between Estonia and Finland for Tallink.

Specifications

Length overall	78.7 m
Length waterline	68.5 m
Beam	23.0 m
Draught	2.4 m
Crew	24
Passengers	600

Vehicles	163 cars or 38 cars and 10 coaches
Fuel capacity	50,000 litres
Water capacity	4,000 litres
Operational speed	34.2 kt
Range	360 n miles

Classification: Det Norske Veritas.

Propulsion: The main propulsion consists of four Ruston 16 RK270 diesel engines each driving a Kamewa 100 S11 water-jet via Reintjes VLJ 4430 reduction gearboxes.

Electrical system: Electrical power is provided by four Caterpillar 3406T/SR4 generators rated at 200 kW each.

Auto Express 72

Highspeed 2

Highspeed 3

Two 72 m Auto Express ferries, *Highspeed 2* and *Highspeed 3*, were delivered to Hellas Flying Dolphins in May and June 2000 to operate on various routes in the Greek Cyclades and Sporades. The 72 m designs feature a single vehicle deck with stern loading and unloading.

Specifications

Length overall	72.0 m
Length waterline	63.5 m
Beam	17.5 m
Draught	2.5 m
Deadweight	190 t
Passengers	620
Vehicles	70 cars
Fuel capacity	60,000 litres
Operational speed	40.0 kt

Classification: Germanisher Lloyd.

Propulsion: The vessel is powered by four MTU 16V 595 TE 70L diesel engines, each rated at 3,866 kW at 1,750 rpm, driving Kamewa 90 S11 water-jets via Reintjes VLJ 2230 reduction gearboxes.

Auto Express 66

Bocayna Express

A 66 m Auto Express vehicle-passenger catamaran was ordered in December 2002 and delivered in 2003 for operation by Fred Olsen SA.

Specifications

Length overall	66.2 m
Length waterline	59.0 m
Beam	18.2 m
Draught	2.5 m
Passengers	450
Vehicles	69 cars
Operational speed	31 kt

Propulsion: The vessel is powered by two Paxman 18VP185 and two Paxman 12VP185 diesel engines. Each 18V engine drives a Kamewa 90 SII water-jet via a Reintjes VLJ 2230 gearbox, and each 12V engine drives a Kamewa 80 SII water-jet via a Reintjes VLJ1130 gearbox.

Auto Express 65 m

Austal's two multipurpose Auto Express 65 m ferries are expected to be delivered in 2007/08 to the Sultanate of Oman.

Specifications

Length overall	64.8 m
Length waterline	61.1 m
Beam	16.7 m
Draught	6.2 m
Passengers	203
Vehicles	56 cars
Operational speed	50 kt

Propulsion: The vessel is powered by four MTU 20V 1163 TB73L diesel engines each rated at 6,500 kW, each driving a Kamewa 90 SII water-jet via a Reintjes gearbox.

65 m Trimaran crewboat (design)

Based on the company's experience of designing and building the 126 m Trimaran ferry, Austal has developed a 65 m trimaran crewboat design. Capable of operating at 40 knots in head seas at seastate 5, the vessel is designed to carry a maximum deadweight of 98 tonnes.

48 m catamaran Opale Express

0045108

42 m Draupner, *one of two identical catamarans built to a Norwegian design for HSD of Norway*
0084075

42 m catamaran Zhao Qing
0038695

Austal 44 m fast cruise catamaran Geka 2000
0084074

Specifications

Length overall	65.1 m
Length waterline	62.3 m
Beam	17.3 m
Draught	1.6 m
Passengers	149
Crew	8
Fuel capacity	39,400 litres
Water capacity	3,785 litres

Propulsion: Three MAN B&W 18VP185 diesels each rated at 3,500 kW and each driving a Kamewa 90 SII water-jet.

Auto Express 60

Turgut Reis 1

Cezayirli Hasan Pasa 1

The 60 m Auto Express vehicle-passenger catamarans *Turgut Reis 1* and *Cezayirili Hasan Pasa 1* were delivered to Istanbul in September 1997 for the Turkish operator Istanbul Deniz Otobüsleri. They operate a 24 n mile route across the sea of Marmara with a 15 minute turnaround time.

Specifications

Length overall	59.9 m
Length waterline	51.2 m
Beam	17.5 m
Draught	3.2 m
Crew	16
Passengers	450
Vehicles	94 cars
Fuel capacity	24,000 litres
Water capacity	2,000 litres
Max speed	35 kt
Operational speed	33 kt

Classification: Det Norske Veritas.

Propulsion: The vessel is powered by two MTU 20V 1163 TB 73L diesel engines, each rated at 6,500 kW at 1,250 rpm, driving Kamewa 125 S11 water-jets via Reintjes VLJ 4431 reduction gearboxes.

Auto Express 56

Fares Al Salam

Aremiti 5

The first vessel was ordered by the Saudi Arabian company Maritime Company for Navigation for operation by the El Salam Maritime Company of Egypt across the Red Sea. This was delivered in 2002. A second vessel was ordered by Aremiti Cruises in 2003 and delivered in 2004. This vessel is similar to the first but has a passenger capacity of 700 and a vehicle capacity of 30 cars.

Specifications

Length overall	56.0 m
Length waterline	49.8 m
Beam	14.0 m
Draught	2.7 m
Crew	12
Passengers	430
Vehicles	43
Fuel capacity	50,000 litres
Water capacity	5,000 litres
Maximum speed	38 kt
Operational speed	34.5 kt

Propulsion: Four MTU 16V 4000M70 diesel engines each rated at 2,320 kW, and each driving a Kamewa 71 SII water-jet via a Reintjes VLJ 930 gearbox.

Auto Express 48

Jade Express

Austal delivered its first 48 m Auto Express car-passenger catamaran in June 1998 to Guadeloupe. It was built for French shipping group Compagnie Chambon and is operated by L'Express Des Iles, Chambon's French Caribbean fast ferry operation.

Unlike Austal's larger Auto Express vessels, which feature either stern or drive-through vehicle access, cars are loaded and discharged from the garage, located aft on the main deck, via a 3.0 m ramp on the port side. This allows the vessel to berth at existing facilities with only simple shore-based ramps required to reach the height of the car deck.

Austal Auto Express 86, Villum Clausen *powered by two GE 2500 gas turbines* 0084077

Austal Auto Express 101 Euroferrys Pacifica 0102270

Austal 52 m fast catamaran ferry Cat No 1 *designed for operation in Germany* 0084081

Auto Express 101 WestPac Express, *currently operated by the US Army* 0121764

Specifications

Length overall	47.6 m
Length waterline	41.6 m
Beam	13.0 m
Draught	1.4 m
Crew	10
Passengers	329
Operational speed	39 kt

Classification: Bureau Veritas.

Propulsion: The vessel is powered by four MTU 16V 396 TE 74L diesel engines each rated at 1,980 kW and driving Kamewa 63 SII water-jets via Reintjes VLJ 930 reduction gearboxes.

52 m Passenger catamaran

Cat No 1

Betico

Two Austal 52 m vessels were delivered in 1999. *Cat No 1* was delivered in May to German operator AG Reederei Norden-Frisia for operation between Norderney, Helgoland and Cuxhaven. *Betico* was delivered in September to New Caledonia operator Compagnie Maritime Des Iles (CMI). *Betico* was fitted out for 366 passengers at a service speed of 34 kt.

Specifications

Length overall	52.4 m
Length waterline	49.2 m
Beam	13.0 m
Draught	1.5 m
Passengers	450
Operational speed	40 kt

Classification: (Cat No 1) Germanischer Lloyd + 100 A5 HSC-A OC3 High-Speed Passenger Craft, Machinery +MC, HSC-A, (*Betico*) Bureau Veritas.

Propulsion: (Cat No1) Four MTU 16V 4000 M70 diesel engines each driving a Kamewa 71 SII water-jet via a Reintjes VLJ 930 reduction gearbox. *(Betico)* Four MTU 12V 4000 M70 each driving Kamwea 63 SII water-jets.

48 m Passenger catamaran

Flying Dolphin 2000

Austal delivered the 48 m passenger catamaran *Flying Dolphin 2000* to Greek operator Ceres Hydrofoil Joint Service in June 1998.

It operates in the Saronic Gulf, connecting the port of Piraeus with Poros, Hydra, Spetses and Porto Heli.

Specifications

Length overall	47.6 m
Length waterline	41.6 m
Beam	13.0 m
Draught	1.4 m
Passengers	516

Classification: Germanischer Lloyd.

Propulsion: The vessel is powered by four MTU 16V 4000 M70 diesel engines, each rated at 2,320 kW and each driving Kamewa 63 SII water-jets via Reintjes 930 VLJ gearboxes.

47.5 m Passenger catamaran

New Ferry LXXXI

New World LXXXII

New World LXXXIII

New Ferry LXXXIV

New Ferry LXXXV

The first three craft were delivered in 2002 and two further craft were delivered in 2004.

Specifications

Length overall	47.5 m
Length waterline	44.0 m
Beam	11.8 m
Draught	1.4 m
Deadweight	55.8 t
Crew	8
Passengers	416
Fuel capacity	20,000 litres
Fresh water capacity	2,000 litres
Operational speed	40 kt

Classification: Det Norske Veritas +1A1 HSLC Passenger R2 EO.

Propulsion: Two MTU 16V 4000 M70 diesel engines each driving a Kamewa 63 SII water-jet via a Reintjes VLJ930HL gearbox.

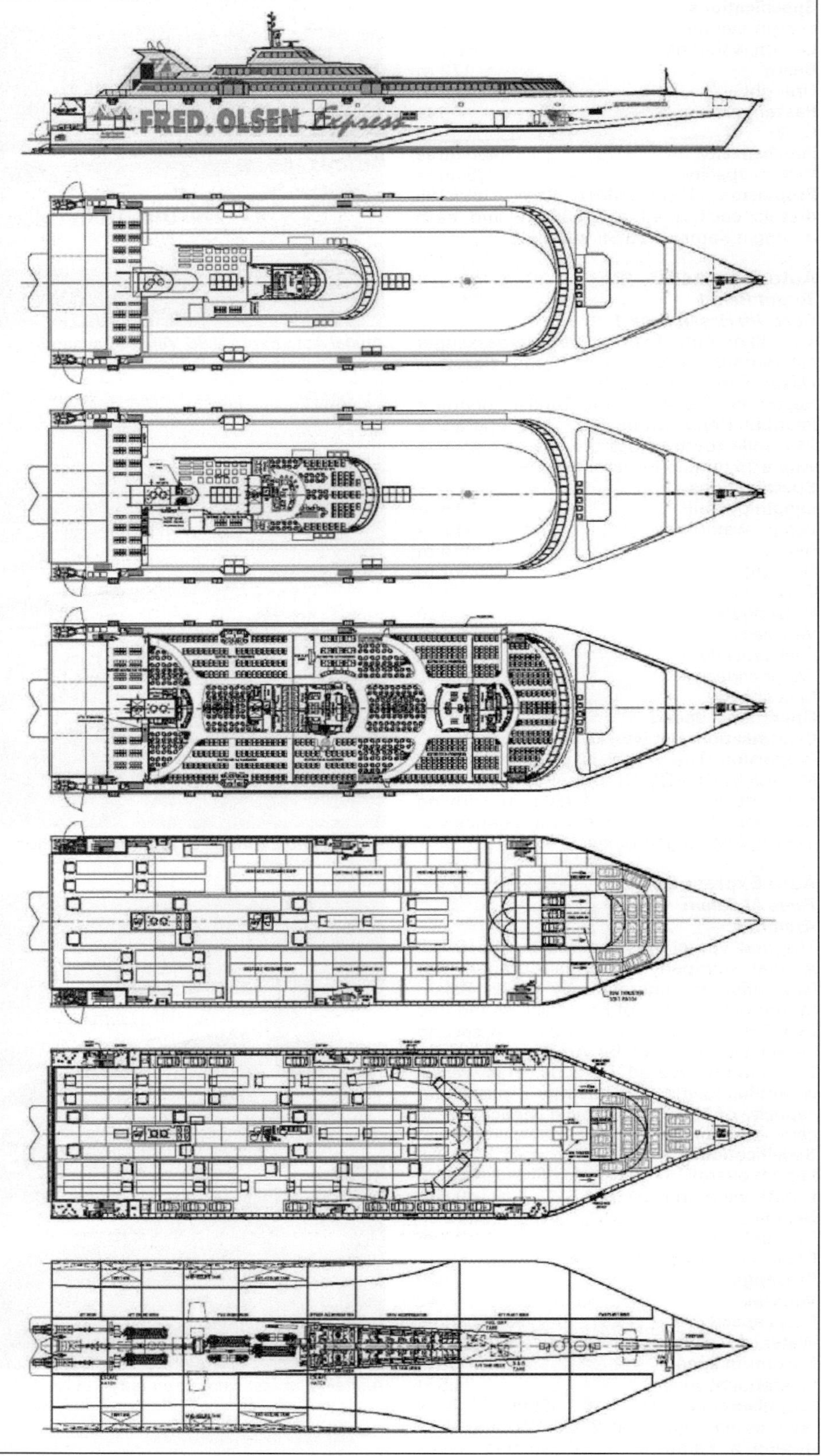

General arrangement of Austal 125 m Trimaran 1120231

47.5 m Passenger catamaran

In 2007 Austal announced orders for ten of these vessels to Venetian Marketing Services to operate from Hong Kong to the Cotai Strip in Macau. A further two, 418 passenger vessels were ordered by New World First Ferry for delivery in 2008.

Specifications

Length overall	47.5 m
Length waterline	43.8 m
Beam	11.8 m
Draught	1.6 m
Deadweight	70 t
Crew	8
Passengers	411
Fuel capacity	20,000 litres
Operational speed	42 kt

Classification: Det Norske Veritas +1A1 HSLC Passenger R2 EO.

Propulsion: Four MTU 16V 4000 M70 diesel engines each rated at 2,320 rpm and each driving a Rolls-Royce/Kamewa 63 SII water-jet via a Reintjes VLJ930 gearbox.

44 m Cruise catamaran

Geka 2000

Delivered to Bounty Cruises of Bali in Indonesia, in July 1999 to operate between the islands of Bali and Lombok. This vessel offers three operational modes including high-speed passenger transport, day cruise operations and night entertainment cruises. With a speed of 30 kt, *Geka 2000* has the capacity to carry 500 passengers.

Specifications

Length overall	44.0 m
Length waterline	38.4 m
Beam	11.8 m
Draught	2.5 m
Passengers	500
Fuel capacity	11,000 litres
Operational speed	30 kt

Classification: Western Australian Department of Transport Uniform Shipping Laws Code.

Propulsion: Two MTU 16V 396 TE74L diesel engines each rated at 1,960 kW and driving fixed-pitch propellers via a Reintjes WVS 930 reverse reduction gearbox.

43 m Catamaran

Speeder

Speeder was delivered to Diamond Ferry, a subsidiary of Mitsui, in May 1995. It was Austal's first sale to Japan and represented the first Australian-built ferry constructed for regular commuter service in Japan.

Specifications

Length overall	43.0 m
Length waterline	37.7 m
Beam	11.2 m
Draught	1.3 m
Crew	6
Passengers	331
Fuel capacity	14,000 litres
Water capacity	1,500 litres
Propulsive power	7,840 kW
Operational speed	42 kt

Classification: JG, Japanese Ministry of Transport.

Propulsion: The vessel is powered by four MTU 16V 396 TE 74L diesels each rated at 1,980 kW and driving through a ZF BU755-D reduction gearbox to a Kamewa 63 S11 water-jet.

42 m Catamaran

Zhao Qing

The Austal 42 m passenger catamaran *Zhao Qing* was delivered to Hong Kong in 1997.

Specifications

Length overall	42.1 m
Length waterline	37.0 m
Beam	11.5 m
Draught	1.35 m
Crew	12
Passengers	352
Fuel capacity	13,400 litres
Water capacity	1,500 litres
Operational speed	34 kt

Classification: China Classification Society.

Propulsion: The vessel is powered by two MTU 16V 396 TE 74L diesel engines, each rated at 1,980 kW and each driving Kamewa 71 SII water-jets via a Reintjes VLJ 930 gearbox.

41.5 m Catamaran

Salten

Steigtind

Both of Austal's 41.5 m passenger catamarans were delivered to OVDS in Norway in 2003.

Specifications

Length overall	41.3 m
Length waterline	36.3 m
Beam	11.6 m
Draught	1.54 m
Crew	6
Passengers	214
Fuel capacity	16,600 litres
Operational speed	33 kt

Classification: Det Norske veritas 1A1 HSLC R2 (nor) Passenger EO Norwegian Flag.

Propulsion: The vessel is powered by two MTU 16V 4000 M70 diesel engines, each rated at 2,320 kW at 2,000 rpm and each driving Kamewa 71 SII water-jets via a ZF 7550 gearbox.

40 m Catamaran

Equator Triangle

Specifications

Length overall	40.1 m
Beam	13.3 m
Draught	2.4 m

Austal Auto Express 92 m catamaran, Highspeed 4 0077478

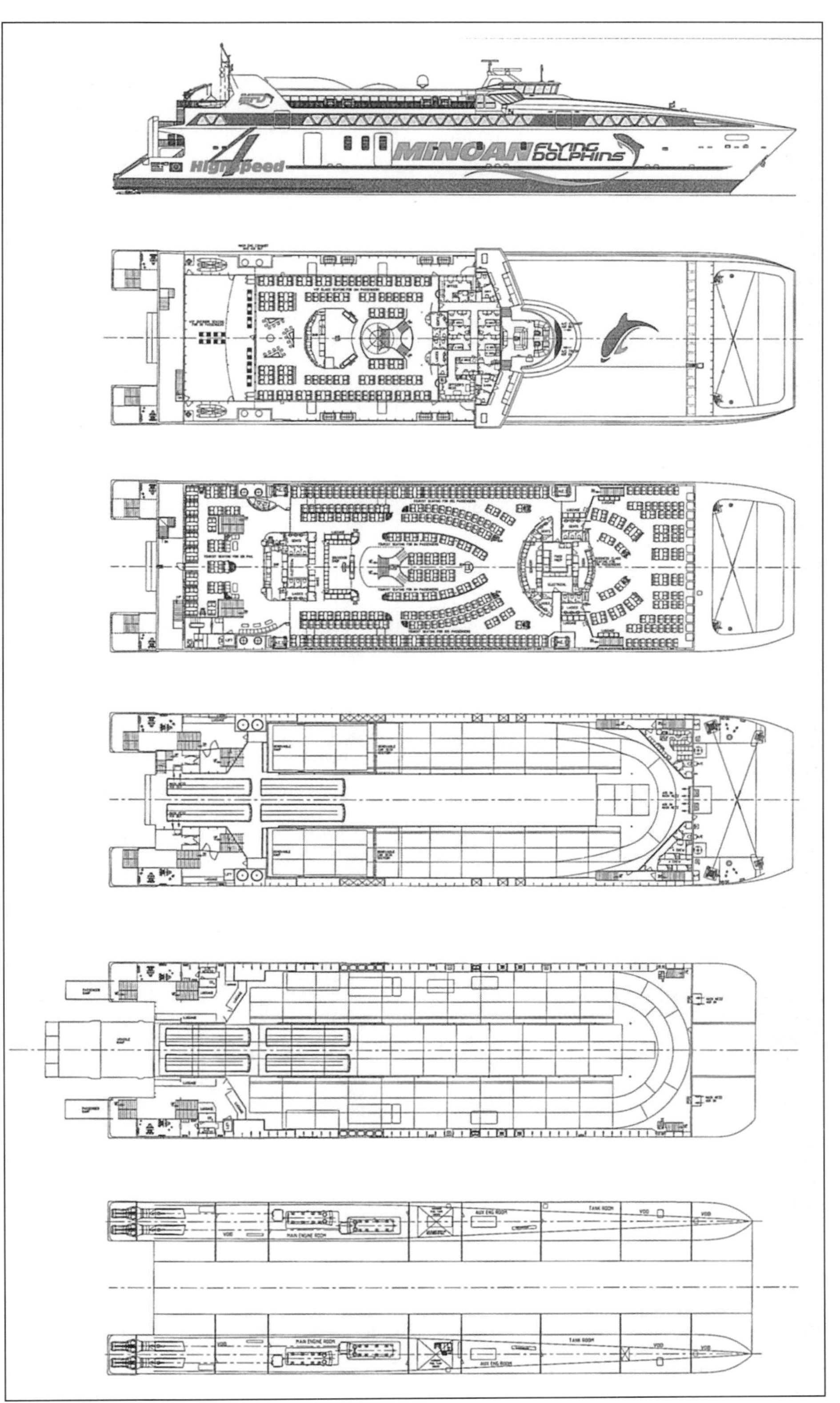

General arrangement of Austal Auto Express 92 m Highspeed 4 0103290

Displacement, max 176 t
Crew 34
Passengers 216
Fuel capacity 20,000 litres
Water capacity 10,000 litres
Propulsive power 2 × 1,430 kW
Max speed 25.5 kt

Structure: Hull, superstructure and deck construction material: marine grade aluminium alloy.

Propulsion: Main engines: two Caterpillar 3516TA diesels, 1,430 kW each at 1,800 rpm, driving fixed-pitch propellers.

40 m Catamaran

Kai Ping Class

There have been 11 of these craft delivered to various owners in China.

Specifications

Length overall 39.9 m
Length waterline 35.0 m
Beam 10.0 m
Draught 1.4 m
Crew 12
Passengers 318–368
Fuel capacity 10,000 litres
Water capacity 1,500 litres
Propulsive power 2 × 1,580 kW
Operational speed 32 kt

Structure: Hull material: aluminium alloy.

Superstructure material: aluminium alloy.

Propulsion: Engines: two MTU 16V 396TE 74 diesels, 1,580 kW each.

Transmissions: two Reintjes VLJ 930 gearboxes.

Thrust devices: two Kamewa 71S water-jets.

Auxiliary systems: Engines: two MTU 6V 183 AA51 diesels, 98 kW each.

Electrical system: 380 V/220 V 50 Hz AC, 24 V DC.

40 m Gas Turbine Catamaran Class

There have been five of these vessels constructed, *Shun Jing, Lian Gang Hu, Yi Xian Hu, Zhong Shan* and *Shun De*, all for Chinese owners.

Specifications

Length overall 40.0 m
Length waterline 35.0 m
Beam 11.5 m
Draught 1.4 m
Crew 12
Passengers 354
Fuel capacity 10,000 litres
Water capacity 1,500 litres
Propulsive power 2 × 2,570 kW
Operational speed 39 kt

Propulsion: These vessels are powered by two Textron Lycoming TF40 gas turbines, rated at 2,570 kW at 15,400 rpm. Each turbine drives through a MAAG MPG-80 gearbox to a Kamewa 71S water-jet. (For *Zhong Shan* and *Shun De*, the gearbox is a Cincinnati Gear MA-107 model.)

Auxiliary systems: Two MTU 6V 183 AA51 diesel generators rated at 98 kW at 1,500 rpm.

40 m Catamaran

Sinan Pasa

Piyale Pasa

These two 40 m craft were delivered in December 1996 and operate routes in the Bosporus and the Sea of Marmara.

Specifications

Length overall 40.1 m
Length waterline 35.0 m
Beam 10.5 m
Draught 1.2 m
Crew 4
Passengers 450
Fuel capacity 9,000 litres
Water capacity 3,000 litres
Operational speed 34.5 kt

Propulsion: The vessel is powered by two MTU 16V 396 TE 74L diesel engines, each rated at 1,980 kW at 1,940 rpm and each driving a Kamewa 71 S11 water-jet.

One of three 47.5 m catamarans built for New World First Ferry 0594976

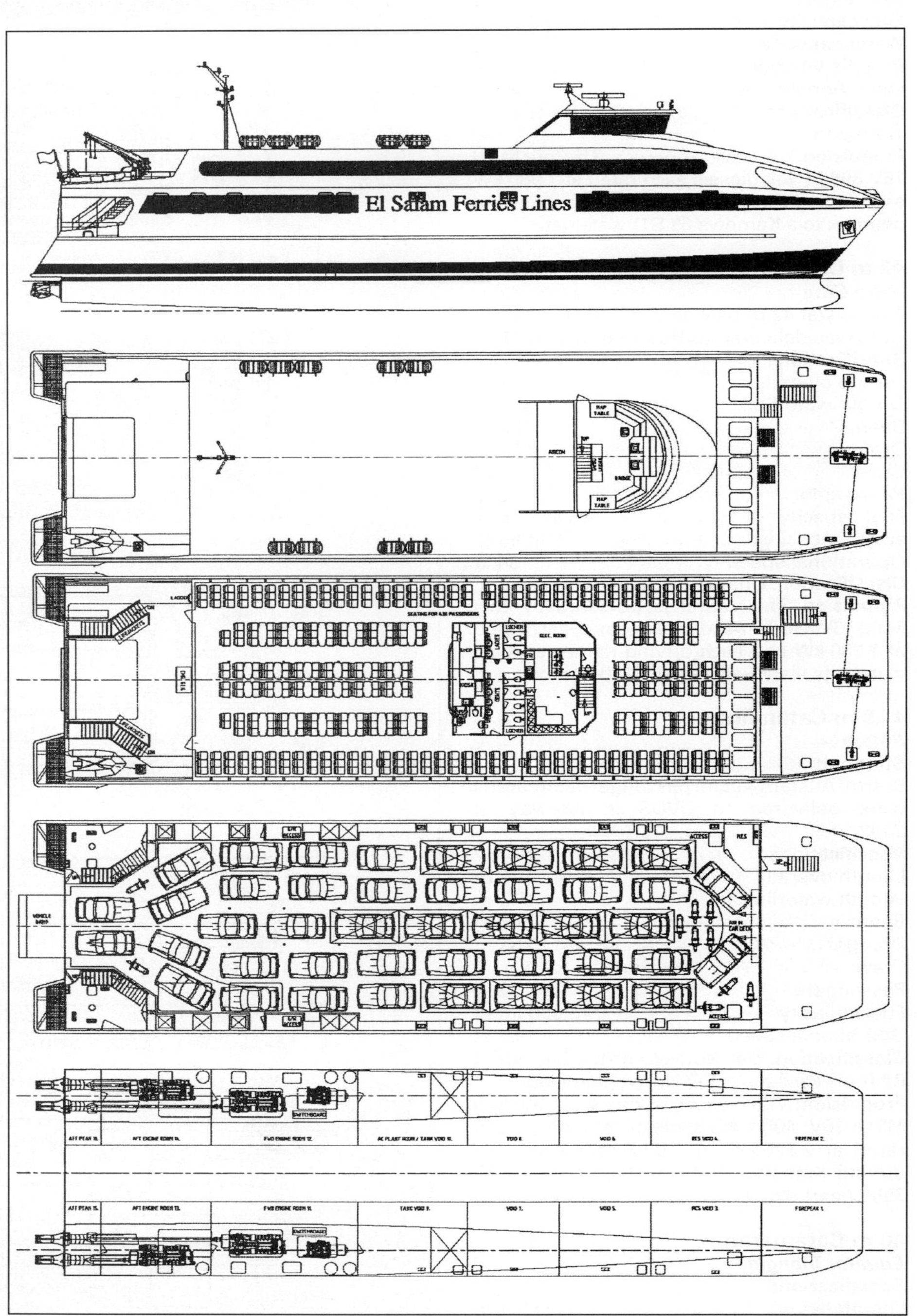

General arrangement of Austal Auto Express 56 0526488

Benchijigua Express *during trials* 1120230

40 m Catamaran

Tai Jian

Nan Hua

These two craft were delivered in July 1997 for operation between Hong Kong and Guangdong Province for operator CKS. These craft represent a radical departure from the earlier 40 m craft, featuring a three-deck arrangement, with the wheel house situated above the passenger decks. (*Tai Jian* is only fitted for 318 passengers).

Specifications

Length overall	40.1 m
Length waterline	35.0 m
Beam	11.5 m
Draught	1.4 m
Crew	12
Passengers	338
Fuel capacity	10,000 litres
Water capacity	1,500 litres
Operational speed	34.5 kt

Propulsion: The vessel is powered by two MTU 16V 396 TE 74L diesels, each rated at 1,980 kW at 1,940 rpm, each driving a Kamewa 71S water-jet via a ZF BU755 reduction gearbox.

38 m Catamaran

Tong Zhou

In June 1989, Austal Ships secured its first Chinese contract for the construction of this 38 m, 430 passenger catamaran ferry for the Nantong High-Speed Passenger Ship Company, China. Delivered in November 1990, it operates on the Yangtse River between the cities of Nantong and Shanghai.

Specifications

Length overall	38.0 m
Length waterline	32.4 m
Beam	11.8 m
Draught	1.3 m
Displacement, max	145 t
Fuel capacity	2 × 7,500 litre tanks
Water capacity	1,200 litres
Crew	12
Passengers	430
Max speed	30 kt

Structure: Hull, superstructure and deck construction material: marine grade aluminium alloy.

Propulsion: Main engines: two MTU 16V 396TB 83 marine diesels developing 1,470 kW each at 1,940 rpm. Two MJP J650R water-jet units.

36 m Catamaran

Bali Hai II

Bali Hai II is the second catamaran to be built for Bali Hai Cruises, to run day cruises from Bali to the islands of Nusa Penida and Lembongan. This vessel was delivered in April 1994.

Specifications

Length overall	35.7 m
Beam	10.5 m
Draught	1.2 m
Crew	8
Passengers	322
Fuel capacity	12,000 litres
Water capacity	6,000 litres

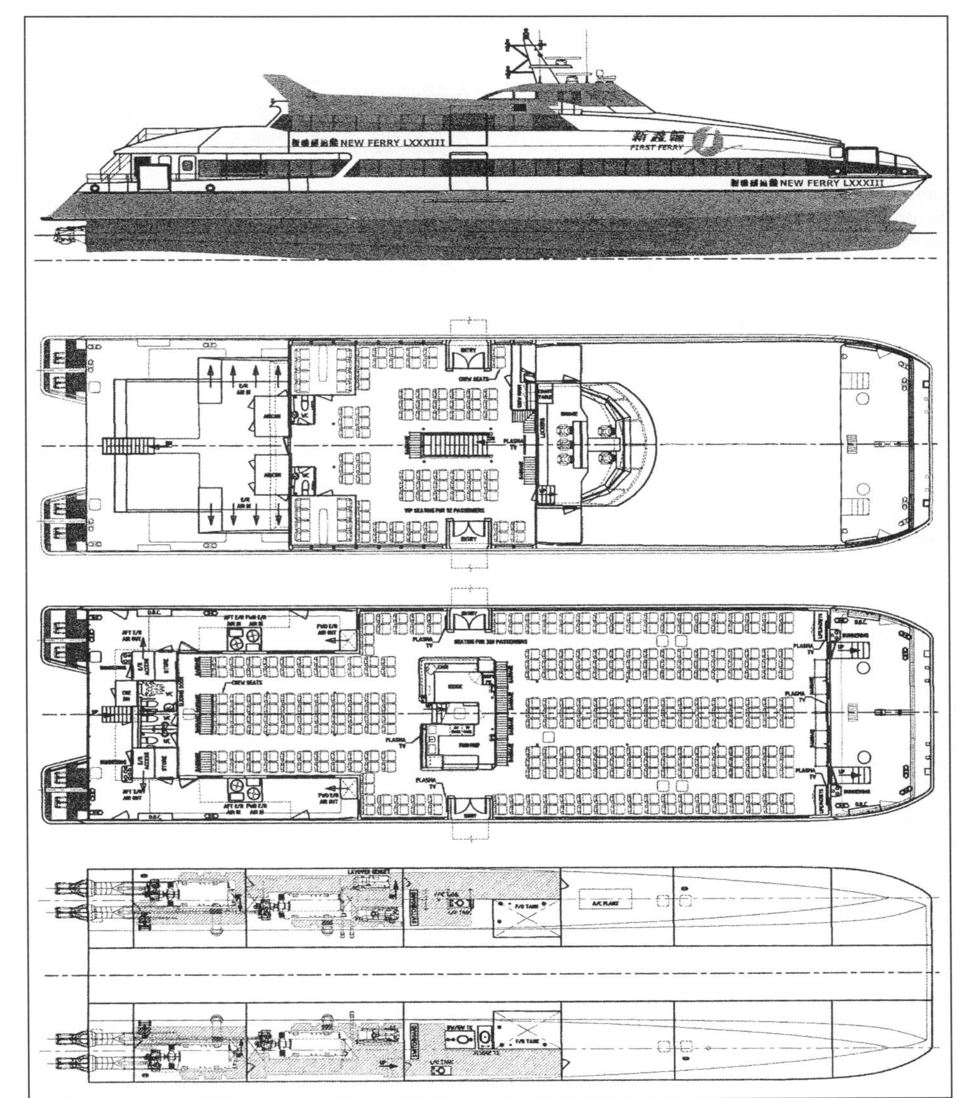

General arrangement of the Austal 47.5 m Passenger Catamaran 0526490

Austal Auto Express 88 Osman Gazi 1120346

Propulsive power 2 × 1,470 kW
Operational speed 29 kt
Structure: Hull, superstructure and deck construction material: aluminium alloy.
Propulsion: Main engines: two MTU 12V 396TE 74L diesels rated at 1,470 kW each at 1,940 rpm, driving two MJP J650R DD water-jets.

34 m Catamaran

Bali Hai

Delivered in March 1990 to run day cruises from Bali to the islands of Nusa Penida and Lembongan. *Bali Hai* is constructed in aluminium and designed to carry 300 passengers at 20 kt.

Specifications

Length overall	33.6 m
Length waterline	30.7 m
Beam	10.8 m
Draught	1.9 m
Crew	8
Passengers	300
Fuel capacity	16,000 litres
Water capacity	4,000 litres
Propulsive power	2 × 735 kW
Operational speed	20 kt

Classification: DnV.
Structure: Hull, superstructure and deck construction material: marine grade aluminium alloy.
Propulsion: Engines: two MAN D284 2LYE diesels developing 735 kW each at 2,300 rpm and driving fixed-pitch propellers via a ZF BW 195S gearbox.

Auto Express 60 Turgut Reis 1 *and* Cezayirili Hasan Pasa 1 0007915

Australus Catamarans

Rippleside, Liverpool Street, North Geelong, Victoria 3215, Australia

Tel: (+61 3) 527 229 09
Fax: (+61 3) 527 720 22
e-mail: australus_cats@bigpond.com

20 m CATAMARAN

Built to a Williams Tech Marine design, the catamaran is used for whale watching tourist trips.

Specifications

Length overall	20.0 m
Length waterline	18.0 m
Beam	7.0 m
Draught	1.7 m
Passengers	130
Maximum speed	25 kt
Operational speed	19 kt

Propulsion: Two Caterpillar 3406E diesels each rated at 518 kW and each driving a fixed pitch propeller via a Twin Disc gearbox.

Vessel type	Vessel name	Yard No	Length (m)	Speed (kt)	Seats	Vehicles	Originally delivered to	Date of build
20 m Catamaran	*Australus III*	–	20.0	25.0	130	none	Whale Planet Australia	1994

BSC Marine Group

Suite 12, 621 Coronation Drive, Toowong, Queensland 4066, Australia

Tel: (+61 7) 38 76 15 22
Fax: (+61 7) 38 76 15 66
e-mail: info@bscship.com.au
Web: www.bscship.com.au

Michael Hollis, *Managing Director*

BSC Marine Group designs and builds aluminium and composite commercial and paramilitary craft.

38 m LOW WASH FERRY

In 2006 Thames Clipper ordered six Aimtek River Runner 200 MIIB catamarans for delivery in 2007.

Specifications

Length overall	38.04 m
Length waterline	35.65 m
Beam	9.3 m
Draught	1.3 m
Passengers	196
Crew	3
Maximum speed	27 kt

Propulsion: Two MTU 10V 2000 M72 diesel engines delivering 720 kW at 2,090 rpm to fixed-pitch propellers via Twin Disc MGX 6598 SC Quickshift gearboxes.

32 m CATAMARAN CREWBOAT

The vessels were designed and built specifically for the Seatrucks Group to operate as crew support boats on the west coast of Africa.

Specifications

Length overall	32.1 m
Length waterline	29.2 m
Beam	9.87 m
Draught	0.9 m
Deadweight	14,700 kg
Crew	6
Passengers	132
Maximum speed	32 kt
Operational speed	29 kt
Range	280 n miles

Propulsion: Two Caterpillar 3412E diesels each rated at 820 kW at 2,300 rpm driving two Hamilton HM 571 water-jets.
Classification: Lloyd's Register +100 A1 + SSC Passenger Catamaran HSC G2.

MV Suzanne 0069439

30 m CATAMARAN

Maggie Cat

Sun Cat

These vessels have been designed by One2Three Naval Architects for Sunferries, Queensland, as part of a AUD7.8 million upgrade of the Townsville to Magnetic Island service. The 239-passenger ferries have a maximum speed of 30 knots and complete the 8 nautical mile crossing in 8 minutes.

Specifications

Length overall	29.5 m
Length waterline	27.3 m
Beam	8.5 m
Draught	1.3 m
Passengers	239
Fuel capacity	4150 l
Freshwater	1,200 l
Operational speed	28 kt

Propulsion: Two Caterpillar 3412E diesels each rated at 895 kW at 2,300 rpm driving Bruntons fixed pitch propellers via Twin Disc MG 6557 SC gearboxes.

28 m LOW WASH FERRY (DESIGN)

The company proposed construction of a 28 m low wash ferry in 2000.

Specifications

Length overall	28.1 m
Length waterline	27.4 m
Beam	7.7 m
Draught	1.2 m
Crew	2
Passengers	146
Fuel capacity	3,127 litres
Water capacity	300 litres
Operational speed	22 kt

Propulsion: The vessel is to be powered by two Caterpillar CAT 3196 engines each rated at 356 kW and each driving an 760 mm diameter fixed-pitch propeller via a Twin Disc 5114 gearbox.

26 m LOW WASH FERRY

MV Suzanne

Initially built for Fantasea Cruises, this vessel is currently owned by Transtejo, operating on the Tejo River in Lisbon, Portugal.

Specifications

Length overall	26.4 m
Length waterline	24.3 m
Beam	7.8 m
Draught	1.5 m
Crew	3
Passengers	150
Fuel capacity	2,300 litres
Water capacity	600 litres
Max speed	26.3 kt
Operational speed	25 kt

Propulsion: The vessel is powered by two Caterpillar CAT 3196TA engines each rated at 356 kW and each driving a 760 mm diameter fixed-pitch propeller via a Twin Disc MG5114A gearbox.

25 m LOW WASH FERRIES

An initial order for six craft for the CityCats service, operated by Brisbane City Council, was completed in 1997. This was followed by a further order for two craft which were delivered in 1998 and 1999. The vessel hulls are fabricated from aluminium and the composite superstructures were adhesively bonded to the hulls and cross-structure.

Specifications

Length overall	25.0 m
Beam	7.2 m
Draught	0.9 m
Crew	2
Passengers	130
Fuel capacity	1,600 litres
Water capacity	500 litres
Max speed	28 kt
Operational speed	24 kt

Propulsion: The initial six vessels were powered by two Scania DSI 11 60M diesels each rated at 280 kW at 1,800 rpm and each driving a four-blade propeller via a Twin

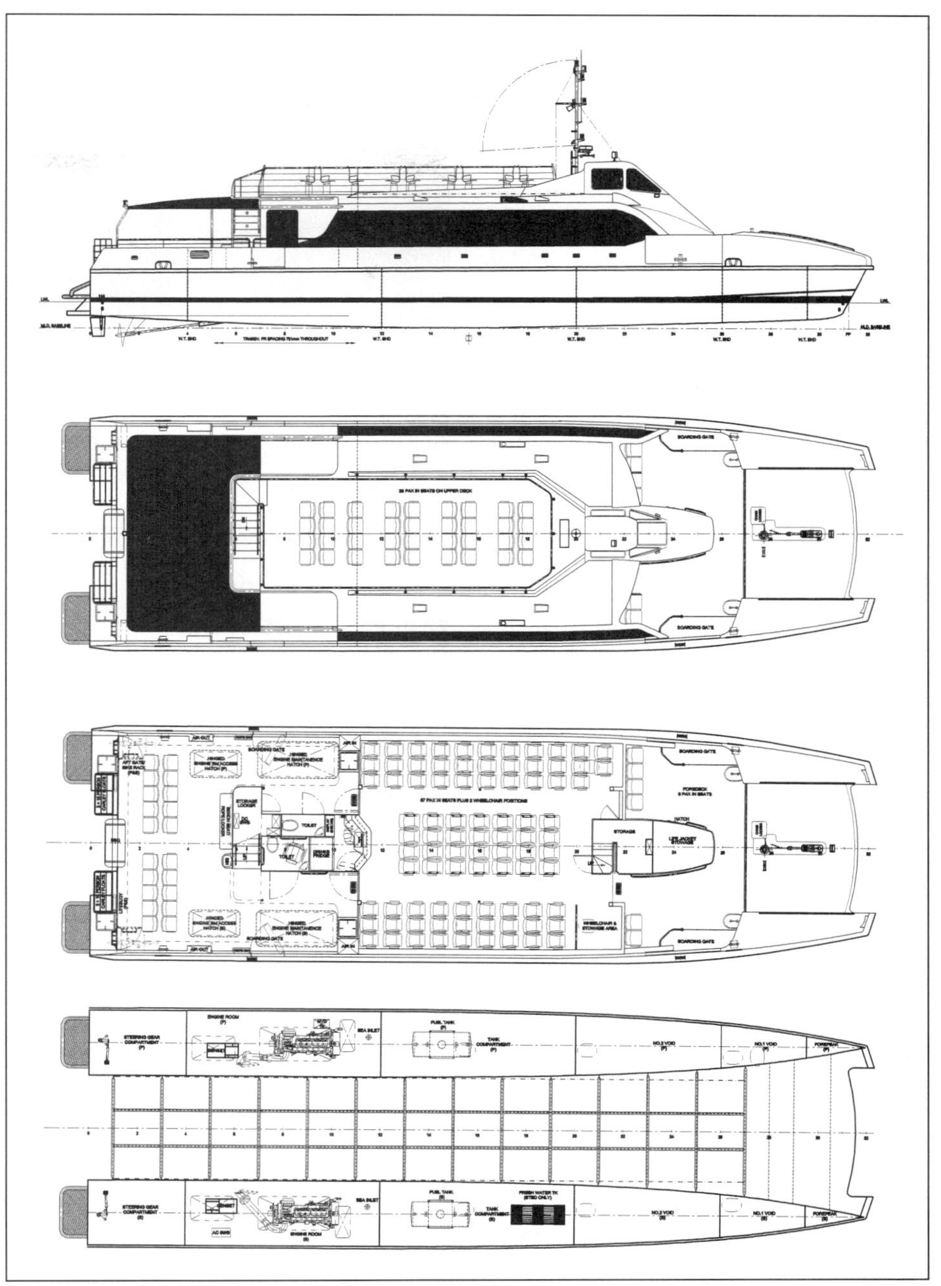

General arrangement of 26 m low wash catamaran Suzanne 0080478

Four 22 m customs boats for Vietnam 0045110

Disc 5114A reverse reduction gearbox. The subsequent two craft were powered by the later model Scania DSI 11 73M.

25 m PASSENGER FERRY

This vessel, designed to Lloyds Register Special Service Craft regulations, was delivered to Fitzroy Island Ferries in June 2000.

Specifications

Length overall	24.4 m
Beam	8.0 m
Draught	1.5 m
Crew	3
Passengers	188
Water capacity	500 litres
Max speed	26 kt
Operational speed	24 kt

Propulsion: The vessel is powered by two Volvo Penta 163P diesels each rated at 500 kW at 2,100 rpm and each driving a four-blade propeller via a ZF IRM 35A reverse reduction gearbox.

22 m CUSTOM BOATS

Four 22.4 m fast customs boats, built to a Stolkraft design, were delivered to the Vietnamese Customs Department in 1998.

Specifications

Length overall	22.4 m
Length waterline	19.4 m
Beam	7.5 m
Draught	1.2 m
Crew	11
Fuel capacity	11,000 litres
Water capacity	500 litres
Operational speed	27 kt
Range	1,500 n miles

Propulsion: Two MTU 12V 183 TE93 diesel engines each rated at 846 kW and each driving a Doen DJ220 water-jet.

The first 25 m ferry Kurilpa 0507106

22 m PASSENGER FERRY

The company commenced construction of a 22 m passenger ferry in 1999. This vessel is designed to operate in coastal and sheltered waters exposed to significant wave heights of 1.2 m or less.

Specifications

Length overall	22.9 m
Length waterline	22.0 m
Beam	8.0 m
Draught	1.6 m
Crew	3
Passengers	172
Fuel capacity	2,300 litres
Water capacity	300 litres

Propulsion: The vessel is powered by two Volvo TAMD163P engines each rated at 510 kW and each driving an 813 mm diameter fixed-pitch propeller via a Twin Disc 5114 gearbox.

Vessel type	Vessel name	Yard No	Length (m)	Speed (kt)	Seats	Vehicles	Originally delivered to	Date of build
25 m Low Wash Ferry	*Tugulawa*	–	25.00	24.0	108	–	Brisbane City Council	1996
25 m Low Wash Ferry	*Bianjin*	–	25.00	24.0	108	–	Brisbane City Council	1996
25 m Low Wash Ferry	*Barambin*	–	25.00	24.0	108	–	Brisbane City Council	1996
25 m Low Wash Ferry	*Migbarpa*	–	25.00	24.0	108	–	Brisbane City Council	1997
25 m Low Wash Ferry	*Mianjin*	–	25.00	24.0	108	–	Brisbane City Council	1997
25 m Low Wash Ferry	*Kurilpa*	–	25.00	24.0	108	–	Brisbane City Council	1997
22 m Custom Boat	*HQ56*	–	22.40	27.0	–	–	Vietnamese Customs Department	1998
22 m Custom Boat	*HQ57*	–	22.40	27.0	–	–	Vietnamese Customs Department	1998
22 m Custom Boat	*HQ58*	–	22.40	27.0	–	–	Vietnamese Customs Department	1998
22 m Custom Boat	*HQ59*	–	22.40	27.0	–	–	Vietnamese Customs Department	1998
25 m Low Wash Ferry	*Beenung–urrung*	–	25.00	24.0	108	–	Brisbane City Council	1999
25 m Low Wash Ferry	*Tunamun*	–	25.00	24.0	108	–	Brisbane City Council	1999
26 m Low Wash Ferry	*MV Suzanne*	–	26.40	25.0	150	–	Fantasea Cruises	June 1999
25 m Passenger Ferry	*Yasawa Flyer* (ex-*Queenslander I*)	–	22.90	26.0	188	–	Fitzroy Island Ferries	June 2000
32 m Crewboat	*Jascon 14*	–	32.10	29.0	132	–	Seatrucks Group	2004
32 m Crewboat	*Jascon 15*	–	32.10	29.0	132	–	Seatrucks Group	2004
32 m Crewboat	*Jascon 16*	–	32.10	29.0	132	–	Seatrucks Group	2004
18 m Catamaran	*Swordfish*	–	17.60	25.0	123	–	Cruise Whitsundays	2006
18 m Catamaran	*Bermuda*	–	17.60	25.0	123	–	Cruise Whitsundays	2006
27 m Passenger Ferry	*Yasawa Flyer II*	–	27.45	26.0	267	–	South Sea Cruises	2006
30 m Catamaran	*Maggie Cat*	–	29.50	–	239	–	Sunferries	2006
30 m Catamaran	*Sun Cat*	–	29.50	–	239	–	Sunferries	2007
38 m Low Wash Ferry	–	–	38.04	27.0	196	–	Thames Clippers	(2007)
38 m Low Wash Ferry	–	–	38.04	27.0	196	–	Thames Clippers	(2007)
38 m Low Wash Ferry	–	–	38.04	27.0	196	–	Thames Clippers	(2007)
38 m Low Wash Ferry	–	–	38.04	27.0	196	–	Thames Clippers	(2007)
38 m Low Wash Ferry	–	–	38.04	27.0	196	–	Thames Clippers	(2007)
38 m Low Wash Ferry	–	–	38.04	27.0	196	–	Thames Clippers	(2007)

Cougar Catamarans Australia Pty Ltd

PO Box 5344, Gold Coast Mail Centre, Bundall 9726, Gold Coast City, Australia

Tel: (+61 7) 55 39 22 44
Fax: (+61 7) 55 38 88 33

e-mail: cougar@cougarcatamarans.com
Web: www.cougarcatamarans.com

Harry Roberts, *Managing Director*

Cougar Catamarans has been in operation for over 30 years, building glass fibre catamarans from 8 to 36 m and aluminium catamarans from 18 to 40 m. It is the largest producer of GRP catamarans in the Asia Pacific region. Vessel applications have included fast ferries, patrol, charter, police and fishing.

San Bei

A 27.2 m catamaran ferry delivered in 1994 to Shanghai, China.

Vessel type	Vessel name	Yard No	Length (m)	Speed (kt)	Seats	Vehicles	Originally delivery to	Date of build
25 m Catamaran	*Ying Bin 5*	–	25.0	25.0	200	–	Hong Kong	1993
32 m Catamaran	*Yun Tong*	–	32.0	28.0	266	–	Shanghai, China	1994
27.2 m Catamaran	*San Bei*	–	27.2	27.0	220	–	Shanghai, China	1994
55 Passenger Ferry	*Wildcat*	–	14.6	45.0	55	–	–	1998
CAT Crowther 24 m	*Te Hukatai*	–	24.0	30.0	150	–	SeaCat, New Zealand	1998
50 Passenger Ferry	*Maloo*	–	16.6	29.0	50	–	–	1998
CAT Crowther 24 m	*Cat Cocos*	–	25.0	32.0	159	–	Interisland Boats, Seychelles	1998

Specifications

Length overall	27.2 m
Passengers	220
Crew	5
Maximum speed	31 kt
Operational speed	27 kt

Propulsion: Engines: two 820 kW Detroit Diesel 16V 9TA.

Auxiliary systems: two 50 kVA Perkins generators.

Yun Tong

A 32 m catamaran ferry delivered in 1994 to Shanghai, China.

Specifications

Length overall	32.0 m
Beam	8.4 m
Passengers	266
Crew	5
Propulsive power	2 × 1,342 kW
Maximum speed	32 kt
Operational speed	28 kt

Propulsion: Engines: two 1,342 kW Detroit Diesel 16V 149TI.

Auxiliary systems: two 67 kVA Perkins generators.

Ying Bin 5

A 25 m catamaran ferry delivered in 1993 to Hong Kong.

Specifications

Length overall	25.0 m
Beam	8.4 m
Passengers	200
Crew	5
Maximum speed	28 kt
Operational speed	25 kt

Propulsion: Engines: two 820 kW Detroit Diesel 16V 92TA.

Auxiliary systems: Engines: two 37 kVA Isuzu generators.

Te Hukatai

This 150 passenger ferry was delivered in February 1998.

Specifications

Length overall	24 m
Crew	5
Passengers	150
Crew	5
Propulsive power	1,575 kW
Operational speed	30 kt

Classification: Bureau Veritas coastal 150+5.

Propulsion: Engines: two MTU 2000V12 M70 diesels each rated at 787 kW.

Cat Cocos

This 160 passenger ferry was delivered in September 1998.

Specifications

Length overall	25 m
Passengers	159
Crew	5
Propulsive power	1,600 kW
Operational speed	32 kt

Classification: Bureau Veritas

Propulsion: Engines: two MTU 2000V12 M70 diesels each rated at 800 kW.

Maloo

This 50 passenger ferry was delivered in August 1998.

Specifications

Length overall	16.6 m
Passengers	50
Crew	3
Propulsive power	720 kW
Operational speed	29 kt

Classification: Bureau Veritas

Propulsion: Engines: two Caterpillar 3406C diesel engines each rated at 360 kW.

Wildcat

This 55 seat passenger ferry was delivered in February 1998.

Specifications

Length overall	14.6 m
Operational speed	45 kt

Cat Cocos 0038849

Te Hukatai 0038848

Maloo 0038850

Yun Tong 0506886

Image Marine Pty Ltd

Trading as Austal
100 Clarence Beach, Henderson, Perth, Western Australia 6166, Australia

Tel: (+61 8) 94 10 11 11
Fax: (+61 8) 94 10 25 64
e-mail: marketing@austal.com
Web: www.austal.com

John Rothwell, *Chairman*
Bob McKinnon, *Managing Director*
Michael Atkinson, *Director, Financial Control*

Image Marine was established in 1986 and is a builder of customised aluminium vessels of both catamaran and monohull design. Image has built in excess of 165 vessels for the domestic and export market including Australasia, the Asia-Pacific, East Africa and the Caribbean.

While vessels for the niche live-aboard dive boat and fast ferry sectors form a major component of Image Marine's portfolio of expertise, the product range also includes cruise and dinner-cruise vessels, crewboats, pilot and patrol vessels and private motor yachts, all in both catamaran and monohull design.

Image Marine was acquired by the Austal Group in July 1998 and is dedicated to the construction of aluminum vessels of up to 50 m in length.

Evercrest 0109642

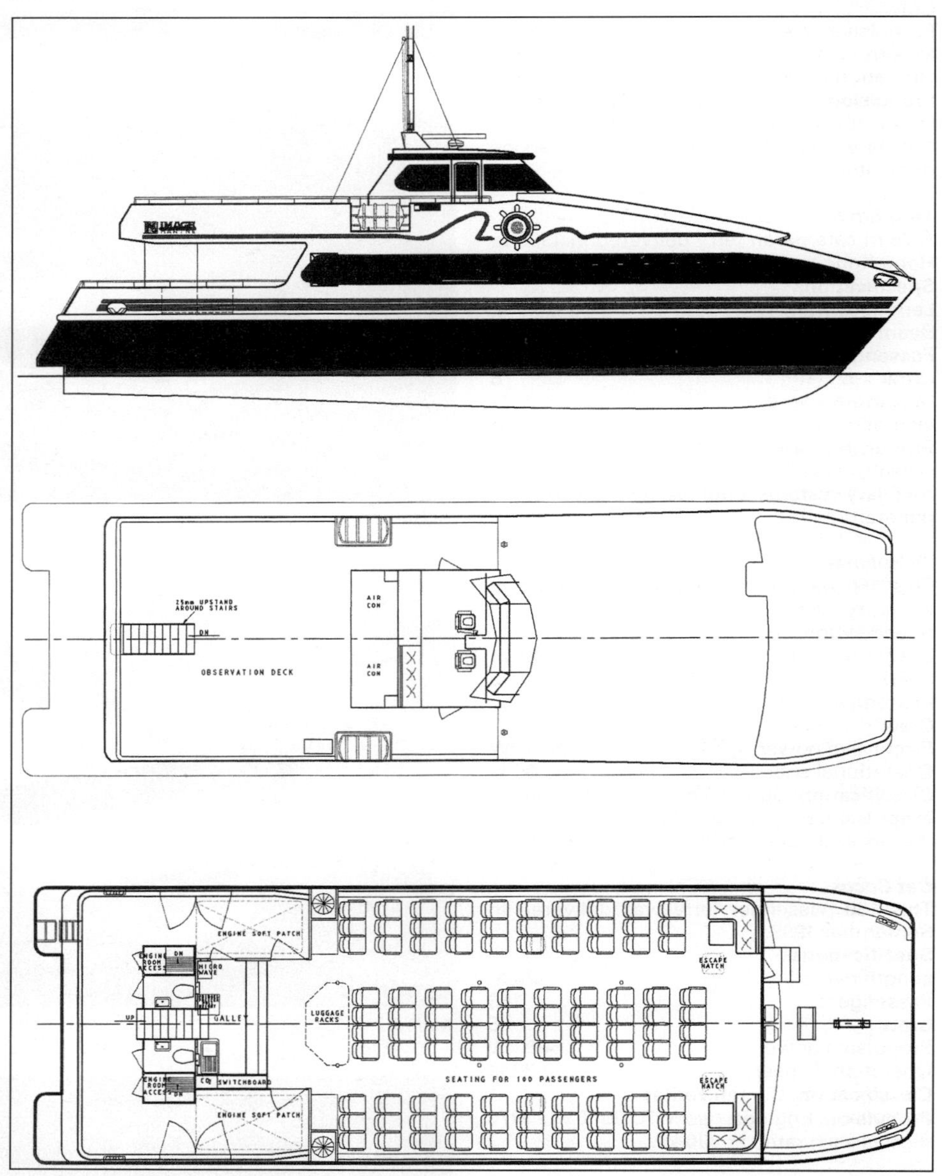

General arrangement of Evercrest 0109643

41.5 m CATAMARAN

Salten
Steigtind

Two of these craft were ordered by OVDS of Norway in 2002 and delivered in 2003.

Specifications

Length overall	41.5 m
Length waterline	36.3 m
Beam	11.6 m
Draught	1.5 m
Passengers	214
Crew	6
Operational speed	33 kt
Range	360 nm

Propulsion: Two MTU 16V 4000 M70 diesel engines each driving a Kamewa 71 SII water-jet via a Reintjes or ZF gearbox.

41.4 m CATAMARAN

First Travel XXXI
First Travel XXXII

Delivered in October 2003 to Hong Kong's New World First Travel Services Ltd, *First Travel XXXI* and *XXXII* are similar to the 41.5 vessels delivered to Norway but will operate at only 16 knots on their twin 788 kW MTU engines.

Specifications

Length overall	41.4 m
Length waterline	39.5 m
Beam	11.8 m
Draught	1.8 m
Passengers	254
Crew	5
Operational speed	16 kt

Propulsion: Two MTU 12V 2000 M70 diesel engines each rated at 788 kW at 2,100 rpm.

37 m RIVER FERRY

Cesario Verde
Pedro Nunes

Two of these craft were ordered by Transtejo of Portugal in 2001 and delivered in 2002. The vessel order was taken by Austal, the parent company of Image Marine.

Specifications

Length overall	37.4 m
Length waterline	34.2 m
Beam	10.7 m
Draught	1.2 m
Passengers	292
Operational speed	27 kt
Maximum speed	29 kt

Classification: Germanischer Lloyd.

Marineview *constructed by Image Marine under contract to Austal* 0007923

Propulsion: Two MTU 16V 2000 M70 diesel engines each rated at 1,050 kW at 2,100 rpm and driving a Hamilton HM 651 water-jet via a Reintjes WVS 430/1 gearbox.

30 m CATAMARAN

Marineview

Marineview was built for Kumamoto Ferry of Japan and entered service in July 1997. It operates five return trips daily between Kumamoto City and Hondo City in Amakusa.

Specifications

Length overall	30.0 m
Length waterline	26.2 m
Beam	8.70 m
Draught	1.1 m
Crew	8
Passengers	140
Fuel capacity	6,000 litres
Water capacity	400 litres
Operational speed	33 kt

Propulsion: Two MTU 12V 396 TE 74L diesel engines each rated at 1,499 kW at 1,915 rpm driving Kamewa 56 S11 water-jets via ZFBW465 reduction gearboxes.

24 m SEABUS CATAMARAN

Seabus I and II

These two passenger catamarans, designed by Crowther Multihulls, were delivered to a Tanzanian operator in 1997 for operation along the east Africa coast.

Specifications

Length overall	23.9 m
Length waterline	21.9 m
Beam	8.0 m
Draught	1.4 m
Passengers	161
Fuel capacity	4,000 litres
Water capacity	1,000 litres
Operational speed	28 kt

Classification: DnV +1A1 HSLC Passenger A3

Structure: All aluminium hull and superstructure.

Propulsion: Powered by two V12 MAN D2842 LE408 diesel engines each rated at 755 kW at 2,100 rpm and each driving a Teignbridge five-bladed propeller via a ZF BW 190 reverse reduction gearbox.

Electrical System: Two Perkins 62 kVA.

24 m FERRY

Evercrest

The 24 m passenger catamaran, *Evercrest*, was delivered in August 1998 to Gotesco Properties of the Philippines.

Specifications

Length overall	24.0 m
Length waterline	21.8 m
Beam	8.0 m
Draught	1.8 m
Passengers	100
Fuel capacity	4,200 litres
Operational speed	31.0 kt

Classification: Lloyd's Register.

Structure: Aluminium hull and superstructure.

Propulsion: Two MTU 12V 2000 M70 diesels each rated at 873 kW and each driving a Teinbridge propeller through a ZF BW 255 gearbox.

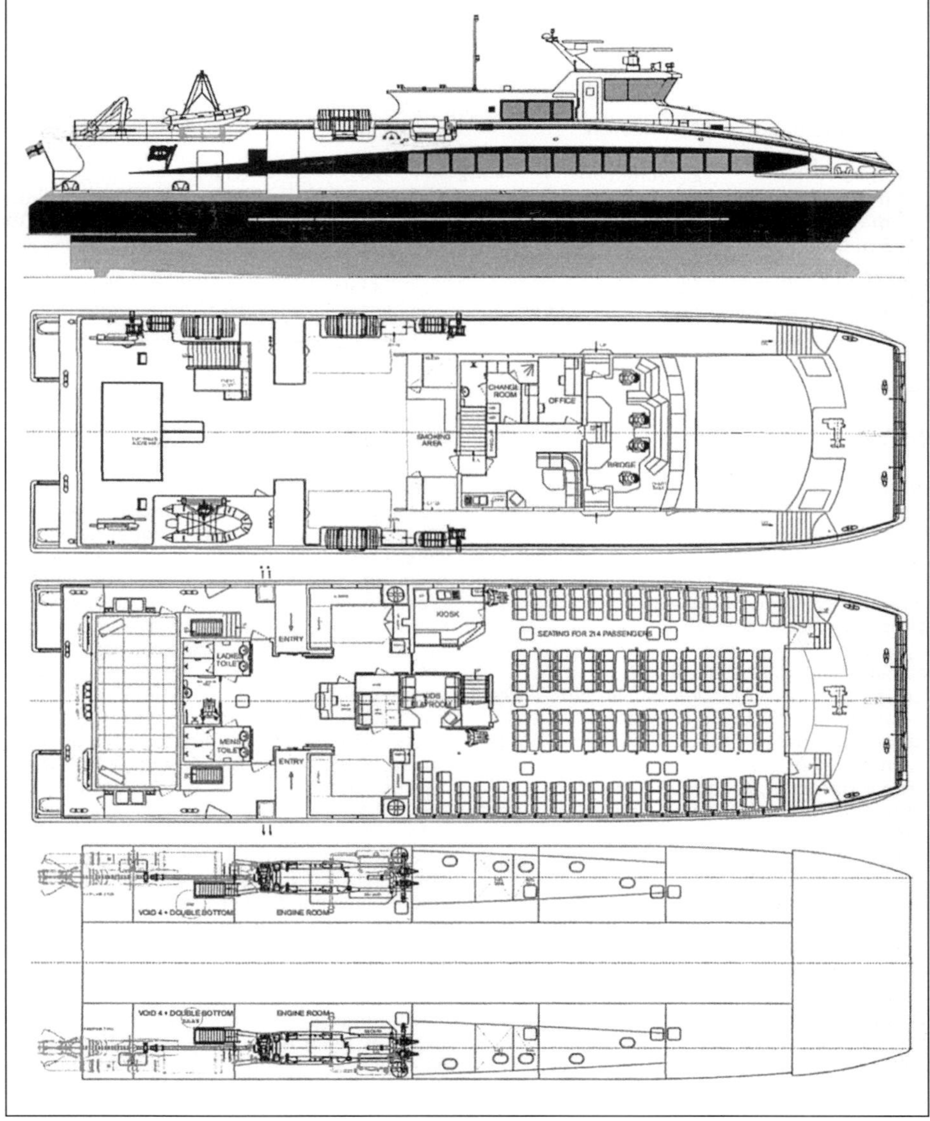

General arrangement of Image 41.5 m catamaran (Austal) 0554533

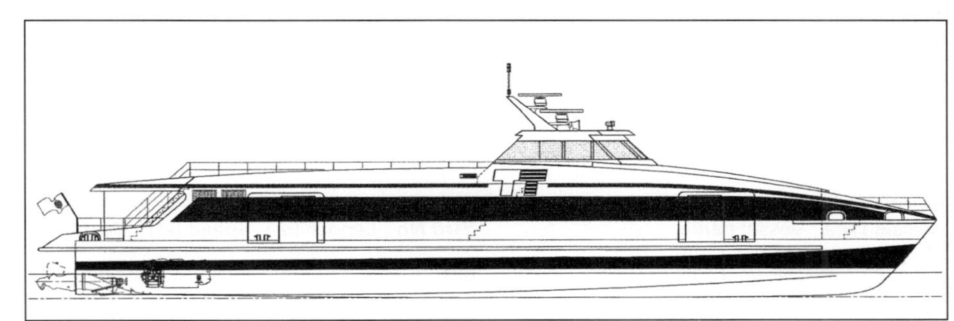

Profile view of the 37 m River Ferry for Transtejo 0121747

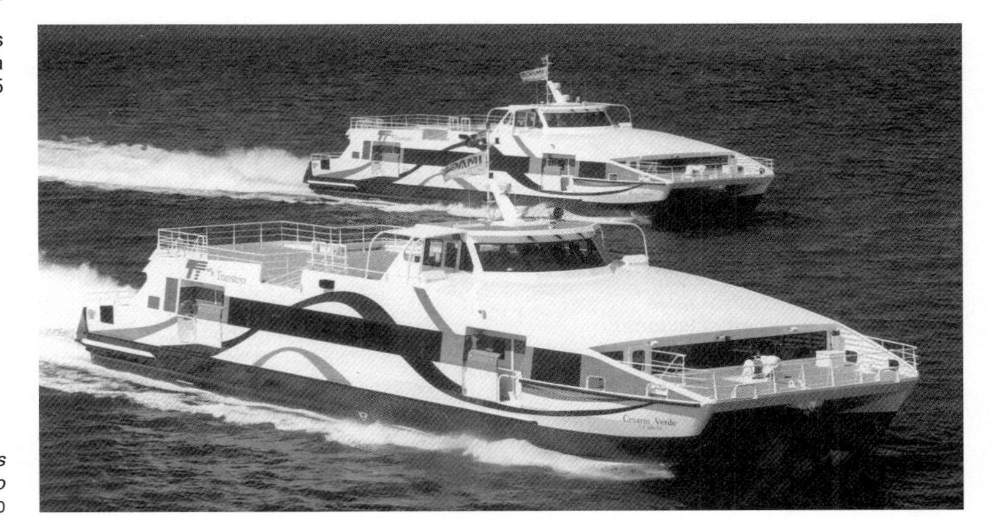

37 m River Ferries built for Transtejo 0594980

Vessel type	Vessel name	Yard No	Length (m)	Speed (kt)	Seats	Vehicles	Originally delivered to	Date of build
24 m Seabus Catamaran	*Seabus I*	139	23.9	28.0	161	none	Tanzania	1997
24 m Seabus Catamaran	*Seabus II*	138	23.9	28.0	161	none	Tanzania	1997
30 m Cartamaran	*Marineview*	–	30.0	33.0	140	none	Kumamoto Ferry, Japan	1997
24 m Ferry	*Evercrest*	142	24.0	31.0	100	none	Gotesco Properties	1998
37 m River Ferry	*Pedro Nunes*	275	37.2	27.0	292	none	Transtejo	2002
37 m River Ferry	*Cesario Verde*	247	37.4	27.0	292	none	Transtejo	2002
41.5 m Catamaran	*Salten*	242	41.5	33.0	214	12 tonnes	OVDS, Norway	2003
41.5 m Catamaran	*Steigtind*	243	41.5	33.0	214	12 tonnes	OVDS, Norway	2003
41.4 m Catamaran	*First Travel XXXI*	–	41.4	16	354	–	First Travel Services Ltd	2003
41.4 m Catamaran	*First Travel XXXII*	–	41.4	16	354	–	First Travel Services Ltd	2003

InCat

18 Bender Drive, Hobart, Tasmania 7009, Australia

Tel: (+61 3) 62 73 06 77
Fax: (+61 3) 62 73 09 32
e-mail: incat@incat.com.au
Web: www.incat.com.au

Robert Clifford, *Chairman*
Craig Clifford, *Managing Director, Incat Chartering Pty Ltd*
Leith Thompson, *Managing Director, Incat Finance Pty Ltd*
Trevor Hardstaff, *Production Director*
Kim Clifford, *Managing Director, Incat Marketing Pty Ltd*
John Harris, *Managing Director, Incat Tasmania Pty Ltd*
Robin Allardice, *Group Company Secretary*

European office
Incat Europe Pty Ltd
Dr Tvaergade 8B, DK-1302 Copenhagen K, Denmark

Tel: (+45) 33 14 50 75
Fax: (+45) 33 14 50 79

Gitte Helgren, *Office Manager*

The company was formed in 1972 and commenced operating passenger ferries across the Derwent River in Hobart, Tasmania. Soon after, the building of ferries became the company's main business.

The wave-piercing concept was conceived in 1983 and an 8.7 m test craft, *Little Devil*, was built and tested. A number of wave-piercing and conventional catamarans followed and in 1990 the first 74 m wave-piercing catamaran passenger car ferry *Hoverspeed Great Britain* was delivered. This craft was the first InCat vessel to hold the Hales Trophy for the fastest crossing of the Atlantic by a passenger vessel, and achieved an average speed of 36.65 kt on its delivery voyage to the UK in June 1990. Two subsequent InCat vessels, both 91 m wave-piercing designs, have since broken the record (*Catalonia*, 3 days 4 hours and 32 minutes at an average speed of 38.877 kt and *Cat-Link V* in 3 days, 2 hours and 20 minutes at an average speed of 41.284 kt).

In 1994, fast ferry demand led to the development of the K Class design; a ferry suited to very high-speed operation on relatively sheltered routes. Two K Class vessels, both of which achieved over 50 kt during trials, were delivered from Hobart in 1995 and continue to be built under licence by Afai Shipyard in China.

Continuing development and demand has now seen nine 74 m, three 78 m, three 81 m, four 86 m, four 91 m, six 96 m and two 98 m wave-piercing catamarans built up to 2001. The shipyard currently has the proven capacity to build five of these large craft per year and is expanding the facilities to build yet larger vessels.

In 2001 the company formed an alliance with Bollinger Shipyard in the US primarily to market Incat craft to the US Navy. In 2002 this new alliance had received a charter agreement for a 96 m WPC (hull 050) with the US military. The vessel, *Joint Venture HSV-X1*, has since been deployed in the Persian Gulf. A second vessel (hull 060) was leased to the US Army Tank-Automotive and Armaments Command (TACOM) as a Theatre Support Vessel (TSV) in 2002. This craft, *TSV-1X Spearhead*, was leased to the US Military Sealift Command to support the US Navy Mine Warfare Commando in 2003.

112 m EVOLUTION ONE12 SEAFRAME

The first Evolution One12, *Natchan Rera*, ordered by Higashinihon Ferry, was delivered in in 2007. Despite having space for more than 1,000 passengers, Higashinihon Ferry have chosen to cater for 800 passengers in more luxurious comfort. The vessel has a capacity for 355 cars or 450 truck lane metres and 193 cars. Constructed as a base vessel or Seaframe, the Evolution One12 craft can be fitted out for numerous purposes including passenger/commercial and military deployment and the vessel is reputed to be the most fuel-efficient diesel-powered high speed catamaran to date.

Specifications

Length overall	112.6 m
Length waterline	105.6 m
Beam	30.2 m
Draught	3.93 m
Deadweight	1,450 t
Passengers	800
Vehicles	355 cars or 589 truck lane m
Maximum speed	45 kt
Operational speed	40 kt

Classification: DnV +1A1 HSLC R1 Car Ferry 'B' EO.
Structure: All aluminium hull and superstructure.

Vessel type	Vessel name	Yard No	Length (m)	Speed (kt)	Seats	Vehicles	Originally delivered to	Date of build
18 m	*Derwent Explorer* (ex-*Jeremiah Ryan*)	001	–	26.0	145	None	–	1977
18 m	*Tropic Princess* (ex-*James Kelly I*)	002	–	28.0	100	None	Ecrolight	1979
20 m	*Fitzroy Flyer* (ex-*Fitzroy*)	004	–	28.0	–	None	Great Adventures	1981
20 m	*Tangalooma*	005	–	28.0	200	None	Tangalooma Island Resort	1981
20 m	*Islander* (ex-*Green Islander*)	007	–	28.0	220	None	–	1982
20 m	*Low Isles Reef Express* (ex-*Quicksilver*)	008	–	–	100	None	Outer Barrier Reef Cruises	1982
29 m	*Spirit of Roylen*	009	–	27.0	250	None	McLeans Roylen Cruises Pty	1982
20 m	*Magnetic Northerner* (ex-*Keppel Cat 2*, ex-*Trojan*)	010	–	–	200	None	Hydrofoil Seaflight Services Pty Ltd	1983
20 m	*Keppel Cat I* (ex-*Tassie Devil*)	011	–	–	195	None	Hydrofoil Seaflight Services Pty Ltd	1984
27.4 m	*Spirit of Paradise* (ex-*Spirit of Victoria*)	016	–	28.0	–	None	Indonesia	1985
30 m	*Our Lady Patricia*	020	–	31.0	452	None	Wightlink Ferries Ltd	1986
30 m	*Our Lady Pamela*	021	–	31.0	452	None	Wightlink Ferries Ltd	1986
31 m	*WPC 2001* (ex-*Tassie Devil 2001*)	017	–	30.0	196	None	InCat Charters	1986
22.8 m	*Spirit of the Bay* (ex-*Starship Genesis*, ex-*Genesis*)	018	–	36.0	200	None	New South Wales	1987

Vessel type	Vessel name	Yard No	Length (m)	Speed (kt)	Seats	Vehicles	Originally delivered to	Date of build
37.2 m WPC	*Tamahine Moorea II* (ex-*Seaflight*)	019	–	30.0	–	None	Seaflight Ltd, New Zealand	1988
31 m WPC	*2000*	019	–	–	231	None	Hamilton Island Cruises	1988
74 m WPC	*Hoverspeed Great Britain* (ex-*SeaCat Great Britain*, ex-*Christopher Columbus*)	025	73.6	35.0	600	90	Sea Containers, Hoverspeed Ltd	1990
74 m WPC	*SeaCat France* (ex-*Croazia Jet*, ex-*Atlantic II*, ex-*SeaCat Calais*, ex-*SeaCat Tasmania*)	023	73.6	42.0	350	84	Sea Containers, Hoverspeed Ltd	1990
74 m WPC	*SeaCat Isle of Man* (ex-*SeaCat Norge*, ex-*Hoverspeed France*)	026	73.6	35.0	483	80	Sea Containers, Hoverspeed Ltd	1991
74 m WPC	*SeaCat Danmark* (ex-*Hoverspeed Boulogne*, ex-*Hoverspeed Belgium*, ex-*Seacatamaran Danmark*)	027	73.6	35.0	420	85	Sea Containers, Hoverspeed Ltd	1991
74 m WPC	*SeaCat Scotland*	028	73.6	35.0	450	80	Sea Containers, Hoverspeed Ltd	1992
74 m WPC	*Patricia Olivia*	024	73.6	37.0	520	92	Los Cipreses SA Buquebus	1992
74 m WPC	*Condor 10* (ex-*Euroferry 26*)	030	73.6	37.0	584	84	Condor Ltd	1993
74 m WPC	*Ocean Flower* (ex-*Avant*, ex-*Stena Lynx*, ex-*Stena Sea Lynx*, ex-*Stena Sea Lynx 1*)	031	73.6	37.0	450	84	Buquebus (under charter to Stena Line)	1993
74 m WPC	*Atlantic III* (ex-*Juan L*)	032	73.6	41.0	600	110	Los Cipreses SA Buquebus	1993
78 m WPC	*Jaume I* (ex-*Thundercat 2*, ex-*Ronda Marina*, ex-*Stena Sea Lynx II*)	033	77.8	35.0	640	150	Stena Sealink Line	1994
K55 CAT	*Juan Patricio*	036	70.4	50.0	450	58	Los Cipreses SA Buquebus	1995
78 m WPC	*Elenacra* (ex-*Euroferrys 1*, ex-*Cat Link I*, *I* ex-*Condor 11*)	034	77.8	37.0	700	150	Condor Int	1995
78 m WPC	*Thundercat 1* (ex-*Cat Link I*)	035	77.8	36.0	640	150	Scandlines Cat-Link AS	1995
K50 CAT	*Sunflower*	037	79.2	50.0	750	32	Dae A Gosok Ferry Company, South Korea	1995
81 m WPC	*Rapide* (ex-*Holyman Rapide*, ex-*Condor 12*)	038	81.1	38.0	700	173	Condor Ltd	1996
81 m WPC	*Stena Sea Lynx III* (ex-*Elite*)	040	81.1	38.0	700	173	Holyman Ports	1996
81 m WPC	*Diamant* (ex-*Holyman Diamant*, ex-*Holyman Express*)	041	81.1	38.0	700	173	Holyman Ports	1996
86 m WPC	*Condor Express*	042	86.3	42.0	800	200	Condor Ltd	1996
86 m WPC	*Sicilia Jet*	043	86.3	42.0	900	200	Jetmarine Ltd	1997
86 m WPC	*Condor Vitesse*	044	86.3	42.0	800	200	Condor Ltd	1997
86 m WPC	*Speed One* (ex-*Winner*, ex-*HMAS Jervis Bay*, ex-*InCat 045*)	045	86.3	42.0	876	200	–	1997
91 m WPC	*T&T Express* (ex-*The Cat*, ex-*Devil Cat*)	046	91.3	43.0	900	240	–	1997
91 m WPC	*Max Mols* (ex-*Cat Link IV*)	048	91.3	43.0	900	240	Scandlines Cat-Link AS	1998
91 m WPC	*Portsmouth Express* (ex-*Catalonia*)	047	91.3	42.0	900	240	Buquebus	1998
91 m WPC	*Master Cat* (ex-*Mads Mols*, ex-*Cat Link V*)	049	91.3	43.0	900	240	Scandlines Cat-Link AS	1998
96 m Ro-Ro-Pax WPC	*HSV-X1 Joint Venture* (ex-*Top Cat*, ex-*Devil Cat*)	050	96.0	42.0	600	105	Fast Cat Ferries	1998
96 m Ro-Ro-Pax WPC	*Alborán* (ex-*Avemar*)	052	96.0	45.0	900	300	Buquebus Espana	1999
96 m Ro-Ro-Pax WPC	*Bonanza Express*	051	96.0	42.0	600	105	Fred Olsen Lines	1999
96 m Ro-Ro-Pax WPC	*Bentago Express* (ex-*Bentayga Express*, ex-*Benchijigua Express*)	053	96.0	42.0	900	260	Fred Olsen SA	1999
96 m Ro-Ro-Pax WPC	*Bencomo Express* (ex-*Benchijigua Express*)	055	96.0	40.0	900	260	Fred Olsen SA	2000
96 m Ro-Ro-Pax WPC	*Milenium Dos*	056	96.0	40.0	900	260	Trasmediterranea	2000
98 m Evolution 10 B WPC	*Normandie Express* (ex-*The Lynx*, ex-*InCat Tasmania*)	057	97.2	38.0	900	260	Australian Trade Commission	2000
98 m Evolution 10 B WPC	*Milenium Dos*	058	97.2	38.0	900	260	Trasmediterranea	2001
98 m Evolution 10 B WPC	*The Cat*	059	97.2	38.0	900	260	–	2002
98 m Evolution 10 B WPC	*T&T Spirit* (ex-*TSV-1X Spearhead*)	060	97.2	38.0	–	–	US Army	2002
98 m Evolution 10 B WPC	*HSV 2 Swift*	061	97.2	38.0	–	–	US Navy	2003
98 m Evolution 10 B WPC	*Milenium Tres*	062	97.2	38.0	900	280	Acciona Trasmediterranea	2006
112 m Evolution One12 WPC	*Natchan Rera*	064	112.6	45.0	800	355	Higashinihon Ferry	2007
112 m Evolution One12 WPC	–	064	112.6	45.0	1,000	312	–	(2008)

Propulsion: Four MAN 20V 28/33D diesel engines each rated at 9,000 kW at 1000 rpm.

98 m EVOLUTION 10 B WPC CATAMARAN

While the last six 96 m Ro-Pax wave-piercing catamarans were identified as Evolution 10s, this mainly referred to their car and passenger layout differences. The Evolution 10 B is a larger vessel, 6 m longer on the waterline, 1.2 m longer overall and with a 75 tonne increase in deadweight. The first craft was used by the Australian Trade Commission at the Sydney Olympic Games in September and October 2000 and was then delivered to New Zealand for operation by Tranz Rail with the name *The Lynx*. The second vessel, *Milenium Dos* (058), was launched in 2001 and the third, *The Cat* (059) in 2002. Two further vessels (060 and 061) were leased to the United States Military in 2002–03 and a similar vessel, Milenium Tres was delivered to Trasmeditteranea in 2006. These ships are capable of maintaining 35 kt or more with 500 tonnes of payload consisting of military personnel and equipment. In 2007 one of the vessels leased to the US Army, TSV-IX Spearhead, was completely refitted for commercial use and renamed *T&T Spirit*. The vessel is now owned by the Ministry of Works and Transport, Trinidad and Tobago.

Specifications

Length overall	97.2 m
Length waterline	92.0 m
Beam	26.6 m
Hull Beam	4.5 m
Draught	3.4 m
Displacement, max	750 t
Passengers	900
Vehicles	260 cars or 80 cars and 380 truck lane m
Fuel capacity	160,000 litres
Water capacity	5,000 litres
Operational speed	38 kt

Classification: DnV +1A1 HSLC R1 Car Ferry B EO.

Propulsion: Four Caterpillar 3618 or four Ruston 20 RK 270 diesels driving LIPS 150D water-jets.

86 m WPC *Sicilia Jet* 0007926

81 m WPC *Stena Sea Lynx III* 0507048

96 m RO-PAX WPC CATAMARAN

A series of six wave-piercing catamarans were constructed and designed in association with freight forwarding companies, based on the technology of the large wave-piercing catamaran ferries.

The first vessel, *Devil Cat*, was delivered in December 1998 and spent four months operating across the Bass Strait under TT Line management. This vessel, renamed *Top Cat,* then operated across the Cook Strait in New Zealand with FastCat Ferries. In 2001 this vessel was refitted, delivered for operation with the US Navy and renamed *Joint Venture HSV-X1.* The second vessel (052) was delivered to Buquebus in 1999, and the third, fourth and fifth vessels, *Bonanza Express, Bentayga Express* and *Benchijigua Express,* to Fred Olsen lines in the Canary Islands.

The freight deck has been designed to carry cars, vans, coaches and trucks, or a combination of these.

The first vessel had a passenger capacity of 600 and a deck layout for 330 truck lane m or 370 car lane m. *Bonanza Express* and subsequent craft had a capacity for 735 passengers with a similar vehicle deck layout. The sixth craft, *Millennium,* was delivered to Transmediterranea in 2000.

Specifications

Length overall	96.0 m
Length waterline	86.0 m
Beam	26.0 m
Draught	3.70 m
Hull beam	4.5 m
Deadweight	800 t

Natchan Rera on tria 1120333

Passengers 600
Vehicles 245 cars
Propulsive power 28,800 kW
Max speed 51 kt
Operational speed 35 kt

Propulsion: *Top Cat* is powered by four Caterpillar 3618 diesel engines each rated at 7,200 kW and each driving a LIPS 150D water-jet. *Bonanza Express* is powered by four Alstom Ruston 20RK270 diesel engines each rated at 7,080 kW driving LIPS 150D water-jets via a Reintjes VLJ6831 gearbox.

91 m WAVE-PIERCING CATAMARAN

The 91 m wave-piercing catamaran is a further development of the InCat 86 m design. The first of these vessels, InCat 046, entered service at the end of 1997 on Bass Strait for a Tasmanian summer charter. The second vessel, InCat 047, was contracted to the South American fast ferry operator, Buquebus, in early 1998. This vessel has an extended third tier passenger lounge for duty-free sales and is powered by four Caterpillar 3618 medium-speed diesels each rated at 7,200 kW. The third and fourth vessels were delivered in 1998 to Scandlines Cat-Link in Denmark.

Specifications

Length overall 91.3 m
Length waterline 81.7 m
Beam 26.0 m
Hull beam 4.3 m
Draught 3.7 m
Deadweight 450 t
Crew 23
Passengers 877
Vehicles 240
Fuel capacity 47,500 litres
Water capacity 5,000 litres
Propulsive power 28,320 kW
Max speed 50 kt
Operational speed 43 kt

Classification: DnV +1A1 HSLC R1 car ferry 'A' EO.

Structure: Constructed from marine grade aluminium with each hull subdivided into eight watertight compartments. An aluminium superstructure is supported on vibration-damping mounts in common with most other company designs.

Propulsion: The vessels are powered by four Ruston V20 RK270 diesels each rated at 7,080 kW or four Caterpillar diesels rated at 7,200 kW each and each driving a LIPS LJ145D water-jet via a Renk ASL60 reduction gearbox.

Ride control: The vessels are fitted with MDI active ride control systems incorporating active twin tabs and provision for bow-mounted T-foils.

86 m WAVE-PIERCING CATAMARAN

The 86 m catamaran design replaced the 74, 78 and 81 m designs. Importantly, the capacity to transport large vehicles and coaches was incorporated into the 86 m range. Four 86 m vessels were completed in the period from December 1996 to July 1997 with an evolving deadweight capacity from 340 tonnes on *Condor Express* to 380 tonnes on InCat 045. The higher deadweight vessels have a capacity for 900 passengers and 200 cars.

Specifications

Length overall 86.3 m
Length waterline 76.4 m
Beam 26.0 m
Hull beam 4.3 m
Draught 3.5 m
Deadweight 340–380 t
Passengers 800
Vehicles 200 cars
Fuel consumption 200 g/kW h
Propulsive power 28,320 kW
Max speed 48 kt
Operational speed 42 kt

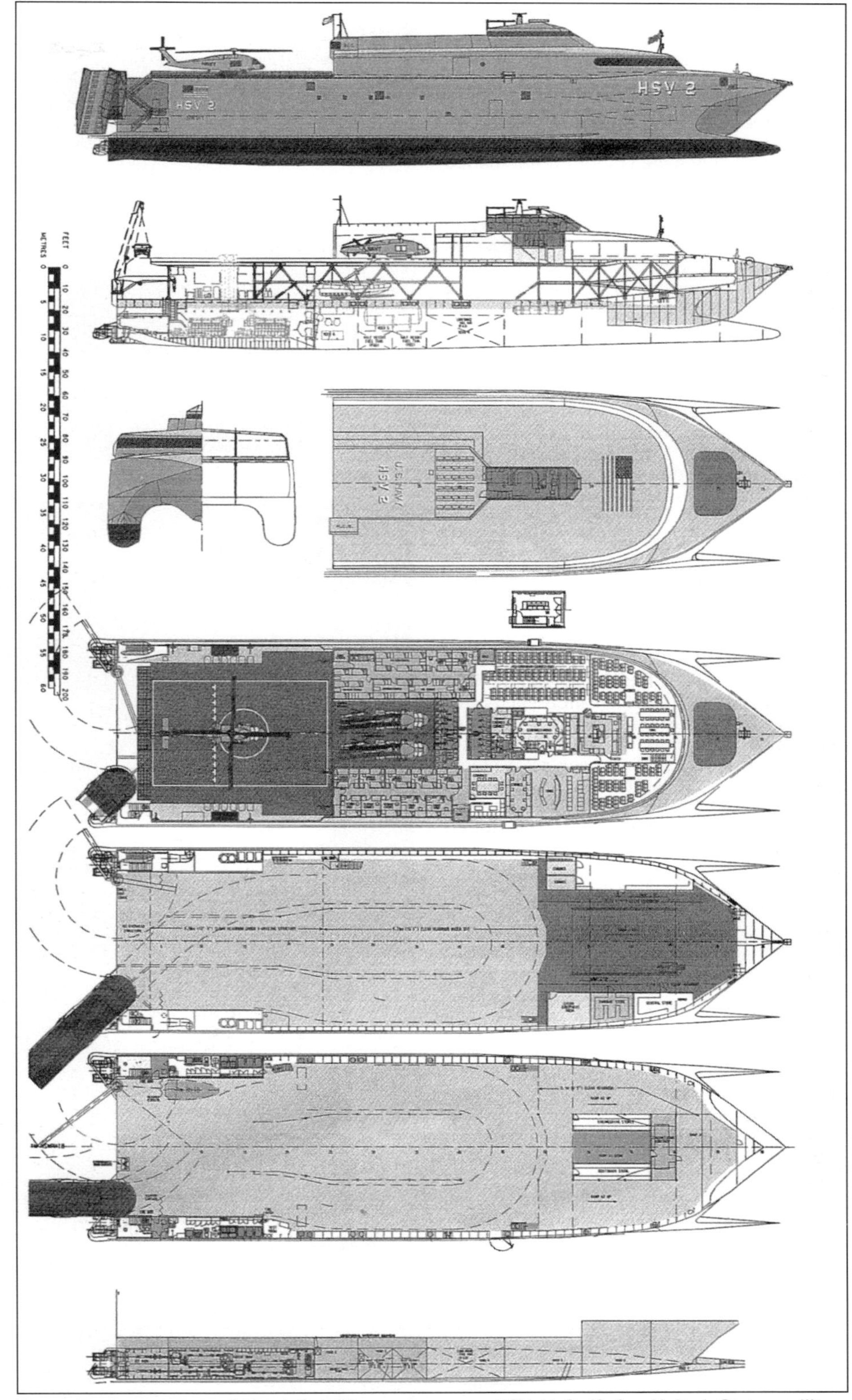

General arrangement of the InCat 98 m catamaran *HSV 2 Swift* for United States military forces 0554530

K50 *Sunflower* on trials 0506964

Classification: Built to the requirements of DnV HSLC +1A1 R1 Car Ferry 'A' EO.
Structure: Constructed from marine grade aluminium alloys. Each waterborne hull is subdivided into eight watertight compartments. These are connected by an arched bridging structure with a central forward hull above the smooth water loaded waterline. An aluminium superstructure supported on vibration-damping mounts provides seating for up to 800 passengers and crew. A full-width wheelhouse is provided with central and wing positions for docking.
Propulsion: The vessel is powered by 2 × 2 conventional medium-speed diesel engines developing 7,080 kW each. Ruston 20V RK270 engines are standard. Each engine drives a transom-mounted water-jet (LIPS IR45D is standard) through a Renk ASL 60 reduction gearbox with internal clutch. The jet control system provides, apart from steering and reversing, the option of thrust vectoring.
Ride control: Maritime Dynamics ride control system is fitted to the vessel, consisting of an active trim tab mounted at the transom of each hull. These provide trim and motion dampening. This vessel is also fitted with structural foundations and hydraulic services for the fitting of optional forward active ride control foils.

81 m WAVE-PIERCING CATAMARAN

The 81 m vessel is a further development of the successful 74 m and 78 m InCat car ferries. A considerable number of design changes have been incorporated into the vessels including the fitting of ride control systems (an active trim tab is fitted as standard and a forward foil system is optional). Three of these vessels were built at InCat's Hobart shipyard during 1996.

Specifications

Length overall	81.1 m
Length waterline	66.3 m
Beam	26.0 m
Hull beam	4.3 m
Draught, fully loaded	3.0 m
Deadweight	320 t
Passengers	700
Vehicles	173 cars
Fuel consumption	200 g/kW h
Propulsive power	22,000 kW
Max speed	44 kt
Operational speed	38 kt

Classification: Built to the requirements of DnV HSLC +1A1 R1 Car Ferry 'A'.
Structure: Constructed from marine grade aluminium alloys. Each waterborne hull is subdivided into seven watertight compartments. These are connected by an arched bridging structure with a central forward hull above the smooth water loaded waterline. An aluminium superstructure supported on vibration-damping mounts provides seating for up to 700 passengers and crew. A full-width wheelhouse is provided with central and wing positions for docking.
Propulsion: Powered by 2 × 2 conventional high-speed diesel engines developing 5,500 kW each. Ruston 16V RK270 engines are standard. Each engine drives a transom-mounted water-jet (LIPS IR135D is standard) through a Reintjes VLJ 4431 reduction gearbox with internal clutch. The jet control system provides, apart from steering and reversing, the option of thrust vectoring.
Ride control: Maritime Dynamics ride control system is fitted to the vessel, consisting of an active trim tab mounted at the transom of each hull. These provide trim and motion dampening. This vessel is also fitted with structural foundations and hydraulic services for the fitting of forward active ride control foils.

78 m WAVE-PIERCING CATAMARAN

The 78 m design evolved directly from developments to the 74 m design.

78 m WPC Condor 11  0506965

The 70 m InCat/AMD K55 on sea trials before delivery 0506979

91 m WPC Hull 046 Devil Cat *on crossing from Nova Scotia to Tasmania* 0102227

86 m WPC Hull 045 on trials 0007928

Specifications

Length overall	77.8 m
Length waterline	64.0 m
Beam	26.0 m
Hull beam	4.3 m
Draught	3.1 m
Deadweight	250 t
Passengers	600
Vehicles	150 cars
Fuel consumption at 35 kt	3,400 litres/h
Propulsive power	4 × 4,320 kW
Max speed	41 kt
Operational speed	35 kt

Classification: Built to the requirements of DnV HSLC +1A1 R1 Car Ferry 'A'.

Structure: Constructed from marine grade aluminium alloys. Each waterborne hull is subdivided into seven watertight compartments. These are connected by an arched bridging structure with a central forward hull above the smooth water loaded waterline. An aluminium superstructure supported on vibration-damping mounts provides seating for up to 600 passengers and crew. A full-width wheelhouse is provided with central and wing positions for docking.

Propulsion: Powered by 2 × 2 conventional high-speed diesel engines (Ruston 16 RK270 or Caterpillar 3616 engines are standard). Each engine drives a transom-mounted water-jet (LIPS IR11DX is standard) providing an arrangement for thrust vectoring and jet reverse and steering.

Ride control: Maritime Dynamics ride control system is fitted to the vessel, consisting of an active trim tab mounted at the transom of each hull. These provide trim and motion dampening. *Condor II* had an additional bow T foil fitted.

74 m WAVE-PIERCING CATAMARAN

Five of these vessels were originally ordered, four for Hoverspeed Ltd, a subsidiary of Sea Containers Ltd, and one for Tasmanian Ferry Services for the Bass Strait crossing. The first two were ordered by Sea Containers on 18 September 1988 and the next two on 26 January 1990. The vessels had an approximate price of AUD20 million each. Of exceptional interest was the decision to employ relatively heavy, medium-speed diesel engines in an advanced lightweight aluminium vessel structure. The aim was to exploit the low fuel consumption and long time between overhaul of these engines as well as avoiding the use of gearboxes. The nominal dry weight, with flywheel, of the 16 RK270 Ruston engine was 25.82 tonnes.

By January 1993, nine 74 m wave-piercing catamarans had been delivered. A considerable number of design changes were made in the more recent vessels including the fitting of ride control systems (an active trim tab is fitted as standard and a forward foil system is optional) and in the sixth craft the use of Caterpillar 3616 medium-speed diesels instead of the Ruston 16 RK270 diesels. Power rose from 3,650 kW in the first vessel to 4,050 kW in the last. In addition, large superstructure and accommodation changes were made in the vessels built for Buquebus, Argentina, and Condor Ltd and Stena, UK. This has allowed passenger numbers to increase to 600 and vehicle numbers to over 100.

Specifications

Length overall	73.6 m
Beam	26.0 m
Hull beam	4.4 m
Draught, fully loaded	3.0 m
Displacement, max	650 t
Deadweight	171 t

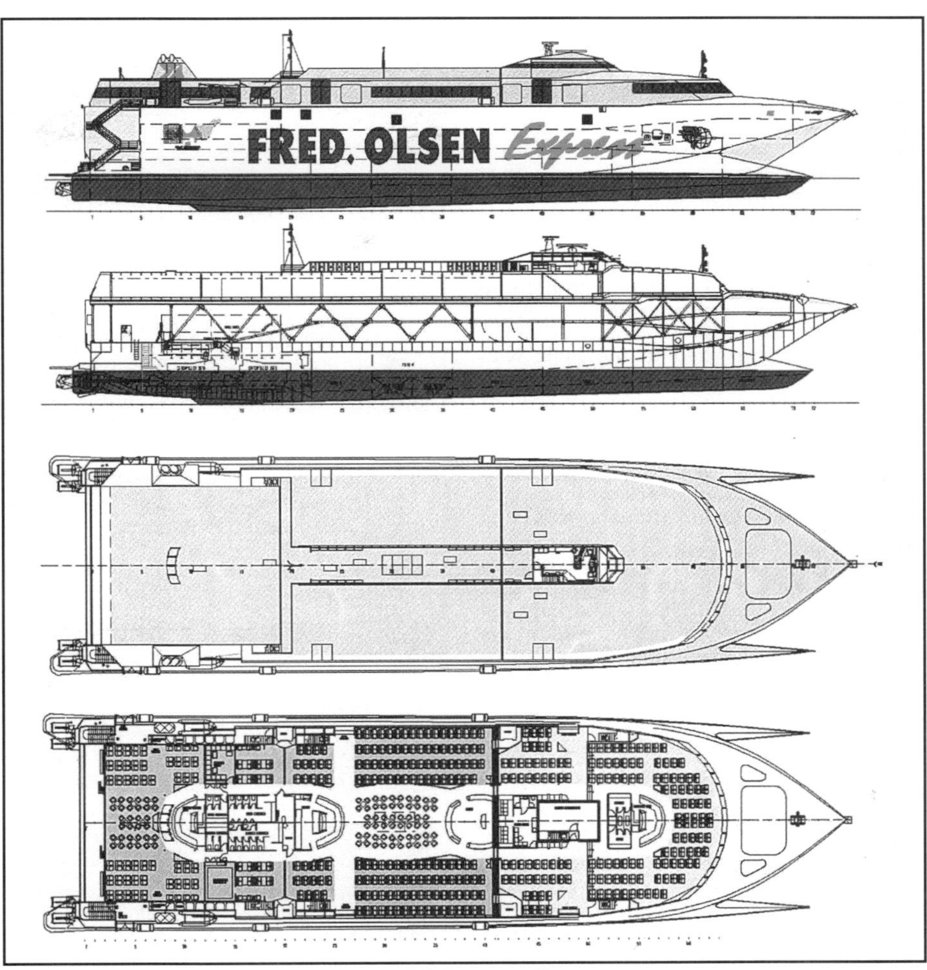

General arrangements of the InCat 96 m WPC, Bentayga Express 0069442

Incat 96 m Bonanza Express *leased to Fred Olsen Lines* 0063220

98 m The Lynx *(ex-* InCat Tasmania*) operating in the Tranz Rail livery in New Zealand* 0580418

Passengers	383
Vehicles	80 cars
Fuel capacity	20,000 litres
Water capacity	3,000 litres
Fuel consumption	200 g/kW h
Propulsive power	4 × 4,050 kW
Operational speed	35 kt

Classification: Built to the requirements of DnV +1A1 LCC + MV R280 passenger ship, EO, Car Ferry 'A'.

Structure: All-welded construction, most plating thickness 6 to 20 mm and up to 50 mm in the transom/water-jet areas. Superstructure (65 tonnes) is built as a separate unit and fitted to the hulls with resilient mountings.

Propulsion: Four Ruston 16 RK270 or Caterpillar 3610 medium-speed diesels, each 4,050 kW at 720 rpm. These directly drive four Riva Calzoni (now LIPS Jet) IRC 115 DX water-jet units, with only one on each side being equipped with steering and reversing systems.

Controls: The manoeuvring of the ship is controlled by a LIPS Ancos 2000/joystick system called LIPS-STICK. The control system is split up into individual controls for each water-jet (two steerable and two boosters) through which rpm, angle of thrust and reversing of thrust are controlled. The LIPS-STICK system combines all individual controls in one single lever.

K50 CLASS CATAMARAN

Sunflower

Afai 08

In August 1994, InCat Australia received an order from the Dae A Gosok Ferry Company of South Korea for a 79.25 m conventional catamaran. The *Sunflower* was designed and built for operation on a 117 n mile route between Pohang on the Korean mainland and Ullung Island in the Sea of Japan. The vessel transports tourists to a resort on this extinct volcanic island. Construction of the vessel commenced in late November 1994 and the vessel was delivered in late July 1995. The vessel has a capacity for 750 passengers and 32+ cars and has a maximum speed of 53 kt. At the time of its sea trials, *Sunflower* was the fastest large ferry in the world. A second vessel was built in a joint venture with Afai Ships of Hong Kong and delivered to European waters in 1998.

Specifications

Length overall	79.2 m
Length waterline	72.3 m
Beam	19.5 m
Hull beam	5.0 m
Draught	2.2 m
Deadweight	174 t
Crew	20
Passengers	815
Vehicles	32 cars
Fuel consumption	200 g/kW h
Propulsive power	4 × 5,420 kW
Max speed	53 kt
Operational speed	50 kt

Classification: The vessel is built to the requirements of DnV HSLC Car Ferry 'A' and the requirements of the Korean Registry.

Structure: Constructed from marine grade aluminium alloys. Each waterborne hull is subdivided into eight watertight compartments. These are connected by an arched bridging structure above the loaded waterline. An aluminium superstructure supported on vibration-damping mounts provides seating for up to 750 passengers and crew. A full-width wheelhouse is provided with central and wing positions for docking.

Propulsion: The vessel is powered by 2 × 2 Caterpillar 3616 conventional high-speed diesel engines (Ruston 16 RK270 are optional). Each engine directly drives a transom-mounted water-jet (KaMeWa 80 water-jets are standard) providing an arrangement for thrust vectoring and jet reverse and steering.

96 m Milenium (Richard Bennett) 0103287

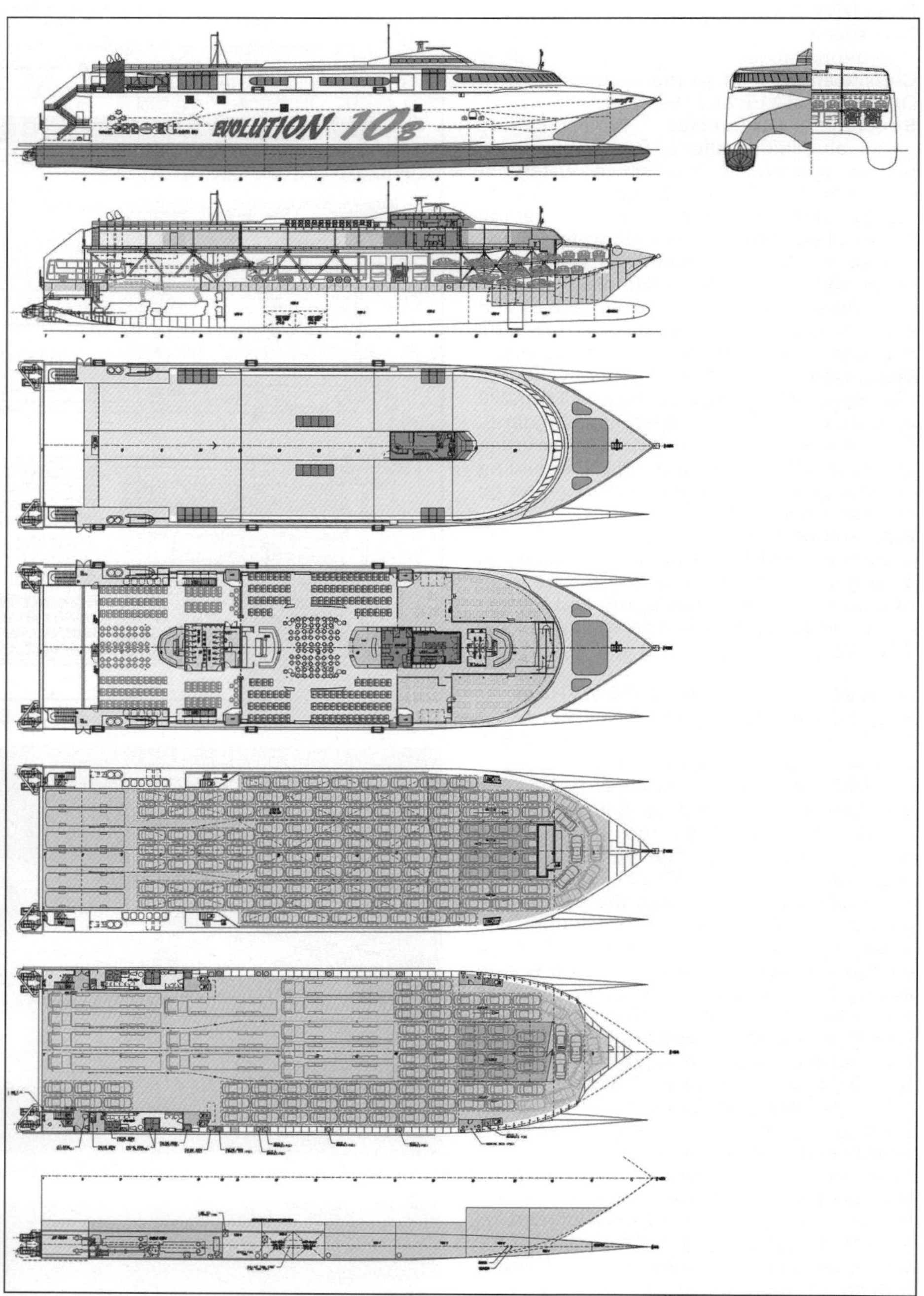

General arrangement of 98 m The Lynx 0097447

96 m HSV-X1 (ex- Top Cat*), (ex-* Devil Cat*) in operation with the US military forces* 0101973

Ride control: A Maritime Dynamics ride control system is fitted to the vessel, consisting of an active trim tab mounted at the transom of each hull. These provide trim and motion dampening.

K55 CLASS CATAMARAN

Juan Patricio

The *Juan Patricio* was the first K Class conventional catamaran built. It was designed for operation across the Rio de la Plata between Uruguay and Argentina. The vessel has a capacity for 436 passengers and 60+ cars or a mix of cars with light trucks on the aft open deck.

Specifications

Length overall	70.4 m
Length waterline	63.9 m
Beam	19.5 m
Hull beam	5.0 m
Draught	2.2 m
Deadweight	125 t
Passengers	436
Vehicles	60+ cars
Fuel consumption	200 g/kW h
Propulsive power	4 × 5,420 kW
Max speed	50 kt

Classification: The vessel is built to the requirements of DnV HSLC +1A1 R4 Car Ferry 'B' EO.

Structure: Constructed from marine grade aluminium alloys. Each waterborne hull is subdivided into seven watertight compartments. These are connected by an arched bridging structure above the loaded waterline. An aluminium superstructure supported on vibration-damping mounts provides seating for up to 436 passengers and crew. A full-width wheelhouse is provided with central and wing positions for docking.

Propulsion: The vessel is powered by 2 × 2 Caterpillar 3616 medium-speed diesel engines (Ruston 16 RK270 are optional). Each engine directly drives a transom-mounted water-jet (Kamewa 80 water-jets are standard) providing an arrangement for thrust vectoring and jet reverse and steering.

Ride control: A Maritime Dynamics ride control system is fitted to the vessel, consisting of an active trim tab mounted at the transom of each hull. These provide trim and motion dampening.

Incat 98 m Milenium Dos 0552701

InCat 98 in Evolution 10B fast wave-piercing catamaran HSV 2 Swift 0569820

112 m WPC Natchan Rera 1120333

K40 CLASS CATAMARAN (DESIGN)

This design, not yet built, is a 58 m conventional catamaran which has a capacity for 350 passengers and 45 cars. The K40 concept has been developed to provide or improve service on shorter routes, (most probably 10 to 30 n miles) where there is a volume of light vehicle and passenger traffic. During the design stage, particular attention was given to fast turnaround. In conjunction with a suitable berth, disembarkation and embarkation of both vehicles and passengers can take as little as 5 minutes.

Specifications

Length overall	58.0 m
Length waterline	52.0 m
Beam	14.5 m
Draught	2.0 m
Deadweight	100 t
Passengers	350
Vehicles	45 cars
Fuel consumption	200 g/kW h
Propulsive power	2 × 4,320 kW
Max speed	39 kt
Operational speed	34 kt

Classification: Built according to DnV HSLC +1A1 R5 Car Ferry 'B' EO. The vessel will be built to comply with the IMO High-Speed Craft Code.

Propulsion: Propulsion machinery will consist of two turbocharged marine diesel engines (one in each hull) each directly driving a water-jet through a flexible coupling. Each water-jet shall be fitted with a thrust vectoring arrangement for steering and reversing. The main engines are Ruston 16 RK270 or Caterpillar 3616, operating at 4,320 kW at 782 rpm as standard.

Lightning Boats

873 Kingsford Smith Drive, Eagle Farm, Queensland 4009, Australia

Tel: (+61 7) 32 68 59 06
Fax: (+61 7) 38 68 16 26

Lightning Boats, in association with Eagle Farm Marine Fabrication, has built a number of small fast ferries.

20 m FAST CATAMARAN

Spirit of 1770 was designed by Crowther Design and delivered by Lightning Boats in 2003 to Captain Cook Great Barrier reef Cruises.

Specifications

Length overall	20.0 m
Length waterline	18.0 m
Beam	8.0 m
Draught	1.4 m
Deadweight	17 t
Crew	10
Passengers	134
Fuel capacity	3,000 litres
Water capacity	200 litres
Operational speed	26 kt

Propulsion: Two Caterpillar 3406E diesel engines each rated at 447 kW and driving a fixed-pitch propeller via a Twin Disc MG 5114 gearbox.

19 m FAST CATAMARAN

Golden Spirit was delivered in 2001 for operation from Palm Beach to Ettalong on the Central Australian East Coast.

Specifications

Length overall	18.4 m
Length waterline	17.28 m
Beam	5.0 m
Draught	1.2 m
Displacement, min	31.7 t
Crew	3
Passengers	160
Fuel capacity	2,800 litres
Water capacity	1,000 litres
Max speed	22.5 kt
Operational speed	17.5 kt

Structure: All aluminium hull and superstructure.

Propulsion: Two Cummins 6 CTA8 3-M diesel engines each rated at 224 kW at 2,500 rpm, driving a fixed-pitch propeller via a ZF 280A gearbox.

New Wave Catamarans Pty Ltd

152 Riverside Place, Morningside 4170, Australia

Tel: (+61 7) 38 99 14 00
Fax: (+61 7) 38 99 54 00
e-mail: admin@newwavecats.com
Web: www.newwavecats.com

Formed in 2002, New Wave Catamarans provides design, build and repair services for aluminium and steel vessels. In 2003 the company launched its first fast ferry, the 25 m *Aqua Spirit*.

25 m CATAMARAN

Aqua Spirit
Crystal Spirit

Designed by CMCS of Australia, the first vessel was delivered in 2002 and the second in 2003, both operated by Palm Beach Ferry Services.

Specifications

Length overall	24.9 m
Length waterline	23.8 m
Beam	7.8 m
Draught	1.4 m
Passengers	192
Fuel capacity	3,000 litres
Water capacity	400 litres
Maximum speed	27 kt
Operational speed	25 kt

Propulsion: *Aqua Spirit:* Two Cummins QSM 11 diesels rated at 433 kW driving fixed-pitch propellers via a reverse reduction gearbox. *Crystal Spirit*: Two Cummins N14M diesels rated at 391 kW driving fixed-pitch propellers via a reverse reduction gearbox.

24.5 m CATAMARAN

Tusa T5

Delivered to Tusa Dive in 2007, this vessel was designed by CMCS of Australia. The vessel will be used for dive trips from Cairns.

Specifications

Length overall	24.5 m
Length waterline	23.1 m
Beam	7.66 m
Draught	1.75 m
Displacement	50 t
Passengers	95
Crew	5
Maximum speed	27 kt

Propulsion: Two MTU Series 60 diesel engines delivering 496 kW each to fixed-pitch propellers via Twin Disc 5224A gearboxes.

24m CATAMARAN

Eclipse (2006)

Designed by CMCS for Calyso Reef Cruises as a dive boat, this vessel was delivered in February 2006 and replaces the 23.5 m catamaran of the same name built in 2004.

Specifications

Length overall	23.5 m
Length waterline	22.6 m
Beam	7.5 m
Draught	1.8 m
Passengers	192
Displacement	58 tonnes
Passengers	107
Crew	5
Maximum speed	27.5 kt

18 m Catamaran Crewboat Bue Tobol 1120284

23.5 m Catamaran Sea Quest 1120285

Propulsion: Two MTU Detroit Diesel Series 60 14L diesel engines each delivering 485 kW to fixed-pitch propellers.

23.5 m CATAMARAN

Eclipse (2004)

Designed by Commercial Marine Consulting Service, the vessel is primarily used as a dive boat for trips to the Great Barrier Reef. An aft hydraulic platform can be lowered down between the hulls.

Specifications

Length overall	23.5 m
Length waterline	22.62 m
Beam	7.5 m
Draught	1.8 m
Displacement	48 t
Maximum speed	27 kt
Operational speed	24 kt
Passengers	95

Propulsion: Two MTU Detroit Series 6014L (each producing 481 kW).

23.5 m CATAMARAN

Sea Quest

Designed by Commercial Marine Consulting Service, the vessel is used as a dive boat for trips to the Great Barrier Reef. The vessel was built to Australia's USL IC regulations.

Specifications

Length overall	23.5 m
Length waterline	21.65 m
Beam	7.5 m
Draught	1.75 m
Displacement, full load	60 t
Maximum speed	31.5 kt
Operational speed	27 kt
Passengers	130
Fuel capacity	5,000 litres
Water capacity	7,000 litres

Propulsion: Two MTU 12V 183 TE72 diesels rated at 735 kW at 2,300 rpm driving fixed-pitch propellers via ZF BW 655A gearboxes.

20 m CATAMARAN

Tasman Venture III

Designed by CMCS of Australia, this 20 m catamaran was delivered in 2004.

Specifications

Length overall	18.8 m
Length waterline	12.8 m
Beam	6.2 m
Draught	1.4 m
Maximum speed	26 kt
Operational speed	22 kt
Passengers	140
Fuel capacity	2000 litres
Water capacity	500 l

Vessel type	Vessel name	Yard No	Length (m)	Speed (kt)	Seats	Originally delivered to	Date of build
CMCS Seacat 25 m	*Aqua Spirit*	V201	25.00	27.0	192	Palm Beach Ferry Service	2002
CMCS Seacat 25 m	*Crystal Spirit*	V301	25.00	27.0	192	Palm Beach Ferry Service	2003
CMCS Seacat 17 m	*Quick Cat II*	–	17.00	22.0	80	Hervey Bay Whale Watch	2003
CMCS Seacat 23.5 m	*Eclipse*	–	23.50	27.0	95	Deep Sea Divers Den	2004
CMCS Seacat 23.5 m	*Sea Quest*	–	23.50	27.0	130	Deep Sea Divers Den	2004
CMCS Seacat 20 m	*Tasman Venture III*	V303	19.40	28.0	120	Tasman Venture	2004
CMCS Seacat 18 m	*Bue Tobol*	–	18.15	30.0	50	Bue Marine	2005
CMCS Seacat 18 m	*Bue Narya*	0520	18.15	30.0	50	Bue Marine	2005
CMCS Seacat 18 m	*Bue Bo*	0521	18.15	30.0	50	Bue Marine	2005
CMCS Seacat 18 m	*Bue Darya*	–	18.15	30.0	50	Bue Marine	2005
CMCS Seacat 19 m	*Fantasea*	–	18.8	26.0	130	Fantasea Cruises	2005
CMCS Seacat 24 m	*Eclipse*	–	23.50	27.5	101	Calypso Reef Cruises	2006

Propulsion: Two Caterpillar 3406E diesel engines driving fixed-pitch propellers via Twin Disc MDG 5114A gearboxes.

19 m CATAMARAN

Fantasea

Launched in July 2005, this vessel was delivered to Fantasea Cruises of Australia and carries 130 passengers at 25 kt.

Specifications

Length overall	19.4 m
Maximum speed	28 kt
Crew	4
Passengers	120
Fuel capacity	8,000 litres
Operational speed	24 kt

Propulsion: Two 330 kW Deutz diesel engines.

18 m CATAMARAN CREWBOATS

Bue Tobol
Bue Narya
Bue Darya
Bue Bo
Bue Tekes

Designed by Sea Speed Design of Australia (formerly Commercial Marine Consulting Services) these 18 m crewboats will be used on the Caspian Sea.

Specifications

Length overall	18.15 m
Length waterline	17.24 m
Beam	6.24 m
Draught	0.9 m
Displacement	30 t
Passengers	50
Crew	2
Maximum speed	30 kt
Operating speed	25 kt

Propulsion: Two Caterpillar C18 diesel engines delivering 510 kW at 2,100 rpm driving Doen Jet 05 220 water-jets via ZF 5505L gearboxes.

18 m CATAMARAN

Four 18 m catamarans were ordered in 2005 for crew boat duties in Eastern Europe. Three of the vessels have a design speed of 25 kt, being powered by a pair of Caterpillar C18 diesel engines driving a Doen DJ 220 water-jet. The fourth has a speed of 35 kt and is powered by Caterpillar 3412E engines.

17 m CATAMARAN

Quick Cat II

This whale watching catamaran, custom designed by CMCS, was launched in 2003 for operations in Hervey Bay, Queensland.

24 m Catamaran operated by Calypso Reef Cruises  1120286

23.5 m Catamaran Eclipse, *built in 2004* 1120287

Specifications

Length overall	17.4 m
Length waterline	16.6 m
Beam	6.5 m
Draught	1.33 m
Displacement, full load	35 t
Maximum speed	26 kt
Operating speed	22 kt
Passengers	80
Fuel capacity	2,000 litres
Water capacity	400 litres

Propulsion: Two 320 kW Cummins diesel engines.

North West Bay Ships Pty Ltd

83 Gemalla Road, Margate, Tasmania 7054, Australia

Tel: (+61 3) 62 67 01 04
Fax: (+61 3) 62 67 01 02
e-mail: info@nwbships.com.au
Web: www.nwbs.com.au

John Fuglsang, *Chairman*

North West Bay Ships Pty Ltd was formed in 2000 with the aim of developing modern high-speed marine transport solutions. In 2001 the company launched its first vessel, a 55 m trimaran fast ferry, closely followed by a 33 m catamaran ferry. The company, which specialises in high quality aluminium ships, offers other fast ferry designs including low wash river ferries.

In 2003 the Government of Bermuda placed an order for two 23 m foil-assisted semi-planing ferries capable of serving any existing dock in Bermuda and thereby opening up new destinations.

In February 2007 the shipyard facilities of North West Bay Ships were acquired by Austal.

55 m trimaran Dolphin Ulsan *(ex-*Triumphant*) at speed* 0526482

55 m TRIMARAN

Dolphin Ulsan

Launched in 2001, this vessel was the subject of considerable research, including testing on an 11 m prototype vessel. The craft is supported by a main central hull, two smaller side hulls and a hydrofoil near amidships spanning the centre hull to side hulls.

Specifications

Length overall	54.5 m
Length waterline	52.1 m
Beam	15.2 m
Draught	2.1 m
Deadweight	57 t
Crew	6
Passengers	473
Fuel capacity	15,000 litres
Water capacity	2,500 litres
Propulsive power	6,960 kW
Max speed	40 kt

Classification: DnV +1A1 HSLC Passenger, R1 EO Cat B.

Structure: All aluminium hull and superstructure.

Propulsion: Three MTU 4000 M70 diesel engines each driving a Kamewa 63 SII water-jet.

Auxiliary Systems: One 300 mm Naiad bow thruster.

39 m CATAMARAN FERRY

Red Jet 4

Ordered in 2002, this vessel was delivered in April 2003.

Specifications

Length overall	39.2 m
Displacement (load)	121 t
Passengers	300
Fuel capacity	5,600 litres
Fresh water capacity	250 litres
Operational speed	35 kt

Propulsion: Two MTU 12V 4000 M70 diesels each driving a water-jet unit.

36 m TRIMARAN (DESIGN)

Having launched the 55 m trimaran *Triumphant* in early 2001, the company is also marketing a 36 m trimaran design with a passenger capacity of 260 and speed of up to 38 kt, depending on propulsion engines selected.

Specifications

Length overall	36.0 m
Length waterline	33.5 m
Beam	10.0 m
Draught	1.5 m
Deadweight	26.5 t
Crew	8

Vessel type	Vessel name	Yard No	Length (m)	Speed (kt)	Seats	Vehicles	Originally delivered to	Date of build
55 m Trimaran	*Dolphin Ulsan* (ex-*Triumphant*)	–	54.50	40.0	473	none	Unknown	2001
33 m Passenger Catamaran	*Ten*	2201	34.90	33.0	273	none	Fantasea Cruises	2001
39 m Catamaran	*Red Jet 4*	–	39.20	35.0	300	none	Red Funnel	2003
34 m Catamaran	*Simply Majestic*	–	34.00	18.0	–	none	Majestic Cruises	2003
23 m Catamaran	*Tempest*	–	23.40	34.0	177	none	Government of Bermuda	2004
23 m Catamaran	*Venturilla*	–	23.40	34.0	177	none	Government of Bermuda	2004
30 m Catamaran	*Silversonic*	–	29.45	31.0	150	none	Quicksilver Connections	2006
30 m Catamaran	*Silverswift*	–	29.45	31.0	150	none	Quicksilver Connections	2006
30 m Catamaran	*Aristocat V*	–	29.45	31.0	150	none	Aristocat Cruises	2006

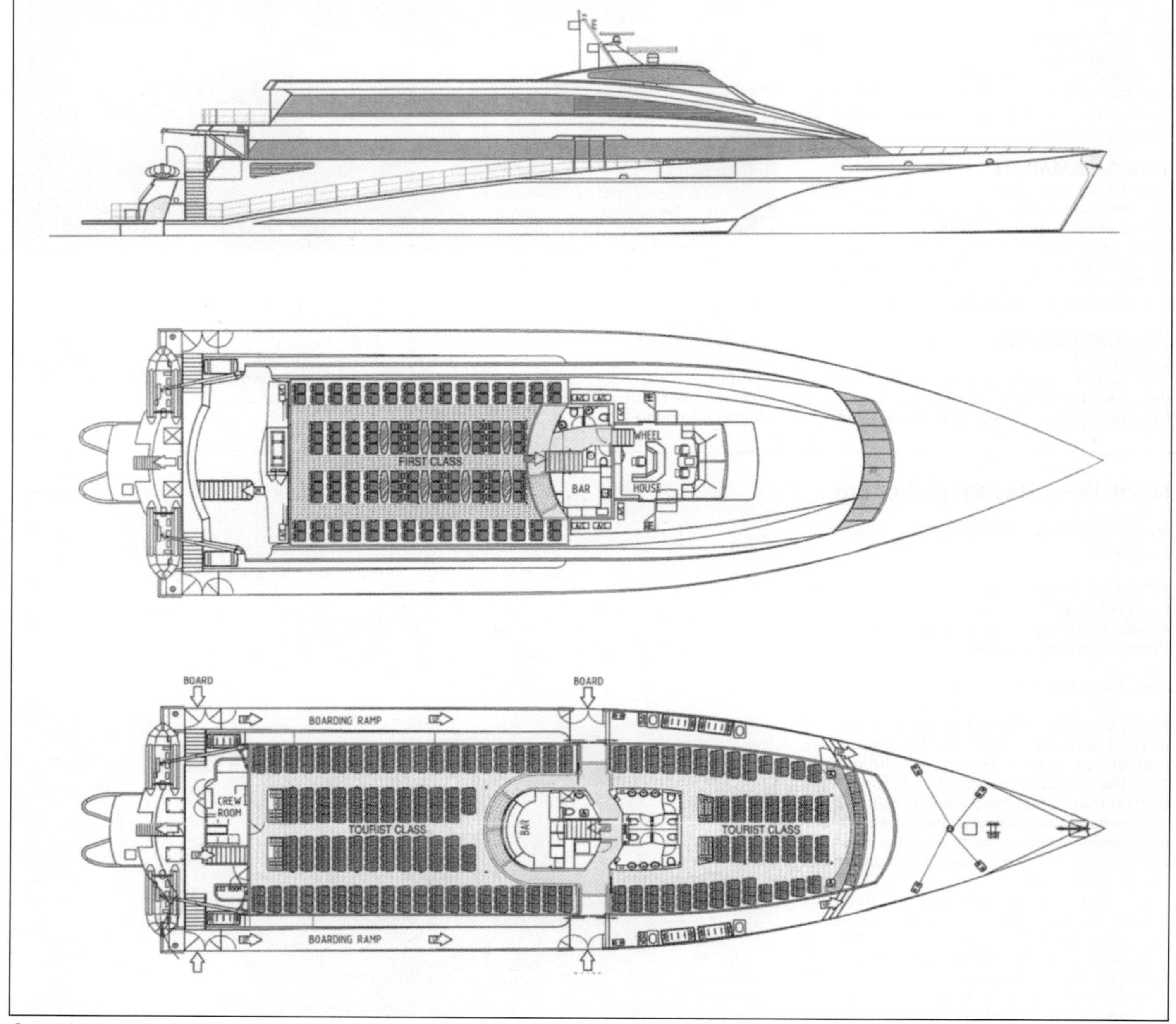

General arrangement of 55 m trimaran

0109644

Passengers	260
Fuel capacity	5,000 litres
Water capacity	1,000 litres
Operational speed	38 kt

Propulsion: Three MTU 2000 M70 diesel engines each rated at 788 kW driving water-jets.

34 m CATAMARAN FERRY

Simply Majestic

Based on the successful *Red Jet 4* design this ferry has a service speed of 18 knots.

Specifications

Length overall	34.0 m
Maximum speed	21.2 kts
Service speed	18 kts

Propulsion: Two Caterpillar 3406E diesel engines each rated at 448 at 2,100 rpm and each driving a five-bladed propeller via a ZF 350 gearbox.

32 m CATAMARAN FERRY

Ten

This vessel was delivered to Australian operator Fantasea Cruises in 2001 for operation to Great Barrier Reef resorts.

Specifications

Length overall	32.0 m
Length waterline	31.6 m
Beam	7.8 m
Draught	1.1 m
Deadweight	26.9 t
Crew	3
Passengers	220
Fuel capacity	5,000 litres
Water capacity	1,000 litres
Max speed	33 kt
Operational speed	31 kt

Propulsion: Two Deutz TBD 616 V16 diesel engines each rated at 934 kW at 2,100 rpm and driving a fixed-pitch propeller.

30 m CATAMARAN FERRY

Silversonic

Aristocat V

Silverswift

Delivered in August 2005 to Quicksilver Connections for service in North Queensland, this vessel was built to Australian USL Class 1C regulations and fitted out for either 100 passengers on daily excursions or up to 150 on charters. Another 30 m catamaran of the same design was delivered to Aristocat Cruises in April 2006.

Specifications

Length overall	29.45 m
Length waterline	27.10 m
Beam	7.80 m
Passengers	100/150
Max speed	31 kt

Propulsion: Two Deutz TBD 616 V16 diesel engines providing 936 kW at 2,100 rpm

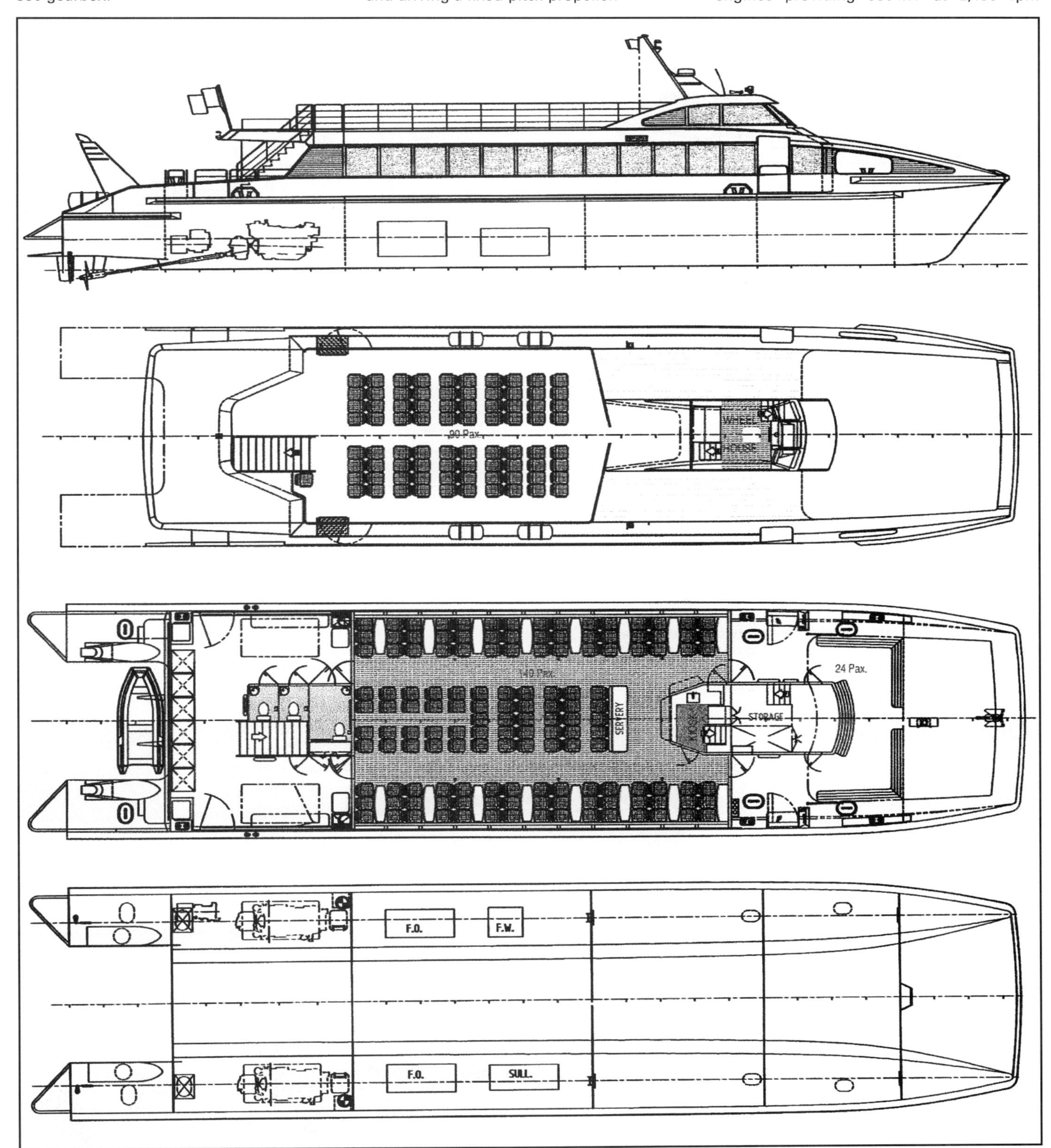

General arrangement of the NWBS 33 m Catamaran Ten

0121748

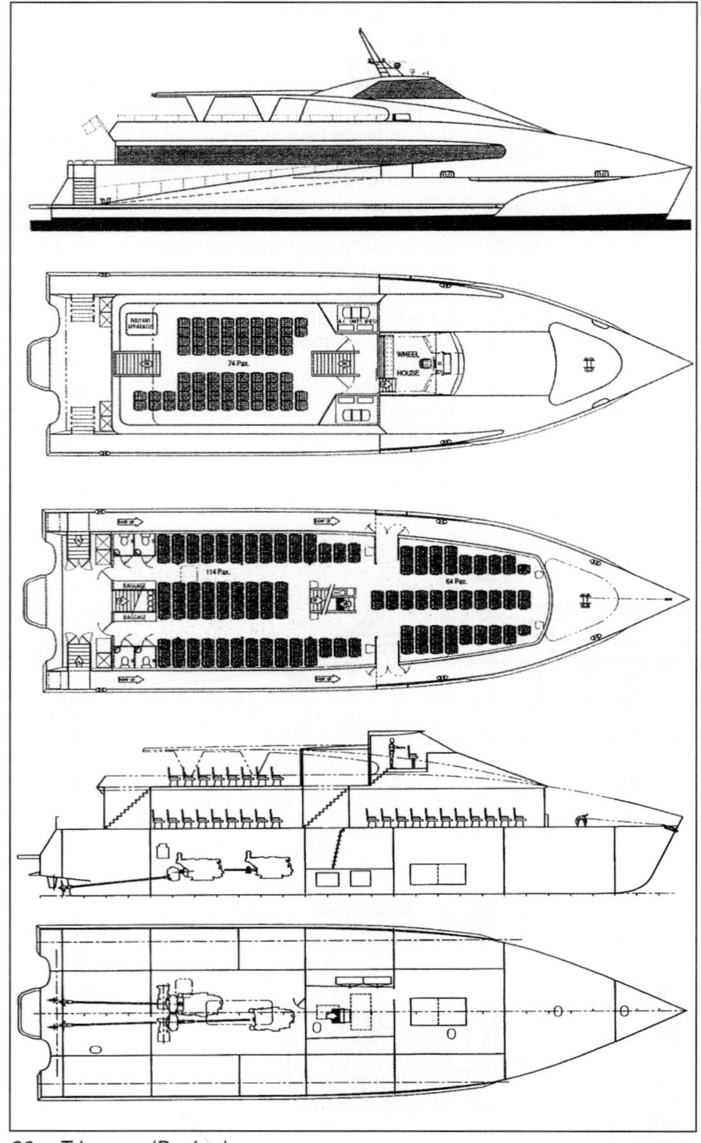

36 m Trimaran (Design) 0121763

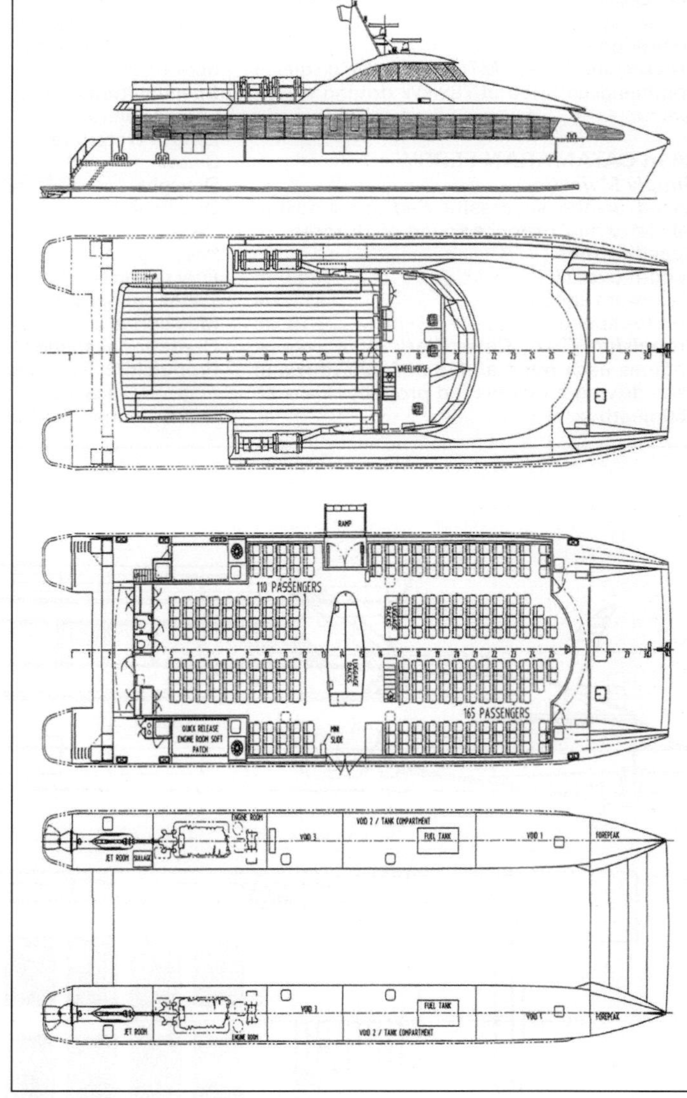

General arrangement of Red Jet 4 0554532

each power five-blade fixed-pitch propellers through ZF BW 2550 gearboxes.

23 m CATAMARAN FERRY

Tempest
Venturilla

Designed by Teknicraft of New Zealand, two of these foil-assisted semi-planing vessels were delivered to the Government of Bermuda in 2004.

Specifications

Length overall	23.3 m
Beam	7.8 m
Draft	1.0 m
Passengers	177
Max speed	35 kt

Classification: Lloyd's Register SSC G2.
Structure: Aluminium.
Propulsion: Two MTU Detroit Diesel 12V2000 diesel engines, each rated at 525 kW driving Hamilton water-jets.

NQEA Australia Pty Ltd

NQEA ceased operations in October 2005 after 60 years of business.

The company started services in 1948 from the residence of its founder with a staff of three, the principal activity being the operation of general engineering agencies, leading to general engineering manufacture. In 1964 it entered the shipbuilding industry with the construction of dumb barges. This was followed by work on many types of vessels, including landing barges, work boats and patrol vessels. In 1977 NQEA was the successful tenderer for 14 Fremantle class 42 m patrol craft for the Australian Navy, the first having been built by Brooke Marine, UK.

Approximately 220 craft of various types have been built to date including 46 for the Australian Department of Defence.

Throughout the 1980s, NQEA built large numbers of high-speed ferry craft for the international market-place as well as some special purpose vessels. Search and rescue vessels, luxury motor yachts, dredge and

Hurricane Clipper 1033986

NQEA low-wash catamarans 0506649

Vessel type	Vessel name	Yard No	Length (m)	Speed (kt)	Seats	Vehicles	Originally delivered to	Date of build
NQEA 22 m Catamaran	*Reef Magic* (ex-*Fantasea Princess* ex-*Telford Reef*)	107	22	29	204	0	Reef Magic Cruises	1982
NQEA 22 m Catamaran	*Reef Adventure III* (ex-*Magnetic Express*)	108	22	29	240	0	Great Adventures	1983
28 m Catamaran	*Fantasea Monarch* (ex-*Telford Capricorn*)	111	29	30	352	0	Fantasea Cruises	1983
NQEA 24 m Catamaran	*Quickcat I*	115	24	30	175	0	Quickcat Cruises	1984
NQEA 22 m Catamaran	*Commander Peak* (ex-*Cougar*)	113	22	29	200	0	Fjordland Travel, NZ	1984
NQEA 22 m Catamaran	*Reef Link*	109	22	29	200	0	Magnetic Link Townsville	1984
NQEA 24 m Catamaran	*Quickcat II*	123	24	30	195	0	Fantasea Cruises	1985
NQEA 30 m Catamaran	*Reef Cat*	126	30	30	310	0	Great Adventures	1986
NQEA 30 m Catamaran	*Bacolod Express* (ex-*Quicksilver III*)	127	30	30	312	0	Philippines	1986
NQEA 24 m Catamaran	*Auto Batam 3* (ex-*Supercat II*)	138	24	28	239	0	Singapore	1986
NQEA 30 m Catamaran	*Sea Princess* (ex-*Reef Link II*)	147	30	30	342	0	Fantasea Cruises	1987
NQEA 24 m Catamaran	*Tiger IV* (ex-*Taupo Cat*)	148	24	29	202	0	Fullers, NZ	1987
NQEA 30 m Catamaran	*Island Jade* (ex-*Quicksilver IV*)	130	30	30	314	0	Philippines	1987
NQEA 24 m Catamaran	*Roylen Sunbird*	151	24	29	217	0	Roylen Cruises Pty Ltd	1987
NQEA 30 m Catamaran	*Diamond* (ex-*Supercat III*)	125	30	30	329	0	IGSA Shipping	1987
NQEA 30 m Catamaran	*Reef King*	152	30	30	324	0	Great Adventures	1988
NQEA 26.9 m Catamaran	*Adaire*	156	24	29	151	0	Kuwait Public Transport Co	1988
NQEA 26.9 m Catamaran	*Na-Ayem*	157	24	29	151	0	Kuwait Public Transport Co	1988
38.6 m WPC	*Quicksilver V*	158	38.6	27.0	252	0	Quicksilver Connections	1988
38.6 m WPC	*Quicksilver Beluga* (ex-*Quicksilver VII*)	159	38.6	27.0	350	0	Beluga Quicksilver, Bali	1989
39 m WPC	*Prince of Venice*	160	38.6	27.0	303	0	Kompas Touristik	1989
38.6 m WPC	*Quicksilver VII*	161	38.6	27.0	350	0	Beluga Quicksilver, Bali	1989
34.8 m Catamaran	*Blue Fin*	163	34.8	35.0	250	0	NSW State Transit Authority	1990
40.17 m WPC	*Seacom I*	170	40.2	29.0	302	0	Secom Corporation of Japan	1990
34.8 m Catamaran	*Sir David Martin*	172	34.8	35.0	250	0	NSW State Transit Authority	1990
34.8 m Catamaran	*Sea Eagle*	173	34.8	35.0	250	0	NSW State Transit Authority	1991
Low Wash Catamaran	*Betty Cuthbert*	181	36.8	22.5	200	0	NSW State Transit Authority	1992
Low Wash Catamaran	*Dawn Fraser*	180	36.8	22.5	200	0	NSW State Transit Authority	1992
NQEA 22 m Catamaran	*Green Island Express*	106	22	29	230	0	Great Adventures	1982
Low Wash Catamaran	*Shane Gould*	184	36.8	22.5	200	0	NSW State Transit Authority	1993
Low Wash Catamaran	*Marlene Mathews*	185	36.8	22.5	200	0	NSW State Transit Authority	1993
Low Wash Catamaran	*Marjorie Jackson*	187	36.8	22.5	200	0	NSW State Transit Authority	1993
Low Wash Catamaran	*Evonne Goolagong*	186	36.8	22.5	200	0	NSW State Transit Authority	1993
24 m Catamaran	*Payar*	154	24	27	162	0	Sriwani Tours, Penang	1994
45 m WPC	*Quicksilver VIII*	192	45.5	35.6	456	0	Quicksilver Connections	1995
29 m Luxury Passenger Ferry	*Sun Eagle*	195	29.2	30.0	156	0	Ansett Transport Industries	1996
24 m Passenger Commuter Ferries	*Zon 2*	198	25.0	22.5	166	0	Kelana Megah, Johor Bahru	1996
24 m Passenger Commuter Ferries	*Zon 1*	197	25.0	22.5	166	0	Kelana Megah, Johor	1996
River Runner 150	*Vavau*	202	30.6	22.0	115	0	Bora Bora Navettes, Frech Polynesia	1998
River Runner 150	*Motu Tapu O Ra*	203	30.6	22.0	115	0	Bora Bora Navettes, Frech Polynesia	1998
Cheetah Patrol Boat	*Wauri*	129	25.3	28	12	0	Queensland Fisheries	1988
River Runner 200	*Piet Hein*	206	37.0	30.0	150	0	High Speed Ferries, Netherlands	1999
River Runner 200	*Witte De With*	207	37.0	30.0	150	0	High Speed Ferries, Netherlands	1999
River Runner 150 Mk 3	*Down Runner*	208	32.0	25	149	0	–	2001
17 m Cat	*Aristocrat 4*	209	17.0	23	58	0	Charter Corporation	–
River Runner 150 Mk 3	*Antrim Runner*	210	32.0	25	149	0	–	2001
River Runner 200 Mk II	*Hurricane Clipper*	212	37.5	27	200	0	Thames Clippers	2002
22 m Transit Catamaran	*PVT Sorensen*	–	22.0	30.0	140	0	Kwajalein Range Services, Marshall Islands	2004
22 m Transit Catamaran	*PVT Anderson*	–	22.0	30.0	149	0	Kwajalein Range Services, Marshall Islands	2005

barge units (capable of being linked using an NQEA-designed Autodock system), hovercraft and patrol boats have all been produced at the Cairns Australia NQEA facilities.

After delivery of the 34.8 m catamarans (Yard Nos 163, 172 and 173) to the New South Wales State Transit Authority as hydrofoil replacements, NQEA was awarded a contract to build six low-wash catamaran passenger ferries for the same operator, which were delivered in 1993.

In April 1995 the shipyard delivered a 45 m InCat-designed WAVE-PIERCING catamaran to Quicksilver Connections and in 1996 delivered a 30 m luxury fast ferry to Hayman Island Resort and two 24 m commuter ferries to a Malaysian client.

By 2001, the company had delivered 12 fast low-wash river ferries and five more were built under license by Damen Shipyards in the Netherlands. Another 37.5 m River Runner catamaran was delivered to the UK in 2002 for Thames Clippers to operate on the River Thames.

SEAJET PV 330-SS-D (DESIGN)

The PV 330-SS-D has a total deadweight capacity of 330 tonnes and an operational speed of 42 kt with gas-turbine propulsion and 38 kt with diesel propulsion.

Specifications

Length overall	81.0 m
Length waterline	75.0 m
Beam	23.4 m
Passengers	600
Vehicles	216 cars
Fuel capacity	50,000 litres
Water capacity	7,000 litres
Operational speed	42 kt

SEAJET PV 250-SS-T

The Seajet PV 250-SS-T is a joint venture between Fry Design and Research, a consultancy division of NQEA Australia and Danyard of Frederikshavn, Denmark. Two of these vessels have recently been constructed at the Danyard shipyard.

The semi-Swath hull form was developed with the Swedish Marine Research Institution, SSPA. The resultant hull form experiences relatively low vertical accelerations compared to a conventional catamaran for a given sea state without the use of a ride control system.

Sea trials on the first vessel for Mols-Linien achieved 50.3 kt in a light load condition and 46.3 kt fully loaded.

The four loading lanes at the bow and twin side exit lanes aft would allow for very fast turnaround times.

Specifications

Length overall	76.1 m
Beam	23.4 m
Draught	3.4 m
Crew	12
Passengers	450
Vehicles	120 cars
Propulsive power	24,800 kW
Operational speed	43.6 kt
Range	240 n miles

Classification: DnV +1A1 HSLC R2 Passenger Car Ferry A EO ICS NAUT.

Propulsion: Engines: two GE LM1600 gas turbines each rated at 12,400 kW, driving Maag Propjet M6-127/2 gearboxes. This CODOG arrangement drives four Kamewa water-jets.

45 m WAVE-PIERCING CATAMARAN

Quicksilver VIII

This is the largest fast ferry delivered by NQEA to date and represents the flagship of the owners, Quicksilver Connections. As with *Quicksilver V*, the vessel operates out of Port Douglas on Queensland's northeast coast, to the Agincourt Reef on the outer section of the Great Barrier Reef.

Quicksilver VIII 0506966

Quicksilver V 0506967

Blue Fin 0506647

Specifications

Length overall	45.5 m
Length waterline	39.6 m
Beam	16.2 m
Hull beam	3.0 m
Draught	1.9 m
Crew	8
Passengers	450
Fuel capacity	12,000 litres
Water capacity	6,000 litres
Max speed	42.4 kt
Operational speed	35.6 kt
Range	220 n miles

Classification: DnV +1A1 R2 HSLC Passenger Ferry EO.

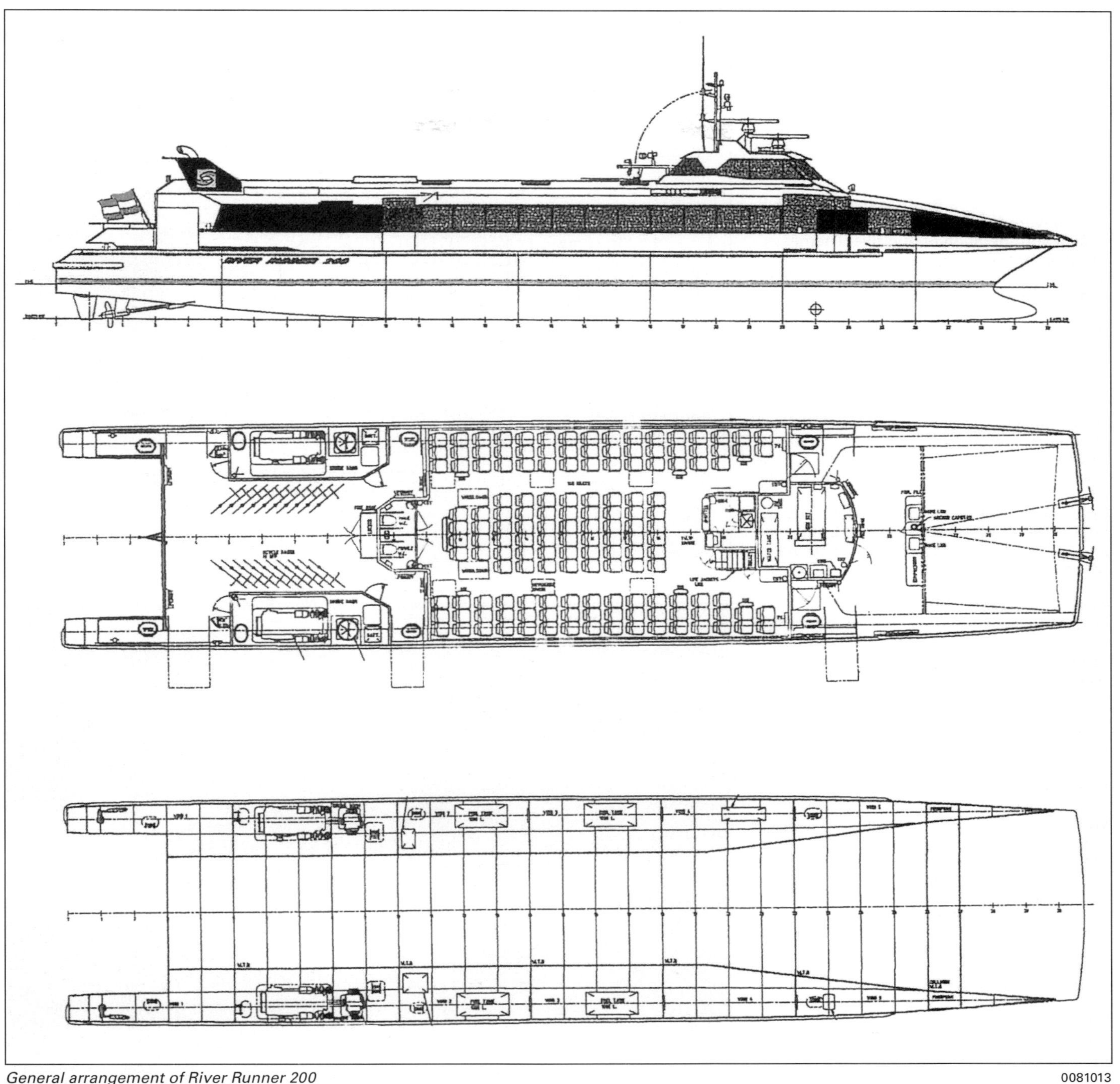

General arrangement of River Runner 200 0081013

Propulsion: Powered by four DDC 16V 149TI diesels each rated at 1,343 kW at 1,900 rpm and each driving a Kamewa 56 S11 water-jet via a reduction gearbox.

40.17 m WAVE-PIERCING CATAMARAN

Seacom I

SeaCom Corporation of Japan (formerly known as the Nisshin Steamship Company) took delivery of this vessel in Cairns in July 1990. The ferry sailed to Japan under its own power and is operating as a ferry around Yokahama.

Specifications

Length overall	40.2 m
Length waterline	31.4 m
Beam	15.6 m
Hull beam	2.5 m
Draught	1.6 m
Crew	5
Passengers	302
Fuel capacity	9,000 litres
Water capacity	3,500 litres
Propulsive power	2 × 1,435 kW
Operational speed	29 kt
Max speed	32 kt

Classification: Nippon Kaiki Kyokai NS (restricted coastal service/aluminium catamaran/passenger ship).

Propulsion: Two GM Detroit diesel engines model 16V-149 TIB, each coupled to ZF BU460 gearboxes. The engines are each rated at 1,435 kW at 2,000 rpm. Propulsion is delivered through two Kamewa 63 water-jets.

Electrical system: Provided by two Hino WO6/DTI engines driving Stamford 100 kVA UCM274D marine alternators.

38.6 m WAVE-PIERCING CATAMARAN

Quicksilver V

The first of these vessels, *Quicksilver V*, was completed by NQEA in November 1988 for Quicksilver Connections, with a further two identical craft delivered in February and September 1989 for the same operator and a fourth vessel delivered to Yugoslavia, also in June 1989.

A 39 m variation of the Quicksilver style was constructed for Kompas Touristik International, underwent an 11,000 mile delivery voyage from the builder's yards to Piran in Yugoslavia, and immediately went into operation, plying between the Istrian ports and Venice. This vessel is fitted with Kamewa water-jet units driven by DDC 16V 149 TA diesel engines.

Specifications

Length overall	38.6 m
Length waterline	31.4 m
Beam	15.6 m
Hull beam	2.6 m
Draught	1.6 m
Passengers	340
Fuel capacity	2 × 2,000 litre tanks
Water capacity	3,000 litres
Propulsive power	2 × 1,230 kW
Max speed	30 kt
Operational speed	27 kt

Structure: all-welded aluminium generally using alloys 5083 H321 and 6061 T6. Some light plates are 5086 H32.

Classification: DnV +1A1 R45 Queensland Department of Harbours and Marine, Class 1G.

Propulsion: Two GM diesel engines Model 16V-149 TIB, each coupled to ZF gearbox Model BU 460. Each engine is rated at 1,230 kW at 1,800 rpm, and drives two Kamewa water-jets Model 63S 62/6.

Electrical system: Main services are supplied by either of two diesel engine (Cummins 6BT5) driven 100 kVA Stamford alternator sets. These supply 415 V three-phase and 415 V single-phase 50 Hz power. Engine starting, auxiliary and emergency services are supplied from 24 V DC battery banks.

Outfit: 350 passengers. Interior seats are individual armchairs with woollen upholstery. Exterior seats are moulded polypropylene shells. A food service area is fitted at the aft end of the lower cabin and a drinks bar in the middle of the lower cabin. Passenger spaces are air conditioned.

RIVER RUNNER 200 MK II

Hurricane Clipper, a River Runner 200 Mk II was delivered to Thames Clippers in London in 2002 for operation on the river Thames.

Specifications

Length overall	37.8 m
Length waterline	35.4 m
Beam	9.3 m
Draught	1.3 m
Deadweight	20.4 t
Crew	5
Passengers	206
Fuel capacity	3,400 litres
Water capacity	500 litres
Maximum speed	27.5 kt
Operational speed	24 kt

Classification: LR +A1 SSC Passenger Catamaran HSC G1 LDC.

Propulsion: Two Caterpillar 3412E diesel engines each rated at 690 kW at 2,100 rpm and driving a fixed-pitch propeller via a reverse reduction gearbox.

RIVER RUNNER 200

Two of these craft were delivered to High Speed Ferries BV in the Netherlands in 1999.

Specifications

Length overall	37.0 m
Beam	8.3 m
Draught	1.3 m
Crew	3
Passengers	150
Freight capacity	0.4t
Fuel capacity	4,000 litres
Water capacity	1,500 litres
Operational speed	30 kt

Classification: DNV +1A1 HSLC Passenger R5EO (NL) and Dutch Shipping Inspectorate.

Structure: Aluminium hull with aluminium and composite superstructure.

Propulsion: Two Deutz TBD 616V16 diesel engines each rated at 1,920 kW at 2,165 rpm driving a five-bladed propeller via a ZF BW255V gearbox.

low-wash CATAMARAN

The New South Wales State Transit Authority, in June 1991, placed an order with NQEA to build two 36.8 m low-wash catamaran ferries to link the city of Parramatta with Sydney's Circular Quay. In 1992 the same client placed an order for a further four vessels with NQEA after the successful operation of the first Rivercats. All six were completed by September 1993.

Designed by Grahame Parker, construction is light aluminium with foam FRP coring in the top cabin structure.

The propulsion system for these vessels is unusual. The designers chose Schottel Rudderpropellers, an uprated system based on the Schottel SRP 132/131 type. The modifications of this unit enable power input to be raised to about 367 kW. The position of these thrust and steering devices is also unusual in that they are positioned beneath the bridging structure of the catamaran towards its aft end with the propellers facing forward.

Specifications

Length overall	36.8 m
Length waterline	35.0 m
Beam	10.5 m
Draught	1.3 m
Displacement, min	45 t
Passengers	200
Fuel capacity	5,500 litres
Water capacity	600 litres
Propulsive power	2 × 373 kW
Operational speed	22.5 kt
Max speed	23.7 kt
Range	660 n miles

Propulsion: Two DDC 8V 92TA engines, 373 kW each at 2,100 rpm; driving two Schottel steerable propeller units SRP 132/131.

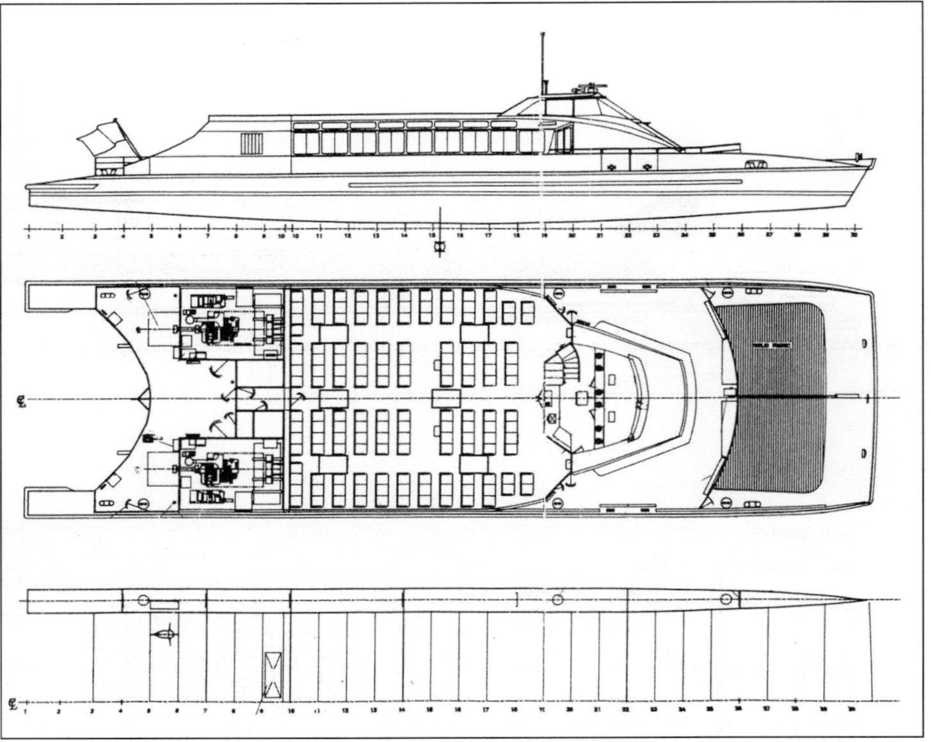

General arrangement of 150-seat low-wash ferry for New South Wales State Transit Authority
0506648

The NQEA Danyard Seajet 250 Mai Mols
0507060

Zon 1 *at launch*
0507061

34.8 m CATAMARAN

Blue Fin

One of three InCat catamarans ordered by the State Transit Authority of New South Wales.

Survey: Maritime Services Board of New South Wales class 1D.

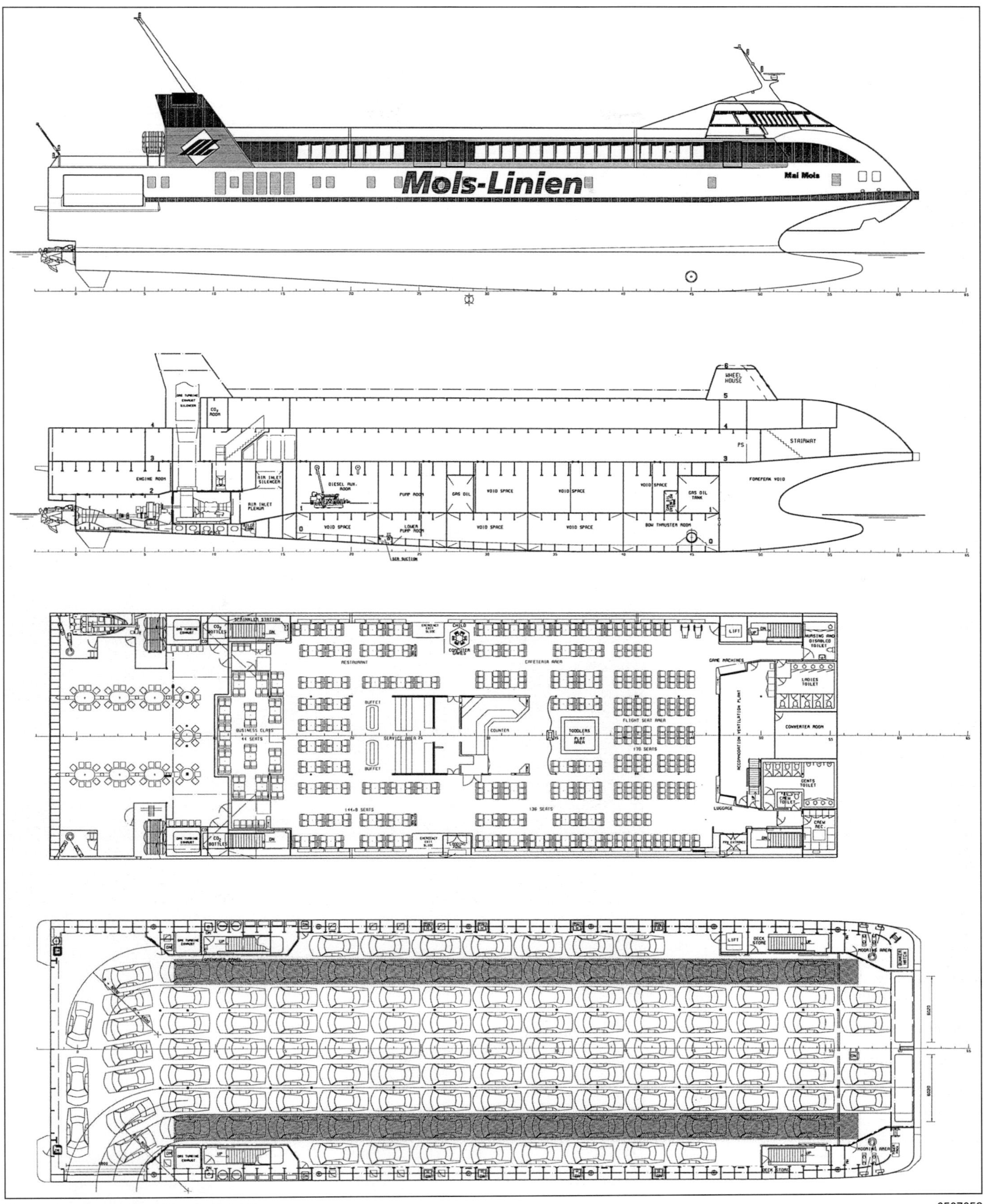

General arrangement of the Seajet 250

0507053

Specifications

Length overall	34.8 m
Length waterline	28.4 m
Beam	9.7 m
Hull beam	2.8 m
Draught	1.0 m
Passengers	250
Fuel capacity	8,830 litres
Water capacity	500 litres
Propulsive power	2 × 1,675 kW
Operational speed	35 kt
Max speed	41 kt

Classification: Lloyd's Register of Shipping@100 A1 Catamaran Passenger Vessel + LMC.

Propulsion: Engines: two MWM TBD 604B V16, 1,675 kW each, driving two Kamewa 63 S11 water-jets.

RIVER RUNNER 150

The company delivered the first pair of river runner low-wash catamarans in mid 1998, to Bora Bora Navettes in French Polynesia. Four further craft were built under license in the Netherlands by Damen Shipyards. Two further vessels were built by NQEA to a MK 3 design in 2001 which were 1.4 m longer and 0.8 m wider that the original River Runner 150 craft.

Specifications

Length overall	30.6 m
Length waterline	29.5 m
Beam	7.0 m
Draught	1.3 m
Deadweight	12.4 t
Displacement	43 t
Crew	3
Passengers	115

Fuel capacity	2,800 litres
Water capacity	410 litres
Operational speed	22 kt

Propulsion: Two Caterpillar 3196 DITA diesel engines each rated at 365 kW at 2,300 rpm and each driving a propeller via a Twin Disc 5114A gearbox.

Sun Eagle

This luxury passenger ferry was delivered in 1996 to the Hayman Island Resort at a cost of A$5 million. The vessel is used for day excursions to and around the Great Barrier Reef.

Specifications

Length overall	29.2 m
Length waterline	24.3 m
Beam	9.5 m
Draught	1.5 m
Displacement, max	110 t
Passengers	156
Operational speed	30 kt

Propulsion: Powered by two MTU 12V 396TE 74 diesels driving Kamewa 56S water-jets via ZF BU 460 gearboxes.

28 m CATAMARAN

InCat design with an all-welded aluminium hull and a resiliently mounted superstructure to minimise noise and vibration.

Specifications

Length overall	29.2 m
Length waterline	25.0 m
Beam	11.2 m
Draught	1.8 m
Fuel capacity	2 × 5,000 litre tanks
Propulsive power	2 × 1,200 shp
Max speed	29 kt
Operational speed	26 kt

Classification: Queensland Marine Board, Class 1C, November 1983.

Propulsion: Two 1,200 shp GM 16V-92 TA high-speed diesels with ZF reverse reduction gearbox 2.4:1; driving a five-blade aluminium-bronze propeller.

Electrical systems: 415 V AC from shore power, or 80 kVA GM diesel alternator set 24 V DC.

Payar

This is a 25 m catamaran built for Sriwani Tours and Travel Snd Bhd, for transporting tourists from Langkawi, Malaysia, to a small island called Pulau Payar.

Specifications

Length overall	26.5 m
Length waterline	23.0 m
Beam	8.7 m
Draught	2.0 m
Passengers	210
Fuel capacity	4,800 litres
Water capacity	2,500 litres
Propulsive power	2 × 736 kW
Max speed	27.2 kt

Classification: DnV +1A1 HSLC R4 Passenger EO.

Propulsion: Main engines are two MAN D2842 LE402 diesels; driving two Veem five-blade series 'B' 1,000 × 1,015 mm propellers; via ZF BW 255 Rev Red gearboxes.

CHEETAH PATROL BOAT

NQEA has designed a patrol boat variant of the 23 m commercial catamaran design. The first vessel, *Wauri*, is currently operated by the Australian Customs Service for surveillance work in northern Australian waters. Normally operated by a six-person crew, the craft has facilities to carry an additional 12-person landing party and has extended range cruising capabilities.

Specifications

Length overall	25.3 m
Beam	8.7 m
Hull beam	2.5 m
Draught	2.0 m
Crew	6
Propulsive power	2 × 605 kW
Operational speed	25 kt
Range	1,000 miles at 20 kt

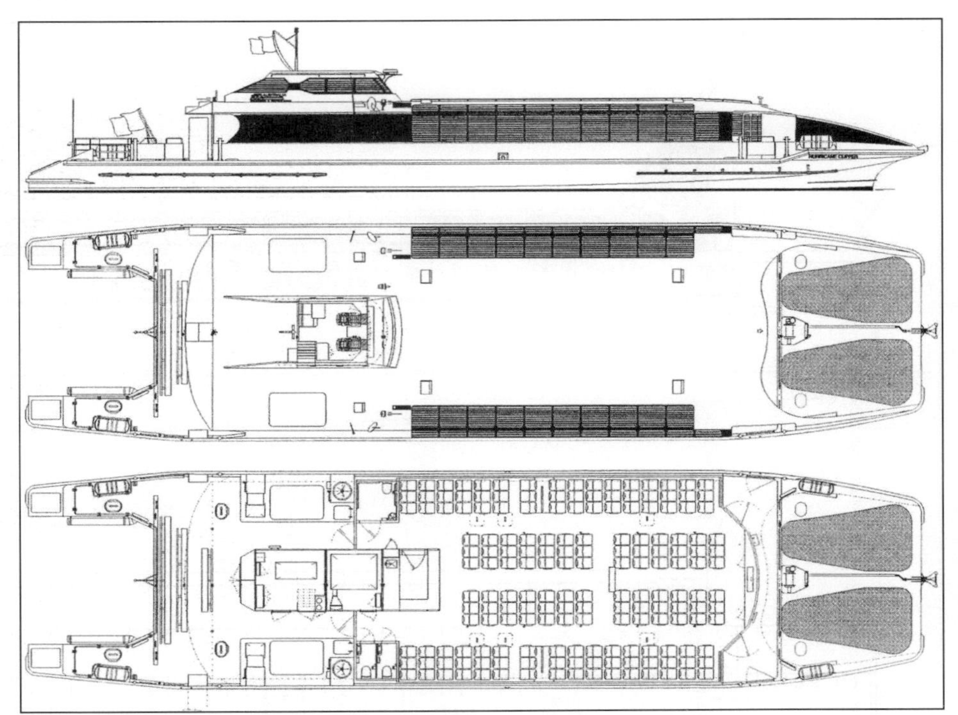

General arrangement of the River Runner 200 Mk II Hurricane Clipper 0137813

River Runner 200 Witte de With 0079976

River Runner 150 on trials at NQEA shipyard 0079977

Propulsion: Two Deutz MWM TBD 234 V12 engines, 605 kW each at 2,200 rpm; driving two ZF BW195 gearboxes, ratio 2.46:1; to two five-blade, 1 m diameter, aluminium-bronze propellers.

Zon 1* and *Zon 2

There are two 24 m passenger commuter ferries which were delivered to Kelana Megah in mid-1996 for operation at the Zon Complex in Johor Bahru. Designed by InCat Designs of Sydney, the vessels are powered by MAN diesels driving a fixed-pitch propeller system.

Specifications

Length overall	25.0 m
Length waterline	23.3 m
Beam	8.4 m
Draught	1.3 m
Displacement, max	70 t

Passengers	166
Fuel capacity	4,900 litres
Water capacity	2,000 litres
Operational speed	22.5 kt

Propulsion: Powered by two MAN D2842 LE402 diesels each rated at 625 kW at 2,100 rpm and driving fixed-pitch propellers.

CROWTHER 22 m TRANSIT CATAMARAN

Under construction in 2004 to a Crowther Design, this 22 m catamaran is being built to USCG Sub Chapter 'T' classification.

Specifications

Length overall	22.0 m
Passengers	149
Operational speed	30.0 kt
Range	270 nm

Propulsion: Powered by four MTU/Detroit Diesel Series 60 engines each rated at 444 kW, each coupled to a Hamilton HJ water-jet.

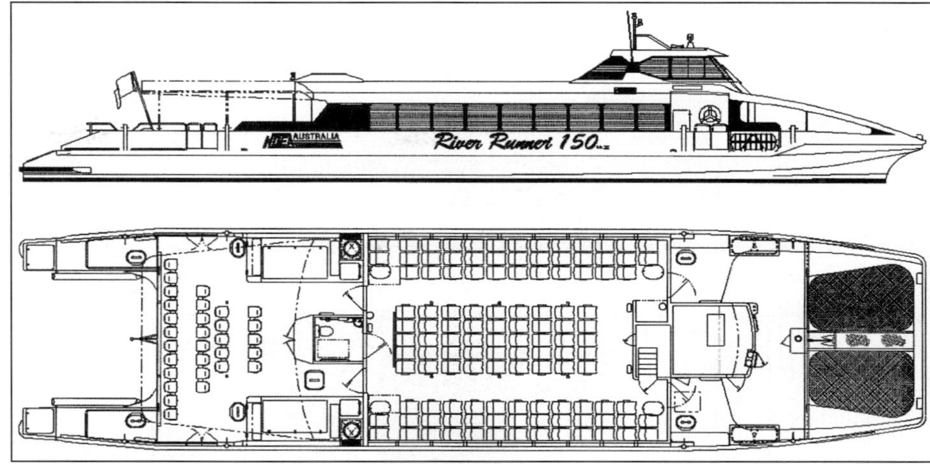

General arrangement of the River Runner 150 0121760

Richardson Devine Marine Constructions Pty Ltd

PO Box 1673, Hobart, Tasmania 7001, Australia

Tel: (+61 3) 62 72 78 00
Fax: (+61 3) 62 72 65 77
e-mail: rdm@rdm.com.au
Web: www.rdm.com.au

Founded in 1989, the company specialises in the production of lightweight, high-performance aluminium catamaran and monohull craft for the commercial and tourist industries.

35 m Catamaran

Designed by InCat Crowther Design, this 35 m catamaran is currently in build for World Heritage Cruises. The design encompasses low wash features developed for low speed operation in ecologically sensitive regions.

Specifications

Length overall	35.0 m
Passengers	221
Operational speed	28 kt

Propulsion: Two MTU 8V 4000 M70 engines each rated at 1,159 kW.

32 m Catamaran

Ordered in 2007 by Ishigaki Dream Tours, this 31 kt, 245 passenger InCat Crowther designed vessel will operate in the Ryukyu Islands off the southernmost tip of Japan.

Specifications

Length overall	32.0 m
Length waterline	28.6 m
Beam	8.5 m
Draught	1.7 m
Deadweight	25.5 t
Passengers	245
Fuel capacity	8,000 l
Freshwater capacity	2000 l
Operational speed	28 kt

Construction: Aluminium.

32 m Catamaran

Lady Jane Franklin II

Designed by Crowther Design, this luxury ferry was delivered to Gordon River Cruises in November 2003.

Specifications

Length overall	32.0 m
Length waterline	29.3 m
Beam	9.0 m
Draught	1.65 m
Passengers	234
Crew	6
Maximum speed	32 kt
Operational speed	29 kt

Propulsion: Two MTU 16V 2000 M70 engines each rated at 1,050 kW at 2,100 rpm.

32 m Catamaran

Patea Explorer

Delivered in December 2005, *Patea Explorer* is the first InCat Crowther design to be launched. The vessel is designed to cope with the harsh conditions of Doubtful Sound, New Zealand.

Specifications

Length overall	31.6 m
Length waterline	26.65 m
Beam	9.0 m
Draught	2.13 m
Deadweight	25 t
Passengers	195
Crew	6
Operational speed	28.5 kt

Propulsion: Two Caterpillar C32 diesel engines delivering 1,045 kW at 2,300 rpm to fixed pitch propellers via Twin Disk MG 6619 SC gearboxes.

29 m Catamaran

Adventurer

Launched in 2002, this vessel is the second Crowther designed catamaran delivered to World Heritage Cruises in Tasmania.

Specifications

Length overall	28.8 m
Length waterline	26.8 m
Beam	8.5 m
Draught	1.4 m
Deadweight	24.9 t
Crew	5
Passengers	232
Operational speed	30 kt

Propulsion: Two DDEC 16V-92TA diesel engines each rated at 962 kW and each driving a five-bladed propeller via a ZF 2050 gearbox.

26 m Catamaran

Wanderer II

Lady Jane Franklin

Wanderer II was delivered in 2000 for operation in the Gordon River in Australia. *Lady Jane Franklin* was delivered in 2001 to Macquarie Harbour. Both vessels were designed by Crowther Multihulls.

Specifications

Length overall	25.6 m
Length waterline	22.8 m
Beam	8.5 m
Draught	1.0 m
Passengers	198
Fuel capacity	6,000 litres
Water capacity	1,000 litres
Max speed	32 kt
Operational speed	30 kt

Classification: DnV +1A1 HSLC R4 Passenger EO.

Propulsion: *Wanderer III* is powered by two DDEC 16V TA DEC 3 diesel engines each rated at 820 kW and driving fixed-pitch propellers. *Lady Jane Franklin* is powered by two Caterpillar 3412E diesel engines each rated at 820 kW.

24 m Catamaran

Marana

This vessel was ordered by Roche Brothers & Sons for Port Arthur Cruises of Tasmania.

Vessel type	Vessel name	Yard no	Length (m)	Speed (kt)	Seats	Vehicles	Originally delivered to	Date of build
26 m Catamaran	*Wanderer II*	–	25.6	30.0	198	none	World Heritage Cruises, Australia	2000
26 m Catamaran	*Wanderer III*	023	25.6	32.0	198	none	World Heritage Cruises, Australia	2000
26 m Catamaran	*Lady Jane Franklin*	–	25.6	30.0	193	none	World Heritage Cruises, Australia	2001
26 m Catamaran	*Peppermint Bay I*	027	25.6	31.0	193	none	Hobart Cruises	2001
17 m Catamaran	*Matthew Flinders*	29	17.0	–	–	none	Inter Island Ferries, Australia	2002
29 m Catamaran	*Adventurer*	30	29.0	30.0	–	none	World Heritage Cruises, Australia	2002
24 m Catamaran	*Marana*	–	24.0	25.0	225	none	Port Arthur Cruises	2002
23 m Catamaran	*Sea Melbourne*	036	23.4	25.0	150	none	Cruise Victoria	2003
32 m Catamaran	*Lady Jane Franklin II*	035	37.0	32.0	234	none	Gordon River Cruises	2003
22 m Catamaran	*Peppermint Bay II*	040	22.0	27.0	203	none	Hobart Cruises	2005
32 m Catamaran	*Patea Explorer*	042	31.6	30.0	200	none	Fiordland Travel	2005
23 m Catamaran	*Luminosa*	043	23.0	28.0	95	none	Real Journeys	2006
35 m Catamaran			35.0	28	221	none	World Heritage Cruises	(2008)
32 m Catamaran	–	–	32.0	28	245	none	Ishigaki Dream Tours	(2008)

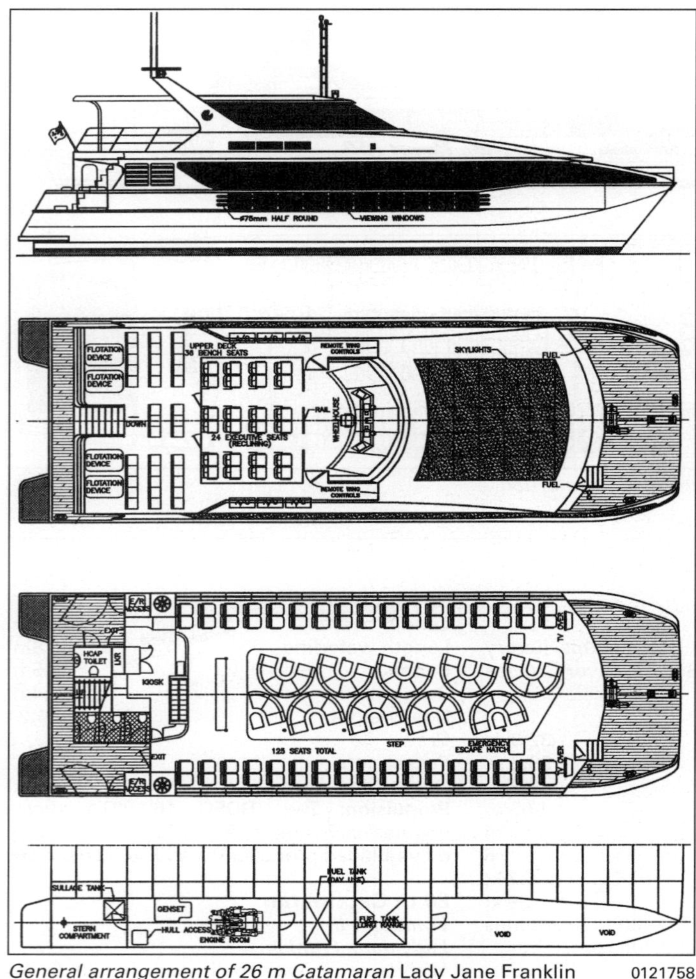

General arrangement of 26 m Catamaran Lady Jane Franklin 0121758

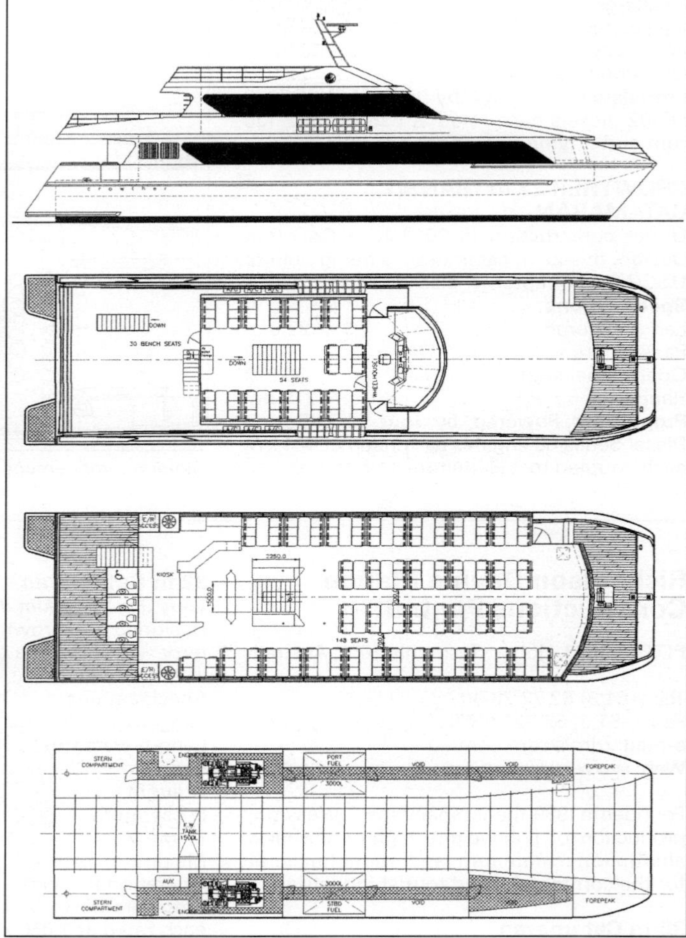

General arrangement of Adventurer 0526481

Specifications

Length overall	24.0 m
Length waterline	22.5 m
Beam	8.5 m
Draught	1.6 m
Deadweight	20 t
Crew	3
Passengers	230
Fuel capacity	4,000 litres
Fresh water capacity	1,000 litres
Operational speed	30 kt

Propulsion: Two Caterpillar 3406E diesel engines each rated at 447 kW and each driving a fixed-pitch propeller.

23 m Catamaran

Luminosa

Ordered by Red Journeys for delivery in mid 2006, this vessel was designed by Incat Crowther.

Specifications

Length overall	23.0 m
Length waterline	20.71 m
Beam	6.8 m
Draught	1.7 m
Deadweight	11.5 t
Passengers	95
Maximum speed	28 kt
Operational speed	25 kt

Propulsion: Two Caterpillar C18 diesel engines each delivering 533 kW at 2,100 rpm.

22 m Catamaran

Pepermint Bay II

Delivered to Hobart Cruises in 2005 the vessel will be used for tourist excursions from Hobart. The vessel was built to Australian USL 1D regulations.

Specifications

Length overall	22.0 m
Beam	7.5 m
Draught	1.75 m
Passengers	167
Maximum speed	27 kt

Propulsion: Two Caterpillar 3406E diesel engines each rated at 520 kW driving fixed-pitch propellers via Twin Disk gearboxes.

23 m Catamaran

Sea Melbourne

This vessel was delivered to Cruise Victoria in July 2003 for excursions within Port Philip Bay.

Specifications

Length overall	23.4 m
Length waterline	19.7 m
Beam	7.8 m
Draught	1.5 m
Deadweight	11 t
Crew	8
Passengers	150
Fuel capacity	4,000 litres
Fresh water capacity	1,910 litres
Maximum speed	28 kt
Operational speed	25 kt

Propulsion: Two Cummins QSM11 diesels each rated at 311 kW and each driving a fixed-pitch propeller via a ZF 350 gearbox.

17.5 m Catamaran

Matthew Flinders

Designed by Crowther Multihulls, this vessel was delivered to owners Inter Island Ferries of Australia in 2002.

Specifications

Length overall	17.5 m
Length waterline	15.7 m
Beam	6.2 m
Crew	3
Passengers	84
Fuel capacity	2,000 litres
Fresh water capacity	200 litres
Max speed	25 kt

Classification: USL Code 1C and 1D.

Propulsion: Two Cummins 6CTA 8.3 m engines each rated at 221 kW and driving four bladed fixed-pitch propellers via a Twin Disc gearbox.

Sabre Catamarans Pty Ltd

156 Barrington Street, Spearwood, Western Australia 6163, Australia

Tel: (+61 8) 94 18 30 00
Fax: (+61 8) 94 34 14 57
e-mail: wgharry@sabrecat.com.au; sabrecat@sabrecat.com.au
Web: www. sabrecat.com.au

Bill Harry, *Principal*

Builder of a number of aluminium fast catamaran fishing and passenger vessels. The most recent builds include: *Stewart Islander*, a Sabre 22 m for an operator in New Zealand and *Star 7*, a 24 m catamaran built as a private motor yacht. Other vessels have included *Aremiti 11*, a 29 m 30 kt catamaran ferry built for Aremiti Pacific Cruises. The two earlier catamarans, *Aremiti* and *Saladin Sabre,* were operated by Stirling Marine Services during the construction of the Saladin oilfield off the coast of Western Australia. *Saladin Sabre* is now working as a Tahiti-Moorea ferry. Two smaller high-speed catamarans have also been built, the *OT Manu* and the *Paia*. These 16.7 m catamarans each have accommodation for 73 passengers and a maximum speed of 25 kt. They operate in Bora Bora lagoon in French Polynesia.

Sabre 55

Aremiti

Saladin Sabre

Specifications

Length overall	16.8 m
Length waterline	14.9 m

Stewart Islander 1146732

Sabre 22 m Naser 1 *delivered to Iran in 2002* 0527036

22 m catamaran Emirates 0103266

Quickcat II 0007930

Beam	7.2 m
Draught	0.7 m
Passengers	100
Fuel capacity	2 × 1,000 litre tanks
Water capacity	500 litres
Max speed	34.6 kt
Operational speed	28 kt

Propulsion: Main engines are two DDC 8V 92T diesels, driving two Levi 800 series surface-piercing drive units.

Auxiliary systems: The auxiliary generator is an Isuzu 17 kVA model.

OT Manu and Paia

These two 16.7 m passenger vessels were built for Bora Bora Navettes to carry Air Tahiti passengers from the airport to the tropical island of Bora Bora in French Polynesia. They were both delivered in 1991.

Specifications

Length overall	16.7 m
Length waterline	13.3 m
Beam	6.5 m
Draught	0.9 m
Passengers	73
Fuel capacity	3,000 litres
Max speed	25 kt
Operational speed	20 kt

Propulsion: Main engines are two Detroit Diesel 6V 92 TA diesels, driving Arneson surface-piercing propellers.

Sabre 22 m catamaran

Makana
Emirates
Ocean 20
Chogogo
Peacemaker
Naser 1
Voyager

Makana was delivered in 1998 for operation between Cape Town and Robben Island. *Emirates* was delivered to the UAE in 2000, *Ocean 20* to the Maldives in 2001, *Chogogo* to the Netherlands in 2001 and *Peacemaker* to the British Virgin Islands and *Naser 1* to Iran, both in 2002.Voyager was delivered to Hook Island Adventure Cruises in 2006.

Specifications

Length overall	22.3 m
Length waterline	19.8 m
Beam	7.9 m
Draught	1.5 m
Passengers	160
Fuel capacity	4,000 litres
Water capacity	1,000 litres
Max speed	30 kt
Operational speed	28 kt

Vessel type	Vessel name	Yard No	Length (m)	Speed (kt)	Seats	Vehicles	Originally delivered to	Date of build
Sabre 23 m	*Quickcat II*	–	23.0	20.0	250	none	Fullers Gulf Ferries, New Zealand	1986
Sabre 55	*Aremiti*	–	16.8	28.0	100	none	Stirling Marine Services	1988
Sabre 20 m	*Foveaux Express*	–	19.9	30.0	67	none	Stewart Marine Services	1991
17.7 m Ferry	*OT Manu*	–	16.7	20.0	73	none	Bora Bora Navettes, French Polynesia	1991
17.7 m Ferry	*Paia*	–	16.7	20.0	73	none	Bora Bora Navettes, French Polynesia	1991
Sabre 30 m	*Aremiti II*	–	29.9	–	288	none	Aremiti Pacific Cruises, Taihiti	1992
Sabre 32 m	*Delta Cat II (ex Flying Cat)*	–	32.0	28.0	300	none	Boat Torque Cruises	1996
Sabre 22 m	*Lucky 2*	–	22	26	129	none	Fil Estate properties	1997
Sabre 22 m	*Ok Ka Ferry I*	–	22	26	200	none	Mt Samat Ferry Express	1997
Sabre 25 m	*Royal Ferry 2*	–	25.5	28.0	169	none	Royal Ferry Service, Phillipines	1998
Sabre 22 m	*Makana*	–	22.3	28.0	160	none	South Africa	1998
Sabre 22 m	Autshumatu	–	22.3	28	172	none	South Africa	1998
Sabre 22 m	*Emirates*	–	22.3	28.0	160	none	UAE	2000
Sabre 22 m	*Ocean 20*	–	22.3	29.0	157	none	Maldives	2001
Sabre 22 m	*Chogogo*	–	22.3	29.0	177	none	Flamingo Fast Ferries	2001
Sabre 22 m	*Peacemaker*	–	22.3	29.0	215	none	Speedy's (BVI)	2002
Sabre 22 m	*Alaleh I (ex Naser I)*	–	22.3	29.0	–	none	Iran	2002
Sabre 24 m	*Seabreeze II*	147	24.0	27.0	164	none	Mackenzies Island Cruises	2003
Sabre 22 m	*Seabreeze II*	–	22	26	164	none	Mackenzies Island Cruises	2003
Sabre 22 m	*Tiger V (ex Stewart Islander)*	–	22.3	26.0	184	none	Stewart Island Experiences, New Zealand	2004
Sabre 24 m	*Star 7*	–	24.1	35.5	–	none	Private Owner, United States	2004
Sabre 22m	*Voyager*	150	22.3	29	139	none	Hook Island Adventure Cruises	2006

Propulsion: *Makana* is powered by two DDEC 16V 92TA diesel engines, each rated at 820 kW and each driving a five bladed fixed-pitch propeller via a Niigata MGN 273-E gearbox. *Emirates, Chogogo, Peacemaker, Naser 1* and *Voyager* are powered by two MTU 12V 2000 M70 diesel engines each rated at 788 kW.

Quickcat II
Owned by Fullers Gulf Ferries of New Zealand, this 23 m vessel has a passenger capacity of 250 on three decks.

Specifications

Length overall	23.0 m
Beam	8.3 m
Draught	1.4 m
Passengers	250
Fuel capacity	20,000 litres
Max speed	23 kt
Operational speed	20 kt

Propulsion: Powered by twin MTU 12V 183 TE72 diesel engines.

Sabre 22 m

Tiger V (ex Stewart Islander)

Specifications

Length overall	22.3 m
Length waterline	19.8 m
Beam	7.9 m
Draught	1.7 m
Passengers	184
Fuel capacity	4,000 litres
Fresh water capacity	900 litres
Maximum speed	28 kt
Operational speed	26 kt

Classification: Lloyd's Register +100 A1 SSC Passenger (A) Catamaran HSC G3 + LMC.
Propulsion: The craft is propelled by two MTU 12V 2000 M70 engines each producing 788 kW at 2,100 rpm driving fixed pitch propellers via ZF BW 250 gearboxes.

Foveaux Express
Owned by Stewart Island Marine Services and delivered in 1991.

Specifications

Length overall	19.9 m
Length waterline	17.6 m
Beam	7.2 m
Draught	1.5 m
Passengers	67
Fuel capacity	8,000 litres
Operational speed	30 kt

Sabre 24m

Star 7
Built as a private motor yacht to be operated on the River Mississippi in the US.

Specifications

Length overall	24.1 m
Length waterline	22.8 m
Beam	8.0 m
Draught	1.0 m
Fuel capacity	10,000 litres
Fresh water capacity	5,000 litres
Maximum speed	35.5 kt
Operational speed	30 kt
Range	600 nmiles

Propulsion: Two MTU 16V 2000 M91 engines, each rated at 1,490 kW at 2,350 rpm and each driving a Hamilton HM 651 water-jet via a ZF 2550 gearbox.

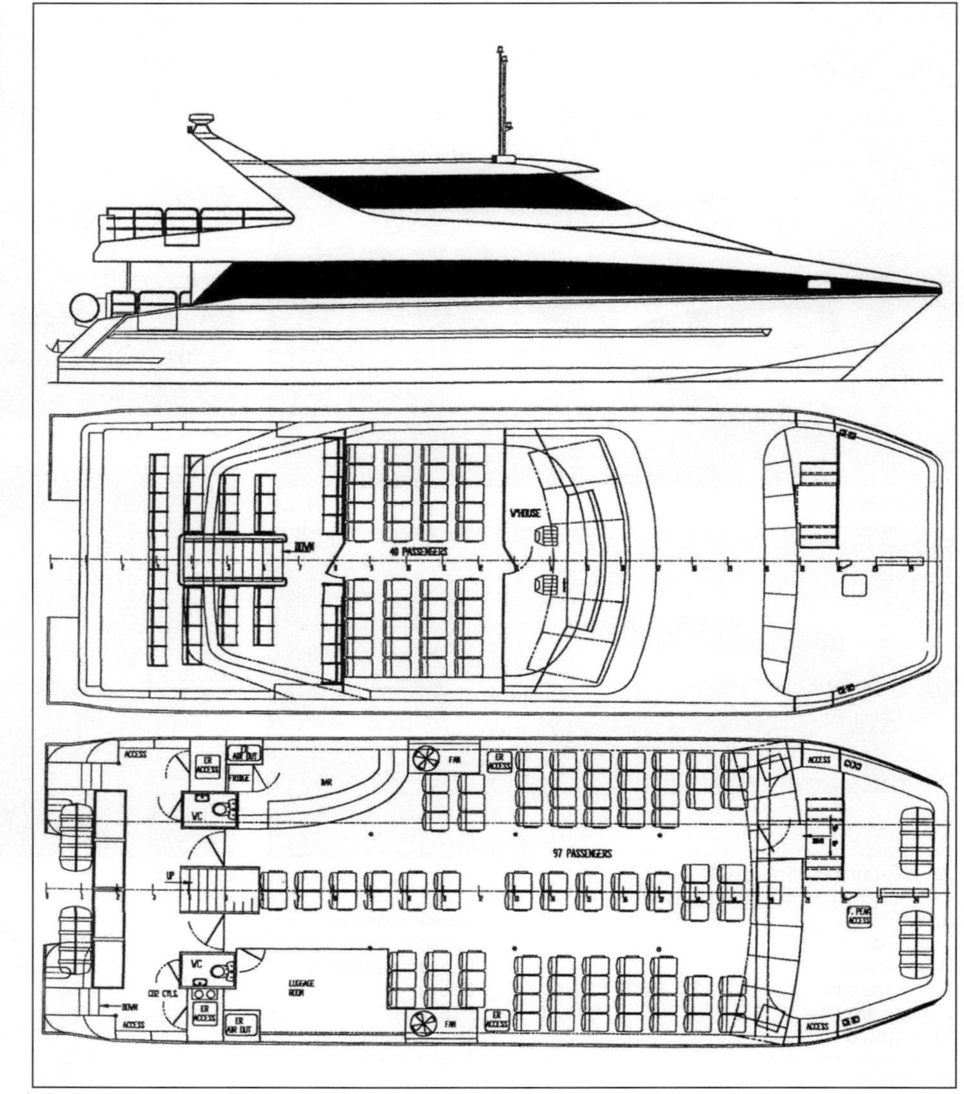

General arrangement of Sabre 22 m Chogogo 0132894

Star 7 1146731

Aremiti II. *The fourth high-speed commuter ferry delivered to Tahiti by Sabre Catamarans* 0007931

Sabre 22 m Peacemaker 0132891

Royal Ferry 2
This 25 m catamaran was delivered to Royal Ferry Services in the Philippines in mid-1998.

Specifications

Length overall	25.5 m
Length waterline	21.9 m
Beam	8.0 m
Draught	1.8 m
Passengers	169
Fuel capacity	6,000 litres
Water capacity	1,000 litres
Propulsive power	2,000 kW
Maximum speed	30 kt
Operational speed	28 kt

Propulsion: Two Detroit Diesel 12 V 149 engines driving a Teignbridge fixed-pitch propeller via a ZF BW 460 gearbox.

OT Manu *for Bora Bora Navettes* 0007932

Aremiti II
Owned by Aremiti Pacific Cruises, this vessel operates from Papeete, Tahiti. Delivered June 1992.

Specifications

Length overall	29.9 m
Length waterline	26.9 m
Beam	9.3 m
Draught	1.5 m
Passengers	288
Fuel capacity	2,000 litres
Propulsive power	2 × 1,360 kW
Max speed	32 kt

Propulsion: Two Detroit Diesel 12V 149TIB diesels rated at 1,360 kW each.

Flying Cat
This is the largest catamaran built by Sabre and was delivered to Boat Torque Cruises in mid-1996. The 32 m vessel, designed by Mark Ellis Design and Management, operates between Rottnest Island and mainland Australia.

Specifications

Length overall	32.0 m
Beam	9.3 m
Draught	1.7 m
Passengers	300
Fuel capacity	20,000 litres
Water capacity	2,000 litres
Operational speed	28 kt

Propulsion: Powered by two MTU 12V 396 TE 74 diesels each rated at 1,260 kW at 1,900 rpm and each driving a four-blade fixed-pitch propeller via a Reintjes WVS 730 reverse reduction gearbox.

SBF Shipbuilders

33 Clarence Beach Road, Henderson, Western Australia 6166, Australia

Tel: (+61 8) 94 10 22 44
Fax: (+61 8) 94 10 18 07
e-mail: info@sbfships.com.au
Web: www.sbfships.com.au

Don Dunbar, *Principal*
Alan McCombie, *Managing Director*
Samantha Dunbar, *Administration Manager*
Brent Dunbar, *Design Engineer*

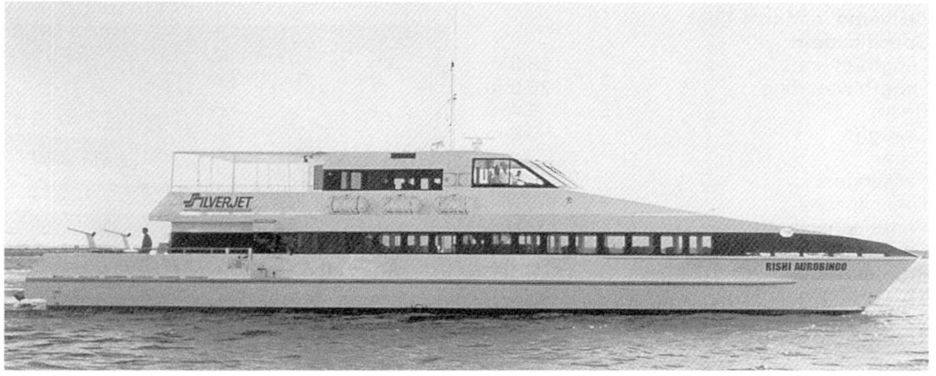

Rishi Aurobindo 0069445

SBF Shipbuilders started trading in 1977, as SBF Engineering, building fishing vessels. The company's first aluminium fast ferry, the monohull *SeaFlyte* was constructed in 1981 and by 2000 the company had delivered over 40 vessels. The shipyard facilities in Henderson cover an area of 3,600 m² and include a 200 tonne boat lift and 300 tonne slipway.

Tropic Sunbird and Quickcat
Tropic Sunbird was first used as a press boat at the 1987 Americas Cup in Perth and then subsequently as a passenger ferry with Sunseeker Cruises in Queensland. *Quickcat* has a similar specification except upper passenger cabin extends the full length and width of the main superstructure.

Specifications

Length overall	33.4 m
Length waterline	30.0 m
Beam	13.0 m
Draught	1.4 m
Passengers	600
Max speed	32 kt
Operational speed	27 kt

Structure: The catamaran hull and superstructure are constructed in all-welded marine grade aluminium alloy. The hulls incorporate bulbous bows which reduce pitching through their extra buoyancy

Propulsion: Main engines are two Deutz MWM TBD 604B V12, driving through two ZF BW 455 gearboxes.

32 m Catamaran

Yukon Queen
This 32 m catamaran was delivered to Alaska in 1999 for operation on the Yukon River between Dawson City in Canada and Eagle in Alaska. The vessel has a passenger capacity of 114. It is powered by four MTU

Vessel type	Vessel name	Yard No	Length (m)	Speed (kt)	Seats	Vehicles	Originally delivered to	Date of build
SBF 33 m Catamaran	*Victory III*	QC31	31.0	32.0	448	None	Tourist Services Pty Ltd	1986
SBF 33.4 m Catamaran	*Tropic Sunbird*	QC34	33.4	27.0	600	None	Sunseeker Cruises, Queensland	1986
SBF 33.4 m Catamaran	*Quickcat*	NZ34	33.4	27.0	600	None	Waiheke Shipping, Auckland	1987
SBF 31.7 m Catamaran	*Lada Tiga*	LK3	31.7	35.0	250	None	Lada Langkawi, Malaysia	1993
SBF 31.7 m Catamaran	*Lada Dua*	LK2	31.7	35.0	250	None	Lada Langkawi, Malaysia	1993
SBF 31.7 m Catamaran	*Lada Satu*	LK1	31.7	35.0	250	None	Lada Langkawi, Malaysia	1993
SBF 19.9 m Catamaran	*Seabreeze*	932	19.9	25.0	116	None	Mcckenzie Marine	mid-1994
SBF 20 m Catamaran	*Lada Lima*	942	20.0	28.0	106	None	Lada Langkawi, Malaysia	late-1994
SBF 31.7 m Catamaran	*Lada Empat*	941	31.7	35.0	250	None	Lada Langkawi, Malaysia	1994
SBF 28 m Catamaran	*Rishi-Aurobindo*	953	28.0	36.0	130	None	Development Consultants, India	March 1996
SBF 26 m Catamaran	*Blue Banter*	961	26.0	28.0	196	None	PT Binaperkasa	1996
SBF 26 m Catamaran	*Lake Express III (formerly MV Sydney)*	964	25.8	32.0	196	None	Geoje Shipping Company, South Korea	September 1997
SBF 22 m Catamaran	*Shelly Taylor Smith*	971	–	20.0	156	None	D.C.M.	1997
SBF 28 m Catamaran II	*Hanryu Cruises*	972	–	20.0	250	None	Hanryeo Sudo Ship	1998
SBF 32 m Catamaran	*Yukon Queen II*	981	32.0	34.0	114	None	Alaska	1999

12V 2000 M70 diesels, each driving a Hamilton HW 461 water-jet via a ZF BW250 reversing gearbox, giving a maximum speed of 34 kt and a cruise speed of 32 kt.

31.7 m Catamarans

Lada Satu
Lada Dua
Lada Tiga
Lada Empat

By the end of 1994 SBF Shipbuilders had delivered all four 250-passenger catamarans under construction following contracts signed in 1992 and 1993 with Lada Langkawi in Malaysia.

Specifications

Length overall	31.7 m
Length waterline	27.0 m
Beam	9.6 m
Draught	1.0 m
Passengers	250
Fuel capacity	5,000 litres (12,000 litres for *Lada Empat*)
Water capacity	1,000 litres
Propulsive power	2 × 1,435 kW
Operational speed	35 kt

Propulsion: Main engines are two MTU 12V 396 TE 74L, 1,435 kW each at 2,000 rpm; driving two Kamewa 56 water-jets, via two ZF BUK455-1 gearboxes.

Rishi-Aurobindo

Built for Development Consultants of India, this ferry operates on the Hooghly river between Calcutta and Haldia. The vessel was delivered in March 1996.

Specifications

Length overall	28.0 m
Length waterline	26.0 m
Beam	9.4 m
Draught	0.8 m
Crew	7
Passengers	130
Fuel capacity	3,600 litres
Water capacity	1,000 litres
Operational speed	36 kt

Propulsion: Powered by two Deutz MWM TBD 616 V16 diesels rated at 1,107 kW at 2,165 rpm, each driving a Kamewa 50 S11 water-jet via a Reintjes WLS 430 reduction gearbox.

26.0 Catamaran

MV Sydney

This fast-commuter ferry was delivered to Geoje Shipping Company of South Korea in 1997.

Specifications

Length overall	25.8 m
Length waterline	23.4 m
Beam	8.6 m
Draught	1.3 m
Passengers	196
Fuel capacity	7,000 litres
Water capacity	1,000 litres
Propulsive power	2 × 1,107 kW
Max speed	35 kt
Operational speed	32 kt

Classification: Korean Register of Shipping
Propulsion: Powered by two MWM TBD616 V16 diesel engines driving fixed-pitch propellers via ZF BW 255 reverse reduction gearboxes
Auxiliary: 2 by Cummins 37 kVA.

Seabreeze *in service* 0507050

Yukon Queen II *in service* 0102135

MV Sydney *on trials* 0007933

Seabreeze

This vessel was delivered to Western Australian operator Mackenzies Marine in mid-1994.

Specifications

Length overall	19.9 m
Length waterline	18.0 m
Beam	7.0 m
Draught	1.4 m
Passengers	116
Fuel capacity	3,000 litres
Water capacity	1,000 litres
Propulsive power	2 × 550 kW
Max speed	25 kt

Structure: Marine grade aluminium
Propulsion: Main engines are two MTU 12V 183 TE 62 diesels, each producing 550 kW at 2,000 rpm; driving two five-blade Stone Marine propellers; via two Twin Disc MG 5141 gearboxes.

Lada Lima

This vessel was delivered to Lada Langkawi Holdings in Malaysia in late 1994.

Specifications

Length overall	20.0 m
Length waterline	18.0 m
Beam	7.0 m
Draught	0.8 m
Passengers	106
Fuel capacity	7,000 litres
Water capacity	1,000 litres
Propulsive power	2 × 610 kW
Max speed	28 kt

Classification: Bureau Veritas
Structure: Marine grade aluminium
Propulsion: Main engines are two MTU 12V 183 TE 72 diesels, each producing 610 kW at 2,100 rpm; driving two Castoldi 07 series water-jets
Auxiliary: 2 Perkins 19 kVA diesel.

South Pacific Marine

PO Box 619, Burpengary 4505, Queensland, Australia

Tel: (+61 7) 54 98 63 26
Fax: (+61 7) 54 98 54 37
e-mail: sales@spmarine.com.au
Web: www.spmarine.com.au

Established in 1992, South Pacific Marine specialises in the construction of aluminium craft of up to 40 m in length.

31 m PASSENGER CATAMARAN

Super Seabus III

Delivered in 1998 to Azam Marine, this vessel was designed by Crowther Designs.

Specifications

Length overall	31.0 m
Beam	9.5 m
Draught	1.9 m
Crew	5
Passengers	292
Fuel capacity	10,000 litres

Propulsion: Four MAN D2842 LE408 diesel engines each rated at 745 kW driving in pairs to two Servogear VD 820B controllable-pitch propellers.

Vessel type	Vessel name	Yard No	Length (m)	Speed (kt)	Seats	Vehicles	Originally delivered to	Date of build
Crowther 31 m Catamaran	*Super Seabus III*	006	31	35	282	–	Azam Marine	1998
Crowther 24 m Catamaran	*Lucky I*	107	24	30	150	–	–	1998
Crowther 24 m Catamaran	*Cat Cocos*	115	24	30	160	–	Inter Island Boats	1998
Crowther 29 m Catamaran	*Eye Spy*	669	29.5	30	338	–	Moreton Bay Whale Watching Tours	2002

29.5 m CATAMARAN

Eye Spy

South Pacific Marine delivered this Crowther designed 29.5 m passenger catamaran to Moreton Bay Whalewatching in 2002.

Specifications

Length overall	29.5 m
Length waterline	26.4 m
Beam	9.5 m
Draught	1.8 m
Deadweight	29.6 t
Crew	6
Passengers	338
Fuel capacity	19,600 litres
Fresh water capacity	4,000 litres
Operational speed	30 kt

Propulsion: Two MTU 16V 2000 M70 diesel engines each rated at 1,050 kW and each driving a fixed-pitch propeller.

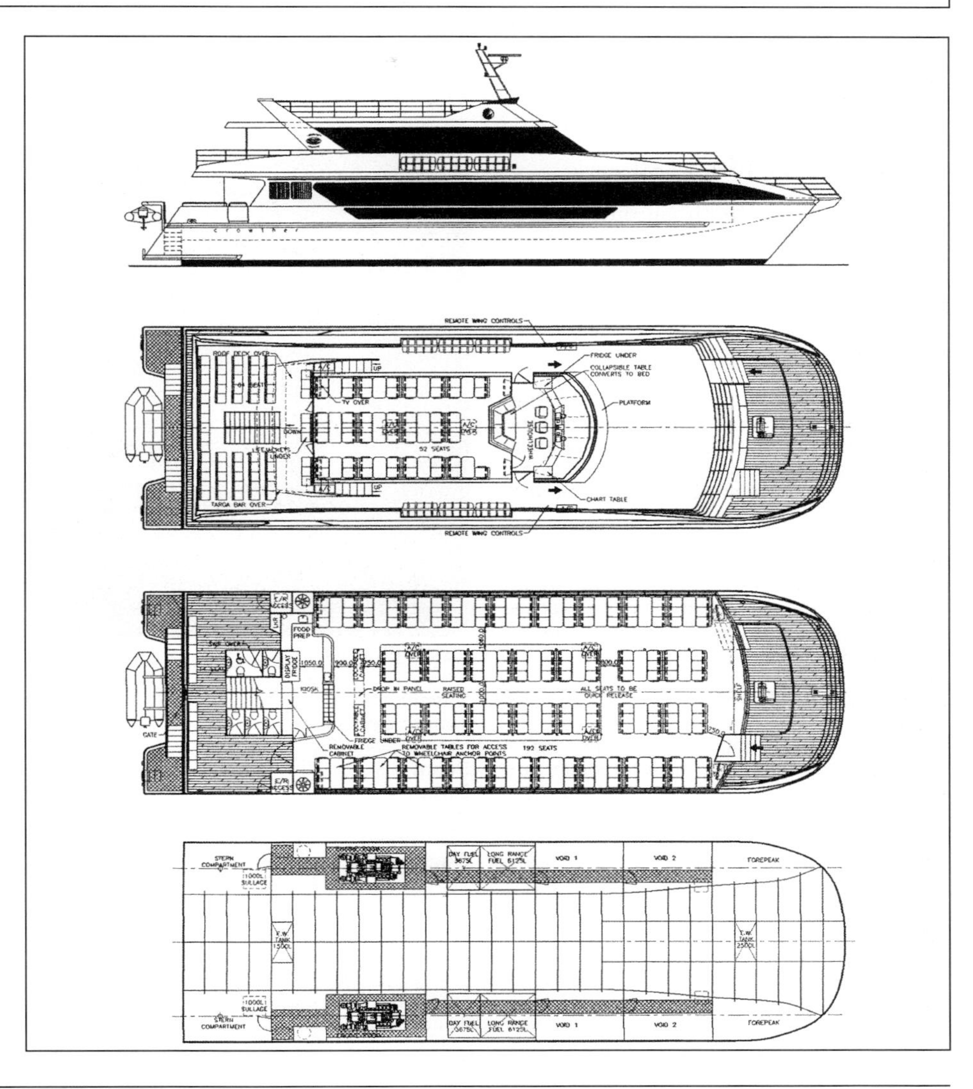

General arrangement of 29 m Eye Spy
0526485

Strategic Marine Pty Ltd

Lot 5 Clarence Beach Road, Henderson, Western Australia 6166, Australia

Tel: (+61 8) 94 37 48 40
Fax: (+61 8) 94 37 48 38
e-mail: info@strategicmarine.com
Web: www.strategicmarine.com

Mark Newbold, *Chairman*
Ron Anderson, *Managing Director, International Business*
John Fitzhardinge, *Director of Design*

Strategic Marine, formerly Geraldton Boat Builders, specialises in building small fast patrol craft and commercial vessels, offering a range of both monohull and catamaran designs, all high-speed planing type. Hulls are manufactured from aluminium or GRP and can be constructed in their facilities in Australia or Singapore.

39.6 m Emergency response vessel

Trumper

The yard completed a conversion project in 2004 to extensively alter *Tamhine Moorea II*. Originally built by Oceanfast in 1995 as a surface effect ship (Ulstein UT928), Strategic Marine converted the vessel into a conventional catamaran for Arco Petroleum Engineering to serve as an emergency response and support vessel for the Nigerian oil and gas industry.

Specifications

Length overall	39.6 m
Beam	11.9 m
Operational speed	30 kt

Propulsion: Two MTU 16V 396 TE74 L engines each rated at 2,000 kW at 2,000 rpm, each driving a Kamewa 63 SII water-jet.

Brazil

Rodriquez Cantieri Navali Do Brazil

Rua da Assembléia, 11-5° and Centro CEP 20010-010 Rio De Janeiro, Brazil

Tel: (+55 21) 25 10 40 30
Fax: (+55 21) 22 15 70 02
e-mail: riodejaneiro@rodriquez.it
Web: www.rodriquez.it

This company was formed in 2001 by the Italian shipyard Rodriquez Cantieri Navali to build a number of catamaran ferries for Brazilian operator Barcas SA of Rio

Vessel type	Vessel name	Yard No	Length (m)	Speed (kt)	Seats	Vehicles	Originally delivered to	Date of build
29 m CityCat	*Zeus I*	–	28.00	27.0	200	–	(Barcas SA, Brazil)	2004
29 m CityCat	*Apolo I*	–	28.00	27.0	200	–	(Barcas SA, Brazil)	2004
29 m CityCat	*Netuno I*	–	28.00	27.0	200	–	(Barcas SA, Brazil)	2005

de Janiero. It is anticipated that the new company will also develop the capabilities to build all the vessels currently marketed by Rodriquez of Italy.

29 m CITY CAT FERRY

Three of these craft were ordered in 2001 for delivery to Barcas SA in 2002. They are scheduled to serve the Rio de Janeiro Charitas operation, a six nautical mile route.

Specifications

Length overall	29.20 m
Beam	9.60 m
Draught	1.44 m
Passengers	237
Operational speed	25 kt
Maximum speed	28 kt

Propulsion: Two MTU 16 V 2000 M70 diesel engines each rated at 1,050 kW and driving a fixed-pitch propeller via a reverse reduction gearbox.

Canada

Allied Shipbuilders Ltd

1870 Harbour Road, North Vancouver, British Columbia V7H 1A1, Canada

Tel: (+1 604) 929 23 65
Fax: (+1 604) 929 53 29
e-mail: asl@alliedship.com
Web: www.alliedship.com

Allied Shipbuilders has been operating for almost 60 years, during which time the primary product has been commercial coastal vessels such as fishing vessels, tugs, ferries and offshore supply vessels ranging in size from 12 to 120 m.

In 1996, the yard delivered its first fast aluminium catamaran *Inkster* and shortly afterwards started to supply aluminium fabrications for the 122 m Pacificat ferries under construction by Catamaran Ferries International.

Inkster on trials 0101520

22 m PATROL CRAFT

Inkster

The vessel was delivered to the Royal Canadian Mounted Police for patrol, inspection and policing functions within coastal waters of British Columbia. The vessel is fitted with a fast rescue/assault boat launched at the aft end between the two main hulls.

Specifications

Length overall	21.6 m
Length waterline	17.8 m
Beam	6.7 m
Draught	0.7 m
Propulsive power	1,225 kW
Max speed	32 kt
Operational speed	25 kt

Structure: All-welded marine grade aluminium.

Propulsion: Two MAN D2840 diesel engines each rated at 610 kW and each driving via a ZF BW165 gearbox to an ASD12 Arneson drive surface piercing propeller.

Electrical system: Two 12 kW diesel generators.

Vessel type	Vessel name	Yard No	Length (m)	Speed (kt)	Seats	Vehicles	Originally delivered to	Date of build
22 m Patrol Craft	*Inkster*	–	21.6	25.0	–	–	Royal Canadian Mounted Police	1996

Chantier Naval Matane Inc

1460 Matane Sur Mer, CP 204 Matane (Quebec), G4W 3NI, Canada

Tel: (+1 418) 368 60 35
Fax: (+1 418) 368 39 68
e-mail: cnfg@globetrotter.net
Web: www.chantier-naval.com

Established in 1987 for fishing vessel construction, the shipyard is owned by Chantier Naval Forillon.

In 1997 the company launched a 43 m aluminium high-speed catamaran *CNM Evolution* arranged for passenger and car transportation.

Specifications

Length overall	43.10 m
Length waterline	40.10 m
Beam	10.40 m
Draught	1.40 m
Crew	8
Passengers	150
Vehicles	30 cars
Operational speed	30 kt

Propulsion: Powered by two Caterpillar 3516B diesel engines each rated at 2,088 kW at 1,925 rpm each driving a Hamilton HM 811 water-jet.

Vessel type	Vessel name	Yard No	Length (m)	Speed (kt)	Seats	Vehicles	Originally delivered to	Date of build
43 m Catamaran	CNM Evolution	–	43.10	30.0	150	30	–	1997

Sylte Shipyards

20076 Wharf Street, Maple Ridge, British Columbia, V2X 1A1, Canada

Tel: (+1 604) 465 1441
Fax: (+1 604) 465 0282

Sylte Shipyards is experienced in producing aluminium sailing yachts. In 2007 the company delivered its first fast ferry.

22 m CATAMARAN

Inside Passage

Delivered to West Coast Launch in British Columbia in August 2007, the vessel, which carries up to 80 passengers, will be used for wilderness cruises.

Specifications

Length overall	22 m
Length waterline	19.6 m
Beam	7.5 m
Draught	1.2 m
Crew	5
Passengers	80
Maximum speed	26 kt

Propulsion: Two Cummins QSK 19-M diesel engines delivering 597 kW each to fixed pitch propellers via ZF 655 gearboxes.

Vessel type	Vessel name	Yard No	Length (m)	Speed (kt)	Seats	Vehicles	Originally delivered to	Date of build
InCat Crowther 22 m catamaran	*Inside Passage*	–	22	26	80	none	West Coast Launch Ltd	2007

Chile

Alwoplast SA

PO Box 114, Valdivia, Chile

Tel: (+56 63) 20 32 00
Fax: (+56 63) 20 32 01
e-mail: info@alwoplast.cl
Web: www.alwoplast.cl

Alexander Wopper, *Managing Director*
Ronald Klingenberg, *Production Manager*
Mauncio Vidal, *Head of Engineering*

Alwoplast SA, established in 1987, is a German managed, Chilean based company producing custom-built catamarans from 10 m to 25 m. The company specialises in composite sandwich construction methods.

NewCat 54 1120259

NewCat 54

Fiordos de Sur

This 60 seat, FASTcc designed foil-assisted catamaran was launched in June 2005. The hull is fabricated using a diving cell foam core construction and is being operated in the Patagonian channels in southern Chile.

Specifications

Length overall	16.24 m
Beam	5.1 m
Draught	0.97 m
Displacement	22 tonnes
Crew	4
Passengers	60
Max speed	35 kts
Operational speed	30 kts

Propulsion: Two Cummins QSM-II diesel engines delivering 432 kW at 2,300 rpm to Hamilton HJ 362 water-jets.

ASENAV

Fidel Oteiza 1956 P 13 Providencia, Santiago 9, Chile

Tel: (+56 2) 274 15 15
Fax: (+56 2) 204 91 18
e-mail: geren.scl@asenav.cl
Web: www.asenav.cl

Andre Krasemann, *Naval Architect*

Astilleros y Servicios Navales (ASENAV) delivered its first catamaran, the *Patagonia Express*, in 1991. Since then it has delivered a further two aluminium craft and one steel version (slow speed). All these designs have been the result of a collaboration between ASENAV and Båtservice Holding, using the standard Sea Lord 28 design as the basis.

29 m CATAMARAN

Patagonia Express
Pacifico Express
Luciano Beta

Although all these catamarans have slightly differing specifications, they are all based on the same standard vessel and are grouped here for convenience. The *Patagonia Express* was originally built to carry 220 passengers but was refitted to accommodate 60 passengers in comfort for 10-hour trips through the channels of southern Chile. *Pacifico Express* was designed to transport smolts (small live fish) from the cultivation centres to the remote breeding grounds in the fiords of southern Chile. The smolts are carried in six removable tanks on the vessel's working deck. *Luciano Beta* was designed to carry 130 passengers for charter, again in the extreme south of the country. All vessels are constructed of aluminium.

Specifications

Length overall	29.0 m
Beam	8.3 m
Draught	1.6 m
Fuel capacity	12,000 litres
Water capacity	2,000 litres
Max speed	30 kt
Operational speed	24 kt

Propulsion: The vessels are all powered by two Detroit Diesel 12V 149TI (*Luciano Beta* DD 16V 92 TA) each driving a LIPS or Kamewa propeller via a Reintjes reduction gearbox.

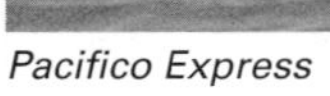
Pacifico Express  0506968

Luciano Beta 0506969

Vessel type	Vessel name	Yard No	Length (m)	Speed (kt)	Seats	Vehicles	Originally delivered to	Date of build
29 m Catamaran	*Patagonia Express*	085	29.0	30.0	60	–	Patagonia Connection	1991
29 m Catamaran	*Luciano Beta*	106	29.0	26.0	150	–	Tolkeyen	1994
29 m Catamaran	*Pacifico Express*	105	29.8	30.0	–	–	Transaldi	1994

Cuba

Damex Shipbuilding & Engineering

Km 7½ Corretera de Punta Gorda, Santiago de Cuba, Cuba

Tel: (+53 22) 68 61 01
Fax: (+53 22) 68 81 01
e-mail: damex@damex.ciges.inf.cu
Web: www.damex.biz

Established in January 1995, this shipyard acts as a licence holder and aftersales service for Damen Shipyards, Netherlands. The yard is able to perform new builds up to 100 m in length and repair and refit for vessels up to 70 m in length.

Damen DFF 3009 catamaran

Rio Las Casas
Rio Jucaro

The first vessel, *Rio Las Casas*, of the pair of DFF 3009 catamarans ordered in 2004, was delivered in late 2005 for operation between Batabano and Isla de la Juventud. The second vessel was delivered in January 2006.

Specifications

Length overall	30.0 m
Beam	9.0 m
Draught	1.6 m
Crew	12
Passengers	237
Fuel capacity	4,500 litres
water capacity	1,800 litres
Operational speed	22 kt

Propulsion: Two Caterpillar 3412 diesel engines each rated at 1,640 KW and driving fixed pitch propellers via a ZF 2050V reverse reduction gearbox.

Vessel type	Vessel name	Yard No	Length (m)	Speed (kt)	Seats	Vehicles	Originally delivered to	Date of build
DFF 3009	*Rio Las Casas*	538,301	30	25	237	–	Intermar Clarion Shipping	2005
DFF 3009	*Rio Jucaro*	538,302	30	25	237	–	Intermar Clarion Shipping	2006

Egypt

Mapso Marine

44 Industrial Zone, Cairo-Ismailia Road, PO Box 11769, Cairo, Egypt

Tel: (+202) 29 68 47 77
Fax: (+202) 26 99 07 80
e-mail: marine@mapso.com
Web: www.mapso.com

The company was established in 1976 and supplies equipment and kits to other companies as well as building its own vessels. Built in 2003, *Hassan Taha* is the company's first high-speed craft.

14.9 m FIREBOAT

Hassan Taha

Designed by Teknicraft of New Zealand and supplied as a kit by Australian company One Steel Aluminium, the vessel is used as a fire and rescue boat. The vessel has a unique propulsion system employing a two-speed gearbox which allows the fire pumps to be run at full capacity with the vessel retaining the ability to drive its water-jets.

Specifications

Length overall	14.9 m
Beam	6 m
Draught	0.7 m
Displacement, min	22 t
Fuel capacity	2,400 litres
Crew	4
Max speed	29.5 kt
Range	330 n miles

Propulsion: Two Scania DI12 diesel engines each rated at 423 kW, driving Hamilton 361 water-jets via ZF two speed gearboxes.

Finland

Aker Finnyards Inc

PO Box 132, FI-00151 Helsinki, Finland

Tel: (+358 10) 67 00
Fax: (+358 10) 670 67 00
e-mail: cruiseferries@akeryards.com
Web: www.akeryards.com

Yrjö Julin, *Managing Director*
Juha Heikinheimo, *Sales Director*

In January 2005, Finland's three largest shipyards combined to form one company. While the yards in Helsinki and Turku specialise in cruise ships, ferries are built at the Rauma shipyard on the southwest coast of Finland. The company also has a division specialising in Arctic technology.

Aker Finnyards is part of the international shipbuilding group Aker Yards, Europe's largest and one of the world's five largest shipbuilders.

Stena HSS 1500

There have been three vessels ordered by Stena AB for operation in the UK. The first vessel was delivered in February 1996 and operates on the Holyhead-Dun Laoghaire route. The second vessel was delivered in June 1996 for operation between Stranraer and Belfast and the third in spring 1997.

The hull design is of a patented narrow-hulled catamaran semi-Swath form designed to DnV certification, with a service speed of 40 kt in a significant wave height of 5 m.

Specifications

Length overall	125.0 m
Beam	40.0 m
Passengers	1,500
Vehicles	375 cars or
–	50 lorries + 100 cars
Operational speed	40 kt

Propulsion: The vessel is fitted with four gas turbines, one General Electric/Kværner

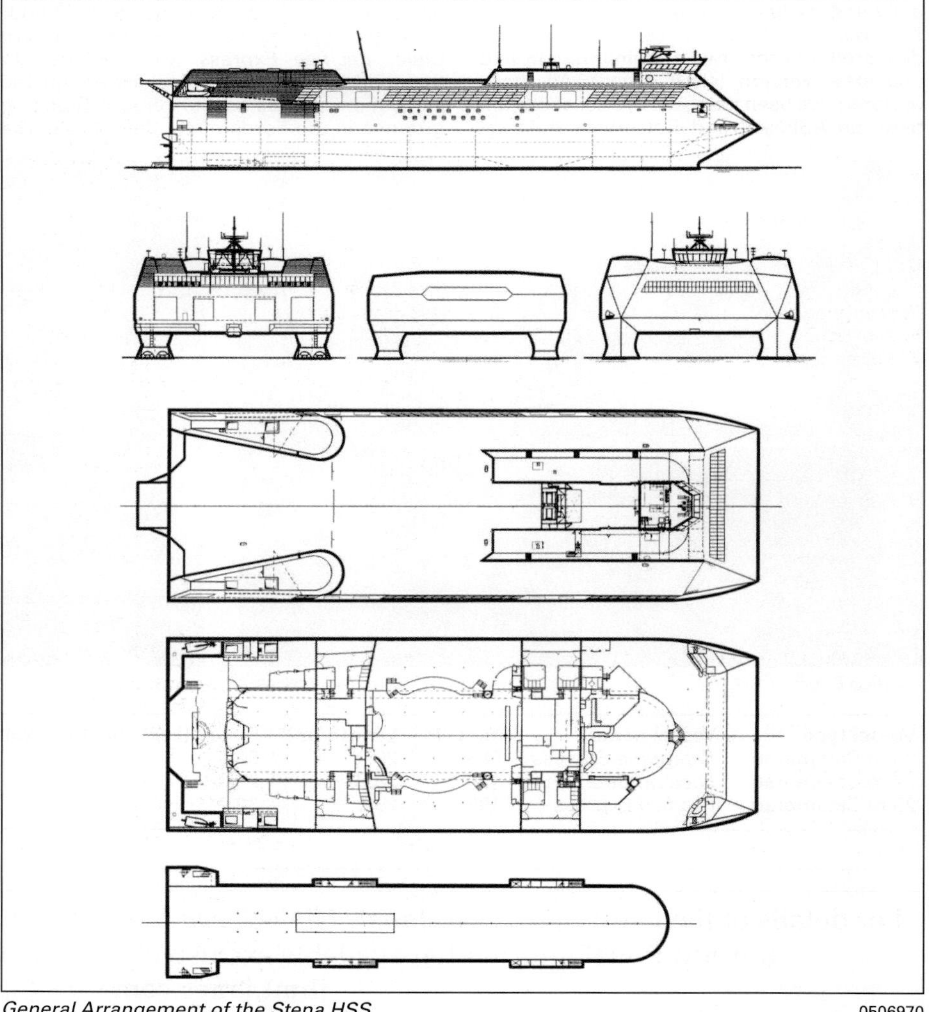

General Arrangement of the Stena HSS 0506970

LM 1600 and one LM 2500 turbine in each hull. The turbines will drive through a MAAG gearbox to two Kamewa 760 S11 water-jets. This arrangement of the gas turbines allows the vessel to cruise at 24 kt on the two LM 1600s, at 32 kt on the two LM 2500s and at 40 kt on all four.

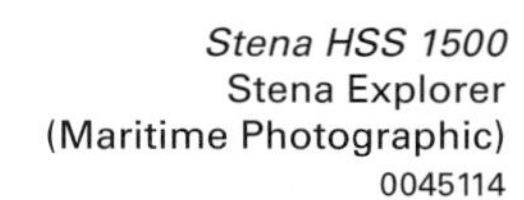

Stena HSS 1500
Stena Explorer
(Maritime Photographic)
0045114

Vessel type	Vessel name	Yard No	Length (m)	Speed (kt)	Seats	Vehicles	Originally delivered to	Date of build
HSS 1500	*Stena Explorer*	404	125.0	40.0	1500	375	Stena AB	1996
HSS 1500	*Stena Voyager*	405	125.0	40.0	1500	375	Stena AB	1996
HSS 1500	*Stena Discovery*	406	125.0	40.0	1500	375	Stena AB	1997

France

Construction Navale Bordeaux (CNB)

Part of the Bénéteau Group
162 quai De Brazza, F-33100 Bordeaux, France

Tel: (+33 5) 57 80 85 50
Fax: (+33 5) 57 80 92 81
e-mail: cnbpro@cnbpro.com
Web: www.cnbpro.com

Builder of composite and aluminium sailing and powered craft, CNB delivered two catamaran ferries in 1999, *Iguana Beach* and *Osprey Shuttle*. The company is marketing similar designs from 16 to 27 m.

VOYAGER CAT 27 m

Iguana Beach
Osprey Shuttle
Caribe Surf

The most recent craft, *Caribe Surf*, differs from the earlier Voyager Cats because it is equipped with two fixed carbon fibre hydrofoils between the hulls to improve performance.

Specifications

Length overall	27.2 m
Length waterline	23.8 m
Beam	8.5 m
Draught	1.7 m
Crew	4
Passengers	206
Fuel capacity	10,000 litres
Water capacity	1,000 litres
Max speed	30 kt
Operational speed	27 to 28 kt

Iguana Beach *in operation in Guadeloupe* 0084083

Structure: All-aluminium hull and superstructure.
Propulsion: *Iguana Beach*: Two Bandouin 12M26SR diesel engines each rated at 870 kW, driving a four-bladed propeller via a ZF BW 255 gearbox. *Osprey Shuttle*: Two Mitsubishi S12R MPTK diesel engines each rated at 1,200 kW driving fixed-pitch propellers via Twin Disc MGN 433 gearboxes.

Vessel type	Vessel name	Yard No	Length (m)	Speed (kt)	Seats	Vehicles	Originally delivered to	Date of build
Voyager Cat 23.8 m	*Voyager 2*	–	23.8	20.0	205	–	Mariteam Caraibes	–
Voyager Cat 27 m	*Osprey Shuttle*	–	27.2	27.0	206	–	–	1999
Voyager Cat 27 m	*Iguana Beach*	–	27.2	27.0	206	–	–	1999
Voyager Cat 27 m	*Caribe Surf*	–	27.2	28.0	226	–	M&M Transportation Services	2003

Iris Catamarans

Avenue du President Wilson, Port-Neuf, La Rochelle F-17000, France

Tel: (+33 5) 46 42 42 18
Fax: (+33 5) 46 42 41 82
e-mail: info@iris-catamarans.com
Web: www.iris-catamarans.com

Jean François Fountaine, *President*
Patrice Desprez, *Managing Director*

The IRIS (InteR Islands Shuttle) project is a new multipurpose catamaran concept for the transportation of passengers, containers, or special cargo services for relatively short voyages. The company now has a complete range of craft including *IRIS 6.2, IRIS 6.1, IRIS Cargo, IRIS 3.1 and IRIS 37.*

The catamaran design consists of an open platform suspended between the two hulls, with the bridge situated at the forward end of this platform. Passenger modules have the dimensions of 40 ft containers, or of course containers, and can be mounted on this platform.

IRIS 6.1

The first IRIS catamaran was launched on 28 November 1997 and operates an inter-island shuttle service between Pointe-à-Pitre on Guadeloupe to the islands of Marie Galante and Les Saintes.

Specifications

Length overall	42.8 m
Length waterline	39.0 m
Beam	10.6 m
Draught	1.6 m
Payload	51 t (incl modules)
Passengers	240
Fuel capacity	9,000 litres
Water capacity	1,200 litres
Propulsive power	3,480 kW
Max speed	37 kt
Operational speed	33 kt
Range	300 n miles

Classification: DnV +1A1 HSLC R3 Passenger EO.

Structure: The main construction material is vacuum-bagged GRP-Sandwich. This method is used to construct the hulls, control station and passenger modules; but all shock-sensitive areas are single-skin laminate. The transverse beams that support the bridgedeck structure are built from aluminium alloy and attached to the hulls on reinforced watertight bulkheads.

Propulsion: Main engines are two MTU 12V 4000 M70 diesels, each producing 1,740 kW; driving two water-jets.

IRIS 6.2

This version can carry six container-sized modules, on two floors. The first vessel of this type was launched in 2000, with one other on order. Two of these craft were delivered in 2001 for operation on Lake Neuchâtel in Switzerland. During the Expo 02 event, these vessels ran for 159 days with seven daily trips each, making a total of 2,226 trips.

In 2007, an IRIS 6.2, *KriloJet* was delivered to Uto Kapetan Luka for operation on the Split-Hvar-Korkula route in Croatia.

Specifications

Length overall	42.8 m
Length waterline	39.0 m
Beam	10.6 m
Draught	1.6 m
Payload	72 t (incl modules)
Passengers	392
Fuel capacity	10,000 litres
Water capacity	1,900 litres
Propulsive power	4,640 kW
Max speed	40.0 kt
Operational speed	35.0 kt
Range	300 n miles

Classification: DnV +1A1 HSLC R3 Passenger EO.

Propulsion: Main engines are two MTU 16V 4000 M70 diesels, each producing 2,320 kW; driving two MJP 750 water-jets.

IRIS 3.1

The first vessel to this design was launched in November 2000 and is designed for protected water services with a passenger capacity of up to 200. The first vessel was fitted with a pair of 350 kW diesels giving a maximum speed of 18 kt. Four further vessels with a speed of 26 kt were ordered in 2001 for use in the Swiss Expo 02 in 2002. During the Expo 02 event these vessels ran for a total of 3,816 trips over 159 days.

Specifications

Length overall	26.0 m
Length waterline	24.9 m
Beam	9.0 m
Draught	1.0 m
Payload	18.1 t
Passengers	200
Fuel capacity	2,500 litres
Water capacity	800 litres
Propulsive power	700 kW
Max speed	18–26 kt
Operational speed	15–23 kt

Propulsion: The first vessel was powered by two 350 kW diesel engines each driving a fixed-pitch propeller.

IRIS CARGO (DESIGN)

This design is based on the IRIS 6.1 and 6.2 hulls with a deck structure which provides for a 70 tonne payload.

An IRIS 6.2 vessel on Lake Neuchâtel 0526486

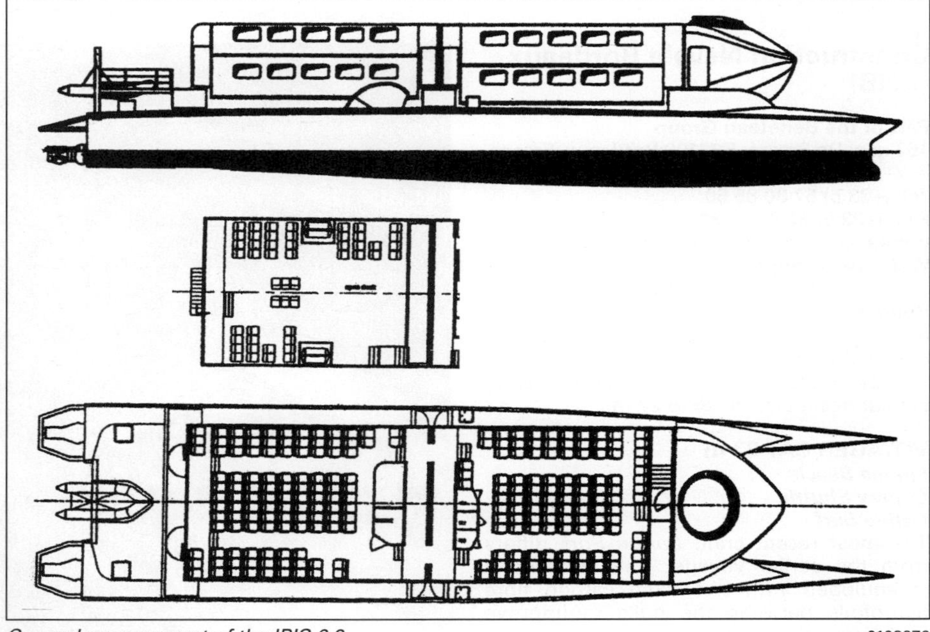

General arrangement of the IRIS 6.2 0103276

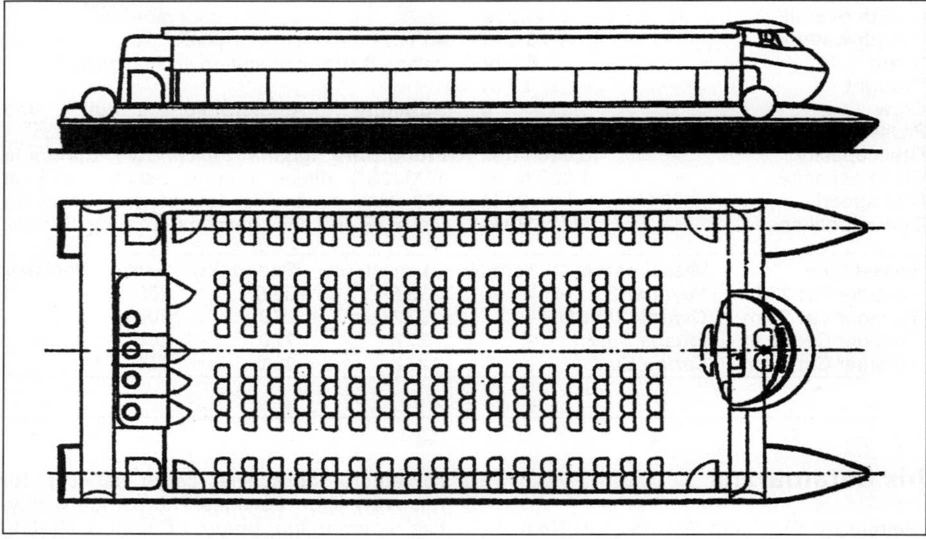

General arrangement of the IRIS 3.1 0103267

Vessel type	Vessel name	Yard No	Length (m)	Speed (kt)	Seats	Vehicles	Originally delivered to	Date of build
IRIS 6.1	*Iris 1*	–	42.8	33.0	240	–	Guadeloupe	28th November 1991
IRIS 6.2	*Iris Jet*	–	42.8	35.0	392	–	–	2000
IRIS 3.1	–	–	26.0	16.0	200	–	–	2000
IRIS 6.2	*Lyon*	–	42.8	35.0	392	–	Swiss Expo 02	2001
IRIS 3.1	–	–	26.0	26.0	200	–	Swiss Expo 02	2001
IRIS 3.1	–	–	26.0	26.0	200	–	Swiss Expo 02	2001
IRIS 3.1	–	–	26.0	26.0	200	–	Swiss Expo 02	2001
IRIS 3.1	–	–	26.0	26.0	200	–	Swiss Expo 02	2001
IRIS 6.2	*KriloJet*	–	42.8	40.0	398	–	Uto Kapetan Luka	2007

IRIS 37 (DESIGN)

This is the smallest of the IRIS range at an overall length of 11.2 m and has a passenger capacity of approximately 50.

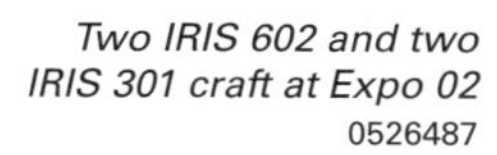

Two IRIS 602 and two IRIS 301 craft at Expo 02
0526487

Hong Kong

Afai Ships Ltd

Suite 1516, Tower Two, Grand Century Place, Mongkok, Kowloon, Hong Kong

Tel: (+852) 23 07 62 68
Fax: (+852) 23 07 51 70
e-mail: afai@afaiships.com
Web: www.afaiships.com

Vitus Szeto, *Managing Director*

Afai began as a ship repairer in 1949. It started shipbuilding in the 1960s and delivered Hong Kong's first aluminium catamaran, a 21 m vessel, in 1982. Today Afai specialises in the construction of aluminium high-speed vessels, and is currently constructing an 80 m high-speed catamaran.

AQUAN2400

Afai delivered the AQUAN2400 to a Hong Kong operator for local usage in December 1996. Its hull is an upgrade of the AMD170 hull form used for the Wuzhou vessel. The uniqueness of the AQUAN2400 is in its appearance, which looks more like a bullet train than any ship on the market today.

Intended to operate in Hong Kong waters, it is designed to withstand wave heights of 1.2 m at speeds of 25 kt. Total passenger seating is for 200. It is powered by two diesels operating at a continuous rating of 700 kW at 1,925 rpm. It has a normal fuel capacity of 3,000 litres, or a maximum capacity of 6,000 litres, and has an endurance of 200 n miles.

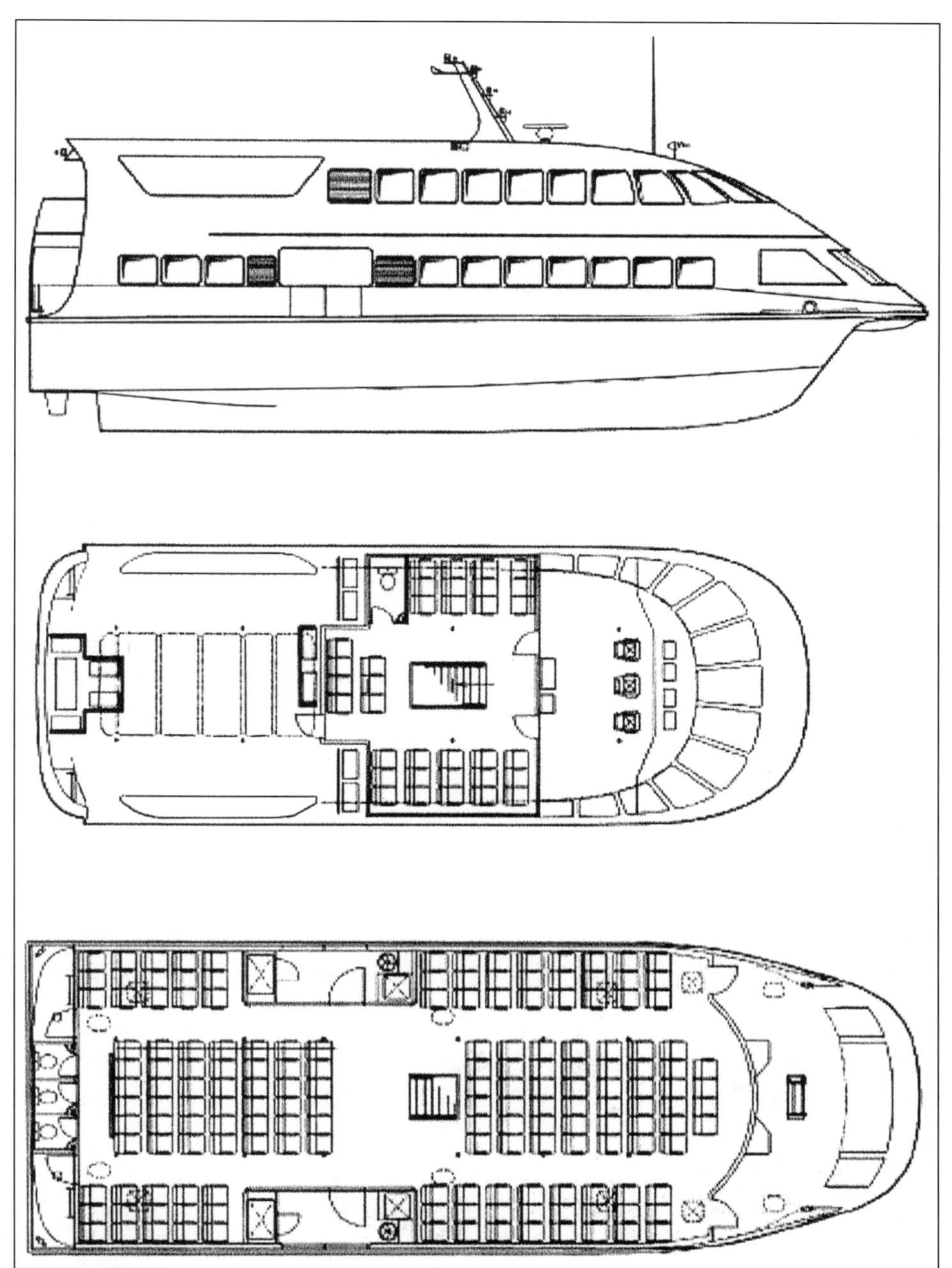

General arrangement of Aquan catamaran 0554534

80 m A50 CAR FERRY

The Cat

Three of these vessels have now been built (one by InCat, Australia and two by Afai). Different configurations of accommodation for passengers and cars were used on all three vessels. Built under DnV classification, the vessels satisfy class +1A1 HSLC R2 Car Ferry B EO, and are registered under the IMO HSC code.

The 80 m vessel *The Cat* can carry 89 vehicles and 400 passengers, while travelling at speeds of 48 kt. It is powered by four diesel engines each with 5,650 kW, and four steerable, reversable water-jets.

Specifications

Length overall	80.1 m
Beam	19.0 m
Draught	2.2 m
Passengers	400
Vehicles	89 cars
Max speed	50 kt
Operational speed	48 kt

Classification: DnV +1A1 HSLC R2 car ferry B EO.

Propulsion: Powered by four Ruston 16 RK 270 marine diesels each rated at 5,650 kW and each directly coupled to a Kamewa S80II steerable water-jet.

AMD200 Jin Xing 0506828

CA3 AIRPORT RESCUE BOAT

Afai delivered an InCat designed 23 m airport rescue catamaran to the Hong Kong government's Civil Aviation Department in July 1990. In addition to providing rescue services at Kai Tak airport, the vessel is also equipped for firefighting duties. Standard crewing is seven persons.

Intended to be normally operated only within class I and II waters, it is designed to withstand minimum wave heights of 1.2 m at speeds of at least 25 kt. Forward there is a wheelhouse with a floor 1 m above the main deck level to give the crew a good all-round view. A firefighting monitor is mounted on the roof of the wheelhouse.

An emergency recess with two stairways is provided on each side of the vessel, the length of the rescue recess platforms being at least 2.5 m to facilitate the easy handling of stretchers. Aluminium fold-up panels with non-skid steps are fitted at each side of the recesses to allow survivors in the water to climb on board.

An awning aft of the wheelhouse is high enough to provide ample headroom for two tiers of stretchers and not restrict rearward visibility from the wheelhouse. Below the awning, there are three floodlights on each side to illuminate the deck area after dark.

Six rows of seats located port and starboard, adjacent the outer awning supports, are used as weathertight lockers for rescue equipment. Total rescue capability of the design is 32 people on stretchers, 64 seating and at least 154 standing. With the two diesels operating at their continuous rating of 550 kW at 2,000 rpm and leaving base with 2,400 litres of fuel on board, endurance is 4.5 hours.

The vessel is the thirteenth InCat design built by Afai at its Hong Kong yard since 1982.

AMD200

Jin Xing

Constructed by Afai to an AMD design for a Chinese customer, this 28 m AMD design was delivered in June 1993.

Specifications

Length overall	28.0 m
Beam	10.3 m
Draught	1.8 m
Passengers	242

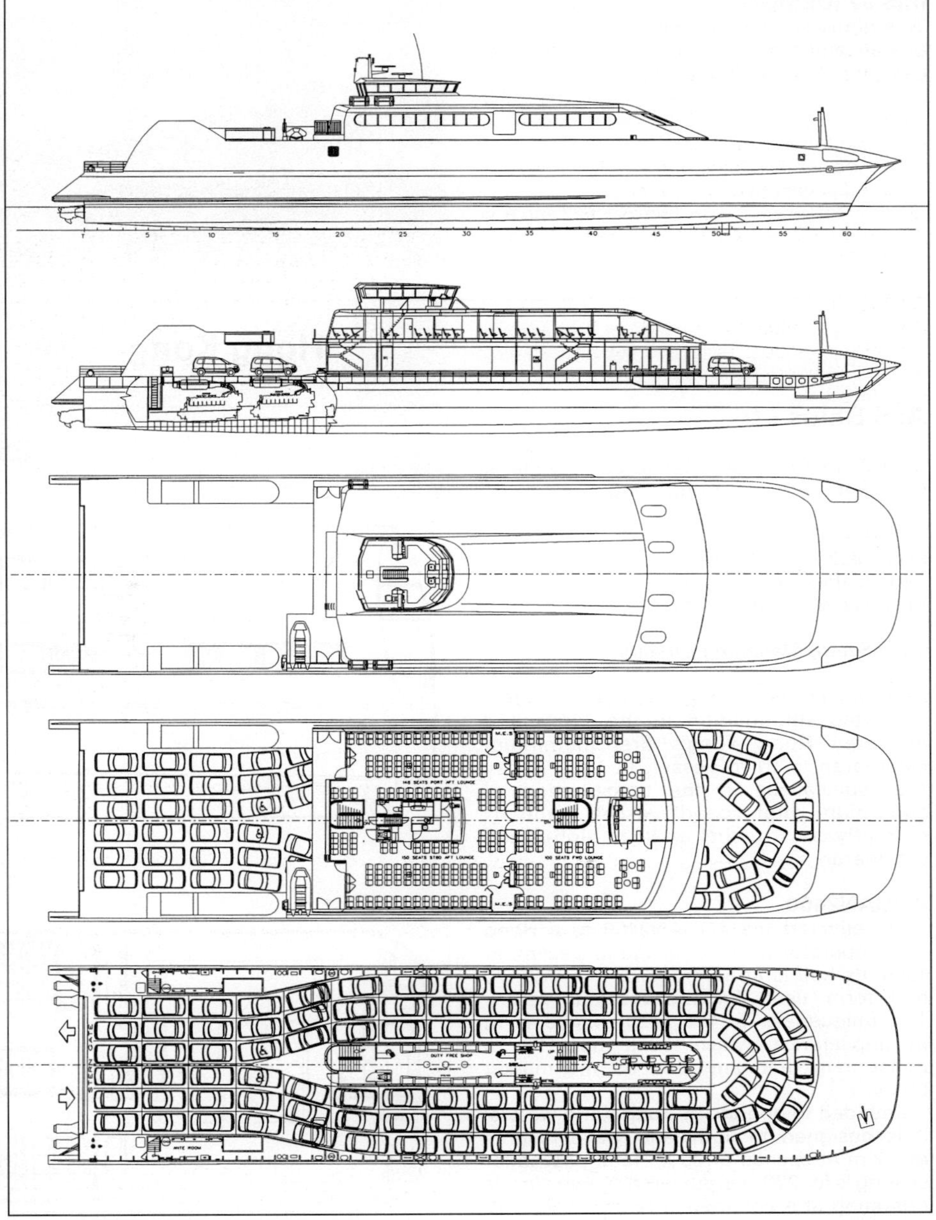

A50 general arrangement 0038692

Vessel type	Vessel name	Yard No	Length (m)	Speed (kt)	Seats	Vehicles	Originally delivered to	Date of build
Afai 21 m	*Mingzhu Hu*	–	–	29.0	150	None	Nig Bo Hua Gang Co Ltd, Zhe Jiang Province, China	1982
Afai 21 m	*Yin Zhou Hu*	–	–	29.0	150	None	Nig Bo Hua Gang Co Ltd, Zhe Jiang Province, China	1982
Afai 21 m	*Liu Hau Hu*	–	–	29.0	150	None	Macao Navigation, Hong Kong	1982
Afai 21 m	*Li Jiang*	–	–	29.0	150	None	Kwai Kong Shipping Co Ltd, Hong Kong	1983
Afai 16 m	*Kwong Fai*	–	–	21.0	40	None	Castle Peak Power Co Ltd, Hong Kong	1984
Afai 21 m	*Yue Hai Chun*	–	–	29.0	169	None	Shen Zhen Shipping, China	1984
Afai 21 m	*Shen Zhen Chun*	–	–	29.0	169	None	Shen Zhen Shipping, China	1985
Afai 21 m	*Zhu Hai Chun*	–	–	29.0	169	None	Shen Zhen Shipping, China	1986
Afai 22 m	*Gong Bian 153*	–	–	33.0	–	None	Guang Zhou Customs Dept, China	1988
Afai 22 m	*Ling Nan Chun*	–	–	28.0	187	None	Shen Zhen Shipping, China	1988
Afai 22 m	*Nan Hai Chun*	–	–	28.0	187	None	Shen Zhen Shipping, China	1989
23 m Rescue Catamaran	*CA3*	–	23.0	25.0	250	None	Hong Kong Government Civil Aviation Department	1990
Afai 23 m	*Dong Feng Chun*	–	–	28.0	197	None	Shen Zhen Shipping, China	1991
AMD200	*Jin Xing*	–	28.0	25.0	235	None	Shen Zhen Shipping, China	1993
AMD170	*Wuzhou*	–	24.0	29.0	150	None	Wuzhou Guangxi Navigation Co	1994
Afai 17 m	*Harbour Jet 10*	–	–	28.0	85	None	Harbour Jet Shipping Co	1994
Afai 14 m	*Harbour Jet 5*	–	–	25.0	60	None	Harbour Jet Shipping Co	1995
Afai 14 m	*Harbour Jet 6*	–	–	25.0	60	None	Harbour Jet Shipping Co	1995
Afai 16 m	*Polly 3*	–	–	24.0	100	None	The Polly Ferry Co Ltd	1996
Afai 16 m	*Polly 4*	–	–	24.0	118	None	The Polly Ferry Co Ltd	1996
Afai 15 m	*Afai 18*	–	–	25.0	108	None	–	1996
AQAN 2400	*Aquan One*	–	–	25.0	200	None	New World First Ferry	1996
A50	*The Cat*	–	80.1	48.0	400	89	Savory Industries Inc	1998
AMD188	*Aquan Two*	–	27.6	23.0	222	None	New World First Ferry	1999
A50	–	–	80.1	48.0	700	80	–	(2008)

Fuel capacity	8,000 litres
Water capacity	1,000 litres
Operational speed	25 kt

Classification: Chinese ZC (CCS).

Propulsion: Engines: two Deutz MWM TBD 604B V8, 840 kW each at 1,800 rpm (other engine options available).

Transmissions: two Reintjes WVS 430 gearboxes.

Thrust devices: two FP Propellers.

Electrical system: 380 V AC 3-phase, 220 V AC single phase, 24 V DC.

Generators: two 75 kVA.

AMD188

Delivered in 1999, the Afai AMD188 is a high-capacity, low cost ferry currently operating in Hong Kong harbour.

Specifications

Length overall	27.6 m
Beam	8.9 m
Draught	1.6 m
Passengers	222

A50 (Afai 08) on trials 0024757

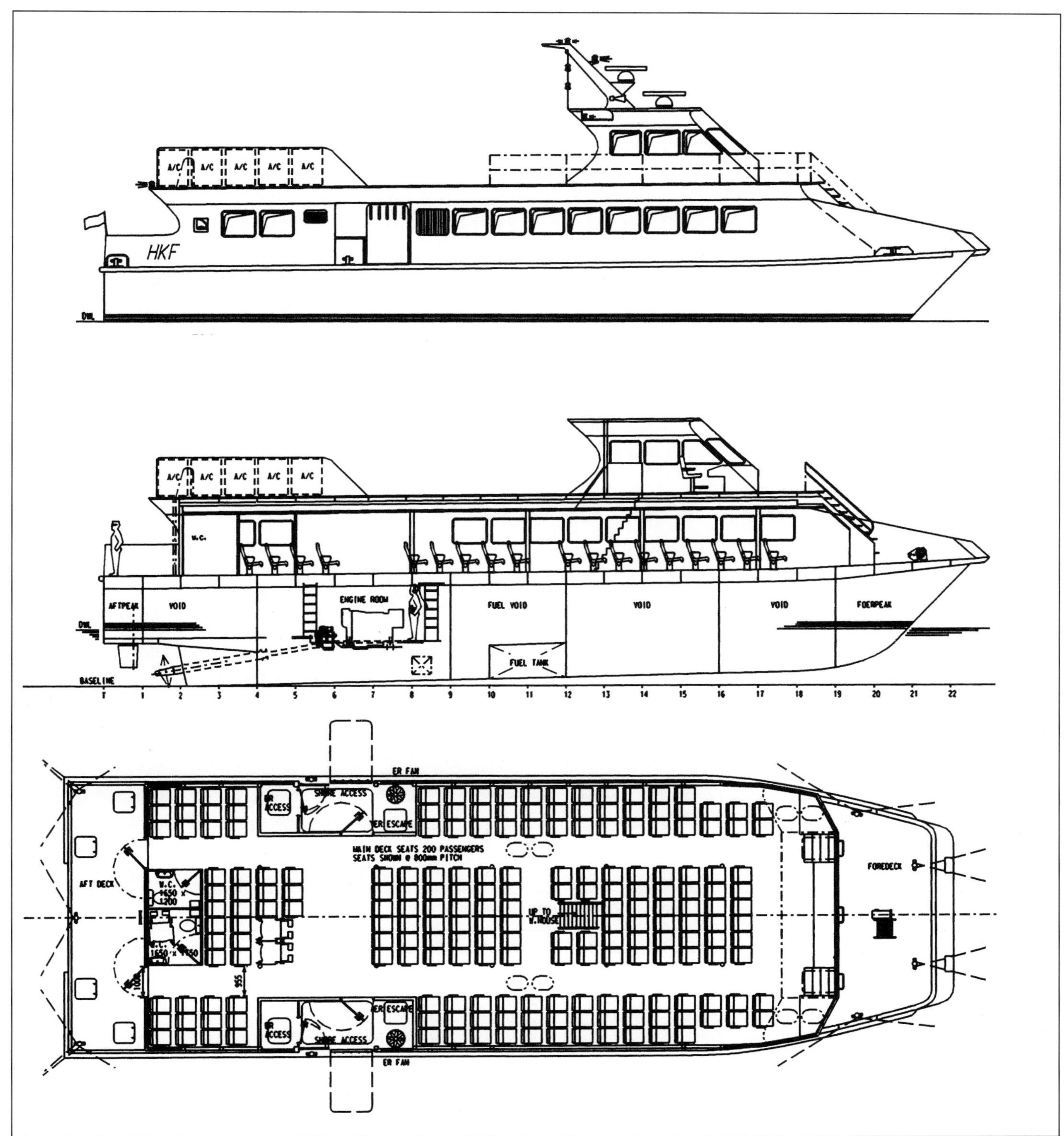

Layout of AMD 188 catamaran 0084071

Afai's AQUAN catamaran 0038693

AMD170 Wuzhou 0506888

Fuel capacity 5,000 litres
Water capacity 600 litres
Operational speed 23 kt
Structure: All-aluminium hull and superstructure.
Propulsion: Powered by twq GM 12V 92 TA DDEC diesel engines each rated at 560 kW at 2,100 rpm driving fixed-pitch propellers via a ZF BW 190 gearbox.
Auxiliary Systems: Two 48 kW generators.

AMD170

Wuzhou

Constructed by Afai engineers to an AMD design for the Wuzhou Guangxi Navigation Co, this vessel was delivered in mid-1994.

Specifications

Length overall 24.0 m
Beam 8.9 m
Draught 1.8 m
Passengers 150
Fuel capacity 8,000 litres
Operational speed 29 kt
Classification: Chinese ZC (CCS).
Propulsion: Two Issotta Fraschini ID 36SS12 diesels each driving a fixed-pitch propeller via a ZF BW460 reverse reduction gearbox.

Cheoy Lee Shipyards Ltd

89-91 Hing Wah Street West, Lai Chi Kok, Kowloon, Hong Kong

Tel: (+852) 23 07 63 33
Fax: (+852) 23 07 55 77
e-mail: info@cheoyleena.com
Web: www.cheoyleena.com

Ken Lo, *Director*
Jonathan Cannon, *Sales Manager*

The company was originally founded in Shanghai over a century ago. It moved to a small yard in Hong Kong in 1937, then to a much larger yard in 1940 and was occupied with building and repairing cargo ships. It diversified into building teak sailing yachts for export in 1956 and acquired a large site for a second yard on Lantau Island in 1960. The company commenced building motor yachts and building in GRP in 1961 and now builds commercial and pleasure craft in steel, composites and aluminium. The company has built over 4,700 vessels since the current records commenced in 1950. The company is currently expanding operations into mainland China.

Cheoy Lee 20 m GRP Catamaran Hung Wai II 0007942

28 m GRP CATAMARAN

Sea Supreme
Park Island 6
Sea Superior

The 28 m catamaran *Sea Supreme* entered service in Hong Kong in 2000 and was the second in a series of 24/28 m vessels designed by Crowther Multihulls of Australia. The first hull from this GRP mould was the shortened 24 m VIP catamaran, *Tin Hau*. Two vessels, *Park Island 6* and *Sea Superior*, were delivered in 2003 and powered with uprated engines developing 970 kW at 1,800 rpm.

Cheoy Lee 28 m catamaran Sea Supreme 0100694

Vessel type	Vessel name	Yard No	Length (m)	Speed (kt)	Seats	Vehicles	Originally delivered to	Date of build
Cheoy 20 m GRP Catamaran	–	–	20.0	24.0	200	none	–	–
Cheoy 20 m GRP Catamaran	–	–	20.0	24.0	200	none	–	–
Cheoy 20 m GRP Catamaran	–	–	20.0	24.0	200	none	–	–
Cheoy 20 m GRP Catamaran	–	–	20.0	24.0	200	none	–	–
Cheoy 20 m GRP Catamaran	–	–	20.0	24.0	200	none	–	–
Cheoy 20 m GRP Catamaran	*Hung Wai II*	–	20.0	24.0	200	none	–	1997
Cheoy 24 m VIP Catamaran	*Tin Hau*	–	24.0	26.0	–	none	–	1998
Cheoy 28 m Catamaran	*Sea Supreme*	–	28.0	25.0	380	none	Hong Kong	2000
Cheoy 28 m Catamaran	*Park Island 6*	–	28.0	25.0	398	none	Park Island	2003
Cheoy 28 m Catamaran	*Sea Superior*	–	28.0	25.0	398	none	Park Island	2003
Cheoy 26 m Catamaran	*Park Island 7*	–	26.0	27.0	223	none	Park Island	2003
Cheoy 26 m Catamaran	*Park Island 8*	–	26.0	27.0	223	none	Park Island	2004

Specifications

Length overall	28.0 m
Beam	8.1 m
Displacement	111 t
Passengers	380
Propulsive power	1,492 kW
Operational speed	25.0 kt

Structure: Single skin GRP hull and GRP foam sandwich superstructure.
Propulsion: Powered by two Cummins KTA-38-M1 each rated at 746 kW and coupled to a Reintjes WVS 430/1M gearbox driving a fixed-pitch propeller.
Auxiliary Systems: Two 30 kW Perkins diesel generators.

26 m CATAMARAN

Designed by Crowther Design of Australia, two of these vessels were delivered to Park Island Transport Company in 2003.

Specifications

Length overall	26.0 m
Length waterline	23.7 m
Beam	8.5 m
Draught	1.8 m
Passengers	223
Crew	5
Operational speed	27.0 kt

Classification: Hong Kong Marine Dept.
Construction: Aluminium.
Propulsion: Powered by two Cummins KTA-38-M2 diesel engines each rated at 895 kW at 1,800 rpm.

24 m VIP CATAMARAN

The 24 m catamaran, *Tin Hau*, was delivered to the Hong Kong government in 1998 for VIP transport duties. Designed by Crowther Multihulls of Australia, the vessel is built of GRP.

Specifications

Length overall	24 m
Beam	8.1 m
Propulsive power	1,940 kW
Max speed	26 kt

Structure: Single-skin GRP hull and GRP foam sandwich superstructure.
Propulsion: Powered by two Cummins KTA38-M2 diesel engines each rated at 970 kW at 1,800 rpm and each driving a fixed-pitch propeller.
Electrical system: Two Onan 45 kW generators.

20 m GRP CATAMARAN

Delivered in 1997, the 20 m *Hung Wai II* is the first of three fast catamaran models that Cheoy Lee has developed with Australian designer Crowther Multihulls, covering the 15 to 28 m range. Classed by DnV, they can all be constructed in GRP or aluminium. Six 20 m vessels are now in service.

Specifications

Length overall	20.0 m
Beam	6.5 m
Passengers	200
Propulsive power	798 kW
Operational speed	24 kt

Classification: Det Norske Veritas.
Structure: GRP hull and superstructure.
Propulsion: Two diesel engines each rated at 399 kW.

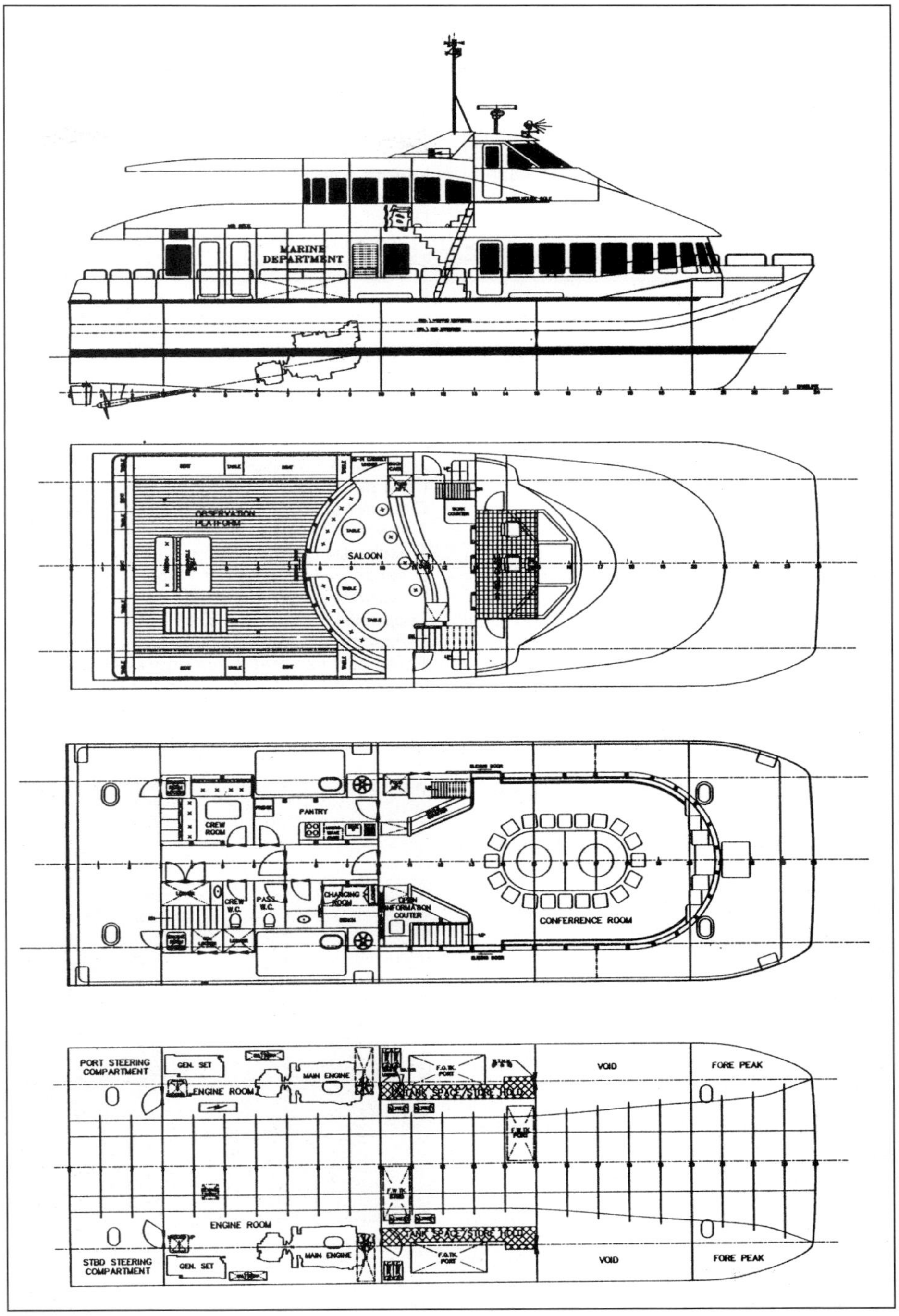

Cheoy Lee 24 m catamaran, general arrangement 0024758

Wang Tak Engineering and Shipbuilding Co Ltd

Wang Tak Building, 85 Hing Wah Street West, Lai Chi Kok, Kowloon, Hong Kong

Tel: (+852) 27 46 28 88
Fax: (+852) 23 07 55 00
e-mail: info@wangtak.com.hk

Feat Szeto, *Director and General Manager*

The Wang Tak Shipyard based in Hong Kong offers shipbuilding and repair services in steel and aluminium materials. In 2005 the company delivered one 33 m catamaran to Shanghai communter operator and three 28 m catamarans to New World First Ferry in Hong Kong in 2003.

Seacat 41 m catamaran

Designed by Sea Speed Design for operation with Shanghai Ya Tong, this vessel is the seventh Seacat catamaran delivered by Wang Tak.

Specifications

Length overall	41.6 m
Length waterline	40.0 m
Beam	11.8 m
Draught	1.4 m
Passengers	398
Max Speed	30 kts

Propulsion: Two MTU 12V4000 M70 diesel engines rated at 1740 kW at 2000 rpm, each driving MJP 650-DD water-jets

SEACAT 33 m CATAMARAN

Xunlong 3

Designed by Commercial Marine Consulting Services to be fuel efficient, the first vessel was delivered to Shenzhen Xunlong Passenger Shipping in 2007.

Specifications

Length overall	33.0 m
Length waterline	27.1 m
Beam	8.0 m
Draught	1.02 m
Passengers	232
Crew	8
Operational speed	28 kt

Propulsion: Two MTU 16V2000 M70 diesel engines delivering 1050 kW each driving MJP 550-DD water-jets via ZF 4540 gearboxes.

33 m CATAMARAN

Liang Yu

Designed by Commercial Marine Consulting Services, this vessel is expected to be launched in late 2005.

Specifications
Length overall 33.0 m
Length waterline 30.7 m
Beam 8.5 m
Draught 1.1 m
Deadweight 22 tonnes
Passengers 252
Fuel capacity 4200 l
Maximum speed 28 kt
Operational speed 26 kt
Propulsion: Two MTU 16V2000 M70 diesel engines delivering 1050 kW each driving MJP 550-DD water-jets.

28 m catamaran

First Ferry IX, X and XI
Specifications
Length overall 28.23 m
Beam 8.5 m
Draught 2.5 m
Passengers 231
Fuel capacity 5,600 litres
Water capacity 1,500 litres
Operational speed 25 kt
Classification: Lloyd's Register of Shipping SSC Passenger catamaran HSC G2.
Propulsion: Two MTU 12V2000 M70 diesel engines each rated at 788 kW at 2,100 rpm and each driving a Teignbridge 5 bladed propeller via a ZF 2500 A gearbox.
Construction: Aluminium alloy hull and superstructure.

Italy

Cantiere Navali di Pesaro

Strada Tra i Due Porti, 48, I-61100 Pesaro, Italy

Tel: (+39) 072 14 25 81
Fax: (+39) 072 12 46 79
e-mail: cnp@cnp.it
Web: www.cnp.it

Founded in the 1950s, Cantieri Navale di Pesaro has been mainly involved in ship repair and cargo craft construction. In 2000 the company delivered two 28 m aluminium catamaran craft for operation on the Italian Lakes. A further four vessels were delivered in 2003.

28 m CATAMARAN

Città di Como
Città di Lecco
Tivano
Freccia del Garda
G Verga
D'Annunzio
Leopardi
G Pascoli
Three of these craft have now been delivered for operations on Lake Como with a further three on order. All have been specified with surface piercing propeller stern drives apart from one, which was delivered with water-jet propulsion for evaluation.
Specifications
Length overall 27.7 m
Length waterline 24.6 m
Beam 6.0 m
Displacement 91.6 t
Passengers 300
Max speed 28 kt
Classification: RINA class 100 A 1.1*.
Propulsion: Two MTU 16V 2000 M70 diesel engines each rated at 1,030 kW and driving a LAME Searider Ti90 surface piercing propeller unit.

Rodriquez Cantieri Navali SpA

Via S Raineri 22, I-98122 Messina, Italy

Tel: (+39 0907) 76 52 83
Fax: (+39 0907) 76 52 93
e-mail: marketing@rodriquez.it
Web: www.rodriquez.it

Luciano La Noce, *President*
Marco Ragazzini, *Managing Director*
Alcide Sculati, *Technical Director*
Ettore Morace, *Sales and Marketing Director*
Sam Crockford, *Commercial Director*

Since 1956, when the shipyard started to build commercial fast ferries, Rodriguez has delivered over 220 high-speed craft and the company continues to be a world leader in the design, engineering and manufacture of such craft. The company has four shipyards in Italy, one at Messina in the south, one at Pietra Liguna in the north, the Intermarine Yard at Sarzana in the north and the fourth is in Naples where the pleasure craft production is managed. It also has a shipyard in Rio de Janeiro, Brazil.

In May 2004, RCN Finanziaria purchased a majority share of Rodriquez Cantieri Navali. The other shareholders are MRS Sviluppo and Ustica Lines.

Achenar 0506830

52 m WAVE-PIERCING FERRIES

In May 2006, the Sultanate of Oman ordered three 52 m wave-piercing ferries designed by AMD Marine Consulting. The vessels, due for delivery in 2008, will be fitted out as fast ferries but will also be used as search and rescue vessels.
Specifications
Length overall 52 m
Length waterline 45.5 m
Beam 15.7 m
Draught 2.0 m
Passengers 106
Vehicles 22
Service speed 40 kt
Propulsion: Engines: four MTU 16V 4000 M71 diesel engines each rated at 2465 kW at 2,000 rpm, driving two Kamewa 71 S11 water-jets.

CITYCAT 43

Achernar
The first vessel of this type was ordered by Caremar SpA and delivered to this Naples-based operator in February 1993. The vessel is operated on the company's route from Naples to Capri.
Specifications
Length overall 43.2 m
Length waterline 36.4 m
Beam 10.9 m
Draught 1.4 m
Displacement, min 115 t
Displacement, max 150 t
Passengers 354
Propulsive power 2 × 2,000 kW
Max speed 34 kt
Range 200 n miles
Construction: Welded and riveted aluminium.
Propulsion: Engines: two MTU 16V 396 TE 74L, 2,000 kW each at 2,000 rpm, driving two Kamewa 71 S11 water-jets.

CITYCAT 40

Maria Celeste Lauro
Maria Sole Lauro
V. Cat 40
In December 2003, the company received orders for three CityCat 40 vessels, one for Bahamas Fast Ferry and two for delivery to Naples-based Alilauro.

The craft destined for the Bahamas has been modified to suit local conditions and is fitted with water-jets and four 610 kW MTU engines. The vessels for Alilauro are powered by twin 2,000 kW engines. The first

Vessel type	Vessel name	Yard No	Length (m)	Speed (kt)	Seats	Vehicles	Originally delivered to	Date of build
CityCat 43	*Achernar*	259	43.2	–	354	none	Caremar SpA	1993
CityCat 40	*Maria Celeste Lauro*	339	39.7	28.0	300	none	Alilauro	2004
CityCat 40	*Maria Sole Lauro*	342	39.7	32.0	320	none	Alilauro	2004
CityCat 40	*V.Cat 40*	313	39.7	27.0	260	none	Bahamas Fast Ferry	2006
AMD 52	–	350	52.0	40	106	22	Sultanate of Oman	(2008)
AMD 52	–	351	52.0	40	106	22	Sultanate of Oman	(2008)
AMD 52	–	352	52.0	40	106	22	Sultanate of Oman	(2008)

vessel to be completed, Maria Celeste Lauro, was delivered in December 2004.

Specifications

Length overall	39.7 m
Length waterline	36.0 m
Beam	10.9 m
Draught	1.6 m
Passengers	260/300
Propulsive power	4 × 610 kW/2 × 2,000 kW
Max speed	27.5/31.5 kt

Classification: RINA HSC class.

Propulsion: Two Caterpillar 3576B/MTU 12V 396 diesel engines each driving a Kamewa 71 SII water-jet. Alilauro vessel or two MTU 16V 396TE74L diesels each driving a Kamewa 63 SII water-jet (Bahamas Fast Ferry).

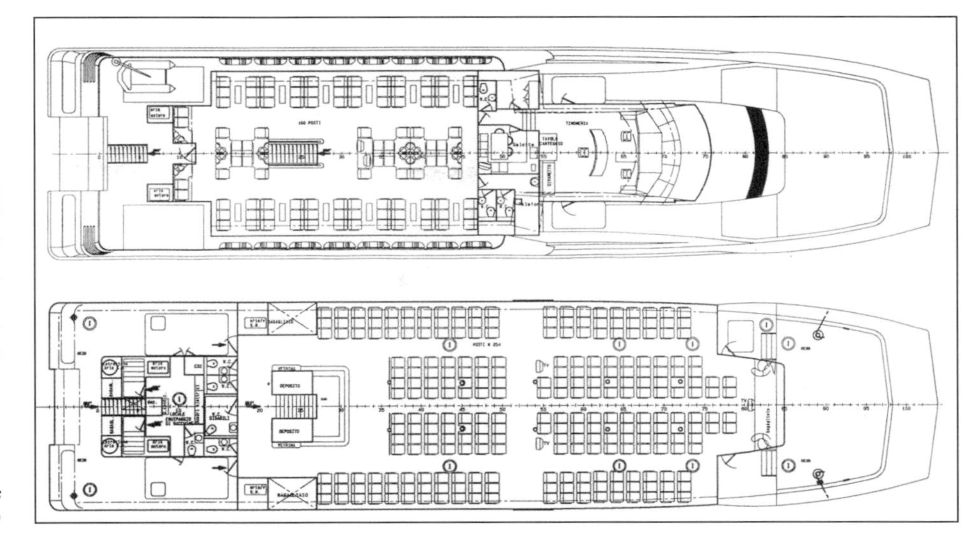

Achenar deck layouts
0506829

Japan

IHI Corporation

22-23, Kaigan 3-chome, Minato-ku, Tokyo 108-0022, Japan

Tel: (+81 3) 62 04 78 00
Fax: (+81 3) 62 04 88 00
Web: www.ihi.co.jp

Hirohiko Sakurai, *Assistant Manager*

IHI has two main shipyards, one in Yokohama and one in Kure.

The Super Slender Twin Hull (SSTH) concept was developed by IHI. In 1991, the first SSTH project was launched, called the SSTH 30, with the SSTH 70 following in 1998.

SSTH 30 0506650

SSTH 30

This craft experiences vertical accelerations less than 0.1 g measured in head sea of 40 m wave length and 1 m significant wave height.

Specifications

Length overall	30.4 m
Beam	5.6 m
Passengers	68
Propulsive power	880 kW
Max speed	28.2 kt
Operational speed	21 kt

Propulsion: Engines: two MTU 8V 183TE 92 diesels, 440 kW (MCR) each.

Ocean Arrow 0024754

SSTH 50 (DESIGN)

Specifications

Length overall	49.9 m
Beam	8.0 m
Draught	1.2 m
Passengers	320
Propulsive power	3,640 kW
Max speed	35 kt
Range	150 n miles

Structure: Hull material: welded aluminium.

Propulsion: Engines: two 1,820 kW (MCR) diesels, driving two water-jets.

SSTH 70

Construction of this 72 m catamaran vessel was started in 1997 and the vessel was delivered to her owner, Kumamoto Ferry Co Ltd, in 1998. The vessel can carry up to nine buses or trucks in place of the 51 cars.

Specifications

Length overall	72.0 m
Length waterline	67.9 m
Beam	12.9 m
Draught	2.0 m
Deadweight	204 t
Passengers	430
Vehicles	51 cars

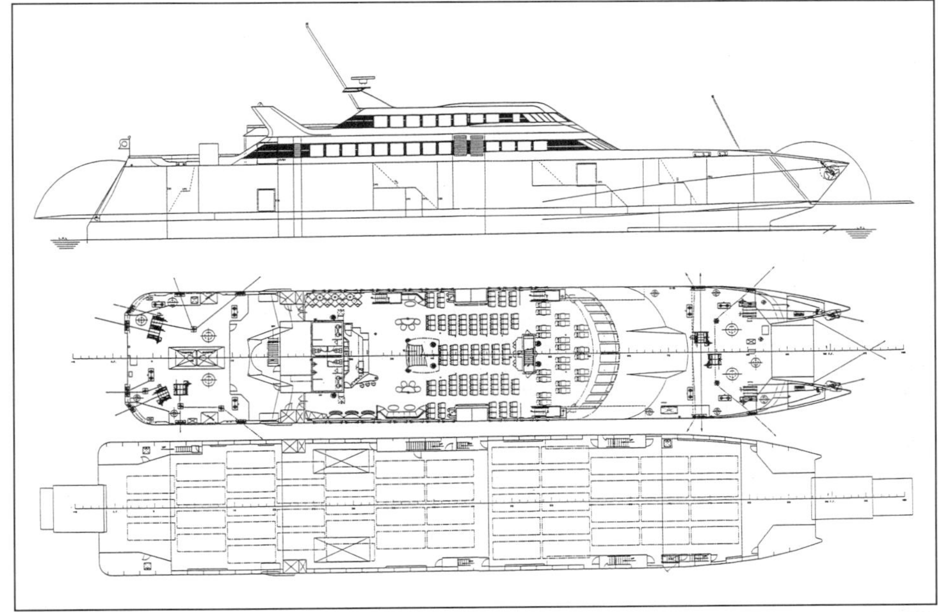

General arrangement of Ocean Arrow 0024755

Propulsive power	7,850 kW
Max speed	34 kt
Operational speed	30 kt

Propulsion: The vessel is to be powered by two MTU 16V 595 TE 70 diesel engines each rated at 3,925 kW and each driving a fixed-pitch propeller.

SSTH 90 (DESIGN)

Specifications

Length overall	92.4 m
Beam	19.4 m
Draught	2.1 m
Payload	170 t
Passengers	400
Vehicles	80 cars
Propulsive power	10,300 kW
Max speed	42 kt
Range	200 n miles

Structure: Hull material: welded aluminium.

Propulsion: Engines: four 5,150 kW high-speed diesels, driving water-jets.

SSTH 150 (DESIGN)

Passenger/car carrying high-speed ferry designed with steel hull and aluminium superstructure.

Specifications

Length overall	153.5 m
Beam	27.5 m
Draught	3.5 m
Passengers	1,000
Vehicles	300 cars
Propulsive power	41,200 kW
Max speed	37 kt
Range	400 n miles

Propulsion: Two gas-turbine engines each driving a water-jet unit.

SSTH 200 (DESIGN)

The SSTH 200 is a new all-steel design arranged for cargo and passenger operations.

Specifications

Length overall	199.9 m
Beam	29.8 m
Draught	4.9 m
Passengers	500
Propulsive power	41,200 kW
Max speed	32 kt
Range	600 n miles

Propulsion: Two LM 2500 19,800 kW gas-turbine engines each driving a water-jet propulsor.

Kawasaki Heavy Industries Ltd

Ship Division Tokyo Head Office:
World Trade Center Building, 4-1 Hamamatsu-cho 2-chome, Minato-ku, Tokyo 105-6116, Japan

Tel: (+81 3) 34 35 21 11
Fax: (+81 3) 34 36 30 37

Kobe Works:
1-1 Higashi Kawasaki-cho 3-chome, Chou-ku, Kobe 50-8670, Japan

Tel: (+81 78) 682 50 01
Fax: (+81 78) 682 55 00
e-mail: ono_h@khi.co.ltd
Web: www.khi.co.jp

Tomokazu Taniguchi, *Managing Director and President*

AMD1500 Kawasaki Jet Piercer, Hayabusa (since refitted and renamed Alkioni) 0506890

Kawasaki Heavy Industries Ltd has two shipyards, one in Kobe and the other in Sakaidi.

The concept design of wave-piercing catamarans was introduced from Advanced Multihull Designs Pty Ltd in 1990. Since then Kawasaki has been developing the design by conducting a wide range of tank tests, structural analysis and fire tests of aluminium structures.

Kawasaki Heavy Industries Ltd executed a construction contract between Maritime Credit Corporation of Japan and Kyushi Ferry Boat Co Ltd for a Kawasaki Jet Piercer AMD1500 in March 1994.

The keel of the vessel was laid at KHI's Kobe yard on 2 April 1994. The vessel, *Hayabusa*, was on trials by November 1994 and was delivered in December 1994. The vessel was introduced on the route between Yawatahama, on the west coast of Shikoku, and Usuki, on the east coast of Kyushi by Kyushi Ferry Boat Company, reducing the trip time from 2 hours 10 minutes to 1 hour 30 minutes.

The design of the AMD1500 allows for two modes of operation. In car/passenger mode the vessel will carry 460 passengers and 94 cars at a maximum speed of 35 kt. In freight mode the single vehicle deck can accommodate up to 24 12-tonne freight trucks or 32 eight-tonne freight trucks. This freight capacity results in a maximum deadweight of 570 tonnes. Such capacity provides flexibility for the operator, allowing low season or night freight services for additional revenue.

In 2006 the vessel was extensively refitted at Salamis shipyard in Greece. The work

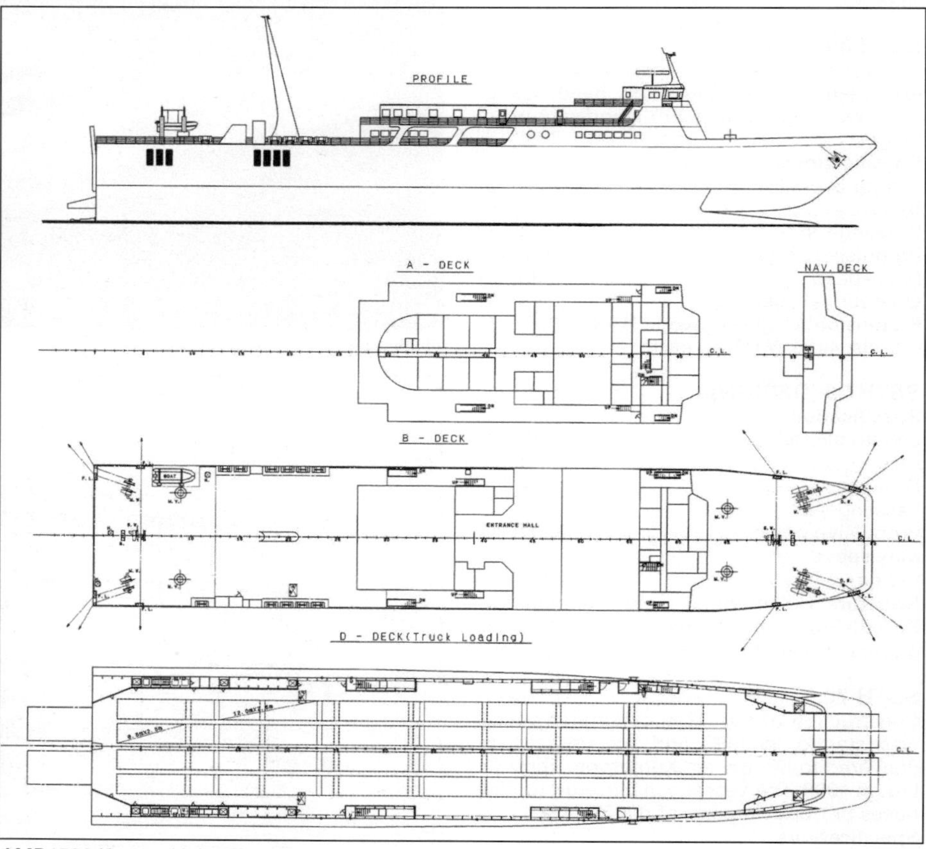

AMD1500 Kawasaki Jet Piercer 0506891

Vessel type	Vessel name	Yard No	Length (m)	Speed (kt)	Seats	Vehicles	Originally delivered to	Date of build
AMD1500 Kawasaki Jet Piercer	*Alkioni* (ex-*Hayabusa*)	–	100.0	30	460	94	Kyushi Ferry Boat Co Ltd	1994

included replacing the superstructure and increasing vehicle and passenger capacity. The vessel, now named *Alkioni*, is currently in service with Eagle ferries in Greece.

AMD1500 KAWASAKI JET PIERCER
Alkioni
Specifications

Length overall	100.0 m
Beam	20.0 m
Draught	3.1 m
Passengers	460
Vehicles	94 cars
Max speed	35.5 kt

Propulsion: Two Caterpillar 3616 and two Caterpillar 3612 engines, driving four Kawasaki KPJ-169A water-jets.

Mitsubishi Heavy Industries (MHI) Ltd

16-5 Konan 2-chome, Minato-ku, Tokyo 108-8215, Japan

Tel: (+81 3) 6716 31 11
Fax: (+81 3) 67 16 58 00
e-mail: ueda@ship.hq.mhi.co.jp
Web: www.mhi.co.jp

Naoki Ueda, *Manager*

MHI has built a number of high-speed craft mainly at the Shimonoseki Shipyard and Machinery Works in Hiroshima.

Kin-in *on trials* 0024756

27.6 m CATAMARAN
Kin-in 1
This 25 kt, 162 passenger catamaran was delivered to Fukuoka City in 1996.
Specifications

Length overall	27.6 m
Beam	8.0 m
Draught	1.2 m
Passengers	162
Propulsive power	1,470 kW
Max speed	25.3 kt

HI-STABLE CABIN CRAFT (HSCC)
Ukishiro
Completed by Mitsubishi in the 3rd quarter of 1987, this prototype catamaran craft features a passenger cabin mounted on hydraulically actuated autostabilising rams. The cabin is centrally mounted on top of a 200 mm diameter central hydraulic jack having a stroke of 1 m. This supporting member is attached by a ball and socket joint to the hulls and positions the cabin completely clear of the hulls. Four other 85 mm hydraulic rams, attached with shock-absorbers, support the cabin's four corners.

When motion of the basic catamaran craft occurs in heave, yaw, surge or sway, all hydraulic rams are activated simultaneously by an onboard computer (a 16 bit personal computer) with a rapid response capability to monitoring sensors, thereby almost eliminating or minimising cabin motion. Trials have demonstrated reductions of cabin motion to one-third of the motion of the basic supporting catamaran structure.

The catamaran type was chosen as the basic vehicle because of its good stability characteristics and cabin width advantages. It is reported that the computer software required for the system was developed in a 5-year study programme initiated by MHI and supported by the semi-governmental Japan Foundation for Shipbuilding Advancement.
Specifications

Length	12.7 m
Beam	5.4 m
Tonnage	17 t (GRT)
Max speed	20 kt

Propulsion: Two Mitsubishi 239 kW diesels.
Operator: Higashi Chugoku Ryoju Kosan.

Voyager 0506651

The Mitsubishi HSCC at sea 0506652

HI-STABLE CABIN CRAFT (HSCC)
Voyager
Following the prototype HSCC *Ukishiro* completed in 1987, Mitsubishi, in July 1990, completed a 200-passenger HSCC delivered to Nishi-Nippon Kaiun Kaisha Ltd.
Specifications

Length overall	26.5 m
Beam	9.0 m
Draught	1.4 m
Tonnage	132 t (GRT)
Passengers	200
Max speed	18 kt

Propulsion: Two DDC 16V 92TA engines, 843 kW each.

Vessel type	Vessel name	Yard No	Length (m)	Speed (kt)	Seats	Vehicles	Originally delivered to	Date of build
HSCC prototype	*Ukishiro*	–	12.7	–	–	–	Higashi Chugoku Ryoju Kosan	1987
HSCC	*Voyager*	–	26.5	–	200	–	Nishi-Nippon Kaiun Kaisha Ltd	July 1990
27.6 m Catamaran	*Kin-in 1*	–	27.6	–	162	–	Fukuoka City	1996

Mitsui Engineering & Shipbuilding Company Ltd

6-4 Tsukiji 5-chome, Chuo-ku, Tokyo 104, Japan

Tel: (+81 3) 35 44 31 47
Fax: (+81 3) 35 44 30 50
e-mail: prdept@mes.co.jp
Web: www.mes.co.jp

Shoichi Yabuki, *Managing Director and General Manager, Ship & Ocean Project HQ*
Mitsuru Kawamura, *General Manager, Government Ship & High-Speed Ship Sales Department*
Kimio Uchida, *Manager, Government Ship & High-Speed Ship Sales Department*

The construction of fast catamarans at Mitsui started in 1975 with the delivery of a Westamarin-designed craft carrying 162 passengers. Subsequently Mitsui designed its own craft and in 1978 delivered the Supermaran CP20HF, seating 195 passengers with a cruising speed of 30 kt. A development of this design, the CP30 was delivered in 1983 carrying 280 passengers at a cruising speed of 28 kt. Mk II versions of these craft had a higher speed of 32 kt.

In 1987 the Supermaran CP10 was delivered followed by CP30s, CP15s and Mightycat 40s which are themselves a development of the CP30 design.

Mitsui Mightycat 40 Neptune 0506971

There are now six craft of the latest version Mightycat 40, with a maximum speed of over 40 kt, completed. The company is currently involved in the development of the Japanese Techno Superliner, a high speed passenger and freight ferry.

SUPERMARAN CP15

Aquajet I, II, III and IV

These catamaran passenger ferries have water-jet propulsion and were built for Kyodo Kisen Company Ltd. They operate on the route between Osaka, Kobe and Awaji Island.

Specifications

Length overall	34.2 m
Beam	8.0 m
Draught	1.2 m
Max speed	35 kt
Operational speed	31 kt

Propulsion: Two MTU 16V 396 TB 83, 1,469 kW each, at 1,940 rpm. These drive two Kamewa S63 water-jets.

Vessel type	Vessel name	Yard No	Length (m)	Speed (kt)	Seats	Vehicles	Originally delivered to	Date of build
Supermaran CP20	*Blue Hawk*	–	26.5	–	111	–	–	1975
Supermaran CP20	*Ying Bin 3* (ex-*Marine Star*)	–	26.5	–	162	–	–	1976
Supermaran CP20	*Nam Hae 7* (ex-*Dong Hae Kosok* ex-*Sun Beam*)	–	26.5	–	180	–	–	1978
Supermaran CP20 HF	*Sun Shine*	1600	–	–	195	–	Tokushima Kosokusen Co Ltd	March 1979
Supermaran CP20 HF	*Blue Sky*	1601	–	–	195	–	Tokushima Kosokusen Co Ltd	June 1979
Supermaran CP30	*Marine Hawk*	1603	40.9	28.1	280	–	Nankai Ferry Co Ltd	July 1983
Supermaran CP30 Mk II	*Marine Shuttle*	1604	41.9	32.0	280	–	Tokushima Kosokusen Co Ltd	February 1986
Supermaran CP10	*Marine Queen*	1607	21.7	21.8	88	–	Kyowa Kisen KK	April 1987
Supermaran CP30 MkII	*Blue Star*	1606	41.0	32.0	280	–	Tokushima Kosokusen Company Ltd	June 1987
Supermaran CP30 MkII	*Sun Rise*	1605	41.0	32.0	280	–	Tokushima Kosokusen Company Ltd	July 1987
Supermaran CP20 Mk II	*Queen Rokko*	1609	33.2	25.0	250	–	Awaji Ferry Boat Co Ltd	June 1988
Supermaran CP30 MkII	*Coral*	1608	41.0	32.0	280	–	Tokushima Kosokusen Company Ltd	July 1988
Supermaran CP5	*Mon Cheri*	1610	19.9	17.0	53	–	Tenmaya Marine Corporation	July 1988
Supermaran CP15	*Aquajet I*	1613	34.2	31.0	198	–	Kyodo Kisen Co Ltd	March 1989
Supermaran CP20	*Wakashio*	1612	–	–	96	–	Chiba Prefecture	March 1989
Supermaran CP25	*New Tobishima*	1611	–	–	300	–	Sakata City	May 1989
Supermaran CP15	*Aquajet II*	1614	34.2	31.0	198	–	Kyodo Kisen Co Ltd	June 1989
Supermaran CP15	*Aquajet III*	1616	34.2	31.0	190	–	Kyodo Kisen Co Ltd	November 1990
Supermaran CP15	*Aquajet IV*	1617	34.2	31.0	190	–	Kyodo Kisen Co Ltd	January 1991
Supermaran CP10	*Fusanami*	1620	21.7	21.8	–	–	Chiba Prefecture	February 1991
Mk IV Mightycat 40	*Sun Shine*	1618	43.2	37.0	300	–	Tokushima Kosokusen Co Ltd	June 1991
Mk IV Mightycat 40	*Soleil*	1619	43.2	37.0	300	–	Tokushima Kosokusen Co Ltd	September 1991
Supermaran CP20	*Yumesaki*	1621	–	–	83	–	Osaka City	March 1992
Mk IV Mightycat 40	*Argo*	1622	43.2	37.0	300	–	Tokushima Kosokusen Co Ltd	December 1992
Mk IV Mightycat 40	*Polar Star*	1623	43.2	37.0	300	–	Tokushima Kosokusen Co Ltd	March 1993
Mk IV Mightycat 40	*Venus*	1624	43.2	37.0	300	–	Tokushima Kosokusen Co Ltd	July 1993
Supermaran CP15	*Marine Bridge*	1626	30.3	24.0	235	–	Bantan Renraku Kisen Co Ltd	March 1995
Mk IV Mightycat 40	*Neptune*	1625	43.2	37.0	300	–	Tokushima Kosokusen Co Ltd	June 1995
Supermaran CP5	*Cosmo*	1627	19.9	17.0	–	–	Japan Ministry of Transport	September 1995
Supermaran CP10	*Inaba 2*	1628	21.7	21.8	120	–	Yokkaichi City LFA	March 1997

SUPERMARAN CP15

Marine Bridge

This is the fifth CP15 passenger ferry although it differs in specification from the other four.

Specifications

Length overall	30.3 m
Beam	8.3 m
Draught	1.3 m
Passengers	235
Max speed	32.8 kt
Operational speed	24 kt

Propulsion: The vessel is powered by two Niigata 12V16Fx diesel engines each rated at 1,324 kW at 1,900 rpm. Each engine drives a fixed-pitch propeller via a reverse reduction gearbox.

SUPERMARAN CP20 Mk II

Queen Rokko

This is the catamaran double-decker cabin cruiser built for Awaji Ferry Boat Company in 1988.

Specifications

Length overall	33.2 m
Beam	9.0 m
Draught	1.5 m
Displacement	217 t
Passengers	250
Max speed	30 kt
Operational speed	25 kt

Propulsion: Two Deutz MWM 604B V12 engines, 1,278 kW (1,714 hp PS) each, at 1,800 rpm. These drive two fixed-pitch propellers through Niigata Converter MGN 433 EW gearboxes with electric variable propeller speed control device (reduction ratio 2.06:1).

SUPERMARAN CP30 Mk III

Sun Rise

This is a catamaran passenger ferry and the fourth one of the CP30 series, built for Tokushima Kosokusen Company Ltd in 1987.

Specifications

Length overall	41.0 m
Beam	10.8 m
Depth	3.4 m
Draught	1.3 m
Passengers	280
Max speed	35 kt
Operational speed	32 kt

Propulsion: Two Fuji Pielstick 12 PA4V 200 VGA engines, 1,961 kW each, at 1,475 rpm. The craft is propelled by two fixed-pitch propellers.

CP30 Mk IV MIGHTYCAT 40

This is the latest catamaran class of passenger ferry of the CP30 series, having the highest speed of all Japanese catamarans.

Six craft were built for Tokushima Kosokusen Company Ltd and Nankai Ferry Company Ltd.

Specifications

Length overall	43.2 m
Beam	10.8 m
Passengers	300
Max speed	41 kt

Propulsion: Two Niigata 16 PA4V 200 VGA engines, driving Kamewa 71 S11 water-jets.

Mitsui Supermaran CP15 Aquajet III

0506653

Mitsui Supermaran CP10 Inaba 2 on trials 0024743

Mitsui Supermaran CP15 Marine Bridge 0506972

Universal Shipbuilding Corporation

Head office:
Muza-Kawasaki Central Tower, 1310, Omiya-cho, Saiwai-ku, Kanagawa 212-8554, Japan

Tel: (+81 4) 45 43 27 00
Fax: (+81 4) 45 43 27 10
Web: www.u-zosen.co.jp

Universal Shipbuilding Corporation, formerly Hitachi Zosen, constructed a large number of PT-50 hydrofoils between 1960 and 1985 based on its technology

Superjet-30 Miyajima 0506889

licence with Supramar of Switzerland. The experience gained with these vessels and the co-operative research with the University of Tokyo led to the development of the foil-assisted Superjet-30 catamarans constructed between 1993 and 1994. In 1996 the company received an order for a 62 m wave-piercing catamaran from Yasuda Ocean Line which was delivered the next year. In 1997 the company received an order for a Superjet-40, based on its experience from the Superjet-30, from Ishizaki Kisen. This vessel, the *Seamax*, was delivered in February 1998.

Hitachi Zosen Superjet-40 Seamax 0038683

Superjet-30

Superjet-30 is a foil-assisted catamaran with a maximum speed of 40 kt. The hull is a hybrid design consisting of a catamaran and fully submerged foils. The lift of hydrofoils supports more than 80 per cent of the ship's weight. The control system of flaps and ailerons on the hydrofoils offers high sea-keeping capability.

There have been eight Superjet-30s constructed at the Hitachi Zosen Kanagawa Works.

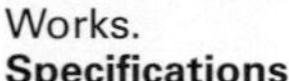

Specifications

Length overall	31.5 m
Beam	9.8 m
Draught	1.9 m
Draught, hullborne	2.8 m (incl foils)
Passengers	160
Propulsive power	3,676 kW
Max speed	40 kt

Propulsion: Main engines are two high-speed diesel engines driving two water-jets.
Control: Motion control system provided.

62 m WPC Sea-Bird 0007943

Superjet-40

Superjet-40 is a larger development of the Superjet-30 craft, with a 25 per cent increase in passenger capacity and a substantially higher speed.

Specifications

Length overall	39.0 m
Beam	11.4 m
Draught	1.9 m (excl foils)
Displacement	300 t
Passengers	200
Propulsive power	8,088 kW
Max speed	45 kt

Propulsion: Main engines and four high-speed diesels driving two water-jets.

62 m WAVE-PIERCING CATAMARAN

This vessel was ordered by Yasuda Ocean Line in 1996 and delivered in March 1997.

Specifications

Length overall	62.0 m
Beam	15.4 m
Draught	2.3 m
Crew	8
Passengers	296
Vehicles	48 cars
Operational speed	30 kt

Propulsion: Powered by four 2,022 kW high-speed diesels.

Vessel type	Vessel name	Yard No	Length (m)	Speed (kt)	Seats	Vehicles	Originally delivered to	Date of build
Superjet-30	*Trident Ace*	7306	31.5	38	160	–	Airport Awaji Aqualine Co Ltd	1993
Superjet-30	*Artemis*	7307	31.5	38	160	–	Airport Awaji Aqualine Co Ltd	1993
Superjet-30	*Zuiko*	7309	31.5	38	156	–	Isizaki Kisen Co Ltd	1993
Superjet-30	*Dogo*	7310	31.5	38	156	–	Setonaikai Kisen Co Ltd	1993
Superjet-30	*Shoko*	7311	31.5	38	156	–	Isizaki Kisen Co Ltd	1994
Superjet-30	*Miyajima*	7312	31.5	38	156	–	Setonaikai Kisen Co Ltd	1994
Superjet-30	*Apollon*	7308	31.5	38	160	–	Airport Awaji Aqualine Co Ltd	1994
Hitachi 62 m WPC	*Seabird*	7320	62.0	30.0	296	48	Yasuda Ocean Line	1997
Superjet-30	*Blue Sonic*	7323	31.5	45	160	–	Kashima Futo Co Ltd	1997
Superjet-40	*Seamax*	–	39.0	45	200	–	Isizaki Kisen Co Ltd	1998

Yamaha Motor Company Ltd

Gamagori Shipyard, 2500 Shingai, Iwata 438, Japan

Tel: (+81 538) 32 11 45
Fax: (+81 538) 37 42 50
Web: www.yamaha.co.jp

Yamaha Motor started in the marine business as a manufacturer of marine engines and FRP boats in 1960. In 1989 three large catamaran ferries were constructed for Tokyo Blue Cruises.

Yamaha Motor Company Ltd catamaran Type 291 Bay Bridge 0506654

29.1 m CATAMARAN TYPE 291

There have been three of these vessels delivered to Tokyo Blue Cruises.

Built in GRP, they are powered by two DDC 16V 92TA 895 kW diesels driving five-blade propellers employing skew blades.

Vessel type	Vessel name	Yard No	Length (m)	Speed (kt)	Seats	Vehicles	Originally delivered to	Date of build
Yamaha Catamaran Type 291	*Bay Dream*	–	29.1	23.0	182	–	Tokyo Blue Cruises	1989
Yamaha Catamaran Type 291	*Bay Bridge*	–	29.1	25.0	230	–	Tokyo Blue Cruises	1989
Yamaha Catamaran Type 291	*Bay Frontier*	–	29.1	25.0	230	–	Tokyo Blue Cruises	1989

Bay Dream
Specifications

Length overall	29.1 m
Beam	8.1 m
Draught	2.1 m
Displacement, min	98.2 t
Crew	5
Passengers	182
Fuel capacity	5,000 litres
Water capacity	2,000 litres
Propulsive power	2 × 895 kW
Max speed	30.7 kt
Operational speed	23 kt (95% MCR, 70% load)
Range	200 n miles

Classification: J G certified for limited coastal water operation.
Propulsion: Main engines are two DDC 16V 92TI, 895 kW each at 100 per cent MCR.
Electrical system: Auxiliary engines: four-stroke diesels, one 48.5 kW at 1,800 rpm, one 61 kW at 1,800 rpm.
Alternator: three-phase AC, one 50 kVA, one 60 kVA.

Bay Bridge
Bay Frontier
Specifications
Details as for *Bay Dream* except as follows:

Displacement, 70% load	82 t
Passengers	230
Max speed	33 kt
Operational speed	25 kt

Electrical system: Auxiliary engines are four-stroke diesels, one 37.5 kW at 1,800 rpm, one 35.2 kW at 1,800 rpm.
Alternator: three-phase AC, one 34.1 kVA, one 20 kVA.

Korea, South

Daewoo Shipbuilding and Marine Engineering Co Ltd

541, 5-GA Namdaemun-ro-Jung-Gu, Seoul, South Korea

Tel: (+82 2) 21 29 01 14
Web: www.dsme.co.kr/eng

Jung Sung Leep, *Chief Executive Officer*
Y S Hans, *Director and Head of Industrial Application R&D Institute*

Daewoo Shipbuilding and Marine Engineering Co Ltd was established in 1978 and has two divisions covering commercial vessels and offshore and special ships. In 1994 Daewoo's Okpo shipyard completed a 40 m 40 kt passenger catamaran in 1994 for operation from Koje island to Pusan in South Korea. The company is now marketing a range of vessels including an 80 m passenger/car ferry.

F-CAT 40

Royal Ferry
This all-aluminium vessel has been developed by Daewoo for Korea's domestic market. In order to achieve the high speed with acceptable sea-keeping qualities, the design incorporates a foil extending between the two hulls. The vessel was delivered in late 1994.

Specifications

Length overall	40.2 m
Length waterline	37.0 m
Beam	9.3 m
Draught	1.5 m
Passengers	350
Vehicles	8
Fuel capacity	2 × 2,500 litres
Water capacity	1,400 litres
Propulsive power	2 × 2,000 kW
Max speed	40 kt
Operational speed	36 kt
Range	400 n miles

Classification: Korean Register.
Propulsion: Two MTU 16V 396TE 74L diesels, producing 2,000 kW each at 2,000 rpm. Each drives a Kamewa 63 S11 water-jet.
Electrical system: Two Cummins 6BT5.9G1(M) generators, two Onan-Newage 75 kW alternators.

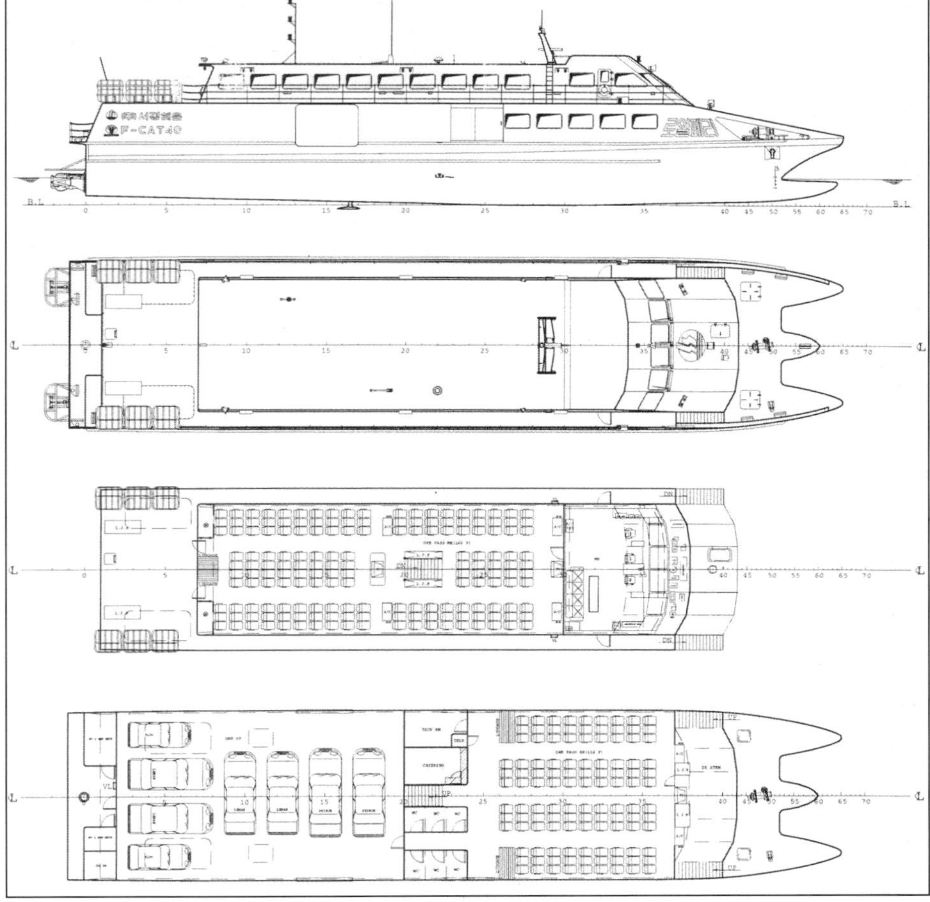

General arrangement of the F-CAT 40 0506892

Royal Ferry 0506893

HSPC 80 DESIGN

This 80 m catamaran design is an all-aluminium craft designed for passenger and car transportation.

Specifications

Length overall	82.0 m
Length waterline	74.0 m
Beam	22.0 m
Draught	2.6 m

Vessel type	Vessel name	Yard No	Length (m)	Speed (kt)	Seats	Vehicles	Originally delivered to	Date of build
F-Cat 40	*Royal Ferry*	–	40.2	36.0	350	8	Korea	1994

Payload	320 t
Crew	20
Passengers	630
Vehicles	167 cars
Fuel capacity	4 × 14,000 litres
Water capacity	6,000 litres
Propulsive power	4 × 7,080 kW
Max speed	41 kt
Operational speed	39 kt

Classification: DnV +1A1 HSLC, Passenger A, R1, EO.

Propulsion: Main engines are four Ruston 20 RK 270 M, each producing 7,080 kW at 1,030 rpm, driving two Lips LJ 140 DL water-jets via a reduction gearbox.

Electrical system: Two 280 kW generators, 450 V AC, 60 Hz, 3PH.

Hyundai Heavy Industries Company Ltd

Head Office and Ulsan Works:
1 Cheonha-Dong, Dong-Ku, Ulsan, South Korea

Tel: (+82 522) 02 92 50
Fax: (+82 522) 02 34 21
e-mail: sbplan@hhi.co.kr
Web: www.hhi.co.kr

Seoul office:
Special and Naval Shipbuilding Division
140-2 Kye-Dong, Chongro-Ku, Seoul, South Korea

Tel: (+82 2) 746 46 71; 2
Fax: (+82 2) 741 11 52
Telex: 28361/27496 HHIYARD K

Hyundai 45.5 m catamaran on trials (without foils fitted) 0506831

Hyundai 35 m Catamaran ferry Han-Ma-Eum II 0068906

45.5 m Catamaran ferry

Texroc (ex-Han Ma Um Ho)

A demand from a domestic customer for a high-speed passenger vessel for a 700 n mile round trip with more than 300 passengers was received in 1990. A comprehensive survey was undertaken of existing catamaran designs, which motivated Hyundai to develop a design which could meet the long-range and corresponding sea-keeping requirements. The sea-keeping capability has been met by incorporating a two-foil ride control system, the foils being positioned beneath the hulls at extreme forward and aft locations and spanning the distance between the hulls. The company has previous experience of foil technology from its involvement in the build of an RHS 70 hydrofoil, *Angel IX* in 1985.

The design work began in August 1990 and was completed in mid-1991, supported by a large number of model tests both within and outside Hyundai Heavy Industries. The vessel with and without foils fitted, underwent instrumented sea trials throughout 1994. Data from this vessel is understood to be the basis of a much larger craft being developed by Hyundai.

Specifications

Length overall	45.5 m
Beam	11.4 m
Draught	1.6 m
Passengers	300
Propulsive power	3,061 kW
Operational speed	35 kt
Range	600–700 n miles

Classification: DnV +1A1 R170, EO, HSLC.

Propulsion: Main engines are two Paxman Valenta 18RP 200CM, 3,061 kW each at 1,540 rpm, driving two water-jets.

35 m Catamaran ferry

Han-Ma-Eum II

Specifications

Length overall	34.6 m
Beam	9.4 m
Draught	1.5 m
Passengers	250
Fuel capacity	12,000 litres
Water capacity	1,600 litres
Operational speed	37 kt
Range	250 n miles

Propulsion: Two Hyundai-Paxman 12 VP 185 m diesel engines each rated at 2,200 kW and each driving a Kamewa 62SII water-jet.

Electrical System: Two 75 kW MTU 6RO99 TA31 diesel generators.

Vessel type	Vessel name	Yard No	Length (m)	Speed (kt)	Seats	Vehicles	Originally delivered to	Date of build
45.5 m Catamaran Ferry	Tezroc (Han Ma Um Ho)	P062	45.5	35.0	300	–	Korea	1994
35 m Catamaran Ferry	Han-Ma-Eum II	P081	34.6	37.0	250	–	Hyundai Transportation	1997

Semo Company Ltd

Shipbuilding Division, 1 Jangiri Donghaemyun, Kosungkun, Kyungnam, South Korea

Tel: (+82 556) 70 35 16
Fax: (+82 556) 70 35 68
e-mail: semo@semo.co.kr
Web: www.semo.co.kr

Jae-Joong Yoon, *Managing Director*

80 m CATAMARAN

It is understood that the company is to build an 80 m catamaran car ferry for delivery in 2003.

Specifications

Length overall	80.0 m
Length waterline	69.1 m
Beam	19.8 m
Draught	2.9 m
Crew	20
Passengers	550
Vehicles	87 cars
Propulsive power	21,000 kW
Operational speed	42 kt

40 m CATAMARAN

Pegasus

Built on the FRP experience gained from the company's SES craft, *Pegasus* is a fast passenger catamaran with a capacity of over 370 passengers and a maximum speed of 38 kt.

Specifications

Length overall	40.0 m
Beam	10.4 m

Vessel type	Vessel name	Yard No	Length (m)	Speed (kt)	Seats	Vehicles	Originally delivered to	Date of build
40 m catamaran	*Pegasus*	212	40.0	35.0	392	–	Semo Ferry Division	1995

Draught	1.35 m
Passengers	392
Max speed	38 kt
Operational speed	35 kt

Classification: Korean Register of Shipping.
Structure: The hull and superstructure are constructed of FRP sandwich with a Divinycel foam core.
Propulsion: Two MTU 16V 396 TE 74L diesel engines each rated at 1,940 kW driving RNG 140K-I reduction gearboxes and KMW 63 S11 water-jets.

Semo 40 m FRP catamaran Pegasus
0007946

Malaysia

Kay Marine SDN BHD

Lot 4961, Telok Pasu, Manir, 21200 Kuala Terengganu, Malaysia

Tel: (+60 9) 615 51 45
Fax: (+60 9) 615 39 84
e-mail: kaym@tm.net my

Kay Marine is a specialist marine fabricator in Malaysia and has built a range of small fast craft. Its most recent delivery is a 24 m oil spill response vessel for the Marine Department of Malaysia.

26 m RESCUE CATAMARAN

Delivered to the Marine Department of Peninsula Malaysia, this vessel is designed for patrol and rescue services. The aluminium vessel is powered by two MTU 12V 2000 M91 diesel engines and Doen waterjets.

Specifications

Length overall	26.0 m
Beam	9.2 m
Draught	1.1 m
Crew	6
Max speed	27.5 kt

Construction: Aluminium
Propulsion: Two MTU 12V diesel engines delivering 1,119 kW at 2,350 rpm to DOEN Dj 260 waterjets via ZF 2550 gearboxes.

24 m OIL SPILL RESPONSE VESSEL

This vessel, the *Al Nilam*, was delivered in 2001 to the Marine Department of Malaysia. Designed by Mark Ellis, Marine Department of Malaysia, the vessel carries tanks of 5,000 litres capacity and 3 tonnes of deck cargo.

Specifications

Length overall	24.0 m
Beam	9.0 m
Crew	8
Fuel capacity	12,500 litres
Max speed	23 kt
Operational speed	19 kt

Construction: Aluminium hull and superstructure.
Propulsion: Two Caterpillar 3412E diesel engines each rated at 820 kW at 2,300 rpm and driving fixed-pitch propellers via Reintjes reverse reduction gearboxes.

Vessel type	Vessel name	Yard No	Length (m)	Speed (kt)	Seats	Vehicles	Originally delivered to	Date of build
CAT 24 m	*Al Nilam*	–	24 m	33	–	–	Marine Dept of Malaysia	2001
CAT 26 m	*Sirius*	–	26 m	27.5	12	–	Marine Dept of Malaysia	2007

NGV Tech SDN BHD

No 508 block D, Pusat Dagangan Phileo Damansara 1, No 9 Jalan 16/11, Off Jalan Damansara, 46350 Petaling Jaya, Selangor, Malaysia

Tel: (+60 3) 7660 1780
Fax: (+60 3) 7660 1781
e-mail: enquiry@ngvtech.com.my

Established in 1992, NGV Tech has developed a varied product range including offshore support vessels, high speed patrol and interceptor craft and tug boats as well as fast catamarans. The company fabricates vessels in aluminium, steel and composites.

32 m CATAMARAN FERRY

Black Marlin
Blue Marlin
Skipjack

Deisgned by BSC Marine, the first three of a series of six 32 m fast passenger catamarans have been built by NGV Tech for operation off India's south-west coast between the city and the Lakshadweep islands. They have a deadweight tonnage of 68 t, a range of 215 nautical miles and seating for 50 passengers.

Specifications

Length overall	32.13 m
Length waterline	29.3 m
Beam	9.0 m
Draught	1.2 m
Displacement	83 t
Deadweight	68 t
Passengers	50
Maximum speed	25 kt
Range	215 n.miles

Propulsion: Two Caterpillar 3412E diesel engines rated at 745 kW at 2,100 rpm each, driving twin HM571B Hamilton waterjets via a Reintjes WVS334 gearbox.

Vessel type	Vessel name	Yard No	Length (m)	Speed (kt)	Seats	Vehicles	Originally delivered to	Date of build
32 m Catamaran Ferry	*Black Marlin*	–	32.13	25	50	none	Union Territory of Lakshadweep Adminitration, India	2006
32 m Catamaran Ferry	*Blue Marlin*	–	32.13	25	50	none	Union Territory of Lakshadweep Adminitration, India	2006
32 m Catamaran Ferry	*Skipjack*	–	32.13	25	50	none	Union Territory of Lakshadweep Adminitration, India	2006

Mauritius

Diogene Marine

Taylor Smith Yard, Old Quay D Road, Port Louis, Mauritius

Tel: (+230) 423 45 70
Fax: (+230) 240 05 47
e-mail: diogene@intnet.mu
Web: www.diogenemarine.com

Reginald de Baritault, *Director*

Diogene Marine was set up to build vessels designed by Craig Loomes, such as the range of fast trimarans. In 2006 Diogene Marine launched the C-Rex 23, a 23 m trimaran yacht designed to operate on the Mauritius to Reunion route.

C-REX 23 m TRIMARAN

Specifications

Length overall	23 m
Length waterline	20 m
Beam	8.3
Max speed	35 kt

Propulsion: Caterpillar C18 diesel engines delivering 750 kW each to Seafury surface drives via Twin-Disc gearboxes.

Netherlands

Damen Shipyards

Damen Fast Ferries Division
Industrieterrein Avelingen West 20, PO Box 1, NL-4200AA, Gorinchem, Netherlands

Tel: (+31 183) 63 99 22
Fax: (+31 183) 63 21 89
e-mail: info@damen.nl
Web: www.damen.nl

E van der Noordaa, *Managing Director*
H O van Herwijnen, *Damen Fast Ferries*

Damen Fast Ferries is the product group of Damen Shipyards supporting the marketing and sales of fast ferries for inland, coastal and international voyages in the 28 to 53 m range.

Damen Shipyards started in 1927 and in 1969 began the construction of commercial vessels to standard designs using modular construction. Since 1969, more than 3,000 vessels have been built and delivered to 117 countries worldwide.

In 1998, Damen Shipyards entered the High Speed Catamaran Ferry market with the construction of a series of five Low Wash Commuter Ferries, built under licence from NQEA, Australia. In 2000, Damen Shipyards purchased the Kvaerner Fjellstrand Singapore facility and, along with it, the design rights for a wide range of catamaran craft.

Marketing of the range of fast ferries is by Damen Shipyards Gorinchem in the Netherlands. Production of the majority of fast ferries is undertaken from the Singapore yard Damen Shipyards Singapore Pty Ltd.

All Damen Fast Ferries have been given a DFF designation for identification of model types based on length and beam, for example DFF 5313.

Since 2000, Damen Fast Ferries has delivered or has under construction, 12 fast ferries: one DFF 3810, two DFF 4010, seven DFF4912 and two SWATH 3717 vessels. Damen Shipyards is accredited to the ISO 9001 Quality Assurance Standard.

Damen DFF 3007 (Damen NL) 1173602

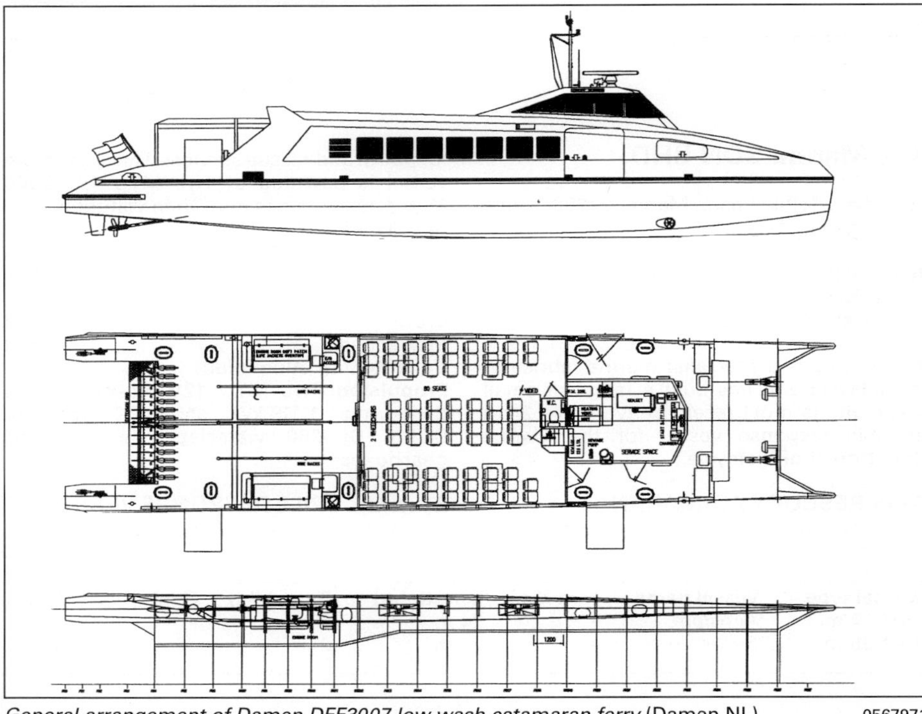

General arrangement of Damen DFF3007 low wash catamaran ferry (Damen NL) 0567971

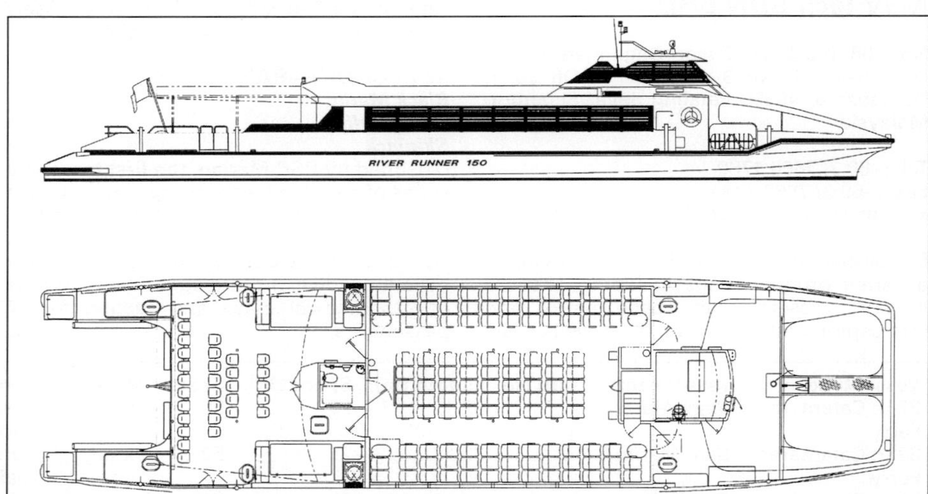

General arrangement of the DFF3007 0116592

RIVER RUNNER 150

The first craft delivered from the Damen Fast Ferry division were five River Runner 150 low wash ferries in 1999. These 20 kt craft were designed by NQEA of Australia and built under licence at Damen Shipyards.

Specifications

Length overall	30.5 m
Beam	7.0 m
Draught	1.3 m
Crew	2
Passengers	130
Freight capacity	1.2 t
Fuel capacity	3,000 litres
Water	250 litres
Max speed	23 kt
Operational speed	20 kt

Classification: DnV +1A1 HSLC Passenger R5EO (NL) and Dutch Shipping Inspection Inland Water Passenger Craft.

Structure: All-aluminium hull with an aluminium and composite superstructure.

Propulsion: Two Caterpillar 3196B D1-TA diesel engines each rated at 650 kW at 90 per cent MCR and driving a five-bladed fixed-pitch propeller via Twin Disc gearboxes.

Electrical System: One 30 kW generator.

Auxiliary Systems: Two electrically driven bow thrusters.

Outfit: Seating for 80 with an additional 50 passengers standing.

DFF 3007

The first vessel was ordered in 2003. Designed by Nigel Gee and Associates, the vessel has a similar power, speed and carrying capacity as the River Runner 150.

Vessel type	Vessel name	Yard No	Length (m)	Speed (kt)	Seats	Vehicles	Originally delivered to	Date of build
River Runner 150	*Devel*	9101	30.00	21.0	130	–	Waterbus BV	1999
River Runner 150	*Wantij*	9102	30.00	21.0	130	–	Waterbus BV	1999
River Runner 150	*Rietbaan*	9103	30.00	21.0	130	–	Waterbus BV	1999
River Runner 150	*Albas*	9104	30.00	21.0	130	–	Waterbus BV	1999
River Runner 150	*Aquarunner*	9107	30.00	21.0	130	–	Aqualiner BV	2005
DFF 3207	*Constanze*	538221	33.64	21.6	130	–	Bodensee Katarmaran	2005
DFF 3207	*Fridolin*	538222	33.64	21.6	130	–	Bodensee Katarmaran	2005

Specifications

Length overall	31.1 m
Beam	7.4 m
Draught	1.3 m
Crew	2
Passengers	130
Fuel capacity	2,000 litres
Water	250 litres
Max speed	22 kt
Operational speed	21 kt

Propulsion: Two Caterpillar 3196 diesel engines each rated at 330 kW and driving a five-bladed fixed-pitch propeller via a Twin Disc MG5114A gearbox.

DFF 3207

The DFF 3207 is 2 m longer than the DFF 3007, allowing the vessel to accommodate 185 passengers and 30 bicycles. This vessel is also designed by Nigel Gee and Associates and shares the low wash attributes of the DFF 3007. Two DFF 3207 vessels have been delivered to an operator on Lake Constance.

Specifications

Length overall	33.64 m
Beam	7.6 m
Draught	1.65 m
Crew	3
Passengers	185
Operational speed	21.6 kt

Propulsion: Twin MAN 8V D2848 LE 422 diesel engines each rated at 551 kW and driving fixed-pitch propellers through ZF 665 A gearboxes.

Luyt BV

Nijverheidsstraat 1, NL-1779 AS Den Oever, Netherlands

Tel: (+31 227) 51 23 41
Fax: (+31 227) 51 21 48
e-mail: info@luyt-denoever.nl
Web: www.luytgroep.nl

G Wiegman, *Manager*

Founded in the 1960s, Luyt BV specialises in the fishing and survey industry sector.

21 m COASTAL CATAMARAN

In May 2000, the 21 m coastal catamaran *Coastal Liner* was handed over to Rederij Waterweg for multirole operations including crew tender, passenger ferry and survey vessel.

Specifications

Length overall	21.0 m
Beam	6.7 m
Draught	1.3 m
Crew	4
Passengers	36
Fuel capacity	6,000 litres
Water capacity	240 litres
Max speed	25 kt

Propulsion: Two MTU/Detroit 12 V 2000 M70 diesel engines driving fixed-pitch propellers via a reverse reduction gearbox.

main deck

below main deck

front view

General arrangement of Coastal Liner
0106430

Coastal Liner *at speed*
0106429

Schelde Naval Shipbuilding (Royal Schelde)

Part of the Damen Shipyards Group
PO Box 555, NL-4380 AN Vlissingen, Netherlands

Tel: (+31 118) 48 21 18
Fax: (+31 118) 48 50 20
e-mail: info@schelde.com
Web: www.schelde.com

Rene Berkvens, *Managing Director*
M Pater, *Marketing and Sales*

Schelde Shipbuilding has developed a range of high-speed vessels, starting 10 years ago with an SES craft, from which was built a 26 m demonstration vessel *Seaswift 23*, which operated for a period between Southampton and the Isle of Wight in the UK. Subsequent to this vessel, Schelde Shipbuilding developed a range of catamarans and monohulls, drawing on its experience from the building of naval frigates.

In 1996, Schelde Shipbuilding delivered a 77 m fast catamaran passenger-car ferry for operation in Greece. This vessel, *Highspeed I*, operates between Piraeus and the Cyclades.

In order to meet market demands, Schelde Shipbuilding continued to enhance the basic designs and further extend the standard product range from 60 m to over 100 m in length and capacities from 450 passengers and 30 cars to 1,400 passengers and 340 cars.

Highspeed 1 (*ex*-Captain George) *at speed* 0507056

70 m CATAMARAN–HEAVY LIFT (HL)

Highspeed I (ex-Captain George)

This design is similar to the standard design but incorporates increased passenger and vehicle loads with a higher superstructure. The first vessel of this design was delivered to Catamaran Lines of Greece in mid-1996 and is now operated by Minoan Lines between Piraeus and the Cyclades. The car deck is arranged with a mezzanine side deck and hoistable centre portion. The free height of the main car deck is 2.5 m and for the mezzanine deck is 2.0 m. The bus lanes have a total of 118.0 m length with a free height of 4.5 m.

Specifications

Length overall	76.6 m
Length waterline	68.0 m
Beam	22.2 m
Draught	3.3 m
Passengers	726
Vehicles	154 cars
Fuel capacity	45,000 litres
Water capacity	4,000 litres
Max speed	42 kt
Operational speed	36 kt
Range	300 n miles

Propulsion: Main engines are four Caterpillar 3616 diesel engines, each driving through reduction gearboxes to a Kamewa 112 S11 water-jet.

Vessel type	Vessel name	Yard No	Length (m)	Speed (kt)	Seats	Vehicles	Originally delivered to	Date of build
70 m Catamaran–Heavy Lift	*Highspeed I* (ex-*Captain George*)	–	76.6	36.0	726	154	Catamaran Lines, Greece	1996

New Zealand

Fitzroy Yachts

Ocean View Parade, Private Bay 2014, New Plymouth, New Zealand

Tel: (+64) 67 69 93 80
Fax: (+64) 67 69 93 81
e-mail: md@fitzroy.com
Web: www.fitzroyyachts.com

Peter White-Robinson, *Managing Director*
Gorden French, *Production Manager*
Hamish Neale, *Procurement Manager*

Fitzroy Yachts was established in 1997 as a subsidiary of Fitzroy Engineering, from where it drew experience of designing, building and managing large projects. Now a stand alone company, Fitzroy Yachts is producing large sailing yachts as well as Teknicraft Design catamarans.

18 m CATAMARAN

Paikea

This Teknicraft Design 18 m foil-assisted catamaran was ordered by Whale Watch Kaikorura and delivered in 2004.

Specifications

Length overall	17.7 m
Beam	6.4 m
Draught	0.8 m
Passengers	48
Maximum speed	32 kt
Operational speed	28 kt

Propulsion: Two MTU 8V 2000 M70 diesel engines each rated at 525 kW at 2,400 rpm drive Hamilton HJ391 water-jets.

16 m CATAMARAN

Yellowcat II

Designed by Teknicraft this foil-assisted catamaran was specified for operation in Alaska.

Specifications

Length overall	16.1 m
Beam	5.7 m
Maximum speed	50 kt

Propulsion: Two MAN diesel engines delivering 772 kW each to water-jets.

Q-West

2A Gilberd Street, PO Box 862, Wanganui, New Zealand

Tel: (+64 6) 344 35 72
Fax: (+64 6) 344 35 92
e-mail: sales@q-west.co.nz
Web: www.q-west.com

Myles Fothergill, *Managing Director*
Colin Mitchell, *Sales and Project Manager*
Tara Hoyle, *Office Manager*
Dave Turner, *Purchasing Officer*

17.7 m catamaran Discovery 4 1176170

Q-West builds a wide range of vessels including pleasure, fishing boats, pilot craft, survey vessels, patrol boats and passenger vessels. The foil-assisted high-speed passenger craft have been designed by Teknicraft of New Zealand. In 2006 Q-West won the tender to build a new 18.4 m, 25 knot foil supported catamaran for the New Zealand Police and Customs service.

17.5 m Foil-assisted catamaran

Three Saints

Designed by Teknicraft, this vessel is used for pleasure trips in Alaska.

Specifications

Length overall	17.5 m
Beam	6.7 m
Draught	0.8 m
Passengers	6
Operational speed	26 kt

Structure: All aluminium hull and superstructure.

Propulsion: Two C12 Caterpillar marine engines with Twin Disc reduction gearboxes driving 403 Hamilton water-jets.

20 m Wave piercing catamaran

Polaris

Designed by Crag Loomes Design, this vessel will be used as a charter vessel in Brisbane, Australia.

Specifications

Length overall	21.3 m
Beam	7.5 m
Draught	1.3 m
Passengers	8
Operational speed	25 kt

Structure: All aluminium hull and superstructure.

Propulsion: Two 3406 Caterpillar diesel engines with Twin Disc remote V drive, driving fixed pitch propellers.

23 m Catamaran

Dolphin Seeker

Designed by Williams Marine Design, this vessel is used as a high volume sightseeing craft.

Specifications

Length overall	23.0 m
Beam	7.5 m
Draught	1.6 m
Passengers	230
Operational speed	23 kt

Structure: All aluminium hull and superstructure.

Propulsion: Two MTU 12V2000 diesel engines each driving a fixed pitch propeller.

20 m Catamaran

Lagi Lagi

This foil-assisted catamaran was delivered to Beachcomber Tours and Cruises of New Zealand in 2001.

Specifications

Length overall	20.0 m
Beam	6.9 m
Draught	0.9 m
Passengers	150
Fuel capacity	7,000 litres
Speed, maximum	34 kt

Propulsion: Four MAN D 2848 LE 405 diesel engines each rated at 485 kW and each driving a 362 Hamilton water-jet.

17.7 m foil assisted catamaran

1066888

16.6 m foil assisted catamaran MV 439

1176172

18 m Catamaran

Tahora

Ordered by Whale Watch Kaikoura, this 18 m catamaran was delivered in April 2006 and was designed by Teknicraft.

Specifications

Length overall	17.6 m
Length waterline	17.0
Beam	6.4 m
Draught	0.75 m
Passengers	52
Crew	2
Operational Speed	28 kt

Propulsion: Two MTU 8V 2000 M70 diesel engines delivering 525 kW at 2100 rpm to Hamilton HJ 403 water-jets.

16.6 m Adjustable foil assisted catamaran

MV 439

Designed by Teknicraft, this vessel operates from Auckland scanning for dredging areas.

Specifications

Length overall	16.6 m
Beam	6.4 m
Draught	0.8 m
Passengers	6
Speed, maximum	47 kt
Speed, operational	40 kt

Structure: Aluminium hull and superstructure.

Propulsion: Two C18 Caterpillar diesel engines each driving a 403 Hamilton water-jet. Blue arrow control system.

Vessel type	Vessel name	Yard No	Length (m)	Speed (kt)	Seats	Vehicles	Originally delivered to	Date of build
CAT 20 m	*Lagi Lagi*	–	20	34	150	none	Beachcomber Tours and Cruises	2001
CAT 17.7 m	*Wheketere*	–	17.7	30	50	none	Whale Watch Kaikoura	–
CAT 17.7 m	*Te Ao Marama*	–	17.7	30	50	none	Whale Watch Kaikoura	–
CAT 17.7 m	*Discovery 4*	–	17.7	30	75	none	Dolphin Discoveries Ltd	–
CAT 17.7 m	*Aoraki*	–	17.7	30	50	none	Whale Watch Kaikoura	2002
CAT 23 m	*Dolphin Seeker*	–	23.0	23	230	none	King's Tours and Cruises	–
Teknicraft 18 m	*Tahora*	–	18.0	32	48	none	Whale Watch Kaikoara	2004
Teknicraft 17.7 m	*Three Saints*	–	17.7	26	–	none	Private Owner	2003
Teknicraft 16.6 m	*MV 439*	–	16.6	47	–	none	Seascan Ltd	2005
Teknicraft 15 m	*Clipper II*	–	15	40	49	none	Pine Harbour	2006
Teknicraft 14.6 m	*White Morph*	–	14.6	36	–	none	Private Customer	2007
Teknicraft 15 m	*Clipper III*	–	15	40	49	none	Pine Harbour	(2008)

17.7 m Catamaran

A number of these vessels, designed by Teknicraft, have been delivered for whale and dolphin watching, with passenger capacities up to 75 people.

Specifications

Length overall	17.7 m
Beam	6.4 m
Draught	0.7 m
Passengers	50–75
Fuel capacity	2,000 litres
Speed, maximum	30 kt

Propulsion: Two Caterpillar E 3406 diesel engines each rated at 520 kW or two MTU 8V 2000M70 diesels rated at 525 kW at 2,100 rpm, each driving 391 Hamilton water-jets.

15 m Catamaran

Clipper II
Clipper III

Built for Pine Harbour Ferries, this 40 knot vessel operates ferry trips to Auckland, Waiheke and Rakino Islands.

Specifications

Length overall	14.95 m
Beam	5.8 m
Passengers	49
Fuel capacity	2,000 litres
Speed, maximum	40 kt

Propulsion: Two MTU 60 series diesel engines each rated at 550 kW and each driving Hamilton 364 water-jets.

Norway

Båtservice Group ASA

Foged Heibergsgate 29, PO Box 113, N-4502 Mandal, Norway

Tel: (+47) 38 27 13 00
Fax: (+47) 38 26 45 80
e-mail: post@batservice.no
Web: www.batservice.no

Bjorn Fjellhaugen, *Managing Director*
Ottar Ramsfjord, *Technical Manager*
Jarl Mydland, *Marketing Manager*

31 m CATAMARAN

Twin City Liner

Designed by Ola Lilloe-Olsen, this vessel operates between Vienna and Bratislava on the River Danube. The vessel has a draught of 0.74 m and a maximum wash height of 0.45 m at 30 kt making it suitable for river use.

Specifications

Length overall	31.2 m
Beam	8.35 m
Draught	0.74 m
Displacement (max)	54,000 kg
Passengers	102
Crew	4
Operational speed	30 kt

Propulsion: Two MTU 8V 2000 M72 diesel engines delivering 780 kW at 2,250 rpm to Hamilton HM 461 waterjets.

Sea Lord 28

Fjorddrott
Fjordbris
Fjordsol

In November 1990, Båtservice Holding A/S launched its first Sea Lord 28 high-speed catamaran. Two further vessels have been delivered; all three are operated by Rogaland Trafikkselskap A/S.

Specifications

Length overall	28.0 m
Beam	8.3 m
Draught	2.3 m
Passengers	150
Fuel capacity	6,000 litres
Water capacity	400 litres
Propulsive power	2 × 1,500 kW
Max speed	37 kt
Operational speed	35 kt

Classification: DnV +1A1 R15 Light Craft–EO.
Propulsion: Main engines are two MTU 12V 396 TE 84, 1,500 kW each at 2,000 rpm (100% MCR), driving Servogear CP propellers, via ZF Type BW 465S gearboxes.
Auxiliary systems: Two MTU generators, 76 kW each.

Sea Lord 32

Beinveien

Delivered to Rutelaget Askoy, Bergen, in December 1992.

Specifications

Length overall	32.0 m
Beam	9.2 m
Draught	1.8 m
Crew	8
Passengers	177

Sea Lord 38 Xunlong 2 0045116

Sea Lord 38 catamaran Fjorddronningen 0506973

Sea Lord 36 Helgeland 0045117

Fuel capacity	12,000 litres
Water capacity	14,000 litres
Max speed	33 kt
Operational speed	27 kt
Range	450 n miles

Propulsion: Main engines are two Caterpillar 16V 3576 diesels, 1,630 kW each, each driving a Servogear CPP via a ZF reduction gearbox.

Sea Lord 36 (Pollution control)

Ain Dar 7
Ain Dar 8

Early in 1991, Båtservice signed a contract with the Saudi Arabian company Aramco for the building of two 36 m Pollution Control catamarans for operation out of the oil terminal at Ra's Tanurah. The contract was valued at USD15 million and was handled by the Aramco Overseas Company BV of the Netherlands.

The two vessels are equipped with MARCO pollution control oil recovery filter belt systems. Produced by MARCO in Seattle, United States, since 1972. These systems employ a continuous mesh belt that allows water to pass through it but retains the collected oil. Oil and debris are carried towards the drive roller and collected in separate containers, the heavier material being scraped off first by a scraper blade.

Specifications

Length overall	36.0 m
Beam	11.8 m
Draught	1.8 m
Displacement, max	210 t
Crew	8
Fuel capacity	12,000 litres
Water capacity	14,000 litres
Propulsive power	2 × 1,630 kW
Max speed	33 kt
Operational speed	27 kt
Range	450 n miles

Structure: Hull material: aluminium.
Superstructure material: aluminium.

Propulsion: Main engines are two Caterpillar 16V 3516 diesels, 1,630 kW each, driving two VD802A Servogear controllable-pitch propellers, via ZF gearboxes.

Auxiliary systems: Engines: two 185 kW.

Sea Lord 36 (Passenger)

Helgeland

A further Sea Lord 36 fitted out as a passenger ferry, was delivered to Norwegian operator Helgeland Trafikkselskap in late 1997. A cargo hold with cooling capacity was arranged for 20 europallets.

Specifications

Length overall	37.2 m
Length waterline	33.0 m
Beam	10.6 m
Draught	1.7 m
Passengers	199
Fuel capacity	8,000 litres
Water capacity	1,000 litres
Propulsive power	2,500 kW
Max speed	35 kt
Operational speed	33 kt

Classification: DnV+1A1 HSLC R3 Passenger, EO.

Propulsion: Four MTU 12V183 TE92 diesel engines each rated at 625 kW driving two Servogear controllable-pitch propellers via Servogear HD 295 PTI gearboxes.

Sea Lord 38

Fjorddronningen
Xunlong 2

The first Sea Lord 38 was delivered to Troms Fylkes Dampskibsselskap in April 1995. The second vessel *Xunlong 2* was delivered to Shen Zhen Xun Long Passenger Transportation in China in 1998 with a slightly larger passenger capacity of 357.

Specifications

Length overall	38.0 m
Beam	11.2 m
Draught	1.6 m

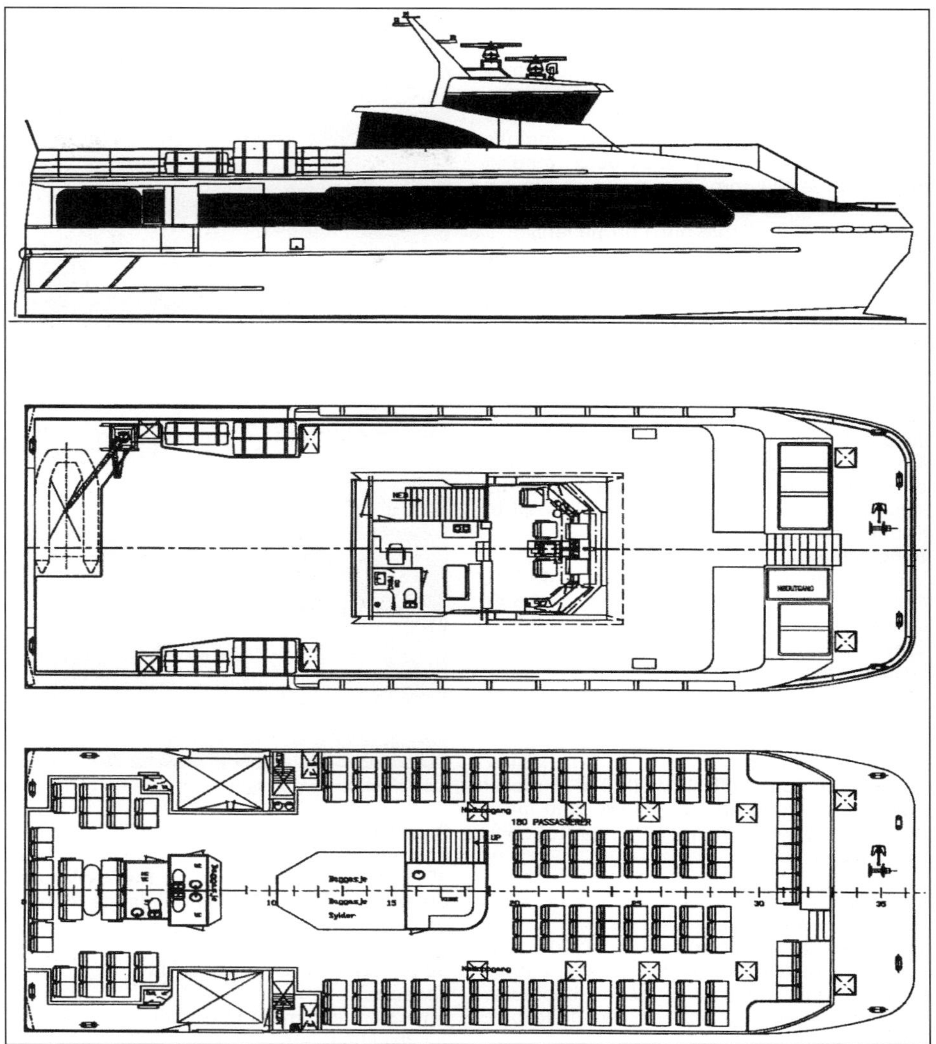

General arrangement of the Båtservice 28 m catamaran 0121756

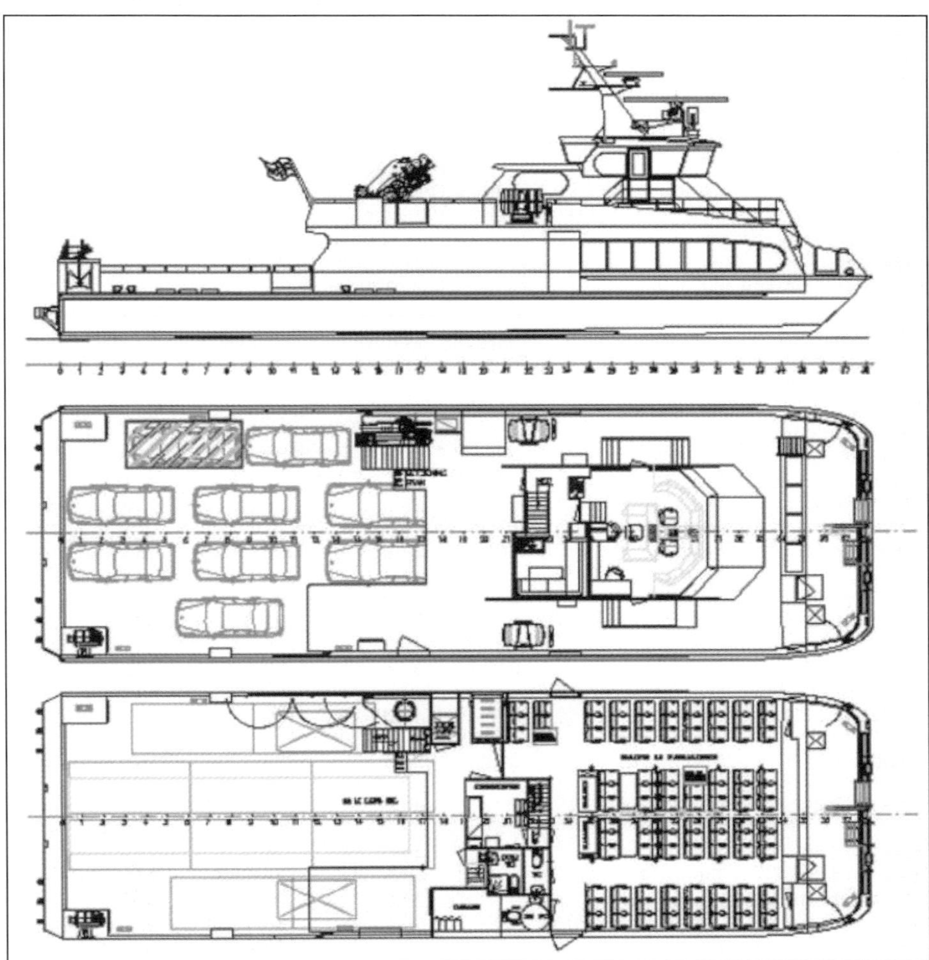

General arrangement of Tansøy 1120332

Crew	5
Passengers	336
Fuel capacity	8,000 litres
Water capacity	1,000 litres
Operational speed	35 kt

Propulsion: *Fjorddronningen*, main engines are four MTU 12V 183 TE 92 each rated at 625 kW at 2,300 rpm. Two pairs of diesels drive through a Servogear 250 gearbox to Servogear controllable-pitch propellers.

Xunlong 2 was powered by four MTU 12V 2000 MTO diesel engines each rated at 788 kW, also driving a pair of Servogear VD820E controllable-pitch propellers.

Sea Lord 40

Dodekanisos Express
Tavous 1
Tavous 2

The first vessel, *Dodekanisos Express*, was delivered to Greece in early 2000, and the second and third vessels were delivered to Iran in late 2001.

Specifications

Length overall	40.0 m
Beam	11.5 m
Draught	1.8 m
Crew	13
Passengers	341
Vehicles	6 cars
Fuel capacity	9,000 litres
Water capacity	3,000 litres
Operational speed	34 kt

Propulsion: Four MTU 12V 2000 M70 diesel engines, each rated at 788 kW and each driving a Servogear VD820F controllable-pitch propeller.

W-4200

Seajet 2(ex-Mirage)

This design, a Westamaran 4200, was originally developed by the Oskashamnas Varv AB yard (now no longer in business) and a single vessel, *Seajet I*, was built for a Greek operator.

A second vessel constructed by Båtservice Holding A/S was delivered to Strintzis Lines in July 1998.

Specifications

Length overall	42.0 m
Length waterline	37.4 m
Beam	10.0 m
Draught	1.8 m
Crew	6
Passengers	386
Fuel capacity	17,600 litres
Water capacity	2,000 litres
Max speed	42 kt
Operational speed	40 kt

Tansøy 1120331

Dodekanisos Express 0103272

Propulsion: Two MTU 16 V 396 TE 74L diesel engines at 2,000 kW each and two MTU 12V 39TE 74L diesel engines at 1,500 kW each, driving Kamewa 56SI water-jets via Reintjes VL7 730 gearboxes.

30 m COMBI CATAMARAN

Tansøy

This 30 m catamaran was delivered to Fjord 1 Fylkesbaatane in January 2007. The vessel was designed for flexible operation and can carry cars, trucks, up to 96 passengers and various goods.

Specifications

Length overall	29.5 m
Length waterline	28.6 m
Beam	9.0 m
Draught	1.85 m
Passengers	97
Crew	3
Vehicles	10 cars or 4 cars and 1 truck
Fuel capacity	5,000 litres
Fresh water capacity	1000 litres
Operating speed	28 knots

Propulsion: Two Caterpillar C32 diesel engines rated at 1045 kW at 2100 rpm, driving twin Servogear 9P810 controllable pitch propellers via Servogear HD295H gearboxes.

28 m catamaran

Vindafjord

Båtservice built one of two identical craft ordered by Torghatten Trafikkelskap from Oma Baatbyggeri which was delivered in 2002.

Specifications

Length overall	27.3 m
Length waterline	27.0 m
Beam	9.0 m
Draught	1.6 m
Crew	4
Passengers	180
Fuel capacity	4,000 litres
Fresh water capacity	500 litres
Operational speed	30 kt

Propulsion: Two Mitsubishi S 12R-MPTA diesel engines each rated at 970 kW and each driving a Servogear VD820G controllable-pitch propeller.

Vessel type	Vessel name	Yard No	Length (m)	Speed (kt)	Seats	Vehicles	Originally delivered to	Date of build
Sea Lord 28	*Fjorddrott*	002	28.0	35.0	150	–	Rogaland Trafikkselskap A/S	1990
Sea Lord 36 (Pollution control)	*Ain Dar 8*	–	36.0	27.0	–	–	Aramco, Saudi Arabia	1991
Sea Lord 36 (Pollution control)	*Ain Dar 7*	–	36.0	27.0	–	–	Aramco, Saudi Arabia	1991
Sea Lord 28	*Nordfolda (ex Fjordsol)*	004	28.0	35.0	150	–	Rogaland Trafikkselskap A/S	1991
Sea Lord 28	*Fjordbris*	003	28.0	35.0	150	–	Rogaland Trafikkselskap A/S	1991
Sea Lord 32	*Beinveien*	008	32.0	27.0	177	–	Rutelaget Askroy	1992
SeaLord 38	*Fjorddronningen*	010	38.0	35.0	336	–	Troms Fylkes Dampskibsselskap	1995
Sea Lord 36 (passenger)	*Helgeland*	016	37.2	33.0	199	–	Helgeland Trafikkselskap	1997
W-4200	*Seajet 2 (ex-Mirage)*	017	42.0	40.0	386	–	Strinzis Lines	1998
SeaLord 38	*Xunlong 2*	018	38.0	35.0	357	–	Shen Zhen Xun Long Passenger Transportation	1998
Sea Lord 40	*Dodekanisos Express*	–	40.0	34.0	341	6	Greece	2001
Sea Lord 40	*Tavous 1*	–	40.0	34.0	341	6	Iran	2001
Sea Lord 40	*Tavous 2*	–	40.0	34.0	341	6	Iran	2001
28 m Catamaran	*Vindafjord*	–	28.0	30.0	160	–	Torghatten Trafillselskap	2002
40 m Catamaran	*Dodekanisos Pride*	049	40.1	30.0	280	–	Dodekanisos Naftiliaki NE	2005
32 m Catamaran	*Twin City Liner*	050	32	30	102	–	Central Danube Region	2006
30 m Catamaran	*Tansøy*	–	29	28	97	10	Fjordfykesbaatane	2007

Båtutrustning Bømlo A/S

Termetangen, N-5420 Rubbestadneset, Norway.

Tel: (+47) 53 42 84 00
Fax: (+47) 53 42 84 02
e-mail: bu@bomlo.as
Web: bu.bomlo.as

Mainly a builder of monohull craft, this company has recently produced its first high-speed catamaran.

16 m Catamaran

Nordic II

Nordic II was constructed in 2004 for Vinnes Skyssbåtservice. The vessel has a composite sandwich structure.

Specifications

Length overall	16.25 m
Beam	6.5 m
Draught (static)	2 m
Passengers	76
Operational speed	25 kt

Propulsion: Two Caterpillar 3406E diesel engines each rated at 450 kW driving fixed pitch propellers through Twin Disk gearboxes.

Brødrene Aa

N-4829 Hyen, Norway

Tel: (+47) 57 86 87 00
Fax: (+47) 57 86 99 14
e-mail: braa@braa.no
Web: www.braa.no

In 2003 Brødrene re-entered the high-speed ferry market with an innovative development. *Rygerkatt*, a 62 passenger, 25 kt ferry was constructed in a new carbon fibre material from Devold AMT. Using multiaxial fabrics over a core material, Brødrene have produced a vessel which has a structural weight reduction of 40 per cent over conventional materials but with an increased strength.

23 m Catamaran ferry

Kvaenangen
Rygerkongen
Rygerfonn

Ordered by Norwegian operator TFDS in 2004, this is the largest carbon fibre catamaran built by Brødrene Aa. Delivered in May 2005, the vessel operates as a ferry and an ambulance boat.

Specifications

Length overall	23.0 m
Length waterline	21.8 m
Beam	8.0 m
Draught	1.0 m
Crew	3
Passengers	50
Fuel capacity	2200 l
Water capacity	400 l
Operational speed	27 kt

Structure: All carbon fibre sandwich construction.

Propulsion: Two MTU 12V 2000 M70 diesel engines each driving a Servogear P805 controllable pitch propeller.

27 m carbon fibre catamaran, Fjordfart 1323689

20.5 m Catamaran ferry

Rygerfjord

Ordered by L Rodne and Sønner, this craft is due to enter service in May 2004.

Specifications

Length overall	20.5 m
Beam	7.2 m
Crew	3
Passengers	97
Operational speed	30 kt

Propulsion: Two MAN D284Z LE 410 diesels rated at 749 kW at 2,100 rpm driving a Servogear HD 220 H gearbox.

Construction: Carbon fibre reinforced sandwich.

18.5 m Catamaran ferry

Rygerkatt

Ordered by L Rodne and Sønner, this craft was delivered in 2003 and operates as a harbour ferry in the Stavanger region.

Specifications

Length overall	18.5 m
Beam	8.0 m
Draught	0.8 m
Crew	3
Passengers	62
Fuel capacity	2,500 litres
Freshwater capacity	200 litres
Maximum speed	29 kt
Operational speed	25 kt

Construction: All composite (carbon fibre) hull and superstructure.

Propulsion: Two Detroit Diesel Series 60 engines each rated at 440 kW and each driving a Servogear VD 80F controllable pitch propeller system.

Vessel type	Vessel name	Yard No	Length (m)	Speed (kt)	Seats	Vehicles	Originally delivered to	Date of build
18.5 m Catamaran	*Rygerkatt*	236	18.5	29	62	–	L Rodne & Sonner	2003
20 m Catamaran	*Rygerfjord*	237	20.5	30	97	–	L Rodne & Sonner	2004
18.5 m Catamaran	*Fjordsol*	241	18.5	29	50	–	Fjordservice	2005
23 m Catamaran	*Kvaenangen*	240	23	27	50	1	Troms Fylkes Dampskipsselship	2005
18.5 m Catamaran	*Nordic Queen*	239	18.5	31	97	–	Hardanger Sunnhordland	2005
23 m Catamaran	*Rygerkongen*	243	23	28	130	–	L Rodne & Sonner	2006
23 m Catamaran	*Cetacea*	245	23	28	100	–	Jakobsen & Sonner	2006
27 m Catamaran	*Fjordkatt*	246	27	32	180	–	Fjordservice	(2006)
23 m Catamaran	*Dagning*	244	23	28	50	4	Radenet Dagning	(2007)
23 m Catamaran	*Rygerfonn*	247-	23	36.5	97	–	L Rodne & Sonner	(2007)
27 m Catamaran	*Fjordfart*	246	27	32	180	–	Fjordservice	(2007)
27 m Catamaran	*Fjorddrott*	246	27	32	180	–	Fjordservice	(2007)
24.5 m Catamaran	–	251	24.5	30	130	–	Fosen Trafikklag	(2007)
24.5 m Catamaran	–	252	24.5	30	130	–	Fosen Trafikklag	(2008)
24.5 m Catamaran	–	253	24.5	30	147	–	Stavangerske	(2008)
27 m Catamaran	–	–	27.0	–	–	–	Stavangerske	(2008)
32.3 m Catamaran	–	–	32.3	–	250	–	Stavangerske	(2008)

Fjellstrand A/S

N-5632 Omastrand, Norway

Tel: (+47) 56 55 41 00
Fax: (+47) 56 55 42 44
e-mail: fjellstrand@fjellstrand.no
Web: www.fjellstrand.no

Asbjørn Tolo, *Managing Director*
Bjarne Børven, *Sales Director*
Stig Oma, *Technical Manager*

Fjellstrand was founded in 1928 in Omastrand, on the west coast of Norway. In January 1989 Fjellstrand merged with Kværner to form Kværner Fjellstrand A/S.

In 2000, the Norwegian yard was subject to a management buyout and re-named Fjellstrand AS. In 2006 the company was acquired by Hans m Gravdel. New yard facilities include a 160 m assembly hall which will allow vessels of up to 120 m in length to be constructed.

Fjellstrand has achieved the most extensive and sustained export penetration of the world high-speed ferry market, both in terms of number of vessels sold and number of countries sold into. By 1999, Fjellstrand had delivered high-speed catamarans to 35 different countries.

Following the delivery of nearly 400 aluminium vessels starting in 1952, Fjellstrand built its first catamaran (25.5 m) in 1976, the Alumaran 165 type. A further four vessels of this size followed, up to 1981. The first 31.5 m passenger catamaran was delivered to a Norwegian operator. Between 1981 and 1985 a further 12 31.5 m catamarans were delivered worldwide for passenger ferry work and crew/supply operations in the offshore oil industry. Design work on a larger 38.8 m type, the Advanced Slender Catamaran (ASC), started in 1983 and by June 1991 Fjellstrand had delivered 33 of this type.

Fjellstrand now offers seven main types of advanced high-speed catamarans to the market, the JumboCat™ 150, JumboCat™ 60, FlyingCat™ 50, FlyingCat™ 45, FlyingCat™ 46, FlyingCat™ 40 and FlyingCat™ 30. Additionally, in collaboration with the Norwegian Public Roads Administration, Fjellstrand has developed an 80 m ferry with a capacity for 120 cars and 400 passengers for river transportation.

In 2007, Fjellstrand built *Miljodronningen II*, a 33.9 m catamaran to be used as a platform for conferences and environmental studies.

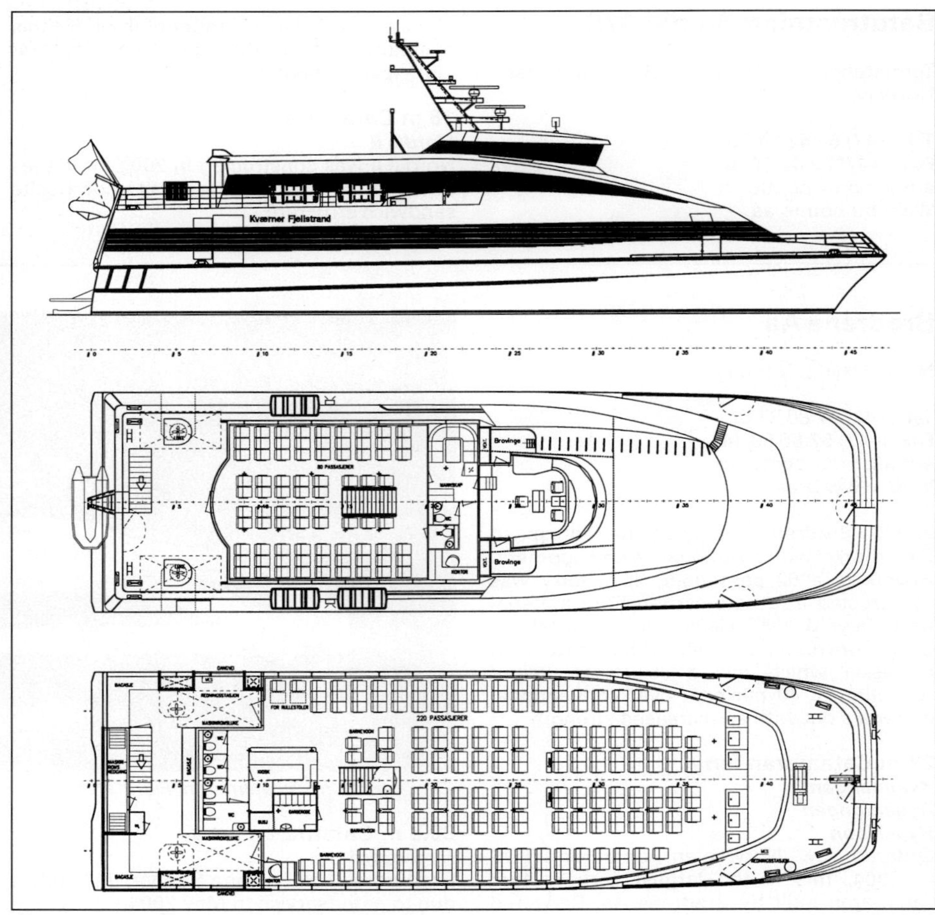

General arrangement of FlyingCat™ 40 Hjorungavag 0058909

ENVIRONMENTAL STUDIES VESSEL

Miljodronningen II

This vessel, built in 2007 for Green Warriors, is a conference and environmental studies platform for operation in Norway.

Specifications

Length overall	33.9 m
Length waterline	33.0
Beam	10.4 m
Passengers	50
Crew	4
Vehicles	1 2-tonne helicopter
Maximum speed	25 kt

Classification: Norwegian Maritime Directorate IHHT; HSC Code 2000, category A for operation in Norway, area 4.

Propulsion: Two 890 kW diesel engines driving water-jets.

Ferrycat 120

Fjellstrand has designed a fast double ended ferry in conjunction with other maritime industry participants, for fast river and fjord operations. It is understood that the first vessel has been ordered by Rogaland for delivery in 2003 with two further vessels ordered by IDO of Turkey for delivery in 2003 to 2004. The vessel is 80 m in length, 20 m in beam with a capacity for 400 passengers and 120 cars. The propulsion system is based on four Ulstein Aquamaster Azipull units, one in each corner of the vessel giving operational speeds up to 25 kt with high levels of manoeuvrability.

JumboCat™ 60

Solidor 3
Nordic Jet
Baltic Jet
Nasser Travel
Solidor 5

The first 60 m Jumbo Cat, *Solidor 3*, was delivered to Emeraude Lines in 1996. The vessel (Yard No 1630) operates between St Malo in France and the Channel Islands. A second vessel, *Nordic Jet* (Yard No 1646), was delivered to Nordic Jet Line in 1998, with a third vessel, *Baltic Jet* (Yard No 1649), to the same operator in 1999. The specifications differ slightly, with the first vessel being powered by MTU engines and having a service speed of 33 kt. A fourth vessel, the *Nasser Travel*, was delivered to Amco

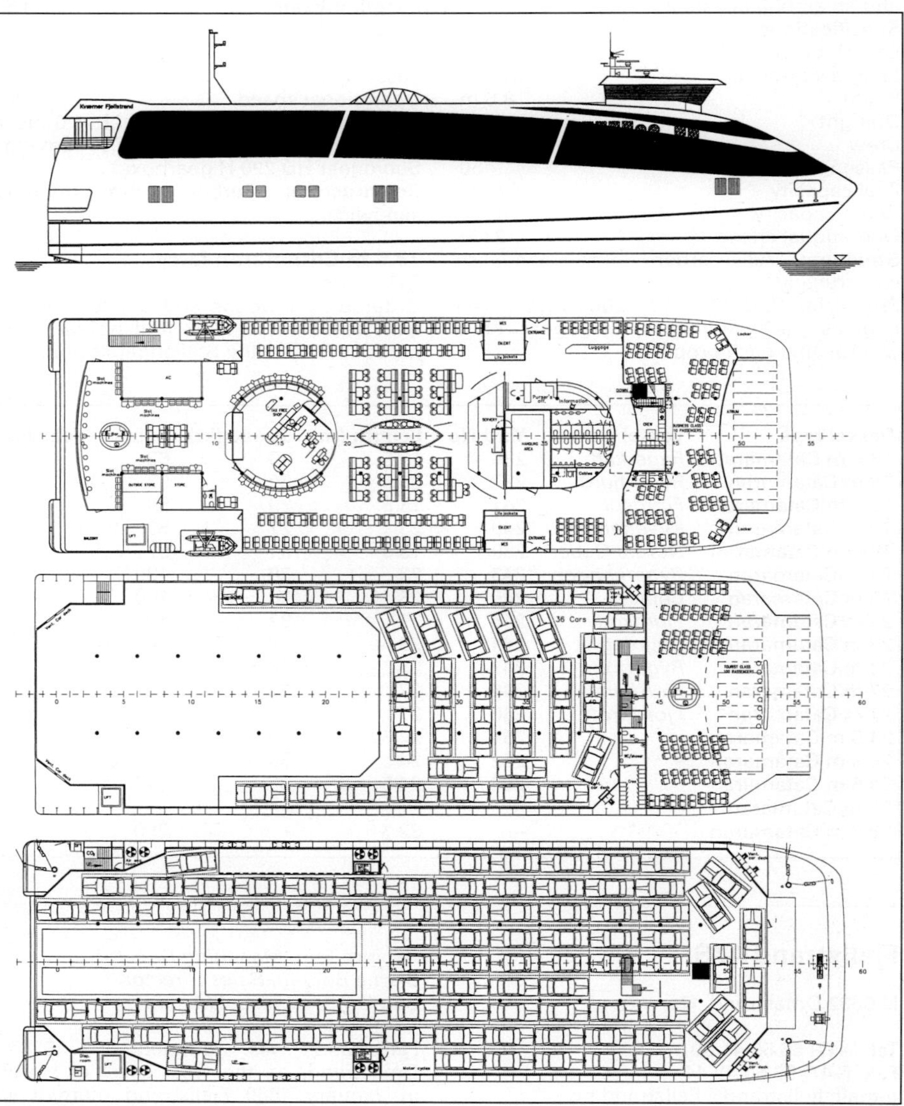

General arrangement of the Fjellstrand JumboCat™ 150 (DESIGN) 0058904

Vessel type	Vessel name	Yard No	Length (m)	Speed (kt)	Seats	Vehicles	Originally delivered to	Date of build
Advanced Slender Catamaran (38.8 m)	*Tian Lu Hu*	–	38.8	35.0	326	–	Zhen Hing Enterprises Co Ltd, China	1985
Advanced Slender Catamaran (38.8 m)	*Yong Xing*	1570	38.8	35.0	312	–	Nigbo Huagang Ltd, China	1985
Advanced Slender Catamaran (38.8 m)	*Mexico* (ex-*Can Cun)*	1571	38.8	35.0	370	–	Cruceros Maritimos del Caribe SA, Mexico	1986
Advanced Slender Catamaran (38.8 m)	*Victoria Clipper*	1572	38.8	35.0	330	–	Clipper Navigation Inc, US	1986
Advanced Slender Catamaran (38.8 m)	*Anne Lise*	1573	38.8	35.0	195	–	Hardanger Sunnhordlanske A/S, Norway	1986
Advanced Slender Catamaran (38.8 m)	*Sevilla 92* (ex-*Caribbean Princess)*	1574	38.8	35.0	310	–	Islena de Navegacion, Spain	1986
Advanced Slender Catamaran (38.8 m)	*Caka Bey*	1576	38.8	35.0	449	–	Istanbul Great City Municipality, Turkey	1987
Advanced Slender Catamaran (38.8 m)	*Rapido de Algeciros* (ex-*Bahamian Princess)*	1575	38.8	35.0	310	–	Islena de Navegacion, Spain	1987
Advanced Slender Catamaran (38.8 m)	*Yeditepe*	1579	38.8	35.0	449	–	Istanbul Great City Municipality, Turkey	1987
Advanced Slender Catamaran (38.8 m)	*Umur Bey*	1578	38.8	35.0	449	–	Istanbul Great City Municipality, Turkey	1987
Advanced Slender Catamaran (38.8 m)	*Sarica Bey*	1581	38.8	35.0	449	–	Istanbul Great City Municipality, Turkey	1987
Advanced Slender Catamaran (38.8 m)	*Fjordprins*	1577	38.8	35.0	201	–	Fylkesbaatane i Sogn og Fjordane, Norway	1987
Advanced Slender Catamaran (38.8 m)	*Ulubatli Hasan*	1582	38.8	35.0	449	–	Istanbul Great City Municipality, Turkey	1987
Advanced Slender Catamaran (38.8 m)	*Sognekongen*	1580	38.8	35.0	201	–	Fylkesbaatane i Sogn og Fjordane, Norway	1987
Advanced Slender Catamaran (38.8 m)	*Uluc Ali Reis*	1583	38.8	35.0	449	–	Istanbul Great City Municipality, Turkey	1988
Advanced Slender Catamaran (38.8 m)	*Nusret*	1584	38.8	35.0	449	–	Istanbul Great City Municipality, Turkey	1988
Advanced Slender Catamaran (38.8 m)	*Karamursel Bey*	1585	38.8	35.0	449	–	Istanbul Great City Municipality, Turkey	1988
Advanced Slender Catamaran (38.8 m)	*Sea Cat/Blue Manta*	1586	38.8	35.0	249	–	–	1988
Advanced Slender Catamaran (38.8 m)	*Jetcat*	1594	38.8	35.0	449	–	–	1988
Advanced Slender Catamaran (38.8 m)	*Hezarfen Celebi*	1587	38.8	35.0	449	–	Istanbul Great City Municipality, Turkey	1988
Advanced Slender Catamaran (38.8 m)	*Cavli Bey*	1588	38.8	35.0	449	–	Istanbul Great City Municipality, Turkey	1988
Advanced Slender Catamaran (38.8 m)	*Draupner*	1590	38.8	35.0	449	–	Flaggruten, Norway	1989
Advanced Slender Catamaran (38.8 m)	*Sleipner*	1589	38.8	35.0	449	–	Flaggruten, Norway	1989
Advanced Slender Catamaran (38.8 m)	*Nordlicht*	1592	38.8	35.0	449	–	AG EMS Emden, Germany	1989
Advanced Slender Catamaran (38.8 m)	*Leopardo*	1593	38.8	35.0	449	–	Cat Lines SA, Spain	1989
Advanced Slender Catamaran (38.8 m)	*Eyra*	1595	38.8	35.0	449	–	Cat Lines SA, Spain	1989
Advanced Slender Catamaran (38.8 m)	*Nam Hae Star*	1596	38.8	35.0	449	–	Nam Hae Express, South Korea	1989
Advanced Slender Catamaran (38.8 m)	*Dae Won Catamaran*	1591	38.8	35.0	449	–	Dae Won Ferry Co Ltd, South Korea	1988
Advanced Slender Catamaran (38.8 m)	*Mercury*	1598	38.8	35.0	449	–	AKP Sovcomflot, CIS	1990
Advanced Slender Catamaran (38.8 m)	*Solovki*	1599	38.8	35.0	449	–	AKP Sovcomflot, CIS	1990
FlyingCat 40	*Kommandoren*	1597	40.0	30.0	252	–	Fylkesbaatane I Sogn og	1990
Advanced Slender Catamaran (38.8 m)	*Nilo* (ex-*Varangerfjord*)	1600	38.8	35.0	449	–	Finmark Fylkesredi og Ruteselskap, Norway	1990
Advanced Slender Catamaran (38.8 m)	*Loberen*	1601	38.8	35.0	449	–	Dampskibsselskabet Oresund, Denmark	1990
FlyingCat 40	*Flying Cat 1*	1602	40.0	30.0	372	–	Fjordane, Norway	1990
FlyingCat 40	*Jet Cat*	1603	40.0	30.0	272	–	Ceres Hellenic Shipping Co	1990
Advanced Slender Catamaran (38.8 m)	*Springaren*	1608	38.8	35.0	449	–	Svenska Rederi AB Oresund, a subsidiary of Dampskibsselskabet Oresund, Denmark	1991
FlyingCat 40	*Aliconer*	1605	40.0	30.0	352	–	Seatran Travel, Malaysia	1991
FlyingCat 40	*Royal Vancouver*	1607	40.0	30.0	302	–	Caremar SpA, Italy	1991
FlyingCat 40	*Ahinora* (ex-*Royal Victoria)*	1609	40.0	30.0	302	–	Royal Sealink Express, Canada	1991
FlyingCat 40	*Soloven*	1611	40.0	30.0	296	–	KatExpress, Denmark	1992
FlyingCat 40	*Orca Spirit*	1610	40.0	30.0	296	–	Nanaimo Express, Canada	1992
FlyingCat 40	*Hai Ou*	1614	40.0	30.0	400	–	Dalian Steamship Co, China	1993
FlyingCat 40	*Victoria Clipper IV*	1615	40.0	30.0	324	–	Clipper Navigation Inc, US	1993
FlyingCat 40	*Soloven II*	1613	40.0	30.0	292	–	Dampskibsselskabet Oresund, Denmark	1993
FlyingCat 40	*Xin Shi Ji*	1612	40.0	30.0	300	–	Yantai Marine Shipping Company, China	1993
FlyingCat 40	*Hai Yan*	1618	40.0	30.0	447	–	Dailan Marine Transport (Group) Company	1993
FlyingCat 40	*Sifka Viking*	1621	40.0	30.0	168	–	DSO, Denmark	1994

Vessel type	Vessel name	Yard No	Length	Speed	Seats	Vehicles	Originally delivered to	Date of build
FlyingCat 40	*Kraka Viking*	1620	40.0	30.0	168	–	DSO, Denmark	1994
FlyingCat 40	*Vargoy*	1623	40.0	30.0	230	–	Fylkesrederi og Ruteselskap, Norway	1994
FlyingCat 40	*Fjordkongen*	1624	40.0	30.0	320	–	Troms Fylkes Dampskibusselskap, Norway	1994
FlyingCat 40	*New Arcardia*	1628	40.0	30.0	300	–	Seo Kyung Shipping Co Ltd, South Korea	1995
FlyingCat 40	*Evridika*	1604	40.0	30.0	374	–	Black Sea Shipping, Bulgaria	1995
FlyingCat 40	*Morro de San Paulo*	1626	40.0	30.0	449	–	Eagle Ridge SA, Panama	1996
JumboCat 60	*Solidor 3*	1630	60.0	36.0	428	52	Emeraude Lines	1996
FlyingCat 40	*Hanse Jet*	1629	40.0	30.0	342	–	Weisse Flotte, Germany	1996
FlyingCat 40	*Baltic Jet*	1644	40.0	30.0	341	–	Weisse Flotte, Germany	1997
FlyingCat 35 m	*Kaptan Pasa*	1635	–	–	358	–	Istanbul Seabus Co	1997
FlyingCat 35 m	*Seidi Ali Reis-I*	1636	–	–	–	–	Istanbul Seabus Co	1997
FlyingCat 35 m	*Oruc Reis-V*	1637	–	–	358	–	Istanbul Seabus Co	1997
FlyingCat 35 m	*Piri Reis-II*	1638	–	–	–	–	Istanbul Seabus Co	1997
FlyingCat 35 m	*Hizir Reis-III*	1639	–	–	358	–	Istanbul Seabus Co	1997
FlyingCat 40 (35 m)	–	–	35.0	33.0	300	–	Istanbul Seabus Co	1997
FlyingCat 24 m	*Rodoyloven*	1643	–	–	358	–	Ofotens og Vesteraalens Dampskibsselskab	1997
FlyingCat 35 m	*Temel Reis-II*	1640	–	–	–	–	Istanbul Seabus Co	1997
FlyingCat 35 m	*Ofoten*	1641	–	–	199	–	Ofotens og Vesteraalens Dampskibsselskab	1997
FlyingCat 40 (35 m)	–	–	35.0	33.0	300	–	Istanbul Seabus Co	1997
FlyingCat 40 (35 m)	–	–	35.0	33.0	300	–	Istanbul Seabus Co	1997
FlyingCat 40 (35 m)	–	–	35.0	33.0	300	–	Istanbul Seabus Co	1997
FlyingCat 40 (35 m)	–	–	35.0	33.0	300	–	Istanbul Seabus Co	1997
FlyingCat 40 (35 m)	–	–	35.0	33.0	300	–	Istanbul Seabus Co	1997
FlyingCat 35 m	*Salten*	1642	–	–	199	–	Ofotens og Vesteraalens Dampskibsselskab	1997
FlyingCat 40 (35 m)	*Salten*	–	35.0	33.0	199	–	Ofotens og Vesteraalens Dampskibsselskap	1997
FlyingCat 50	*Eid Travel*	1647	48.2	39.5	308	12	Amco Travel	1998
JumboCat 60	*Nordic Jet*	1646	60.0	36.0	428	52	Nordic Jet Line	1998
FlyingCat 40	*Locmaria*	1650	40.0	30.0	–	–	Companie Morbilhannaise	1998
FlyingCat 36 m	*Prinsen*	1645	–	–	290	–	As Nesodden-Bunderfjord Dampskipsselskap	1998
FlyingCat 38 m	*Hjornungavag*	1648	–	–	300	–	More og Fylkesbatar	1998
FlyingCat 40 (37.6 m)	–	–	37.6	35.0	300	–	MRF, Norway	1999
JumboCat 60	*Baltic Jet*	1649	60.0	36.0	428	52	Nordic Jet Line	1999
FlyingCat 30	*Grip*	–	29.2	33.5	150	–	Møre Og Fylkesbatar	2000
JumboCat 60	*Solidor 5*	–	60.0	36.0	450	60	Emeraude Lines	2000
JumboCat 60	*Nasser Travel*	1652	60.0	36.0	450	49	Amco Travel	2000
Flying Cat 50	*San Gwann*	1658	52.0	37.5	420	20	Virtu Ferries	2001
Flying Cat 40	*Millenium Star*	1659	37.6	35.0	–	–	Albania	2001
Flying Cat 40	*Ladejarl*	1663	37.6	35.0	276	–	Møre Og Fylkesbatar	2002
Flying Cat 40	*Mørejarl*	1669	37.6	35.0	276	–	Møre Og Fylkesbatar	2002
Flying Cat 45	*Flying Viking*	–	44.0	32.5	354	–	Førde Reederei Seetouristik	2002
FerryCat 120	–	–	80.0	25.0	400	120	Rogaland	2003
Ferry Cat 52	*Halunder Jet*	–	51.0	36.0	587	none	Førde Reederei Seetouristik	2003
Ferry Cat 120	–	–	80.0	25.0	400	120	IDO, Turkey	–
Ferry Cat 120	–	–	80.0	25.0	400	120	IDO, Turkey	–
Flying Cat 40	*Atlantic Jet*	–	41.1	31.0	264	–	SPM Express SA, France	2004
Flying Cat 46	*Fjordkongen*	–	45.2	36.5	356	–	TFDS, Norway	2004
Flying Cat 46	*Pont D'Yeu*	–	45.2	32.0	442	6	Compagnie Yeu Continent	2005
Flying Cat 46	*Fjorddronningen*	–	45.2	36.5	356	–	TFDS, Norway	2005
33 m Catamaran	*Teislen*	1675	33.0	35.0	180	–	HSD, Norway	2006
33 m Catamaran	Tranen	1676	33.0	35.0	180	–	HSD, Norway	2006
Flying Cat 46	*Le Châtelet*	1674	45.2	32.0	442	6	Compagnie Yeu Continent	2006
Flying Cat 35	*Miljodronningen II*	–	33.9	25	50	Helicopter	Green Warriors, Norway	2007

Travel in 2000, and a fifth was delivered to Emeraude Lines in 2000, in a contract worth NOK185 million.

Specifications

Length overall	60.0 m
Beam	16.8 m
Draught	2.9 m
Deadweight	135 t
Passengers	428
Vehicles	52
Fuel capacity	30,000 litres
Water capacity	3,000 litres
Propulsion power	14,400 kW
Operational speed	36 kt

Classification: DnV +1A1 HSLC Car Ferry A, R2, EO.

Propulsion: Powered by two Caterpillar 3618 diesel engines each rated at 7,200 kW at 1,050 rpm driving through Rientjes VLJ6831 reduction gears to Kamewa 112 S11 waterjets.

Control: Fitted with a Kværner motion damping system.

Fjellstrand FlyingCat 50 San Gwann *during delivery to Virtu Ferries* 0110416

Flying Cat™ 52

Based on the success of the Flying Cat 45 and 50, this all passenger vessel was ordered by FRS in 2002 and delivered in 2003. Specification details are currently unavailable.

Flying Cat™ 50

The first Flying Cat™ 50 passenger/car catamaran was ordered in 1997 by Amco Travel in a contract worth NOK85 million. The vessel was delivered in May 1998 for operation in the Red Sea between Safaga in Egypt and Deba in Saudi Arabia. A second slightly larger Flying Cat™ 50, the *San Gwann,* was ordered by Virtu Ferries of Malta and was delivered in 2001. This vessel has a capacity of 420 passengers and 20 cars.

Specifications

Length overall	48.2 m
Beam	12.5 m
Crew	8
Passengers	308
Vehicles	12 cars
Fuel capacity	16,000 litres
Water capacity	1,500 litres
Propulsive power	8,000 kW
Maximum speed	41.5 kt
Operational speed	39.5 kt

Classification: GL +100 A5 HSC A 0C2 High Speed Passenger Craft +MC HSC-A.

Propulsion: The vessel is powered by four MTU 16V 396 74L diesel engines each rated at 2,000 kW at 2,000 rpm and each driving a KMW 63 S11 water-jet via a Reintjes gearbox.

Control: The vessel is fitted with a Kvaerner foil and flap motion damping system.

Flying Cat™ 46
Pont d'Yeu

Two vessels were ordered by the Department de la Vendée. The first vessel was delivered in 2005 and the second is due for delivery in October 2006.

Specifications

Length overall	45.5 m
Beam	11.2 m
Draught	1.5 m
Passengers	431
Vehicles	6 cars
Operational speed	32 kt

Propulsion: Four Cummins KTA 50 M-Z diesel engines each delivering 1399 kW at 1950 rpm to Kamewa 63SII water-jets via ZF 4650 gearboxes.

Flying Cat™ 46
Fjordkongen
Fjorddronningen

Two vessels were ordered by Troms Fylkes Dampskibsselskap for delivery in late 2004 and early 2005. The vessels were designed and built primarily for business commuters.

Specifications

Length overall	45.2 m
Beam	11.5 m
Passengers	356
Operational speed	36.5 kt

Classification: DnV +1A1 HSLC Passenger R3 (Nor) EO.

Propulsion: Four diesel engines driving four water-jets.

Ride control: Stern Interceptors.

Flying cat™ 45

Based on the experience of the Flying Cat 40 and 50 craft, this 45 m craft provides for a passenger capacity of 350 with an operating speed of 35 to 40 kt with four independent propulsion systems. The first vessel was completed in 2002.

Specifications

Length overall	44.0 m
Length waterline	40.5 m
Beam	11.6 m
Draught	1.6 m
Passengers	354

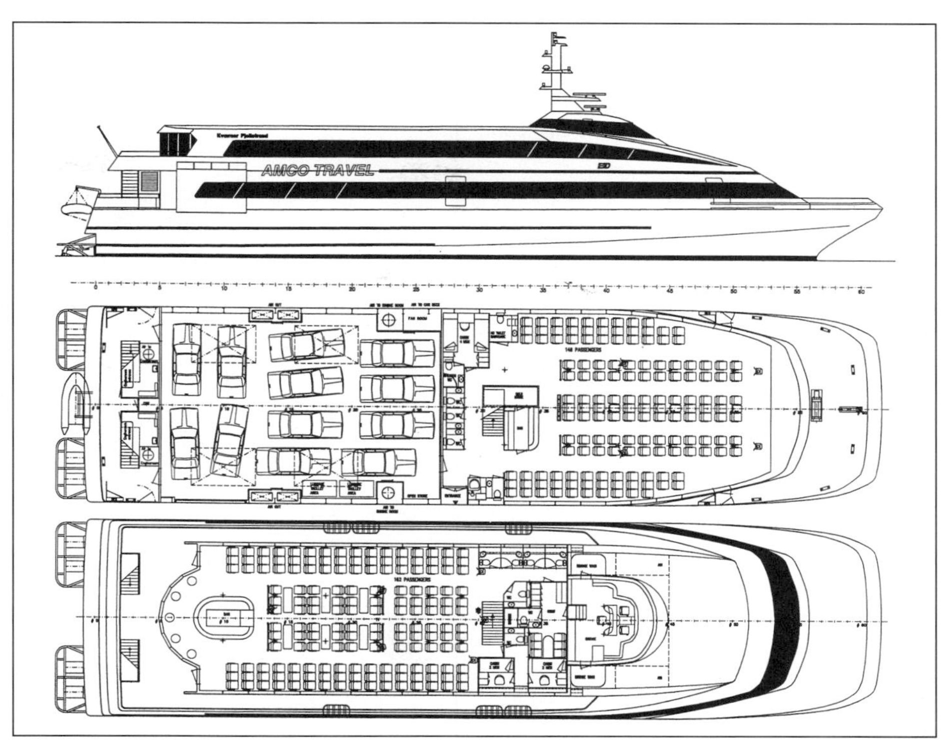

General arrangement of Fjellstrand FlyingCat™ 50 0024751

FlyingCats™ 40 (35 m) Ofoten *and* Salten 0010510

FlyingCat™ 50 Eid Travel 0024746

Flying Viking *on trials* 0132898

Fuel capacity	18,400 litres
Water capacity	2,230 litres

Propulsion: Four Caterpillar 3508 BTA diesel engines each rated at 1,020 kW and each driving a Kamewa A50 water-jet via a ZF 4540 NR gearbox.

Flying Cat™ 40

This design was introduced in 1997 based on an order for six 35 m craft from Istanbul Seabus Company in Turkey. An additional two vessels were also ordered by Ofotens og Vesteraalens Dampskibsselskap of Norway. These latter two craft have seating capacity of only 199 passengers but can carry 14 cargo pallets in addition. Their engines are TE 92 rated at 640 kW each giving a service speed of 33 kt. A 37.6 m version was ordered by MRF of Norway and delivered in 1999, followed by two further craft in 2002. In 2004 a Flyingcat 40 was ordered by French operator SPM Express for delivery in the same year. The hull, built at the former Kvaerner Fjellstrand yard in Singapore is the second of two shipped from Singapore to Norway when the management of the Omastrans yard purchased Fjellstrand from Kvaerner in 2000.

Specifications

Length overall	37.6 m
Beam	10.4 m
Draught	1.6 m
Passengers	300
Fuel capacity	7,000 litres
Water capacity	800 litres
Operational speed	35 kt

Classification: DnV +1A1 HSLC Passenger R3, EO.

Propulsion: Powered by two MWM TBD 620 V16 diesel engines each rated at 2,012 kW at 2,000 rpm driving a Servogear CPP system.

Flying Cat™ 40 (standard)

This Kvaener Fjellstrand design was launched in 1989. By 1999, 59 craft had been built, 23 from the Omastrand yard and 36 from the Singapore yard.

Baltic Jet

This 40 m Flying Cat was ordered by Weisse Flotte of Germany in 1996 and was delivered in 1997. This is a version of the M2 modified 40 m Flying Cat now known as the FlyingCat 40 (standard).

Specifications

Length overall	40.0 m
Beam	10.1 m
Draught	1.6 m
Crew	6
Passengers	341
Fuel capacity	12,000 litres
Water capacity	1,800 litres
Propulsive power	4,000 kW
Operational speed	30 kt

Classification: GL +100 A5 HSC.B.0C2.HSPC +MC HSC-B.

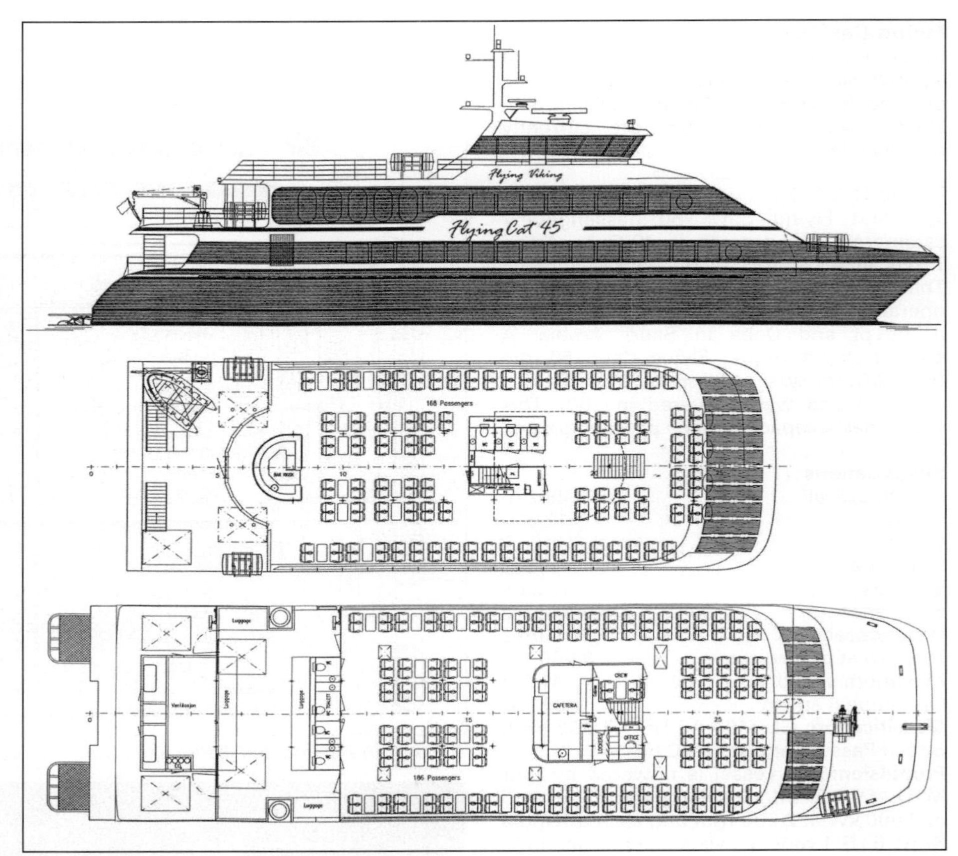

General arrangement of the Flying Cat™ 45 0132899

Artist's impression of the JumboCat™ 150 (DESIGN) 0058905

FlyingCat™ 40 (38 m) Hjorungavåg 0103281

Fjorddronningen *and* Fjordkongen *side by side at sea* 1141278

Propulsion: Powered by two MTU 16V 396 TE 74L diesel engines each rated at 2,000 kW at 2,000 rpm. Each engine drives through a Reintjes VLJ 930C reduction gearbox to a Kamewa S6311 water-jet.
Control: A Kvæner motion damping system is installed on the vessel.

Flying Cat™ 30

Grip

This is the smallest of the Kværner Fjellstrand fast ferry range with a capacity for 150 passengers.

Specifications

Length overall	29.2 m
Length waterline	25.9 m
Beam	9.0 m
Passengers	150
Fuel capacity	3,000 litres
Water capacity	500 litres
Operating speed	33.5 kt

Propulsion: Two Deutz MWM diesel engines each rated at 1,135 kW.

Holen MEK Verksted A/S

PO Box 20, N-6030 Langevaag, Norway

Tel: (+47) 70 19 25 78
Fax: (+47) 70 19 35 84

Capt S Gudmundset, *Director*

Holen mek Verksted A/S specialises in the construction of high-speed vessels in welded aluminium, of up to 60 m in length. Comprehensive facilities for conversion and repair work on both aluminium and steel craft are also available at the yard, which is located close to Ålesund on Norway's west coast.

New buildings in aluminium include two 26 m catamaran passenger tenders carrying 400 passengers.

MM 24 PC

Baronessa

Baronen

These two identical high-speed catamarans were delivered to Hardanger Sunnhordlandske Dampskipsselskap A/S on the west coast of Norway, in May and August 1994. They have been designed for operation at high speed in adverse weather conditions.

Specifications

Length overall	24.0 m
Length waterline	22.0 m
Beam	8.2 m
Draught	1.2 m
Passengers	125
Fuel capacity	5,000 litres
Water capacity	1,000 litres
Propulsive power	2 × 735 kW
Max speed	32 kt

Classification: DnV +1A1 HSLC Passenger R3 EO, IMO Resolution A373(X), Norwegian Maritime Directorate.

Baronessa *at speed* 0506900

Structure: The hull and superstructure are constructed from aluminium.
Propulsion: Main engines are two MTU 12V 183 TE 92 diesels; driving two Servogear variable-pitch propellers.

MM 21 PC

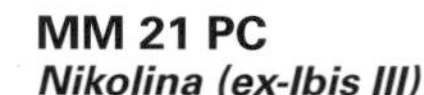

Nikolina (ex-Ibis III)

Three of these craft have been constructed for Scavfer International SA, to be operated by Interoil Services Limited as crew boats on the west coast of Nigeria. The vessels are all-aluminium with a passenger saloon for 62 and crew facilities for four.

Specifications

Length overall	21.0 m
Length waterline	19.0 m
Beam	7.7 m
Draught	1.2 m
Passengers	62
Fuel capacity	5,000 litres
Water capacity	2,000 litres
Max speed	34 kt

Classification: DnV +1A1 HSLC Passengers R3.
Propulsion: Powered by two MAN D2842 LE 402 diesel engines each driving a Servogear variable-pitch propeller via a Mekanord V gearbox.

Vessel type	Vessel name	Yard No	Length (m)	Speed (kt)	Seats	Vehicles	Originally delivered to	Date of build
MM 24 PC	*Cosmos* (ex-*Baronessa*)	–	24.0	32	125	–	Hardanger Sunnhordlandske Dampskipsselskap A/S	1994
MM 24 PC	Seka (ex-*Baronen*)	133	24.0	32	125	–	Hardanger Sunnhordlandske Dampskipsselskap A/S	1994
MM 21 PC	*Nikolina* (ex- *Ibis III*)	144	21.0	34	62	–	Scavfer International SA	1995
MM 21 PC	–	–	21.0	–	62	–	Scavfer International SA	–
MM 21 PC	–	–	21.0	–	62	–	Scavfer International SA	–

Lindstøls Skips- & Båtbyggeri A/S

PO Box 250, Risoer Aust Agder, N-4953, Norway

Tel: (+47) 37 150 344
Fax: (+47) 37 15 20 60
e-mail: post@lindstol-ship.no

Einar K Lindstøl, *Director*
Arne Lindstøl, *Manager*

Lindstøls Skips- & Båtbyggeri A/S was founded in 1870 by Erik K Lindstøl. Through the years Lindstøls Skips has acquired a broad experience in the construction of various types of vessels in the size range up to 60 m in length. A large number of sailing cargo vessels, rescue vessels, research, arctic, fishing and pleasure vessels were delivered to customers in several countries.

One of the most famous of these vessels was delivered in 1917 and named *Quest*; it travelled the oceans with Captain Shackleton on Antarctic expeditions for many years.

Erik K Lindstøl's son, Arne K Lindstøl, headed the yard for 40 years, up until 1952, when Einar K Lindstøl changed the building material from wood to steel. The company began using aluminium in construction in 1960.

Since that time Lindstøls Skips has built a number of fast catamaran craft. The yard signed an agreement with Abeking and Rasmussen of Germany in 1995 for the joint construction of two 27.5 catamarans designed by Lindstøls Skips. These were delivered in June 1996.

33 m PASSENGER FERRY

Hansestar

Specifications

Length overall	33.3 m
Beam	10.3 m
Draught	1.7 m
Passengers	322
Fuel capacity	7,400 litres
Water capacity	1,000 litres
Maximum speed	31 kt
Operational speed	28 kt

Propulsion: Two MTU 12V 396 TE74L diesel engines rated at 1,500 kW @ 2,000 rpm each, driving two controllable-pitch propeller via ZF Bu 465 gearboxes.

32 m PASSENGER FERRY

Fjordtroll

This 32 m catamaran ferry was designed by Ola Lilloe-Olsen and ordered by Fylkesbaatane i Sogn Og Fjordane in 2001. The vessel entered service on the Selje to Bergen route in April 2002.

Specifications

Length overall	32.1 m
Length waterline	29.8 m
Beam	9.5 m

Passengers	170
Operational speed	36 kt

Propulsion: Two MTU 12V 396 TE74L diesel engines each rated at 1,500 kW and driving a Servogear VD 820F controllable-pitch propeller.

27.5 m PASSENGER FERRY

Bairro Alto (ex-Hanseblitz)
Parque das Nacoes (ex-Hansepfeil)

These vessels were originally delivered to Elbe City Jets in 1996.

Specifications

Length overall	27.5 m
Beam	9.0 m
Draught	1.7 m
Passengers	202
Crew	5
Fuel	5,200 litres
Maximum speed	35 kt

Propulsion: Two MTU 8V 396 TE74L diesel engines rated at 1,000 kW at 2,000 rpm each and each driving a controllable-pitch propeller via Servogear HD250 gearboxes.

27.5 m CATAMARAN

Namdalingen

This vessel was delivered in 2000 to Namsos Traikkselskap based in Gutvik, Norway, for operation to Rørvik and Namsos.

Specifications

Length overall	27.5 m
Beam	9.3 m
Draught	1.4 m
Crew	3
Passengers	144
Fuel capacity	5,000 litres
Water capacity	500 litres
Operational speed	28 kt

Propulsion: Two MTU 16 V 2000 M70 diesel engines each rated at 1,050 kW at 2,100 rpm and driving a Servogear VD 810B controllable-pitch propeller system.

25 m FAST VEHICLE FERRY

Designed by Ola Lilloe-Olsen, these vessels were delivered in 2000 to Norwegian operator, Fylkesbaatane I Sogn Og Fjordane.

Specifications

Length overall	25.5 m
Length waterline	23.8 m
Beam	8.0 m
Draught	1.6 m
Crew	3
Passengers	81
Vehicles	5 cars
Fuel capacity	4,000 litres
Water capacity	500 litres
Max speed	28 kt

Propulsion: Two Mitsubishi S6R2 diesel engines each rated at 610 kW and each driving a Servogear VD810E controllable-pitch propeller.

LIGHT 22.5 m CATAMARAN

The first Light 22.5 m catamaran built by Lindstøl Skips A/S, *Namdalingen*, was delivered from the yard in August 1990.

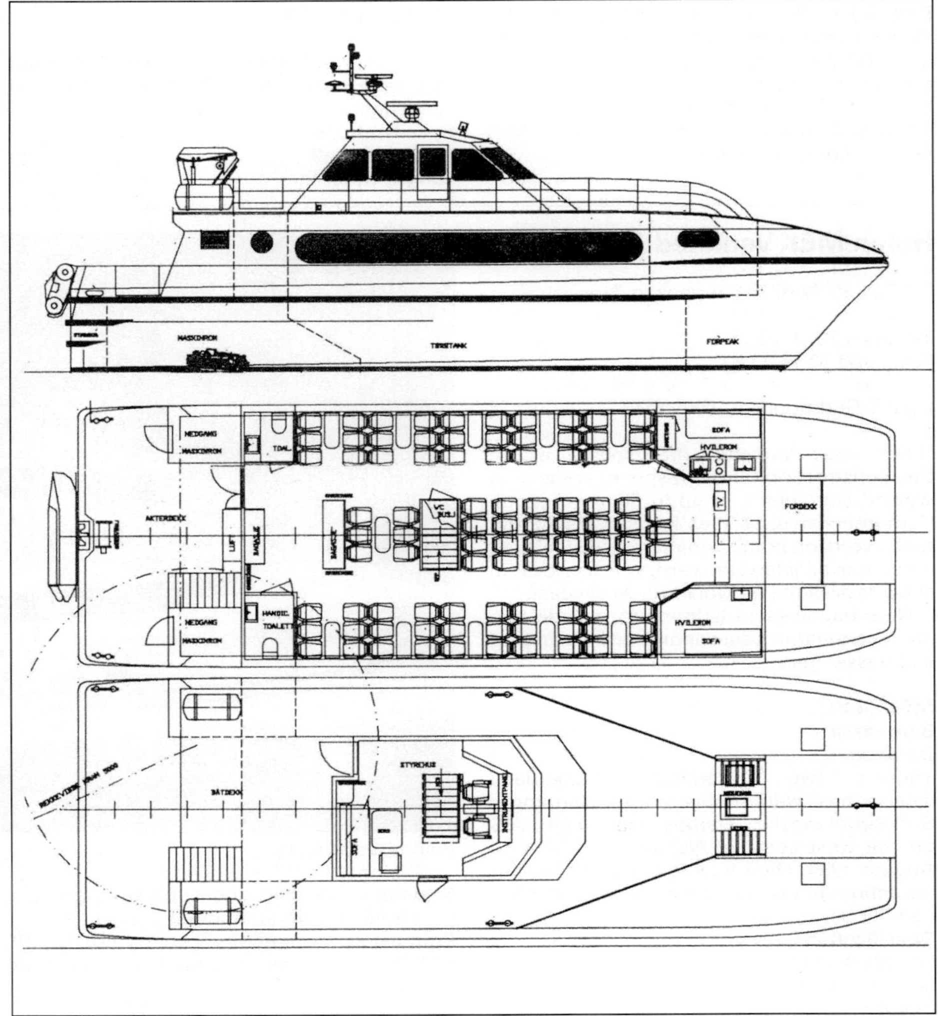

General arrangement of Lindstøl's Light 22.5 m catamaran 0506656

Fosningen 0506657

Vessel type	Vessel name	Yard No	Length (m)	Speed (kt)	Seats	Vehicles	Originally delivered to	Date of build
Light 22.5 m Catamaran	*Namdalingen II* (ex-*Namdalingen*)	296	22.5	30.0	110	–	A/S Namsos Trafikkselskap, Norway	August 1990
Light 22.5 m Catamaran	*Fosningen*	298	22.5	30.0	110	–	Fosen Trafikkselag A/S, Trondheim	25 June 1991
25 m Fast Vehicle Ferry	*Sylvarnes*	–	25.5	–	81	5	Fylkesbaatane I sogn Og Fjordane	2000
25 m Fast Vehicle Ferry	*Fjordglytt*	–	25.5	–	81	5	Fylkesbaatane I sogn Og Fjordane	2000
27.5 m Catamaran	*Namdalingen*	–	27.5	28	144	–	A/S Namsos Trafillselskap, Norway	2002
32 m Passenger Ferry	*Fjordtroll*	320	32	36	170	–	Fylkesbaatane I sogn Og Fjordane	2002
33 m Catamaran	*Hansestar*	308	33.3	31.0	322	–	Elbe City Jet	1997
27.5 m Catamaran	*Bairro Alto* (ex-*Hanseblitz*)	305	27.5	35	202	–	Elbe City Jet	1996
27.5 m Catamaran	*Parque das Nacoes* (ex-*Hansepfeil*)	305	27.5	35	202	–	Elbe City Jet	1996

Ordered by A/S Namsos Trafikkselskap, the vessel has entered the rough coastal service route between Namsos and Rörvik/Leka in mid-Norway, north of Trondheim.

The Light 22.5 m is a 100 per cent Lindstøl-designed concept based on the first Norwegian-built slender catamarans with symmetric hulls developed and built by Lindstøl Skips A/S in 1983 to 1985.

Namdalingen is the latest design in a continuous line of deliveries since the first sailing ship was built in 1870.

Lindstøl Skips A/S has been in aluminium constructions since the beginning of the 1960s and is today equipped with the most advanced CAD/CAM systems available, in order to offer the customer the best service during the projecting and building period.

Consideration has been given to ensure the possibility of changing a main engine within 8 hours. The reason for this is that *Namdalingen* is the only high-speed passenger vessel covering the Namsos – Rörvik and Leka area, and the vessel has to be in traffic 360 days a year. The owner, A/S Namsos Trafikkselskap, therefore has one spare engine in stock for replacing when routine overhauls are necessary.

Model tests indicated that when operating at 30 kt in a significant wave height of 1 m (maximum wave height of 2 m), accelerations in the passenger saloon will be 0.1 *g*. The wash at 30 kt is measured to 0.3 m, 30 m from the side of the vessel.

Special attention has been given to passenger and crew visibility; the wheelhouse has 360° visibility.

Namdalingen II (ex-Namdalingen), Yard No 296

Specifications

Length overall	22.5 m
Beam	7.6 m
Draught	1.1 m
Crew	2
Passengers	110
Fuel capacity	2 × 1,700 litre tanks
Water capacity	400 litres
Propulsive power	2 × 672 kW
Operational speed	30 kt
Range	440 n miles

Classification: DnV +1A1 R5, EO Light-Craft Passenger Catamaran.

Structure: Aluminium alloy, decks and superstructure of Lindstøl aluminium alloy sandwich profile.

Propulsion: Main engines are two MTU 12V 183TE 92, 672 kW each, at 2,300 rpm; driving Servogear CP propellers type 800 A, V drive gear VD 250 B.

Electrical system: Auxiliary engine: Mitsubishi/Mustang A 372-22, 16 kW, 20 kVA.

Fosningen, Yard No 298

This is the second Light 22.5 m catamaran built by Lindstøls Skips and was delivered on 25 June 1991. This vessel has 118 seats and is in service with Fosen Trafikklag A/S of Trondheim on an 8.3 n mile route between Skansen and Vanvikan. *Fosningen* is similar in many respects to *Namdalingen,* the main differences being large 1.1 m diameter four-blade CP propellers to give good starting, stopping and manoeuvring qualities, and a change of saloon layout with the inclusion of a kiosk and office section. Some changes have also been made to the hull form below the waterline.

27.5 m CATAMARAN

Two of these vessels were completed in 1996 at a total contract value of £5 million. They were produced jointly by Lindstøls Skips and the German yard Abeking and Rasmussen. The crafts seat 210 passengers each and are powered by two MTU 8V 936 diesels at speeds of up to 35 kt. The vessels operate from Hamburg in Germany.

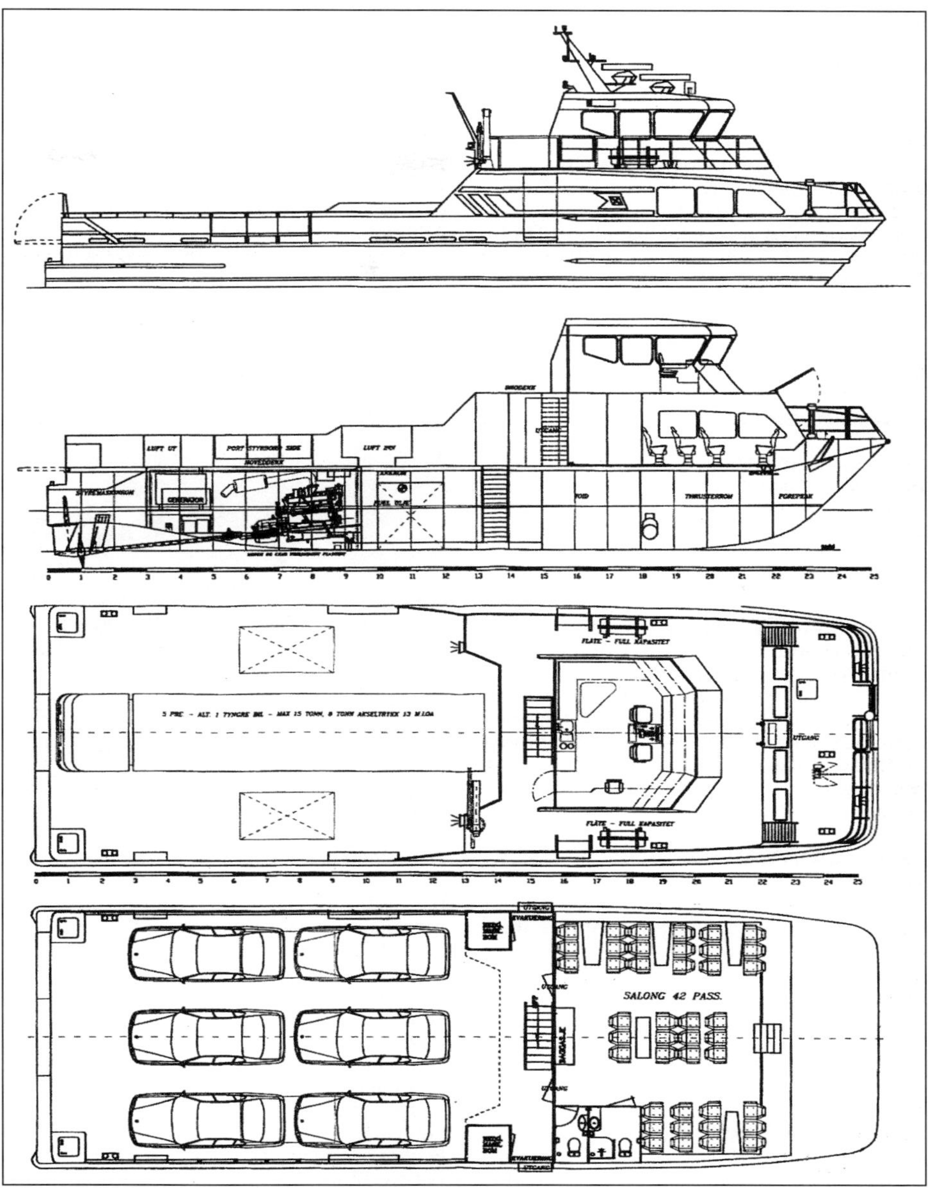

General arrangement of Lindstøls 25 m vehicle ferry 0103274

Profile view of the 32 m Passenger Ferry for FSO 0129050

32 m Fjordtroll 1 1120249

Måløy Verft

6718 Deknepolien, Norway

Tel: (+47 57) 84 98 00
Fax: (+47 57) 84 98 01
e-mail: firmapost@maloy-verft.no
Web: www.maloy-verft.no

The Måløy yard was established in 1978 by the Saetremyr family as a service facility for their fleet of trawlers. In 1995 a new building hall was built for the production of composite vessels. The company offers a range of passenger, cargo, patrol, search and rescue vessels designed by Ola Lilloe-Olsen.

22 m Catamaran

Fjord Aalesund

This vessel, constructed from composites, was delivered to 62° Nord Cruises in 2006. Additional vessels with a similar specification are being built forTalisman Energy to operate in the Orkney Islands.

Specifications

Length overall	21.75 m
Length waterline	19.45 m
Beam	7.6 m
Draught	1.2 m
Passengers	90
Max speed	30 kt

Propulsion: Two Scania DI 1643 diesel engines delivering 580 kW at 2,250 rpm to Servogear CP propellers.

Oma Baatbyggeri

PO Box 200, Stord N-5402, Norway

Tel: (+47) 53 40 98 00
Fax: (+47) 53 40 98 01
e-mail: post@oma.no
Web: www.oma.no

Gustav Oma, *Managing Director*

Situated at the foot of Hordangerfjord on the island of Stord in Norway, Oma Baatbyggeri is a ship building and repair yard which specialises in aluminium and FRP materials. Having designed over 500 vessels since its establishment in 1909, the yard is currently involved in building a number of high-speed catamaran craft.

26 m Snarveien 1178311

32.5 m CATAMARAN

Lysefjord

Ordered by KS Lysefjord, this vessel was delivered in late 2006 and operates from Stavanger.

Specifications

Length overall	32.5 m
Beam	10.6 m
Passengers	85
Vehicles	12 cars
Operational speed	28 kt

29 m CATAMARAN

Jernøy

Designed for operation between Alta and Hammerfest on Norway's northwest coast, the vessel is equipped with a quadruple propulsion system and capacity for 3 cars and 145 passengers.

Specifications

Length overall	29.2 m
Length waterline	28.5 m
Beam	9.0 m
Draught	1.5 m
Deadweight	27 tonnes
Passengers	145
Vehicles	3
Fuel capacity	6,000 litres
Freshwater capacity	1,200 litres
Operational speed	35 kt

Propulsion: Four MTU 8V 2000 M72 diesel engines each rated at 720 kW at 2,250 rpm, each driving a Kamewa A50 waterjet via a ZF 2050 gearbox.

28 m CATAMARAN

Strandafjord

Two of these craft were ordered by Torghatten Trafikkselskap in 2001 and delivered in 2002. Due to the short time for delivery, Oma Baatbyggeri and Båtservice co-operated in building one craft each.

Specifications

Length overall	27.3 m

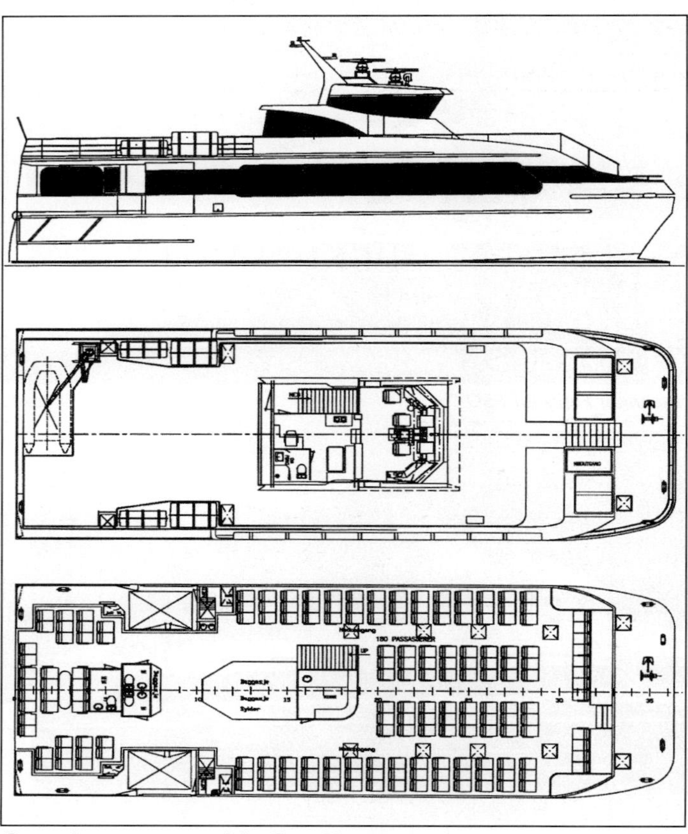

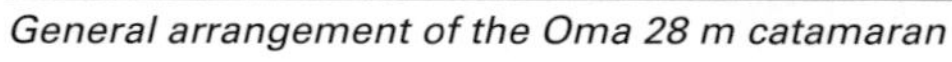

General arrangement of the Oma 28 m catamaran 0121756

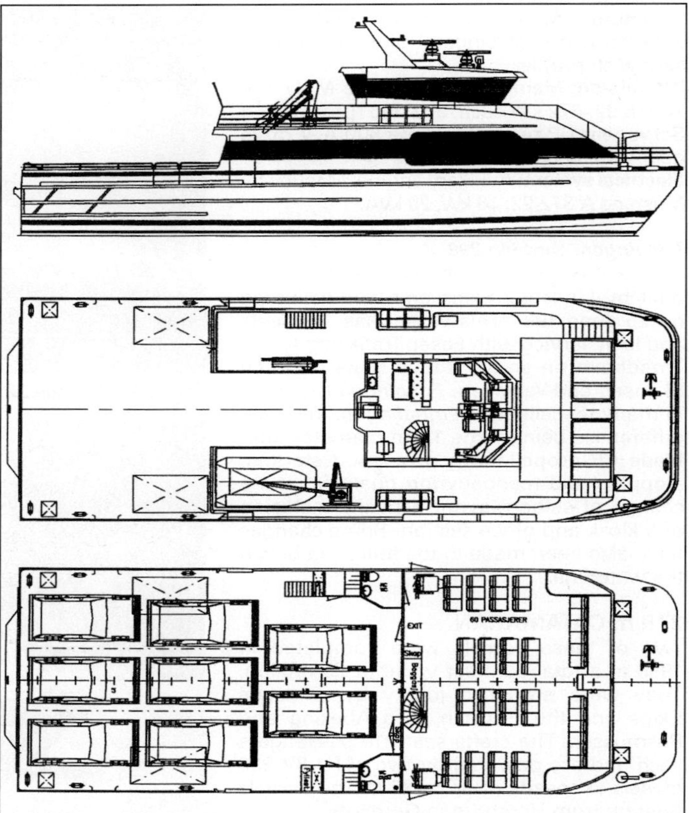

General arrangement of the Oma 26 m catamaran 0121757

Vessel type	Vessel name	Yard No	Length (m)	Speed (kt)	Seats	Vehicles	Originally delivered to	Date of build
20 m Catamaran	*Vøringen*	507	19.8	30.0	86	none	HSD	1995
24 m Catamaran	*Mårøy*	510	24.5	28.5	132	none	Finnmark	1997
21 m Catamaran	*Rypløy*	509	21.0	29.5	90	none	Finnmark	1997
24 m Catamaran	*Rødøyløven*	508	24.5	28.5	132	none	Ofotens Og Vesteraalens Dampskibsselskap	1997
24 m Catamaran	*Kobbøy*	511	24.5	28.5	84	none	Finnmark Fylkesrederi	1998
30 m Catamaran	*Renøy*	514	30.0	34.0	185	none	Finnmark Fylkesrederi	1999
30 m Catamaran	*Fjorprincessen*	513	30.0	33.5	198	none	Troms Fylkes Campskibsseskap	1999
24 m Catamaran	*Vågsfjord*	512	24.5	28.5	149	none	Troms Fylkes Campskibsseskap	1999
26 m Catamaran	*Kostervåg*	515	25.8	20.0	270	none	Kostermarin AB	2000
28 m Catamaran	*Strandafjord*	516	28.0	30.0	180	none	Torghatten Trafikkselskap	2002
26 m Catamaran	*Fjordlys*	517	26.0	28.0	60	8 cars	Rogaland Trafikkselskap	2002
26 m Catamaran	*Donna*	520	26.0	28.0	60	8 cars	Helgelandske	2003
26 m Catamaran	*Snarveien*	521	26.0	27.0	180	none	BNR Fjordcharter	2005
32.5 m Catamaran	*Lysefjord*	522	32.5	28.0	85	12	KS Lysefjord	2006
22.5 m Catamaran	*Tedno*	523	22.5	35.0	85	none	HSD and Fjord 1	2007
29.2 m Catamaran	Jernøy	524	29.2	35.0	147	3	Finnmark Fylkestederi og	(2007)

Length waterline	27.0 m
Beam	9.0 m
Passengers	180
Operational speed	30 kt

26 m CATAMARAN

Fjordlys
Donna

The first vessel, *Fjordlys* was ordered by Rogaland Trafikkselskap in 2001 for operation with Rodne and Sonner, an associate company. The hull of this craft was built by Båtservice with the final outfitting and delivery undertaken by Oma in 2002. A second vessel, *Donna,* was ordered by Helgelandske and delivered in 2003. This vessel has a capacity for 50 passengers and 6 cars.

Specifications

Length overall	25.8 m
Length waterline	24.0 m
Beam	9.0 m
Passengers	60
Vehicles	8 cars
Operational speed	28 kt

Propulsion: Two Volvo Penta D30A-AT diesels each rated at 610 kW.

26 m CATAMARAN

Snarveien

Ordered by BNR Fjordcharter this vessel was delivered in 2005 to be used on operations out of Hardanger.

Specifications

Length overall	26 m
Beam	9.0 m
Passengers	180
Operational speed	27 kt

Propulsion: Two Scania DI 16 diesel engines each rated at 550 kW driving controllable-pitch propellers.

22.5 m CATAMARAN

Jointly ordered by HSD and Fjord 1, and delivered in 2007, this vessel, styled by Torup Design, is capable of carrying 85 passengers and up to 1.5 t of cargo.

Specifications

Length overall	22.5 m
Beam	7.6 m
Draught	1 m
Passengers	85
Additional payload	1.5 tonnes cargo
Operational speed	35 kt

Propulsion: Two MTU 12V2000 diesel engines delivering 1080 kW each to Servogear CP propellers.

Philippines

FBMA Marine Inc

Arpili, Balamban, Cebu, Philippines

Tel: (+63 32) 465 00 01; 00 07
Fax: (+63 32) 465 00 08; 00 10
e-mail: dougb@fbma.com.ph
Web: www.fbma.com.ph

D Border, *General Manager*
T Redfern, *Sales Manager*
Tim Liu, *Finance Director*

FBMA launched the first 50 m Tricat fast passenger ferry in mid-1998 and the second at the end of 1999, for service in Asia. The 50 m Tricat hull design is a derivative of the successful 45 m Tricat that is in service on routes out of Hong Kong. The yard recently constructed three Australian-designed 25 m and 28 m fast passenger ferries, which entered service in the region during the first half of 2000.

57 m TRICAT

This BMT Nigel Gee design has been specially developed for SAS Sudiles of New Caledonia in the South Pacific. The vessel has been designed to operate in the frequently challenging sea conditions between Noumea and the surrounding islands, and features a hull form based on the Modcat design developed for the US Navy.

Specifications

Length overall	57.8 m
Length waterline	51.6 m
Beam	14.0
Draught	2.0 m
Passengers	356
Vehicles	10
Speed	32 kt

52 m TRICAT

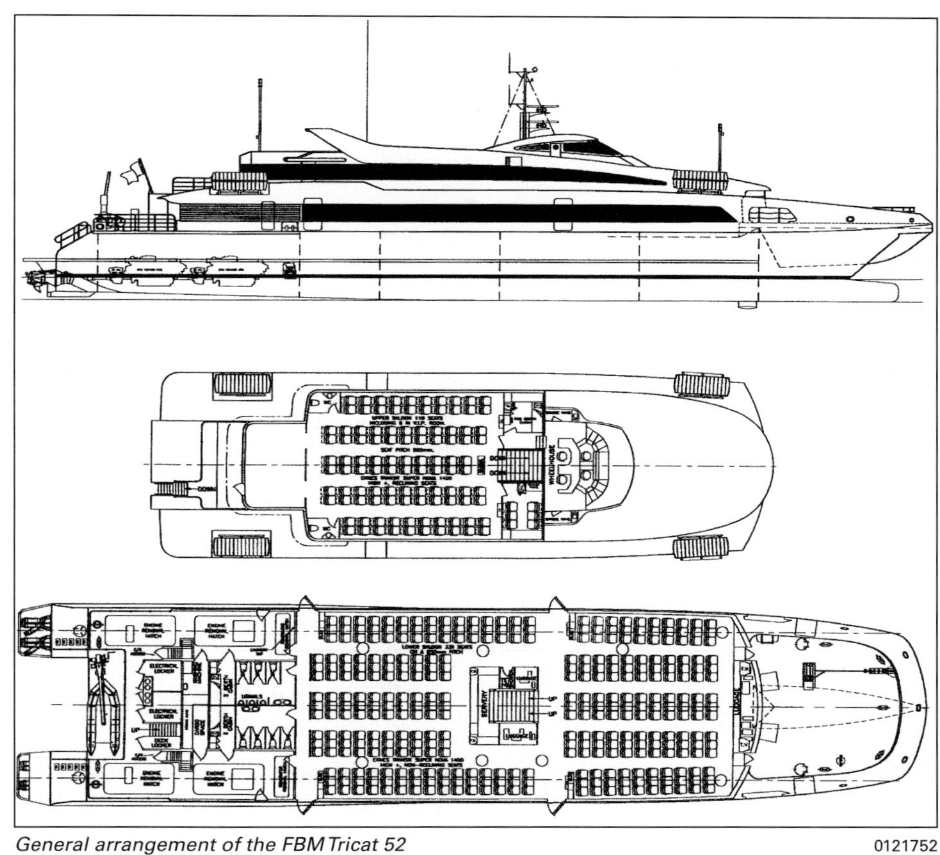

General arrangement of the FBM Tricat 52 0121752

This vessel was ordered by Dae-A Express of South Korea in 2000 for delivery in July 2001. The vessel is a derivative of the Tricat 45 m, but is fitted with a quadruple diesel engine arrangement.

Specifications

Length overall	52.5 m
Length waterline	47.5 m
Draught	1.6 m
Deadweight	59.2 t
Crew	9
Passengers	455
Fuel capacity	17,650 litres
Water capacity	2,000 litres
Max speed	43 kt
Range	350 n miles

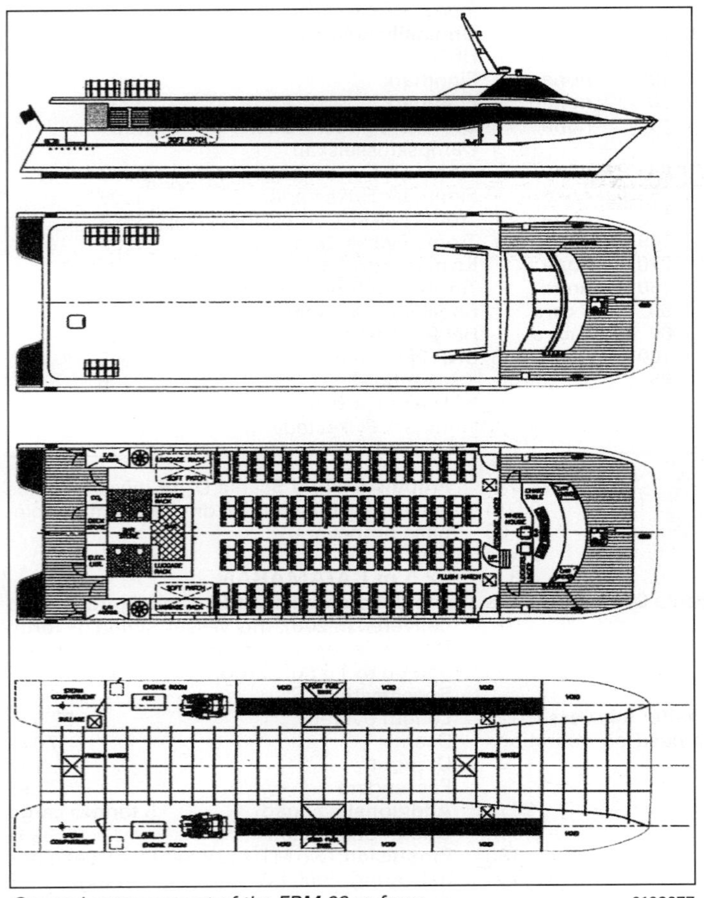

General arrangement of the FBM 28 m ferry 0103277

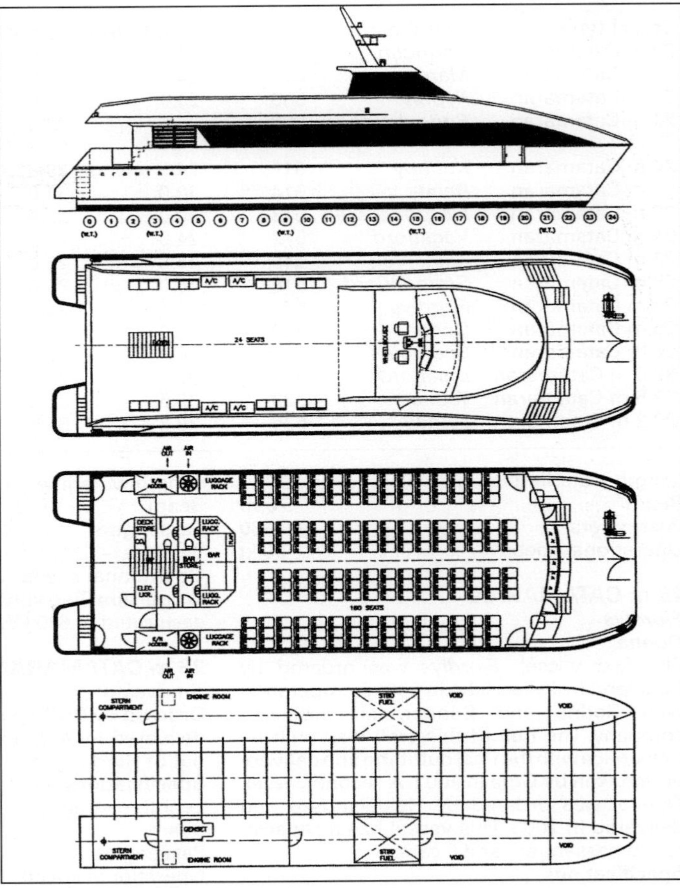

General arrangement of the FBM 25 m ferry 0103268

Vessel type	Vessel name	Yard No	Length (m)	Speed (kt)	Seats	Vehicles	Originally delivered to	Date of build
50 m Tricat	–	–	–	–	–	–	Asia	1998
50 m Tricat	–	–	–	–	–	–	Asia	1999
Australian design	–	–	–	–	–	–	–	2000
Australian design	–	–	–	–	–	–	–	2000
Australian design	–	–	–	–	–	–	–	2000
25 m Ferry	MV West Frontier	–	25.0	25.0	204	None	Shin Han Shipping	2000
28 m Ferry	*Mt Samat Ferry 3*	–	28.0	28.0	180	None	Mt Samat Ferry Express	2000
28 m Ferry	*Mt Samat Ferry 5*	–	28.0	28.0	180	None	Mt Samat Ferry Express	2000
52 m Tricat	*Hankyoreh*	–	52.5	43.0	455	None	Dae-A Express	2001
57 m Ferry	*Betico II*	–	57.8	32.0	356	10	SAS Sudiles	(2007)

Classification: Korean Register + KR SOC Passenger Cat (HSLC SA 2) ICR VMA.
Structure: All aluminium hull and superstructure with GRP fairings on the superstructure.
Propulsion: Four MTU 16V 4000 M70 diesel engines each rated at 2,320 kW and each driving a Kawema 63 SII water-jet.

28 m FERRY

Mt Samat Ferry 3
Mt Samat Ferry 5

Two of these vessels were built in 2000 to a Crowther Multihull of Australia design for the Philippine operator Mt Samat Ferry Express.

Specifications

Length overall	28.0 m
Length waterline	25.2 m
Beam	8.5 m
Draught	1.6 m
Deadweight	20.5 t
Crew	8
Passengers	180
Fuel capacity	6,000 litres
Water capacity	1,500 litres
Propulsive power	1,640 kW
Operational speed	28 kt

Propulsion: Two Caterpillar 3412 E diesel engines each rated at 820 kW and driving a fixed-pitch propeller via a Reintjes WVS 334P gearbox.

25 m FERRY

One of these vessels was built in 2000 also to a Crowther Multihull design for South Korean operator Shin Han Shipping.

Specifications

Length overall	25.0 m
Length waterline	22.9 m
Beam	8.5 m
Draught	1.5 m
Deadweight	20.5 t
Crew	4
Passengers	204
Fuel capacity	6,000 litres
Water capacity	1,500 litres
Operational speed	25 kt

Propulsion: Two Caterpillar 3412 C diesel engines each driving a fixed-pitch propeller via a reverse reduction gearbox.

Russian Federation

Almaz Marine Yard

19, Uralskaya Street, 199155 St Petersburg, Russia

Tel: (+7 812) 350 27 08
Fax: (+7 812) 350 77 50
e-mail: market@shipconstruction.ru
Web: www.shipconstruction.ru

Almaz Marine Yard was founded in 1931 and specialises in the construction of fast ferries, patrol craft, work boats and sailing and motor yachts in aluminium and steel.

The 42,000 m² facilities include a 355 m moorage quay, building areas for vessels up to 75 m in length and launching weight of 1,500 tonnes.

Vessel type	Vessel name	Yard No	Length (m)	Speed (kt)	Seats	Vehicles	Originally delivered to	Date of build
40 m Superfoil-40	*Shi Ji Kuai Hang* (ex-*Linda Express*)	600	40.0	55.0	294	none	Linda Line	2002

SUPERFOIL 40

Delivered in July 2002 to Linda Lines for operation between Estonia and Finland, this particularly fast foil-assisted catamaran was designed by the St Petersburg-based Marine Technology Development company. It is understood that the vessel has been designed to maintain operation in up to Sea State 5 (2.5 m significant wave height).

Specifications

Length overall	40.2 m
Beam	11.2 m
Draught (foilborne)	1.2 m
Draught (hullborne)	2.0 m
Lightship displacement	123 t
Crew	5
Passengers	294
Fuel capacity	10,000 litres
Fresh water capacity	2,000 litres
Maximum speed	55 kt
Range	140 n miles

Propulsion: Four MTU 12V 4000 M70 diesel engines each rated at 1,740 kW and each driving an MJP 550 water-jet.

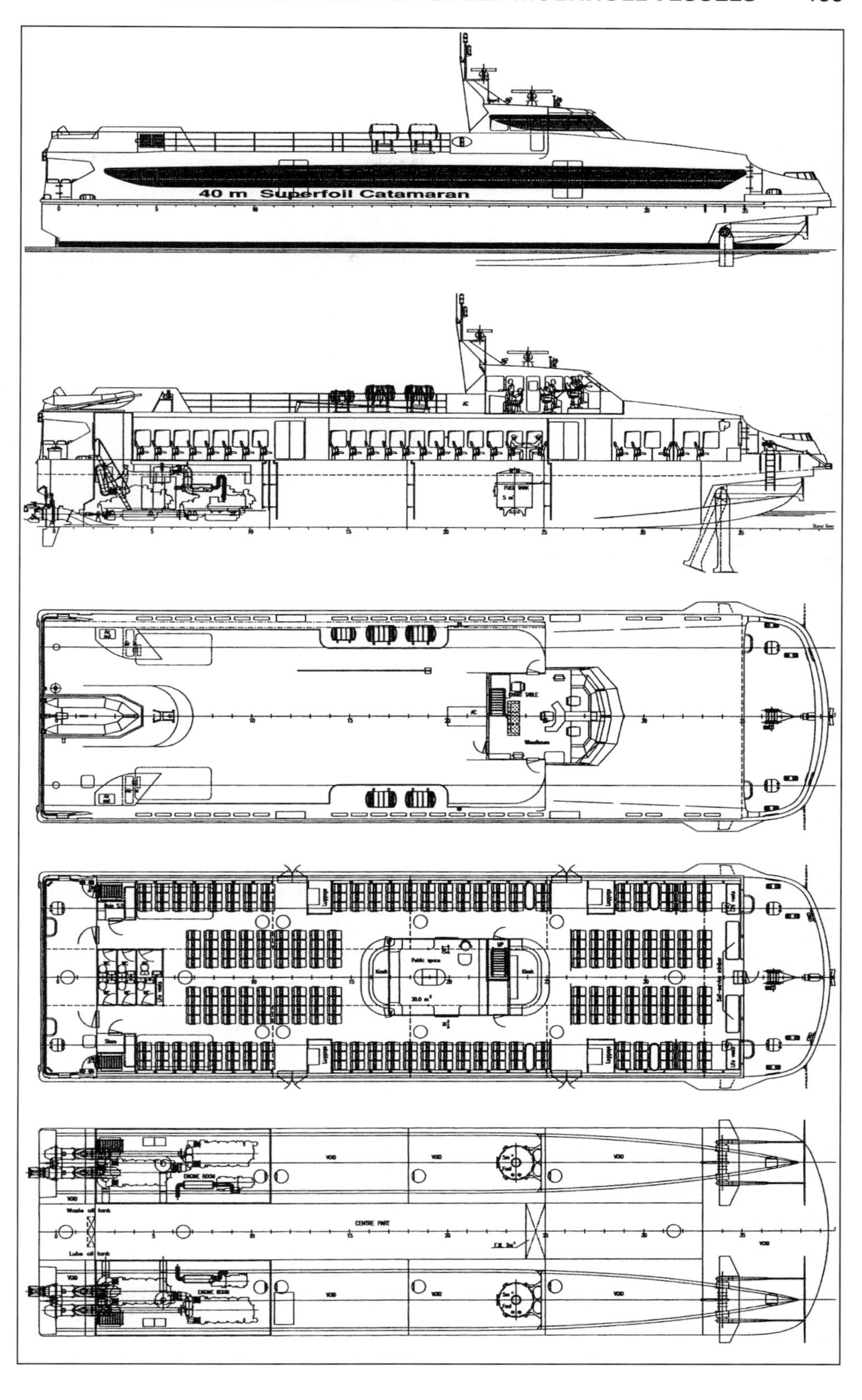

General Arrangement of Superfoil 40
0526489

Almaz Shipbuilding Company Ltd

26 Petrovsky Pr, St Petersburg 197110, Russia

Tel: (+7 812) 235 51 48
Fax: (+7 812) 235 70 69
e-mail: market@almaz.spb.ru
Web: www.almaz.spb.ru

Alexander Shliakhtenko, *President*
Nikolai Ivakin, *Shipyard Director*
Leonid Grabovets, *General Director*
Sergei Galichenko, *Marketing Manager*

Almaz Shipbuilding Company was founded in 1901 in St Petersburg by Alexander Zolotoff. Since that time over 1,000 vessels have been delivered including fast patrol

Artist's impression of Almaz 80
0593908

craft, passenger ferries and the world's largest hovercraft, the *Pomoznik*.

ECOPATROL

This is a specialised ecological monitoring catamaran built in aluminium and fitted with over 60 different ecological sensors including remote-controlled aircraft equipment for aerial video observations. The vessel was designed by Engineering Shipbuilding Centre in St Petersburg. Two vessels were delivered during 1996.

Specifications

Length overall	30.2 m
Beam	6.6 m
Draught	0.8 m
Crew	6
Operational speed	22 kt
Range	150 n miles

Propulsion: The vessel is conventionally propelled with two ducted fixed-pitch propellers each driven by a 590 kW Zvezda diesel engine.

The first Ecopatrol vessel on trials 0507051

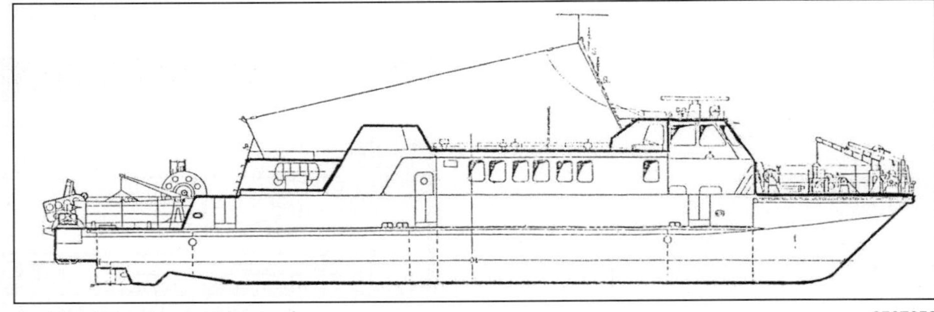

Profile of the Ecopatrol vessel 0507052

ALMAZ 80 (DESIGN)

Designed to operate on the St Petersburg to Kaliningrad route, this 74 m ferry design has a wide range of propulsion options, using either MTU, Ruston, Wartsila or Caterpillar engines.

Specifications

Length overall	74 m
Beam	21.4 m
Draught	3 m
Deadweight	340 t
Crew	18
Passengers	450
Vehicles	158 cars or 112 cars and 4 buses
Operational speed	22 kt
Range	550 n miles

Vessel type	Vessel name	Yard No	Length (m)	Speed (kt)	Seats	Vehicles	Originally delivered to	Date of build
Ecopatrol	–	–	30.2	22.0	–	–	–	1996
Ecopatrol	–	–	30.2	22.0	–	–	–	1996

Singapore

Damen Shipyards Singapore Pte Ltd

29 Tuas Crescent, Singapore 638720

Tel: (+65) 68 61 41 80
Fax: (+65) 68 61 41 81
e-mail: central@damen.com.sg
Web: www.damen.com.sg

Dennis Carroll, *Managing Director*
Karl Nilssen, *Engineering Manager*
Lester Mak, *Projects Manager*

Damen Shipyards Singapore is a company belonging to the Damen Group of the Netherlands. The company occupies a modern purpose-built facility that specialises in the construction of aluminium vessels.

Damen Shipyards Singapore provides a range of catamaran and monohull fast ferries in the 28 to 60 m range. The company also constructs a range of pilot boats, launches and crew tenders.

After sales support and servicing for operators of aluminium vessels in the Asia Pacific region is also provided from the Singapore facility.

The shipyard operates under a certified ISO 9001 Quality Assurance system.

52 m WAVE-PIERCING CATAMARAN

To be delivered in 2007, the 52 m wave-piercing vessel was designed by AMD Marine Consulting with capacity to carry 337 passengers and 10 cars or 36 tonnes of cargo.

Specifications

Length overall	52.0 m
Length waterline	45.5 m
Beam	13.8 m
Draught	1.9 m
Passengers	337
Cars	10
Crew	10
Operational speed	35 kt

Propulsion: Four MTU 12V 4000 M70 diesel engines each rated at 1740 kW at 2000 rpm, driving Kamewa 63 SII water-jets.

31.5 m FLYING CAT (DFF 3209)

Due to be delivered in December 2006, this 31.5 m catamaran will be operated between Cape Town and Robben Island.

Specifications

Length overall	31.5 m
Beam	9.3 m
Draught	1.45 m
Passengers	297
Operational speed	27 kt

Propulsion: Two MTU 16V 2000 M70 diesel engines delivering 1,050 kW each to a Teignbridge fixed-pitch propeller via a ZF gearbox.

38 m FLYING CAT (DFF 3810)

A DFF 3810 vessel was ordered in 2001 by Bintan Resort Ferries and delivered in July 2002. The ferry will be used for guest transfers from Bintan Island Resort to Singapore, a distance of 25 miles.

Specifications

Length overall	37.6 m

DFF 4912, *delivered to SOFLUS SA* 0587795

Beam	10.1 m
Draught	1.6 m
Passengers	268
Fuel capacity	12,000 litres
Water capacity	1,500 litres
Operational speed	31.5 kt

Classification: ABS + A1 HSC Ferry Service +AMS.

Propulsion: Four MTU 12V 2000 M70 diesel engines each rated at 788 kW and driving a steerable water-jet.

Vessel type	Vessel name	Yard No	Length (m)	Speed (kt)	Seats	Vehicles	Originally delivered to	Date of build
Flying Cat 40	*Perdana Express*	01	40.0	30.0	352	–	Inlandpark Sdn Bhd	1992
Flying Cat 40	*Mabua Express*	02	40.0	30.0	248	–	PT Mabua Intan Express	1992
Flying Cat 40	*Universal Mk II*	05	40.0	30.0	259	–	Woolaston Holdings Ltd	1992
Flying Cat 40	*Universal Mk I*	04	40.0	30.0	259	–	Woolaston Holdings Ltd	1992
Flying Cat 40	*HKF II*	09	40.0	30.0	449	–	HKY Ferry Co Ltd	1993
Flying Cat 40	*HKF I*	08	40.0	30.0	449	–	HKY Ferry Co Ltd	1993
Flying Cat 40	*Universal Mk III*	10	40.0	30.0	258	–	Woolaston Holdings Ltd	1993
Flying Cat 40	*Universal Mk V*	06	40.0	30.0	266	–	Universal Mk V Ltd	1994
Flying Cat 40	*Universal Mk IV*	03	40.0	30.0	266	–	Universal Mk IV Ltd	1994
Flying Cat 40	*Nam Hae Queen*	07	40.0	30.0	350	–	Nam Hae Express Co	1994
Flying Cat 40	*Paradise*	12	40.0	30.0	380	–	Won Kwang Shipping	1994
Flying Cat 40	*Indera Bupula*	11	40.0	30.0	312	–	Bintan Resort Ferries	1994
Flying Cat 40	*Damania I*	13	40.0	30.0	392	–	Damania Shipping Ltd	1994
Flying Cat 40	*Aremiti*	16	40.0	30.0	449	–	Arimiti Pacific Cruises	1994
Flying Cat 40	*HKF III*	14	40.0	30.0	433	–	HKY Ferry Co Ltd	1994
Flying Cat 40	*Arai Bupula*	15	40.0	30.0	270	–	Bintan Resort Ferries	1995
Flying Cat 40	*Jumbo Cat II*	23	40.0	30.0	441	–	Transtur-Aerobarcos do Brasil	1995
Flying Cat 40	*Jumbo Cat I*	22	40.0	30.0	441	–	Transtur-Aerobarcos do Brasil	1995
Flying Cat 40	*Stella Maris*	17	40.0	30.0	346	–	Calm Water Approach Investment Ltd	1995
Flying Cat 40	*Baekk Ryeong*	21	40.0	30.0	333	–	Jindo Transportation Company Ltd	1996
Flying Cat 40	*Dong Yang New Gold*	24	40.0	30.0	324	–	Dong Yang Express Ferry Co Ltd	1996
Flying Cat 40	*Water Jet 2*	20	40.0	30.0	370	–	Water Jet Netherland Antilles N.V.	1996
Flying Cat 40	*Water Jet 1*	18	40.0	30.0	370	–	Water Jet Netherland Antilles N.V.	1996
Flying Cat 40	*St Raphael*	19	40.0	30.0	361	–	Negros Navigation Company	1996
Flying Cat 34	*Continental*	26	34.0	–	250	–	Jindo Transportation Company Ltd	1996
Flying Cat 40	*Princess*	25	40.0	30.0	426	–	Won Kwang Shipping	1996
Flying Cat 40	*Tinian Express*	27	40.0	30.0	333	–	Tinian Shipping Co Inc.	1996
Flying Cat 40	*Saipan Express*	28	40.0	30.0	333	–	Tinian Shipping Co Inc.	1996
Flying Cat 40	*Angel of Hope*	29	40.0	30.0	312	–	Negros Navigation Company	1996
Flying Cat 40	*Angel of Freedom*	30	40.0	30.0	312	–	Negros Navigation Company	1997
Flying Cat 40	*St Gabriel*	31	40.0	30.0	274	–	Negros Navigation Company	1997
Flying Cat 40	*Baltic Jet*	32	40.0	30.0	341	–	Warnow Schiffahrtsgesellschaft GmbH & Co FG	1997
Flying Cat 40	*Water-jet 3*	33	40.0	30.0	268	–	Sembawang Mulpha Pte Ltd	1997
Flying Cat 40	*Water-jet 5*	34	40.0	30.0	268	–	Sembawang Mulpha Pte Ltd	1997
Flying Cat 40	*Sembferries 19*	38	40.0	30.0	300	–	Auto Batam	1998
Flying Cat 40	*Locmaria*	35	40.0	30.0	376	–	CMNN, France	1998
Flying Cat 50	*Aremiti 4*	42	49.0	37.5	495	20	Aremiti Pacific Cruises	1999
Flying Cat 40	*Lanka Rani*	39	40.0	30.0	300	–	Ceylon Shipping Corp.	2000
Flying Cat 40	*Lanka Devi*	40	40.0	30.0	300	–	Ceylon Shipping Corp.	2000
DFF 4010	*New Ferry V*	41	40.0	32.5	409	–	New World First Ferry Services	2000
DFF 4010	*New Ferry VI*	44	40.0	32.5	405	–	New World First Ferry Services	2001
DFF 3810	*Arung Mendara*	52	37.6	31.5	268	–	Bintan Resort	2002
DFF 4912	*Damiao De Goes*	53	49.2	34.0	600	–	Sociedade Fluvial de Transportes SA (SOFLUS)	2003
DFF 4912	*Augusto Gil*	54	49.2	34.0	600	–	Sociedade Fluvial de Transportes SA (SOFLUS)	2003
DFF 4912	*Miguel Torga*	55	49.2	34.0	600	–	Sociedade Fluvial de Transportes SA (SOFLUS)	2003
DFF 4912	*Fernando Namore*	56	49.2	34.0	600	–	Sociedade Fluvial de Transportes SA (SOFLUS)	2003
DFF 4912	*Gil Vicente*	57	49.2	34.0	600	–	Sociedade Fluvial de Transportes SA (SOFLUS)	2004
DFF 4912	*Jorge de Sena*	58	49.2	29.0	600	–	Sociedade Fluvial de Transportes SA (SOFLUS)	2004
DFF 4912	*Almeida Garret*	59	49.2	29.0	600	–	Sociedade Fluvial de Transportes SA (SOFLUS)	2004
DFF 4912	*Fernando Pessoa*	71	49.2	29.0	600	–	Sociedade Fluvial de Transportes SA (SOFLUS)	2004
DFF 4912	*Antero de Quental*	72	49.2	29.0	600	–	Sociedade Fluvial de Transportes SA (SOFLUS)	2004
DFF 3007	*Vlij*		31.3	22.0	80	–	Waterbus BV	2004
DFF 3007	*Schie*		31.3	22.0	80	–	Waterbus BV	2006
DFF 4212	Burak Reis 3	89	42.9	35.8	449	–	Istanbul Deniz Otobusleri	2007
DFF 4212	Salih Reis 4	90	42.9	35.8	449	–	Istanbul Deniz Otobusleri	2007
DFF 4212	Kemal Reis 5	91	42.9	35.8	449	–	Istanbul Deniz Otobusleri	2007
DFF 4212	Mehmet Reis 11	92	42.9	35.8	449	–	Istanbul Deniz Otobusleri	2007
DFF 4212	Murat Reis 7	93	42.9	35.8	449	–	Istanbul Deniz Otobusleri	2007
AMD 428	–	538901	52.0	35.0	300	10 cars	Kuwait & Gulf Link Transport Company	2007
DFF 4212	–	–	47.9	37.0	300	–	Bintan Resort	(2009)

40 m FLYING CAT (DFF 4010)

The designation DFF 4010 represents the 40 m Flying Cat craft. Over 35 such vessels have been built in the Singapore shipyard to date.

Specifications

Length overall	40.0 m
Beam	10.1 m
Draught	1.7 m
Crew	6
Passengers	409
Fuel capacity	12,000 litres
Water capacity	1,500 litres
Propulsive power	4,000 kW
Operational speed	32.5 kt

Classification: DnV +1A1 HSLC R3 Passenger EO, Category A.

Propulsion: Powered by two MTU 16V 396 TE 74L diesel engines each rated at 2,000 kW at 2,000 rpm. Each engine drives through a ZF BW755 reduction gearbox to a Kamewa S6311 water-jet.

Control: A motion damping system is installed on the vessel.

DFF 4912

Portuguese operator Sociedade Fluvial de Transportes SA (SOFLUSA) ordered seven DFF 4912 catamarans in 2002 and a further two in 2003.

Specifications

Length overall	49.2 m
Beam	12.3 m
Draught	1.55 m
Crew	4
Passengers	600
Fuel capacity	22,000 litres
Water capacity	1,500 litres
Propulsive power	4,640 kW
Operational speed	29 or 34 kt

Classification: Bureau Veritas and RINAVE.

Propulsion: Two MTU 16V 4000 M70 diesel engines each rated at 2,320 kW at 2,000 rpm, and each driving Hamilton HM 811 water-jet via a Reintjes gearbox.

50 m FLYING CAT (DFF 5012)

The designation DFF 5012 refers to the previously named 50 m Flying Cat vessels. Damen Shipyards Singapore Pte Ltd delivered this 49 m car-passenger fast ferry *Aremiti 4* to Aremiti Pacific Cruises on 30 April 1999. The vessel is fitted with an MDS Motion-Damping System and is classified by Germanicher Lloyd as a Category B vessel.

Specifications

Length overall	49.0 m
Beam	12.0 m
Draught	2.1 m
Crew	5
Passengers	445
Vehicles	20 cars
Fuel capacity	20,000 litres
Water capacity	1,500 litres
Propulsive power	4,640 kW
Operational speed	37.5 kt

DFF 3810 Arung Mendora 1066433

DFF 5012 Aremeti 4 0142349

DFF 4010 New Ferry V 0142347

Classification: Germanicher Lloyd, HSC Category B.

Propulsion: Four MTU 16V 4000 M70 diesel engines each driving a Kamewa 63 water-jet via a ZFBW 755 gearbox.

Auxiliary systems: Two Cummins 135 kW generators, Active MDS ride control system.

DFF 4212

Burak Reis 3
Salih Reis 4
Kemal Reis 5
Mehmat Reis 11
Murat Reis 7

Five Damen DFF 4212 catamarans were delivered to IDO in Italy in 2007 to operate in the Bosphorus and the Sea of Marmara. The vessels, which can accommodate 449 passengers, are powered by four Caterpillar C32 diesel engines giving a service speed of over 30 knots.

Specifications

Length overall	42.9 m
Beam	12.4 m
Draught	1.55 m
Passengers	449
Crew	4
Operational speed	30.9 kt

Propulsion: Four Caterpillar C32 diesel engines delivering 1093 kW each to two Servogear CP propellers, via Servogear HD450 PT1R gearboxes.

Greenbay Marine Pte Ltd

27 Pioneer Road, Jurong, Singapore 628500

Tel: (+65) 68 61 41 78
Fax: (+65) 68 61 81 09
e-mail: greenbay@singnet.com.sg
Web: www.greenbay.com.sg

Raymond Toh, *Chief Executive Officer*

Greenbay Marine was incorporated in 1980 as a shipbuilding company. Since that time it has built a wide range of vessels in steel and aluminium.

The company also provides design, consultancy and support for the vessels it supplies. These have covered offshore supply vessels, cargo vessels, tugs, patrol craft and catamaran ferries.

35 m Rescue Command Vessel 0038690

Vessel type	Vessel name	Yard No	Length (m)	Speed (kt)	Seats	Vehicles	Originally delivered to	Date of build
35 m Rescue Vessel	–	–	35.5	25.0	–	–	Hong Kong International Airport Authority	1997/98
35 m Rescue Vessel	–	–	35.5	25.0	–	–	Hong Kong International Airport Authority	1997/98

35 m RESCUE VESSEL

Two of these rescue command vessels, designed by Incat Designs of Sydney, Australia, and delivered in 1997–98, are used by the new Hong Kong International Airport Authority.

Specifications

Length overall	35.5 m
Length waterline	32.8 m
Beam	12.0 m
Hull Beam	3.0 m
Draught	1.8 m
Fuel capacity	10,000 litres
Water capacity	2,000 litres
Propulsive power	2,910 kW
Max speed	28 kt
Operational speed	25 kt

Classification: DnV HS&LC Patrol R3 at 1.75 g.

Propulsion: Two MTU 12V 396TE 74L diesel engines each rated at 1,455 kW at 2,000 rpm and each driving a Kamewa 71 SII water-jet via a ZF BW 465 gearbox.

Marinteknik Shipbuilders (S) Pte Ltd

31 Tuas Road, Singapore 638493

Tel: (+65) 68 61 17 06
Fax: (+65) 68 61 42 44
e-mail: msspl@pacific.net.sg
Web: www.marinteknikdesign.se

Hong Kong
Suite 1805, Lippo Centre, Tower 11, 89 Queensway, Admiralty, Hong Kong

Tel: (+852) 25 23 10 54
Fax: (+852) 25 96 09 19
e-mail: david@marinteknik.com

David Liang, *Group Chairman*
Priscilla Lim, *Financial Director*
Hans Erikson, *Director, Marketing and Sales*
Andrew Yeo, *Director, Project and Design*

The Marinteknik Group has over 25 years experience in the construction of high-speed aluminium vessels, patrol boats and car/cargo/truck ferries. To date, the yard has built more than 100 craft of both monohull and catamaran type varying in length from 27 to 63 m and with a speed range of 18 to 55 kt.

Marinteknik Verkstads AB of Sweden, an associated company, ceased building in 1993. Details of its craft are included in this entry. Marinteknik Design Pte in Singapore works closely with Marinteknik Design (E) AB in Sweden to develop more advanced designs for production in the Singapore yard. Marinteknik designs are available to shipbuilders worldwide under licence.

CPV 32 CATAMARAN

Auto Jet

This vessel was delivered in 1998 to Auto Batam.

Turquoise, *a CPV 41 delivered to Antilles Trans-Express in 1995* 0506975

Rubis Express *in service with Antilles Trans-Express* 0507057

Specifications

Length overall	32.0 m
Beam	8.0 m
Draught	1.2 m
Passengers	200
Propulsive Power	2940 kW
Operational speed	34 kt
Range	300 n miles

Classification: DnV.

Propulsion: Four MAN B&W D2842 LE 408 diesel engines each rated at 735 kW at 2,100 rpm and each driving an MJPJ 450-DD water-jet.

CPV 35 CATAMARAN

Jet Caribe

This vessel was delivered in 2001 although the hull was laid down some 10 years earlier from a contract that was subsequently discontinued.

Vessel type	Vessel name	Yard No	Length (m)	Speed (kt)	Seats	Vehicles	Originally delivered to	Date of build
34 CCB	*Hakeem*	101	34.0	27.0	50	–	–	1985
34 CCB	*Layar Sinar*	103	34.0	29.0	70	–	–	1986
34 CCB	*Layar Sentosa*	105	34.0	29.0	70	–	–	1986
34 CPV	*Jiu Zhou* (ex-*Shun de*)	112	34.0	27.5	250	–	Shun Gang	1987
36 CPV	*Condor 8*	115	36.0	35.0	300	–	Condor Ltd	1988
41 CPV	*Camoes*	119	41.0	37.0	306	–	Hong Kong Macao Hydrolfoil Co. Ltd	1989
36 CPV	*Selesa Express*	117	36.0	30.0	350	–	Kuala Perlis-Langkawi Ferry Services Bhd	1990
41 CPV	*Estrella do Mar*	120	41.0	37.0	306	–	Hong Kong Macao Hydrolfoil Co. Ltd	1990
41 CPV	*Lusitano*	121	41.0	37.0	306	–	Hong Kong Macao Hydrolfoil Co. Ltd	1990
27 CPV	*Marine Star II*	126	27.0	30.0	215	–	Samavest Sdn Bhd	1991
41 CPV	*Vasco da Gama*	123	41.0	37.0	306	–	Hong Kong Macao Hydrolfoil Co. Ltd	1991
41 CPV	*Magellan*	128	41.0	37.0	306	–	Hong Kong Macao Hydrolfoil Co. Ltd	1991
41 CPV	*St Malo*	129	41.0	35.0	250	–	France	1991
41 CPV	*Santa Cruz*	127	41.0	37.0	306	–	Hong Kong Macao Hydrolfoil Co. Ltd	1991
41 CPV	*Nam Hae Prince*	133	41.0	34.0	359	–	Nam Hae Express Co. Ltd	1992
41 CPV	*Saphir Express*	130	41.0	39.0	398	–	Antilles Trans-Express	1994
42 CPV	*Discovery Bay 1*	137	42.0	33.0	500	–	Hong Kong Resort Co	1994
42 CPV	*Discovery Bay 2*	138	42.0	33.0	500	–	Hong Kong Resort Co	1994
CPV 41	*Turquoise Express*	140	42.1	38.5	395	–	Antilles Trans-Express	1995
CPV 45	*Rubis Express*	136	45.0	40.0	450	–	Antilles Trans-Express	1996

Vessel type	Vessel name	Yard No	Length (m)	Speed (kt)	Seats	Vehicles	Originally delivered to	Date of build
42 CPV	*Discovery Bay 3*	139	42.0	33.0	500	–	Hong Kong Resort Co	1996
42 CPV	*Discovery Bay 6*	153	42.0	33.0	500	–	Hong Kong Resort Co	1997
42 CPV	*Discovery Bay 5*	152	42.0	33.0	500	–	Hong Kong Resort Co	1997
36 CPV	*Airone Jet*	111	36.0	35.0	318	–	Med Mar Srl, Naples	1988
32 CAT	*Auto Jet*	148	32.0	44.0	222	–	Auto Batam	1998
55 m Fast Ro-Ro Ferry	*La Vikinga II*	156	55.0	36.0	400	45	Albadria	1998
42 CPV	*Discovery Bay 7*	157	42.0	33.0	500	–	Hong Kong Resort Co	1998
42 CPV	*Discovery Bay 8*	159	42.0	33.0	500	–	Hong Kong Resort Co	1999
43 CPV	*Passion*	162	43.0	40.0	387	–	Brudey Ferries	2000
35 CPV	*Jet Caribe*	118	35.0	34.0	313	–	Venezuela	2000
36 CPV	*First Ferry 3*	170	36.0	25.0	403	–	New World First Ferry	2001
36 CPV	*First Ferry 5*	171	36.0	25.0	403	–	New World First Ferry	2001
36 CPV	*First Ferry 6*	172	36.0	25.0	403	–	New World First Ferry	2001
36 CPV	*First Ferry 7*	173	36.0	25.0	403	–	New World First Ferry	2001
36 CPV	*First Ferry 8*	175	36.0	25.0	403	–	New World First Ferry	2001
38 CPV	*Park Island 1*	180	38.5	25.5	402	–	Park Island Transportation, Hong Kong	2002
38 CPV	*Park Island 2*	181	38.5	25.5	402	–	Park Island Transportation, Hong Kong	2002
38 CPV	*Park Island 3*	182	38.5	25.5	402	–	Park Island Transportation, Hong Kong	2003
38 CPV	*Park Island 4*	183	38.5	25.5	402	–	Park Island Transportation, Hong Kong	2003
63 m Fast Ro-Ro Ferry	*Nixe*	185	63.0	32.0	536	70	Balearia	2004
63 m Fast Ro-Ro Ferry	*Nixe II*	186	63.0	32.0	536	70	Balearia	2004
65 m Fast Ro-Ro Ferry	To be named	–	63.0	30.0	606	34	–	(2006)
65 m Fast Ro-Ro Ferry	To be named	–	63.0	30.0	606	34	–	(2006)

Fast Catamaran craft built by Marinteknik Verkstads AB of Sweden which ceased building in 1993

Vessel type	Vessel name	Yard No	Length (m)	Speed (kt)	Seats	Vehicles	Originally delivered to	Date of build
29 m CPV (JC-F1)	*Aegean Jet* (ex-*Formentera Jet*, ex-*Jaguar* ex-*Jaguar Prince* ex-*Mavi Hall*)	42	30.0	29.0	185	–	Spain	1980
29 m CPV (JC3000)	*Hercules Jet*	47	30.0	30.0	215	–	Hong Kong Macao Hydrofoil Company Ltd	1982
29 m CPV (JC3000)	*Janus Jet*	48	30.0	30.0	215	–	Hong Kong Macao Hydrofoil Company Ltd	1982
29 m CPV (JC3000)	*Apollo Jet*	46	30.0	30.0	215	–	Hong Kong Macao Hydrofoil Company Ltd	1982
29 m CPV (JC3000)	*Duan Zhou Hu* (ex-Triton Jet)	50	30.0	30.0	215	–	Hong Kong Macao Hydrofoil Company Ltd	1982
33 CPV (PV 2400)	*Alize Express* (ex-*Nettuno Jet*)	51	30.0	30.0	218	–	SURF, Congo	1984
33 CPV (PV 2400)	*Jet Kat Express* (ex-*Jetkat I*)	54	33.0	30.0	223	–	ATE	1984
34 CPV-D (PV3100)	*Lomen*	56	33.0	35.0	235	–	Dampskibssellskabet Oresund A/S	1985
33 CPV (PV 2400)	*Giove Jet*	55	33.0	33.0	276	–	Alilauro SpA	1985
34 m CCB (CV3400)	*Emeraude Express*	60	34.0	38.0	242	–	Chambon (SURF)	1986
34 CPV-D (PV3100)	*Ornen*	59	33.0	35.0	235	–	Dampskibssellskabet Oresund A/S	1986
34 CPV-D	*Giunone Jet*	62	34.0	40.0	300	–	Alilauro SpA	1987
41 CPV-SD	*Oregrund*	74	41.0	37.0	306	–	Hongkong Macao Hydrofoil Company Ltd	1988
34 m CPV	*Nettuno Jet*	70	34.0	40.0	300	–	Alilauro SpA	1988
34 m CPV-D	*Acapulco Jet*	69	34.0	41.0	298	–	Alilauro SpA	1989
34 m CPV	*Saud*	73	34.0	35.0	252	–	Yasmine Line	1990
41.5 m CPV	*Jet Kat Express II*	82	41.5	40.0	380	–	Compagnie Maritime des Carabies	1991
34 m CPV	*Antilles Express*	86	34.0	38.0	297	–	Antilles Trans-Express	1992

Specifications

Length overall	35.0 m
Beam	9.4 m
Draught	1.2 m
Displacement, max	122 t
Deadweight	30.7 t
Crew	8
Passengers	313
Fuel capacity	7,000 litres
Water capacity	2,000 litres
Operational speed	34 kt
Range	300 n miles

Propulsion: Two Caterpillar 3516 B DITA diesel engines each rated at 1,939 kW at 1,835 rpm and driving an MJPJ 650R-DD water-jet via a Reintjes VLJ 930 gearbox.

CPV 36 CATAMARAN

Five vessels *First Ferry 3, 5, 6, 7* and *8* were delivered to New World Fast Ferry in 2001.

Specifications

Length overall	36.0 m
Length waterline	33.4 m
Beam	10.0 m
Draught	1.9 m
Displacement, max	135 t
Deadweight	36 t
Crew	5
Passengers	403
Fuel capacity	10,000 litres
Water capacity	1,000 litres
Operational speed	26 kt
Range	480 n miles

Classification: DNV +1A1 HSLC R3 (HKG) Passenger EO.

Propulsion: Two MTU 16V 2000 M70 diesel engines each rated at 1,050 kW at 2,100 rpm and driving a fixed-pitch propeller via a ZF BW 460-4 gearbox.

CPV 38 CATAMARAN

Park Island 1

Park Island 2

Park Island Transportation was established by Sun Hung Kai Properties to manage public transportation on the MA Wan Island shopping and industrial complex. Two CPV 38 catamaran ferries were delivered in 2002 with a further two due for delivery in 2003.

Specifications

Length overall	38.5 m
Length waterline	32.0 m
Beam	10.2 m
Draught	1.35 m
Displacement	150 t
Deadweight	36.5 t
Crew	10
Passengers	402
Fuel capacity	10,000 litres
Fresh water capacity	2,000 litres
Operational speed	29 kt

Classification: DnV

Propulsion: Two Cummins KTA 50-MZ diesel engines each rated at 1,343 kW and each driving an MJP 650R-DD water-jet via a ZF 4650 gearbox.

CPV 41 CATAMARAN

Saphir Express

Turquoise

Saphir Express is a 42 m catamaran of Marinteknik standard and a sister vessel to *Jet Cat Express II*, earlier delivered from Marinteknik Verkstads in Sweden. *Saphir Express* was delivered to Antilles Trans-Express, Guadeloupe, in September 1994.

Turquoise, of the same size but with different superstructure styling, was delivered to Antilles Trans-Express in mid-1995.

Specifications

Length overall	42.1 m
Beam	11.0 m
Draught	1.2 m
Crew	10
Passengers	395
Propulsive power	5,920 kW
Operational speed	40 kt

Classification: Bureau Veritas.

Propulsion: Main engines are four MTU 12V 396 TE 74L diesel engines, each rated 1,480 kW at 2,000 rpm and each driving a Kamewa 63 S11 water-jet via ZF 465 gearboxes.

CPV 42 CATAMARAN

Five craft were ordered in 1994 by the Hong Kong Resort Co, all of the same type. They were designed to offer a 20 minute shuttle service from Discovery Bay to central Hong Kong. The first three vessels were delivered in 1995–96 and the fourth and fifth in 1997.

Specifications

Length overall	42.0 m
Beam	11.5 m
Draught	1.3 m
Passengers	500
Operational speed	33 kt
Range	300 n miles

Classification: Det Norske Veritas.

Propulsion: Powered by two MWM TBD 620 V16 diesel engines each rated at 1,935 kW and each driving an MJP 650 water-jet.

CPV 43 CATAMARAN

Passion

This 43 m vessel was delivered to Brudey Freres of Guadaloupe in April 2000.

Specifications

Length overall	43.5 m
Beam	11.0 m
Draught	1.2 m
Displacement, max	201 t
Passengers	398
Max speed	40 kt
Operational speed	38 kt
Range	275 n miles

Classification: DnV.

Propulsion: Four Caterpillar 3512 B diesel engines each rated at 1,454 kW at 1,835 rpm and each driving a MJP 650 water-jet.

CPV 45 CATAMARAN

Rubis Express

Antilles Trans-Express ordered one CPV 45 catamaran in 1995. The craft *Rubis Express* was delivered in November 1996 to the French West Indies.

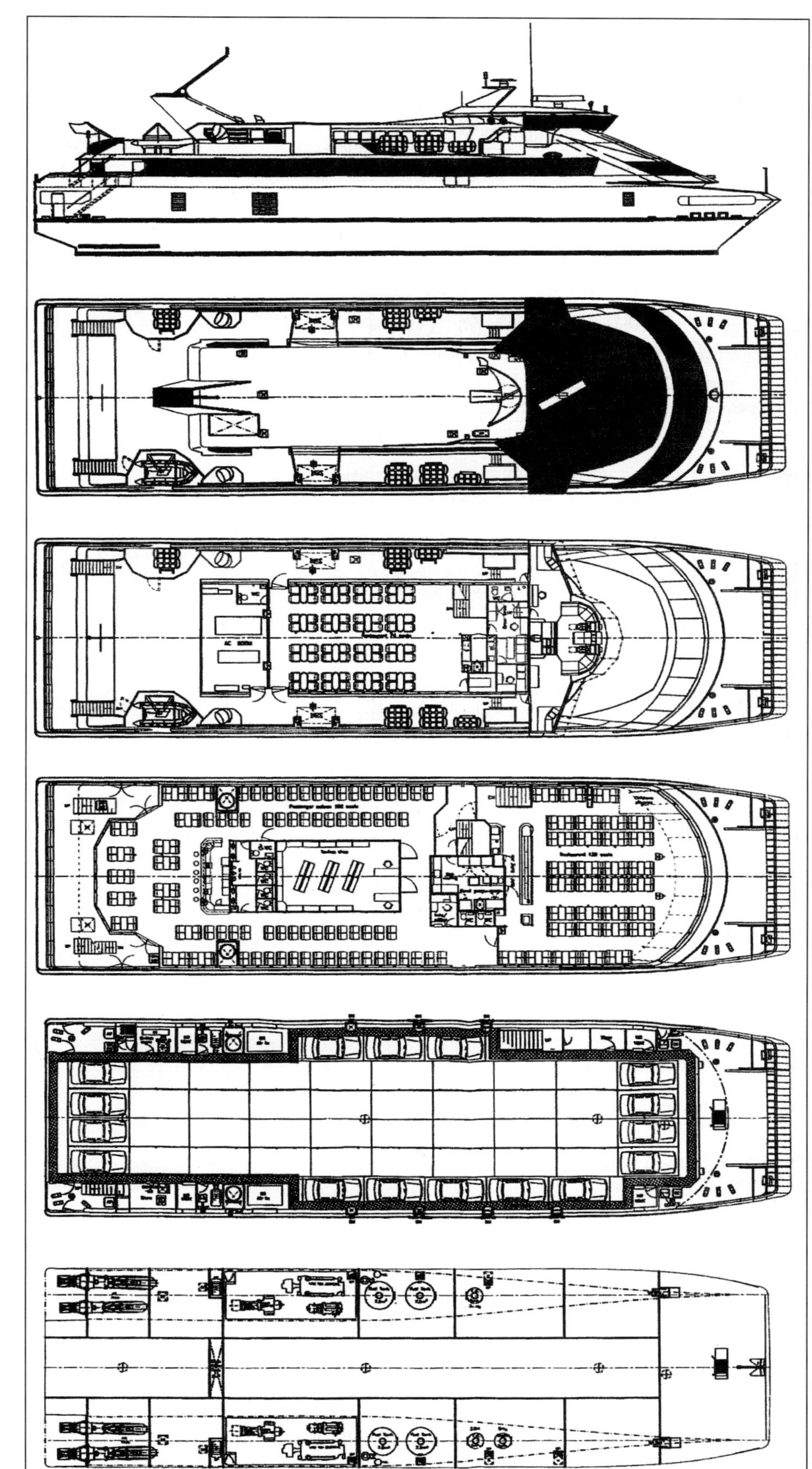

General arrangement of 55 m Fast Ro-ro Ferry La Vikinga II

0103269

Discovery Bay 6 *on trials*

0007955

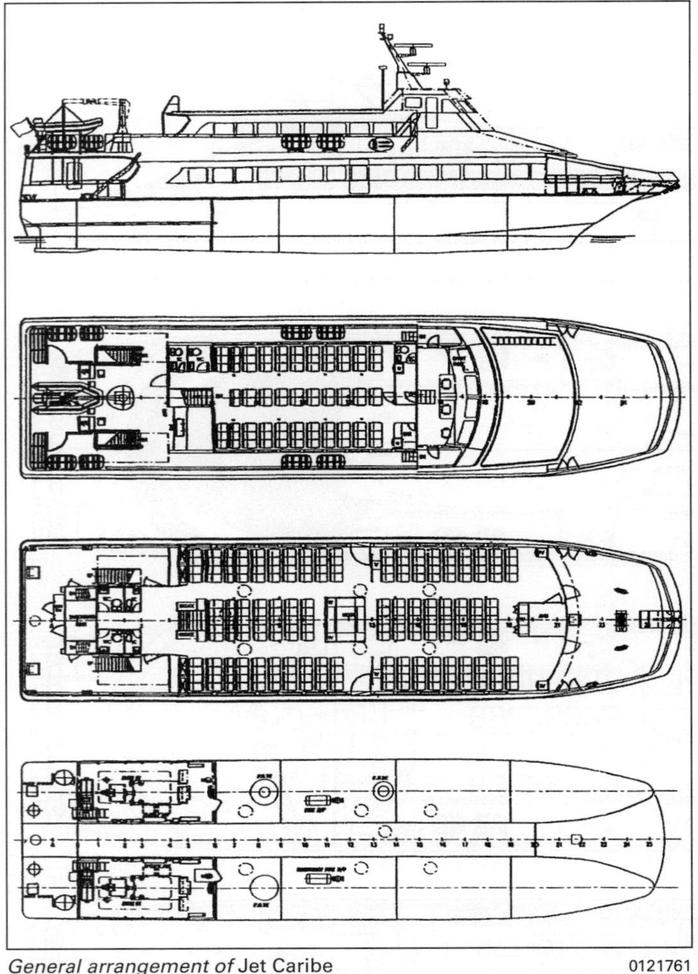
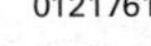

General arrangement of Jet Caribe 0121761

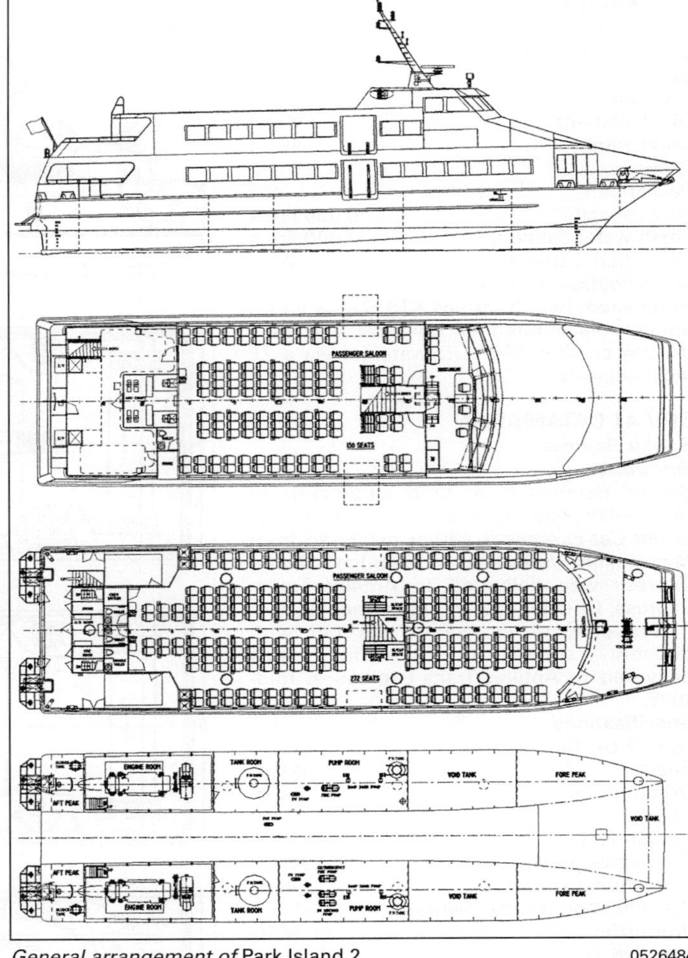

General arrangement of Park Island 2 0526484

Specifications

Length overall	45.0 m
Length waterline	37.0 m
Beam	11.0 m
Draught	1.3 m
Displacement, min	180 t
Crew	8
Passengers	445
Operational speed	42 kt
Range	290 n miles

Propulsion: Powered by four MTU 16V 396 TE 74L diesels each rated at 1,970 kW, each driving a Kamewa 63 S11 water-jet unit via a reduction gearbox.

43 CPV catamaran Passion 0103270

55 m FAST RO-RO FERRY

La Vikinga II

Marinteknik announced its entry into the fast ro-ro ferry market in 1996 with its 55 m design. The first order was obtained in March 1997 and delivered in 1998.

Specifications

Length overall	55.0 m
Beam	15.0 m
Draught	2.0 m
Crew	8
Passengers	400
Vehicles	48 cars
Fuel capacity	12,000 litres
Operational speed	38 kt
Range	180 n miles

Propulsion: Powered by two AlliedSignal TF40 gas turbines each rated at 2,840 kW, two MTU 16V 396 TE 74L diesels each rated at 2,000 kW, and each driving a water-jet via a reduction gearbox.

63 m FAST RO-RO FERRY

Nixe

Nixe II

Two 63 m catamarans originally for Gran Caique of Venezuela, have been ordered by Spanish operator Balearia.

Specifications

Length overall	63.2 m
Beam	16.1 m
Draught	2.0 m
Deadweight, maximum	273 t
Crew	10
Passengers	606
Vehicles	70 cars
Fuel capacity	20,000 litres
Water capacity	4,000 litres
Max speed	32 kt
Range	450 n miles

Classification: DNV +1A1 HS LC R1 Passenger Car Ferry A Cargo A EO.

Propulsion: Four MTU 16V 4000 M70 diesels each rated at 2,320 kW and each driving an MJP 850 water-jet.

Penguin Shipyard International Ltd

A subsidiary of Penguin Boat International
11 Tuas Crescent, 638705 Singapore

Tel: (+65) 68 62 83 22
Fax: (+65) 6892 31 34
e-mail: andy@penguin.com.sg
Web: www.penguin.com.sg

Dana Lui, *Business Development/Marketing Manager*
Andy Heng, *Commercial Manager*

Penguin Shipyard International has two shipyards, one with a covered area of 7,500 m², including a 250-tonne mobile straddle carrier, for shipbuilding and repair and the other a 12,000 m² area for superyacht repair and painting. The latter facility includes a 500-tonne mobile straddle carrier. The company is a builder of offshore vessels, crewboats and survey vessels and high-speed passenger catamarans.

35 m CATAMARAN

Parali

Cheriyapani

Valiyapani

Three vessels have been delivered to the Shipping Corporation of India. The vessels, delivered in 2007, were designed by Nigel

Gee and Associates. The vessels, capable of operating in rough weather conditions, are used on inter-island routes.

Specifications

Length overall	35 m
Length waterline	33.5 m
Beam	9.6 m
Draught	4.3 m
Deadweight	43.5 t
Passengers	154
Payload	5 t
Operational speed	25 kt

Propulsion: Two Cummins KTA 50 MZ diesels each rated at 1,343 kW at 1,900 rpm and each driving a Hamilton HM721 water-jet via a Twin Disc MGX 6846 gearbox.

Precision Craft(88) Pte Ltd

55 Gul Road, Singapore 629 353

Tel: (+65) 862 48 00
Fax: (+65) 862 48 03
email: precision@sbg.com.sg

Aluminium specialist, Precision Craft (88) Pte Ltd has accumulated over 24 years of experience in the building and repair of aluminium craft and industrial structures. Apart from its own proven monohull it also offers other hull designs such as SWATH (Small-Waterplane-Area Twin Hull), SES (Surface Effect Ship) and catamarans.

Island Pearl 0506658

Island Pearl
Sea Pearl
These semi-displacement hull catamarans were delivered in 1991 for service between Singapore and Tioman Island. Air conditioning is fitted.

Specifications

Length overall	34.5 m
Beam	9.8 m
Draught	1.5 m
Crew	6
Passengers	228
Fuel capacity	9,000 litres
Water capacity	3,000 litres
Propulsive power	2 × 1,260 kW
Speed	25 kt

Classification: GL + 100A4K.
Propulsion: Engines: two Deutz MWM TBD 604B V12, MCR 1,260 kW each, driving two nickel aluminium-bronze five-blade fixed-pitch propellers, left- and right-handed.
Auxiliary systems: Two Perkins/Stamford 80 kVA.

Dong Qu Er Hao 0506896

Tai An
Dong Qu Er Hao
Ordered by Humen Lungwei Passenger Transportation of Guandong in mid-1992, *Tai An* was delivered in July 1993. The craft was designed by Lock Crowther of Australia.

A repeat order was contracted in late 1993 and *Dong Qu Er Hao* was delivered in mid-1994.

Specifications

Length overall	35.0 m
Beam	11.0 m
Draught	1.5 m
Crew	8
Passengers	270
Fuel capacity	10,000 litres
Propulsive power	2 × 1,260 kW
Max speed	29 kt

Propulsion: Main engines are two MTU 12V 396 TE 74L each driving conventional fixed-pitch propellers via a reverse reduction gearbox.
Auxiliary systems: Generators: two 100 kVA MTU 6R 099 TE 51.

38 m CATAMARAN
Tai King
Ordered by Humen Transportation Co in 1994, this 360-passenger, 15-crew vessel is similar to *Tai An* and was delivered in mid-1995.

Vessel type	Vessel name	Yard No	Length (m)	Speed (kt)	Seats	Vehicles	Originally delivered to	Date of build
34.5 m Catamaran	*Sea Pearl*	–	34.5	25.0	228	–	–	1991
34.5 m Catamaran	*Island Pearl*	–	34.5	25.0	228	–	–	1991
35 m Catamaran	*Tai An*	–	35.0	–	270	–	Humen Lungwei Passenger Transpotation, Guandong	mid-1992
35 m Catamaran	*Dong Qu Er Hao*	–	35.0	–	270	–	Humen Lungwei Passenger Transpotation, Guandong	mid-1994
38 m Catamaran	*Tai King*	–	38.0	–	360	–	Humen Transportation Co	mid-1995

Singapore Technologies Marine Ltd

7 Benoi Road, Singapore 629882

Tel: (+65) 861 22 44
Fax: (+65) 861 30 28
e-mail: mktg.marine@stengg.com
Web: www.stengg.com

Teh Yew Shyan, *Senior Vice-President, Tuas Yard*
Lim Nian Hua, *Vice-President, Benoi Yard*

Singapore Technologies Marine (STM), the marine arm of Singapore Technologies Engineering Ltd, announced in 1990 a contract to build a 282-passenger, 34.6 m high-speed catamaran ferry for operation between Hong Kong and the Pearl River Delta of Kwangtung Province, China.

Completely designed by STM engineers, this vessel marked a significant move into the high-speed catamaran ferry business.

The company has, to date, built and on order, more than 100 aluminium high-speed vessels (with a length of 12 m or more) with speeds in excess of 20 kt.

34 m CATAMARAN
Tai Ping

Specifications

Length overall	34.6 m
Beam	10.5 m
Draught	2.1 m
Crew	10
Passengers	282
Fuel capacity	7,000 litres

Vessel type	Vessel name	Yard No	Length (m)	Speed (kt)	Seats	Vehicles	Originally delivered to	Date of build
34 m Catamaran	*Tai Ping*	–	34.6	28.0	282	–	–	–

Water capacity 1,000 litres
Propulsive power 2 × 1,412 kW
Max speed 29 kt
Operational speed 28 kt

Structure: Each hull is divided into seven separate watertight compartments by means of watertight bulkheads and has a forepeak, store void space, main engine room, auxiliary engine room and steering compartment. The accommodation cabin is sited above the main deck. The wheelhouse is arranged at the forward end of the upper deck.

Outfit: Immediately aft of the foredeck is the passenger saloon which also houses the toilets, a pantry and a kiosk.

The passenger saloon accommodates 272 passengers in aircraft-style seats (non-reclinable type) with a fold-down plastic table. The VIP saloon, sited at the aft of the wheelhouse on the upper deck, accommodates 10 VIPs in settee-seating. There are two cabins, provided on the upper deck, one for two officers and the other for eight crew.

All accommodation spaces are air conditioned.

Propulsion: Main engines are two MTU 12V 396 TE 74L diesels, each producing 1,412 kW at 2,000 rpm MCR, driving fixed-pitch propellers through stern tube and shaft bearings via ZF BW 465 gearboxes.

Electrical system: The primary system is two 100 kVA, 380 V, 50 Hz, three-phase; the secondary system is 230 V, 50 Hz, single-phase by 380/230 V transformers.

The emergency and engine starting use 24 V DC battery banks.

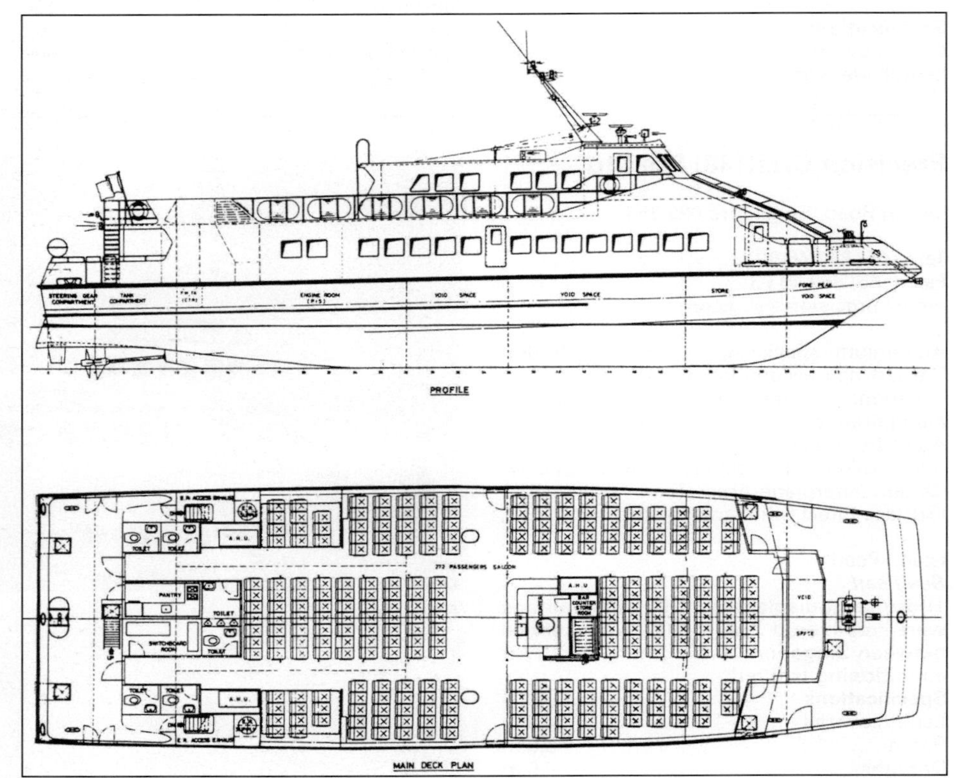

General arrangement of Tai Ping 0506659

Tai Ping
0506660

South Africa

Farocean Marine

PO Box 6075, Roggebaai, 8012, South Africa

Tel: (+2721) 447 17 14
Fax: (+2721) 447 86 55
e-mail: jendo@farocean.co.za

The company, established in 1988, has received an order to build a new Damen 3209 catamaran for Robben Island Museum. The 297 passenger vessel is due for delivery in 2007.

DFF 3209 CATAMARAN

Specifications

Length overall 31.5 m
Beam 9.3 m
Draught 1.45 m
Crew 3
Passengers 297
Operational speed 27 kt

Propulsion: Two MTU 16V 2000 M70 diesel engines delivering 1050 kW each to Teignbridge fixed pitch propellers via ZF gearboxes.

Vessel type	Vessel name	Yard No	Length (m)	Speed (kt)	Seats	Vehicles	Originally delivered to	Date of build
DFF 3209 Catamaran	–	–	31.5	27	297	none	Robben Island Museum	(2007)

Spain

Navantia

Formerly known as Izar.

Puerto Real Shipyard, San Fernando, Carretera de la Carraca s/n, PO Box 18, 11080 San Fernando, Spain

Tel: (+34 91) 335 8400
Fax: (+34 91) 335 8652
e-mail: navantia@navantia.es
Web: www.navantia.es

Navantia is a state-owned shipbuilding company undertaking naval and commercial ship design and construction. Current collaboration, however, also includes joint design and construction programmes with Germany (frigates), the Netherlands (frigates), Norway (frigates) and France (submarines and minehunters), and the design and construction of the Royal Thailand Navy carrier. Navantia also continues to design and build a range of commercial vessels, many exported worldwide. The company has three shipyards, at Ferrol, Cartagena and San Fernando. San Fernando shipyard, near Cadiz, is where the high-speed aluminium craft are constructed.

B60 FAST CATAMARAN

Luciana Federico L

The first B60 was ordered by Buquebus in July 1996 and delivered in September 1997. The vessel is based on the Advanced Multihull Design (AMD) 1130 design. The Australian company provided the design with Navantia contributing to the detail design. This fast passenger car ferry operates across the River Plate.

Specifications

Length overall	77.3 m
Length waterline	70.0 m
Beam	19.0 m
Draught	2.1 m
Passengers	450
Vehicles	52 cars
Propulsive power	32,200 kW
Operational speed	57 kt
Range	200 n miles

Classification: DnV +1A1 HSLC 4 Car Ferry B EO.

Propulsion: Powered by two ABB GT35 gas turbines, each driving a steering and reversing water-jet.

Auxiliary systems: Electrical power is provided by two main generator sets each rated at 400 kW.

PENTAMARAN 175 (DESIGN)

Designed to carry up to 1,000 tonnes of deadweight at speeds of up to 40 knots in moderate to rough seas, the Pentamaran 175 is powered by four medium-speed diesel engines more usually associated with slower conventional vessels. The ship is constructed entirely in NV 36 high-tensile steel and consists of a long slender monohull carrying 99.5 per cent of the ship's displacement, and is stabilised by four steel sponsons port and starboard.

Specifications

Length overall	175.25 m
Beam	31.3 m
Crew	30
Passengers	1200
Vehicles	280 cars and 28 trucks
Propulsive power	52,200 kW
Operational speed	38 kt
Range	450 n miles

Classification: DnV +1A1 HSLC Car Ferry (Pentamaran) A R1 EO.

Propulsion: Four Wartsila 18V 28B engines, each rated at 13 MW driving four water-jets.

B60 catamaran Luciana Federico L 0109641

Vessel type	Vessel name	Yard No	Length (m)	Speed (kt)	Seats	Vehicles	Originally delivered to	Date of build
B60 Fast Catamaran	*Luciana Federico L*	–	77.3	57.0	450	52	Buquebus	September 1997

Thailand

Asimar

Asian Maritime Services

128 Mu 3 Suksawad Rd, Prasamutjedee, Samutprakarn 10290, Thailand

Tel: (+66 281) 520 60-7
Fax: (+66 245) 372 14
e-mail: mkd@asimar.com
Web: www.asimar.com

Asimar shipyard is situated in the port area of Chaa Phaya, with a 20,000 m² shipyard area. Asimar undertakes ship repair, building and engineering contracts. In 2003 the company provided a 20.0 m fast catamaran for a local operator. The vessel was designed by Crowther Design of Australia for a capacity of 120 passengers and a speed of 25 kts.

20 m CATAMARAN

Lom Phraya 2000

Specifications

Length overall	20.0 m
Length waterline	17.0 m
Beam	8.0 m
Draught	1.4 m
Passengers	120
Fuel capacity	2,000 litres
Water capacity	1,000 litres
Operational speed	25 kt

Classification: DnV +1A1 HSLC R4 Passenger EO.

Propulsion: Two MTU 8V 183TE 72 diesel engines each driving a fixed-pitch propeller.

Crowther 25 m CATAMARAN

Namuang

Specifications

Length overall	25.0 m
Beam	8.5 m
Draught	1.7 m
Passengers	228
Crew	4
Operational speed	26 kt

Propulsion: Two MTU 12V 2000 M70 diesel engines rated at 788 kW each driving fixed pitch propellers via ZF 2150 gearboxes.

Italthai Marine Ltd

389 Soi.Italianthai, Taiban Road, Tambon Taiban, Amphur Muang, Samutprakarn 10280, Thailand

Tel: (+66 2) 387 10 59
Fax: (+66 2) 387 10 63
e-mail: info@italthaimarine.com
Web: www.italthaimarine.com

Preamchai Karnasuta, *Chairman*
Wirat Chanasit, *General Manager*
Arayant Ummartyotin, *Marketing Manager*

***Dou Men* (Yard No 78) *Zhong Shan Hu* (Yard No 80)**
Two catamaran ferries were delivered in June and July 1990 to Yuet Hing Marine Supplies, Hong Kong, for the Hong Kong to Tao Mun, China route. These vessels are powered by two MWM TBD 604B V12 1,250 kW high-speed diesel engines driving Kamewa 63 S water-jet units.

Dou Men 0506661

Vessel type	Vessel name	Yard No	Length (m)	Speed (kt)	Seats	Vehicles	Originally delivered to	Date of build
Catamaran Ferry	*Dou Men*	78	–	–	–	–	Yuet Hing Marine Supplies, Hong Kong	1990
Catamaran Ferry	*Zhong Shan Hu*	80	–	–	–	–	Yuet Hing Marine Supplies, Hong Kong	1990

Turkey

Pendik Shipyards

Kayana, 801504 Pendik, Istanbul, Turkey

Tel: (+90 212) 390 03 10
Fax: (+90 212) 354 49 95

Pendik Shipyard delivered three FlyingCat 35 catamarans to IDO of Istanbul. The build of these craft was based on a co-operation agreement between the Turkish Shipbuilding Industry (TSI) and Kvaerner Fjellstrand (now Fjellstrand AS).

35 m FLYING CAT

Specifications
Length overall 35 m
Beam 10.1 m
Draught 1.6 m
Passengers 400
Operational speed 32 kt
Classification: DnV +1A1 HSLC Passenger R3 EO, passenger.
Propulsion: Four MTU 12V 183 TE 72 diesel engines driving, in pairs, two Servogear HD 295PTI gearboxes and Servogear VD 10131A propellers.

Vessel type	Vessel name	Yard No	Length (m)	Speed (kt)	Seats	Vehicles	Originally delivered to	Date of build
35 m FlyingCat	*Temel Reis II*	1640	35.0	32.0	400	none	IDO, Istanbul	1997
35 m FlyingCat	*Barbaros Hayreddin Pasa*	1653	35.0	32.0	400	none	IDO, Istanbul	2000
35 m FlyingCat	*Sokullu Mehmed Pasa*	1654	35.0	32.0	400	none	IDO, Istanbul	2000

Ukraine

Feodosia Shipbuilding Association (Morye)

Desantrnikov Str 1, 98176 Feodosia, Crimea, Ukraine

Tel: (+380 65) 626 37 22
Fax: (+380 65) 626 99 05
e-mail: kgp@morye.kafa.crimea.ua
Web: www.morye.crimea.ua

Leninar Bogatov, *Chairman*
Boris Kozlov, *Technical Director*
Gregory Klebanov, *Head of Marketing and Sales*

Well known for its production of hydrofoil craft, the company developed a foil assisted catamaran design in 2000.

SUPERFOIL 30 MK1 (DESIGN)

This aluminium catamaran passenger ferry incorporates a patented foil and interceptor system and is powered by diesel engines with fixed-pitch ventilating propellers or water-jets.

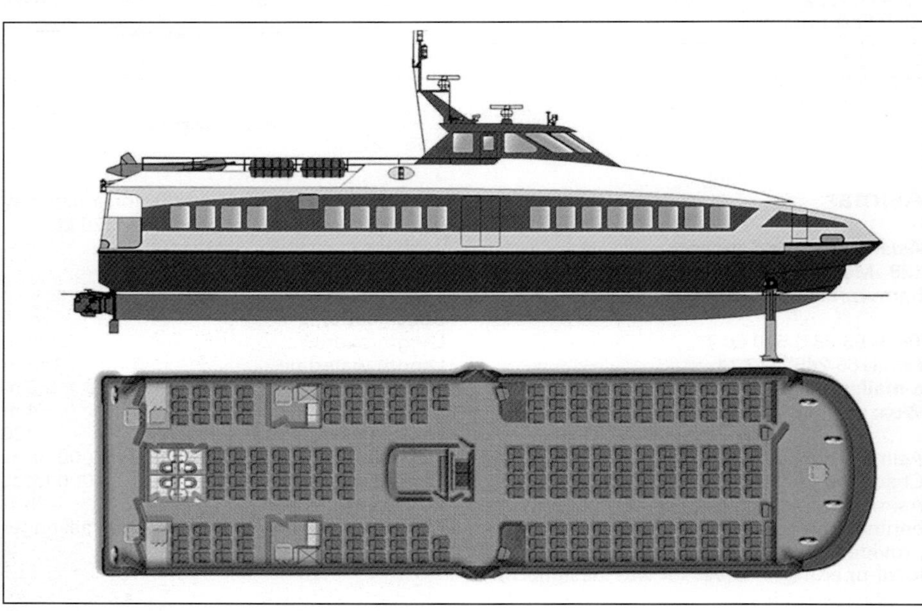
General arrangement of the Superfoil 30 (Design) 0589521

Specifications

Length overall	30.9 m
Length waterline	26.7 m
Beam	8.6 m
Draught	2.7 m
Displacement, min	63.8 t
Displacement, max	85.0 t
Passengers	200
Fuel capacity	6,000 litres
Water capacity	1,500 litres
Operational speed	48 kt
Range	250 n miles

Classification: Bureau Veritas/DnV.
Structure: All aluminium hull and superstructure.
Propulsion: Two MTU 16V 4000 M70 diesel engines driving MJP JSSOR water-jets or fixed pitch propellers.

SUPERFOIL 27 (DESIGN)

Specifications

Length overall	27.7 m
Beam	7.3 m
Displacement, min	42.0 t
Displacement, max	58.0 t
Passengers	150
Fuel capacity	5,000 litres
Water capacity	1,000 litres
Operational speed	42kt
Range	200 n miles

Classification: Bureau Veritas/DnV.
Structure: All aluminium hull and superstructure.
Propulsion: Two diesel engines each driving a MJP 450R DD water-jet via a ZF 2550 gearbox.

United Kingdom

Alnmaritec Ltd

Alnmarin House, Willowburn Industrial Estate, Alnwick, Northumberland, NE66 2PQ, United Kingdom

Tel: (+44 1665) 60 29 17
Fax: (+44 1665) 60 53 99
e-mail: sales@alnmaritec.co.uk
Web: www.alnmaritec.co.uk

Chris Milman, *Managing Director*
Dave Slater, *Operations Manager*
Pat Hume, *Sales Manager*

Alnmaritec specialises in aluminium fabrication for the marine industry, providing workboats, pontoons and specialist aluminium structures including high-speed catamaran craft.

16.25 m CATAMARAN

Braye Spirit

A variation of Alnmaritec's successful Wave-Train 1,400 design, the first 16.25 m catamaran was built for IT Ferries to operate between the Channel Islands and France.

Specifications

Length overall	16.25 m
Passengers	12
Cars	4

Propulsion: Two Cummins 430 kW diesel engines coupled to Ultrajet 376 water-jets.

Aluminium Shipbuilders Ltd

The Shipyard, Ashlake Copse Road, Fishbourne, Isle of Wight PO33 4EY, United Kingdom

Tel: (+44 1983) 88 22 00
Fax: (+44 1983) 88 47 20

J A Davies, *Managing Director*
M J Love, *Chairman*

Aluminium Shipbuilders was formed in 1983 by the current Managing Director, John Davies. Initially, ASL undertook a design and build contract for specialist craft.

In early 1985, ASL was instrumental in the sale of two Australian designed and built 30 m catamarans to Sealink (UK) Ltd for service on the Portsmouth to Isle of Wight route. These craft, designed by International Catamaran Designs Pty Limited (InCat) were built at its yard in Hobart, Tasmania.

The first InCats built in Europe were the result of an initiative in late 1985 whereby, in association with McTay Marine Limited, the company responded to a Ministry of Defence (Navy) MoD(N) enquiry for the supply of Towed-Array Recovery/Deployment Vessels (TARVs). In December 1985 this joint venture resulted in the award of a contract for three TARVs. ASL undertook the fabrication and assembly of these vessels for fitting out by McTay Marine on Merseyside.

In 1986 ASL was awarded a contract for a 16 m Catamaran Riverbus from Thames Line plc. Subsequently, and after successful proving trials with the first craft, named the *Daily Telegraph*, a further seven slightly larger 17.5 m craft were built in the succeeding two years, all for Thames Line.

Early in 1989 ASL obtained a GBP5 million contract from Condor Limited to build a 49 m wave-piercing catamaran of InCat design.

Condor 9 was delivered in August 1990 and operated a daily service from March to November on the arduous Western Channel route from St Malo to Weymouth and return, via the Channel Islands. This vessel, which carries 450 passengers in two air conditioned saloons at a service speed of 35 kt, was built to the classification requirements of Det Norske Veritas Light Craft rules.

Condor 9 built by Aluminium Shipbuilders Ltd

0506662

One of eight Aluminium Shipbuilders InCat 17.5 m catamarans 0506668

During the last quarter of 1991 the company received a contract for the fabrication of a novel high-speed trimaran passenger ferry. The craft is the prototype of a 70-seat variant which, with its inherent low-wash characteristics and economical propulsion arrangement, is intended as the next generation of smooth-water fast river ferries.

In 1992 ASL expanded its high-speed craft repair facilities to cater for the increasing volume of specialised repair work associated with the entry into service of further SeaCats.

The company is currently in the process of acquiring construction facilities large enough to accommodate the next generation of SeaCat (of 78 m and above) for an ever-growing marketplace.

RIVER 50

In August 1987, the first of a series of eight lightweight, high-speed passenger catamarans was delivered to Thames Line plc for operation between Charing Cross Pier and West India Dock on the River Thames. The vessel is powered by twin Volvo Penta TAMD 71A engines driving Riva Calzoni water-jets. Delivery was completed in 1988.

Specifications

Length overall	16.3 m
Beam	5.4 m
Draught	0.6 m
Fuel capacity	2 × 320 litres
Water capacity	100 litres
Propulsive power	2 × 228 kW
Operational speed	23–25 kt

Classification: UK DoT Class V smooth water limits.

Propulsion: Main engines are two Volvo Penta TAMD 71A, 228 kW each at 2,500 rpm, driving Riva Calzoni IRCL 39 D water-jets via MPM IRM 301 PL-1 gearboxes.

17.5 m CATAMARAN

A total of seven vessels was ordered by Thames Line plc for delivery during 1988. The vessels are 17.5 m in length and have revised interior arrangements which allow for a spacious cabin for 62 passengers.

Specifications

Length overall	17.5 m
Beam	5.4 m
Draught	0.6 m
Passengers	50–70
Max speed	30 kt

Propulsion: Water-jets.

49 m WAVE-PIERCING CATAMARAN

Condor 9

The order by Condor Ltd for a 49 m InCat wave-piercing catamaran was announced on 18 April 1989. Designed to carry 450 passengers, *Condor 9* replaced two hydrofoil vessels. The vessel was handed over in late August 1990. In 2003 and 2004 the vessel was extensively refitted at Eastern Shipbuilding in Panama City. The vessel was stripped back to a bare hull, re-engines and completely refitted. The craft was relaunched and now operates as *Jessica W.*

Specifications

Length overall	48.7 m
Length waterline	40.5 m
Beam	18.2 m
Hull beam	3.3 m
Draught	1.9 m
Crew	15
Passengers	450
Fuel capacity	2 × 7,200 litres
Water capacity	2 × 1,000 litre tanks
Operational speed	35 kt

Classification: DnV +1A1 Light Craft (CAT) R45 Passenger Ship EO. Also surveyed under UK Department of Transport category Class 2, Short International Voyage vessel.

Structure: Welded aluminium.

Propulsion: Main engines are four MWM TBD 604BV 16, 1,682 kW each at 1,800 rpm, driving four MJP J650R-DD water-jet units.

If one engine should break down, a speed of 30 kt can be maintained; each engine with its water-jet unit is entirely independent of the other three in its operation.

Vessel type	Vessel name	Yard No	Length (m)	Speed (kt)	Seats	Vehicles	Originally delivered to	Date of build
17.5 m River Cat	*Barclays Bank*	–	17.5	–	51	–	Thames Line Plc	1987
17.5 m River Cat	*London Docklands*	–	17.5	–	62	–	Thames Line Plc	April 1988
17.5 m River Cat	*Chelsea Harbour*	–	17.5	–	62	–	Thames Line Plc	May 1988
InCat River 50	*Le Premier* (ex-*Daily Telegraph*)	–	16.3	25.0	62	–	Thames Line Plc	May 1988
17.5 m River Cat	*Debenham Tewson and Chinnocks*	–	17.5	–	62	–	Thames Line Plc	August 1988
17.5 m River Cat	*Harbour Exchange*	–	17.5	–	62	–	Thames Line Plc	August 1988
17.5 m River Cat	*Daily Telegraph*	–	17.5	–	62	–	Thames Line Plc	1988
17.5 m River Cat	*London Broadcasting Company*	–	17.5	–	62	–	Thames Line Plc	1989
Incat 49 m WPC	*Jessica W* (ex-*Condor 9*)	–	48.7	35.0	450	–	Condor Ltd	August 1990

FBM Babcock Marine Group

The Courtyard, St Cross Business Park, Monks Brook, Newport, Isle of Wight, PO30 5BF, United Kingdom

Tel: (+44 1983) 82 57 00
Fax: (+44 1983) 82 41 80
e-mail: fbm@babcock.co.uk
Web: www.fbmuk.com

Mark Graves, *General Manager*

FBM builds a wide range of monohull, catamaran and SWATH vessels for both commercial and paramilitary applications. In 2000, Babcock International purchased the FBM interests, adding a further shipyard in Rosyth.

THAMES CLASS CATAMARAN

(RTL HYDROCAT)

There have been three of these 62-seat minimum-wash high-speed ferries delivered. The concept originated with Robert Trillo Ltd (RTL) and was introduced to Fairey Marine Ltd (now FBM Babcock Marine Ltd) in 1986. It combines a number of features specifically aimed at optimising a design of a relatively high-speed vessel for river use. Of particular importance in such applications is the minimisation of wash disturbance and noise. Both of these aspects are inherent in the RTL concept, coupled with minimum water and air draughts.

The remarkably low wash of the concept is of particular value in rivers such as the Thames where very shallow conditions occur at low tide in many areas. The Riverbus Partnership took delivery of three Thames Class catamarans in January 1992. In 1996 they were refitted by FBM for delivery to Serco Denholm for use as passenger ferries for naval staff.

Specifications

Length overall	25.0 m
Length waterline	23.0 m
Beam	5.7 m
Draught	0.7 m
Crew	2
Passengers	62
Fuel capacity	1,000 litres
Max speed	25 kt

Structure: The hull structure is an FRC sandwich laminate with the bridge platform, box beams and saloon constructed in low-maintenance marine grade aluminium alloy.

Propulsion: Two six-cylinder in-line, turbocharged, heat exchange-cooled Scania DSI 11 marine diesel engines fitted

45 m TriCat for Goutos Lines 0038742

with flanged-mounted MPM 320 reverse gearboxes, drive the two Riva Calzoni IRC 390 water-jet units, via Aquadrive Cardan shafts.

30 m CITY SLICKER CATAMARAN (DESIGN)

The 30 m City Slicker is an ultra low wash aluminium catamaran specifically designed for high-speed operation on riverine and harbour ferry routes.

Designed for medium- and high-speed passenger routes, this craft is capable of carrying 100 to 135 passengers at speeds up to 35 kt.

Specifications

Length overall	32.4 m
Length waterline	30.0 m
Beam	8.0 m
Draught	1.0 m
Passengers	100–135
Operational speed	30 kt

Structure: Aluminium hull and superstructure.

Propulsion: Main engines: two MWM TBD 234 V12, driving a pair of water-jets.

SUBMARINE SUPPORT VESSEL

Adamant

The vessel was delivered in 1983 to the UK Ministry of Defence and is based on the symmetrical hulls of the Solent Class range of passenger catamarans. This water-jet-propelled aluminium catamaran craft features a constant tension brow/gangway arrangement and large hydraulically operated fenders to facilitate 'at sea' transfers to and from submarines lying offshore.

The transfer system consists of an 8.1 m long aluminium brow pivoted at the inboard end to the base of an Effer Model 9600/3S hydraulic crane. A constant tension winch fitted on the crane supports the brow, allowing its outboard end, which is fitted with wheels and rubber tyres, to rest on the submarine. Relative movement between the two vessels is compensated by the automatic operation of the constant tension winch.

To prevent any possibility of damage to the submarine hull during such operations, two large foam-filled fenders, each 2.5 m long × 1.25 m diameter, can be deployed on each side of the SSV. These fenders are normally stowed in deck cradles and are lowered and maintained in position by locally controlled hydraulic winches and an endless belt which passes through an eye located at water level.

Specifications

Length	30.8 m
Beam	7.8 m
Draught	1.1 m
Propulsive power	1,014 kW
Max speed	23 kt
Range	250 n miles

Propulsion: Engines: two Cummins KTA19M diesels.

Transmissions: two ZF gearboxes.

Thrust device: two MJP water-jets.

31.5 m SOLENT CLASS CATAMARANS

Red Jet 1* and *Red Jet 2

Another low wash catamaran design, the 31.5 m Solent Class is in service with Red Funnel; the first of two was delivered in February 1991, following its launch on 4 December 1990.

The Solent Class catamarans are specifically designed for operations on coastal routes, rivers and urban environments.

During trials, *Red Jet 1* in full-load condition (that is, 120 passengers, baggage, full fuel, water and three crew) achieved the following results:

Speed, maximum (measured mile): 38.3 kt at 100 per cent MCR

Speed, contract: 32.5 kt at 63 per cent MCR

Speed, single engine: 24 kt

Acceleration: 35 kt in 47 s in 520 m

Fuel consumption: 375 litres/h at 32.8 kt

FBM Marine Thames Class catamaran (RTL HydroCat) in service on the Thames 0506663

FBM Marine Solent Class catamaran ferry Red Jet 1 *operated by Red Funnel* 0506976

FBM 45 m River Cat Algés 0506977

53 m TriCat Class catamaran Sea Speed 1 *during sea trials* 0081006

The 30 m Solent Class catamaran can maintain a cruising speed of 38 kt in a 1 m head sea and is well within the vertical acceleration limits prescribed for normal passenger comfort. The craft responds very quickly to the steering control at all speeds and the maximum rate of turn at 1,800 rpm is 6.6°/s. The turning circle under full helm becomes progressively tighter and the speed falls away to about 20 kt after a full circle. The diameter of this circle is approximately 325 m.

The craft can rotate about its own axis at the rate of 7.5°/s.

Specifications

Length overall	31.5 m
Beam	8.4 m
Draught	1.1 m
Displacement, max	65 t
Passengers	120
Fuel capacity	1,685 litres
Water capacity	250 litres
Propulsive power	2,720 kW
Operational speed	32.5 kt

Classification: DnV +1A1, R15, EO and UK Department of Transport Class IV for Solent use only.

Structure: Hull: marine grade aluminium, BS 1470 N8 plate, H30TF extrusions.

Propulsion: Engines: two MTU 12V 396 TE 84, 1,360 kW each, at MCR, 1,940 rpm, driving two MJP 650 water-jet units, driven via ZF gearboxes.

35 SOLENT CLASS CATAMARANS

Red Jet 3

Bo Hengy

In July 1998 FBM Marine delivered the third Red Jet to the Red Funnel Group. *Red Jet 3* is an extended version of the Solent Class *Red Jets 1* and *2*, has a passenger capacity of 188 and a speed of 33.7 kt while maintaining the same propulsion system with a slightly upgraded rating. In 1999 a slightly stretched sistership vessel, *Bo Hengy*, was built under licence by Pequot River Shipworks in the US and delivered to Bahamas Fast Ferries.

Specifications

Red Jet 3

Length overall	32.0 m
Beam	8.4 m
Draught	1.25 m
Passengers	188
Fuel capacity	2,500 litres
Water capacity	100 litres
Propulsive power	3,000 kW
Operational speed	33.7 kt

Specifications

Bo Hengy

Length overall	35.0 m
Length waterline	29.5 m
Beam	8.3 m
Draught	1.25 m
Passengers	177
Fuel capacity	4,000 litres
Water capacity	250 litres
Propulsive power	3,480 kW
Operational speed	35 kt

Propulsion: Two MTU/DDC 12V 4000 diesel engines each driving a KMW 63 water-jet.

35 m TRICAT CLASS CATAMARAN

Delivered in January 1994, this development of the well proven Solent Class catamaran transports passengers on the increasingly popular route from Hong Kong to the new airport at Shenzhen on the Chinese mainland.

This 35 m catamaran is designed for use in coastal and sheltered waterways and is particularly suitable for operation in busy harbours due to its manoeuvrability, shallow draught and low wash characteristics.

Specifications

Length overall	35.1 m
Length waterline	29.5 m
Beam	8.3 m
Draught	1.2 m

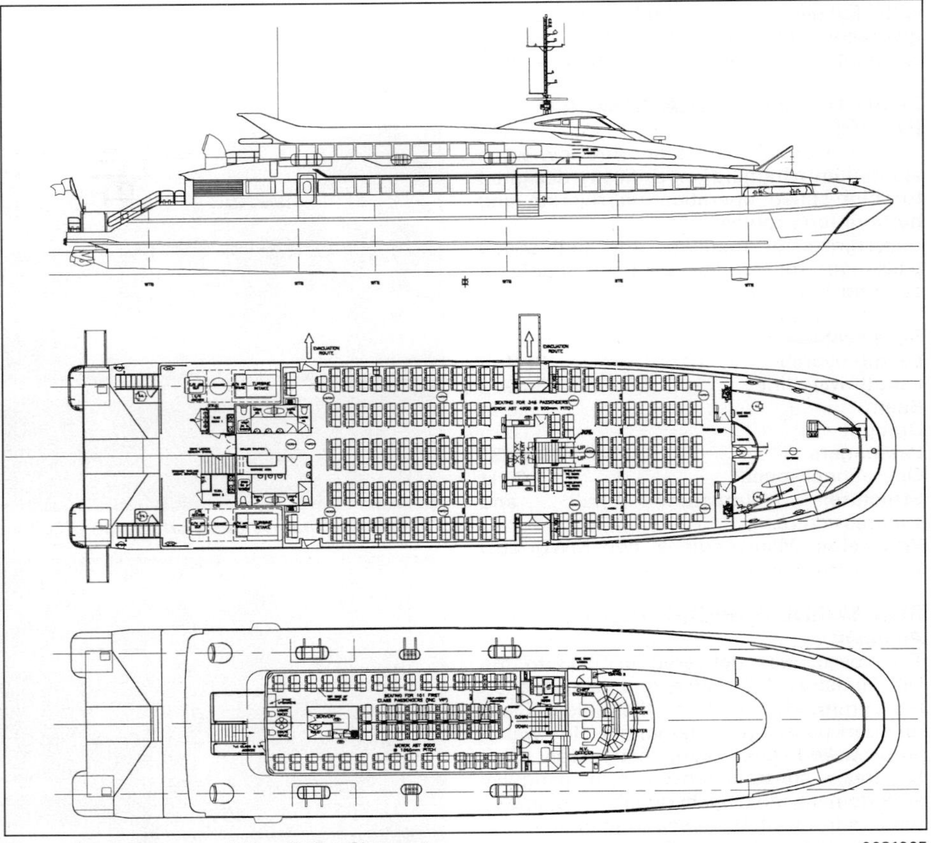

General arrangement of 53 m TriCat Class catamaran 0081005

45 m TriCat Class catamaran universal 2003 (Maritime Photographic) 0045118

FBM 30.8 m submarine support vessel Adamant 0506823

Passengers	160
Max speed	34 kt

Structure: Hull and superstructure material is aluminium.

Propulsion: Engines: two MTU 12V 396 TE 74. Thrust devices: two MJP water-jets.

45 m TRICAT CLASS CATAMARAN

FBM's 45 m TriCat fast passenger ferry has been specifically designed in association with one of the world's leading designers of luxury yachts and mini cruise liners, Terry Disdale.

This 312-seat catamaran is the result of FBM's extensive research and development programme to build a passenger catamaran which can achieve very high speeds with good passenger comfort without recourse to foils or air cushioned systems.

This 45 m fast passenger catamaran is designed to meet the requirements of the Hong Kong Marine Department for short international voyages out of Hong Kong under a Hong Kong flag and DnV Classification R1 for High-Speed Light Craft.

There have been eight TriCats delivered to CTS-Parkview Ferry services for the route between Hong Kong and Macau. One TriCat was built under licence by Babcock Rosyth (Fabricators) Ltd at the Rosyth Royal Dockyard. The first TriCat was launched on 8 September 1994 and delivered to Hong Kong in January 1995.

The eighth TriCat for Hong Kong was built at the company's shipyard in the Philippines.

In June 1998 a further 45 m TriCat was delivered to the Athens-based operator Goutos Lines. Although the vessel has been modified to accommodate 375 passengers, the specification is similar to the Hong Kong craft. Two further craft were built under licence in the US by Pecquot River Shipworks and delivered in 1997 and 1999 respectively.

Specifications

Length overall	45.0 m
Length waterline	40.0 m
Beam	11.8 m
Draught	1.5 m
Payload	36.1 t
Passengers	312
Fuel capacity	7,500 litres
Propulsive power	8,400 kW
Max speed	47 kt
Operational speed	44 kt
Range	125 n miles

Classification: DnV Classification R1 for High-Speed Light Craft.

Structure: The hull is constructed with aluminium and the superstructure is aluminium with lightweight FRP cladding.

Propulsion: The TriCat is powered by two Caterpillar Solar Taurus gas turbines each rated at 4,200 kW at 13,000 rpm. These engines drive, via separate rigidly mounted reduction gearboxes, two Kamewa water-jets. The turbines and gearboxes are mounted on a raft rigidly bolted to the ship's structure.

FBM 45 m TRANSCAT

This passenger ferry is intended for intensive commuter passenger services within shallow harbours and sheltered routes. The TransCat specifically addresses the requirements for low fuel consumption and low wash. This catamaran incorporates the latest requirements of the IMO code for safety and incorporates the FBM TriCat unique futuristic styling.

Eight of these craft are currently operated by Transtejo Transportes on the River Tagus in Lisbon, Portugal. Four have been constructed at the company's shipyard in Cowes and four have been subcontracted to a Portuguese yard. The first craft was delivered in 1995.

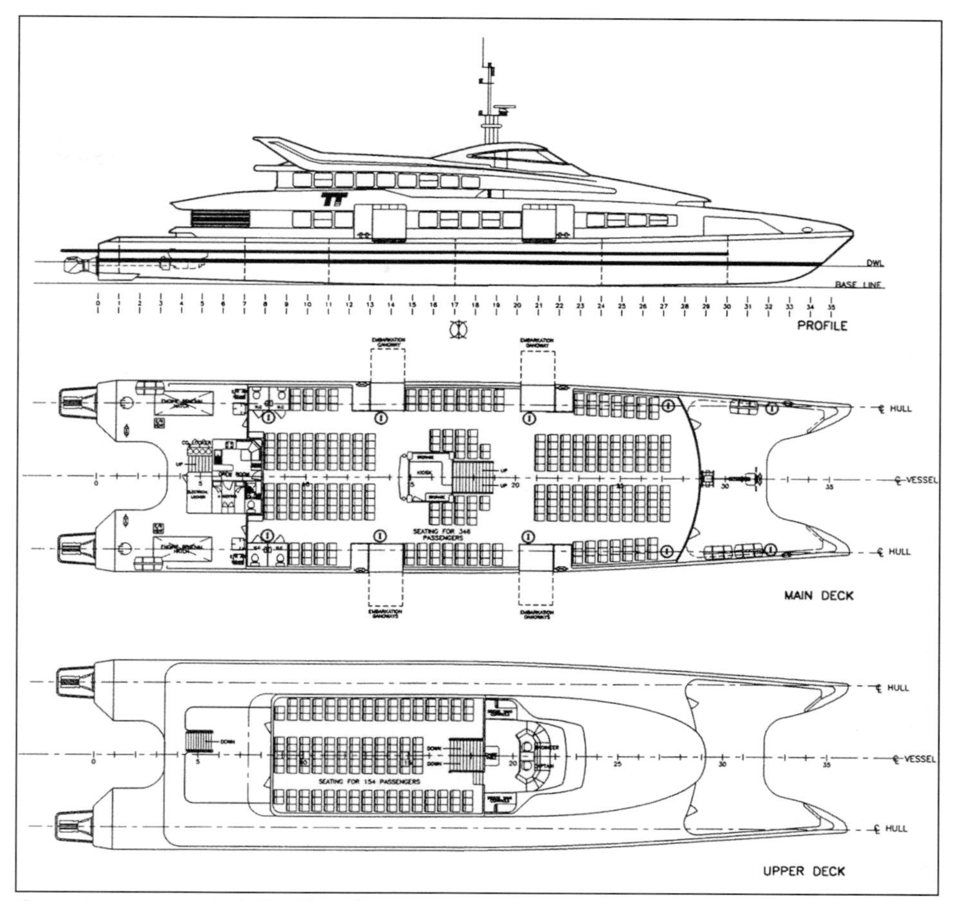

General arrangement of 45 m TransCat 0506898

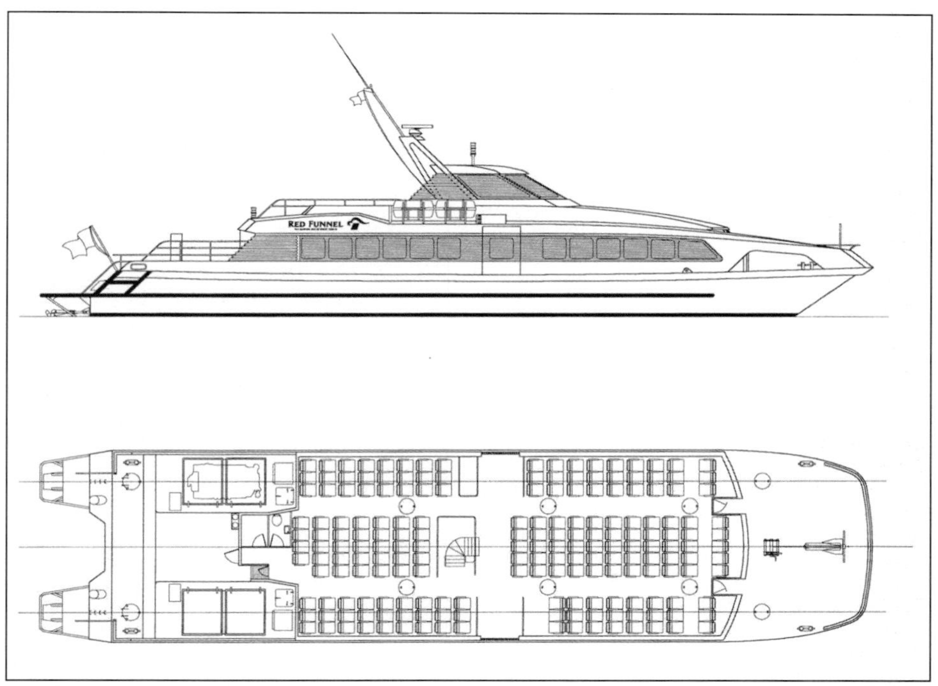

Solent Class Red Jet 3 *general arrangement* 0007957

Solent Class Red Jet 3 *on trials* (Maritime Photographic) 0045119

Vessel type	Vessel name	Yard No	Length (m)	Speed (kt)	Seats	Vehicles	Originally delivered to	Date of build
Submarine Support Vessel	*Adamant*	–	30.8	–	–	–	UK Ministry of Defence	1983
31 Solent Class Catamaran	*Red Jet 2*	–	31.5	32.5	120	–	Red Funnel	February 1991
31 Solent Class Catamaran	*Red Jet 1*	–	31.5	32.5	120	–	Red Funnel	February 1991
Thames Class Catamarans	–	–	25.0	–	62	–	The Riverbus Partnership	January 1992
Thames Class Catamarans	–	–	25.0	–	62	–	The Riverbus Partnership	January 1992
Thames Class Catamarans	–	–	25.0	–	62	–	The Riverbus Partnership	January 1992
35 m TriCat	–	–	35.1	–	160	–	Hong Kong/China	January 1994
45 m TriCat	*Universal Mk 2001*	–	45.0	44.0	312	–	CTS Parkview Ferry Services	January 1995
45 m TriCat	*Universal Mk 2002*	–	45.0	44.0	312	–	CTS Parkview Ferry Services	1995
45 m TransCat	*Castelo*	–	45.0	25.0	500	–	Transtejo Transportes	1995
45 m TransCat	*Alges*	–	45.0	25.0	500	–	Transtejo Transportes	1995
45 m TriCat	*Universal Mk 2004*	–	45.0	44.0	312	–	CTS Parkview Ferry Services	1995
45 m TriCat	*Universal Mk 2003*	–	45.0	44.0	312	–	CTS Parkview Ferry Services	1995
45 m TriCat	*Universal Mk 2006*	–	45.0	44.0	312	–	CTS Parkview Ferry Services	1996
45 m TriCat	*Universal Mk 2007*	–	45.0	44.0	312	–	CTS Parkview Ferry Services	1996
45 m TriCat	*Universal Mk 2005*	–	45.0	44.0	312	–	CTS Parkview Ferry Services	1996
45 m TransCat	*S Juliao*	–	45.0	25.0	500	–	Transtejo Transportes	1997
45 m TriCat	*Universal Mk 2009*	–	45.0	44.0	312	–	CTS Parkview Ferry Services	1997
35 m Solent Class Catamaran	*Red Jet 3*	–	32.0	33.7	188	–	Red Funnel	July 1998
45 m TransCat	*Carnide*	–	45.0	25.0	500	–	Transtejo Transportes	1998
45 m TriCat	*Athina 2004*	–	45.0	44.0	312	–	Goutos Lines, Greece	1998
53 m TriCat	*Sea Speed 1*	–	52.8	40.0	450	–	Agapitos, Greece	1999
56 m TriCat	*Universal Mk I*	–	56.5	42.0	450	–	Naviera Parkview Espanola	2001

Specifications

Beam	11.8 m
Draught	1.4 m
Passengers	500
Fuel capacity	4,000 litres
Water capacity	1,000 litres
Operational speed	25 kt

Classification: IMO code for safety of High-Speed Craft. Lloyd's Register of Shipping.

Structure: The hull is constructed of aluminium and the superstructure is aluminium with lightweight FRP cladding.

Propulsion: Main engines are twin MWM TBD 616 V16s. These drive twin LIPS water-jets LJ76DL.

FBM 53 m TRICAT

Specifications

Length overall	52.8 m
Length waterline	47.5 m
Beam	13.0 m
Draught	1.62 m
Deadweight	68 t
Crew	10
Passengers	450
Fuel capacity	32,000 litres
Water capacity	1,500 litres
Operational speed	40 kt

Classification: DNV +1A1, HSLC, R3, EO Passenger HSC Category A.

Structure: All aluminium hull and superstructure (with GRP fairing).

Propulsion: Twin Caterpillar SOLAR TAURUS gas turbines each rated at 4,444 kW at 13,000 rpm, each driving a Kamewa S90 water-jet via a reduction gearbox.

Electrical system: Two Caterpillar 3306 generators each rated at 100 kW.

Auxiliary systems: An MDI ride control system consisting of two forward T foils is fitted.

FBM 56 m TRICAT

The first of this series, *Universal Mk I*, was constructed at FBM Babcock's Rosyth shipyard in 2000/2001 and delivered to Naviera Parkview Espanola in June 2001 for operation from Barcelona to Majorca.

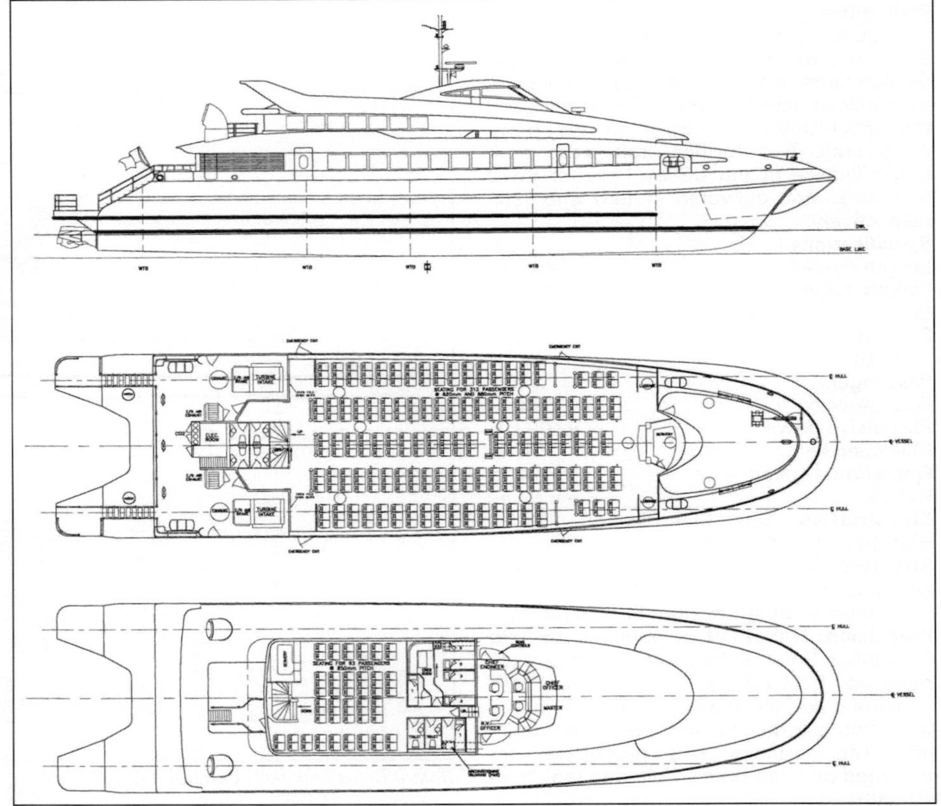

General arrangement of 45 m TriCat Class for Greek Goutos Line

0007958

FBM 56 m TriCat Universal Mk 1

0134113

Specifications

Length overall	56.5 m
Length waterline	49.16 m
Beam	13.0 m
Draught	1.62 m
Passengers	450
Fuel capacity	32,000 litres
Water capacity	1,500 litres
Operational speed	42 kt

Classification: DNV +1A1, HSLC, R3, EO Passenger HSC Category A.

Structure: All aluminium hull and superstructure (with GRP fairing).

Propulsion: Four MTU 16V 4000 diesel engines each rated at 4,640 kW and driving a Kamewa 63 water-jet.

VT Shipbuilding

Fleet Way, Portsmouth, PO1 3AQ, UK

Tel: (+44 23) 92 85 72 00
Fax: (+44 23) 92 85 74 00
e-mail: shipsales@vosperthornycroft.com
Web: www.vtplc.com

VT Shipbuilding continues the shipbuilding business established over a century ago by two separate companies, Vosper Ltd and John I Thornycroft and Company Ltd. These companies merged in 1966, were nationalised in 1977, returned to the private sector in 1985 and floated on the London Stock Exchange in 1988. The company has designed, built and repaired warships of all sizes and has always specialised in high-speed craft. Since the early 1970s it has also developed the use of Fibre-Reinforced Plastic (FRP) for warships, particularly mine countermeasures vessels.

Vosper Thornycroft was awarded the contract to build a trimeran demonstrator for the UK Defence and Evaluation Research Agency (DERA) in 1998. The vessel was launched in May 2000. With an overall length of 98.7 m, the two side hulls have a length of 34.2 m and a draught of 2.4 m. The vessel is being used for performance evaluation during 2001 and 2002.

HYDROCRUISER 27

Three of these craft were ordered by Fast Ferry Leasing of the UK, having been designed by Teknicraft Design of New Zealand and were delivered in 2004.

Specifications

Length overall	27 m
Beam	9.0 m
Draught	1.0 m
Passengers	150
Operational speed	39 kt

Classification: DnV.

Propulsion: Four Cataerpillar 3412E diesel engines each rated at 830 kW and driving a Hamilton 391 water-jet.

100 m TRIMARAN

R.V. TRITON

Specifications

Length overall	98.7 m
Length waterline	90.0 m
Beam	22.5 m
Hull beam	6.0 m
Draught	3.4 m
Displacement, min	1,010 t
Displacement, max	1,400 t
Crew	12
Passengers	12
Max speed	20 + kt
Operational speed	12 kt
Range	3,000 n miles

Classification: DnV High-Speed Light Craft.

Structure: All steel hull and superstructure.

Propulsion: Diesel electric propulsion. Two Paxman 12VP185 generators each rated at 2,085 kW driving a single HMA 3,500 kW electric motor, itself driving a fixed pitch propeller.

Auxiliary systems: Two 350 kW electric motors driving a Schottel SRP 200 thruster, one in each side hull.

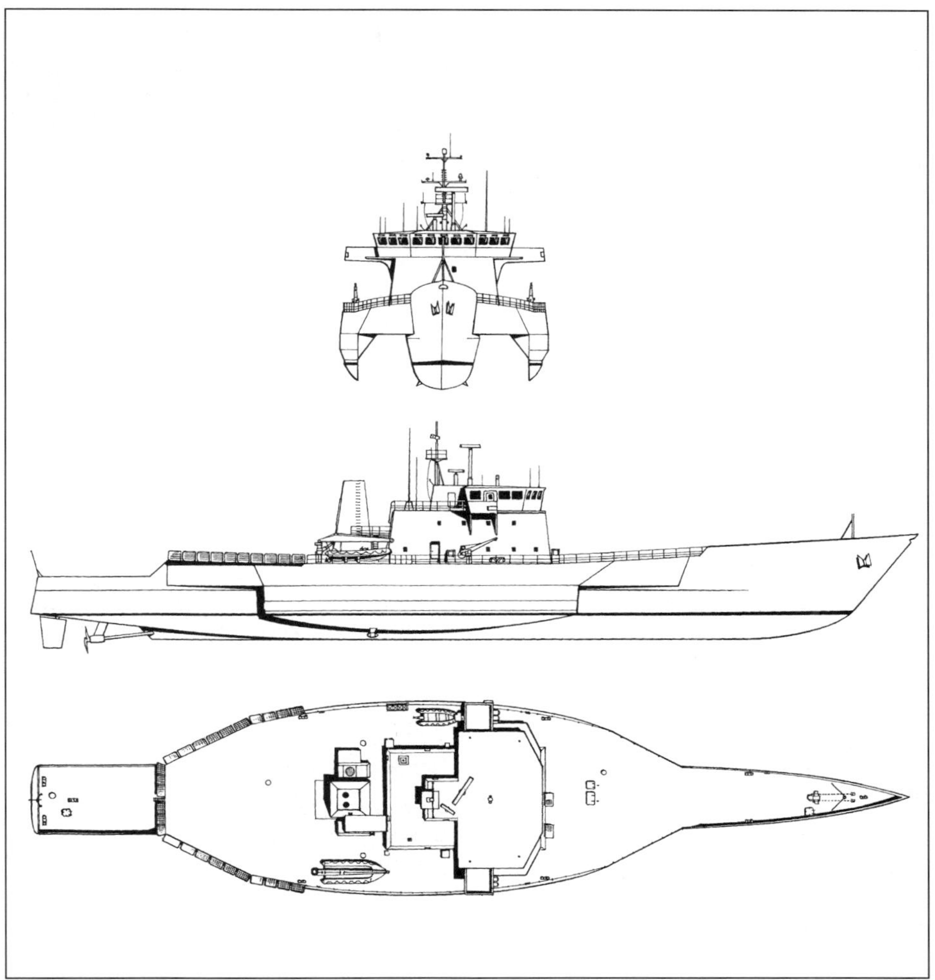

General arrangement of Triton

0100697

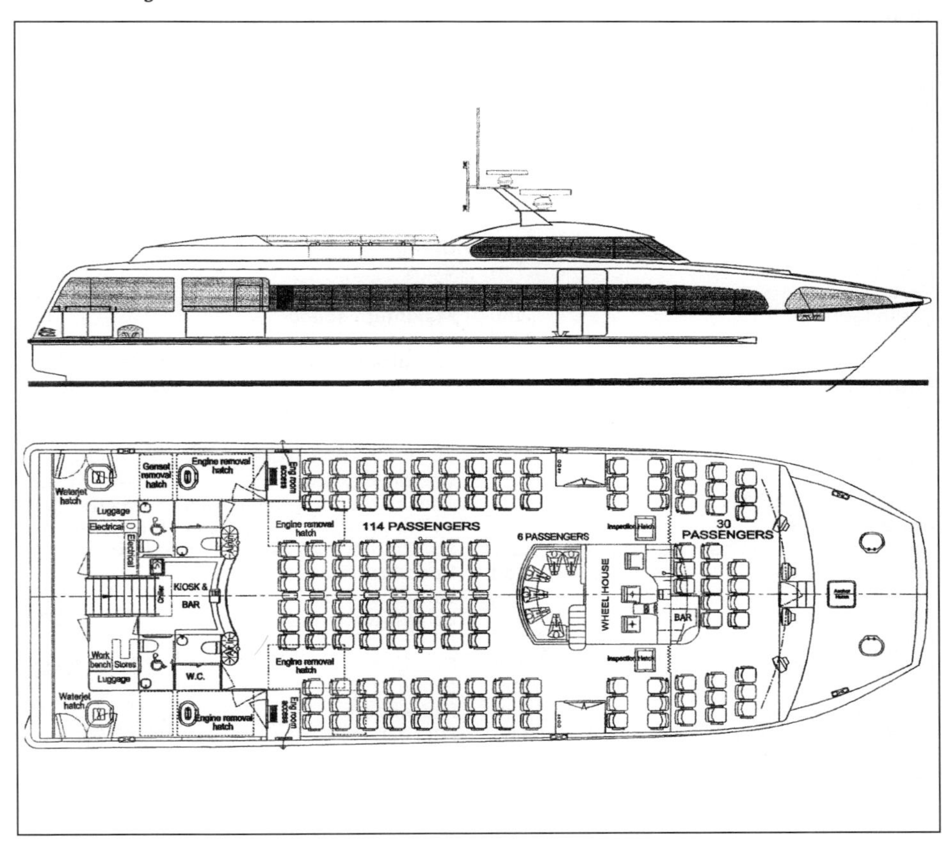

General arrangement of the Hydrocruiser 27
0121750

Bow view of Triton 0100696

Stern view of Triton 0100695

Vessel type	Vessel name	Yard No	Length (m)	Speed (kt)	Seats	Vehicles	Originally delivered to	Date of build
100 m Trimaran	*RV Triton*	–	98.7	12.0	12	none	DERA, UK	May 2000
CAT Hydrocruiser 27	*HC Katia*	04323	27.0 m	39	150	none	Fast Ferry Leasing	2004
CAT Hydrocruiser 27	*HC Milancia*	04324	27.0 m	39	150	none	Fast Ferry Leasing	2004
CAT Hydrocruiser 27	*HC Olivia*	04328	27.0 m	39	150	none	Fast Ferry Leasing	2004

United States

All American Marine Inc

200 Harris Avenue, Bellingham, Washington, 98225, United States

Tel: (+1 360) 647 76 02
Fax: (+1 360) 647 76 07
e-mail: sales@allamericanmarine.com
Web: www.allamericanmarine.com

Matt Mallett, *Chief Executive Officer*

All American Marine is a builder of lightweight fast aluminium craft for commercial and pleasure use. In 1998 the company started to build fast catamaran craft to the designs of Teknicraft of New Zealand.

25 m CATAMARAN

Aialik Voyager

Delivered in March 2007 to Kenai Fjords Tours, this 25.2 m catamaran was designed by Teknicraft. The vessel features the Teknicraft hull shape and patented foil system. This system was designed to impreove vessel motion and passenger comfort.

Specifications

Length overall	25.2 m
Beam	8.9 m
Draught	1.1 m
Crew	4
Passengers	150
Max speed	30 kt
Operational speed	27 kt

Propulsion: Two Caterpillar C32 diesel engines delivering 1,044 kW at 2,300 rpm to FB propellers via ZF gearboxes.
Classification: USCG Subchapter T.

23 m CATAMARAN

Condor Express

Certified under US Coast Guard Subchapter T for 149 passengers, this vessel was delivered in 2002.

Specifications

Length overall	22.7 m
Length waterline	19.7 m
Beam	7.8 m
Draught	1.0 m
Displacement (load)	53.6 t
Deadweight	20 t
Crew	3
Passengers	149
Fuel capacity	7,600 litres
Fresh water capacity	570 litres
Max speed	35 kt
Operational speed	30 kt
Range	800 n miles

Propulsion: Four Detroit Diesel engines each rated at 552 kW and each driving a Hamilton HM 362 water-jet.

22.0 m CATAMARAN

Specifications

Length overall	22.0 m
Beam	7.8 m
Draught	0.95 m
Passengers	149
Maximum speed	35 kt

Propulsion: Four Detroit Diesel Series 60 14L engines each rated at 550 kW at 2,300 rpm and each driving a Hamilton water-jet.

21.8 m CATAMARAN

Spirit

An aluminium catamaran built to a Teknicraft design using their hydrofoil support technology.

Specifications

Length overall	21.8 m
Beam	7.7 m
Draught	0.9 m
Crew	3
Passengers	149
Maximum speed	35 kt
Operational speed	28 kt

Propulsion: Four Detroit Diesel Series 60 14L engines each rated at 550 kW at 2,300 rpm and each driving a Hamilton HJ 362 water-jet.

20 m CATAMARAN

Chilkat Express

This all aluminium catamaran was designed by Teknicraft of New Zealand for operation in Alaska between Skagway and Haines.

Specifications

Length overall	19.3 m
Length waterline	16.3 m
Beam	6.7 m
Draught	0.85 m
Crew	2
Passengers	63
Fuel capacity	2,000 litres
Max speed	48 kt
Operational speed	42 kt
Range	180 n miles

Propulsion: Four Caterpillar 3406E diesels each rated at 597 kW and driving a Hamilton HJ362 water-jet.

19.8 m CATAMARAN

Fulmar

Designed by Teknicraft of New Zealand for the Monterey Bay National Marine Sanctuary, this foil assisted vessel is based on the Shearwater research craft delivered in 2002.

Specifications

Length overall	19.8 m
Beam	7.5 m

Propulsion: Two MTU series 60 engines each rated at 500 kW at 2,300 rpm.

19.7 m CATAMARAN

Expeditions IV

An all aluminium foil supported catamaran built for the Hone Hebe Corporation as an expedition boat.

Specifications

Length overall	19.7 m
Beam	7.1 m
Draught	1.7 m
Crew	3
Passengers	149
Operational speed	28 kt

Propulsion: Two Detroit Diesel Series 60 engines each rated at 550 kW at 2,300 rpm and each driving a fixed-pitch propeller.

17 m CATAMARAN

Peter Gladding

This vessel was designed by Teknicraft of New Zealand for the Florida Keys Law Enforcement Agency.

Specifications

Length overall	17 m
Beam	6.3 m

Propulsion: Two MTU series 60 engines each rated at 600 kW at 2,300 rpm, driving Hamilton 362 water-jets.

16.8 m CATAMARAN

Expedition 5

Following on from the 19.7 m catamaran Expedition IV, in 2007 Hone Heke ordered a 16.8 m version for operation between Lahaina (Maui) and the island of Lanai, Hawaii.

Vessel type	Vessel name	Yard No	Length (m)	Speed (kt)	Seats	Vehicles	Originally delivered to	Date of build
20 m Catamaran	*New Golden Eye*	–	20.0	30.0	49	–	–	1999
17 m Catamaran	*IKE*	–	17.0	29.0	–	–	–	2000
20 m Catamaran	*MV Islander*	–	20.0	25.0	149	–	Channel Island Cruises	2001
20 m Catamaran	*Chilkat Express*	–	20.0	46.0	63	–	Chilkat Cruises	2001
23 m Catamaran	*Condor Express*	–	23.0	30.0	149	–	Condor, US	2002
19 m Catamaran	*Noaa Shearwater*	–	19.0	24.0	22	–	Noaa, US	2002
20 m Catamaran	*Island Adventure*	–	20.0	24.0	149	–	Superboats Inc	2003
19.7 m Catamaran	*Expeditions IV*	–	19.7	28.0	149	–	Hone Heke Corporation	2003
21.8 m Catamaran	*Spirit*	–	21.8	35.0	149	–	Four Seasons Marine	2004
Teknicraft 19.8 m Catamaran	*Fulmar*	–	19.8	22.0	24	–	Monterey Bay National Marine Sanctuary, US	2005
Teknicraft 15 m Catamaran	*Sbums*	–	15.0	24.0	14	–	Stellwagen Bank National Marine Sanctuary, US	2005
Teknicraft 17 m Catamaran	*Auk*	–	17.0	42.0	4	–	Florida Keys Law Enforcement, US	2005
Teknicraft 17 m Catamaran	*Hula Kai*	–	17.0	–	51	–	Fair Wind Cruises	2006
Teknicraft 20 m Catamaran	*Valdez Spirit*	–	26.0	25.0	151	–	Stan Stephens Cruises	2006
Teknicraft 17 m Catamaran	*Expeditions Five*	–	17.0	24.0	107	–	Home Henk Corporation	2006
Teknicraft 25 m Catamaran	*Aialik Voyager*	25	25.2	28.0	154	–	Kenai Fjords Tours	2007

Specifications

Length overall	16.8 m
Beam	6.5 m
Draft	1.68
Maximum speed	30 kt
Operational speed	24 kt

Classification: USCG Subchapter T.
Structure: Marine grade aluminium.
Propulsion: Two Caterpillar C-18 engines rated at 715 hp at 2,100 rpm.

15 m CATAMARAN

Auk

This vessel was designed for scientific research and is equipped with an a-frame and winch. The vessel will be operated by the Stellwagon Bank National Marine Sanctuary in the US.

Specifications

Length overall	15 m
Beam	5.5 m

Propulsion: Two Cummins QSC 8.3 M490 engines each rated at 370 kW at 2,600 rpm.

Allen Marine Inc

1512 Sawmill Creek Road, PO Box 1049, Sitka, Alaska 99835-1049, United States

Tel: (+1 907) 747 81 00
Fax: (+1 907) 747 81 99
e-mail: sale@allenmarine.com
Web: www.allenmarineinc.com

Ken Baker, *Project Manager*

30 m CATAMARAN

Alaskan Dream

Built in 5 months in 1988 for transport of miners to an island off Juneau, Alaska.

Specifications

Length overall	30.0 m
Displacement, max	72 t
Crew	3
Passengers	150
Operational speed	33 kt

Propulsion: Main engines: four Caterpillar 3412 560 kW at 2,100 rpm, driving four Hamilton 422 water-jet units, driven via a pneumatic clutch.

24 m CATAMARAN

St Gregory
Sit'Ku
St Herman
Keety
St Phillip
St Nicholas
Cape Aialik
Yogi Berra
Fiorello La Guardia
Frank Sinatra
Vince Lombardi
Giovanni de Verrazono
Fairweather Express II
Brooklyn
Frank R Lautenberg
Hobeken
Congressman Robert A Roe
Queens
Jersey City
Bayonne

Vessel type	Vessel name	Yard No	Length (m)	Speed (kt)	Seats	Vehicles	Originally delivered to	Date of build
30 m Catamaran	*Alaskan Dream*	4	30.0	33	150	none	Alaska	1988
20 m Catamaran	*St Tatiana*	9	20.0	26	110	none	Allen Marine Inc	1995
20 m Catamaran	*St Michael*	10	20.0	26	110	none	Allen Marine Inc	1995
20 m Catamaran	*St Eugene*	11	20.0	26	110	none	Allen Marine Inc	1995
24 m Catamaran	*St Gregory*	12	23.8	28	150	none	Allen Marine Inc	1996
24 m Catamaran	*Sit'Ku*	13	23.8	28	150	none	Allen Marine Inc	1996
24 m Catamaran	*St Herman*	16	23.8	28	150	none	Allen Marine Inc	1997
24 m Catamaran	*St Phillip*	15	23.8	28	150	none	Allen Marine Inc	1997
20 m Catamaran	*Majestic Fjord*	18	20.0	30	100	none	Allen Marine Inc	1998
24 m Catamaran	*Keet*	17	23.8	28	150	none	Allen Marine Inc	1998
24 m Catamaran	*St Nicholas*	19	23.8	28	150	none	Allen Marine Inc	1999
24 m Catamaran	*Cape Aialik*	20	23.8	28	150	none	Allen Marine Inc	1999
24 m Catamaran	*Yogi Berra*	21	23.8	28	150	none	NY Waterway	2000
24 m Catamaran	*Fiorello La Guardia*	22	23.8	28	150	none	NY Waterway	2000
24 m Catamaran	*Frank Sinatra*	23	23.8	28	150	none	NY Waterway	2000
24 m Catamaran	*Vince Lombardi*	–	23.8	28	150	none	NY Waterway	2000
24 m Catamaran	*Giovanni de Verrazono*	25	23.8	28	150	none	NY Waterway	2001
24 m Catamaran	*Fairweather Express II*	26	23.8	28	150	none	Chilkat Cruises	2001
24 m Catamaran	*Brooklyn*	31	23.8	28	150	none	NY Waterway	2002
24 m Catamaran	*Frank R Lautenberg*	32	23.8	28	150	none	NY Waterway	2002
24 m Catamaran	*Hoboken*	33	23.8	28	150	none	NY Waterway	2002
24 m Catamaran	*Congressman Robert A Roe*	36	23.8	28	150	none	NY Waterway	2003
24 m Catamaran	*Queens*	37	24.0	28	150	none	NY Waterway	2003
24 m Catamaran	*Jersey City*	38	24.0	28	150	none	NY Waterway	2003
24 m Catamaran	*Bayonne*	39	24.0	28	150	none	NY Waterway	2003
24 m Catamaran	*Admiral Richard E Bennis*	40	24.0	28	150	none	NY Waterway	2003
27 m Catamaran	*St Aquilina*	–	27.0	36	150	none	Allen Marine Tours	2004
24 m Catamaran	*St John*	48	23.9	28	149	none	Allen Marine Tours	2005
24 m Catamaran	*St Nona*	49	23.9	28	149	none	Allen Marine Tours	2005

Admiral Richard E Bennis
St John
St Nona

A total of 23 of these craft have been launched since 1996. The first 16 craft had a twin passenger deck arrangement whereas the next five have a single deck layout, designed to minimise loading and unloading times.

Specifications

Length overall	23.8 m
Beam	8.5 m
Draught	0.8 m
Crew	4
Passengers	150
Fuel capacity	8,500 litres
Water capacity	420 litres
Propulsive power	1,800 kW
Max speed	33 kt
Operational speed	28 kt

St Philip *at speed*
0038687

Alumacat Corporation

PO Box 1207, Calvert Marina Building B, Solomons, Maryland 20688, United States

Tel: (+1 800) 676 0653
Fax: (+1 410) 326 6067
e-mail: alumacat@tqci.net
Web: www.tqci.net/~alumacat

Randy Payne, Sr, *Owner*

Established in 1987, Alumacat recently completed the *Bay King III* for charter operation in Maryland.

Bay King III

Specifications

Length overall	25.9 m
Beam	8.23 m
Passengers	149
Operational speed	35 kt

Structure: Aluminium
Propulsion: Two 1100 hp MAN diesel engines each driving an Ultra 450 water jet.

Vessel type	Vessel name	Yard No	Length (m)	Speed (kt)	Seats	Originally delivered to	Date of build
CAT	*Bay King III*	–	25.9	35	149	Dale Scheibe, US	2006

Austal

Vessels built at Austal's United States shipyard

100 Dunlap Drive, Mobile, Alabama 36602, United States

Tel: (+1 334) 434 80 00
Fax: (+1 334) 434 80 80
e-mail: usasales@austal.com
Web: www.austal.com

Bob Browning, *Chief Executive Officer*

Austal USA was established in Mobile, Alabama, in December 1999. In partnership with leading US shipbuilder Bender Shipbuilding and Repair Co Inc, Austal's US joint venture makes available its market leading aluminium vessels to the US market and surrounding regions.

Austal's product range extends from fast passenger ferries and vehicle-passenger ferries to cruise vessels, patrol and offshore craft, luxury yachts and defence platforms. It is expected that the full product range will be built at the Mobile, Alabama shipyard. In October 2005, it was announced that Austal had been awarded a contract for the construction of a seaframe for *Flight* , the General Dynamics Littoral Combat Ship (LCS) for the US Navy. The design is based on the 127 m trimaran hull *Benchijigua Express* and construction commenced in 2006.

Lake Express during sea trials
0580421

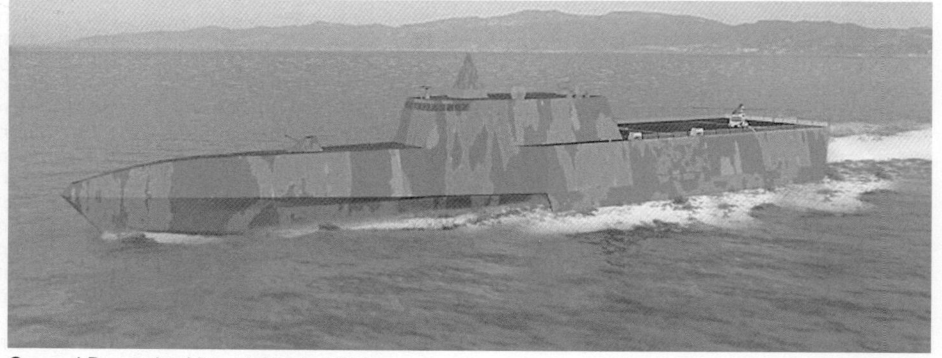

General Dynamics Littoral Combat Ship
1120269

LITTORAL COMBAT SHIP (LCS)

In 2002 the US Navy began a programme to design, develop and build a new class of small stealth ships for operation in littoral waters. In 2002, six concept studies were awarded, one of which was to the General Dynamics consortium which included Austal as builder. The consortium was successful in reaching the final stage, and construction of *Flight 0* began in 2006.

The General Dynamics LCS design features a high-speed trimaran hull, based on the successful *Benchijigua Express* passenger and vehicle ferry. The high stability provided by the trimaran hull will allow helicopter operations to take place in heavy sea states. Due to the wide beam of the vessel, a large flight deck is possible, allowing the use of two helicopters or multiple unmanned aeriel vehicles. The design will provide the largest payload volume of any US Navy surface combatant allowing the vessel to carry interchangeable mission modules.

The first General Dynamics vessel and any successive vessels will be constructed at Austal's Mobile shipyard in Alabama. Delivery of the vessel is scheduled for 2007.

Specifications

Length overall 127.1 m
Beam 30.4 m
Draught 4.5 m
Maximum speed 45 kt

Propulsion: CODAG arrangement of two GE LM2500 gas turbines and two diesel engines, driving four Kamewa waterjets.

AUTO EXPRESS 107

Alakai

This is the first of two 107 m vehicle-passenger catamarans being built for Hawaii Superferry. The second vessel is due for completion in early 2009. The vessels are intended for service between the islands of Honolulu, Maui and Kahului with a service speed of 38 knots.

Specifications

Length overall 106.5 m
Length waterline 92.4 m
Beam 23.8 m
Draught 3.65 m
Passengers 866
Vehicles 282 cars
Fuel capacity 215,000 l
Maximum speed 40 kt
Operational speed 38 kt

Classification: Germanischer Lloyd.

Propulsion: Four MTU 20V 8000 M70 engines each rated at 8,200 kW at 1150 rpm, each driving a Kamewa 125 SII water-jets via ZF 53 000 gearboxes.

AUTO EXPRESS 58

This vessel was delivered to Milwaukee-based operator Lake Express in 2004. It was the first high-speed vehicle-passenger vessel to operate in the US.

Lake Express

Specifications

Length overall 58.4 m
Length waterline 52.2 m
Beam 17.6 m
Draught 2.5 m
Passengers 248
Vehicles 46 cars
Fuel Capacity 53,000 litres
Maximum speed 34 kt

Propulsion: Four MTU 16 V 4000 M70 diesel engines each rated at 2,320 kw at 2,000 rpm and each driving a Kamewa 80 SII water-jet via a ZF 7500 gearbox.

43.5 m PASSENGER CATAMARAN

This vessel was delivered to Circle Line Statue of Liberty Ferry in 2003 for service between Lower Manhattan, Ellis Island and the Statue of Liberty.

Zephyr

Specifications

Length overall 43.5 m
Length waterline 37.4 m
Beam 11.5 m
Draught 1.4 m
Crew 6
Passengers 600
Fuel capacity 23,700 litres
Freshwater capacity 3,800 litres
Operational speed 30 kt

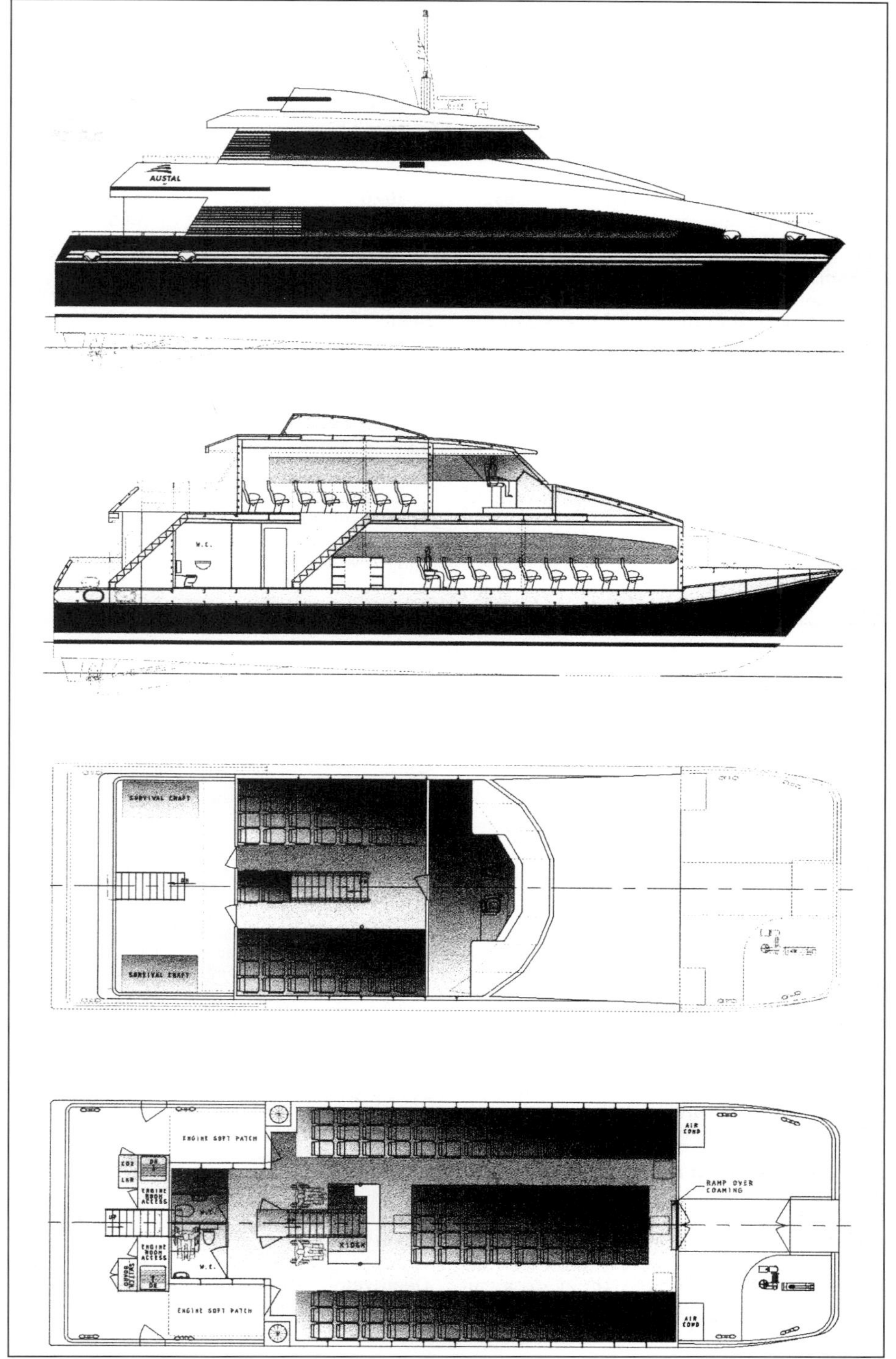

General arrangement of the 26 m Passenger Catamaran 0121749

Auto Express 107 *Alakai* 1120348

Vessel type	Vessel name	Yard No	Length (m)	Speed (kt)	Seats	Vehicles	Originally delivered to	Date of build
26 m Catamaran	*Patriot*	101	26.3	26	189	none	Circle Line, Statue of Liberty Ferry	2002
43.5 m CAT	*Zephyr*	611	43.5	29	600	none	Circle Line, Statue of Liberty Ferry	2003
AutoExpress 58	*Lake Express*	614	58.4	36	248	46	Lake Express	2004
Air Ride Craft 31 m SESS	–	612	31.2	50+	–	none	American Marine Holdings, US	2004
107 m Catamaran	*Alakai*	615	107.0	35	866	282	Hawaii Superferry	2007
107 m Catamaran	–	616	107.0	35	866	282	Hawaii Superferry	(2009)

Classification: US Coast Guard 46 CFR Subchapter K-Lakes, Bays and Sounds.
Propulsion: Four Cummins KTA 38-M2 diesel engines each rated at 895 kW at 1,800 rpm, each driving a Hamilton HM 571 water-jet via a Reintjes WVS 440 gearbox.

32 m SEACOASTER

Austal USA are currently building a 32 m version of American Marine Holdings' Seacoaster SES vessel. The project is funded by the ONR and once the ship completes trials it will be converted to a 149 passenger fast ferry.

Specifications

Length overall	31.2 m
Beam	9.75 m
Operational speed	56 kt

Propulsion: Four Caterpillar diesel engines each rated at 1,045 kW and each driving a ZF surface piercing propeller.

26.3 m PASSENGER CATAMARAN

This vessel was delivered in 2003 to Circle Line-Statue of Liberty Ferry.

Specifications

Length overall	26.3 m
Length waterline	24 m
Beam	8.0 m
Draught	1.75 m
Deadweight	18 t
Crew	12
Passengers	189
Fuel capacity	22,710 litres
Water capacity	800 litres
Operational speed	25 kt

Propulsion: Two MTU 16V 2000 M70 diesel engines each rated at 960 kW and driving a fixed-pitch propeller.

Blount Boats Inc

PO Box 368, 461 Water Street, Warren, Rhode Island 02885, United States

Tel: (+1 401) 245 83 00
Fax: (+1 401) 245 83 03
e-mail: info@blountboats.com
Web: www.blountboats.com

Luther H Blount, *President*

Blount Boats is a full service shipyard specialising in innovative boat design, engineering, craftsmanship and repair. Founded as Blount Marine Corporation in 1949 by Luther H Blount, a leading innovator in the shipbuilding industry, Blount Boats designs and builds steel, aluminium and composite vessels and to date has built over 300 vessels ranging in size from 5 to 80 m.

40 m Whale Watching Catamaran Atlanticat 1027300

40 m PASSENGER CATAMARAN

Atlanticat

In 2002 the company received an order from Arcadian Whale Adventures in Maine, US. The vessel, designed by Crowther Multihulls of Sydney, Australia is a three deck, all aluminium vessel equipped with a sonar system to detect whales and a ride control system.

Specifications

Length overall	39.8 m
Length waterline	32.6 m
Beam	11.0
Draught	1.4 m
Deadweight	51 t
Passengers	442
Fuel capacity	11,350 litres
Fresh water capacity	6,425 litres
Operational speed	35 kt

Propulsion: Four Cummins KTA 50-M2 diesel engines each rated at 1,340 kW and each driving a Hamilton HM 651 water-jet via a ZF 4600D gearbox. Generators are two Cummins Onan 956 kW.

Dakota Creek Industries

820 Fourth Street, PO Box 218, Anacortes, Washington 9822, US

Tel: (+1 360) 293 95 75
Fax: (+1 360) 293 64 32
e-mail: dci@fidalgo.net
Web: www.dakotacreek.com

Richard Nelson, *President*
David Longdale, *Sales and Marketing*

Dakota Creek Industries, founded in 1975, is a licensed builder of Australia's AMD catamaran designs and the company has delivered six such vessels to date.

AMD 360

Intintoli
Mare Island
Del Norte (360B)
Solano (360C)

Two AMD 360 craft were ordered for operation between the City of Vallejo and San Francisco, California. A third craft, *Del Norte* was ordered by the Golden Gate District Transportation Authority and is powered by four Detroit 16V 149 DDEC diesels driving MJP J550 water-jets via a Reintjes VLJ 730 gearbox giving a speed of 36 kt. This craft operates between Larkspur and San Francisco. A fourth vessel, *Solano*, was ordered by City of Vallejo in 2003 and specified with four MTU 16V 4000 M70 diesels driving Hamilton HM811 water-jets. This vessel was delivered in 2004.

Specifications

Length overall	41.5 m
Beam	12 m
Draught	1.5 m
Passengers	325
Fuel capacity	14,000 litres
Water capacity	3,800 litres
Max speed	38 kt
Operational speed	38 kt

Vessel type	Vessel name	Yard No	Length (m)	Speed (kt)	Seats	Vehicles	Originally delivered to	Date of build
AMD 360	*Intintoli*	28	41.5	34.0	325	–	Baylink Ferries	1997
AMD 360B	*Del Norte*	31	41.5	36.0	325	–	Golden Gate Transportation Authority	1997
AMD 360	*Mare Island*	29	41.5	34.0	325	–	Baylink Ferries	1997
AMD 385	*Chinook*	32	43.6	34.0	350	–	Washington State Ferries	1998
AMD 385	*Snohomish*	–	43.6	34.0	350	–	-	1999
AMD 360C	*Solano*	44	41.3	34.0	325	–	Baylink Ferries	2004

The first AMD 360 Intintoli *during trials* 0507109

AMD 385 0038698

Propulsion: The vessel is powered by two MTU 16V 396 TE 74L diesels each firing on Bird Johnson (MJP) 750 water-jet via a ZF BU 755 gearbox.

AMD 385

Chinook

Snohomish

Chinook was delivered to Washington State Ferries in 1998 and *Snohomish* followed in 1999.

Specifications

Length overall	43.6 m
Beam	12.0 m
Draught	1.5 m
Deadweight	44 t
Passengers	350
Fuel capacity	17,000 litres
Water capacity	1,100 litres
Propulsive power	5,370 kW
Max speed	40 kt
Operational speed	37 kt

Propulsion: Four Detroit 16V 149TI DDEC diesel engines each driving a Bird Johnson MJP650 water-jet via a Reintjes VLJ 730 gearbox.

Electrical system: Two Northern Lights 125 kW generators.

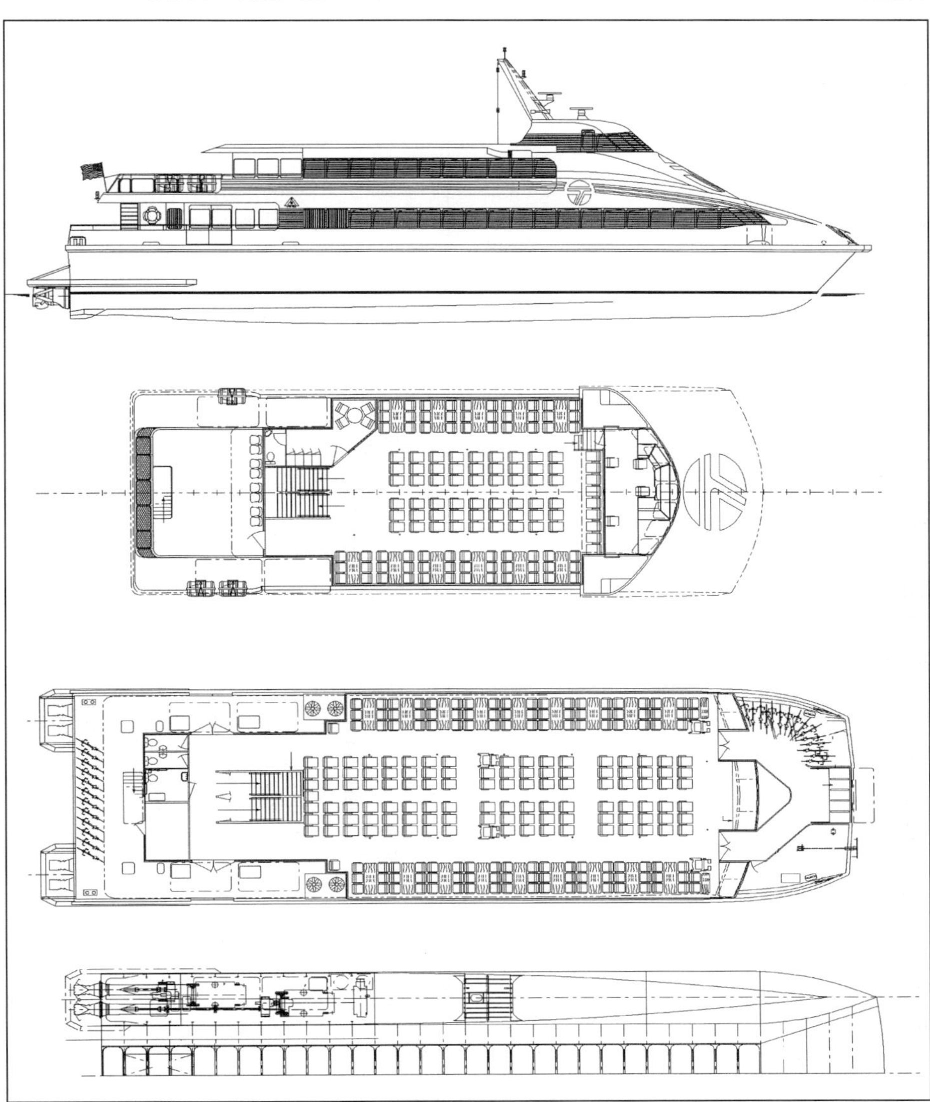

Dakota AMD 385 general arrangement 0038701

Derecktor Shipyards

311 East Boston Post Road, Mamaroneck, New York 10543, United States

Tel: (+1 914) 698 50 20
Fax: (+1 914) 698 46 41
e-mail: general@derecktor.com
Web: www.derecktor.com

Paul Derecktor, *President*
Krista Stanley, *Marketing Manager*

Primarily a builder of fast monohull luxury yachts, the company also builds a wide range of other craft. In 1997 two fast catamaran ferries for operation in New York Harbour were delivered. The company then delivered a larger 45 m gas-turbine propelled catamaran in 1998–99 and a 41 m fast catamaran for Woods Hole SSA in 2000. In early 2006 the company announced an order for a 38 m, 350 passenger commuter ferry for the Bermudan government, with delivery in June 2007. Designed and constructed to Lloyd's survey, the

Derecktor 38 m catamaran Finest 0507058

Ernest Hemmingway *(now* Patricia Olivia II*) on trials* 0045120

ferry will be powered by four MTU diesel engines with ZF gearboxes and Hamilton water-jets and will be an all aluminium construction. The company has two shipyards, one in Florida and one at the above address.

73 m CAR FERRY

Two 73 m fast passenger/car ferries were built for the Alaska Marine Highway System (AMHS) in 2004. The vessels were designed by Nigel Gee and Associates of the UK. *Fairweather* operates from the port of Juneau in south eastern Alaska and *Chenega* connects the ports of Valdez, Cordova and Whittier in south central Alaska.

Specifications

Length overall	73.4 m
Length waterline	63.8 m
Beam	18.0 m
Draught	2.7 m
Deadweight	200 t
Crew	10
Passengers	250
Vehicles	35 cars
Operational speed	35 kt

Classification: DNV +1A1 HSLC R3 EO.

Structure: All aluminium hull and superstructure.

Propulsion: Four MTU 16V 595TE 70L diesel engines each rated at 3,600 kW at 1,750 rpm and each driving a Kamewa 90 SII water-jet.

38 m PASSENGER CATAMARAN

Warbaby Fox

J.L. Cecil Smith

The Government of Bermuda ordered a BMT Nigel Gee and Associates 38 m catamaran to be delivered by Derecktors in June 2006. The 349 passenger, low wake catamaran has a resiliently mounted superstructure. A second vessel was ordered in January 2007 and delivered in September 2007.

Specifications

Length overall	37.8 m
Length waterline	34.0 m
Beam	9.0 m
Draught	1.4 m
Deadweight	34 t
Passengers	349
Crew	5
Operational speed	33 kt

Propulsion: Four MTU 12V 2000 M70 diesel engines, each delivering 788 kW to Hamilton HM 512 water-jets.

38 m PASSENGER CATAMARAN

Built for New York Fast Ferry and designed by N Gee and Associates Ltd, these 35 kt craft feature slender hulls in order to reduce wash generation at high and intermediate speeds. The first vessel, *Finest*, was launched in September 1996 and the second, *Bravest*, in October 1996.

Specifications

Length overall	37.7 m
Length waterline	33.0 m
Beam	10.0 m

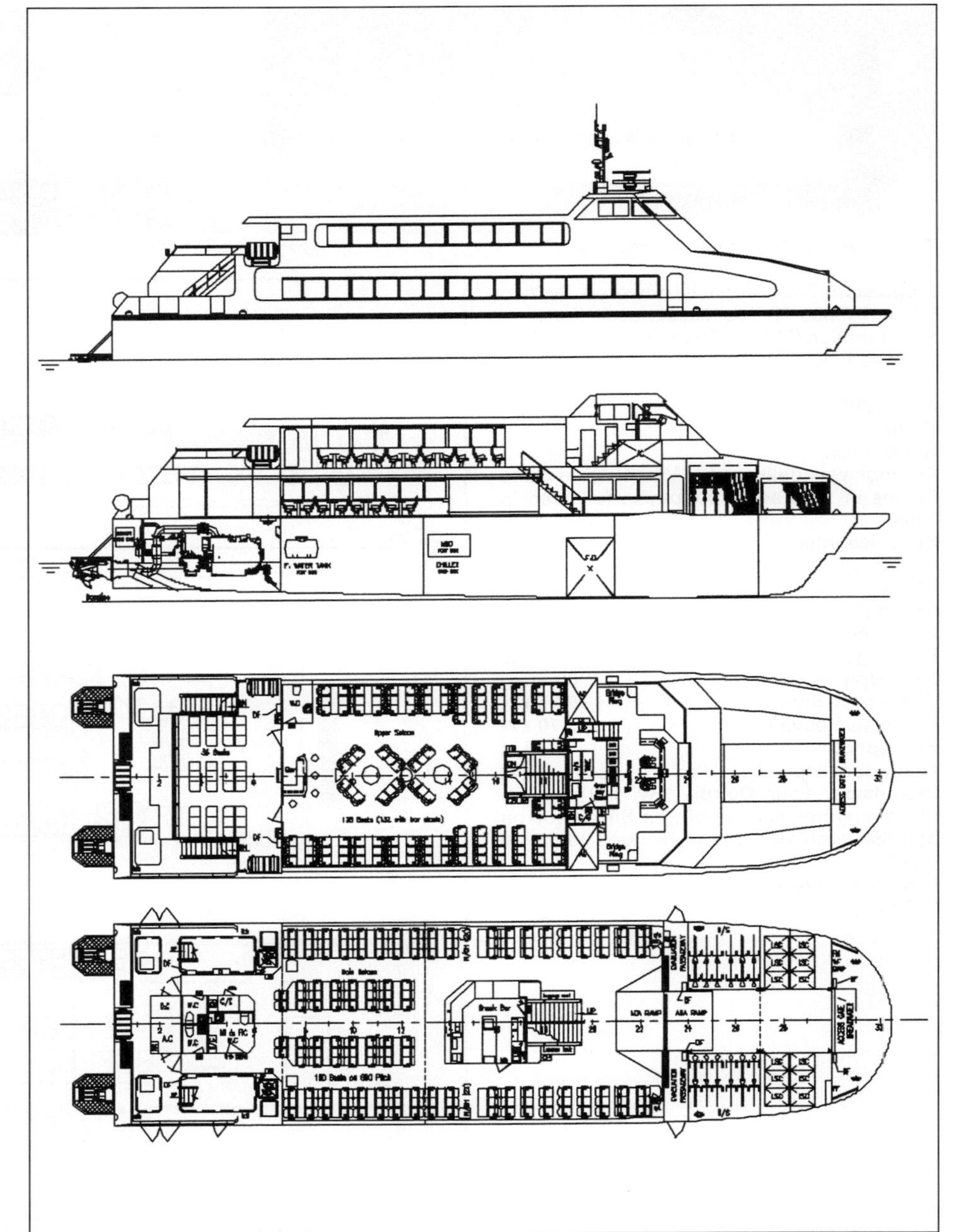

Derecktor 41 m catamaran general arrangement 0080470

Fairweather, *the 73 m fast car ferry for AMHS, Alaska* 1114409

Flying Cloud *on trials* 0098001

16 m water taxi Mickey Murphy 0593078

Draught	1.6 m
Displacement, max	140 t
Passengers	350
Propulsive power	4,000 kW
Operational speed	35 kt

Classification: American Bureau of Shipping.
Structure: All-aluminium hull and superstructure.
Propulsion: Two MTU 16V 396TE 74L diesels, one in each hull, each rated at 2,000 kW at 1,900 rpm and driving an MJP J650R-DD water-jet via a ZF BW 755 reduction gearbox.

41 m PASSENGER CATAMARAN

Flying Cloud

This vessel was ordered by the Woods Hole Steamship Authority and was delivered in May 2000. Based on the 38 m design, Nigel Gee and Associates also designed this vessel which is to be operated to Martha's Vineyard in Massachusetts, US.

Specifications

Length overall	41.0 m
Length waterline	35.2 m
Beam	10.5 m
Draught	1.5 m
Displacement, max	170 t
Payload	38.5 t
Passengers	300
Propulsive power	5,000 kW
Operational speed	36 kt

Structure: All aluminium hull and superstructure.
Propulsion: Two Paxman 12V P185 diesel engines each driving a Kamewa 71 SII water-jet via a reverse reduction gearbox.

Vessel type	Vessel name	Yard No	Length (m)	Speed (kt)	Seats	Vehicles	Originally delivered to	Date of build
38 m Passenger Catamaran	*Finest*	156	37.7	35.0	350	none	New York Fast Ferry	September 1996
38 m Passenger Catamaran	*Bravest*	157	37.7	35.0	350	none	New York Fast Ferry	October 1996
45 m Fast Passenger Ferry	*Patricia Olivia II* (ex-*Ernest Hemmingway*)	161	45.6	53.0	304	none	Buquebus	1998
41 m Passenger Catamaran	*Flying Cloud*	162	41.0	36.0	300	none	Woods Hole Steamship Authority	May 2000
16 m Water Taxi	*Mickey Murphy*	–	16.3	25.0	54	none	New York Water Taxi	2002
16 m Water Taxi	*Mickey Mann*	–	16.3	25.0	54	none	New York Water Taxi	2002
16 m Water Taxi	*Curt Berger*	–	16.3	25.0	54	none	New York Water Taxi	2002
16 m Water Taxi	*Schyler Meyer Jr*	–	16.3	25.0	54	none	New York Water Taxi	2003
16 m Water Taxi	*John Keith*	–	16.3	25.0	54	none	New York Water Taxi	2003
16 m Water Taxi	*Ed Rogowsky*	–	16.3	25.0	54	none	New York Water Taxi	2003
73 m Car Ferry	*Fairweather*	–	73.4	35.0	250	35 cars	AMHS, Alaska	2004
73 m Car Ferry	*Chenega*	–	73.4	35.0	250	35 cars	AMHS, Alaska	2004
Crowther 28 m Catamaran	*Whaling City Express*	–	28.0	30.5	186	none	New England Fast Ferry	2004
Crowther 28 m Catamaran	*Martha's Vineyard Express*	–	28.0	30.5	186	none	New England Fast Ferry	2005
BMT Nigel Gee 38 m Catamaran	*Warbaby Fox*	–	37.8	33.0	349	none	Government of Bermuda	2006
BMT Nigel Gee 38 m Catamaran	*J L Cecil Smith*	–	37.8	33.0	349	none	Government of Bermuda	2007

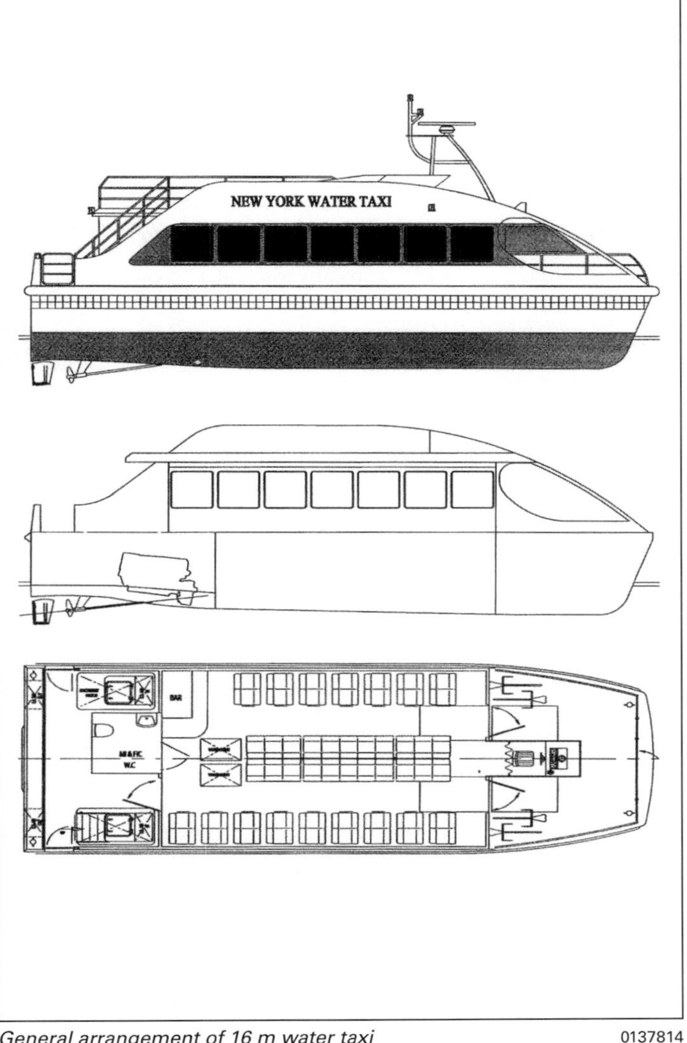

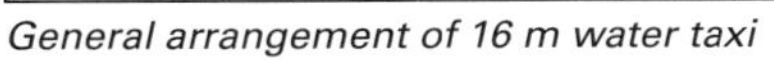
General arrangement of 16 m water taxi 0137814

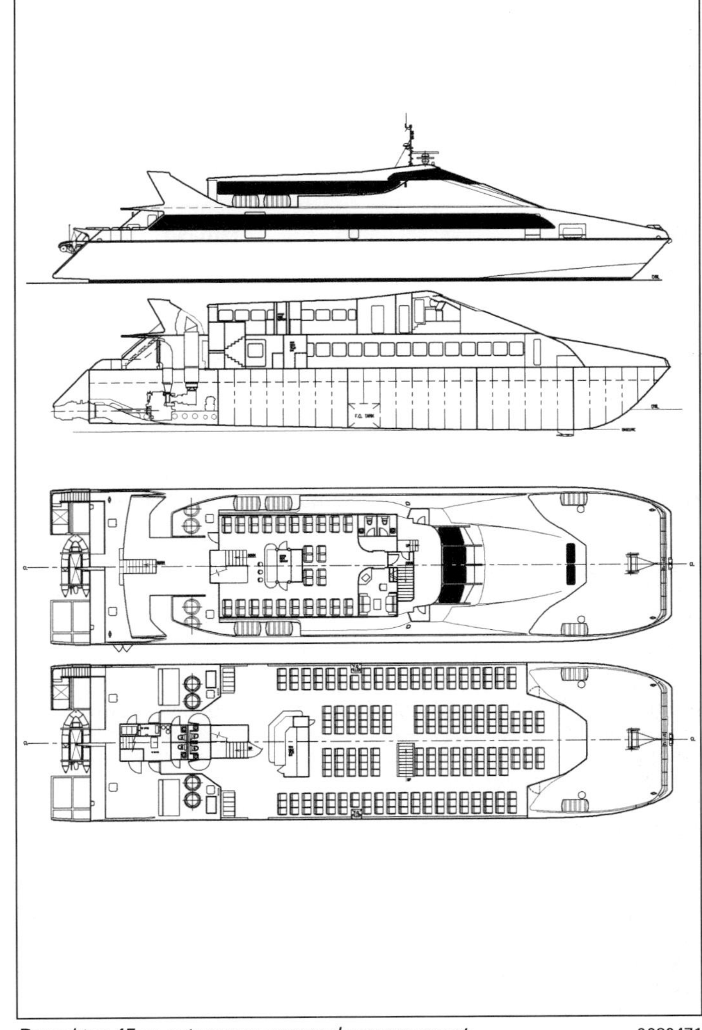
Derecktor 45 m catamaran general arrangement 0080471

45 m FAST PASSENGER CATAMARAN

***Patricia Olivia II* (ex-*Ernest Hemmingway*)**

Derecktor Shipyards delivered this particularly fast passenger catamaran to Buquebus in Argentina in 1998. Designed by Nigel Gee and Associates, the vessel operates from Uruguay.

Specifications

Length overall	45.6 m
Length waterline	40.2 m
Beam	11.8 m
Draught	1.5 m
Displacement	202 t
Crew	12
Passengers	304
Propulsive power	11,940 kW
Max speed	57 kt
Operational speed	53 kt

Classification: American Bureau of Shipping.

Propulsion: Powered by two AlliedSignal TF 80 gas turbines each rated at 5,970 kW and each driving an MJP 950 water-jet.

28 m CATAMARAN

Whaling City Express

Martha's Vineyard Express

Dereckor Shipyards built the superstructures and fitted out two Crowther-designed catamarans for US operator New England Fast Ferry. The hulls were built by Seattle-based Kvichak Marine.

The vessels carry 186 passengers at 30.5 kt powered by twin MTU 16V 2000 diesels.

16 m WATER TAXI

Six of these were ordered in 2002 from New York Water Taxi for operation throughout New York Harbour.

Specifications

Length overall	16.3 m
Length waterline	15.8 m
Beam	5.6 m
Draught	1.0 m
Passengers	54
Maximum speed	24 kt

Classification: USCG Subchapter T, Rivers.

Construction: All aluminium hull and superstructure.

Propulsion: Two Detroit Diesel Series 60 engines each rated at 335 kW and each driving a fixed-pitch propeller.

Gambol Industries

1825 Pier D Street, Long Beach, California 90802, United States

Tel: (+1 562) 901 2470
Fax: (+1 562) 901 2472
e-mail: jeffopsuper@earthlink.net
Web: www.gambolindustries.com

Founded in 1992, Gambol Industries specialise in building, servicing and refitting yachts up to 50 m in length. In 2006 Semisub Inc ordered a 26 m semi submersible catamaran from Maillion Associates design to be built by Gambol.

26 m SEMI SUBMERSIBLE CATAMARAN

Ordered by Semisub Inc this vessel, featuring large windows, will be used for reef cruises. This novel catamaran has large ballast tanks that allow it to increase its draft significantly upon reaching a reef.

Specifications

Length overall	25.9 m
Beam	7.6 m
Passengers	125
Operational speed	40 kts

Propulsion: Two Caterpillar C32C diesel engines delivering 1,045 kW each to Hamilton water-jets.

Geo Shipyards Inc

4817 South Lewis Street, PO Box 9622, New Iberia, Louisiana 70562-9622, USA

Tel: (+1 877) 991 6137
Fax: (+1 337) 364 7493
e-mail: information@geoshipyard.com
Web: www.geoshipyard.com

Geo Shipyards was established in 1979 and specialises in commercial aluminium boats for passenger tours, fishing, crewboat and survey duties.

25 m FOIL ASSISTED CATAMARAN

Catalina Adventure

Designed by Viking Fast Craft Solutions, in conjuction with South Africa's Foil Assisted Ship Technologies, Catalina Adventure is a fast ferry built for Pacific Adventure Cruises for operation in the Marina del Rey to Catalina Island region near Los Angeles. The foil system consists of a fixed foil forward and an adjustable aft trim foil.

Specifications

Length overall	24.7 m
Length waterline	22.0 m
Beam	7.4 m
Draught	1.2 m
Passengers	199
Fuel capacity	5675 litres
Fresh water capacity	1500 litres
Operational speed	29 kt

Classification: USCG Subchapter T.

Structure: Aluminium.

Propulsion: Two Caterpillar C32 diesel engines each rated at 1045 kW at 2100 rpm and each driving a five-bladed fixed propeller via ZF2550 gearboxes.

Vessel type	Vessel name	Yard No	Length (m)	Speed (kt)	Seats	Vehicles	Originally delivered to	Date of build
25 m Foil Assisted Catamaran	*Catalina Adventure*	–	24.7	29	199	none	Pacific Adventure Cruises	2007

Gladding-Hearn Shipbuilding

PO Box 300, One Riverside Avenue, Somerset, Massachusetts 02726-0300, United States

Tel: (+1 508) 676 85 96
Fax: (+1 508) 672 18 73
e-mail: sales@gladding-hearn.com
Web: www.gladding-hearn.com

George R Duclos, *Chairman and CEO*
John F Duclos, *President, Director of Operations*
Peter J Duclos, *President, Director of Business Development*

This shipyard is a member of the Duclos Corporation and is well known for its range of commercial workboat and fast ferry craft. The yard has built a large number of small pilot craft as well as its well-known InCat catamaran ferries, (it is a licensed builder of International Catamarans Designs Pty Ltd, Australia).

47 m CATAMARAN

Iyanough

This 47 m, 350 passenger fast catamaran was delivered to Woods Hole, Martha's Vineyard and Nantucket Steamship Authority in December 2006. The vessel's arrangement is optimised for bow loading to match existing ferries and the propulsion system was designed to allow the vessel to stay on schedule, even with one engine shut down.

Specifications

Length overall	46.54 m
Length waterline	41.07
Beam	11.9 m
Draught	1.62 m
Deadweight	51 t
Passengers	389
Fuel capacity	14,000 litres
Water capacity	950 litres
Operational speed	38 kt

Classification: USCG sub chapter 1C.

Construction: Marine grade aluminium.

Propulsion: Four MTU 12V 4000 M70 diesel engines each rated at 1,688 kW at 2,000 rpm and each driving a Hamilton HM 721 water-jet via a ZF BU 4650 gearbox.

Ride control: MDI Active Interceptors.

44 m CATAMARAN

Grey Lady

Delivered in 2002, this was the third catamaran ordered from the company by Hy-Line Cruises.

Specifications

Length overall	43.7 m
Length waterline	39.1 m
Beam	10.6 m
Draught	2.0 m
Passengers	312
Fuel capacity	14,000 litres
Operational speed	36.6 kt

Propulsion: Four Cummins KTA 50 M2 diesel engines each rated at 1,350 kW and driving a Hamilton HM 651 water-jet.

44 m CATAMARAN

Salacia

This vessel was ordered by Boston Harbor Cruises and delivered in 2000.

Specifications

Length overall	44.65 m
Length waterline	41.10 m
Beam	11.9 m
Draught	1.8 m
Deadweight	42 t
Passengers	417
Fuel capacity	17,000 litres
Operational speed	35 kt
Range	425 n miles

Propulsion: Four Caterpillar 3512 B diesel engines each rated at 1,450 kW and each driving a Kamewa A56 water-jet via a ZF BW 465D gearbox.

43 m CATAMARAN

Seastreak New York
Seastreak New Jersey
Seastreak Wall Street
Seastreak Highlands

Two 43 m catamaran vessels were ordered in 2000 for operation by Seastreak of New York. The first vessel was delivered in 2001 and operates on the Highland to Lower Manhattan route.

Specifications

Length overall	42.9 m
Length waterline	37.9 m
Beam	10.4 m
Draught	2.0 m
Crew	7
Passengers	384
Fuel capacity	10,600 litres
Water capacity	1,100 litres
Max speed	41 kt
Operational speed	39 kt

Propulsion: Four Cummins KTA 50 M2 diesel engines each rated at 1,400 kW at 1,950 rpm driving a Kamewa A50 water-jet via a Reintjes VLJ 730D gearbox.

37 m CATAMARAN

Millennium
Nora Vittoria
Aurora

Delivered to Boston Harbor Cruises in 1998 and 1999 for commuter and whale watching services.

Specifications

Length overall	37 m
Length waterline	34 m
Beam	10.2 m
Draught	1.5 m
Crew	4
Passengers	350
Max speed	38 kt
Operational speed	34 kt

Structure: All-aluminium hull and superstructure.

Propulsion: Powered by four Cummins diesel engines each rated at 970 kW and each driving an MJP J500S-DD Bird-Johnson water-jet.

34 m CATAMARAN

Voyager III

Delivered in mid-1999 to the New England Aquarium, the vessel is fitted out for 325 passengers, has an operational speed of 31 kt and is used for whale watching activities.

Specifications

Length overall	34.1 m
Length waterline	30.1 m
Beam	9.2 m
Draught	1.6 m
Crew	5
Passengers	325
Max speed	35 kt
Operational speed	31 kt

Structure: All-aluminium hull and superstructure.

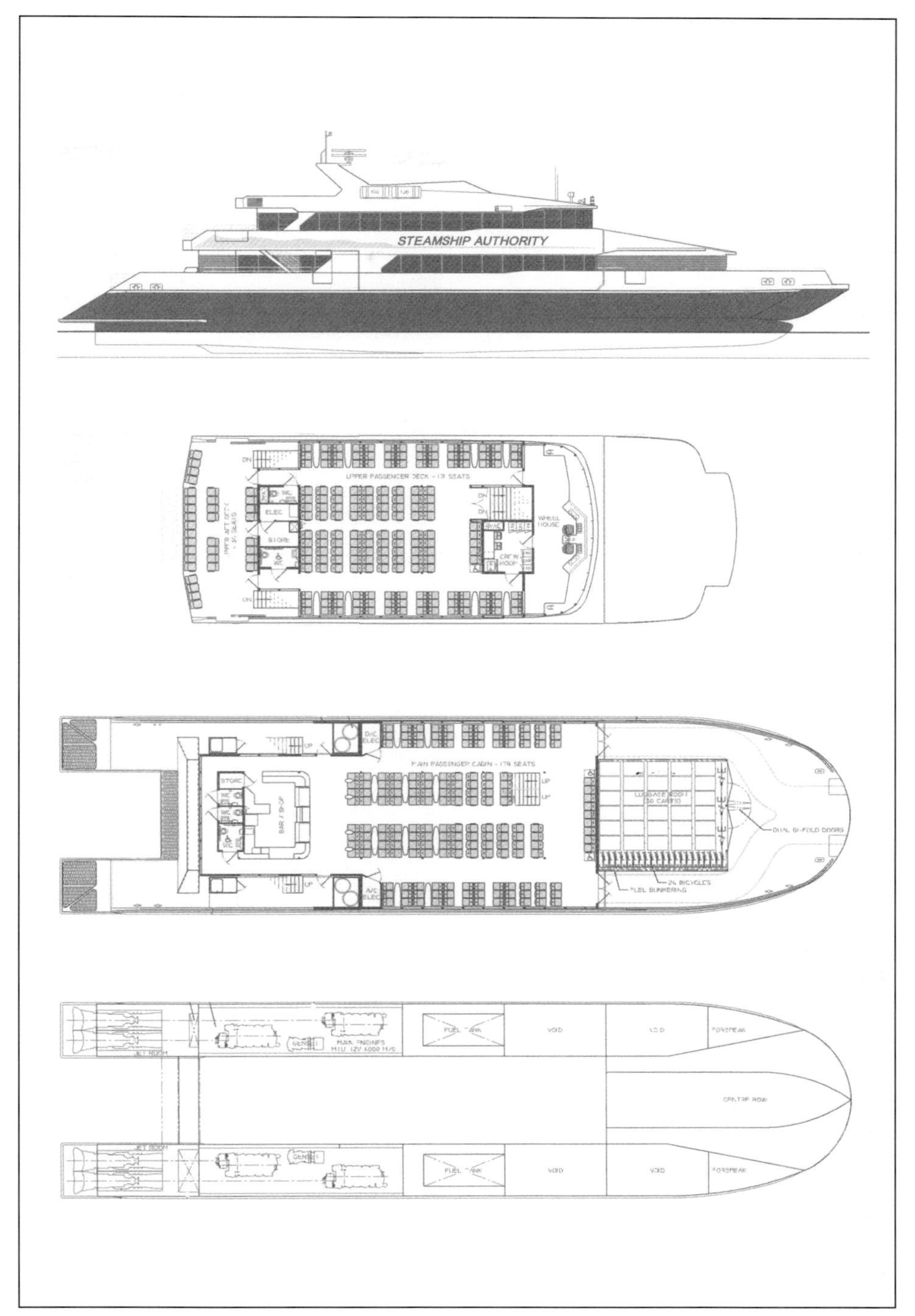

47 m catamaran general arrangement  1120245

43 m Seastreak New York 0129587

Propulsion: Four MTU/DDEC 12V 2000 M70 diesel engines each rated at 700 kW and each driving a Hamilton HM461 water-jet.

34 m CATAMARAN

Lady Martha (ex-Grey Lady II)

This InCat-designed vessel was delivered to Hy-Line Cruises in November 1997 for service to Nantucket Island. The vessel was repowered and renamed in 2005 for service to Martha's Vineyard.

Specifications

Length overall	32.4 m
Beam	9.2 m
Draught	1.25 m
Crew	3
Passengers	149
Max speed	38 kt
Operational speed	35 kt

Structure: All-aluminium hull and superstructure.

Propulsion: Powered by four Detroit diesel 16V 92TA DDEC engines each driving an MJPJ450-QD water-jet via a ZFBW250 reduction gearbox.

32 m CATAMARAN

Friendship V

This all-aluminium vessel delivered in May 1996, is certified to carry up to 325 passengers with a fully loaded speed of 29 kt and a maximum speed of 35 kt. The vessel is used for whale watching tours in Florida.

Specifications

Length overall	32.0 m
Length waterline	28.4 m
Beam	8.9 m
Hull beam	2.7 m
Draught	1.3 m
Passengers	325
Fuel capacity	10,000 litres
Water capacity	890 litres
Propulsive power	2,430 kW

Propulsion: Powered by four high-speed diesel engines each rated at 608 kW and each driving a water-jet.

30 m CATAMARAN

Provincetown

This vessel was built to an InCat design to operate on a service between Boston and Cape Cod, US.

Specifications

Length overall	29.74 m
Length waterline	28.8 m
Beam	9.1 m
Draught	1.85 m
Passengers	149
Operational speed	30 kt

Propulsion: Two Cummins KTA 38-M2 each rated at 1,003 kW at 195 rpm driving five-bladed fixed propellers via ZF 2550 gearboxes.

30 m CATAMARAN

Jet Cat Express

Delivered on 15 April 1991 for service with Catalina Channel Express.

Specifications

Length overall	31.2 m
Beam	8.7 m
Draught	1.0 m
Crew	4
Passengers	368
Fuel capacity	7,570 litres
Water capacity	757 litres
Propulsive power	2 × 1,298 kW
Operational speed	31 kt

Propulsion: Main engines are two DDC 16V 149TA, 1,298 kW each; driving two Kamewa 63 S2 water-jet units; via two ZF BUK485 gearboxes.

30 m CATAMARAN

Designed by InCat Crowther, two vessels are expected to be delivered in Summer 2006.

44 m Salacia 0129597

Provincetown III 1139552

44 m Grey Lady (Gladding-Hearn Shipbuilding) 1120260

Specifications

Length overall	29.8 m
Length waterline	27.7 m
Beam	9.15 m
Draught	1.68 m
Passengers	285

Propulsion: Two Caterpillar C32 diesel engines.

30 m CATAMARAN

Jet Express II

This craft is powered by two 1,298 kW MWM diesel engines driving Kamewa water-jets and can reach 32 kt fully loaded. The draught is 1 m.

29 m CATAMARAN Z-BOW

Athena

This vessel entered service in July 2001 with Hi-Speed Ferry, operating between Galilee and Block Island at Rhode Island, US.

Specifications

Length overall	30.4 m
Length waterline	27.2 m
Beam	9.4 m
Hull beam	2.8 m
Draught	2.0 m
Deadweight	29 t
Passengers	256
Fuel capacity	7,600 litres
Water capacity	1,100 litres

26 m catamarans Serenity *and* Resolute (Roland Skinner) 1120261

Vessel type	Vessel name	Yard No	Length (m)	Speed (kt)	Seats	Vehicles	Originally delivered to	Date of build
InCat 24 m	*Mackinac Express*	263	25.2	31.0	365	none	Arnold Transit Company	1987
InCat 24 m	*Express II* (ex-*Vineyard Spray*)	266	25.2	31.0	300	none	Hydro Lines Inc	September 1988
InCat 24 m	*Island Express*	267	25.2	31.0	300	none	Arnold Transit Company	1988
InCat 24 m	*Express I*	268	25.2	31.0	260	none	Hydro Lines Inc	February 1989
InCat 30 m	*Jet Express*	270	28.2	–	380	none	Put-in-Bay Boat Line Co	May 1989
InCat 37 m	*Nora Vittoria*	318	37.0	34.0	350	none	Boston Harbour Cruises	October 1998
InCat 31 m	*Victoria Clipper III* (ex-*Audubon Express*)	272	31.1	–	368	none	Clipper Navigation Inc	1990
InCat 30 m	*Jet Cat Express*	273	31.2	31.0	368	none	Catalina Channel Express	15 April 1991
InCat 30 m	*Jet Express II*	281	30.0	–	395	none	Put-in-Bay Boat Line Co	May 1992
InCat 28 m Z bow	*Friendship IV*	292	28.0	25.0	149	none	Bar Harbour Whale Watch	June 1994
InCat 28.2 m	*Theodore Roosevelt* (ex-*Grey Lady*)	301	28.2	30.0	89	none	Hyannis Harbour Tours	November 1995
InCat 39 m	*Friendship V*	303	32.0	–	325	none	Bar Harbour Whale Watch	June 1996
InCat XP300	*Flying Cloud*	307	26.8	25.0	149	none	Water Transport, Boston	November 1996
InCat XP300	*Lightning*	308	26.8	25.0	149	none	Water Transport, Boston	December 1996
InCat 34 m	*Lady Martha* (ex- *Grey Lady II*)	301	32.4	35.0	149	none	NYWW	November 1997
InCat 37 m	*Millenium*	317	37.0	34.0	350	none	Boston Harbour Cruises	August 1998
InCat 34 m	*Voyager III*	320	34.1	31.0	325	none	New England Aquarium	1999
InCat 28 m Z bow	*Yankee Freedom II*	–	28.0	25.0	250	none	Yankee Roamer	1999
InCat 37 m	*Aurora*	323	37.0	34.0	350	none	Boston Harbour Cruises	1999
InCat 44 m	*Salacia*	326	43.5	35.0	500	none	Boston Harbour Cruises	2000
InCat 29 m Z bow	*Athena*	330	30.4	34.0	256	none	Island Hi-Speed Ferry	2001
InCat 43 m	*Seastreak New York*	327	42.9	39.0	384	none	Seastreak	2001
InCat 43 m	*Seastreak New Jersey*	328	42.9	39.0	384	none	Seastreak	2002
InCat 26 m	*Serenity*	331	25.0	–	–	none	Bermuda Government	2002
InCat 26 m	*Resolute*	332	25.0	–	–	none	Bermuda Government	2002
InCat 27 m	*Jet Cat Express III*	337	25.0	35.0	149	none	Put-in-Bay Boat Line	2002
InCat 44 m	*Grey Lady*	339	44.0	38.0	350	none	Hy-Line Cruises	2003
InCat 39 m	*Millenium*	340	39.0	38.0	350	none	Boston Harbor Cruises	2003
InCat 39 m	–	341	39.0	38.0	350	none	Boston Harbor Cruises	2003
InCat 43 m	*Seastreak Wall Street*	342	42.9	39.0	384	none	Seastreak	2003
InCat 43 m	*Seastreak Highlands*	343	42.9	39.0	384	none	Seastreak	2004
InCat 30 m	*Provincetown III*	349	30.4	33.0	250	none	Bay State Cruise Company	2004
InCat 22 m	*Seymour B Durst*	351	22.0	26.0	147	none	New York Water Taxi	2005
InCat 22 m	*Sam Holmes*	352	22.0	26.0	147	none	New York Water Taxi	2005
Crowther 47 m	*Iyanough*	357	46.0	37.0	452	none	Woods hole, Marthas Vineyard and Nantucket Steamship Authority	2006
InCat Crowther 22 m	*Ed Rogowsky*	360	22.0	26.0	147	none	New York Water Taxi	(2006)
InCat Crowther 22 m	*Marian S Heiskell*	–	22.0	26.0	147	none	New York Water Taxi	(2006)
InCat Crowther 35 m Catamaran	–	–	35.0	–	199	none	San Francisco Bay Area Water Transit Authority	(2008)

Operational speed	34 kt
Range	300 n miles

Propulsion: Four Caterpillar 3412E diesel engines each rated at 820 kW and driving a Hamilton HM 521 water-jet.

28.2 m CATAMARAN

This vessel, delivered in November 1995, operates a fast ferry service between Hyannis and Nantucket taking less than 1 hour.

Specifications

Length overall	28.2 m
Length waterline	23.6 m
Beam	8.7 m
Hull beam	2.7 m
Draught	1.0 m
Crew	2
Passengers	89
Fuel capacity	5,000 litres
Water capacity	350 litres
Operational speed	30 kt

Propulsion: Powered by two DDEC 16V 92 diesel engines each driving an MJP 550 water-jet via a reduction gearbox.

28 m CATAMARAN Z-BOW

Friendship IV

Yankee Freedom II

Friendship IV is designed and certified to carry up to 149 passengers for distances of up to 100 miles offshore. Operating from Bar Harbor Maine, the vessel carries passengers on whale watching/sightseeing excursions which last for approximately 3 hours. The craft was designed by InCat in 1993 and delivered by Gladding-Hearn in June 1994. *Yankee Freedom II* has a similar specification with an increased passenger capacity of 250, and was delivered to Yankee Roamer of Florida in 1999.

Specifications

Length overall	28.0 m
Beam	9.0 m
Draught	1.7 m
Passengers	149–250
Fuel capacity	6,000 litres
Max speed	27 kt
Operational speed	25 kt

Propulsion: *Friendship IV* is powered by two 600 kW diesel engines each driving a fixed-pitch propeller. *Yankee Freedom II* has a similar arrangement using two Caterpillar 3412C diesels each rated at 780 kW.

27 m CATAMARAN

Jet Express III

Delivered in July 2002 to Put-in-Bay Boat Line in Lake Erie, US.

Specifications

Length overall	26.6 m
Length waterline	23.2 m
Beam	8.4 m
Hull beam	2.3 m
Draught	1.2 m
Passengers	149
Fuel capacity	5,500 litres
Operational speed	30 kt

Propulsion: Two Caterpillar 3412E diesel engines each rated at 820 kW at 2,300 rpm and each driving a Kamewa FF 550 water-jet.

XP 300 STANDARD CATAMARAN

Flying Cloud

Lightning

This catamaran design is arranged for series production. The hulls are manufactured in epoxy FRP and the deck and superstructures in aluminium. The ferries are certified by the USCG for coastal operations.

Specifications

Length overall	26.8 m
Beam	8.4 m

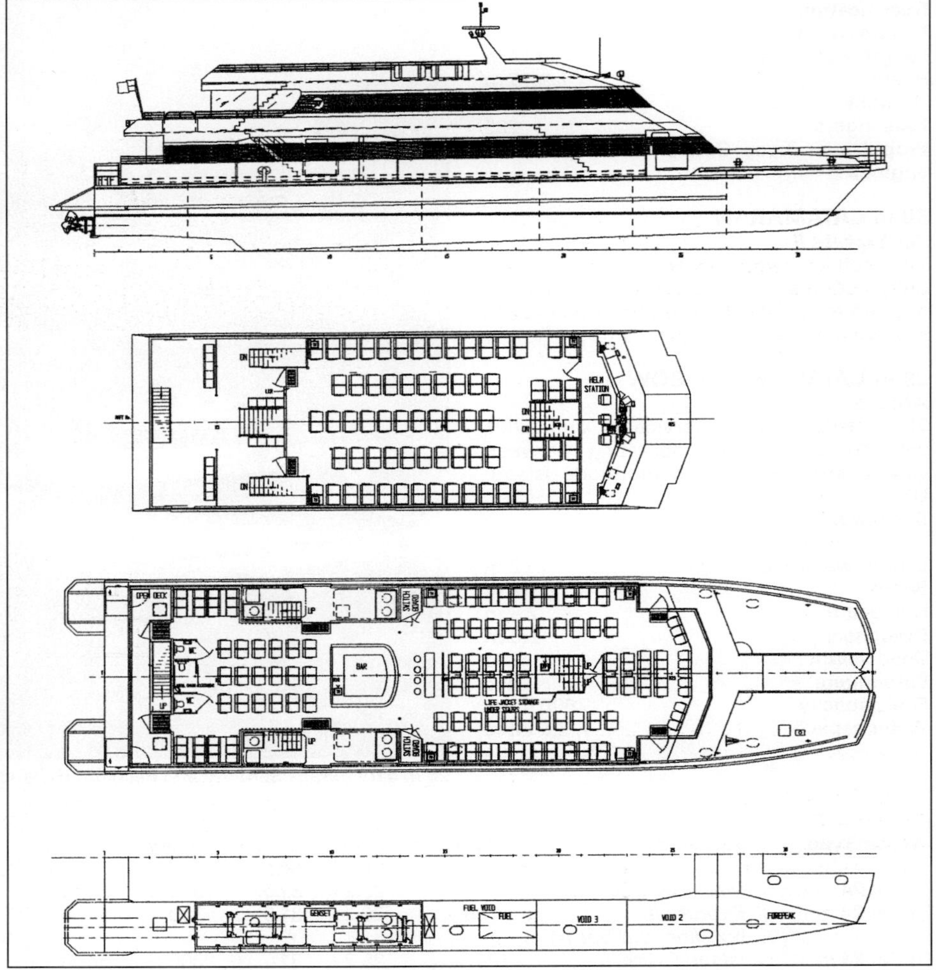

General arrangement of InCat 43 m Seastreak New York 0121759

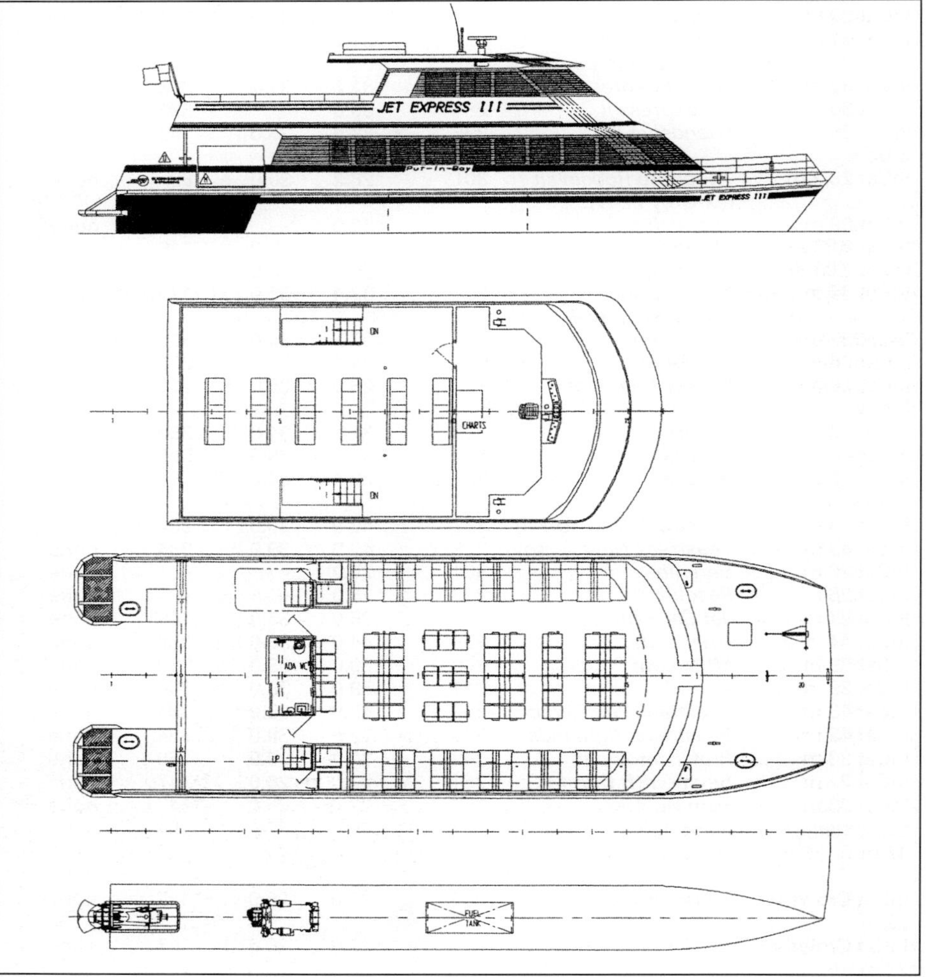

General arrangement of Jet Express III 0526483

Draught 0.9 m
Crew 3
Passengers 149
Fuel capacity 3,000 litres
Propulsive power 1,800 kW
Max speed 26 kt
Operational speed 25 kt

Structure: Epoxy FRP hulls and aluminium superstructure.

Propulsion: These vessels are powered by two DDEC 16V 92TA diesels driving via reverse reduction gearboxes to MJP J550S-DD water-jets. Each engine is rated at 900 kW at 2,100 rpm. The craft have been designed to accommodate larger engines and/or propellers for alternative design speeds.

26 m CATAMARAN

Serenity
Resolute

These harbour ferries were delivered to the Government of Bermuda in 2002.

Specifications

Length overall 25.7 m
Beam 9.5 m
Draught 1.9 m
Passengers 250
Fuel capacity 5,100 litres
Operational speed 25 kt
Range 345 n miles

Propulsion: Two MTU Detroit Diesel 12V 2000 diesel engines each rated at 720 kW.

24 m CATAMARAN

Mackinac Express
Island Express

Two InCat catamaran ferries were built for the Arnold Transit Company for service from the Upper and Lower Peninsulas of Michigan to the resort island in the Mackinac Straits. The vessels are powered by two MWM TBD 604B V8 diesel engines.

Specifications

Length overall 25.2 m
Length waterline 21.5 m
Beam 8.7 m
Hull beam 2.5 m
Draught 2.1 m
Passengers 365 (*Mackinac Express*)
300 (*Island Express*)
Operational speed 31 kt

24 m CATAMARAN

Express II
(***ex-Vineyard Spray***)

Delivered September 1988.

Specifications

Length overall 25.2 m
Length waterline 21.5 m
Beam 8.7 m
Hull beam 2.5 m
Draught 2.1 m
Passengers 300
Fuel capacity 3,974 litres
Propulsive power 2 × 1,290 kW
Operational speed 31 kt

Propulsion: Main engines are two MWM 1,290 kW diesels.

Electrical system: Two 35 kW Lister generators.

22 m CATAMARAN

Seymour B Durst
Sam Holmes
Ed Rogowsky
Marian S Heiskell

Four InCat designed 22 m catamarans have been built for the New York Water Taxi service. The vessels have been designed for operation in New York Harbour with low wake and sound levels and are also optimised for bow loading and quick turn arounds.

Specifications

Length overall 22 m
Beam 8.3 m

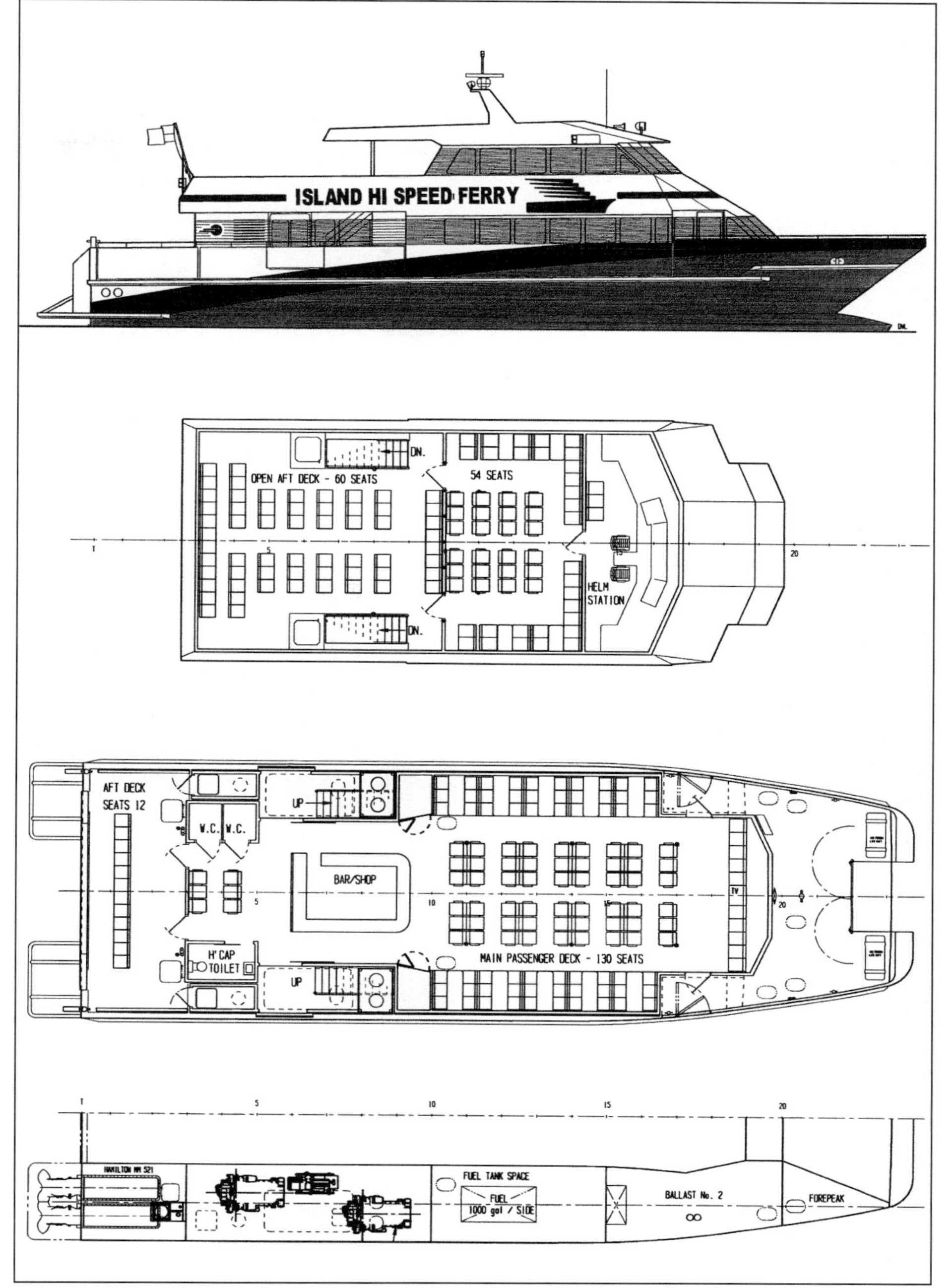

General arrangement of 29 m catamaran Athena 0121754

22 m New York water taxi Seymour B Durst (G Justin Zizes) 1120262

Draught 1.3 m
Passengers 147
Operational speed 26 kt
Propulsion: Two Cummins QSK 19M diesel engines rated at 597 kW at 2,100 rpm each driving Bruntons five-blade fixed-pitch propellers via Twin Disk gearboxes.

Athena 0129594

Gulf Craft

3904 Highway 182 W, Patterson, Louisiana 70392, United States

Tel: (+1 985) 395 52 54; 62 59
Fax: (+1 985) 395 36 57
e-mail: info@gulfcraft.com
Web: www.gulfcraft.com

Kevin Tibbs, *President*

Gulf Craft Inc. is a builder of high-speed aluminium passenger ferries and work boats. It is one of the licensees of InCat Crowther Design (Australia).

M/V *Big Cat Express* 1048773

46 m PASSENGER CATAMARAN
Big Cat Express

Designed by Crowther Design, this vessel was delivered in 2004 to Key West Ferry, an operator in Florida.

Specifications
Length overall 46.04 m
Length waterline 39.02 m
Beam 10.37 m
Draught 1.37 m
Deadweight 55.90 t
Passengers 378
Maximum speed 37 kt
Propulsion: Four Cummins KTA 50-M2 diesel engines each rated at 1,342 kW at 1,900 rpm and each driving a Hamilton HM 651 water-jet via a ZF 4600D gearbox.

46 m PASSENGER CATAMARAN
Galaxy Wave

Designed by InCat Crowther Design, this vessel was delivered to Safe Way Marine Transportation in Bay Island, Honduras in August 2006. The vessel features a higher than normal freeboard and special measures for operating in tropical climates.

Specifications
Length overall 46.3 m
Length waterline 38.6 m
Beam 10.4 m
Draught 1.75 m
Displacement 65 tonnes
Passengers 447
Maximum speed 37 kt
Propulsion: Four Caterpillar 3512B diesel engines each delivering 1,340 kW at 1,785 rpm driving Hamilton HM 651 water-jets.

50 m CATAMARAN CREWBOAT
Seacor Cheetah

Specifications
Length overall 50.29 m
Length waterline 43.89 m
Beam 11.58 m
Draught 2.13 m
Deadweight 152 t
Passengers 150
Crew 10
Fuel capacity 52,314 litres
Water capacity 14,000 litres
Operational speed 41.5 kt
Classification: USCG Sub Chapter T-Ocean Service, ABS +A1 HSC Crewboat +AMS+OPS-2.
Construction: Marine grade aluminium.
Propulsion: Four MTU 16V 4000 diesel engines each driving a Hamilton HM 811 water-jet.

52 m PASSENGER CATAMARAN
Key West Express

Designed by Crowther Design, this vessel was delivered to Key West Express in 2006.

Specifications
Length overall 51.8 m
Beam 11.6 m
Draught 11.4 m
Passengers 513
Maximum speed 45 kt
Range 785 n mile
Propulsion: Four MTU 16V 4000 M71 diesel engines each delivering 2,470 kW at 2,000 rpm driving Hamilton HM 811 water-jets.

Vessel type	Vessel name	Yard no	Length (m)	Speed (kt)	Seats	Vehicles	Originally delivered to	Date of build
Crowther 46 m Catamaran	*Big Cat Express*	452	46.04	31	378	none	Key West Ferry	2004
Crowther 52 m	*Key West Express*	460	51.80	45	513	none	Key West Ferry	2006
InCat Crowther 46 m	*Galaxy Wave*	461	46.30	37	452	none	Safe Way Maritime Transportation	2006
InCat Crowther 50 m Crewboat	*Seacor Cheetah*	–	50.00	41.5	150	none	Seacor Marine	(2007)

Island Boats Inc

5109E Old SpanishTrail, Jeanerette, Louisiana 70544, United States

Tel: (+1 337) 560 44 83
Fax: (+1 337) 560 44 73
e-mail: mail@islandboats.com
Web: www.islandboats.com

Island Boats design and build a wide range of small aluminium craft.

22 m CATAMARAN

Ocean Discovery

In late 2006 Island Boats delivered a 22 m catamaran to the Pacific Whale Foundation in Hawaii. The One2three-designed vessel was built to Det Norske Veritas rules and features thick plating and high wet deck clearance to cope with the conditions off Hawaii.

Specifications

Length overall	22.4 m
Length, waterline	19.75 m
Beam	8.2 m
Draught	1.34 m
Passengers	149
Operational speed	25 kt

Propulsion: Two Cummins KTA-19 diesel engines delivering 522 kW each to fixed-pitch propellers via Twin Disc Max 5135 SC gearboxes.

21 m CATAMARAN

Four vessels were ordered by Royal Caribbean Cruise Lines in 2003 for delivery in 2004 to operate in Belize.

Specifications

Length overall	21.0 m
Beam	6.6 m
Draught	1.5 m
Passengers	230
Fuel capacity	7,600 litres
Water capacity	950 litres
Operational speed	25 kt

Construction: All aluminium hull and superstructure.

Propulsion: Two Caterpillar 3406E diesels each rated at 520 kW and driving fixed-pitch propellers.

20 m CATAMARAN

Three vessels were delivered to Norwegian Cruise Lines in 2001 as ship tenders operating at Fanning Island in the Republic of Kiribati.

Specifications

Length overall	20.0 m
Beam	6.0 m
Draught	1.4 m
Passengers	200
Fuel capacity	2,700 litres
Maximum speed	22.4 kt
Operational speed	20 kt

Construction: All aluminium hull and superstructure.

Propulsion: Two John Deere 6081 diesels each rated at 280 kW and each driving a fixed-pitch propeller.

17 m CATAMARAN

Quicksilver

Designed by Morelli & Melvin (USA) as a tourist vessel, *Quicksilver* is built to USCG SubchapterT classification.

Specifications

Length overall	17.0 m
Length waterline	16.8 m
Beam	7.0 m
Draught	1.0 m
Passengers	149

Propulsion: Twin Lugger 6140AL2 diesels each rated at 446 kW, driving fixed-pitch propellers via Twin Disc MG 5144 gearboxes.

Kitsap Catamarans

Kitsap Catamarans, 201 E 11th Street, Tacoma, Washington 98421, United States

Tel: (+1 253) 383 87 88
Fax: (+1 253) 404 01 79

Based in Tacoma, Washington, Kitsap Catamarans offers a range of commercial catamarans constructed using an epoxy infusion process.

T-85 catamaran ferry

Marina Flyer

The T-85 catamaran, *Marina Flyer*, was ordered by Catalina Ferries for delivery in April 2006.

Specifications

Length overall	26.3
Beam	11.2
Passengers	180
Maximum speed	40 kt
Operational speed	35 kt

Propulsion: Two Caterpillar C32 diesel engines each delivering 1,045 kW at 2,300 rpm to Hamilton HM521 waterjets via ZF 2550 gearboxes.

Vessel type	Vessel name	Yard No	Length (m)	Speed (kt)	Seats	Vehicles	Originally delivered to	Date of build
T-85 Catamaran Ferry	*Marina Flyer*	–	26	40	149	–	Catalina Ferries	2006

Knight and Carver Yachtcenter

San Diego Bay, 1313 Bay Marina Drive, National City, California 91950, United States

Tel: (+1 619) 336 41 41
Fax: (+1 619) 336 40 50
e-mail: info@knightandcarver.com
Web: www.knightandcarver.com

Established in 1972, the company is well known for the production and repair of FRP composite motor yachts. In 2001 the company produced its first commercial high-speed catamaran ferry and in 2004 it was awarded a contract to construct a 24.4 m demonstrator of the m Hull technology for the US Navy. This vessel was launched in January 2006.

Fjordland 0527037

M 80 STILETTO

The hull developed by M Ship Co (www.mshipco.com) consists of a main centre hull and rigid skirts either side. The hull has been proven in a Venice water taxi design, *The Vaparetto,* and is now being considered as a future hull form for the US Navy. A 26.8 m demonstrator vessel was launched by Knight and Carver in January 2006. Built of carbon fibre, and regarded as the largest carbon fibre vessel ever built, the M 80 Stiletto is characterised by its sleek appearance, high speeds (up to 51 kt in recent sea trials) and a patented Double-M hull design which gives the vessel aerodynamic lift and noticeably reduced wake, especially at higher speeds.

M 80 Stiletto 1198842

Vessel type	Vessel name	Yard No	Length (m)	Speed (kt)	Seats	Vehicles	Originally delivered to	Date of build
CAT 20 m	*Fjordland*	–	20.0	30.0	49	–	Alaska Fjordlands	2001
M-Hull 80	–	–	24.0	40.0-50.0	–	–	M Ship Co	(2005)

The vessel is currently undergoing a series of tests by the US Navy.

Specifications

Length overall	26.8 m
Beam	12.2 m
Draught	0.9 m
Displacement	24.5 tonnes
Fuel capacity	22,750 litres
Max speed	51 kt

Construction: Carbon fibre.

Propulsion: Four Caterpillar C32 diesel engines driving Arneson Surface piercing drives.

20 m CATAMARAN

The company's first vessel, a 20 m, 30 kt sightseeing catamaran was delivered to Alaska Fjordlands in 2001 for operation between Juneau, Skagway and Haines in Alaska.

Specifications

Length overall	19.8 m
Length waterline	18.8 m
Beam	5.9 m
Draught	1.3 m
Displacement	24.5 tonnes
Crew	3
Passengers	54
Fuel capacity	1,900 litres
Fresh water	570 litres
Max speed	30 kt
Operational speed	27 kt

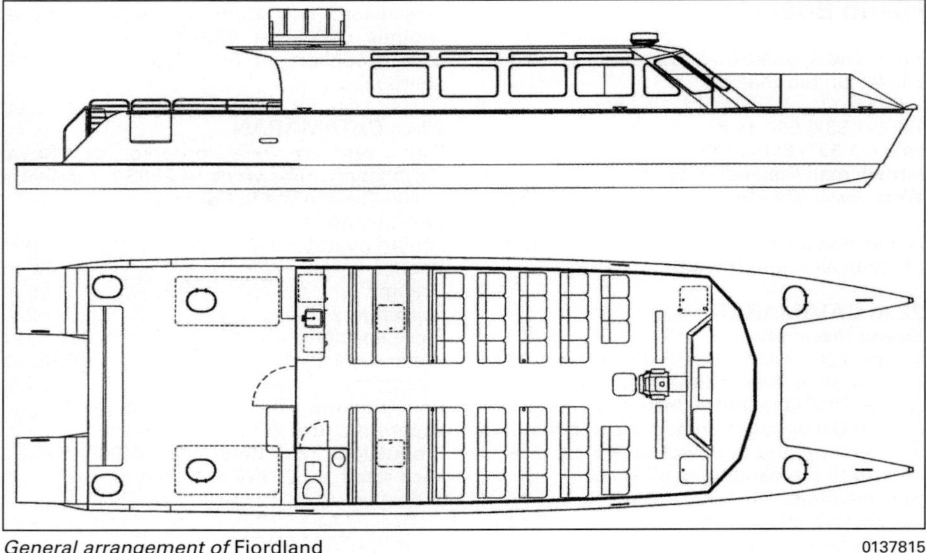

General arrangement of Fjordland 0137815

Construction: Fibre reinforced plastic.

Propulsion: Two Lugger 6140 AL diesel engines each rated at 447 kW and driving a fixed-pitch propeller via a Twin Disc MG 5114 gearbox.

Kvichak Marine Industries

469 NW Bowdoin Place, Seattle, Washington 98017, United States

Tel: (+1 206) 545 84 85
Fax: (+1 206) 545 35 04
e-mail: sales@kvichak.com
Web: www.kvichak.com

Founded in 1981, Kvichak Marine specialises in marine aluminium fabrication, covering a wide range of commercial craft including oil response, fishing, dive and passenger ferry vessels.

20 m Patrol catamaran

Cama'i

This vessel was delivered to The Alaska Department of Public Safety (ADPS) in 2002. Designed by Australian Crowther Multihulls of Sydney, the vessel is to patrol the waters around Kodiak Island, the Alaska Peninsula and Bristol Bay.

Specifications

Length overall	19.8 m
Beam	7.6 m
Displacement	–
Crew	6
Maximum speed	25 kt
Operational speed	20 kt

Propulsion: Twin Caterpillar 3196 engines, 660 BHP at 2,300 rpm driving 76 cm propellers through Twin Disc MG5114A marine gears.

18 m Catamaran ferry

Designed by Morelli and Mervin of the United States, this craft was delivered to Lake Mead Cruises of Boulder City, Nevada in 2001.

Specifications

Length overall	17.4 m
Beam	5.2 m
Draught	1.4 m
Displacement	17.2 t
Passengers	68
Fuel	1,500 litres
Fresh water	570 litres
Maximum speed	32 kt
Operational speed	27 kt

Propulsion: Two Caterpillar 3196 TA diesel engines each rated at 492 kW and driving a fixed-pitch propeller via a ZF 350 A gearbox.

36 m Catamaran ferry

Two 25 knot vessels have been ordered by the San Francisco Water Transit Authority for operation beginning in 2008, providing a ferry service between south San Francisco and Oakland. The vessels are being completed in conjunction with Nichols Brothers Boatbuilders Inc. and will be equipped with up to date emission control devices.

Specifications

Length overall	36.0 m
Length waterline	34.6 m
Beam	8.7 m
Draught	1.9 m
Passengers	200
Operational speed	25 kt

Propulsion: Twin MTU 16V 2000 M70 diesels rated at 1050 kW at 2100 rpm, driving fixed pitch propellerss.

22 m Catamaran ferry

Two vessels have been ordered by Luxury Resorts to InCat Crowther designs for delivery in December 2006.

Specifications

Length overall	22.1 m
Length waterline	19.6 m
Beam	7.5 m
Draught	1.2 m
Passengers	118
Operational speed	27 kt

Propulsion: Two Cummins Q5K 19M diesel engines delivering 596 kW each at 2,100 to fixed-pitch propellers.

20 m Catamaran ferry

Designed by Crowther Multihulls of Australia, one craft was delivered to Andy's Sailing Adventures of Hawaii in 1999 with a second to Long Beach Transport (US) which

Vessel type	Vessel name	Yard No	Length (m)	Speed (kt)	Seats	Vehicles	Originally delivered to	Date of build
20 m Catamaran Ferry	–	–	19.5	25.0	100	none	Andy's Sailing Adventures, Hawaii	1999
20 m Catamaran Ferry	*Aqualink*	–	19.5	24.0	75	none	Long Beach Transport	2001
18 m Catamaran Ferry	*Velocity*	–	17.4	32.0	68	none	Lake Mead Cruises	2001
20 m Tourist Catamaran	*Ocean Odyssey*	–	19.5	27.0	149	none	Pacific Whale Foundation	2001
20 m Patrol Catamaran	*Cama'i*	–	19.8	20	–	none	Alaska Department of Public Safety	2002
Crowther 20 m Catamaran	–	–	20.0	25	180	none	Eco Tours	2004
Crowther 19 m Catamaran	*Ocean Voyager*	–	19.0	27	149	none	Pacific Whale Foundation	2004
Crowther 20 m Catamaran	–	–	19.5	23	149	none	–	2005
InCat Crowther 22 m Catamaran	Taina Dancer	–	22.1	27	118	none	Luxury Resorts	2006
InCat Crowther 22 m Catamaran	Yunque Princess	–	22.1	27	118	none	Luxury Resorts	2006
InCat Crowther 36 m Catamaran	–	–	36.0	25	200	none	San Francisco Water Transit Authority	(2007)
InCat Crowther 36 m Catamaran	–	–	36.0	25	200	none	San Francisco Water Transit Authority	(2007)

is operated by Catalina Express. In 2005 a further vessel to this design was delivered to a Californian whale tour operator. Unlike its sister vessels, this craft is powered by two Cummins KTA 19 engines driving Nibral propellers through ZF 350A gearboxes (2,077:1 reduction ratio).

Specifications

Length overall	19.5 m
Beam	8.5 m
Displacement	27 t
Passengers	75–100
Operational speed	25 kt

Propulsion: Two Cummins N-14 marine diesel engines each driving a fixed-pitch propeller.

20 m Tourist catamaran

Ocean Odyssey

This vessel was delivered to the Pacific Whale Foundation in Hawaii in 2001. Designed by Australian Crowther Multihulls of Sydney, the vessel is operated by PWF's Eco-Adventurer Cruises.

Specifications

Length overall	19.5 m
Beam	8.5 m
Displacement	43 t
Passengers	149
Fuel capacity	3,330 litres
Water capacity	760 litres
Maximum speed	27 kt
Operational speed	23 kt

Construction: All aluminium hull and superstructure.

Propulsion: Two Cummins KTA-19M4 diesel engines each rated at 520 kW at 2,100 rpm and driving a fixed-pitch propeller via a ZF IRM 350-A2 gearbox. N-14 marine diesel engines each driving a fixed-pitch propeller.

Ocean Odyssey 0121762

19 m Tourist catamaran

Ocean Voyager

Built to a Crowther Multihulls Design, the vessel was delivered to the Pacific Whale Foundation in 2004. The engines use Maui Bio Fuel which is made from recycled cooking oil.

Specifications

Length overall	19.0 m
Beam	8.5 m
Passengers	149
Maximum speed	27 kt
Operational speed	23 kt

Propulsion: Two Cummins KTA-19M4 diesel engines each rated at 520 kW at 2,100 rpm coupled to ZF IRM 350-A2 gearboxes.

Marinette Marine Corporation

Part of the Manitowoc Marine Group
1600 Ely Street, Marinette, Wisconsin 54143-2434, United States

Tel: (+1 715) 735 93 41
Fax: (+1 715) 735 35 16
e-mail: fcharrier@marinettemarine.com
Web: www.manitowocmarine.com

Robert Herre, *President & General Manager, Manitowoc Marine Group*
Floyd Charrier Jr, *Vice President Sales & Marketing, Manitowoc Marine Group*

Marinette Marine was incorporated in 1942 and proceeded to build a large number of slow and fast craft for the US Navy and other US government organisations. The company currently handles large marine fabrications and launched its first fast catamaran ferry, *Straits Express* in 1995.

30 m CATAMARAN

Straits Express

This is the first fast catamaran constructed by Marinette Marine. The vessel was delivered to Arnold Transit Co in May 1995 for operation on the Great Lakes.

Straits Express 0507059

Specifications

Length overall	30.0 m
Length waterline	27.0 m
Beam	9.1 m
Draught	0.9 m
Crew	4
Passengers	400
Fuel capacity	7,600 litres
Max speed	35 kt
Operational speed	35 kt

Structure: All-aluminium construction.

Propulsion: Powered by two CAT 3512 diesel engines each driving a MJP 650 water-jet via a ZF BW 465 reduction gearbox.

Vessel type	Vessel name	Yard No	Length (m)	Speed (kt)	Seats	Vehicles	Originally delivered to	Date of build
30 m Catamaran	*Straits Express*	401	30.0	35.0	400	–	Arnold Transit Co	1995

Midship Marine Inc

Midship Marine, Inc, P.O. Box 125 Harvey, Louisiana 70058, United States

Tel: (+1 504) 341 43 59
Fax: (+1 504) 340 89 97
e-mail: midboats@aol.com
Web: www.midshipmarine.net

Michael Hinojosa, *President*

Midship Marine was established in 1989 and builds custom aluminum vessels for a wide range of customers. The company's founder and president, Michael Hinojosa, has been designing aluminum ships and boats for over 27 years.

Midship Marine builds a wide range of aluminum water craft including: passenger vessels and ferries, power catamarans, sailing catamarans, semi-submersibles, crew boats and supply vessels. These vessels range in size from 7 to 60 m in length and can be built from in-house designs or custom built according to the client's specifications.

In 2007 Midship Marine received an order for three Incat Crowther 27 m catamaran ferries for a Mexican operator in Cancun.

27 m CATAMARAN FERRY

In 2007 Midship Marine received an order for three 27 m catamarans design by Incat

Crowther. The vessels will be operated from Cancun in Mexico and feature full height windows for the 125 seats in the main cabin, 124 seats on the upper deck and luggage carts for overnight travellers. The vessels will be built from an aluminium plate and extrusion kit provided by Incat Crowther.

Specifications

Length overall	27.8 m
Length waterline	23.4 m
Beam	7.75 m
Draught	1.2 m
Passengers	249
Fuel capacity	2,000 litres
Max speed	30
Operational speed	28

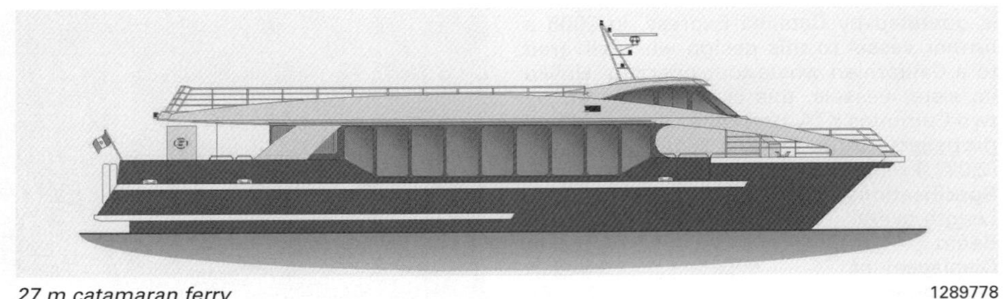

27 m catamaran ferry 1289778

Propulsion: 2 x Caterpillar 3412E diesel engines delivering 895 kW each to fixed-pitch propellers.

Vessel type	Vessel name	Yard No	Length (m)	Speed (kt)	Seats	Originally delivered to	Date of build
27 m Catamaran Ferry	To be named	–	27.8	28.0	249	Ultramar	2007
27 m Catamaran Ferry	To be named	–	27.8	28.0	249	Ultramar	2007
27 m Catamaran Ferry	To be named	–	27.8	28.0	249	Ultramar	2007

Nichols Brothers Boat Builders Inc

5400 S Cameron Road, Freeland, Whidbey Island, Washington 98249, United States

Tel: (+1 360) 331 55 00
Fax: (+1 360) 331 74 84
e-mail: nichols@whidbey.com
Web: www.nicholsboats.com

Matt Nichols, *Chief Executive Officer*
Bryan Nichols, *President*
Ken Schoonover, *Yard Supervisor*

Nichols Brothers Boat Builders is a licensed builder of International Catamaran Designs Pty Ltd, Australia, and has built a wide range of craft to its own design. In 2003 the company was contracted by the Titan Corporation to build a 50 kt, 1,230 t experimental catamaran (X-Craft) for the US Office of Naval Research.

It is understood that in November 2007 the company filed for bankruptcy protection.

X-Craft on sea trials 1198535

80 m X-CRAFT

Sea Fighter

Designed by BMT Nigel Gee and Associates, and ordered by the Titan Corporation of California in 2003, the X-Craft was delivered in mid-2005. The X-Craft is designed for coastal region warfare but has the payload capacity and performance to cross oceans at high speed.

Specifications

Length overall	80.0 m
Length waterline	73.3 m
Beam	22.0 m
Draught	3.0 m
Displacement, min	950 t
Deadweight	721 t
Crew	26
Fuel capacity	545,000 litres
Water capacity	5,000 litres
Maximum speed	55+ kt
Operational speed	50 kt
Range	4,000 n miles at 20 kt

Classification: ABS +A1 HSC Naval Craft OE +AMS +ACCU R2-S.

Structure: All aluminium construction.

Propulsion: Two GE LM 2500 gas turbines each rated at 22,380 kW and two MTU 16V 595 TE90 diesel engines each rated at 4,320 kW. One gas turbine and one diesel in each hull driving two KMW 125 SII water-jets via a single RENK BS 140/2 gearbox.

The 37 m InCat wave-piercing catamaran Nantucket Spray, *renamed* SeaJet 1 0506665

44 m CATAMARAN

Jet Cat Express

This 44 m passenger catamaran is the largest completed by Nichols Brothers Boat Builders and was delivered to Catalina Express Lines in California in early 2001.

The 400-passenger, 26 m Dolphin 0506664

Specifications

Length overall	44.5 m
Length waterline	39.0 m
Beam	10.5 m
Hull beam	2.8 m
Draught	2.0 m
Passengers	381
Fuel capacity	15,000 litres
Water capacity	900 litres
Operational speed	37 kt

Propulsion: Four Cummins KTA50 diesel engines, each rated at 14,000 kW and driving a Hamilton Jet 651 through a ZF reduction gearbox.

44 m CATAMARAN

Catalina Jet

This catamaran was delivered to Catalina Cruises in Long Beach, California in May 1999. The vessel is operated to Catalina Island alongside larger conventional craft. With a length of 44 m this vessel has a capacity for 450 passengers and is powered by four Caterpillar 3512B diesel engines each rated at 1,475 kW and driving an MJP J550DD water-jet, giving an operational speed of 36 kt.

42 m CATAMARAN

Mendocino

This 42 m catamaran entered service between Larkspur and San Francisco, US in mid 2001. It is understood that the funding provided for the vessel included USD8.5 million for construction and USD1.5 million for spares and training.

Specifications

Length overall	42.1 m
Length waterline	37.5 m
Beam	10.4 m
Hull beam	2.8 m
Draught	1.5 m
Passengers	400

Structure: All aluminium.

Propulsion: Four Cummins KTA50 diesel engines, each driving a HamiltonJet 571 water-jet through a ZF reduction gearbox.

38.6 m INCAT WPC

Seajet I (ex-Metro Atlantic, ex-Nantucket Spray)

This USD4 million vessel was ordered in June 1988 and entered service with Bay State Cruises, Boston.

42 m catamaran Mendocino

1198536

34 m catamaran Peralta 1198539

Vessel type	Vessel name	Yard No	Length (m)	Speed (kt)	Seats	Vehicles	Originally delivered to	Date of build
InCat 22 m	*Great Rivers II* (ex-*Klondike*)	–	22.0	26.0	210	–	Yukon River Cruises Inc	1984
InCat 22 m	*Spirit of Alderbrook*	–	22.0	26.0	240	–	Seattle Harbour Tours	1984
InCat 26 m	*Encinal* (ex-*Catamarin*)	–	26.1	28.0	400	–	Crowley Marine Corporation	1985
InCat 26 m	*Gold Rush*	–	26.1	28.0	400	–	Glacier Bay Yacht Tours	1985
InCat 26 m	*Klondike II* (ex-*Victoria Clipper* ex-*Glacier Express* ex-*Baja Express*)	–	26.1	28.0	400	–	Clipper Navigation Inc	1985
InCat 30 m	*Executive Explorer*	–	30.0	22.0	49	–	Glacier Bay Yacht Tours Inc	1986
InCat 26 m	*Dolphin*	–	26.1	28.0	400	–	Crowley Marine Corporation	1986
InCat 23.17 m	*Jera FB-816*	–	23.2	30.0	232	–	US Army	1988
InCat 23.17 m	*Jelang K FB-817*	–	21.9	30.0	232	–	US Army	1988
InCat 36 m	*Catalina Flyer*	–	36.0	27.0	500	–	Catalina Passenger Service	1988
InCat 22 m	*San Geronimo*	–	22.0	26.0	167	–	Port of Puerto Rico	1989
InCat 37 m WPC	*Seajet I* (ex-*Metro Atlantic*)	–	38.6	32.0	367	–	Bay State Cruises, Boston	1989
InCat 22 m	*Viejo San Juan*	–	22.0	26.0	167	–	Port of Puerto Rico	1989
InCat 22 m	*Cristobal Colon*	–	22.0	26.0	167	–	Port of Puerto Rico	1989
InCat 22 m	*Martin Pena*	–	22.0	26.0	167	–	Port of Puerto Rico	1989
InCat 22 m	*Amelia*	–	22.0	26.0	167	–	Port of Puerto Rico	1989
InCat 22 m	*Covadonga*	–	22.0	26.0	167	–	Port of Puerto Rico	1989
InCat 37 m	*Kona Aggressor II*	–	37.0	–	10	–	Alaska Dive Boat Inc	1992
InCat 29 m	*Bay Breeze*	–	29.0	28.0	250	–	City of Alameda	1994
InCat 32 m Z-Bow	*Palau Aggressor II*	–	32.0	23.0	49	–	Alaska Dive Boat Inc	1994
InCat 30 m	*Fiji Aggressor*	–	–	–	49	–	Alaska Dive Boat Inc	1997
InCat 44 m	*Catalina Jet*	–	44.0	36.0	450	–	Catalina Cruises	1999
InCat 40 m	*Klondike Express*	–	40.0	–	400	–	Phillips Cruises and Tours	1999
InCat 44 m	*Jet Cat Express*	140	44.5	37.0	379	–	Catalina Express	2001
InCat 42 m	*Mendocino*	139	42.1	36.0	400	–	Golden Gate Bridge	2001
InCat 34 m	*Peralta*	141	34.0	–	315	–	Alameda Oakland Ferry Service	2001

Bay Breeze 0506897

44 m catamaran Catalina Jet 1198537

Specifications

Length overall	38.6 m
Beam	15.6 m
Hull beam	2.6 m
Draught	1.3 m
Passengers	367
Fuel capacity	2 × 14,130 litres
Water capacity	7,570 litres
Propulsive power	2 × 1,768 kW
Max speed	36.4 kt
Operational speed	32 kt

Classification: USCG, SOLAS.
Propulsion: Main engines are two MWM TBD 604 V16 1,768 kW each, at 1,800 rpm, driving two Kamewa 63 S62/6 water-jet units.
Electrical system: Two John Deere 55 kW generators.

36 m CATAMARAN

Catalina Flyer

The largest capacity high-speed catamaran to be built in the US, the 500-seat *Catalina Flyer* (delivered May 1988) is in service between Newport Harbor and Catalina Island in Southern California with the Catalina Passenger Service.

Specifications

Length overall	36.0 m
Beam	12.2 m
Draught	2.4 m
Passengers	500
Fuel capacity	11,355 litres
Water capacity	1,514 litres
Propulsive power	2 × 1,490 kW
Max speed	30 kt
Operational speed	27 kt

Propulsion: Main engines are two Caterpillar 3516 TA, 1,490 kW each, specially lightened diesels; driving three-blade CuNiAl Bronze, 1.3 m diameter propellers; via two Reintjes WS-1023, 2.538:1 gearboxes.
Electrical system: John Deere generators, two 40 kW units.
Navigation and communications: Ross DR 600D flasher, two ICOM VHF radios, Furuno 1510D and 8030D radars, Sperry 8T autopilot, Furuno LC-90 Loran.

34 m CATAMARAN

Peralta

Alameda/Oakland Ferry Service ordered this 34 m, 315 passenger ferry in 2000 for delivery in the second half of 2001. The vessel is powered by two Cummins KTA 38 diesel engines each rated at 890 kW and each driving a fixed-pitch propeller.

General arrangement of the 44 m catamaran for Catalina Express Lines 0121755

30 m CATAMARAN

Executive Explorer

A 31.7 m cruising catamaran, built in 1986 for Alaskan and Hawaiian islands cruising.

Specifications

Length overall	30.0 m
Beam	11.2 m
Draught	2.5 m
Crew	20
Passengers	49
Fuel capacity	63,588 litres
Water capacity	19,682 litres
Propulsive power	2 × 1,004 kW
Operational speed	22 kt

Propulsion: Main engines are two Deutz BAM 816, 1,004 kW continuous, each at 1,800 rpm, driving ISO class 1, five-blade, Nicel 1.346 × 1.422 m Columbia Bronze propellers via two Reintjes 842 2.96:1 gearboxes.
Electrical system: The generator is a 120 kW, PDC 120 MB Mercedes-Benz W/OM-421.

26 m CATAMARAN

Catamarin

Dolphin

Gold Rush
Klondike II (ex-Victoria Clipper, ex-Glacier Express, ex-Baja Express)

Specifications

Length overall	26.1 m
Beam	9.4 m
Draught	2.4 m
Fuel capacity	20,440 litres
Water capacity	1,893 litres (*Catamarin*) 1,515 litres (*Klondike II*)
Propulsive power	2 × 1,004 kW
Operational speed	28 kt

Propulsion: Main engines are two Deutz BAM 16M 816C diesels, each 1,004 kW continuous at 1,800 rpm; driving Coolidge five-blade, 1.169 m × 1.194 m propellers via Reintjes WVS 832 gearboxes, ratio 2.29:1.

Electrical system: Generators: two John Deere 4275 engines and Northern Lights 50 kW generators. *Klondike II* has Pacific diesel units.

22 m CATAMARAN

Great Rivers II (ex-Klondike)

There have been eight of these craft delivered since 1984, the main order coming in 1988 for six identical craft for the Port of Puerto Rico. These vessels operate on a five mile route between old San Juan and the newer metropolitan area of Puerto Rico city.

Specifications

Length overall	22.0 m
Beam	8.7 m
Passengers	210
Fuel capacity	1,000 US gallons
Water capacity	1,000 US gallons
Operational speed	26 kt

Propulsion: *Great Rivers II* is powered by two Caterpillar 3412TA diesels, 522 kW each; *Spirit of Alderbrook* by two 12V-92TA diesels; and the Puerto Rico boats by two DDC GM 12V71 diesels.

Gearboxes: Niigata MGN-80.

Propellers: Coolidge five-blade.

Electrical system: Northern Lights 40 kW generator.

Control: The rudders are the curved dipping InCat type.

Jera FB-816
Jelang K FB-817

There were two 23.17 m InCat catamarans built in 1988 for ferry service at the US Army's missile test range in the Marshall Islands. *Jera* was handed over to the Nichols Brothers Boatyard on 9 April 1988 and *Jelang K* on 7 October 1988.

Specifications

Length overall	23.2 m (*Jera*) 21.9 m (*Jelang K*)
Beam	8.9 m (*Jera*) 8.7 m (*Jelang K*)
Draught	1.8 m
Passengers	232
Fuel capacity	5,300 litres (*Jera*) 4,164 litres (*Jelang K*)
Water capacity	946 litres
Propulsive power	2 × 715 kW
Operational speed	31 kt

Propulsion: Main engines are two DDC GM 16 V92 TA, 715 kW each driving five-blade Osborne bronze propellers, 940 × 927 mm via two ZF BW 250 2.03:1 ratio gearboxes.

Electrical system: Two 50 kW Northern Lights generators, John Deere 4276 generators.

Bay Breeze

Designed by International Catamarans, this craft was delivered in April 1994.

Specifications

Length overall	29.0 m
Beam	9.0 m
Draught	1.2 m
Passengers	250
Fuel capacity	4,166 litres
Water capacity	2,840 litres
Propulsive power	2 × 768 kW
Max speed	30 kt
Operational speed	28 kt

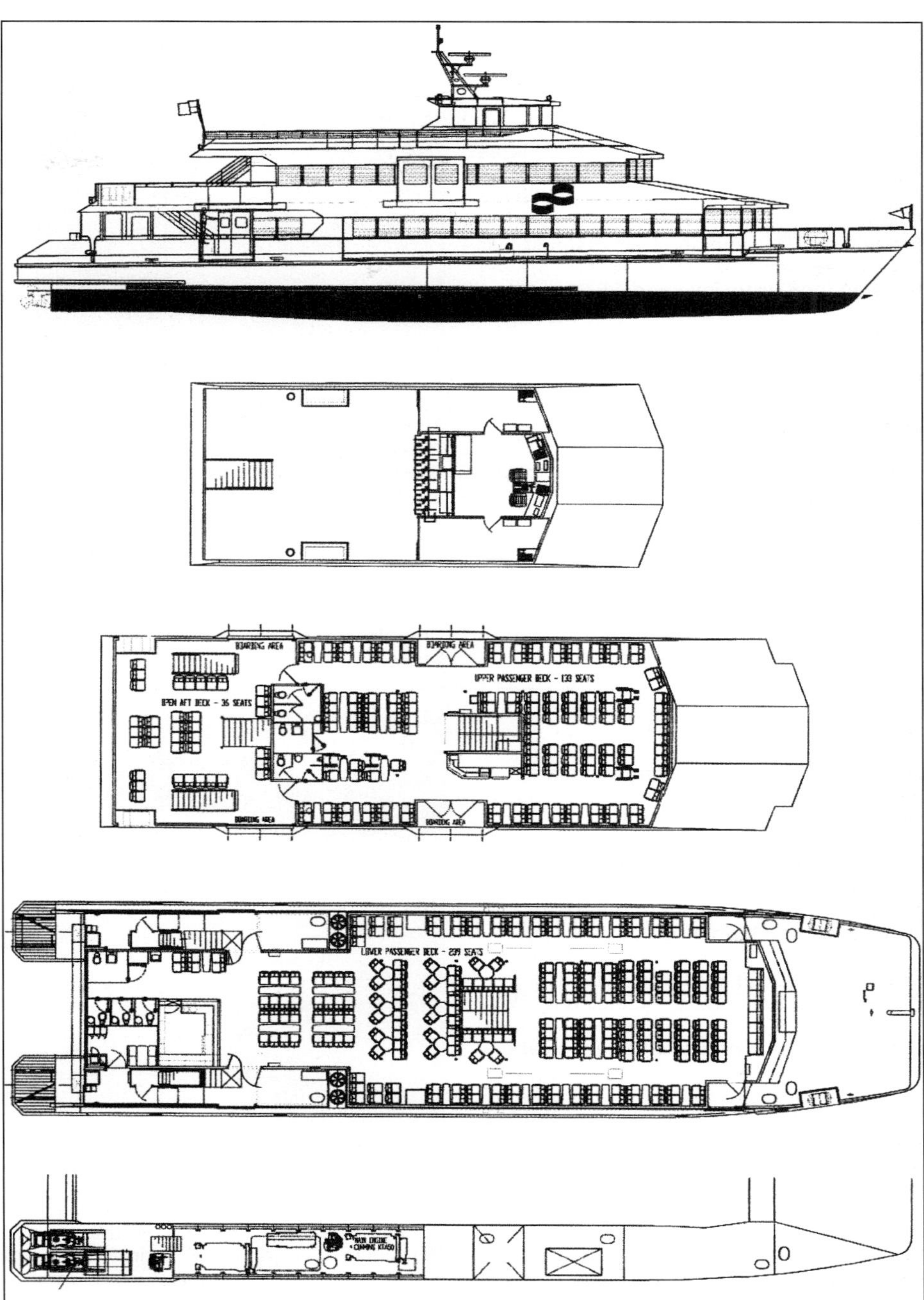

General arrangement of the 42 m catamaran for Golden Gate Bridge 0121751

44 m catamaran Jet Cat Express 1198538

Propulsion: Main engines are two Detroit Diesel 16V 92TA DDEC–768 kW; driving two Kamewa 50 water-jet units.

Electrical system: Two John Deere 4276 50 kW generators.

Palau Aggressor II

This second vessel purchased by the Alaska Dive Boat Company carries 16 overnight passengers and 49 day passengers. It was designed by InCat Designs Pty Ltd using the company's Z-Bow hull form.

Specifications

Length overall	32.0 m
Beam	9.0 m
Draught	1.3 m
Passengers	49
Fuel capacity	17,000 litres
Water capacity	7,600 litres
Propulsive power	1,570 kW
Operational speed	23 kt

Propulsion: Main engines are two Caterpillar 3412 diesels each rated at 785 kW, driving two Hamilton water-jets model 5711.

InCat Jelang K
0506667

Royal Crown Yachts

5411 West Tyson, Tampa, Florida 33611, United States

Tel: (+1 813) 805 24 45
Fax: (+1 813) 902 85 65

Best known for its production of fast luxury recreational craft, the company has built a 30 m catamaran surface effect ferry.

30 m CATAMARAN SES

Purrseaverance

This vessel was built for Fast Cat Boat Works to a novel catamaran design incorporating air lubrication.

Specifications

Length overall	30.5 m
Beam	8.5 m
Draught, on-cushion	0.5 m
Draught, off-cushion	1.1 m
Crew	6
Passengers	183
Fuel capacity	7,570 litres
Water capacity	750 litres
Operational speed	42 kt

Propulsion: Four Caterpillar 3412E diesel engines each rated at 820 kW and each driving a France Helice surface piercing propeller, and two Caterpillar 3208C diesels each rated at 235 kW and driving a lift fan.

VT Halter Marine Inc

A Vision Technologies Systems company
900 Bayou Casotte Parkway, Pasagoula, Mississippi 39581, United States

Tel: (+1 228) 696 68 88
Fax: (+1 228) 696 68 99
e-mail: vthmsales@vthaltermarine.com
Web: www.vthaltermarine.com

Boyd E King, *Chief Executive Officer*
Richard T McCreary, *Executive Vice President*
Cynthia Borries, *Corporate Communications Manager*

VT-Halter Marine, better known for its monohull vessels, has completed a development project for an advanced catamaran design. The development work was undertaken in conjunction with the Advanced Research Projects Agency with MARITECH funding. The E-Cat design has been arranged for operation on rivers and waterfront areas where the low wash characteristics of the vessel will be advantageous.

VT Halter E-Cat prototype on trials 0043933

E-CAT

This prototype craft was launched in October 1998 for initial trials. Further trials in 1999/2000 with additional hydrofoil support have indicated substantial performance improvements in speed, fuel consumption and wash. The final superstructure will be fitted once owner's requirements are ascertained.

Specifications

Length overall	45.0 m
Beam	11.6 m
Draught	1.3 m
Passengers	400
Operational speed	35 kt

Vessel type	Vessel name	Yard	Length (m)	Speed (kt)	Seats	Vehicles	Originally delivered to	Date of build
E-Cat	–	–	45.0	35.0	400	–	–	October 1998

Yank Marine

Mosquito Landing Road, Tuckahoe, New Jersey 08250, United States

Tel: (+1 609) 628 29 28
Fax: (+1 609) 628 26 28
e-mail: BJWJCY@hotmail.com
Web: www.yankmarine.com

This company launched a 31 m fast catamaran, *Royal Miss Belmar,* in 2001 which was subsequently chartered to Acadian Whale Adventures in Bar Harbour and also fast ferry operators in the New York area during 2001–02.

Vietnam

Tam Bac

A division of Vietnam Shipbuilding Industry Corporation (Vinashin)
157 Durong Ha Ly, Quan Hong Bang, Thanh pho Hai Phong, Vietnam

Tel: (+84 31) 84 27 75
Fax: (+84 31) 84 28 48
e-mail: tambacs@vinashin.com.vn
Web: www.vinashin.com.vn/tambac

The company was formed in 1955 and in 2004 started construction of an aluminium catamaran designed by the Australian company Commercial Marine Consulting Services for a Chinese operator.

SEACAT 29

Vinashin Rose

Specifications

Length overall	30.0 m
Beam	8.5 m
Crew	8
Passengers	211
Fuel capacity	3,000 litres
Water capacity	100 litres
Operational speed	31 kt

Propulsion: Two Cummins KTA 50M2 engines each rated at 1,395 kw at 1,950 rpm driving Hamilton HM651 water-jets via ZF gearboxes.

SMALL-WATERPLANE-AREA TWIN-HULL (SWATH) VESSELS

Company listing by country

Germany
Abeking and Rasmussen Shipyard

Japan
Mitsui Engineering & Shipbuilding Company Ltd

Korea, South
Hyundai Heavy Industries Company Ltd

Netherlands
Damen Shipyards

Philippines
FBMA Marine Inc

Russian Federation
Almaz Shipbuilding Company

United Kingdom
FBM Babcock Marine Group

United States
Eastern Shipbuilding Group
Navatek Ships Ltd
Nichols Brothers Boat Builders Inc
Swath Ocean Systems Inc

Germany

Abeking and Rasmussen Shipyard

PO Box 1160, D-27805 Lemwerder, Germany

Tel: (+49 421) 673 30
Fax: (+49 421) 673 31 12
e-mail: info@abeking.com
Web: www.abeking.com

Herman H Schaedla, *President*
Klaas Spethmann, *Managing Director, Sales and Technical*
Hans M Schaedla, *Managing Director, Yacht Division*
Erich Bischoff, *Managing Director, Commercial*

50 m SWATH pilot ship on station (vessel on right hand side) 0129183

The Abeking and Rasmussen shipyard has developed a range of SWATH ships offering levels of seaworthiness and response motions comparable with much larger conventional vessels. The first three vessels operate as pilot vessels in the Elbe Approach and have proved themselves in adverse weather conditions in the North Sea.

25 m SWATH PILOT TENDER

These vessels have been designed for frequent pilotage operations in exposed coastal waters and tidal estuaries with a design wave height of 3.5 m. Two vessels, *MV Döse* and *Duhnen* were delivered in mid-1999.

Specifications

Length overall	25.7 m
Beam	14.3 m
Draught	2.7 m
Deadweight	13.8 t
Crew	2
Passengers	8
Fuel capacity	8,500 litres
Water capacity	1,000 litres
Propulsive power	1,580 kW
Operational speed	18 kt

Classification: GL +100 K OC3, Pilot Tender.
Structure: Aluminium hull and superstructure.
Propulsion: Two MTU 12V 2000 M70 diesel engines each rated at 790 kW and driving a fixed-pitch propeller via a Reintjes WVS 430 reverse reduction gearbox.
Electrical system: Two STN ATLAS generators rated at 1,000 kW each.

25 m SWATH pilot tender 0129182

50 m SWATH PILOT STATION SHIP

This large SWATH craft, *MV Elbe*, has been designed for operation in rough coastal waters and tidal estuaries as a pilot station vessel, being able to remain on station for extended periods. The vessel was commissioned in mid-2000.

Specifications

Length overall	49.9 m
Beam	22.6 m
Draught	5.9 m
Deadweight	155 t
Fuel capacity	8,500 litres
Water capacity	20,000 litres
Propulsive power	2,000 kW
Operational speed	14 kt

Classification: GL +100 A5, K+MC AUT Pilot Station Ship.
Structure: Steel hull and superstructure.
Propulsion: Diesel-Electric drive with two 1,000 kW electric motors each driving fixed-pitch propellers and powered by four MTU 8V 396 TE54 diesel generators.
Auxiliary systems: Bow thruster in one hull.

Vessel type	Vessel name	Yard No	Length (m)	Speed (kt)	Seats	Vehicles	Originally delivered to	Date of build
25 m SWATH pilot tender	*Döse*	6427	25.2	18.0	8	–	–	1999
25 m SWATH pilot tender	*Duhnen*	6428	25.2	18.0	8	–	–	1999
50 m SWATH pilot station ship	*Elbe*	6429	49.8	14.0	20	–	–	1999
25 m SWATH	*Borkum* ex-*Explorer*	6467	25.0	–	–	–	–	2004
25 m SWATH pilot tender	*Wangeroog*	6468	25.0	18.0	8	–	–	2004
25 m SWATH pilot tender	*Cetus*	6469	25.0	18.0	8	–	–	2005
25 m SWATH pilot tender	*Perseus*	6470	25.0	18.0	8	–	–	2000
60 m SWATH pilot station ship	*Elbe*	6484	60.0	14.0	20	–	–	(2009)
60 m SWATH pilot station ship	*Elbe*	6485	60.0	14.0	20	–	–	(2009)

Japan

Mitsui Engineering & Shipbuilding Company Ltd

6-4 Tsukiji 5-chome, Chuo-ku, Tokyo 104, Japan

Tel: (+81 3) 35 44 31 47
Fax: (+81 3) 35 44 30 50
e-mail: prdept@mes.co.jp
Web: www.mes.co.jp

Takao Motoyama, *President*
Tamiyoshi Iwasaki, *Managing Director and General Manager, Ship and Ocean Project HQ*
Kiyoshi Hara, *Deputy Director and Deputy General Manager, Ship and Ocean Project HQ*
Mitsuru Kawamura, *General Manager, Governmental Ship and HSS Sales*
Kimio Uchida, *Manager, Governmental Ship and HSS Sales*
Masakata Hashimoto, *General Manager of Research and Development HQ*

Mitsui began its high-speed Semi-Submerged Catamaran (SSC) development programme in 1970. Since 1976, the programme has been operated in conjunction with the Japanese Marine Machinery Development Association (JMMDA). In 1977 Mitsui built the experimental 18.37 tonne *Marine Ace* in order to obtain practical experience with this hull form. In 1979 the first SSC high-speed passenger vessel was launched under the provisional name *Mesa 80*. After extensive trials it was completed in 1981 and renamed *Seagull*. It has since been operated by Tokai Kisen Company Ltd on a passenger ferry service between Atami and Oshima island.

The company has also developed and built an SSC hydrographic survey vessel, *Kotozaki*, for the Fourth District Port Construction Bureau of the Japanese Ministry of Transport. This vessel was completed in 1981.

Marine Ace

Mitsui's first experimental SSC, *Marine Ace*, is built in marine grade aluminium alloy and can operate in Sea States 2 to 3.

Specifications

Length overall	12.4 m
Beam	6.5 m
Draught	1.6 m
Displacement, max	18.4 t*
Fuel capacity	1.5 m^3
Operational speed	18 kt

*after modification in 1978

Propulsion: Main engines are two V-type four-cycle petrol engines, each developing 150 kW at 3,700 rpm. Each drives, via a vertical intermediate transmission shaft and bevel gear, a three-blade fixed-pitch propeller.

Control: Four sets of fin stabilisers, driven by hydraulic servo motors, reduce ship motion in heavy seas.

Seagull

Developed jointly by Mitsui Engineering & Shipbuilding Company Ltd and the Japanese Marine Machinery Development Association (JMMDA), the 27 kt *Seagull* was the world's first commercial SSC. Despite its small size, the overall length is just under 36 m, the vessel provides a stable ride in seas with 3.5 m waves.

During the first 10 month commercial run in a service between Atami and Oshima, *Seagull* established an operating record of 97 per cent availability.

Specifications

Length overall	35.9 m
Beam	17.1 m
Draught	3.2 m
Crew	7
Passengers	446
Max speed	27 kt

Structure: Marine grade aluminium alloy.

Propulsion: Main engines are two Fuji-SEMT marine diesels, each developing 3,000 kW max continuous at 1,475 rpm. Each drives, via a vertical transmission shaft and bevel gear, a four-blade fixed-pitch propeller.

Electrical system: Two 206.25 kVA generators provide electrical power.

Control: Four sets of fin stabilisers driven by hydraulic servo motors reduce ship motion in heavy seas.

Mitsui SSC high-tech cruiser SSC 15 type, Marine Wave, *built in glass and carbon-reinforced plastic* 0506670

SSC 40 Seagull 2 0506671

SSC 20 Diana 0506672

SSC 15

Marine Wave

The first SSC 15 type cruiser *Marine Wave* built by MES for Toray Industries Inc was delivered in July 1985.

Marine Wave, which is only about 15 m in length, is relatively free from rolling and pitching by virtue of its SSC design which also allows a spacious deck to be provided and facilitates comfortable cruising. One

Vessel type	Vessel name	Yard No	Length (m)	Speed (kt)	Seats	Vehicles	Originally delivered to	Date of build
Experimental vessel	*Marine Ace*	–	12.4	18.0	20	–	–	1977
SSC passenger craft	*Seagull*	–	35.9	–	446	–	Tokai Kisen Company Ltd	1979
Survey vessel	*Kotozaki*	–	27.0	–	–	–	–	1981
SSC 15	*Marine Wave*	–	15.1	16.0	15	–	–	1985
SSC 15	*Sun Marina*	–	15.1	17.0	30	–	San Marina Hotel, Okinawa	1987
SSC 20	*Bay Queen*	–	18.0	–	40	–	Tokyo Bay	1989
SSC 40	*Seagull 2*	–	39.3	–	410	–	–	1989
SSC 20	*Bay Star*	–	19.5	–	40	–	–	1991
SSC 30	*Cosmos*	–	29.2	–	96	–	Japanese Shipping Company	1995

of its most interesting features is the combination of its unusual shape with Toray's newly developed hull material incorporating carbon fibre composites. The SSC 15 has two sets of computer-controlled stabilising fins and two fixed fins.

Marine Wave is certificated by the Japan Craft Inspection Organisation for use in coastal waters. By August 1986 *Marine Wave* had operated over 860 hours including a voyage to West Japan in which it experienced waves 4.5 m in height.

Specifications

Length overall	15.1 m
Beam	6.2 m
Draught	1.6 m
Crew	2
Passengers	15
Fuel capacity	2,000 litres
Water capacity	300 litres
Max speed	18 kt
Operational speed	16 kt

Structure: Glass-reinforced plastic and carbon-reinforced plastic.

Propulsion: Two high-speed marine diesel Ford Sabre 5,950 cc engines, 200 kW each at 2,500 rpm, driving fixed-pitch propellers via Twin Disc MG 506 gearboxes, ratio 2.03:1.

Electrical system: Onan MDJJF-18R diesel unit.

SSC 30 Cosmos 0506980

SSC Bay Star 0506673

SSC 15

Sun Marina

The second of the SSC 15 series, *Sun Marina* was built by Mitsui for San Marina Hotel, opened as a grand resort hotel in Okinawa in March 1987.

Sun Marina has a large luxurious party cabin which can accommodate 30 guests of the hotel, and sails round many coral reefs from the privately owned marina of the hotel.

Specifications

Length overall	15.1 m
Beam	6.4 m
Draught	1.6 m
Crew	3
Passengers	30
Fuel capacity	2 × 900 litres
Water capacity	400 litres
Max speed	20.5 kt
Operational speed	17 kt

Propulsion: The propulsive power is provided by two 170 kW marine diesels.

Electrical system: 15 kW generator.

SSC 20

Bay Queen

Bay Queen is the seventh SSC vessel built by Mitsui since 1977. It can carry a maximum of 40 passengers and has entered service for operation in inspection tours, sightseeing, crew transportation and many other purposes in Tokyo Bay.

Bay Queen is the first and smallest commercialised SSC vessel in the world, made of aluminium to a design based on the concept of *Marine Wave* which was made of FRP.

Specifications

Crew	4
Passengers	40
Fuel capacity	2,400 litres
Water capacity	1,000 litres

Propulsion: Two high-speed marine diesel engines, 350 kW each at 2,000 rpm.

Electrical system: Yanmar 4JHL-TN diesel generator unit.

SSC 20

Diana

Diana is the second vessel of the Mitsui SSC-20 series and was completed in the middle of March 1990. It is operated for day and night cruises as a party boat in Osaka Bay, a good application of the Mitsui SSC type, which can provide spacious deck area and a comfortable ride in rough sea conditions. It has an automatic motion control system to minimise ship motion.

Specifications

Length overall	20.7 m
Beam	6.8 m
Draught	1.6 m
Crew	3
Passengers	40
Max speed	19.2 kt

Structure: Marine grade aluminium alloy.

Propulsion: Main engines are two high-speed diesel engines, each developing 276 kW max continuous at 2,250 rpm.

Control: Maintenance-free sensors and a sophisticated fin control system are applied. One pair of canard fin stabilisers is automatically controlled by an electric motor.

SSC 30

Cosmos

Mitsui delivered the first SSC 30, *Cosmos,* to the Japanese Shipping Company in October 1995. The vessel is operated as a crew support ship for offshore transportation.

The SSC 30 has been developed as a medium-sized standard SSC vessel. SSC 30 can give a comfortable ride in a 2 to 2.5 m wave height with a cruising speed of 20 kt.

Specifications

Length overall	29.2 m
Length waterline	24.3 m
Beam	11.3 m
Payload	5 t deck cargo
Crew	4
Passengers	96

Structure: Marine grade aluminium alloy.

Propulsion: The engines are two high-speed marine diesels, 1,120 kW.

Control: One set of forward fin stabilisers driven by hydraulic servo motors, and one set of fixed fins aft.

SSC 40

Seagull 2

Seagull 2 has a capability of running at 30.6 kt at maximum continuous rating and 27.5 kt service speed with 410 passengers. According to the analysis of log book records, *Seagull 2* has better speed sustainability than *Seagull* in wave heights of 2.5 m and higher.

Specifications

Length overall	39.3 m
Beam	15.6 m
Draught	3.5 m
Crew	7
Passengers	410
Max speed	30.6 kt

Propulsion: Main engines are four MTU 16V 396 TB 84, each developing 2,000 kW max continuous at 1,940 rpm. Each pair of engines drives a four-blade fixed-pitch propeller, through a long straight tube shaft with two ZF reversible reduction gearboxes. A microcomputer-based remote-control system is applied.

Control: The ship control system is composed of a display of nautical information and ballast control. Nautical information includes draught, trim, heel, rudder angle, fin angle, ship speed, wind data, ship course and so on. A ballast control system displays ballast line layout and can operate ballast pumps and valves remotely.

Four fin stabilisers are automatically controlled by hydraulic rotary actuators.

Korea, South

Hyundai Heavy Industries Company Ltd

1 Cheonha-Dong, Dong-Ku, Ulsan, South Korea

Tel: (+82 52) 202 29 50
Fax: (+82 52) 202 34 21
e-mail: sbplan@hhi.co.kr
Web: english.hhi.co.kr/Business/Shipbuilding.asp

Choi, Kil-Seon, *President/CEO*
Min, Keh-Sik, *President/CEO*
C S Kim, *General Manager Ship Sales Planning Department*

35 m SWATH vessel

Building of the 35 m SWATH vessel started in 1991 and it was delivered in April 1993.

Specifications

Length overall	34.5 m
Length waterline	32.5 m
Beam	15.0 m
Draught	3.5 m
Propulsive power	2 × 2,000 kW
Operational speed	21.6 kt

Propulsion: Main engines are two MTU 16V 396TE 84, each developing 2,000 kW at 1,940 rpm.

35 m Hyundai SWATH vessel 0506832

Vessel type	Vessel name	Yard No	Length (m)	Speed (kt)	Seats	Vehicles	Originally delivered to	Date of build
35 m SWATH vessel	–	–	34.5	21.6	–	–	–	1993

Netherlands

Damen Shipyards

Damen Fast Ferries Division
Industrieterrein Avelingen West 20, PO Box 1, NL-4200AA, Gorinchem, Netherlands

Tel: (+31 183) 63 99 22
Fax: (+31 183) 63 21 89
e-mail: info@damen.nl
Web: www.damen.nl

E van der Noordaa, *Managing Director*
H O van Herwijnen, *Damen Fast Ferries*

Damen Fast Ferries is the product group of Damen Shipyards, supporting the marketing and sales of fast ferries for inland, coastal and international voyages in the 28 to 60 m range.

Damen Shipyards started in 1927 and in 1969 began the construction of commercial vessels to standard designs using modular construction. Since 1969, more than 3,000 vessels have been built and delivered to 117 countries worldwide.

In 1998, Damen Shipyards entered the high-speed catamaran ferry market with the construction of a series of five low wash commuter ferries, built under licence from NQEA, Australia. In 2000, Damen Shipyards purchased the Kvaerner Fjellstrand Singapore facility and, along with it, the design rights for a wide range of catamaran craft.

Fast ferries are marketed by Damen Shipyards Gorinchem in the Netherlands. They are constructed at Damen Shipyards Singapore or Damen Shipyards Gorinchem, or may be licensed to other yards.

Princess Maxima *during trials* (Nigel Gee & Associates) 0572483

SWATH 3717

Princess Maxima
Prins Willem Alexander

Two SWATH 3717 vessels were ordered in 2002 by Province of Zeeland of the Netherlands and delivered in 2004. The ferries are used to transport passengers and bicycles across the Westerschelde estuary. The craft are capable of operating in significant wave heights up to 2.5 m. The conceptual design was undertaken by Sovereign Marine Services NV and the detail design by Nigel Gee and Associates Ltd, UK.

Specifications

Length overall	37.7 m
Beam	17.3 m
Draught	4.2 m
Crew	4
Passengers	181
Fuel capacity	22,000 litres
Water capacity	4,000 litres
Operational speed	16.5 kt

Classification: BV1+Hull+ Mach Light Ship/ SWATH/Fast Passenger Ferry (NL).

Propulsion: The vessel will be powered by a diesel electric drive.

Vessel type	Vessel name	Yard No	Length (m)	Speed (kt)	Seats	Vehicles	Originally delivered to	Date of build
SWATH 3717	*Princess Maxima*	–	37.7	16.5	181	none	Province of Zeeland	2004
SWATH 3717	*Prins Willem Alexander*	–	37.7	16.5	181	none	Province of Zeeland	2004

Philippines

FBMA Marine

Arpili, Balamban, Cebu, Philippines

Tel: (+63 32) 465 00 01
Fax: (+63 32) 465 00 08
e-mail: dougb@fbma.com/ph
Web: www.fbma.com.ph

D Border, *General Manager*
C Patrick, *Sales Manager*

Operating since 1997, FBMA Marine Inc has delivered a number of high-speed ferries and patrol boats. In 2005, FBMA Marine Inc received an order from Hotelena Servicos Petroleos for two Lockheed Martin SLICE swath crewboats.

LOCKHEED MARTIN SLICE

Lider
Tenaz

Based on the Lockheed SLICE swath hull design, these vessels transport up to 150 passengers together with light cargo, and offer improved seakeeping and operational efficiency over conventional crewboats.

Specifications

Length overall	28 m
Passengers	150
Max speed	30 kts
Operational speed	22 kts

Structure: Aluminium.

Russian Federation

Almaz Shipbuilding Company

26 Petrovsky Pr, St Petersburg 197110, Russia

Tel: (+7 812) 235 51 48
Fax: (+7 812) 235 70 69
e-mail: office@almaz-kb.sp.ru
Web: www.almaz.spb.ru

Alexander V Shliakhtenko, *President*
Nikolai Ivakin, *Shipyard Director*
Leonid Grabovets, *General Director*
Sergei Galichenko, *Marketing Manager*

Almaz Shipbuilding Company was founded in 1901 in St Petersburg by Alexander Zolotoff. Since that time over 1,000 vessels have been delivered including fast patrol craft, passenger ferries and the world's largest hovercraft, the *Pomoznik*.

In 1993 Almaz announced that a contract had been signed to construct a SWATH craft designed by the AGAT Design Bureau for Marine Systems of Sukhoi. It is understood that this craft was delivered in 2000.

Sukhoi A4 Class SWATH 1120281

SUKHOI A4 CLASS SWATH

Specifications

Length overall	32.3 m
Beam	10.5 m
Draught	2.3 m
Displacement, max	154 t
Payload	27.5 t
Passengers	186
Fuel capacity	9,700 litres
Water capacity	2,000 litres
Operational speed	28 kt
Range	250 n miles

Classification: DnV.
Propulsion: Powered by two MTU 12V 396 TE 74L diesel engines each rated at 1,500 kW driving fixed-pitch propellers.

Vessel type	Vessel name	Yard No	Length (m)	Speed (kt)	Seats	Vehicles	Originally delivered to	Date of build
Sukhoi A4 class SWATH	–	–	32.3	28.0	186	–	Marine Systems of Sukhoi	1993

United Kingdom

FBM Babcock Marine Group

The Courtyard, St Cross Business Park, Monks Brook, Newport, Isle of Wight PO30 5BF, United Kingdom

Tel: (+44 1983) 82 57 00
Fax: (+44 1983) 82 41 80
e-mail: fbm@babcock.co.uk
Web: www.fbmuk.com

Mark Graves, *General Manager*

FBM builds a wide range of monohull, catamaran and SWATH vessels for both commercial and paramilitary applications. In 2000 Babcock International purchased the FBM interests, adding a further shipyard in Rosyth.

Atlantic Class FDC Patria *in service between Funchal and Porto Santo* 0506675

23 m PASSENGER TRANSFER CRAFT

In May 1996, FBM Babcock Marine was awarded a contract by the UK Ministry of Defence worth GBP6.5 million for the design and construction of two 23 m SWATH passenger transfer craft. The craft are used by the MoD for the rapid transfer of staff to and from warships in Plymouth harbour. The transfer system is designed around a stabilised and telescopic brow allowing teams of up to 30 people to transfer within a very short time period. These all-aluminium craft are designed to have a maximum speed enabling them to be classified under the IMO HSC Code of Safety.

Specifications

Length overall	23.9 m
Length, waterline	21.0 m
Beam	11.1 m
Draught	2.2 m
Crew	5
Passengers	75
Max speed	15.5 kt
Operational speed	12 kt

Classification: Lloyds Register of Shipping.
Structure: All-aluminium hull and superstructure.

37 m ATLANTIC CLASS FAST DISPLACEMENT CATAMARAN (FDC)

Patria

In August 1988 Fairey Marinteknik, now FBM Marine, announced the award of a contract for the supply of a 400-seat fast displacement catamaran to the regional government of Madeira, Portugal. Valued at GBP4 million the craft was delivered in early 1990.

First trials started in October 1989. On 5 October 1989 the vessel achieved a speed of 32.1 kt at two-thirds load, some four to five kt higher than the previous maximum recorded speed for a SWATH type, therefore establishing a record.

In late January 1990 *Patria* went on trial in storm conditions. It spent a total of 7 hours at sea averaging 28 kt in a fully loaded condition in wave heights of 2.5 to 3.5 m, and wind speeds that seldom dropped below 70 kt and gusted, at times, above 90 kt. On 19 August 1990 *Patria* went into service.

Specifications

Length overall	36.5 m
Beam	13.0 m
Draught	2.7 m
Displacement, max	approx 180 t
Crew	8–10
Passengers	400
Fuel capacity	17,750 litres
Propulsive power	2 × 2,040 kW
Operational speed	31.7 kt
Range	472 n miles
Operational limitations	wave height 3.5 m

Classification: DnV Passenger Ship Light Craft.
Propulsion: Main engines: two MTU 16V 396 TB 84 rated at 2,040 kW (25°C air, 25°C sea) at 1,940 rpm, driving twin shaft and three-blade fixed-pitch Lips propellers via 3.23:1.0 ZF gearboxes.
Auxiliary systems: A ballast tank system is operated to adjust craft attitude: Pump: 50 m³/h, 1.5 kW, 380 V on each of four ballast tanks in the platform, each of 7.5 t capacity.

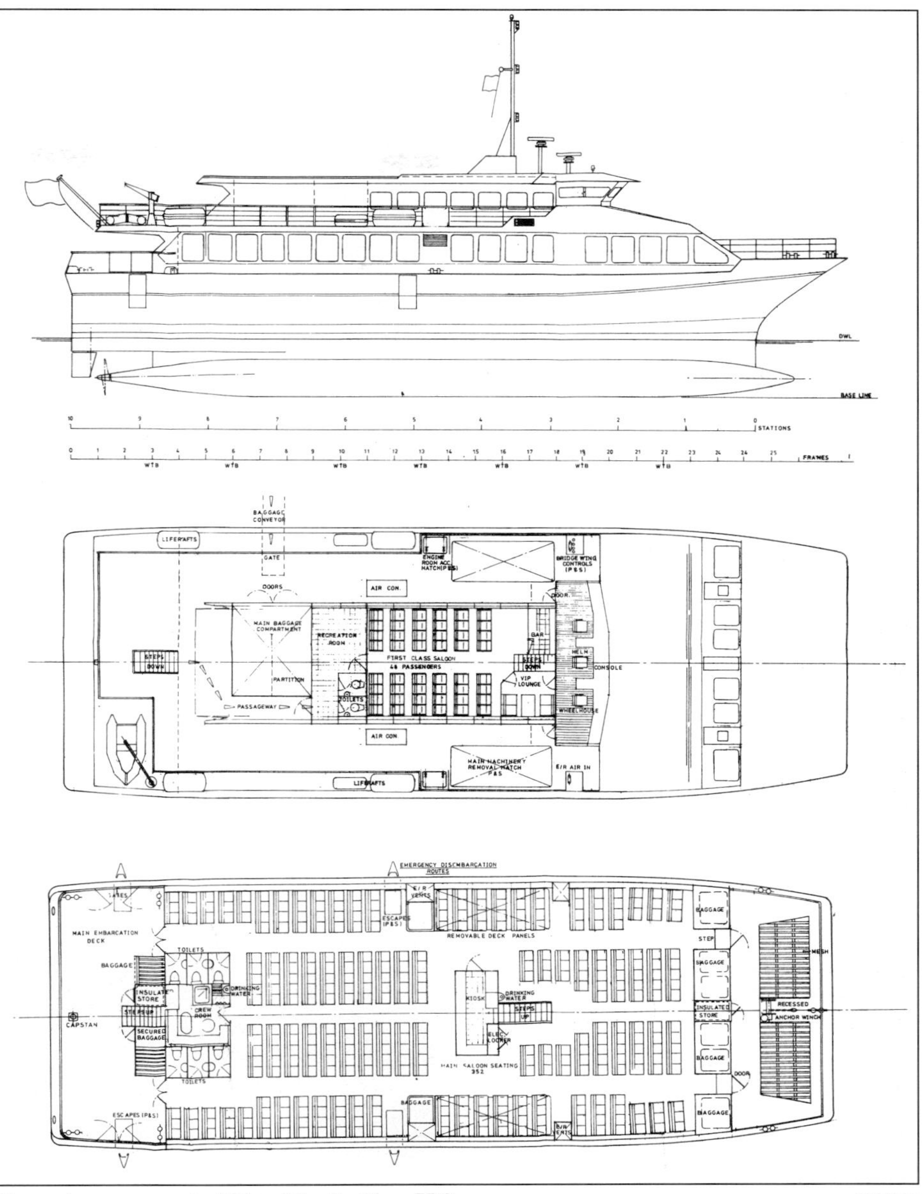

General arrangement of 37 m Atlantic Class FDC 0506674

23 m Passenger transfer craft, Cawsand (Maritime Photographic) 0045121

Vessel type	Vessel name	Yard No	Length (m)	Speed (kt)	Seats	Vehicles	Originally delivered to	Date of build
Atlantic Class FDC	*Patria*	–	36.5	31.7	400	–	Regional Government of Mariera, Portugal	1990
23 m Passenger Transfer Craft	*Cawsand*	–	23.9	12.0	75	–	UK Ministry of Defence	1996
23 m Passenger Transfer Craft	*Bovisand*	–	23.9	12.0	75	–	UK Ministry of Defence	1996

United States

Eastern Shipbuilding Group

2200 Nelson Street, Panama City, Florida 32401, US

Tel: (+1 850) 763 19 00
Fax: (+1 850) 763 79 04
e-mail: info@easternshipbuilding.com
Web: www.easternshipbuilding.com

K Munroe, *Vice-President*

Known for the design and production of steel supply vessels and offshore service craft, Eastern Shipbuilding also has proven capabilities in the construction of lightweight aluminium craft.

Stillwater River *on trials* 0045232

Stillwater River

Designed by Alan C McClune Associates Inc, the *Stillwater River* was built by Eastern Shipbuilding Group for TRICO Marine Services for operation in Brazil's Campos Basin area as a crewboat. The vessel was delivered in November 1998.

Specifications

Length overall	36.6 m
Beam	13.1 m
Draught	3.5 m
Crew	6
Passengers	250
Propulsive power	7,200 kW
Max speed	31 kt
Operational speed	25 kt

Classification: ABS+A1+AMS (USCG Subchapter K)

Structure: All aluminium.

Propulsion: Two Allison 501 KFS gas turbines driving Lips controllable-pitch propellers via a Philadelphia HP-25 gearbox with a ratio of 28:1.

Electrical system: Two Cummins NTA-855-G3 generator sets each rated at 315 kW.

Auxiliary systems: Ride control system, MDI, bow thruster rated at 150 kW.

Vessel type	Vessel name	Yard No	Length (m)	Speed (kt)	Seats	Vehicles	Originally delivered to	Date of build
37 m SWATH	*Stillwater River*	–	36.6	25.0	250	–	–	November 1998

Navatek Ships Ltd

A subsidiary of Pacific Marine
Suite 1880, 841 Bishop Street, Honolulu, Hawaii 96813, United States

Tel: (+1 808) 531 70 01
Fax: (+1 808) 523 76 68
e-mail: schmicker@navatekltd.com
Web: www.navatekltd.com

Steven Loui, *President*
Gary Shimozono, *Vice President, R&D*
Todd Peltzer, *Director of Programs*
Michael Schmicker, *Vice President, Corporate Communications*

In 1977, Pacific Marine and Dr Ludwig Seidl of the University of Hawaii formed Pacific Marine Engineering Science Co to develop a commercial SWATH ship design. In 1979 the company was granted a patent on its SWATH design.

Navatek Ships Ltd researches, designs, tests and builds advanced marine hull forms for both commercial and military clients. These hull forms include small waterplane area craft (SWATH, SLICE) and HYbrid Small WAterplane Craft (HYSWAC, Midfoil, HDV 100) employing lifting bodies (see Hydrofoils). The company holds multiple patents on its proprietary hull forms.

Navatek I 00506676

Navatek I

Pacific Marine designed and built its first SWATH vessel, *Navatek I*. After being outfitted at Northwest Marine, Portland, Oregon with a partial superstructure and pilot-house, sea trials commenced in May 1989. During those trials, *Navatek I* sailed 600 n miles from Portland to San Francisco, California, then 2,100 n miles across the Pacific to Honolulu, Hawaii. During the six day trip it averaged 15 kt, its design cruising speed, through seas of 1.8 to 3 m. It also demonstrated the ability to maintain 93 per cent of its speed through Sea State 5. *Navatek I* achieved a top speed of over 17 kt and its fuel consumption was 265 litres/h at 1,600 rpm cruising speed. Its reduced ships motions were extremely good; waves up to 2.4 m produced almost no motion.

With the successful completion of sea trials in September 1989, Royal Hawaiian Cruises Ltd of Honolulu, Hawaii, signed a long-term lease on the vessel for use as a day excursion/dinner cruise boat serving the Hawaiian tourist trade. In October 1989, *Navatek I* sailed 6,500 miles via the Panama Canal to New Orleans, Louisiana, where Trinity Marine installed the rest of the two-deck superstructure.

In early 1990, the US Coast Guard certified *Navatek I* to carry 400 passengers on an ocean route. It thus became the first SWATH to receive US Coast Guard approval as a commercial, passenger-carrying vessel.

Navatek I began commercial service in March 1990, and currently operates three scheduled cruises a day, seven days a week, in ocean waves ranging from 1.5 to 4.5 m.

In September 1991, the Coastguard increased the passenger certification of *Navatek I* to 430 persons.

Specifications

Length overall	43.0 m
Beam	16.0 m
Draught	2.4-4.3 m
Passengers	430
Propulsive power	2 × 1,007 kW
Operational speed	15 kt
Max speed	18 kt

Propulsion: Powered by twin Deutz MWM 16V-816CR diesels, continuous rating of 1,007 kW each, driving Ulstein reduction

Vessel type	Vessel name	Yard No	Length (m)	Speed (kt)	Seats	Vehicles	Originally delivered to	Date of build
43 m SWATH	*Navatek I*	–	43.0	15.0	430	None	Royal Hawaiian Cruises Ltd	September 1989
25 m SWATH	*NavatekII*	–	25.0	19.0	150	None	Hawaii	April 1994
32 m SLICE	*SLICE*	–	32.0	27.0	–	None	–	November 1996

gears and Ulstein controllable-pitch four-blade propellers.

Electrical system: Twin Detroit Diesel 6 to 7 litre generators rated at 99 kW each.

Navatek II

In 1991, Navatek Ships Ltd completed design, development and tank model testing in California of a proprietary, second-generation SWATH design developed by engineers at Navatek Ships Ltd. In contrast to the original PAMESCO design, the new design features twin-canted struts for which Navatek Ships Ltd holds patents.

Navatek's patented canted strut design offers several advantages over the vertical strut design employed in *Navatek I*. Inherent in any SWATH design is the tendency of the bow to dive, requiring fin stabilisation to overcome this phenomenon. Fin systems, however, add resistance, requiring increased horsepower to move the vessel through the water. The twin-canted hull design of Navatek Ships Ltd produces a smaller bow down trimming moment, thus allowing smaller fins which result in less resistance and more speed per horsepower installed. Other advantages of canted struts include a larger damping effect in pitch, roll and heave, resulting in significantly better motion characteristics.

In November 1992, Navatek Ships Ltd began construction of its first twin-canted strut SWATH, the *Navatek II,* at its Honolulu shipyard.

Navatek II was launched in January 1994, and began commercial service in April 1994 as a day cruise/adventure boat in the Hawaii tourist trade. Variations of this design for US Coast Guard duties have also been studied.

Specifications

Length overall	25.0 m
Beam	11.6 m
Draught	1.7 m
Passengers	150
Operational speed	19 kt
Range	750 n miles

Propulsion: Engines: two MTU 12V 183.
Transmissions: two ZF BW 250 gearboxes.
Thrust devices: two four-blade workboat-style propellers.

Auxiliary systems: Generators: two 45 kW Northern Lights.

SLICE prototype on trials

0024759

Navatek II *in operation* 0506902

32 m SLICE prototype under construction 0002640

SLICE technology partnership with Lockheed Martin

In September 1993, Navatek Ships and Lockheed Martin Company announced that they had teamed up to commercialise SWATH ship technology originally developed by Lockheed for the defence industry. Lockheed Martin Company, a leading defence contractor, has been involved in the development of many unique, high-technology marine vehicles. In addition to the world-recognised Polaris, Poseidon and Trident missile programmes, Lockheed developed one of the world's first deep diving manned vehicles, *Deep Quest,* followed by development of the Navy's submarine rescue submersibles, Deep Submergence Rescue Vehicles I and II. Lockheed's patented motion control system has been installed in the US Navy's T-AGOS 19 and T-AGOS 23 SWATH surveillance ships. The company's high-technology SWATH hull designs and computer-aided ship control systems continue to push the state of the art. In April 1993, the US Navy unveiled its SWATH stealth ship *Sea Shadow,* designed and built by Lockheed. In addition to their collaboration on SWATH design, Navatek and Lockheed are exploring other hull forms, first developed by Lockheed for the US defence industry, which may offer commercial potential. One of these hull forms, called SLICE, is a high-speed, 35 kt variant of SWATH technology. By minimising wave-making resistance at high speeds, it produces significantly higher speeds for constant horsepower when compared to conventional SWATH hull technology, yet it retains the sea-keeping of SWATH technology. Navatek is licensed to market commercial applications of this technology. In late 1993 the two companies received a USD10 million advanced technology development grant from the US Navy to design and build a 177 tonne, 32 m SLICE prototype.

Design and tank testing work on the SLICE prototype was conducted at Lockheed's California facilities. Construction of the prototype began in early 1995. The prototype was launched on 26 November 1996 and underwent sea trials in Hawaiian waters. It is currently being used as a test and evaluation platform.

32 m SLICE

Specifications

Length overall	32.0 m
Beam	16.8 m
Draught	4.3 m
Displacement	max 180 t
Fuel capacity	26,000 litres
Water capacity	3,000 litres
Max speed	30 kt
Operational speed	27 kt
Range	450 n miles

Structure: All-aluminium (5083 H111) hull and superstructure.

Propulsion: Powered by two MTU 16V 396 TB94 main diesels driving CP propellers via reduction gearboxes.

Nichols Brothers Boat Builders Inc

5400 S Cameron Road, Freeland, Whidbey Island, Washington 98249, United States

Tel: (+1 360) 331 55 00
Fax: (+1 360) 331 74 84
e-mail: nichols@whidbey.com
Web: www.nicholsboats.com

Matt Nichols, *Chief Executive Officer*
Bryan Nichols, *President*
Ken Schoonover, *Vice President, Production*
Kevin Tallman, *Sales and Marketing*

Nichols Brothers received an order for a 37 m SWATH vessel in late 1992, to be operated by Party Line Cruises of Miami. This vessel was constructed in the same facilities used for the shipyard's catamaran ferries.

37 m SWATH FERRY

Cloud X

Designed by Swath International Limited, this craft was launched in May 1995. Trials were undertaken in 1997 and final delivery took place in 2002. The craft finally started operations from Florida to Freeport, Grand Bahamas in August 2005. It is understood that the vessel was withdrawn from service in 2006.

Cloud X on trials 0526360

Specifications

Length overall	37.2 m
Beam	18.0 m
Draught	3.4 m
Passengers	365
Fuel capacity	16,000 litres
Water capacity	3,000 litres
Propulsive power	2 × 2,870 kW
Max speed	30 kt
Operational speed	28 kt

Propulsion: Main engines are two Textron Lycoming TF40 gas turbines rated at 2,870 kW each, driving two Kamewa CPP propellers.

Vessel type	Vessel name	Yard No	Length (m)	Speed (kt)	Seats	Vehicles	Originally delivered to	Date of build
37 m SWATH Ferry	*Cloud X*	–	37.2	28.0	365	–	Party Line Cruises	1995

Swath Ocean Systems LLC

997 G Street, Chula Vista, California 91910, United States

Tel: (+1 619) 409 78 88
Fax: (+1 619) 409 78 91
e-mail: info@swathocean.com
Web: www.swathocean.com

Paul Kotzebue, *General Manager*
Robert Thompson, *Sales Manager*

Swath Ocean Systems (SOS) moved to a 5.5 acre facility in July 1993. This new facility offers indoor construction with overhead crane handling equipment, outdoor yard with bayside launch capabilities and materials storage area. SOS has a complete in-house design and engineering department using the latest hydrostatics, resistance and CAD software.

MHS-1

EODMU Seven

This vessel was delivered to the US Navy as an explosive ordinance disposal minehunter craft.

Specifications

Length overall	12.0 m
Beam	5.5 m
Draught	1.4 m
Displacement, full load	25 t
Fuel capacity	3,200 litres
Max speed	15 kt

Construction: All aluminium hull and superstructure.

Propulsion: Two Caterpillar 3116 DITA diesel engines each rated at 190 kW at 2,600 rpm and driving a five-bladed fixed-pitch propeller.

Chubasco

Chubasco was launched on 28 March 1987 by James Betts Enterprises of San Diego for Leonard Friedman for drift fishing. The design and building are under patents of Dr Thomas G Lang of the Semi-Submerged Ship Corporation.

It served as the official committee boat for the judges for the 1988 America's Cup races held at San Diego and in 1995 was sold to the New Zealand Group, the winners of the Cup in that year.

Specifications

Length overall	21.9 m
Beam	9.4 m
Draught	3.0 m
Displacement, max	70 t
Fuel capacity	18,920 litres
Water capacity	1,890 litres
Propulsive power	2 × 559 kW
Max speed	21 kt
Operational speed	20 kt

Structure: Marine grade aluminium.

Propulsion: Main engines are two DDC 8V 92 TI diesels, 559 kW each, turbocharged.

Control: Gyro-activated stabiliser fin system.

Steering station at stern, steering station at each wing outside pilot-house, bow thruster and controls at three stations.

***Betsy* (*ex-Suave Lino*)**

Completed in 1981 for SOS as a fishing boat, *Betsy* was used as a tender by the Sail America Syndicate during the America's Cup defence off Fremantle, Australia.

The vessel is constructed of aluminium with a single-strut configuration for each hull. Two high-speed diesels of 425 hp each, mounted at deck level, drive fixed-pitch propellers through bevel gears. Automatic fin control fitted.

Specifications

Length overall	19.2 m
Beam	9.1 m
Draught	2.1 m
Displacement, min	53 t
Payload	14 t
Max speed	18 kt

Propulsion: Main engines are two DDC 8V 71, 317 kW each.

1500 CLASS

Lonestar

This vessel was delivered to the Houston Pilots.

Specifications

Length overall	15.3 m
Beam	9.2 m
Draught	2.1 m
Displacement, full load	65 t
Operational speed	23 kt

Construction: All aluminium hull and superstructure.

Vessel type	Vessel name	Yard No	Length (m)	Speed (kt)	Seats	Vehicles	Originally delivered to	Date of build
MHS-1	*EODMU Seven*	–	12.0	15	–	–	US Navy	–
1500 Class Pilot	*Lonestar*	–	15.25	23.0	–	–	Houston Pilots	–
19 m SWATH	*Betsy* (*ex-Suave Lino*)	–	19.2	–	–	–	SWATH Ocean Systems Inc	1981
22 m SWATH	*Chubasco*	–	21.9	20.0	–	–	James Betts Enterprises	1987
2000 Class	*Frederick G Creed*	–	20.0	25.0	125	–	Canadian Department of Fisheries and Oceans	1989
2000 Class	*Houston*	–	20.4	23.0	–	–	Houston Pilots	1993
2000 Class	*Navigante*	–	20.4	–	–	–	–	–
4000 Class	*Chubasco I*	–	27.5	–	–	–	Chubasco Charters	1993
6000 Class	*Western Flyer*	–	30.0	8.0	–	–	Monterey Bay Aquarium Research Institute	1996
12 m SWATH	–	–	12.0	21	US Navy	–	US Navy	2006

2000 CLASS

Frederick G Creed

A 20.4 m 80 tonne Swath vessel, launched in November 1989. It was made available for the Canadian Department of Fisheries and Oceans under a lease purchase contract for oceanographic and hydrographic service off the eastern coasts of Canada and the US.

Specifications

Length overall	20 m
Beam	10.7 m
Draught	2.6 m
Displacement, min	57.9 t
Displacement, max	80.3 t
Passengers	125
Propulsive power	2 × 805 kW
Max speed	29 kt
Operational speed	25 kt

Propulsion: Main engines are two DDC 12V 92TA 805 kW diesels.

2000 CLASS

Houston

A 20.4 m 80 tonne Swath vessel, based on the SOS 2000 class craft, was delivered to the Houston Pilots in early 1993. By December 1995 the vessel had logged over 14,000 hours and 24,000 pilot transfers.

Specifications

Length overall	20.4 m
Beam	11.3 m
Draught	2.4 m
Displacement, min	57 t
Displacement, max	78 t
Propulsive power	2 × 750 kW
Operational speed	23 kt

Propulsion: Main engines are two Cat 3412 diesels rated at 750 kW each.

4000 CLASS

Chubasco 1

The vessel was launched in October 1993 and delivered to her owner Chubasco Charters of San Diego. It is currently used for long-range fishing charters.

Specifications

Length overall	27.5 m
Beam	13.7 m
Displacement, max	182.8 t

Propulsion: Two Caterpillar 3516TA diesels each rated at 2,100 kW at 1,880 rpm and each driving a fixed-pitch propeller via a ZF BWK 755 reverse reduction gearbox.

6000 CLASS

RV Western Flyer

This 30 m Swath was completed in 1996 for the Monterey Bay Aquarian Research Institute (MBARI) to serve as a mother ship for Remotely Operated Vehicle (ROV) operations. Deployment and recovery of the ROV is designed to take place via a central 'moonpool' and the arrangement of the vessel is built around this concept. Propulsion of the vessel is diesel-electric which allows effective control of the craft at very low speeds whilst providing a maximum speed of 14.5 kt. The vessel was originally designed as an 18 kt craft. The two diesel engines are Caterpillar 3512 type driving 850 kW generator sets. The main propulsion motors are GE 930 kW DC electric motors with a maximum of 400 rpm. At a cruise speed of 8 kt the vessel has a range of 5,200 n miles and at 14 kt, 2,400 n miles.

SOS 4000 Class Chubasco 1 0506982

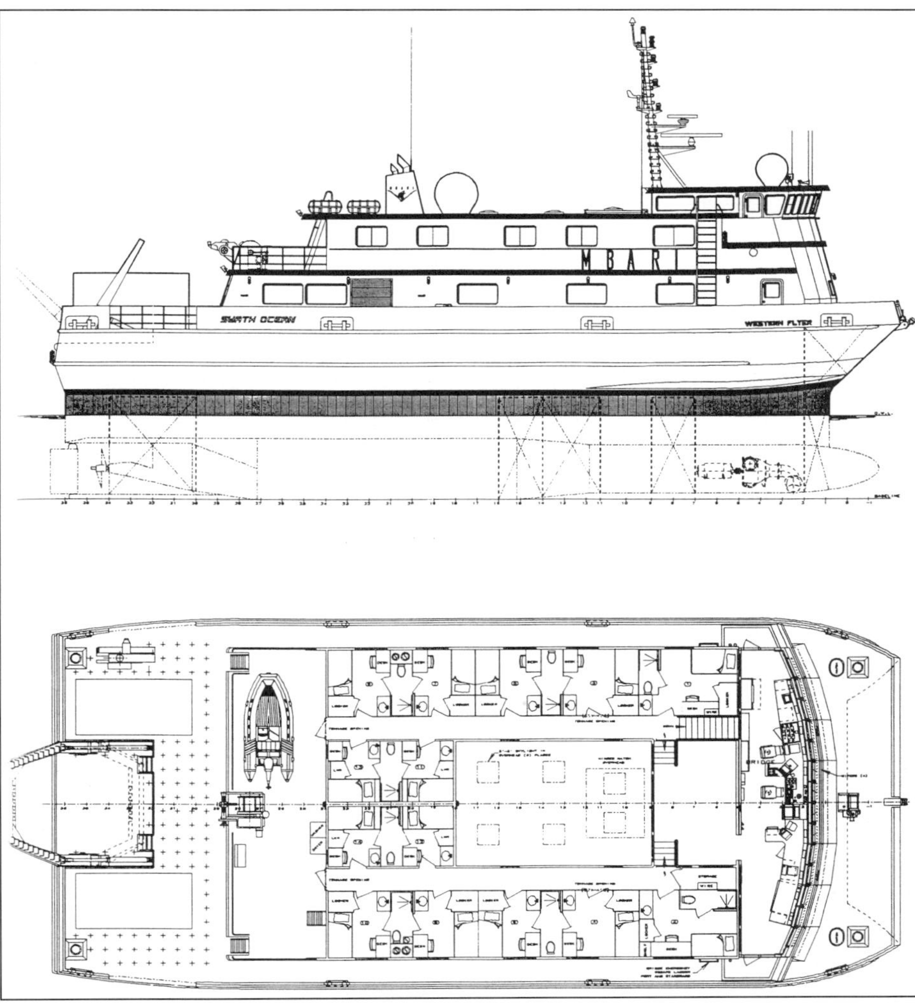

General arrangement of RV Western Flyer 0002641

Houston Pilot Cutter 0506834

HIGH-SPEED MONOHULL CRAFT

Company listing by country

Australia
Austal
BSC Marine
Image Marine
SBF Shipbuilders
Strategic Marine Pty Ltd

Chile
Asmar Shipbuilding and Shiprepairing Company

France
Aker Yards SA
Chantiers Navale Gamelin
Construction Navale Bordeaux
Constructions Mécaniques de Normandie (CMN)
OCEA

Germany
Lürssen Werft GmbH & Co

Hong Kong
Cheoy Lee Shipyards Ltd

Italy
Azimut-Benetti SpA
Cantieri Navale Foschi
De Poli Cantieri Navali
Fincantieri Cantieri Navali Italiani SpA
Industrie Navales Meccaniche Affini SpA (INMA)
Moschini Cantieri Ing SpA
Riva
Rodriquez Cantieri Navali SpA

Japan
Kiso Shipyard
Mitsubishi Heavy Industries Ltd
Mokubei Shipbuilding Company
Yamaha Motor Company Ltd

Korea, South
Semo Company Ltd

Malaysia
Malaysia Marine & Heavy Engineering
Yong Choo Kui Shipyard

Monoco
Wally Yachts

Netherlands
Damen Shipyards
Heesen Shipyards

Norway
Bergen Mekaniske Verksted AS
Brødrene Aa
GS Marine AS
Maloy Werft
Rubb MarineA/S

Russian Federation
Central Hydrofoil Design Bureau (CHDB)

Singapore
Damen Shipyards Singapore Pte Ltd
Greenbay Marine Pte Ltd
Marinteknik Shipbuilders (S) Pte Ltd
Penguin Shipyard International Ltd
Precision Craft (88) Pte Ltd
Strategic Marine Pte Ltd

Spain
Astillerod de Huelva
Navantia

Sweden
Boghammar International AB

Taiwan
Dragon Yacht Building

United Arab Emirates
Dubai Drydocks
Seaspray Marine Services & Engineering

United Kingdom
Fairline Boats plc
FBM Babcock Marine Group
McTay Marine

United States
Allen Marine
Austal US
Blount Boats Inc
Breaux Brothers Enterprises Inc
Breaux's Bay Craft Inc
Christensen Shipyards
Derecktor Shipyards
Gladding-Hearn Shipbuilding
Gulf Craft Inc
Magnum Marine Corporation
Marinette Marine
Navatek Ships Ltd
Nordlund Boat Company
Seaark Marine Inc
Seacraft Shipyard Corporation
Swiftships Shipbuilders, LLC
Trinity Yachts Inc
VT Halter Marine Inc
Westport Shipyard Inc

Australia

Austal

100 Clarence Beach, Henderson, Perth, Western Australia 6166, Australia

Tel: (+61 8) 94 10 11 11
Fax: (+61 8) 94 10 25 64
e-mail: marketing@austal.com
Web: www.austal.com

John Rothwell, *Executive Chairman*
Stephen Murdoch, *Chief Operating Officer*
Michael Atkinson, *Director, Financial Control*

Austal is a leading world supplier of fast commercial ferries and aluminium patrol craft.

48 m HIGH-SPEED MONOHULL PASSENGER FERRY

Ono-Ono

This vessel was delivered in June 1994 to SPI Maritime to operate from Tahiti's main port Papeete, to the neighbouring islands of Huahine, Raiatea, Bora Bora and Tahea.

Specifications

Length overall	48.0 m
Length waterline	41.3 m
Beam	9.0 m
Draught	1.2 m
Crew	11
Passengers	450
Fuel capacity	20,000 litres
Water capacity	2,000 litres
Propulsive power	3 × 1,960 kW
Max speed	35 kt

Propulsion: Main engines are three MTU 16V 396 TE 74L diesels rated at 1,960 kW at 1,940 rpm, driving three Kamewa 63 water-jets via three Reintjes VLJ930 gearboxes.

Ono-Ono *on trials* 0506903

30 m MONOHULL FERRY

Ertugrul Gazi

Aksemseddin

These two 30 m monohull vessels were sold to the Turkish operator Istanbul Deniz Otobüsleri in December 1994. The vessels were purchased to operate between Istanbul and ports in the Sea of Marmara and the Bosphorus.

Specifications

Length overall	30.0 m
Length waterline	25.8 m
Beam	7.0 m
Draught	1.2 m
Crew	4
Passengers	155
Fuel capacity	4,000 litres
Water capacity	400 litres
Propulsive power	2 × 822 kW
Operational speed	25 kt

Propulsion: Powered by two MTU 8V 396 TE 74 diesels rated at 822 kW each. Each diesel drives through a ZF BU255 gearbox to an FFJet 550 water-jet.

Vessel type	Vessel name	Yard No	Length (m)	Speed (kt)	Seats	Vehicles	Originally delivered to	Date of build
48 m High Speed Monohull Ferry	*Ono-Ono*	–	48.0	–	450	none	SPI Marine	1994
30 m Monohull Ferry	*Ertugrul Gazi*	–	30.0	25.0	155	none	Istanbul Deniz Otobusleri	1994
30 m Monohull Ferry	*Aksemseddin*	–	30.0	25.0	155	none	Istanbul Deniz Otobusleri	1994

BSC Marine Group

Suite 12, 621 Coronation Drive, Toowong, Queensland 4066, Australia

Tel: (+61 7) 38 76 15 22
Fax: (+61 7) 38 76 15 66
e-mail: info@bscship.com.au
Web: www.bscship.com.au

Michael Hollis, *Managing Director*

Known for its low wash catamaran craft, BSC has also produced a range of monohull commercial vessels including the MV *Delphi*, a 19 m, 28–30 kt workboat which is designed to Lloyd's Register SSC G3.

20 m PATROL VESSEL

MV Flinders

The first vessel was delivered to Queensland Boating and Fisheries Patrol in 2002. The all aluminium vessel has an overall length of 20 m and is powered by two Caterpillar 3406E diesel engines each driving a fixed-pitch propeller via a ZF 305A gearbox.

Vessel type	Vessel name	Yard No	Length (m)	Speed (kt)	Seats	Vehicles	Originally delivered to	Date of build
20 m Patrol vessel	*MV Flinders*	–	20	22	–	none	Queensland Boating & Fisheries	2002
19 m Workboat	*MV Delphi*	–	19	28	–	none	Queensland National Parks	–

Image Marine PTY Ltd

Part of the Austal Group
100 Clarence Beach, Henderson, Perth, Western Australia 6166, Australia

Tel: (+61 8) 94 10 11 11
Fax: (+61 8) 94 10 25 64
e-mail: marketing@austal.com
Web: www.austal.com

Image Marine was established in 1986 and is a builder of customised aluminium vessels of both catamaran and monohull design. Image has built in excess of 165 vessels for the domestic and export market including Australasia, the Asia-Pacific, East Africa and the Caribbean.

While vessels for the niche live-aboard dive boat and fast ferry sectors form a major component of Image Marine's portfolio of expertise, the product range also includes cruise and dinner-cruise vessels, crewboats, pilot and patrol vessels and private motor yachts, all in both catamaran and monohull design.

Image Marine was acquired by the Austal Group in July 1998 and is dedicated to the construction of aluminium vessels of up to 50 m in length.

Vessel type	Vessel name	Yard No	Length (m)	Speed (kt)	Seats	Vehicles	Originally delivered to	Date of build
17 m Patrol Boat	*R J Hawke*	–	17.0	26.0	–	none	–	–
22 m Image Marine Seasbus	*Subic Eagle 1*	–	22.0	29.0	150	none	Eagle Ferries	1996
22 m Image Marine Seasbus	*Subic Eagle 2*	–	22.0	29.0	150	none	Eagle Ferries	1996
30 m Ferry	*Mary D*	145	30.0	29.0	184	none	Sarl Aquarius, New Caladonia	1998

30 m FERRY

Mary D

This vessel was delivered to tourist resort operator Sarl Aquarius of Noumea, New Caledonia, in 1998.

Specifications

Length overall	30.0 m
Length waterline	26.8 m
Beam	7.3 m
Draught	1.0 m
Deadweight	18.6 t
Passengers	189
Fuel capacity	6,000 litres
Max speed	29 kt

Classification: Lloyds Register.

Structure: All aluminium hull and superstructure.

Propulsion: Three MTU 12V 183 TE 72 diesel engines each rated at 660 kW and each driving a Hamilton 461 water-jet via a ZF BW190 gearbox.

30 m Ferry Mary D 0106089

SBF Shipbuilders

33 Clarence Beach Road, Henderson, Western Australia 6166, Australia

Tel: (+61 8) 94 10 22 44
Fax: (+61 8) 94 10 18 07
e-mail: info@sbfships.com.au
Web: www.sbfships.com.au

Don Dunbar, *Principal*
Alan McCombie, *Managing Director*

SBF Shipbuilders builds high-speed aluminium crew boats and passenger ferries, designed to meet the operational needs of individual shipping companies.

35 m PASSENGER FERRY

Eagle Express

This vessel was delivered to Rottnest Express of Freemantle in 2000.

Specifications

Length overall	35.0 m
Length waterline	31.7 m
Beam	7.0 m
Draught	1.9 m
Passengers	350
Fuel capacity	10,000 litres
Water capacity	1,000 litres
Max speed	32 kt
Operational speed	30 kt

Classification: Australian Dept of Transport Class 1D.

Structure: Aluminium hull and superstructure.

Propulsion: Three MTU 12V 2000 M70 diesel engines each rated at 790 kW at 2,100 rpm each driving a five-bladed Stone Marine fixed-pitch propeller via a Twin Disc MG6557-00-SC gearbox.

32 m CREWBOAT

Abeer Thirty Nine

Designed and built for Abeer Marine Services, the crewboat is equipped with a fire monitor and chemical spray dispersant system.

Specifications

Length overall	32.3 m
Beam	7.2 m
Draught	1.53 m
Passengers	116

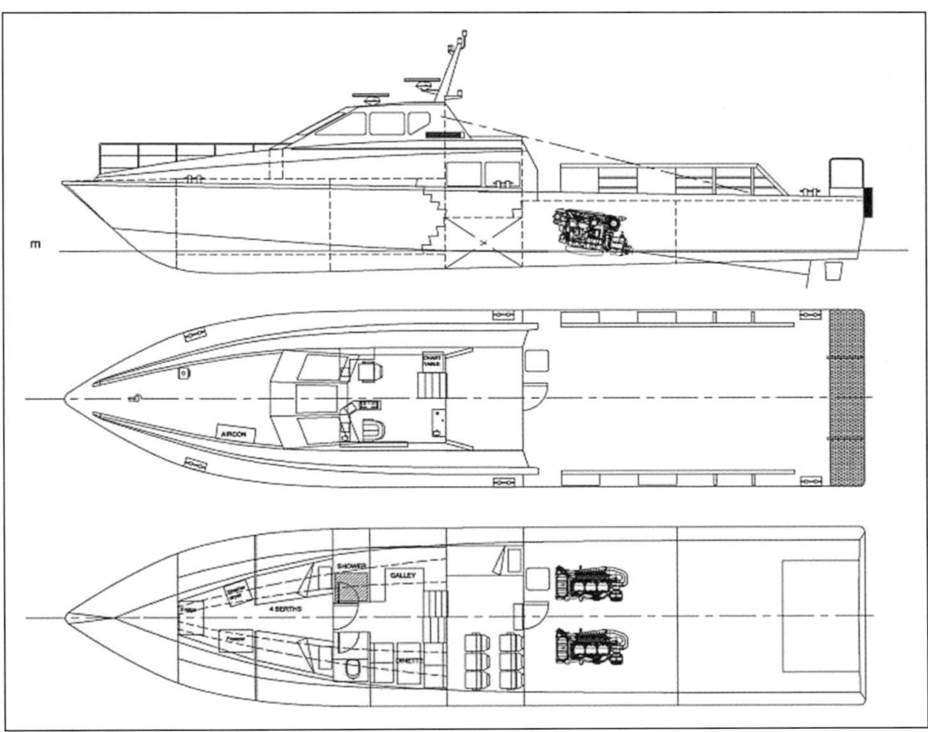

General arrangement of Abeer Forty One *and* Abeer Forty Two 1112088

Auto Batam 7 0506677

35 m Eagle Express 0102136

Tenggara Satu 1036533

Crew	8
Fuel capacity	24,000 litres
Water capacity	10,000 litres
Operational speed	28 kt

Classification: ABS +A1 HSC Crewboat AMS.

Propulsion: Two Cummins KTA 38-M2 diesel engines, each rated at 1,044 kW at 1,950 rpm and each driving a fixed-pitch propeller via a ZF 4540A gearbox.

Auxillary: Two Cummins 68MXDGDA 58 kW generators.

Najade

This 31 m high-speed monohull ferry was delivered to Rederij Doeksen in Holland in 1999 for operation from Harlingen to the Friesland Islands.

Specifications

Length overall	31.0 m
Length waterline	27.0 m
Beam	6.7 m
Draught	0.9 m
Passengers	180
Fuel capacity	8,000 litres
Water capacity	1,000 litres
Propulsive power	2,370 kW
Max speed	37.5 kt
Operational speed	32 kt

Classification: DnV +1A1 HSLC Passenger R4.

Structure: All-aluminium hull and superstructure.

Propulsion: Three MTU 12V 2000 M70 diesel engines each rated at 790 kW and each driving an MJP 450 water-jet via a ZF BW 250 gearbox.

Strait Runner

This vessel was ordered by Auto Batam Ferry Services of Singapore and delivered in July 1995.

Specifications

Length overall	31.0 m
Length waterline	27.6 m
Beam	6.5 m
Draught	1.8 m
Passengers	152
Propulsive power	1,830 kW
Max speed	32 kt

Propulsion: Powered by three MTU 12V 183 TE 72 diesels, each rated at 610 kW at 2,100 rpm and each driving a fixed-pitch propeller via a reverse reduction gearbox.

Auto Batam 7

This monohull ferry (Yard No KP07) was delivered in March 1991 for service with Kalpin Shipping and Trading Company of Singapore, for the Singapore to Batam Island route. *Auto Batam 7* is a sister vessel of *Auto Batam 6*, delivered in 1990.

Specifications

Length overall	30.0 m
Length waterline	24.6 m
Beam	6.5 m
Draught	1.8 m
Displacement	54 t
Passengers	200
Fuel capacity	5,500 litres
Operational speed	27 kt

Classification: Bureau Veritas.

Structure: Hull material: aluminium.

Superstructure material: aluminium.

Propulsion: Main engines are three MTU 12V 183 TE 62 diesels, driving three five-blade propellers via three ZF BW755 gearboxes.

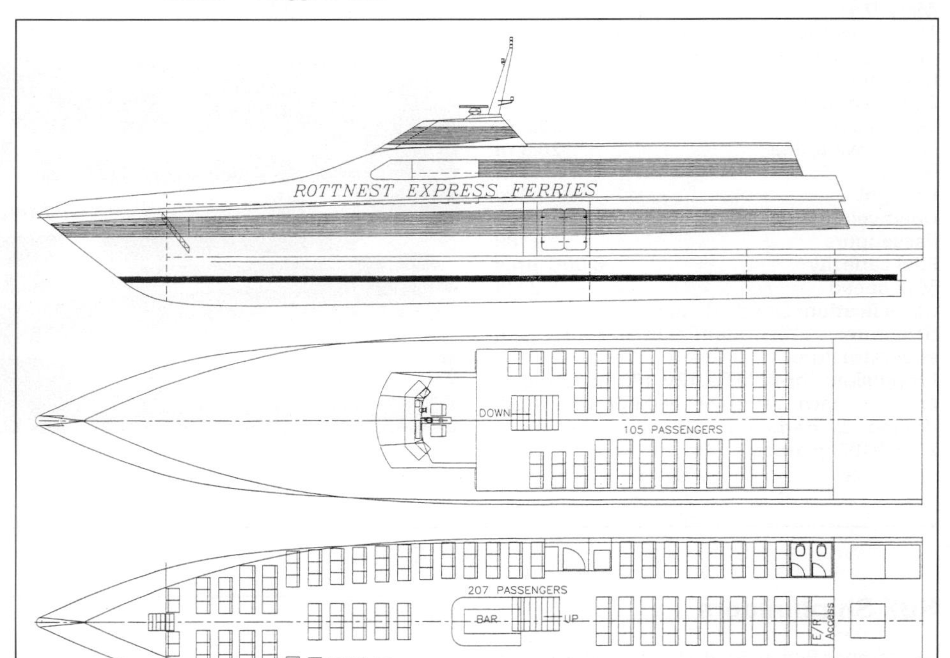

General arrangement of 35 m passenger ferry 0103308

Abeer Thirty Nine 1036534

Electrical system: One Cummins 4B3.9 37 kVA 50 Hz 415/250 V generator, one Lister-Petter 19 kVA air-cooled generator.

Sing Batam 1

This triple screw vessel (Yard No SB28) was ordered at the same time as *Sea Dragon 2* by the same owners but under the company name of Sing Batam Ferries Pte Ltd. This vessel is based upon *Auto Batam 7* delivered to Kalpin Shipping and Trading Company Pte Ltd earlier in 1991. The difference is that the upper deck is fully enclosed and air conditioned, with the same interior seating as the main deck. This vessel was launched in Fremantle in December 1991 and delivered the same month, it took seven days under its own power to arrive in Singapore.

Specifications

Length overall	30.0 m
Length waterline	24.6 m
Beam	6.5 m
Draught	1.8 m
Displacement	54 t
Passengers	226
Fuel capacity	5,500 litres
Operational speed	27 kt

Classification: Bureau Veritas.

Safety standards: Singapore Marine Department.

Structure: Hull material: aluminium.

Superstructure material: aluminium.

Propulsion: Main engines are three MTU 12V 183TE 62 diesels, driving three Stone Marine five-blade propellers via three Niigata MGN 232 gearboxes.

Electrical system: Engines: two Cummins 4B3.9 37 kVA 50 Hz 415/250 V generators.

Vessel type	Vessel name	Yard No	Length (m)	Speed (kt)	Seats	Vehicles	Originally delivered to	Date of build
30 m Ferry	*Sea Flyte*	SF103	30.00	28.0	240	none	Singapore	1981
29 m Ferry	*James Kelly II*	TF90	27.00	27.0	200	none	Tasmania	1983
26 m Ferry	*Sundancer V*	SB90	24.00	34.0	160	none	Singapore	1983
20 m Ferry	*Wilderness Seeker*	TF20	20.00	80.0	100	none	Tasmania	1985
30 m Crewboat	*Sea Phantom* (ex-*Satrya Express)*	SE103	31.00	30.0	62	none	Indonesia	1985
30 m Ferry	*Mary D Princess*	NC28	30.00	23.0	226	none	Australia	1989
30 m Ferry	*Auto Batam 6*	SP28	30.00	30.0	200	none	Kalpin Shipping and Trading Company of Singapore	1990
30 m Ferry	*Auto Batam 7*	KP07	30.00	30.0	200	none	Kalpin Shipping and Trading Company of Singapore	1991
23 m Workboat	*Sea Dragon 2*	SD23	23.00	22.0	12	none	Kalpin Shipping and Trading Company, Singapore	1991
30 m Ferry	*Auto Batam 1* (ex-*Sing Batam 1)*	SB28	30.00	27.0	226	none	Sing Batam Ferries Pte Ltd	1991
23 m Workboat	*Sea Dragon 3*	SD25	23.00	22.0	12	none	Sea Dragon Marine Services Pty Ltd	1992
27 m Ferry	*Tung Hsin*	TW27	27.00	31.0	193	none	Taiwan	1993
30 m Ferry	*Osprey V*	931	30.00	30.0	257	none	Rottnest Express Pty Ltd	1994
31 m Ferry	*Strait Runner*	943	31.00	30.0	152	none	Auto Batam Ferry Services, Singapore	1995
30 m Ferry	*Sea Eagle III*	962	30.00	30.0	260	none	Rottnest Express Pty Ltd	1996
31 m Ferry	*Mistral*	963	31.00	30.0	205	none	Italy	1997
31 m Ferry	*Najade*	982	31.00	32.0	191	none	Rederi Doeksen, Holland	1999
35 m Ferry	*Eagle Express*	991	35.00	30.0	312	none	Rottnest Express	2000
20 m Dive Boat	–	201	20.00	26.0	50	none	–	2002
30 m Ferry	*Tenggara Satu*	227	30.00	25.0	200	none	Asia Tenggara Marine	2002
17 m Crewboat	*Abeer Forty*	211	17.00	25.0	16	none	Barber Dubai Shipping, UAE	2002
32 m Crewboat	*Abeer Thirty Nine*	232	32.00	28.0	113	none	Abeer Marine, Indonesia	2004
21 m Security Boat	*Abeer Forty One*	241	20.83	30.0	4+	none	Abeer Marine, Indonesia	2004
21 m Security Boat	*Abeer Forty Two*	242	20.83	30.0	4+	none	Abeer Marine, Indonesia	2004
21 m Crewboat	*Abeer Forty Three*	251	21.00	25.0	12	none	UAE	2005

Tenggara Satu

This vessel was built for Asia Tenggara Marine and delivered in 2002. The vessel is used for the transfer of gold mining personnel between islands in Indonesia.

Specifications

Length overall	30.0 m
Beam	6.7 m
Draught	1.8 m
Deadweight	20 t
Passengers	200
Fuel capacity	10,000 litres
Water capacity	1,000 litres
Operational speed	25 kt

Classification: Bureau Veritas.

Propulsion: Two Caterpillar 3412E diesel engines each rated at 745 kW and each driving a fixed-pitch propeller via a ZF 2500 A reverse reduction gearbox.

Osprey V

This vessel was delivered in September 1994 to Rottnest Express Pty Ltd for operation from Perth to Rottnest Island.

Specifications

Length overall	30.0 m
Length waterline	26.6 m
Beam	6.5 m
Draught	1.8 m
Passengers	257
Propulsive power	3 × 610 kW
Max speed	32 kt

Classification: Surveyed for Australian Department of Transport (Marine) 1D.

Structure: Marine grade aluminium.

Propulsion: Main engines are three MTU 12V 183 TE 72 diesels, driving three five-blade Stone Marine propellers via three Twin Disc MG 5141 reversing gearboxes.

Sea Eagle

Delivered in December 1996, the *Sea Eagle* is a sister ship to MV *Osprey* although modified for whale watching off the Western Australian coast.

Sea Dragon 2

A 23 m workboat for ship delivery and support work in Singapore harbour anchorage, delivered to Kalpin Shipping and Trading Company in late 1991.

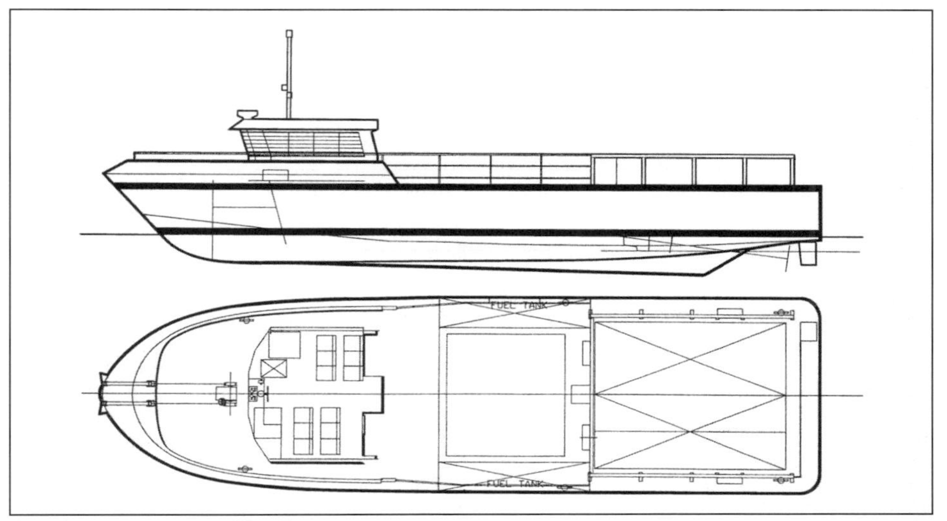

General arrangement of Sea Dragon 2 

0506904

MV Osprey V 0506905

Najade 0069451

Specifications

Length	23.0 m
Beam	6.2 m
Draught	1.3 m
Operational speed	22 kt

Propulsion: Main engines are two Cummins KT19, 380 kW at 2,100 rpm.

Sea Dragon 3

This vessel was ordered when the owners, Sea Dragon Marine Services Pte Ltd, took delivery of her sister vessel *Sea Dragon 2*. Launched and delivered in March 1992, the vessel was delivered under its own power.

Specifications

Length overall	22.5 m
Beam	6.2 m
Draught	1.3 m
Passengers	12
Propulsive power	2 × 317 kW

Structure: Hull material: aluminium.
Superstructure material: aluminium.

Propulsion: Main engines are two Cummins KT19-M diesels, 317 kW each, driving two four-blade propellers, through two Twin Disc 5111A gearboxes.

21 m SECURITY BOAT

Abeer Forty One
Abeer Forty Two

The boats were designed for Abeer marine services to patrol oil and gas fields for BP Indonesia.

Specifications

Length overall	20.83 m
Beam	4.8 m
Draught	1.15 m
Crew	4
Fuel	9,000 litres
Fresh water	2,000 litres
Operational speed	30 kt

Propulsion: Two Caterpillar 3196 diesel engines each rated at 425 kW at 2,300 rpm, each driving two fixed-pitch propellers via Twin Disc 5114 SC gearboxes.

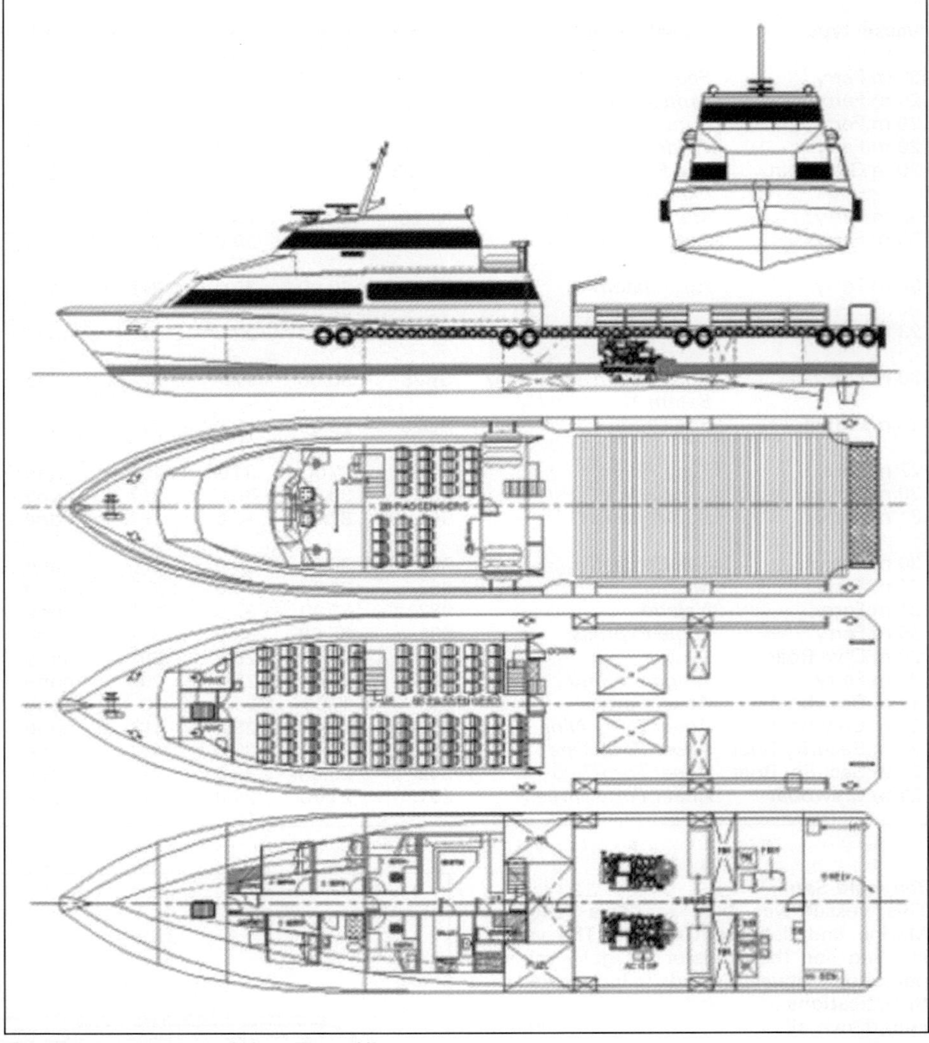

General arrangement of Abeer Thirty Nine 1048775

Strategic Marine PTY Ltd

Lot 5 Clarence Beach Road, Henderson, Western Australia 6166, Australia

Tel: (+61 8) 9437 48 40
Fax: (+61 8) 9437 48 38
e-mail: info@strategicmarine.com
Web: www.strategicmarine.com

Mark Newbold, *Chairman*
Mark Schiller, *Managing Director, Australia*
Ron Anderson, *Managing Director, International Business*

Strategic Marine, formerly Geraldton Boat Builders, specialises in building small fast patrol craft and commercial vessels, offering a range of both monohull and catamaran designs, all of a high-speed planing type. Hulls are manufactured from aluminium or GRP and can be constructed in their facilities in Australia or Singapore.

38 m PASSENGER FERRY

Jacque Cartier

In 2007 Strategic Marine completed the build of a 38 m monohull ferry originally started by Wavemaster International. The vessel was puchased by French operator Societe D'Armement De L'Ouest.

Specifications

Length overall	38.0 m
Length waterline	31.6 m
Beam	7.0 m
Draught	2.3 m
Passengers	210
Max speed	30 kt
Operational speed	28 kt
Range	250 n miles

Propulsion: Powered by three Caterpillar 3412C diesel engines each delivering 820 kW at 2,300 rpm to fixed-pitch propellers via ZF BW 190 gearboxes.

Vessel type	Vessel name	Yard No	Length (m)	Speed (kt)	Seats	Vehicles	Originally delivered to	Date of build
38 m Ferry	*Jacque Cartier*	207	38.0	29.0	210	none	Societe D'Armement De L'Ouest	2006

Chile

Asmar Shipbuilding and Shiprepairing Company

Edifico Rapa Nui, Prat 856, Piso 13, Valparaiso, Chile

Tel: (+56 32) 26 00 00
Fax: (+56 32) 26 01 57
e-mail: secgen@asmar.cl
Web: www.asmar.cl

Claudio Pelaez, *Public Relations Manager*

ASMAR Shipbuilding and Shiprepairing Company has four main facilities in Chile, one each at Valparaiso and Punta Arenas and two at Talcahuano. The company builds a range of military and commercial high-speed vessels such as the Protector Class pilot launches. Recently, four 42.5 m patrol craft have been delivered to the Chilean Navy with two more on order. A 75 m patrol vessel for the government of Mauritius Isles was launched in December 1995.

The company was awarded ISO 9001 quality approval in 1997.

PROTECTOR CLASS PILOT LAUNCHES

Lep Hallef
Lep Alacalufe

In total 16 Protector 33 vessels have been built under licence from FBM for the Chilean Navy. *Iquique,* the last vessel of

Vessel type	Vessel name	Yard No	Length (m)	Speed (kt)	Seats	Vehicles	Originally delivered to	Date of build
Protector Class pilot launch	*Lep Hallef*	–	32.7	18.0	–	None	–	September 1989
Protector Class pilot launch	*Lep Alacalufe*	–	32.7	18.0	–	None	–	September 1989

the production run spanning six years was named in 2004.

Specifications

Length overall	32.7 m
Length waterline	29.0 m
Beam	6.7 m
Draught	2.1 m
Displacement, max	100 t
Crew	14
Fuel capacity	24,000 litres
Water capacity	5,000 litres
Propulsive power	2 × 900 kW
Max speed	20 kt
Operational speed	18 kt
Range	1,100 n miles

One of two FBM Marine-designed Chilean pilot boats
0024762

France

Aker Yards SA

Formerly Alstom Leroux Naval, now part of the Aker Finnyards Group

BP 90180, F-44613 Saint Nazare Cedex, France

Tel: (+33 2) 51 10 91 001
Fax: (+33 2) 51 10 97 97
e-mail: france@akeryards.com
Web: www.akeryards.com

F Sieur, *Commercial Manager*
P Neri, *Naval Architect*

Aker Yards SA is composed of Chantiers de l'Atlantique and Alstom Leroux Naval, both specialised in high added value ship building (cruise ships, military ships, high speed ferries, LNG tankers and other complex ship types) and Ateliers de Montoir, specialised in maritime furniture.

Alstom Leroux Naval delivered its first fast ferry, Corsaire 6000 in 1994 followed by two Corsaire 11000s in 1996, one Corsaire 11500 in 1999, one Corsaire 12000 for NEL Lines and one Corsaire 13000 for SNCM in 2000, and a further two vessels (Corsaire 14000 and 10000) in 2001 for NEL Lines.

Corsaire 6000

Moorea 8

The first vessel of this design was delivered to Emeraude Lines in Spring 1994 for service on the St Malo to Channel Islands route. The vessel is now in service in Polynesia.

Specifications

Length overall	66.0 m
Length waterline	58.0 m
Beam	10.9 m
Draught	2.0 m
Deadweight	92 t
Passengers	450

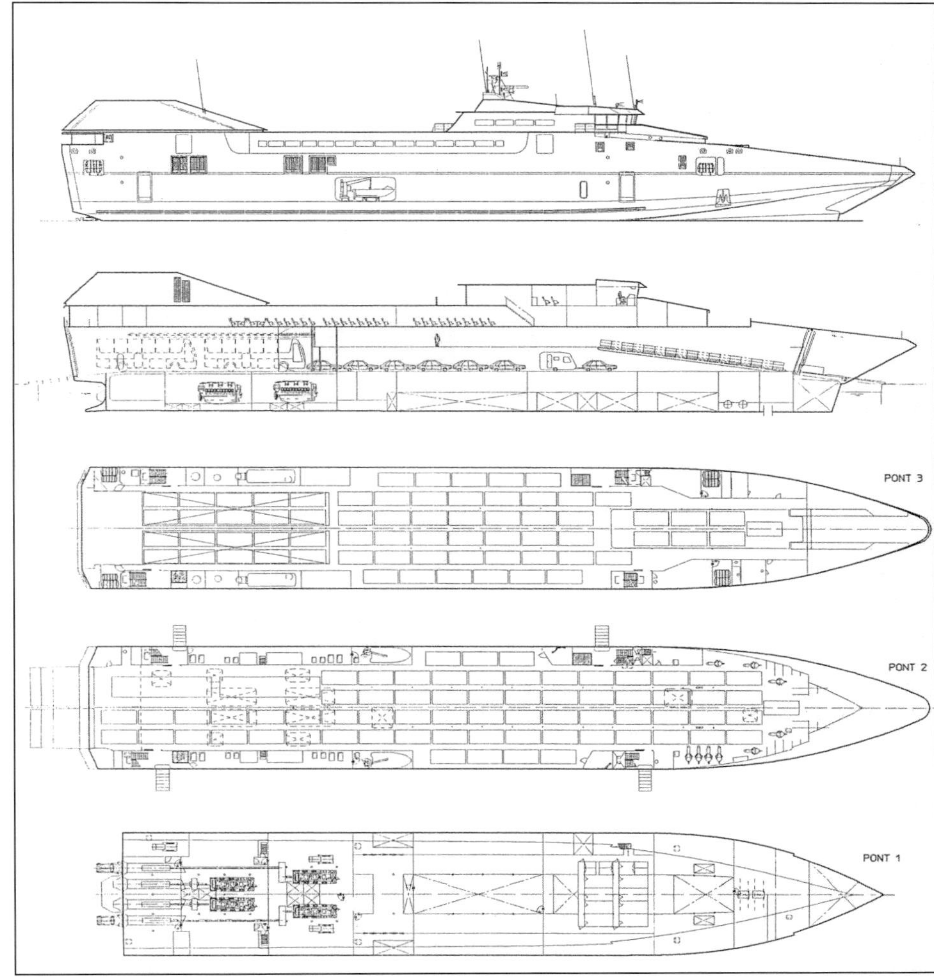

General arrangement of Corsaire 11000
0507025

Vessel type	Vessel name	Yard No	Length (m)	Speed (kt)	Seats	Vehicles	Originally delivered to	Date of build
Corsaire 6000	*Moorea 8*	–	66.0	32.0	450	42	Emeraude Lines	1994
Corsaire 11000	*NGV Asco*	–	102.0	37.0	600	150	SNCM	1996
Corsaire 11000	*NGV Aliso*	–	102.0	37.0	600	150	SNCM	1996
Corsaire 11500	*Destination Gotland*	–	112.0	35.0	700	140	Rederi AB Gotland	1999
Corsaire 12000	*Aeolos Express*	–	119.0	36.0	1,000	210	Lesvos SA	2000
Corsaire 13000	*NGV Liamone*	–	134.0	42.0	1,116	250	Societe Nationale Corse-Mediterranee	2000
Corsaire 14000	*Aeolos Kenteris*	–	140.0	40.0	1,800	450	Lesvos SA	2001
Corsaire 10000	*Aeolos Express II*	–	104.0	35.0	800	190	Lesvos SA	2001

Vehicles	42 cars
Fuel capacity	30,000 litres
Water capacity	8,000 litres
Propulsive power	7,200 kW
Max speed	34 kt
Operational speed	32 kt
Range	150 n miles

Classification: Bureau Veritas 1 3/3 AUT MS-F Lightship.
Propulsion: Main engines are four MWM 620B V16 diesels rated at 2,245 kW each; three Kamewa water-jets, two S80 and one B90 units.
Control: The vessel is fitted with four roll stabilisation fins.
Electrical system: Three 250 kVA generators.

Corsaire 8000 (Design)

A larger design, the Corsaire 8000 has also been introduced based on the first Alstom Leroux Naval fast ferry Corsaire 6000.

Specifications

Length overall	77.0 m
Beam	11.5 m
Deadweight	140 t
Passengers	450/500
Vehicles	45/50 cars
Propulsive power	12,600/15,000 kW
Operational speed	36/38 kt
Range	500 n miles

Propulsion: Main engines can be two Ruston 20 RK 270, two MTU 20V 1163 or two Pielstick 20 PAB STC driving two Kamewa 112 SII water-jets.

Corsaire 10000

Aeolos Express II

The Greek maritime company of Levos SA (NEL Lines) ordered a Corsaire 10000 in 1999 for delivery in 2001. The vessel is in service on the Peloponese to Greek island routes with a capacity for 800 passengers and 190 cars.

Specifications

Length overall	104.0 m
Length waterline	95.5 m
Beam	13.9 m
Draught	2.6 m
Deadweight, max	450 t
Crew	14
Passengers	800
Vehicles	190
Propulsive power	24,300 kW
Operational speed	35 kt
Operational limitation	4.5 m sig wave height

Classification: Bureau Veritas.
Structure: Steel hull with aluminium superstructure.
Propulsion: Three SEMT Pielstick 20 PA6 B STC diesel engines each rated at 8,100 kW and each driving a Kamewa 125SII water jet.
Electrical system: Three 418 kW generator units.
Auxiliary systems: Two 300 kW bow thrusters. Active Ride Control system comprising stern flaps, fins and bow T-foil.

Corsaire 11000

NGV Asco
NGV Aliso

Construction of these vessels commenced in mid-1994 and was completed in early 1996. The craft were purchased by SNCM to operate on the Nice to Calvi (Corsica) route.

Specifications

Length overall	102.0 m
Beam	15.4 m
Draught	2.5 m
Deadweight	360 t
Passengers	600
Vehicles	150 cars or 126 cars + 4 coaches
Fuel capacity	120,000 litres
Water capacity	10,000 litres
Propulsive power	26,000 kW
Max speed	40 kt
Operational speed	37 kt
Range	300 n miles

Classification: Bureau Veritas I 3/3 E.

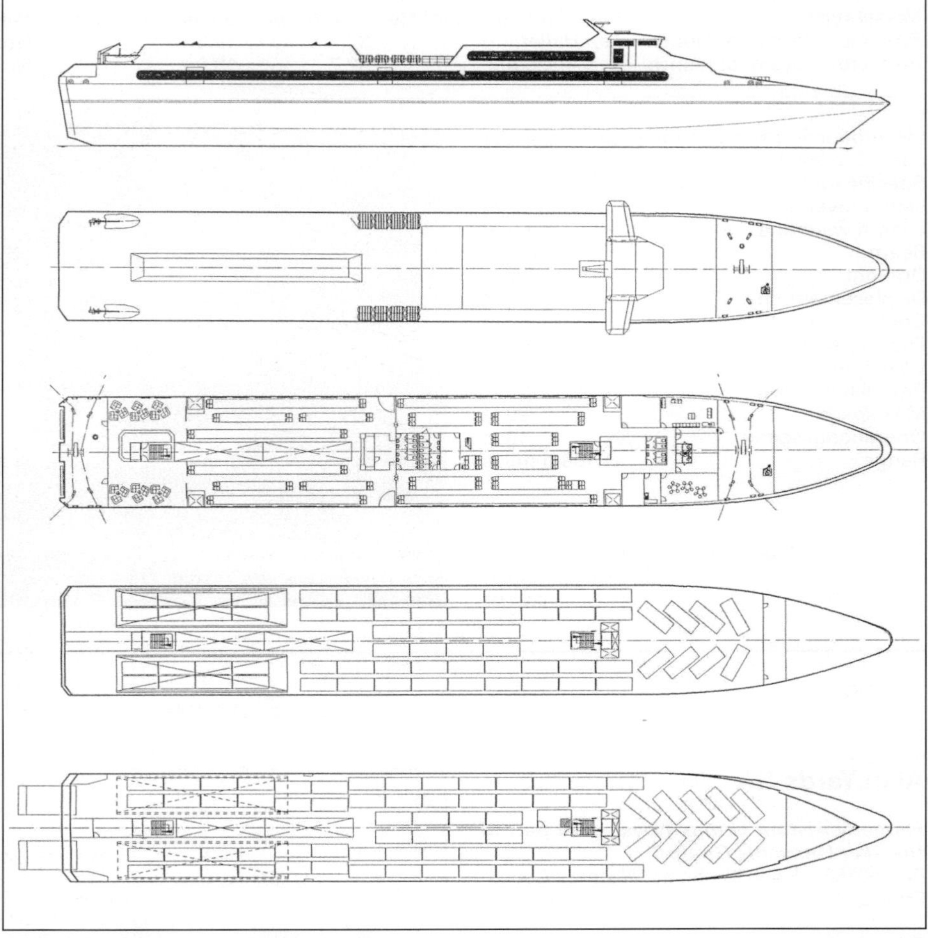

General arrangement of Corsaire 11500 0002645

Corsaire 11000 NGV Asco 0507069

Corsaire 6000 0507068

Structure: All-aluminium hull and superstructure.
Propulsion: Main engines are four MTU 20V 1163TB 73L diesels, each producing 6,500 kW; driving two Kamewa S112 and two Kamewa B112 water-jets.
Electrical system: Three 418 kW generator sets.
Auxiliary system: One 250 kW bow thruster. Active Ride Control System comprising one T-Foil, two fins and two stern flaps supplied by MDI.

Corsaire 11500

Destination Gotland

The Corsaire 11500 was launched in September 1998 for the Swedish operator Rederi AB Gotland and delivered in February 1999. The vessel's engines are fitted with Siemens SCR units to reduce NO_x emissions.

Specifications

Length overall	112.0 m
Length waterline	100.0 m
Beam	15.7 m
Draught	2.6 m
Deadweight	450 t
Passengers	700
Vehicles	140 cars
Operational speed	35 kt
Operational limitation	3.1 m sig wave height

Classification: Lloyds +100A1 SSC Passenger (B) Mono-HSC LDCG3 +LMS UMS 1BS.
Structure: The hull is constructed of EH36 high-tensile steel with a superstructure fabricated from aluminium extrusions of grade 5383.
Propulsion: The vessel is powered by four ALSTOM Ruston 20RK 270 diesel engines each rated at 7,080 kW at 1,030 rpm, each driving a Kamewa 125 SII water-jet.
Electrical systems: Three 418 kW generator sets providing 380 V/50 Hz/3 phase and 220 V/50 Hz single-phase networks. Four 24 V networks are provided for engine start, emergency supply and radio power supplies.
Auxiliary systems: The vessel is fitted with two 350 kW bow thrusters and is equipped with an active ride control system based on stabiliser fins and transom flaps.

Alstom Leroux Naval Corsaire 11500 Gotland *in service with Rederi AB Gottland* 0063224

NGV Liamone 0077908

Aeolos Express 0077909

Corsaire 12000

Aeolos Express

The Greek maritime company of Lesvos SA, (NEL Lines) ordered a Corsaire 12000 in 1998 for delivery in 2000. The vessel is on service on the Piraeus to Greek island routes with a capacity for 210 cars or 62 cars and 12 lorries.

Specifications

Length overall	119.0 m
Length waterline	105.0 m
Beam	15.7 m
Draught	2.6 m
Deadweight	500 t
Crew	22
Passengers	1,000
Vehicles	210
Propulsive power	32,400 kW
Operational speed	36 kt
Operational limitation	4.5 m sig wave height

Classification: Bureau Veritas.
Structure: Steel hull and aluminium superstructure.
Propulsion: Four SEMT Pielstick marine diesels each rated at 8,100 kW and each driving a Kamewa 125 SII water-jet.
Electrical system: Three 418 kW generator units.
Auxiliary systems: Two 300 kW bow thrusters. Active Ride Control system comprising stern flaps, fins and bow T-foil.

Corsaire 13000

NGV Liamone

Delivered to Société Nationale Corse-Méditerranée in 2000, this vessel was ordered in September 1998 for operation between France and Corsica at a cost of USD69 million.

Specifications

Length overall	134.0 m
Beam	18.9 m
Draught	3.2 m
Passengers	1,116
Vehicles	250 cars or 112 cars and 14 lorries
Propulsive power	63,000 kW
Operational speed	42 kt
Operational limitation	5.5 m sig wave height

Classification: Bureau Veritas.
Structure: Steel hull and aluminium superstructure.
Propulsion: Two GE LM2500 gas turbines each rated at 25,000 kW driving Kamewa 180 SII water-jets and two MTU 20V 1163 TB73L diesel engines each rated at 6,500 kW driving Kamewa 140 SII units.
Electrical system: Five 432 kW generator sets.
Auxiliary systems: Two 480 kW bow thrusters, MDI active Ride Control system comprising one T-Foil, four fins and two stern flaps.

Corsaire 14000

Aeolos Kenteris

In 1999 the Greek maritime company of Levos SA (NEL Lines) ordered a Corsaire 14000 for delivery in 2001. The vessel is in service on the Piraues to Greek island routes with a capacity for 450 cars or a combination of cars, coaches and lorries.

Specifications

Length overall	140.0 m
Length waterline	126.0 m
Beam	21.8 m
Draught	3.6 m
Deadweight, max	1,300 t
Crew	22
Passengers	1,742
Vehicles	450
Propulsive power	66,200 kW
Operational speed	40 kt
Operational limitation	6.0 m sig wave height

Classification: Bureau Veritas.

Structure: Steel hull and aluminium superstructure.

Propulsion: Two SEMT Pielstick 20 PA6 B STC diesels each rated at 8,100 kW and each driving a Kawema 140 SII water-jet and two GE LM2500 + gas turbines each rated at 25,000 kW driving Kawema 200 SII water-jets.

Electrical system: Four 500 kW generator units.

Auxiliary systems: Two 480 kW bow thrusters. Active Ride Control system comprising stern flaps, fins and bow T-foil.

Corsaire 16000 (DESIGN)

A 160 m vessel with capacity for approximately 500 cars, 2,000 passengers and speeds of 42 to 50 kt is currently under development.

Mistral

Alstom are also researching a SES assisted monohull with capacity for 2,000 to 3,000 passengers and 500 to 700 cars and a target speed of 50 to 60 kt.

Chantiers Navale Gamelin

Commerce de la Pallice 17000, La Rochelle, France

Tel: (+35 5) 46 43 97 77
Web: www.chantiers-gamelin.fr

Formed in 1980, Gamelin produce aluminium vessels fom 11 to 40 m.

37 m MONOHULL

In 2006, operator Manche-Îles-Express ordered a 37 m monohull for service on its routes between Normandy and the Channel Islands. The vessel, designed by Mer 8 Design was delivered in 2007.

Specifications

Length overall	37.2
Beam	8.6 m
Draught	2
Passengers	260
Operational speed	27 kts

Propulsion: Two MTU 12V 4000 M71 diesel engines delivering 1,850 kW at 2,000 rpm to FP propellers.

Vessel type	Vessel name	Yard No	Length (m)	Speed (kt)	Seats	Vehicles	Originally delivered to	Date of build
37 m Ferry	*Tocqueville*	–	37.2	27	260	none	Manche-Îles-Express	2007

Construction Navale Bordeaux (CNB)

Part of the Bénéteau Group
162 quai De Brazza, F-33100 Bordeaux, France

Tel: (+33 0) 557 80 85 50
Fax: (+33 0) 557 80 85 51
e-mail: cnb@cnbpro.com
Web: www.cnb.fr

Builders of composite and aluminium sailing and powered craft, CNB delivered a fast monohull ferry to Guadeloupe in 2000.

Specifications

Length overall	19.1 m
Length waterline	14.5 m
Beam	5.5 m
Draught	0.8 m
Crew	3
Passengers	105
Fuel capacity	2,500 litres
Water capacity	400 litres
Max speed	28 kt
Operational speed	25 kt

Propulsion: Two MAN 2842 LE 401 diesel engines each rated at 600 kW at 2,100 rpm and driving a Hamilton HM 422 water-jet.

Vessel type	Vessel name	Yard No	Length (m)	Speed (kt)	Seats	Vehicles	Originally delivered to	Date of build
19 m ferry	*Iguana Sun*	–	19.1	28	105	none	Groupe Jacques, Guadeloupe	2000

Constructions Mecaniques DE Normandie (CMN)

51 rue de la Brettionnière, BP 539, F-50105 Cherbourg Cedex, France

Tel: (+33 2) 33 88 30 00
Fax: (+33 2) 33 44 01 09
e-mail: info@cmn-cherbourg.com
Web: www.cmn-group.com

M Eric Berthelot, *Commercial Director*

CMN is well known for its pioneering work in the design and construction of lightweight high-speed military craft. Combined with its experience in the construction of large yachts and other craft such as SES, the company has designed a large monohull fast ferry for passenger, car and freight transportation, the *Voyager 70.*

MARLIN 38 m

Amporelle

This vessel was built to service the population of Yeu Island off the coast of France in 1992.

Specifications

Length overall	38.0 m
Beam	7.75 m
Draught	1.35 m
Passengers	367
Maximum speed	32 kt

Propulsion: Powered by two diesel engines rated at 1730 kW each driving a water-jet.

OCEA SA

Quai de la Cabaude, F-85100 Les Sables d'Olonne, France

Tel: (+33 2) 51 21 05 90
Fax: (+33 2) 51 21 20 06
e-mail: commercial@ocea.fr
Web: www.ocea.fr

Roland Joassard, *Chief Executive Officer*
Fabrice Epaud, *Commercial Director*

Founded in 1987, OCEA specialises in the design and construction of aluminium craft up to 50 m long.

EXPLORER 100-240

Queen of Aran II

This vessel was delivered in 2001 to Inishmore Ferries for operation between Roseaville and the Aran Islands in Connemara Bay on the west coast of Ireland.

Specifications

Length overall	30.0 m
Length waterline	27.6 m
Beam	7.0 m
Draught	1.9 m
Passengers	235
Fuel capacity	7,350 litres
Water capacity	1,500 litres
Max speed	22 kt

Construction: All aluminium hull and superstructure.

Propulsion: Two MTU 12V 2000 M70 diesel engines, each rated at 788 kW and driving a fixed-pitch propeller.

TP 75 MK MONOHULL FERRY

Marcus Garvey I

The first vessel of this design was delivered to La Saintannaise in the Caribbean in 1998.

Specifications

Length overall	24.55 m
Length waterline	20.0 m
Beam	6.5 m
Draught	1.2 m
Deadweight	15.8 t

Vessel type	Vessel name	Yard No	Length (m)	Speed (kt)	Seats	Vehicles	Originally delivered to	Date of build
TP 75 Mk 1	*Marcus Garvey I*	–	24.6	27	135	–	La Saintannaise	1997
TP 98 Mk II	*Le Sabia*	–	30.0	30.0	110	–	Vendée, France	2000
Explorer 100–240	*Cupidon*	–	23.0	22.0	140	–	Brittany, France	2000
Explorer 100–240	*Queen of Aran II*	–	30.0	22.0	235	–	Inishmore Ferries	2001

Crew	4
Passengers	135
Fuel capacity	5,000 litres
Water capacity	1,000 litres
Propulsive power	1,830 kW
Max speed	27 kt
Range	280 n miles

Classification: Bureau Veritas.
Propulsion: Three MTU 183 TE 72 diesel engines, each rated at 610 kW and each driving a Kamewa A40 water-jet.

TP 78 MK II MONOHULL FERRY

La Sabia

Specifications

Length overall	36.0 m
Length waterline	27.2 m
Beam	5.5 m
Draught	2.1 m
Deadweight	12 t
Passengers	97
Fuel capacity	4,400 litres
Water capacity	1,000 litres
Operational speed	30 kt
Range	275 n miles

Propulsion: Two MTU 16V 2000 diesel engines, each rated at 960 kW at 2100 rpm, each driving a fixed pitch propellers via a ZF BW255A gearbox.

Type 98 MKII monohull ferry La Sabia 1179491

Explorer 100–240 Queen of Aran II 1310554

Germany

Lürssen Werft GmbH & Co kg

Zum Alten Speicher 11, 28759 Bremen, Germany

Tel: (+49 421) 66 03 34
Fax: (+49 421) 660 43 95
e-mail: kt@luerssen.de
Web: www.luerssen.de

Detlef Schlichting, *Sales Director*

Lürssen Werft is a member of the Lürssen group of companies. In 1980 the parent company acquired Kröger Werft in Rendsburg and in 1997 the Bremer Vulkan Marineschiffbau which incorporates a large covered dry dock, allowing the support of ships up to 170 m in length. Fast motor boats, motor cruisers and motor yachts have been built by Lürssen Werft since 1890. Recent construction includes fast attack craft, corvettes, mine combat vessels, megayachts, search and rescue boats and fast ferries.

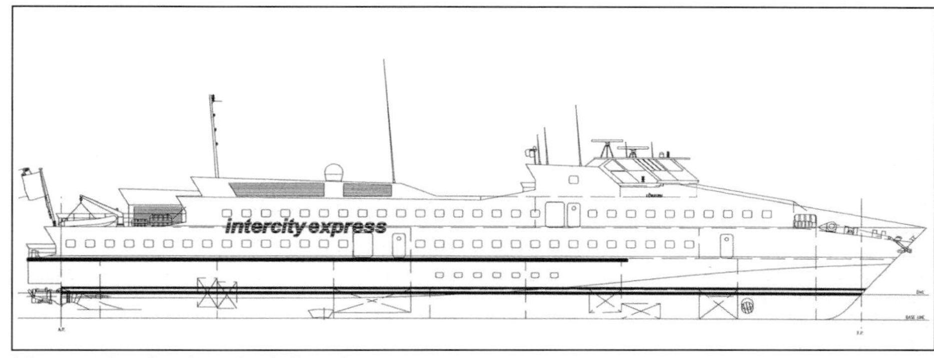

Lürssen 70 m fast ferry for Indonesia 0507071

70 m FAST FERRY

Ambulu
Cisadane
Serayu
Mahakam
Barito

Lürssen Werft has constructed five 70 m fast monohull ferries for the transportation of passengers within the Indonesian archipelago. The order was received from the Indonesian Ministry of Transportation and the first three craft were delivered in 1998, with the remaining two in 1999.

Specifications

Length overall	70.0 m
Beam	10.4 m

Vessel type	Vessel name	Yard No	Length (m)	Speed (kt)	Seats	Vehicles	Originally delivered to	Date of build
70 m Fast Ferry	*Serayu*	–	70.0	–	925	none	Indonesia	1998
70 m Fast Ferry	*Cisadane*	–	70.0	–	925	none	Indonesia	1998
70 m Fast Ferry	*Ambulu*	–	70.0	–	925	none	Indonesia	1998
70 m Fast Ferry	*Mahakam*	–	70.0	–	925	none	Indonesia	1999
70 m Fast Ferry	*Barito*	–	70.0	–	925	none	Indonesia	1999

Draught	2.0 m
Crew	30
Passengers	925
Speed	38 kt
Range	550 n miles

Classification: BKI (Indonesia) and GL as HSC.

Structure: Aluminium hull and superstructure.

Propulsion: Four MTU 16V 595 TE 70L marine diesels each rated at 3,800 kW and driving four Kamewa 90 water-jet units (two steerable).

Safety equipment: Four MES stations and two rescue boats.

The ferries are equipped with stabilisers, bow thruster and standard nautical equipment including GPS, GMDSS and SATCOM.

Hong Kong

Cheoy Lee Shipyards Ltd

89-91 Hing Wah Street West, Lai Chi Kok, Kowloon, Hong Kong

Tel: (+852) 23 07 63 33
Fax: (+852) 23 07 55 77
e-mail: info@cheoyleena.com
Web: www.cheoyleena.com

Ken Lo, *Director*
Jonathan Cannon, *Sales Manager*

The company was originally founded in Shanghai over a century ago. It moved to a small yard in Hong Kong in 1937, then to a much larger yard in 1940, and was occupied with building and repairing cargo ships. It diversified into building teak sailing yachts for export in 1956 and acquired a large site for a second yard on Lantau Island in 1960. The company commenced building motor yachts and building in GRP in 1961 and now builds commercial and pleasure craft in steel, composites and aluminium. The company has built over 4,700 vessels since the current records commenced in 1950. The yard is currently extending its operations into mainland China.

Oceanjet 3 *on trials* 0134198

Sea Falcon *on trials* 0134197

32 m PASSENGER VESSEL

Three vessels were delivered to Socor Shipping Lines in Cebu, Philippines in 2001 to 2003. The first vessel has an increased passenger capacity of 296.

Specifications

Length overall	32.0 m
Length waterline	29.8 m
Beam	7.1 m
Draught	2.0 m
Displacement	98 t
Crew	6
Passengers	287
Fuel capacity	5,000 litres
Water capacity	500 litres
Propulsive power	3,972 kW
Maximum speed	30 kt
Operational speed	28 kt

Classification: BV + Hull + Mach Light Ship Fast Passenger Carrier.

Structure: All aluminium hull and superstructure.

Propulsion: Two Cummins KTA50-MZ diesel engines each rated at 1,324 kW and each driving propellers via ZF BW465-1A gearboxes.

28 m PASSENGER FERRY

Lamma IV

Lamma IV is a fast passenger vessel designed by Naval Consult of Singapore and was delivered to the Hong Kong Electric Company by Cheoy Lee in 1996.

Specifications

Length overall	28.0 m
Beam	6.7 m
Draught	1.8 m
Passengers	222
Propulsive power	1,492 kW
Max speed	23.5 kt
Operational speed	22 kt

Propulsion: Powered by two Caterpillar 3412 TA diesel engines each rated at 746 kW driving fixed-pitch propellers.

22.8 m PASSENGER FERRY

Discovery Bay 17

Fast passenger vessel delivered in October 1988 and followed by a second, delivered to Penguin Boat Services Pte Ltd, Singapore, in 1991.

Owner: Hong Kong Resort Company Ltd.

Specifications

Length overall	22.8 m
Length waterline	20.0 m
Beam	5.4 m
Draught	1.7 m
Displacement	36.1 t
Crew	4
Passengers	163
Fuel capacity	5,100 litres
Water capacity	450 litres

Vessel type	Vessel name	Yard No	Length (m)	Speed (kt)	Seats	Vehicles	Originally delivered to	Date of build
22.8 m passenger vessel	*Discovery Bay 17*	–	22.8	23.0	163	none	Hong Kong Resort Company Ltd	1988
22.8 m passenger vessel	–	–	22.8	23.0	163	none	Penguin Boat Services Ltd	1991
20 m surveillance vessel	–	–	19.8	28.0	–	none	Environment Protection Council of Kuwait	1993
44 m motor yacht	–	–	44.1	–	–	none	–	1995
28 m passenger vessel	*Lamma IV*	–	28.0	22.0	222	none	Hong Kong Electric Company	1996
28 m passenger vessel	*Sea Falcon*	–	28.0	27.0	200	none	Tian San Shipping	2001
28 m passenger vessel	*Sea Osprey*	–	28.0	27.0	200	none	Tian San Shipping	2001
32 m passenger vessel	*Ocean Jet 3*	–	32.0	28.0	296	none	Socor Shipping Lines	2001
32 m passenger vessel	*Ocean Jet 5*	–	32.0	28.0	287	none	SOCOR Shipping Lines	2003
32 m passenger vessel	*Ocean Jet 6*	–	32.0	28.0	287	none	SOCOR Shipping Lines	2003

Propulsive power	2 × 737 kW
Max speed	25 kt
Operational speed	23 kt

Classification: American Bureau of Shipping 100 A1 + AMS.

Structure: Deep vee constant deadrise construction in solid GRP. It has a two-tier superstructure with a wheelhouse and short passenger saloon on the upper deck forward half, after half and open under awning.

Propulsion: Main engines are two Stewart and Stevenson Detroit Diesel Model 16V 92MCTAB, 737 kW each driving propellers through Nico MGN 273 gearboxes.

Electrical system: One Mercedes-Benz/Stamford generator set, 32.5 kVA, 380 V, 50 Hz, three-phase, two 24 V 200 Ah starting battery sets, one 24 V 120 Ah starting battery set.

Lamma IV 0507073

20 m SURVEILLANCE VESSEL

This vessel was constructed for the Environment Protection Council of Kuwait in 1993.

Specifications

Length overall	19.8 m
Length waterline	17.6 m
Beam	5.5 m
Draught	1.6 m
Displacement	28 t
Crew	8
Fuel capacity	6,000 litres
Water capacity	400 litres
Propulsive power	2 × 820 kW
Max speed	30 kt
Operational speed	25–28 kt

Structure: Hull and superstructure material is GRP.

Propulsion: Main engines are two MAN D2842 LZE diesels, 820 kW each, driving two propellers via two ZF 195 2:1 reduction gearboxes.

Electrical system: Two Onan MDGBA 30 kW generators.

Wang Tak Engineering and Shipbuilding Co Ltd

Wang Tak Building, 85 Hing Wah Street West, Lai Chi Kok, Kowloon, Hong Kong

Tel: (+852) 27 46 28 88
Fax: (+852) 23 07 55 00; 23 07 55 22
e-mail: info@wangtak.com.hk

Feat Szeto, *Director and General Manager*

The Wang Tak Shipyard based in Hong Kong offers shipbuilding and repair services in steel and aluminium materials. In 1998 the company delivered a fast monohull firefighting vessel to the Fire Services Department of HKSAR (Hong Kong Special Administrative Region).

28 m FIREBOAT

Fireboat 3

Fireboat 3 carries a UK "Counterfire" firefighting system.

Specifications

Length overall	28.5 m
Beam	6.0 m
Max speed	23 kt
Operational speed	20 kt

Propulsion: Two MTU 12V 396 TE 74L diesel engines each rated at 1,455 kW at 1,900 rpm and driving a 1.4 m Teignbridge propeller via a ZF BW465 reverse reduction gearbox.

Fireboat 3 0077062

Vessel type	Vessel name	Yard No	Length (m)	Speed (kt)	Seats	Vehicles	Originally delivered to	Date of build
Firefighting Vessel	*Fireboat 3*	Y971	28.5	20.0	–	–	Fire Services Dept of Hong Kong	1998

Italy

Azimut-Benetti SpA

Via Martin Luther King 9-11, I-10051 Avigliana, Turin, Italy

Tel: (+39 011) 931 61
Fax: (+39 011) 93 16 72 70
e-mail: proffice@azimutyachts.net
Web: www.azimutyachts.com

Dr Ing Paolo Vitelli, *President and Managing Director*
Ing Marco Simonetti, *Chief Executive and Managing Director, Azimut Yachts*
Ing Vincenzo Poerio, *Chief Executive and Managing Director, Benetti Dvision*
Mrs Carla Demaria, *Chief Executive Azimut Capital Division and President of Gobbi SpA*
Dr Mary Brayda Bruno, *Public Relations Manager*

Azimut SpA was founded in 1969 by Dr Paolo Vitelli and began by importing distinguished yachting marques. Later Vitelli decided to design his own boats and commission yards to build them. As the company grew so did its sales network. Between 1978 and 1980 it became the first company in Italy to manufacture glass fibre boats and, in 1982, launched the first ever yacht of over 30 m to be built in that material, the 105′ *Failaka*.

In 1985 the company took over the Fratelli Benetti yard in Viareggio, known for its 40 to 50 m steel-hulled megayachts. In 1988 it also acquired a yard for the production of 10 to 18 m glass fibre boats in Avigliana near Turin.

By 2003 the company had acquired additional boatyards in Sariano (Gobbi Boatyard) and Fano (Benetti Boatyard), and had set up two service centres, one in Savona and the other in Viareggio. With an output of 120 boats a year, Azimut is one of Europe's major boat yards.

The company's production is split into three distinct lines to meet the demands of a wide section of the market:

AZ 10 to 15 m glass fibre boats (AZ 36', 40', 43', 46', 50').

AZIMUT 18 to 30 m boats (58' Full, 70' Seajet, 78' Ultra, 90'–100' Jumbo).

BENETTI 40 m+ aluminium-steel yachts.

The company markets its boats through a sales network of about 35 dealers, covering the European, American, Southeast Asian and Japanese markets. The group has also invested considerably in the setting up of an aftersales service network.

Today the company is divided into four areas:

Azimut Yachts – fibreglass motorcruises and motoryachts from 12 to 35 m.

Benetti – megayachts in steel over 50 m and in fibreglass from 30 to 45 m.

Azimut Capital – company acquisitions, Gobbi management.

Marinas and Real estate – management of tourist real estate and private marinas.

AZIMUT103S

Specifications

Length overall	30.9 m
Beam	7.1 m
Draught	1.34 m
Displacement, max	105 t
Crew	4
Passengers	8
Fuel capacity	12,350 litres
Water capacity	2,500 litres
Maximum speed	34 kt
Operational speed	30 kt

Construction: Fibreglass hull and superstructure.

Propulsion: Two MTU V16 2000 M93 diesel engines each rated at 1,790 kW and driving Kamewa waterjets.

Azimut 98 Leonado 0538413

AZIMUT 98 LEONADO

This vessel was launched in 2002.

Specifications

Length overall	30.0 m
Beam	7.1 m
Draught	2.1 m
Displacement, max	90 t
Crew	4
Passengers	8
Fuel capacity	10,000 litres
Water capacity	2,000 litres
Maximum speed	33 kt
Operational speed	28 kt

Construction: Fibreglass hull and superstructure.

Propulsion: Two MTU 2000 M91 diesel engines each rated at 1,471 kW and driving fixed-pitch propellers.

Cantiere Navale Foschi

Via Toscanelli 12, I-47042 Cesenatico, Italy

Tel: (+39 05) 478 05 10
Fax: (+39 05) 477 55 62
e-mail: info@cantierefoschi.it
Web: www.cantierefoschi.it

38 m FAST FERRY

Eolian Queen

This vessel, the fourth built for Taranto Navigazione was delivered in 2007.

Specifications

Length overall	38 m
Length waterline	34.4 m
Beam	8.1 m
Passengers	400
Fuel capacity	18,000 litres
Water capacity	2,000 litres

Propulsion: Two MTU 16V 4000 M70 diesel engines each driving a Bisognani five-bladed propeller via a ZF 7549C gearbox.

35 m FAST FERRY

Neocastrum

Delivered in May 1993, this 35 m fast ferry was constructed at Cantiere Foschi's facilities in Cesenatico, Italy, for Foderaro Navigazione.

Magic Panarea 0583005

Vessel type	Vessel name	Yard No	Length (m)	Speed (kt)	Seats	Vehicles	Originally delivered to	Date of build
30 m Fast Ferry	*Hipponion*	54	30.3	26.0	350	none	–	1990
24 m Fast Ferry	*Golfo Di Arzachena*	61	23.7	22.0	200	none	–	1992
35 m Fast Ferry	*Neocastrum*	58	35.4	27.0	350	none		1993
21 m Fast Ferry	*Lady M*	67	20.7	–	130	none	Taranto Navigazione	1997
23 m Fast Ferry	*Positano Jet*	71	23.0	33.0	200	none	Positano Jet	1998
34 m Fast Ferry	*Citrarium*	74	34.4	–	350	none		1999
29 m Fast Ferry	*Stromboli Express*	75	28.8	25.0	250	none	Comerci Navigazione	2000
23 m Fast Ferry	*La Conia*	76	23.4	22.0	180	none	Cantiere Foschi, Italy	2000
30 m Fast Ferry	*Eolian Princess*	77	30.0	30.0	274	none	Taranto Navigazione	2002
27 m Fast Ferry	*Eolian Star*	78	27.1	–	256	none	Taranto Navigazione	2003
28.8 m Fast Ferry	*Magic Panarea*	80	28.8	–	285	none	Savadori Navigazione	2004
22 m Fast Ferry	*Lipari Island*	81	22.0	–	157	none	Merenda Navigazione	2006
25.5 m Fast Ferry	*Delfino Verde AS*	82	25.5	–	208	none	Delfino Verde Navigazione	2007
38 m Fast Ferry	*Eolian Queen*	83	38.0	–	400	none	Taranto Navigazione	2007

The vessel is equipped with a pair of Mitsubishi 1,632 kW diesels for a top speed of 31 kt. Passenger capacity is 350 with about two-thirds seated in the main deck saloon and one-third in the upper deck saloon. The captain and crews' cabins are located just aft of the bridge. Forward of the main deck saloon, a crew mess and galley have direct access to the fore deck.

The vessel is operated between the southwest coast of Italy and the islands off the north coast of Sicily. The daily round trip is approximately 160 n miles with passage taking approximately 3 hours each way.

Specifications

Length overall	35.4 m
Beam	7.6 m
Displacement, min	85 t
Passengers	350
Fuel capacity	12,000 litres
Water capacity	2,000 litres
Propulsive power	3,264 kW
Max speed	31 kt

Propulsion: Two Mitsubishi S16R MPTK diesels.

34 m FAST FERRY

Citrarium

This 34 m triple water-jet fast ferry was launched in mid 1999 and is operated by Foderaro between the south coast of Italy and the Eolian Islands off Sicily.

Specifications

Length overall	34.4 m
Beam	7.4 m
Displacement, min	75 t
Passengers	350
Fuel capacity	10,000 litres
Water capacity	1,500 litres
Propulsive power	3,600 kW

Propulsion: Three Deutz MWN 616V1+ diesel engines each rated at 1,200 kW and each driving a Hamilton HM 571 water-jet via a ZF 255 gearbox.

30 m FAST FERRY

Hipponion

Delivered in May 1990.

Specifications

Length overall	30.3 m
Beam	7.1 m
Displacement, min	52.1 t
Passengers	350
Propulsive power	2,085 kW
Operational speed	26 kt

Propulsion: Two diesels.

29 m FAST FERRY

Stromboli Express

Delivered in June 2000, this fast ferry was constructed at Cantiere Foschi's facilities in Cesenatico, Italy, for Comerci Navigazione. Equipped with a pair of Caterpillar 3412 diesel engines which give a cruising speed of 25 kt and carrying 250 passengers, the vessel operates along the coast of Calabria and the islands offshore.

Specifications

Length overall	28.8 m
Beam	6.5 m
Displacement, min	47.5 t
Displacement, max	75.0 t
Passengers	250
Fuel capacity	8,000 litres
Water capacity	1,000 litres
Propulsive power	1,875 kW
Max speed	25 kt

Propulsion: Two caterpillar 3412 diesel engines.

28.8 m FAST FERRY

Magic Panarea

This vessel was delivered in June 2004 to operator Savadori Navigazione.

Specifications

Length overall	28.87 m
Beam	6.85 m
Passengers	285
Fuel capacity	10,000 litres
Water capacity	2,000 litres

Propulsion: Three Caterpillar 3412E diesel engines driving fixed-pitch propellers.

Neocastrum 0506838

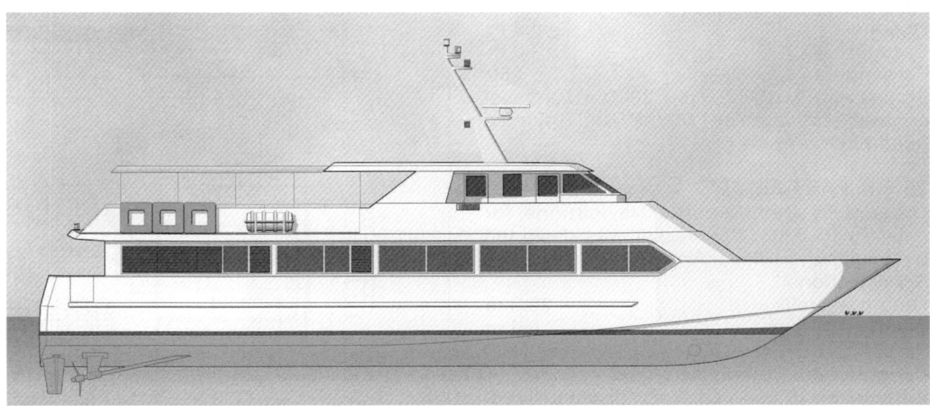

Eolian Star 0583006

Hipponion 0506690

Citrarium 0068946

Positano Jet 0038852

20 m fast ferry Lady M 0038851

27 m FAST FERRY

Eolian Star

This vessel was ordered in 2002 for operation as a tourist vessel for a Sicilian operator.

Specifications

Length overall	27.1 m
Length waterline	23.3 m
Beam	6.8 m
Draught	1.0 m
Passengers	250

Propulsion: Two MTU 16V 2000 M70 diesel engines each rated at 1,050 kW and driving a fixed-pitch propeller.

25.5 m FAST FERRY

Ordered by Delfino Verde Navigazione for delivery in 2006, this vessel was designed by Sciomachen Naval Architects.

Specifications

Length overall	25.5 m
Beam	6.4 m
Passengers	250

Construction: Composite wood.

Propulsion: Diesel engines driving fixed-pitch propellers via Twin Disc gearboxes.

24 m FAST FERRY

Golfo Di Arzachena

Delivered in May 1992.

Specifications

Length overall	23.7 m
Beam	6.1 m
Displacement, min	29.3 t
Passengers	200
Propulsive power	672 kW
Speed	22 kt

Propulsion: Two diesels.

23 m FAST FERRY

Positano Jet

La Conia

Positano Jet was completed in 1998 for Positano Jet for service between Positano and Capri in the Gulf of Naples. La Conia was constructed at Cantieri Foschi, Cesenatico and was delivered in 2000 for operation along the north coast of Sardinia with capacity for 180 passengers and equipped with a pair of Volvo TAMD 122 A main engines.

Specifications

Length overall	23.0 m
Length waterline	19.4 m
Beam	6.1 m
Draught	1.9 m
Displacement, min	50 t
Passengers	200
Fuel capacity	5,000 litres
Water capacity	1,000 litres
Propulsive power	2,014 kW
Max speed	33 kt

Propulsion: Powered by two Caterpillar 3412 diesel engines each rated at 1,007 kW and each driving propellers via a ZF V drive gearbox (*Positano Jet*), two Volvo TAMD 122A diesels (*La Conia*).

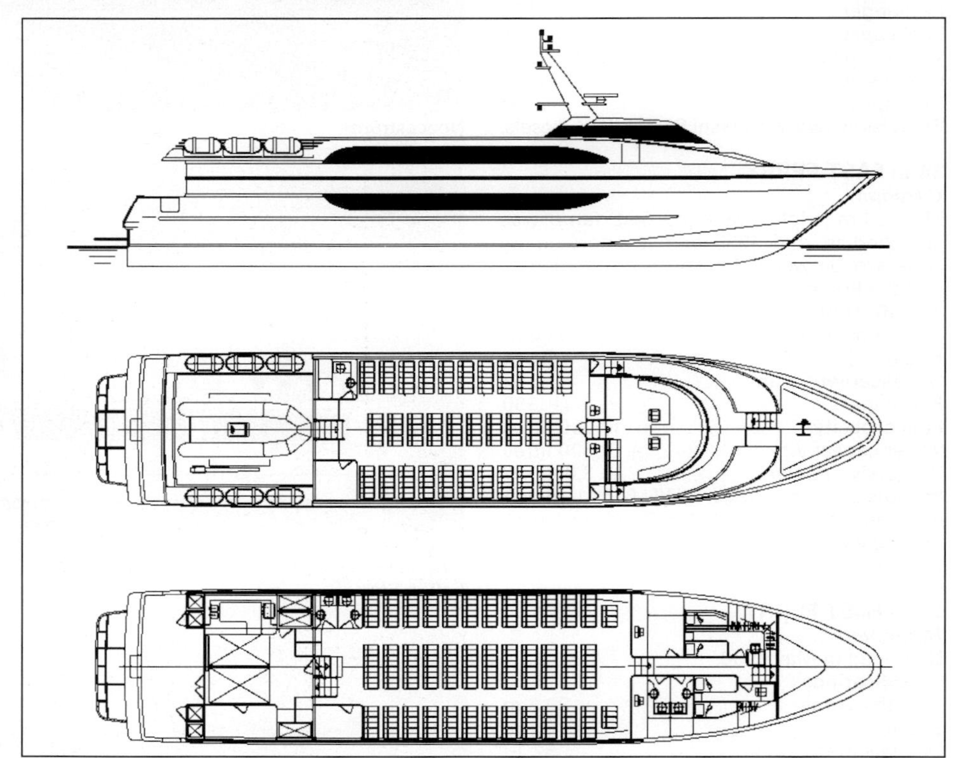

General arrangement of the Sciomachen *33 m fast ferry (design)* 0038853

22 m FAST FERRY

Lipari Island

Designed by Sciomachen, this vessel was ordered by Merenda Navigazione. The vessel was delivered in 2006.

Specifications

Length overall	22 m
Construction	Wood Composite
Passengers	149

Propulsion: Two Daewoo V158 diesel engines each driving fixed-pitch propellers through Twin Disk 5114A gearboxes.

21 m FAST FERRY

Lady M

A 21 M ferry with a top speed of 25 kt, *Lady M* was launched in 1997 and is operated in the south of Italy. *Lady M* can carry 130 passengers, including 80 seats in the enclosed saloon; there is a crew galley forward and the bridge is elevated above the main deck. Propulsion is by two Volvo Penta TAMD 122D engines, each rated at 331 kW at 2,000 rpm.

Specifications

Length overall	20.7 m
Length waterline	17.5 m
Beam	5.3 m
Displacement, min	21 t
Passengers	130
Fuel capacity	3,000 litres
Water capacity	500 litres
Propulsive power	662 kW
Max speed	25 kt

De Poli Cantieri Navali

Sestiere Scarpa, Via dei Murazzi 1216, I-30010 Venezia Pellestrina, Italy

Tel: (+39 041) 527 81 11
Fax: (+39 041) 96 77 30
e-mail: depoli@depoli.com
Web: www.depoli.com/eng

Davino De Poli, *Director*

Known for its steel shipbuilding and repair activities, Cantieri Navali De Poli constructed its first fast ferry, *Isoli Di Stromboli,* in 1998/99.

TMV 95

Isola Di Stromboli

This 95 m vessel was designed in conjunction with and built under licence from Rodriquez Cantieri Navali and delivered to Siremar in 1999.

Specifications

Length overall	95.0 m
Length waterline	86.5 m
Beam	16.8 m
Draught	2.6 m
Deadweight	400 t
Crew	22
Passengers	800
Vehicles	170 cars
Propulsive power	15,600 kW
Max speed	30 kt

Vessel type	Vessel name	Yard No	Length (m)	Speed (kt)	Seats	Vehicles	Originally delivered to	Date of build
95m Fast Monohull	Isola Di Stromboli	–	95.0	28.0	800	170	SIREMAR	1999

Operational speed 28 kt
Range 500 n miles
Classification: RINA 100A 1.1 – NavS-HSC (Cat B) – 1AQ2 –TpTr (Ro-Ro).
Structure: High tensile steel hull with aluminium superstructure.
Propulsion: Three diesel engines each rated at 5,200 kW with the wing engines driving propellers and the centre engine driving a water-jet.

Isola Di Stromboli
0097417

Fincantieri Cantieri Navali Italiani SpA

Head office
Via Genova 1, I-34121 Trieste, Italy

Merchant ship business unit
Passeggio S Andrea 6B, I-34123 Trieste, Italy

Tel: (+39 040) 319 31 11
Fax: (+39 040) 319 23 05
e-mail: info@fincantieri.it
Web: www.fincantieri.com

Giuseppe Bono, *Chief Executive Officer*
Corrado Antonini, *Chairman*
Giovanni Romano, *Merchant Ship Business Unit Executive Vice President*
Vincenzo Farinetti, *Auxiliary Naval Vessels and High Speed Craft Business Manager*

Fincantieri is one of the major and most diversified shipbuilding groups in Europe.

In May 1996 Fincantieri delivered the first of three MDV 1200 Pegasus to Ocean Bridge Investments. The vessel was chartered to Stena Line operated on the Newhaven to Dieppe route between the United Kingdom and France for one year. *Pegasus One* is now in operation in Venezuela. In the same year Seacontainers ordered six similar vessels also for operation in Europe.

In May 1997 *Pegasus Two* was delivered to Ocean Bridge Investments and chartered to Color Line (Skagen-Larvik route). *Superseacat One* and *Superseacat Two* were also delivered to Seacontainers for the Gothenburg-Frederikshaven and Dover-Calais routes.

MDV 3000 Taurus *on trials*
0038710

MDV 1200 Pegasus Two *operated by Color Line*
0002650

Vessel type	Vessel name	Yard No	Length (m)	Speed (kt)	Seats	Vehicles	Originally delivered to	Date of build
Blue Riband Challenger	*Destriero*	–	67.7	40.0	–	None	–	1991
MDV 1200 (steel)	*Pegasus One*	C5965	95.0	36.0	600	170	Ocean Bridge Investments	1996
MDV 1200 (steel)	*Pegasus Two*	C6005	95.0	36.0	600	170	Ocean Bridge Investments	1997
MDV 1200 (Aluminium)	*Superseacat One*	C5799	100.0	38.0	800	None	Sea Containers	1997
MDV 1200 (Aluminium)	*Superseacat Two*	C6000	100.0	38.0	800	None	Sea Containers	1997
MDV 1200 (Aluminium)	*Superseacat Three*	C6001	100.0	38.0	800	None	Sea Containers	1997
MDV 1200 (Aluminium)	*Superseacat Four*	C6002	100.0	38.0	800	None	Sea Containers	1997
MDV 3000 Jupiter	*Aries*	C6003	146.0	42.0	1,800	460	Tirrenia, Italy	1998
MDV 3000 Jupiter	*Taurus*	C6009	146.0	42.0	1,800	460	Tirrenia, Italy	1998
MDV 3000 Jupiter	*Capricorn*	C6046	146.0	42.0	1,800	460	Tirrenia, Italy	1999
MDV 3000 Jupiter	*Scorpio*	C6047	146.0	42.0	1,800	460	Tirrenia, Italy	1999
MDV 2000	*Gotlandia II*	C6128	122.0	40.0	800	160 cars and 8 coaches	Rederi AB Gotland	2006

In 1998 two MDV 3000 Jupiters were delivered to Tirrenia and in August 1997 Fincantieri received an order for two more MDV 3000 Jupiter vessels for delivery in early 1999.

Destriero

Destriero, which is owned by the Yacht Club Costa Smeralda, was built specifically to challenge the Blue Riband reward.

Destriero was built at the Fincantieri Yards of Riva Trigoso (Genova) and Muggiano (La Spezia).

The design of *Destriero* started in March 1990. The contract was signed in May 1990, the vessel was launched in March 1991 and the sea trials were completed by the end of May 1991. On trials *Destriero* has reached 62.8 kt.

Capacity is provided for 750 tonnes of fuel; each of the three engines burns up to 3.5 t/h.

Design and engineering of the vessel was undertaken by Fincantieri, hull forms by American naval architect Donald L Blount and styling by Pininfarina.

On 9 August 1992 *Destriero* broke the Blue Riband record for the fastest crossing of the Atlantic by more than 21 hours, completing the crossing from Ambrose lighthouse (New York) to Bishop Rock (Scilly Isles, UK), a distance of 3,106 n miles, in 58 hours 34 minutes and 50 seconds at an average speed of 53.1 kt.

It has been reported that on the initial westbound crossing of *Destriero* the starting displacement was 1,070 tonnes with a corresponding speed of 43 kt. As fuel was consumed speed increased, rising eventually to its maximum. The reported eastbound crossing time of 62 hours 7 minutes, together with the record westbound crossing time, also set a new double-crossing record of 159 hours 48 minutes and 15 seconds, a feat no other marine vessel has ever achieved.

Specifications

Length	67.7 m
Beam	12.9 m
Displacement, max	1,070 t
Fuel capacity	750 t
Propulsive power	44,742 kW
Max speed	65 kt
Operational speed	40 kt
Range	3,000 n miles

Classification: The vessel is classified according to DnV Light Craft Rules, DnV having co-operated closely with the designers during design and construction; also Registro Italiano Navale.

Propulsion: The craft is powered by three GE LM 1600 gas turbines, 14,914 kW each, packaged in resiliently mounted MTU modules, driving three Kamewa 125 SII water-jets.

Reduction gears: Renk-Tacke Bus 255.

Structure: Aluminium alloy.

Control: Reverse thrust and steering affected by the two outer water-jets.

SuperseacatTwo *in service* (Maritime Photographic) 0045179

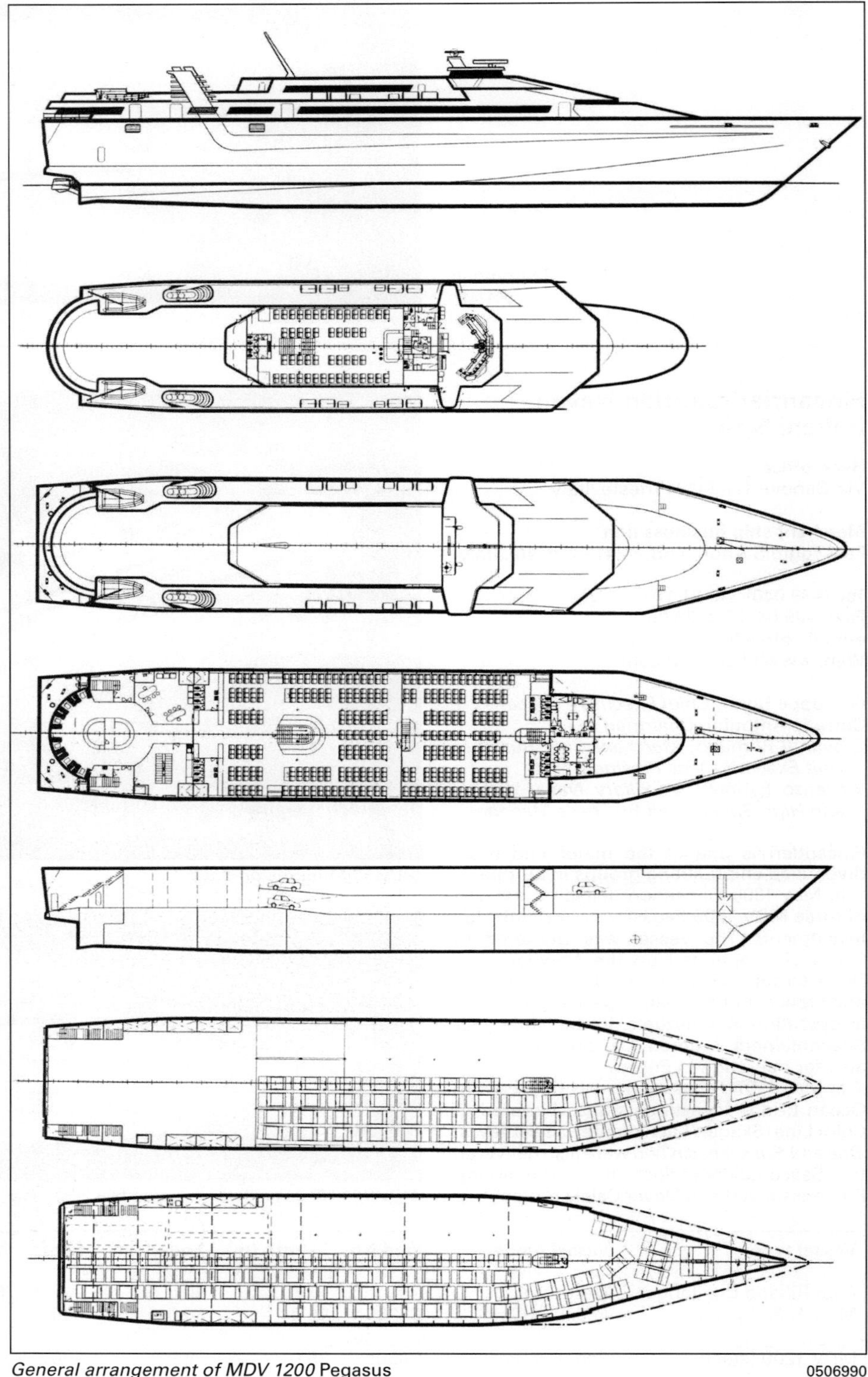

General arrangement of MDV 1200 Pegasus 0506990

MDV 1200 (STEEL)

Tallink Autoexpress 4 (ex Pegasus One)

Tallink Autoexpress 3 (ex Pegasus Two)

The MDV 1200 *Pegasus* is a deep vee monohull fast ferry designed to maintain high speeds in exposed sea conditions. The first MDV 1200 was delivered on 29 May 1996 and the second in May 1997, the contract for *Pegasus Three* is still pending. The hull, tested at the Danish Maritime Institute, was specifically optimised to ensure excellent sea-keeping performance.

The craft is capable of carrying 600 passengers and up to 170 cars at a speed of 36 kt.

Styling was defined jointly with Pininfarina.

The passenger and crew areas are subdivided into three main zones and are protected by a sprinkler system. The hull is made of high-tensile steel and the superstructures of light alloy.

The propulsion system and generating sets are split into two compartments in order to give the ship a return capability in case of failure of one engine room.

Specifications

Length overall	95.0 m
Length waterline	82.0 m
Beam	16.5 m
Draught	10.3 m
Passengers	600
Vehicles	170 cars
Propulsive power	24,000 kW
Operational speed	36 kt
Range	300 n miles

Classification: This design was developed in line with Category B of the new IMO regulations for Fast Ferries, and classed by Registro Italiano Navale.

Propulsion: Four MTU 20V 1163TB 73L diesel engines, each driving a Kamewa water-jet via a reduction gearbox.
Electrical system: Three main diesel generators and one emergency generator are specified.

100 m MDV (Aluminium) 1200

Superseacat
Superseacat One
Superseacat Two
Superseacat Three
Superseacat Four

Four of these craft have been ordered by Sea Containers. Two were delivered in 1997 and the other two in 1999. These craft are slightly larger than the three MDV 1200 Pegasus, and have the hull constructed in aluminium.

Specifications

Length overall	100.0 m
Length waterline	88.0 m
Beam	17.1 m
Passengers	800
Vehicles	175
Propulsive power	27,500 kW
Operational Speed	38 kt

Structure: All-aluminium construction with vehicles entering and leaving by the rear doors. There are no bow doors fitted to these craft.
Propulsion: Four Ruston 20 RK 270 diesel engines each rated at 875 kW and each driving a Kamewa 112S steerable water-jet via a Renk reduction gearbox.

MDV 2000

Gotlandia II

Gotlandia II was delivered to Rederi AB, Gotland in 2006. The hull material is high-tensile steel with aluminium alloy superstructure and the design is sized between the MDV 1200 Pegasus and the MDV 3000 Jupiter.

Specifications

Length overall	122.0 m
Length waterline	112.2 m
Beam	16.05 m
Deadweight	580 tonnes
Passengers	780
Vehicles	160 cars
Max speed	35 kt

Propulsion: Four MAN B&W 20RK280 diesel engines each rated at 9000 kW at 1000 rpm and each driving a Kamewa 140 SII waterjet via a ZF 6000 NRH gearbox.

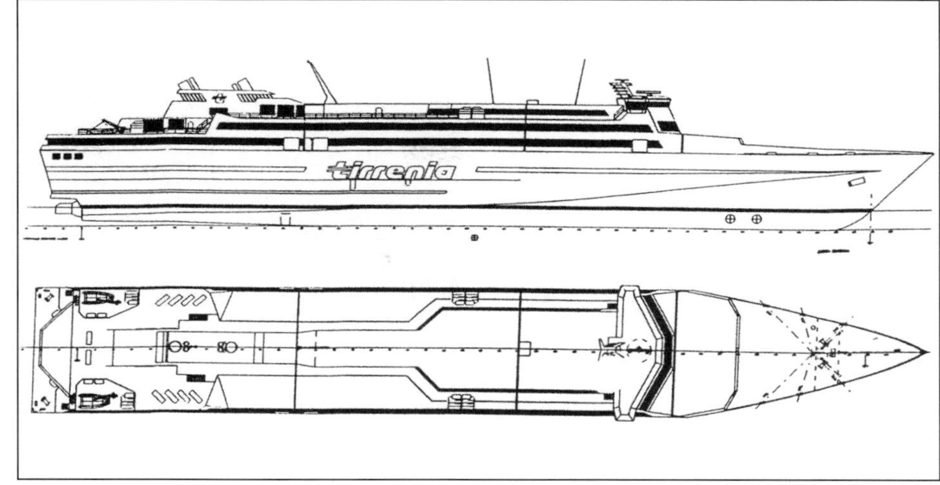

Profile and plan of the MDV 3000 Jupiter 0002649

MDV 3000 JUPITER

Aries
Taurus
Capricorn
Scorpio

The MDV 3000 is a deep vee monohull fast ferry designed to transport passengers, cars and trucks at high speed. Two of these craft were ordered by Tirrenia of Italy in 1996 and delivered in 1998 with a further two ordered in 1997 and delivered in 1999.

The vessel can be used in two different operating conditions: in a tourist service, where the cargo capability of the vessel is 460 cars and 1,800 passengers (deadweight 800 tonnes) with a speed of 42 kt, and in a freight service, where the cargo capability of the vessel is 120 cars and 420 m of freight lanes (deadweight 1,200 tonnes) with a speed of 40 kt.

Specifications

Length overall	146.0 m
Beam	22.0 m
Passengers	1,800
Vehicles	460 cars
Propulsive power	72,000 kW

Classification: This design has been developed in compliance with Category B of new IMO regulations.
Structure: The hull is constructed in high-tensile steel and the superstructure is aluminium.
Propulsion: The propulsion system and generating sets are split into two compartments and consist of two LM 2500 gas turbines and four diesel engines with a total power installed of 72,000 kW. The two LM 2500 gas turbines, rated at 22 MW, each drive a water-jet whereas the four MTU 20V 1163TB73L diesels, each rated at 6.7 MW are paired, each pair driving a water-jet unit.

Industrie Navales Meccaniche Affini SpA (INMA)

Viale F Bartholomeo, 88-19138 La Spezia, Italy

Tel: (+39 0187) 544 11
Fax: (+39 0187) 52 41 81
e-mail: inmatec@iol.it
Telex: 270297 IMAI

V Solinas, *Marketing Manager*

Industrie Navales Mechaniche Affini (INMA) is a licensee of Rodriquez Cantieri Navali and has delivered two FFM 1035 passenger/vehicle monohulls, this being the designation given by INMA for Aquastrada craft built under licence from Rodriquez.

INMA FFM Aquastrada Corsica Express 0069458

FFM 1035 (AQUASTRADA)

Three of these craft were delivered to Corsica Ferries in 1996.

Specifications

Length overall	103.5 m
Length waterline	86.5 m
Beam	14.5 m
Draught	2.2 m
Passengers	540
Vehicles	150 cars
Operational speed	37 kt
Range	370 n miles

Propulsion: Powered by four MTU 20V 1163 TB 74L diesel engines, two arranged to power the central booster water-jet and the other two driving individual wing water-jets.

Vessel type	Vessel name	Yard No	Length (m)	Speed (kt)	Seats	Vehicles	Originally delivered to	Date of build
FFM 1035	*Corsica Express II*	–	103.5	37.0	540	150	Corsica Ferries	1996
FFM 1035	*Corsica Express III*	–	103.5	37.0	540	150	Corsica Ferries	1996
FFM 1035	*Sardinia Express*	–	103.5	37.0	540	150	Corsica Ferries	1996

Moschini – Cantieri Ing Moschina SpA

Via de Nicolas 5, I-61032 Fano, Pesaro, Italy

Tel: (+39 0721) 85 42 36
Fax: (+39 0721) 85 49 34
e-mail: moshini@mobilia.it

Cantieri del Porto
Lungomare Mediterraneo 2-4, I-61032 Fano, Pesaro, Italy

Tel: (+39 0721) 80 99 88

L Doglioni, *Technical Manager*

Cantieri Ing Moschini SpA has been building GRP boats in the Pesaro area of Italy since 1970. The company operates three yards and has 100 employees.

36 m FERRY

Ischiamar II

Launched in 1992 following experience gained with the 35 m ferry, this vessel is of all-GRP construction.

35 m ferry built by Cantieri Ing Moschini SpA and SIAR SpA 0506683

Specifications

Length overall	36.0 m
Beam	7.1 m
Displacement, max	134 t
Passengers	350
Operational speed	28 kt

Propulsion: Main engines are two MTU 12V 396TE 94 diesels.

35 m FERRY

The hull and main deck of this vessel are built in GRP with the superstructure built in aluminium alloy. The passenger capacity is 350 and speed is 25 kt.

21 m SUPPLY VESSEL

With an overall length of 22 m and beam of 6 m, this 25 kt vessel was delivered in 1992.

Vessel type	Vessel name	Yard No	Length (m)	Speed (kt)	Seats	Vehicles	Originally delivered to	Date of build
21 m Supply vessel	–	–	22.0	25.0	–	none	–	1992
36 m Ferry	*Ischiamer II*	–	36.0	–	350	none	–	1992
35 m Ferry	–	–	35.0	25.0	350	none	–	–

Riva

Part of the Ferretti Group
Via Predore 30, I-24067 Sarnico, Italy

Tel: (+39 035) 91 02 02
Fax: (+39 035) 91 10 59
e-mail: info@riva-yacht.com
Web: www.riva-yacht.com

Based on Lake Iseo in northern Italy, but with an additional yard at La Spezia for construction of larger vessels, Riva is a manufacturer of luxury powerboats in the 10 to 35 m range. Its customers include many heads of state, senior industrialists and celebrities.

Riva has the capability to build 85 powerboats each year. Its principal markets are in Europe and the Middle East and the sales network is being expanded to take advantage of opportunities worldwide.

RIVA 20.8 m EGO SUPER

Specifications

Length overall	20.8 m
Length waterline	17.31 m
Beam	5.45 m
Displacement min	36.2 t
Displacement max	40.0 t
Passengers	16
Fuel capacity	3600 litres
Water capacity	720 litres,
Max speed	38 kt

Propulsion: Two MAN V12 1550 CRM diesel engines, rated at 1,140 kW each at 2300 rpm.

RIVA 26 m OPERA SUPER

Specifications

Length overall	26.02 m
Length waterline	20.85 m
Beam	6.2 m
Max speed	33.5 kt
Displacement min	62.0 t
Displacement max	72.73 t
Passengers	20
Fuel capacity	8500 litres
Freshwater capacity	1500 litres
Operational speed	30.0 kt

Propulsion: Two MTU 16V2000 M91 diesel engines each rated at 2029 kW at 2350 rpm.

RIVA 35.5 m ATHENA

Specifications

Length overall	35.4 m
Length waterline	29.5 m
Beam	7.08 m
Displacement, min	136 t
Displacement, max	162 t
Fuel capacity	23,450 litres
Water capacity	4000 litres
Passengers	20
Max speed	27 kt

Propulsion: Two MTU 12V 4000 diesel engines rated at 2040 kW each at 2100 rpm.

Vessel type	Vessel name	Yard No	Length (m)	Speed (kt)	Seats	Vehicles	Originally delivered to	Date of build
Riva19.5 Vertigo	Series Production	–	19.5	34.0	12	none	–	–
Riva 20.8 Ego Super	Series Production	–	20.8	38.0	16	none	–	–
Riva 23 Venere	Series Production	–	23.0	32.5	16	none	–	–
Riva 26 Opera Super	Series Production	–	26.0	33.5	20	none	–	–
Riva 35.5 Athena	Series Production	–	35.4	27.0	20	none	–	–

Rodriquez Cantieri Navali SpA

Via S Raineri 22, I-98122 Messina, Italy

Tel: (+39 090) 776 52 83
Fax: (+39 090) 776 52 93
e-mail: marketing@rodriquez.it
Web: www.rodriquez.it

Luciano La Noce, *President*
Marco Ragazzini, *Managing Director*
Alcide Sculati, *Technical Director*
Ettore Morace, *Sales and Marketing Director*
Sam Crockford, *Commercial Director*

Since 1956, when the shipyard started to build commercial fast ferries, Rodriguez has delivered over 220 high-speed craft and the company continues to be a world leader in the design, engineering and manufacture of such craft. The company has four shipyards in Italy, one at Messina in the south, one at Petra Liguna in the north, the Intermarine yard in Sarzana close to La Spezia and a fourth in Naples where the Pleasure craft production is managed. Rodriquez also has a yard in Rio de Janeiro, Brazil.

In May 2004 RCN Finanziaria purchased a majority share of Rodriquez Cantieri Navali. The other shareholders are MRS Sviluppo and Ustica Lines.

TMV 47 monohull

The first vessel *Marconi* was delivered to Adriatica di Navigazione SpA in December 1991 and the second *Pacinotti* in June 1992. The third vessel *Isola di San Pietro* was delivered to Saremar, a state-owned company, in June 1993. A fourth vessel, *Princess of Dubrovnik* was delivered to Croatian operator Atlas Travel Agency in July 1998. Following the introduction of the 37 m foil-assisted monohull, these larger vessels accommodate 400 passengers. Interior design is by Ennio Cantu who has been responsible for many Adriatica di Navigazione SpA vessels. The two passenger saloons are air conditioned. The third and fourth boats do not have aft stabilising foils but four Vosper Thornycroft stabilising fins, two forward and two aft.

Specifications

Length overall	46.0 m
Length waterline	37.2 m

Beam 7.6 m
Draught 1.2 m
Displacement, max 145 t
Passengers 340
Propulsive power 2 × 2,000 kW
Operational speed 34 kt
Range 200 n miles

Propulsion: Engines: two MTU 16V 396 TE 74L, 2,000 kW each at 2,000 rpm. The *Princess of Dubrovnik* is powered by two MWM TBD 604B V16 engines giving an operational speed of 30 kt.

Thrust devices: two Kamewa water-jets (first two boats); two Ulstein Propeller Speed-Z CPZ 60/42-125 CPP (third boat).

TMV 50

Segesta Jet
Tindari Jet
Selinunte Jet

Three TMV 50 craft were delivered to the Ferrovia dello Stato (Italian State Railways) in 1999: March, May and June respectively.

Specifications

Length overall 50.46 m
Length waterline 43.0 m
Beam 9.2 m
Draught 1.35 m
Displacement, max 183 t
Deadweight 56 t
Crew 7
Passengers 511
Propulsive power 4,700 kW
Operational speed 30 kt
Range 280 n miles

Classification: RINA HSC.

Structure: All-aluminium structure.

Propulsion: Two MTU 16V 4000 diesels each rated at 2,350 kW at 1,975 rpm and each driving a Kamewa 71 SII water-jet.

TMV 50 NLG

Vesuvio Jet

A derivative of the TMV 50, *Vesuvio Jet* was delivered in June 2003 to Navigazione Libera del Golfo.

Specifications

Length overall 50.5 m
Length waterline 43.0 m
Beam 8.0 m
Draught 1.4 m
Displacement 190 t
Deadweight 56 t
Crew 7
Passengers 444
Operational speed 36 kt

Construction: All aluminium hull and superstructure.

Propulsion: Three MTU 396 16V diesel engines each rated at 2,000 kW driving two Kamewa 71 steerable water-jets and one Kamewa 63 booster water-jet.

TMV 70

Isola di Capri
Isola de Capria
Isola di Vulcano
Isola di Procida

Rodriquez Shipyard delivered the *Isola di Capri* to Caremar in Italy in September 1998. Three further vessels were delivered in 1999, one each to Caremar, Siremar and Toremar.

Specifications

Length overall 70.9 m
Length waterline 64.0 m
Beam 12.4 m
Draught 2.45 m
Displacement, min 425 t
Displacement, max 625 t
Passengers 550
Vehicles 57 cars
Fuel capacity 40,000 litres
Propulsive power 9,400 kW
Max speed 32 kt
Operational speed 29 kt
Range 280 n miles

Classification: RINA HSC.

Structure: High-tensile steel hull and aluminium superstructure.

TMV 50 Vesuvio Jet 0580426

The 47 m monohull Princess of Dubrovnik 0097291

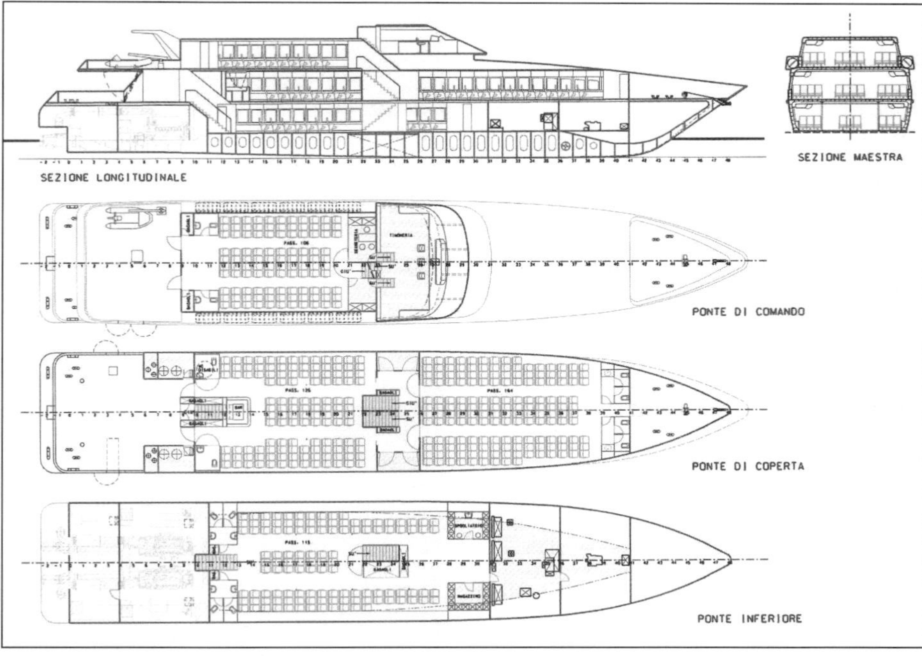

General arrangement of TMV 50 0097293

Aquastrada Guizzo *in service with Tirrenia* 0506908

Propulsion: Four MTU 16V 4000 diesel engines each rated at 2,350 kW at 2,100 rpm. The central engines drive a single Kamewa 112 BIC booster water-jet and the side engines drive fixed-pitch propellers.
Auxiliary systems: Rodriquez SMS ride control system composed of four non-retractable fins. A 100 kW hydraulic bow thruster is also fitted.

TMV 84

Ramon Llull
Princess

Two vessels were ordered in 2002, one for operation in Spain (*Ramon Llull*) having a capacity for 1,246 passengers and 56 cars, and one for Arab Bridge of Jordan (*Princess*), having a capacity for 710 passengers and 58 cars. Both vessels have all-aluminium hulls and superstructures and are powered by diesel driven water-jets. *Ramon Llull* is fitted with four anti-roll fins and T-foils fore and aft for motion control. *Princess* is fitted with anti-roll fins only but has two retractable bow thrusters for low speed manoeuvring.

Specifications
(Spanish Vessel)

Length overall	83.4 m
Beam	13.6 m
Draught	1.8 m
Passengers	462
Vehicles	56
Maximum speed	39 kt
Operating speed	37 kt
Range	300 n miles

Structure: All-aluminium construction.
Propulsion: Four MAN B&W VP 185 diesel engines each rated at 3,700 kW and each driving a water-jet unit.

Specifications
(Arab Bridge Vessel)

Length overall	83.4 m
Beam	13.5 m
Draught	2.0 m
Payload	247 t
Passengers	710
Vehicles	58 cars
Maximum speed	35 kt
Operating speed	33 kt
Range	300 n miles

Structure: All-aluminium hull and superstructure.
Propulsion: Four MTU 16V595 diesel engines each rated at 3,925 kW and each driving a LIPS LJ 91E water-jet via a Reintjes VLJ 2230 gearbox.

100 AND 103 m Aquastrada (TMV 103)

Guizzo
Scatto
Sardinia Express

There have been two 100 m fast monohull passenger/car ferries constructed for Tirrenia of Italy *Guizzo* and *Scatto*. They both have a similar arrangement although *Scatto* carries a slightly increased payload. The vessels' hulls are of a hard chine form and constructed of high-tensile steel. The superstructure for the passenger accommodation is constructed of aluminium. A further 103 m was subsequently constructed for Corsica Ferries. Two similar 103 m craft were constructed for the same operator by INMA of Italy.

Specifications

Length overall	101.8 m
Length waterline	85.3 m
Beam	14.5 m
Draught	2.1 m
Displacement, min	800.5 t
Displacement, max	1,057.5 t
Passengers	450
Vehicles	126 cars
Propulsive power	24,065 kW
Operational speed	43 kt
Range	370 n miles

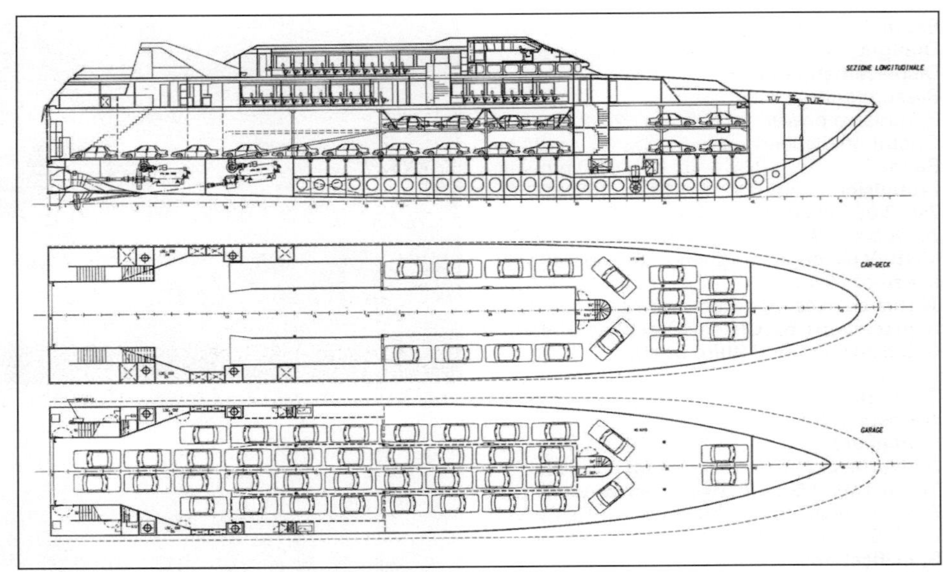

General arrangement of TMV 70 Isola di Capri 0097296

TMV 70 Isola de Capria *in service* 0097295

TMV 50 Segesta *Jet* 0069459

TMV 115 0097300

Propulsion: Engines: two MTU 16V 595TE 70, 3,565 kW each, one GE LM 2500 gas turbine, 20,500 kW; driving two Kamewa 100 SII, one Kamewa 180 SII.

Three more Aquastrada-type vessels, one constructed by Rodriquez and the other two under licence at INMA of La Spezia, were delivered to Corsica Ferries in 1996.

Specifications

Length overall	103.5 m
Length waterline	86.5 m
Beam	14.5 m
Draught	2.2 m
Displacement, max	1,400 t
Passengers	540
Vehicles	150 cars
Propulsive power	24,000 kW
Operational speed	37 kt

Propulsion: Powered by four MTU 20V 1163 TB 74L diesel engines each rated at 6,000 kW. Two engines are arranged to power a central booster water-jet and the other two drive individual wing water-jets.

TMV 114

Volcan de Tauro

Rodriquez Cantieri Navali delivered a TMV 114, hull number 274, for Naviera Armas (Spain) in 2000. The vessel has a Rodriguez SMS (Seaworthiness Management System) composed of four lateral fins and two T-foils fitted.

Specifications

Length overall	113.5 m
Length waterline	96.2 m
Beam	16.5 m
Draught	2.5 m
Deadweight	547 t
Crew	16
Passengers	880
Vehicles	200 cars
Fuel capacity	50,000 litres
Water capacity	3,000 litres
Max speed	45.0 kt
Operational speed	40.0 kt

Classification: Bureau Veritas 13/3 (–) +HSC-CAT B/Passenger and Vehicles carrier, Deep Sea + Mach. AUT-MS.

Structure: High-tensile steel hull and aluminium superstructure.

Propulsion: Six Caterpillar 3616 diesel engines each rated at 6,000 kW driving in pairs, three LIPS water-jets.

TMV 115

Frederico Garcia Lorca

The TMV 115 is an evolution of the TMV 114, the primary difference has been the change from high tensile steel hull to an aluminium hull. The weight savings allowed the propulsion system to be changed from 6 to 4 diesels with a negligible change in service speed.

Specifications

Length overall	115.3 m
Length waterline	96.2 m
Beam	17 m
Draught	2.5 m
Displacement, max	1,500 t
Deadweight	540 t
Passengers	884
Vehicles	200 cars
Propulsive power	28,800 kW
Operational speed	40 kt
Range	650 n miles

Classification: RINA +100A1 NAVS HSC-CAT B + 1AQ-2TpTr (RoRo).

Structure: All-aluminium hull and superstructure.

Propulsion: Four Caterpillar 3618 diesel engines each rated at 7,200 kW and each driving a LIPS water-jet.

Auxiliary Systems: Four roll stabiliser fins, two T-foils (one fore and one aft) and an interceptor train control system.

TMV 130 (Design)

This mid-sized monohull vessel has been designed to operate efficiently at a range

Rodriguez TMV 114 Volcan de Tauro 0059114

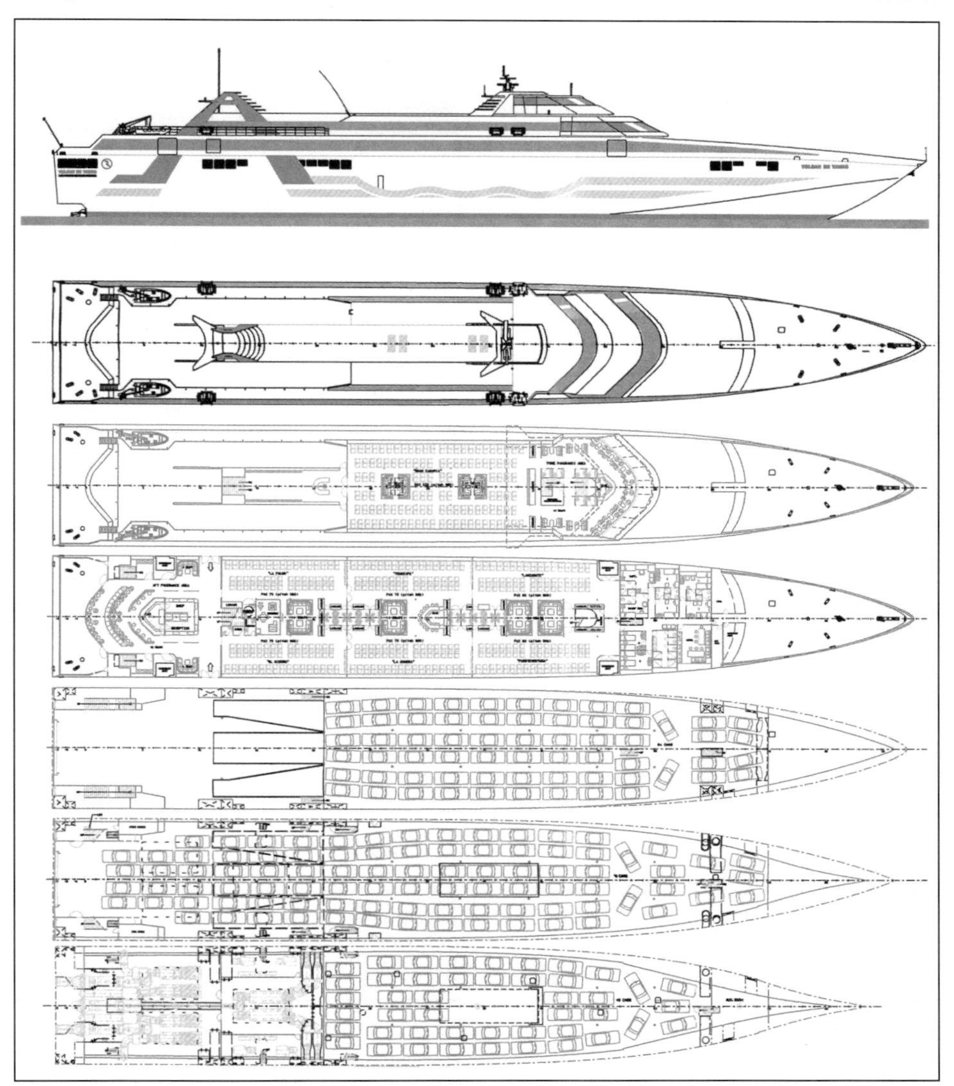

TMV 114 general arrangement 0093275

TMV 84 Ramon Llull (Rodriquez) 0567969

of operational speeds, from 36 to 40 kt depending on the deadweight carried.

Specifications

Length overall	130.0 m
Beam	19.6 m
Passengers	1,400
Vehicles	300 cars
Propulsive power	28,800 kW
Operational speed	36 kt
Range	530 n miles

Propulsion: Four Caterpillar 3618 diesel engines each rated at 7,200 kW and each driving a Kamewa water-jet.

TMV 84 Princess (Rodriquez)
0567970

Vessel type	Vessel name	Yard No	Length (m)	Speed (kt)	Seats	Vehicles	Originally delivered to	Date of build
47 m foil-assisted monohull	*Marconi*	–	46.0	34.0	340	None	Adriatica di Navigazione SpA	December 1991
47 m foil-assisted monohull	*Pacinotto*	–	46.0	34.0	340	None	Adriatica di Navigazione SpA	June 1992
47 m foil-assisted monohull	*Isola Di San Pietro*	–	46.0	34.0	340	None	Saremar	June 1993
100 m Aquastrada	*Guizzo*	–	101.8	43.0	450	126	Tirrenia, Italy	1993
100 m Aquastrada	*Scatto*	–	101.8	43.0	450	126	Tirrenia, Italy	1994
103 m Aquastrada	*Sardinia Express*	–	103.5	37.0	540	150	Corsica Ferries	1996
47 m foil-assisted monohull	*Princess of Dubrovnik*	–	46.0	30.0	340	None	Atlas Travel Agency	July 1998
TMV 70	*Isola di Capri*	–	70.9	29.0	550	57	Caremar	September 1998
TMV 50	*Segesta Jet*	–	50.5	30.0	511	None	Ferrovia dello Stato (Italian State Railway)	March 1999
TMV 50	*Selinunte Jet*	–	50.5	30.0	511	None	Ferrovia dello Stato (Italian State Railway)	May 1999
TMV 50	*Tindari Jet*	–	50.5	30.0	511	None	Ferrovia dello Stato (Italian State Railway)	June 1999
TMV 70	*Isola di Vulcano*	–	70.9	29.0	550	57	Siremar	1999
TMV 70	*Isola di Procida*	–	70.9	29.0	550	57	Toremar	1999
TMV 70	*Isola Capria*	–	70.9	29.0	550	57	Caremar	1999
TMV 114	*Volcan de Tauro*	274	113.5	40.0	880	200	Armas Ocean Jet SA	April 2000
TMV 115	*Frederico Garcia Lorca*	–	115.3	40	884	200	Baleria Eurolineas Maritimes	2001
TMV 84	*Ramon Llull*	–	83.4	37.0	462	58	Baleria Eurolineas Maritimes	2003
TMV 84	*Princess*	–	83.4	37.0	462	58	Arab Bridge Maritime	2003
TMV 50	*Vesuvio Jet*	306	50.5	30.0	511	none	Navigazione Libera del Golfo	2003
TMV 42	*Agostino Lauro*	339	42.6	31	320	none	Lauro Shipping	2006

Japan

Kiso Shipbuilding

906 Mukaihigashi-Cho, Onomichi-Shi, Hiroshima, 722-0062 Japan

Tel: (+81 848) 44 56 00
Fax: (+81 848) 44 56 01
e-mail: info@kisoship.jp
Web: www.kisoship.jp

26 m FAST FERRY

New Noshima

New Noshima, a 26 m monohull fast ferry was delivered in 1998.

Specifications

Length overall	25.8 m
Beam	5.9 m
Draught	0.8 m
Crew	4
Passengers	95
Maximum speed	30 kt
Operational speed	28 kt

Propulsion: Two Detroit Diesel 16V 92TA DDEC each rated at 820 kW and driving a fixed pitch propeller via a reverse reduction gearbox.

23 m FAST FERRY

Hayabusa No 1
Hayabusa No 2

Two 23 m monohull fast ferries were delivered to Sanyo Shosen for operation on the Seto Inland Sea.

Specifications

Length overall	23.4 m
Beam	4.9 m
Draught	2.2 m
Crew	2
Passengers	95
Fuel capacity	5,000 litres
Water capacity	200 litres
Maximum speed	36 kt
Operational speed	33 kt

Propulsion: Two MTU 16V 2000 M70 diesel engines each rated at 1,050 kW and each driving a fixed pitch propeller via a NICO MGN 353 reverse reduction gearbox.

22.8 m FAST FERRY

Taiyu 2

The vessel was designed by Katsuhiko Kiso for the Tanaka Transportation Company.

Specifications

Length overall	22.8 m
Length waterline	20.9 m
Beam	5.2 m
Draught	0.87 m
Displacement	32 t
Crew	3
Passengers	85
Fuel capacity	2,500 litres
Water capacity	500 litres

Propulsion: Two Isuzu UM6RBITCK diesel engines each producing 374 kW driving fixed pitch five-bladed propellers.

New Noshima *ferry* 0580423

Kagayaki No 1 *fast ferry* 0580420

21 m FAST FERRY

Yunagi

Ordered by Kumashima City as an ambulance boat, this vessel was delivered in early 2006.

Specifications

Length overall	20.8 m
Beam	5.0 m
Draught	0.8 m
Crew	2
Displacement	19 tonnes
Passengers	12
Fuel capacity	2,200 l
Maximum speed	30.5 kt
Operational speed	25 kt

Propulsion: Two Mitsubishi 56B5-MTK25 diesel engines delivering 537 kW to fixed pitch propellers.

20 m FAST FERRY

Kagayaki No 1

Kagayaki No 1, a 20 m monohull fast ferry was delivered in 1999.

Specifications

Length overall	19.3 m
Beam	4.2 m
Draught	0.7 m
Crew	2
Passengers	56
Maximum speed	33 kt
Operational speed	30 kt

Hayabusa No 2 0580415

Propulsion: Two MTU 12V 183 TE72 diesel engines each rated at 560 kW and each driving a fixed pitch propeller via a reverse reduction gearbox.

Vessel type	Vessel name	Yard No	Length (m)	Speed (kt)	Seats	Vehicles	Originally delivered to	Date of build
26 m Fast Ferry	*New Noshima*	–	25.8	28	95	none	–	1998
20 m Fast Ferry	*Kagayaki No 1*	–	19.3	30	56	none	–	1999
23 m Fast Ferry	*Hayabusa No 1*	–	23.4	33	95	none	Sanyo Shosen	2003
23 m Fast Ferry	*Hayabusa No 2*	–	23.4	33	95	none	Sanyo Shosen	2003
22.8 m Fast Ferry	*Taiyu 2*	–	22.8	23.5	85	none	Tanaka Transportation	2004
21 m Fast Ferry	*Yunagi*	–	20.8	30.5	12	1	Kumishima City	2006

Mitsubishi Heavy Industries (MHI) Ltd

16-5 Konan 2-chome, Minato-ku, Tokyo 108-8215, Japan

Tel: (+81 3) 6716 31 11
Fax: (+81 3) 6716 58 00
e-mail: ueda@ship.hq.mhi.co.jp
Web: www.mhi.co.jp

Naoki Ueda, *Manager*

MHI has built a number of high-speed craft mainly at the Shimonoseki Shipyard and Machinery Works in Hikoshima, Shimonoseki, Japan.

Seagrace 0506995

90 m FAST RO-RO FERRY

Unicorn

The company received an order from Higashi Nihon for a 90 m fast ro-ro ferry in 1996 which was delivered in 1997. The vessel operates between Aomori on Honshu and Hakodate on Hokkaido.

Specifications

Length overall	101.0 m
Length waterline	90.0 m
Beam	14.9 m
Crew	11
Passengers	423
Vehicles	106 cars
Propulsive power	26,000 kW
Operational speed	35 kt
Max speed	42.4 kt

90 m fast Ro-Ro ferry Unicorn 0009743

Vessel type	Vessel name	Yard No	Length (m)	Speed (kt)	Seats	Vehicles	Originally delivered to	Date of build
48.5 m Fast Ferry	*Marine Star*	–	48.5	–	351	None	Oki Kisen Company Ltd	1983
48.5 m Fast Ferry	*Seagrace*	–	48.5	–	286	None	Kyushu Shosen Ltd	1993
90 m Fast RoRo Ferry	*Unicorn*	–	101.0	35.0	423	106	Higashi Nihon	1997
48.5 m Fast Ferry	*Seahawk*	–	48.5	–	301	None	Koshikishima Shosen Ltd	–
46 m Fast Ferry	*Oniyozu*	–	45.95	23	–	None	Mishimi Island	–

Propulsion: Powered by four MTU 20V 1163 TB 73L diesels each rated at 6,500 kW and each driving a Kamewa water-jet.

48.5 m FAST FERRY

Seahawk

Specifications

Length overall	48.5 m
Beam	8.2 m
Passengers	301
Propulsive power	2 × 2,023 kW
Max speed	29.2 kt

Propulsion: Two diesel engines, 2,023 kW each, 1,450 rpm.
Owner: Koshikishima Shosen Ltd.

Marine Star

Built in 1983 and originally in service with Oki Kisen Company Ltd, this craft has now been sold to a foreign owner.

Specifications

Length overall	48.5 m
Beam	8.2 m
Passengers	351
Propulsive power	2 × 1,800 kW
Max speed	30.7 kt

Propulsion: Two 1,800 kW high-speed diesels.

Seagrace

This vessel was delivered to its owner, Kyushu Shosen Ltd, in March 1993.

Specifications

Length overall	48.5 m
Beam	8.2 m
Draught	3.9 m
Passengers	286
Max speed	29.8 kt
Propulsive power	2 × 2,023 kW

Propulsion: Two diesels, 2,023 kW each at 1,450 rpm.
Control: Fin stabiliser, anti-pitching fin.

Mokubei Shipbuilding Company

1-2-20 Imakatata, Otsu, Shiga Prefecture, 520 0241 Japan

Tel: (+81 775) 72 21 01
Fax: (+81 775) 72 21 11

Mamoru Nakano, *President*
Masaru Ikeda, *Naval Architect*

Mokubei Lansing

0506691

Kaiyo 0506692

22 m TOURIST BOAT

Lansing

A water-jet-propelled sightseeing tourist boat, built in 1982 for operation by the Biwa Lake Sightseeing Company on the very shallow Biwa Lake. Built in aluminium, the vessel has seating for 86 passengers and two crew. The fully laden power-to-weight ratio is 41.5 hp/t.

Specifications

Length overall	22.0 m
Length waterline	19.8 m
Beam	4.0 m
Draught	0.6 m
Displacement, max	26 t
Crew	2
Passengers	86
Propulsive power	2 × 400 kW
Max speed	27 kt
Operational speed	23 kt

Propulsion: Main engines are MAN 254 MLE V12 diesels rated at 400 kW each at 2,230 rpm; driving two Hamilton Model 421 type water-jets directly driven from the engine flywheel via a torsionally flexible coupling and Cardan shaft.

Control: The vessel is fitted with an auxiliary diesel AC generator which enables AC motor-driven hydraulic power packs to be used for the water-jet unit controls. A tandem-pump hydraulic power pack is used for the reverse ducts while a single-pump hydraulic power pack operates the steering. At the helm, control for steering is via a wheel, the reverse is operated via a single electric joystick and two position indicators are fitted, one for the steering deflector angle and the other for the reverse duct position.

25.5 m ESCORT BOAT

Kaiyo

A 21.5 kt steel hull firefighting boat, equipped to carry foam liquid, launched on 19 February 1986 and owned by Sanyo-kaiji Company Ltd and Nihon-kaiji-kogyo Company Ltd.

Specifications

Length overall	25.5 m
Beam	5.6 m
Draught	1.1 m
Displacement, min	54.9 t
Crew	9
Propulsive power	2 × 746 kW

Propulsion: Main engines are two Yanmar 12LAAK-UT1, 746 kW each at 1,850 rpm, driving two fixed-pitch, three-blade, aluminium bronze propellers, 1,000 mm diameter, 1,000 mm pitch, developed blade-area ratio: 0.90:1. Propeller shaft diameter: 99 mm, length 6,500 mm; via NICO-MGN 332 gearboxes, shaft output MCR: 1,000 hp at 907 rpm.

23.5 m RESEARCH BOAT

Mizusumashi II

Water quality research boat delivered 31 March 1989.

Specifications

Length overall	23.5 m
Beam	4.8 m
Displacement, max	39.1 t
Crew	30
Propulsive power	2 × 1,007 kW
Max speed	28 kt

Propulsion: Two DDC 16V 92TA, 1,007 kW each, MCR, 2,300 rpm; driving two five-blade propellers, 700 mm diameter, 884 mm pitch, developed area ratio: 1.0; via NICO MGN 332E reduction gearboxes, ratio: 1.42:1.

Vessel type	Vessel name	Yard No	Length (m)	Speed (kt)	Seats	Vehicles	Originally delivered to	Date of build
22 m Tourist Boat	*Lansing*	–	22.0	23.0	86	None	Biwa Lake Sightseeing Company	1982
22.5 m Escort Boat	*Kaiyo*	–	25.5	–	–	None	Sanyo-kaiji Co Ltd and Nihon-kaiji-kogyo Co Ltd	1986
23.5 m Research Boat	*Mizumashi II*	–	23.5	–	–	None	–	1989

Sumidagawa Shipyard Co Ltd

2-1-16 Shiomi, Koto-ku, Tokyo, 135-0052, Japan

Tel: (+81 3) 3647 6111
Fax: (+81 3) 3647 5210
e-mail: info@sumidagawa.co.jp
Web: www.sumidagawa.co.jp

Horoshi Ishiwata, *President*

Formed in 1913, Sumidagawa Shipyard produced its first fast ferry in 2005.

Monohull Fast Ferry

Ibis

Designed in house by Sumidagawa, this high speed monohull ferry features a very long slender bulbous bow which is intended to improve seakeeping qualities. The vessel is of aluminium construction, has a service speed of 25 knots and is powered by diesel engines.

Fast Ferry *Ibis* 1120343

Yamaha Motor Company Ltd

2500 Shingai, Iwata, Shizuoka 438-3501, Japan

Tel: (+81 538) 32 11 45
Fax: (+81 538) 37 42 50
Web: www.yamaha.co.jp

Yamaha Motor started in the marine business as a manufacturer of marine engines and FRP boats in 1960. The company has manufactured a number of large fast monohull and catamaran craft since that time.

The Tsugaru

This boat was designed to engage in supervision and survey operations in Matsu Bay, Aomori Prefecture.

Specifications

Length overall	19.6 m
Beam	4.4 m
Draught	0.8 m
Passengers	22
Fuel capacity	1,900 litres
Water capacity	200 litres
Propulsive power	2 × 410 kW
Max speed	33.5 kt
Operational speed	29 kt
Range	240 n miles

Structure: Hull material is FRP, and is single skin.

Propulsion: Main engines are two GM 8V 92TA diesels, 410 kW at 2,240 rpm, 373 kW continuous at 2,170 rpm; driving two surface step drives.

Electrical system: One 18.2 kW generator.

The Tsugaru *in Matsu Bay* 0506694

Yamaha 24 m motor yacht 0506986

24 m MOTOR YACHT

This GRP motor yacht (Yard No S-314) was delivered to a Yokohama owner in July 1995.

Specifications

Length overall	24.0 m
Beam	5.0 m
Draught	1.9 m
Displacement, min	36 t
Fuel capacity	4,000 litres
Max speed	31.5 kt

Classification: JG certification for coastal water operation.

Propulsion: Powered by two MAN 12V diesel engines each rated at 800 kW and each driving a fixed-pitch three-blade propeller via a reverse reduction gearbox.

Vessel type	Vessel name	Yard No	Length (m)	Speed (kt)	Seats	Vehicles	Originally delivered to	Date of build
24 m Motor Yacht	–	S-314	24.0	–	–	none	–	1995
20 m Vessel	*The Tsugaru*	–	19.6	29.0	22	none	Aomori Prefecture	–
Yamaha 65	*Lumba Lumba*	–	19.5	30	50	none	–	1989
Yamaha 91	*Superjet Kasumi*	–	27.3	30	153	none	Kasumigaura Jet Line	1985

Korea, South

Semo Company Ltd

Part of the Semo Kimchi Group

Shipbuilding Division: 1 Jank-ri, Tonghae-Myon, Kosung-gun, KyongSangnam, South Korea

Tel: (+82 556) 672 35 35
Fax: (+82 556) 672 35 70
e-mail: semo@semo.co.kr

28 m PASSENGER FERRY

Baram Tara
Sewol Tara

Two of these GRP craft were delivered in 1997 for operation on the Pusan to Keojedo route.

Specifications

Length overall	28.0 m
Beam	7.2 m
Draught	1.2 m
Crew	6
Passengers	200
Propulsive power	2,942 kW
Max speed	37 kt
Operational speed	32 kt

Classification: Korean Register of Shipping.

Propulsion: Two Niigata 12V-16FX diesel engines each rated at 1,471 kW driving Niigata/MJP 650 water-jets via Niigata 2RGN 130K reduction gearboxes.

Vessel type	Vessel name	Yard No	Length (m)	Speed (kt)	Seats	Vehicles	Originally delivered to	Date of build
28 m Passenger Ferry	*Sewol Tara*	–	28.0	32.0	200	None	Japan	1997
28 m Passenger Ferry	*Baram Tara*	–	28.0	32.0	200	None	Japan	1997

Semo 28 m ferry *Baram* Tara 0007178

Malaysia

Malaysia Marine and Heavy Engineering

P.O. Box 77, 81707 Pasir Gudang, Johor, Malaysia

Tel: (+60 7) 251 21 11
Fax: (+60 7) 251 38 90
e-mail: enquires@mmhe.com.my
Web: www.mmhe.com.my

Wan Yusoff Wan Hamat, *Chief Executive Officer*

Jana Utama *on trials* 0093271

Malaysia Shipyard and Engineering Sdn Bhd (MSE) marine and industrial complex is located by the Straits of Johor at the southern tip of Malaysia, adjacent to the island of Singapore. It was incorporated on the 18 May 1973 and privatised on 8th July 1991. Present shareholding equity is held by a consortium comprising Malaysia International Shipping Corporation Bhd (43 per cent), Kuok Brothers Sdn Bhd (28.5 per cent) and IMC Enterprise Incorporated (28.5 per cent). All MSE's operating units are accredited to ISO 9000 series certification.

Employing 1,625 people, over 125 ha of land in the industrial area of Pasir Gudang has been developed to cover MSE's core activities of ship repair, shipbuilding and engineering (offshore and onshore). The ship building division has to date delivered more than 85 vessels of various designs to both international and local clients.

34 m crewboat

Jana Utama
Karang Putih

Specifications

Length overall	34.0 m
Length waterline	31.8 m
Draught	1.25 m
Fuel capacity	45,000 litres
Operational speed	25 kt

Classification: Lloyds register.
Structure: Aluminium hull and superstructure.
Propulsion: Four Cummins KTA 38-M2 diesel engines each rated at 1,120 kW at 2,050 rpm driving Stone Marine fixed-pitch propellers via ZF BW365 gearboxes.
Auxiliary systems: Two Cummins 6BT5.9 engines each rated at 50 kW.

Vessel type	Vessel name	Yard No	Length (m)	Speed (kt)	Seats	Vehicles	Originally delivered to	Date of build
34 m crewboat	*Jana Utama*	–	34.0	25.0	–	None	Syarikat Borcos Shipping	1999
34 m crewboat	*Karang Putih*	–	34.0	25.0	–	None	Syarikat Borcos Shipping	1999

Yong Choo Kui Shipyard Sdn Bhd

13 Chengal Road, 96000 Sibu, Sarawak, Malaysia.

Tel: (+60 8) 431 99 22
Fax: (+60 8) 313 37 06
e-mail: chookui@tm.net.my
Web: www.yckshipyard.com

Yong Choo Kui, *Managing Director*

Established in 1976, the company has constructed a number of 40 m fast commercial craft such as Weesam 3 and 5 in 1999. It is understood that various local subcontractors are used to assist in the building of the company's fast monohull ferries.

47 m MONOHULL FERRY

There were three of these craft delivered in 1996, two vessels for Chinese operations and one for Indonesian operations between Pontianak and Jakarta.

Specifications

Length	46.8 m
Beam	6.1 m
Passenger (approx)	300
Operational speed	28 kt

Propulsion: The two vessels for China are powered by two Mitsubishi S16R-MPTK diesels rated at 1,610 kW at 1,800 rpm, and the Indonesian vessel by three Yanmar 16LAK STI diesels each rated at 1,120 kW at 1,900 rpm.

Weesam No 5

Built in 1999 for operations in Zamboanga in the Philippines.

Specifications

Length overall	43.0 m
Length waterline	37.4 m
Beam	5.5 m
Passengers	284
Operational speed	33 kt

Vessel type	Vessel name	Yard No	Length (m)	Speed (kt)	Seats	Vehicles	Originally delivered to	Date of build
39.5 m monohull ferry	*Bullet Express III*	–	39.5	20.0	139	None	Malaysia	1995
42 m monohull ferry	*Jambo Jet 218*	–	42.1	30.0	231	None	Singapore	1995
32.8 m monohull ferry	*Jambo Jet 118*	–	32.8	26.0	96	None	Malaysia	1995
42 m monohull ferry	*Labuan Express Tiger*	–	42.1	32.0	207	None	Malaysia	1995
47 m monohull ferry	–	–	46.8	28.0	300	None	China	1996
47 m monohull ferry	–	–	46.8	28.0	300	None	China	1996
47 m monohull ferry	–	–	46.8	28.0	300	None	Indonesia	1996
41 m monohull ferry	*MS Express*	–	41.0	30.0	197	None	Phillipines	1998
41.7 m monohull ferry	*Weesam No 3*	–	41.7	32.0	197	None	Phillipines	1999
43 m monohull ferry	*Weesem No 5*	–	43.0	33.0	284	None	Phillipines	1999
39.5 m monohull ferry	–	–	39.5	28.0	120	None	–	2000
41.7 m monohull ferry	–	–	41.7	32.0	197	None	–	2000
41.7 m monohull ferry	–	–	41.7	32.0	197	None	–	2000
40.8 m monohull ferry	*Desa Intan*	–	40.8	32.0	282	None	Malaysia	–

Classification: CM (Malaysian Government).
Structure: Mild steel hull and superstructure.
c
Propulsion: Two Caterpillar 3516B diesels, each rated at 1,790 kW and driving fixed-pitch propellers.

Jambo Jet 218
Built in 1995 for operation from Singapore and P Tioman.

Specifications

Length overall	42.1 m
Beam	5.5 m
Passengers	231
Operational speed	30 kt

Classification: NKK.
Structure: The main hull and superstructure are of mild steel construction.
Propulsion: Powered by two DDC 16V 149TI diesel engines, each rated at 1,340 kW at 1,900 rpm.

Labuan Express Tiger
This vessel was built in 1995 with a very similar specification to *Jambo Jet*. However, with passenger numbers reduced to 207, the operational speed increases to 32 kt with the same engine power. The vessel operates from Labuan to Kota Kinabalu.

Weesam No 3
Weesam No 3 was built in 1999 for operations in Zamboanga in the Philippines. Two further craft were built to this design in 2000, one with Yanmar 16 LAK-STE engines and the other with Mitsubishi S12R-MTK engines.

Specifications

Length overall	41.7 m
Length waterline	37.7 m
Beam	4.7 m
Passengers	197
Operational speed	32 kt

Classification: KM (Malaysian Government).
Structure: Mild steel hull and superstructure.
Propulsion: Two Caterpillar 3512B diesels, each rated at 1,120 kW, driving fixed-pitch propellers.

MS Express
Built in 1998 for operation in Zamboanga in the Philippines.

Specifications

Length overall	41.0 m
Length waterline	38.0 m
Beam	5.2 m
Passengers	197
Propulsive power	2,310 kW
Operational speed	30 kt

Structure: The hull material is mild steel.
Propulsion: Main engines are two Mitsubishi diesel engines each rated at 1,155 kW and each driving a fixed-pitch propeller via a ZF BW450 gearbox.

Weesam No 3 0081016

Weesam No 5 0081017

Desa Intan
Operating area: Melacca and Dumai.

Specifications

Length overall	40.8 m
Beam	5.5 m
Passengers	282
Propulsive power	3 × 902 kW
Max speed	35 kt
Operating speed	32 kt

Classification: NKK.
Structure: The hull material is mild steel.
Propulsion: Main engines are three Yanmar 12LAK ST2, 902 kW each at 1,910 rpm.

Bullet Express III
Built in 1995 for operation between Cebu, Ormocand and Tagbilaran. Another vessel was built in 2000, having a beam of 3.8 m, passenger capacity of 120 and a speed of 38 kt.

Specifications

Length overall	39.5 m
Beam	4.4 m
Passengers	139
Operational speed	20 kt

Propulsion: Powered by two Yanmar 12LAK ST2 diesels each rated at 900 kW.

Jambo Jet 118
Built in 1995 for operation between Mersing and P Tioman.

Specifications

Length overall	32.8 m
Beam	4.2 m
Passengers	96
Operational speed	26 kt

Propulsion: Powered by two MTU 12V 183 TC60 diesel engines.

35.6 m MONOHULL

Delivered to Ton Guan Hock and Sons Realty, this 35.6 m monohull is driven by water-jets and has a maximum speed of 40 kt.

Specifications

Length overall	35.6 m
Length waterline	31.5 m
Beam	3.9 m
Displacement	25 t
Passengers	131
Maximum speed	40 kt

Propulsion: Powered by two MAN diesels each rated at 809 kW at 2300 rpm.

Monaco

Wally Yachts

8 Avenue des Ligures, M-98000, Monaco

Tel: (+337) 931 00 93
Fax: (+337) 931 00 94
Web: www.wally.com

Luca Bassini Antivari, *President and Founder*

Wally is the world's most prolific builder of advanced composite large yachts. The company was founded in 1993 by Luca Bassani Antivari to produce a cruising yacht with the performance of a racing yacht by using advanced composites. In 2001, Wally entered the motor yacht market, using the same concept of combining technology and design to obtain improved performance and style.

The holding company of the group is based in Monte Carlo, with shipyards located in Massa Carrara, Italy, for the construction of the Wally Power production boats, and Fano, Italy, for advanced composite construction with a post-cure oven for 50 m yachts, and technical office with designers, engineers, naval architects, interior designers and structural engineers.

Wally Yachts facilities include 265 employees, 35,500 m² (382,118 ft²) of building area, four ovens to cure the advanced composites and a 120 m long mooring dock.

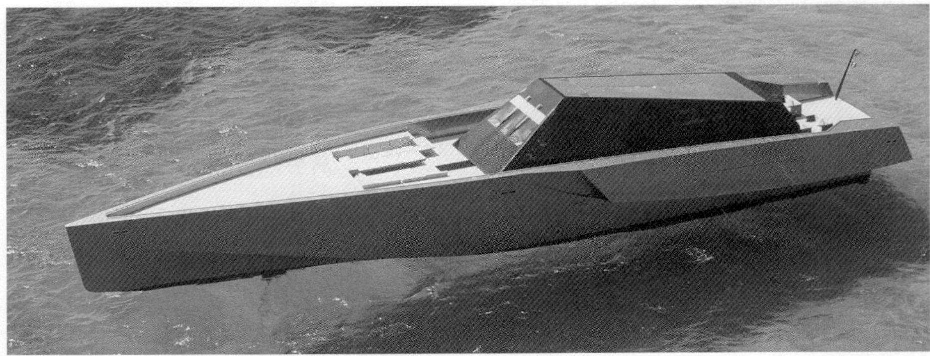

Wally Power 118 (Gilles Martin-Raget, Wally Yachts) 1120257

Wally Power 70 (Philip Plisson, Wally Yachts) 1120258

WALLY POWER 118

Wally Yachts first large power boat is a high performance superyacht with characteristics such as a vertical bow, large air inlets, and a unique deck superstructure. The vessel is constructed from carbon fibre composites and is powered by three gas turbines generating 12,000 kW, driving water-jets, two steerable outboard and a booster on the centreline. For manoeuvring and long deliveries, the steerable water-jets are powered by two diesel engines of 280 kW each. The exhaust system is made of titanium, saving weight, while being resistant to high temperatures generated by the gas turbines.

Specifications

Length overall	36 m
Beam	9 m
Draught	1.36 m
Displacement	95 t
Speed	60 kt

Propulsion: CODAG arrangement of three Detroit Diesel Company TF50 gas turbines each delivering 4,000 kW to Kamewa water-jets via Maag gearboxes or two Cummins diesel engines delivering 28 kW to two of the water-jets.

WALLY POWER 90 (DESIGN)

In 2007 Wally Yachts introduced the Wally Power 90 design.

Specifications

Length overall	27.9 m
Beam	23.5 m
Draught	7.08 m
Displacement	70 t
Max Speed	40 kt

Propulsion: Two MTU 16V 2000 M93 CR diesel engines delivering 1,790 kW each to Kamewa 56-SII waterjets via ZF BW 3060 gearboxes.

WALLY POWER 70

Launched in 2005, the Wally Power 70, is built in a fibreglass sandwich with carbon fibre structural reinforcements while the superstructure combines carbon fibre with laminated glass. The propulsion system consists of two MTU diesel engines generating 2,250 kW, driving KaMeWa water-jets, providing a maximum speed of 40 kt.

Specifications

Length overall	21.9 m
Beam	5.48 m
Draught	0.94 m
Displacement	34 t
Speed	40 kt

Propulsion: Two MTU 10V 2000 M93 diesel engines delivering 1,120 kW each to Kamewa A45 water-jets via ZF 2050 gearboxes.

Netherlands

Damen Shipyards

Industrieterrein Avelingen West 20, PO Box 1, NL-4200 AA, Gorinchem, Netherlands

High Speed and Naval Craft

Tel: (+31 183) 63 99 22
Fax: (+31 183) 63 21 89
e-mail: hsnc@damen.nl
Web: www.damen.nl

J L Gelling, *Product Director*

Damen Shipyards specialises in the design and construction of a wide range of work boats, dredgers and patrol craft up to 90 m in length and with installed powers up to 10,000 kW.

The shipyard started in 1927 and in 1969 began the construction of commercial vessels to standard designs by modular construction. Since 1969 more than 3,000 ships have been built and delivered worldwide to 117 countries.

Damen Fast Crew Supplier 3507 *Silni* 1120344

Vessel type	Vessel name	Yard No	Length (m)	Speed (kt)	Seats	Vehicles	Originally delivered to	Date of build
Stan Tender 4709	*Mauricio*	547101	46.50	29.0	121	none	Navierra Integral Mexico	2001
Damen AluCat 1605	*Malembo Tide*	532006	16.45	30.0	20	none	Tidewater Marine	2001
Stan Tender 4709	*Juan Pablo II*	547102	46.50	25.0	–	none	Naviera Integral, Mexico	2003
Alucat 1605	*Rubani*	532204	16.45	28.0	–	none	Kenya Ports Authority	2003
Alucat 1104	*Tiskj4*	533410	11.85	34.0	–	none	Nigeria	2003
Alucat 1605	*Lucola Tide*	532209	16.05	28.0	–	none	Egypt	2003
Alucat 1605	*Bakama Tide*	532208	16.05	28.0	–	none	Tidewater Marine Services	2003
Alucat 1605	*Luvassa Tide*	532206	16.05	28.0	–	none	Tidewater Marine Services	2003
Alucat 1605	*Nyota*	532210	16.45	28.0	–	none	Kenya Ports Authority	2003
Alucat 1604	*Com. N'Dozi*	532213	16.45	28.0	–	none	Angola	2003
Alucat 1604	*Com. Orlog*	532212	16.45	28.0	–	none	Angola	2003
Alucat 1604	*Com. Gika*	532211	16.45	28.0	–	none	Angola	2003
Fast Crew Supplier 1405	*Maria*	532014	14.50	25.0	–	none	Germany	2005
Fast Crew Supplier 1405	*Stephanie*	532015	14.50	25.0	–	none	Germany	2005
Fast Crew Supplier 1405	–	532017	–	–	–	none	Belgium	–
Fast Crew Supplier 1605	*Zimbo Tide*	532207	16.04	26.0	30	none	US	2005
Fast Crew Supplier 1605	*Smit Oloma*	532214	16.55	26.0	30	none	US	2005
Fast Crew Supplier 1605	–	532215	–	–	–	none	Nigeria	–
Fast Crew Supplier 1605	–	532216	–	–	–	none	Qatar	–
Fast Crew Supplier 1605	–	532217	–	–	–	none	Nigeria	–
Fast Crew Supplier 1605	*Lubambi Tide*	532218	16.04	26.0	30	none	US	2005
Fast Crew Supplier 1605	*Loge Tide*	532219	16.04	26.0	30	none	US	2005
Fast Crew Supplier 1605	*Lucunga Tide*	532220	16.04	26.0	30	none	US	2005
Fast Crew Supplier 1605	*Belize Tide*	532221	16.04	26.0	30	none	US	2005
Fast Crew Supplier 1605	*Landana Tide*	532222	16.04	28.0	30	none	US	2005
Fast Crew Supplier 1605	*Lobito Tide*	532223	16.04	28.0	30	none	US	2006
Fast Crew Supplier 1605	*Zhenis*	532224	16.55	–	30	none	Kazakhstan	2006
Fast Crew Supplier 1605	*Zambeze Tide*	532225	16.55	–	29	none	US	2006
Stan Tender 1605	*Oeud Chouly 1*	541704	16.25	23.0	–	none	Algeria	2005
Stan Tender 1605	*Oeud Chouly 2*	541705	16.25	23.0	–	none	Algeria	2005
Stan Tender 1605	*Oeud Saf-Saf 1*	541706	16.25	23.0	–	none	Algeria	2005
Stan Tender 1605	*Oeud Saf-Saf 2*	541707	16.25	23.0	–	none	Algeria	2005
Stan Tender 1605	*Visura*	541710	16.25	22.0	–	none	Germany	2006
Stan Tender 1605	–	541711	16.25	22.0	–	none	Germany	–
Stan Tender 4709	–	547104	–	–	–	none	Congo-Brazzaville	–
Stan Pilot 1505	*Fourth Leopard*	541801	15.40	–	–	none	UK	2006
Stan Pilot 1505	–	541802	–	–	–	none	China	–
Stan Pilot 1505	–	541803	–	–	–	none	China	–
Fast Crew Supplier 3307	*Axe-Bow 101*	544801	33.25	–	–	none	Netherlands	2006
Fast Crew Supplier 3307	*Axe-Bow 102*	544802	33.25	–	–	none	Netherlands	2006
Fast Crew Supplier 3507	*Silni*	544803	35.95	29.0	30	none	Brodospas	2007
Fast Crew Supplier 5009	Doña Diana	547101	51.3	25	80	none	Naviera Intergal SA	2007
Fast Crew Supplier 5009	–	–	51.3	25	80	none	Naviera Intergal SA	2007

The High Speed and Naval Craft division produces a wide range of high-speed craft for commercial, patrol, surveillance and rescue operations. From 1969, more than 700 high-speed craft have been built by Damen Shipyards Gorinchems' High Speed and Naval Craft department.

FAST CREW SUPPLIER 3507

Silni

Delivered to Brodospas in January 2007, *Silni* is the third Damen vessel to feature the Axe bow.

Specifications

Length overall	35.95 m
Beam	7.35 m
Draught	2.0
Passengers	30
Maximum speed	29 kt
Operating speed	26 kt

Propulsion: Three Caterpillar C32 C diesel engines delivering 1045 kW each to FP propellers via Reintjes WVS 730 gearboxes.

FAST CREW SUPPLIER 5007

Doña Diana

Delivered to Naviera Intergral SA in October 2007, this vessel features the Damen Axe bow and a steel hull construction.

Damen Fast Crew Supplier 5007 *Doña Diana* 1120345

Specifications

Length overall	51.3 m
Beam	9.2 m
Draught	3.2
Passengers	80
Deadweight	350 tonnes
Maximum speed	25 kt
Operating speed	21 kt

Propulsion: Four Caterpillar 3512B diesel engines delivering 1118 kW each to FP propellers via Reintjes WVS 730 gearboxes.

Heesen Shipyards BV

Rijnstraat 2, NL-5347-KL Oss, Netherlands

Tel: (+31 412) 66 55 44
Fax: (+31 412) 66 55 66

email: info@heesenshipyards.nl
Web: www.heesenshipyards.nl

Frans Heesen, *Managing Director*
Jan Gremmen, *General Manager*
Nancy Jansen, *Publicity Officer*

Heesen Shipyards was founded in 1963 and has focused on the construction of high-performance motor yachts since that time. The company currently employs 165 staff at Oss and a further 55 at Winterwijk where interior outfitting takes place.

Norway

Bergen Mekaniske Verksted AS

PO Box 14, Laksevåg, N-5847 Bergen, Norway.

Tel: (+47 55) 54 24 00
Fax: (+47 55) 54 24 01
Web: www.bergenmek.no

Trygve Aarlan, *Managing Director*
Terje Sjumarken, *Technical Manager*

Bergen Mekaniske Verksted AS (BMV), established in 2002, took over the full resources of Mjellem & Karlsen Verft AS. BMV undertakes shipbuilding and conversion, ship repair, engine overhaul and repair and testing. BMV is now the Royal Norwegian Navy's main civilian partner for maintenance of the RNN's fleet. It is understood that this company no longer produces fast ferries.

The yard's present facilities include:
Two floating docks 207 m long by 32.5 m wide by 8.75 m deep
Two dry docks 109.7 m long by 14.7 m wide by 5.1 m deep
Two slipways 92 m long by 21 m wide and 7.2 m deep
A floating crane with lifting capacity of 130 t
Three shipbuilding construction halls
A diesel engine overhaul centre.

95 m PASSENGER/CAR FERRY

Kattegat
Jetliner

Based on the company's experience in the design and construction of military vessels, the 95 m ferries were constructed in a series of modules that were assembled at the Halsnøy Verft AS yard.

The design is equipped with bow and stern ramp system, is compatible with conventional ferry terminals and allows quick unloading of cars, buses and trucks. The six vehicle lanes run the full length of the vessel. A combination of fixed and hoistable tween-decks ensures maximum flexibility. With the hoistable tween-decks in the upper position, two lanes with six buses in each lane are available together with 52 cars.

Specifications

Length overall	95.0 m
Length waterline	86.5 m
Beam	17.4 m
Draught	3.7 m
Crew	12
Passengers	600
Vehicles	160 cars + 12 buses
Deadweight	650 t
Fuel capacity	196,000 litres
Water capacity	27,000 litres
Propulsive power	4 × 5,800 kW
Max speed	38 kt
Operational speed	3 kt

Structure: Hull and superstructure are constructed from seawater-resistant aluminium. The hull structure combines the use of plates and stiffeners with extruded profiles, depending on local and general strength requirements.

Propulsion: Main engines are four MTU 20V 1163 TB 73 diesels, each producing 5,800 kW at 1,200 rpm; each driving Kamewa 112 SII water-jets.

Control: All four water-jets are steerable, and two powerful bow thrusters have been fitted. Four computer-controlled fins are fitted for stabilisation. The vessel is controlled from an integrated bridge system.

Vessel type	Vessel name	Yard No	Length (m)	Speed (kt)	Seats	Vehicles	Originally delivered to	Date of build
95 m passenger car ferry	*Kattegat*	–	95.0	35.0	600	160	European Ferries Denmark	1995
95 m passenger car ferry	*Jetliner*	–	95.0	35.0	600	160	P&O European Ferries	1996

Brødrene Aa

N-4829 Hyen, Norway

Tel: (+47 57) 86 87 00
Fax: (+47 57)86 99 14
e-mail: braa@braa.no
Web: www.braa.no

In 2003 Brødrene Aa re-entered the high-speed ferry market with an innovative new ferry. The 62 passenger, 25 knot vessel, *Rygerkatt*, is constructed in a new carbon fibre material from Deold AMT. Using multiaxial fabrics over a core material, Brødrene has produced a vessel which has a structural weight reduction of 40 per cent over conventional materials, but with increased strength. In 2002, Brødrene produced its first monhull for more than 10 years, a water ambulance, for ferry operator Rødne and Sønner AS.

19 m MONOHULL

Rygerdoktoren

Ambulance boat *Rygerdoktoren* is built in carbon composite materials using closed mould methods.

Specifications

Length overall	19 m
Beam	4.5 m
Draught	0.8 m
Passengers	12
Maximum speed	44 kt

Propulsion: 2 × MAN D248LE 413 diesel engines delivering 750 kW to Rolls Royce FF Jet 450s water-jets.

Vessel type	Vessel name	Yard No	Length (m)	Speed (kt)	Seats	Vehicles	Originally delivered to	Date of build
TMS 16 m	*Rygerskyss*	–	16	37	52	–	Rødne and Sønner AS	1986
Otto Scheen 16 m	*Dagring*	185	16	34	53	–	Rederiet Dagning	1986
Cirrus M27 P/LC	*Klipperfjord*	182	27	26	49	–	Finnmark Fylkesrederi	1986
Otto Scheen 18 m	*Rognsøy Junior*	181	17	33	50	–	Trygve Rognsøy	1986
TMS 18 m	*Sea Runner*	179	18 m	38	50	–	Øksfjord Slip & Mek	1986
TMS 20 m	*Alden*	186	20 m	32	55	–	PR Traust	1986
TMS 22 m	*Romdalsfjord*	188	22 m	33	–	–	Romdalsfjord	1987
TMS 18 m	*Sjøvakt*	193	18 m	32	50	–	Sjøvakt AS	1987
TMS 19 m	*Frøysprint*	–	19	28	68	–	Hammerwid Shipping	1989
19 m Monohull	*Rygerdoktoren*	–	19	44	12	–	Rødne & Sonner AS	2002
19 m Monohull	*Rygerstril*	–	19	39	25	–	Rødne & Sonner AS	2004

GS Marine AS

N-6082 Gursken, Norway

Tel: (+47) 70 02 52 42
Fax: (+47) 70 02 53 15
e-mail: gsmarine@online.no
Web: www.gsmarine.no

Established in 1997, GS Marine is a Norwegian boat builder specialising in composite construction.

SCIROCCO 50 Monohull Ferry

The company is building three new Scirocco 50 Shuttle ferries, *Fjordbuen*, *Helgøy Princess* and a third, *Nordic Sky*, based on the original Scirocco 50 ferry *Jon Arne Helgøy Skyssbåt*. The first shuttle ferry, *Fjordbuen*, was delivered in April 2005.

Specifications

Length overall	16.5 m
Beam	4.7 m
Draught	1.5 m
Crew	2
Passengers	50
Fuel capacity	2000 litres
Fresh water capacity	600 litres
Operational speed	25 kt

Structure: Carbon fibre construction.

Propulsion: Two Volvo D12 diesels each rated at 370 kW and each driving a fixed pitch propeller via a ZF325 reverse reduction gearbox.

Måløy Verft

6718 Deknepolien, Norway

Tel: (+47 57) 84 98 00
Fax: (+47 57) 84 98 01
e-mail: firmapost@maloy-verft.no
Web: www.maloy-verft.no

The Måløy yard was established in 1978 by the Saetremyr family as a service facility for their fleet of trawlers. In 1995 a new building hall was built for the production of composite vessels. The company offers a range of passenger, cargo, patrol, search and rescue vessels designed by Ola Lilloe-Olsen.

16 m Monohull

Designed for operations in the Faroe Islands, this vessel is constructed from a glass fibre reinforced plastic sandwich structure.

Specifications

Length overall	15.7 m
Length waterline	13.75
Beam	4.65
Draught	1.25
Crew	2
Passengers	48
Max speed	28 kts

Propulsion: Two Scania DI 14 diesel engines rated at 480 kW at 2,250 rpm driving fixed pitch propellers.

Rubb Marine AS

Formerly Båtutrustning Bømlo A/S

Termetangen, N-5420 Rubbestadneset, Norway

Tel: (+47) 53 42 84 00
Fax: (+47) 53 42 84 02
e-mail: admin@rubbmarine.no
Web: www.rubbmarine.no

Askepott

Designed by Teknisk Modellcenter A/S, this three-engine 118 passenger 40 kt ferry was delivered to Per Vold in 1989. The vessel has an unusually high maximum speed capability of 47 kt.

Specifications

Length overall	25.3 m
Length waterline	21.5 m
Beam	6.0 m
Passengers	118
Fuel capacity	6,000 litres
Propulsive power	2,616 kW
Max speed	47 kt
Operational speed	40 kt
Range	420 n miles

Structure: Hull of GRP sandwich construction.

Askepott (Alan Bliault) 0506695

Propulsion: Main engines are three MWM TBD 234 V16, each 872 kW max, 780 kW continuous; driving three Kamewa water-jet units, via Type 40S three ZF BU250 gearboxes.

Builder of Monohull Vessels

Vessel type	Vessel name	Yard No	Length (m)	Speed (kt)	Seats	Vehicles	Originally delivered to	Date of build
25.3 m Monohull Ferry	*Askepott*	–	25.3	40.0	118	None	Per Vold	1989
Otto Sheen 19 m	*Delfin Junior II*	88	19.2	33	117	none	Trygve Roynsoy	1988
Otto Sheen 20 m	*Hennøy*	90	20	38	77	none	Arvid Hopen	1989
Båtutrustning 15 m	*Fjord Express*	70	15.6	28	58	none	Nils Agasoster	1984
Otto Sheen 18 m	*Lyse Express*	86	18.1	33	58	none	PR Concorde	1988
Båtutrustning 18 m	*Kingen*	91	17.8	35	50	none	Kvinherad Batservice	1990
Båtutrustning 19 m	*Clipperskyss*	92	20.1	35	88	none	Rødne & Sonner	1990
Båtutrustning 19 m	*Øyvon*	98	19.4	38	50	none	Sandøy Kommune	1996
Båtutrustning 15 m	*Respons*	97	15	37	50	none	Kvinnhead Batservice	1996

Russian Federation

Central Hydrofoil Design Bureau (CHDB)

51 Svoboda Street, Nizhny Novgorod 603602, Russian Federation

Tel: (+7 8312) 73 10 37
Fax: (+7 8312) 73 02 48
e-mail: rostislav@sinn.ru

I M Vasilavsky, *Director*
M A Suslov, *Technical Director*
Sergey V Patonov, *General Director*

In 1991 Sudoexport announced the availability of two high-speed monohull ferry designs, one for river use and one for open sea use. Since that time over 10 of the Linda craft have been delivered.

In 1993 details of the 25 m landing craft Serna were released and it is understood that two of these craft have been constructed for operation on the Azov and Baltic seas.

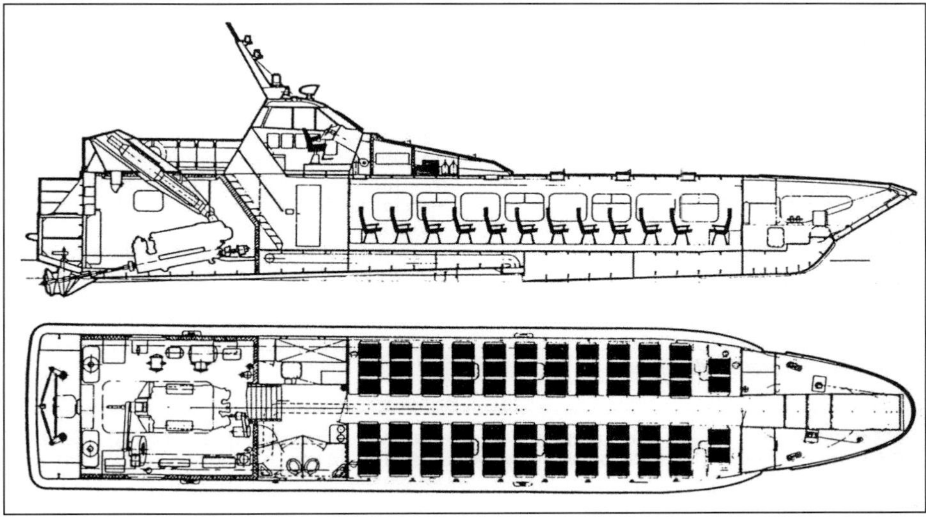

General arrangement of Linda 0506699

All craft benefit from a patented air lubrication system. The CHDB designs and builds prototype craft such as these and also a number of wing-in-ground effect craft.

It is understood that the design bureau was subject to complete reorganisation in 2003 and some vessels are now produced by Zelenodolsk Shipyard.

LINDA

Designed for passenger use on shallow rivers, the vessel takes two basic forms: for long routes accommodation is provided for 50 passengers with a bar and baggage space of 5 m^3; on shorter routes the vessel is equipped with 70 seats with no bar and the same 5 m^3 baggage space. Over 15 of these craft have now been constructed with a continuing production planned for 30 craft. The high, calm water speed is achieved by a patented hull air lubrication system.

Specifications

Length overall	24.1 m
Beam	4.6 m
Draught	1.0 m
Passengers	60
Propulsive power	800 kW
Max speed	38 kt
Operational speed	33 kt
Range	300 n miles

Propulsion: Engine is a single M401A-2 Zvezda diesel rated at 736 kW.

SERNA

These craft are intended for the fast transportation of containers, heavy vehicles and other cargo. It is understood that at least two have been constructed.

The vessel incorporates an air cavity within the hull which allows a significantly higher speed than would normally be available for a conventional craft of this size and installed power.

Specifications

Length overall	25.4 m
Beam	5.9 m
Draught	1.6 m
Displacement, max	105 t
Payload	45 t
Crew	3
Operational speed	30 kt
Range	600 n miles

Propulsion: Two M503A-Z Russian-built Zvezda diesels drive two tunnelled surface propellers.

Serna *at speed* 0007180

Linda *showing bow unloading ramp* 0104153

Linda *at speed* 0507067

Vessel type	Vessel name	Yard No	Length (m)	Speed (kt)	Seats	Vehicles	Originally delivered to	Date of build
Linda	*Series production (15)*	–	24.1	30.0	70	None	–	Series production (1991)
Serna	–	–	25.4	32.0	–	None	Russia	1993
Serna	–	–	25.4	32.0	–	None	Russia	1993

Singapore

Damen

29 Tuas Crescent, Singapore 638720

Tel: (+65) 68 61 41 80
Fax: (+65) 68 61 41 81
e-mail: central@damen.com.sg
Web: www.damen.com.sg

Dennis Carroll, *Managing Director*
Karl Nilssen, *Engineering Manager*
Lester Mak, *Projects Manager*

Damen Shipyards Singapore is a company belonging to the Damen Group of the Netherlands. The company occupies a modern, purpose-built facility that specialises in the construction of aluminium vessels.

Damen Shipyards Singapore provides a range of catamaran and monohull fast ferries in the 28 to 60 m range. The company also constructs a range of pilot boats, launches and crew tenders.

DAMEN 4108 FERRY

Bornholm Express

In 2006 Damen Shipyards constructed a 41 m monohull ferry at the Singapore yard. The vessel will be operated by Danish company Christiansøfarten between the islands of Christiansø, Bornholm and Simrishamn.

Specifications

Length overall	41.0 m
Beam	8.3 m
Deadweight	26 t
Passengers	311

Vessel type	Vessel name	Yard No	Length (m)	Speed (kt)	Seats	Vehicles	Originally delivered to	Date of build
DFF 4108	*Bornholm Express*	538101	41.0	25.0	311	–	Christiansofarten	2006
Fast Crew Supplier 3307	*Axebow 101*	544801	33.5	28.0	70	–	Oceanteam	2006
Fast Crew Supplier 3307	*Axebow 102*	544802	33.5	28.0	70	–	Oceanteam	2006

Fuel capacity	21,000 litres
Water	1,000 litres
Operational speed	25 kt

Classification: BV 1 + Hull HSC Cat-A Coastal Area + AUT–UMS

Propulsion: Three Caterpillar C32 diesel engines each rated at 1,645 kW at 2,300 rpm driving fixed-pitch propellers via reverse reduction gearboxes.

FAST CREW SUPPLIER 3307

The first two Damen Fast Crew Supplier 3307 vessels, *Axebow 101* and *Axebow 102* were delivered by Damen Shipyards Singapore to Oceanteam in July 2006. The vessels' axe-shaped bow maximises the waterline length and prevents the bow from lifting out of the water, reducing slamming and improving fuel consumption.

Specifications

Length overall	33.5 m
Beam	6.8 m
Passengers	70
Crew	8
Maximum speed	28 kt

Construction: Aluminium.

Damen 4108 monohull ferry 1173604

Greenbay Marine PTE Ltd

27 Pioneer Road, Jurong, Singapore 628500

Tel: (+65) 68 61 41 78
Fax: (+65) 68 61 81 09
e-mail: greenbay@singnet.com.sg
Web: www.greenbay.com.sg

Raymond Toh, *Chief Executive Officer*

Greenbay Marine was incorporated in 1980 as a shipbuilding company with facilities in Singapore and China. The company has ISA 9002 certification and by 2004 had built over 150 vessels.

The company also provides design, consultancy and full aftersales service and support for the vessels it supplies. These have covered offshore supply vessels, cargo vessels, tugs, patrol craft and fast passenger monohull and catamaran ferries.

20 m CREWBOAT

Zara

Zada

Two 20 m crewboats were delivered to the Gulf region in 2005.

Specifications

Length overall	20.0 m
Beam	5.3 m
Draught	1.3 m
Crew	4
Passengers	12
Fuel capacity	4,000 litres
Water capacity	1,000 litres
Maximum speed	21 kt
Operational speed	18 kt

Propulsion: Two Caterpillar 3406 diesel engines each rated at 447 kW at 2,100 rpm and driving two five-bladed propellers via a ZFBW2500 reverse reduction gearbox.

28 m CREWBOAT

The 28 m utility crewboat *MV Sanergy Landok* was delivered in 1999.

20 m Crewboat Zara  1139051

Wavemaster 10 *37 m Monohull* 1097732

Specifications

Length overall	28.0 m
Length waterline	26.9 m
Beam	6.8 m
Draught	1.3 m
Crew	6
Passengers	45
Fuel capacity	8,000 litres
Water capacity	3,000 litres
Maximum speed	25 kt
Operational speed	23 kt

Propulsion: Two Caterpillar 3412 C diesel engines each rated at 750 kW at 2,000 rpm and driving a Hamilton HM 571 water-jet via a ZFBW190 reduction gearbox.

Vessel type	Vessel name	Yard No	Length (m)	Speed (kt)	Seats	Vehicles	Originally delivered to	Date of build
32 m fast passenger ferry	*Auto Batam 9*	89	32.0	26.0	200	none	Auto Batam	1992
32 m fast passenger ferry	*Auto Batam 13*	–	32.0	26.0	200	none	Auto Batam	1995
32 m fast passenger ferry	*Auto Batam 10*	93	32.0	26.0	200	none	Auto Batam	1995
32 m fast passenger ferry	*Auto Batam 12*	109	32.0	26.0	200	none	Auto Batam	1995
36 m fast passenger ferry	*Auto Tioman*	113	36.0	34.0	220	none	Singapore	1996
37 m fast passenger ferry	*Wavemaster 10*	–	37.2	32.0	200	none	Berlian Ferries	–
28 m Utility Crewboat	*Sanergy Landok*	127	28.0	25.0	46	none	–	1999
20 m Crewboat	*Zada*	148	20.0	–	–	none	Gulf Region	2005
20 m Crewboat	*Zara*	148	20.0	–	–	none	Gulf Region	2005

32 M FAST PASSENGER FERRY

There have been four of these vessels constructed for *Auto Batam* ferry services of Singapore.

Specifications

Length overall	32.0 m
Length waterline	28.7 m
Beam	6.4 m
Draught	1.3 m
Passengers	200
Fuel capacity	5,000 litres
Water capacity	2,000 litres
Operational speed	26 kt
Range	250 n miles

Classification: Singapore Marine Department.

Structure: All-aluminium hull and superstructure.

Propulsion: Powered by three MTU 12V 183 TE72 diesel engines each rated at 610 kW, two driving fixed-pitch propellers and one driving a water-jet.

36 m FAST PASSENGER FERRY

Auto Tioman

This 36 m vessel is a development of the well-proven 32 m craft. The craft is designed for more exposed sea conditions and incorporates an active ride control system using stern trim tabs.

Specifications

Length overall	36.0 m
Length waterline	34.0 m
Beam	6.8 m
Draught	1.4 m
Passengers	220
Operational speed	34 kt

Classification: Singapore Marine Department.

Structure: All-aluminium.

Propulsion: Powered by three marine diesels each rated at 970 kW and driving an FF550 water-jet.

Auto Tioman 0507075

Greenbay Marine's 32 m Auto Batam 13 0507074

37 m FAST PASSENGER FERRY

Wavemaster 10

One vessel is currently under construction for Berlian Ferries of Singapore.

Specifications

Length overall	37.2 m
Beam	7.0 m
Draught	1.12 m
Passengers	171
Fuel capacity	12,000 litres
Water capacity	2,000 litres
Operational speed	30 kt

Classification: Bureau Veritas.

Structure: All-aluminium hull and superstructure.

Propulsion: Three MTU 12V 2000 M70 diesel engines each rated at 788 kW, two driving fixed-pitch propellers and one driving a water-jet.

Marinteknik Shipbuilders (S) PTE Ltd

31 Tuas Road, Singapore 638493

Tel: (+65) 861 17 06
Fax: (+65) 861 42 44
e-mail: david@marinteknik.com
Web: www.marinteknikdesign.se

David Liang, *Group Chairman*
Andrew Yeu, *Director and Technical Manager*
Hans Erikson, *Executive Manager Marketing and Sales*

The Marinteknik Group has over 25 years experience in the construction of high speed aluminium vessels, patrol boats and car/cargo.truck ferries. To date, the yard has built more than 100 craft of both monohull and catamaran type varying in length from 27 to 55 m and with a speed range of 18 to 55 kt.

Marinteknik Verkstads AB of Sweden, an associated company, ceased building in 1993. Details of its craft are included in this entry. Marinteknik Design Pte in Singapore works closely with Marinteknik Design (E) AB in Sweden to develop more advanced designs for production in the Singapore yard. Marinteknik designs are available to shipbuilders worldwide under licence.

35 m PASSENGER FERRY

A number of 35 m passenger ferries has been built to different operational requirements. Details for *Auto Batam 16* are as follows:

Auto Batam 16 0007181

Specifications

Length overall	35.0 m
Beam	7.5 m
Draught	1.2 m
Passengers	200
Propulsive power	3 × 735 kW
Operational speed	32 kt

Classification: DnV.

Structure: Deep vee forward and flat vee with flat chines aft. Constructed in marine grade aluminium.

Propulsion: Main engines are three MAN D2842 LE408 marine diesels rated at 735 kW each; driving three MJP J450-TB water-jet propulsion units.

45 m PASSENGER FERRY

Superjet

Superjet was constructed in 1999 for Navigazione e Libera di Golfo of Italy.

Specifications

Length overall	45.0 m
Beam	9.0 m
Draught	1.3 m
Passengers	550
Propulsive power	2× 2,000 kW
Operational speed	32 kt
Range	850 nm

Craft built (high-speed monohull)

Vessel type	Vessel name	Yard No	Length (m)	Speed (kt)	Seats	Vehicles	Originally delivered to	Date of build
30 MCB	*Hamidah*	102	300	19.5	50	none	Ocean Tug Services	1985
30 MCB	*Zakat*	104	300	18.5	38	none	Black Gold (M) Sdn Bhd, Malaysia	1986
30 MCB	*Amal*	106	300	18.5	38	none	Black Gold (M) Sdn Bhd, Malaysia	1988
41 MPV	*Europa Jet*	68	41.0	30.0	400	none	Alilauro SpA	1987
41 MPV	*Cinderella West*	65	42.0	25.0	450	none	City Jet Line Rederi AB	1987
MPV	*Discovery Bay 12*	107	35.0	26.0	250	none	Discovery Bay	1987
35 MPV	*Discovery Bay 15*	108	35.0	26.0	250	none	Discovery Bay Co	1987
35 MPV	*Discovery Bay 16*	109	35.0	26.0	250	none	Discovery Bay Co	1987
35 MPV	*Zhen Jiang Hu (ex-Wu Yi Hu)*	110	35.0	28.0	262	–	Jiangmen Jiang Gang Passenger Traffic CO, China	1987
41 MPV-D	*Celestina*	116	41.0	29.0	300	none	Alilauro SpA	1988
42 MPV	*Rosario Lauro (ex-Aurora Jet)*	71	42.0	31.0	400	none	Alilauro SpA	1988
40 MPV	*Cosmopolitan Lady*	66	40.0	27.0	20	none	Private Cruise International I Ltd	1989
42 MPV	*Cinderella II*	72	42.0	31.0	450	none	City Jet Line	1989
42 MPV	*Iris*	76	42.0	31.0	351	none	Kvaerner Express	1989
42 MPV	*Cinderella*	78	42.0	30.0	450	none	City Jet Line	1990
35 MPV	*Discovery Bay 19*	122	35.0	25.0	250	none	Hong Kong Resort Co Ltd	1990
35 MPV	*Discovery Bay 20*	125	35.0	25.0	250	none	Hong Kong Resort Co Ltd	1990
42 MPV	*Blue Crystal*	77	42.5	30.0	19	none	Fyneside Shipping Caraibes	1991
42 MPV	*Diamant Express*	79	42.0	31.0	405	none	Compagnie Maritime des	1991
43 MPV	*Northern Cross*	80	43.0	29.0	19	none	Lillbacka Shipping	1991
41 MPV	*Napoli Jet*	83	41.5	30.0	400	none	Navigazione Libera del Golfo	1992
35 MPV	*Discovery Bay 21*	131	35.0	24.0	300	none	Hong Kong Resort Co Ltd	1992
41.5 m MPV	*Saint Malo*	B88	41.5	36.0	350	–	France	1993
35 MPV	*Discovery Bay 22*	132	35.0	24.0	300	none	Hong Kong Resort Co Ltd	1993
43 MPV	*Lamda Mar*	84	43.0	45.0	500	none	BM Marine	1993
41 MPV	*Adler Express*	87	41.5	27.0	427	none	Nallia-und Inselreederi	1993
35 MPV	*Auto Batam 16*	147	35.0	32.0	200	none	Auto Batam	1997
45 m	*Superjet*	161	45.0	30.0	550	none	Navigazione Libera di Golfo	1999
50 m	*SNAV Orion*	188	50.0	36.0	688	none	Aliscafi SNAV	2004

Classification: RINA.
Propulsion: Two MTU 16V 396TE 74L diesel engines each rated at 2000 kW at 2,000 rpm, each driving an MJP J 650R-DD water-jet.

50 m PASSENGER FERRY

SNAV Orion

Designed and built to RINA classification rules, this vessel was delivered to Aliscafi SNAV in 2004.

Specifications

Length overall	49.9 m
Length waterline	44.9 m
Beam	9.0 m
Draught	1.4 m
Passengers	688
Crew	10
Fuel capacity	20,000 l
Water capacity	2,000 l
Maximum speed	36 kt

Classification: RINA.
Propulsion: Four MTU 16V 396TE 74L diesel engines each rated at 2000 kW at 2,000 rpm, each driving a Kamewa 63 SII waterjet via a reintjes VLJ 930 K31 gearbox.

Penguin Shipyard International Ltd

A subsidiary of Penguin Boat International
11 Tuas Basin Link, 638784 Singapore

Tel: (+65) 68 68 06 26
Fax: (+65) 98 22 75 60
e-mail: andy@penguin.com.sg
Web: www.penguin.com.sg

David Lui, *Business Development/Marketing Manager*
Andy Heng, *Commercial Manager*

Penguin Shipyard International Ltd has two shipyards in Singapore and an additional yard in Batam Island, Indonesia. The comprehensive facilities include two undercover shipbuilding halls and the yards are equipped with a 250 tonne and a 500 tonne mobile straddle carrier for easy docking of vessels.

Penguin are builders of multipurpose offshore vessels, fast supply vessels, crew boats, patrol craft, ferries and survey vessels as well as carrying out maintenance and repair of super-yachts. All crew boats are classified by ABS and are fully compliant with ISM, MARPOL, STCW and SOLAS.

36 m CREW BOAT

In 2007 four 36 m crewboats were ordered by offshore company Access Charter Sdn Bhd of Malaysia. The 25 kt vessels will carry 60 passengers and 70 t of rear deck cargo, and will be designed to operate in wave heights up to 3 m.

Specifications

Length overall	36.0 m
Passengers	50
Operating speed	25 kt

34 m CREW BOAT

Poksay

This crew boat was delivered in 2003 with a cargo deck capacity of 20 tonnes.

Specifications

Length overall	34.0 m
Length waterline	31.0 m
Beam	7.6 m
Draught	1.6 m
Crew	8
Passengers	50
Fuel capacity	50,000 litres
Water capacity	15,000 litres
Max speed	28 kt
Operating speed	25 kt

Classification: ABS.
Propulsion: Three MTU 16V2000 M70 diesel engines each driving a fixed-pitch propeller via a ZF 4540 reverse reduction gearbox.

39 m CREW BOAT

Specifications

Length overall	39.6 m
Beam	7.5 m
Draught	1.85 m
Passengers	75
Max speed	30 kt

Propulsion: Three Caterpillar 3512B diesels each rated at 1305 kW at 1600 rpm, each driving a fixed-pitch five bladed propeller.

47 m CREW BOAT

Pelican Challenge
Pelican Glory
Pelican Pride
Pelican Venture
Pelican Vision

Designed by Pelmatic Neg Pte Ltd for Penguin Offshore Services, these craft were delivered between 1998 and 2000.

Specifications

Length overall	45.8 m
Beam	9.1 m
Draught	2.5 m
Displacement	425 t
Crew	8
Passengers	80
Fuel capacity	76,500 litres
Water capacity	140,000 litres
Max speed	30 kt

Structure: All aluminium hull and superstructure.
Propulsion: *Pelican Glory* and *Pelican Challenge* are powered by four Detroit Diesel 16V-92 DDEC III diesels each driving a fixed-pitch propeller via a ZF BW 255 gearbox. *Pelican Pride* and *Pelican Venture* are powered by four MTU 8V 4000 M70 diesel engines each driving a fixed-pitch propeller,

Vessel type	Vessel name	Yard No	Length (m)	Speed (kt)	Seats	Vehicles	Originally delivered to	Date of build
47 m crew boat	*Pelican Glory*	–	45.8	25	100	None	Penguin Offshore Services	1998
47 m crew boat	*Pelican Challenge*	–	45.8	28	100	None	Penguin Offshore Services	1998
47 m crew boat	*Pelican Venture*	–	45.8	30	80	None	Penguin Offshore Services	1999
47 m crew boat	*Pelican Pride*	–	45.8	28	80	None	Penguin Offshore Services	1999
47 m crew boat	*Pelican Vision*	–	45.8	30	80	None	Penguin Offshore Services	1999
50 m crew boat	*Pelican Champion*	–	50.0	32	200	None	Penguin Offshore Services	2000
50 m crew boat	*Pelican Chaser*	–	50.0	31	100	none	Penguin Offshore Services	2002
50 m crew boat	*Pelican Pursuer*	–	50.0	31	100	none	Penguin Offshore Services	2002
34 m crew boat	*Poksay*	–	34.0	25	50	none	Penguin Offshore Services	2003
39 m Crew Boat	*Sea Mirage*	–	39.6	30	75	none	Middle East	2003
40 m Crew Boat	*Hull 134*	134	40.0	28	70	none	–	2004
36 m Crew Boat	*Hull 135*	135	36	27	80	none	–	2004
18 m Patrol Craft	*Hull 136*	136	18	30	12	–	Indonesia	2005
18 m Patrol Craft	*Hull 137*	137	18	30	12	–	Indonesia	2005
18 m Patrol Craft	*Hull 138*	138	18	30	12	–	Indonesia	2005
18 m Patrol Craft	*Hull 139*	139	18	30	12	–	Indonesia	2005
36 m Crew Boat	–	–	36.0	25	60	70 t cargo	Access Charter Sdn Bhd	(2007)
36 m Crew Boat	–	–	36.0	25	60	70 t cargo	Access Charter Sdn Bhd	(2008)
36 m Crew Boat	–	–	36.0	25	60	70 t cargo	Access Charter Sdn Bhd	(2008)
36 m Crew Boat	–	–	36.0	25	60	70 t cargo	Access Charter Sdn Bhd	(2008)

and *Pelican Vision* is powered by four CAT 3512B diesel engines driving four fixed-pitch propellers via a ZFBW365 gearbox.

50 m CREW BOAT

Pelican Champion
Pelican Chaser
Pelican Pursuer

This vessel is an extended version of the 47 m crew boat, designed to accommodate up to 200 passengers.

Specifications

Length overall	50.0 m
Length waterline	45.6 m
Beam	9.1 m
Draught	2.5 m
Displacement	555 t
Deadweight	365 t
Crew	8
Passengers	100–200
Fuel capacity	21,000 litres
Water capacity	150,000 litres
Max speed	31 kt
Operating speed	25 kt

Classification: ABS +A1 HSC+AMS.

Propulsion: Four Caterpillar 3516 B SCAC DITA diesel engines each rated at 1,770 kW, each driving a fixed-pitch propeller via a ZF BW 755 gearbox.

Precision Craft (88) PTE Ltd

55 Gul Road, Singapore 629 353

Tel: (+65) 862 48 00
Fax: (+65) 862 48 03
e-mail: precision@sbg.com.sg

Aluminium specialist, Precision Craft (88) Pte Ltd has accumulated over 24 years of experience in the building and repair of aluminium craft and industrial structures. Apart from its own proven monohull it also offers other hull designs such as SWATH (Small-Waterplane-Area Twin Hull), SES (Surface Effect Ship) and catamarans.

Borcos 112 *crew boat* 0506696

27.5 m CREW BOAT

Borcos 112
Borcos113

Specifications

Length overall	27.5 m
Length waterline	25.0 m
Beam	6.2 m
Draught	1.0 m
Crew	6
Passengers	68
Payload	6 t (additional)
Fuel capacity	6 t
Water capacity	2,000 litres
Operational speed	18 kt
Range	400 n miles

Classification: Lloyd's@100A1 +LMC for local trade limits, personnel carrier.

28 m FERRY

Delivered in the first quarter of 1993 for a Chinese owner, this 138-passenger vessel was classed and surveyed by the China Classification Society (ZC). A service speed of 28 kt is achieved with 1,700 kW total output from two marine diesel engines.

Vessel type	Vessel name	Yard No	Length (m)	Speed (kt)	Seats	Vehicles	Originally delivered to	Date of build
28 m Ferry	–	–	28.0	28.0	138	None	China	1993
27.5 m Crew Boat	*Borcos 112*	–	27.5	18.0	68	None	Syarikat Borcos Shipping	1991
27.5 m Crew Boat	*Borcos 113*	–	27.5	18.0	68	None	Syarikat Borcos Shipping	1993

Strategic Marine Pte Ltd

No 4, Pioneer Sector 1, Singapore, 624816

Tel: (+65 6558) 7877
Fax: (+65 6558) 7277
Web: www.strategicmarine.com

Mark Newbold, *Chairman*
John Fitzhardinge, *Director of Design*

Strategic Marine specialises in the construction of patrol boats and commercial vessels in aluminium. The company has recently expanded and now has yards in Australia, Singapore and Vietnam.

31 m OFFSHORE SERVICE VESSEL

Baruna Raya Logistics ordered three 31 m offshore service vessels in 2007 for delivery in February 2009. Capable of carrying 50 passengers and 134 tonnes of cargo, the vessels are powered by three Cummins KTA 38 engines and will have a top speed of 27 knots.

Specifications

Length overall	31 m
Beam	7.5 m
Draught	1.5 m
Passengers	50
Maximum speed	27 kt

Propulsion: Three Cummins KTA 38 diesel engines delivering a total of 2650 kW.

40 m OFFSHORE SERVICE VESSEL

Sgarikat Borcos Shipping of Malaysia ordered three 40 m Offshore Utility vessels for delivery in August 2008. The vessels will be capable of carrying 80 passengers at up to 28 knots.

Specifications

Length overall	40 m
Beam	7.5 m
Draught	2.3 m
Passengers	80
Maximum speed	28 kt
Operational speed	25 kt

Propulsion: One Cummins KTA 50 and two Cummins KTA 38 diesel engines delivering a total of 3350 kW.

Spain

Astilleros de Huelva

Glorieta Norte S/N, E-21001 Huelva, Spain

Tel: (+34) 959 28 09 77
Fax: (+34) 959 26 21 33
e-mail: comercial@astihuelva.es
Web: www.astihuelva.es

Established in 1971, Astilleros de Huelva is a shipbuilding and ship repair yard located near the Straits of Gibraltar on Spain's southwest Atlantic Coast. With two building berths of 150 × 23 and 160 × 24 m, the shipyard specialises in steel vessels up to 160 m in length.

In 2000, the shipyard reached an agreement with Rodriguez Cantieri Navali SpA for licensed production of the Rodriguez TMV range of high-speed monohull ferries. In 2007 the company formed a new entity with Sevilla Shipyard (formerly part of the Izar group).

Navantia

Formerly known as Izar.
Puerto Real Shipyard, San Fernando, Carretera de la Carraca s/n, PO Box 18, 11080 San Fernando, Spain

Tel: (+34 956) 59 98 97
Fax: (+34 956) 59 98 98
Web: www.navantia.es

Navantia is the new name for Izar which was formed in 2000 by the merger of Empresa Nacional Bazan and Astilleros Espanoles. The company has a staff of over 11,000 with 12 yards and factories located on the Spanish Atlantic and Mediterranean coasts. Their activities comprise four business areas: Shipbuilding; Repairs and conversions; Propulsion and energy; and Systems and weapons. Naval forces from several countries such as Norway, Chile and Thailand and shipowners from Belgium, Finland, France, Norway, Sweden, UK and the US rank among the customers of Navantia. In the fast ferry business, Navantia developed and built a series of high-speed small- to medium-size monohull and multihull vessels.

Silvia Ana L *at speed* 0594389

The Mestral passenger/vehicle ferry Almudaina (Maritime Photographic) 0045182

MESTRAL CAR FERRY

Albayzin
Alcántara
Almudaina

Late in 1992, two 96 m monohull car ferries were ordered by Compania Trasmediterranea from Izar (formerly Bazan). The first vessel, *Albayzin*, was delivered from the San Fernando shipyard in October 1994 and was immediately transferred to Buquebus. This Uruguayan company subsequently leased the vessel to the New Zealand operator Sea Shuttle in late 1994. The second vessel, *Alcántara*, was delivered in May 1995 to Compania Transmediterranea to be operated initially on the route between Palma de Mallorca and the Spanish mainland and most recently in the Gibraltar Strait between Algeciras and Ceuta. The vessel is constructed entirely of aluminium with a deep vee hull form providing the sea-keeping and powering performance specified for the Trasmediterranea ferry routes.

A third vessel, *Almudaina* was delivered to Transmediterranea in May 1996 being deployed between Palma de Mallorca and the Spanish mainland. All vessels are of aluminium construction with a deep vee hull form.

Specifications

Length overall	96.2 m
Length waterline	84.0 m
Beam	14.6 m
Draught	2.1 m
Crew	16
Passengers	534
Vehicles	80 cars
Propulsive power	22,600 kW
Max speed	39 kt
Operational speed	38 kt
Range	300 n miles

Classification: DnV +1A1 HSLC R1 Car Ferry A EO.

Structure: All aluminium hull and superstructure.

Propulsion: Powered by four Caterpillar 3616 diesel engines each rated at 5,650 kW and each driving a size 100 Kamewa water-jet, the outer two jets being of a steering and reversing type.

Electrical system: Power is provided by three Caterpillar 3408 diesels each rated at 270 kW.

Auxiliary systems: An active ride control system is provided on all vessels, the *Albayzin* having four Vosper Thornycroft stabiliser fins and the other vessels having a system based on two active fins forward and two active transom flaps aft.

Vessel type	Vessel name	Yard No	Length (m)	Speed (kt)	Seats	Vehicles	Originally delivered to	Date of build
Mestral car ferry	*Albayzin*	–	96.2	38.0	450	80	Buquebus	October 1994
Mestral car ferry	*Alcantara*	–	96.2	38.0	450	80	Compania Transmediterranea	May 1995
Mestral car ferry	*Almudaina*	–	96.2	38.0	450	80	Compania Transmediterranea	May 1996
Alhambra	*Silvia Ana L*	–	125.0	38.0	1250	240	Buquebus	1996

ALHAMBRA DF-110 FAST RO-RO VESSEL (DESIGN)

This mid-sized vessel has been designed to provide a craft with a capacity between that of the Mestral and Alhambra range.

Specifications

Length overall	110.0 m
Length between perpendiculars	96.2 m
Moulded Breadth	17.0 m
Depth to deck level 1	11.0 m
Full load draught	2.6 m
Crew	16
Passengers	800
Vehicles/Coaches	180 cars/4 coaches
Power output	28,800 kW
Speed	38 kt
Max deadweight	370 t
Range	300 n miles
Fuel	60 t
Fresh water	6 t

Classification: DnV +1A1 HSLC Car Ferry AR1 EO.

Structure: All aluminium hull and superstructure.

Propulsive: Powered by four diesel engines developing a total power of 4 × 7,200 kW driving four steerable water-jets.

Electrical system: Power is provided by three generators rated at 406 kW.

Auxiliary systems: An active ride control system is provided based on two active fins forward and two active transom flaps aft. Manoeuvrability capacity is strongly enhanced by the azimuthal retractable bow thruster.

ALHAMBRA DF-125

Silvia Ana L

The first Alhambra was ordered by Buquebus in 1995 and was delivered in late 1996.

This vessel is an enlarged version of the Mestral type in which all the shipyard experience has been utilised. Initially operating on the River Plate, the vessel as then chartered to Color Line for operation between Norway and Denmark. A new version of this vessel has been designed to meet the requirements of the HSC Code 2000 (for category B vessels).

Specifications

Length overall	125.0 m
Length waterline	110.0 m
Beam	18.7 m
Draught	2.5 m
Crew	27
Passengers	1,250
Vehicles	238 cars
Propulsive power	33,900 kW
Max speed	40 kt
Operational speed	38 kt
Range	300 n miles

Classification: DnV +1A1 HSLC R1 Car Ferry A EO.

Propulsive: Powered by six Caterpillar 3616 diesel engines each rated at 5,650 kW. The two central diesel engines drive a booster Kamewa 140 water-jet and each of the four lateral diesel engines drives a Kamewa 112 water-jet propulsor with steering and reversing capacity.

Electrical system: Power is provided by three diesel generators, each rated at 350 kW.

Auxiliary systems: An active ride control system is provided based on two active fins forward and two active transom flaps aft.

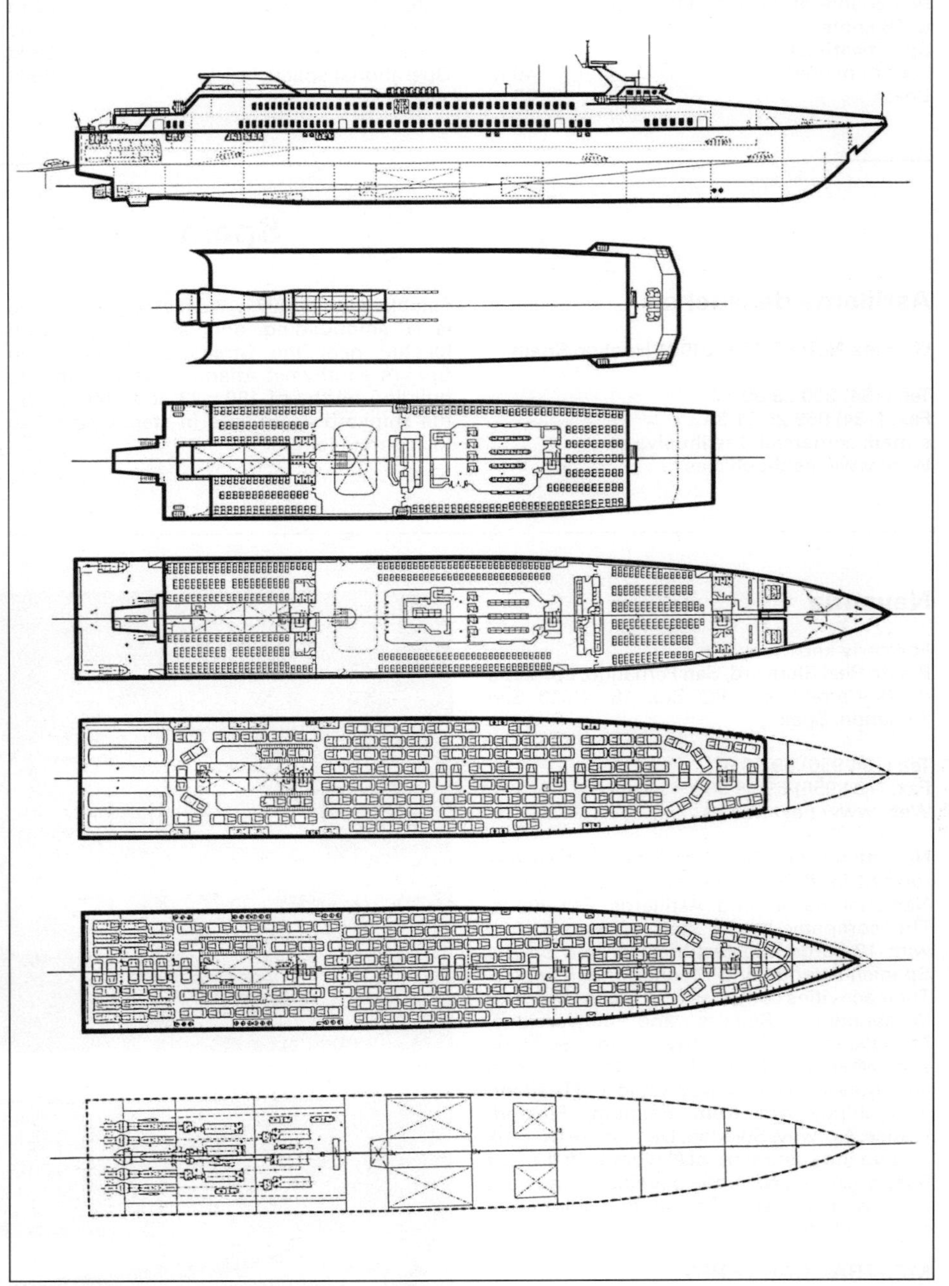

General arrangement of Alhambra *car ferry* 0506987

Artists impression of Alhambra DF-140 (design) 0594385

ALHAMBRA DF-140 FAST RO-RO VESSEL (DESIGN)

The Alhambra DF-140 is designed to carry passengers, cars, trucks and coaches at high speed.

Specifications

Length overall	143.6 m
Length waterline	128.0 m
Beam	21.0 m
Crew	50
Passengers	1,290
Cars/Truck lanes	295 cars/460 m
Propulsive power	48,600 kW
Max speed	40.0 kt
Range	300 n miles
Fuel	100 t
Fresh water	10 t

Classification: DnV +1A1 HSLC R1 Car Ferry AR1 EO.

Structure: All aluminium hull and superstructure.

Propulsion: Six diesels engines developing a total propulsion power of 6 × 8,100 coupled via gearboxes to four steerable and one booster water-jets, distributed in three compartments located aft.

Electrical system: Power is provided by three generators each rated at 560 kW.

Auxiliary systems: An active ride control system is provided, based on two active fins forward and two active transom flaps aft. Manoeuvrability is strongly enhanced by the twin bow thrusters.

Sweden

Boghammar International AB

Nysaetravagen 6-8, S-181 61 Lidingo, Sweden

Tel: (+46 8) 766 01 90
Fax: (+46 8) 766 18 55
e-mail: international@boghammar.se
Web: www.boghammar.se

Anders Boghammar, *Executive Director*
Lars Boghammar, *Director*

The company was formed in 1984 and is the international selling company for Boghammar Marine AB. The yard was formed in 1906, it is family owned and is today headed by the third generation. Boghammar International AB is situated at Lidingo just outside Stockholm.

Nearly 1,200 boats have been built by the yard to date, with 90 per cent being of the yard's own design. It currently specialises in building all-aluminium constructed boats for commercial use, such as high-speed monohull craft for patrol duty and day passenger transports.

The latest design of high-speed monohull craft for day passenger transports is the Kungsholm series. Three have now been built, *Gripsholm* in 1989, *Kungsholm* in 1990 and *August Lindholm* in 1992. The first two vessels are almost identical with some minor superstructure differences. The interior design of *Kungsholm* and its machinery and bridge are different from those of *Gripsholm*. *August Lindholm* has two funnels instead of one.

Gripsholm

Delivered in August 1989 for operation on Lake Malaren.

Specifications

Length overall	26.5 m
Beam	5.6 m
Draught	1.8 m
Crew	5
Passengers	125
Propulsive power	3 × 744 kW
Max speed	36 kt
Operational speed	33 kt

Propulsion: Main engines are three MAN V12 D2842 LXE-LYE, 744 kW each at 2,200 rpm; Servogear CP propellers.

August Lindholm 0506697

Kungsholm

This vessel was delivered in June 1990 for use on a high speed route between Stockholm and Manfred on Lake Malaren.

Specifications

Length overall	27.5 m
Beam	5.6 m
Draught	1.8 m
Propulsive power	3 × 723 kW
Max speed	35 kt
Operational speed	30 kt

Propulsion: Main engines are three MWM TBD 234 V12, 723 kW each at 2,300 rpm, driving two Servogear CP propellers.

August Lindholm

This vessel (Yard No 1154) was delivered in 1991 to Bore Lines for inter island traffic in the Stockholm archipelago.

Specifications

Length overall	28.1 m
Beam	5.6 m
Draught	1.2 m
Displacement, max	41.5 t
Passengers	173
Fuel capacity	4,800 litres
Water capacity	600 litres
Propulsive power	2 × 723 kW
Max speed	30 kt
Operational speed	25 kt

Propulsion: Main engines are two MWM TBD 236 V12 diesels, 723 kW each at 2,300 rpm, driving two Michigan-Wheel fixed-pitch propellers.

Vessel type	Vessel name	Yard No	Length (m)	Speed (kt)	Seats	Vehicles	Originally delivered to	Date of build
Kungsholm Series	*Gripsholm*	1148	23.5	33.0	125	None	–	1989
Kungsholm Series	*Kungsholm*	1153	27.5	30.0	–	None	–	1990
Kungsholm Series	*August Lindholm*	1154	28.1	25.0	173	None	Stockholm Sightseeing	1991

Taiwan

Lung Teh Shipbuilding Co Ltd

No.15, De-Shing 5th Road, Dung-Shan Country, ILAN, Taiwan

Tel: (+886) 39 90 11 66
Fax: (+886) 39 90 56 50
e-mail: sales@ms.lts.com.tw
Web: www.lts.com.tw

Mr Seldon Huang, *President*
Mr Chin-Hwa Chang, *Sales Vice President*
Mr Solon Huang, *Production Vice President*

Founded in 1979 the company has considerable experience in the design and construction of a wide range of patrol boats, military landing craft, support craft, pilot boats, fire and rescue boats as well as tugs and leisure boats.

The company has experience of building vessels in aluminium, steel and GRP up to 40 m in length and up to speeds of 60 kt. Recent deliveries have included 28 and 40 m passenger/cargo ferries.

28 m PASSENGER/CARGO FERRY

Glory

This vessel was delivered in April 2007 for operation in the Liuchiu and TungGang areas of Taiwan.

Specifications

Length overall	28.0 m
Beam	6.5 m
Deadweight	30 t
Max speed	34.5 kt
Range	400 nm

Construction: All aluminium hull and superstructure.

Propulsion: Three Caterpillar C32 diesels driving fixed pitch propellers via MG 6620 gearboxes.

30.5 m MONOHULL FERRY

Gi-Shan-Ru-Yi

This vessel, which was ordered in March 2003, was delivered in October 2003 to the Liu Qiu county government in Taiwan.

Specifications

Length overall	30.5 m
Beam	6.8 m
Draught	1.8 m
Displacement	100 t
Fuel capacity	6,000 litres
Freshwater capacity	2,000 litres
Max speed	34 kt
Operational speed	24 kt

Construction: All aluminium hull and superstructure.

Classification: CR100+ Ferry, Coasting Service, CMS.

Propulsion: Four MAN 2848 LE410 diesel engines each driving an MJP 450 QD water-jet via a reduction gearbox.

40.0 m PASSENGER/CARGO FERRY

Star of Southern Sea

Delivered in January 2007 to the Penghu County Government for operation on the Magong-Wangan-Cimei route in the Penghu islands, this vessel can carry 30 t of cargo.

Specifications

Length overall	40.0 m
Beam	8.5 m
Max speed	28.8 kt
Range	400 nm

Construction: All aluminium hull and superstructure.

Classification: CR100+ Ferry, Coasting Service, CMS+, Coastal Service.

Propulsion: Two Caterpillar 3516C HD diesels driving fixed pitch propellers via ZF 7650 gearboxes.

Dragon Yacht Building

129 Angi Road, Anping District, Tainan City 708, Taiwan

Tel: (886 6) 295 1686
Fax: (+886 6) 295 2990
e-mail: dragonyacht@dragonyacht.com.tw
Web: www.dragonyacht.com.tw

Established in 1986, Dragon Yacht Building specialises in fibreglass vessel construction.

29 m FAST FERRY

Fei Ma

Delivered to Fei-Ma Steamship Company in 2007 this vessel is capable of carrying 185 passengers at a maximum speed of 36 knots.

Specifications

Length overall	29.05 m
Length waterline	23.77 m
Draught	1.1 m
Crew	8
Passengers	185
Vehicles	12 motorcycles
Fuel capacity	10,000 litres
Fresh water capacity	1000 litres
Maximum speed	36 kt

Structure: Fibre-reinforced plastic

Propulsion: 3 × MTU 16V 2000 M90 diesels engines delivering 1400 kW to fixed pitch propellers via 2 × ZF 2555A and 1 × ZF 2555V gearboxes.

Vessel type	Vessel name	Yard No	Length (m)	Speed (kt)	Seats	Vehicles	Originally delivered to	Date of build
29 m Passenger Ferry	*Fei Ma*	–	29.05	36	185	none	Fei-Ma Steamship Company	2007

United Arab Emirates

Dubai Drydocks

PO Box 8988, Juneirah Beach Road, Dubai, United Arab Emirates

Tel: (+971 4) 345 06 26
Fax: (+971 4) 345 01 16
e-mail: drydocks@drydocks.gov.ae
Web: www.drydocks.gov.ae

Dubai Drydocks is the largest ship repair facility in the Middle East although it also has a small shipbuilding division.

27 m CREWBOAT

In 2001 the company launched a new 27 m, 25 kt aluminium crewboat, *Khulood*, designed by Robert Allan Ltd of Canada.

Specifications

Length overall	27.4 m
Beam	6.7 m
Draught	2.0 m
Deadweight	35.8 t
Passengers	31
Fuel capacity	11,400 litres
Water capacity	2,000 litres

Construction: All aluminium hull and superstructure.

Classification: +100 A1 SSC Passenger Mono HSC G3.

Propulsion: Two Cummins KTA 38-MZ diesel engines driving Teignbridge fixed-pitch propellers via a Reintjes WVS 430 gearbox.

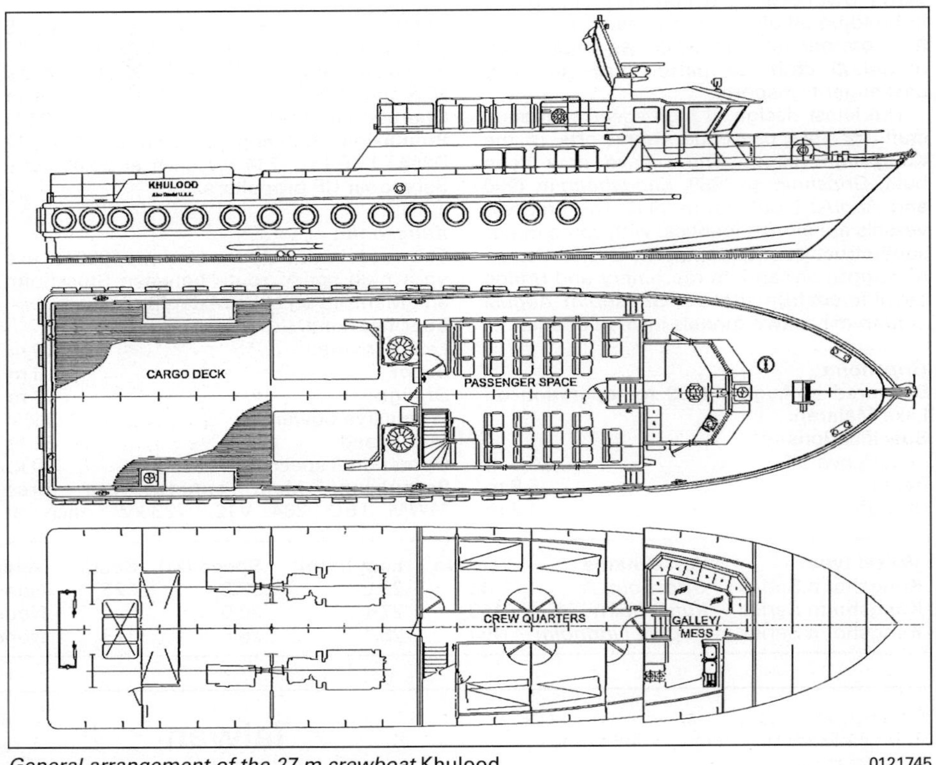

General arrangement of the 27 m crewboat Khulood 0121745

Vessel type	Vessel name	Yard No	Length (m)	Speed (kt)	Seats	Vehicles	Originally delivered to	Date of build
27 m Crewboat	*Khulood*	–	27.4 m	25	31	none	–	2001

Seaspray Marine Services and Engineering FZC

PO Box 42406, Hamriyah Free Zone, Sharjah, United Arab Emirates

Tel: (+971 6) 526 06 77
Fax: (+971 6) 526 06 88
e-mail: seaspray@emirates.net.ae
Web: www.seasprayboats.com

21 m CREWBOAT

The VSA-1 REEM was delivered to V-Ships Agency in December 2002.

Specifications

Length overall	21.0 m
Length waterline	18.0 m
Beam	6.0 m
Draught	1.2 m
Crew	5
Passengers	30
Fuel capacity	4,000 litres
Water capacity	1,000 litres
Max speed	25 kt
Operational speed	19 kt

Construction: All aluminium.

Propulsion: Two Caterpillar 3412 diesel engines each rated at 560 kW and each driving a fixed-pitch propeller.

20 m CREW BOAT

This vessel was delivered to Barwil Saudi Arabia in 2002 for crewboat duties. Two further vessels were delivered to GAC Saudi in 2003 and Bemat in 2006.

Specifications

Length overall	20.0 m
Length waterline	18.6 m
Beam	6.0 m
Draught	1.2 m
Crew	4
Passengers	12
Additional payload	15 t
Fuel capacity	10,000 litres
Water capacity	4,000 litres
Max speed	20 kt
Operational speed	18 kt
Range	800 n miles

Propulsion: Two Daewoo V10 diesel engines each rated at 440 kW and each driving a fixed-pitch propeller via a Nico gearbox.

United Kingdom

Fairline Boats PLC

Oundle, Northamptonshire PE8 5PA, United Kingdom

Tel: (+44 1832) 27 36 61
Fax: (+44 1832) 27 34 32
e-mail: sales@fairline.com
Web: www.fairline.com

Derek Carter, *Chairman and Managing Director*

Fairline produces a range of fast luxury yachts in three main ranges: the Targa range, from 10 to 18 m long; the Phantom range, from 12 to 15 m; and the Squadron range, from 16 to 22 m. The largest craft is the 22 m Squadron 74.

Squadron 74

Specifications

Length overall	22.7 m
Beam	5.7 m
Draught	1.6 m
Displacement, min	41 t
Fuel capacity	5,900 litres
Propulsive power	2 × 1,100 kW
Speed	31 kt

Propulsion: Two MAN diesel engines each rated at 1,100 kW, driving a four bladed fixed-pitch propeller via a reverse reduction gearbox.

Vessel type	Vessel name	Yard No	Length (m)	Speed (kt)	Seats	Vehicles	Originally delivered to	Date of build
Squadron 74	(Series Production)	–	22.7	–	–	none	–	–
Squadron 78	(Series Production)	–	24.38	33	–	none	–	–

FBM Babcock Marine Group

The Courtyard, St Cross Business Park, Monks Brook, Newport, Isle of Wight, PO30 5BF, United Kingdom

Tel: (+44 1983) 82 57 00
Fax: (+44 1983) 82 41 80
e-mail: fbm@babcock.co.uk
Web: www.fbmuk.com

M Graves, *General Manager*

FBM builds a wide range of monohull, catamaran and SWATH vessels for both commercial and paramilitary applications. In 2000 Babcock International purchased the FBM interests, adding a further shipyard in Rosyth.

AIR CREW TRAINING VESSEL

Dee
Don
Yare
Towy
Spey
Dart

Designed by FBM Babcock Marine these training vessels were delivered to SMIT (Scotland) in May 2003 following a contract awarded to SMIT by the UK Ministry of Defence. Construction was split between Babcock BES's Rosyth Dockyard and the FBMA Babcock Marine yard in the Philippines.

Specifications

Length overall	27.6 m
Length waterline	24.0 m
Beam	6.6 m
Draught	1.5 m
Displacement, full load	55 t
Passengers	12
Propulsive power	2 × 518 kW
Max speed	21 kt
Operational speed	18 kt

27.6 m SMIT air crew training vessel 0576686

Sorrento Jet 0506700

30 m FAST PASSENGER FERRY (DESIGN)

This all-aluminium planing vessel has been designed for passenger transportation on rivers and in estuarial sea areas.

Specifications

Length overall	32.0 m
Length waterline	30.0 m
Beam	7.0 m
Draught	1.8 m
Max speed	32 kt
Operational speed	30 kt

Classification: DnV +1A1 HSLC R3 EO.
Propulsion: Twin or triple diesel engines driving propellers or water-jets.

41 m PASSENGER FERRY

Capri Jet
Sorrento Jet

Designed by Marinteknik Verkstads AB and FBM Marine Ltd, *Capri Jet*, a 41 m passenger ferry, was completed in 1988 and operates on the Naples to Capri route.

A second similar vessel, *Sorrento Jet*, was ordered in May 1989 and was delivered in 1990.

Specifications

Length overall	41.0 m
Beam	7.8 m
Draught	1.1 m
Payload	27 t
Passengers	350

Vessel type	Vessel name	Yard No	Length (m)	Speed (kt)	Seats	Vehicles	Originally delivered to	Date of build
41 m Passenger Ferry	*Sorrento Jet*	–	41.0	29.0	350	none	Italy	1990
41 m Passenger Ferry	*Capri Jet*	–	41.0	–	350	none	Italy	1988
Air Crew Training Boat	*Dee*	–	27.6	21.0	12	–	SMIT, Scotland	2003
Air Crew Training Boat	*Don*	–	27.6	21.0	12	–	SMIT, Scotland	2003
Air Crew Training Boat	*Yare*	–	27.6	21.0	12	–	SMIT, Scotland	2003
Air Crew Training Boat	*Towy*	–	27.6	21.0	12	–	SMIT, Scotland	2003
Air Crew Training Boat	*Spey*	–	27.6	21.0	12	–	SMIT, Scotland	2003
Air Crew Training Boat	*Dart*	–	27.6	21.0	12	–	SMIT, Scotland	2003

Fuel capacity	7,000 litres
Water capacity	1,000 litres
Propulsive power	2,360 kW
Max speed	33.5 kt
Service speed	29 kt

Classification: Built to rules of the Code of Safety for Dynamically Supported Craft comparable to the SOLAS and Load Line Conventions, class notation RINA is *100-A (UL) 1.1-Nav. S-TP.

Structure: The superstructure and hull are built in welded marine grade aluminium, subdivided into seven watertight compartments. A semi-planing hull with a deep vee forward progressing to a shallow vee aft with hard chines.

The extruded profiles are of quality Alcan B51 SWP. The sheets are of quality Alcan B54 S½.

Propulsion: Two MTU 12V 396 TB 83 marine diesels giving minimum 1,180 kW each at MCR (1,045 kW for *Sorrento Jet*), 1,940 rpm at ambient air temperature of 27° C and sea water temperature at 27° C. Engines are coupled to ZF marine reduction gearboxes, and drive two MJP water-jet units, steering deflection angle is 30° port and starboard with an estimated reverse thrust of approximately half of the forward gross thrust.

Electrical system: Two Perkins 4.236M diesel generators each driving two 30 kVA alternators, Stamford MSC 234A.

Electrical supplies: 380 V AC, 50 Hz three-phase, 220 V AC, 50 Hz single-phase, 24 V DC, shore supply connection.

Control: All controls and machinery instruments are within reach of the helmsman's seat. Steering and reversing buckets for the water-jets are controlled electrohydraulically from the wheelhouse. A retractable bow thruster unit is fitted.

McTay Marine

The Magazines, Port Causeway, Bromborough, Wirral CH62 4YP, United Kingdom

Tel: (+44 151) 346 13 19
Fax: (+44 151) 334 00 41
e-mail: robertmcburney@mctaymarine.com

M R Brodie, *Managing Director*
R McBurney, *Commercial Director*

Although high-speed craft have not in the past been the main output of the shipyard, the company has completed two fast monohull craft: a patrol vessel and a high-speed survey vessel.

Chartwell

A high-speed survey vessel for hydrographic survey duties in the Thames and Thames estuary. The vessel has a longitudinally framed steel hull with an aluminium superstructure. The hull form is based on the NPL high-speed, round bilge series. The survey duties include echo-sounding, sonar sweeping, wire sweeping, marking obstructions, tidal stream observations and search and rescue.

Chartwell *on the River Mersey* 0007184

Specifications

Length overall	26.6 m
Beam	5.8 m
Propulsive power	2 × 1,074 kW
Max speed	22.8 kt

Propulsion: Main engines are two Paxman diesels 12 SETCWM, 1,074 kW each at 1,500 rpm. Two fixed-pitch three-blade Brunton propellers, driven via ZF BW 460 gearboxes, ratio 1.509:1.0. One PPJet PP170 water-jet unit (for improved control at low speed) driven by Volvo Penta AMD121D diesel, 283 kW, 1,800 rpm.

Vessel type	Vessel name	Yard No	Length (m)	Speed (kt)	Seats	Vehicles	Originally delivered to	Date of build
High-Speed Survey Vessel	*Chartwell*	–	26.6	–	–	–	Port of London Authority	–

United States

Allen Marine

1512 Sawmill Creek Road, PO Box 1049, Sitka, Alaska 99835-1049, United States

Tel: (+1 907) 747 81 00
Fax: (+1 907) 747 81 99
e-mail: saledept@allenmarine.com
Web: www.allenmarine.com

Ken Baker, *Project Manager*

Well known for the design, construction and operation of fast catamaran craft, Allen Marine also produce a range of fast monohull craft.

20 m JET OTTER FERRY

Six of these vessels were delivered to NY Waterway of New York in 2001 to 2002 for operation in Manhattan.

Specifications

Length overall	19.8 m
Beam	5.2 m
Draught	0.7 m
Crew	2
Passengers	97
Fuel capacity	3,600 litres
Maximum speed	34 kt
Operational speed	31 kt

Propulsion: Three Caterpillar 3406E diesel engines each rated at 450 kW and each driving a Hamilton 362 water-jet.

Vessel type	Vessel name	Yard No	Length (m)	Speed (kt)	Seats	Vehicles	Originally delivered to	Date of build
19.5 m Jet Otter	*Sea Otter Express*	5	19.5	36	49	none	Allen Marine Tours	1993
19.5 m Jet Otter	*Sea Lion Express*	6	19.5	36	49	none	Allen Marine Tours	1993
19.5 m Jet Otter	*Katlion Express*	7	19.5	36	49	none	Allen Marine Tours	1993
19.5 m Jet Otter	*Kalinin Express*	8	19.5	36	49	none	Allen Marine Tours	1993
20 m Jet Otter	*Austin Tobin*	27	19.8	34	87	none	NY Waterway	2001
20 m Jet Otter	*Fr Mychal Judge*	28	19.8	34	87	none	NY Waterway	2001
20 m Jet Otter	*Moira Smith*	29	19.8	34	87	none	NY Waterway	2002
20 m Jet Otter	*Douglas Gurian*	30	19.8	34	87	none	NY Waterway	2002
20 m Jet Otter	*Fred Morrone*	35	19.8	34	87	none	NY Waterway	2002
20 m Jet Otter	*Enduring Freedom*	34	19.8	34	87	none	NY Waterway	2002

Austal

100 Dunlap Drive, Mobile, Alabama 36602, US

Tel: (+1 251) 434 80 00
Fax: (+1 251) 434 80 80
e-mail: usasales@austal.com
Web: www.austal.com

Austal established a US yard in Mobile, Alabama, in December 1999. In partnership with leading US shipbuilder Bender Shipbuilding & Repair Co Inc, Austal's US joint venture now makes available its market leading aluminium vessels to the domestic US market and surrounding regions.

Austal's product range extends from fast passenger ferries and vehicle-passenger ferries to cruise vessels, patrol and offshore craft, luxury yachts and defence platforms. It is expected that the full product range will be built at the Mobile shipyard.

46 m CREWBOATS

Veronica
Anapaula

In 2001 the company received a contract for the design and build of two crewboats for Otto Candies LLC of the US. The vessels were delivered in 2002.

Specifications

Length overall	45.8 m
Length waterline	40.6 m
Beam	8.2 m
Draught	2.2 m
Crew	4
Passengers	80
Fuel capacity	28,500 litres
Water capacity	76,000 litres
Maximum speed	30 kt

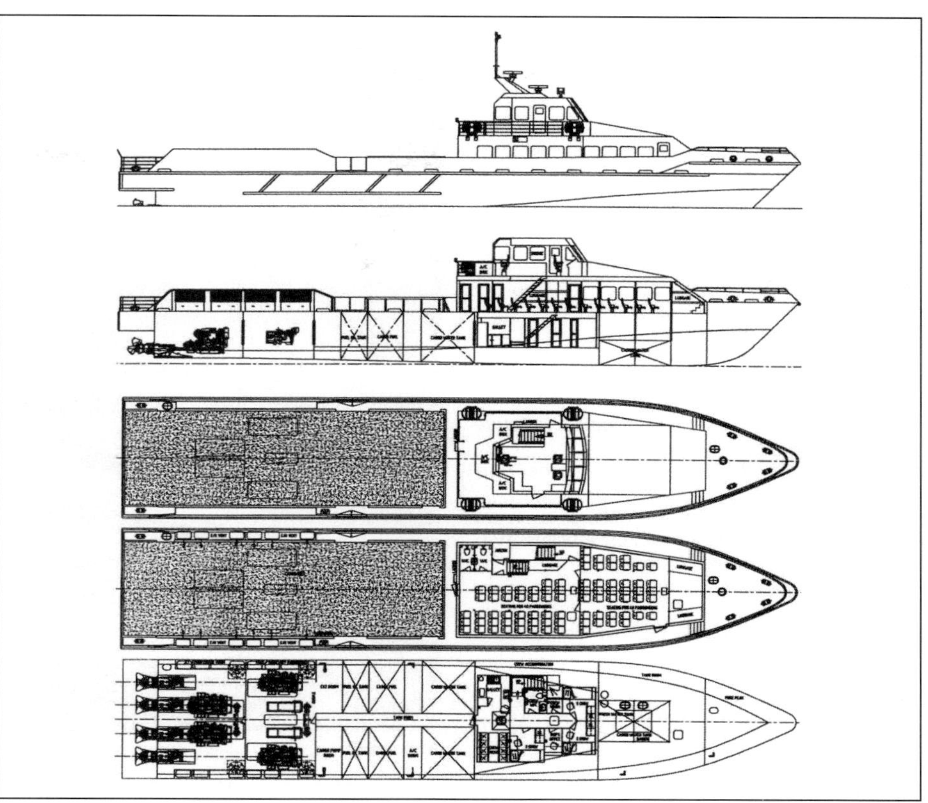

General arrangement of the 46 m crewboat 0132893

Structure: All-aluminium hull and superstructure.

Propulsion: Four Caterpillar 3508B diesel engines each rated at 783 kW and each driving a Hamilton HM571 water-jet.

Blount Boats Inc

461 Water Street, Warren, Rhode Island 02885, US

Tel: (+1 401) 245 83 00
Fax: (+1 401) 245 83 03
e-mail: info@blountboats.com
Web: www.blountboats.com

Luther H Blount, *President*

Blount Boats Inc (formerly Blount Marine Corporation) was formed in 1952 and, has designed and built over 300 vessels ranging in size from 5 to 80 m. Among the many types built has been the world's first small stern trawler and a significant number of passenger/commuter vessels including mini-cruise ships. The Hitech composite hull was invented by Luther Blount. A development of this hull concept, using an aluminium frame supporting a plastic or composite fabric skin, was patented in 1999 primarily for catamaran construction.

26 m FERRY

Fire Island Flyer

Delivered to Fire Island Ferries in 2001, the craft operates from Bay Shore New York to Fire Island.

Specifications

Length overall	25.9 m
Beam	6.3 m
Draught	1.6 m
Crew	5
Passengers	400
Fuel capacity	3,750 litres
Max speed	28 kt
Operational speed	23 kt

Blount Marine Hitech Express 0506702

Construction: All aluminium hull and superstructure.

Propulsion: Three DD C/MTU 8V 2000 diesel engines each rated at 481 kW at 2,100 rpm and driving fixed-pitch propellers via a Twin Disc MG5141 gearbox.

Hitech Express

Hull No 251, Design No P452.

A multipurpose craft which was designed in 1983 and built in 1984. The vessel has been granted USCG Certification for 149 passengers. Two smaller versions have also been built.

Hitech Express has been engaged in demonstrating a reliable, fast, economical commuting service in the US. It has made five runs to New York City from Warren, Rhode Island, a distance of 150 miles, in just over 5 hours (November 1984, May 1985, June 1985). It has made the run from the Battery, New York to Staten Island fully loaded in 10 minutes, against 28 minutes via conventional ferry. It also crossed the Hudson at mid-Manhattan in 2 minutes 46 seconds in November 1984 and June 1985.

Specifications

Length overall	23.5 m
Length waterline	22.7 m
Beam	6.1 m
Draught	1.1 m
Displacement, max	56.4 t
Displacement, min	23.9 t
Payload	15 t
Crew	2
Passengers	149
Fuel capacity	1,923 litres
Propulsive power	2 × 380 kW
Max speed	27.8 kt
Operational speed	24.3 kt
Operational limitation	wave height 1.5 m Beaufort 4–5
Range	250 n miles

Structure: Aluminium structural frames, bulkheads and decks with polyurethane foam sprayed over and a glass fibre skin $^1/_4$ to $^1/_8$ in thick forming the outer hull covering.

Propulsion: Two GM 12V-71TI, 380 kW at 2,300 rpm (each), driving two 711 mm diameter, 622 mm pitch Colombian Bronze propellers.

Electrical system: Two 60 A alternators, one on each main engine providing 32 V DC power throughout vessel. Engine room area available for optional generator.

Vessel type	Vessel name	Yard No	Length (m)	Speed (kt)	Seats	Vehicles	Originally delivered to	Date of build
Multipurpose craft	*Hitech Express*	251	23.5	24.3	149	none	US	1984
26 m Ferry	*Fire Island Flyer*	508	25.9	28.0	400	none	Fire Island Ferries	2001

Breaux Brothers Enterprises Inc

PO Box 1100, Highway 86, Loreauville, Los Angeles 70552, US

Tel: (+1 337) 229 42 32
Fax: (+1 337) 229 49 51
Web: www.breauxboats.com

Vance Breaux, *Director*
Ward Breaux, *Director*

Founded in 1983, the company builds a range of small aluminium vessels including a number of fast crewboats and catamarans.

Breaux's Bay Craft Inc

PO Box 370, Loreauville, Louisiana 70552, United States

Tel: (+1 337) 229 42 46
Fax: (+1 337) 229 83 32

Roy Breaux Jr, *President*
Royce Breaux, *Chairman and Executive Vice-President*
Hub Allums, *Chief of Design and Engineering*

Breaux's Bay Craft manufacture a wide range of passenger craft and crewboats, two of which are described below.

Genesis
This all-aluminium crew boat was delivered in 1998.
Specifications

Length overall	46.3 m
Beam	9.1 m
Payload	250 t
Crew	3
Passengers	52
Fuel capacity	77,000 litres
Water capacity	115,000 litres
Propulsive power	4,625 kW

Mexico III
This all-aluminium passenger ferry was delivered in April 1993 for use in rivers and protected waters.
Specifications

Length overall	45.7 m
Beam	9.1 m
Draught	2.1 m
Passengers	800
Fuel capacity	23,000 litres
Operational speed	24 kt

Propulsion: Main engines are four DDC 16V 92TA diesels, driving via four Twin Disc MGN273EV gearboxes.

Mexico III 0506842

Genesis 0045183

Vessel type	Vessel name	Yard No	Length (m)	Speed (kt)	Seats	Vehicles	Originally delivered to	Date of build
46 m Passenger Ferry	*Mexico III*	–	45.7	24.0	800	None	Cruceros Maritimos	1993
46 m Crew Boat	*Genesis*	–	46.3	–	52	None	Jade Marine	1998

Christensen Shipyards

4400 South East Columbia Way, Vancouver, Washington 98661, US

Tel: (+1 360) 695 32 38
Fax: (+1 360) 695 32 52
e-mail: inquiry@christensenyachts.com
Web: www.christensenyachts.com

David H Christensen, *Chief Executive Officer*
Joe Foggia, *President*
Henry Luken, *Principal*
John Lance, *Marketing Director*

The Christensen Shipyard, located in Vancouver, Washington, consists of 14,000 m^2 of climate controlled manufacturing space and a seven-acre marina. Nine large manufacturing and assembly bays, one of which houses an expandable mold for hulls from 33 to 53 m, are the heart of the facility. One bay is fully enclosed to create a clean controlled environment in which to paint yacht exteriors. The dry dock can accommodate a 350 t yacht. Christensen yachts are built to full ABS +1A1-AMS and MCA classification standards for hull, superstructure and machinery.

Royal Oak 0506703

Royal Oak
This triple-deck motor yacht was launched in October 1988 for a Japanese owner.
Specifications

Length overall	39.6 m
Length waterline	35.7 m
Beam	8.2 m
Draught	1.8 m
Displacement, min	131.5 t
Displacement, max	172.3 t
Fuel capacity	30,280 litres
Water capacity	5,677 litres
Propulsive power	1,732 kW
Operational speed	20 kt

Propulsion: Engines are two Mitsubishi Model S12A2-MPTIC each rated at 866 kW at 2,100 rpm driving propellers through a ZF BW-256 reduction gearbox.

155 ft MOTOR YACHT

Silver Lining
Specifications

Length overall	47.3 m
Beam	8.4 m

Vessel type	Vessel name	Yard No	Length (m)	Speed (kt)	Seats	Vehicles	Originally delivered to	Date of build
40 m Motor Yacht	*Royal Oak*	–	39.6	20.0	–	None	Japan	October 1988
155 ft Motor Yacht	*Silver Lining*	–	47.3	18.0	–	None	–	–
110 ft Motor Yacht	*Rendezvous*	–	33.5	22.0	–	–	–	–
124 ft Motor Yacht	*Wehr Nuts*	–	37.8	24.0	–	–	–	–
141 ft Motor Yacht	*Walkabout*	–	42.9	22.0	–	–	–	–
155 ft Motor Yacht	*Liquidity*	–	47.2	20	–	–	–	–
145 ft Motor Yacht	*Primadonna*	–	44.2	20	–	–	–	–

Draught 2.1 m
Fuel capacity 48,000 litres
Propulsive power 2,900 kW
Operational speed 18 kt
Propulsion: Engines: two Deutz MWM TBD 604B V12 diesels, 1,450 kW each at 1,800 rpm.
Transmissions: two ZF reduction gearboxes.
Electrical system: Generators: two Northern Lights 99 kW.

Christensen 155 ft motor yacht 0007275

Derecktor Shipyards

311 East Boston Port Road, Mamaroneck, New York 10543, US

Tel: (+1 914) 698 50 20
Fax: (+1 914) 698 46 41
e-mail: general@derecktor.com
Web: www.derecktor.com

Paul Derecktor, *President*
George Best, *Marketing Manager*

MIT sea AH 0506843

Primarily a builder of fast monohull luxury yachts, the company also builds a wide range of other craft. In 1997 two fast catamaran ferries for operation in New York Harbour were delivered. The company then delivered a larger 45 m gas-turbine propelled catamaran in 1998–99 and a 41 m fast catamaran for Woods Hole SSA in 2000. In 2002/3 six 16 m water taxis were delivered and in 2004 two 73 m car ferries for Alaska and two Crowther 28 m catamarans for New England Ferry Company were on order. The company has three shipyards, one in Florida, one in New York and a yard in Bridgeport, Connecticut which opened in 2001.

19 m RESEARCH FISHERY VESSEL

This vessel, *First State,* was designed by Nigel Gee and Associates and delivered in July 2002 to the State Department of Fish and Wildlife.

Specifications
Length overall 19 m
Length waterline 17.5 m
Beam 5.3 m
Draught 1.4 m
Propulsion: Two Daewoo T180VIM diesel engines each rated at 485 kW and each driving a fixed pitch propeller.

Dillinger 0506706

Dillinger 74

This vessel, completed in Autumn 1990, was designed by Frank Mulder, the Dutch naval architect designer of *Octopussy. Dillinger 74* is built in advanced carbon fibre pre-impregnated materials.

Specifications
Length overall 22.7 m
Length waterline 18.5 m
Beam 5.3 m
Draught 1.1 m
Fuel capacity 7,200 litres
Water capacity 2,100 litres
Max speed 56 kt
Propulsion: Main engines are two MTU 12V 396 TB 93; two Kamewa S45 water-jet units.

MIT sea AH

This vessel was designed by Richard Liebowitz Design with Sparkman & Stephens

Vessel type	Vessel name	Yard No	Length (m)	Speed (kt)	Seats	Vehicles	Originally delivered to	Date of build
Derecktor 66	*Wicked Witch III*	128	20.1	–	–	None	US	1985
S&S 105	*Lady Frances*	144	32	25	–	None	US	1987
31 m Motor Yacht	*Eyeball*	149	31	–	–	None	US	1989
22 m Fexas Express Motor Yacht	*Transition*	150	22.6	30	–	None	US	1990
23 m Motor Yacht	*Dillinger 74*	152	22.7	–	–	None	US	1990
35 m Motor Yacht	*MIT sea AH*	153	34.7	26	–	None	US	1993
20 m Sport Fishing Boat	*The Boat III*	154	20	35	–	None	US	1995
19 m Research Fishing Vessel	*First State*	167	19	22	–	–	State of Delaware	2002
22 m Passenger Ferry	*Fire Island Empress*	–	22	22	260	–	Sayville Ferries	2004

and is the third and largest *sea AH* for the same owner.

Specifications

Length overall	34.7 m
Length waterline	29.9 m
Beam	7.0 m
Draught	1.9 m
Propulsive power	2 × 1,700 kW
Operational speed	26 kt
Max speed	29 kt

Classification: American Bureau of Shipping.

Structure: Aluminium hull structure.

Propulsion: Main engines are two MTU 12V 396 TE 94, 1,700 kW each; driving two propellers via ZF V-drive gearboxes, one Arneson surface drive.

S&S 105

Lady Frances

A high-speed water-jet-propelled motor yacht built in 1987, to American Bureau of Shipping@A1 for Yachting Service.

Designers: Sparkman & Stephens Inc, 79 Madison Avenue, New York 10016, US.

Specifications

Length overall	32.0 m
Length waterline	27.7 m
Beam	7.0 m
Draught	1.2 m
Fuel capacity	22,710 litres
Water capacity	6,056 litres
Propulsive power	2 × 1,462 kW
Max speed	31 kt
Operational speed	25 kt

Structure: Aluminium alloy, 5086 series welded aluminium, 6061 aluminium extrusions, teak main deck and boat deck.

Propulsion: Engines: Main engines are two MTU 12V 396TB 93 diesel engines 1,462 kW each, driving two Kamewa Series 63 water-jet units.

DERECKTOR 66

Wicked Witch III

High-speed motor yacht.

Specifications

Length overall	20.1 m
Length waterline	16.4 m
Beam	5.4 m
Draught	1.7 m
Displacement, max	29.51 t
Propulsive power	2 × 649 kW
Max speed	33 kt

Structure: Aluminium alloy, stepped surface hull form, skeg on centreline.

Propulsion: Main engines are two GM 12V-71 T marine diesels 649 kW each, coupled to 195 ZF, V-drive gearboxes; propeller shaft angle: 7°.

31 m MOTOR YACHT

Eyeball

Built in early 1989 this 31 m luxury yacht was designed by DIANA Yacht Design BV, Netherlands.

Specifications

Length overall	31.0 m
Length waterline	26.0 m
Beam	7.5 m
Draught	1.4 m
Displacement, max	120 t
Fuel capacity	25,000 litres
Water capacity	6,000 litres
Propulsive power	2 × 2,040 kW
Max speed	28 kt

Structure: All-aluminium fully planning deep vee hull with transom stern, trim wedge, three spray rails and chine rail each side on bottom, flared bow, three watertight bulkheads. Fuel, water, oil and sanitary tanks built in double bottom.

Material of hull plating and superstructure: aluminium alloy grade 5083; remaining parts of construction to be aluminium alloy grade 6082.

Propulsion: Main engines are two diesel engines MTU 16V 396 TB 84, rating to ISO 3046/i, 2,040 kW each, driving two Kamewa water-jet units Series 71.

Auxiliary systems: Two 40 hp thrusters, by Cramm (Holland) mounted in engine room for low-speed propulsion and manoeuvring, Schottel Bow Thruster, two deck cranes by Cramm, two Steen Anchor Windlass, windows supplied by WIGO, Holland.

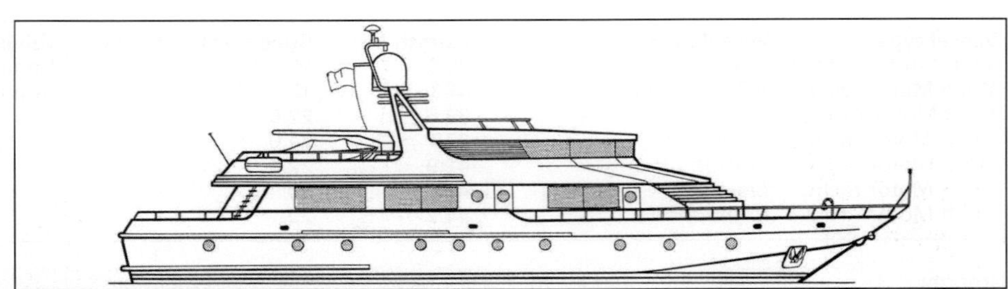

31 m (DIANA Yacht Design BV) water-jet luxury motor yacht Eyeball, *built by R E Derecktor Shipyard*
0506911

Transition
0506707

22 m FEXAS EXPRESS MOTOR YACHT

Transition

Designed by Tom Fexas Yacht Design Inc, Florida, this vessel was delivered towards the end of 1990.

Specifications

Length overall	22.6 m
Length waterline	19.5 m
Beam	5.8 m
Draught	0.9 m
Displacement, max	34.5 t
Fuel capacity	10,598 litres
Water capacity	2,498 litres
Propulsive power	2 × 746 kW
Max speed	35 kt
Operational speed	30 kt
Range	1,200 n miles

Propulsion: Main engines are two MAN D2842 LYE, 746 kW each.

20 m SPORT FISHING BOAT

The Boat III

Designed by Tripp Design, this all-aluminium vessel was delivered from the New York yard in 1995.

Specifications

Length overall	20.0 m
Length waterline	17.7 m
Beam	5.4 m
Draught	1.7 m
Propulsive power	2 × 950 kW
Max speed	40 kt
Operational speed	35 kt
Range	900 n miles

Propulsion: Two MTU 8V 396 TE 94 rated at 950 kW each driving through a ZF BW 255S gearbox.

22 m PASSENGER FERRY

This vessel was delivered to Sayville Ferry Service Inc in 2003.

Specifications

Length overall	21.6 m
Length waterline	20.7 m
Beam	5.8 m
Draught	2.4 m
Passengers	260
Operational speed	22 kt

Construction: All aluminium hull and superstructure.

Propulsion: Three Detroit Diesel Series 60 engines.

Gladding-Hearn Shipbuilding

One Riverside Avenue, PO Box 300, Somerset, Massachusetts 02726-0300, United States

Tel: (+1 508) 676 85 96
Fax: (+1 508) 672 18 73
e-mail: sales@gladding-hearn.com
Web: www.gladding-hearn.com

George R Duclos, *Chairman and CEO* John F Duclos, *President, Director of Operations*
Peter J Duclos, *President, Director of Business Development*

This shipyard is a member of the Duclos Corporation and is known for its range of commercial workboats and fast ferry craft. The yard has also built a large number of small pilot craft (not covered within the scope of this database) as well as its well-known InCat catamaran ferries.

19 m pilot boat

This all-aluminium Hunt-designed pilot boat was delivered to the Lake Charles Pilots in September 1995.

Specifications

Length overall	19.3 m
Beam	5.6 m
Draught	1.7 m
Fuel capacity	3,800 litres
Propulsive power	1,216 kW
Max speed	25 kt

20 m pilot boat

Delivered in July 1993 to the San Francisco Bar Pilots, the boat was designed jointly with C Raymond Hunt Associates.

The deep vee, twin screw vessel is designed and equipped to run 24 hours a day at speeds over 25 kt. The rugged, all-aluminium hull and superstructure was built with heavy duty, low-maintenance equipment, including keel-coolers and a 3,800 litre remote-controlled ballast tank.

Specifications

Length overall	20.4 m
Beam	5.3 m
Draught	1.8 m
Max speed	25.5 kt

23 m pilot boat

Two of these all aluminium Hunt-designed pilot boats were delivered to Charleston Navigation Co in 2001.

Specifications

Length overall	23.1 m
Beam	6.1 m
Draught	1.8 m
Crew	2
Passengers	4
Fuel capacity	10,425 litres
Water capacity	380 litres
Operational speed	27 kt

Propulsion: Two MTU 16 V 2000 DDEC diesel engines each rated at 960 kW at 2,100 rpm and driving Bruntons five-bladed fixed-pitch propellers via Twin Disc 6619A gearboxes.

24.5 m ferry

Delivered in May 1996, this all-aluminium, triple screw vessel is operated by the Sayville Ferry company carrying over 400 passengers at speeds of up to 22 kt.

Specifications

Length overall	24.5 m
Beam	6.4 m
Draught	0.9 m
Passengers	415
Propulsive power	1,823 kW
Operational speed	22 kt

Propulsion: Powered by three Detroit Diesel 12V 92TA diesel engines each driving a fixed-pitch propeller via a reverse reduction gearbox.

30 m monohulled ferry

This is a new generation of monohulled fast ferries designed for fast offloading of passengers.

The first four all-aluminium vessels of this class, *Henry Hudson*, *Robert Fulton*, *Empire State* and *Garden State* were delivered to New Jersey-based Port Imperial Ferry Company for passenger service between New Jersey and New York City.

The vessel's unique bow-loading system safely offloads 100 passengers per minute.

Specifications

Length overall	29.6 m
Beam	7.8 m
Draught	1.8 m
Passengers	400

24.5 m South Bay Clipper 0507078

Monohulled passenger ferry Robert Fulton 0506845

23 m pilot boat Fort Moultrie *operated by the Charleston Navigation Co* 0121746

Vessel type	Vessel name	Yard No	Length (m)	Speed (kt)	Seats	Vehicles	Originally delivered to	Date of build
20 m Pilot Boat	*Golden Gate*	282	20.4	25.0	–	none	San Francisco Bar Pilots	1993
19 m Pilot Boat	*Lake Charles Pilot*	297	19.3	25	–	none	Lake Charles Pilots	1995
20.5 m Ferry	*South Bay Clipper*	302	24.5	22.0	415	none	Sayville Ferry Company	1996
30 m Monohull Ferry	*Garden State*	293	29.4	–	400	none	Imperial Ferry Company	1994
30 m Monohull Ferry	*Empire State*	289	29.4	–	400	none	Imperial Ferry Company	1999
30 m Monohull Ferry	*Henry Hudson*	285	29.4	–	400	none	Imperial Ferry Company	1993
30 m Monohull Ferry	*Robert Fulton*	286	29.4	–	400	none	Imperial Ferry Company	1993
23 m Pilot Boat	*Fort Moultrie*	225	23.1	27.0	–	none	Charleston Navigation Co	2000
23 m Pilot Boat	*Fort Sumter*	224	23.1	27.0	–	none	Charleston Navigation Co	2000

Propulsive power	2 × 500 kW
Max speed	18 kt

Propulsion: Main engines are twin Caterpillar 3412 diesels, each rated at 500 kW at 1,800 rpm. The engines drive two 1.07 m bronze propellers via ZF 2.57:1 reverse reduction gears.

Gulf Craft Inc

3904 Highway 182, Patterson, Louisiana 70392, United States

Tel: (+1 985) 395 52 54
Fax: (+1 985) 395 36 57
e-mail: info@gulfcraft.com
Web: www.gulfcraft.com

Kevin Tibbs, *President*

Gulf Craft Inc is a builder of high-speed aluminium conventional hull passenger ferries and crew boats.

58 m CREWBOAT

The 58 m *Granville McCall* was delivered in 2002 to McCall Boat Rental for Seacor of Houston for use in the Gulf of Mexico.

Specifications

Length overall	57.9 m
Length waterline	53.7 m
Beam	10.7 m
Draught	3.1 m
Crew	8
Passengers	97
Fuel capacity	174,000 litres
Water capacity	310,000 litres
Maximum speed	26 kt
Operational speed	23 kt

Propulsion: Five Cummins KTA50-M2 diesel engines each rated at 1,340 kW and each driving a fixed-pitch propeller via a Twin Disc MG 8858 gearbox.

56 m CREWBOAT

Gulf Craft recently delivered the first of an order for three 56 m crewboats. The *John B Martin McCall* is currently operating from Chevron's Leeville/Port Fuchon bases.

Keith G McCall 0121744

Specifications

Length overall	56.4 m
Beam	10.7 m
Fuel capacity	120,000 litres
Water capacity	240,000 litres
Propulsive power	6,040 kW

Propulsion: Six Cummins V-12 KTA38-M diesel engines each driving 1.2 m diameter fixed-pitch propellers.

54 m CREWBOAT

Gilbert McCall
Seth McCall

Two of these vessels were delivered to Seacor Marine in 2003.

Specifications

Length overall	54 m
Length waterline	48 m
Beam	9.1 m
Draught	2.4 m
Max speed	33 kt
Operational speed	25 kt

Propulsion: Four Cummins KTA50-M2 diesel engines each rated at 1,340 kW and each driving a Hamilton HM 811 water-jet via a Twin Disc MG 6848 reduction gearbox.

53 m CREWBOAT

Joyce McCall

Delivered to Seacor Marine in September 2004, this vessel is fitted with an active stabilisation system designed Maritime Dynamics.

Specifications

Length overall	53 m
Beam	9.1 m
Max speed	32 kt
Operational speed	20 kt
Fuel capacity	84,500 litres
Water capacity	165,000 litres

Propulsion: Four Cummins KTA50-M2 diesel engines each rated at 1,340 kW at 1,900 rpm and each driving a Hamilton HM 811 water-jet.

50 m CREWBOAT

This vessel was delivered in 2001.

Specifications

Length overall	50.3 m
Length waterline	44.4 m
Beam	9.1 m
Draught	2.2 m
Crew	14
Passengers	84
Fuel capacity	10,000 litres

Vessel type	Vessel name	Yard No	Length (m)	Speed (kt)	Seats	Vehicles	Originally delivered to	Date of build
19 m	*Nicolet*	290	18.80	25.0	150	none	Star Line	1985
24 m	*Radisson*	313	23.60	30.0	350	none	Star Line	1988
49 m Crew Boat	*Annabeth McCall*	–	48.80	25.0	–	none	McCall Boat Rental	1989
49 m Crew Boat	–	–	48.80	25.0	–	none	McCall Boat Rental	1989
49 m Crew Boat	*Norman McCall*	–	48.80	25.0	–	none	McCall Boat Rental	1989
49 m Crew Boat	–	–	48.80	25.0	–	none	McCall Boat Rental	1989
47 m Crew Boat	*Blair McCall*	–	47.30	–	92	none	–	–
56 m Crew Boat	*John B Martin McCall*	–	56.40	–	–	none	–	–
44 m Crew Boat	*Caleb McCall*	–	44.20	23.0	75	none	McCall Boat Rental	–
20 m	*Cadilac*	349	19.70	28.0	150	none	Star Line	1990
30 m Whale Watch Ferry	*Atlantis*	*Whale Watcher*	30.50	27.0	–	none	–	1998
37 m Whale Watch Ferry	–	–	36.50	28.0	395	none	–	1999
44 m Crew Boat	*Abuja Eagle*	439	44.20	23.0	75	none	Fymak Marine	2000
40 m Crew Boat	*Savannah*	433	40.00	23.0	–	none	Texas Crewboats	2000
44 m Crew Boat	*Ms Adrienne*	435	44.00	23.0	75	none	Gulf Logistics	2000
44 m Crew Boat	*Mr Jake*	434	40.00	23.0	–	none	Texas Crewboats	2000
44 m Crew Boat	*Delta Princess*	439	44.20	23.0	75	none	Fymak Marine	2001
44 m Crew Boat	*Omambala River*	437	44.20	23.0	75	none	Fymak Marine	2001
44 m Crew Boat	*Qua Iboe River*	438	44.20	23.0	75	none	Fymak Marine	2001
58 m Crew Boat	*Granville McCall*	–	57.90	26.0	97	none	McCall Boat Rental	2002
46 m Crew Boat	*Gilbert C McCall*	443	46.00	23.0	96	none	Seacor Marine	2002
46 m Crew Boat	*French Broad River*	447	46.00	23.0	78	none	Trico Marine	2002
46 m Crew Boat	*Holston River*	448	46.00	23.0	78	none	Trico Marine	2002
46 m Crew Boat	*Tennessee River*	449	46.00	25.0	78	none	Trico Marine	2003
54 m Crew Boat	*Seth McCall*	451	54.00	25.0	76	none	Seacor Marine	2003
47 m Crew Boat	*Elizabeth Anne McCall*	454	47.20	27.0	76	none	Seacor Marine	2004
53 m Crew Boat	*Joyce McCall*	–	53.00	32.0	50	none	Seacor Marine	2004
47 m Crew Boat	*Roxanne B McCall*	–	47.24	27.0	76	none	Seacor Marine	2004
55 m Crew Boat	*Jenny McCall*	–	55.32	26.0	80	none	Seacor Marine	2005
55 m Crew Boat	*Ingrid McCall*	–	55.32	26.0	60	none	Seacor Marine	2005
25 m Fast Ferry	*Marquette II*	458	25.20	25.0	330	none	Star Line	2005
40 m Fishing Vessel	*Big E*	457	39.6	25	149	none	Capt. Elliot's Party Boats	2005
58 m Crew Boat	*John G McCall*	–	57.9	26	38	none	Seacor Marine	2007
51 m Crew Boat	*Harlan S McCall*	–	50.6	27	80	none	Seacor Marine	2007

Water capacity	12,500 litres
Max speed	33 kt
Operational speed	26 kt

Classification: USCG Subchapter T.
Construction: All aluminium hull and superstructure.
Propulsion: Four Cummins KTA50-M2 diesel engines each rated at 1,324 kW at 1,900 rpm and driving a Hamilton HM 811 water-jet via a Twin Disc MG 6848 gearbox.

Annabeth McCall
Norman McCall
These are two of four very large crew boats delivered in 1989 to McCall Boat Rental of Cameron, Louisiana. These boats combine increased size (48.78 m length) for improved sea-keeping with a reasonably high speed of around 25 kt.
Specifications

Length overall	48.8 m
Beam	9.1 m
Propulsive power	6 × 507 kW

Propulsion: Main engines are six Cummins KTA19-M2 diesels, each 507 kW, giving a total power of 3,042 kW.

47 m CREWBOAT
Elizabeth Anne McCall
This vessel was delivered to Seacor Marine in June 2004.
Specifications

Length overall	47.24 m
Beam	8.23 m
Draught	1.98 m
Fuel capacity	77,000 litres
Max speed	27 kt
Operational speed	24 kt

Propulsion: Four Cummins KTA38-M2 diesel engines each driving a fixed-pitch propeller via a Twin Disc reduction gearbox.

46 m CREWBOAT
Three of these vessels were delivered to Trico Marine in 2002.
Specifications

Length overall	46 m
Length waterline	41.7 m
Beam	8.2 m
Draught	2.7 m
Fuel capacity	77,000 litres
Freshwater capacity	106,000 litres
Max speed	28 kt
Operational speed	23 kt

Propulsion: Four Cummins KTA38-M2 diesel engines each rated at 1,000 kW and each driving a fixed-pitch propeller via a MG 6850 reverse reduction gearbox.

44 m CREWBOAT
Owned by McCall Boat Rental, Cameron, Louisiana, US, the *Caleb McCall* is a 44 m crew boat with a maximum speed of 27 kt. An interesting feature of this craft is that there are five engines which the owner claims give better manoeuvrability and make engine replacement or repair easier.
Specifications

Length overall	44.2 m
Beam	8.5 m

John B Martin McCall 0043934

Seth McCall 1048774

Draught	2.4 m
Crew	6
Passengers	75
Fuel capacity	45,420 litres
Water capacity	75,700 litres
Max speed	27 kt
Operational speed	23 kt

Classification: USCG approved Gulf of Mexico, 200 miles offshore.
Structure: Superstructure and hull are built in aluminium. Hull plating is in 5086 grade, 15.8 mm thick over the propellers, the remainder of the bottom is 12.7 mm thick with 8.4 mm side plating, superstructure is in 4.8 mm plate.
Propulsion: Five Cummins KTA19-M diesel engines coupled to Twin Disc MG 518, 2.5:1 reduction gearboxes, driving five three-blade Columbian Bronze FP propellers.
Electrical system: Two Cummins 6B5.8, 56 kW generators.
Navigation and communications: Raytheon/JRC colour and Furuno FR 8100 radars, Si-Tex Koden Loran, Datamarine sounder, Decca 150 autopilot.
Motorola Triton 20 SSB, Raytheon 53A VHF.
Control: Orbitrol/Charlynn steering and Kobelt engine controls. Duplicate engine controls also fitted on the upper bridge allowing operator to face aft with a clear view while working cargo at rigs and oil platforms.
Outfit: Three two-berth cabins, galley and mess room.

40 m CREWBOAT
MV Savannah
MV Savannah was delivered in 2000 to Texas Crewboats.
Specifications

Length overall	40 m
Beam	7.9 m
Fuel capacity	50,000 litres
Freshwater capacity	80,000 litres
Operational speed	23 kt

Propulsion: Four Cummins KTA19 diesels each driving a fixed-pitch propeller via a ZF BW190 reverse reduction gearbox.

37 m WHALE WATCH FERRY
Specifications

Length overall	36.5 m
Beam	8.5 m
Draught	1.5 m
Passengers	395
Fuel capacity	22,000 litres
Water capacity	2,000 litres
Maximum speed	38 kt
Operational speed	33 kt

Propulsion: Five MTU 16V 2000 diesels each rated at 960 kW and each driving a Hamilton HM571 water-jet via a Twin Disc 6619 SC reduction gearbox.

30 m WHALE WATCH FERRY
Specifications

Length overall	30.5 m
Beam	7.6 m
Draught	1.8 m
Fuel capacity	20,000 litres
Water capacity	1,000 litres
Maximum speed	29 kt
Operational speed	27 kt

Propulsion: Four Caterpillar 3412 diesel engines each rated at 750 kW and each driving a fixed pitch propeller via a ZF BW190 reverse reduction gearbox.

Magnum Marine Corporation

2900 Northeast 188th Street, North Miami Beach, Aventura, Florida 33180, United States

Tel: (+1 305) 931 42 92
Fax: (+1 305) 931 00 88
e-mail: sales@magnummarine.com
Web: www.magnummarine.com

Katrin Theodoli, *President and Chief Executive Officer*

Magnum Marine specialises in the construction of high-performance, patrol and pleasure craft from 8 to 24 m in length.

The company's patrol craft are in service with the FBI, US Customs, US Coast Guard, US Navy and many other non-US agencies.

Magnum Marine offers both armed and unarmed patrol craft, all built to RINA classification.

Magnum Marine and the former McDonnell Douglas (now Boeing) have jointly developed a range of patrol craft based on the proven Magnum race boats and incorporating state-of-the-art surveillance and targeting systems and weapons from

McDonnell Douglas. This range of craft, named Barbarian, is available in sizes up to 24 m overall length.

MAGNUM 44

A follow-on to the prototype Magnum 40 Barbarian originally prepared by McDonnell Douglas and Magnum Marine, this GRP, 24° deadrise model is now Magnum's smallest diesel-powered patrol craft with cruise and top speeds suitable for coastal patrol and drug interdiction operations.

Specifications

Length overall	13.4 m
Beam	3.7 m
Draught	1.1 m
Fuel capacity	1,514 litres
Water capacity	260 litres
Max speed	50 kt

Propulsion: Two Caterpillar or MAN diesel engines each rated at 600 kW with Twin Disc/Arneson ASD-10 surface drives.

MAGNUM 50

Specifically designed for rapid response to coastal threats and rapid offensive actions, this GRP, 24° deadrise craft has low freeboard, low profile, quick acceleration, high manoeuvrability and a top speed of 60 kt.

Specifications

Length overall	15.2 m
Beam	4.2 m
Draught	1.1 m
Displacement	20 t
Crew	3
Passengers	5
Fuel capacity	2,365 litres
Water capacity	565 litres
Max speed	60 kt
Operational speed	48 kt
Range	385 n miles

Structure: GRP hull.

Magnum Barbarian 40 prototype 0093270

Propulsion: Two MTU 10 cylinder common rail engines each rated at 1100 kW with two Twin Disc/Arneson ASD-12 Surface Drives.

MAGNUM 60

This vehicle was built and tested in 2000.

Specifications

Length overall	18.4 m
Beam	4.8 m
Draught	1.1 m
Fuel capacity	4,160 litres
Water capacity	565 litres
Max speed	59 kt
Operational speed	43 kt

Propulsion: Two MTU common rail diesel engines each rated at 1,800 kW, each driving an Arneson ASD-15 surface drive.

MAGNUM 70

Can be used for long-range, high-performance transport of special forces for insertion/extraction operations. Fully air transportable.

Specifications

Length overall	21.35 m
Beam	5.18 m
Draught	1.06 m
Displacement	36 t
Crew	3
Passengers	10
Fuel capacity	4,540 litres
Water capacity	567 litres
Max speed	47 kt
Operational speed	38 kt

Construction: Aramid/carbon fibre.

Propulsion: Two MTU common rail diesel engines each rated at 1,800 kW, each driving two Twin Disc Arneson ASD15 surface drives.

MAGNUM 80

Specifications

Length overall	24.0 m
Beam	5.8 m
Draught	1.4 m
Fuel capacity	6,190 litres
Water capacity	1,590 litres
Max speed	46 kt
Operational speed	39 kt

Propulsion: Two MTU common rail diesel engines each rated at 1,800 kW.

Vessel type	Vessel name	Yard No	Length (m)	Speed (kt)	Seats	Vehicles	Originally delivered to	Date of build
Magnum 44	*Banzai*	–	13.4	50	–	None	–	–
Magnum 50	*Bestia*	–	15.2	60.0	5	None	–	–
Magnum 60	*Furia*	–	18.4	60	–	None	–	2000
Magnum 70	–	–	23.4	50	10	None	–	–
Magnum 80	–	–	24.0	46.0	–	None	–	–

Marinette Marine Corporation

A Subsidiary of Manitowoc Marine Corporation
1600 Ely Street, Marinette, WI 54143-2434, United States

Tel: (+1 715) 735 9341
Fax: (+1 715) 735 3516
e-mail: fcharrier@marinettemarine.com
Web: www.manitowocmarine.com

Robert Herre, *President & General Manager, Manitowoc Marine Group*
Floyd Charrier Jr, *Vice President Sales & Marketing, Manitowoc Marine Group*

Since incorporation in 1942, Marinette has built over 1300 vessels. It is a full service shipyard with in-house capabilities to design and construct technologically advanced vessels.

Lockheed Martin Littoral Combat Ship 1290776

LITTORAL COMBAT SHIP (LCS)

In 2002 the US Navy began a programme to design, develop and build a new class of small stealth ships for operation in littoral waters. In 2002 one of six concept studies was awarded to Lockheed Martin, which included Marinette Marine as builders. Lockheed Martin was successful in reaching the final stage and in 2005 Marinette Marine began construction of the first LCS.

The Lockheed Martin design is based on a proven semi-planing monohull which provides agility and high-speed manoeuvrability with known seakeeping characteristics to support launch and recovery operations, mission execution and for crew comfort. The vessel features a steel hull and aluminium superstructure and is understood to be capable of supporting air operations up to seastate 5. The first vessel was launched in 2007, however the second vessel was cancelled by the US Navy in the same year.

Specifications

Length overall	115 m
Beam	13 m
Draught	4 m
Displacement	3000 tonnes
Crew	50
Max speed	50 kts

Propulsion: CODAG arrangement of two Rolls-Royce MT30 gas turbines and two Fairbanks Morse diesel engines.

Midhip Marine Inc

P.O. Box 125 Harvey, Louisiana 70058, United States

Tel: (+1 504) 341 43 59
Fax: (+1 504) 340 89 97
e-mail: midboats@aol.com
Web: www.midshipmarine.net

Michael Hinojosa, *President*

Midship Marine was established in 1989 and builds custom aluminum vessels for a wide range of customers. The company's founder and president, Michael Hinojosa, has been designing aluminum ships and boats for over 27 years. Midship Marine builds a wide range of aluminum water craft including: passenger vessels and ferries, power catamarans, sailing catamarans, semi-submersibles, crew boats and supply vessels. These vessels range in size from 7 to 60 m in length and can be built from in-house designs or custom built according to the client's specifications. In 2004 and 2005 Midship Marine delivered three 39 m monohull ferries to Mexican operator Ultramar.

Vessel type	Vessel name	Yard No	Length (m)	Speed (kt)	Seats	Vehicles	Originally delivered to	Date of build
39 m Paseenger Ferry	*Ultramar II*	–	39 m	27	450	none	Ultramar	2004
39 m Paseenger Ferry	*Ultramar III*	–	39 m	27	450	none	Ultramar	2004
39 m Paseenger Ferry	*Ultramar IV*	–	39 m	27	450	none	Ultramar	2005

Navatek Ships Ltd

A subsidiary of Pacific Marine
Suite 1880, 841 Bishop Street, Honolulu, Hawaii 96813, United States

Tel: (+1 808) 531 70 01
Fax: (+1 808) 523 76 68
e-mail: pacmar@aloha.net
Web: www.navatekltd.com

Steven Loui, *President*
Gary Shimozono, *Vice President, R&D*
Robert Gornstein, *Chief Scientist*
Michael Schmicker, *Vice President, Marketing and Corporate Communications*

Waverider profile 0536817

Pacific Marine subsidiary Navatek Ltd, a developer of advanced hull forms for the US Navy, conducted research on both high- and low-flying hydrofoil and hybrid hydrofoil concepts, with the programme resulting in a high-speed, low-wake, hybrid hydrofoil design called *Waverider*. Advantages of a hybrid hydrofoil design versus a conventional hydrofoil design include a simpler foil system (one forward foil only) resulting in reduced complexity and greater mechanical reliability as well as reduced construction costs and greater inherent stability.

In 2001 the company finished converting the existing 85 ft, 85 ton high-flying hydrofoil *Westfoil*, built in Westport, Washington US in 1992 into the first prototype *Waverider*. While multi-hull hybrid hydrofoil designs exist, the *Waverider* appears to be the first commercial monohull partial hydrofoil to enter the market. The *Waverider* prototype vessel features a single titanium foil forward, an interceptor ride control system aft and four Detroit Diesel 12V 92 engines coupled to two MJP650 water-jets.

In 2002 sea trials were conducted in waters off Hawaii, with the vessel reaching a speed of 44 kt and demonstrating a superior ride in rough seas. The measured wake wash was low compared to a conventional vessel of the same size and speed. In winter 2002, following commercial outfitting, the prototype *Waverider* vessel was leased to SCX Corporation for use as a demonstration commuter ferry in San Diego, US. The vessel entered service as *Oceanside Wave* in March 2003.

The company has also sea trialled the base *Waverider* design with a Navatek underwater lifting body which significantly increases vessel payload while also allowing the vessel to operate more comfortably in Sea State 4/5. The vessel is currently in service and being operated as a ferry between the US Virgin Islands and the British Virgin Islands by Caribbean Maritime Excursions.

WAVERIDER

Specifications

Length overall	26.0 m
Beam	7.3 m
Draught	1.2 m
Passengers	149
Max speed	44 kt
Operational speed	38 kt
Propulsive power	2,983 kW

Propulsion: Four Detroit Diesel 12V 92 diesel engines driving in pairs to MJP650 water-jets.

Vessel type	Vessel name	Yard No	Length (m)	Speed (kt)	Seats	Vehicles	Originally delivered to	Date of build
Waverider	*Oceanside Wave*	–	26.0	40	149	none	SCX Inc	2001

Nordlund Boat Company

1626 Marine View Drive, Tacoma, Washington 98422, United States

Tel: (+1 253) 627 06 05
Fax: (+1 253) 627 07 85
e-mail: info@nordlundboat.com
Web: www.nordlundboat.com

23 m PILOT VESSEL

This pilot vessel was delivered to Puget Sound Pilots in 2000.

Specifications

Length overall	22.7 m
Beam	6.0 m
Draught	0.9 m
Max speed	26 kt

Propulsion: Two Caterpillar 3412E diesel engines each rated at 671 kW at 2,300 rpm and each driving a Hamilton 571B water-jet via a Reintjes WFS 334 gearbox.

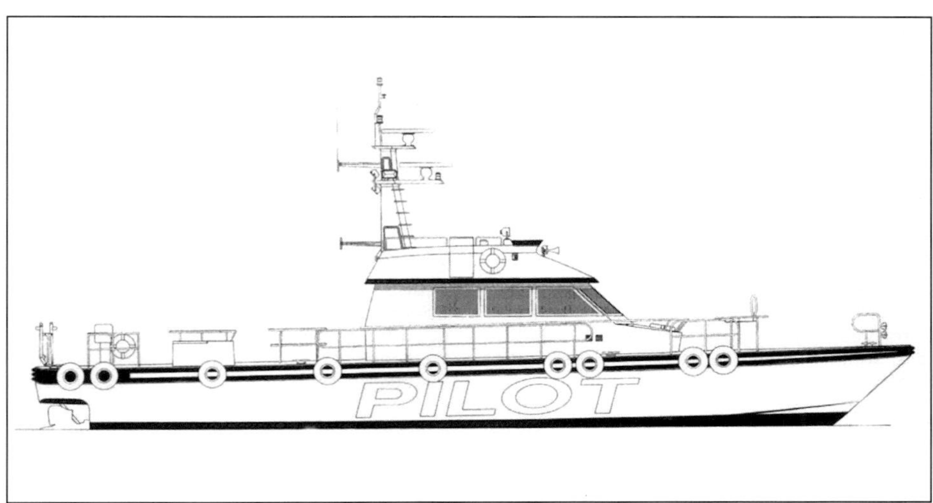

Profile of the 23 m Pilot Vessel 0103298

Vessel type	Vessel name	Yard No	Length (m)	Speed (kt)	Seats	Vehicles	Originally delivered to	Date of build
30 m Motor Yacht	*Shadowfax*	–	30.5	26.0	–	–	–	1993
30 m Motor Yacht	*Southern Way*	–	30.5	20.0	–	–	–	1999
22 m Motor Yacht	*Alexa C*	–	22.3	27.0	–	–	–	1999
27 m Motor Yacht	*Victor E*	–	22.3	27.0	–	–	–	1999
23 m Pilot Vessel	–	–	22.7	–	–	–	Puget Sound Pilots	2000
25 m Motor Yacht	*Huntress*	–	25.3	25.0	–	–	–	2001
32 m Motor Yacht	*lauren C*	–	32.3	21.0	–	–	–	2001
32 m Motor Yacht	*Mixer*	–	32.3	22.0	–	–	–	2002
35 m Motor Yacht	*Alexa C2*	–	34.7	24.0	–	–	–	2003
29 m Motor Yacht	*Phantasma*	–	28.6	25.0	–	–	–	2003
35 m Motor Yacht	*Southern Way*	–	34.7	21.0	–	–	–	2005
33 m Motor Yacht	*Victorious*	–	33.5	25.0	–	–	–	2006
26 m Motor Yacht	*Elvato*	–	26.5	27.0	–	–	–	2006

Seaark Marine Inc

PO Box 210, 404 North Gabbert Street, Monticello, Arkansas 71657, US

Tel: (+1 870) 367 97 55
Fax: (+1 870) 367 21 20
e-mail: sales@seaark.com
Web: www.seaark.com

J McClendon, *President*
K McFalls, *Vice-President Sales*
C Kirby, *Engineering Manager*

SeaArk Marine, formed in 1959 as the Mon-Ark Boat Company, constructs all-welded aluminium craft for patrol, firefighting, rescue, pollution control and passenger transport. Fast passenger ferries up to 27 m in length carrying 149 passengers are currently being marketed.

26 m FAST PASSENGER FERRY

The 26 m craft is offered as a medium-speed, 149-passenger ferry classed under US Coast Guard Subchapter T for operation in protected or partially protected waters. The vessel operates tows on Lake Powell.

Specifications

Length overall	26.0 m
Beam	6.1 m
Crew	5
Fuel capacity	3,200 litres
Water capacity	700 litres
Propulsive power	1,060 kW

SeaArk 26 m fast ferry
0045184

Classification: US Coast Guard Subchapter T.
Structure: All-aluminium.
Propulsion: Powered by two Detroit Diesel 12V92TA engines each rated at 530 kW at 2,100 rpm and each driving a five-blade fixed-pitch propeller via Twin Disc MG 518 reverse reduction (2.5:1) gears.

20 m SURVEY VESSEL

Two of these vessels were delivered to the US Army Corp of Engineers in 2001–02.

Specifications

Length overall	19.8 m
Beam	5.5 m
Draught	1.7 m
Displacement	27.2 tonnes
Fuel capacity	2,000 litres
Maximum speed	24 kt

Propulsion: Two Cummins KTA 19 M4 diesels each rated at 540 kW and each driving a fixed pitch propeller.

Vessel type	Vessel name	Yard No	Length (m)	Speed (kt)	Seats	Vehicles	Originally delivered to	Date of build
26 m Fast Passenger Ferry	*Desert Shadow*	–	26.0	–	149	None	US	–
20 m Survey Vessel	*Teche*	–	19.8	24	–	–	US Army	2001
20 m Survey Vessel	*La Fouche*	–	19.8	24	–	–	US Army	2002

Seacraft Shipyard Corporation

PO Box 1550, 3820 Lake Palourde Road, Amelia, Louisiana 70340-1550, United States

Tel: (+1 985) 631 26 28
Fax: (+1 985) 631 35 13
e-mail: seacraft@seacraftshipyard.com
Web: www.seacraftshipyard.com

Known for conversions and repair of aluminium vessels, Seacraft delivered a 35 m, 30 kt crewboat to Tidewater in 1999.

35 m CREW BOAT

Contender

This vessel was designed as a crew boat and support vessel to be used in the Gulf of Mexico.

Specifications

Length overall	42.7 m
Passengers	50
Fuel capacity	29,400 litres
Water capacity	5,670 litres
Operational speed	25 kt

Classification: US Coast Guard.
Structure: All-aluminium hull and superstructure.
Propulsion: Four Cummins KTA 19-M engines each rated at 520 kW at 2,100 rpm, each driving fixed-pitch propellers via ZF gearboxes.

35 m CREW BOAT

Laurie Tide

Specifications

Length overall	35 m
Passengers	45
Fuel capacity	16,940 litres
Water capacity	15,400 litres
Max speed	30 kt
Operational speed	28 kt

Classification: US Coast Guard.
Structure: All-aluminium hull and superstructure.
Propulsion: Three Cummins KTA 38-M2 engines each rated at 1,000 kW driving fixed-pitch propellers via a Reintjes WVS 430 gearbox.
Electrical System: Two Cummins 4BT3.9 engines coupled to a Newage Stanford 50 kW generator.

Vessel type	Vessel name	Yard No	Length (m)	Speed (kt)	Seats	Vehicles	Originally delivered to	Date of build
35 m Crew Boat	*Laurie Tide*	–	35.0	28.0	45	none	Tidewater	1999
42 m Crew Boat	*Contender*	–	42.7	25	50	none	–	2003

Swiftships Shipbuilders, LLC

PO Box 2869, 1105 Levee Road, Morgan City, Louisiana 70831, United States

Tel: (+1 985) 384 17 00
Fax: (+1 985) 384 09 14
e-mail: swiftships@swiftships.com
Web: www.swiftships.com

Calvin J LeLeux, *President*
Jeffery Perin, *Military Products*
Jeff LeLeux, *Commercial Products and Special Projects*

Swiftships produces a wide range of aluminium craft including crewboats and patrol craft. These craft are in service with US armed forces, naval coast guard and police forces in some 20 countries throughout the world. In 1999, Swiftships and Rodriguez Cantieri Navali of Italy signed a licence agreement for the construction of the Italian fast ferry designs in the US. It is understood that initial construction will focus on the 54 and 76 m designs.

115 ft CREW BOAT

Fast supply/crew boat.

Specifications

Length overall	35.0 m
Beam	7.6 m
Draught	2.2 m
Payload	75.5 t
Crew	6
Passengers	49
Fuel capacity	14,000 litres
Propulsive power	3 × 596 kW
Speed	23 kt
Range	750 n miles

Propulsion: Three MTU 8V 396 TC 82 diesels providing 596 kW at 1,845 rpm each, driving a four-blade bronze propeller via ZF BW 255 2:1 reduction gearboxes and a 101 mm diameter Aquamet 17 shaft.

135 ft CREW BOAT

Specifications

Length overall	41.1 m
Beam	7.9 m
Payload	100 t
Fuel capacity	38,600 litres
Max speed	24 kt

Propulsion: Four MTU 12V 183 TE 72 diesel engines.

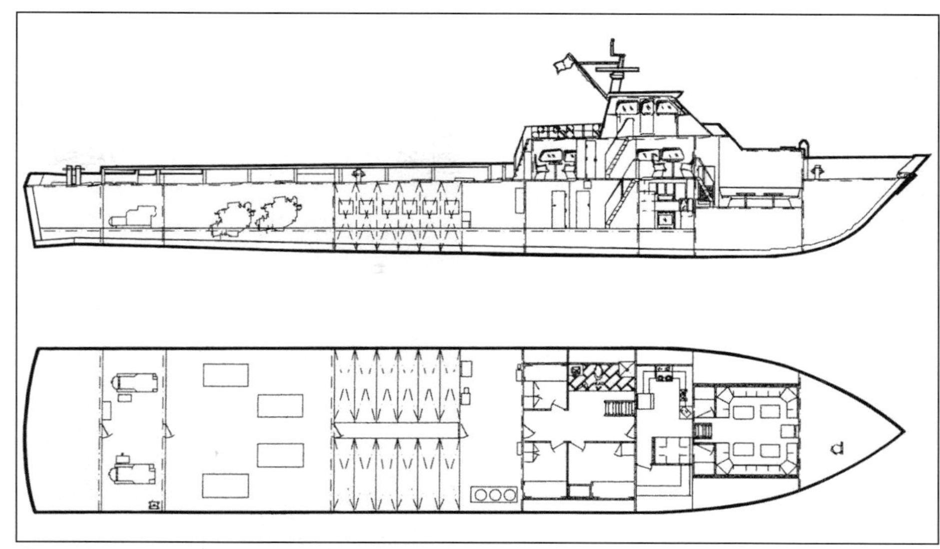

General arrangement of Swiftships 135 ft crew boat 0506913

51.8 m crew boat Paula Kay 0045185

145 ft CREW BOAT

This is a quadruple water-jet propelled all-aluminium crewboat.

Specifications

Length overall	44.2 m
Beam	8.2 m
Payload	122 t
Passengers	79
Fuel capacity	17,200 litres
Water capacity	54,000 litres
Speed	25 kt

Propulsion: Four Caterpillar 3412 DITA diesel engines driving Hamilton HM 571 water-jets.

170 ft CREW BOAT

Paula Kay
Mr Stephen
Miss Julie

A series of these vessels has been constructed, the first delivered in May 1997 and the most recent in 2002.

Specifications

Length overall	51.8 m
Beam	9.1 m
Payload	200 t
Passengers	64
Fuel capacity	67,000 litres
Water capacity	94,000 litres
Speed	31.5 kt

Propulsion: Four V12 Cummins KTA 38 M1 (*Paula Kay*), Four Caterpillar 3412 DITA (*Mr Stephen*) each driving Hamilton HM 571 water-jets.

185 ft CREWBOAT

This is a quadruple water-jet propelled all aluminium crewboat.

Specifications

Length overall	56.4 m
Beam	9.1 m
Crew	5
Passengers	64
Fuel capacity	116,000 litres
Freshwater capacity	108,500 litres
Maximum speed	28 kt

Propulsion: Four Caterpillar 3512 engines each rated at 1,000 kW and each driving a Hamilton HM 651 water-jet.

Vessel type	Vessel name	Yard No	Length (m)	Speed (kt)	Seats	Vehicles	Originally delivered to	Date of build
170 ft Crew Boat	*Paula Kay*	–	51.8	31.5	64	none	–	1997
170 ft Crew Boat	*Mr Stephen*	–	51.8	31.5	64	none	–	1998
115 ft Crewboat	–	–	35.0	23.0	49	none	–	–
135 ft Crew Boat	–	–	41.1	–	–	none	–	–
170 ft Crew Boat	*Miss Julie*	–	51.8	31.5	80	none	Diamond Services	2002
185 ft Crew Boat	–	–	56.4	28	64	none	–	–

Trinity Yachts Inc

13058 Seaway Road, Gulfport, Mississippi, 39503, United States

Tel: (+1 228) 276 10 00
Fax: (+1 228) 276 10 01
e-mail: info@trinityyachts.com
Web: www.trinityyachts.com

Trinity Yachts Inc was established in 1990 and specialises in the design and construction of large motor yachts. In September 2005 the company purchased a second shipyard in Gulfport, Mississippi, effectively doubling the company's capacity. The new yard has 120,000 m² of covered building area and will allow the company to build 8 to 10 mega-yachts per year. The original shipyard in New Orleans is situated on the deepwater Inner Harbour Navigation canal connecting Lake Pontchartrain and the Mississippi river. Large undercover fabrication and outfitting bays allow a number of large yachts to be built concurrently at both yards.

22 m SPORT YACHT

This sport fishing yacht was delivered in 1994.

Specifications

Length overall	22.1 m
Beam	6.0 m
Draught	1.4 m
Fuel capacity	10,500 litres
Water capacity	1,000 litres
Max speed	32 kt
Operational speed	27 kt
Range	750 n miles

Propulsion: Two Caterpillar 3412 diesels.

22 m sport yacht
0506917

Vessel type	Vessel name	Yard No	Length (m)	Speed (kt)	Seats	Vehicles	Originally delivered to	Date of build
22 m Sport Fishing Boat	*Cetacea* (ex-*Contigo*)	–	22.10	27.0	–	none	–	1994
46 m Motor Yacht	*Magic* (ex-*Noble House*)	–	46.00	24.0	–	none	–	1997
48 m Motor Yacht	*Themis* (ex-*Victory Lane*)	–	48.00	24.0	–	none	–	1998
46 m Motor Yacht	*Nova Spirit*	–	46.00	20.0	–	none	–	1999
46 m Motor Yacht	*Utopia III* (ex-*Bellini*)	–	46.00	23.0	–	none	–	1999
38 m Motor Yacht	*Marlena*	–	42.00	31.0	–	none	–	–
46 m Motor Yacht	*My Iris* (ex-*Sea Hawk*)	–	45.70	22.0	–	none	–	2003
49 m Motor Yacht	*Zoom Zoom Zoom*	–	49.07	24.0	–	none	–	2005
47 m Motor Yacht	*Lady Linda*	–	47.90	23.0	–	none	–	2006

VT Halter Marine Inc

A Vision Technologies Systems Company
900 Bayou Casotte Parkway, Pascagoula, Missisippi 39581-2552, United States

Tel: (+1 228) 696 68 88
Fax: (+1 228) 696 68 99
e-mail: vthmsales@vthaltermarine.com
Web: www.vthaltermarine.com

Key personnel

Sid Mizell, *Senior Vice President, Business Development*

More than 2,600 vessels have been built at VT Halter Marine facilities. The company has three shipyards located in Pascagoula, Moss Point and Escatawpa, Mississippi.

The following craft descriptions represent some of the principal types of high-speed craft designed and built at VT Halter Marine Inc facilities, having lengths of approximately 20 m or over with speeds over 20 kt.

HALMAR 65 (CREW BOAT TYPE)

This is a crew boat design but is also available in the following variants: customs

Halmar 101 Fraih Al-Fraih
0506989

Vessel type	Vessel name	Yard No	Length (m)	Speed (kt)	Seats	Vehicles	Originally delivered to	Date of build
Halmar 65 Crew Boat	*Series production (120)*	–	19.6	19.0	–	none	–	Series production (1964)
Halmar 101 Crew Boat	*Series production (38)*	–	31.0	20.0	55	none	–	Series production (1978)
34 m Crewboat	*Hebah*	–	33.7	–	25	none	Marine and Transport Services, Saudi Arabia	1988
34 m Crewboat	*Daina*	–	33.7	–	25	none	Marine and Transport Services, Saudi Arabia	1989
34 m Crewboat	*Marwa*	–	33.7	–	25	none	Marine and Transport Services, Saudi Arabia	1989

launch, tanker service launch, pilot boat, ferry, communications launch, workboat and ambulance launch. The first of this type was designed and built in 1964 and over 120 of these craft have now been built. The craft have been classified by various authorities including USCG, ABS and Lloyd's. Most are in service with the offshore oil industry.

Specifications

Length overall	19.6 m
Beam	5.9 m
Draught	1.1 m
Crew	6
Fuel capacity	3,596 litres
Propulsive power	380 kW
Max speed	21 kt (aluminium hull)
–	18 kt (steel hull)
Operational speed	19 kt (aluminium hull)
–	16 kt (steel hull)
Operational limitation	Sea State 3–4

Structure: Corten steel, longitudinally framed; aluminium 5086, transverse framed.
Propulsion: Various options, the most popular being two GM 12V-71 TI, producing 380 kW each at 2,100 rpm, driving two Columbian Bronze propellers, 813 mm diameter, in Nibral.
Electrical system: 110/220 V AC, 60 Hz, generator 2-71 GM, 20 kW.
Control: Morse MD-24, manual, two stations. Steering hydraulic, 600 lb/sq in.
Outfit: 49 passengers and two crew (one captain, one deckhand/engineer).

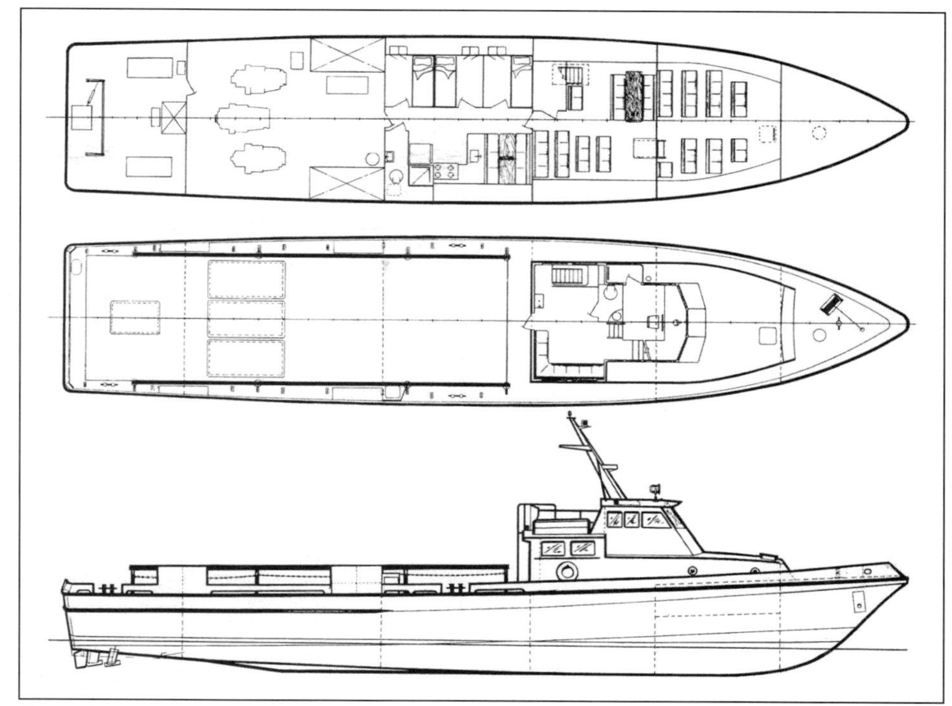

General arrangement of Halmar 101 0506912

HALMAR 101 (CREW BOAT TYPE)

A crew boat design available in a full range of variants. Designed and first built during 1977 to 1978, 38 had been delivered by December 1985, classified either by USCG or ABS. The latest vessel, *Fraih Al Fraih*, was delivered to Kuwait in 1995 for security services for offshore oil rigs.

Specifications

Length overall	31.0 m
Beam	6.5 m
Draught	1.7 m
Displacement, min	55.9 t
Payload	30.5 t
Crew	6
Passengers	55
Fuel capacity	9,092 litres
Propulsive power	3 × 380 kW
Max speed	22 kt
Operational speed	20 kt
Operational limitation	Sea State 5
Range	500 n miles

Structure: Aluminium 5086, transverse framed.
Propulsion: Main engines are three GM 12V-71 TI, 380 kW each at 2,100 rpm, driving via Twin Disc 2:1 reduction gears to three Columbian Bronze propellers, 864 mm diameter, in Nibral (one vessel is fitted with three Rocketdyne water-jet units, two 406.4 mm diameter driven by two GM 16V-92TI diesel engines and one 609.6 mm diameter driven by one Allison 501 gas turbine).
Electrical system: Two generators, GM 3-71, 30 kW each 110/220 V AC, 60 Hz.

The ATCO Hebah, *25 passenger, 25 tonne cargo, crew boat built by Halter Marine Inc* 0506709

Navigation and communications: VHF, Sailor 144, SSB, Motorola, Radar, Decca D-150 36 mile range.
Control: Kobelt, pneumatic, two stations.

34 m CREWBOAT

Hebah
Marwa
Daina

Three of these all-aluminium crew boats were built for Marine and Transportation Services, Saudi (Ltd) Damman, Saudi Arabia, and delivered in 1988 and 1989.

Specifications

Length overall	33.7 m
Beam	6.6 m
Passengers	25
Fuel capacity	12,870 litres
Water capacity	3,785 litres

Propulsion: Main engines are Detroit Diesel 12V-71TI, 380 kW each, at 2,100 rpm; driving through Twin Disc MG 514, ratio 2:1.

Westport Shipyard Inc

1807 Nyhus Street, PO Box 308, Westport, Washington 98595, United States

Tel: (+1 954) 316 6364
Fax: (+1 954) 316 6365
e-mail: info@westportyachtsales.com
Web: www.westportshipyard.com

Randy Rust, *General Manager*

Westport Shipyard was established in 1964 and sold to the present owners in 1977. The yard has a workforce of 170 and consists of 11,100 m² (120,000 sq ft) covered building

Kenai Explorer 0506919

sheds and workshops, plus four acres of fenced storage space. The range of craft built is from 10.4 to 36.6 m in length. Hulls of the larger craft are built with a sandwich-type construction of Airex PVC and GRP, which the builder claims yields a tougher and lighter hull than an all-GRP type. Hull moulds are adjustable for length and beam enabling craft to be built in the following length ranges: 16.15 to 19.8 m, 19.8 to 29 m and 27.4 to 36.6 m. The company has built over 50 vessels to USCG 'S' classification since 1977.

Westship Lady

Designed by Jack Sarin, this vessel was delivered in early 1993. Five other vessels of this design have been delivered: *Norwegian Queen, Lady Lee, Rain Maker, Primadonna* and *Black Sheep. Black Sheep* was 2 m longer overall.

Specifications

Length overall	32.3 m
Beam	6.9 m
Draught	1.7 m
Displacement, max	95 t
Crew	14
Fuel capacity	18,950 litres
Water capacity	2,840 litres
Max speed	28 kt
Operational speed	25 kt
Range	2,000 n miles

Structure: The hull is FRP/Airex core and the superstructure is FRP/composite.

Propulsion: Main engines are two MTU 8V 396TE 94.

Kenai Explorer

Delivered Spring 1993, *Kenai Explorer* operates as a tour vessel in the Kenai Fjords area of Alaska and is certified to operate within 20 miles of a harbour of safe refuge. The hull is built with Airex core and fire-retardant resin with plywood bulkheads and stringers of glass fibre and foam. The superstructure is of glass fibre and core-type construction.

Specifications

Length overall	27.4 m
Beam	6.7 m
Passengers	149
Fuel capacity	11,355 litres
Water capacity	1,135 litres

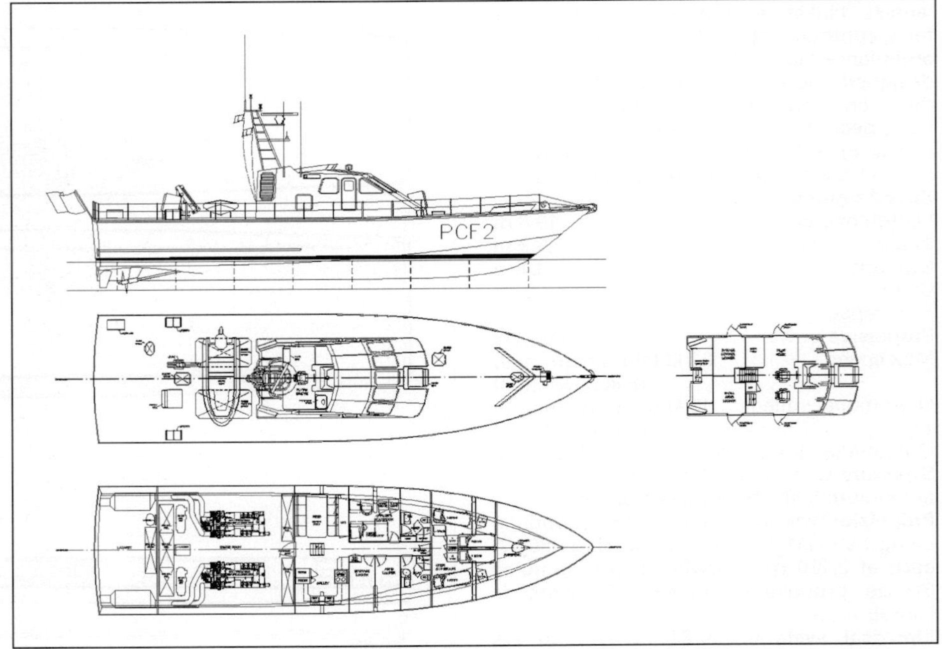

General arrangement of the Guardian 85 patrol craft 0100702

Catalina Express 0506918

Vessel type	Vessel name	Yard No	Length (m)	Speed (kt)	Seats	Vehicles	Originally delivered to	Date of build
32.7 m Motor Yacht	*Northstream*	–	32.7	22.0	–	None	–	1990
Westship 95	*Tahiti*	–	29.1	18.0	10	None	–	1991
19.8 m Motor Vessel	*Bluefin*	–	19.8	–	9	None	California Fish and Game	1991
24.4 m Passenger Ferry	*Chugach*	–	24.4	18.0	149	None	Stan Stephens Charters, US	1992
Westship 98	*Lady Kathryn*	–	29.9	19.0	10	None	–	1992
32 m Motor Yacht	*Westship Lady*	–	32.3	25.0	–	None	–	1993
32 m Motor Yacht	*Rain Maker*	–	32.3	25.0	–	None	–	1993
27.4 m Tour Vessel	*Kenai Explorer*	–	27.4	26.0	149	None	US	1993
32 m Motor Yacht	*Lady Lee*	–	32.3	25.0	–	None	–	1993
32 m Motor Yacht	*Primadonna*	–	32.3	25.0	–	None	–	1993
32 m Motor Yacht	*Norwegian Queen*	–	32.3	25.0	–	None	–	1993
32 m Motor Yacht	*Black Sheep*	–	34.3	25.0	–	None	–	1993
Westport 95	*Catalina Express*	–	30.0	32.0	149	None	Catalina Express Lines	1994
Westport 95	*Islander Express*	–	30.0	32.0	149	None	Catalina Express Lines	1994
Westport 95	*Alaskan Explorer*	–	30.0	28.0	149	None	Kenai Fjord Tours	1995
Westport 95	*Glacier Explorer*	–	30.0	28.0	149	None	Kenai Fjord Tours	1996
Westport 95	*Nunatak*	–	30.0	28.0	149	None	Kenai Fjord Tours	1996
Westport 108	–	–	32.9	26.0	–	None	–	1997
Westport 95	*Coastal Explorer*	–	30.0	28.0	149	None	Kenai Fjord Tours	1998
Westport 95	*Admiralty Wind*	–	30.0	28.0	149	None	Four Seasons Marine	1998
Westport 95	*Tanaina*	–	30.0	28.0	149	None	Prince William Sound Cruises and Tours	1999
Westport 112	–	–	34.1	26.0	–	None	–	1999
Westport 112	–	–	34.1	26.0	–	None	–	1999
Westport 112	–	–	34.1	26.0	–	None	–	1999
Westport 112	–	–	34.1	26.0	–	None	–	1999
Westport 112	–	–	34.1	26.0	–	None	–	1999
Guardian 85 Patrol Craft	–	–	25.9	–	–	None	–	2000
29.9 m Motor Yacht	*Heritage*	–	29.9	22.5	–	None	–	–
Westport 130	*Design*	–	39.6	20.0	–	None	–	–
Westport 164	–	–	50.0	24.0	–	none	–	2006

Propulsive power	2 × 969 kW
Operational speed	26 kt

Propulsion: Main engines are two DDC 16V 92TA diesels, 969 kW each at 2,300 rpm.

Chugach

Delivered in May 1992 this vessel, designed by Jack Sarin, is owned and operated by Stan Stephens Charters, Valdez, Alaska.

Specifications

Length overall	24.4 m
Beam	6.4 m
Draught	1.5 m
Passengers	149
Fuel capacity	7,570 litres
Water capacity	416 litres
Propulsive power	626 kW
Max speed	22.6 kt
Operational speed	18 kt

Propulsion: Main engines are two Lugger L6170A diesels, 626 kW each at 2,100 rpm.

WESTSHIP 98

Lady Kathryn

Designed by Jack Sarin, this vessel was delivered at the end of 1992.

Specifications

Length overall	29.9 m
Beam	6.9 m
Draught	1.7 m
Displacement, min	74.8 t
Passengers	10
Fuel capacity	17,000 litres
Water capacity	2,840 litres
Max speed	23 kt
Operational speed	19 kt
Range	1,720 n miles

Structure: Hull: FRP/Airex core.
Superstructure: FRP/composite.
Propulsion: Main engines are DDC 12V 92.

Northstream

Delivered in 1990, this Sports fisherman type was designed by Jack Sarin, with acoustic engineering by Joe Smullin.

Specifications

Length overall	32.7 m
Beam	7.3 m
Displacement, min	80 t
Displacement, max	93 t
Fuel capacity	22,330 litres
Max speed	27 kt
Operational speed	22 kt

Structure: Hull: double-cored Airex bottom, single-cored Airex sides; superstructure: composite with Airex core.
Propulsion: Main engines are two DDC 16V 92TA with DDEC system.
Control: Stabilisers: Naiad 301s with 0.56 m^2 fins. Bow thruster: Wesmar 12.2 kW.

Heritage

An ocean-going motor yacht, built in advanced plastics. The hull and all structural bulkheads, lower soles, exterior decks and superstructure are constructed in glass fibre composite.

Specifications

Length overall	29.9 m
Beam	6.7 m
Displacement, min	65.3 t
Displacement, max	83 t
Propulsive power	2 × 805 kW
Max speed	26 kt
Operational speed	22.5 kt

Propulsion: Main engines are two DDC 12V 92TAB, 805 kW each at 2,300 rpm, driving two Michigan Wheel Dynaquad, four-blade, 1,067 × 91.4 mm Nibral Alloy propellers.

Tahiti

Delivered early 1991, this is the second in the Westship Series and was designed by Jack Sarin.

Specifications

Length overall	29.1 m
Beam	6.9 m
Draught	1.7 m
Displacement	74.8 t

Northstream

0506712

Primadonna 0507080

Lady Kathryn 0506713

Westport 30 m Ferry Tanaina 0081018

Passengers	10
Fuel capacity	13,250 litres
Water capacity	2,840 litres
Max speed	24 kt
Operational speed	18 kt

Structure: The hull is FRP/Airex core, and the superstructure is FRP/composite.
Propulsion: Main engines are Caterpillar 3412.

Bluefin
This boat was delivered to California Fish and Game in 1991.
Specifications

Length overall	19.8 m
Beam	5.8 m
Passengers	9
Fuel capacity	9,463 litres
Water capacity	1,135 litres
Max speed	30 kt

Structure: The hull is FRP/Airex core, and the superstructure is wood/glass fibre.
Propulsion: Main engines are two DDC 12V 92TAB.

30 m FERRY

Catalina Express
Islander Express
Delivered in 1994, these two vessels operate off the coast of Southern California for Catalina Express Lines. Licensed for coastwise service (20 miles from harbour of safe refuge), the boats will carry 149 passengers at speeds of up to 32 kt.

The hullform features modified propeller tunnels for increased propeller efficiency.
Specifications

Length overall	30.0 m
Beam	6.7 m
Passengers	149
Fuel capacity	9,500 litres
Water capacity	418 litres
Propulsive power	2 × 1,491 kW
Operational speed	32 kt

Structure: The vessels are constructed from Fibre-Reinforced Plastic (FRP) sandwich with an Airex core and fire-retardant resin. The vessels were the first to be built in a new Westport adjustable hull mould.
Propulsion: Main engines are two 16V 149TIB DDEC Detroit diesels, driving propellers. (Re-engineered in 2002 with two MTU 12V 4000 M70 diesels and Twin Disc gearboxes).
Electrical system: The electrical power is supplied by two Northern Lights generators.

Alaskan Explorer
Glacier Explorer
Nunatak
Coastal Explorer
Admiralty Wind
Tanaina
Alaskan Explorer was delivered to Kenai Fjords Tours in 1995. The vessel is based in Seward, Alaska, for day tours to the scenic Kenai Fjords national park area.

The hull form incorporates propeller tunnels, and was designed by Jack Sarin based on Westport's 8500 series high-speed hull mould.

Glacier Explorer and *Nunatak* were both delivered in mid-1996 and *Coastal Explorer* and *Admiralty Wind* in 1998. *Tanaina* was delivered to Prince William Sound Cruises and Tours in 1999.
Specifications

Length overall	30.0 m
Beam	6.7 m
Passengers	149
Fuel capacity	11,360 litres
Water capacity	1,135 litres
Propulsive power	2 × 1,305 kW
Operational speed	28 kt

Propulsion: Main engines are two Caterpillar 3512 DITA diesels, each producing 1,305 kW at 1,800 rpm; driving propellers via a ZF460 gearbox. *Admiralty Wind* and *Tanaina* use Caterpillar 3518B engines with ZF BW465 gearboxes.

General arrangement of the Westport 130 0100701

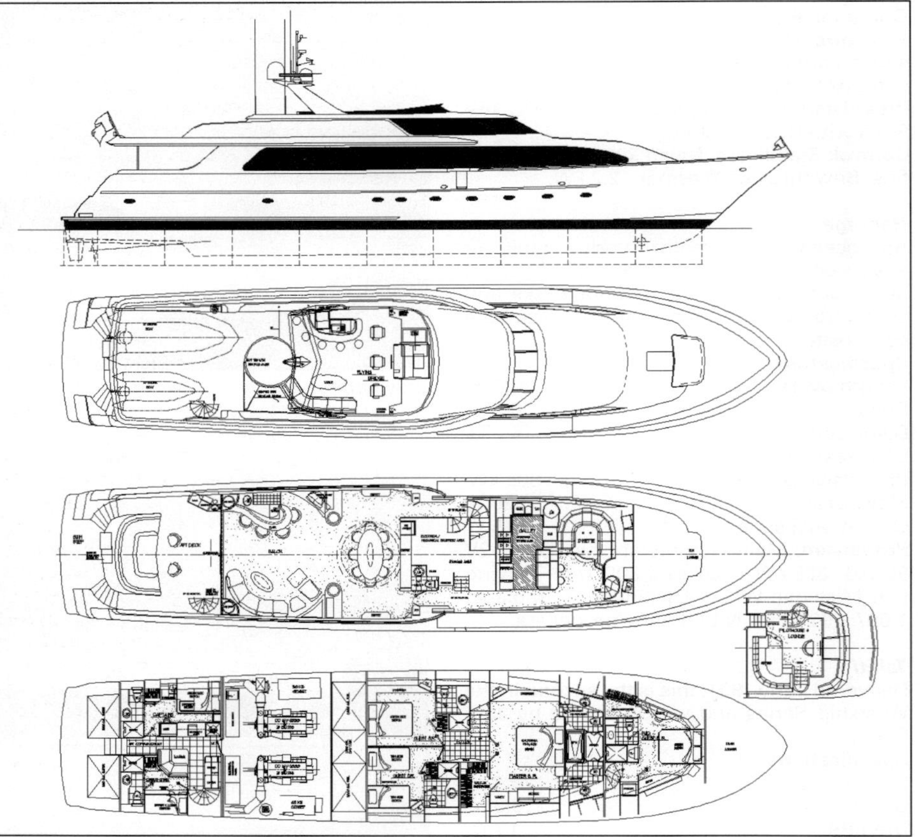

General arrangement of the Westport 108 0100699

Electrical system: Power is supplied by 2 Northern Lights generators, 40 kW and 20 kW.

WESTPORT 108

The first of this new class of fast motor yacht was delivered in 1997.

Specifications

Length overall	32.9 m
Length waterline	29.0 m
Beam	7.2 m
Draught	1.7 m
Displacement, max	113.4 t
Fuel capacity	20,000 litres
Water capacity	4,000 litres
Propulsive power	2,685 kW
Max speed	28 kt
Operational speed	26 kt
Range	2,500 n miles

Propulsion: Powered by two MTU 16V 2000 diesel engines each driving a five-blade fixed-pitch propeller rated at 1,340 kW.

WESTPORT 112

Five of these vessels have been delivered since 1998.

Specifications

Length overall	34.1 m
Length waterline	29.9 m
Beam	7.2 m
Draught	1.7 m
Displacement, max	131 t
Fuel capacity	20,800 litres
Water capacity	5,000 litres
Propulsive power	2,684 kW
Max speed	28 kt
Operational speed	26 kt
Range	2,500 n miles

Structure: GRP sandwich.

Propulsion: Two MTU/DDC 16V 2000 diesel engines each rated at 1,342 kW and each driving a five-bladed fixed-pitch propeller.

WESTPORT 130

Specifications

Length overall	39.6 m
Length waterline	34.7 m
Beam	8.2 m
Draught	1.8 m
Displacement, max	216 t
Fuel capacity	37,500 litres
Water capacity	7,000 litres
Propulsive power	4,080 kW
Max speed	24 kt
Operational speed	20 kt
Range	2,300 n miles

Structure: GRP sandwich.

Propulsion: Two MTU/DDC 12V 4000 diesel engines each rated at 2,040 kW and each driving a five-bladed fixed-pitch propeller.

GUARDIAN 85 PATROL CRAFT

Designed by Jack Sarin in 1998 the first vessel was delivered in 2000.

Specifications

Length overall	25.9 m
Length waterline	22.4 m
Beam	7.0 m
Draught	1.0 m
Fuel capacity	13,250 litres
Water capacity	1,900 litres
Propulsive power	3,480 kW

Structure: GRP.

Propulsion: Two MTU/DDC 12V 4000 diesel engines each rated at 1,740 kW and each driving a three-bladed fixed-pitch propeller.

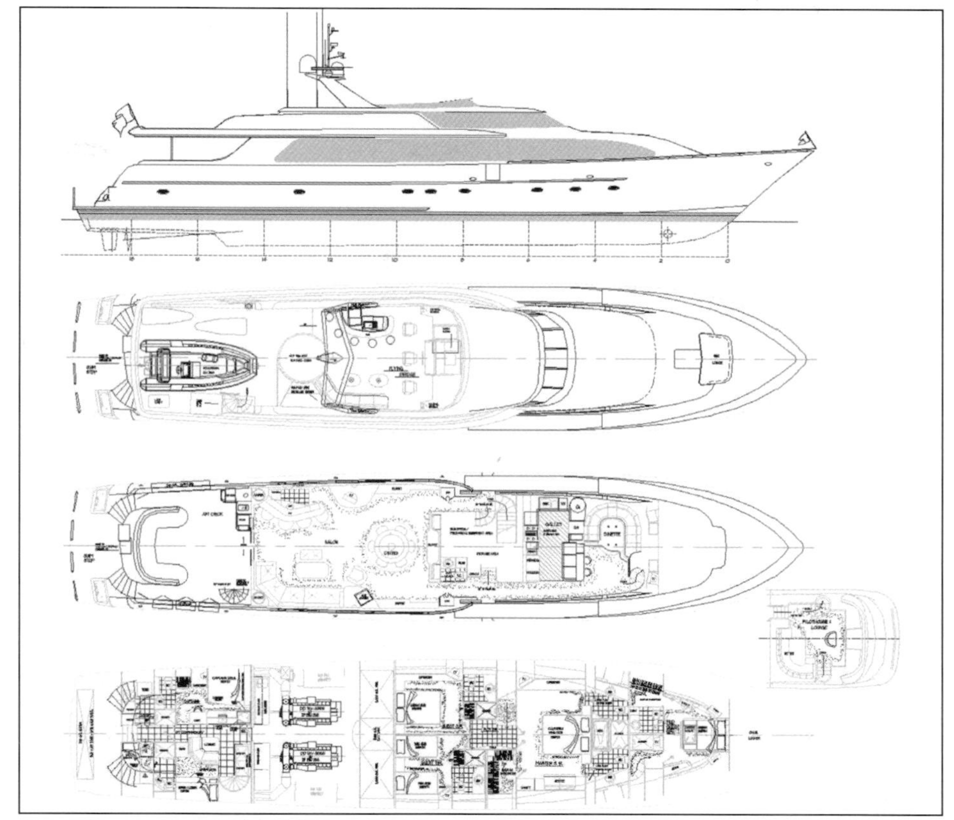

General arrangement of the Westport 112 0100700

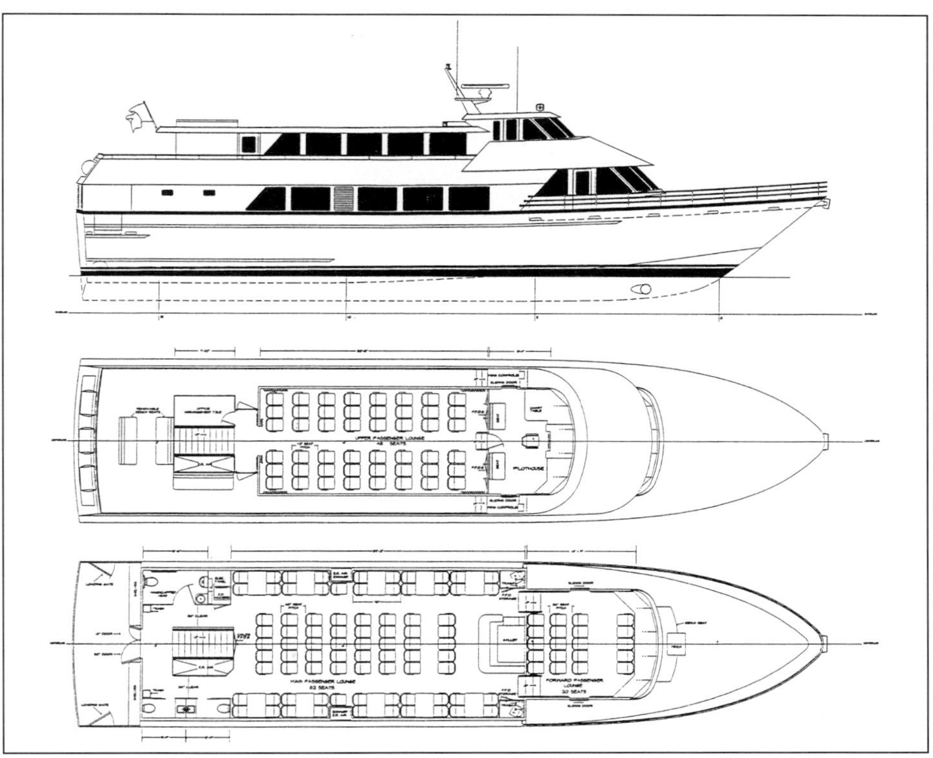

General arrangement of the Westport 95 0104154

Admiralty Wind 0045186

WING-IN-GROUND-EFFECT CRAFT

Company listing by country

China
CSSRC
MARIC

Germany
Botec Ingenieursozietät GmbH
Flyship GmbH
RFB
MTE GmbH

Japan
Tokushima Buri University
Tottori University

Korea, South
Hanjin Heavy Industries

Russian Federation
Alsin
Arctic Trade and Transport
Central Hydrofoil Design Bureau

Ukraine
Feodosia Shipbuilding Company (MORYE)

United States
Boeing Integrated Defense Systems

China

China Ship Scientific Research Centre–CSSRC

PO Box 116, Wuxi, Jiangsu, 214082 China

Tel: (+86) 555 53 17
Fax: (+86) 555 51 93
Web: www.cssrc.com

Guo-Jun Zhou, *Head of International Technical Co-operation*

RAM-WING VEHICLE 902

The research plan behind the development of the 902 began in 1979. A programme of theoretical and wind tunnel experimental research was carried out, which provided solutions for the particular problem of longitudinal stability encountered by such craft, when operating in and out of ground-effect. The 902 was built in 1983 and first flew on 12 November 1984.

Take-off can be achieved in waves up to 0.4 or 0.5 m with winds of Beaufort 6 to 7. The craft can be flown in and out of ground-effect and in the surface-effect region of 0.6 to 8 m altitude, it can fly safely without any automatic stabilising device.

Specifications

Length overall	9.5 m
Span	5.8 m
Weight, min	0.3 t
Weight, max	0.4 t
Payload	0.1 t
Take-off distance	150 m
Take-off speed	40.5 kt
Operational speed	65 kt
Operational limitation	wave height 0.5 m Beaufort 6–7

Propulsion: Main engines are two HS-350A, 15 kW each, aircraft piston engines, driving two fixed-pitch, aircraft-type propellers.

XTW-1

Following the success of the 902 a larger ram-wing craft was built, the XTW-1, for lake and river communications duties. This craft has a weight of 950 kg and is powered by two 30 kW piston engines giving it a cruise speed of 70 kt, with four people on board. A retractable undercarriage is incorporated for slipway handling.

XTW-2

With the experience gained from the 902 and XTW-1, a much larger project was embarked upon in 1988. The XTW-2 was built in 1990 and performance testing was completed in 1992. Operations commenced in 1993. The main centre hull is fitted with a passenger cabin with 14 seats.

Specifications

Length overall	18.5 m
Span	12.7 m
Weight, max	3.6 t
Payload	1.2 t
Propulsive power	440 kW (take-off)
Operational speed	130–150 kt
Flight altitude	1.0–1.5 m
Range	485 n miles
Operational limitation	Sea State 3

Propulsion: Main engines are two IO-540 K1B5 aircraft piston engines.

XTW-3

XTW-3 is a new model of the XTW series. Using aluminium for the main hull structure,

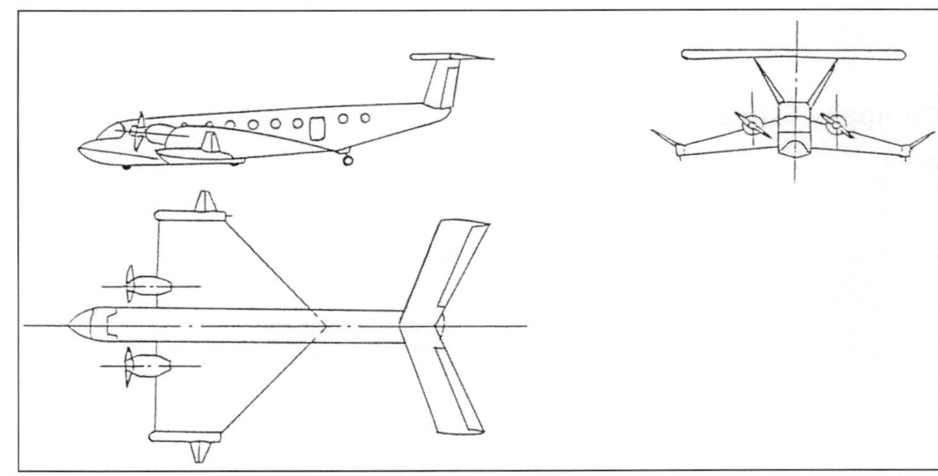

Wing-in-ground-effect craft built by China Ship Scientific Research Centre (XTW-2) 0506996

XTW-2 in flight over the China Sea 0506997

XTW-3 during performance testing 0024766

Model tests of the SDJ 1 at the long tank of CSSRC 0038713

Vessel type	Vessel name	Yard No	Length (m)	Speed (kt)	Seats	Vehicles	Originally delivered to	Date of build
Ram Wing Vehicle	902	–	9.5	65.0	–	None	–	November 1984
XTW-1	–	–	–	70.0	4	None	–	1986
XTW-2	–	–	18.5	130.0	14	None	China	1990
XTW-3	–	–	17.9	77.8	–	None	–	1996
XTW-4	–	–	21.7	81.0	–	None	–	September 1999
SDJ 1 catamaran WIG	Prototype	–	14.5	65.0	6	None	–	2000

the craft was built in 1996 and carried out performance testing in early 1997.

Specifications

Length overall	17.9 m
Span	11.8 m
Weight, max	4.0 t
Payload	0.92 t
Propulsive power	440 kW (take-off)
Operational speed	78 kt
Flight altitude	1.5–2.0 m
Range	200 n miles
Operational limitation	Sea State 3

Propulsion: Main engines are two IO-540 K1B5 aircraft piston engines.

XTW-4

XTW-4 is a new model of the XTW series. Using aluminium alloy for the main hull structure, this construction was completed in September 1999 and the river trials had been successfully completed by 2000.

Specifications

Length overall	21.7 m
Span	14.5 m
Weight, max	6.0 t
Payload	1.8 t
Propulsive power (take-off)	1,014 kW
Operational speed	81 kt
Flight altitude	0.1–2.0 m
Range	162 to 270 n miles
Operational limitation	Sea State 3

Propulsion: Two PT6A–15AG turbo propeller engines.

XTW-4 in cruising flight 0077022

SDJ 1 (PROTOTYPE)

The SDJ 1 is a WIG of a catamaran configuration. Model tests for this design were undertaken at CSSRC in 1997 and a six-seat prototype craft is currently under construction for completion in 2000.

Specifications

Length overall	14.5 m
Beam	12.4 m
Weight, max	2.0 t
Crew	1
Passengers	6
Propulsive power	220 kW
Operational speed	65 kt
Range	130 n miles
Operational limit	Sea State 3

Maric

Marine Design and Research Institute of China
1688 Xizhangnan Road, Shanghai 200011, People's Republic of China

Tel: (+86 21) 63 78 80 98
Fax: (+86 21) 63 77 97 44
Web: www.maric.com.cn

Mrs Hu Ankang, *President*
Liang Qikank, *Director*
Yun Liang, *Deputy Chief Naval Architect*

MARIC PAR WIG Type 750 coming ashore 0506714

PAR WIG TYPE 750

This two-seat experimental Power-Augmented Ram-Wing-In-Ground-effect (PAR WIG) craft was completed in 1985, and over the period March 1985 to June 1987 over 40 trial flights were performed, exploring various operating conditions.

Over-water behaviour: can take-off and fly steadily in wave heights of 0.5 to 0.7 m with Beaufort 4 to 5 and gust 6 wind conditions. In horizontal flight, average pitch is 0.3° and roll, 0.53°.

Specifications

Length overall	8.5 m
Span	4.8 m
Weight, max	0.8 t
Payload	0.2 t
Propulsive power	88 kW
Take-off distance	160 m
Take-off speed	24 kt
Max speed	71 kt
Flight altitude	0.5 m
Range	70 n miles

Structure: Built in GRP with aircraft construction techniques.

Propulsion: Four 22 kW piston engines driving four ducted propellers Type DT-30.

Swan *landing* 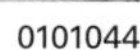 0101044

PAR WIG TYPE 751 (AF-1)

This 20-passenger PAR WIG is understood to have been completed in 1995.

Swan *in flight* 0101045

DACWIG TYPE (SWAN)

A Dynamic wing-in-ground-effect craft (DACWIG), also known in China as an

Vessel type	Vessel name	Yard No	Length (m)	Speed (kt)	Seats	Vehicles	Originally delivered to	Date of build
PAR WIG Type 750	*Experimental craft*	–	8.5	–	–	None	–	March 1985
PAR WIG Type 751	*AF-1*	–	–	–	20	None	China	1995
DACWIG	*Swan*	–	19.4	70.0	20	None	–	–

amphibious wing-in-ground-effect craft (AWIG) has been recently developed. The vessel, *Swan*, can operate, take off and land safely in sea state 3.

Specifications

Length overall	19.4 m
Span	13.4 m
Weight, max	8 t
Passengers	20
Operational speed	70 kt
Range	300 n miles
Operational limit	Sea state 3

Structure: Built in GRP with aircraft construction techniques.

Propulsion: Four 22 kW piston engines driving four ducted propellers Type DT-30.

Germany

Botec Ingenieursozietät GmbH

Odenwaldring 24, D-64401 Gross-Bieberau, Germany

Tel: (+49 61) 62 10 13
Fax: (+49 61) 62 10 14
e-mail: kontakt@airfoil.de
Web: www.airfoil.de

Ingrid Schellhaas, *Chairman*
Reiner A Jörg, *Technical Director*

The first Günther Jörg ram-wing, a two-seater powered by a modified Volkswagen engine, was designed in 1973 and first flew in 1974. After an extensive test programme, Jörg II was designed, performing its first flight in 1976. This incorporated more than 25 design improvements and has travelled more than 10,000 km. During 1978 and 1979, Jörg designed a glass fibre-hulled four- to six-seater Jörg III. The first prototype was completed in 1980. Designs are being prepared for larger and faster craft capable of carrying heavier loads over greater distances.

Stabilisation about the pitch and roll axes, and maintaining the flying height of the Aerodynamic Ground Effect Craft (AGEC) can be regulated independently by the ground/water surface reaction and therefore requires only a simple steering control for movement around the vertical axis. This basic requirement led to the development of a tandem-wing configuration, the handling characteristics of which were first tested on models and then on two-seat research craft. Tests showed that the tandem wings had good aerodynamic qualities, even above turbulent water, with or without water contact.

AGECs have ample buoyancy and are seaworthy, even while operating at low speed (cruising in displacement condition). After increasing speed, the craft lifts off the surface of the water and the resulting air cushion reduces the effects of the waves. After losing water contact the boat starts its ram-wing flight at a height of 4 to 8 per cent of the profile depth. A two-seater with a wing span of 3.2 m and a profile length of 3.1 m will have a flying height above the water surface of up to 0.9 m at a speed of 140 km/h. The low-power requirement of this type of craft is achieved through the improved lift-to-drag ratio of a wing-in-ground-effect as compared to free flight.

A craft of more than 300 tonnes would have a wing chord of 36 m and fly under ground-effect conditions between 3.5 and 7 m. It would have a wing span of 40 m and a length of 100 m, requiring an engine thrust of 40,000 kW. Cruising speed would be 250 km/h from 7,350 kW, and craft of this

TAF VIII-2 in service in Asia 0093267

One-in-ten scale model of Flairboat TAF VIII-4 under test 0058913

TAF VIII-4S during ground test 0506715

Vessel type	Vessel name	Yard No	Length (m)	Speed (kt)	Seats	Vehicles	Originally delivered to	Date of build
Flairboat TAF VIII-3	–	–	14.0	76.0	–	None	–	August 1991
TAF VIII-4S	*Jorg 5*	–	17.2	90.0	5	None	–	September 1986
Flairboat TAF VIII-4	–	–	19.8	95.0	14	None	–	1992
Flairship TAF VIII-5/25	–	–	25.9	90.0	25	None	–	1999
Flairship TAF VIII-7/3	*DESIGN*	–	45.7	–	80	None	–	–
Flairship TAF VIII-10	*DESIGN*	–	70.0	100.0	400	None	–	–
Flairboat TAF VIII-2B	–	–	12.65	94.0	–	None	–	(2006)

size could travel all year round over 90 per cent of the world's sea areas.

The following operational possibilities are foreseen:

Inland waterways and offshore areas: Suitable for rescue boats, customs, police, coastguard units, patrol boats, high-speed ferries, leisure craft.

Coastal traffic: Large craft would be operated as fast passenger and passenger/car ferries and mixed traffic freighters.

Overland: Suitable for crossing swamps, flat sandy areas, snow and ice regions. Another possible application is as a high-speed tracked skimmer operating in conjunction with a guide carriage above a monorail.

The basic advantage of the AGEC is its ability to transport passengers and freight far quicker than conventional ship, rail or road transport. The ratio between empty weight and service load is approximately 2:1 but can certainly be improved. Fuel consumption is 80 to 85 per cent lower than that of a boat of similar construction. A high degree of ride comfort is achieved with these craft.

The Airfoil-Flairboats of Jörg design are designed in accordance with international marine standards.

Flairboat Taf-VIII-2B

Intended for coastguard and rescue units, the TAF-VIII-2B is currently under construction.

Specifications

Length overall	12.65 m
Span	5.6 m
Weight, max	2.3 t
Payload	1 tonne
Operational speed	94 kt
Maximum speed	108 kt
Range	550 n miles

JÖRG V

TAF VIII-4S

This new Jörg craft underwent its first ground tests in September 1986. Open sea tests in the North Sea started in September 1987.

Specifications

Length overall	17.2 m
Weight, max	3.5 t
Passengers	5–8
Propulsive power	485 kW
Operational speed	45–90 kt
Flight altitude	0.6 m
Range	269 n miles

Propulsion: Main engine is one 485 kW Textron Lycoming LTX 101 turboshaft with water-methanol injection available for 10 per cent power increase at Take-off. One diesel engine driving a retractable marine propeller, air propeller is a MT-P four-blade of 2.9 m diameter, cable/hydraulically controlled; via nine V-belts transmissions (Optibelt, Höxter, Germany).

FLAIRBOAT TAF VIII-4

Intended for river and inter-island traffic. The first TAF VIII-4 was designed in 1992.

Specifications

Length overall	19.8 m
Span	8.5 m
Draught, hullborne	0.3 m
Weight, max	8.2 t
Payload	1.5 t
Crew	2
Passengers	13–14
Propulsive power	1,100 kW
Max speed	108 kt
Operational speed	95 kt
Range	400 n miles

Structure: Aluminium and composite plastics.

Propulsion: One turbo prop pusher engine at 1,100 kW.

Jörg TAF VIII-3 with engine cover raised 0506716

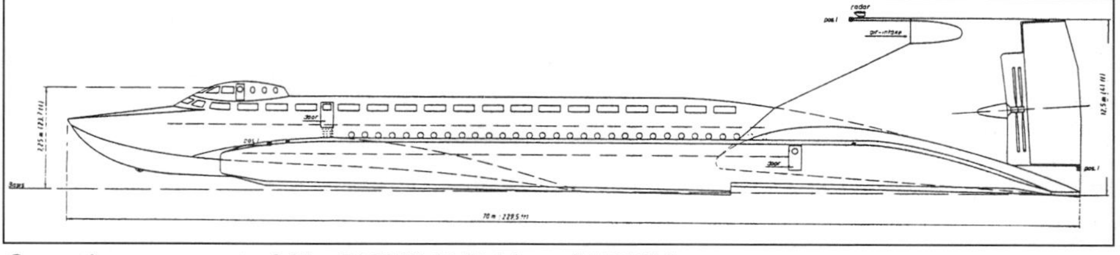

General arrangement of Jörg TAF VIII-10 Flairboat (DESIGN) 0058914

An early Jörg Airfoil Flairboat TAF VII-5 at high speed 0069455

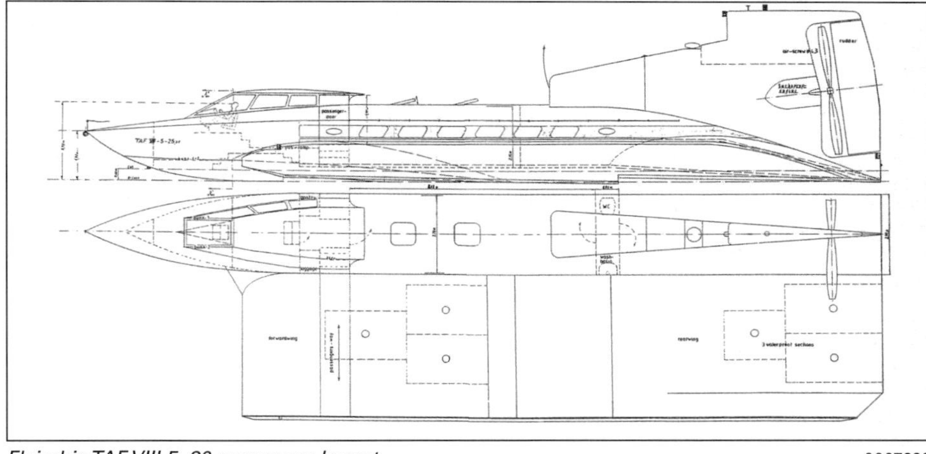

Flairship TAF VIII-5, 30 passenger layout 0007292

Water propulsion: Two retractable propellers for harbour manoeuvring.

FLAIRBOAT TAF VIII-3

Construction of this project began in Germany at the beginning of 1991. It is being built as an eight-seat patrol boat and is due to go into series production. The first craft was launched at the end of August 1991 and high-speed tests on the Oder River followed in September 1991 and in the Baltic Sea between 1992 and 1996.

Specifications

Length overall	14.0 m
Span	5.8 m
Hull beam	1.6 m
Weight, min	2.2 t
Take-off speed	39 kt
Operational speed	76 kt
Max speed	84 kt
Flight altitude	0.3 m

Propulsion: The engine is a Wankel or turboprop engine of approximately 380 kW.

FLAIRSHIP TAF VIII-5/25

This design is for a luxury craft (equivalent to a motor yacht specification) carrying up to 10 people. The craft was scheduled for delivery in 1999, equipped for 25 passengers.

Specifications

Length overall	25.9 m
Length waterline	24.2 m
Span	12.9 m
Weight, max	15 t
Crew	3
Propulsive power	1,600 kW
Operational speed	90 kt
Flight altitude	0.7–0.9 m
Range	432 n miles

Propulsion: Twin turboprop engine rated at 1,600 kW, driving one 4.3 m diameter air propeller and two steerable, retractable water propellers.

FLAIRSHIP TAF VIII-7/3 (DESIGN)

The TAF VIII-7 has been designed to carry 80 passengers and 3 crew at speeds up to 124 kt in sea conditions up to 2.5 m significant wave height. With two of the three engines running the speed is reduced to 119 kt and a 1.5 m wave height limitation. With only one of the three engines running the speed and wave limitation are 92 kt and 0.5 m respectively. With a turboprop engine it is possible to increase passenger numbers from 80 to 100.

Specifications

Length overall	45.7 m
Span	18.6 m
Crew	3
Passengers	80
Fuel capacity	6,000 litres
Max speed	124 kt
Range	750 n miles
Operational limitation	2.5 m significant wave height

Propulsion: Powered by three turboprop-pusher engines each rated at 1,700 kW and each driving a variable-pitch air propeller.

FLAIRSHIP TAF VIII-10 (DESIGN)

The TAF VIII-10 is Botec's largest design and has been conceived to carry 400 passengers or a payload of 40 tonnes with a total weight of 125 tonnes. The 70 m craft is powered by three engines giving a cruising speed of over 100 kt with a range of approximately 1,000 n miles.

RFB

Rhein-Flugzeugbau GmbH
Krefelder Str, 840 D, 11066 Münchengladbach, Germany

While RFB no longer has an interest in wing-in-ground-effect craft, this entry remains as a record of the early development of these craft.

As a result of Dr Lippisch's work on the wing-in-ground-effect machine X-112 in the USA, RFB bought several of his patents and further developed the wing-in-ground-effect technology in close co-operation with Dr Lippisch.

After elaborate towing and wind tunnel tests supplemented by radio control model tests, the X-113 was derived, a 1:1.7 scaled-down version of a four-seat craft. This single-seat trimaran-type craft, built of composite materials in a special sandwich construction developed by RFB, successfully underwent tests on Lake Constance in October 1970.

During additional trials, which were performed in Autumn 1972 in the Weser estuary and the North Sea coastal region over rough water, the X-113's design and the chosen composite construction method proved fully effective.

The flight performance measurements confirmed the analytically expected improvements with respect to lift and drag resulting in gliding angles of 1:23 close to the surface. Even flights out of ground-effect were successfully performed with gliding angles of 1:7 because of the low aspect ratio during this flight phase.

The X-113 showed inherent stabilities even with respect to altitude keeping. Speeds of 97 kt were achieved with a power of 28 kW in ground-effect. The ability to achieve bank altitudes permits very small curve radii and excellent manoeuvrability. Remarkably good sea behaviour was shown from the outset. Take-offs and landings in wave heights of about 0.75 m presented no problem.

RFB X-113 Am during flight demonstration over Wattenmeer 0506717

RFB X-114 Aerofoil boat with hydrofoils extended 0506719

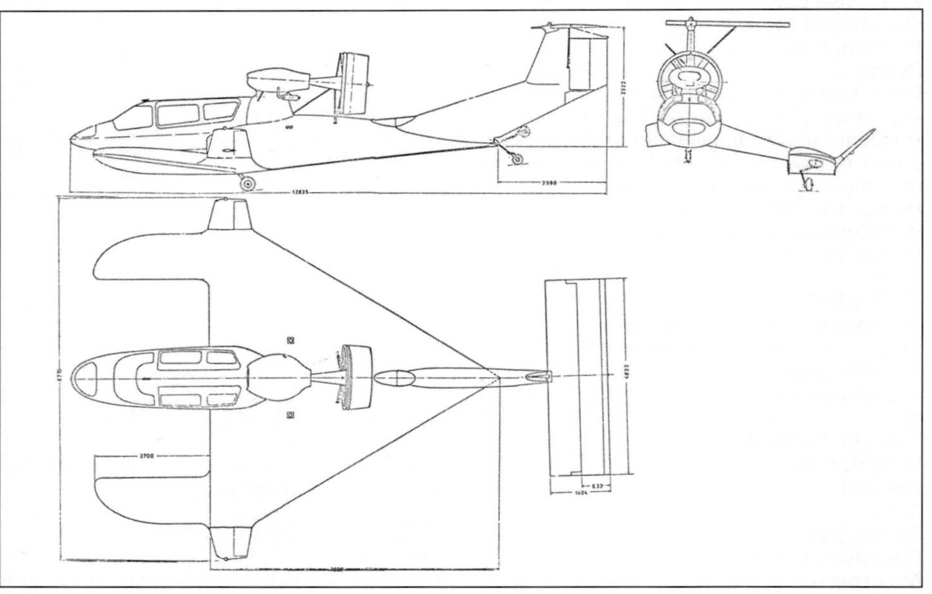
General arrangement of RFB X-114 0506720

RFB X-114 AND -114H AEROFOIL BOATS

Evolved from the X-113, this six- to seven-seater has a maximum take-off weight of 1,500 kg and is fitted with a retractable wheel undercarriage, enabling it to operate from land or water.

Power is provided by a 150 kW Lycoming IO-360 four-cylinder, horizontally opposed air-cooled engine driving a specially designed Rhein-Flugzeugbau ducted fan. Range, with 100 kg of fuel, is more than 1,000 km. Operational speed is 75 to 200 km/h.

An initial trials programme was successfully completed in 1977. A new series of trials was undertaken after modifications which included the fitting of hydrofoils beneath the sponsons and raising maximum take-off weight to 1,750 kg. In this configuration it is known as the X-114H.

The vehicle is designed to operate over waves up to 1.5 m in ground-effect and can therefore be used without restriction during 80 per cent of the year in the Baltic Sea area and 60 per cent of the year in the North Sea. In high seas of more than 1.5 m, take-off and landing takes place in waters near the coast. Flying is virtually unrestricted, providing due allowance is made for the loss in economy.

Fuel consumption costs, while flying in ground-effect, are lower than those for cars. RFB states that its economics cannot be matched by any other form of transport aircraft.

Although built primarily as a research craft to extend the experience gained with the X-113 Am single seater, Aerofoil boats of the size of the X-114 are suitable for air taxi work along coastlines, the supervision

Vessel type	Vessel name	Yard No	Length (m)	Speed (kt)	Seats	Vehicles	Originally delivered to	Date of build
RFB X-114	Research craft	–	12.8	81.0	–	None	–	1977

of restricted areas, patrol, customs and coastguard purposes, and search and rescue missions.

Without any significant new research the construction of a vehicle with a take-off weight of approximately 18,000 kg is possible. On a vehicle of this size, the ratio of empty weight to take-off weight is less than 50 per cent.

Specifications

Length overall	12.8 m
Span	7.0 m
Weight, max	1,500 kg (X-114)
	1,750 kg (X-114H)
Payload	500 kg
Max speed	108 kt
Operational speed	81 kt
Range	1,160 n miles

MTE GmbH

Schonenfahrerstr. 7, D-18057 Rostock, Germany

Tel: (+49 381) 260 54 51
Fax: (+49 381) 808 98 42
e-mail: poste@mte-germany.de
Web: www.mte-germany.de

Dieter Puls, *President*
Reinhard Krause, *Vice-President*

Techno Trans Hydrowing VT01 0009744

MTE was founded in 1993 as TechnoTrans eV, a non-commercial institution dealing with development and technology transfer in marine engineering, shipbuilding and ocean technology, laser Doppler anemometry and environmental technology. One of the major projects is the design of wing-in-ground-effect craft. Initial experience in this area was gained with two tandem-airfoilboats of the BOTEC Jörg type.

Under funding from the German Ministry of Research and Science, a completely new design with a hydrofoil-based take-off aid has been developed to prototype stage; the Hydrowing VT01.

Hydrowing VT01

This manned test craft is a 1:3.35 scale prototype of an 80-seat ferry and was successfully tested in 1997. Further developments and modifications were to be undertaken in 1999.

Specifications

Length overall	9.9 m
Beam	7.8 m
Weight max	1.1 t
Passengers	2
Propulsive power	90 kW
Operational speed	65 kt
Operational limitation	0.5 m wave height

Vessel type	Vessel name	Yard No	Length (m)	Speed (kt)	Seats	Vehicles	Originally delivered to	Date of build
Hydrowing VT01	Test craft	–	9.9	65.0	2	None	–	1997

Japan

Tokushima Buri University

School of Engineering,
Shido-Town, Kagawa Prefecture, 769-21 Japan

Fax: (+81 878) 94 42 01

Dr Shigenori Ando
Motoige 17-75, Iwasaki, Nissin-City, Aichi Prefecture, 470-01 Japan

Tel: (+81 561) 73 32 46

PAR WIG model development by Dr Shigenori Ando 0506998

Dr Ando has developed a number of designs for Power-Assisted-Ram Wing-In-Ground-effect (PAR WIG) craft. Radio-controlled models have been built in several configurations, these developments following his earlier work with Nagoya University.

A 25-person PAR WIG design was presented at the WIG Conference (RINA) in London 1997.

Tottori University

Department of Applied Mathematics and Physics, Faculty of Engineering, Koyama, Tottori 680, Japan

Tel: (+81 857) 31 56 30
Fax: (+81 857) 31 57 47
e-mail: kubo@damp.tottori-u.ac.jp
Web: www.damp.tottori-u.ac.jp

Syozo Kubo, *Professor*

μSKY I AND μSKY II (MARINE SLIDER PROGRAMME)

This ram wing project was first announced by Mitsubishi on 1 February 1990. Both prototypes were aimed at the eventual production of a commercial craft for leisure

Vessel type	Vessel name	Yard No	Length (m)	Speed (kt)	Seats	Vehicles	Originally delivered to	Date of build
μSky-I	Test craft	–	4.4	35.0	–	None	–	December 1988
μSky-II	Test craft	–	5.9	–	–	None	–	February 1990
Kaien	Test craft	–	8	–	–	None	–	–

use but it is understood that this work has now been suspended. The department remains active in the field of computational fluid dynamics applied to Wing- in-Ground effect craft.

μSky I 0506722

μSky I

μSky I first flew in December 1988. The structure is rigid; an earlier craft, Rameses I (US-built), employed a flexible structure.

Specifications

Length overall	4.4 m
Span	3.5 m
Weight, min	0.2 t
Weight, max	0.3 t
Crew	1
Propulsive power	48 kW
Take off speed	35 kt
Max speed	44 kt

Propulsion: Engine is a water-cooled petrol engine 48 kW (Rotax 532), driving a four-blade, fixed-pitch propeller.

Structure: Hull: GFRP, wing: GFRP, tail: aluminium plus cloth.

μSky II 0506723

μSky II

μSky II first flew on 8 February 1990 at Kobe, Hyogo Prefecture.

The prototype craft obtained the Ship Inspection Certificate on 17 July 1990, and five pre-production craft on 11 September 1991 from the Japanese government.

Specifications

Length overall	5.9 m
Span	4.3 m
Weight, min	0.3 t
Weight, max	0.4 t
Crew	2
Propulsive power	48 kW
Take off speed	33.5 kt
Max speed	46 kt
Flight altitude	0.5 m

Propulsion: The engine is a water-cooled petrol 48 kW (Rotax 532), driving a fixed-pitch, three-blade propeller made of GFRP, 1.54 m in diameter.

Structure: Hull: GFRP, wing: aluminium plus cloth, tail: aluminium plus cloth.

Control: Air and water rudders are fitted. The vehicle may be dismantled for transportation.

Korea, South

Hanjin Heavy Industries

Shipbuilding Division
29, 5-Ga, Bongnae-dong, Youngdo-ku, Puran, South Korea

Tel: (+82 51) 410 32 30
Fax: (+82 51) 410 84 75
e-mail: shipsale@hhic.co.kr
Web: www.hanjinsc.com

Choong-Hoon Cho, *Chairman*
Chan Yong Ah, *Shipsales Marketing Team*
Ung Sub Chung, *Shipsales Marketing Team*
In Kuk Chun, *Shipsales Marketing Team*

Hanjin KTMI 2 m remote-control WIG model 0007294

In 1999 Hanjin Heavy Industries merged with Korea Tacoma Marine Industries. Research on the next generation of wing-in-ground effect ships continues. The company has had experience testing three remote-control models and construction of a manned test craft is under review.

Russian Federation

Alsin

CENTRE OF EKRANOPLAN TECHNOLOGIES
PO Box 309, Moscow-62, 103062, Russia

e-mail: alsin@aquaglide.ru
Web: www.aquaglide.ru

D N Sinitsyn, *General Designer*
G L Radovitsky, *Chief Designer*

Alsin was formed to provide a centre for ekranoplan technologies in Russia. The company manufactures the Aquaglide five-seater ekranoplan, which is based on the successful Volga 2 craft, of which 10 have now been built, all certified by the Russian Maritime Register of Shipping.

Alsin has the capability and experience to develop much larger ekranoplan craft and is in the process of negotiating licences for joint development projects. Designs are available for second generation ekranoplans with displacements ranging from 10 to 500 tonnes, passenger capacities of 20 to 500, seaworthiness up to 3.5 m and range of up to 6,000 km for commercial operation.

AQUAGLIDE

Specifications

Length overall	10.4 m
Width	5.9 m
Weight	2.5 t
Passengers	5
Operating speed	92 kt
Range	240 n miles

Classification: Russian Maritime Register of Shipping.

Propulsion: Single Mercedes Benz M119 engine rated at 243 kW driving two air propellers.

Arctic Trade and Transport Company

21a Obraztsova Str, Moscow 127018, Russia

Tel: (+7 95) 737 31 41
Fax: (+7 95) 737 31 41
e-mail: attk@mail.ru
Web: www.attk.ru

Ruben A Nagapetyan, *President*

Arctic Trade and Transport Company (ATTK) develops and promotes innovative maritime technologies and, in particular, designs and builds wing-in-ground effect vehicles. The company staff have previously worked for the Central Hydrofoil Design Bureau (CHDB) and on the Amphistar development programme at the Joint Stock Company Technologies and Transport. At present, ATTK has designs and patents for aero-hydrodynamic configurations of ekranoplans of both A-type and B-type with take-off weights ranging from 2 to 400 tons. According to IMO classification, the A-type vehicles are those designed to operate in ground effect, for example, relatively close to an underlying surface, whereas B-type vehicles are able to climb temporarily to avoid obstacles. Type-B ekranoplans are designed for passenger transportation at a speed ranging from 90 to 260 kt.

The company's design and production bureau is situated in Nizhniy Novgorod, Russia's third largest city, about 400 km east of Moscow.

Having built an Aquaglide-5 in 1997, it is understood that the company is proposing to build a larger version, the Aquaglide-50, capable of seating up to 48 passengers.

In 2004 a joint venture between Companhia Câmara Construções Navals of São Paulo, Brazil and Arctic Trade and Transport was announced. A proposal, to build 15 Aquaglide-5 ekranoplans to operate on the rivers of the Amazon in northern Brazil, was valued at up to USD80 million.

Detail of Aquaglide-5 propulsion and lift system 1120265

Aquaglide-5 in cruising flight over Gorkovsky lake 1120264

Vessel type	Vessel name	Yard No	Length (m)	Speed (kt)	Seats	Vehicles	Originally delivered to	Date of build
Type-A								
Aquaglide-5 (T	–	–	10.6	92	4	–	–	1997
Aquaglide-40 (design)	–	–	21.5	110	40	–	–	–
Aquaglide-50 (design)	–	–	30.0	110	48	–	–	–
Aquaglide-60 (design)	–	–	23.0	65	60 (or 10 tons of cargo)	–	–	–
Aquaglide-200 (design)	–	–	43.0	60–80	200 or 35 tons of cargo	–	–	–
Type-B								
MPE-10	–	–	21.0	135	18	–	–	–
MPE-23	–	–	30.0	165	48	–	–	–
MPE-55	–	–	40.0	175	70	–	–	–
MPE-200	–	–	57.0	205	250	–	–	–
MPE-400	–	–	73	260	450	–	–	–

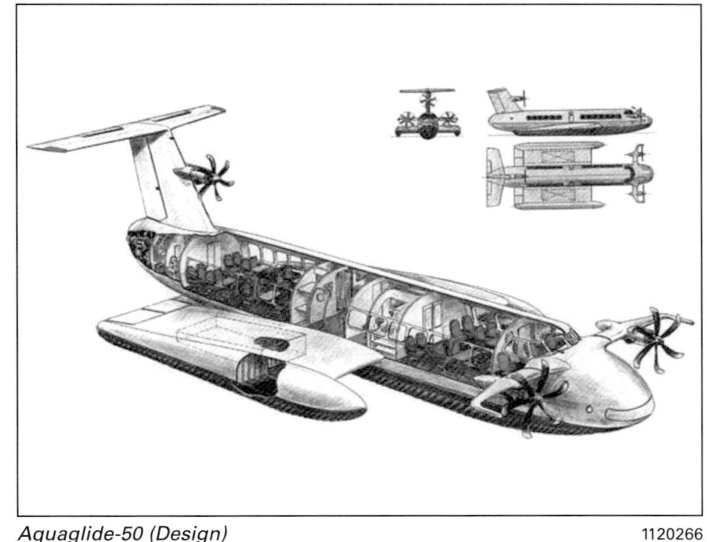

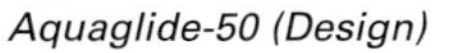

Aquaglide-50 (Design) 1120266

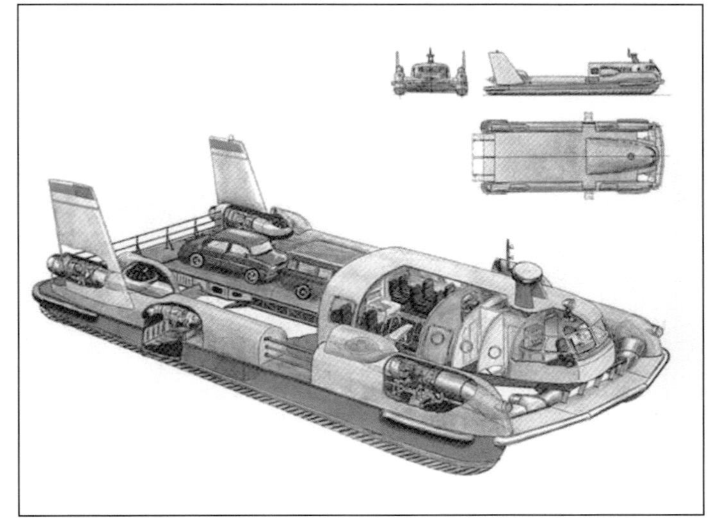

Aquaglide-60 (Design) 1120267

Aquaglide-5

The first vehicle of the Type-A series, Aquaglide-5 was designed, built, extensively tested and certified as a small ship by the Russian Maritime Register of Shipping. The cruising speed of about 90 kt is comparable to a helicopter or a fast train, and fuel consumption is similar to a large truck.

Specifications

Length overall	10.66 m
Span	5.9 m
Height	3.35 m
Weight, normal take-off	2.4 t
Crew	1
Passengers	4
Fuel capacity	180 l
Operational speed	92 kt
Range	400 km

Classification: Russian Maritime Register of Shipping.

Structure: Cantilever monoplane with a multispar wing made of composite material. The sections made up by the wing's spars and ribs are watertight. The lower parts of the endplates, which generate extra buoyancy, are made of soft material capable of absorbing shocks when hitting the water surface.

Propulsion: Water-cooled Mercedes-Benz V8 car engine driving two variable tilt propellers via reduction gearboxes and elongated shafts.

Central Hydrofoil Design Bureau

51 Svoboda Street, Nizhny Novgorod 603003, Russian Federation

Tel: (+7 8312) 73 10 37
Fax: (+7 8312) 73 02 48
e-mail: rostislav@sinn.ru

I M Vasilevsky, *Director*
M A Suslov, *Technical Director*
Sergey V Platonov, *General Director*

SDVPs

SDVP is a Russian abbreviation for Dynamic Air Cushion Vehicle.

In 1990 first details were published in the Soviet Union, on a series of designs of wing-in-ground-effect craft developed under the direction of Dr R Ye Alekseev of the Central Design Bureau of Hydrofoils. The need for these craft arises from the speed limitations of hydrofoils and the appreciable draught and limited speed of catamarans in long-distance river operations. The type of wing-in-ground-effect craft under development employ the power-assisted-lift principle, or PAR WIG (Power-Assisted-Ram Wing-In-Ground-effect) as it is sometimes known. The designs have over hard surface as well as over water capability.

STRIZH

This vessel was designed as a WIG pilot training craft for the Russian Navy in 1989. The design has since been developed to include the Strizh 3, Strizh 4 and Strizh 5 for between 7 and 12 passengers.

Specifications

Length overall	11.4 m
Span	6.7 m
Speed	150 kt
Range	300 n miles

VOLGA-2

This craft can overcome uneven surfaces and obstacles up to 0.4 m in height. The engines for Volga-2 are two VAZ-413 'rotor-piston' engines, believed to be rotary engines. Experimental data gained during trials of Volga-2 show propulsion economics to be on a level with existing hydrofoils. Further details of this craft are given in the table that follows.

A.90.150 Ekranoplan loading cargo (Avico Press) 0507082

An interesting aspect of the SDVP designs is the use of soft balloon-type structures to give the craft amphibious capabilities; the main structures are built in light alloys.

The Gorky Institute of Water Transportation Engineers (GIIVTOM), together with the Scientific Industrial Society (NPO) of the Central Design Bureau of Hydrofoils and the United Volga River Shipping Line (VORP), have carried out research appraising the technical possibilities and economic expediency of using these wing-in-ground-effect craft. As an example it was found that the route Nizhni Novgorod to Kazan, a distance of 432 km, could show positive economic results if wing-in-ground-effect craft were to be used for passenger-carrying services. This craft was awarded the gold medal at the World Invention Exhibition in Brussels in 1995.

In the accompanying table the Raketa, Meteor, Kometa and Vykhr craft are design studies; Volga-2 has been built.

A.90.150 EKRANOPLAN ORLYONOK

It is understood that only three of these craft were built. Working prototypes of cargo versions have also been built enabling the economics of future passenger-carrying versions to be predicted. The fuselage is of a relatively simple girder and stringer design and, like the wings, is divided into watertight compartments so that the craft will float while at rest. The fuselage is divided into three parts; the nose section with the flight deck and crew service area, a middle section containing the cargo or passenger area and the tail section which houses auxiliary machinery.

The nose-mounted jet engines have pivoted exhaust nozzles; during take-off the jet exhaust streams are directed beneath the wing to boost the ram-air pressure beneath the wing. On changing to cruising flight the nozzles are redirected to provide horizontal thrust, accelerating the craft until cruising speed is reached; the take-off jet units are then shut down. The same procedure is used when landing the craft to reduce hydrodynamic loading. The fuselage nose location of the jet units allows their intakes to be positioned in the contours of the nose in such a way as to minimise aerodynamic resistance.

In cruising flight the craft is propelled by the NK-12 turboprop installation mounted

Volga-2 eight passenger, 65 kt wing-in-ground-effect craft (Avico Press) 0507081

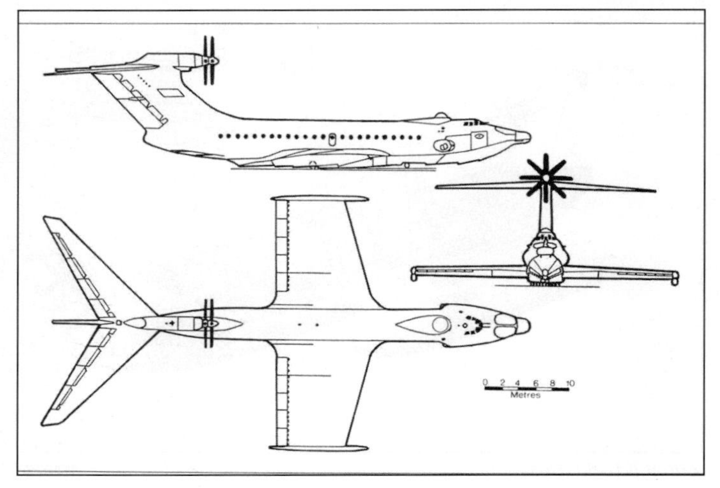

General arrangement of A.90.150 Ekranoplan 0506922

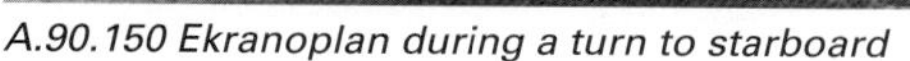

A.90.150 Ekranoplan during a turn to starboard 0507084

Caspian Monster *at speed* (A Belyaev) 0506923

high at the fin and tailplane intersection in order to keep the intake away from sea spray as far as it is feasible.

Pitching stability is more difficult to achieve for a craft operating in ground-effect mode than in normal free flight, because it is influenced by the flight altitude as well as the angle of attack. The Ekranoplan has a very large tailplane to counter this problem, which is greater increased in efficiency by the propeller slipstream.

Specifications

Length overall	58.0 m
Span	31.5 m
Height	15.0 m
Weight, normal take-off	110 t
Weight, max	125 t
Passengers	100–150 (single-deck version) 350 (twin-deck version)
Fuel capacity	33,000 litres (max) 17,650 litres (normal)
Propulsive power	11,000 kW
Operational speed	216 kt
Range	1,080 n miles
Operational limitation	Beaufort 4–5

Propulsion: Main engines are two Kuznetsov NK-8 turbofan engines, up to 10.5 t thrust, for take-off, and one Kuznetsov NK-12, 11,000 kW, turboprop for sustained cruising.

Outfit: Passenger cabin length: 25 m, width: 3.3 m, height: 3 m and volume: 240 m^3.

A.90.150 EKRANOPLAN, CARGO VERSION

During the development of the cargo model, a version was designed to have the fuselage split in two vertically, in order to facilitate loading. It is a unique project, with no equivalent known of anywhere else.

Specifications

Payload	30 t
Range	540 n miles

Outfit: Cargo section length: 25 m, width: 3.3 m and height: 3 m.

CASPIAN MONSTER

The Russians built the first large-scale WIG craft in the 1960s. The 500 t, so called, *Caspian Monster,* remained secret to the West until the late 1980s. There have been at least eight modifications to the single craft built, the number on the tail indicating the modification number. The *Caspian Monster* prototype is reported to have crashed and sunk in the Caspian Sea in 1980 after 14 years of service.

The craft design is ideally suited for rapid transportation of forces, with a greater payload capacity, greater range and less fuel consumption than conventional aircraft, whilst operating at a comparable speed.

Specifications

Length overall	92.4 m
Span	37.8 m
Max weight	544 t
Max speed	270 kt
Operational speed	230 kt
Range	1,000 n miles

A.90.150 Ekranoplan in flight (Avico Press) 0507083

Caspian Monster *SM-8 development craft* (A Belyaev) 0507085

Spasatel-2 (Avico Press) 0507086

Strizh *training craft in ground-effect* 0507087

Vessel type	Vessel name	Yard No	Length (m)	Speed (kt)	Seats	Vehicles	Originally delivered to	Date of build
WIG pilot training craft	*Strizh*	–	11.4	150.0	–	–	Russian Navy	1989
LUN	–	–	75.0	–	–	–	Russian military	1989
Volga-2	–	–	11.6	65.0	8	–	–	1995
A.90.150 Ekranoplane	*Orlyonok*	–	58.0	216.0	350	–	–	–
A.90.150 Ekranoplane	–	–	58.0	216.0	350	–	–	–
A.90.150 Ekranoplane	–	–	58.0	216.0	350	–	–	–
A.90.150 Ekranoplane, Cargo version	Prototype	–	58.0	216.0	–	–	–	–
92 m WIG	*Caspian Monster*	–	92.4	230.0	–	–	–	–
LUN (rescue craft version)	*Spasatel-2*	–	75.0	–	600	–	–	–
Vykhr-2	Design	–	54.0	150.0	250	–	–	–
Meteor-2	Design	–	36.0	92.0	120	–	–	–
Rakata-2.2	Design	–	34.8	80.0	90	–	–	–
Rakata-2	Design	–	26.2	80.0	50	–	–	–
Kometa-2	Design	–	25.0	100.0	150	–	–	–
Strizh 3	Design	–	13.4	200.0	7	–	–	–
Strizh 4	Design	–	15.4	210.0	9	–	–	–
Strizh 5	Design	–	17.0	220.0	12	–	–	–

LUN

A development of the *Caspian Monster*, the LUN design apparently went into military service in 1989, but this craft is now reported to have been withdrawn from military activities.

The craft is propelled by eight NK-87 turbofan engines, and the craft is armed with six anti-ship missiles.

Specifications

Length overall	75.0 m
Span	41.0 m
Weight	400 t
Flight altitude	4.0 m
Max speed	300 kt
Range	1,620 n miles

A number of non-military versions of the LUN craft are currently proposed. It is understood that the LUN craft currently under construction was originally conceived as a military craft but is being converted to a rescue craft named the *Spasatel-2*. This has similar performance characteristics to the original LUN craft with a displacement of 400 tonnes, a cruising speed of 243 to 297 kt and a range of 1,620 n miles. The carrying capacity of this craft is up to 600 people. Other versions of the *Spastel-2* are also proposed – a passenger-carrying craft, the *LUN-P*, and a cargo carrying craft, the *MTER*.

Ukraine

Feodosia Shipbuilding Company (MORYE)

Desantnikov Str 1, 98176 Feodosia, Crimea, Ukraine

Tel: (+380 6562) 325 56
Fax: (+380 6562) 323 73
e-mail: fsa@morye.kafa.crimea.ua
Web: morye.kafa.crimea.ua

Sergei Ostapenko, *Technical Director*
Grigory Klebanov, *Head of Marketing and Sales*

SHMEL

Feodosia have built the Shmel wing-in-ground-effect craft to exploit inland waters and rivers at any surface state, in up to 3m seastates and on flat unprepared land surfaces.

Vessel type	Vessel name	Yard No	Length (m)	Speed (kt)	Seats	Vehicles	Originally delivered to	Date of build
WIGE	*Shmel*	–	16	100	10	None	–	–

FAST INTERCEPTORS

Company listing by country

Australia
Gemini Inflatables
Inflatables RIBS Marine (IRM)
Strategic Marine Pty Ltd

Bahrain
Voyager Marine

Canada
Metalcraft Marine Inc
Reflex Advanced Marine Corporation
Zodiac Hurricane Technologies Inc

Chile
Asmar

China
Shenzhen Jianghua Marine

Croatia
Diokom ANC

Denmark
Viking Lifesaving Equipment

Finland
Astra-Marine Oy
Boomeranger Boats Oy
Marine Alutech OY AB

France
CMN
Plascoa
Sillinger

Germany
DSB (Deutsche Schlauchboot Gmbh & Amp Co)
Fassmer
Peene-Werft

Greece
Motomarine SA
Watercraft Hellas SA

India
ABG Shipyard
Goa Shipyard

Indonesia
North Sea Boats

Ireland
Safehaven Marine

Israel
Israel Aerospace Industries
Israel Shipyards

Italy
Cantieri Capelli srl
Cantieri Navali del Golfo
FB Design
Intermarine SpA
Lomac Nautica Srl
Novamarine SpA
Novurania SpA

Lebanon
Lenco Marine

Malaysia
Hong Leong Lurssen Sdn Bhd
Kay Marine Sdn Bhd
NGV Tech Sdn BHD

Netherlands
Damen Shipyards
Madera Ribs BV

New Zealand
AMF Boats
Aquapro International Ltd
Lancer Industries Ltd
Naiad Inflatables (NZ) Ltd
Rayglass Boats
Sensation Yachts

Norway
H Henriksen Mekaniske Verksted A/S
Maritime Partner A/S
Norsafe A/S

Portugal
Valiant

Russian Federation
Almaz Marine Yard
Central Hydrofoil Design Bureau
Feodosia Shipbuilding Company
Vympel Shipyard

Singapore
Lita Ocean Pte Ltd
Singapore Technologies Marine Ltd

South Africa
Boat Loft CC
Eraco Boat Builders
Stingray Marine

Spain
Duarry (Astilleros Neumaticos Duarry SA)
Rodman Polyships SA

Sri Lanka
Blue Star Marine Ltd
Colombo Dockyard

Sweden
Capo Marin AB
Dockstavarvet
Kockums AB
Ring Powercraft
Storebro Bruks AB
Swede Ship Marine AB

Taiwan
Lung Teh Shipbuilding Co Ltd

Turkey
Altinel Shipyard Inc
Yonca Onuk

United Arab Emirates
Abu Dhabi Ship Building

United Kingdom
Avon Inflatables Ltd
Delta Power Services
Fusion Marine Ltd
Hoyhead Marine
Humber Inflatable Boats
Marine Specialised Technology Ltd
Ocean Dynamics
Redbay Boats Ltd
Ribeye Ltd
RTK Marine
Seaspray Marine Services and Engineering
Tiger Marine Ltd
Tornado Boats International Ltd
VT Halmatic
XS-Ribs

United States
Aluminium Chambered Boats Inc
Almar Boats
Boston Whaler Inc
Brunswick Commercial and Government Products
Demaree
Moose Boats
Nautica International Inc
Northwind Marine Inc
Oregon Iron Works
Otech
Seaark Marine Inc
Silverships
Swiftships Shipbuilders, LLC
United States Marine Inc
USA Avenger Inc
VT Halter Marine
Willard Marine Inc

Australia

Gemini Inflatables

Shop 7, Gold Coast City, Waterways Drive, Coomera 4209, Queensland, Australia

Tel: (+617) 55 73 72 65
Fax: (+617) 55 73 78 34
e-mail: greg@gemini-inflatables.com
Web: www.gemini-inflatables.com

Formed in 1979, Gemini Inflatables offers range of full inflatable and rigid hull inflatable boats from 2.4 to 8.5 m. The range of Gemini inflatables includes luxury tenders, racing boats, commercial RIBs, dive boats, rescue craft, SOLAS vessels and military craft.

The company has sold inflatable craft in over 35 countries and has a network of agents around the world.

Vessel type or class name	Length (m)	Max beam (m)	Max speed (kt)	Hull material	Engines	Propulsion
SOLAS 380	3.8	1.7	–	GRP	Outboards	–
SOLAS 500	5.0	2.0	–	GRP	Outboards	–
SOLAS 600	6.0	2.5	–	GRP	Outboards	–
Dive 470	4.7	2.1	–	GRP	Outboards	–
Dive 530	5.3	2.1	–	GRP	Outboards	–
Waverider 310	3.1	1.5	–	GRP	Outboards	–
Waverider 550	5.5	2.3	–	GRP	Outboards	–
Waverider 730	7.3	2.7	–	GRP	Outboards	–
Waverider 850	8.5	2.8	–	GRP	Outboards	–

Gemini SOLAS 600 RIB 0583200

Gemini Waverider 730 RIB 0583201

Inflatables RIBs Marine

2/1356 Lytton Road, Hemmant, Queensland 4174, Australia

Tel: (+617) 33 90 71 33
Fax: (+617) 33 90 72 33
e-mail: irm@aribs.com.au
Web: www.inflatablesribsmarine.com, www.airib.com.au

J Ferguson, *Director of Marketing*
E Boast, *Director of Production*

Inflatables RIBs Marine produce the AIRIB range of aluminium-hulled rigid inflatable craft for tourist, patrol, diving and rescue operations.

Vessel type or class name	Length (m)	Max beam (m)	Max speed (kt)	Hull material	Engines	Propulsion
AIRIB 430	4.3	1.9	–	Aluminium	Outboard	–
AIRIB 4.4 XL	4.8	2.15	–	Aluminium	Outboard	–
AIRIB 5.0 RV	5.0	2.4	–	Aluminium	Outboard	–
AIRIB 5.3 RV	5.3	2.4	–	Aluminium	Outboard	–
AIRIB 5.8 RV	5.8	2.4	–	Aluminium	Outboard	–
AIRIB 6.2 RV	8.0	2.6	–	Aluminium	Outboard	–
AIRIB 6.2 SW	6.2	2.4	–	Aluminium	Outboard	–
AIRIB 6.5 SW	6.5	2.5	–	Aluminium	Outboard	–
AIRIB 7.0 LS	7.0	2.75	–	Aluminium	Outboard	–
AIRIB 7.5 LR	7.5	2.75	–	Aluminium	Outboard	–

Strategic Marine PTY Ltd

Lot 5, Clarence Beach, Henderson, Western Australia 6166, Australia

Tel: (+61 8) 94 37 48 40
Fax: (+61 98) 94 37 48 38
e-mail: info@strategicmarine.com
Web: www.strategicmarine.com

Mark Newbold, *Chairman*
Mark Schiller, *Managing Director, Australia*
Ron Anderson, *Managing Director, International Business*

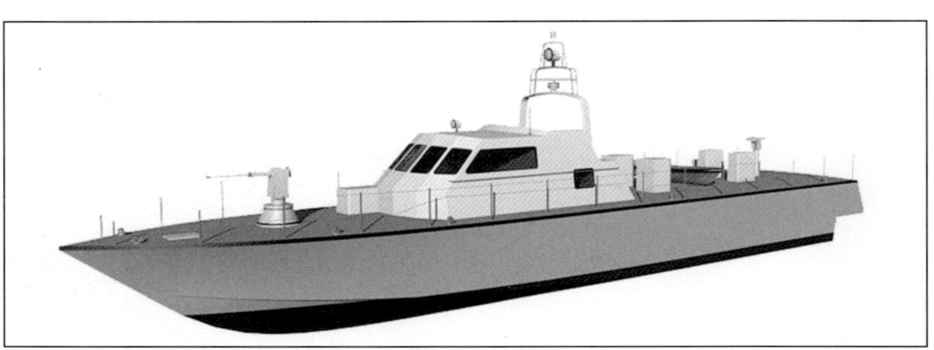

Strategic Marine Stingray 25 m Fast Interceptor Boat 0589528

Strategic Marine 18 m Patrol Boat 0589526

Strategic Marine 22.5 m Police Command Boat 0589527

Vessel type or class name	Length (m)	Max beam (m)	Max speed (kt)	Hull material	Engines	Propulsion
Stingray 10 OBM	10.5	3.2	45	Aluminium	3 × inboard diesel	–
18 m police pursuit boat	18	5.4	40	–	2 × MTU 16V2000M90	–
22.5 m police/coast guard vessel	22.5	7.5	40	–	2 × MTU 16V2000M90	–
25 m patrol boats (Stingray class)	25	5.7	48	–	2 × inboard diesel	Twin water-jet
20 m police/coast guard patrol boat	19.8	6.5	40	Aluminium	2 × MTU 16V2000M90	Twin water-jet

Strategic Marine (formerly Geraldton Boat Builders) offer a wide range of aluminium craft up to 65 m in length. To date they have built over 270 boats consisting of fast interceptors, command and coast guard craft, pursuit craft, fisheries patrol vessels, crew boats, fishing vessels and kit boats.

Swift Marine

19 Reichert Drive, Molendinar, Queensland, Australia

Tel: (+61 7) 5594 6266
Fax: (+61 7) 5594 7886
e-mail: info@swiftmarine.com.au
Web: www.swiftmarine.com.au

Swift Marine has been established for over 20 years and specialises in building high quality, durable RIBS using Orca Hypalon rubber fabric for the inflatable tubes, rather than the more usual PVC-coated fabric, giving enhanced wear resistance and durability.

Vessel type or class name	Length (m)	Max beam (m)	Max speed (kt)	Hull material	Engines	Propulsion
6.5 m SR Rescue	6.5	2.5	30+	aluminium	Twin outboard	outboard
7.0 m SR Pursuit	7.0	2.85	30+	aluminium	Twin outboard	outboard
SR 750	7.5	3.10	30+	aluminium	Twin outboard	outboard

Woody Marine Fabrication Pty Ltd

34 Milenium Place, Tingalpa, Queensland 4173, Australia

Tel: (+61 7) 3393 5995
Fax: (+61 7) 3393 5994
e-mail: sales@wmf.com.au
Web: www.woodymarinefabrication.com.au

Woody Marine Fabrication has built over 500 of its Naiad class hulls since 1990. The Naiad class ranges from 2.5 to 15 m and can be customised for search and rescue, government work or pleasure. The Naiad 8.5 m is used by the New Zealand Coast Guard, Alaskan Fisheries protection and by the New Zealand Ministry of Agriculture and Fisheries.

Vessel type or class name	Length (m)	Max beam (m)	Max speed (kt)	Hull material	Engines	Propulsion
Naiad 5.8 m Offshore	5.8	2.4	–	Aluminium	175 hp inboard/outboard	–
Naiad 6.8 m Passenger	6.8	2.6	–	Aluminium	–	–
Naiad 8.5 m Passenger	8.5	3.0	–	Aluminium	450 hp inboard	–
12.6 m Pilot Boat	12.6	4.0	24	Aluminium	2 × Caterpillar diesel	Henley stern drive
14.6 m Rescue vessel	14.6	–	29	Aluminium	2 × Caterpillar inboard diesel	2 × henley drive shaft and props.

Bahrain

Voyager Marine

A BFG International Company

PO Box 26197, Manama, Bahrain

Tel: (+973) 727 003
Fax: (+973) 727 615
e-mail: info@voyager.com.bh
Web: www.voyagermarine.com.bh

Voyager Marine produces a large range of commercial and pleasure GRP craft. The company has a number of modern factories and recent customers have included the US Navy.

PCH 3500 1120293

Viper 330 SCM 1120294

Vessel type or class name	Length (m)	Max beam (m)	Max speed (kt)	Hull material	Engines	Propulsion
PCH 3500	10.97	3.05	40	GRP	2 × 250 Hp Mercury outboards	–
Viper 330 SCM	10.05	–	60	GRP	2 × 250 Hp Mercury outboards	–

Canada

Metal Craft Marine Inc

347 Wellington Street, Kingston, Ontario K7K 6N7, Canada

Tel: (+1 613) 542 1810
Fax: (+1 613) 542 6515
e-mail: emily.r@metalcraftmarine.com
Web: www.metalcraftmarine.com

Henry Copestake, *General Manager*
Emily Roantree, *Sales and Marketing*

Metalcraft Marine specialises in building high speed patrol and search and rescue craft. The company has become a leader in water jet driven craft with the Kingston design being selected by the US Navy for the Force Protection Program.

Kingston 32 Patrol RHIB 1120340

Kingston 11 m 1120341

Vessel type or class name	Length (m)	Max beam (m)	Max speed (kt)	Hull material	Engines	Propulsion
Kingston 32 Patrol RHIB	9.78	2.74	40	Aluminium	Twin diesel inboard	Twin Hamilton waterjets
Kingston 11 m	10.97	3.66	31	Aluminium	Twin diesel inboard	Twin Hamilton waterjets
Kingston 54–60 Patrol/Crew	16.5	5.6	38	Aluminium	Twin or Triple diesel inboard	Triple Hamilton 362 waterjets

Reflex Advanced Marine Corporation

35-7 Innovation Drive Dundas Ontario, L9H 7H9, Canada

Tel: (+1 905) 690 21 79
Fax: (+1 905) 690 21 81
e-mail: sales@ramboats.com
Web: www.ramboats.com

Ian Taylor, *President and CEO*

Reflex Advanced Marine Corp (RAM) is a design and project management company set up to market the benefits of its unique, extensively tested and patented "Reflex™" tri-hull craft. The first vessel with this hull form has recently been produced.

40 patrol boat 1180181

Vessel type or class name	Length (m)	Max Beam	Max Speed (kt)	Hull Material	Engines	Propulsion
40' Patrol Boat	12.1	4.2	41	Kevlar™/Glass Advanced Composites	2 × 313 kW Inboard Diesel	Twin Hamilton waterjets

Zodiac Hurricane Technologies Inc

3160 Orlando Drive, Unit A, Mississauga, Ontario L4V 1R5, Canada

Tel: (+1 905) 677 42 11
Fax: (+1 905) 677 76 18
e-mail: solassales@zodiac.ca
Web: www.zodiac.ca

Tim Flemming, *Manager*

The company produces a very large range of rigid inflatable craft from 1.8 to 10 m in length. The majority of commercial sales centre on the 7 to 10 m craft for coast guard and military applications. These vessels feature an aluminium hull and a full wrap around inflatable or foam filled collar.

Vessel type or class name	Length (m)	Max beam (m)	Max speed (kt)	Hull material	Engines	Propulsion
H300 OB	3.0	1.5	–	Aluminium	7.5 kW outboard	–
H380 OB	3.9	1.8	–	Aluminium	30 kW outboard	–
H472 OB	4.7	2.0	–	Aluminium	52 kW outboard	–
H472 SOLAS	4.7	2.0	–	Aluminium	52 kW outboard	–
H533 SOLAS	5.3	2.4	–	Aluminium	67 kW outboard	–
H540 OB	5.6	2.2	–	Aluminium	67 kW outboard	–
H590 OB	5.8	2.4	–	Aluminium	2 × 52 kW outboard	–
H630 OB	6.7	2.6	–	Aluminium	2 × 67 kW outboard	–
H733 OB	7.22.7	–		Aluminium	2 × 112 kW outboard	–
H920 OB	9.1	3.2	–	Aluminium	3 × 150 kW outboard	–
H1010 OB	10.1	3.5	–	Aluminium	3 × 168 kW outboard	–

Chile

Asmar

PO Box 150-V, Avenida Altamirano, 1015 Valparaiso, Chile

Tel: (+56 32) 27 20 00
Fax: (+56 32) 27 20 99
e-mail: asmaris@asmar.cl
Web: www.asmar.cl

Captain Duberly Mena, *General Manager*

ASMAR Shipbuilding and Docking Company manufacture a range of fast rescue and workboats from GRP with inflatable collars. The company is currently developing a 5.8 and 9 m version for fast rescue purposes.

ASMAR 1160 0024764

Vessel type or class name	Length (m)	Max beam (m)	Max speed (kt)	Hull material	Engines	Propulsion
R 420W	4.2	1.73	32	GRP	40 hp outboard	–
PUMAR 530	5.3	2.14	–	GRP	75 hp outboard	–
PUMAR 585	5.85	2.48	–	GRP	115 hp outboard	–
PUMAR 775	7.75	2.75	–	GRP	2 × 115 hp outboard	–
RR 1160	12.66	3.9	28	GRP	–	Twin water-jet

China

Shenzhen Jianghua Marine

Gangwan, 2nd Road, Shekou, Shenzhen, China

Tel: (+86 755) 26 82 91 58
Fax: (+86 755) 26 69 53 59
e-mail: sales@integritytrawlers.com
Web: www.jh-boat.com

Founded in 1981, The Shenzhen Jianghua Marine & Engineering Co, Ltd, manufactures yachts, passenger ferries, and commercial power boats. To date the company has supplied over 300 patrol and interceptor craft to the Chinese Customs Service and Marine Police units in south China.

Vessel type or class	Length (m)	Max beam (m)	Max Speed (kt)	Hull Material	Engines	Propulsion
7.5 m Patrol Boat	7.4	2.3	36	Composite	193 kW Mercruiser Diesel Inboard	single water-jet
9.5 m Patrol Boat	9.5	2.55	45	Composite	2 × Mercruiser D42L D-tronic	twin stern drive
11.8 m Work Boat	11.8	3.15	51	Composite	4 × Mercury 250 Hp Outboards	quadruple outboards
11.8 m Patrol Boat	11.8	3.15	45	Composite	3 × 190 kW Mercruiser Diesels	triple stern drive
16 m Work Boat	16.0	4.25	34	Composite	2 × Volvo Penta TAMD 74C EDC	–

Croatia

Diokom

Put Brodarice 6, HR-21000, Split, Croatia

Tel: (+385 21) 30 87 77
Fax: (+385 21) 38 38 59
e-mail: galanterija@diokom-novi.hr

Diokom ANC, which is part of the Diokom group of companies based in Split, manufactures a range of inflatable fast tenders for sport and commercial use.

Vessel type or class name	Length (m)	Max beam (m)	Max speed (kt)	Hull material	Engines	Propulsion
ANC 300	3.00	1.60	–	–	–	–
ANC 350	3.40	1.70	–	–	–	–
ANC 420	4.20	1.98	–	–	–	–
ANC 470	4.68	2.10	–	–	–	–

Denmark

Viking Lifesaving Equipment

PO Box 3060, Saedding Ringvej 13–16, DK-6710, Esbjerg V, Denmark

Tel: (+45) 76 11 81 00
Fax: (+45) 76 11 81 01
e-mail: viking@viking-life.com
Web: www.viking-life.com

Viking produce two fast rescue craft, the MOB (Man Over-Board) and the FRB (Fast Rescue Boat), both SOLAS approved.

Vessel type or class name	Length (m)	Max beam (m)	Max speed (kt)	Hull material	Engines	Propulsion
Viking FRB	6.90	2.6	–	GRP	Outboard engine(s)	–

Finland

Astra-marine OY

Teollisuustie 11, FIN-21250, Masku, Finland

Tel: (+358 2) 433 96 60
Fax: (+358 2) 433 96 70
e-mail: info@astramarine.com
Web: www.astramarine.com

Timo Jarvinen, *Managing Director*
Petteri Lehtovaara, *Sales Manager*
Anton Nojonen, *Sales Director (Russia)*

Astra-Marine OY was established in 1992 to continue the boat-building traditions of Turku Boatyard OY and Waterman OY. The company specialises in work boats such as fast rescue boats, pilot boats and patrol boats but also builds a range of leisure boats.

Masmar 475 Fast Patrol Boat 1120274

Vessel type or class name	Length overall (m)	Max beam (m)	Max speed (kt)	Hull material	Engines	Propulsion
Masmar 33 FPB	10.10	2.60	40	GRP	1 × Volvo D6-350	Stern Drive
Masmar 47J FPB	14.97	3.20	38	GRP	1 × Caterpillar 3,406	Water-jet

Boomeranger Boats

PO Box 95, FIN-07901, Loviisa, Finland

Tel: (+358) 19 51 58 05
Fax: (+358) 19 51 58 25
e-mail: boomeranger@boomeranger.fi
Web: www.boomeranger.fi

Boomeranger Boats Ltd was founded in 1991 and specialises in building high quality Rigid Inflatable Boats (RIBs) primarily for the professional market. The current range varies in size from 5 to 14 m and includes vessels with and without cabins.

11.5 m RIB with triple outboards 1120276

Recent deliveries have included two large cabin RIBs to the Finnish Lifeboat Society, a 6 m RIB to the Norwegian Sea Rescue Society and four specialist RIBs to the Swedish Armed Forces for use onboard the Visby Class stealth corvettes.

6 m fast rescue boat for the Swedish Navy 1120278

Marine Alutech Oy AB

Telakkatie, FIN-25570, Teijo, Finland

Tel: (+358 2) 728 8100
Fax: (+358 2) 728 8129
e-mail: info@marinealutech.com
Web: www.marinealutech.com

Niko Haro, *Managing Director*
Heikki Haro, *Chairman*
Jari Panu, *Naval Architect*

Marine Alutech was set up in 1985 and has built over 350 aluminium commercial and military vessels. These have included passenger ferries, pilot and patrol/SAR-boats and military transport vessels including the Watercat M12/Jurmo troop carrier of which over 38 were built. The vessels feature water-jet propulsion, hydraulically operated bow doors and a lightweight composite superstructure.

Watercat 1100 Patrol Craft 1120297

Vessel type or class name	Length (m)	Max beam (m)	Max speed (kt)	Hull material	Engines	Propulsion
Watercat M12 Jurmo	14.2	3.65	35	Aluminium	2 × 331 kW Caterpillar diesel inboards	2 × water-jets
Watercat M11 Uisko	10.5	3.50	35	Aluminium	490 kW diesel inboard	1 × water-jet
Watercat M8 G-Boat	7.2	2.10	35	Aluminium	179 kW Yanmar 4LHAM-STP diesel inboard	1 × water-jet
Watercat 1100 Patrol Craft	12.1	3.50	30	Aluminium	2 × Caterpillar 312 6TA diesel inboards	2 × water-jets
Watercat 1450 Rescue Boat	15.9	4.50	32	Aluminium	2 ×478 kW Scania DII6 diesel inboards	2 × water-jets

Watercat M12 Jurmo Troop Carrier 1120295

Watercat M8 G-Boat Fast Landing Craft 1120296

France

CMN

52 rue du la Brettonniere, BP 539, F-50105 Cherbourg Cedex, France

Tel: (+33 2) 33 88 30 00
Fax: (+33 2) 33 88 31 98
e-mail: info@cmn-cherbourg.com
Web: www.cmn-group.com

The CMN shipyard covers over 120,000 m^2, where the workforce is able to build ships entirely under cover. Spacious and well equipped outfit workshops with adjacent

Vessel type or class name	Length (m)	Max beam (m)	Max speed (kt)	Hull material	Engines	Propulsion
Interceptor DV-15	15.6	3.0	50	Composite	2 × diesel inboard	Twin surface drive
Interceptor DV-33	33.0	6.7	45	Composite	2 × diesel inboard	Twin waterjets

storage facilities, enable CMN to effectively undertake complete machinery installation and outfitting before the ships leave for the test and trials facility. CMN produce a range of military ships, commercial ships and luxury yachts and the repair, refit and upgrading of these vessels is a significant part of the shipyard's activities.

DV-15 Interceptor
1195340

Plascoa

Couach's Professional Boats Department
Rue De L'Yser, F-33470 Gujan-Mestras, France

Tel: (+33 5) 56 22 35 55
Fax: (+33 5) 56 66 08 20
e-mail: couach@couach.com
Web: www.couach.com

Couach shipyard has been building yachts in glass fibre since 1962 and is well known for its production of fast vessels. The company has diversified into fast patrol craft and currently offers a range from 11 m to 32 m. The company also offers a range of custom made yachts from 20 m to 45 m.

Vessel type or class name	Length (m)	Max beam (m)	Max speed (kt)	Hull material	Engines	Propulsion
Plascoa 1500 HS	15	3.8	41	GRP	2 × MTU 8V 2000 CR	–
Plascoa 1100	11	2.9	36	GRP	2 × Volvo KAD 43 BP	Twin submerged shaft
Plascoa 2000	20	4.85	31	GRP	2 × MTU 8V 2000 CR	Twin submerged shaft
Plascoa 2300	23.47	5.55	35	GRP	2 × MTU 16V 2000 CR	Twin submerged shaft

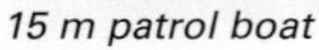

15 m patrol boat 0580411

11 m fast patrol boat 1150 0580409

Sillinger

Z-I de Buray, 41500 Mer, France

Tel: (+33 2) 54 81 08 32
Fax: (+33 2) 54 81 39 95
e-mail: contact@sillinger.com
Web: www.sillinger.com

Sillinger is a specialist in foldable and semi-rigid inflatable boats for military use, with over 40 years of experience throughout the world. The semi-rigid UM range features glass fibre hulls with Hypalom-Neoprene tubes, and can be fitted with a range of modular accessories such as consoles, seats, tables and ballistic protection.

525 RIB UM
1120288

Vessel type or class name	Length (m)	Max beam (m)	Max speed (kt)	Hull material	Engines	Propulsion
330 RIB UM	3.33	1.60	–	GRP	15 kW outboard	–
380 RIB UM	3.82	1.78	–	GRP	33 kW outboard	–
425 RIB UM	4.25	1.78	–	GRP	41 kW outboard	–
470 RIB UM	4.65	1.98	–	GRP	55 kW outboard	–
490 RIB UM	4.88	2.12	–	GRP	60 kW outboard	–
525 RIB UM	5.25	2.15	–	GRP	86 kW outboard	–
650 RIB UM	6.40	2.55	45	GRP	150 kW outboard	–
765 RIB UM	7.65	3.04	55	GRP	2 × 150 kW outboards	–

Germany

DSB (Deutsche Schlauchboot GmbH & Co)

PO Box 1155, Angerweg 5, D-37632, Eschershausen, Germany

Tel: (+49 5534) 30 10
Fax: (+49 5534) 30 12 00
Web: www.deutsche-schlauchboot.de

Günter Boomgaarden, *Chief Executive Officer*
Dirk Hebeler, *Managing Director, Sales*

The DSB rescue boats have been developed in accordance with the SOLAS/IMO regulations and are approved by a number of shipping authorities for use as ships fast rescue boats.

DSB Zephyr 5.1SR 0507030

Vessel type or class name	Length (m)	Max beam (m)	Max speed (kt)	Hull material	Engines	Propulsion
DSB 420 IRB	4.2	1.8	–	–	outboard engine(s)	–
DSB 470 IRB	4.7	1.98	–	–	outboard engine(s)	–
DSB Zephyr 3.9 SR	3.9	1.94	–	–	outboard engine(s)	–
DSB Zephyr 4.3 SR	4.3	1.95	–	–	outboard engine(s)	–
DSB Zephyr 5.1 SR	5.1	2.05	–	–	outboard engine(s)	–
DSB 6.5 SR	6.5	2.51	–	Aluminium	outboard engine(s)	–

Fr Fassmer GmbH & Co

Industriestrasse 2, D-27804 Berne/Motzen, Germany

Tel: (+49 4406) 94 20
Fax: (+49 4406) 94 21 00
e-mail: info@fassmer.de
Web: www.fassmer.de

Joachim Hellrung, *Technical Sales*

Johannes Fassmer set up the company in 1850 and is still family owned five generations later. Today, Fassmer has over 300 employees worldwide in four divisions – shipbuilding, lifeboat construction, deck equipment and FRP components.

The engineers and technical personnel that make up an in-house design office are equipped with the advanced IT and CAD workstations and the production facilities include: heated workshops ranging in size up to 20,000 m², a 250 m fitting pier with a 26 ton crane and a ship elevator with a capacity of 1,500 tonnes.

Vessel type or class	Length (m)	Max beam (m)	Max Speed (kt)	Hull material	Engines	Propulsion
FRR 6.1	6.1	2.23	34	GRP	Single 140 hp outboard	–
FRR 7.0 ID-SF	6.96	2.6	33	GRP	158 kW diesel inboard	Single water-jet
20 m Fast Patrol Boat	20	5.6	30	Aluminium	2 × MTU 12V 183TE 92	Twin submerged shaft
21 m Fast Patrol Boat	21	5.8	34	Aluminium	2 × 1,100 kW MTU diesel inboards	Twin submerged shaft
28.5 m Customs Cruiser	28.62	6.2	26	Aluminium	2 × MWM TBD 604 BV 8	Triple screw

Peene Werft

17438 Wolgast, Schiffbauerdamm 1, Germany

Tel: (+49) 38 36 25 00
Fax: (+49) 38 36 25 01 53
e-mail: info.pw@peene-werft.de
Web: www.peene-werft.de

A member of the Hegemann Group, the Peene Werft shipyard has built more than 650 new ships including 257 naval vessels.

The yard also repairs and refits vessels for a large number of clients.

Vessel type or class name	Length (m)	Max beam (m)	Max speed (kt)	Hull material	Engines	Propulsion
Light Torpedo Attack Craft "Libelle"	18.96	4.42	48	Aluminium	3 × 880 kW diesel engines	Twin screw

Greece

Motomarine SA

27th km Korpiou-Varis Avenue, Koropi, GR-19400, Attica, Greece

Tel: (+30 210) 66 23 29 12
Fax: (+30 210) 602 08 65
e-mail: info@motomarine.gr
Web: www.motomarine.gr

A Baloglou, *President*
A Zoulias, *Vice-President and Managing Director*

Motomarine (using Lambro as a brand name) was established in 1962 to build wooden pleasure craft. It has since specialised in the design and construction of composite military, rescue and commercial vessels up to 22.5 m in length.

The company has an in-house design, research and development office with naval architects collaborating with several marine laboratories such as the National Technical University of Athens.

Recent major orders have included 39 RIBs (9.8 m) for the Hellenic Special Forces, 19 high speed patrol boats (17.4 m) for the Hellenic Coast Guard and two high speed craft (13.4 m) for the Hellenic Navy.

Vessel type or class name	Length (m)	Max beam (m)	Max speed (kt)	Hull material	Engines	Propulsion
Motomarine Lambro 57 PB	17.40	4.65	48.0	Composite	2 × MTU 12V 2000M90	Twin water-jet
Magna RIB 960	9.60	3.25	50.0	Composite	2 × Mercruiser 502 (310 kW each)	Twin stern drive
Magna RIB 44	14.50	–	70.0	Composite	2 × Seatek (560 kW)	Twin surface drive
Guardian 53	16.80	4.68	33.0	Composite	2 × MAN 12V	Twin submerged shaft
Javelin 74	22.65	5.00	40.0	Composite	2 × 1,450 kW diesels	Twin surface drive

Magna RIB 960 high-speed army troop carrier 1120205

Magna RIB 44 special forces vessel (Motomarine) 1120244

Lambro 57 PB Class high-speed patrol craft 1120207

Guardian 53 1198645

Watercraft Hellas

34 Askilipiou Street, GR-18545, Piraeus, Greece

Tel: (+301) 413 29 32
Fax: (+301) 417 47 37
e-mail: info@watercraft.gr
Web: www.watercraft.gr

John Georgiadis, *Manager*

The company produces three main sizes of rigid hull inflatable fast rescue craft, the 6.3, 6.5 and 6.8 m. All are SOLAS approved for this purpose. The vessels are constructed of GRP and powered by a single 48 kW outboard which gives a speed of 25 kt. The company also produces 5.3 and 6.1 m sport diving boats of a similar design.

India

ABG Shipyard

5th Floor, Bhupati Chambers, 13 Matthew Road, Opera House, Mumbai 400 004, India

Tel: (+91 22) 56 56 30 00
Fax: (+91 22) 364 92 36
e-mail: shipyard@abgindia.com
Web: www.abgindia.com

ABG Shipyard, the largest private shipyard in India, is known for conventional steel shipbuilding. In 2004 it was awarded a contract to build three 94 m Coast Guard ships for the Indian government. The vessels will primarily be used as pollution control vessels but additional duties will include surveillance, fishery protection and law enforcement, anti smuggling and search and rescue.

26 m PATROL CRAFT

In 2000, the company developed a fast lightweight aluminium patrol craft with Australian consultants Thornycraft Maritime and Associates. Two vessels were delivered in 2001.

Specifications

Length overall	26.0 m
Length waterline	27.8 m

Beam	6.6 m
Draught	1.2 m
Displacement	75 t
Deadweight	13 t
Crew	10
Fuel capacity	11,000 litres
Water capacity	500 litres
Max speed	45 kt
Operational speed	40 kt

Classification: LR 100AI SSC Patrol Mono LMC HSC G5.
Structure: All aluminium hull and superstructure.
Propulsion: Two MTU 12V 4000 M90 diesel engines each rated at 2,040 kW and driving a Kamewa 63 water-jet via a ZF BW 465-1 gearbox.

26 m Patrol Craft 1170875

Goa Shipyard

Vasco-Da-Gama, Goa, 403802, India

Tel: (+91 832) 251 21 52
Fax: (+91 832) 251 38 70
e-mail: contactus@goashipyard.com
Web: www.goashipyard.co.in

Radm. A K Handa, *Managing Director and Chairman*

Goa Shipyard Limited was established in 1957 and is today a leading shipyard on the West Coast of India, meeting the requirements of varied customers in the field of design, development, construction, repair, modernisation, testing and commissioning of ships. Goa Shipyard's facilities and resources include a CAD/CAM centre, modern manufacturing facilities including a number of building bays and a 180 m fit-out pier, a work force of over 2,000 skilled personnel and over 175 qualified engineers and naval architects. The shipyard can build traditional military ships and marine crafts, surface effect ships, hovercrafts, high-speed aluminium-hulled vessels, pollution control vessels, advance deep sea trawlers, fish factory vessels, catamarans, pleasure submarines and offers a whole range of marine products and services.

Vessel type or class name	Length (m)	Max beam (m)	Max speed (kt)	Hull material	Engines	Propulsion
Fast Attack Craft	25.4	5.67	45	–	2 × MTU 12V396 TE94 diesel engines	2 × Arneson surface drives

Indonesia

North Sea Boats

Jl Letzen S Parman No 1, Singajurah, Bunguwangi, 68464, East Java, Indonesia

Tel: (+62 333) 633 422
Fax: (+62 333) 633 104
e-mail: allan@northseaboats.com
Web: www.northseaboats.com

North Sea Boats is part of an investment company operating in Indonesia. The company offers three models suitable for military, rescue or special operations work. The X2K Interceptor, capable of over 50 knots, is designed for rapid response or regular patrol work. Various engine and equipment options are available including single or twin outboards, enhanced survivability features and the potential for gun installments.

Vessel type or class name	Length (m)	Max beam (m)	Max speed (kt)	Hull material	Engines	Propulsion
X2K Interceptor	11.0	2.33	52	GRP	Single or twin outboard	Outboard
X2K RIB	11.5	2.93	50	GRP	Single or twin outboard	Outboard
X-38 Combat Catamaran	11.8	4.59	40	GRP	Single or twin outboard	Outboard

Ireland

Safehaven Marine

Ashgrove, Cobh, County Cork, Ireland

Tel: (+353 214) 813 726
Fax: (+353 214) 812 483
e-mail: safehavenmarine@eircom.net
Web: www.safehavenmarine.com

Frank Kowalski, *Managing Director*

Safehaven Marine designs and builds the Interceptor range of pilot boats, intended for use is all weathers and featuring exceptional seakeeping and survivability. Interceptor vessels are currently on order for the Port of Portland Pilot and the Port of Saint Malo Pilot.

Interceptor 42 Cork Pilot 1120338

Interceptor 42 1120337

Vessel type or class name	Length (m)	Max beam (m)	Max speed (kt)	Hull material	Engines	Propulsion
Interceptor 42	13.47	4.16	27	FRP	2 × Volvo D12 diesels	Twin 5-blade propellers
Interceptor 46	14.0	4.2	25	FRP	2 × Volvo D12 diesels	Twin screw
Interceptor 55	16.5	5	25	FRP	2 × Scania 429 kW diesels	Twin screw

Israel

Israel Aerospace Industries

RAMTA Division, PO Box 323, Industrial Zone, Beer Sheva, 84102 Israel

Tel: (+972 8) 627 22 31
Fax: (+972 8) 640 22 52
e-mail: cbendov@iai.co.il
Web: www.iai.co.il

M Gilboa, *Deputy General Manager*

Israel Aerospace Industries is a large military equipment manufacturer. Its main business concerns are aircraft production and the supply of advanced avionics and electronic weapon systems around the world.

The company is responsible for the design and production of the Super Dvora Mk II patrol boat which has been chosen by the Israeli Navy as its next generation coastal patrol and interdiction vessel.

Super Dvora Mk III 0590154

Vessel type or class name	Length (m)	Max beam (m)	Max speed (kt)	Hull material	Engines	Propulsion
Super Dvora Mk (II and III)	25	5.7	45.0	Aluminium	–	–

Israel Shipyards

PO Box 10630, Haifa Bay, IL-26118, Israel

Tel: (+972 48) 46 02 46
Fax: (+972 48) 41 87 44
e-mail: marketing@israel-shipyards.com
Web: www.israel-shipyards.com

Israel Shipyards is one of the largest privately owned shipbuilding and repair facilities in the East Mediterranean. Established over 30 years ago the company was government owned until 1995. The manufacturing and repair plant is spread over 400,000 m² with a total wharf length of 900 m. The company has the ability to build both large and small military vessels, tugs and commercial vessels. The company was responsible for producing the successful Shaldag Class of high-speed patrol boats.

Recent deliveries have included three offshore patrol boats to the Hellenic Coast Guard, two fast patrol boats to the Israeli navy and two to an African navy.

Vessel type or class name	Length (m)	Max beam (m)	Max speed (kt)	Hull material	Engines	Propulsion
Shaldag Mk II	24.8	6.0	45.0	Aluminium	2 × MTU 12V 396TE	Twin water-jet Kamewa/Lips/MJP
Shaldag Mk III	27.0	6.0	45.0	Aluminium	2 × MTU/Caterpillar	Twin water-jet Kamewa/Lips/MJP
Seagull	32.0	7.0	40.0	Aluminium	2 × MTU 16V 4000 M90	Twin water-jet Kamewa/
OPV 58	58.0	7.6	33.0	Steel hull, aluminium superstructure	4 × MTU	4 × FPP
OPV 62	62.0	7.6	30.0	Steel hull, aluminium superstructure	4 × MTU	4 × FPP
OPV 75	75.0	9.0	26.0	Steel hull, aluminium superstructure	2 × MTU 1163	4 × CPP

Shaldag Mk II 0589534

Shaldag Mk III 1175912

Italy

Cantieri Capelli srl

Via delle Industrie, 19 26020 Spinadesco, Cremona, Italy

Tel: (+39 0372) 49 13 99
Fax: (+39 0372) 49 21 15
e-mail: info@cantiericapelli.it
Web: www.cantiericapelli.com

Cantieri Capelli, established in 1974, manufactures a wide range of RIBs and boats up to 10 m. The product range includes the Tempest 750 WTD RIB, the official boat of the "Raid Tourquoise" nautical rally challenge.

Vessel type or class name	Length (m)	Max beam (m)	Max speed (kt)	Hull material	Engines	Propulsion
Tempest 750 WTD	7.5	2.95	30	GRP	Yamaha outboard	Outboard

Cantieri Navali Del Golfo

Lungomare Caboto, I-04024 Gaeta (LT), Italy

Tel: (+39) 77 17 12 81
Fax: (+39) 771 31 00 69
e-mail: infocantieri@cantieridelgolfo.it
Web: www.cantieridelgolfo.it

Cantieri Navali del Golfo produces a range of patrol boats and motor yachts. The company recently merged with Cantieri Italcraft, a builder of pleasure craft between 9 and 30 m.

Vessel type or class name	Length (m)	Max beam (m)	Max speed (kt)	Hull material	Engines	Propulsion
Squalo Class patrol boat	13.5	4.2	35	GRP	2 × 450 kW Diesel Inboard	Twin submerged shaft
Class 800 patrol boat	17	5	37	Composite	2 × 820 kW Diesel Inboard	–
Class 500 patrol boat	9.73	3.75	40	Composite	2 × 324 kW Diesel Inboard	Twin water-jet
PS 2 Class patrol boat	11.6	3.8	34	Composite	2 × 275 kW Diesel Inboard	–
P58 Class patrol boat	17	5	37	Composite	2 × 890 kW Diesel Inboard	Twin submerged shaft

FB Design Srl

Via Provinciale 73, I-23841, Annone Brianza, Italy

Tel: (+39 0341) 26 01 05
Fax: (+39 0341) 26 01 08
e-mail: mail@fbdesign.it
Web: www.fbdesign.it

FB Design was established in 1971 by Fabio Buzzi to design and construct high performance motor boats and drive systems. The company has grown to occupy a 12,000 m^2 site. With its roots firmly established in the boat racing industry the company is now a leading supplier of very high-speed commercial and military craft. The current range of RIBs and conventional boats covers a range of sizes from 7.3 to over 24 m.

Customers have included the British MoD, Italian Customs Guardi di Finanza, US Coast Guard, Hellenic Coast Guard, Hong Kong Coast Guard as well as Rwandan and Ugandan forces.

FB 55 ft SC 0589540

FB RIB 24 ft 0589546

FB RIB 55 ft SC 0589520

FB RIB 33 ft SC 0589547

FB RIB 40 ft 0589514

FB RIB 42 ft SC 0589550

Vessel type or class name	Length (m)	Max beam (m)	Max speed (kt)	Hull material	Engines	Propulsion
FB 38′ STAB	11.85	2.89	65	Composite	2 × Yanmar 6LY2 STE (625 kw total)	Twin Surface Drive
FB 38′	11.85	2.32	65	Composite	2 × Yanmar 6LY STE (536 kW total)	Twin Surface Drive
FB 40′	12.94	2.59	68	Composite	2 × CAT C9 (760 kW total)	Twin Surface Drive
FB 55′ (Built as V6000 Levriero)	16.55	2.85	–	Balsa Sandwich	2 × Caterpillar 3406 E (1,190 kW total)	Twin Surface Drive
FB 55′ SC	16.4	2.84	65	Composite	2 × CAT 3406 E (1,190 kW total)	Twin Surface Drive
RIB 55′ SC	16.52	3.89	66	Composite	2 × Isotta Franschini V 1312T2 MS (1,786 kW total)	Twin Surface Drive
RIB 42′ SC	13.23	3.57	66	Composite	2 × CAT C12 (1056 kW total)	Twin Surface Drive
RIB 42′	13.2	3.56	55	Composite	2 × Yanmar 6LY2-STE (625 kW total)	Twin Surface Drive
RIB 38′	11.5	2.74	65	Composite	2 × Yanmar (654 kW total)	Twin Surface Drive
RIB 36′	11	2.65	60	–	2 × Yamaha Outboards (372 kW total)	Twin Outboard
RIB 33′ SC	10.43	2.72	63	Composite	2 × Yanmar (625 kW total)	Twin Surface Drive
RIB 33′	10.37	2.67	54	Composite	2 × Yanmar (446 kW)	Twin Surface Drive
RIB 24′	7.5	2.6	55	Composite	–	Single water-jet
FB 55′ STAB	16.41	3.58	70	Composite	2 × Isotta Fraschini 1312 V12	Twin surface drive
FB 42′ STAB	12.93	3.13	55	Composite	2 × Caterpillar C9	Twin surface drives
RIB 24′	7.5	2.6	55	Composite	2 × Yanmar 6LPA-STP	Single water-jet
RIB 38′ STAB	11.58	2.89	65	Composite	2 × Yanmar 5TE	Twin surface drives
FB 48′ STAB	14.54	2.55	70	Composite	2 × Caterpillar C12	Twin surface drives

Intermarine SpA

Via Alta, I-19038, Sarzana, Italy

Tel: (+39 01) 87 61 71
Fax: (+39 01) 87 67 42 49
e-mail: mail@intermarine.it
Web: www.intermarine.it

Oswaldo Facchinetti, *Business Development Manager*

Intermarine was founded in the 1970's as a shipyard specialising in the design and manufacture of FRP (Fibre Reinforced Plastic) marine craft. The company is historically linked with the development of mine countermeasure vessels sold to many of the world's navies.

In 2002, the company was acquired by Rodriquez Cantieri Navali, itself a world leader in the design and construction of steel and aluminium fast ferries, military craft, special craft and megayachts.

Since the merger of companies, Intermarine is responsible for the design, production and sales of all military products for the group.

Intermarine's main shipyard is located in Sarzana (near La Spezia) and has a total area of 76,000 m². The company's second yard is used primarily for outfitting boats and is located in La Spezia.

V-5000 0130143

V-6000 0104897

Vessel type or class name	Length (m)	Max beam (m)	Max speed (kt)	Hull material	Engines	Propulsion
V-6000 Levriero	16.43	2.84	75	Composite	4 × Seatek 6-4V-10D	Four surface drives
V-5000	16.5	4.5	58	Composite	2 × MTU 12V 2000 M93	Twin surface drive
CP 25 fast patrol boat	25	5.76	34	Composite	3 × 735 kW diesel inboard	Twin submerged shaft + booster water-jet
MV85 fast patrol boat	28.2	7.04	40	Composite	2 × MTU 16V 4000 M90 diesel engines (2720 kW each)	–
V2000	13.2	3.44	45	Composite	2 × Seatek 600Plus (2 × 620 hp)	2 × Rolls-Royce Kamewa FF375S waterjets
22 m	28.2	5.4	30	Composite	2 × MTU 12V 2000 M93 (1340 kW each)	2 × shafts and FP propellers

Lomac Nautica SRL

Via Bruno Buozzi 26, I-20097, S.Donato, Milanese (MI), Italy

Tel: (+39 02) 51 80 05 38
Fax: (+39 02) 51 80 05 31
e-mail: info@lomac.it
Web: www.lomac.it

Lomac, a manufacturer of leisure RIBs, has an associated company, Nautica 90, which produces a 12 m RIB for pilot and other commercial operations. With two 240 kW diesel engines the craft has a speed of 35 kt, although speeds of over 50 kt are understood to be possible with this design.

Nautica 90 Pilotina jet boat
1120322

Vessel type or class name	Length (m)	Max beam (m)	Max speed (kt)	Hull material	Engines	Propulsion
Nautica 90 Pilotina	12	4.2	35	GRP	2 × Inboard Diesel (480 kW total)	Twin water-jet

Novamarine 2 SpA

Via Dei Lidi, Zona Industriale, I-07026, Olbia, Italy

Tel: (+39 0789) 585 38
Fax: (+39 0789) 588 56
e-mail: novamarine@tin.it

Formed in 1983, Novamarine produces a large range of rigid inflatable craft from 2.5 to 18 m in length. Together with traditional RIBs, Novamarine also produces T.U.G. foam collar system non-inflatable RIBs in the same length range with various propulsion systems, for both leisure and heavy duty work and military use.

Vessel type or class name	Length (m)	Max beam (m)	Max speed (kt)	Hull material	Engines	Propulsion
HD Seven	7.00	2.56	–	GRP	1/2 Outboards Max 150 kW	Inboard + Jet/stern drive also available.
HD Seven-Fifty TUG	7.60	2.56	–	GRP	1/2 Outboards Max 190 kW	Inboard + Jet/stern drive also available.
HD Nine	9.20	3.25	–	GRP	1/2 Outboards Max 300 kW	Inboard + Jet/stern drive also available.
HD Ten	9.99	3.5	–	GRP	1/2 Outboards Max 445 kW	Inboard + Jet/stern drive also available.
HD Ten T.U.G.	9.99	3.5	–	GRP	1/2 Outboards Max 445 kW	Inboard + Jet/stern drive also available.
HD Twelve	13.50	4.0	–	GRP	2 Inboard Diesel Max 670 kW	Jet Drives.
HD Fifteen	15.00	4.1	–	GRP	2/3 Inboard Diesel Max 1,045 kW	Stern/Jet drives (3 outboards also available).
HD Fifteen T.U.G.	15.00	4.1	–	GRP	2/3 Inboard Diesel Max 1,045 kW	Stern/Jet drives (3 outboards also available).
HD Sixteen	15.70	4.0	–	GRP	2 × Inboard Diesel Max 893 kW	Jet Drives.

Novurania

Via Circonvallazione 3, I-38079, Tione Di Trento, Italy

Tel: (+39 0465) 33 93 11
Fax: (+39 0465) 32 15 51
e-mail: novuraniaspa@novurania.it
Web: www.novurania.it

Flavia Pellegrini, *Marketing Director*

Novurania MOB 570 SOLAS
0507031

Novurania supplies a range of inflatable rescue and workboat craft. The rescue craft are SOLAS approved (Whitecap range and MOB SOLAS). The MOB 570 SOLAS and Mx 650 FRB are powered by inboard engines and water-jets. The Mx 650 FRB is certified for use in cruise liners as a fast rescue boat. Sales are made for both commercial craft and charter yachts.

Vessel type or class name	Length (m)	Max beam (m)	Max speed (kt)	Hull material	Engines	Propulsion
Whitecap 330	3.30	1.58	–	–	Outboard engine(s)	–
Whitecap 360	3.60	1.63	–	–	Outboard engine(s)	–
Whitecap 390	3.90	1.75	–	–	Outboard engine(s)	–
420SOLAS	4.22	1.84	15	–	Outboard engine(s)	–
MOB 500	5.05	2.07	–	–	Outboard engine(s)	–
MOB 570	5.7	2.30	–	–	Outboard engine(s)	–
MOB 650	6.50	2.76	–	–	Outboard engine(s)	–
MOB 500 SOLAS	5.02	2.10	25	–	Outboard engine(s)	–
MOB 570 SOLAS	5.70	2.30	28	–	Outboard engine(s)	–
Mx 650 FRB	6.70	2.50	31	–	Outboard engine(s)	–

Novurania SpA

Via Circonvallazione 3, I-38079 Tione Di Trento, Italy

Tel: (+39 0465) 33 93 11
Fax: (+39 0465) 32 11 51
e-mail: novuraniaspa@novurania.it
Web: www.novurania.it

Flavia Pellegrini, *Marketing Director*

Novurania MOB 570 SOLAS 0507031

Novurania supplies a range of inflatable rescue and workboat craft. The rescue craft are SOLAS approved (Whitecap range and MOB SOLAS). The MOB 570 SOLAS and Mx 650 FRB are powered by inboard engines and water-jets. The Mx 650 FRB is certified for use in cruise liners as a fast rescue boat. Sales are made for both commercial craft and charter yachts.

Model	Length overall (m)	Beam overall (m)	Tube dia m (mm)	No of compartments	Boat weight (kg)	Capacity (weight kg)	Capacity (persons)	Max engine power (kW)	Approx speed (kt)
Whitecap 330	3.30	1.58	420	3	70	–	4	20	–
Whitecap 360	3.60	1.63	430	3	78	–	5	20	–
Whitecap 390	3.90	1.75	450	3	85	–	6	30	–
420 SOLAS	4.22	1.84	–	–	235	685	–	–	15
MOB 500	5.05	2.07	500	5	220	1,100	8	50	–
MOB 570	5.70	2.30	540	5	351	1,500	8	100	–
MOB 650	6.50	2.76	580	5	583	2,050	12	170	–
MOB 500 SOLAS	5.02	2.10	500	5	560	–	6	50	25
MOB 570 SOLAS	5.70	2.30	540	5	1,335	–	6	85	28
Mx 650 FRB	6.70	2.50	560	5	1,850	2,500	6	215	31

Lebanon

Lenco Marine S.A.L.

Halate, North Highway, PO Box 175, Byblos, Lebanon

Tel: (+961 9) 478 847
Fax: (+961 9) 740 032
e-mail: lencomarine@lenco-marine.com
Web: www.lenco-marine.com

Lenco Marine is a family owned business that designs and constructs high performance sports boats and has recently entered the defence market with the Phenix 55 FPB.

All models are built from composites, using Kevlar honeycomb for stiffeners and comply with international standards. Vessels are distributed in France, the Caribbean, Germany, Norway, UAE and Jordan.

Phenix 55 FPB 1120300

Vessel type or class name	Length (m)	Max beam (m)	Max speed (kt)	Hull material	Engines	Propulsion
Phenix 55 FPB	16.76	3.5	55	Composite	2 × Catperpillar diesels (595 KW)	Twin surface drives

Malaysia

Hong Leong Lurssen Shipyard Sdn Bhd

PO Box 43, 12700 Butterworth, Penang, Malaysia

Tel: (+60 4) 332 99 55
Fax: (+60 4) 332 99 66
e-mail: hlls@po.jaring.my

Chuah Chuan Yong, *General Manager*

The company was established in 1969 as a joint venture between Malaysia's Hong Leong Company, German shipbuilder Fr. Lurssen Werft and Pernea Securities. Since then HLLS has built a large number of craft for the Malaysian armed forces, police force and port authorities.

All craft have been built in the shipyard based on the banks of the Prai River in Butterworth. The shipyard is equipped for new construction and repairs including a 750 tonne syncrolift. It is also equipped for the repair and overhaul of marine engines.

Deliveries have included 23 × 14 m aluminium patrol boats designed by Simonneau Marine International and 15 PZ class patrol boats designed by Lurssen Werft.

Vessel type or class name	Length (m)	Max beam (m)	Max speed (kt)	Hull material	Engines	Propulsion
14 m patrol boat	14	4	37	Aluminium	2 × MTU 12V 183	Twin submerged shaft

Kay Marine SDN BHD

Lot 4961, Telok Pasu, Manir, 21200 Kuala Terengganu, Malaysia

Tel: (+60 9) 615 51 45
Fax: (+60 9) 615 39 84
e-mail: kaym@tm.net my

Kay Marine is a specialist marine fabricator in Malaysia and has built a range of small fast craft.

Vessel type or class name	Length (m)	Max beam (m)	Max speed (kt)	Hull material	Engines	Propulsion
9.0 m RIB	8.85	3.21	55	Marine aluminium	–	–
Bullet Proof Patrol Boat	11.5	3.57	30	Aluminium	Twin outboard	–

NGVTech SDN BHD

No 508 Block D, Pusat Dagangan Phileo Damansara 1, No 9 Jalan 16/11, off Jalan Damansara, 46350 Petaling Jaya, Selangor, Malaysia

Tel: (+60 3) 7660 1780
Fax: (+60 3) 7660 1781
e-mail: enquiry@ngvtech.com.my
Webl: www.ngvtech.com.my

Zulkifli Shariff, *Executive Director*

Established in 1992, NGVTech has developed a varied product range including offshore support vessels, high speed patrol and interceptor vessels as well as fast catamarans. The company fabricates vessels in aluminium, steel and composites and its scope includes design, consultancy, communications and navigational aids as well as manufacture.

NGVTech 18 m interceptor capable of 45 knots 1120306

Vessel type or class name	Length (m)	Max speed (m)	Max speed (kt)	Hull material	Engines	MTU	Propulsion
Fast Interceptor Craft	18.1	4.2	45	Aluminium	[Enter value]		waterjets

Netherlands

Damen Shipyards

Industrieterrein Avelingen West 20, PO Box 1, NL-4200 AA, Gorinchem, Netherlands

Tel: (+31 183) 63 94 36
Fax: (+31 183) 63 95 05
e-mail: info@damen.nl
Web: www.damen.nl

J L Gelling, *Product Director*

Damen Shipyards specialises in the design and construction of a wide range of work boats, dredgers and patrol craft up to 90 m in length and with installed power up to 10,000 kW. The Shipyard started in 1927 and in 1969, began the construction of commercial vessels to standard designs by modular construction. Since 1969 more than 3,000 ships have been built and delivered worldwide to 117 countries.

The High Speed and Naval Craft division produces a range of high-speed craft for commercial, patrol, surveillance and rescue operations. From 1969, more than 700 high-speed craft have been built by Damen Shipyards Gorinchems' High Speed and Naval Craft department.

Damen Interceptor at speed 1120200

Fast Crew Supplier 1605 vessels 1120309

Vessel type or class name	Length (m)	Max beam (m)	Max speed (kt)	Hull material	Engines	Propulsion
Interceptor DI 1503	15	2.7	60	Aluminium	3 × innovation sledgehammer	Triple stern drives
Fast Crew Supplier 1605	16.15	5.4	28	Aluminium	–	–
Fast Crew Supplier 3307	33.5	6.8	28	Aluminium	–	–
SAR 903	9.3	3.04	35	GRP	–	–
Stan Tender 1905	19.05	5.3	25	Steel	–	–

Madera Ribs B.V.

Plomperstraat 6, 3087 BD, Rotterdam, Netherlands

Tel: (+31 10) 429 4511
Fax: (+31 10) 428 0599
e-mail: info@m-ribs.com
Web: www.m-ribs.com

Madera Ribs specialise in the design and construction of large ribs of between 8 and 20 metres for commercial and military customers. The Ribs are designed to carry high payloads over huge distances. The rigid hulls are constructed in GRP and feature a high freeboard and multiple watertight compartments.

Madera Ribs MRC-1450 1120351

Madera Ribs MRC-1250 1120352

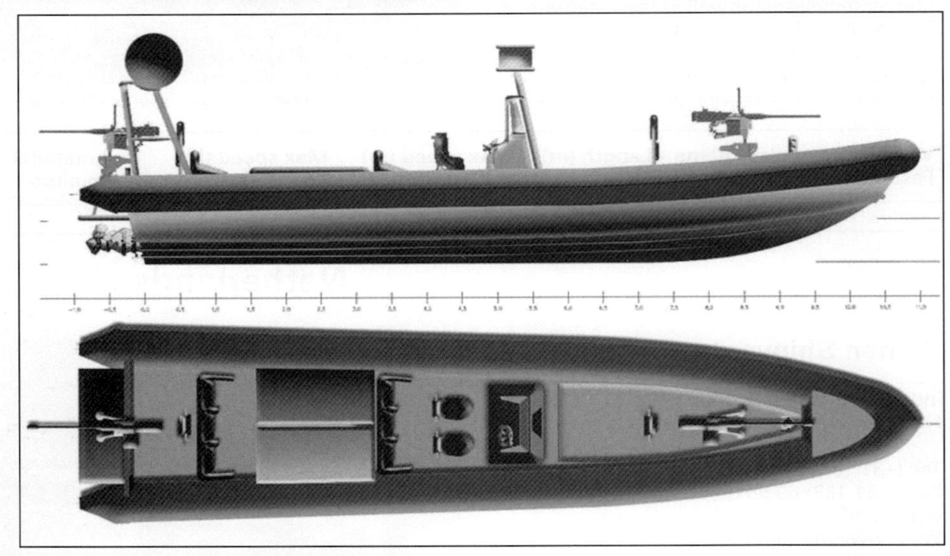

Madera Ribs MRD-1130 1120353

Madera Ribs MRC-1050 1120354

Vessel type	Length (m)	Beam (m)	Speed (kt)	Hull material	Engines	Propulsion
MR-1000	10	2.67	55	GRP	2 × 350 Hp Outboards	–
MR-1130	11.3	2.67	55	GRP	3 × 250 Hp Outboards	–
MR-1230	12.3	2.76	50	GRP	2 × 250 Hp Outboards	–
MRC-1050	10.5	3.21	60	GRP	4 × 250 Hp Outboards	–
MRC-1250	11.89	3.58	60	GRP	4 × 250 Hp Outboards	–
MRC-1450	14.26	3.59	58	GRP	5 × 350 Hp Outboards	–
MRD-1000	10	2.67	50	GRP	2 × 223 kW Diesels	Stern drives
MRD-1130	11.3	2.67	55	GRP	2 × 357 kW Diesels	Stern drives
MRCD-1050	10.5	3.21	60	GRP	2 × 440 kW Diesels	Stern drives
MRCD-1250	11.89	3.58	60	GRP	2 × 533 kW Diesels	Stern drives
MRCD-1450	14.38	4.0	53	GRP	2 × 820 kW Diesels	Waterjets
MRCD-1500	15	4.05	50	GRP	2 × 820 kW Diesels	Waterjets
MRCD-1950	19.5	4.1	35	GRP	2 × 820 kW Diesels	Surface Drives

New Zealand

AMF Boats

125 Hewletts Road, Mount Maunganui, New Zealand

Tel: (+64 7) 574 9377
Fax: (+64 7) 574 6959
e-mail: amfboatsco@xtra.co.nz
Web: www.amfboats.co.nz

AMF builds a variety of aluminium hulled fast monohulls including boats from 7 to 13 m. Although primarily targeted at the fishing industry, the vessels are also suitable for search and rescue, port authority and military use.

Vessel type or class name	Length (m)	Max beam (m)	Max speed (kt)	Hull material	Engines	Propulsion
Commercial 820	8.2	2.8	–	Aluminium	Volvo AD41P 200 hp diesel	Stern drive
Commercial 960	9.6	3.3	–	Aluminium	Volvo D6 310 hp diesel	Stern drive
Commercial 1120	11.2	3.5	–	Aluminium	Volvo D6 350 hp diesel	Stern drive
Commercial 1300	13.0	4.4	–	Aluminium	Twin Volvo 450 hp diesels	Twin Hamilton 292 waterjets
GRV 900 RIB	9.0	2.88	41.7	Aluminium	Yanmar 440 hp diesel	Hamilton 274 waterjets

Aquapro International Ltd

30 Hargreaves Street, Ponsonby, Auckland, New Zealand

Tel: (+649) 373 20 19
Fax: (+649) 307 10 12
e-mail: sales@aquapro.co.nz
Web: www.aquapro.co.nz

Peter Spear, *Sales*

Aquapro build a wide range of RIBs for both the leisure and commercial markets. They have experience in building vessels with and without cabins in a range of sizes up to 14 m in length.

Aquapro Raider 790
0583202

Vessel type or class name	Length (m)	Max beam (m)	Max speed (kt)	Hull material	Engines	Propulsion
RDC 605	6.05	2.50	–	–	–	–
RDC665	6.65	2.75	–	–	–	–
RDC790	7.90	2.90	–	–	–	–
RDC900	9.00	3.00	–	–	–	–
RDC1000	10.00	3.47	–	–	–	–
RDC1400	14.00	4.57	–	–	–	–
Seataxi	9.00	3.30	–	–	–	–

Lancer Industries Ltd

PO Box 83-136, Edmonton, Auckland, New Zealand

Tel: (+649) 837 12 06
Fax: (+649) 837 12 08
e-mail: info@lancer.co.nz
Web: www.lancer.co.nz

R Winstone, *Technical Director*

Lancer Industries manufactures inflatable and rigid inflatable boats for defence forces in New Zealand and overseas. They also supply a wide range of leisure craft and SOLAS /USCG approved rescue craft.

The largest vessel in the range is the 13.9 m offshore RIB which can be supplied in a wide range of configurations.

Lancer 1300RD
0077914

Vessel type or class name	Length (m)	Max beam (m)	Max speed (kt)	Hull material	Engines	Propulsion
390P	3.90	1.76	–	GRP	Outboards	–
450P	4.50	1.93	–	GRP	Outboards	–
520P	5.20	2.14	–	GRP	Outboards	–
600P	5.20	2.14	–	GRP	Outboards	–
620RD	6.20	2.58	–	GRP	Outboards	–
750RD	7.50	2.80	–	GRP	Outboards	–
850RD	8.50	2.75	–	GRP	Outboards	–
940RD	9.40	3.60	–	GRP	Outboards	–
1300RD	13.00	3.45	–	GRP	Outboards	–

Naiad Inflatables

PO Box 252, Picton, New Zealand

Tel: (+64 3) 573 72 46
Fax: (+64 3) 573 72 74
e-mail: enquiries@naiad.co.nz
Web: www.naiad.co.nz

Naiad designs and builds a wide range of aluminium hulled rigid inflatable boats from 2.5 to 15 m in length. The boats have proven very popular with the commercial market and also the Americas Cup Teams.

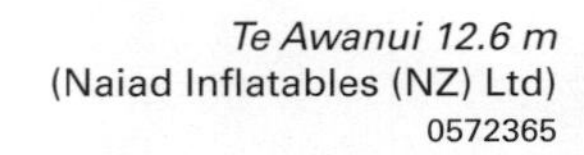

Te Awanui 12.6 m
(Naiad Inflatables (NZ) Ltd)
0572365

Vessel type or class name	Length (m)	Max beam (m)	Max speed (kt)	Hull material	Engines	Propulsion
4.3	4.3	2.2	–	Aluminium	90 hp outboard	–
4.8	4.8	2.2	–	Aluminium	90 hp outboard	–
5.3	5.3	2.2	–	Aluminium	115 hp outboard	–
5.8	5.8	2.2	–	Aluminium	2 × 90 hp outboard	–
6.8	6.8	2.5	–	Aluminium	2 × 140 hp outboard	–
8.5	8.5	2.9	45	Aluminium	2 × 225 hp outboard	–
12.6	12.6	3.8	–	Aluminium	2 × diesel inboard	–

Rayglass Boats

7 Paisley Place, Mt Wellington, Auckland, New Zealand

Tel: (+649) 57 3 7979
Fax: (+649) 57 3 7978
e-mail: enquiry@rayglass.co.nz
Web: www.protectorribs.com

Lyndsay Turner, *Manager*

Known for its range of sportsboats, Rayglass has recently produced a number of rigid inflatable craft of up to 19 m in length. The 8.5 m version is used by the New Zealand Coast Guard for patrol and interception duties.

9.5 m RIB

0583187

11 m cabin RIB 0583188

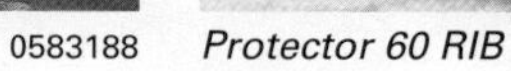

Protector 60 RIB 0583189

Vessel type or class name	Length (m)	Max beam (m)	Max speed (kt)	Hull material	Engines	Propulsion
8.5 m RIB	8.85	2.9	50	GRP	2 × 168 kW outboards	–
10 m RIB	10.0	3.6	45	GRP	2 × 168 kW outboards	–
11 m RIB	11.0	3.6	45	GRP	2 × 168 kW outboards	–
12 m RIB	12.0	3.6	45	GRP	2 × 168 kW outboards	–
Protector 60	19	–	–	GRP	Up to 890 kW	–

Sensation Yachts

11 Selwood Road, Henderson, PO Box 79020,
Auckland 8, New Zealand

Tel: (+649) 837 22 10
Fax: (+649) 836 17 75
e-mail: sensation@sensation.co.nz
Web: www.sensation.co.nz

Carl A Morris, *Sales Manager*

Sensation Yachts' 12 m Interceptor 1120236

Sensation Yachts primarily manufacture luxury motor and sailing yachts. This company has been expanding in recent years and the premises now include specialist workshops for engineering, electrical, interior and exterior woodwork, painting and finishing as well as storage and load-out facilities. Expansion has also been a catalyst for the company to promote its expertise in aluminium construction techniques as well as in steel and composite materials.

Due to the Americas Cup Yachting Regatta in 1999–2000 and 2002–03 the New Zealand Police Maritime Unit needed a fast patrol boat to police the event and awarded a contract to Sensation Yachts to build eleven vessels. Sensation Yachts are now marketing the design as a high-speed patrol boats for military use.

Vessel type or class name	Length (m)	Max beam (m)	Max speed (kt)	Hull material	Engines	Propulsion
12 m RHIB Interceptor	12.17	3.55	40	Aluminium	2 × 225 Hp Outboards	Twin water-jet

Norway

H Henriksen Mekaniske Verksted A/S

Traeleborgveien 15, P.O Box 2404, N-3104 Tonsberg, Norway

Tel: (+47 33) 37 84 00
Fax: (+47 33) 37 84 30
e-mail: postbox@hhenriksen.com
Web: www.hhenriksen.com

The company was founded in 1856 to supply equipment to the marine industry. The current range of products includes lifting hooks, aluminium workboats, man overboard boats and specialist machinery.

The current boat range includes three rescue boats and a larger 42 kt fast patrol boat.

This fast aluminium patrol vessel is designed for military or search and rescue use. The bridge accommodates 4 and the aft cabin has space for 10 passengers or 6 stretchers, with additional space for 4 passengers in the bow. The vessel is designed with a deep-Vee hull giving good seakeeping abilities and is capable of a top speed of 42 knots.

HS 1365 2VD fast patrol vessel 1120321

Specifications

Length overall	13.65 m
Beam	4.1 m
Draught	0.9 m
Crew	18
Max speed	42 kt

Propulsion: Twin Volvo Penta D12 650 hp engines driving Kamewa K32 waterjets.

Vessel type or class name	Length (m)	Max beam (m)	Max speed (kt)	Hull material	Engines	Propulsion
Henriksen HP 1040 2 VD	10.4	3.4	38	Aluminium	2 × Yanmar 6LY (M)-STE	Twin water-jet
HS 1365 2VD	13.65	4.10	42	Aluminium	2 × Volvo Penta D12 (967 kW total)	Twin water-jet
GTC 700	7.55	2.85	35	Aluminium	1 Inboard Diesel (171 kW)	Single water-jet
GTC 900	8.75	3.3	32	Aluminium	1 × Inboard Diesel (253 kW)	Single water-jet

Maritime Partner A/S

Notenesgata 1, PO Box 776 Sentrum, N-6001 Aalesund, Norway

Tel: (+47) 70 11 65 65
Fax: (+47) 70 11 65 55
e-mail: office@maritime-partner.com
Web: www.maritime-partner.com

Peder R Myklebust, *Managing Director*
John Ivar Olsen, *Marketing Director*
Arne K Dybvik, *Sales Director*

Maritime Partner AS was founded in May 1994. In 1999 it acquired the brands Alusafe, Seabear and Weedo. Maritime Partner AS can now supply 50 boats of different types ranging for 5 to 25 m. Their boats are built to comply with IMO regulations as well as DNV and other classification societies.

Vessel type or class name	Length (m)	Max beam (m)	Max speed (kt)	Hull material	Engines	Propulsion
Sjobjorn 25	7.8	2.9	33	Aluminium	Yanmar 6LYA-STE	Single water-jet
MP 741 Springer	7.4	2.72	33	Aluminium	Diesel inboard (up to 250 kW)	Single water-jet
Mp 911 FRDC	9.2	3.5	33	Aluminium	2 × Volvo AD31P/DP	Twin Stern Drive
MP 1000 FRDC	10	3.5	32	Aluminium	2 × Volvo KAD 32 P	Twin Stern Drive
MP 1211 FPC	12.35	3.5	35	Aluminium	2 × Yanmar 6LYA-STE	Twin water-jet
Alusafe 770	7.7	2.7	32	Aluminium	Inboard diesel up to 250 kW	Single water-jet
Alusafe 1290	12.9	3.5	42	Aluminium	2 × Volvo Penta TAMD 74 EDC	Twin water-jet
Alusafe 1500 MkII	14.8	4.6	36	Aluminium	2 × Caterpillar 425 kW	Twin water-jet
Alusafe 2300	23	5.75	33	Aluminium	2 × 820 kW inboard diesels	Twin water-jet

Norsafe A/S

PO Box 115, N-4852, Faervik, Norway

Tel: (+47) 37 05 85 00
Fax: (+47) 37 05 85 01
e-mail: mail@norsafe.no
Web: www.norsafe.no

Oskar Havnelid, *Sales Manager*
Bjarte Skaala, *Technical Manager*

Norsafe A/S is a company was founded in 1992 although its origins date back to 1864 when it was producing wooden lifeboats. The company has supplied more than 15,000 lifeboats to the global shipping and offshore markets and also produces a range of fast rescue craft (FRC). The FRC range in length from 4.65 m to nearly 12 m and certain models achieve speeds of up to 35 kt.

Munin 1200 (Norsafe) 1120211

Vessel type or class name	Length (m)	Max beam (m)	Max speed (kt)	Hull material	Engines	Propulsion
Mako 655	6.55	2.70	25	GRP	Steyr 180 kW diesel inboard	Single water-jet
Magnum 750	7.65	2.90	32	GRP	Yanmar 250 kW diesel inboard	Single water-jet
Munin 1000	10.00	3.51	30	GRP	2 × 186 kW diesel inboard	Twin water-jets
Munin 1200	11.65	3.75	32	GRP	2 × 310 kW diesel inboard	Twin water-jets
Maya 850	9.19	3.29	25	GRP	2 × 120 kW diesel inboard	Twin waterjets

Maya 850 fast rescue craft 1120210

Norsafe Magnum 750 fast rescue craft 1120240

Portugal

Valiant

A subsidiary of the Brunswick Corporation

PO Box 38, P-4920 909, Vila Nova De Cerveira, Portugal

Tel: (+351 251) 70 80 60
Fax: (+351 251) 70 80 69
e-mail: valiant@valiant-boats.com
Web: www.valiant-boats.com

Valiant produces a wide range of rigid hull inflatable craft for both the leisure and commercial sectors. The four main commercial craft are manufactured with a GRP hull and a Neoprene/Hypalon or polyurethane inflatable tubes.

Valiant DR 750
0589524

Vessel type or class name	Length (m)	Max beam (m)	Max speed (kt)	Hull material	Engines	Propulsion
PT550	5.5	2.4	–	GRP	–	–
PT600	6.0	2.5	–	GRP	–	–
PT750	7.5	2.85	–	GRP	–	–
PT850	8.5	3.0	–	GRP	–	–

Russian Federation

Almaz Marine Yard

19 Uralskaya Str, 199155, St. Petersburg, Russian Federation

Tel: (+7 812) 350 28 66
Fax: (+7 812) 350 77 50
e-mail: market@shipconstruction.ru
Web: www.shipconstruction.ru

ALMAZ Marine Yard was founded in 1931 and specialised in repairing Russian Border Guard and Navy high-speed craft. In 1970 the production facilities of ALMAZ Marine Yard were modernised and reconstructed to start the production of the high-speed craft for the Russian Navy. Today ALMAZ Marine Yard produces a wide range of custom craft up to 70 m in length and 30 m in breadth and employs 250 staff.

Vessel type or class name	Length (m)	Max beam (m)	Max speed (kt)	Hull material	Engines	Propulsion
A-125 Fast Patrol Boat	17.6	4.2	48	Aluminium	2 × MTU 8V 2000 M90	Twin water-jet
L2000	8.3	3.5	45	Aluminium	2 × 170 kW diesel inboard	Twin stern drive
A-66 Fast Patrol Boat	6.2	2.16	40	Aluminium	150 hp outboard	Single outboard
A-67 Fast Patrol Boat	8.6	2.4	40	Composite	200 hp outboard	Single outboard
A-99 Fast Patrol Boat	7.9	2.4	43	Aluminium	250 hp outboard	Single outboard

Almaz Shipbuilding Company

26 Petrovsky Pr, St Petersburg 197110, Russia

Tel: (+7 812) 235 51 48
Fax: (+7 812) 235 70 69
e-mail: market@almaz.spb.ru
Web: www.almaz.spb.ru

Alexander Shliakhtenko, *President*
Nikolai Ivakin, *Shipyard Director*
Leonid Grabovets, *General Director*
Sergei Galichenko, *Marketing Manager*

Almaz Shipbuilding occupy a 165 000 m² production facility and, amongst other naval projects, produce fast patrol boats.

Sobol Patrol Boat

Recent deliveries include the Sobol patrol boat for the Russian Federal Security Service.

Specifications

Length overall	30.0 m
Beam	5.82 m
Draught	1.34 m
Displacement max	57 t
Maximum speed	47 kt

Propulsion: Two Deutz TVD 616V16 diesels connected to two surface drives.

Vessel type or class name	Length (m)	Max beam (m)	Max speed (kt)	Hull material	Engines	Propulsion
Sobol	30.0	5.82	47.0	–	2 × Deutz TVD 616V16	Surface drives

Central Hydrofoil Design Bureau (CHDB)

51 Svoboda Street, Nizhny Novgorod 603602, Russian Federation

Tel: (+7 8312) 73 10 37
Fax: (+7 8312) 73 02 48
e-mail: rostislav@sinn.ru

I M Vasilavsky, *Director*
M A Suslov, *Technical Director*
Sergey V Patonov, *General Director*

CHDB design and build prototype craft such as air cushioned vehicles and wing-in-ground effect craft. The Serna craft is intended for fast transportation of containers, heavy vehicles and other cargo. It is understood that at least two have been constructed. The vessel incorporates an air cavity within the hull which allows a significantly higher speed than would normally be available for a conventional craft of this size and installed power.

Serna 0007180

Vessel type or class name	Length (m)	Max beam (m)	Max speed (kt)	Hull material	Engines	Propulsion
Serna	25.4	5.9	30	–	2 × Zvezda M503A-Z	Twin tunnelled surface drives

Feodosia Shipbuilding Company (Morye)

Desantnikov Str 1, Feodosia, Crimea 98176, Ukraine

Tel: (+380 65) 626 37 22
Fax: (+380 65) 626 99 05
e-mail: kgp@morye.kafa.crimea.ua

Boris Kozlov, *Chairman*
Sergei Ostapenko, *Technical Director*
Grygory Klebanov, *Head of Sales*

Well known as builders of hydrofoils and hovercraft, Feodosia Shipbuilding Company also produce patrol boats and luxury yachts.

Kalkan M 0573698

Vessel type or class name	Length (m)	Max beam (m)	Max speed (kt)	Hull material	Engines	Propulsion
Kalkan M	11.75	3.36	36	–	Volvo Penta 426 kW	Single water-jet
Foros 1350	13.8	3.78	37	–	2 × Volvo Penta 331 kW	Twin water-jet
Kafa 1350	13.3	3.7	50	–	2 × TAMD 122P EDC	Twin water-jet

Vympel Shipyard

4 Novaya Street, Rybinsk, 152912, Yaroslavl Reg, Russian Federation

Tel: (+7 855) 20 23 03
Fax: (+7 855) 21 18 77
e-mail: aovympel@yaroslavl.ru
Web: www.vympel-rybinsk.ru

Dmitry A. Belyakov, *Marketing Director*

Throughout its history, spanning over seventy years, the company has manufactured a range of vessels including torpedo boats, utility boats for the engineering troops, torpedo retrieving craft, fuel replenishment boats, ambulance vessels, firefighting vessels, tugs and pontoon bridge erecting boats, leisure craft and attack craft both for the Russian Naval Forces and for export.

Vessel type or class name	Length (m)	Max beam (m)	Max speed (kt)	Hull material	Engines	Propulsion
Mangust Patrol Boat	19.45	4.4	53	Aluminium	2 × Zvezda M470k diesel	2 × ASD-14 Arneson surface drives

Singapore

Lita Ocean PTE Ltd

30/32 Tuas Basin Link, Lita Building, Jurong, Singapore 638786

Tel: (+ 65 861) 36 63
Fax: (+ 65 861) 51 34
e-mail: litama@singnet.com.sg
Web: www.litaocean.com

Lita Ocean Private Ltd was formed in the 1970s primarily to offer services to the shipping industry. As well as owning tankers, tugs and barges for port services (including handling of explosive cargoes) the company has entered the high-speed craft market. Lita Ocean designed and built the V-Marlin class of high-speed patrol boats for the Singapore Customs and Excise Department.

V-Marlin Patrol Boat 0583169

Vessel type or class name	Length (m)	Max beam (m)	Max speed (kt)	Hull material	Engines	Propulsion
V-Marlin	12.0	3.05	52	Aluminium	4 × Mercury 250 hp outboards	
Island Bus	12.6	3.3	28	Aluminium	2 × Cummins 6CTA (255 kW each)	Twin waterjet
Taskmaster	12	4.1	26	Aluminium	2 × Cummins 6BTA-59M (235 kW each)	Twin waterjet
Island class	12.6	3.3	28	Aluminium	2 × Cummins 6CTA-8.3M	Hamilton HJ274 waterjet
Yellow Eye	6.5	2.4	30	Aluminium	Petrol Outboard	Outboard
Sonic Cruiser *Kenta*	12.0	4.1	26	Aluminium	2 × Cummins 6BTA 5.4m	2 × Hamilton HJ 274 waterjets

Singapore Technologies Marine Ltd

Part of Singapore Technologies Engineering Ltd
7 Benoi Road, Singapore, 629882.

Tel: (+65) 6860 9254
Fax: (+65) 6861 1601
e-mail: mkg.marine@stengg.com
Web: www.stengg.com

Singapore Technologies Marine is predominantly a builder of large patrol craft, min counter-measure vessels and missile corvettes, however they also manufacture a small variety of fast landing craft and interceptors.

Vessel type or class name	Length (m)	Max beam (m)	Max speed (kt)	Hull material	Engines	Propulsion
FCEP	13.6	3.66	25	Aluminium	2 × MAN D2866 LXE 40 diesel	2 × Hamilton 362 waterjets
FCU	23.0	6.0	25	Aluminium	2 × MAN D2842 LE 402 diesel	2 × waterjets
High Speed Inshore Patrol	14.5	4.23	38	Aluminium	2 × MAN D2848 LE 401	2 × Hamilton 362 waterjets

South Africa

Boatloft CC

Tamarix, Main Road, Hout Bay, 7806 South Africa

Tel: (+27) 217 90 27 45
Fax: (+27) 217 90 27 45
e-mail: info@boatloft.co.za
Web: www.boatloft.co.za

Based in Cape Town South Africa, Boatloft have been designing Catamarans for more than 30 years. Their team of engineers and technicians have specialised in powered catamarans designed to provide space, stability and comfort. Demanding sea conditions off of the Cape coast has led Boatloft to design a hull with unique seakeeping qualities. Boatloft provides a consultancy service, basic hulls in various configurations, as well as fully fitted boats.

Vessel type or class name	Length (m)	Max beam (m)	Max speed (kt)	Hull material	Engines	Propulsion
Bobkat 8660 patrol craft	8.6	3.2	35	GRP	Outboard Engine(s)	Twin outboard
Bobkat 1012 patrol boat	10	4	37	GRP	2 × 220 kW Diesel Inboard	Twin water-jet
Bobkat 2202 patrol craft	22	6	55	GRP	2 × MTU/DDC V16	Twin submerged shaft
Bobkat 2516 patrol boat	25	7	37	GRP	2 × 12V MTU 193TE92	Twin water-jet

Eraco Boats

34 Plantation Road, Wetton, Cape Town 7790, South Africa

Tel: (+27 21) 762 9284
Fax: (+27 21) 797 3181
e-mail: eraco@lano.co.za
Web: www.eracoboatbuilders.co.za

Deon Erasmus, *Principal*

Eraco boat builders was formed in 1997 specifically to design and build aluminium boats for the defence and commercial industries. Since then the company has built numerous boats ranging in size from 4 to 22 m for clients worldwide, including environmental agencies, navies, police, construction and fishing companies. Eraco Boats produce a standard range of vessels as well as specific customised designs.

14 m interceptor 1120305

Vessel type or class name	Length (m)	Max beam (m)	Max speed (kt)	Hull material	Engines	Propulsion
5.5 m Workhorse	5.5	2.30	33.0	aluminium	90 kW diesel inboards	sterndrive
8.5 m Workhorse	8.5	2.72	36.0	aluminium	2 × 135 kW outboards	–
10 m Assault Craft	10.3	2.80	41.0	aluminium	2 × 235 kW diesel inboards	2 × water-jets
12 m Coastal Patrol Vessel	12.7	3.80	40.0	aluminium	2 × 370 kW diesel inboards	2 × water-jets
14 m Interceptor	14.0	3.25	65.0	aluminium	2 × 610 kW diesel inboards	2 × surface drives

Stingray Marine

P.O. Box 1534, Durbanville, Cape Town 7551, South Africa

Tel: (+27 21) 987 11 90
Fax: (+27 21) 988 69 86
e-mail: alba@iafrica.com
Web: www.stingraymarine.com

Stingray Marine manufactures fast patrol boats, workboats, landing craft, inflatable and rigid inflatable boats.

Stingray has a range of catamaran RIBs, as well as the conventional monohull models, and has built vessels for a wide range of customers including the South African Navy.

Stingray Marine 12 m Horizon Interceptor 0589544

Vessel type or class name	Length (m)	Max beam (m)	Max speed (kt)	Hull material	Engines	Propulsion
12 m Horizon Interceptor	13.42	4.4	35	GRP	2 × Caterpillar 3208	Outboards
4.4 m Predator	4.4	2.06	–	GRP	2 × 22.5 kW	Outboards
4.9 m Predator	5	2.06	–	GRP	2 × 30 kW	Outboards
5.1 m Predator	5.1	2.34	–	GRP	2 × 45 kW	Outboards
5.4 m Predator	5.4	2.36	–	GRP	2 × 45 kW	Outboards
6.3 m Predator	6.3	2.56	–	GRP	2 × 67 kW	Outboards
6.8 m Predator	6.8	2.82	–	GRP	2 × 85 kW	Outboards
7.5 m Predator	7.5	2.82	–	GRP	2 × 97 kW	Outboards

Spain

Duarry (Astilleros Neumaticos Duarry SA)

Pasaje Rosers, S/N-08940, Cornella de llobregat, Barcelona, Spain

Tel: (+34 93) 471 45 00
Fax: (+34 93) 375 76 03
e-mail: duarry@duarry.com
Web: www.duarry.com

F Fargas, *Commercial Director*
J Fargas, *Export Director*

Duarry manufactures a wide range of SOLAS approved rescue craft and rigid inflatable boats for the military, workboat and leisure markets. The most recent delivery was a 13.5 m cabin RIB to a Turkish commercial operator.

Duarry Sportech 12 1027240

Duarry Brio 520 fast rescue craft 0097277

Vessel type or class name	Length (m)	Max beam (m)	Max speed (kt)	Hull material	Engines	Propulsion
SR 5	3.8	1.74	–	GRP	11 kW outboard	–
Master 400	4.00	1.90	–	GRP	30 kW outboard	–
Brio 450	4.5	2.1	–	GRP	37 kW outboard	–
Brio 520	5.2	2.2	–	GRP	75 kW outboard	–
Brio 520 MFB	5.2	2.1	40	GRP	75 kW outboard	–
Supercat 600	6.00	2.40	–	GRP	112 outboard	–
Brio 620	6.2	2.2	–	GRP	2 × 75 kW outboards	–
Cormoran 730	7.3	2.75	35	GRP	2 × 112 kW outboards	–
Cormoran 850	8.50	2.86	–	GRP	370 kW diesel inboard	–
Sportech 12	11.95	3.80	42	GRP	2 × 320 kW diesel inboards	–
Megatech 13.5 Cabin	13.5	3.9	–	GRP	2 × Inboard Diesel (469 kW total)	Twin submerged shaft

Rodman Polyships SAU

Poligono Industrial de la Borna s/n, 36955 Meira-Moana, Pontevedra, Spain

Tel: (+34 986) 81 18 18
Fax: (+34 986) 81 18 17
e-mail: int.sales@rodman.es
Web: www.rodman.es

Manuel Rodriquez, *Chairman*
Rafael Guzman, *Vice Chairman*
Francisco Rivas, *Managing Director*

Rodman Polyships was founded in 1974. The company specialises in the design and production of GRP high-speed craft, principally patrol craft and sports boats. The factory in Meira-Moana is one of the largest such facilities in Europe with 95,000 m² of production facilities.

In early 2001 the company was awarded a contract to build 14 vessels for the Spanish Ministry of Agriculture and Fisheries with delivery to commence in June 2003. In 2002 Rodman was awarded a contract to construct 14 Patrol boats for the Royal Oman Police, Coastguard Division. Three of these 18 m vessels have been delivered with the remainder under construction.

Rodman 101 (RSS Spari) 0097547

Rodman 58, 18 m patrol boat built for the Royal Oman Police, Coastguard Division 0595403

Rodman 55, 17 m patrol boat for the Cyprus Navy 0595404

Rodman 82 for the Spanish Civil Guard 0116886

Vessel type or class name	Length (m)	Max beam (m)	Max speed (kt)	Hull material	Engines	Propulsion
Rodman 101	30.00	6.00	35	GRP	2 × MTU 12V 2000 M90	Twin water-jet
Rodman 55	17.00	3.8	48	GRP	–	–
Rodman 58	18.00	4.9	30	GRP	–	–
Rodman 82	26.00	6	30	GRP	–	–

Sri Lanka

Blue Star Marine Ltd

129, Carmel Mawatha, Elakanda, Hendala, Wattala, Sri Lanka

Tel: (+94 11) 293 62 27
Fax: (+94 11) 293 36 84
e-mail: bluestarmarine@sltnet.lk
Web: www.bluestarmarinelanka.com

Established in 1979, Blue Star Marine is today one of the larger boat builders in Sri Lanka. They are currently producing the Penguin 44 for a variety of military and civilian applications.

Blue Star Marine Penguin 44 *Inshore Patrol Craft* 0130144

Vessel type or class name	Length (m)	Max beam (m)	Max speed (kt)	Hull material	Engines	Propulsion
Penguin 44	13.5	3	32	GRP	2 × Cummins 6 BTA 5.9M	Twin water-jet

Colombo Dockyard Ltd

PO Box 906, Port of Colombo, Colombo 15, Sri Lanka

Tel: (+94 112) 522 46 15
Fax: (+94 112) 44 64 41
e-mail: pricel@cdl.lk
Web: www.cdl.lk

M Prince Lye, *Manager*

Colombo Dockyard Ltd has a large drydock complex and a small building berth facility. A wide range of merchant vessels are built and repaired in these facilities as well as the construction of high-speed aluminium vessels.

Fast attack craft series III 0120174

Vessel type or class name	Length (m)	Max beam (m)	Max speed (kt)	Hull material	Engines	Propulsion
17 m Fast attack craft	17.32	–	32	Aluminium	2 × MAN D2842 LZE	Twin water-jet
Fast attack craft series I	24	5.9	45	Aluminium	2 × MTU 12V 396 TE94	Twin water-jet
Fast attack craft series II	24	5.9	45	Aluminium	2 × MTU 12V 396 TE94	Twin water-jet
Fast attack series III	24.4	5.9	50	Aluminium	2 × Deutz TBD 620 V16	Twin water-jet

Sweden

Capo Marin AB

Vikingavägen, 133 33 Saltsjöbaden, Sweden

Tel: (+46 87)17 07 30
Fax: (+46 87)17 07 31
e-mail: goran@capomarin.se
Web: www.capomarin.se

Capo Marin AB build 9 m and 12 m boats for commercial and pleasure applications. The vessel hulls and superstructures are built in Poland and then shipped to Sweden for fitting out. The vessels are built with a robust aluminium structure and have strong side fenders making them suitable for commercial applications.

PRO 93 fast utility craft 1120243

Vessel type or class name	Length (m)	Max Beam (m)	Max Speed (kt)	Hull material	Engines	Propulsion
Capo 11J	11.2	3.2	38	Aluminium	2 × 246 kW Cummins 6BTA 5.9M	Twin Water-jet
Capo 12J	12.2	3.2	40	Aluminium	2 × 360 kW Cummins QSC	Twin Water-jet
Capo Pro 93	9.3	2.6	35	Aluminium	2 × 186 kW Steyr Diesels	Twin Stern Drive
Capo 12 FC	12.2	3.2	40	Aluminium	2 × 360 kW Cummins QSC	Twin Water-jet

Dockstavarvet

PO Box 93, SE-87033, Docksta, Sweden

Tel: (+46 613) 71 16 06
Fax: (+46 613) 404 66
e-mail: ms@dockstavarvet.se
Web: www.dockstavarvet.se

K-A Sundin, *Managing Director*
Magnus Sorenson, *Sales and Marketing Director*
Torbjorn Larsson, *Technical Director*

Dockstavarvet is a leading producer of aluminium boats for professional use in Scandinavia and on the international market. The first aluminium boat was delivered in 1969 and since then the company has delivered over 200 boats to countries such as Sweden, Finland, Turkey, Denmark, UK, Malaysia, Norway, Greece, Mexico and Russia.

Dockstavarvet has been engaged in the development and serial production of advanced, high-speed patrol boats for military and non-military applications for over 15 years and this has resulted in the development of the highly successful range of CB90 boats which at one stage were being produced at a rate of one every 3 weeks.

More recent developments have included the IC 16 m, IC 18 m, HSPC 11.3 m, and designs for the IC 20 m patrol craft.

IC 16 m security vessel 0589545

CB 90 as built for the Swedish Royal Navy 0587796

CB 90 as delivered to Malaysia 0589542

Vessel type or class name	Length (m)	Max beam (m)	Max speed (kt)	Hull material	Engines	Propulsion
IC 16 m	17.20	3.80	50	Aluminium	2 × inboard diesel	Twin water-jet
CB 90 H	16.10	3.80	45	Aluminium	2 × Skania DSI 14	Twin water-jet
IC 18 M MPHSC	19.40	4.16	50	Aluminium	2 × 1,100 kW	Twin water-jet
IC CG 20 m	19.95	4.60	34	Aluminium	2 × 735 kW	Twin water-jet
HSPC 11.3 m	11.30	2.94	42	Aluminium	2 × Volvo Penta 368 kW	Twin water-jet
ALUPILOT 14000	14.7	4.65	30	Aluminium	2 × inboard diesel	Twin water-jet

Kockums AB

SE-205 55, Malmo, Sweden

Tel: (+46 40) 34 80 00
Fax: (+ 46 40) 97 32 81
e-mail: information@kockums.se
Web: www.kockums.se

Kockums design, build and maintain submarines and naval surface vessels that incorporate advanced stealth technology. Other products include the Stirling Air Independent Propulsion (AIP) system, the Kockums submarine rescue systems and mine clearance systems. Kockums are based in Malmo and Karlskrona, where all production facilities are Kockums is part of ThyssenKrupp Marine Systems.

KBV 301
0106594

Vessel type or class name	Length (m)	Max beam (m)	Max speed (kt)	Hull material	Engines	Propulsion
KBV 301	19.95	4.6	34	Composite	2 × 735 kW diesel inboard	Twin water-jet

Ring Powercraft

Skogsövagen 22, SE-133 33 Saltsjöbaden, Sweden

Tel: (+46 8) 611 88 88
e-mail: info@ringpowercraft.se
Web: www.ringpowercraft.se

Janis Petrov, *Sales Manager*

Ring Powercraft has been manufacturing commercial vessels since 1968 based on racing boat hull forms. Today the company produces a range of RIBs from 6–16 m in length. The vessels are constructed from Lloyds approved GRP materials and feature a unique extruded foam profile stringer system. All hulls feature extra thick transoms and multiple watertight chambers and all vessels above 6.5 m can be fitted with outboard or inboard engines.

Fantasma 10.5 m RIB 1120290

Vessel type or class name	Length (m)	Max beam (m)	Max speed (kt)	Hull material	Engines	Propulsion
Fantasma 6 m	6.0	2.64	50	GRP	2 × 75 Hp outboards	–
Fantasma 6.5 m	6.5	2.66	52	GRP	149 kW diesel inboard	–
Fantasma 7.5 m	7.5	2.72	57	GRP	235 kW diesel inboard	–
Fantasma 8 m	8.0	2.76	55	GRP	235 kW diesel inboard	–
Fantasma 8.6 m	8.6	2.80	53	GRP	235 kW diesel inboard	–
Fantasma 9.5 m	9.5	3.05	68	GRP	2 × 235 kW diesel inboard	–
Fantasma 10.5 m	10.5	3.05	65	GRP	2 × 235kW diesel inboard	–
Fantasma 12 m	12.0	3.29	65	GRP	3 × 235 kW diesel inboard	–
Fantasma 13 m	13.0	3.30	65	GRP	3 × 235 kW diesel inboard	–
Fantasma 15 m	15.0	3.50	65	GRP	3 × 330 kW diesel inboard	–
Fantasma 16 m	16.0	3.50	65	GRP	3 × 330 kW diesel inboard	–

Fantasma 13 m RIB with quadruple outboards 1120291

Fantasma 7.5 m RIB 1120292

Storebro Bruks AB

SE-590 83, Storebro, Sweden

Tel: (+46) 492 195 11
Fax: (+46) 492 303 00
e-mail: lars.bjureus@storebro.se
Web: www.storebro.se

Lars Bjuréus, *Sales Manager*

Storebro, part of the Nimbus Group is primarily a producer of pleasure boats between 12 and 15 m. However, they also produced over 50 Storebro 90E fast-patrol boats, which were developed in co-operation with the Swedish Administration of Defence Material, Shipbuilding Division. The boat is built in accordance with Royal Swedish Navy Standards and has an ability to beach because of highly ductine HD-material in the bottom. The boat, capable of carrying 10 passengers or four stretchers, is used by the Swedish Marines as a command and control or medical platform. Over sixty of the vessels have now been built for Sweden, Norway, Lithuania and China, and most recently for the Danish Navy for search and rescue duties along the coast of Greenland.

Vessel type or class name	Length (m)	Max beam (m)	Max speed (kt)	Hull material	Engines	Propulsion
Storebro 90E	11.9	2.9	40	Composite	Single Scania V8 diesel inboard	Single water-jet

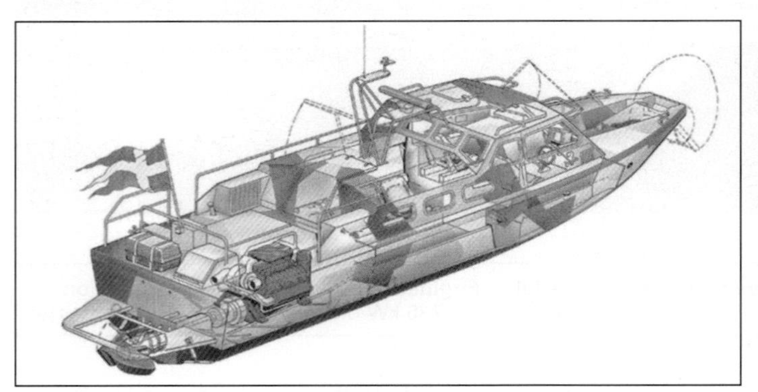

Storebro SRC_90E patrol boat cross section 1120220

Storebro SRC_90E patrol boat 1289655

Swede Ship Marine AB

Djupvik, SE-471 98, Fagerfjall, Sweden

Tel: (+46 304) 67 95 00
Fax: (+46 304) 66 25 00
e-mail: ssm@swedeship.se
Web: www.swedeship.se

Swede Ship Marine AB is a group of three midsize shipbuilding yards, Djupviks Shipyard, Gotlands Shipyard and Swede Ship Composite. The group of yards offer expertise in aluminium and composite construction of patrol, surveillance, transport, pilot and small passenger vessels.

Recent deliveries have included 22 m patrol boats, 24 m fast supply vessels, 13.7 m fast patrol boats and Transport Boat 2000.

Swedeship 45 ft high speed patrol craft 1120233

Vessel type or class name	Length (m)	Max beam (m)	Max speed (kt)	Hull material	Engines	Propulsion
Transport Boat 2000	23.85	5.10	34	Aluminium	3 × Volva Penta TAMD 153P	Triple water-jet
Pilot Boat	17.0	4.90	25	Aluminium	2 × SCANIA DSI 14	Twin submerged shaft
54 ft Rescue Vessel	16.25	5.20	34	Composite	2 × Scania DSI 14	Twin submerged shaft
45 ft High Speed Patrol Craft	13.55	3.05	70	Composite	3 × Seatek 6-4V-9	Sterndrives
Fast Supply Vessel	26.0	5.4	32	Aluminium	2 × MTU 16V 2000 M70	Twin waterjets

Taiwan

Lung Teh Shipbuilding Co Ltd

No.15, De-Shing 5th Road, Dung-Shan Country, ILAN, Taiwan

Tel: (+886) 39 90 11 66
Fax: (+886) 39 90 56 50
e-mail: sales@ms.lts.com.tw
Web: www.lts.com.tw

Mr Seldon Huang, *President*
Mr Chin-Hwa Chang, *Sales Vice President*
Mr Solon Huang, *Production Vice President*

Founded in 1979 the company has considerable experience in the design and construction of a wide range of patrol boats, military landing craft, support craft, pilot boats, fire and rescue boats as well as tugs and leisure boats.

The company has experience of building vessels in aluminium, steel and GRP up to 40 m in length and up to speeds of 60 kt.

20 m FAST PATROL BOAT
The first 20 m vessel was delivered to the Taiwan Coast Guard in 1994. Since then 14 vessels have been delivered to this design.

Specifications

Length overall	20.0 m
Length waterline	18.0 m
Beam	4.8 m
Draught	0.7 m

13.5 m patrol boat 0589516

Crew	8
Propulsive power	2,400 kW
Max speed	45.5 kt
Range	480 n miles

Classification: CR100+, CMS.
Propulsion: Triple Servogear controllable-pitch propellers driven via a U-drive reduction gear.

9.5 m ULTRA FAST PATROL BOAT
The first 9.5 m vessel was delivered to the Taiwan Coast Guard in 2004. To date nine craft have been built.
Specifications

Length overall	9.5 m
Length waterline	8.33 m
Beam	3.0 m
Draught	0.55 m
Crew	4
Propulsive power	540 kW
Max speed	38.0 kt
Range	100 n miles

Classification: CR
Propulsion: Two Yanmar 6LYA-STP diesel main engines driving water-jets via reversible reduction gears.

20 m fast patrol craft 0097813

Vessel type or class name	Length (m)	Max beam (m)	Max speed (kt)	Hull material	Engines	Propulsion
9.5 m Ultra Fast Patrol Boat (PP-601)	9.5	3.00	45	Composite	2 × Yanmar 6LY-STP	Twin water-jet
10 m Coastal Patrol Boat	10.0	3.00	36	GRP	2 × Mercury 225 Hp outboard	Twin outboard
13.5 m Ultra Fast Patrol Boat (15031)	13.5	3.33	56	Composite	2 × DDC 8V 92TA	Twin surface drive
15 m Fast Patrol Boat (Water-Jet)	15.0	4.76	40	GRP	2 × MAN 2480 LE 401	Twin water-jet
16 m Patrol Boat	16.0	4.50	32	Composite	2 × MAN inboard diesel	Twin water-jet
21 m Ultra Fast Patrol Boat (PP-7XX)	20.0	4.80	45	Composite	3 × MAN 2822 LZE 401	Triple submerged shaft
Fishery Patrol Boat	13.6	4.00	38	Composite	2 × MAN inboard diesel	Twin water-jet

Turkey

Altinel Shipyard Inc

Ankara Mercan Yuvasy Sitesi Sahil Yolu No 20, Tuzla, Istanbul, Turkey

Tel: (+90) 212 727 4010
Fax: (+90) 212 727 4011
e-mail info@altinelshipyard.com
Web: www.altinelshipyard.com

Established in 1983, the shipyard has built over 500 boats. The company has been granted the ISO 9002 International Quality Certificate and the Production Certificate from Rina (Registrano Italiano Navale). Altinel builds leisure, commercial and military boats. A new shipyard is currently under construction.

AM 42 FAST PATROL BOAT
Specifications

Length overall	12.65 m
Beam	3.6 m
Draught	0.75 m
Fuel capacity	1,700 litres
Maximum speed	39 kt
Operational speed	35 kt

AM 42 fast patrol boat 1120238

Vessel type or class name	Length (m)	Max beam (m)	Max speed (kt)	Hull material	Engines	Propulsion
AM 42 Fast Patrol Boat	12.65	3.6	45	GRP	2 × Caterpillar 3196 TA	Twin surface drive

Yonca Onuk

Tersaneler Caddesi, 50/3, Tuzla, 34.940, Istanbul

Tel: (+90 216) 392 99 70
Fax: (+90 216) 392 99 69
e-mail: info@yonca-onuk.com
Web: www.yonca-onuk.com

Yonca Onuk MRTP 15TCG 0583197

Yonca Onuk MRTP 33 1148040

Yonca Onuk MRTP 29 1148039

Formed in 1986, the company specialises in the design and production of fast boats from advanced composite materials. The current range includes pleasure, commercial and naval craft up to 33 m in length. The company has built a considerable number of craft for the Turkish Coast Guard.

Vessel type or class name	Length (m)	Max beam (m)	Max speed (kt)	Hull material	Engines	Propulsion
Onuk MRTP 15 (KAAN 15 Class)	16.7	4.04	54–65	Composite	2 × MTU 12V 183TE93 ECU	Twin Surface Drive
Onuk MRTP 29 (KAAN 29 Class)	31.60	6.70	49	Composite	2 × MTU 16V 4000M90	Twin water jet
Onuk MRTP 33	35.6	6.7	47	Composite	2 × MTU Diesel + 1	Twin water jet
Onuk MRTP 15 S.O.C	16.80	4.04	64	Composite	2 × MTU 12V183TE94 ECU	Twin Surface Drive
Onuk MRTP15/U	16.8	4.04	35	Composite	2 × MTU 8V183TE93ECU	Twin water jet
Onuk MRTP21	22.75	5.02	34	Composite	2 × MTU 12V183TE92	Twin water jet
Onuk MRTP 16 (KAAN 16 Class)	17.75	4.19	60	Composite	2 × MTU 8V2000	Twin Surface Drive

United Arab Emirates

Abu Dhabi Ship Building

PO Box 8922, Musafah Industrial Area, Abu Dhabi, UAE

Tel: (+971) 25 02 80 00
Fax: (+971) 25 51 04 55
Web: www.adsb.net

William Saltzer, *CEO*

Abu Dhabi Ship Building began operations in 1996 to serve naval forces, coast guards and local operators of tugs, barges, offshore supply and support vessels. The company employs 750 staff to work in a 17,500 m² shipyard with a ship lift capacity of 2,000 tons. Abu Dhabi Ship Building is capable of working with steel and aluminium and is expanding to produce composite craft.

Fast troop carrier 1120209

Vessel type or class name	Length (m)	Max beam (m)	Max speed (kt)	Hull material	Engines	Propulsion
Fast Troop Carrier	24.25	5.10	38	Aluminium	2 × MTU 12V 2000 M90	Twin water-jet
Assault Boat	9.5	3.4	50	Aluminium	2 × Mercury 250 hp outboard	Twin outboard

Seaspray Marine Services and Engineering

PO Box 42406, Hamriyah, Sharjah, United Arab Emirates

Tel: (+971 6) 526 0677
Fax: (+971 6) 526 0688
e-mail: seaspray@emitates.net.ae
Web: www.seasprayboats.com

Seaspray has been established in the Middle East since 1995 and before that operated in Australia. The company specialises in constructing fast vessels for the military and commercial marine sectors and has the capacity to handle vessels up to 60 m in length. All of Seaspray's products are built in aluminium to international classification society standards using the latest equipment. Recent projects include the building of 42 9.5 m fast patrol boats for the UAE Navy and UAE coastguard.

Seaspray 9.5 m fast patrol vessel 0563487

Vessel type or class name	Length (m)	Max beam (m)	Max speed (kt)	Hull material	Engines	Propulsion
9.5 m Fast Patrol Vessel	9.5 m	3.45	50	Aluminium	2 × 250 hp outboard	Twin outboard
11 m Fast Interception Vessel	11.0	3.34	50	Aluminium	3 × 250 hp outboard	3 × outboard
12 m Fast attack vessel	12.0	3.8	45	Aluminium	3 × 250 hp outboard	3 × outboard

United Kingdom

Avon Inflatables Ltd

Dafen, Llanelli, Carmarthenshire SA14 8NA, United Kingdom

Tel: (+44 1554) 88 20 00
Fax: (+44 1554) 88 20 39
e-mail: workboat@avon-inflatable.com
Web: www.avon-workboats.com

Steve Lang, *Military and Commercial Sales*

Established in 1959, the company pioneered the development of inflatable craft for military and commercial applications and have since produced over 200,000 boats. The current range of commercial and military boats are in service worldwide with navies, coastguards, rescue organisations and as ships' boats.

Avon SR7.4M in use with the US Navy 0507032

Vessel type or class name	Length (m)	Max beam (m)	Max speed (kt)	Hull material	Engines	Propulsion
–	4.05	1.80	30	GRP	50 hp outboard	–
SR4.7M	4.70	2.03	36	GRP	70 hp outboard	–
SR5.4M	5.43	2.03	38	GRP	90 hp outboard	–
SR6M	6.05	2.34	40	GRP	2 × 75 hp outboards	–
SR6.4M	6.40	2.39	36	GRP	2 × 130 hp outboards	–
SR6.4MD	6.40	2.39	30	GRP	170 kW diesel inboard	Single stern drive
SR7.4M	7.47	2.62	50	GRP	2 × 200 hp outboards	–
SR7.4MD	7.47	2.62	34	GRP	220 kW diesel inboard	Single stern drive
SR8.4M	8.38	2.84	50	GRP	2 × 225 hp outboards	–
SR8.4MD	8.38	2.84	38	GRP	320 kW diesel inboard	Single stern drive

Delta Power Services

Newby Road Industrial Estate, Hazel Grove, Stockport, Cheshire SK7 5DR, UK

Tel: (+44 161) 456 65 88
Fax: (+44 161) 456 66 86
e-mail: sales@deltapower.co.uk
Web: www.deltapower.co.uk

Delta Power Services is one of the largest special operations rigid inflatable boat manufacturers in Europe, supplying craft for military, special forces, coast guards, police, customs, rescue and fire services. The company is ISO 9001.2000 approved and builds to the following standards: DnV, SOLAS, Maritime and Coastguard Agency, Nordic Standards and Lloyds Special Service Craft.

The company specialises in developing craft for specific operations as well as having an extensive range of craft which can be configured to customers specifications. The current range includes open boats from 5.5 to 12 m and enclosed cabin boats, including self-righting models from 8 to 20 m.

Recent projects include the delivery of four Delta 10 m TX vessels to the Middle East, a 10.5 m, 45 knot patrol/interceptor for the Finnish Frontier Guard, a 14 m, 40 knot vessel for the Middle East and a range of 12 m fast interceptors. This table details the popular SX, TX, wheelhouse and SOLAS approved fast rescue craft ranges.

Delta Power 10 m craft 0101735

Vessel type or class name	Length (m)	Max beam (m)	Max speed (kt)	Hull material	Engines	Propulsion
Lightnening	6.75	2.5	30	GRP	2 × 60 hp outboards	Single submerged shaft
Seahawk	6.75	2.5	30	GRP	170 kW diesel inboard	Single water-jet
MOR	8.0	2.6	30	GRP	170 kW diesel inboard	Single water-jet
FRDC9.5	10.0	3.3	30	GRP	–	–
FRDC 11.5	11.7	3.3	40	GRP	–	–
FSC	15.3	5.0	38	GRP	–	–

Fusion Marine Ltd

Marine Resource Centre, Barcaldine by Oban, PA37 1SE, Scotland, United Kingdom

Tel: (+44 1631) 72 07 30
Fax: (+44 1631) 72 07 31
e-mail: info@fusionmarine.com
Web: www.fusionmarine.co.uk

Fusion Marine's range of hand built workboat, personnel carrier and rescue craft offer durability, impact resistance and an enhanced lifespan. The Polycraft FFB:1 and FRC:1 are manufactured entirely from polyethylene. A unique feature of the Polycraft is its deep V-shaped hull, which, through its natural planing prism at high and low speeds, brings stability in rough weather conditions without compromising on handling. The vessel, which can be powered by a variety of units, is produced in sizes up to 6.75 m and can be adapted to meet specific customer requirements.

Fusion Marine FRC 1 0589522

Vessel type or class name	Length overall (m)	Max beam (m)	Max speed (kt)	Hull material	Engines	Propulsion
FRC 1	6.75	2.6	–	Polypropylene	Outboard	Single submerged shaft

Holyhead Marine

Newry Beach Yard, Holyhead, Anglesea, LL65 1YB, United Kingdom

Tel: (+44 1407) 760111
Fax: (+44 1407) 764531
e-mail: marine.services@holyhead.co.uk

Holyhead Marine was established in 1962, and since 1994 has been constructing commercial vessels up to 24 m in steel, aluminium and GRP.

Vessel type or class name	Length (m)	Max bean (m)	Max speed (kt)	Hull material	Engines	Propulsion
6.4 m Fire & Rescue Boat	6.4	2.5	25	Aluminium	Steyr 164	HJ741 Hamilton waterjet
9 m Raiding Craft	9.1	2.9	40	Aluminium	2 × Steyr M256	2 × Rolls-Royce Kamewa FF270 waterjets

Humber Inflatable Boats

246 Wincolmlee, Hull, HU2 OPZ, United Kingdom

Tel: (+44 1482) 22 61 00
Fax: (+44 1482) 21 58 84
e-mail: sales@humberboats.co.uk
Web: www.humberinflatables.co.uk

Frank Roffee, *Technical Director*

Humber Inflatable boats is one of the largest producers of inflatable boats and RIBs in the UK, manufacturing a wide range of craft for private and commercial use. The company has produced approved craft for offshore safety and rescue craft. The Ocean Pro range of RIBs is available in a range of sizes from 7 to 11 m, with a beam of 2.82 m.

Humber RH550 rescue craft 0507112

Vessel type or class name	Length (m)	Max beam (m)	Max speed (kt)	Hull material	Engines	Propulsion
Ocean Pro 7.0	7.0	2.82	–	GRP	2 × Outboard (260 kw total)	Outboard
Ocean Pro 7.5	7.5	2.82	–	GRP	2 × Outboard (298 kW total)	Outboard
Ocean Pro 8.0	8.0	2.82	–	GRP	2 × Outboard (335 kW Total)	Outboard
Offshore Pro 8.5	8.5	2.82	–	GRP	2 × Outboard (372 kW total)	Outboard

Marine Specialised Technology Ltd

Unit 1, Atlantic Way, Brunswick Business Park, Liverpool, L3 4BE, United Kingdom

Tel: (+44 151) 708 41 12
Fax: (+44 151) 708 41 13
e-mail: sales@mstltd.com
Web: www.mstltd.com

Ben Kerfoot, *Managing Director*
Philip Hilbert, *Sales Director*
Andy Phillips, *Technical Director*

MST 750 workboat RIB 0583170

MST 750 wheelhouse RIB 0583171

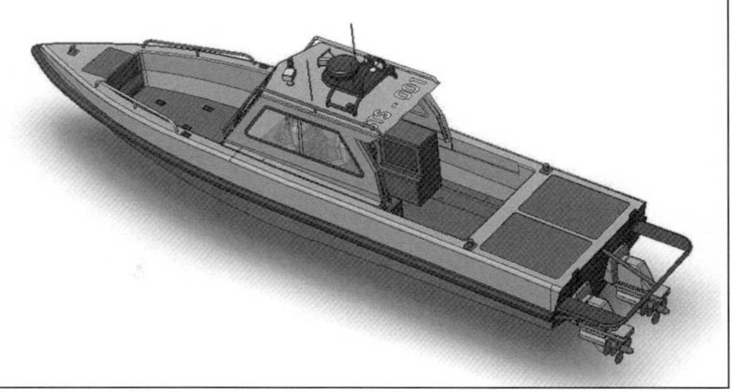
MST T1100 fast patrol craft (design) 0583172

Marine Specialised Technology Limited (MST) is a commercial rigid inflatable and high speed work boat builder based in Liverpool, UK, serving the global specialist boat markets with both standard and custom designed craft. MST craft are widely used by rescue organisations, law enforcement agencies and military teams across the world.

Vessel type or class name	Length (m)	Max beam (m)	Max speed (kt)	Hull material	Engines	Propulsion
MST 680	6.8	2.6	–	GRP	Inboard Diesel or Petrol Outboards	Single water-jet
MST 750	7.5	2.6	–	GRP	Inboard diesel or petrol outboard.	Single water-jet
MST 850	8.5	3.1	–	GRP	Inboard diesel or petrol outboards.	Twin water-jet
MST 1200	12.0	3.5	–	GRP	Inboard diesel or petrol outboards.	Twin water-jet
MST T1100 Interceptor (Design)	11.0	2.6	–	GRP	Inboard diesel or Petrol Outboards	Twin submerged Shaft

Ocean Dynamics

A subsidiary of Mustang Marine (Wales) Ltd
The Dockyard, Pembroke Dock, Pembrokeshire SA72 6TE, United Kingdom

Tel: (+44 1646) 681117
Fax: (+44 1646) 686414
e-mail: sales@mustangmarine.com
Web: www.mustangmarine.com

Ian Strugnell, *Commercial Director*

Ocean Dynamics was formed in the 1960s to develop and produce large commercial aluminium rigid-hulled inflatable craft. The company produces tailor-made vessels to client specification: the main output of the company is in the 6 to 20 m length range with speeds up to 60 knots. The range of models include lifeboats, assault vessels, passenger and expedition craft and the durability of the vessels has been proven over many years in extreme conditions. Propulsion tends to be via water-jets but all other means are utilised.

Ocean Dynamics pioneered the design and development of aluminium hulled RIBs over thirty years ago and has gained an enviable reputation in the workboat world for staright-forward ruggedness, longevity, ease of repair and flexibility of design. The company uses the very latest design software and welding techniques to ensure that its range of RIBs is at the forefront of RIB technology.

Vessel type or class name	Length (m)	Max beam (m)	Max speed (kt)	Hull material	Engines	Propulsion
Ribworker E-900	10.1	3.4	33	Aluminium	Yanmar 6LYA-STP	Hamilton HJ292 Jet
Ribworker D-1100	10.5	3.84	31	Aluminium	Scania DI-9	Hamilton 322 Jet
Ribworker)-1250	13.0	4.5	38	Aluminium	2 × Caterpillar 3126 B	2 × Hamilton 322 Jets

Parker Ribs

A H Parker & Sons Ltd
Suite 8, Laura House, Jengers Mead, Billingshurst, West Sussex RH14 9NZ, United Kingdom

Tel: (+44 1403) 780470
Fax: (+44 1403) 780490
e-mail: info@parkerribs.com
Web: www.parkerribs.com

Parker Ribs has recently received orders from the Bulgarian SAR for six 10 m Parker Baltic cabin RIBs and from the Icelandic Search and Rescue for one 900 Baltic RIB. The Icelandic boat, delivered in September 2007, will be equipped with patrol console, four single jockey seats and self righting equipment. The boats for the Bulgarian SAR will be driven by twin MCD 4.2-250 hp inboard diesel engines and Hamilton 241 waterjet drives.

Parker 900 Baltic 1120328

Parker 750 Baltic 1120329

Parker 900 Baltic with cabin 1120330

Vessel type or class name	Length (m)	Max beam (m)	Max speed (kt)	Hull material	Engines	Propulsion
750 Baltic	7.5	3.12	30+	GRP	Twin inboard or outboard diesel	Propeller or waterjet
900 RS	9.0	2.83	50+	GRP	Twin inboard or outboard diesel	Propeller or waterjet
900 Baltic	9.0	3.16	50+	GRP	Twin inboard or outboard diesel	Propeller or waterjet

Redbay Boats Inc

Coast Road, Cushendall, Co Antrim, Northern Ireland, BT44 0TE, United Kingdom

Tel: (+44 28) 217 713 31
Fax: (+44 28) 217 714 74
e-mail: info@redbayboats.com
Web: www.redbayboats.com

Tom McLaughlin, *Managing Director*

Redbay boats has been an established boat builder since 1977 but has now diversified into the RIB market. The Redbay Stormforce range of RIBs, ranging from 6 to 11 m, is particularly suitable for rough sea conditions. Boats have been exported to Russia, the Faroe Islands and Norway and the South African Life Saving Organisation has recently announced that it is replacing its entire fleet with Redbay Stormforce 11 m RIBs.

Vessel type or class name	Length (m)	Max beam (m)	Max speed (kt)	Hull material	Engines	Propulsion
8.4 m Expedition	8.4	2.76	35	GRP	300 hp Yamaha diesel	Stern drives
9.1 m Stormforce	9.1	3.3	35+	GRP	Twin diesel	Stern drives
11 m Stormforce	11.0	3.3	35+	GRP	Twin diesel	Stern drives

Ribeye Ltd

Collingwood Road, Townstal, Dartmouth, Devon, TQ6 9JY, United Kingdom

Tel: (+44 1803) 832060
Fax: (+44 1803) 839090
e-mail: info@ribeye.co.uk
Web: www.ribeye.co.uk

Charles Chivers, *Export Director*

Ribtec Ltd supplies fast semi-rigid inflatable craft for military, commercial and pleasure use.

The company was formed in 1998 and in 2001 it took over Ribtec Ltd.

Vessel type or class name	Length (m)	Max beam (m)	Max speed (kt)	Hull material	Engines	Propulsion
Ribtec 455	4.55	2.05	–	GRP	45 kW outboard	–
Ribtec 535	5.37	2.05	–	GRP	60 kW outboard	–
Ribtec 585	5.90	2.28	–	GRP	75 kW outboard	–
Ribtec 655	6.50	2.30	–	GRP	112 kW outboard	–
Ribtec 740	7.35	2.55	–	GRP	2 × 86 kW outboard	–

Tiger Marine Ltd

Prospect Road, West Cowes, Isle of Wight, PO31 7AD, United Kingdom

Tel: (+44 19) 83 20 09 88
Fax: (+44 19) 83 20 09 95
e-mail: enquiries@tigermarine.co.uk
Web: www.tigermarine.co.uk

Mark Gulesserian, *Managing Director*

Tiger Marine Ltd currently markets various configurations of two fast rescue craft, the T7000 and T9000. Built of all-welded marine grade aluminium within closed cell foam buoyancy sponsons. The vessels are powered by inboard STEYR diesel engines of up to 190 kW driving Hamilton water-jet units. The craft are manufactured to ISO 9001 and DnV rules.

Tiger Marine T9000 FRC 0583196

Vessel type or class name	Length (m)	Max beam (m)	Max speed (kt)	Hull material	Engines	Propulsion
T7000	7.0	2.6	35	Aluminium	1 × Steyr Diesel	Single water-jet
T9000	9.0	2.8	35	Aluminium	2 × Steyr Diesel	Twin water-jet

Tornado Boats International Ltd

Wiltshire Road, Hull, HU4 6PA, United Kingdom

Tel: (+44 1482) 35 39 72
Fax: (+44 1482) 57 24 75
e-mail: info@tornado-boats.com
Web: www.tornado-boats.com

Tornado Boats was formed 20 years ago and was one of the first to design and develop the modern Rigid Inflatable Boat (RIB). Tornado Boats cover commercial and marine engineering, diving and surveying. It manufactures a range of RIB's from 4.5 to 11 m, marketed by Barnet Marine Centre Ltd.

Tornado Boats 7 m RIB 1120215

Vessel type or class name	Length (m)	Max beam (m)	Max speed (kt)	Hull material	Engines	Propulsion
Tornado 6.5	6.6	2.65	50	GRP	2 × 90 kW outboards	–
Tornado 7.	7.2	2.85	55	GRP	2 × 115 kW outboards	–
Tornado 8.5	8.55	2.7	60	Composite	2 × 200 kW outboard or 300 hp inboard	–
Tornado 9.5	9.56	2.7	60	GRP	2 × 300 kW outboard or 313 kW inboard	–
Tornado 10	10.05	3.0	50	GRP	2 × 300 kW outboard or twin inboard	–
Tornado 11m	11.5	3.0	50	GRP	3 × 300 kW outboard or twin inboard	–

Tornado Boats 8.5 m RIB 1120216

Tornado Boats 9.5 m RIB 1120217

VT Halmatic

Portchester Shipyard, Hamilton Road, Portsmouth, Hampshire, United Kingdom

Tel: (+44 23) 92 53 96 00
Fax: (+44 23) 92 53 96 01
e-mail: simon.thomson@vtplc.com
Web: www.vtplc.com/halmatic

David Hobbs, *Managing Director*
Bob Cripps, *Technical Director*
Simon Thomson, *Export Sales and Marketing Manager*

VT Halmatic has been a supplier of fast rescue craft, patrol boats and commercial craft since 1952 and is now regarded as a world leader for manufacturing specialist GRP or composite boats in the 6 to 40 m range. The current range includes:
Fast patrol boats from 9 to 40 m
Ultra-fast patrol boats from 9 to 23 m
Rigid Inflatable boats from 6 to 12 m.

Beach Raider craft 6.5 m 1149905

Enforcer 33 0560075

Combat Support Boat  1149912

Vessel type or class name	Length (m)	Max beam (m)	Max speed (kt)	Hull material	Engines	Propulsion
Enforcer 33	10.88	2.84	45.0	Composite	2 × Yanmar 6 cylinder diesel	Twin water-jet
VSV 16 m	16.00	2.82	50.0	Composite	2 × inboard diessl	Twin surface drive
VSV 22 m (Design)	22.86	4.35	60.0	Composite	2 × inboard diesel	Twin surface drives
Cat 900	9.20	2.89	50.0	Composite	–	–
Enforcer 40	12.19	2.74	60.0	Composite	2 × inboard diesel	Twin surface drive
Enforcer 46	13.90	2.80	60.0	Composite	2 × inboard diesel	Twin surface drive
Arctic 22	6.96	2.54	45.0	GRP	2 × 130 Hp outboards	–
Pacific 22 Mk II	7.10	2.44	35.0	GRP	diesel inboard	–
Pacific 24	8.10	2.86	45.0	GRP	220 kW diesel inboard	–
Pacific 28/30	8.70	3.02	45.0	GRP	375 kW diesel inboard	–
Pacific 32	10.20	3.02	45.0	GRP	375 kW diesel inboard	–

As well as supplying vessels to organisations around the world they have also supplied the UK MoD, RNLI and HM Customs & Excise.

In October 2007 the company delivered the first of a new generation of advanced rigid inflatable boats to the French Marine commandos. Based on the proven Pacific 28 design, the new variant has a carbon composite hull and is designed for improved performance in such areas as marine assault, fire support and as a command and control platform.

The structures of all vessels are built under Lloyd's Register of Shipping Inspection and completion can be to Lloyd's Register, Germanischer Lloyd, ABS or other classification society rules.

XS-Ribs

14 West Burrowfield, Welwyn Garden City, Hertfordshire AL7 4TW, United Kingdom

Tel: (+44 1707) 33 13 89
Fax: (+44 1707) 33 13 83
e-mail: info@ribs.co.uk
Web: www.xs-ribs.co.uk

XS-Ribs manufacture a range of RIB's from 4.6 m to 9.9 m which are marketed by Barnet Marine Centre Ltd.

XS-600 RIB
1120222

XS 750 RIB
1120327

Vessel type or class name	Length (m)	Max beam (m)	Max speed (kt)	Hull material	Engines	Propulsion
XS-550	5.5	2.6	45	GRP	75 kW outboard	Single outboard
XS-600	6	2.6	50	GRP	112 kW outboard	Single outboard
XS-650	6.5	2.6	50	GRP	167 kW outboard	Single outboard
XS Diesel rib	7.2	2.6	32	Composite	Mercruiser 1.7 diesel inboard	Single stern drive
XS-750	7.5	2.9	50	GRP	Single Mercruiser Outboard	Single outboard
XS-850	8.5	2.9	50	GRP	275 Hp Verado Outboard	Single outboard
XS-950	9.9	2.9	50	GRP	Twin outboard/inboard	Twin stern drive

United States

ACB

Aluminium Chambered Boats Inc
355 Harris Avenue, Suite 107, Bellingham, WA 98225, United States

Tel: (+1 360) 647 0345
Fax: (+1 360) 255 2295
e-mail: info@acbboats.com
Web: www.acbboats.com

ACB manufactures high performance aluminium boats for commercial, governmental and recreational markets. Models range from 7 to 9 m and all feature ACB's sealed chambered hull design, designed to increase life expectancy compared to fibreglass boats. Clients include the US Navy, US Marines and US Air Force.

Riverine Demonstrator
ACB and Northrop Grumman have jointly developed this vesselin response to the

Vessel type or class name	Length (m)	Max beam (m)	Max speed (kt)	Hull material	Engines	Propulsion
Riverine Demonstrator	12.4	3.28	39	Aluminium	2 × Cummins QSC 8.3 Quantum	2 × Ultra Dynamics Ultrajet 340 waterjets

US Navy's programme to recapitalise its riverine force elements. The boat is based on a 12.4 m multi-mission aluminium vessel with a low draft (0.66 m) and is able to navigate class 5 rapids. Two Cummins QSC 8.3 Quantum engines powering two Ultra Dynamic Ultrajet 340 waterjets will give the demonstrator a top speed of 39 knots.

Almar Boats

117 Puyallup Avenue E, Tacoma, Washington 98421, United States

Tel: (+1 253) 572 28 77
Fax: (+1 253) 572 00 65
e-mail: info@almarboats.com
Web: www.almarboats.com

Established in 1975 Almar specialises in aluminium commercial, patrol and leisure boats up to 11 m in length. A recent contract involved the construction of a 35 kt RIB for the National Oceanographic and Atmospheric Administration (US).

Vessel type or class name	Length (m)	Max beam (m)	Max speed (kt)	Hull material	Engines	Propulsion
Almar 32 NOAA	9.8	3.6	35	Aluminium	2 × Yanmar 6LP	Twin water-jet

Brunswick Commercial and Government Products

A subsidiary of the Brunswick Corporation

420 Megan Z Avenue, Edgewater, Fl 32132, United States

Tel: (+1 386) 423 2900
Fax: (+1 386) 423 9187
e-mail: bdimiternko@whaler.com
Web: www.brunswickcgboats.com

Brunswick Commercial and Government Products was previously the government and commercial division of Boston Whaler. The company specialises in special operations/combat boats, fire and rescue craft and workboats, all based on Boston Whaler designs. Vessels are available from 4.6 m to 10 m and are all constructed from GRP.

Vessel type or class name	Length (m)	Max beam (m)	Max speed (kt)	Hull material	Engines	Propulsion
22′ Guardian	6.78	2.26	–	GRP	Twin outboard	Outboard
25′ Challenger	7.49	2.43	–	GRP	Twin outboard	Outboard
27′ Vigilant	8.10	3.04	–	GRP	Twin outboard	Outboard
27′ Justice	8.22	2.59	–	GRP	Twin outboard	Outboard

Demaree

410 Oak Street, Freindsville, Maryland 21531, United States

Tel: (+1 301) 746 58 15
Fax: (+1 301) 746 50 19
e-mail: dib@gcnetmail.net
Web: www.dibboats.com

Demaree Inflatable Boats Inc manufactures a range of fast inflatable craft for general industrial use covering geophysical uses, product harvesting, oil spill recovery, salvage and diving companies, and research and survey groups as well as for patrol and rescue operations.

Chesapeake
0129329

Vessel type or class name	Length (m)	Max beam (m)	Max speed (kt)	Hull material	Engines	Propulsion
Chesapeake	5.2	2.4	–	–	–	–
Rapid Response	3.7	1.8	–	–	–	–
Rescue Sled	4.3	2.1	–	–	–	–

Moose Boats

274 Sears Point Road, Port Sonoma Marina, Petaluma, California 94954, United States

Tel: (+1 866) 466 66 73
Fax: (+1 707) 778 98 17
e-mail: info@mooseboats.com
Web: www.mooseboats.com

Moose Boats specialises in waterjet-driven catamarans from 10–14 m, mainly for law enforcement agencies, military bodies, fire and rescue and port security authorities.

All models have a symmetrical planing hull constructed in marine grade aluminium, twin inboard diesels and twin waterjets, although outboard motors or stern drive options are available.

Vessel type or class name	Length (m)	Max beam (m)	Max speed (kt)	Hull material	Engines	Propulsion
M2-33′	10.06	4.12	40	Aluminium	Twin diesel inboard	Twin waterjets
M2-35′	10.80	4.12	38	Aluminium	Twin diesel inboard	Twin waterjets
M2-37′	11.4	4.12	39	Aluminium	Twin diesel inboard	Twin waterjets
M2-44′	13.4	4.72	42	Aluminium	Twin diesel inboard	Twin waterjets

Nautica International Inc

1500 S.66 Avenue, Pembroke Pines, Florida 33023, United States

Tel: (+1 954) 986 16 00
Fax: (+1 954) 986 16 31
e-mail: nautica@nauticaintl.com
Web: www.nauticaintl.com

Howard Rogers, *Manager*
Sonia Sousa, *Marketing*

Formed in 1982, the company has sold a wide range of rigid inflatable craft to the US Coast Guard, Customs and Marine Corps as well as to the leisure market. The smallest design is a 3.2 m diesel powered vessel and the largest a 12.5 m Cabin RIB. The company has recently launched a range of catamaran RIBs (5.9 to 8.5 m) complete with bow loading ramps.

X-41 RIB 0589518

Vessel type or class name	Length (m)	Max beam (m)	Max speed (kt)	Hull material	Engines	Propulsion
X-36	10.97	3.40	–	GRP	2 × Yanmar 6LY2A-STP (327.5kW each)	Twin water-jet
X-41	12.49	3.40	–	GRP	2 × Yanmar 6LY2A-STP (327.5 kW each)	Twin water-jet
X-25	7.79	3.15	–	GRP	Yanmar 6LPA-STP	Single water-jet

X-36 Cabin RIB 0589519

X-41 Cabin RIB 0589525

Northwind Marine Inc

605 South Riverside Drive, Seattle, Washington 98108, United States

Tel: (+1 206) 767 44 97
Fax: (+1 206) 767 67 24
e-mail: info@northwindmarine.com
Web: www.northwindmarine.com

Scott Tucker, *Programme Manager*

21X rescue RIB 0583175

Vessel type or class name	Length (m)	Max beam (m)	Max speed (kt)	Hull material	Engines	Propulsion
15DXSJ	4.73	1.68	45	Aluminium	–	–
17XOBCC	5.13	2.39	40	Aluminium	–	–
17XSJCC	5.13	2.39	40	–	–	–
19XOBCC	5.69	2.29	40	Aluminium	–	–
19XSJCC	5.69	2.29	40	Aluminium	–	–
21XOBCC	6.40	2.39	40	Aluminium	–	–
23XOBTT	7.01	2.59	40	Aluminium	–	–
26XDJ315CC	7.95	2.79	40	Aluminium	–	–
26XE20BFCF	8.10	3.12	45	Aluminium	–	–
27XE2OB	8.20	3.30	45	Aluminium	–	–
28XDJ315CC	8.48	2.69	40	Aluminium	–	–
28XDJ315PM	8.48	2.69	35	Aluminium	–	–
29.5X2DJ315PM	8.99	3.05	40	Aluminium	–	–
34X20BCC	10.3	3.66	40	Aluminium	–	–
36X2DJ315PM	11.02	3.66	30	Aluminium	–	–
Seafox 21 (Unmanned vehicle)	6.40	2.39	–	–	–	–

23X RIB 0583176

25X RIB 0583177

28X RIB 0583180

34X RIB 0583183

Northwind Marine Inc supplies a range of fast aluminium-hulled rigid inflatable rescue craft from 5 to 15 m, the most recent of which is the 8.3 m SAFE™ boat.

Seafox 21 Unmanned RIB 0583184

Oregon Iron Works

9700 S.E. Lawnfield Road, Clackamas, Oregon, 97015, United States

Tel: (+1 503) 653 6300
Fax: (+1 503) 653 5870
e-mail: sales@oregoniron.com
Web: www.oregoniron.com

Oregon Iron Works offers engineering design and build to a wide range of industries and US Government Departments. In 2003 they delivered the Sealion Technology demonstrator to Navsea for research work into seakeeping ability and littoral operations.

SEALION DEMONSTRATOR

The monohull Sealion Technology Demonstrator has the following characteristics.

Specifications

Length overall	21.6 m
Beam	4.4 m
Displacement, min	30 tonnes
Payloads	2.2 tonnes
Maximum speed	40+ kt
Range	300 n.miles

Vessel type or class name	Length (m)	Max beam (m)	Max speed (kt)	Hull material	Engines	Propulsion
Sealion	21.6	4.4	40+	–	–	–

Otech

1140 Peters Road, Harvey, Louisiana 70058, United States

Tel: (+1 800) 783 74 42
Fax: (+1 504) 362 59 49
e-mail: jane@oceantech.com
Web: www.oceantech.com

Esteban Fernandez, *President and CEO*
Jorge Fernandez, *Vice President, Operations*

Ocean Technical Services, Inc is an ISO-9002 certified service shipyard and manufacturer of the Ocean Sprint RIBs. The hull has a deep vee design that flattens to a conic planing surface at the transom and a series of concave spray chines that run the length of the boat. This hull design was developed by Crompton Marine Ltd (UK) after 18 years of experience with such craft in North Sea conditions. Licensed by Crompton and in conjunction with Sintes Fibreglass Designs, Otech has modified the series to incorporate a moulded deck, a quick tube replacement system and the SCRIMP method of fibreglass construction.

This design is currently in use by Ensco, Semco, the Royal Swedish Navy, Spanish Red Cross, Anglian Water, Swedish Rescue Society, North African Customs and Excise Department and various others. The hulls carry a 5 year commercial hull warranty. Inboard, outboard and water-jet propulsion packages are available along with various topside configurations.

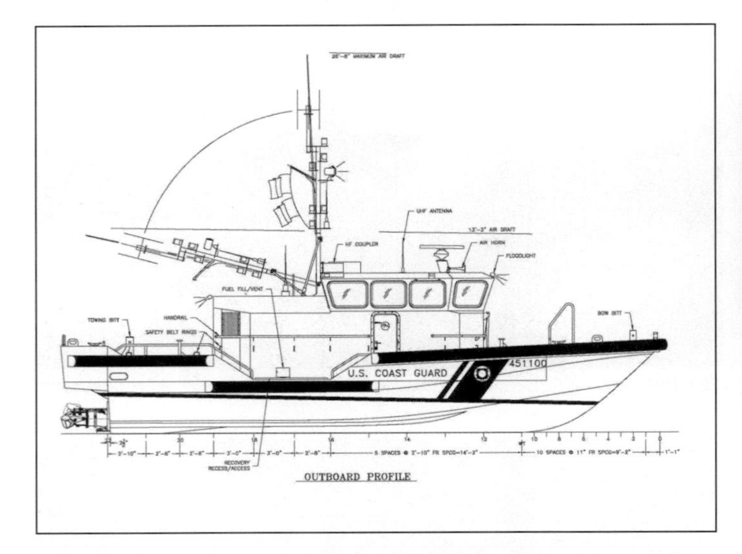

13.7 m fast response boat 0583185

General arrangement of 13.7 m fast response boat 0583186

OTECH have designed and built a new 13.7 m fast response boat as a demonstrator for the US Coast Guard. This boat began an extensive testing program in 2003 and was competing against two other boats for a potential contract to build 180 boats.

Vessel type or class name	Length (m)	Max beam (m)	Max speed (kt)	Hull material	Engines	Propulsion
4.8 m	4.8	2.30	–	GRP	41 kW outboard	–
5.6 m	5.6	2.30	–	GRP	67 kW outboard	–
7.0 m	7.0	2.90	–	GRP	2 × 112 kW outboard	–
Response Boat	13.7	4.17	40.0	Aluminium	2 × Yanmar 6LY2A	Twin water-jet

SeaArk Marine

PO Box 210, 404 North Gabbert Street, Monticello, Arkansas 71657, United States

Tel: (+1 870) 367 97 55
Fax: (+1 870) 367 21 20
e-mail: sales@seaark.com
Web: www.seaark.com

John McClendon, *President and CEO*
Ken McFalls, *Vice President Sales*

In January 2003, SeaArk delivered the first of 36 10.5 m Dauntless fast boats destined for the Naval Coastal Warfare units in the US Atlantic and Pacific fleets. The vessel reached 36 kt during trials. In April 2003, the company delivered an 11 m Dauntless patrol/rescue boat to the New York Police Department Harbor Unit. The vessel is to be used in patrol, search and rescue, port security, diving and anti-terrorism activities in New York Harbour and the waters surrounding New York City.

11 m Dauntless built by SeaArk for the New York Police Department 0553216

Vessel type or class name	Length (m)	Max beam (m)	Max speed (kt)	Hull material	Engines	Propulsion
23′ (7.01 m) Commander Series	7.01	2.59	–	Aluminium	2 × Outboard (224 kW total)	Twin submerged shaft
34′ (10.36 m) Dauntless RAM	10.36	3.66	–	Aluminium	Inboard Diesel or Outboard (632 kW total)	Twin submerged shaft

Seaark 23′ V Commander 0583190

Seaark 34′ Dauntless 0583191

Silverships

PO Box 1266, Theodore, Alabama 36582, United States

Tel: (+1 888) 564 56 67
e-mail: slvr_ship@silverships.com
Web: www.silverships.com

Mike McCarty, *President*
Jason Powers, *General Manager*
Jim Sorenson, *Production Manager*

The company was formed in 1985 and has since designed and built over 1,000 aluminium boats. The facility includes over 8,000 m^2 of covered work space and is used to produce a wide range of patrol boats, fire boats, workboats and rescue boats. In 2000, Silver Ships purchased the established aluminium boat builder Ambar Marine.

Silverships 34 ft patrol boat 1120239

Silverships AM900 fast rescue craft 0507113

Vessel type or class name	Length (m)	Max beam (m)	Max speed (kt)	Hull material	Engines	Propulsion
34 ft (10.36 m) patrol boat	10.36	2.9	40	Aluminium	2 × Outboard (375 kW total)	Twin submerged shaft
44 ft (13.4 m) patrol boat	13.4	3.15	40	Aluminium	2 × Caterpillar 3196 (848 kW total)	Twin water-jet
48 ft (14.6 m) patrol boat	14.6	3.66	40	Aluminium	2 × Caterpillar 3196 (848 kW total)	Twin surface drive
AM800	8.3	2.8	45	Aluminium	2 × Outboard (335 kW total)	Twin submerged shaft
AM900	8.6	3.4	47	Aluminium	2 × Outboard (372 kW Total)	Twin submerged shaft
AM1100	10.97	3.15	38	Aluminium	2 × Outboard (335 kW total)	Twin submerged shaft
30 ft patrol boat	9.2	3.1	35	Aluminium	2 × diesel inboard	Twin submerged shaft
27 ft patrol boat	8.2	2.6	40	Aluminium	2 × 200 hp outboards	Twin outboard
26 ft patrol boat	8	2.6	45	Aluminium	2 × 225 hp outboard	Twin outboard
25 ft patrol boat	7.6	2.6	30	Aluminium	Single Volvo AD41	Single stern drive

Swiftships Shipbuilders, LLC

PO Box 2869, 1105 Levee Road, Morgan City, Louisiana 70380, United States

Tel: (+1 504) 384 17 00
Fax: (+1504) 384 09 14
e-mail: swiftships@swiftships.com
Web: www.swiftships.com

Established in 1969 Swiftships is owned by the Swift Group LLC holding company. It has built over 550 vessels in the 10 to 68 m range and is well known for delivering 200 Swift patrol boats for the Vietnam conflict. The company's current range covers fast patrol boats and utility vessels primarily for the US market.

11 m Riverine Assault Craft 0106808

Vessel type or class name	Length (m)	Max beam (m)	Max speed (kt)	Hull material	Engines	Propulsion
Swiftships 38PB 1098	11.6	3.81	45	Aluminium	2 × Caterpillar 3406	–
14 m Patrol Craft Riverine	13.71	3.57	30	Aluminium	2 × Detroit Diesel 6V92TA	Twin water-jet
11 m Riverine Assualt Craft	10.9	2.71	39	Aluminium	2 × Cummins GBTA5.9Ms	Twin water-jet
Swiftships 65PA1090	19.81	5.48	35	Aluminium	2 × MTU 16V2000	Twin submerged shaft
Swiftships 42PA1184	12.8	4.42	30	Aluminium	2 × Diesel Inboard	Twin submerged shaft

United States Marine Inc

10011 Lorraine Road, Gulfport, MS-39503, United States

Tel: (+1 228) 679 10 05
Fax: (+1 228) 679 10 10
e-mail: shawn@usmi.com
Web: www.usmi.com

Shawn Killeen, *President*
Barry T Dreyfus, *CEO*

United States Marine Inc designs, manufactures, tests and delivers high-speed, lightweight composite craft and logistical support systems to military and commercial industries. By 2003 the company had delivered over 380 specialised craft to the Department of Defence (DoD) and clients worldwide. The current range includes the 10 m SOCR (Special Operations Craft Riverine) vessel and the 11 m NSW (Naval Special Warfare) rigid inflatable boat. The company also offers a 25 m eXtra Fast Patrol Boat (XFPB).

United States Marine Inc are licensed builders of the VSV™ fast patrol craft designed by Paragon Mann Ltd of the UK. The company is a fully integrated manufacturer capable of designing, building and testing boats totally in house.

In 2005 United States Marine Inc moved to a new production site in Gulfport, Mississippi following Hurricane Katrina which destroyed their previous site in New Orleans.

11 m NSW rigid inflatable boat at speed 0079941

United States Marine Inc 25 m XFPB 1020762

XFPB

Capable of speeds of up to 50 kt, the XFPB is propelled by two Kamewa water-jets.

Specifications

Length overall	25.5 m
Beam	5.64 m
Crew	12
Operational speed	45 kt

Propulsion: Two diesel engines rated at 2040 kW driving two Kamewa water-jets.

Vessel type or class name	Length (m)	Max beam (m)	Max speed (kt)	Hull material	Engines	Propulsion
SOCR	10.05	2.74	42	Aluminium	2 × 260 kW Inboard Diesel	Twin water-jet
NSW	11.0	3.2	44	GRP	2 × 350 kW Inboard Diesel	Twin water-jet
XFPB	25.5	5.64	50	GRP	2 × 2,040 kW Inboard Diesel	Twin water-jet
VSV	14.6	2.74	60	Composite	2 × 750 ho Inboard Diesel	–

USA Avenger Inc

Part of The Superyacht Technologies Group (Fort Lauderdale)
1300 SE 17th Street, Suite 219, Fort Lauderdale, Florida 33315, United States

Tel: (+1 954) 761 79 34
Fax: (+1954) 761 79 21
e-mail: info@superyachts.com
Web: www.super-yachts.com

Peter Baker, *Managing Director*

The company produces all alloy boats built to military specifications. The Avenger Fleet Designs were developed in the 1990s for use in a number of applications worldwide ranging from patrol boats to troop carriers (11 to 23 m). The Avenger Patrol Boats have been delivered to the governments of Greece and the Bahamas.

The company has manufacturing agreements with Kennedy Ships and Textron Marine and Land Systems.

10 m Avenger 0007898

Vessel type or class name	Length (m)	Max beam (m)	Max speed (kt)	Hull material	Engines	Propulsion
Avenger 11 m	11	2.74	50	Aluminium	Range of engine options	–
Avenger 12 m	12.3	3.0	50	Aluminium	Range of engine options	–
Avenger 17.5 m	17.5	4.8	50	Aluminium	Range of engine options	–
Avenger 23 m Offshore	23	6.25	35	Aluminium	Range of engine options	Twin water-jet

VT Halter Marine

A Vision Technologies Systems Company
900 Bayou Casotte Parkway, Pasacagoula, Mississippi 39581-2552, United States

Tel: (+1 228) 696 68 88
Fax: (+1 228) 696 68 90
e-mail: vthsales@vthaltermarine.com
Web: www.vthaltermarine.com

Boyd E King, *Chief Executive Officer*
Richard T McCreary, *Executive Vice President*
Cynthia Borries, *Corporate Communications Manager*

More than 2,600 vessels have been built by VT Halter Marine including 20 Mark V Special Operations Craft (SOC) for the US Navy's Special Warfare Command.

Mark V SOC 1120289

Vessel type or class name	Length (m)	Max beam (m)	Max speed (kt)	Hull material	Engines	Propulsion
Mark V SOC	25	5.3	50	Aluminium	2 × 1700 kW MTU diesel inboards	2 × waterjets

Willard Marine Inc

1250 N Grove St., Anaheim, California 92806, United States

Tel: (+1 714) 666 21 50
Fax: (+1 714) 632 81 36
e-mail: webmaster@willardmarine.
Web: www.willardmarine.com

Willard Marine, Inc is a manufacturer of fibreglass boats for the US Navy, US Coast Guard and US commercial operators. The company specialises in the construction of Rigid Inflatable Boats (4.8 to 16.5 m), personnel boats (6.7 to 12.2 m) and utility boats (6.7 to 15.2 m).

In recent years Willard Marine has also added aluminium construction to its portfolio.

Willard Marine SOLAS 5.4 m craft 0045332

Vessel type or class name	Length (m)	Max beam (m)	Max speed (kt)	Hull material	Engines	Propulsion
Willard 11 m RIB	11	3.66	40	GRP	2 × Inboard Diesel	Twin water-jet
490 (SOLAS)	4.9	2.0	–	GRP	70 Hp	–
540 (SOLAS)	5.4	2.3	–	GRP	100 Hp	–
670 (SOLAS)	6.7	2.7	–	GRP	200 Hp	–
Sea Force 730 RIB	7.3	2.7	–	GRP	2 × 112 kW	–
Sea Force 730	7.3	2.7	–	Aluminium	2 × Inboard Diesel	–
Sea Blazer 820	8.2	2.5	58	Aluminium	1 × Outboard	–
32' Kingston Patrol Boat	9.7	2.7	40	Aluminium	2 × Cummins 6BTA	Hamilton 274 waterjet
22' Utility Boat	6.7	2.7	40	GRP	2 × 150 hp Outboard	Twin Outboard

HIGH-SPEED PATROL CRAFT

Company Listing by Country

Australia
Austal
BSC Marine
Sea Chrome Marine Eritrea Share Company
Strategic Marine Pty Ltd
Tenix Defence Pty Ltd

Chile
Astilleros Y Maestrangas de la Armada

China
Guangzhou Huangpu

Croatia
Adria-Mar Shipbuilding Ltd

Denmark
Danyard Aalborg

Estonia
BLRT Laevaehitus

Finland
Aker Yards

France
CMN
DCN International
OCEA SA

Germany
Fr Fassmer GmbH
Lürssen Werft
Peene Werft

Greece
Elefsis Shipbuilding
Hellenic Shipyards

Hong Kong
Cheoy Lee Shipyard Ltd
Wang Tak Engineering

India
Garden Reach
Goa Shipyard Ltd

Indonesia
PT Pal
PT Palindo

Israel
Israel Shipyards

Italy
Cantieri del Golfo
Fincantieri
Intermarine SpA

Japan
Mitsubishi Heavy Industries Ltd
Mitsui Engineering and Shipbuilding
Sumidagawa Shipyard
Universal Shipbuilding Corporation

Malaysia
PSC Naval Dockyard

Netherlands
Damen Shipyard

Norway
Umoe Mandal AS

Russian Federation
Almaz Shipbuilding Co
Vympel Shipyard

Singapore
Singapore Technologies Marine Ltd

Spain
Astilleros de Murueta
Navantia

Sri Lanka
Colombo Dockyard

Taiwan
Jong Shyn Shipbuilding Co Ltd
Lung Teh Shipbuilding Co Ltd

United Kingdom
FBM Babcock Marine
McTay
VT Shipbuilding

United States
Bender Shipbuilding
Bollinger Shipyards Inc
Swiftships Shipbuilders, LLC
VT Halter Marine Inc

Australia

Austal

100 Clarence Beach, Henderson, Perth, Western Australia 6166

Tel: (+61 8) 94 10 11 11
Fax: (+61 8) 94 10 25 64
e-mail: marketing@austal.com
Web: www.austal.com

John Rothwell, *Executive Chairman*
Michael Atkinson, *Director*

Austal is a leading supplier of fast commercial ferries and aluminium patrol craft. Austal is able to provide patrol vessels to address a broad cross-section of needs for heightened security and surveillance in ports and coastal waters. The company has received sizeable orders from clients such as the Australian Customs Service, the Royal Australian Navy, Kuwait Coast Guard and the Republic of Yemen. Austal offers a significant range of craft dedicated to specialised patrol needs, incorporating the following:

- coast guard/customs/Navy patrol
- search and rescue
- EEZ surveillance
- fisheries protection
- fast attack craft

Armidale Class 56 m Patrol Boat 1123535

ARMIDALE CLASS 56 m PATROL BOAT

This semi-planning monohull was designed to replace the Royal Australian Navy's 'Freemantle' Class patrol boats. The first vessel, Armidale was delivered in May 2005 with all twelve vessels due for delivery within 48 months of the project commencement. In June 2006 Austal received an order for a further two vessels, which were launched in 2007.

37.5 m PATROL BOAT

Austal Ships received an order for 10 of these craft in June 2003 for delivery to the Republic of Yemen Ministry of Defence. The vessels were all transported to Yemen on a heavy lift ship in 2005.

16 m PATROL BOAT

Seven 16 m patrol boats were delivered to the New South Wales Police in time for the Sydney Olympics in 2000. A further six vessels were ordered in 2005 for delivery in 2006.

Vessel type	Dimensions L × B × T (m)	Disp. (tonnes)	Main machinery	Max speed (kt)	Complement	No. built	Date of build
Bay Class 38 m Patrol Boat	38.2 × 7.2 × 2.4	30	2 × MTU 16V 2000M70 (1,492 kW each); 2 × submerged shafts	24	16	8	1999
16 m Patrol Boat	17 × 4.88 × 1.2	–	2 × Scania DI 12 42 (485 kW each); 2 × shafts and FP propellers	28	2	13	2000
22 m Patrol Boat	21.2 × 5.5 × 1.7	–	2 × MAN D2842 LE408 (735 kW each); 2 × shafts and FP propellers	28	4	2	2000
37.5 m Patrol Boat	37.5 × 7.2 × 2.2	–	2 × Caterpillar 3512 (1,305 kW each); 2 × shafts and FP propellers	29	16	10	2004
22 m Patrol Boat	21.6 × 5.96 × 1.5	9.8	2 × MTU 12V 183TE92 (735 kW each); 2 × shafts and FP propellers	25	3	3	2004
Armidale Class 56 m Patrol Boat	56.8 × 9.5 × 2.7	–	2 × MTU 16V 4000 M70 (2,320 kW each); 2 × shafts & FP propellers	25	29	14	2005–2007
16 m Patrol Boat	17 × 4.88 × 1.2	–	2 × Caterpillar C12 (492 kW each); 2 shafts and FP propellers	28	2	1	2006

BSC Marine Group

Suite 12, 621 Coronation Drive, Toowong, Queensland 4066, Australia

Tel: (+61 7) 38 76 15 22
Fax: (+61 7) 38 76 15 66
e-mail: info@bscship.com.au
Web: www.bscship.com.au

Michael Hollis, *Managing Director*

Known for its low wash catamaran craft, BSC also produce a range of patrol vessels.

20 m PATROL VESSEL

The first vessel was delivered to Queensland Boating and Fisheries Patrol in 2002. The all aluminium vessel has an overall length of 20 m and is powered by two Caterpillar 3406E diesel engines each driving a fixed-pitch propeller via a ZF 305A gearbox.

23 m PATROL VESSEL

In 1998 BSC completed the delivery of four 23 m, 30 kt all aluminium patrol vessels for the Vietnam Department of Customs. The vessels were built under survey to both Bureau Veritas and USL 2B. All four vessels were built simultaneously and were completed within 14 months.

Vessel type	Dimensions L × B × T	Disp. (tonnes)	Main machinery	Max speed (kt)	Complement	No. built	Date of build
20 m Patrol Boat	–	–	2 × Caterpillar 3406E diesel engines, FP propellers, ZF 305A gearboxes	–	–	–	2002
23 m Patrol Boat	–	–	–	30		4	1998

Strategic Marine Pty Ltd

Lot 5 Clarence Beach Road, Henderson, Western Australia 6166

Tel: (+61 8) 9437 48 40
Fax: (+61 8) 9437 48 38
e-mail: info@strategicmarine.com
Web: www.strategicmarine.com

Mark Newbold, *Chairman*
Mark Schiller, *Managing Director, Australia*
Ron Anderson *Managing Director, International Company*

Strategic Marine, formerly Geraldton Boat Builders, specialises in building small fast patrol craft and commercial vessels, offering a range of both monohull and catamaran designs, all of a high speed planing type. Hulls are manufactured from aluminium or GRP and can be constructed at their facilities in Australia or Singapore.

20 m Command Vessel 0093272

22.5 m Patrol Boat 0589527

22.5 m PATROL BOAT

Designed for police and coast guard duties the 22.5 m Patrol Boat is capable of speeds up to 40 kt.

20 m PATROL BOAT

There have been five of these vessels delivered to the Australian defence forces.

20 m COMMAND VESSEL

Delivered in 1999 and 2000, these vessels were part of an order for 20 craft, at 18 and 20 m in length, for the Singapore Police Coast Guard. While they were assembled at the Asia-Pac Geraldton facilities in Singapore, much of the prefabrication was undertaken in the Geraldton shipyard in Australia.

Vessel type	Dimensions L × B × T (m)	Disp. (tonnes)	Main machinery	Max speed (kt)	Complement	No. built	Date of build
20 m Patrol Boat	22 × 6.6 × 1.6	43	2 × MTU 12V 183TF 92 diesel engines (1,500 kW each); 2 × shafts and FP propellers	28	–	1	1993
20 m Command Vessel	20 × 6.0 × 1.5	–	2 × MTU 16V 2000 M90 diesel engines (1,300 kW each); 2 × Hamilton 521HM water-jets	30	7	2	1999
20 m Patrol Boat	20 × 5.7 × 1.5	–	2 × MTU 8V 183TE diesel engines (485 kW each); 2 × shafts and propellers	25	7	5	1999
22.5 m Patrol Boat	22.5 × 7.5 × 1.0	–	2 × MTU 16V 2000 M90 diesel engines (1,350 kW each); 2 × shafts and FP propellers	40	–	1	–
10 m Patrol Boat	10.5 × 3.2 × 0.9	–	3 × 190 kW diesel engines	45	–	50	2006

Sea Chrome Marine Eritrea Share Co

PO Box 109, North Fremantle, Western Australia 6159, Australia

Tel: (+61 8) 93 35 11 55
Fax: (+61 8) 94 30 52 45

e-mail: info@seachrome.com.au
Web: www.seachrome.com.au

N Wilhelm, *Managing Director*
J Wilhelm, *Sales and Contracts*

Sea Chrome, a builder of high-speed commercial vessels, has entered the patrol craft market with its 24 m patrol boat.

Vessel type	Dimensions L × B × T	Disp. (tonnes)	Main machinery	Max speed (kt)	Complement	No. built	Date of build
24 m Patrol Boat	24 × 5.8 × 1.5	–	2 × D.DEC V12 diesels, FP and reverse reduction gearbox	30	–	–	–

Tenix Defence Pty Ltd

100 Arthur Street, North Sydney, New South Wales 2060, Australia

Tel: (+61 2) 99 63 98 00
Fax: (+61 2) 94 60 22 78

e-mail: marine.division@tenix.com
Web: www.tenix.com

David Miller, *Executive General Manager*
Jock Thornton, *Engineering Manager*
Hector Donohue, *Business Development Manager*

Tenix Defence Marine Division is one of Australia's leading designers and builders of naval, paramilitary and commercial vessels. The Division has particular skills in naval systems design and integration and brings together sensor and communications systems and weapons from different

Vessel type	Dimensions L × B × T (m)	Disp. (tonnes)	Main machinery	Max speed (kt)	Complement	No. built	Date of build
31.5 m Patrol Boat	31.5 × 6.5 × 2.0	150	2 × MTU 16 V 396 TB 94 diesels engines (2,100 kW each); 2 × shafts and FP propellers. 1 × MTU 8V 183 TE 62 diesel engine (550 kW); Hamilton 422 water-jet	28	11	4	1993
33 m Police Patrol Boat	32.6 × 8.2 × 1.6	145	2 × Caterpillar 3516 TA diesels engines (2,134 kW each); 2 × shafts and FP propellers. 1 × Caterpillar 3412 diesel engine (750 kW); single water-jet	30	19	6	1993
31.5 m Pacific Patrol Boat	31.5 × 8.1 × 2.1	162	2 × Caterpillar 3516 TA diesel engines (1,050 kW each); 2 × shafts and FP propellers	23	17	23	1987–1997
San Juan Class 56 m Patrol Boat	56 × 10.5 × 2.5	500	2 × Caterpillar 3612 diesels (3,530 kW each); 2 × shafts and CP propellers	25	38	4	2000

suppliers, integrating them into fully operational combat systems. Tenix owns major dockyards at Williamstown (Victoria, Australia) and Henderson (Western Australia) and a module construction site at Whangerei, New Zealand. In addition to these shipbuilding facilities, Tenix Marine has several smaller sites around Australia providing specialist technical support services. With Australia's largest shiplift and rotary ship turntable, each capable of lifting 8,065 tonnes, the division provides ship repair, dry-docking, maintenance, refit and support services for commercial ship operators and for the Royal Australian Navy. Tenix Defence marine Division has designed, built and delivered more than 200 naval, paramilitary and technically advanced commercial vessels and has the full capability to design, construct, outfit, set-to-work, trial and support major surface combatants.

31.5 m Pacific Patrol Boat 0017831

SAN JUAN CLASS 56 m PATROL BOAT

The San Juan Class patrol boats are primarily used for search and rescue, with facilities for 300 survivors. The vessels also carry firefighting and pollution control equipment, have a platform for one light helicopter and can launch and retrieve a 6.5 m rigid inflatable from a transom ramp and four 4.5 m rigid inflatables from davits. The vessel is capable of maintaining 12 knots in Sea State 4.

31.5 m PACIFIC PATROL BOAT

Tenix won the contract to build patrol boats for the Pacific Patrol Boat Project. The craft undertake surveillance and enforcement of the Exclusive Economic Zones of the countries concerned, namely Papua New Guinea, Vanuatu, Western Samoa, Solomon Islands, Cook Islands, Tonga, Marshall Islands, Fiji, Kiribati, Tuvalu and the Federated States of Micronesia.

San Juan Class 56 m Patrol Boat 0105228

Chile

Astilleros y Maestranzas de la Armada

ASMAR
Edificio Rapa Nui, Prat 856, Valparaíso, Chile

Tel: (+56 32) 26 00 00
Fax: (+56 32) 26 01 57
e-mail: gcomercial@asmar.cl
Web: www.asmar.cl

Claudio Pelaez, *Public Relations Manager*

ASMAR Shipbuilding and Shiprepairing Company has four main facilities in Chile, one each at Valparaiso and Punta Arenas and two at Talcahuano. The Valparaiso yard, which was founded in the 1890s has two dry-docks, of 180 m and 240 m long, a 160 m long slipway a 300 m fit out berth and is capable of building in steel and aluminium. The company builds a range of military and commercial high-speed vessels such as the Protector Class patrol vessels. Recently, six 42.5 m medium-speed patrol craft have been delivered to the Chilean Navy with two more on order. A 75 m patrol vessel for the government of Mauritius Isles was launched in December 1995.

Vessel type	Dimensions L × B × T (m)	Disp. (tonnes)	Main machinery	Max speed (kt)	Complement	No. built	Date of build
Protector Class 33 m Coastal Patrol Boat	33.1 × 6.6 × 2	110	2 × MTU diesel engines (1,900 kW each); 2 × CP propellers	24	10	18	2004

PROTECTOR CLASS 33 m COASTAL PATROL BOAT

Built to an FBM design under licence these vessels were completed between 1989 and 2004.

Protector Class 33 m Coastal Patrol Boat
0056733

China

Guangzhou Huangpu

Changzhou Town, Huangpu District, Guangzhou, Guangdong 510336, China

Tel: (+86 20) 220 17 29
Fax: (+86 20) 220 13 87
Web: www.cssc.net.cn

Part of the China State Shipbuilding Corporation. Established in 1851, Guangzhou Huangpu Shipyard specializes in military and civil service craft, cargo ships, yachts, work boats, high-speed aluminum ships, offshore structures, large-sized steel structures for land-based construction and heavy duty harbour machinery.

HOUJIAN CLASS 65 m FAST ATTACK CRAFT

Seven Houjian Class vessels were built for the Chinese Navy from 1991 to 2000.

Houjian Class 65 m Fast Attack Craft
0012767

Vessel type	Dimensions L × B × T (m)	Disp. (tonnes)	Main machinery	Max speed (kt)	Complement	No. built	Date of build
Houjian Class 65 m Fast Attack Craft	65.4 × 8.4 × 2.4	520	3 × SEMT-Pielstick 12 PA6 280 diesel engines (3,900 kW each)	32	75	7	1991

Croatia

Adria-Mar Shipbuilding Ltd

10000 Zagreb, Trg kralja Tomislava 8, Croatia

Tel: (+385 1) 4886 560
Fax: (+385 1) 4870 430
e-mail: office@adria-mar.hr
Web: www.adria-mar.hr

Adria-Mar offer desig, refit and new build services at their shipyard in Zagreb. In 2006 they delivered two 30 m patrol craft to the Libyan Coast Guard. Four more vessels were built in 2007 and another four are expected in 2008.

30 m Patrol Boat for Libyan Coast Guard 1292027

Vessel type	Dimensions L × B × T	Disp. (tonnes)	Main machinery	Max speed (kt)	Complement	No. built	Date of build
30 m Patrol Craft	31.3 × B × T	116 t	2 × Deutz TBD630 diesels, FP propellers	30 kt	14	10	2006

Denmark

Danyard Aalborg

Danyard Holding, Speditørvej 1, PO Box 660, DK-9100 Aalborg, Denmark

Tel: (+45 70) 24 36 01
Fax: (+45 70)10 05 16
e-mail: info@danyard.dk
Web: www.danyard.dk

Danyard Aalborg is a subsidiary of Danyard Holding A/S Frederikshavn. The company has specialised in composites material shipbuilding and fabrication of outfitting components. The technology and skills base is founded on construction of composite ships and craft for the Royal Danish Navy.

Flyvefisken Class 54 m Offshore Patrol Boat 1123507

FLYVEFISKEN CLASS 54 m OFFSHORE PATROL BOAT

Fourteen Flyvefisken Class multi-role vessels were built for the Royal Danish Navy between 1987 and 1996. Based on the Standard Flex 300 design, the Flyvefisken Class is based on a modular concept – using a standard hull with containerised weapon systems and equipment, which allows the vessel to change role quickly for surveillance, surface combat, anti-submarine warfare, mine countermeasures/minehunter, minelayer or pollution control. The hull is constructed of a fibre-reinforced plastic sandwich with a cellular core between outer and inner laminates. This gives the vessel a low magnetic signature, important for minehunting operations. The vessel features a combined diesel and gas turbine (CODAG) propulsion arrangement. A General Electric LM 500 gas turbine drives a fixed pitch centre-line propeller, providing 4,100 kW. Two MTU 16V 396 TB94 diesels providing 2,130 kW each, drive controllable-pitch propellers. The vessel also uses a roll stabilisation system which acts on the rudders and by passive stabilisation tanks.

Vessel type	Dimensions L × B × T (m)	Disp. (tonnes)	Main machinery	Max speed (kt)	Complement	No. built	Date of build
Flyvefisken Class 54 m Offshore Patrol Boat	54 × 9 × 2.5	480	CODAG; 1 × GE LM 500 gas turbine (4,100 kW); 2 × MTU 16V 396TB94 diesel engines (2,130 kW each); 1 × FP propeller (centre); 2 × CP propellers (outer)	30	29	14	1989

Estonia

BLRT Laevaehitus

48 Tööstuse Street, 10416 Tallinn, Estonia

Tel: (+372) 610 20 85
Fax: (+372) 610 21 69
e-mail: shipbuilding@bsr.ee
Web: www.bsr.ee

Fjodor Berman, *CEO, BLRT Group*

BLRT Laevaehitus was founded in 2003 when the BLRT Group made a strategic decision to develop its shipbuilding capabilities by re-launching and modernising the Noblessner shipyard in Tallin. At present, technical facilities at the yard allow the building of vessels up to 110 m length with a 1,750 t launch weight. There are 11 slipways, two of which are indoor slipways equipped with 2 × 10 t cranes and two-roofed slipways equipped with 100 t cranes. Since 1997 the yard has produced a patrol boat for the Estonian Coast Guard Service, a pilot boat for the Estonian Maritime Administration, a high-speed patrol boat for the Estonian Coast Guard Service and a multi-purpose work boat for the Port Authority of Kaliningrad.

PIKKER II CLASS 31 m COASTAL PATROL BOAT

This vessel is designed for patrolling the territorial waters of Estonia in rough sea conditions and to withstand small broken ice. Launched in 2000 the vessel has a top speed of 27 kt and is capable of carrying an RIB tender.

PIKKER I CLASS 30 m COASTAL PATROL BOAT

Built in 1996 this vessel can carry an RIB tender and was designed for the challenging conditions off the Estonian coast.

Vessel type	Dimensions L × B × T (m)	Disp. (tonnes)	Main machinery	Max speed (kt)	Complement	No. built	Date of build
Pikker 1 Class 30 m Coast Patrol Boat	30 × 5.8 × 1.5	90	2 × diesel engines (1,000 kW each); 2 × propellers	23	5	1	1996
Pikker II Class 31 m Coastal Patrol Boat	31.4 × 6.0 × 1.8	114	2 × Duetz TBD 620 V12 diesel engines (1,524 kW each); 2 × propellers	27	7	1	2000

Finland

Aker Yards

PO Box 132, FI-00151, Helsinki, Finland

Tel: (+358 10) 67 00
Fax: (+358 10) 670 67 00
e-mail: akerfinnyards@akeryards.com
Web: www.akerfinnyards.com

Yrjö Julin, *Managing Director*
Juha Heikinheimo, *Sales Director*

Aker Finnyards is part of Aker Yards, the largest shipbuilding group in Europe and one of the world's five largest shipbuilders. Aker Finnyards has thee shipyards situated in Helsinki, Rauma and Turku producing cruise ships, ferries, technically advanced vessels and military vessels.

HAMINA CLASS 51 m FAST ATTACK CRAFT

A modernised version of the Rauma Class, four Hamino Class vessels were delivered to the Finnish Navy in 1998, 2003, 2005 and

2006. With a length of 51 m, a beam of 8.5 m, draught of 1.7 m and displacement of 240 t, the vessel is of aluminium construction and also features lightweight composite materials in the superstructure as well as enhanced stealth characteristics. The propulsion system consists of two diesel engines driving water-jets, which give the vessel a maximum speed of 32 kt.

Hamina Class 51 m Fast Attack Craft 0094528

RAUMA CLASS 48 m FAST ATTACK CRAFT

Aker Finnyards built a series of four Rauma Fast Attack Craft for the Finnish Navy in the early 1990s. The vessels are of an all aluminium construction and have a water-jet propulsion system powered by 2 × 2,760 kW diesel engines. They are designed with moderate stealth technology and armed for fast combat missions.

Rauma Class 48 m Fast Attack Craft 0076823

Vessel type	Dimensions L × B × T (m)	Disp. (tonnes)	Main machinery	Max speed (kt)	Complement	No. built	Date of build
Rauma Class 48 m Fast Attack Craft	48 × 8 × 1.5	215	2 × MTU 16V 538 diesel engines (2,760 kW each); 2 × Riva Calzoni IRC 115 water-jets	30	19	4	1990
Hamina Class 51 m Fast Attack Craft	50.8 × 8.3 × 2	240	2 × MTU 16V 538TB93 diesel engines (2,760 kW each); 2 × Kamewa 90SOII water-jets	32	21	4	1998–2006

France

CMN

52 rue du la Brettonniere, BP 539, F-50105, Cherbourg Cedex, France

Tel: (+33 2) 33 88 30 00
Fax: (+33 2) 33 88 31 98
e-mail: info@cmn-cherbourg.com
Web: www.cmn-group.com

The CMN shipyard extends to over 120,000 m² of covered space. Spacious and well-equipped outfit workshops with adjacent storage facilities, enable CMN to effectively undertake complete machinery installation and outfitting before the ships leave for the test and trials facility. CMN produces a range of military ships, commercial ships and luxury yachts, and the repair, refit and upgrading of these vessels is a significant part of the shipyard's activities.

P-400 Class 55 m Large Patrol Boat 0104463

P-400 CLASS 55 m LARGE PATROL BOAT

Built between 1986 and 1988 for the French Navy, the P-400 Class have a steel structure, are capable of transporting up to 20 passengers and can be quickly converted to fast missile craft.

Vessel type	Dimensions L × B × T (m)	Disp. (tonnes)	Main machinery	Max speed (kt)	Complement	No. built	Date of build
P-400 Class 55 m Large Patrol Boat	54.5 × 8 × 2.5	477	2 × SEMT-Pielstick 16 PA4 200 VGDS diesel engines (2,940 kW each); 2 × propellers	25	26	10	1986
Flamant Class 55 m Offshore Patrol Boat	54.8 × 10 × 2.8	390	2 × Deutz 16V TBD 620 (1,740 kW each); 2 × MWM 12V TBD 234 (455 kW each); 2 × CP propellers	23	19	3	1995
Al Bushra Class 55 m Offshore Patrol Boat	54.5 × 8.0 × 2.7	475	2 × MTU 16V 538TB93 diesel engines (2,940 kW each); 2 × FP propellers	24	43	3	1995
Um Al Maradim Class 42 m Missile Attack Craft	42 × 8.2 × 1.9	245	2 × MTU 16V 538TB93 diesel engines (1,470 kW each); 2 × Kamewa water-jets	30	29	8	1998

DCN International

rue Choiseul, 56311 Lorient Cedex, France

Tel: (+33 2) 97 12 10 00
Fax: (+33 2) 97 12 26 90
Web: www.dcn.fr

Jean-Marie Poimbœuf, *CEO*

DCN is one of the few naval defence companies to propose an integrated approach to warships and services. As a prime contractor, shipbuilder and systems integrator, the company offers resources and expertise spanning the entire naval defence value chain and product lifecycles, from design concept to decommissioning. The company's four main shipyards are in Lorient, Cherbourg, Brest and Toulon and are capable of building vessels in steel or fibre re-enforced plastics.

Stellis Class 25 m Coastal Patrol Boat 0104484

GERANIUM CLASS 32 m PATROL BOAT

Six vessels of this class were built in 1997. They feature a triple engine installation with two conventional shafts with propellers and a central water-jet for loiter or boost operations.

STELLIS CLASS 25 m COASTAL PATROL BOAT

Two vessel were built in 1993 for service in French Guiana. The vessels feature a triple diesel installation with two conventional shafts and propellers and a central booster water-jet for high speeds in conjunction with the propellers or for low-speed operations on its own.

Vessel Type	Dimensions L × B × T (m)	Disp. (tonnes)	Main machinery	Max speed (kt)	Complement	No. built	Date of Built
Stellis Class 25 m Coastal Patrol Boat	24.9 × 6.1 × 1.7	60	3 × diesel engines (1,800 kW in total); 2 × propellers; 1 × water-jet	28	8	2	1993
Geranium Class 32 m Patrol Boat	32.2 × 6.1 × 1.9	96	2 × Duetz TBD 516 V16 (1,092 kW each); 1 × Duetz TBD 516 V12 (724 kW); 2 × propellers; 1 × Hamilton 422 water-jet	30	15	6	1997

OCEA SA

Quai de la Cabude, F-85100, Les Sables d'Olonne, France

Tel: (+33 2) 51 21 05 90
Fax: (+33 2) 51 21 20 06
e-mail: commercial@ocea.fr
Web: www.ocea.fr

Roland Joassard, *Chief Executive Officer*
Fabrice Epaud, *Commercial Director*

OCEA specialises in the design and construction of both monohull and catamaran aluminium craft in the 20 to 45 m range. The company has recently delivered a number of high speed patrol craft to Kuwait.

Subahi Class FPB 110 Patrol Boat 1179477

SUBAHI CLASS FPB 110 PATROL BOAT

Ten of these patrol vessels were ordered by the Kuwait Ministry of the Interior in 2002 for delivery over a 3-year period. These vessels are an improved and lengthened version of the FPB 100 Mk 1 vessels.

AL SHAHEED CLASS FPB 100 MK 1 PATROL BOAT

Three vessels were delivered to the Kuwait Coast Guard between 1997 and 2001. The vessels are constructed entirely of aluminium and feature twin water-jets providing a top speed of 30 kt.

Al Shaheed FPB 100 Mk 1 Patrol Boat 1179481

Vessel type	Dimensions L × B × T (m)	Disp. (tonnes)	Main machinery	Max speed (kt)	Complement	No. built	Date of build
Al Shaheed Class FPB 100 Mk 1 Patrol Boat	33.3 × 7.0 × 1.2	–	2 × MTU 12V 396TE 94 diesel engines (1,600 kW each); 2 × Kamewa water-jets	30	11	3	1997
Subahi Class FPB 110 Patrol Boat	35.2 × 6.8 × 1.2	116	2 × MTU 12V 396TE 94 diesel engines (1,715 kW each); 2 × LIPS LJ65E water-jets	32	17	10	2003

Germany

Fassmer GmbH & Co

Industriestr 2, D-27804 Berne Motzen (Weser), Germany

Tel: (+49 4406) 94 20
Fax: (+49 4406) 94 21 00
e-mail: info@fassmer.de
Web: www.fassmer.de

Harald Fassmer, *Director*
Peter Schmidt, *Sales Director*

Established in 1850, Fassmer specialises in the design and construction of fast steel and aluminium monohull craft. With a building shed facility of up to 80 m in length, the largest craft launched by the shipyard to date is a 58 m ferry.

30 m PATROL BOAT

Delivered in 2000 to the Lower Saxony Water Police, the vessel is used for patrol and surveillance.

26 m POLICE BOAT

Delivered in 2001, *Warnaw* is operated by the Rostock Water Police in Germany, patrolling water in the 12 mile sea zone.

30 m Patrol Craft 0097104

Vessel type	Dimensions L × B × T	Disp. (tonnes)	Main machinery	Max speed (kt)	Complement	No. built	Date of build
28.5 m Customs Boat	28.62 × 6.2 × 1.65	95	3 × MWM TBD 604 BV8	26	6	5	1991
29.5 m Police Boat	29.5 × 6.4 × 2.0	95	3 × MWM TBD 234 V 12	23	6	2	1992
25.0 Surveillance Craft	25 × 6 × 1.5	59	2 × MAN V12 diesel engines	23	4	1	1997
21 m Patrol Boat	21 × 5.8 × 1.3	45	2 × MTU diesel engines	34.0	13	3	1997
24.5 m Patrol Boat	24.5 × 6.25 × 1.5	58	2 MAN D2842 LE408	24.0	4	1	1997
30 m Police Boat	30.5 × 6.9 × 2.1	120	2 × MTU 12V 396TE84	22	6	1	2000
46.0 m Rescue Vessel	46 × 10.6 × 2.8	400	2 × MTU 12V 4000 M90, 1 × MTU 16V 4000 M90	25	14	1	2002

Lurssen Werft GmbH & Co Kg

Zum Alten Speicher 1, 28759 Bremen, Germany

Tel: (+49 421) 66 03 34
Fax: (+49 421) 66 03 95
e-mail: kt@lurssen.de
Web: www.lurssen.de

Detlef Schlichting, *Sales Director*

Lürssen Werft is a member of the Lürssen group of companies. In 1980 the parent company acquired Kröger Werft in Rendsburg and in 1997 the Bremer Vulkan Marineschiffbau which incorporates a large covered dry dock, allowing the support of ships up to 170 m in length. Fast motor boats, motor cruisers and motor yachts have been built by Lürssen Werft since 1890. Recent construction includes fast attack craft, corvettes, mine combat vessels, patrol craft, megayachts, search and rescue boats and fast ferries.

AHMAD EL FATEH CLASS 45 m FAST ATTACK CRAFT

Four vessels of this type were delivered between 1984 and 1989. The vessels have a quadruple diesel installation driving four shafts and propellers, giving the vessels a top speed of 40 kt.

MUBARRAZ CLASS 45 m FAST ATTACK CRAFT

Two vessels of the class were delivered to the UAE in 1990.

28 m CUSTOMS PATROL BOAT

Five of these vessels were delivered to the Indonesian Customs Service in 1999. The first two were built by Lürssen and the last three by PT Pal Indonesia.

Vessel type	Dimensions L × B × T (m)	Disp. (tonnes)	Main machinery	Max speed	Complement (kt)	No. built	Date of build
Ahmad El Fateh Class 45 m Fast Attack Craft	44.9 × 7 × 2.5	228	4 × MTU 16V 538TB92 diesel engines (2,500 kW each); 4 × propellers	40	36	4	1984
Mubarraz Class 45 m Fast Attack Craft	44.9 × 7 × 2.2	260	2 × MTU 20V 538TB93 diesel engines (3,450 kW each); 2 × propellers	40	40	2	1990
28 m Customs Patrol Boat	28.2	85	2 × MTU 16VTE94 diesel engines (1,070 kW each); 2 × propellers	40	11	5	1999

Peene Werft

17438 Wolgast, Schiffbauerdamm 1, Germany

Tel: (+49) 38 36 25 00
Fax: (+49) 38 36 25 01 53
e-mail: info.pw@peene-werft.de
Web: www.peene-werft.de

A member of the Hegemann Group, the Peene Werft shipyard has built more than 650 new ships including 257 naval vessels. The yard also repairs and refits vessels for a large number of clients.

Grajau Class 47 m Large Patrol Boat 0104228

GRAJAU CLASS 47 m LARGE PATROL BOAT

Six Grajau Class vessels were built for Brazil by Peene Werft to a Vosper QAF design between 1994 and 1999. The vessels can be used for patrol duties and diver support work and carry a RIB tender.

Vessel type	Dimensions L × B × T (m)	Disp. (tonnes)	Main machinery	Max speed (kt)	Complement	No. built	Date of build
Grajau Class 47 m Large Patrol Craft	46.5 × 7.5 × 2.3	263	2 × MTU 16V 296TB94 diesel engines (2,130 kW each); 2 × propellers	26	29	6	1994
Sassnitz Patrol Boat	48.9 × 8.65 × 2.15	348	3 × M520 diesel engines	37	33	–	–
Patrol Boat for Brazilian Navy	46.5 × 7.5 × 2.3	217	2 × 2469 kW diesel engines	25	31	–	–

Greece

Elefsis Shipbuilding & Industrial Enterprises SA

192 00 Elefsis, Greece

Tel: (+30 210) 553 51 11
Fax: (+30 210) 554 60 16
e-mail: cm@mail.elefsis-shipyards.gr
Web: www.elefsis-shipyards.gr

Located in Elefsis bay, close to the port of Piraeus, Elefsis Shipyard carries out commercial new buildings, repair and conversion work and industrial work. The yard covers an area of 250,000 m², of which 56,000 m² is under cover. It has three drydocks with a maximum capacity of 120,000 t deadweight.

Roussen Class 62 m Fast Attack Craft 1155336

ROUSSEN CLASS 62 m FAST ATTACK CRAFT

Designed by Vosper Thornycroft, these 62 m vessels feature a steel hull and aluminium superstructure. The main propulsion system consists of four MTU 16V 595 TE 90 engines with ZF type BW 1556666/1557 S gear boxes driving four fixed-pitch propellers. The craft can achieve a speed of 35 kt and at a cruise speed of 12 kt has a range of 1,800 n miles. The steering system and stabiliser system were supplied by Vosper Thornycroft Marine Products.

Vessel type	Dimensions L × B × T (m)	Disp. (tonnes)	Main machinery	Max speed (kt)	Complement	No. built	Date of build
Roussen Class 62 m Fast Attack Craft	61.9 × 9.5 × 2.6	580	4 × MTU MD 16V 538TB 90 diesel engines (2,205 kW each); 4 × propellers	34	45	5	2004

Hellenic Shipyards

PO Box 3480, Athens GR-10233, Greece

Tel: (+30 1) 557 83 15
Fax: (+30 1) 557 07 19
e-mail: marketing@hellenic-shipyards.gr
Web: www.hellenic-shipyards.gr

Hellenic Shipyards SA is the largest shipyard in the Eastern Mediterranean. Founded over 50 years ago, Hellenic Shipyards SA is now part of the TKMS Group, and has moved into complex ship construction as well as maintaining its traditional business in ship repair. With the capacity to repair up to 18 ships concurrently and the ability to service most repair requirements for its customers, the shipyard has repaired and converted well over 9,000 commercial and navy vessels of all types and sizes, including over 130 ships of the US Navy fleet. In the mid 1970s, the Shipyard entered the naval new buildings market by constructing a series of Delos Class Coastal Patrol Vessels for the Hellenic Navy. It has since expanded into the construction of fast attack craft, fast patrol and missile boats, frigates, gunboats and most recently submarines, the first to be constructed in Greece.

Vessel type	Dimensions L × B × T (m)	Disp. (tonnes)	Main machinery	Max speed (kt)	Complement	No. built	Date of build
SAAR 4 Class 58 m Large Patrol Boat	58 × 7.8 × 2.4	415	4 × MTU 12V 331TC92 diesel engines (2,750 kW each); 2 × propellers	32	30	4	2003
Machitis Class 57 m Patrol Boat	56.5 × 10 × 3.6	575	2 × Wartsila 16V 25 diesel engines (3700 kW each); 2 × propellers	22	50	4	2003–2005
Pyroplitis Class 57 m Patrol Boat	56.5 × 10 × 2.7	563	2 × Wartsila 16V 25 diesel engines	24	36	2	1993

SAAR 4 CLASS 58 m LARGE PATROL BOAT

In 2002 the Greek Coast Guard ordered four SAAR 4 Class vessels. The first two vessel were constructed by Israel Shipyards and the following vessels were built by Hellenic Shipyards. The vessels are used for patrol duties and search and rescue missions.

MACHITIS CLASS 57 m LARGE PATROL BOAT

Four Machitis Class Large Patrol Craft were launched between 2002 and 2003. The vessels feature a steel hull and aluminium superstructure and are an updated version of the Pyrpolitis Class.

PYRPOLITIS CLASS 57 m LARGE PATROL BOAT

Two vessels were built in 1992 and 1993 respectively for the Hellenic Navy. The vessels use modularity so that weapon and sensor fits can be interchanged for different missions. The vessels, propulsion system consists of resiliantly mounted diesel engines driving propellers, giving the vessel a top speed of 24 kt.

Machitis Class 57 m Large Patrol Boat 1120275

Hong Kong

Cheoy Lee Shipyard Ltd

Tel: (+825) 23 07 63 33
Fax: (+825) 23 07 55 77
e-mail: info@cheoyleena.com
Web: www.cheoyleena.com

Ken Lo, *Director*
Jonathon Cannon, *Sales Manager*

The company commenced building motor yachts and building in GRP in 1961 and now builds commercial and pleasure craft in steel, composites and aluminium. The company has built over 4,700 vessels since 1950 and is currently expanding its operations into mainland China.

KEKA CLASS 30 m PATROL BOAT

Two of these 30 m aluminium vessels were delivered to the Hong Kong Police in 2002 and a further two were delivered in 2005. The vessel features a triple engine installation, two driving FP propellers and one driving a low-speed manoeuvring or high speed booster water-jet. The round bilge hull form has a fine bow entry which offers good performance in a wide range of sea conditions.

Keka Class 30 m Patrol Boat 1109895

13 m PATROL LAUNCH

Six vessels were delivered in 2000 to the Hong Kong Police, based on a design by Peterson Shipbuilders used by both the American and Egyptian Navies. Fast and manoeuverable, this jet powered aluminum design is ideally suited to operations in congested or confined waters.

13 m Patrol Launch 1109904

Vessel type	Dimensions L × B × T (m)	Disp. (tonnes)	Main machinery	Max speed (kt)	Complement	No. built	Date of build
13 m Patrol Launch	13 × 3.96 × 0.7	–	2 × MAN D2842 LE 403 diesel engines (530 kW each); 2 × Hamilton water-jets	35	4	6	2000
Keka Class 30 m Patrol Boat	30 × 6.3 × 1.7	114	2 × MTU 12V 396TE84 diesel engines (1,520 kW each); 2 × FP propellers; 1 × MTU 8V 2000 M60 diesel engine (405 kW); 1 × Hamilton HM 571 booster water-jet	28	14	4	2002

Wang Tak Engineering

Wang Tak Building, 85 Hing Wah Street West, Lai Chi Kok, Kowloon, Hong Kong

Tel: (+852) 27 46 28 88
Fax: (+852) 23 07 55 00
e-mail: info@wangtak.com.hk

Feat Szeto, *Director and General Manager*

The Wang Tak Shipyard based in Hong Kong, offers shipbuilding and repair services in steel and aluminium materials. In 2000 the company delivered two 32 m patrol launches to the Hong Kong Customs and Excise department.

32 m Patrol Launch 0101463

32 m PATROL LAUNCH

Two of these patrol craft were delivered to Hong Kong Customs and Excise in 2000. The vessels feature steel hulls with aluminium superstructures and meet Lloyds Special Service Craft classification.

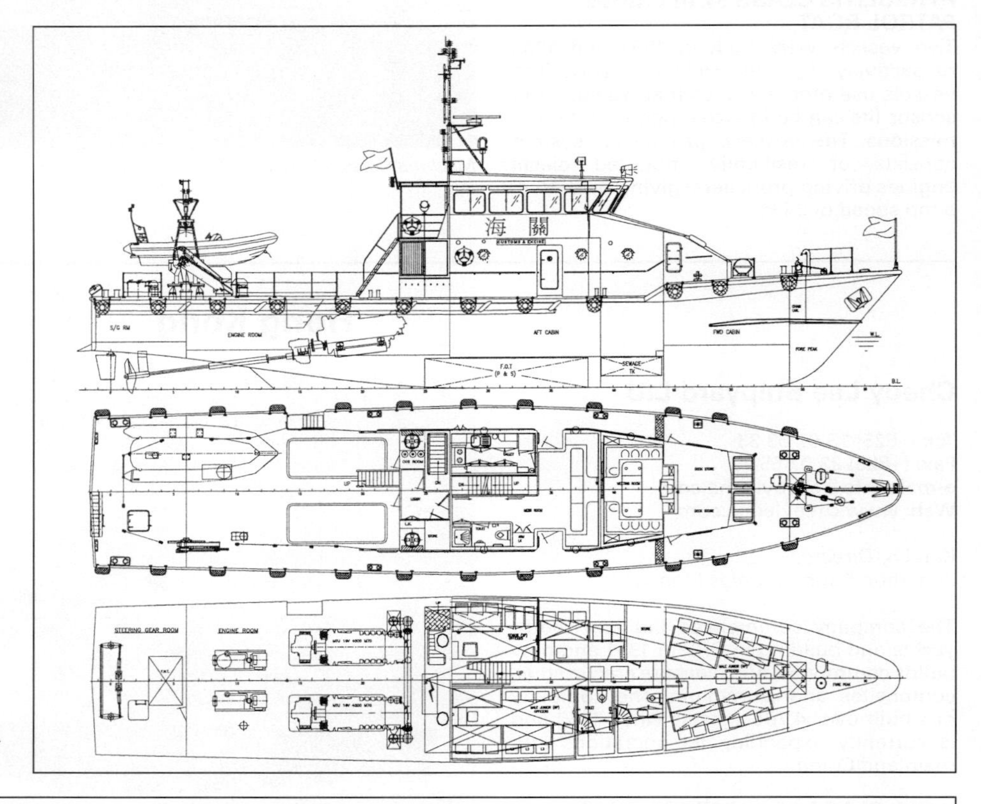

32 m Patrol Launch, general arrangement 0116595

Vessel type	Dimensions L × B × T (m)	Disp. (tonnes)	Main machinery	Max speed (kt)	Complement	No. built	Date of build
32 m Patrol Launch	32 × 6.8 × 2	180	2 × MTU 16V 4000 M70 diesel engines (2,320 kW each); 2 × Teignbridge FP propellers	29	–	2	2000

India

Garden Reach

Garden Reach Road, Kolkatta 700024, India

Tel: (+33) 246 961 20
Fax: (+33) 246 981 44
e-mail: cgmsm@vsnl.net
Web: www.grse.nic.in

Rear Admiral T.S Ganeshan, *Chairman and MD*

The company was taken over by the Government of India in 1960 and has gradually expanded and modernised to meet growing maritime needs, particularly those of the Indian Navy and Coast Guard.

SDB Mk 5 Trinkat Class 46 m Patrol Boat 0096069

Vessel type	Dimensions L × B × T (m)	Disp. (tonnes)	Main machinery	Max speed (kt)	Complement	No. built	Date of build
Tara Bai Class 45 m Coastal Patrol Boat	44.9 × 7 × 1.9	195	2 × MTU 12V 538 diesel engines (1,480 kW each); 2 × propellers	26	34	6	1987
Priyadarshini Class 46 m Coastal Patrol Boat	46 × 7.5 × 1.9	306	2 × MTU 12V 538 diesel engines (1,480 kW each); 2 × propellers	23	34	4	1992
SDB Mk 5 Trinkat Class 46 m Patrol Boat	46 × 7.5 × 2.5	260	2 × MTU 16V 538 TB92 diesel engines (2,500 kW each); 2 × propellers	30	34	8	2000

Today Garden Reach is one of the leading shipyards in India, building a wide range of ships from modern warships to sophisticated commercial vessels and small harbour craft to fast patrol boats. It is among the few shipyards in the world with its own engine manufacturing division. Between 2000 and 2006, the yard delivered 8 SDB Mk 5 Trinkat Class patrol craft.

SDB Mk 5 TRINKAT CLASS 46 m PATROL BOAT

Four Trinkat Class vessels were built between 2000 and 2002 with a further four being built by 2006. In 2005 the Indian Navy predicted that a further ten vessels could be required. The vessels which are designed for EEZ patrol duties feature a propulsion system consisting of two diesel engines rated at 2,500 kW each driving conventional shafts and propellers giving the vessels a top speed of 30 kt and a range of 2,000 nm at 12 kt.

PRIYADARSHINI CLASS 46 m COASTAL PATROL BOAT

Developed from the Tara Bai Class, the Priyadarshini Class, of which eight vessels were built between 1992 and 1998, has a range of 2,400 nm at 12 kt and a top speed of 23 kt. Garden Reach built four of these vessels with Goa Shipyard producing the remaining vessels.

TARA BAI CLASS 45 m COASTAL PATROL BOAT

Six Tara Bai Class vessels were built between 1987 and 1990 for the Indian Coast Guard. The first two vessels were built in Singapore with the final four being built by Garden Reach.

Goa Shipyard Limited

Vasco-Da-Gama, Goa, 403802, India

Tel: (+91 832) 251 21 52
Fax: (+91 832) 251 38 70
e-mail: contactus@goashipyard.com
Web: www.goashipyard.co.in

Radm. Sampath Pillai, *Managing Director and Chairman*

Goa Shipyard Limited was established in 1957 and is today a leading shipyard on the West Coast of India, meeting the requirements of varied customers in the field of design, development, construction, repair, modernisation, testing and commissioning of ships. Goa Shipyard's facilities and resources include a CAD/CAM centre, modern manufacturing facilities including a number of building bays and a 180 m fit-out pier, a work force of over 2,000 skilled personnel and over 175 qualified engineers and naval architects. The shipyard can build traditional military ships and marine crafts, surface effect ships, hovercrafts, high-speed aluminium-hulled vessels, pollution control vessels, advance deep sea trawlers, fish factory vessels, catamarans, pleasure submarines and offers a whole range of marine products and services.

SAROJINI NAIDU CLASS 48 m FAST PATROL BOAT

The Sarojini Class is a sea going, high speed, armed surveillance platform, capable of shallow water operations. The vessel is primarily designed for anti-smuggling operations, anti-terrorist deployment, fisheries protection and search and rescue operations. Powered by three 2,720 kW MTU diesel engines, driving independent Kamewa water jets, this vessel is designed for good manoeuvrability and is capable of operating in up to Sea State 4 and can withstand Sea State 6. The vessel has an operational range of 1,500 nm.

PRIYADARSHINI CLASS 46 m COASTAL PATROL CRAFT

Developed from the Tara Bai Class, the Priyadarshini Class, of which eight vessels were built between 1992 and 1998, has a range of 2,400 nm at 12 kt and a top speed of 23 kt. Goa Shipyard built four of these vessels with Garden Reach producing the remaining vessels.

Vessel type	Dimensions L × B × T (m)	Disp. (tonnes)	Main machinery	Max speed (kt)	Complement	No. built	Date of build
Priyadarshini Class 46 m Coastal Patrol Craft	46 × 7.5 × 1.9	306	2 × MTU 12V 538 diesel engines (1,480 kW each); 2 × propellers	23	34	4	1992
Sarojini Naidu Class 48 m Fast Patrol Boat.	48.1 × 7.5 × 2	270	3 × MTU 16V 4,000 M90 diesel engines (2,720 kW each); 3 × Kamewa 71SII waterjets	35	35	7	2002

Indonesia

PT Pal Indonesia

Ujung Surabaya, PO Box 1134, Indonesia

Tel: (+62) 31 3292275
Fax: (+62) 31 3292530
e-mail: palsub@pal.co.id
Web: www.pal.co.id

Adwin Suryohadiprojo, *President*
Mochamad Moenir, *Shipbuilding Director*
Kusimawati Rukminto, *Business Development Director*

PT Pal Indonesia's current naval vessel products include fast patrol boats of 57 m, 28 m, and 14 m, corvettes and minehunters.

Singa Class 58 m Large Patrol Boat 0546342

Vessel type	Dimensions L × B × T (m)	Disp. (tonnes)	Main machinery	Max speed (kt)	Complement	No. built	Date of build
Singa Class 58 m Large Patrol Boat	58.1 × 7.6 × 2.8	447	2 × MTU 16V 956TB92 diesel engines (3,250 kW each); 2 × propellers.	27	42	4	1988
Kakap Class 58 m Large Patrol Boat	58.1 × 7.6 × 2.8	423	2 × MTU 16V 956TB92 diesel engines (3,250 kW each); 2 × propellers.	28	49	4	1988
Todak Class 58 m Large Patrol Boat	58.1 × 7.6 × 2.8	447	2 × MTU 16V 956TB92 diesel engines (3,250 kW each); 2 × propellers.	27	53	4	2000

PT Pal Indonesia also produce a wide range of merchant vessels including general cargo vessels, container ships, bulk carriers, tankers, tugs, survey vessels and conventional ferries. The yard's facilities include two large dry docks, a 5,000 t floating dock, a 1,500 t ship lift, hull construction and outfitting workshops.

SINGA CLASS 58 m LARGE PATROL BOAT

Designed by Lürssen Werft, the first two vessels of the class of four were partially built at Lurssen and shipped to PT Pal Indonesia for fitting out. The remaining vessels were built by PT Pal Indonesia and delivered by 1993. The vessels are fitted with Vosper Thornycroft stabiliser fins.

KAKAP CLASS 58 m LARGE PATROL BOAT

Four vessels of this class were built between 1988 and 1995. The first two vessels were partially built by Lürssen Werft and completed by PT Pal Indonesia. The third and fourth vessels were built entirely by PT Pal Indonesia. The vessels are can be used for patrol duties, search and rescue and firefighting. They are fitted with Vosper Thornycroft fin stabilisers and carry a fast tender with launch davit.

TODAK CLASS 58 m LARGE PATROL BOAT

Developed from the Singa and Kakap Class vessels, four of this class were delivered between 2000 and 2004. The vessels feature a twin diesel installation driving standard shafts and propellers, giving a top speed of 27 kt, which can be maintained for 2,200 n miles.

Kakap Class 58 m Large Patrol Boat

0546346

Todak Class 58 m Large Patrol Boat 0093587

PT Palindo

Jl.Mt.Haryono No.8, Tanjung Pinang, Indonesia

Tel: (+62) 77 12 76 88
Fax: (+62) 77 12 74 56
e-mail: wisnu_kn@yahoo.com
Web: www.palindo-shipyard.com

PT Palindo, located on Bintan Island, Indonesia, has built a number of GRP ferries for operators in Indonesia, Malaysia and Singapore. The yard has also produced various patrol vessels for the Indonesian Navy and other government authorities.

36 m Patrol Boat 1310441

Vessel type	Dimensions L × B × T (m)	Disp. (tonnes)	Main machinery	Max speed (kt)	Complement	No. built	Date of build
36 m Patrol Boat	36 × 7 × 1.35	90	2 × MAN D2842 LE 410 diesel engines (820 kW each); 2 × propellers	31	18	13	2003–2005
40 m Patrol Boat	40 × 7.3 × 1.35	105	3 × MAN D2842 LE410 diesel engines (820 kW each)	30	20	3	–

36 m PATROL BOAT

Thirteen vessels of this type were built between 2003 and 2005 by a number of yards including PT Palindo. The vessels feature a GRP construction and some have a stern ramp for launch and recovery of a RIB. The vessels have a triple diesel engines driving fixed-pitch propellers giving the vessels a top speed of 38 kt.

40 m Patrol Boat
1310768

Israel

Israel Shipyards

PO Box 10630, Haifa Bay, IL-26118, Israel

Tel: (+972 48) 46 02 46
Fax: (+972 48) 41 87 44
e-mail: marketing@israel-shipyards.com
Web: www.israel-shipyards.com

Israel Shipyards is one of the largest privately owned ship building and repair facilities in the East Mediterranean. Established over 30 years ago the company was government owned until 1995. The manufacturing and repair plant is spread over 400,000 m^2 with a total wharf length of 900 m. The company has the ability to build both large and small military vessels, tugs and commercial vessels. Israel Shipyards was responsible for producing the successful Shaldag Class of high-speed interceptor and has exported a number of patrol boats, including the SAAR 4 Class to Greece.

SAAR 58 m Large Patrol Boat

The SAAR 4 represents the newest generation of small, fast-attack missile craft designed and built by Israel Shipyards. The vessel is capable of accepting a full range of sensors and weapons.

SAAR Class 58 m Fast Attack Craft
0075861

Vessel type	Dimensions L × B × T (m)	Disp. (tonnes)	Main machinery	Max speed (kt)	Complement	No. built	Date of build
Hetz Class 62 m Fast Attack Craft	61.7 × 7.6 × 2.5	488	4 × MTU 16V 538TB93 diesel engines (3,050 kW each); 4 × FP propellers	31	53	8	1981
SAAR 4 Class 58 m Large Patrol Boat	58 × 7.8 × 2.4	415	4 × MTU 12V 331TC92 diesel engines (2,750 kW each); 2 × propellers	32	30	4	2002

Italy

Cantieri Navali del Golfo

Lungomare Caboto, I-04024 Gaeta (LT), Italy

Tel: (+39) 77 17 12 81
Fax: (+39) 771 31 00 69
e-mail: infocantieri@cantieridelgolfo.it
Web: www.cantieridelgolfo.it

Cantieri Navali del Golfo produces a range of patrol boats and motor yachts.

Vessel type	Dimensions L × B × T	Disp. (tonnes)	Main machinery	Max speed (kt)	Complement	No. built	Date of build
Corrubia Patrol Boat	26.4 × 7.4 × –	–	2 × diesel inboard	43	–	10	–

Fincantieri

Naval Vessel Business Unit, Via Cipro 11, 16129 Genoa, Italy

Tel: (+39 010) 599 51
Fax: (+39 010) 599 53 79
e-mail: info@fincantieri.it
Web: www.fincantieri.it

Saettia Class 53 m Patrol Boat
0104899

Corrado Antonini, *Chairman*
Giuseppe Bono, *Chief Executive Officer*

Fincantieri is one of the largest and most diversified shipbuilding groups in Europe, designing and building cruise ships, ferries, surface combatants, auxiliary ships and submarines. Fincantieri's Naval Vessel Business Unit has built over 2,000 ships for the Italian and foreign Navies. Vessels designed and built have included patrol vessels, frigates, aircraft carriers, auxiliary ships, oceanographic vessels and submarines. The unit has a design centre in Genoa and shipyards in Muggiano and Riva Trigoso (Genoa).

SAETTIA CLASS 53 m PATROL BOAT

The first vessel of this class was built in 1985 and then a further five vessels were built between 2002 and 2004 for the Italian Coast Guard. The later vessels feature a propulsion system consisting of 4 x diesel engines and CP propellers giving the vessels a top speed of 31 kt.

Antonio Zara Class 51 m Patrol Boat 0012620

ANTONIO ZARA CLASS 51 m PATROL BOAT

Two vessels of this class were built in 1990 with a further vessel being completed in 1998. The final vessel has a propulsion system consisting of four diesel engines driving two shafts and propellers giving the vessel a top speed of 34.5 kt.

Vessel type	Dimensions L × B × T	Disp. (tonnes)	Main machinery	Max speed (kt)	Complement	No. built	Date of build
Saettia Class 53 m Patrol Boat	52.8 × 8.1 × 2	427	4 × Isotta Fraschini V1716T2 MSD diesel engines (2,360 kW each); 4 × CP propellers	31	30	6	1985
Antonio Zara Class 51 m Patrol Boat	51 × 7.5 × 1.9	340	4 × MTU 16 V 369TB94 diesel engines (2,395 kW each); 2 × propellers	34	33	3	1990

Intermarine SpA

Via Alta, I-19038 Sarzana, Italy

Tel: (+39 0187) 61 75 55
Fax: (+39 0187) 61 75 50
e-mail: mail@intermarine.it
Web: www.intermarine.it

Osvaldo Facchinetti, *Business Development Manager*

A subsidiary of Rodriquez Cantieri Navali SpA and internationally known as a leading designer and builder of minehunting vessels, Intermarine also manufactures fast patrol craft from 9 to 40 m in length.

MV100 30 m Patrol Boat 0114501

MV115 36 m MAZZEI CLASS PATROL BOAT

These are stretched versions of the MB85 Bigliani Class. Two of these vessels, in a training configuration, were delivered to the Italian Guardia di Finanza (Customs) in 1998. Five further vessels, in an operational configuration, were delivered between 2003 and 2005.

MV85 27 m BIGLIANI CLASS PATROL BOAT

A composite fast patrol boat of which 22 boats in four series have been built for the Guardia di Finanza between 1987 and 2006. The latest vessels include Kevlar armour and are fitted with a remote-controlled Breda 12.7 mm gun and 40 mm grenade launcher instead of the Breda 30 mm gun. All are being fitted with infra-red search and surveillance sensors. A fifth series is currently under construction consisting of five enlarged vessels with different configurations for both the propulsion and the mission systems.

MV115 36 m Mazzei Class Patrol Boat 0103304

CP25 Patrol Boat

The CP25 Class has been designed to provide high speed and excellent seaworthiness. 28 vessels have been built for the Italian Coast Guard. The CP25 fast patrol boats have a crew of 12. Propulsion is by two diesel engines driving fixed-pitch propellers and by one central diesel engine driving a "booster" water-jet giving the CP25 a maximum speed of 34 kt.

Vessel type	Dimensions L × B × T	Disp. (tonnes)	Main machinery	Max speed (kt)	Complement	No. built	Date of build
MV85 27 m Bigliani Class Patrol Boat	27 × 7.0 × 1.2	95	2 × MTU 16V 396TB 94 diesel engines (2,560 kW each); 2 × shafts and FP propellers	42	12	22	1987–2007
MV85 27 m Bigliani Class – VIII Series	28.2 × 7.0 × 1.2	100	2 × MTU 16V M90 diesel engines (2720 kW each); 2 × shafts and FP propellers	40	15	5	2006–2009
MV115 36 m Mazzei Class Patrol Boat	35.5 × 7.5 × 1.2	115	2 × MTU 16V 396TB 94 diesel engines (2,560 kW); 2 × shafts and FP propellers	35	34	7	1998
MV115 36 m Mazzei Class Patrol Boat VII Series	35.5 × 7.5 × 1.2	140	2 × MTU 16V 4000 M90 diesel engines (2720 kW each); 2 × shafts and FP propellers	35	19	2	2006–2009
22 m Buratti Class Patrol Boat	22 × 5.4	56	2 × MTU 12V 2000 M93 diesel engines (1340 kW each); 2 × shafts and FP propellers	30	9	23	2007–2010
CP25/CP200 Patrol Boat	25 × 5.76 × 1	53	3 × diesel engines (735 kW each); 2 × shafts and FP propellers, central water-jet booster	34	12	28	2003

Japan

Mitsubishi Heavy Industries Ltd

16-5 Konan 2-chome, Minato-ku, Tokyo 108-8215, Japan

Tel: (+81 3) 67 16 31 11
Fax: (+81 3) 67 16 58 00
e-mail: ueda@ship.hq.mhi.co.jp
Web: www.mhi.co.jp

Naoki Ueda, *Manager*

Mitsubishi Heavy Industries has three main shipyards in Japan, Kobe, Nagasaki and Shimonoseki with an additional dockyard at Yokohama. The company employs a total of over 10,000 people in all four facilities. Patrol boats are built at the Shimonoseki yard.

TSURUUGI CLASS 50 m PATROL BOAT

Six Tsuruugi Class vessels were built for the Japanese Coast Guard between 2001 and 2005. The vessels were built by Universal Shipbuilding, Mitsubishi and Mitsui.

HAYABUSA CLASS 50 m FAST ATTACK CRAFT

Six Hayabusa Class vessels were built between 2001 and 2003 for the Japanese Maritime Self-Defence Force. The vessels feature a propulsion system consisting of three LM500 gas turbines driving three water-jets giving the vessel a top speed of 44 kt.

MIHASHI CLASS 43 m PATROL BOAT

Twelve Mihashi Class vessels were built between 1988 and 2004 by Mitsubishi, Universal Shipbuilding and Mitsui. The vessels feature a triple diesel installation. Two diesels drive conventional shafts and propellers whilst the central engine drives a Kamewa 80 water-jet. On water-jet alone the vessel is capable of 15 kt, with all propellers and water-jet the vessel has a top speed of 35 kt.

Mihashi Class 43 m Patrol Boat 0104968

Hayabusa Class 50 m Fast Attack Craft 0132700

Vessel type	Dimensions L × B × T	Disp. (tonnes)	Main machinery	Max speed (kt)	Complement	No. built	Date of build
Mihashi Class 43 m Patrol Boat	43 × 7.5 × 1.7	195	2 × SEMT Pielstick 16 PA4 V200 VGA diesel engines (2,600 kW each), 1 × SEMT Pielstick 12 PA4 V 200 VGA diesel engine (2,000 kW); 2 × propeller, 1 × Kamewa 80 water-jet	35	35	5	1988
Takatsuki Class 35 m Patrol Boat	35 × 6.7 × 1.3	180	2 × MTU 16V 396TB94 diesel engines (1,910 kW each); 2 × Kamewa 71 water-jets	35	13	1	1992
Matsunami Class 35 m Patrol Boat	35 × 8 × 3.3	165	2 × diesel engines (1,910 kW each); 2 × water-jets	25	30	1	1995
Amami Class 56 m Patrol Boat	56 × 7.5 × 2	230	2 × Fuji 8S40B diesel engines (2,985 kW each); 2 × CP propellers	25	–	2	1998
Hamayuki Class 32 m Patrol Boat	32 × 6.5 × 3.5	100	2 × diesel engines (1,910 kW each)	36	10	1	1999
Tsuruugi Class 50 m Patrol Boat	50 × 8 × 4	220	3 × diesel engines; 3 × water-jets.	45	–	1	2001
Hayabusa Class 50 m Fast Attack Craft	50.1 × 8.4 × 4.2	200	3 × LM 500-G07 gas turbines (4,025 kW each); 3 × water-jets	44	18	6	2001
Tokara Class 56 m Patrol Boat	56 × 8.5 × 4.4	335	3 × diesel engines; 3 × water-jets	30	–	1	2003

Mitsui Engineering & Shipbuilding Company Ltd

6-4 Tsukiji 5-chome, Chuo-ku, Tokyo 104-8439, Japan

Tel: (+81 3) 35 44 31 47
Fax: (+81 3) 35 44 30 50
e-mail: prdept@mes.co.jp
Web: www.mes.co.jp

Tamiyoshi Iwasaki, *Managing Director and General Manager, Ship and Ocean Project Headquarters*

Kiyoshi Hara, *Deputy Director and Deputy General Manager, Ship & Ocean Project Headquarters*

Mitsui Engineering & Shipbuilding Company Ltd has two main shipyards and a dockyard in Japan, at Chiba, Tamano and Yura respectively. Both the Chiba and Tamano shipyards have built lightweight high-speed craft, with the Tamano yard producing patrol craft.

TOKARA CLASS 56 m PATROL BOAT
Two Tokara class vessels were built in 2003 by Universal Shipbuilding and Mitsubishi and a third was built in 2004 by Mitsui.

HAMAYUKI CLASS 32 m PATROL BOAT
Based on the Asogiri class, the design has been enlarged and uses water-jet propulsion giving the vessels a higher top speed. Three vessel of this class were built by Mitsubishi, Universal Shipbuilding Corporation and Mitsui respectively and were delivered between 1999 and 2003.

Vessel type	Dimensions L × B × T	Disp. (tonnes)	Main machinery	Max speed (kt)	Complement	No. built	Date of build
Mihashi Class 43 m Patrol Boat	43 × 7.5 × 1.7	195	2 × SEMT Pielstick 16 PA4 V200 VGA diesel engines (2,600 kW each), 1 × SEMT Pielstick 12 PA4 V 200 VGA diesel engine (2,000 kW); 2 × propeller, 1 × Kamewa 80 water-jet	35	35	2	1994
Hamayuki Class 32 m Patrol Boat	32 × 6.5 × 3.5	100	2 × diesel engines (1,910 kW each)	36	10	1	2003
Tokara Class 56 m Patrol Boat	56 × 8.5 × 4.4	335	3 × diesel engines; 3 × water-jets	30	–	1	2004
Tsuruugi Class 50 m Patrol Boat	50 × 8 × 4	220	3 × diesel engines; 3 × water-jets	45	–	2	2005

Sumidagawa Shipyard

1-16, Shiomi 2-chome, Koto-ku, Tokyo, Japan

Tel: (+81 3) 36 47 61 11
Fax: (+81 3) 36 47 52 10
Web: www.sumidagawa.co.jp

Hiroshi Ishiwata, *President*

Established in 1913, Sumidagawa Shipyard constructs and repairs patrol boats, firefighting boats and passenger vessels. Construction is of high tensile steel, aluminium light alloy or fibre reinforced plastic.

YODO CLASS 37 m PATROL/ FIREFIGHTING BOAT
Three Yodo Class vessels were built for the Japanese Coast Guard by Sumidagwa and one by Ishihara. The vessels can perform as a coastal patrol craft or as a firefighting vessel, for which they are equipped with four fire monitors, two of which are mounted on extendable supports.

AKAGI CLASS 35 m PATROL BOAT
Four Akagi Class patrol boats were built between 1980 and 1988 by Sumidagawa and a further three vessels were built by Ishihara and Yokahoma Yacht Co. in 1987 and 1988.

HAYANAMI CLASS 35 m PATROL BOAT
Sumidagawa have built nine Hayanami Class vessel for the Japanese Coast Guard. The later vessels have firefighting equipment fitted.

Akagi Class 35 m Patrol Boat 0104971

Hayanami Class 35 m Patrol Boat 0104977

Vessel type	Dimensions L × B × T	Disp. (tonnes)	Main machinery	Max speed (kt)	Complement	No. built	Date of build
Akagi Class 35 m Patrol Boat	35 × 6.3 × 1.3	115	2 × Pielstick 16 PA4 V 185 diesel engines (1,310 kW each); 2 × propellers	28	22	4	1980
Natsugiri Class 27 m Patrol Boat	27 × 5.6 × 1.2	68	2 × diesel engines (1,105 kW each); 2 × propellers	27	10	2	1990
Hayanami Class 35 m Patrol Boat	35 × 6.3 × 2.3	110	2 × diesel engines (1,470 kW each); 2 × propellers	25	–	9	1993
Yodo Class 37 m Patrol/Firefighting Boat	37 × 6.7 × 3.4	125	2 × diesel engines; 2 × waterjets	25	–	3	2002

Universal Shipbuilding Corporation

Muza-Kawasaki Central Tower, 1310, Omiya-cho1310, Omiya-cho, Saiwai-ku, Kawasaki, Kanagawa 212-8554, Japan

Tel: (+81 44) 543 27 00
Fax: (+81 44) 543 27 10
Web: www.u-zosen.co.jp

Universal Shipbuilding Corporation, formed by the merger of Hitachi Zosen and HKK Corporation, has five main shipyards in Japan, the Ariake Works, Maizuru Works, Keihin Works, Innoshima Works and Tsu Works. While the Maizuru Works has built naval vessels, the Keihin Works specialises in commercial high-speed craft. Hitachi Zosen built a large number of patrol craft at their Kanugawa Works before the merger.

AMAMI CLASS 56 m PATROL BOAT

Four Amami class patrol boats were built between 1992 and 1998 by Universal Shipbuilding and Mitsubishi. The vessels have a stern ramp for the launch and recovery of a RIB tender.

Amami Class 56 m Patrol Boat 0104970

TAKATSUKI CLASS 35 m PATROL BOAT

Two Takatsuki class vessels were built in 1992 and 1993 by Mistubishi and Universal Shipbuilding.

Vessel type	Dimensions L × B × T	Disp. (tonnes)	Main machinery	Max speed (kt)	Complement	No. built	Date of build
Amami Class 56 m Patrol Boat	56 × 7.5 × 2	230	2 × Fuji 8S40B diesel engines (2,985 kW each); 2 × CP propellers	25	–	2	1992
Takatsuki Class 35 m Patrol Boat	35 × 6.7 × 1.3	180	2 × MTU 16V 396TB94 diesel engines (1,910 kW each); 2 × Kamewa 71 water-jets	35	13	1	1993
Tsuruugi Class 50 m Patrol Boat	50 × 8 × 4	220	3 × diesel engines; 3 × water-jets	45	–	1	2001
Hamayuki Class 32 m Patrol Boat	32 × 6.5 × 3.5	100	2 × diesel engines (1,910 kW each)	36	10	1	2002
Tokara Class 56 m Patrol Boat	56 × 8.5 × 4.4	335	3 × diesel engines; 3 × water-jets	30	–	1	2003

Malaysia

PSC Naval Dockyard

Pangkalan TLDM, 32100 Lamut, Perak, Malaysia

Tel: (+60 5) 683 57 01
Fax: (+60 5) 683 57 08
e-mail: pscnd@pscnd.com.my

The PSC Naval Dockyard of Malaysia delivered two fast troop vessels in 2001 based on a proven monohull design from Wavemaster of Australia.

38 m FAST TROOP CARRIER

These craft are designed to carry up to 32 passengers and cargo at speeds of up to 25 kt.

Vessel type	Dimensions L × B × T	Disp. (tonnes)	Main machinery	Max speed (kt)	Complement	No. built	Date of build
38 m Fast Troop Carrier	37.5 × 7.2 × 1.1	116.5	4 × MAN D2842 LE 408 diesel engines (735 kW each); 4 × Hamilton 521 water-jets	25	–	2	2001

Netherlands

Damen Shipyards

Industrieterrein Avelingen West 20, PO Box 1, NL-4200AA , Gorinchem, Netherlands

Tel: (+31 183) 63 99 22
Fax: (+31 183) 63 21 89
e-mail: info@damen.nl
Web: www.damen.nl

E van der Noordaa, *Managing Director*

Damen Shipyards specialises in the design and construction of a wide range of work boats, dredgers and patrol craft up to 90 m in length and with installed powers up to 10,000 kW. The shipyard started in 1927 and in 1969 began the construction of commercial vessels to standard designs by modular construction. Since 1969 more than 3,000 ships have been built and delivered worldwide to 117 countries. The High Speed and Naval Craft division produces a wide range of high-speed craft for commercial, patrol, surveillance and rescue operations. From 1969, more than 700 high-speed craft have been built by Damen Shipyards' Gorinchem High Speed and Naval Craft department. The largest vessel currently in the standard range is a 53 m patrol vessel. The Damen Offshore Surveillance Vessels range up to 95 m. In the patrol craft range, three Damen Stan Patrol 4207 were delivered to the Dutch Navy in 1999, and in 2001 four vessels were delivered to the UK Customs and two to the Dutch Customs. In 2004 three Search and Rescue vessels were delivered to the Vietnamese government and in 2005 the Jamaican Coast Guard took delivery of a Damen Stan Patrol 4207. The Dutch police have received 13 Damen Stan Patrol 1800, delivered in the period 1996 to 2002. These all-aluminium vessels operate on the rivers, seaports and Ijselmeer in the Netherlands. In 2002 the first of five sea-going patrol vessels of 27 m in length were delivered to the Dutch police. In the period 2003 to 2004 the Dutch Customs replaced a further four vessels and now their fleet is made up of only Damen built vessels.

STAN PATROL 4207

Based on the Stan Patrol 4207 design supplied to the UK Customs service, three vessels were supplied to the Jamaican Coast Guard in late 2005 and early 2006. The vessels feature a stern ramp launched RIB that can reach speeds of 30 kt. The boats will be used for a number of tasks including search and rescue, fisheries protection, pollution control and anti-smuggling activities. The ships, which can travel at speeds in excess of 24 kt, will also carry out evacuation duties if hurricanes are predicted.

STAN PATROL 4100

Ordered from Damen shipyards in March 1997 and delivered to the Netherlands Coastguard in late 1998.

Stan Patrol 4207 Searcher built for the UK Customs Service
0522410

Stan Patrol 4708 built for South Africa
1120299

Vessel type	Dimensions L × B × T	Disp. (tonnes)	Main machinery	Max speed (kt)	Complement	No. built	Date of build
Stan Patrol 4100	42.8 × 6.8 × 2.5	205	2 × Caterpillar 3516B diesels (2,090 kW); 2 × shafts and CP propellers	26	17	3	1998
Damen Stan Patrol 4207	42.8 × 7.11 × 2.52	–	2 × Caterpillar 3516 DI-TA elec (2,100 kW); 2 × shafts and CP propellers	26	12	4	2001
Stan Patrol 2606	26.5 × 6.2 × 1.8	55	2 × MTU 12V 396TE94 (1,650 kW); 2 × shafts and FP propellers	28	3	6	2003
Stan Patrol 4708	46.8 × 8.11 × 2.9	–	2 × MTU 16V 4000 M90 (2,720 kW each); 2 × shafts and CP propellers	24	16	1	2004
Stan Patrol 4100	42.8 × 6.8 × 2.5	205	2 × Caterpillar 3516B diesels (2,090 kW); 2 × shafts and CP propellers	28	17	3	2004
Stan Patrol 4207	42.8 × 7.11 × 2.52	–	2 × Caterpillar 3516 DI-TA elec (2,100 kW); 2 × shafts and CP propellers	26	12	3	2005

Norway

Umoe Mandal AS

Gismerøya, PO Box 902, N-4509 Mandal, Norway

Tel: (+47) 38 27 92 00
Fax: (+47) 38 26 03 88
e-mail: mandal@umoe.no
Web: www.mandal.umoe.no

Peter Klemsdal, *Managing Director*
Ole Ihme, *Contracts Manager*
Are Soreng, *Purchasing Manager*

Umoe Mandal AS specialises in the design and construction of advanced marine craft built from FRP composite materials and has

Skjold Class 47 m Fast Patrol Boat
0111029

built a number of lightweight high-speed craft including patrol craft, minehunters and minesweepers. The company has developed specialist expertise in EMC/EMI, stealth technology, titanium welding, ship signatures, naval architecture and system integration.

SKJOLD CLASS 47 m FAST PATROL BOAT

A prototype SES vessel was delivered to the Royal Norwegian Navy in April 1999. The vessel is currently being reconfigured with a new gas turbine propulsion system and weapons suite. Five additional vessels are under construction under a programme which will last until 2009. Umoe Mandal AS is working together with Armaris and Kongsberg Defence and Aerospace in a consortium for the delivery of six Skjold Class vessels.

Vessel type	Dimensions L × B × T	Disp. (tonnes)	Main machinery	Max speed (kt)	Complement	No. built	Date of build
Skjold Class 47 m Fast Patrol Boat	47 × 13.5 × 2.2	270	CODOG 2 × Alison KF571 gas turbines (6,000 kW each); 2 × MTU 183 diesel engines (350 kW each); 2 × water-jets	53	15	6	1998

Russian Federation

Almaz Shipbuilding Company

26 Petrovskij Prospect, St Petersburg, 197042
Russian Federation

Tel: (+7 812) 235 48 20
Fax: (+7 812) 350 11 64
e-mail: market@almaz.spb.ru
Web: www.almaz.spb.ru

Alexander Shliakhtenko, *President*
Nikolai Ivakin, *Shipbuilding Director*
Leonid Grabovets, *General Director*

Svetlyak Class 50 m Patrol Boat 0081676

Almaz Shipbuilding Company was founded in 1901 in St Petersburg by Alexander Zolotoff. Since that time over 1,000 vessels have been delivered including fast patrol craft, passenger ferries and the world's largest hovercraft; the Pomoznik. The company is situated in the central part of St Petersburg on Petrovskij island, in the immediate vicinity of the Gulf of Finland. The 165,000 m² yard, employing more than 1,000 people, includes 50,000 m² of workshops fitted with modern equipment, modern-style shipbuilding halls and two floating docks of 3,000 and 2,500 tonnes capacity.

SVETLYAK CLASS (PROJECT 10412) 50 m PATROL BOAT

The Svetlyak Class patrol boats were designed to carry out a variety of missions and to be economical and easy to operate. The triple diesel propulsion layout provides a maximum speed of 30 kt and range of up to 2,200 miles at 13 kt. The boats can effectively operate in moderate and hot climates, employ weapons in up to Sea State 5 and can safely navigate in Sea State 7. Up to 23 vessels are in service with the Russian Border Guard and three are in service in the Russian Navy, which were delivered in the late 1990s. A further two vessels were delivered in 2002 for the Vietnamese Navy.

Vessel type	Dimensions L × B × T	Disp. (tonnes)	Main machinery	Max speed (kt)	Complement	No. built	Date of build
Svetlyak Class (Project 10412) 50 m Patrol Boat	49.5 × 9.2 × 2.2	375	3 × M550 diesel engines (3,530 kW each); 3 × CP propellers	30	28	22	2002

Vympel Shipyard

4 Novaya Street, Rybinsk, 152912, Yaroslavl Reg, Russian Federation

Tel: (+7 855) 20 23 03
Fax: (+7 855) 21 18 77
e-mail: aovympel@yaroslavl.ru
Web: www.vympel-rybinsk.ru

Dmitry A. Belyakov, *Marketing Director*

Throughout its history spanning over seventy years, the company has manufactured a range of vessels including torpedo boats, utility boats for the engineering troops, torpedo retrieving craft, fuel replenishment boats, ambulance vessels, firefighting vessels, tugs and pontoon bridge erecting boats, leisure craft and attack craft both for the Russian Naval Forces and for export.

MIRAZH CLASS 34 m PATROL BOAT

Designed by Almaz, the Mirazh Class features an aluminium-magnesium alloy structure. Two vessels have been delivered to the Russian Border Guard. The vessel has a top speed of 48 kt.

Vessel type	Dimensions L × B × T	Disp. (tonnes)	Main machinery	Max speed (kt)	Complement	No. built	Date of build
Mirazh Class 34 m Patrol Boat	34 × 6.6 × 2.7	126	2 × Zvezda M521 diesel engines (5,950 kW each); 2 × propellers	48	12	2	1998

Singapore

Singapore Technologies Marine Ltd

7 Benoi Road, Singapore 629882

Tel: (+65) 68 61 22 44
Fax: (+65) 68 61 30 28
e-mail: mktg.marine@stengg.com
Web: www.stengg.com

See Leong Teck, *President*
Han Yew Kwang, *Chief Operating Officer*
Larry Loon, *President (Naval Business Group)*
Lim Siew Koon, *Vice-President Commercial Marketing & Business Development*

Singapore Technologies Marine (STM), the marine arm of Singapore Technologies Engineering, was established in 1968 as a specialist shipyard. Current areas of expertise include the building of specialised commercial vessels, military engineering equipment fabrication and the reconstruction and modernisation of old vessels. STM is located in a 30 acre site at the mouth of the Benoi Basin. In addition to the Swift class patrol boat, described below, STM has built over 30 of its PT class patrol boats (14.54 m, 30 kt), including 19 for the Singapore Marine Police, two for Singapore Customs and Excise Department and seven for the Royal Brunei Police Force; this last order was completed by the end of 1987.

SWIFT CLASS PATROL BOAT

Twelve of these vessels were built between 1979 and 1980 for the Singapore Navy.

Vessel type	Dimensions L × B × T	Disp. (tonnes)	Main machinery	Max speed (kt)	Complement	No. built	Date of build
Swift Class Patrol Boat	22.7 × 6.2 × 1.6	47	2 × Deutz SBA 16M 816 diesels	33 kt	12	12	1980

Spain

Astilleros de Murueta

E-48394, Murueta, Bizkaia, Spain

Tel: (+34 94) 625 20 00
Fax: (+34 94) 625 52 44
e-mail: mail@astillerosmurueta.com
Web: www.astillerosmurueta.com

Agustin Oritz de Eribe, *Managing Director*

Astilleros de Murueta was founded in 1943 on the River of Gernika and has built 170 vessels since then. The yard produces fishing vessels, dredgers, Ro-Ro ferries, and general cargo ships and is now producing patrol craft. The yard has two enclosed dry docks of 100 m and 90 m in length and has an in-house design office, which is equipped with CAD and CAM software enabling effective use of the yard's CNC machines and automatic welding facilities.

VIGILANTE CLASS 45 m OFFSHORE PATROL BOAT

Based on the FBM Marine Protector 45 design, three of these vessels will be built under licence. The steel hulled vessels will feature a triple engine installation with two engines for normal operation and a smaller central engine for slow-speed operations. Facilities are provided to carry a 5 m tender for boarding operations.

Vessel type	Dimensions L × B × T	Disp. (tonnes)	Main machinery	Max speed (kt)	Complement	No. built	Date of build
Vigilante Class 45 m Offshore Patrol Boat	45 × 9.8 × 2.5	300	2 × MTU 16V 4000 M90 diesel engines (2,720 kW each); 1 × MTU 12V 4000 M80 (1,740 kW); 3 × propellers	25	27	3	2006

Navantia

Puerto Real Shipyard, San Fernando, Carretera de la Carraca s/n, PO Box 18, 11080 San Fernando, Spain

Tel: (+34 956) 59 98 97
Fax: (+34 956) 59 98 98
Web: www.navantia.com

Navantia is the new name for Izar which was formed in 2000 by the merger of Empresa Nacional Bazan and Astilleros Espanoles. The company has a staff of over 11,000 with 12 yards and factories located on the Spanish Atlantic and Mediterranean coasts. Their activities comprise four business areas: shipbuilding; repairs and conversions; propulsion and energy; and systems and weapons. Naval vessels have beed exported to countries such as Norway, Chile and Thailand and commercial vessels to Belgium, Finland, France, Norway, Sweden, UK and US.

CORMORAN CLASS 57 m FAST ATTACK CRAFT

Built for the Spanish Navy in 1985, this vessel has a top speed of 32 kt and a range of 2,500 nm at 15 kt. The vessel was transferred to Colombia in 1995.

Vessel type	Dimensions L × B × T	Disp. (tonnes)	Main machinery	Max speed (kt)	Complement	No. built	Date of build
Cormoran Class 57 m Fast Attack Craft	56.6 × 7.5 × 2	358	3 × MTU 16V 956TB91 diesel engines (2,755 kW each); 3 × propellers	32	31	1	1985

Sri Lanka

Colombo Dockyard

PO Box 906, Port of Colombo, Colombo 15, Sri Lanka

Tel: (+94 112) 522 46 15
Fax: (+94 112) 47 13 35
e-mail: pricel@cdl.lk
Web: www.cdl.lk

M. Prince Lye, *Manager*

Colombo Dockyard Ltd has a large drydock complex and a small building berth facility. A wide range of merchant vessels are built and repaired in these facilities and they are also used for the construction of high-speed aluminium vessels.

ISKANDHAR CLASS 25 m COASTAL SURVEILLANCE VESSEL

This vessel was built for the Maldivian Coast Guard to carry out patrol missions in the coastal waters of the Maldives. As the island nation's general mode of transportation is by sea, the CSV, in addition to its surveillance role, is also suitably furnished for the high-speed transfer of VIPs. Powered by twin Paxman main engines, each developing 1,564 kW, driving Kamewa water-jets, the vessel is capable of achieving a maximum speed of 40 kt.

Vessel type	Dimensions L × B × T	Disp. (tonnes)	Main machinery	Max speed (kt)	Complement	No. built	Date of build
Iskandhar Class 25 m Coastal Surveillance Vessel	24.4 × 5.8 × 1.3	58	2 × Paxman diesel engines (1,564 kW each)	40	18	2	1997

Taiwan

Jong Shyn Shipbuilding Co Ltd

92, 108 Lane, Chung Chow 3rd Road, Chi Chin District, Kaohsiung, Taiwan

Tel: (+886 7) 5714 651
Fax: (+886 7) 5717 002
e-mail: cp@jade-yachts.com.tw
Web: www.jssc.com.tw

Jong Shyn Shipbuilding Co is one of Taiwan's main builders of steel and aluminium boats, specialising in patrol boats and rescue boats for the Taiwanese Navy and Coast Guard.

Vessel type	Dimensions L × B × T	Disp. (tonnes)	Main machinery	Max speed (kt)	Complement	No. built	Date of build
Patrol Boat PP-5039	26.2 × 6.2 × 1.2	63.36	4 × MAN D2842 LE410 diesels	43.15	11	1	2007

Lung Teh Shipbuilding Co Ltd

No.15, De-Shing 5th Road, Dung-Shan Country, ILAN, Taiwan

Tel: (+886) 39 90 11 66
Fax: (+886) 39 90 56 50
e-mail: sales@ms.lts.com.tw
Web: www.lts.com.tw

Seldon Huang, *President*
Chin-Hwa Chang, *Sales Vice President*
Solon Huang, *Production Vice President*

Founded in 1979 the company has considerable experience in the design and construction of a wide range of patrol boats, military landing craft, support craft, pilot boats, fire and rescue boats as well as tugs and leisure boats.

The company has experience of building vessels in aluminium, steel and GRP up to 40 m in length and up to speeds of 60 kt. Recent deliveries have included three 34 m patrol boats for the Taiwan Navy.

30.5 m FAST PATROL BOAT

The first 30.5 m vessel was delivered to the Taiwan Coast Guard in 1994 with a total of 14 boats produced to date.

Specifications

Length overall	30.5 m
Length waterline	28.0 m
Beam	6.8 m
Draught	1.5 m
Crew	16
Propulsive power	4,600 kW
Max speed	33.5 kt
Range	1,000 n miles

28 m ultra fast patrol boat 0589523

Classification: DnV 1A1; R2 patrol.
Propulsion: Two MTU 16 V 396 diesel engines each rated at 2,290 kW and driving a fixed-pitch propeller via a U-drive reduction gear.

26.5 m FAST PATROL BOAT

The first 26.5 m patrol boat was delivered to the Taiwan Coast Guard in 2004. To date 4 vessels have been built.

Specifications

Length overall	26.5 m
Length waterline	23.05 m
Beam	6.2 m
Draught	1.2 m
Crew	11
Propulsive power	3,200 kW
Max speed	36.25 kt
Range	600 n miles

Classification: CR 100+, CMS+, Coastal Service, Patrol, DnV +1A1, HSLC, R2, patrol.
Propulsion: Four MAN D2482 LE 410 diesel engines driving water-jets via a reversible reduction gear.

Vessel type	Dimensions L × B × T	Disp. (tonnes)	Main machinery	Max speed (kt)	Complement	No. built	Date of build
19 m self-righting SAR boat	19 × 5.6 × 2.4	43	MAN 2842 LE406	25	–	–	–
21 m fast patrol boat	21 × 5 × 1.5	42	MTU 12V 183	31	–	–	–
25 m fast patrol boat	25 × 5.8 × 1.1	–	MTU 12V 196	32.5	16	13	1990
26.5 m fast patrol boat	26.5 × 6.2 × 1.2	–	MAN D2482 LE410	36.25	11	4	2004
28 m Ultra Fast patrol boat	28 × 6.2 × –	–	Paxman VP185	42	–	–	–
30.5 m Ultra Fast patrol boat	30.5 × 6.8 × 1.5	140	MTU 16V 396 T894	33.5	16	14	1994
34 m patrol boat	34 × 7.0 × 1.3	–	–	36	14	3	2007

United Kingdom

FBM Babcock Marine

The Courtyard, St. Cross Business Park, Monks Brook, Newport, Isle of Wight, PO30 5BF, United Kingdom

Tel: (+44 19) 83 82 57 00
Fax: (+44 19) 83 82 41 80
e-mail: fbm@babcock.co.uk
Web: www.fbmuk.com

M Graves, *General Manager*

FBM builds a wide range of monohull, catamaran and SWATH vessels for many applications at its yards in the UK and the Philippines. In 2000, Babcock International purchased the FBM interests, adding a shipyard in Rosyth as well as the Lairsdside yard in Birkenhead, UK.

PROTECTOR CLASS 33 m PATROL BOAT

Two 33 m Protector class vessels were ordered from FBM Marine in 1998 for the

Vessel type	Dimensions L × B × T	Disp. (tonnes)	Main machinery	Max speed (kt)	Complement	No. built	Date of build
Protector Class 33 m Patrol Boat	33 × 6.7 × 1.9	294	2 × MTU 16V 396 TE94 diesel engines (2,173 kW each); 2 × FP propellers	33	14	2	1999
Protector Class 26 m Customs and Excise Cutter	25.7 × 6.2 × 1.7	–	2 × Paxman 1,074 kW diesel engines, ZF BW 460S gearboxes	25	8	4	1988

UAE Coast Guard. The vessels feature a steel hull and aluminium superstructure.

PROTECTOR CLASS 26 m CUSTOMS AND EXCISE CUTTER

The UK Customs and Excise Marine Branch ordered three 26 m Protector class fast patrol cutters which entered service during 1988. The craft are a development of the 33 m 'Protector' class, designed to meet the needs of HM Customs and Excise for operation anywhere round the coastline of the UK. A fourth 26 m Protector for HM Customs and Excise was built under licence by Babcock Thorn at the Rosyth Royal Dockyard in 1993.

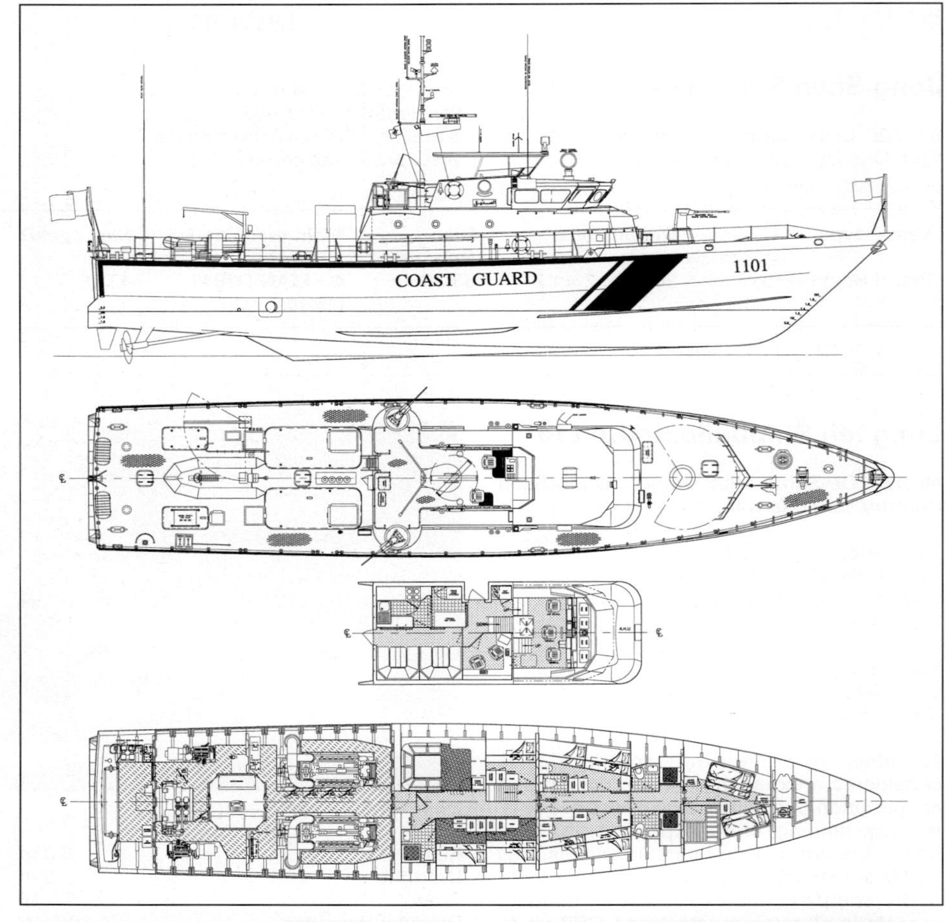

Protector class 33 m patrol boat, general arrangement
0081015

Protector class 33 m patrol boat
0081014

McTay Marine

The Magazines, Port Causeway, Bromborough, Wirral CH62 4YP, United Kingdom

Tel: (+44 151) 346 13 19
Fax: (+44 151) 334 00 41
e-mail: robertmcburney@mctaymarine.com

M R Brodie, *Managing Director*
R McBurney, *Commercial Director*

Although high-speed vessels are not the main output of the McTay yard, the company built the Europatrol 250 1994.

EUROPATROL 250

This 47 m customs craft, the result of a collaboration between McTay and Vosper International, was delivered to the Greek Customs service in 1994.

Europatrol 250 on trials 0506920

Vessel type	Dimensions L × B × T	Disp. (tonnes)	Main machinery	Max speed (kt)	Complement	No. built	Date of build
Europatrol 250	47.3 × 7.5 × 2.5	240	3 × GEC Paxman diesels, 3 × FP	25	21	1	1994

VT Shipbuilding

Fleet Way, Portsmouth, PO1 3AQ, United Kingdom

Tel: (+44 2392) 85 72 00
Fax: (+44 2392) 85 74 00
e-mail: shipsales@vtplc.com
Web: www.vtplc.com/shipbuilding

Part of the VT Group, VT Shipbuilding is a major supplier to the Royal Navy and, through VT Shipbuilding International, is one of the world's leading exporters of naval vessels and technology transfer. The company produces advanced surface warships including frigates, corvettes, fast attack craft and mine counter measure solutions, offshore patrol vessels and patrol craft for EEZ management and fishery protection duties. As an established prime contractor in the UK and overseas, VT can offer a total solutions approach including vessel design and system integration plus comprehensive training and support packages. VT also offers complete in-country build support and technical assistance through technology transfer agreements and by working with customer's local supplier bases to ensure local industrial participation for major naval programmes.

Dhofar Class 57 m Fast Attack Craft 0010822

Barzan (Vita) Class 56 m Fast Attack Craft 0004369

DHOFAR CLASS 57 m FAST ATTACK CRAFT

Four Dhofar Class vessels were built between 1981 and 1988 for the Royal Navy of Oman.

BARZAN (VITA) CLASS 56 m FAST ATTACK CRAFT

In 1992 Qatar ordered four Barzan (Vita) Class fast strike craft for the Qatar Emiri Navy. The first two ships, Barzan and Huwar were handed over in 1996 and the final two ships, Al Udeid and Al Deebel, were delivered in 1998. The design is a derivative of the 56 m patrol craft built for Oman and Kenya, with a redesigned internal arrangement and superstructure to provide a second deck at tank level. Modifications to the hull included increasing the freeboard amidships and flaring the topsides to widen the beam at the weatherdeck. The hull and superstructure design has incorporated features for reduction of radar cross section. Active roll-damping fin stabilisers have been added for operation in high Sea States.

AL HUSSEIN CLASS 31 m FAST ATTACK CRAFT

Three Al Hussein Class vessels were delivered in 1991 to Jordan. The vessels are of GRP construction.

Vessel type	Dimensions L × B × T	Disp. (tonnes)	Main machinery	Max speed (kt)	Complement	No. built	Date of build
Ramadan Class 52 m Fast Attack Craft	52 × 7.6 × 2.3	307	4 × MTU 20V 538TB91 diesel engines (2,825 kW each); 4 × FP propellers	40	30	6	1981
Dhofar Class 57 m Fast Attack Craft	56.7 × 8.2 × 2.4	394	4 × Paxman Valenta 18CM diesel engines (2,800 kW each); 4 × propellers	38	45	4	1981
Nyayo Class 57 m Fast Attack Craft	56.7 × 8.2 × 2.4	–	4 × Paxman Valenta 18CM diesel engines (2,800 kW each); 4 × propellers	40	40	2	1986
Al Hussein Class 31 m Fast Attack Craft	30.5 × 6.9 × 1.5	124	2 × MTU 16V 396TB94 diesel engines (2,130 kW each); 2 × propellers	32	16	3	1989
Barzan (Vita) Class 56 m Fast Attack Craft	56.3 × 9 × 2.5	376	4 × MTU 20V 538TB93 diesel engines (3,450 kW each); 4 × propellers	35	35	4	1995

United States

Bender Shipbuilding & Repair Co. Inc

265 South Water Street, Mobile, Alabama 36603, United States

Tel: (+1 251) 625 6213
Web: www.bendership.com

T. B. Bender Jr, *President*
Frank Terrell, *Vice President*
Kirk Burgamy, *Engineering Manager*

For more than 75 years, Bender has been a leading ship repair facility on the central Gulf of Mexico. The company has over 2,000 m of deep water frontage and has a lifting capacity of more than 24,000 t. The yard has a Panamax dry dock for large vessel repair. Bender builds a wide range of vessels including offshore supply vessels, tug boats and patrol craft. There are currently more than 800 Bender-built ships operating worldwide.

Vessel type	Dimensions L × B × T	Disp. (tonnes)	Main machinery	Max speed (kt)	Complement	No. built	Date of build
Toledo Class 35 m Patrol Boat	35.4 × 7.6 × 2.1	140	2 × MTU 12V 396 TE94 diesel engines (3,050 kW each); 2 × propellers	25	25	2	1994

TOLEDO CLASS 35 m PATROL BOAT

Built in 1994, these vessels have a maximum speed of 25 kt and a range of 1,500 n miles at 15 kt.

Toledo Class 35 m Patrol Boat
0114511

Bollinger Shipyards Inc

P.O. Box 250, 8365 Hwy 308, Lockport, LA-70374, United States

Tel: (+1 985) 532 25 54
Fax: (+1 985) 532 72 25
e-mail: sales@bollingershipyards.com
Web: www.bollingershipyards.com

Donald Bollinger, *CEO*
Chris Bollinger, *Executive Vice-President*
René Leonard, *Director of Engineering*

Family owned and operated since 1946, the company specializes in the repair, conversion and construction of a wide variety of small to medium-sized offshore and inland vessels. The company currently operates thirteen shipyards, all of which are ISO 9001 certified and strategically located throughout South Louisiana and Texas with direct access to the central Gulf of Mexico and the Mississippi River. With 40 dry-docks, ranging in capacity from 100 tons to 22,000 tons, Bollinger provides a wide variety of services for both shallow and deepwater vessels and rigs.

Cyclone Class 55 m Coastal Patrol Boat

0131288

CYCLONE CLASS 55 m COASTAL PATROL BOAT

Built between 1993 and 2000, these vessels were built to a Vosper Thornycroft 'Ramadam' design, which was modified to US specifications for coastal patrol, drug enforcement duties and special forces transport and support. The vessels have a quadruple diesel installation driving four fixed shafts and propellers giving the vessels a top speed of 35 kt.

ISLAND CLASS 38 m PATROL BOAT

The Island Class project which was completed in 1992, began in 1982 as an effort to procure "off the shelf" patrol boats in response to congressional and administration imperatives to increase law enforcement resources in southeast US. The US Coast Guard was so satisfied with these vessels that their deployment was extended as far west as Guam and as far north as Alaska. The vessels are based on an existing patrol craft–the Vosper Thornycroft 33 m patrol boat, which has been operated in several countries. Bollinger modified the design to suit US Coast Guard requirements. The vessel possesses the same hull form and dimensions, underwater appendages and propulsion configuration as the parent craft. The hull is a semi-displacement type monohull made of high-strength steel; the main deck and superstructure are aluminium

Island Class 38 m Patrol Boat
1120272

and active fin stabilisation is added to enhance sea keeping.

MARINE PROTECTOR CLASS 27 m PATROL BOAT

This vessel was designed as a multi-mission platform capable of performing search and rescue, law enforcement and fisheries patrols, as well as drug interdiction and illegal immigrant interdiction duties up to 200 miles off shore. The design is based on the Damen STAN 2600, which was originally developed for the Hong Kong Police. Bollinger modified the design to meet the specific requirements of the US Coast Guard.

Marine Protector Class 27 m Patrol Boat
1120273

Vessel type	Dimensions L × B × T	Disp. (tonnes)	Main machinery	Max speed (kt)	Complement	No. built	Date of build
Island Class 38 m Patrol Boat	37.5 × 6.4 × 2.2	168	2 × Paxman Valenta 16RP 200M diesel engines (2,500 kW each); 2 × propellers	29	16	49	1986
Cyclone Class 55 m Coastal Patrol Boat	54.6 × 7.9 × 2.4	360	4 × Paxman Valenta 16RP 200M diesel engines (2,500 kW each); 4 × CP propellers	35	27	13	1993
Marine Protector Class 27 m Patrol Boat	26.5 × 5.8 × 1.6	91	2 × MTU 8V 396TE94 diesel engines (1,923 kW each); 2 × FP propellers	25	10	75	1998

Swiftships Shipbuilders, LLC

PO Box 2869, 1105 Levee Road, Morgan City, Louisiana 70831, United States

Tel: (+1 985) 384 17 00
Fax: (+1 985) 384 09 14
e-mail: swiftships@swiftships.com
Web: www.swiftships.com

Calvin J LeLeux, *President*
Jeffery Perin, *Military Products*
Jeff LeLeux, *Commercial and Special Products*

Swiftships produces a wide range of aluminium craft including crewboats and patrol craft. These craft are in service with US armed forces, naval coast guard and police forces in some 20 countries throughout the world.

SWIFTSHIPS 35 m CLASS PATROL BOAT

The Swiftships 35 m class is a triple-waterjet diesel patrol craft designated as a coastal capable craft. The vessel is capable of operating in Sea State 4 and surviving Sea State 5 at night and under all weather conditions. The vessel incorporates a stern launching ramp for a 4.7 m RIB.

CANOPUS CLASS 34 m PATROL BOAT

Two aluminium vessels were built in 1984. In 2003 and 2004, the two vessels were extensively refitted by Swiftships.

JOSE MARIA PALAS CLASS 34 m PATROL BOAT

These vessels feature a quadruple diesel propulsion system powering conventional shafts and propellers. The vessels were acquired by the Colombian Coast Guard with US assistance.

Vessel type	Dimensions L × B × T	Disp. (tonnes)	Main machinery	Max speed (kt)	Complement	No. built	Date of build
Canopus Class 34 m Patrol Boat	33.5 × 7.3 × 1.8	93.5	3 × Detroit 12V92TA diesel engines (760 kW each); 3 × propellers	23	19	2	1984
Jose Maria Palas Class 34 m Patrol Boat	33.5 × 7.5 × 2	99	4 × Detroit 12V71TI diesel engines (450 kW each)	25	19	2	1989
Swiftships 35 m Class Patrol Boat	35.1 × 7.3 × 1.5	95	3 × Caterpillar 3412 diesel engines (735 kW each); 3 × Hamilton HM 651 water-jets	25	–	2	2003

VT Halter Marine Inc

900 Bayou Casotte Parkway, Pascagoula, Mississippi 39581-2552, United States

Tel: (+1 228) 696 68 88
Fax: (+1 228) 696 68 90
e-mail: vthmsales@vthaltermarine.com
Web: www.vthaltermarine.com

Boyd E King, *Chief Executive Officer*
Sid Mizell, *Senior Vice President, Business Development*

Bahamas Class 61 m Patrol Boat
0130420

A Vision Technologies Systems Company, VT Halter Marine has built more than 2,600 vessels. The company has three shipyards, located in Pascagoula, Moss Point and Escatawpa, Mississippi. VT Halter Marine has expertise in the design and construction of displacement and planning monohulls. The company also builds catamarans, trimarans, SWATH vessels, Surface Effects Ships (SESs) and Very Slender Vessels (VSVs). The yards use modern, efficient construction techniques including numerically controlled cutting, modular construction and zone outfitting. VT Halter Marine designs and builds to the requirements of the US Coast Guard, US Navy, regulatory bodies and classification societies.

BAHAMAS CLASS 61 m PATROL BOAT

Based on the Vosper International Europatrol 250 design, the Bahamas Class features a steel hull with an aluminium superstructure. The vessel carries a RIB and davit system on the stern and has a triple diesel and submerged shaft propulsion layout.

GAVION CLASS 24 m PATROL BOAT

Ordered in 1998, 12 of these vessels were delivered between 1999 and 2000. The vessels have an aluminium construction and a diesel propulsion system that gives the vessels a top speed of 25 kt and a range of 1,000 n miles at 12 kt.

HALTER 24 m PATROL BOAT

Seventeen of these vessels were built between 1991 and 1993, in aluminium, for Saudi Arabia. A further 22 vessels were built for the Philippines. The vessels have two Detroit Diesel 16V-92TA diesel engines driving the vessel to a top speed of 28 kt.

Vessel type	Dimensions L × B × T	Disp. (tonnes)	Main machinery	Max speed (kt)	Complement	No. built	Date of build
Halter 24 m Patrol Boat	23.8 × 6.1 × 1.8	56	2 × Detroit 16V-92TA diesel engines (1,040 kW each); 2 × propellers	28	8	17	1991
Gavion Class 24 m Patrol Boat	24.4 × 5.2 × 1.5	45	2 × Detroit 12V-92TA diesel engines (805 kW each); 2 × propellers	25	10	12	1999
Bahamas Class 61 m Patrol Boat	60.6 × 8.9 × 2.6	375	3 × Caterpillar 3516B diesel engines (1,620 kW each); 3 × propellers	24	35	2	2000

CIVIL OPERATORS OF HIGH-SPEED CRAFT

Company listing by country

Argentina
Buquebus

Australia
Aristocat Reef Cruises
Australian Customs Service
Australian Whale Watching
Bali Hai Cruises
Big Cat Green Island Cruises
Brisbane City Council
Calypso Reef Charters
Captain Cook Great Barrier Reef Cruises
Cruise Victoria
Cruise Whitsundays
Darwin Hovercraft Tours
Deep Sea Divers Den
Down Under Cruise and Dive
Fantasea Cruises Pty Ltd
Freedom Fast Cats
Gordon River Cruises
Great Adventures
Haba Dive and Snorkel
Hayman Resort
Hervey Bay Whale Watch
Hillary's Fast Ferries
Hobart Cruises
Inter Island Ferries
Lady Musgrave Barrier Reef Cruises
Mackay Island and Reef Cruises
Mackenzies Island Cruises
Moreton Bay Whalewatching
Navigators
Palm Beach Ferry Service
Peel's Tourist and Ferry Services Pty Ltd
Poseidon Cruises
Queensland Government
Quick Cat Cruises
Quicksilver Connections Ltd
Raging Thunder Pty Ltd
Reefjet
Reef Magic Cruises
Rottnest Express Pty Ltd
Sea-Cat Ferries and Charters P/L
Spirit of the Bay Cruises
Sunferries
Sunlover Cruises
Sydney Ferries
Tangalooma Wild Dolphin Resort Ltd
Tasman Venture
Voyages
World Heritage Cruises

Bahamas
Bahamas Ferries

Belgium
Eurosense Belfgotop NV

Bermuda
Government of Bermuda

Bolivia
Crillon Tours SA

Brazil
Barcas Transportes Maritimos
Siem Consub SA
Transtur

Bulgaria
Black Sea Shipping plc

Canada
Bay Ferries
Canadian Coast Guard Hovercraft Unit
West Coast Launch Ltd

Cape Verde
Moura Company

Chile
Catamaranes del sur
Hotelsa Patagonia Expedition
Patagonia Connection

China
Dalian Steamship Company
Dalian Yuan Feng Ferry Company
Jiang Men Hong Kong Macau Transportation
Ming Zhu Passenger Transport
Nantong High-Speed Passenger Ship Company
Ningbo Fast Ferries Ltd
Shanghai Free Flying Transport
Shanghai Huandoo Passenger Shipping
Shanghai Ya Tong Co Ltd
Shenzhen Hengtong Shipping
Shenzhen Shipping Co Ltd
Shenzhen Xunlong Transportation Shipping Co Ltd

Croatia
Alpex
Atlas Turistička Plovidba
Commodore Cruises
G&V Line
Ivante
Jadrolinija
Mia Tours
Split Tours dd
Uto Kapetan Luka

Cuba
Cubanacan Jet
Intermar Clarion Shipping Co Ltd

Cyprus
Fergun Deniz Cilik

Denmark
Bornholms Trafikken
Christiansøfarten
CT Offshore ApS
Mols-Linien

Egypt
El Salam Maritime Company
International Fast Ferries
United Company for Marine Lines

Estonia
Nordic Jet Line AS
Silja Line Superseacat

Fiji
Beachcomber Island Resort and Cruises Ltd
South Sea Cruises

Finland
Linda Line Express

France
Bourbon Offshore
Brittany Ferries
Compagnie Corsaire
Compagnie Yeu Continent
Corsica Ferries/Sardinia Ferries
Manche Iles Express
SNCM Société National Corse Mediterranée
Société Morbihannaise de Navigation

French Polynesia
Aremiti Pacific Cruises
Bora Bora Navette Ltd
Société de Développement de Moorea
Tahaa Transport Services

Germany
AG EMS
Elbe City Jet
Förde Reederei Seetouristik GmbH
Hallig und Inselreederei (Adler-Schiffe)
Wilhelmshaven Helgoland Linie

Greece
Aegean Shipping Company
Aegean Speed Lines N.E.
Agistri Piraeus
Anes-Symi Lines
C-Link Ferries
Dodekanisos Naftiliaki NE
Eagle Ferries
Euroseas
GA Ferries
Hellenic Seaways
Iason Jets Shipping
Ionian Cruises
Kallisti Ferries
Kiriacoulis Maritime
Laumzis Sun Cruises
Nel Lines
Paxos Hydrofoil Shipping
Planet Seaways
Samos Hydrofoils
Saronic Dolphins
Tilos 21st Century Shipping
Vasilopoulos Flying Dolphin

Guadeloupe
Comatrile
L'Express Des Iles
Société de Transports Maritimes Brudey Frères

Honduras
Safeway Maritime Transportation Co

Hong Kong (Special Administrative Region)
China Merchant Development Company
Chu Kong Passenger Transport Co Ltd
Chu Kong Shipping Development Company Ltd
Customs and Excise Department
Discovery Bay Transportation Services Ltd
Hong Kong & Kowloon Ferry Ltd
Hong Kong Police Force, Marine Region
Nansha Ferry Company Ltd
New World First Ferry Services Ltd
Park Island Transportation Company
Turbojet Sea Express

Hungary
Mahart Passnave Passengers Shipping Ltd

India
Development Consultants
Samudra Link Ferry Shipping
Shipping Corporation of India

Indonesia
Aquaria Shipping
ASDP
Bali Bounty Cruises
Bali Cruises Nusantara
Pelni
Quicksilver

Iran
Valfajr Shipping Company

Ireland
Aran Islands Direct Ltd
Irish Ferries
Island Ferries

Italy
Alilauro SpA
Alimar SpA
Aliscafi SNAV SpA

Caremar
Delfino Verde Navigazione
Foderado Navigazione
Larivera Lines
Navigazione Golfo dei Poeti
Navigazione Lago di Como
Navigazione Lago Maggiore
Navigazione Libera Del Golfo SpA
Navigazione Navigargano Srl
Navigazione Sul Lago di Garda
RFI Bluvia
Righetti Navi
Siremar
Snav Croazia Jet
Taranto Navigazione
Tirrenia Navigazione
Toremar
Ustica Lines
Venezia Lines
Vetor Srl

Japan
Awashima Kisen Company Ltd
Biwako Kisen Company Ltd
City of Kamishima
Cosmo Line
Fuke Kaiun Company Ltd
Goto Sangyu Kisen
Ishizaki Kisen Company Ltd
Japan Coast Guard
JR Kyushu Railway Company
Kagoshima Shosen Company Ltd
Kashima Futo Co Ltd
Koshikijima Shosen Ltd
Kumamoto Ferry Company
Kyushu Shosen Company Ltd
Kyushu Yusen Company Ltd
Oita Hover Ferry Company Ltd
Oki Kisen KK
Sado Kisen Kaisha
Sanyo Shosen
Setonaikai Kisen Company Ltd (Seto Inland Sea Lines)
Tokai Kisen
Tokushima Shuttle Line Company Ltd
Yaeyama Kanko Ferry Company Ltd
Yasuda Sangyo Kisen Company Ltd (Yasuda Ocean Group) Airport Line

Jordan
Arab Bridge Maritime Co

Korea, South
Chonghaejin Marine Company
Dae-A Express Shipping Company Ltd
Dong Yang Express
Jindo Transportation
Mirejet
Nam Hae Express Company
Onbado
Seogyeong Marine
Sung Woo Ferry
Uri Express Ferry Co Ltd

Kuwait
Falaika Heritage Village

Malaysia
Ampangship Marine
Fast Ferry Ventures
Lada Langkawi Holdings
Langkawi Ferry Service
Sriwani Tours and Travel

Malta
Virtu Ferries Ltd

Mexico
Cruceros Maritimos del Caribe SA De CV
Ultramar

Netherlands
Connexxion Fast Flying Ferries
Rederij G Doeksen en Zonen BV
Veolia Transport
Waterbus

Netherlands Antilles
Rapid Explorer

New Caledonia
Compagnie Maritime des Iles
Mary-D Enterprises

New Zealand
Dolphin Discoveries
Fullers Bay of Islands Ltd
Fullers Group Ltd
Pine Harbour Ferries
Real Journeys
Stewart Island Experience
Whale Watch Kaikoura

Northern Mariana Islands
Tinian Shipping and Transportation

Norway
62° Nord Cruises
Arctic Voyager
Bergen Norhordland Rutelag A/S
BNR Fjordcharter
FFR
Fjord1 Fylkesbaatane A/S
Fosen Trafikklag A/S
Gjendebåtene
Helgeland Ska A/S
Hurtgruten Group ASA
Jon Arne Helgoy A/S
L Rødne and Sønner A/S
Master Ferries
Namsos Trafikkselskap A/S
Nesodden-Bundefjord Dampskipsselskap A/S
North Cape Minerals A/S
SeaAction AS
Stavangerske AS
Tide ASA
Torghatten Trafikkselskap A/S

Pakistan
Pakistan Water and Power Development Authority

Philippines
Aquajet Maritime
Delta Fast Ferries
El Greco Jet
Mt Samat Ferry Express Inc
Oceanjet
Supercat Fast Ferry Corporation
Weesam Express

Poland
Zegluga Gdanska

Portugal
Sociedade Fluvial de Transportes
Transtejo

Puerto Rico
Luxury Resorts
Puerto Rico Ports Authority

Russian Federation
Alien Shipping

Saudi Arabia
Saudi Aramco
United Company for Marine Lines

Seychelles
Inter Island Boats

Sierra Leone
Diamond Airlines Hovercraft Services

Singapore
Batam Fast Ferry Pte Ltd
Berlian Ferries
Bintan Resort Ferries
Lewek Shipping Pte Ltd
Pelican Offshore Services
Penguin Ferry Services
Tian San Shipping (Pte) Ltd
Waterfront City Resorts

Slovakia
Slovak Shipping and Ports

Slovenia
Kompas dd

South Africa
Robben Island

Spain
Acciona Trasmediterranea SA
Balearia Eurolineas Maritimes
Buquebus Espana
Cape Balear
Europa Ferrys SA
Fred Olsen SA
FRS Iberia
Garajonay Expres
Iscomar Ferrys
Mediterranea Pitiusa
Nautus Al Maghreb
Trasmapi

Sri Lanka
Ceylon Shipping Corporation

Sweden
Destination Gotland
Hoverline AB
Stena Line AB
Stromma

Switzerland
Compagnie Générale de Navigation sur le Lac Léman

Taiwan
Tung Hsin Steamship Company Ltd
Yeuam Shian Shipping

Tanzania
Azam

Thailand
Andaman Wave Master
Kon-tiki Diving and Snorkelling Co Ltd
Lomprayah High Speed Ferry
Seatran Ferry Co

Trinidad and Tobago
Port Authority of Trinidad and Tobago

Turkey
Akgunier Shipping
Bodrum Express Lines
Istanbul Deniz Otobusleri
Yesil Marmaris Shipping

Ukraine
Black Sea Shipping
Ukrainian Danube Shipping Company

United Arab Emirates
Danat Dubai Cruises

United Kingdom
F H Bertling Ltd
Condor Ferries Ltd
HD Ferries
HM Revenue and Excise
Hoverspeed Ltd
Hovertravel Ltd
Isle of Man Steam Packet Company
P&O Irish Sea
Red Funnel Ferries
Sea Containers Ltd
Speed Ferries
Stena Line
Thames Clippers
Wightlink Ltd

United States
Alameda Harbor Bay Ferry
Alameda Oakland Ferry
Alaska Fjordlines
Alaska Hovercraft Ventures
Alaska Marine Highway System
Aleutians East Borough
Allen Marine Tours
Arnold Transit Company
Auk Nu Tours
Bar Harbor Whale Watch Co

Bay State Cruise Company
Billybey Ferry Company
Blue and Gold Fleet, L.P.
Boston Harbor Cruises
Catalina Channel Express
Catalina Ferries
Catalina Passenger Services
Chilkat Cruises & Tours
Circle Line
Clipper Navigation Inc
Condor Cruises
Cross Sound Ferry Inc
Expeditions
Four Seasons Tours
Golden Gate Ferry
Harbor Express
Hawaii Superferry
Holland America Westours
Hy-Line Cruises
Interstate Navigation
Island Hi Speed Ferry
Island Packers Cruises Inc
Jet Express
Kenai Fjords Tours
Key West Express
Kitsap Ferry Co
Lake Express LLC
Miss Barnegat Light
New England Aquarium
New England Fast Ferry Company
New York Water Taxi
NY Waterway Tours
Otto Candies LLC
Pacific Whale Foundation
Phillips Cruises & Tours (Yukon River Cruises)
Prince William Sound Cruises and Tours
Royal Caribbean Cruise Lines
Salem Ferry
Sayville Ferry Service Inc
Seastreak
Shepler's Mackinac Island Ferry
Star Line
Trico Marine Operations Inc
Vallejo Baylink Ferries
Vineyard Fast Ferry
Yankee Fleet
Washington State Ferries
Woods Hole and Martha's Vineyard Steamship Authority

Uruguay
Ferrylineas

Venezuela
Conferry
Ferryven
Gran Cacique

Vietnam
Duong Dong Express

Virgin Islands (UK)
M & M Transportation Services
Speedy's
Road Town Fast Ferry

Virgin Islands (United States)
Smith´s Ferry Services Ltd
Transportation Services of St John
VI Sea Trans

CIVIL OPERATORS OF HIGH-SPEED CRAFT

Abbreviations

The following abbreviations are used in this section:

ALH	Air-Lubricated-Hull craft
CAT	CATamaran vessel
F-CAT	Foil-CATamaran vessel
HOV	HOVercraft, amphibious capability
HYD	HYDrofoil, surface-piercing foils, fully immersed foils
MH	MonoHull vessel
SES	Surface-Effect Ship or sidewall hovercraft (non-amphibious)
SWATH	Small-Waterplane-Area Twin-Hull vessel
TRI	TRImaran
WPC	Wave-Piercing Catamaran

Argentina

Buquebus

Av Antártida Argentina 821, Pto Madero, Buenas Aires, Argentina

Tel: (+54 11) 43 16 65 00
Fax: (+54 11) 43 16 64 05
e-mail: atclients@buquebus.com
Web: www.buquebus.com

High-speed craft operated

Type	Name	Seats	Additional payload	Delivered
WPC InCat 74 m	*Atlantic III**	600	110 cars	1993
CAT InCat Australia/ AMD K50	*Juan Patricio*	450	55 cars	1995
MH Bazan Mestral 96 m	*Albayzin*	450	84 cars	1995
MH Bazan 127 m	*Silvia Ana*	1065	200 cars	1996
CAT AMD/Bazan B60	*Luciano Federico*	450	55 cars	1997
CAT Derektor/N.Gee	*Patricia Olivia II*	300	–	1998
WPC InCat 91 m	*Catalonia**	900	240 cars	1998
CAT FBM/Pecquot 45 m	*Thomas Edison*	302	–	1999

*These vessels are currently on charter to other operators.

Operations
Montevideo (Punta del Este) to Buenos Aires, 100 n miles.

Catalonia *is chartered to P&O European Ferries to operate between Larne and Troon during the summer months in the northern hemisphere*
0045234

Australia

Aristocat Reef Cruises

Marina Mirage Shop 18, 4 Wharf Street, Port Douglas, QLD 4877, Australia

Tel: (+61 7) 40 99 47 27
Fax: (+61 7) 40 99 45 65
e-mail: reef@aristocat.com.au
Web: www.aristocat.com.au

31 m catamaran Aristocat V 1324789

High-speed craft operated

Type	Name	Seats	Delivered
CAT NWBS 31 m	*Aristocat V*	100	2006

Operations
Day tours to wilderness habitats on the Great Barrier Reef.

Australian Customs Service

Tel: (+61 2) 62 75 66 66
e-mail: information@customs.gov.au
Web: www.customs.gov.au

High-speed craft operated

Type	Name	Delivered
CAT NQEA InCat 22 m Cheetah Class	*Wauri*	1988
MH Austal 38 m	*Botany Bay*	1999
MH Austal 38 m	*Holdfast Bay*	1999
MH Austal 38 m	*Roebuck Bay*	1999
MH Austal 38 m	*Corio Bay*	2000
MH Austal 38 m	*Hervey Bay*	2000
MH Austal 38 m	*Arnhem Bay*	2000
MH Austal 38 m	*Storm Bay*	2000
MH Austal 38 m	*Dame Roma Mitchell*	2000

Operations
National civil surveillance and response around Australia's 37,000 km coastline.

Roebuck Bay *on trials* 0084067

Australian Whale Watching

PO Box 386, Albion, QLD 4010, Australia

Tel: (+ 61 7) 36 30 26 66
Fax: (+61 7) 36 30 24 44
e-mail: info@australianwhalewatching.com.au
Web: www.australianwhalewatching.com.au

High-speed craft operated

Type	Name	Seats	Delivered
CMCS Seacor 24 m	*Eclipse*	95	2004

Operations
Whale watching from Gold Coast, Brisbane and Moreton Bay.

Bali Hai Cruises

267 Rokeby Road, Subiaco, Western Australia 6008, Australia

Tel: (+61 8) 93 88 98 98
Fax: (+61 8) 93 88 98 99
e-mail: balihai@q-net.net.au
Web: www.balihaicruises.com

Bali Office:
PO Box 3548, Denpasar 80001, Bali, Indonesia.

Colin Beeck, *Group General Manager*

High-speed craft operated

Type	Name	Seats	Delivered
CAT Austal 36 m	*Bali Hai II*	316	1994

Operations
Lembongan Island reef cruise.
Note: *Bali Hai II* is operated in Indonesia by PT Bali Hai Cruises Nusantara.

Operations
Benoa to Lembongan Island.

Big Cat Green Island Cruises

Reef FleetTerminal, 1 Spence Street, Cairns, Queensland 4870, Australia

Tel: (+61 7) 40 51 04 44
Fax: (+61 7) 40 51 88 96
e-mail: info@bigcat-cruises.com.au
Web: www.bigcat-cruises.com.au

Paula Wallace, *Managing Director*

The Wallace family who own Big Cat Green Island Reef Cruises have over 20 years experience in operating cruises to the Great Barrier Reef.

High-speed craft operated

Type	Name	Seats	Delivered
CAT N.Wright 24 m	*Reef Rocket*	123	1999
CAT NQEA InCat 30 m	*Big Cat* (ex-*Quicksilver III*)	300	1986

Operations
Cairns to Green Island.

Big Cat 0142155

Brisbane City Council

GPO Box 1434, Brisbane, Queensland 4001, Australia

Tel: (+61 7) 3403 8888
Fax: (+61 7) 3403 6905
Web: www.brisbane.qld.gov.au

High-speed craft operated

Type	Name	Seats	Delivered
CAT BSC/Grahame Parker 25 m	*Kurilpa*	150	1996
CAT BSC/Grahame Parker 25 m	*Mirburpa*	150	1996
CAT BSC/Grahame Parker 25 m	*Barrambin*	150	1996
CAT BSC/Grahame Parker 25 m	*Tugulawa*	150	1996
CAT BSC/Grahame Parker 25 m	*Mianjin*	150	1996
CAT BSC/Grahame Parker 25 m	*Binkinba*	150	1996
CAT BSC/Grahame Parker 25 m	*Mooroolbin*	150	1998
CAT BSC/Grahame Parker 25 m	*Banerabu*	150	1998
CAT BSC/Grahame Parker	*Beenung-urrung*	162	2004
CAT BSC/Grahame Parker	*Tunamun*	162	2005

Operations
Brisbane River from Brett's Wharf to University of Queensland via the Central Business District.

Calypso Reef Charters

PO Box 653, Port Douglas, QLD 4877, Australia

Tel: (+61 7) 40 99 69 99
Fax: (+61 7) 40 99 67 77
e-mail: info@calypsocharters.com.au
Web: www.calypsocharters.com.au

High-speed craft operated

Type	Name	Seats	Delivered
CAT CMCS Seacat 24 m	*Eclipse*	60	2006

Operations
Great Barrier Reef trips from Port Douglas.

Captain Cook Great Barrier Reef Cruises

1770 Great Barrier Reef Cruises
Shop 23, Endeavour Plaza, Captain Cook Drive, Agnes Water, Queensland 4677, Australia

email: info@1770reefcruises.com
Web: www.1770reefcruises.com

High-speed craft operated

Type	Name	Seats	Delivered
CAT Lightning/Crowther 20 m	*Spirit of 1770*	134	2003

Operations
Great Barrier Reef cruises from Town of 1770 to Lady Musgrave Island.

Cruise Victoria

Apt 1612/100, Docklands, Victoria 2008, Australia

Tel: (+61 408) 82 36 33
Fax: (+61 3) 96 70 89 06
e-mail: info@cruisevictoria.com.au
Web: www.cruisevictoria.com.au

High-speed craft operated

Type	Name	Seats	Delivered
CAT RDM/Crowther 23 m	*SeaMelbourne*	112	2003

Operations
Yarra River and Port Phillip Bay.

Cruise Whitsundays

Shingley Drive, Abel Point Marina North, PO Box 1268, Airlie Beach, Queensland 4802, Australia

Tel: (+61 7) 49 46 72 40
Fax: (+61 7) 49 46 43 82
e-mail: info@cruisewhitsundays.com
Web: www.cruisewhitsundays.com

Rod Gluth, *Chief Executive Officer*

High-speed craft operated

Type	Name	Seats	Delivered
WPC InCat 37 m*	*Seaflight*	300	1988
CAT BSC 18 m	*Swordfish*	120	2006
CAT BSC 18 m	*Barracuda*	120	2006

*This craft was rebuilt by NQEA in 2004.

Operations
Airlie Beach, South Molle Island, Daydream Island, Knuckle Reef, Hayman Island and Hamilton Island Airport.

Darwin Hovercraft Tours

2 Larrakeyha Terrace, Darwin, Northern Territory 800, Australia

Tel: (+61 8) 89 81 68 55
Fax: (+61 8) 89 81 57 44

High-speed craft operated

Type	Name	Seats	Delivered
HOV	–	20	–

Operations
Darwin Harbour Tours.

Deep Sea Divers Den

319 Draper Street, Cairns, Queensland 4870, Australia

Tel: (+61 7) 40 46 73 33
Fax: (+61 7) 40 31 12 10
e-mail: info@diversden.com.au
Web: www.diversden.com.au

High-speed craft operated

Type	Name	Seats	Delivered
CMCS Seacat 25	*Reef Quest*	120	2001
CMCS Seacat 23.5	*Sea Quest*	130	2004

Operations
Specialising in diving and snorkelling trips to the Great Barrier Reef from Cairns.

Down Under Cruise and Dive

287 Draper street, Cairns, QLD 4870, Australia

Tel: (+61 7) 40 52 83 07
Fax: (+61 7) 40 31 13 73
e-mail: sales@downunderdive.com.au
Web: www.downunderdive.com.au

High-speed craft operated

Type	Name	Seats	Delivered
MH SBF 30 m	*Osprey V*	240	1994

Operations
Great Barrier reef dive and snorkel trips from Cairns.

Fantasea Cruises Pty Ltd

A subsidiary of Riverside Marine
PO Box 616, Airlie Beach, Queensland 4802, Australia

Tel: (+617) 49 46 51 11
Fax: (+617) 49 46 55 20
e-mail: info@fantasea.com.au
Web: www.fantasea.com.au

D G Hutchen, *Managing Director*

High-speed craft operated

Type	Name	Seats	Delivered
CAT NQEA InCat 30 m	*One* (ex-*Fantasea One*, ex-*Reef Link II*)	405	1988
CAT NQEA InCat 29 m	*Two* (ex-*Fantasea Monarch*, ex-*South Molle* and *Telford Capricorn*)	332	1983
CAT NQEA InCat 24 m	*Three* (ex-*Quickcat II*)	200	1985
CMCS 24 m	*Eleven*	149	1998
CAT NWBS 35 m	*Ten*	276	2001
CAT Crowther 35 m	*Five*	350	2005
CAT 18 m	*Four*	150	2005

Operations
Specialising in cruises to the Great Barrier Reef daily from Hamilton, Daydream, South Molle, Club Med Lindeman and Long Islands as well as mainland resorts and whale watching. Also fast ferry shuttle service (Blue Ferries) between Shute Harbour and Hamilton, South Molle, Daydream and Long Islands.

Fantasea One 0038873

Freedom Fast Cats

Rosslyn Bay, Capricorn Coast, Queensland 4703, Australia

Tel: (+61 7) 49 33 62 44, (+61 7) 49 33 67 70
email: info@keppelbaymarina.com.au
Web: www.keppelbaymarina.com.au

Max Allen, *Managing Director*

Freedom Fast Cats operating from Keppel Bay Marina transport passengers to Great Keppel Island.

Type	Name	Seats	Delivered
InCat 22 m	*Freedom Adventurer*	216	1983

Operations
Keppel Bay Marina to Great Keppel Island, three services daily.

Gordon River Cruises

The Esplanade, Strahan, Tasmania 7468, Australia

Tel: (+61 3) 64 71 43 00
Fax: (+61 3) 64 71 42 65
e-mail: alan.gifford@strahanvillage.com.au
Web: www.federalresorts.com.au/

Greg Astell, *General Manager*
Alan Gifford, *Gordon River Cruises Manager*

High-speed craft operated

Type	Name	Seats	Delivered
CAT 32 m	*Lady Jane Franklin II*	212	2003

Operations
High-speed cruising in Macquarie Harbour and Gordon River World Heritage area.

Great Adventures

Part of the Daikyo Group
Fleet reef Terminal, 1 Spence Street, Cairns, Queensland 4870, Australia

Tel: (+61 7) 40 44 99 44
Fax: (+61 7) 40 44 99 55
e-mail: info@greatadventures.com.au
Web: www.greatadventures.com.au

Tony Baker, *General Manager*

High-speed craft operated

Type	Name	Seats	Delivered
CAT NQEA InCat 22 m	*Green Island Express*	203	1982
CAT NQEA InCat 30 m	*Reef King*	390	1987
CAT Lloyd's 38 m	*Reef Prince*	393	1995

Operations
Great Barrier Reef (Outer Reef) direct or via Green Island.
Carrying over 300,000 passengers annually.

Haba Dive and Snorkel

PO Box 122, Port Douglas, QLD 4871, Australia

Tel: (+61 7) 40 99 52 54
Fax: (+61 7) 48 99 53 85
e-mail: enquiries@habadive.com.au
Web: www.habadive.com.au

High-speed craft operated

Type	Name	Seats	Delivered
CAT SFM 28 m	*Evolution*	108	2005

Operations
Trips to great Barrier Reef from Port Douglas.

Hayman Island Resort

A division of BT Hotel Ltd
Hayman Island, Queensland 4801, Australia

Tel: (+61 7) 49 40 12 34
Fax: (+61 7) 49 40 15 67

e-mail: sales@hayman.com.au
Web: www.hayman.com.au

P Stacher, *General Manager*
D Miers, *Director of Engineering and Marine*

Hayman Resort operates a fleet of luxury passenger ferries from Hayman Island to Hamilton Island, the mainland (Shute Harbour) and various destinations throughout the Whit Sundays. Hayman carries over 70,000 passengers per year.

High-speed craft operated

Type	Name	Seats	Delivered
CAT NQEA Seajet P35	*Sun Eagle*	190	1996
MH Riverboat 34 m	*Hayman Sun Goddess*	200	1984

Operations
Hayman Island to Great Barrier Reef
Hayman Island to Hamilton Island
Hayman to Shute Harbour.

Sun Eagle *operated by Hayman Resort* 0507088

Hervey Bay Whale Watch

PO Box 7334, Urangan, Hervey Bay 4655, Queensland, Australia

Tel: +61 7 4128 9611
Fax: +61 7 4124 9111
Web: www.herveybaywhalewatch.com.au

High-speed craft operated

Type	Name	Seats	Delivered
CAT CMCS 17 m	*Quickcat II*	80	2003

Operations
Whale watching trips from Hervey Bay.

Hillary's Fast Ferries

Shop 56, Southside Drive, Hillary's Boat Harbour, WA 6025, Australia

Tel: (+61 8) 92 46 10 39
Fax: (+61 8) 94 48 64 55
e-mail: reservations@hillarysfastferries.com.au
Web: www.hillarysfastferries.com.au

High-speed craft operated

Type	Name	Seats	Delivered
MH Wavemaster 35 m	*Sea Flyte*	264	1995

Operations
Hillary's Boat Harbour to Rottnest Island, and whale watching cruises.

Hobart Cruises

Brook Street Pier, Sullivan's Cove, Hobart, Tasmania, Australia

Tel: (+61 1300) 137 919
e-mail: info@hobartcruises.com
Web: www.hobartcruises.com

High-speed craft operated

Type	Name	Seats	Delivered
CAT RDMC 22 m	*Peppermint Bay II*	173	2005

Operations
Cruises from Hobart to Peppermint bay and Recherche Bay.

Inter Island Ferries

PO Box 11, Parkdale, Victoria 3194, Australia

Tel: (+61 3) 95 85 57 30
e-mail: query1@interislandferries.com.au
Web: www.interislandferries.com.au

High-speed craft operated

Type	Name	Seats	Delivered
CAT Aluminium Marine 17.5 m	*Matthew Flinders*	112	2005
CAT RDM 16 m	*George Bass*	92	1998

Operations
Stony Point, Victoria to French Island, Phillip Island and Sea Rocks.

Lady Musgrave Barrier Reef Cruises

Tel: (+61 7) 494 9077
e-mail: info@lmcruises.com.au
Web: www.lmcruises.com.au

High-speed craft operated

Type	Name	Seats	Delivered
CAT 24 m	*Spirit of Musgrave*	150	–
TRI 24 m	*Spirit of Musgrave*	75	–
MH SBF 15 m	*Discovereef*	50	1987

Operations
Bundaberg and town of 1770 to Lady Musgrave Island.

Mackay Island and Reef Cruises

PO Box 11126, Mackay Canelands, Queensland 4740, Australia

Tel: (+61 7) 49 55 36 33
Fax: (+61 7) 49 55 64 03
e-mail: info@islandreefcruises.com
Web: www.islandreefcruises.com

High-speed craft operated

Type	Name	Seats	Delivered
MH Precision 22 m	*Reef Goddess*	60	1987

Operations
Mackay Habour to Scawfell and Brampton Island National Park.

Mackenzies Island Cruises

71 The Esplanade, Esperance, Western Australia 6450, Australia

Tel: (+61 8) 90 71 57 57
Fax: (+61 8) 90 71 49 93
e-mail: info@woodyisland.com.au
Web: www.woodyisland.com.au

High-speed craft operated

Type	Name	Seats	Delivered
CAT Sabre 24 m	*Seabreeze II*	164	2003

Operations
Esperance to Woody Island and tours around the Recherche Archipelago.

Moreton Bay Whalewatching

Redcliffe Jetty, Anzac Avenue, Moreton Bay, Queensland, Australia

Tel: (+61 7) 38 80 04 77
Fax: (+61 7) 38 80 11 22
e-mail: info@whalewatching.net
Web: www.whalewatching.net

High-speed craft operated

Type	Name	Seats	Delivered
CAT South Pacific 29 m	*Eye Spy*	320	2002

Navigators

Brooke Street Pier, Sullivan's Cove, Hobart 7000, Tasmania, Australia

Tel: (+61 3) 62 23 19 14
Fax: (+61 3) 62 24 83 33
e-mail: info@derwentrivercruises.com.au

High-speed craft operated

Type	Name	Seats	Delivered
CAT Crowther/RDM 24 m	*Marana*	225	2002
CAT Crowther/RDM 17.5 m	*Excella*	120	2001

Operations
Leisure cruises from Hobart to Port Arthur, Moorilla Estate Winery, Upper Harbour and Outer Harbour.

Palm Beach Ferry Service

PO Box 622, Avalon Beach, New South Wales 2107, Australia

Tel: (+61 2) 99 74 24 11
Fax: (+61 2) 99 74 32 79
e-mail: info@palmbeachferry.com.au
Web: www.palmbeachferry.com.au

High-speed craft operated

Type	Name	Seats	Delivered
CAT Lightning 20 m	*Golden Spirit*	160	2001
CAT NWC 25 m	*Aqua Spirit*	172	2002
CAT NWC 25 m	*Crystal Spirit*	199	2003

Operations
Palm Beach to Ettalong and Wagstaff
Palm Beach to Mackerel Beach.

Peel's Tourist and Ferry Services Pty Ltd

PO Box 197, Lakes Entrance, Victoria 3909, Australia

Tel: (+61 3) 51 55 12 46
Fax: (+61 3) 51 55 17 38
e-mail: info@peelcruises.com
Web: www.peelcruises.com

Barrie Peel, *Managing Director*

High-speed craft operated

Type	Name	Seats	Delivered
CAT InCat (Hobart) 20.4 m Bulls Marine Pty Ltd	*Thunderbird*	190	1984

Operations
Two routes are operated on Gippsland Lakes, ranging from 35 to 70 km.

Peel's InCat Thunderbird 0506725

Poseidon Cruises

PO Box 431, Port Douglas, Queensland 4877, Australia

Tel: (+61 7) 40 99 47 72
Fax: (+61 7) 40 99 41 34
e-mail: info@poseidon-cruises.com.au
Web: www.poseidon-cruises.com.au

High-speed craft operated

Type	Name	Seats	Delivered
CAT Aluminium Marine 24 m	*Poseidon III*	90	2001

Operations
Trips from Port Douglas to Outer Great Barrier Reef.

Queensland Government

Department of Primary Industries, Queensland Boating and Fisheries Patrol, PO Box 7453, Cairns, Queensland, Australia

Web: www.dpi.qld.gov.au/fishweb

Patrol craft New Investigator 0507089

High-speed craft operated

Type	Name	Delivered
MH NQEA Wasp Pacific, 25 m	*New Investigator*	1989
MH Western 23 m	*K I Ross*	2001

Operations
Fisheries protection out of Cairns and Gladstone.

Quick Cat Cruises

PO Box 208, Mission Beach, Queensland 4852, Australia

Tel: (+61 7) 40 68 72 89
Fax: (+61 7) 40 68 71 85
e-mail: qcat@bigpond.net.au
Web: www.quickcatcruises.com.au

High-speed craft operated

Type	Name	Seats	Delivered
CAT InCat/NQEA 24 m	*Quick Cat*	195	1985

Operations
Great Barrier Reef diving excursions and Dunk Island Cruises.

Quicksilver Connections Ltd

Marina Mirage, PO Box 171, Port Douglas, North Queensland 4871, Australia

Tel: (+61 7) 40 87 21 00
Fax: (+61 7) 40 99 55 25
e-mail: enquiries@quicksilver-cruises.com
Web: www.quicksilver-cruises.com

Tony Baker, *Managing Director*
Matthew Hurley, *General Manager*
Travis Clarke, *Operations Manager*

High-speed craft operated

Type	Name	Seats	Delivered
WPC InCat 37.2 m	*Quicksilver V*	315	November 1988
WPC NQEA/InCat 45 m	*Quicksilver VIII*	442	May 1995
CAT Grahame Parker 24 m	*Silver Swift II*	100	June 2000
CAT North West Bay Ships 30 m	*Silversonic*	150	2005
CAT North West Bay Ships 30 m	*Silver Swift*	150	(2006)

Operations
Port Douglas to Agincourt Reef North, 39 n miles, 1 hour 30 minutes
Port Douglas to Low Isles.

45 m wave-piercing catamaran operated by Quicksilver Connections 0506999

Raging Thunder Pty Ltd

PO Box 1109, Cairns 4870, Australia

Tel: (+61 7) 40 30 79 90
Fax: (+61 7) 40 30 79 11
email: info@ragingthunder.com.au
Web: www.ragingthunder.com.au

High-speed craft operated

Type	Name	Seats	Delivered
CAT InCat (Hobart) 20 m	*Fitzroy Flyer*	169	1981

Operations
Marlin Marina, Cairns to Fitzroy Island.

Reefjet

Airlie Beach, Queensland, Australia

Tel: (+61 7) 4946 5366
Fax: (+61 7) 4946 6033
e-mail: info@reefjet.com.au
Web: www.reefjet.com.au

High-speed craft operated

Type	Name	Seats	Delivered
MH Nexcastle 24 m	*Reefjet*	100	1999

Operations
Trips to the Great Barrier Reef and Whitsunday Islands from Airlie Beach.

Reefjet 1310669

Reef Magic Cruises

Reef Fleet Terminal, 1 Spence Street, PO Box 905, Cairns, Queensland 4870, Australia

Tel: (+61 7) 40 31 15 88
Fax: (+61 7) 40 51 19 20
email: enquiries@reefmagiccruises.com
Web: www.reefmagic.com

Tin North, *Director*
Don Cowie, *Director*

Located in Cairns, Reef Magic Cruises runs dining cruises to the Great Barrier Reef.

High-speed craft operated

Type	Name	Seats	Delivered
CAT InCat 22 m	*Reef Magic*	214	1982

Operations
Cruises to the Outer Great Barrier Reef.

Rottnest Express Pty Ltd

1 Emma Place, North Fremantle, Western Australia, Australia

Tel: (+61 8) 94 30 58 44
Fax: (+61 8) 94 30 40 36
e-mail: bookings@rottnestexpress.com.au
Web: www.rottnestexpress.com.au

High-speed craft operated

Type	Name	Seats	Delivered
MH SBF 31 m	*Sea Eagle III*	243	1996
MH SBF 35 m	*Eagle Express*	329	1999
MH Wavemaster 41 m	*Star Flyte*	506	1988

Eagle Express 1033513

Eagle Express 0103301

Operations
Swan River cruises between Perth and Fremantle
Day trips and tours to Rottnest Island
Whale watching excursions from mid-September to November.

Sea-Cat Ferries and Charters P/L

Cullen Bay Ferry Terminal, Darwin, Northern Territory, Australia

Tel: (+61 8) 89 41 19 91
e-mail: seacat@bigpond.com
Web: www.seacat.com.au

High-speed craft operated

Type	Name	Seats	Delivered
CAT 18 m	*Sea Cat 1*	150	1998
CAT 18 m	*Sea Cat 2*	150	1998
CAT Aluminium Marine 23.5 m	*Arafura Pearl*	200	2004

Operations
Darwin to Mandorah and the Tiwi islands.

Spirit of the Bay Cruises

PO Box 4911, Bundall, Queensland 4217, Australia

Tel: (+61 7) 55 27 77 55
Fax: (+61 7) 55 27 64 11
e-mail: spirit1@bigpond.net.au
Web: www.goldcoastwhalewatching.com

High-speed craft operated

Type	Name	Seats	Delivered
CAT InCat 23 m	*Spirit of Gold Coast*	200	1988

Operations
Whale watching trips from the Gold Coast and Port Stephens.

Sunferries

Breakwater Terminal, Sir Leslie Thiess Drive, Townsville, Queensland 4810, Australia

Tel: (+61 7) 47 71 38 55
Fax: (+61 7) 47 71 39 55
e-mail: info@sunferries.com.au
Web: www.sunferries.com.au

High-speed craft operated

Type	Name	Seats	Delivered
CAT InCat (Hobart) 29 m	*The Spirit of Magnetic* (ex-*Spirit of Roylen*)	240	1982
CAT InCat 25 m	*Sunbird*	123	1987
CAT InCat 22 m	*Islander IV*	215	1983
CAT BSC 29 m	*Maggie Cat*	239	2006
CAT BSC 29 m	*Sun Cat*	239	2006

Operations
Palm Island, Magnetic Island and John Brewer Reef.

Sunlover Cruises

A subsidiary of Ocean Capital
PO Box 835, Cairns, Queensland 4870, Australia

Tel: (+61 7) 40 50 13 33
Fax: (+61 7) 40 50 13 13
e-mail: res@sunlover.com
Web: www.sunlover.com.au

Terry Russell, *Managing Director*

High-speed craft operated

Type	Name	Seats	Delivered
CAT Aluminium Craft (88) 34 m	*Island Pearl*	200	1998

Operations
Cairns and Port Douglas to the Great Barrier Reef.

Sydney Ferries

PO Box R1799, Royal Exchange, New South Wales 1225, Australia

Tel: (+61 2) 92 46 83 00
Fax: (+61 2) 92 46 83 61
Web: www.sydneyferries.info

Sue Sinclair, *Chief Executive Officer*

High-speed craft operated

Type	Name	Seats	Delivered
CAT NQEA 36 m InCat	*Blue Fin*	280	Mar 1990
CAT NQEA 36 m InCat	*Sir David Martin*	280	Dec 1990
CAT NQEA 36 m InCat	*Sea Eagle*	280	April 1991
CAT NQEA 35 m RiverCat	*Dawn Fraser*	230	May 1992
CAT NQEA 35 m RiverCat	*Betty Cuthbert*	230	May 1992
CAT NQEA 35 m RiverCat	*Shane Gould*	230	Jan 1993
CAT NQEA 35 m RiverCat	*Marlene Matthews*	230	Jan 1993
CAT NQEA 35 m RiverCat	*Evonne Goolagong*	230	Sep 1993
CAT NQEA 35 m RiverCat	*Marjorie Jackson*	230	Sep 1993
CAT WaveMaster 35 m RiverCat	*Nicole Livingston*	230	Oct 1995
CAT N.Wright 29.6 m	*Ann Sargeant*	150	1998
CAT N.Wright 29.6 m	*Pam Burridge*	150	1998
CAT ADI 37.8 m SuperCat	*Mary Mackillop*	250	2000
CAT ADI 37.8 m SuperCat	*Louise Sauvage*	250	2001
CAT ADI 37.8 m SuperCat	*Susie O'Neil*	250	2000
CAT ADI 37.8	*SuperCat 4*	250	2001

Operations
Sydney (Circular Quay) to Manly, 7 n miles, 15 minutes
Sydney (Circular Quay) to Parramatta, 14 n miles, 1 hour.

36 m Jetcat Blue Fin 0506924

35 m Low-Wash RiverCat Evonne Goolagong 0506925

36 m Jetcat Sea Eagle 0010476

34 m SuperCat Mary Mackillop 0097642

Tangalooma Wild Dolphin Resort Ltd

PO Box 1102, Eagle Farm, Queensland 4009, Australia

Tel: (+61 7) 36 37 21 39
Fax: (+61 7) 32 68 62 99
e-mail: admin@tangalooma.com
Web: www.tangalooma.com

Jeff Osborne, *Director*

High-speed craft operated

Type	Name	Seats	Delivered
CAT InCat 20 m	*Tangalooma Flyer*	200	1981
WPC NQEA/ Incat 40 m	*Tangalooma Jet* (ex-*Seacom I*)	400	1990
WPC Aluminium Boats Australia 22 m	*Tangalooma Express* (ex-*Seacom I*)	120	2004

Operations
Brisbane to Moreton Island, 26 n miles, 1 hour 15 minutes.

Tangalooma Express 1178211

Tasman Venture

26 Gilston Road, Torquay, Hervey Bay, Queensland 4655, Australia

Tel: (+61 7) 4124 3222
Fax: (+61 7) 4128 1847
e-mail: tasmanventure@bigpond.com
Web: www.tasmanventure.com.au

High-speed craft operated

Type	Name	Seats	Delivered
CAT CMCS 20 m	*Tasman Venture*	120	2004

Operations
Whale watching, fishing and tours from Hervey Bay.

Voyages

GPO Box 3589, Sydney, New South Wales 2001, Australia

Tel: (+61 2) 82 96 80 00
Fax: (+61 2) 82 96 80 52
e-mail: travel@voyages.com.au
Web: www.voyages.com.au

High-speed craft operated

Type	Name	Seats	Delivered
MH Austal 35 m	*Heron Spirit*	162	2003
CAT InCat Crowther 33 m	*Reef Voyager*	175	2007

Operations
Gladstone to Heron Island, daily, 1 hour 40 minutes.

World Heritage Cruises

PO Box 93, Strahan, Tasmania 7468, Australia

Tel: (+61 3) 64 71 71 74
Fax: (+61 3) 64 71 74 31
e-mail: enquiries@worldheritagecruises.com.au
Web: www.worldheritagecruises.com.au

World Heritage Cruises operates a number of fast as well as slow-speed ships. Only the high-speed craft are shown here.

High-speed craft operated

Type	Name	Seats	Delivered
CAT RDM 35 m	–	221	(2008)

Bahamas

Bahamas Ferries Ltd

PO Box N3709, Nassau NP, Bahamas

Tel: (+1 242) 323 21 66
Fax: (+1 242) 393 74 51
e-mail: info@bahamasferries.com
Web: www.bahamasferries.com

High-speed craft operated

Type	Name	Seats	Delivered
CAT FBM/Pequot 35 m	*Bo Hengy*	177	1999

Belgium

Eurosense Belfotop NV

Main office: Nerviërslaan 54, B-1780 Wemmel, Belgium

Tel: (+32 2) 460 70 00
Fax: (+32 2) 460 49 58
e-mail: info.be@eurosense.com
Web: www.eurosense.com

E Maes, *Managing Director*
E Leys, *Manager, Coastal Surveys & Hydrography*

High-speed craft operated

Type	Name	Delivered
HOV ABS M10	*Beasac IV*	February 1996
SES Vosper Hovermarine 218	*Beasac V*	September 1983

Operations
Eurosense Belfotop NV employs a modified ABS M10 hovercraft and a Hovermarine 218 surface effect ship for Belgian coast hydrographic survey and remote sensing work. These craft are based at Zeebrugge.

Beasac IV 0507042

Bermuda

Government Of Bermuda

Ministry of Transport, Magnolia Place, 2nd Floor, 45 Victoria Street, Hamilton HM12, Bermuda

Tel: (+1 441) 295 65 75
e-mail: info@seaexpress.bm
Web: www.seaexpress.bm

Two fast ferries were introduced in 2002 in order to reduce traffic on the island. A further two vessels are on order for delivery in 2004.

High-speed craft operated

Type	Name	Seats	Delivered
CAT Gladding Hearn 26 m	*Serenity*	201	2002
CAT Gladding Hearn 26 m	*Resolute*	201	2002
CAT North West Bay Ships	*Venturilla*	177	2004
CAT North West Bay Ships	*Tempest*	177	2004
CAT NGA 38 m	*Warbaby Fox*	349	2006

Operations
Commuter and tourist services from Hamilton to Rockaway to Dockyard to St Georges.

Bolivia

Crillon Tours SA

PO Box 4785, Avenida Camacho 1223, La Paz, Bolivia

Tel: (+591 2) 233 75 33
Fax: (+591 2) 211 64 82
e-mail: titicaca@entelnet.bo
Web: www.titicaca.com

Darius Morgan Sr, *President*
Elsa Morgan, *General Manager*

US office:
1450 South Bayshore Drive, Miami, Florida 33133, United States

Tel: (+1 305) 358 53 53
Fax: (+1 305) 372 00 54
e-mail: darius@titicaca.com

Crillon Tours SA was founded in 1958 by Darius Morgan and started its hydrofoil services on Lake Titikaka in 1966.

High-speed craft operated

Type	Name	Delivered
HYD Ludwig Honold Manufacturing Co	*Inca Arrow*	1966
HYD Ludwig Honold Manufacturing Co	*Copacabana Arrow*	1964
HYD Ludwig Honold Manufacturing Co	*Andes Arrow*	1965
HYD Ludwig Honold Manufacturing Co	*Titikaka Arrow*	1963
HYD Ludwig Honold Manufacturing Co	*Bolivia Arrow*	1976
HYD Seaflight SpA H.57	*Sun Arrow*	1979
HYD Batumi Ship Building Plant, RFAS	*Glasnost Arrow*	1990

Operations
Lake Titikaka, serving La Paz, Huatajata, Sun Island, Moon Island, Copacabana, Urus Floating Islands Juli-Puno (Peru), Tiahuanacu-Guaqui, Suriqui and Kalauta Island. Daily national and international itineraries (Bolivia/Peru).

Crillon transports approximately 20,000 passengers per year.

Bolivia Arrow *operated by Crillon Tours SA* 1135692

Brazil

Barcas Transportes Maritimos

Rua Miguel de Lemos 80, Ponta D´Arela, Niterói, Rio de Janeiro, RJ 24040-260, Brazil

Tel: (+55 21) 22 06 42 22
Fax: (+55 21) 22 06 42 24
e-mail: sac@barcas-sa.com.br
Web: www.barcas-sa.com.br

High-speed craft operated

Type	Name	Seats	Delivered
CAT Fjellstrand 40 m	*Morro de São Paulo*	125	1995
SES Cirrus 120P	*Express Brasil*	330	1988
SES Cirrus 120P	*Express Macae*	330	1989
CAT Rodriquez 29 m	*Zeus 1*	200	2005
CAT Rodriquez 29 m	*Apollo 1*	200	2005
CAT Rodriquez 29 m	*Netuno 1*	200	2005

Operations
Rio de Janeiro to Niteról.
Rio de Janeiro to Charitas.
Barcas transports over 80,000 passengers daily.

Siem Consub SA

Av Rio Branca, 108-28andar, Centro, Rio de Janeiro, RJ 20040-001, Brazil

Tel: (+55 21) 3515 9700
Fax: (+55 21) 3515 9790
e-mail: siemconsub@siemconsub.com.br
Web: www.siemoffshore.com

Celso Costa, *Chief Executive Officer*

High-speed craft operated

Type	Name	Seats	Delivered
MH Swiftships	*Parintins*	70	1972
MH Breaux's Bay	*Piracicaba*	56	1975
MH Halter Marine	*Atalaia*	70	1980
MH Halter Marine	*Capela*	70	1980
MH Inace	*Norsul Paracuru*	60	1986
MH Inace	*Norsul Parnaiba*	60	1986
MH Inace	*Norsul Propria*	60	1986

Operations
Offshore support along the Brazilian coast with bases in Rio, Macaé, Natal, Fortaleza and Aracaju.

Atalaia *operated by Consub SA* 0506727

Transtur

Aerobarcos do Brasil, Transportes Maritimos e Turismo SA
Praça Iaia Garcia 3, Rio de Janeiro, RJ 21930-040, Brazil

Tel: (+55 21) 33 96 85 41
e-mail: transtur@ism.com.br

Rodriquez PT 20 operated by Transtur 0506726

High-speed craft operated

Type	Name	Seats	Delivered
HYD Rodriquez PT 20	*Flecha do Rio*	85	1969
HYD Rodriquez PT 20	*Flecha de Niterói*	85	1970
HYD Rodriquez PT 20	*Flecha das Ilhs*	85	1966
HYD Rodriquez PT 20	*Flecha de Itaipú*	85	1971
HYD Rodriquez PT 20	*Flecha Fluminense*	85	1962
HYD Rodriquez PT 20	*Flecha de Ipanema*	83	1977
HYD Rodriquez PT 20	*Flecha de Icarai*	83	1964
HYD Rodriquez PT 20	*Flecha da Ribeira*	83	1964
CAT Fjellstrand(s) 40 m	*Jumbo Cat I*	420	1996
CAT Fjellstrand(s) 40 m	*Jumbo Cat II*	420	1996

Operations
Rio de Janeiro to Niterói, 2.8 n miles, 5 minutes
Rio de Janeiro to Paquetá Island, 9.2 n miles, 20 minutes.

Canada

Bay Ferries Ltd

PO Box 634, 94 Water Street, Charlottetown, Prince Edward Island, C1A 7L3, Canada

Tel: (+1 902) 566 38 38
Fax: (+1 902) 566 15 50
e-mail: mitch@nfl-bay.com
Web: www.catferry.com

High-speed craft operated

Type	Name	Seats	Additional payload	Delivered
WPC InCat 98 m	*The Cat**	900	240 cars	2002

*Leased to Port Authority of Trinidad and Tobago (November to April).

Operations
Bar Harbour (US) to Yarmouth (Nova Scotia), (June to October)
Yarmouth to Portland, Maine.

Canadian Coast Guard Hovercraft Unit

Laurentian Region ACV Unit: 7100 DuFleuve, Trois Rivierer Ouest, Québec G9A 6M2, Canada

Tel: (+1 819) 379 73 38
Fax: (+1 819) 379 99 65
e-mail: lheureuxd@dfo-mpo.gc.ca
Web: www.ccg-gcc.gc.ca

D L'Heureux, *Officer in Charge*

High-speed craft operated

Type	Name	Delivered
HOV BHC AP1-88/200 Serial No 201, registration CH-CGC	*Waban-Aki*	September 1987
HOV BHC AP1-88/400	*Sipu Muin*	April 1998
HOV BHC AP1-88/400	*Siyay*	1998
HOV AP1-88/100S	*Penac*	–

Operations
Waban-Akir is designed with a forward well-deck and equipped with an easily removable crane, together with a hydraulic capstan and winch. As well as ice-breaking in winter, *Waban-Aki* maintains marine aids for navigation along 350 miles of the St Lawrence river and also responds to search and rescue incident calls.

AP1-88/400 Sipu Muin 0043236

API-88/200 Waban Aki 0098039

AP1-88/400 Siyay 0543301

In 1998 the Coast Guard took delivery of a second craft, a BHC AP1-88/400, named *Sipu Muin*, for service on the St Lawrence river. The craft is larger than *Waban Aki* and has a greater payload. Both craft operate from the same maintenance facility in Trois Rivières, Quebec and are used extensively all year round for Coast Guard duties which include: icebreaking for flood prevention and safe commercial navigation, oil pollution clean up operations and transportation of material and equipment to remote areas difficult to reach by conventional marine craft.

The most recent vessels, *Siyay* and *Penac*, operate from the Sea Island base on the West Coast.

West Coast Launch Ltd

207-3rd Avenue East, Prince Rupert, BC V8J 1K4, Canada

Tel: (+1 250) 627 9166
Fax: (+1 250) 624 3151
e-mail: mail@westcoastlaunch.com
Web: www.westcoastlaunch.com

High-speed craft operated

Type	Name	Seats	Additional payload	Delivered
CAT 22 m	*Inside Passage*	100	–	2007

Operations
Wilderness cruises from Prince Rupert.

Cape Verde

Moura Company

Praia, Santiago, Cape Verde, CP 49-A

Tel: (+238) 260 30 97
Fax: (+238) 264 76 91
Web: www.mouracompany.net

High-speed craft operated

Type	Name	Seats	Delivered
CAT Marinteknik 32 m	*Auto Jet*	222	1998
CAT Marinteknik 35 m	*Jet Caribe*	313	2000

Operations
São Vicente to Antão.
São Vicente to São Nicolau.
Praia to Fogo and Brava.
Praia to Maio, Boa Vista and Sal.

Chile

Catamaranes Del Sur

Av Pedro de Valdivia Norte 0210, Providencia, Santiago, Chile

Tel: (+56 2) 231 19 02
Fax: (+56 2) 231 19 93
e-mail: info@catamaranesdelsur.cl
Web: www.catamaranesdelsur.cl

High-speed craft operated

Type	Name	Seats	Delivered
CAT CMCS Fastcat 23	*Campo Hielo Sur*	90	1997

Operations
Purto Chacabuco to San Rafael Lagoon.

Hotelsa Patagonian Expedition

Santa Magdalena 75 Of. 705, Providencia, Chile

Tel: (+56) 23 35 05 79
email: operaciones@hotelsa.cl
Web: www.hotelsa.cl

High-speed craft operated

Type	Name	Seats	Delivered
CAT Fjellstrand 38.8 m	*Aysen Express* (ex-*Sog Nekongen*)	201	1987

Operations
San Rafael Glacier.

Patagonia Connection

Fidel Oteiza 1921 Of.1006, Providencia, Santiago, Chile

Tel: (+00 56 2) 225 64 89
Web: www.patagonia-connection.com

High-speed craft operated

Type	Name	Seats	Delivered
CAT AYSN Båtservice Sea Lord 28	*Patagonia Express*	60	1991

Operations
Puerto Chacabuco to San Rafael Lagoon.

China

Dalian Steamship Company

1 Minzhu Plaza, Zhongshan Qu, Dalian, Liaoning, People's Republic of China

High-speed craft operated

Type	Name	Seats	Delivered
CAT Fjellstrand (S) Flying Cat 40 m	*Hai Ou*	400	January 1993
CAT Fjellstrand (S) Flying Cat 40 m	*Hai Yan*	400	January 1993

Operations
Dalian to Nantai.

Dalian Yuan Feng Ferry Company

7 You Hso Guang Chang, Zhong Shan, Dalian 116001, People's Republic of China

Tel: (+86 411) 280 87 38
Fax: (+86 411) 263 45 00

High-speed craft operated

Type	Name	Delivered
CAT Fjellstrand 38.8 m	*Fei Yu*	1993

Operations
Dalian to Yantai.

Ming Zhu Passenger Transport Co-op Co Ltd

Yan Hai Road, Gao Ming City, Guandong, China

Tel: (+86 757) 882 56 57
Fax: (+86 750) 882 56 57

High-speed craft operated

Type	Name	Seats	Delivered
CAT Austal 39.9 m	*Gao Ming*	338	1993

Jiang Men Hong Kong Macau Joint Passenger Transportation Co Ltd

New Advanced Technology Property, Development Zone, Waihai Jiangmen, Guangdong, People's Republic of China

Fax: (+86 750) 377 32 20
e-mail: jgpt@ecok.com
Web: www.jgpt.com

High-speed craft operated

Type	Name	Seats	Delivered
CAT Wavemaster 39	*Peng Lai Hu*	354	1992
CAT Wavemaster 39	*Wu Yi Hu*	354	1992
CAT Austal 40 m	*Gang Zhou*	318	1993

Operations
Jiangmen to Hong Kong.

Nantong High-speed Passenger Ship Company

Nantong, Jiangsu Province 226000, People's Republic of China

High-speed craft operated

Type	Name	Seats	Delivered
CAT Austal Ships 36 m	*Tong Zhou*	430	1990

Operations
Shanghai to Nantong.

Ningbo Fast Ferries Ltd

305 Jiangdong Road North, Ningbo, People's Republic of China

Tel: (+86 574) 775 74 93
Fax: (+86 574) 770 84 00

This company also operates travel agencies, fast craft repair yards and other tourist cruise organisations.

High-speed craft operated

Type	Name	Seats	Delivered
CAT Fjellstrand 38.8 m	*Yong Xing*	312	1985
CAT Fjellstrand 31.5 m	*Yong Wang* (ex-*Lygra*)	306	1995

Operations
Ningbo to Shanghai, Suzhou and Putuo.

Yong Xing *operated by Ningbo Fast Ferries* 0506728

Yong Wang *operated by Ningbo Fast Ferries* 0507001

Shanghai Free Flying Transport

Room 458, No 7 Lane 622, Huai Hai Road C, Shanghai, People's Republic of China

Tel: (+86 21) 63 27 64 27
Fax: (+86 21) 62 47 79 17

High-speed craft operated

Type	Name	Seats	Delivered
SES MARIC 719 II	–	257	1988
CAT Austal 40 m	*Free Flying*	–	1994

Operations
Shanghai to Ningbo.

MARIC 719 II 257-seat, 28 kt SES built by Wu-Hu Shipyard 0506729

Shanghai Huandoo Passenger Ship

3/F Passenger Transportation Station, Luchaogang, Shanghai

Tel: (+86 215) 828 2178
Fax: (+86 215) 828 2852

High-speed craft operated

Type	Name	Seats	Delivered
CAT Fjellstrand 38.8	*Fei Hai*	326	1985
CAT Fjellstrand	*Fei Ren*	289	1984

Shanghai Ya Tong Co Ltd

No 9 Bayi Road, Chongming County, Shanghai 202150, China

Tel: (+86 216) 595 4641

High-speed craft operated

Type	Name	Seats	Delivered
Cat Fjellstrand 40 m	*Xin Shi Ji Hao*	324	1993
CAT CMCS/Wang Tak 33 m	*Liang Yu*	250	2005

Shenzhen Hengtong Shipping

13A Nan Sha Ge, Hua Cai Gar, Shekou, Guangdong Province, People's Republic of China

High-speed craft operated

Type	Name	Seats	Delivered
SES Karlskronarvet Jet Rider 3400	*Heng Yi*	292	1987
CAT Afai/AMD 170	*Peng Xing* (ex-*Wuzhou*)	150	1994

Shenzhen Pengxing Shipping Co Ltd

9/F Ming Wah International Convention Centre, No 8 Guishan Road, Shekou Industrial Zone, Nanshan District, Shenzhen, People's Republic of China

Tel: (+86 755) 26 88 60 20
Fax: (+86 755) 26 88 60 21
e-mail: pwzx@pengzingshipping.com
Web: www.pengzingshipping.com

High-speed craft operated

Type	Name	Seats	Delivered
CAT Afai InCat 21 m	*Yue Hai Chun*	169	1984
CAT Afai InCat 21 m	*Shen Zhen Chun*	169	1985
CAT Afai InCat 21 m	*Ling Nan Chun*	169	1987
CAT Afai InCat 21 m	*Nan Hai Chun*	169	1989
CAT Afai InCat 21 m	*Dong Fang Chun*	169	1991
CAT Afai InCat 21 m	*Zhu Hai Chun*	169	1986
CAT Afai AMD 28 m	*Peng Xing*	242	1993
CAT Fjellstrand 31.5 m	*Yin Xing*	291	1985

Operations
Shenzhen to Zhu Hai.
Shenzhen to Skypier (Hong Kong Airport).

Shenzhen Xunlong Shipping Co Ltd

2/F, Shekou Harbour Passenger Terminal, Harbour Road 1, Shekou Shenzhen 518067, China

Tel: (+86 755) 26 68 94 71
Fax: (+86 755) 26 67 52 15
e-mail: shulin.lai@163.com
Web: www.xunlongferry.com

High-speed craft operated

Type	Name	Seats	Delivered
CAT Rosendal Verft 35 m	*Xunlong 1*	312	1996
CAT Batservice Sea Lord 38 m	*Xunlong 2*	350	1998
CAT Austal 39.9 m	*Tai Shan*	326	1993
CAT Wavemaster 34.6 m	*Peng Jiang*	223	1993
CAT Atlay Catamarans 24.5 m	*Ying Bin*	104	1993
CMCS Seacat 33 m	*Xunlong 3*	232	(2007)
CMCS Seacat 33 m	*Xunlong 4*	232	(2007)

Operations
She Kou to Hong Kong, Macaua and Hong Kong International Airport.

Croatia

Alpex

Splitska 2, HR-51000 Rijeka, Croatia

Tel: (+385 51) 32 54 60
Fax: (+385 51) 32 37 78
e-mail: alpex@alpex.hr
Web: www.alpex.hr

High-speed craft operated

Type	Name	Seats	Delivered
CAT Harding 28 m	*Nona Ana* (ex-*Lauparen*, ex-*Steutdik*)	202	1992
CAT Holen 21 m	*Nikolina*	96	1995

Operations
Dubrovnik to Polače
Dubrovnik to Mljet.

Atlas Turisticka Plovidba

Vukovarska 19, 20,000 Dubrovnik, Croatia

Tel: (+385 20) 44 22 22
Fax: (+385 20) 41 11 00; 44 26 45
e-mail: atlas@atlas.hr
Web: www.atlas-croatia.com

Owned by Croatian travel agency Atlas, Turisticka Plovidba operates two Russian Kometa and two Kolkhida hydrofoils and an Italian Monostab fast ferry for passenger and tourist services on the Adriatic Sea.

High-speed craft operated

Type	Name	Seats	Delivered
HYD Ord Kometa 35 m	*Krila Hvara*	116	1982
HYD Ord Kometa 35 m	*Krila Braca*	116	1982
HYD Ord Kolkhida 34.5 m	*Krila Istre* (ex-*Kamelija*)	145	1994
HYD Ord Kolkhida 34.5 m	*Krila Dalmacije* (ex-*Magnolija*)	145	1994
MH Rodriquez 46 m	*Princess of Dubrovnik*	330	1998
HYD Kometa	Krila Dubrovnica	116	1971
HYD Kometa	*Krila Kostrene*	116	1971
HYD Volgograd Katran	*Venice Carneval*	150	1996
HYD Volgograd Katran	*Adriatic Joy*	150	1995

Commodore Cruises

Riva 14, 52100 Pula, Croatia

Tel: (+385 52) 21 16 31
Fax: (+385 52) 21 90 91
e-mail: commodore-travel@pu.htnet.hr
Web: www.commodore-travel.hr

High-speed craft operated

Type	Name	Seats	Delivered
MH Marinteknik 42 m	*Dora*	351	1989

Operations
Pula – Venice, Trieste, Brijuni, Istratour.

G&V Line

Metohijska 2, 20000 Dubrovnik, Croatia

Tel: (+385 20) 313 401
Fax: (+385 20) 418 186
e-mail: gv-line@gv-line.hr
Web: www.gv-line.hr

High-speed craft operated

Type	Name	Seats	Delivered
CAT Yamaha 2910	*Paula*	200	1989
CAT Harding Werft 29 m	*Nona Ana*	200	1990

Operations
Zverinac, Bozava, Sestrunj, Rivanj, Zadar.
Zadar, Iz, Rava.
Dubrovnik, Sobra, Polace, Korcula, Lastovo.

Ivante

Obala Hrvatske mornarice bb, 2200 Sibenik, Croatia

Tel: (+385 22) 20 04 33
Fax: (+385 22) 20 04 34
e-mail: info@ivante.hr
Web: www.ivante.hr

High-speed craft operated

Type	Name	Seats	Delivered
SES Hovermarine 221	*Mislav*	108	1985

Jadrolinija

PO Box 123, Riva 16, 51000 Rijeka, Croatia

Tel: (+385 51) 66 61 11
Fax: (+385 51) 21 31 16
e-mail: passdept_e@adrolinija.hr
Web: www.jadrolinija.hr

High-speed craft operated

Type	Name	Seats	Delivered
CAT Westamaran 86	*Mediteran*	140	1978
CAT Westamaran 95	*Olea*	250	1994
CAT K.Fjellstrand 40 m	*Adriana* (ex-*Perdana Ekspress*)	356	1998
CAT Wavemaster 36 m	*Silba*	310	1998
CAT Marinteknik 41 m	*Judita* (ex-*Supercat 5*, ex-*Lusitano*)	306	1990
CAT Marinteknik 41 m	*Dubravka* (ex-*Supercat 8*, ex-*Magellan*)	306	1991
CAT Marinteknik 41 m	*Karolina* (ex-*Supercat 3*, ex-*Estrala Je Mer*)	306	1995
CAT Marinteknik 41 m	*Navolja* (ex-*Supercat 7*, ex-*Vasco Di Gama*)	306	1996

Operations
Zadar to Adriatic Islands (Ist and Molat Islands)
Split to Ancona
Zadar to Ancona.

Miatours

Vrata Sv Krsevana, HR-23000 Zadar, Croatia

Tel: (+385 23) 25 43 00
Fax: (+385 23) 25 44 00
e-mail: info@miatours.hr
Web: www.miatours.hr

The company has recently acquired a Fjellstrand 38.8 m catamaran to join its existing hydrofoils operating on routes between Italy and Croatia.

High-speed craft operated

Type	Name	Seats	Delivered
CAT Fjellstrand 38.8 m	*Prince Zadra*	272	1989
HYD Kometa	*Zman*	116	1978
HYD Kometa	*Zverinac*	116	1978

Operations
Božava and Zadar (Croatia) to Ancona (Italy)
Zadar to Sali to Zaglav.

Split Tours dd

Boktuljin Put BB, Croatia

Tel: (+385 21) 35 24 81
Fax: (+385 21) 35 24 82
e-mail: info@splittours.hr
Web: www.splittours.hr

High-speed craft operated

Type	Name	Seats	Delivered
CAT Westamaran 86	*Komiza*	136	1973
CAT Westamaran 86	*For*	136	1975
CAT Westamaran 86	*Broc* (ex-*Mons Calpe*)	136	1975
CAT Westamaran 86	*Milna*	140	1978

Operations
Split to Rogac, Milna, Stomorska and Vis
Excursions from Split.

Uto Kapetan Luka

Polijcka Cesta br 28, 21314 Krilo Jeseice, Croatia

Tel: (+385 21) 87 28 77
Fax: (+385 21) 87 28 77
e-mail: luka.tomic@st.htnet.hr
Web: www.krilo.hr

High-speed craft operated

Type	Name	Seats	Delivered
MH 36 m	*Krilo*	253	2006
CAT Cirrus 27	*Mala Lara*	184	1985
CAT Iris 6.2	*Krilo Jet*	398	2007

Operations
Split-Hvar-Korcula.

Cuba

Cubanacan Jet

Cubacan Jet is a joint venture between the state holding company Cubacan and the private Argentine-based Dodero Cruceros (Del Bene Group).

High-speed craft operated

Type	Name	Seats	Delivered
CAT Marinteknik 41 m	*Jet Kat Express II*	380	1991

Operations
Havana to Varadero.

Intermar Clarion Shipping Co Ltd

c/o Practicos de Cuba, Muelle Sierra Maestra No 2, San Pedro 1, Havana Vieja, Havana, Cuba

High-speed craft operated

Type	Name	Seats	Delivered
CAT Damex 30 m	*Rio Jucaro*	237	2006
CAT Damex 30 m	*Rio Las Casas*	237	2006

Operations
Batabano to Isla de Juventud.

Cyprus

Fergun Deniz Cilik

Yeni Linman Yolu Fergün, Apt No 1, Girne, Northern Cyprus

Tel: (+90 392) 815 17 70
Fax: (+90 392) 815 19 89
Web: www.fergun.net

High-speed craft operated

Type	Name	Seats	Delivered
CAT Westamaran 100	*Prince of Girne*	256	1981
CAT Westamaran 95	*Fergün Express III*	210	1982
CAT Marinteknik 34 m	*Emeraude Express*	243	1986

Operations
Girne to Tasucu.
Girne to Alanya.

Denmark

Bornholms Trafikken

Dampskibskajen 3-5, DK-3700, Ronne, Denmark

Tel: (+45) 56 95 18 66
Fax: (+45) 56 91 07 66
Web: www.bornholmstrafikken.dk

High-speed craft operated

Type	Name	Seats	Additional payload	Delivered
CAT Austal Auto Express 86 m	*Villum Clausen*	1,037	186 cars	1999

Operations
Ystad to Ronne.

Christiansøfarten

Ejnar Mikkelsensvej 25, DK-3760 Gudhjem, Denmark

Tel: (+455) 648 51 76
Fax: (+455) 648 55 77
e-mail: info@christiansoefarten.dk
Web: www.bornholmexpress.dk

Peter Strange Jensen, *Director*

High-speed craft operated

Type	Name	Seats	Delivered
MH Damen 41 m	*Bornholm Express*	245	2006

Operations
Bornholm to Christiansø and Simrishamn.

Mols-linien A/S

Faergehavnen, DK-8400 Ebeltoft, Denmark

Tel: (+45) 89 52 52 00
Fax: (+45) 89 52 52 90
e-mail: administration@mols-linien.dk
Web: www.mols-linien.dk

High-speed craft operated

Type	Name	Seats	Additional payload	Delivered
CAT Danyard Seajet 250	*Mai Mols*	450	120 cars	1996
CAT Danyard Seajet 250	*Mei Mols*	450	120 cars	1996
WPC InCat 91 m	*Max Mols** (ex-*Catlink V*)	900	240 cars	1998

*WPC InCat 91 m *Max Mols* (ex-*CatLink V*) leased to P&O Ferries.

Operations
Northern Denmark, Ebeltoft to Odden
Århus to Kalundborg.

Egypt

EL Salam Maritime Company

30 Yakoub Artten Street, Heliopolis, Cairo, Egypt

Tel: (+20 202) 290 85 35
Fax: (+20 202) 291 70 32
e-mail: info@elsalammaritime.com
Web: www.elsalammaritime.com

High-speed craft operated

Type	Name	Seats	Additional payload	Delivered
CAT Austal 56 m	*Fares Al Salam*	450	43 cars	2002
CAT InCat 78 m	*Elanora*	674	146 cars	1995

Operations
Safaga, Hurghada and Sharm El Sheik in Egypt to Duba, Saudi Arabia.

Fares Al Salam 0143598

International Fast Ferries

Hurghada, Egypt

Tel: (+20 65) 44 75 71
Fax: (+20 65) 44 75 73
e-mail: info@internationalfastferries.com
Web: www.internationalfastferries.com

High-speed craft operated

Type	Name	Seats	Additional payload	Delivered
CAT Fjellstrand 46 m	*Eid Travel*	308	12 cars	1998
CAT Jumbo Cat 60	*Nasser Travel*	450	49 cars	2000

Operations
Hurghada to Duba in Saudi Arabia
Hurghada to Sharm El Sheikh.

The Kværner Fjellstrand 46 m Flying Cat Eid Travel *for International Fast Ferries* 0045255

Nasser Travel 0081027

United Company for Marine Lines

2 Alsagadla Building Apartment, 802 15th May Road, Smoutha, Alexandria, Egypt

Tel: (+20 10) 510 28 40
Fax: (+20 20) 35 85 42 26

High-speed craft operated

Type	Name	Seats	Additional payload	Delivered
CAT Austal Auto Express 82	*Al Mattohedah I*	600	175 cars	1996
SES Cirrus 120P	*Red Sea Jet*	330	–	1990

Estonia

Nordic Jet Line AS

Sadama 25-4, EE-0001 Tallinn, Estonia

Tel: (+372) 613 70 00
Fax: (+372) 613 72 42
e-mail: njl@njl.ee
Web: www-eng.njl.fi

Götz Becker, *Managing Director*

High-speed craft operated

Type	Name	Seats	Additional payload	Delivered
CAT K. Fjellstrand Jumbo Cat	*Nordic Jet*	428	52 cars	1998
CAT K. Fjellstrand Jumbo Cat	*Baltic Jet*	428	52 cars	1999

Operations
Helsinki to Tallinn.

Nordic Jet 0024750

SuperSeaCat

A Sea Containers company
Narva Mnt 7, 10117 Tallinn, Estonia

Tel: (+372) 610 00 01
Fax: (+372) 610 00 10
Web: www.superseacat.com

Sea Containers began operating a SeaCat service between Sweden and Denmark in 1993. The service now operates between Finland and Estonia. In 2005 0.9 million passengers were carried on the Helsinki-Tallinn fast ferry.

High-speed craft operated

Type	Name	Seats	Additional Payload	Delivered
MH Fincantieri MDV 100 m	*Superseacat Three*	726	139 cars and 3 buses	1999
MH Fincantieri MDV 100 m	*Superseacat Four*	726	139 cars and 3 buses	1999

Operations
Helsinki to Tallinn (100 minutes).

Fiji

Beachcomber Island Resort & Cruises Ltd

1 Walu Street, Lautoka, PO Box 364, Fiji

Tel: (+679 6) 66 15 00
Fax: (+679 6) 66 44 96
e-mail: info@beachcomberfiji.com.fj
Web: www.beachcomberfiji.com

High-speed craft operated

Type	Name	Seats	Delivered
CAT Teknicraft 18 m	*Lagilagi*	150	2000

Operations
Port Denarau to Beachcomber Island.

South Sea Cruises Ltd

A subsidiary of Cruise Whitsundays
PO Box 718, Nadi, Fiji

Tel: (+679) 67 50 500
Fax: (+679) 675 05 01
e-mail: info@ssc.com.fj
Web: www.ssc.com.fj

Mark Fifield, *Proprietor*

High-speed craft operated

Type	Name	Seats	Delivered
CAT NQEA InCat, 26 m	*Tiger IV* (ex-*Taupo Cat*)	240	1993
MH BSC 25 m	*Yasawa Flyer*	188	2000
MH BSC 25 m	*Yasawa Flyer II*	267	2006

Operations
Nadi to Mamanuca Islands.

Finland

Linda Line Express

Part of the Lindaliini AS Group
Makasiiniterminaal, South Port, FIN-00140 Helsinki, Finland

Tel: (+358 9) 668 97 00
Fax: (+358 9) 66 89 70 70
e-mail: sales@lindaline.fi
Web: www.lindaline.fi

Enn Rohula, *Proprietor*

High-speed craft operated

Type	Name	Seats	Delivered
HYD Morye Olympia 36.6 m	*Jaanika*	186	1994
CAT Austal 52.6 m	*Merilin* (ex-*Cat No 1*)	432	1999

Operations
Tallinn to Helsinki, 1 hour 30 min.

Austal 52 m Merilin 0084081

Hydrofoil Jaanika *operated by Linda Line Express* 0038865

France

Bourbon Offshore

148 rue Sainte, F-13007 Marseilles, France

Tel: (+33 4) 91 13 08 00
Fax: (+33 4) 91 13 08 61
e-mail: boubon-offshore@bourbon-online.com
Web: www.bourbon-online.com

High-speed craft operated

Type	Name	Seats	Delivered
MH 53.6 m	*Bourbon Express*	50	–
MH 53.6 m	*Bourbon Ocean*	50	–
MH 53.6 m	*Surf Express*	50	–
MH 53.6 m	*Kemba*	50	–
MH 53.4	*Bourbon Bora*	80	–
MH Surfer 322	*Fifi*	90	–
MH Surfer 323	*Fifi*	90	–
MH Surfer 254	*Fifi*	70	–
MH Surfer 255	*Fifi*	70	–
MH Surfer 321 Combi	–	60	–
MH Surfer 200	*Fifi*	43	–
MH Surfer 320	*Fifi*	90	–

Operations
Offshore support, off Nigeria, Congo, Angola, Brazil and other areas.

Brittany Ferries

Port du Bloscon, F-29688 Roscoff, France

Tel: (+33 8) 25 82 88 28
Web: www.brittany-ferries.co.uk

High-speed craft operated

Type	Name	Seats	Additional payload	Delivered
WPC InCat 98 m	*Normandie Express*	850	235 cars	2000
WPC InCat 98 m	*Normandie Vitesse*	718	185 cars	1998

Operations
Portsmouth to Caen, 3 h 45 min
Portsmouth to Cherbourg, 3 h
Poole to Cherbourg, 2 h 15 min.

Compagnie Corsaire

Gare Maritime de la Bourse, St Malo, France

Tel: (+33 825) 138 110
Fax: (+33 825) 180 297
e-mail: info@compagniecorsaire.com
Web: www.compagniecorsaire.com

High-speed craft operated

Type	Name	Seats	Delivered
MH	*Jacques Cartier*	2003	2007

Operations
St Malo (France) to Jersey, 2 hours.

Compagnie Yeu Continent

3 av de l'Estacade, Fromentine, F-85550, La Barre-de-Monts, France

Fax: (+33 2) 51 49 59 70
e-mail: l.burgaud@compagnie-yeu-continent.fr
Web: www.compagnie-yeu-continent.fr

High-speed craft operated

Type	Name	Seats	Delivered
CAT Fjellstrand 46 m	*Pont D-Yeu*	431	2005
CAT Fjellstrand 46 m	*Le Châtelet*	431	2006

Operations
Fromentine to Iles d'Yeu.

Corsica Ferries/Sardinia Ferries

A member of the Tourship Group
5 Bis rue Chanoine Leschi, PO Box 239, F-20296 Bastia, Corsica, France

Tel: (+33 4) 95 32 95 95
Fax: (+33 4) 95 32 14 71
Web: www.corsicaferries.com

High-speed craft operated

Type	Name	Seats	Additional payload	Delivered
MH INMA/ Rodriquez 103 m	*Corsica Express II*	535	150 cars	1996
MH INMA/ Rodriquez 103 m	*Corsica Express III*	535	150 cars	1996
MH Rodriquez Aquastrada 103 m	*Sardinia Express*	535	150 cars	1996

Leased to Kallisti Ferries, Greece

Operations
Ajaccio, Bastia and Calvi to Nice
Golfo Aranci to Livorno
Savona to Bastia.

Manche Iles Express

Granville, rue des Iles, France

Tel: (+33 825) 13 30 50
e-mail: info@marine-iles-express.com
Web: www.manche-iles-express.com

High-speed craft operated

Type	Name	Seats	Delivered
CAT Flying Cat 35	*Victor Hugo* (ex-*Salten*)	199	1997
CAT Flying Cat 40	*Marin Marie* (ex-*Aremiti 3*)	350	1994

Operations
France to Channel Islands
Dielette to Guernsey, 1 hour
Dielette to Alderney, 45 minutes
Guernsey to Alderney, 1 hour.

SNCM (Société National Corse Mediterranée)

61 boulevard des Dames, F-13002 Marseille, France

Tel: (+33 441) 563200
Fax: (+33 441) 563 636
Web: www.sncm.fr

High-speed craft operated

Type	Name	Seats	Additional payload	Delivered
MH Corsaire 13,000	*NGV Liamone*	1,116	250 cars	2000

Operations
Nice to Ajaccio, 5 h 50 min
Nice to Balagne, 4 h.

Société Morbihannaise de Navigation

BP 1, F-56360 le Palais, Belle Ile-en-Mer, Brittany, France

Tel: (+33 297) 35 02 00
Fax: (+33 297) 31 56 81
e-mail: contact@snm-navigation.fr
Web: www.smn-navigation.fr

High-speed craft operated

Type	Name	Seats	Delivered
CAT Fjellstrand 40 m	*Locmaria 56*	376	1998
CAT Naval Aluminium Atlantique	*Dravanteg*	182	1991

Operations
Quiberon – Belle-Ile-en-Mer, Ile de Croix, Honat and Hoedic.

French Polynesia

Aremiti Pacific Cruises

BP 9274 Motu Uta, Papeete, Tahiti, French Polynesia

Tel: (+689) 50 57 57
Fax: (+689) 50 57 58
e-mail: aremiticruise@mail.pf
Web: www.aremiti.pf

High-speed craft operated

Type	Name	Seats	Additional payload	Delivered
CAT Fjellstrand 50 m	*Aremiti 4*	450	20 cars	2000
CAT Austal 56 m	*Aremiti 5*	650	30 cars	2004

Operations
Papeete to Moorea.

Aremiti 4 0103302

Bora Bora Navette Ltd

An Air Tahiti Company
French Polynesia

High-speed craft operated

Type	Name	Seats	Delivered
CAT NQEA RR150	*Motu Tapu o Ra*	115	1998
CAT NQEA RR150	*Vavau*	115	1998

Operations
Connecting Bora Bora airport with Vaitape, 15 mins.

Motu Tapu o Ra *and* Vavau 0045256

Société de Développement de Moorea

BP 635, Papeete 98713, Tahiti, French Polynesia

Tel: (+689) 50 11 11
Fax: (+689) 42 10 49
e-mail: reservation@mooreaferry.pf
Web: www.mooreaferry.pf

High-speed craft operated

Type	Name	Seats	Additional payload	Delivered
CAT	*Moorea Express*	306	–	2004

Operations
Papeete to Moorea.

Tahaa Transport Services

PO Box 612, 98730 Vaitape, Bora Bora, French Polynesia

Tel: (+689 67) 66 69
Fax:(+689 67) 66 69
e-mail: maupitiexpress@mail.pf

Gerald Sachet, *Principal*

High-speed craft operated

Type	Name	Seats	Delivered
MH 20 m	*Maupiti Express*	82	–

Operations
Bora Bora to Tahaa 1 hour 20 min
Bora Bora to Raiatea 1 hour 35 min
Bora Bora to Maupiti 1 hour 45 min.

Germany

AG EMS

PO Box 1154, Am Borkumkai, D-26691 Emden 1, Germany

Tel: (+49 21) 890 70
Fax: (+49 21) 890 74 05
e-mail: info@ag-ems.de
Web: www.ag-ems.de

High-speed craft operated

Type	Name	Seats	Delivered
CAT Fjellstrand 38.8 m	*Nordlicht*	272	March 1989
CAT Wavemaster 44 m	*Polarstern**	402	2001

*In 2007 this vessel was operated by Wilhelmshaven Helgoland Linie.

Operations
Emden to Borkum.

Nordlicht *operated by AG Ems* 0506731

Elbe City Jet

A Schnellfahren GmbH & Co KG
Ritscherstrasse 8, D-21706 Drochtersen, Germany

Tel: (+49 4148) 61 02 76
Fax: (+49 4148) 61 60 63
e-mail: info@elbe-city-jet.de
Web: www.elbe-city-jet.de

High-speed craft operated

Type	Name	Seats	Delivered
CAT Lindstøl 33 m	*Hansestar*	322	1997

Operations
Hamburg to Stade, Germany.

Förde Reederei Seetouristik GmbH

Postfach 26 26, D-24916 Flensburg, Germany

Tel: (+49 461) 86 40
Fax: (+49 461) 86 444
e-mail: info@frs.de
Web: www.frs.de

Götz Becker, *Managing Director*

High-speed craft operated

Type	Name	Seats	Additional payload	Delivered
CAT Fjellstrand 52 m	*Halunder Jet*	587	–	2003

Insel – und Halligreederei (Adler-Schiffe)

Boysenstrasse 13, D-25980 Westerland, Sylt, Germany

Tel: (+49 4651) 987 00
Fax: (+49 4651) 263 00
e-mail: info-sylt@adler-schiffe.de
Web: www.adler-schiffe.de

High-speed craft operated

Type	Name	Seats	Delivered
MH Marinteknik 42 m	*Adler Express*	420	1993

Wilhelmshaven Helgoland Linie GmbH & Co KG

Fax: (+49 21) 890 74 05
e-mail: info@helgolandlinie.de
Web: www.helgolandlinie.de

High-speed craft operated

Type	Name	Seats	Delivered
CAT Oceanfast 47 m	*Polarstern*	402	2001

Leased from AG Ems.

Operations
Borkum to Nordeney to Helgoland.

Greece

Aegean Shipping Company

46 Gr Lambrakis Str, Rhodes 85100, Greece

Tel: (+30 2241) 07 65 35
Fax: (+30 2241) 07 36 65
e-mail: info@seadreams.gr
Web: www.seadreams.gr

Paris Kakas, *Managing Director*

The Aegean Shipping Company is the longest established fast ferry operator in the Dodecanese Islands.

High-speed craft operated

Type	Name	Seats	Delivered
CAT Crowther 35 m	*King Saron*	500	1989
HYD Kometa	*Eolos*	116	1976

Operations
Rhodes- Marmaris, Symi, Lindos.

Aegean Speed Lines NE

A joint partnership between Sea Containers Ltd and the Eugenides Group
85 Vouliagmenis Avenue & Antheon Streets, GR-16674 Glyfada, Attiki, Greece

Tel: (+30 210) 969 09 50
Fax: (+30 210) 969 09 59
e-mail: info@aegeanspeedlines.gr
Web: aegeanspeedlines.gr

High-speed craft operated

Type	Name	Seats	Additional Payload	Delivered
WPC InCat 74 m	*Speedrunner I* (ex- *Hoverspeed Great Britain*)*	573	80 cars	1990
MH Fincantieri MDV 1200	*Speedrunner II* (ex- *Auto Express IV*)	600	170 cars	1996

*Leased from Sea Containers.

Operations
Piraeus to Sérifos, Sifnos, Mílos, Kythnos and Paros.
Piraeus to Paros and Santorini.

Agistri Piraeus Lines

Piraeus, Attiki, Greece

Tel: (+30 210) 411 31 08
Fax: (+30 22970) 91163

High-speed craft operated

Type	Name	Seats	Delivered
Westamaran 86	*Keravnos*	148	1972
Westamaran 100	*Keravnos II*	250	1982

Operations
Piraeus to Agistri Island, 55 minutes.

Anes-symi Lines

Afstralias 88T.K., Rodus, GR-85100, Greece

Tel: (+30 241) 03 77 69
Fax: (+30 241) 03 44 93
e-mail: anesymis@otenet.gr
Web: www.anes.gr

High-speed craft operated

Type	Name	Seats	Delivered
CAT Fjellstrand 38.8 m	*Panormitis* (ex-*Tallink Express I*, ex-*Sleipner*)	300	1989
HYD Meteor 35 m	*Aegli*	116	1980

Operations
Rhodes to Symi.

C-link Ferries

130 Kolokotroni Street, 18536 Pireaus, Greece

Tel: (+30 210) 459 40 96
Fax: (+30 210) 451 03 86

High-speed craft operated

Type	Name	Seats	Additional Payload	Delivered
MH Corsaire 11000	*Panagia Parou* (ex-*NGV Asco*)	530	148 cars	1988

Operations
Lavrio (Athens) to the Cyclades Islands
Chios and other islands in the Eastern Aegean.

Dodekanisos Naftiliaki NE

3 Aysralias, Rhodes, GR-85100, Greece

Tel: (+30 224) 705 90
Fax: (+30 224) 705 91
e-mail: info@12ne.gr
Web: www.12ne.gr

Christos Spanos, *Managing Director*

Launched in 2000, Dodekanisos Naftiliaki NE operates the only fast passenger/vehicle route between the Dodecanese islands.

High-speed craft operated

Type	Name	Seats	Additional payload	Delivered
CAT Båtservice Sea Lord 40 m	*Dodekanisos Express*	341	6 cars	2000
CAT Båtservice Sea Lord 40 m	*Dodekanisos Pride*	280	9 cars	2005

Operations
Rhodes to Symi, Panormitis, Kos, Kalymnos, Leros, Lipsi, Patmos and Kastelorizo.

Eagle Ferries

9 Himara St and Amarousiou Ave, 151 21 Marousii, Greece

Tel: (+30 210) 610 59 01
Fax: (+30 210) 610 48 88

High-speed craft operated

Type	Name	Seats	Additional payload	Delivered
WPC Kawaski/ AMD 1500	*Alkioni* (ex- *Hayabusa*)	460	94 cars	1995

Operations
Rafina to Andros, Tinos, Mykonos, Paros, Santorini and Heraklion.

Euroseas Shipping

Akti Tzeleph 3, Piraeus, Greece

Tel: (+30 210) 413 21 05
e-mail: enquiries@euroseas.com
Web: www.euroseas.com

High-speed craft operated

Type	Name	Seats	Delivered
CAT Ulstein UT 905	*Eurofast 1* (ex- *Ulstein Turbiner*)	267	1995

Operations
Piraeus to Poros, Spetses and Hydra.

GA Ferries

2 Etolikou Street, GR-18545 Piraeus, Greece

Tel: (+30 210) 419 91 00
Fax: (+30 210) 422 51 19
e-mail: ga@ferries.gr
Web: www.ferries.gr/gaferries

High-speed craft operated

Type	Name	Seats	Additional payload	Delivered
MH Mek 95 m	*Jetferry 1*	600	160 cars	1996

Hellenic Seaways

2 Aetolikou Street & Akti Kondili Avenue, GR-18545 Piraeus, Greece

Tel: (+30 210) 419 90 00
Fax: (+30 210) 413 11 11
e-mail: booking@dolphins.gr
Web: www.dolphins.gr/english/index.stm

Gerasimos Strintzis, *CEO*

High-speed craft operated

Type	Name	Seats	Additional Payload	Delivered
HYD S Ord Kometa-M	*Flying Dolphin X*	132	–	1978
HYD S Ord Kometa-M	*Flying Dolphin XV*	132	–	1981
HYD S Ord Kolkhida	*Flying Dolphin XVII*	155	–	1986
HYD S Ord Kolkhida	*Flying Dolphin XVIII*	155	–	1986
HYD S Ord Kometa	*Flying Dolphin XXI*	132	–	1988
HYD S Ord Kolkhida	*Flying Dolphin XIX*	155	–	1990
CAT K Fjellstrand Flying Cat	*Flyingcat 1*	336	–	1990
HYD S Ord Kolkhida	*Flying Dolphin XXIX*	155	–	1993
CAT Royal Schelde 70 HL	*Highspeed 1* (ex-*Captain George*)	620	156 cars	1996
CAT Austal 48 m	*Flyingcat 2* (ex-*Flying Dolphin 2000*)	516	–	1998
CAT FBM 53 m Tricat	*Flyingcat 4* (ex-*Sea Speed 1*)	447–	–	1999
CAT FBM 45 m Tricat	*Flyingcat 3* (ex-*Athina 2004*)	375	–	June 1998
CAT Austal 72 m	*Highspeed 2*	620	70 cars	2000
CAT Austal 72 m	*Highspeed 3*	620	70 cars	2000
CAT Austal 95 m	*Highspeed 4*	1,050	188 cars	2000
CAT Austal 85 m	*Highspeed 5*	810	154 cars	2005
FlyingCat 40 m	*Flyingcat 5* (ex-*Hansejet*)	342	–	1990
CAT FlyingCat 40 m	*Flyingcat 6*	341	–	1997

Operations
N.E. Aegean: From Piraeus to Syros, Paros, Mykonos, Eydilos, Karlovassi, Vathi, Clios and Myrtilini
Saronic Islands: From Pireaus to Methana, Poros, Hydra, Hermioni, Souvala, Aegina, Spetses and Porto Heli
Sporadic Islands: From Ag, Constantinos and Volos, to Skiathos, Skopelos, Glossa and Alonissos
Cyclades: From Rafina to Andros, Tinos and Mykonos;
from Heraklion (Crete) to Santorini, Ios, Paros, Naxos and Mykonos;
from Piraeus to Chania (Crete);
from Piraeus to Tinos, Syros, Mykonos, Paros, Serifos, Sifnos, Milos, Ios and Santorini.

Fjellstrand Flyingcat 1 *in service 1990 with Hellinic Seaways* 0506732

Iason Jets Shipping

71 Notara Street, GR-18535 Piraeus, Attiki, Greece

Tel: (+30 210) 412 10 01
Fax: (+30 210) 412 19 12
e-mail: superjet@hot.gr

High-speed craft operated

Type	Name	Seats	Delivered
Westamaran 4200	*Superjet*	400	1995
Westamaran 4200	*Superjet 2*	386	1998

Operations
Piraeus to Milos, Folgandos, Los and Santorini
Rafina to Tinos, Mykonos and Paros.

Ionian Cruises

4 Eth Antistaseos, 49100 Corfu, Greece

Tel: (+30 266) 103 86 90
Fax: (+30 266) 103 87 87
e-mail: info@ionian-cruises.com
Web: www.ioniancruises.com

High-speed craft operated

Type	Name	Seats	Delivered
HYD S Ord Kometa	*Santa 2*	124	1981
HYD S Ord Kometa	*Santa III*	124	1981

Operations
Corfu to Paxos, 1 hour.

Kallisti Ferries

25 Akti Miaouli Street, 185 35 Piraeus, Greece

e-mail: info@kallistiferries.gr
Web: www.kallistiferries.gr

High-speed craft operated

Type	Name	Seats	Additional payload	Delivered
MH INMA/ Rodriquez 103 m	*Corsica Express III**	535	150 cars	1996

*Leased from Corsica Ferries.

Operations
Piraeus to Ikaria and Samos.

Kiriacoulis Maritime

7 Alimou Avenue, Alimos, Athens, Greece

Tel: (+ 30 210) 875 52 46
Fax: (+30 210) 985 52 44
e-mail: maritime@kiriacoulis.com
Web: www.kiriacoulis.com

High-speed craft operated

Type	Name	Seats	Delivered
HYD Kometa	*Tzina II*	116	1973
HYD Kometa	*Georgias M*	116	1973
HYD Kometa	*Aristea M*	116	1973
HYD Kometa	*Marilena II*	116	1978
HYD Kometa	*Samos Flying Dolphin 2*	116	1980
HYD Kometa	*Marilena*	116	1981
HYD Kometa	*Tzina I*	116	1981

Operations
Inter Dodecanese Islands and in the Saronic Gulf
Piraeus to islands of the Saronic Gulf.

Laumzis Sun Cruises

62 Alikarnasson Street, GR-85300 Kos, Greece

Tel: (+30 22) 42 02 39 76
Fax: (+30 22) 42 04 81 63
e-mail: info@laumzis-cruises.gr

High-speed craft operated

Type	Name	Seats	Delivered
HYD Kometa	*Christos L*	116	1978
HYD Kometa	*Petros L*	116	1982

Operations
Kos-Bodrum, Rhodes, Symi.

Kometa Petros L 1120298

Nel Lines

47 Kountouriotou Street, GR-81100 Mytilini, Lesbos, Greece

Tel: (+30 22) 51 02 62 12
e-mail: nellines@internet.gr
Web: www.nel.gr

High-speed craft operated

Type	Name	Seats	Additional payload	Delivered
MH Corsaire 12000	*Aeolos Kenteris I*	1,000	210 cars	2000
MH Corsaire 10000	*Aeolos Kenteris II*	854	190 cars	2001
MH Corsaire 11000	*Panagia Thalassini*	574	148 cars	1996

Operations
Aeolos Kenteris I: Piraeus to Lesbos, Syros, Tinos, Mykonos; Piraeus to Kea, Ikarina, Samos
Aeolos Kenteris II: Piraeus to Paros, Ios, Santorini, Amargos; Piraeus to Syros, Mykonos
Panagia Thalassini: Lavrio to Syros, Tenos and Mykonos.

Paxos Hydrofoil Shipping

144 Kolokotroni Street, Piraeus, Greece

Paxos Hydrofoil Shipping has introduced a Kometa Hydrofoil on services between the islands of Paxos and Corfu. Nine return crossings are scheduled each week.

High speed craft operated

Type	Name	Seats	Delivered
HYD S Ord Kometa	*Paxos Flying Dolphin*	132	1974

Operations
Paxos to Corfu.

Planet Seaways

Tel: (+30 237) 405 1115
Fax: (+30 237) 405 1116
e-mail: info@planetseaways.gr
Web: www.planetseaways.gr

High-speed craft operated

Type	Name	Seats	Delivered
HYD Kolkhida Flying Dolphin	*Vardaris* (ex- *Delfini V*)	140	1991

Samos Hydrofoils

4 Sofouli Themistoki, 83100 Samos, Greece

Tel: (+30 273) 25065
e-mail: nkatr2@otenet.gr
Web: www.diavlos.gr/samos/by-ship/hydrofoil.html

High-speed craft operated

Type	Name	Seats	Delivered
HYD Kometa	*Samos Flying Dolphin I*	116	1980

Operations
Samos to Ephesus, Ikaria, Fourni, Patmos, Lipsi, Leros, Kalimnos, Kos and Agathonisi.

Saronic Dolphins

2 Gounari Street, GR-18531 Piraeus, Greece

Tel: (+30 210) 422 49 80
Fax: (+30 210) 417 35 59
Web: www.ferries.gr/saronic

High-speed craft operated

Type	Name	Seats	Delivered
HYD Kometa	*Delfini II*	102	1992

Operations
Daily departures from Heraklion (Crete).

Tilos 21st Century Shipping

Megalo Chorio, GR-85002 Tilos, Greece.

Tel: (+30 22) 46 04 40 00
Fax: (+30 22) 46 04 40 44
e-mail: sea-star@otenet.gr

Tilos 21st Century Shipping operates the fastest ferry in the Dodecanese islands for the local government of Tilos.

High-speed craft operated

Type	Name	Seats	Delivered
SES Cirrus 120P	*Sea Star*	320	1990

Operations
Tilos – Rhodes, Nissiros and Fethige.

Vasilopoulos Hydrofoils

5 Akti Miaouli, 185 31 Piraeus, Greece

Tel: (+30 210) 417 2561
Fax: (+30 210) 417 2562
email: vlines1@otenet.gr

High-speed craft operated

Type	Name	Seats	Delivered
HYD Kometa	*Samos Flying Dolphin III*	116	1980
HYD Kometa	*Samos Flying Dolphin IV*	116	1978

Operations
Athens to Aegina.

Guadeloupe

Comatrile

8 Residence Darse, Wuai Gatine, 97100 Pointe-a-pitre, Guadeloupe

Tel: (+590 590) 91 02 45
Fax: (+590 590) 82 57 73
e-mail: comatrile@wanadoo.fr

High-speed craft operated

Type	Name	Seats	Delivered
MH CNB VoyagerTM 18W5	*Iguana Sun*	105	2000
MH CNB Voyager BAT 27	*Iguana Beach*	206	1999

Operations
Saint Louis to Terre de Haut.

L'Express Des Iles

9 Immeuble Darse, Quai Gatine, F-97110 Pointe à Pître, Guadeloupe

Tel: (+590) 91 52 15
Fax: (+590) 91 11 05
e-mail: rbellemare@express-des-iles.com
Web: www.express-des-iles.com

R Bellemare, *President*
B Moeson, *Vice-President*

High-speed craft operated

Type	Name	Seats	Additional	Delivered
CAT Austal 45 m	*Gold Express*	446	–	2005
CAT Austal 45 m	*Silver Express*	360	10	2005

Operations
Guadeloupe to Dominica, Martinique, Marie-Galante, Les Saintes and Saint Lucia
The company carries over 600,000 passengers per year.

Gold Express *and* Silver Express 1120263

Société de Transports Maritimes Brudey Frères

78 Centre St John Perse, F-97110 Pointe à Pître, Guadeloupe

Tel: (+590) 590 90 04 48
Fax: (+590) 590 82 15 62
e-mail: brudey-freres@wanadoo.fr
Web: www.brudey-freres.fr

Brudey Doenis, *Shipowner*
Vala Claude, *Commercial Director*

High-speed craft operated

Type	Name	Seats	Delivered
MH CN 34.6 m	*Tropic*	253	1988
CAT Westamarin 3700S 37 m	*Maria*	320	1990

CAT Westamarin 27 m	*Flycat* (ex-*Kogelwieck*)	140	1992
SES UT 928	*Atlantica*	340	1995
CAT Westamarin 95	*Acajou* (ex-*Cythere*, ex-*Pegasus*)	220	1997
CAT Marinteknik	*Acacia*	400	1993

Operations
Pointe à Pître to Dominique to Port de France, 3 hours 30 minutes
Pointe à Pître to Les Saintes, 21 n miles, 45 minutes
Pointe à Pître to Marie Galante, 25 n miles, 55 minutes.
The company carries over 400,000 passengers each year.

Honduras

Safeway Maritime Transportation Company

c/o John McNab, 624 Hesper Ave, Metaire, Los Angeles 70005, United States

Tel: (+504) 445 17 95
e-mail: galaxy@yahoo.com

High-speed craft operated

Type	Name	Seats	Delivered
CAT InCat Crowther 46 m	*Galaxy Wave II*	460	2006

Operations
La Ceiba to Isla de Roatán, 1 hour 15 minutes.

Hong Kong

China Merchant Development Company

152 Connaught Road, Central, Hong Kong

High-speed craft operated

Type	Name	Seats	Delivered
SES Hovermarine HM 218	*Yin Bin 1*	84	1986
SES Hovermarine HM 218	*Yin Bin 2*	84	1986
CAT Mitsui CP 20	*Yin Bin 3*	250	1987
SES Huangpu MARIC 7211	*Yin Bin 4*	162	1993
CAT Cougar 32 m	*Yin Bin 5*	–	1993
CAT Fjellstrand 38.8 m	*Heng He*	–	1994
SES Maric	*Yin Bin 6*	–	1998

Operations
Shekou to Hong Kong Central.
Fu Yong to Hong Kong Central.

Chu Kong Passenger Transport Company Ltd

10/F Chu Kong Shipping Tower, 143 Connaught Road, Central, Hong Kong

Tel: (+852) 28 59 15 81
Fax: (+852) 25 40 19 29
e-mail: enquiry@cksp.com.hk
Web: www.cksp.com.hk

Chu Kong Passenger Transport Company Ltd acts as the General Manager in Hong Kong for many transportation companies in the ports around Guangdong Province. These transportation companies include:
Humen Longwei Passenger Transport Co Ltd
Jiangmen Hong Kong Macau Joint Passenger Transportation
Pan Yu Lian Hua Shan Port Passenger Transport Co Ltd
Shenzhen Shekou Port
Shenzhen Xunlong Shipping Co
Shunde Harbour
Zhaoquing Hong Kong Transportation Co Ltd
Zhong Shan Hong Kong Passenger Shipping Co Ltd
Zhuhai Rapid Passenger Liner Company Ltd

High-speed craft operated

Type	Name	Seats	Delivered
CAT Fjellstrand 31.5 m	*Quong Zhou Yi Hau* (ex- *Li Wan Hu*)	289	1984
CAT Fjellstrand 31.5 m	*Hai Tian* (ex-*Peng Lai Hu*)	291	1985

Shun Jing *operated by CKS in Hong Kong* (H & L Van Ginderen) 0058916

Xin Duan Zhou *operated by CKS in Hong Kong* (H & L Van Ginderen) 0058917

CAT Precision Marine 40 m	*Shun Feng*	352	1988
CAT Italthai Marine 32.5 m	*Dou Men*	252	1990
CAT Austal 38 m	*Hai Wei* (ex-*Shun Shui*)	354	1991
CAT Austal 40 m	Qi Jiang (ex-*Xin Duan Zhou*)	338	1992
CAT Austal 40 m	*Hai Bin* (*ex-Shun De*)	354	1992
CAT Austal 40 m	*Hai Zhu*	338	1992
CAT Austal 40 m	*Nan Gui*	338	1992
CAT Austal 40 m	*Lian Xing* (ex-*Kai Ping*)	368	1992
CAT Austal 40 m	*Lian Shan Hu*	338	1992
CAT Austal 40 m	*Cui Heng Hu*	338	1992
CAT Austal 40 m	*Hai Chang*	338	1993
CAT Crowther 35 m	*Dong Tai An*	250	1993
CAT Austal 40 m	*Xing Zhong* (ex-*Hui Yang*)	368	1993
CAT Austal 40 m	*Tai Shan*	338	1993
CAT Austal 40 m	*Gang Zhou*	318	1993
CAT Austal 40 m	*Gao Ming*	338	1993
CAT Austal 40 m	Shun Shui (ex-*Gui Feng*)	313	1993
CAT Austal 40 m	*San Bu*	368	1993
CAT Wave Master 34.5 m	*Peng Jiang*	193	1993
CAT Austal 40 m	*Shun Jing*	355	1993
CAT Austal 40 m	*Lian Gang Hu*	355	1994
CAT Austal 40 m	*Yi Xian Hu*	355	1994
CAT Austal 40 m	*Xin He Shan*	300	1994
CAT Austal 40 m	*Zhong Shan*	355	1994
CAT Austal 40 m	*Shun De*	–	1995
CAT Austal 40 m	*Hai Yang*	–	1995
CAT Austal 40 m	*Zeng Cheng Yi Hao*	268	1996
CAT Austal 40 m	*Tai Jian*	318	1997
CAT Austal 40 m	*Nan Hua*	–	1997
CAT Austal 42 m	*Zhao Quin*	352	1997

Vessels are ordered through the subsidiary company Yuet Hing Marine Supplies Company Ltd.

Customs and Excise Department

8/F Harbour Building, 38 Pier Road, Central, Hong Kong

Tel: (+852) 28 15 77 11
Fax: (+852) 25 42 33 34
e-mail: customsenquiries@customs.gov.hk
Web: www.customs.gov.hk

Richard Yuen, *Commissioner*

High-speed craft operated

Type	Name
MH Chung Wah Shipbuilding and Engineering Co Ltd Damen Mk III Class Yard No 204	*Customs 6 Sea Glory*

MH Chung Wah Shipbuilding and Engineering Co Ltd Damen Mk III Class Yard No 205	*Customs 5 Sea Guardian*
MH Chung Wah Shipbuilding and Engineering Co Ltd Damen Mk III Class Yard No 206	*Customs 2 Sea Leader*
MH Wang Tak Challenge	*Customs 8 Sea Reliance*
MH Wang Tak Challenge	*Customs 9 Sea Fidelity*

Discovery Bay Transportation Services Ltd

Subsidiary of Hong Kong Resort International Ltd
Discovery Bay Sales Office, Shop 140, DB Plaza, Discovery Bay, Lantau, Hong Kong

Tel: (+852) 29 87 62 33
Fax: (+852) 29 87 91 11
e-mail: corporate.affairs@hkri.com
Web: www.hkri.com

Eric Chu, *Executive Director*

High-speed craft operated

Type	Name	Seats	Delivered
SES Hovermarine 218	*RTS 201*	104	1981
SES Hovermarine 218	*RTS 203*	104	1981
MH Marinteknik (S) 35	*Discovery Bay 19*	306	1990
MH Marinteknik (S) 35 MPV	*Discovery Bay 20*	300	1990
MH Marinteknik (S) 35	*Discovery Bay 21*	306	1992
MH Marinteknik (S) 35	*Discovery Bay 22*	306	1993
CAT Marinteknik (S) 42 m	*Discovery Bay 1*	500	1995
CAT Marinteknik (S) 42 m	*Discovery Bay 2*	500	1995
CAT Marinteknik (S) 42 m	*Discovery Bay 3*	500	1996
CAT Marinteknik (S) 43 m	*Discovery Bay 5*	500	1997
CAT Marinteknik (S) 43 m	*Discovery Bay 6*	500	1997
CAT Marinteknik (S) 43 m	*Discovery Bay 7*	500	1998
CAT Marinteknik (S) 26 m	*DB Support*	60	1999
CAT Marinteknik (S) 43 m	*Discovery Bay 8*	500	1999

Operations
Discovery Bay, Lantau to Central, Hong Kong, 9 n miles, 25 minutes
Approximately 39,000 sailings per year.

Discovery Bay 1, 2 *and* 3 0507092

Discovery Bay 8 1048840

DB Support 1048839

Hong Kong & Kowloon Ferry Ltd

Pier 4, New Reclamation, Central District, Hong Kong

Tel: (+852) 28 15 60 63
Fax: (+852) 28 15 62 63
email: info@hkkf.com.hk
Web: www.hkkf.com.hk

Incorporated in 1998, the company is licensed to operate a wide range of routes in the Hong Kong area.

High-speed craft operated

Type	Name	Seats	Delivered
CAT 12 m	*Sea Smooth*	100	–
CAT 22 m	*Sea Splendid*	180	–
CAT 18 m	*Sea Sprint*	170	–
CAT Cheoy Lee/Crowther 28	*Sea Supreme*	388	2003
CAT Cheoy Lee/Crowther 28	*Sea Superior*	–	–

Operations
Central to Yung Shue Wan
Central to Sok Kwa Wan
Aberdeen to Yung Shue Wan.

Hong Kong Police Force, Marine Region

Marine Police Regional HQ, Tai Hong Street, Sai Wan Ho, Hong Kong

Tel: (+852) 28 03 62 00
Fax: (+852) 28 86 49 95
e-mail: marinesb@police.gen.gov.hk
Web: www.police.gov.hk

Chang Mo-See, *Regional Commander*
A Kerrigan, *Deputy Regional Commander*
Lau Yan-Sang, *Superintendent Support Bureau*

High-speed craft operated

Type	Name
MH Cheoy Lee Shipbuilding 30 m Keka Class	*PL 60*
MH Cheoy Lee Shipbuilding 30 m Keka Class	*PL 61*
MH Cheoy Lee Shipbuilding 30 m Keka Class	*PL 62*
MH Cheoy Lee Shipbuilding 30 m Keka Class	*PL 63*
MH Cheoy Lee Shipbuilding 30 m Keka Class	*PL 64*
MH Cheoy Lee Shipbuilding 30 m Keka Class	*PL 65*
MH Chung Wah Shipbuilding 26.5 m Mk III	*PL 70 King Lai*
MH Chung Wah Shipbuilding 26.5 m Mk III	*PL 71 King Yee*
MH Chung Wah Shipbuilding 26.5 m Mk III	*PL 72 King Lim*
MH Chung Wah Shipbuilding 26.5 m Mk III	*PL 73 King Hau*
MH Chung Wah Shipbuilding 26.5 m Mk III	*PL 74 King Dai*
MH Chung Wah Shipbuilding 26.5 m Mk III	*PL 75 King Chung*
MH Chung Wah Shipbuilding 26.5 m Mk III	*PL 76 King Shun*
MH Chung Wah Shipbuilding 26.5 m Mk III	*PL 77 King Tak*
MH Chung Wah Shipbuilding 26.5 m Mk III	*PL 78 King Chi*
MH Chung Wah Shipbuilding 26.5 m Mk III	*PL 79 King Tai*
MH Chung Wah Shipbuilding 26.5 m Mk III	*PL 80 King Kwan*
MH Chung Wah Shipbuilding 26.5 m Mk III	*PL 82 King Yan*
MH Chung Wah Shipbuilding 26.5 m Mk III	*PL 83 King Yung*
MH Chung Wah Shipbuilding 26.5 m Mk III	*PL 84 King Kan*
MH Transfield Shipbuilding 32.6 m	*PL 51 Protector*
MH Transfield Shipbuilding 32.6 m	*PL 52 Guardian*
MH Transfield Shipbuilding 32.6 m	*PL 53 Defender*
MH Transfield Shipbuilding 32.6 m	*PL 54 Preserver*
MH Transfield Shipbuilding 32.6 m	*PL 55 Rescuer*
MH Transfield Shipbuilding 32.6 m	*PL 56 Detector*

26.5 m Chung Wah Patrol Boat Mk III King Shun 0507002

High Speed Interceptor, Damen Sea Stalker 1500 Class 1109896

This organisation also operates a smaller range of high-speed craft including five anti-smuggling 60 kt craft (PL85–89), 11 Seaspray Inshore Patrol Launches, four Seaspray Logistics Launches and seven Boston whalers. An additional six high-speed inshore patrol craft were delivered by Cheoy-Lee Shipyards Ltd in 2000 (PL 40 to 45) with an additional six launches for 2001/2003 (PL 60, 61, 63, 65, 67 and 68). The KEKA Patrol craft were delivered in 2002 and 2003 replacing Damen Mk 1 class patrol craft.

New World First Ferry Services Ltd

71 Hing Wah Street, Lai Chi Kok, Kowloon, Hong Kong

Tel: (+852) 21 31 88 38
Fax: (+852) 21 31 88 77
Web: www.nwff.com.hk

John CY Hui, *Director and General Manager*
Donna Lai, *Communications*

In 2003 the company operated a service co-operation with Turbojet on the Macau route.

High-speed craft operated

Type	Name	Seats	Delivered
CAT AFAI	*First Ferry I*	223	1996
CAT AFAI	*First Ferry II*	220	1999
CAT Marinteknik	*First Ferry III*	403	2001
CAT Marinteknik	*First Ferry V*	403	2001
CAT Marinteknik	*First Ferry VI*	403	2001
CAT Marinteknik	*First Ferry VII*	403	2001
CAT Marinteknik	*First Ferry VIII*	403	2002
CAT Kvaerner Fjellstrand 40 m	*New Ferry II* (ex-*HKF II*)	432	1993
CAT Kvaerner Fjellstrand 40 m	*New Ferry III* (ex-*HKF III*)	432	1994
CAT Damen DDF 4010	*New Ferry V*	411	2000
CAT Damen DDF 4010	*New Ferry VI*	405	2001
Austal 47.5 m	*New Ferry LXXX1*	414	2002
CAT Austal 47.5 m	*New Ferry LXXX11*	414	2002
CAT Austal 47.5 m	*New Ferry LXXX111*	414	2002
CAT Austal 47.5 m	*New Ferry LXXXV*	418	2004
CAT Austal 47.5 m	*New Ferry LXXXVI*	418	2004
CAT Wang Tak/ Crowther 28 m	*First Ferry IX*	231	2003
CAT Wang Tak/ Crowther 28 m	*First Ferry X*	231	2003
CAT Wang Tak/ Crowther 28 m	*First Ferry XI*	231	2003
CAT Austal 47.5 m	–	418	(2008)
CAT Austal 47.5 m	–	418	(2008)

First Ferry IX, X and XI 1020477

First Ferry VII 0143594

Operations
Domestic routes: *First Ferry I–XI*
Hong Kong (China Ferry Terminal) to Macau, 1 hour 15 minutes: *New Ferry I–LXXXVI.*

Nansha Ferry Co Ltd

58th Floor, Bank of China Tower, 1 Garden Road, Central, Hong Kong

High-speed craft operated

Type	Name	Seats	Delivered
HYD Supramar PTS 75 Mk III	*Nan Sha 1*	100	1977
CAT WaveMaster 39 m	*Nan Sha No 11*	300	1992
CAT WaveMaster 42 m	*Nan Sha No 18*	386	1994
CAT WaveMaster 42 m	*Nan Sha No 28*	380	1994
CAT NewTech 42 m	*Nan Sha No 38*	380	1996
CAT NewTech 42 m	*Nan Sha No 68*	380	1998

Operations
Kowloon to Nansha
Nansha to Hong Kong Central.

Park Island Transportation Company

1/F Park Island Ferry Pier, Pak Lai Road, Ma Wan, NT, Hong Kong

Tel: (+852) 29 46 88 88
Fax: (+852) 25 25 55 56
e-mail: enquiry@pitcl.com.hk
Web: www.pitcl.com.hk

This company was set up in 2001 by Sun Hung Kai Properties to manage transportation at the Ma Wan Island residential and retail complex.

High-speed craft operated

Type	Name	Seats	Delivered
CAT Marinteknik 38 m	*Park Island 1*	402	2002
CAT Marinteknik 38 m	*Park Island 2*	402	2003
CAT Marinteknik 38 m	*Park Island 3*	402	2003
CAT Marinteknik 38 m	*Park Island 5*	402	2003
CAT Cheoy Lee/Crowther 28 m	*Park Island 6**	398	2003
CAT Cheoy Lee/Crowther 26 m	*Park Island 7*	–	2004
CAT Cheoy Lee/Crowther 26 m	*Park Island 8*	–	2004

*Leased from Hong Kong & Kowloon Ferry Company.

Operations
Park Island to Central
Park Island to Tsuen Wan.

Turbojet

Turbojet Sea Express
Member of Shun Tak-China Ship Management Ltd
83 Hing Wah Street West, Lai Chi Kok, Hong Kong

Tel: (+852) 23 07 08 80
Fax: (+852) 27 86 51 25
e-mail: com@turbojet.com.hk
Web: www.turbojet.com.hk, www.turbojetseaexpress.com.hk

Stanley Ho, *Group Executive Chairman*
Shujian Che, *Vice Chairman*
Pansy Ho, *CEO and Executive Director*
Jenning Wang, *Executive Director*

Anna Hong, *Deputy General Manager*
Alan Wong, *Commercial Director*
Tsang Man Ching, *Operations Director*

Shun Tak–China Travel Ship Management Ltd is a joint venture which operates one of the largest high-speed ferry operations in Asia. The company provides a ferry service for over 11 million passengers per annum between Hong Kong and key Pearl River Delta destinations. It is also the only operator to provide a 24-hour ferry service between Hong Kong and Macau. In 2003 the company introduced Turbojet Sea Express, which connects three major airports: Hong Kong International Airport, Shenzhen Baoan International Airport and Macau International Airport. Passengers can choose to transit into or out of the Pearl River Delta region at the cross-boundary passenger ferry terminal "Sky Pier" located inside Hong Kong International Airport without the need to pass through Hong Kong immigration and customs formalities.

The company operates a fleet of 33 high-speed passenger vehicles including 17 Jetfoils, two Foilcats, 10 Tricats and four Flying Cats.

High-speed craft operated
STEC

Type	Name	Seats	Delivered
HYD Boeing Jetfoil 929-100	*Madeira*	240	1974
HYD Boeing Jetfoil 929-100	*Santa Maria*	240	1975
HYD Boeing Jetfoil 929-100	*Flores* (ex-*Kalakaua '78*)	240	1974
HYD Boeing Jetfoil 929-100	*Corvo* (ex-*Kamehameha*)	240	1975
HYD Boeing Jetfoil 929-100	*Pico* (ex-*Kuhio*)	240	1975
HYD Boeing Jetfoil 929-100	*Saõ Jorge* (ex-*Jet Caribe I '80* , ex-*Jet de Oriente '78*)	240	1976
HYD Boeing Jetfoil 929-100	*Acores* (ex-*Jet Caribe II*)	240	1980
HYD Boeing Jetfoil 929-100	*Ponte Delgada* (ex-*Flying Princess II*)	240	1981
HYD Boeing Jetfoil 929-115	*Terceira* (ex-*Normandy Princess*)	240	1979
HYD Boeing Jetfoil 929-100	*Urzela* (ex-*Flying Princess*)	240	1976
HYD Boeing Jetfoil 929-115	*Funchal* (ex-*Jetferry One '83*)	240	1979
HYD Boeing Jetfoil 929-115	*Horta* (ex-*Jetferry Two*)	240	1980
HYD Boeing Jetfoil 929-115	*Lilau* (ex-*HMS Speedy*)	240	1979
HYD Boeing Jetfoil 929-100	*Guia* (ex-*Okesa*)	240	1976
HYD Boeing Jetfoil 929-115	*Taipa* (ex-*Princesa Guaeimara*)	240	1981
HYD Boeing Jetfoil 929-115	*Cacilhas* (ex-*Princesa Guayarmina*)	240	1980
HYD China Shipbuilding PS-30	*Balsa*	260	1994
F-CAT Fjellstrand Foilcat 35 m	*Barca*	419	1995
F-CAT Fjellstrand Foilcat 35 m	*Penha*	419	1995
CAT K Fjellstrand (S) FlyingCat 40 m	*Universal Mk I*	303	1992
CAT K Fjellstrand (S) FlyingCat 40 m	*Universal Mk II*	303	1993
CAT K Fjellstrand (S) FlyingCat 40 m	*Universal Mk III*	303	1993
CAT K Fjellstrand (S) FlyingCat 40 m	*Universal Mk IV*	303	1994
CAT FBM Tricat 45 m	*Universal Mk 2001*	303	1995
CAT FBM Tricat 45 m	*Universal Mk 2002*	303	1995
CAT FBM Tricat 45 m	*Universal Mk 2003*	303	1995
CAT FBM Tricat 45 m	*Universal Mk 2004*	303	1996
CAT FBM Tricat 45 m	*Universal Mk 2005*	303	1996
CAT FBM Tricat 45 m	*Universal Mk 2006*	333	1996
CAT FBM Tricat 45 m	*Universal Mk 2007*	333	1997
CAT FBM Tricat 45 m	*Universal Mk 2009*	328	1997
CAT FBM Tricat 45 m	*Universal Mk 2008*	333	1998
CAT FBM Tricat 45 m	*Universal Mk 2010*	328	2005

A Flying Cat 40 m operated by Turbojet 1143961

Barca, *Fjellstrand Foilcat 35 m* 1143962

Horta, *Boeing JetFoil 929-115* 1143963

Operations
Turbojet service routes:
Hong Kong to Macau, 24-hour service departing every 15 minutes.
Hong Kong to Shenzhen, 15 return sailings per day and 13 trips at weekends.
Shenzen to Macau, five return sailings per day, increasing to seven trips at weekends.
TurboJet Sea Express routes:
Hong Kong International Airport to Macau, nine sailings per day.
Hong Kong International Airport to Shenzhen, 18 sailings per day and 16 at weekends.

Hungary

Mahart Passnave Passenger Shipping Ltd

1056 Budapest, Belgrád rakpart, Nemzetközi Hajóállomás, Hungary

Tel: (+36 1) 484 40 08
Fax: (+36 1) 266 63 70
e-mail: hydrofoil@mahartpassnave.hu
Web: www.mahartpassnave.hu

Zoltán Kovács, *Managing Director*
István Kautz, *Marketing Manager*
Viktória Pálffy, *Hydrofoil Team Manager*

MAHART Passnave Ltd introduced three Raketa hydrofoils on the River Danube between Hungary and Austria in 1962. Since the mid-1970s the Raketas have been replaced by Meteor and Voskhod hydrofoils. In 1993, four Polesye-type vessels were introduced on the same routes, these craft have a very low draught of 0.40 m on foils. The cruise speed of all vessels is between 38 and 42 kt. MAHART carries between 30,000 and 40,000 passengers by hydrofoil each year. Hydrofoils and traditional boats are also available for private hire.

High-speed craft operated

Type	Name	Seats	Delivered
HYD Sormovo Voskhod	*Vöcsök II*	71	1985
HYD Sormovo Voskhod	*Vöcsök III*	71	1986
HYD Sormovo Voskhod	*Vöcsök IV*	71	1987
HYD Sormovo Meteor	*Solyom II*	108	1988
HYD Polesye	*Bíbic I*	54	1992
HYD Polesye	*Bíbic II*	54	1992
HYD Meteor	*Solyom III*	108	1992
HYD Polesye	*Bíbic III*	26	1993

Vöcsök III *operated by MAHART* 0507003

Type	Name	Seats	Delivered
HYD Polesye	*Bíbic IV*	54	1993
–	(ex-*Prins William Alexander*)	–	–

Operations
Between Budapest and Vienna (via Bratislava), 282 km, 5½ or 6½ hours (downstream or upstream), April to November, daily.
Between Budapest and Visegrád, May to September, weekends.
Between Budapest and Esztergom via Vác and Nagymaros, May to September, weekends.
Special cruises to Komárom, Solt, Kalocsa and Mohács.

India

Development Consultants Ltd

24B Park Street, Calcutta 700 016, India

Tel: (+91 33) 226 76 01
Fax: (+91 33) 22 49 23 40
e-mail: dcl@dclgroup.com
Web: www.dclgroup.com

High-speed craft operated

Type	Name	Seats	Delivered
CAT SBF 28 m	*Rishi Aurobindo*	132	1996

Operations
Calcutta to Haldia return.

Samudra Link Ferry Shipping

3-A, 6th Floor, New Excelsior Building, Wallace Street, Fort, Mumbai 400001, India

Tel: (+91 22) 56 33 69 45
Fax: (+91 22) 22 07 13 96
e-mail: samlink@usnl.net
Web: www.sam-link.com

Kirti Verdhman Gupta, *Managing Director*

A subsidiary of the US-based Sam Lire Group, the company has introduced a Marinteknik Verkstads 33 m catamaran to link Mumbai with Panajim (Goa).

High-speed craft operated

Type	Name	Seats	Delivered
CAT Marinteknik 33 m	*Arabian Sea King* (ex *Ciudad de Coloma*)*	216	1985

*This vessel is currently not in operation

Operations
Mumbai to Panajim (Goa), 8 hours.

Shipping Corporation of India

Shipping House, 13th Floor, Mumbai 400021, India

Tel: (+91 22) 202 66 66
Fax: (+91 22) 202 69 05
Web: www.shipindia.com

High-speed craft operated

Type	Name	Seats	Delivered
CAT Tille Shipyards 31.9 m	*Khadeeja Beevi*	100	1990
CAT Tille Shipyards 31.9 m	*Hameedath Bee*	100	1990
CAT NGA 35	*Parali I*	154	2007
CAT NGA 35	*Cheriyapani I*	154	2007
CAT NGA 35	*Valiapani I*	154	2007

Operations
Lakshadweep area of Southwest India.

Indonesia

Aquaria Shipping PT

A subsidiary of Svitzer Wijsmuller
Jl. Tarakan No 35, Balikpapan 76111, Indonesia

Tel: (+62 542) 42 46 24
Fax: (+62 542) 42 17 07

High-speed craft operated

Type	Name	Seats	Additional payload	Delivered
MH Strategic 30 m	*Kaltim Maju*	50	40 t	2002
MH Strategic 30 m	–	50	40 t	2002
MH Strategic 30 m	–	50	40 t	2002

Operations
Offshore support in the Makassar Strait.

ASDP Angkutan Sungai Danau dan Penyeberangen

PO Box 2997, Jin Jendri, Ahmad, Yani Kav 52A, Jakarta 10 510, Indonesia

Tel: (+62 21) 420 88 11
Fax: (+62 21) 421 05 44
Web: www.ferry-asdp.co.id

For financial reasons, the five 70 m monohulls operated by ASDP have been transferred to the Indonesian Navy. The company currently does not operate high-speed vessels.

High-speed craft operated

Type	Name	Seats	Delivered
MH Lürssen 70 m	*Ambulu*	925	1998
MH Lürssen 70 m	*Cisadane*	925	1998
MH Lürssen 70 m	*Serayu*	925	1998
MH Lürssen 70 m	*Manakham*	925	1999
MH Lürssen 70 m	*Barito*	925	1999

These vessels are currently operated by the Indonesian Navy.

Ambulu *on trials prior to delivery to ASDP* 0045260

Bali Bounty Cruises

Bounty Cruises Private Jetty, Benoa Harbour, Bali, Indonesia

Tel: (+62 361) 72 66 66
Fax: (+62 361) 72 66 88
email: info@balibountycruises.com
Web: www.balibountycruises.com

High-speed craft operated

Type	Name	Seats	Delivered
CAT Austal 44 m	*Geka 2000*	400	1999

Operations
Bali to Nusa Lembongan.

Bali Cruises Nusantara

PO Box 3548, Denpasar 80001, Bali, Indonesia

Tel: (+62 361) 72 03 31
Fax: (+62 361) 72 03 34
e-mail: sales@balihaicruises.com
Web: www.balihaicruises.com

Pande Ardika, *Senior Manager, Sales and Customer Service*

High-speed craft operated

Type	Name	Seats	Delivered
CAT Austal 36 m	*Bali Hai II*	316	1994

Operations
Benoa to Lembongan Island.

Pelni

Jalan Angkasa 18, Jakarta, Indonesia

Tel: (+62 21) 421 19 21
Fax: (+62 21) 49 16 23
Web: www.pelni.com

Pelni (the National Indonesian Shipping Company) operates a large fleet of passenger vessels in Indonesian waters.

High-speed craft operated

Type	Name	Seats	Additional payload	Delivered
MONO Mjellem & Karlsen 95 m	*HSC Jetliner*	577	160 cars	1996

Leased to the Sri Lanka Navy.

Quicksilver Beluga

Jl Raya Kerthadalem No 96 Sidakarya, Denpasar 80226, Bali, Indonesia

Tel: (+62 361) 72 87 86
Fax: (+62 361) 72 87 86
e-mail: info@quicksilver-bali.com
Web: www.quicksilver-bali.com

High-speed craft operated

Type	Name	Delivered
WPC NQEA InCat 37 m	*Quicksilver Beluga*	1992
WPC NQEA InCat 37 m	*Quicksilver VII*	1996

Operations
Tanjung Marina Nusa Dua to Nusa Penida Island.

Iran

Valfajr Shipping Company

A subsidiary of Islamic republic of Iran Shipping Lines
Abyar Alley, Shahid Azodi St, Karimkhan-Zand Avenue, 15875-4646 Tehran, Iran

Tel: (+98 21) 880 03 69
Fax: (+98 21) 890 24 09
e-mail: valfajr@vesc.net
Web: www.irisl.net

E Jamali, *Technical Director*

High-speed craft operated

Type	Name	Seats	Additional payload	Delivered
CAT WaveMaster 44 m	*Negeen*	242	18 t	1994
CAT WaveMaster 44 m	*Brelian*	236	10 cars	1997
CAT WaveMaster 44 m	*Zomorrod*	236	10 cars	1997
CAT Miho 34.9 m	*Gowhar* (ex *Aquajet Super II*)	235	–	1998
CAT Miho 34.9 m	*Firouzeh* (ex *Aquajet Super I*)	235	–	1998

Operations
Khoramshahr to Kuwait City
Bandar Abbas to Kish Island and Qeshm Island
Bandar Abbas to Sharjah and Dubai.
Busehr to Kuwait, Qatar, Bahrain and Saudi Arabia
Bandar Lengeh to Dubai
Busehr to Khark Island.

Ireland

Aran Islands Direct Ltd

29 Forster Street, Galway City, Ireland

Tel: (+353 91) 56 65 35
Fax: (+353 91) 53 43 15
e-mail: info@aranidirect.com
Web: www.arandirect.com

High-speed craft operated

Type	Name	Seats	Delivered
MH OCEA 30 m	*Queen of Aran II*	235	2001
CAT Fjellstrand 40 m	*Flying Viking III*	312	1996

Operations
Roseaville to the Aran Islands
Inis Oírr to Inis Meaín to Inis Mór.

Irish Ferries

PO Box 19, Ferryport, Alexandra Road, Dublin 1, Ireland

Tel: (+353 1) 818 30 04 00
Fax: (+353 1) 819 39 42
e-mail: info@irishferries.com
Web: www.irishferries.com

High-speed craft operated

Type	Name	Seats	Additional payload	Delivered
CAT Austal 86 m	*Jonathan Swift*	800	200 cars	1999

Operations
Dublin to Holyhead.

Jonathan Swift (Maritime Photographic) 0116591

Island Ferries Teo

4 Foster Street, Galway, Republic of Ireland

Tel: (+353 91) 56 89 03
Fax: (+353 91) 56 85 38
e-mail: island@iol.ie
Web: www.aranislandferries.com

Sally O'Brien, *Managing Director*
Denis Walsh, *Chief Engineer*

High-speed craft operated

Type	Name	Seats	Delivered
MH Wavemaster 37 m	*Draiocht Na Farraige*	294	1999
MH Wavemaster 37 m	*Ceol Na Farriage*	294	2001

Operations
Rossaveal to Inismor
Rosaveal to Inis Meaín and Inis Oírr.

Draiocht Na Farraige 0081028

Italy

Alilauro SpA

Via Caracciolo 11, I-80122 Naples, Italy

Tel: (+39 081) 761 10 04
Fax: (+39 081) 761 42 50
e-mail: informationi@alilauro.it
Web: www.alilauro.it

Rosaria Lauro, *Director*

A sister company also operates:
Alilauro-Gru SO N SpA
Piazza Marinai d'Italia, Sorrento, Naples, Italy

Tel: (+39 081) 878 14 30
Fax: (+39 081) 807 12 21
e-mail: info@alilaurogruson.com
Web: www.alilaurogruson.it

Captain Salvatore Di Leva, *Chairman*

High-speed craft operated

Type	Name	Seats	Delivered
CAT Marinteknik Marinjet 33 CPV	*Giove Jet*	276	1985
MH Marinteknik 35 m	*Citta di Amalfi*	306	1987
HYD S Ord Kolkhida	*Alieolo*	150	1986
HYD Kolkhida	*Aligea*	150	1986
CAT FBM Marinteknik Shipbuilder (S) Pte Ltd Marinjet 34 CPV	*Acapulco Jet*	355	1987
HYD Kolkhida	*Alikenia*	153	1989
MH Marinteknik 42 m	*Citta di Sorrento*	300	1987
MH Marinteknik 42 m	*Citta di Forio*	312	1987
CAT Marinteknik Verkstads 36 CPV	*Airone Jet*	300	1988
MH FBMM (S)	*Celestina*	400	1988
CAT Marinteknik 34 m	*Giunone Jet*	365	1988
MH Marinteknik 34 m	*Europa Jet*	400	1987
CAT Marinteknik Verkstads AB	*Nettuno Jet*	352	1988
MH Marinteknik 41 m	*Rosaria Lauro*	400	1988
MH Rodriquez TMV 50	*Tindan Jet*	500	1999
HYD Kolkhida	*Aliflorida*	150	1989
HYD Kolkhida	*Aliantares*	150	1991
MH Wavemaster 45 m	*Super Flyte*	525	1992
MH Intermarine	*Flash*	300	1999
MH Wavemaster 50 m	*AnnaMaria Lauro*	324	2002
MH Rodriquez City Cat 40	*Maria Celeste Lauro*	320	2004
MH Rodriquez City Cat 40	*Maria Sole Lauro*	320	2005
MH Rodriquez TMV 42	*Agostino Lauro Jet*	320	2007
CAT Marinteknik 43 m	*Angelino Lauro Jet*	348	2000

Operations
Naples to Ischia Porto
Naples to Forio
Capri to Ischia.
Salerno to Amalfi to Positano to Capri to Napoli
Naples to Capri and the ports of the Cilento coast
Naples to Positano
Formia to Ischia to Capri to Sorrento to Napoli
Sorrento to Capri
Sorrento to Ischia
Naples to Eolie
Pozzuoli to Procisda
Fiumicino to Ponza to Ventotene to Ischia to Capri to Sorrento.

Nettuno Jet 0506735

Alimar SpA

Calata Zingari, I-16126 Genova, Italy

Tel: (+39 010) 25 67 75
Fax: (+39 010) 25 29 66
e-mail: alimarsrl@tin.it
Web: www.alimar.ge.it

High-speed craft operated

Type	Name	Seats	Delivered
CAT Westamaran 88	*Marexpress*	181	1981

Operations
Sightseeing excursions from Genova.

Aliscafi SNAV SpA

Società di Navigazione Alta Velocita
A subsidiary of the Aponte Group
Terminal Napoli, Stazione Marittima, Molo Angioino, 80133 Napoli, Italy

Tel: (+39 081) 428 5211
Fax: (+39 081) 428 5209
Web: www.snav.it

High-speed craft operated

Type	Name	Seats	Additional payload	Built
HYD Rodriquez RHS 200	*Superjumbo*	250	–	1981
HYD Rodriquez RHS 160	*Fast Blu*	180	–	1986
HYD Rodriquez RHS 160	*Sinai*	180	–	1986
HYD Rodriquez RHS 150	*Salina*	161	–	1990
HYD Rodriquez RHS 150	*Panarea*	210	–	1990
CAT Marinteknik 36 m	*SNAV Aries* (ex-*Waterways 3*, ex-*Condor 8*)	300	–	1988
CAT Westamaran 3700	*SNAV Andromeda* (ex-*Pilen 3*, ex-*Vindile*)	300	–	1989
CAT Fjellstrand 38.8 m	*SNAV Auriga* (ex-*Ørnen* ex-*Solovkiy*)	292	–	1990
CAT Fjellstrand 38.8 m	*SNAV Alfa* (ex-*Løberen*)	255	–	1990
CAT Fjellstrand 40 m	*SNAV Alcione*	357	–	1990
CAT Westamaran W4100S	*SNAV Altair*	360	–	1990
CAT Westamaran W4100S	*SNAV Antares*	360	–	1990
CAT Fjellstrand 38.8 m	*SNAV Aurora* (ex-*Svalan*)	281	–	1990
CAT Fjellstrand 40 m	*SNAV Aquila* (ex-*Saelen*, ex-*Soloven II*)	288	–	1992
CAT Wavemaster 42 m	*SNAV Alexa**	–	–	1995
MH Marinteknik 50 m	*SNAV Orion*	700	–	2004
CAT Fjellstrand Jumbo Cat 60	*Don Francesco***	450	60 cars	2000

*This vessel was damaged by grounding in 2001.
**This vessel is leased from Emeraude Ferries.

Operations
Eolie to Napoli to Eolie
Milazzo to Isole Eolie
Isole Eolie to Messina to Reggio C
Messina to Reggio C to Isole Eolie
Messina to Isole Eolie Bis
Milazzo to Isole Eolie to Palermo
Palermo to Cefalu' to Eolie
Isole Eolie to Milaai Bis
Napoli to Capri
Napoli to Procida to Casamicciola
Napoli to Ventotene to Ponza
Napoli to Sardinia
Zadar to Ancora
Zadar to Civitanova
Naples to Golfo Arancia.

Alijumbo Messina *Rodriquez RHS 160F* 0043243

Sicilia Jet 0043253

Caremar

Campania Regionale Marittima SpA
Molo Beverello 2, I-80133 Naples, Italy

Tel: (+39 081) 580 51 11
Fax: (+39 081) 551 45 51
e-mail: caremar@infovoce24.it
Web: www.caremar.it

Dr Giampiero Romiti, *Managing Director*

High-speed craft operated

Type	Name	Seats	Additional payload	Delivered
HYD Rodriquez RHS 160F	*Alnilam*	210	–	1986
HYD Rodriquez RHS 160F	*Aldebaran*	210	–	1986
CAT Rodriquez Seagull 400	*Achernar*	350	–	1993
MH Rodriquez 47 m	*Isola Di San Pietro*	350	–	1993
MH Rodriquez TMV 70	*Isola Di Capri*	520	57 cars	1998
MH Rodriquez TMV 70	*Isola Di Procida*	520	57 cars	1999
HYD Rodriquez RHS 160F	*Monte Gargano*	210	–	1989

Operations
Naples to Capri to Sorrento
Naples to Ischia Porto
Naples to Procida
Formia to Ponza/Ventotene
Pozzuoli to Casamicciola.

Isola Di Capri 0038858

Delfino Verde Navigazione

Piazza della Borsa 7, 34100 Trieste TS, Italy

Tel: (+39 335) 548 13 27
Fax: (+39 040) 31 19 13
e-mail: info@delfinoverde.it
Web: www.delfinoverde.it

High-speed craft operated

Type	Name	Seats	Delivered
MH Sciomachen 25.5 m	*Delfino Verde*	50	2000

Operations
Trieste to Muggia and Grado.

Foderaro Navigazione

Via Diaz 10/16, I-88046 Lamezia Terme (CZ), Italy

Tel: (+39 0968) 238 27
Fax: (+39 0968) 44 85 45
e-mail: info@foderaro.it
Web: www.foderaro.it

High-speed craft operated

Type	Name	Seats	Delivered
MH C N Foschi, Sciomachen 290	*Neocastrum*	350	1993
MH C N Foschi, Sciomachen 34 m	*Citrarium*	356	1999

Operations
Vibo Marina to Tropea to Salina to Vulcano to Lipari to Panarea to Stromboli to Tropea to Vibo Marina.

Larivera Lines

Via Egadi 6, Termoli (CB), Italy

Tel: (+39 0875) 822 48
Fax: (+39 0875) 70 91 83
e-mail: info@lariveralines.com
Web: www.lariveralines.com

Giacomo Guidotti, *Managing Director*

High-speed craft operated

Type	Name	Seats	Delivered
CAT Fjellstrand 45 m	*Termoli Jet* (ex-*Flying Viking*)	354	2002

Operations
Termoli (Italy) to Croatian islands.

Navigazione Golfo dei Poeti

Via Don Minzoni 13, 19121 La Spezia, Italy

Tel: (+39 0187) 73 29 87
Fax: (+39 0187) 73 03 36
e-mail: navinfo@navigazionegolfodeipoeti.it
Web: www.navigazionegolfodeipoeti.it

High-speed craft operated

Type	Name	Seats	Delivered
MH SBF 31 m	*Mistral*	238	1998
MH Cantieri Trieste 33 m	*Beluo*	350	1997

Operations
La Spezia to Bocca di Magra, Lerici, Portovenere, Riomaggiore, Manarola, Corniglia, Vernazza, Monterosso, Levanto, Bonassola, Deiva Marina, Monteglia and Portofino.

Navigazione Lago Di Como

Gestione Governativa Navigazione Laghi
Via Per Cernobbio 18, I-22100 Como, Italy

Tel: (+39 31) 57 92 11
Fax: (+39 31) 57 00 80
e-mail: infocomo@navigazionelaghi.it
Web: www.navigazionelaghi.it

Dott Ing Franze Piunti, *Director*

High-speed craft operated

Type	Name	Seats	Delivered
HYD Rodriquez RHS 70	*Freccia dei Gerani*	80	1977
HYD Rodriquez RHS 150SL	*Freccia delle Valli*	176	1980
HYD Rodriquez RHS 150SL	*Guglielmo Marconi*	198	1983
HYD Rodriquez RHS 150SL	*Voloire*	200	1989
CAT Pesaro 28 m	*Cittá Di Como*	187	2002
CAT Pesaro 28 m	*Cittá Di Lecco*	187	2002
CAT Pesaro 28 m	*Tivano*	187	2003

Operations
Como to Argegno to Lezzeno to Lenno to Tremezzo to Bellagio to Menaggio to Varenna to Bellano to Dongo to Gravedona to Domaso to Colico.

Freccia delle Valli (Aerea I Buga) 0506737

Navigazione Lago Maggiore

Gestione Governativa Navigazione Laghi
Viale F Baracca 1, I-28041 Arona, Italy

Tel: (+39 0322) 23 32 00
Fax: (+39 0322) 24 95 30
e-mail: navimaggiore@navigazionelaghi.it
Web: www.navigazionelaghi.it

Massimo Checcucci, *Director*

High-speed craft operated

Type	Name	Seats	Delivered
Rodriquez RHS 150SL	*Freccia dei Giardini*	170	1980
Rodriquez RHS 150SL	*Enrico Fermi*	180	1984
Rodriquez RHS 150FLS	*Lord Byron*	186	1990

Operations
Arona to Angera to Belgirate to Stresa to Verbania to Luino to Cannobio to Brissago to Ascona to Locarno.

Rodriquez RHS 150SL Freccia dei Giardini 0506850

Navigazione Libera Del Golfo SpA

A subsidiary of the Aponte Group
Molo Beverello, I-80133 Naples, Italy

Tel: (+39 081) 552 07 63; 72 09
Fax: (+39 081) 552 55 89
e-mail: info@navlib.it
Web: www.navlib.it

Nello Aponte, *Manager*

High-speed craft operated

Type	Name	Seats	Delivered
MH FBM Marine 41 m	*Capri Jet*	394	1988
MH FBM Marine 41 m	*Sorrento Jet*	394	1990
MH Marinteknik 41 m	*Napoli Jet*	394	1992
MH Rodriquez TMV 50	*Salerno Jet* (ex-*Marconi*)	400	1992
MH Marinteknik 41 m	*Amalfi Jet*	350	1993
MH Marinteknik 41 m	*Ischia Jet*	350	1994
MH Marinteknik 45 m	*Superjet*	500	1999
MH Rodriquez TMV 50	*Vesuvio Jet*	444	2003
MH Austal 48	*Tremiti Jet*	450	1994

Operations
Naples to Capri, 40 minutes
Sorrento to Capri, 20 minutes.

Superjet 0103300

Napoli Jet 0506738

Navigazione Navigargano Srl

Via Brig Sollitto 7, 71019 Vieste FG, Italy

High-speed craft operated

Type	Name	Seats	Delivered
SES Ulstein UT 904	*Iris*	354	1992
SES Cirrus 120P	*Zenit*	330	1990

Operations
Brindisi to Corfu
Brindisi to Paxi.

Navigazione Sul Lago Di Garda

Direzione di Esercizio Navigazione Laghi
Piazza Matteotti 2, I-25015 Desenzano del Garda, Italy

Tel: (+39 030) 914 95 11
Fax: (+39 030) 914 95 20
e-mail: infogarda@navigazionelaghi.it
Web: www.navigazionelaghi.it

Dott Ing Marcello Coppola, *Director*

High-speed craft operated

Type	Name	Seats	Delivered
HYD Rodriquez RHS 150SL	*Freccia delle Riviere*	174	1981
HYD Rodriquez RHS 150SL	*Galileo Galilei*	186	1982
HYD Rodriquez RHS 150FL	*Goethe*	188	1988
CAT Conavi 22 m	*Catullo*	100	1993
CAT Conavi 22 m	*Parini*	100	1993
CAT Clemna 22 m	*Virgilio*	140	1998
CAT Pesaro 27 m	*Freccia del Garda*	193	2002
CAT Siman 27 m	*Verga*	197	2003
CAT Siman 27 m	*D'Annunzio*	197	2003

Operations
Peschiera del Garda to Desenzano to Sirmione to Bardolino to Garda to Salò to Gardone to Torri to Maderno to Gargnano to Malcesine to Limone to Torbole to Riva del Garda.

Rodriquez RHS 150FL Goethe 0506739

RFI Bluvia

Rete Ferroviaria Italiana
Piazza della Croce Rossa 1, 00161 Rome, Italy

Tel: (+39 064) 4101
e-mail: dircomu@rfi.it
Web: www.rfi.it

High-speed craft operated

Type	Name	Seats	Delivered
MH FBMM (S)	*Celestina*	350	1988
MH Rodriquez TMV 50	*Segesta Jet*	500	1999
MH Rodriquez TMV 50	*Selinunte Jet*	500	1999

Operations
Messina (Sicily) to Reggio Calabria.

Righetti Navi Srl

Via Dino Ricci 2b, I-47042 Cesenatico FC, Italy

Tel: (+39 0547) 67 51 84
Fax: (+38 0547) 67 43 64
e-mail: info@righettinavi.it
Web: www.righettinavi.it

Righetti Navi operate a fleet of crew supply boats to the offshore industry in the Adriatic.

High-speed craft operated

Type	Name	Seats	Delivered
MH 26.9 m	Destriero Secundo	–	–
MH 30.41 m	Lupo	–	–
MH 30.7 m	Panther Primo	–	–
MH 38.0 m	Puma Primo	–	–

Operations
Crewboat support to the offshore industry.

Siremar

Sicilia Regionale Marittima SpA (a Tirrenia company)
Calata Marinai d'Italia, I-90139 Palermo, Sicily, Italy

Tel: (+39 091) 749 3111
Fax: (+39 091) 58 22 67
e-mail: uff.commerciale.siremar@siremar.it
Web: www.siremar.it

High-speed craft operated

Type	Name	Seats	Additional payload	Delivered
HYD Rodriquez RHS 140	*Duccio*	140	–	1977
HYD Rodriquez RHS 160	*Algol*	175	–	1978
HYD Rodriquez RHS 160	*Alioth*	180	–	1979
HYD Rodriquez RHS 160	*Donatello*	180	–	1980
HYD Rodriquez RHS 160	*Botticelli*	183	–	1980
HYD Rodriquez RHS 160F	*Masaccio*	210	–	1987
HYD Rodriquez RHS 160F	*Mantegna*	210	–	1989
HYD Rodriquez RHS 160F	*Giorgione*	210	–	1989
HYD Rodriquez Foilmaster	*Tiziano*	242	–	1994
MH Rodriquez TMV101	*Guizzo*	450	100 cars	1993
MH Rodriquez TMV 70	*Isola Di Vulcano*	520	57 cars	1999
MH De Poli 95m	*Isola Di Stromboli*	802	168 cars	1999
HYD Rodriquez Foilmaster	*Antioco*	240	–	2005
HYD Rodriquez Foilmaster	*Calypso*	240	–	2005
HYD Rodriquez Foilmaster	*Eraclide*	240	–	2005
HYD Rodriquez Foilmaster	*Eschilo*	240	–	2005
HYD Rodriquez Foilmaster	*Atanis*	240	–	(2006)
HYD Rodriquez Foilmaster	*Platone*	240	–	(2006)

Operations
Milazzo to Eolie (Vulcano to Lipari to Salina to Panarea to Stromboli to Filicudi to Alicudi)
Palermo to Ustica
Trapani to Egadi (Favignana to Levanzo to Marettimo).

Sicilia Regionale Marittima's Giorgione 0506740

Snav Croazia Jet

A subsidiary of Aliscafi Snav Spa

Web: www.snav.com

Snav Croazia Jet is a joint venture between Snav and Sea Containers Italy

High-speed craft operated

Type	Name	Seats	Additional payload	Delivered
CAT Austal 82 m	*Croazia Jet* (ex-*Felix*)	650	156	1996
CAT InCat 74 m	*Zara Jet* (ex-*Seacat Danmark*)	450	88 cars	1991

Operations
Ancona to Split (4 hours 15 mins)
Zara to Civitanova via Sibenik (seasonal)
Split Hvara and Pescara.

Sea Cat Danmark (Maritime Photographic) 0114493

Taranto Navigazione

Via dei Mille 40, I-98057 Milazzo (ME), Italy

Tel: (+39 0909) 22 36 17
Fax: (+39 0909) 28 52 10
Web: www.tarantonavigazione.it

High-speed craft operated

Type	Name	Seats	Delivered
MH Sciomachen 21 m	*Lady M*	130	1997
MH Sciomachen 30 m	*Eolian Princess*	294	2002
MH Sciomachen 27 m	*Eolian Star*	250	2003

Operations
Tourist excursions to the Eolian Islands.

Tirrenia Navigazione

Rione Sirignano 2, I-80121 Naples, Italy

Tel: (+39 081) 720 11 11
Fax: (+39 081) 720 14 41
Web: www.gruppotirrenia.it

Franco Pecorini, *Manager*

High-speed craft operated

Type	Name	Seats	Additional payload	Delivered
MH Rodriquez Aquastrada	*Scatto*	450	126 cars	1994
MH Fincantieri MDV 3000	*Aries*	1,800	450 cars	1998
MH Fincantieri MDV 3000	*Taurus*	1,800	450 cars	1998
MH Fincantieri MDV 3000	*Capricorn*	1,800	450 cars	1998
MH Fincantieri MDV 3000	*Scorpio*	1,800	450 cars	1999
MH Rodriquez Monostab 47	*Pacinotti*	390	–	1982

Operations
Fiumicino to G.Aranci/Arbatax (Sardinia).

MDV 3000 Aries *and* Taurus 0038857

Toremar

Toscana Regionale Marittima SpA
Via Calatati 6, I-57123 Livorno, Italy

Tel: (+39 0586) 22 45 11
Fax: (+39 0586) 22 46 24
e-mail: toremar@toremar.it
Web: www.toremar.it

High-speed craft operated

Type	Name	Seats	Additional payload	Delivered
HYD Rodriquez RHS 160F	*Fabricia*	210	none	1987
MH Rodriquez TMV 70	*Isola Di Capraia*	522	57 cars	1998

Operations
Portoferraio to Cavo to Piombino (Elba to mainland).

Ustica Lines

Via Amm Staiti 23, I-91100 Trapani, Italy

Tel: (+39 0923) 87 38 13
Fax: (+39 0923) 59 32 00
e-mail: info@usticalines.it
Web: www.usticalines.it

Vittorio Morace, *President and Managing Director*

High-speed craft operated

Type	Name	Seats	Additional payload	Delivered
HYD Rodriquez RHS 160	*Calarossa*	180	–	1974
HYD Rodriquez RHS 160F	*Fiammetta M* (ex-*Tintorera*)	204	–	1989
HYD Rodriquez RHS 160F	*Cris M* (ex-*Marrajo*)	204	–	1990
HYD Rodriquez RHS 160F	*Citti Ships**	210	–	1990
HYD Rodriquez RHS 160F	*Alijumbo Eolie**	210	–	1991
HYD Rodriquez RHS 160F	*AlijunZibibbo**	210	–	1991
HYD Rodriquez RHS 160F	*AlijunMessina**	210	–	1991
HYD Rodriquez RHS 160F	*Moretto 1**	210	–	1991
CAT Flying Cat 38 m	*SNAV Aquarius**	255	–	1991
CAT Flying Cat 40 m	*SNAV Orion**	288	–	1992
HYD Rodriquez Foilmaster	*Eduardo M*	224	–	1996
CAT Flying Cat 40 m	*Federica M*	382	–	1996
CAT Flying Cat 40 m	*SNAV Aldebaran**	308	–	1996
CAT Flying Cat 40 m	*Gabriele M* (ex-*St Gabriel*)	274	–	1997
CAT Austal Autoexpress 48	*Ale M* (ex -*Jade Express*)	330	10 cars	1998
HYD Rodriquez Foilmaster	*Adriana M*	240	–	1999
CAT Flying Cat 40 m	*Vittoria M*	331	–	2001
HYD Rodriquez Foilmaster	*Natalie M*	240	–	2002
MH Wavemaster 38 m	*Gianluca M*	170	–	2003
HYD Rodriquez Foilmaster	*Ettore M*	240	–	2003
HYD Rodriquez Foilmaster	*Misella M*	240	–	2006
HYD Rodriquez Foilmaster	–	240	–	(2007)

*Operated by Aliscafi SNAV SpA, Italy.

Operations
Naples to Ustica to Egadi Islands to Trapani
Trapani to Egadi Islands
Trapani to Pantelleria
P Ehpedocle to Lampeduia to Linosa
Croatian Ports to Venice to Trieste
Mazara de Vaslo (Sicily) to Sousse (Tunisia) via islands of Panterleria.

Venezia Lines SpA

A wholly owned subsidiary of Virtu Ferries Ltd
Isola di Tronchetto 21, I-30135 Venice, Italy

Tel: (+39 041) 242 40 00
Fax: (+39 041) 272 26 45
e-mail: sales@venezialines.com
Web: www.venezialines.com

Venezia Lines operates a high-speed service connecting Venice and Trieste to the Istrian coast from May to September. In December 2006 it commenced a new service from Bari in Italy to Durres in Albania.

High-speed craft operated

Type	Name	Seats	Additional payload	Delivered
SES Ulstein CIRR 120P	*San Frangisk*	316	–	1990
SES Ulstein CIRR 120P	*San Pawl*	316	–	1990
CAT Fjellstrand Flying Cat 52	*San Gwann*	427	20 cars	2001

Operations
Venice, Italy to Slovenia and Croatia. Ports of call: Piran, Umag, Porec, Rovinj, Pula, Rabac, Lussino and Rimini
Bari, Italy to Durres, Albania, 3 hours 30 minutes.

Vetor Aliscafi Srl

Casella Postale Ni 144, I-00042 Anzio, Italy

Tel: (+39 06) 984 50 83
Fax: (+39 06) 984 50 04
e-mail: helpdesk@vetor.it
Web: www.vetor.it

High-speed craft operated

Type	Name	Seats	Delivered
HYD Kometa	*Vetor 944*	102	1984
HYD Kometa	*Freccia Pontina*	102	1985
HYD Kolkhida	*Gabri*	120	1989
HYD Kolkhida	*Vemar*	120	1991

Operations
Anzio to Naples
Anzio to Ponza to Ventotene
Formia to Ponza to Ventotene.

Japan

Awashima Kisen Company Ltd

3 Awashimaura-mura, Iwafune, Niigata, Japan

Tel: (+81 25) 455 21 31
Fax: (+81 25) 455 20 46

High-speed craft operated

Type	Name	Speed	Seats	Delivered
MH Sumidagawa Zosen Co Ltd	*Iwayuri*	24 kt	144	May 1979
MH Sumidagawa Zosen Co Ltd	*Asuka*	24 kt	173	May 1989

Operations
Awashima to Iwafune.

Asuka operated by Awashima Kisen 0506742

Biwako Kisen Company Ltd

Biwa Lake Sightseeing Company
5-1 Hamaotsu, Otsu-City 520, Shiga Prefecture, Japan

Tel: (+81 775) 22 41 15
Fax: (+91 775) 24 78 96

e-mail: info@biwakokisen.co.jp
Web: www.biwakokisen.co.jp

High-speed craft operated

Type	Name	Speed	Seats	Delivered
MH Mokubei Shipbuilding Co Ikeda 22 m	*Lansing*	27 kt	78	April 1982

Operations
Biwa Lake, sightseeing.

The Mokubei Shipbuilding Company water-jet-propelled Lansing 0506927

City of Kamishima

City of Kamishima, Japan

High-speed craft operated

Type	Name	Seats	Additional payload	Delivered
MH 21m	*Yunagi*	12	1 Ambulance	2006

Operations
Operates as an ambulance ferry.

Cosmo Line

23-4 Kinko-cho, Kagoshima, Kagoshima Prefecture, Japan

Tel: (+81 6) 65 72 35 74
Fax: (+81 6) 65 72 49 84
e-mail: osaka.bay.line@nifty.com
Web: www.ichimaru-grp.jp/cosmo

High-speed craft operated

Type	Name	Seats	Delivered
HYD Boeing jetfoil 929-117	*Rocket 2* (ex-*Al Azizah*)	280	1983
HYD Kawasaki jetfoil 929-117	*Rocket*	230	1985
HYD Kawasaki jetfoil 929-117	*Rocket3* (ex-*Princesa Dacil*)	286	1990

Operations
Ferry services in Osaka Bay.

Goto Sangyo Kisen

Kimigoto, Minami Matsuwa, Nagasaki, Japan

Tel: (+81 665) 73 05 30
Web: www.goto-sangyo.co.jp

High-speed craft operated

Type	Name	Seats	Delivered
MH Yatsushiro 35 m	*Elegant II*	230	1999
CAT Mitsui Mighty Cat 40	*Big Earth* (ex- *Neptune*)	300	1995
CAT Mitsui Mighty Cat 40	*Big Earth II* (ex- *Venus*)	300	1993

Operations
Nagasaki to Sangoto (90 minutes).

Higashi Nihon Ferry Company

Nishi -11, Minami 4-jo, Chuo-ku, Sapporo-shi Hokkaido, Japan

Tel: (+81 115) 182 722
Fax: (+81 115) 182 725
Web: www.higashinihon-ferry.com

Higashi Nihon ferry received delivery of the 112 m InCat vessel *Natchan Rera* in July 2007. At the time of her launch she was the largest passenger catamaran ever built.

High-speed craft operated

Type	Name	Seats	Delivered
WPC InCat 112 m	*Natchan Rera*	800	2007

Operations
Across the Tsugaru strait between the islands of Honshu and Hokkaido.

Ishizaki Kisen Company Ltd

1-4-9 Mitsu, Matsuyama-City 791, Ehime Prefecture, Japan

Tel: (+81 89) 951 01 28
Fax: (+81 89) 951 01 29
e-mail: information@ishizakikisen.co.jp
Web: www.ishizakikisen.co.jp

High-speed craft operated

Type	Name	Seats	Delivered
F-Cat Hitachi Superjet 30	*Zuiko*	156	1993
F-Cat Hitachi Superjet 30	*Shoko*	156	1994
F-Cat Hitachi Superjet 40	*Seamax*	200	1998

Operations
Matsuyama to Hiroshima (Superjet 30s)
Matsuyama to Mojii (Superjet 40).

Seamax 0038872

Japan Coast Guard

2-1-3 Kasumigaseki, Chiyoda-ku, Tokyo, Japan

Tel: (+81 3) 35 91 63 61
Fax: (+81 3) 35 97 94 20
Web: www.kaiho.mlit.go.jp

High-speed craft operated

Type	Name	Delivered
23 m patrol craft		
MH Mitsubishi Heavy Industries Ltd	*Hayagiri*	1985
MH Hitachi Zosen	*Shimagiri*	1985
MH Hitachi Zosen	*Setogiri*	1985
MH Sumidagwa Zosen Co Ltd	*Natsugiri*	1990
MH Sumidagwa Zosen Co Ltd	*Suganami*	1990
180 ton patrol vessel		
MH Mitsubishi Heavy Industries Ltd	*Mihashi*	1988
MH Hitachi Zosen	*Saroma*	1989
MH Mitsubishi Heavy Industries Ltd	*Inasa*	1990
MH Hitachi Zosen	*Kirishima*	1991
MH Mitsubishi Heavy Industries Ltd	*Takatsuki*	1992
MH Hitachi Zosen	*Nobaru*	1993
MH Hitachi Zosen	–	1994
MH Mitsubishi Heavy Industries Ltd	–	1994
MH Mitsubishi Heavy Industries Ltd	*Kamui*	1994
MH Hitachi Zosen	*Bizan*	1994

Only those craft with a maximum speed of over 30 kt have been listed here.

Operations
The JMSA was established in May 1948 for the protection of life and property at sea.

Suganami *23 m patrol craft* 0506743

JR Kyushu Railway Company

3-25-21 Hakataekimae, Hakata-Ku, Fukuoka 812 8566, Japan

Tel: +81 92 474 2501
Web: www.jrkyushu.co.jp

This company started Jetfoil operations in 1991.

High-speed craft operated

Type	Name	Seats	Delivered
HYD Kawasaki Jetfoil 929-117	*Beetle 3*	180	1989
HYD Kawasaki Jetfoil 929-117	*Beetle 2*	236	1990
HYD Kawasaki Jetfoil 929-117	*Beetle*	233	1991

Operations
Hakata to Busan (South Korea).

Kagoshima Shosen Company Ltd

Part of the Iwasaki Corporation
12-2 Kamoikeshinmachi, Kagoshima 892, Japan

Tel: (+81 992) 56 77 71
Fax: (+81 992) 51 83 71
Web: www.iwasaki-group.com

K Yamamoto, *Chief of Marine Department*

High-speed craft operated

Type	Name	Seats	Delivered
HYD Kawasaki Jetfoil 929-117	*Toppy*	244	June 1989
HYD Kawasaki Jetfoil 929-117	*Toppy 2*	244	1992
HYD Kawasaki Jetfoil 929-117	*Toppy 3*	244	1995
HYD Boeing Jetfoil 929-115	*Toppy 4*	280	1978

Operations
Kagoshima to Ibusuki to Tanegashima to Yakushima.

Kashima Futo Co Ltd

Dewa 4186-19, Okunoya, Kamisu-Machi, Kashima-Gun, Ibraki, 314-0116, Japan

Tel: +81 299 97 0684
Fax: +81 299 96 8344
e-mail: marine@kashimafuto.co.jp
Web: www.kashimafuto.co.jp

High-speed craft operated

Type	Name	Seats	Delivered
F-CAT Hitachi Superjet 30	*Blue Sonic*	160	1997

Koshikijima Shosen Ltd

Tel: (+81 996) 32 82 32

High-speed craft operated

Type	Name	Seats	Delivered
MH Mitsubishi 48 m	*Sea Hawk*	301	1990

Operations
Kushikino to Koshikijima.

Kumamoto Ferry Company

Tel: (+81 96) 311 41 00
e-mail: info@kumamotoferry.co.jp
Web: www.kumamotoferry.co.jp

High-speed craft operated

Type	Name	Seats	Additional payload	Delivered
CAT IHI SSTH70	*Ocean Arrow*	430	51 cars	1998
CAT Image 30 m	*Marine View*	140	–	1997

Operations
Kumamoto to Shimabara.

Kyushu Shosen Company Ltd

16-12 Motofune-Cho, Nagasaki-City 850, Japan

Tel: (+81 958) 22 91 53
Fax: (+81 958) 23 26 64
email: nga-ksk@kyusho.co.jp
Web: www.kyusho.co.jp

High-speed craft operated

Type	Name	Seats	Delivered
HYD Kawasaki Jetfoil 929-117	*Pegasus*	282	1990
HYD Kawasaki Jetfoil 929-117	*Pegasus 2*	233	1990
MH Mitsubishi 48.5 m	*Sea Grace*	286	1993

Operations
Nagasaki to Narao to Fukue
Sasebo to Narao.

Kyushu Yusen Company Ltd

1-27 Kamiya-Machi, Fukuoka City, Fukuoka 812-0022, Japan

Tel: (+81 92) 281 08 31
Fax: (+81 92) 281 04 44
email: t-nishiyama@kyu-you.co.jp
Web: www.kyu-you.co.jp

T Nishiyama, *Technical Officer*

High-speed craft operated

Type	Name	Seats	Delivered
HYD Kawasaki Jetfoil 929-117	*Venus*	263	1991
HYD Boeing Jetfoil 929-117	*Venus 2*	260	1996
–	(ex-*Falcon*, ex-*Jet 8*)	–	–

Operations
Hakata to Gounoura
Hakata to Ashibe
Hakata to Izuhara to Hitakatsu.

Jetfoil Venus 2 1326747

Oita Hover Ferry Company Ltd

1-14-1 Nishi-shinchi, Oita 870-0901, Japan

Tel: (+81 97) 558 71 80
Fax: (+81 97) 556 62 47
e-mail: ga@oitahover.co.jp
Web: www.oitahover.co.jp

Sakui Kayajima, *Managing Director*
Kouti Wanaka, *Marketing Director*
Akihro Kurokotwa, *Operations Director*

Oita Hover Ferry Company Ltd was established in November 1970 and began operating in October 1971. Annual traffic is approximately 500,000 people.

High-speed craft operated

Type	Name	Seats	Delivered
HOV Mitsui MV-PP5	*Hobby No 6* (ex-*Akatombo 51* and stretched)	75	1971
HOV Mitsui MV-PP5	*Angel No 5*	75	April 1975
HOV Mitsui PP-10	*Dream Aquarmarine* (ex-*Dream No 1*)	105	March 1990
HOV Mitsui PP-10	*Dream Emerald* (ex *Dream No 2*)	105	March 1991
HOV Mitsui PP-10	*Dream Ruby* (ex *Dream No 3*)	105	October 1995
HOV Mitsui PP-10	*Dream Sapphire*	105	2002

Oita Hover Ferry Mitsui PP-10 Dream No 2 0506744

Operations
Oita city to Oita Airport, 15.6 n miles, 25 minutes.

OKI Kisen KK

11-2, Aza Menuki 4-Chome, Oaza Naki-machi, Saigo, Okigun, Shimane Prefecture 685-0013, Japan

Tel: (+81 85) 122 11 22
Fax: (+81 85) 122 11 26

High-speed craft operated

Type	Name	Seats	Delivered
F-CAT Mitsubishi Super Shuttle 400	*Rainbow*	341	April 1993
F-CAT Mitsubishi Super Shuttle 400	*Rainbow II*	341	1998

Operations
Sakai to Okinoshima.

Sado Kisen Kaisha

9-1 Bandaijima, Niigata City 950-0078, Japan

Tel: (+81 25) 245 72 55
Fax: (+81 25) 247 88 30
e-mail: jetfoil@sadokisen.co.jp

M Iwashita, *Jetfoil Manager*

High-speed craft operated

Type	Name	Seats	Delivered
HYD Boeing Jetfoil 929-115	*Ginga* (ex-*Cu na Mara*)	260	July 1986
HYD Kawasaki Jetfoil 929-117	*Tsubasa*	260	April 1989
HYD Kawasaki Jetfoil 929-117	*Suisei*	260	April 1991

Operations
Niigata, Honshu Island, to Ryotsu, Sado Island, 34.5 n miles, 1 hour.

Kawasaki Jetfoil Suisei *in service with Sado Kisen Kaisha* 0507007

Sanyo Shosen

1030-23 Takehara-Cho, Takehara, Hiroshima Prefecture 725, Japan

Tel: (+81 84) 622 21 33

High-speed craft operated

Type	Name	Seats	Delivered
MH Kiso 23 m	*Hayabusa No 1*	95	2003
MH Kiso 23 m	*Hayabusa No 2*	95	2003

Operations
Seto Island Sea.

Setonaikai Kisen Company Ltd (Seto Inland Sea Lines)

1-12-23 Ujinakaigan, Minami-ku, Hiroshima 734, Japan

Tel: (+81 82) 255 33 44
Fax: (+81 82) 251 67 43
e-mail: info@setonaikaikisen.co.jp
Web: www.setonaikaikisen.co.jp

Nobue Kawai, *Tour Co-ordinator of Akinada Line*

Setonaikai Kisen Company has been operating transport services linking the islands in the Inland Sea for many years. It now has a fleet of vessels, operating regular services along nine routes between Hiroshima and Shikoku, including a main route between Hiroshima and Matsuyama as well as Hiroshima to Kobe and Osaka.

High-speed craft operated

Type	Name	Seats	Delivered
HYD Hitachi PT50	*Otori 1*	126	1968
HYD Hitachi PT50	*Otori 2*	113	1970
HYD Hitachi PT50	*Otori 3*	126	1972
HYD Hitachi PT50	*Otori 4*	126	1973
CAT Miho 21 m	*Oyashio*	80	1973
MH Mitsui	*Saphia*	71	1987
MH Miho Zosenjo	*Marine Star 2*	103	November 1979
MH Miho Zosenjo	*Marine Star 3*	120	January 1983
Setouchi Craft MH	*Akinada*	125	May 1989
MH Miho 22 m	*Waka*	84	June 1984
MH Miho 22 m	*Seto*	84	July 1984
F-CAT Hitachi Superjet 30	*Dogo*	200	1993
F-CAT Hitachi Superjet 30	*Miyajima*	200	1994

Operations
Miyajima to Hiroshima to Omishima Island (Port of Inokuchi) to Setoda *(Akinada)*
Hiroshima to Matsuyama 1 hour (*Ohtori No 2, 3 and Ohtori No 5*), some services stopping at Kure, then 1 hour 10 minutes (*Hikari No 2*)
Mihara to Imabari 1 hour (*Marine Star 2, Marine Star 3*), some services stopping at Setoda, then 1 hour 7 minutes.

Akinada, *introduced into service by Setonaik Kisen in May 1989* 0506745

Tokai Kisen

Fujita Kanko Building, 15-9 Kaigan, Minato-ku, Tokyo 105, Japan

Tel: (+81 3) 34 36 11 31
e-mail: annai.tokaikisen.co.jp
Web: www.tokaikisen.co.jp

High-speed craft operated

Type	Name	Seats	Delivered
SWATH Mitsui SSC 40	*Seagull* (ex-*Seagull 2*)	420	December 1989
HYD Boeing Jetfoil 929-115	Seven Island Ai	280	1980
HYD Boeing Jetfoil 929-115	Seven Island Niji	280	1981
HYD Boeing Jetfoil 929-115	Seven Island Yume	280	1981

Operations
Atami to Ohshima
Tokyo to Ohshima*
Tokyo to Niijima*
Inatori to Ohshima
Ito to Ohshima.

*Summer services only

Tokushima Shuttle Line Company Ltd

3-1-3 Sannomiya-Cho, Chao-Ku, Kobe 650, Japan

Tel: (+81 734) 31 41 91

Hiromu Harada, *President*

High-speed craft operated

Type	Name	Seats	Delivered
CAT MES CP30 Mk II	*Marine Shuttle*	–	February 1986
CAT MES CP30 Mk III	*Coral*	250	1988

Operations
Wakayama to Tokushima, Shikoku Island, 1 hour.

Yaeyama Kanko Ferry Company Ltd

1 Misaki-cho, Isigaki-City, Okinawa 907-0012, Japan

Tel: (+81 98) 082 50 10
Fax: (+81 98) 082 35 59
e-mail: hanashiro@yaeyama.co.jp
Web: www.yaeyama.co.jp

High-speed craft operated

Type	Name	Seats	Delivered
MH Suzuki	*Hirugi 2*	96	1986
MH Gouriki	*Niinufabushi*	188	1990
MH Kowa	*Southern Cross No 1*	62	1992
MH Kowa	*Southern Cross No 3*	62	1993
MH Kowa	*Southern Cross No 8*	90	1996
MH Etho	*Southern Cross No 5*	90	1997
MH Etho	*Southern Dream*	150	1997
MH Etho	*Southern Queen*	90	1998
MH Etho	*Southern King*	90	1998
MH Etho	*Southern Eagle*	90	2000
MH Etho	*Southern Paradise*	90	2001
MH Etho	*Southern Coral*	88	2003
MH Etho	*Churasan*	90	2004
MH Etho	*Churasan 2*	144	2006

Operations
Ishigaki to surrounding islands.

Southern Dream 0038864

Yasuda Sangyo Kisen Company Ltd

(Yasuda Ocean Group) Airport Line

5-35 Matsugae-machi, Nagasaki-City 850, Japan

Tel: (+81 95) 826 01 88
Fax: (+81 95) 824 21 82
Web: http://kisen.yasudo-gp.net

Sakichi Yasuda, *President*

High-speed craft operated

Type	Name	Seats	Additional payload	Delivered
MH Nankai	*Erasmus*	123	–	October 1984
MH Nankai	*Airport Liner 7*	66	–	March 1986
MH Nankai	*Airport Liner 8*	66	–	March 1986
MH Nankai	*Airport Liner 10*	66	–	December 1986
MH Nankai	*Airport Liner 11*	66	–	February 1987
MH Nankai	*Airport Liner 13*	66	–	April 1987
MH Nankai	*Airport Liner 15*	66	–	June 1987
MH Nankai	*Airport Liner 17*	66	–	June 1987
MH Nankai	*Nagasaki 8*	39	–	June 1987
MH Nankai	*Nagasaki 10*	39	–	August 1987
MH Nankai	*Taiyo*	231	–	March 1988
MH Uehara	*Ocean Liner 1*	88	–	April 1988
MH Euhara	*Ocean Liner 3*	88	–	September 1988
MH Uehara	*Ocean Liner 5*	97	–	February 1989
MH Nankai	*Ocean Liner 7*	97	–	February 1989
MH Nankai	*Ocean Liner 8*	97	–	February 1989
MH Euhara	*Ocean Liner 10*	–	–	1988
MH	*Ocean Liner 11*	–	–	1988
CAT Incat/ Hitachi 62 m	*Sea Bird*	296	48 cars	1997
MH Okishin	*Tayo 2*	75	–	April 1997

Operations
Nagasaki Airport to Nagasaki (Togitsu), 17 km, 20 minutes
Mogi to Tomioka, 33.7 km, high-speed passenger ship, 35 minutes
The total number of passengers carried per year is 600,000
Nagasaki Airport to Huis Ten Bosch, 26 km, 50 minutes.

Jordan

Arab Bridge Maritime Co

PO Box 989, Aqaba 77 110, Jordan

Tel: (+962 3) 209 20 00
Fax: (+962 3) 209 20 01
e-mail: sales@abmaritime.com.jo
Web: www.abmaritime.com.jo

High-speed craft operated

Type	Name	Seats	Additional payload	Delivered
MH Rodriquez TMV 84	*Princess*	654	58 cars	2003

Operations
Aqaba (Jordan) to Nuwaibeh (Egypt), 1 hour.

Korea, South

Chonghaejin Marine Company

South Korea

Tel: (+82 61) 663 28 24
Web: www.cmcline.co.kr

In 1998 Chonghaejin purchased four vessels from Semo Marine Company.

High-speed craft operated

Type	Name	Seats	Delivered
SES Ulstein CIRR 120P	*Perestroika*	346	1990
SES Semo 36	*Democracy I*	336	1992
SES Semo 36	*Democracy III*	340	1994
WPC AMD 350	*Ogago*	350	1995

DAE-A Express Shipping Company Ltd

58-54 Hang-gu dong, Bukgu, Pohang City, South Korea

Tel: (+82 562) 42 51 11
Tel: (+82 562) 42 51 14
Web: www.daea.com

C K Kim, *Technical Manager*

High-speed craft operated

Type	Name	Seats	Additional payload	Delivered
CAT Incat 74 m	*Mandarin* (ex-*Ocean Flower*, ex-*Avant*, ex-*Stena Sea Lynx*)	450	89 cars	1993
CAT Fjellstrand	*Dae Won Catamaran*	392	–	1988
SWATH FBM	*Sea Flower II* (ex-*Patria*)	400	–	1989
SES Ulstein CIRR 120P	*Sea Flower*	345	–	1991
CAT Incat K50	*Sun Flower*	835	32 cars	1995
CAT FBMA Tricat 50	*Sea Flower I* (ex-*Supercat 2001*)	400	50	1999
CAT FBMA Tricat 52	*Hankyoreh*	455	–	2001
CAT FBMA Tricat 50	*Supercat 2002*	400	–	2001
TRI North West Bay 55 m	*Dolphin Ulsan*	450	–	2001

Operations
Mukho to Ullung Do, 86 n miles, 3 hours, USD25
Pohang to Ullung Do, 117 n miles, 3 hours, USD48
Hupo to Ullung Do, 86 n miles, 3 hours, USD25
Busan to Tsushima.

Dong Yang Express Ferry Co

11th Floor, Daewang Building, 45-1 Kwangchul Dong, Jongno-gu, Seoul 110-111, South Korea

Tel: (+82) 27 35 24 21
Fax: (+82) 27 30 73 17
Web: www.ihongdo.co.kr

High-speed craft operated

Type	Name	Seats	Delivered
SES Samsung 37 m	*Gold Star*	352	1994
CAT Fjellstrand 40 m	*New Gold Star*	324	1996

Jindo Transportation

Ferry Terminal 213, 5 Hang-Dong, Mokpo, Chun Nam, South Korea

Tel: (+82 61) 244 9577
e-mail: jindo@jindotr.co.kr
Web: www.jindotr.co.kr

High-speed craft operated

Type	Name	Seats	Delivered
CAT Fjellstrand 40 m	*Baekk Ryeong Island*	333	1995
CAT Fjellstrand 34 m	*Continental*	250	1996
CAT Mitsui MES CP15	*Auckland* (*ex Marine Bridge*)	235	1995
CAT MES CP30 MKIV	*Marine Bridge* (ex *Soleil*)	300	1991

Operations
Incheon to Baekk Ryeong.

Mirejet

15-3 Chung-ang-dong, 4-ga Chung-gu, Pusan, South Korea

Tel: (+82 51) 465 81 02
Fax: (+82 51) 441 93 09
Web: www.mirejet.com

High-speed craft operated

Type	Name	Seats	Delivered
HYD China Shipbuilding PS-30	*Kobee* (ex-*Praia*)	276	1995
HYD Boeing Jetfoil 929-100	*Kobee III*	240	1976
HYD Boeing Jetfoil 929-100	Kobee IV	240	1977
CAT Sabre 22 m	*Kerdonis*	141	2001

Operations
Busan to Fukuoka (Japan).

Nam Hae Express Company

201 Suite, Ferry Terminal 6, Hang-dong, Mokpo City, Chollanam-do 530-140, South Korea

Tel: (+82 61) 24 49 91 56
Fax: (+82 61) 243 01 53
Web: www.namhaegosok.co.kr

Seong Gi Sun, *President*

High-speed craft operated

Type	Name	Seats	Delivered
CAT MES CP20	*Nam Hae No 7*	–	1978
CAT Fjellstrand 38.8 m	*Nam Hae Star*	350	1989
CAT Marinteknik (S) 41 CPV	*Nam Hae Prince*	354	January 1993
CAT Fjellstrand 40 m	*Nam Hae Queen*	–	1994
CAT Wavemaster 49 m	*New Nam Hae Queen* (ex-*Fei Long*)	512	1994

Operations
Mokpo City to Hong-do Island, Heuksando and Gegeodo.

Onbada Co Ltd

South Korea

Tel: (+82 51) 466 50 50
Fax: (+82 51) 467 50 50
Web: www.onbadaro.co.kr

High-speed craft operated

Type	Name	Seats	Delivered
HYD Hyundai PT 20	*Angel IX*	71	1985
SES Semo 40	*Democracy V*	400	1995
CAT Semo 40	*Pegasus*	400	1996
MH Semo 28 m	*Sewol Tara*	–	1996
SES Ulstein UT 904	*Gagoogo*	341	1992

Seogyeong Marine Co

937 Chine Po-Ri, Llun-myon Koje-shi, Kyongsangnam-do 656-890, South Korea

Tel: (+82 51) 469 59 78

High-speed craft operated

Type	Name	Seats	Additional payload	Delivered
F-CAT Daewoo 40 m	*Royal Ferry*	350	8 cars	1994
CAT Fjellstrand 37 m	*New Arcadia*	300	–	1995
CAT FBM 35 m	*Gold Coast* (ex-*Supercat 5*, ex-*Universal 5*)	160	–	1993

Operations
Pusan to Geo Je
Pusan to Jangsengpo
Pusan to Okpo
Pusan to Gohan.

Sung Woo Ferry

823-9 Angol-dong, Tinhae City, Kyungnam, South Korea

High-speed craft operated

Type	Name	Seats	Delivered
CAT Holen MM 24 PC	*Cosmos* ex-*Baronessa*	125	1994

Uri Express Ferry Co Ltd

A subsidiary of Won Kwang Shipping
7-88, Hang-Dong, Jung-fu, Incheon, Korea

Tel: (+82 32) 88 72 89 15
Fax: (+82 32) 882 17 14
e-mail: uri@urief.co.kr
Web: www.urief.co.kr

High-speed craft operated

Type	Name	Seats	Additional payload	Delivered
CAT Fjellstrand Flying Cat 40 m	*Seaplane*	230	none	1994
CAT Fjellstrand Flying Cat 40 m	*Paradise*	380	none	1994
CAT Fjellstrand Flying Cat 40 m	*Princess*	426	none	1996

Kuwait

Failaka Heritage Village

Faraika Island, Kuwait

Tel: (+965 22) 449 88
e-mail: chartley@safirhotels.com
Web: www.failakaheritagevillage.com

High-speed craft operated

Type	Name	Seats	Delivered
CAT Wavemaster 24 m	*Umm Al Khair* (ex-*Royal Star*)	96	1997
CAT Crowther 26 m	*Bint Al Khair* (ex-_*Peppermint Bay I*)	149	2001

Bint Al Khair 1325084

Operations
Trips to Failaka Island.

Malaysia

Ampangship Marine

B-3A-6 Block B, Tingat 3, Megan Avenue II, 12 Jalan Yap Kwan Seng, 50450 Kuala Lumpur, Malaysia

Tel: +60 3 2171 2944
Fax: +60 3 2171 2943
e-mail: ampang@ampangship.com.my
Web: www.ampangship.com.my

High-speed craft operated

Type	Name	Seats	Delivered
MH SFCN Marlin 33	*Ruby Express*	90	1986
MH SFCN Marlin 33	*Delinia Express*	90	1986

Fast Ferry Ventures Sdn Bhd

No 11 Lebuh Light, 10200 Penang, Malaysia

Tel: (+60 4) 263 13 98
Fax: (+60 4) 263 13 99
e-mail: ffv@tm.net.my

High-speed craft operated

Type	Name	Seats	Delivered
MH Wavemaster 33	*Langkawi II*	200	1994
MH Wavemaster 33	*Langkawi III*	200	1994
MH Wavemaster 27	*Senang Ekspres*	140	1989
MH Wavemaster 27	*Suka Ekspres*	140	1988

Operations
Penang to Medan
Penang to Langkawi.

Langkawi II 0007300

Lada Langkawi Holdings

Tingkat 4/6, Kompleks Lada, Jalan Persiaran Putra, 07000 Kuah, Langkawi, Kedah Darul Aman, Malaysia

Tel: (+60 4) 966 30 88
Fax: (+60 4) 966 30 66
e-mail: ritzalian@hotmail.com

High-speed craft operated

Type	Name	Seats	Delivered
CAT SBF 32 m	*Lada Satu*	250	1993
CAT SBF 32 m	*Lada Dua*	250	1993
CAT SBF 32 m	*Lada Tiga*	250	1994
CAT SBF 32 m	*Lada Empat*	250	1994
CAT SBF 20 m	*Lada Lima*	100	1994

Operations
Kuala Kedah and Kuala Perlis, Malaysia to Langkawi Island.

Langkawi Ferry Service SDN BHD

A subsidiary of Nauticalink Berhad
37/39 Jalan Pandak Mayah 5, Pusat Bandar Kuah, 07000 Langkawi, Kedah, Darul Aman, Malaysia

Tel: (+60 4) 966 63 16
Fax: (+60 4) 966 71 90
e-mail: info@langkawi-ferry.com
Web: www.langkawi-ferry.com

Ooi Cheng Choon, *Managing Director*

High-speed craft operated

Type	Name	Seats	Delivered
MH PT Palindo 30 m	*Impian 2*	164	1997
MH PT Palindo 25 m	*Impian 3*	93	1997
MH PT Palindo 30 m	*Lambaian 1*	162	1997
MH PT Palindo 30 m	*Lambaian 2*	164	1998
MH PT Palindo 30 m	*Lambaian 3*	155	1998
MH PT Palindo 28 m	*Kenangan 1*	117	1999
MH PT Palindo 31 m	*Kenangan 2*	191	1999
MH PT Palindo 34 m	*Kenangan 3*	192	2000
MH PT Palindo 34 m	*Kenangan 6*	218	2001
MH PT Palindo 28 m	*Alaf Baru 1*	116	2000
MH PT Palindo 31 m	*Alaf Baru 2*	152	2000
CAT Marinteknik 36 m	*Selesa Ekspres*	350	1990
MH	*Satun I*	110	–

Operations
Langkawi Island to Kuala Kedah, Kuala Perlis, Penang and Satun (Thailand)
Penang to Medan (Indonesia), Tanjung Gemuk to Tioman, Kukup to Tj. Balai (Indonesia).

Sriwani Tours and Travel

418 Lebuh Chulia, Penang, Malaysia

Tel: (+60 604) 262 85 35
Fax: (+60 604) 261 40 76

High-speed craft operated

Type	Name	Seats	Delivered
CAT InCat/NQEA 24 m	*Payar*	210	1994

Operations
Langkawi to Pulau Payar reef.

Malta

Virtu Ferries Ltd

A subsidiary of Virtu Holdings Ltd
Ta'Xbiex Terrace, Ta'Xbiex, MSD 11, Malta

Tel: (+356) 21 31 67 66
Fax: (+356) 21 31 45 33
e-mail: admin@virtuferries.com
Web: www.virtuferries.com

F A Portelli, *Managing Director*
S Attard Montalto, *Director*
H Saliba, *Director and General Manager*

MV Maria Dolores 1120254

High-speed craft operated

Type	Name	Seats	Additional payload	Delivered
CAT Austal 67 m	*Maria Dolores*	600	65	2006

Operations
Valetta, Malta to Pozzallo, Sicily, 52 n miles, 1 hour 30 minutes
Valetta, Malta to Catania, Sicily, 113 n miles, 3 hours.
(All operations are ISM certified).

Mexico

Cruceros Maritimos Del Caribe SA DE CV

Calle 6 Nte No 14, Cozumel, Quintana Roo 77600, Mexico

Tel: (+52 987) 872 15 88
Fax: (+52 987) 872 19 79
e-mail: info@crucerosmaritimos.com.mx
Web: www.crucerosmaritimos.com.mx

Lorenzo Molina Caceres, *Chairman of the Board*
Mario Arturo Molina Caceres, *Chief Executive Officer*
Captain Manuel Pirez, *Operations Manager*
Eduardo Collado, *Technical Manager*

High-speed craft operated

Type	Name	Seats	Delivered
CAT Fjellstrand 38.8 m	*Mexico*	450	1986
MH Breaux's Baycraft 30 m	*Mexico II*	300	1988
MH Breaux's Baycraft 43 m	*Mexico III*	800	1993
CAT Fjellstrand 40 m	*Mexico IV* (ex-*Sifka Viking*)	450	1994
CAT Fjellstrand 40 m	*Mexico V* (ex-*Kraka Viking*)	450	1994

Mexico II 0506758

Mexico 0043237

Mexico IV 1048744

Mexico III 0043238

Operations
Mexican Caribbean cruise ship support services
Cozumel to Playa del Carmen and Cancun.

Ultramar

Av. Jose Lopez Portillo, S.M. 84 M-5 L-6, Pto Juarez, CP 77520, Cancun, Mexico

Tel: (+52 998) 843 2011
e-mail: info@granpuerto.com.mx
Web: www.granpuerto.com.mx

High-speed craft operated

Type	Name	Seats	Delivered
MH	*Ultramar*	300	–
CAT Fjellstrand Flying Cat 40 m	*Ultramar 1* (ex-*Calypso 1*)	500	1995
MH Midhipmarine 39 m	*Ultramar II*	450	2004
MH Midhipmarine 39 m	*Ultramar III*	450	2004
MH Midhipmarine 39 m	*Ultramar IV*	450	2005
CAT	*Ultramar IV*	150	(2007)
CAT	*Ultramar X*	150	(2007)
CAT	*Ultramar XI*	150	(2007)
CAT	*Ultramar XII*	150	(2007)
CAT	*Ultramar XIII*	150	(2007)

Operations
Cancun to Isla Mujeres
Playa del Carmen to Cozumel.

Netherlands

Connexxion Fast Flying Ferries

PO Box 224, NL-1200 AE Hilversum, Netherlands

Tel: (+31 251) 26 20 10
Fax: (+31 251) 22 53 72
Web: www.connexxion.nl

Fast Flying Ferries started operations in 1998.

High-speed craft operated

Type	Name	Seats	Delivered
HYD Voschod-2	*Anne Marie**	79	1979
HYD Voschod-2	*Archimedes**	79	1985
HYD Poleseye	*Meteoor II*	57	1996
HYD Voskhod-2M	*Karla*	79	2002
HYD Voskhod-2M	*Catharina Amalia*	79	2002
HYD Voskhod-2M	*Rosanna*	79	2002

*These vessels are for sale.

Operations
Ijmuiden to Amsterdam
Velsen to Amsterdam.

Veolia Transport

Mastbosstraat 12, Postbus 3306, NL-4800 DH Breda, Netherlands

Tel: (+31 76) 528 10 00
Fax: (+31 76) 522 11 91
Web: www.bba.nl

High-speed craft operated

Type	Name	Seats	Delivered
SWATH Damen 3717	*Princess Maxima*	181	2004
SWATH Damen 3717	*Prince Willem Alexander*	181	2004

Operations
Passenger and bicycle ferry across Westerschelde estuary.

Rederij G Doeksen En Zonen Bv

Waddenpromenade 5, 8861 NT Harlingen, 1900-Doeksen, Netherlands

Tel: (+31 562) 44 20 02
Web: www.rederij-doeksen.nl

P J M Melles, *Managing Director*

High-speed craft operated

Type	Name	Seats	Delivered
CAT FBM Tricat 50 m	*SuperCat 2002*	415	1998

Operations
Harlingen to Terschelling, 50 minutes
Harlingen to Vlieland, 45 minutes
Terschelling to Vlieland, 30 minutes.

Koegelwieck *operated by BV Rederij G Doeksen en Zonen* 0507008

Waterbus

Voor de Waterbus gelden de Vervoervoorwaarden van het KNV, Deze zijn verkrijgbaar bij, Koninklijk Nederelands Vervoer, Postbus 55, NL-2501 CB Den Haag, Netherlands

e-mail: info@waterbus.nl
Web: www.waterbus.nl

High-speed craft operated

Type	Name	Seats	Delivered
CAT Damen/NQEA RR 150	–	130	1999
CAT Damen/NQEA RR 150	–	130	1999
CAT Damen/NQEA RR 150	–	130	1999
CAT Damen/NQEA RR 150		130	1999
CAT NQEA PR 200	*Piet Hein*	150	1999
CAT NQEA PR 200	*Witte de With*	150	1999
CAT IRIS 3.1	*Maarten Tromp*	144	2002

Operations
Dordrecht to Zwijndrecht, Papendrecht, Hollandse Biesbosch, Sliedrecht and Rotterdam.

Netherlands Antilles

Rapid Explorer

Dock Maarten, Great Bay Marina, J Yrausquin Boulevard, Philipsburg, St Maarten, Netherlands Antilles

Tel: (+599) 542 97 62
Fax: (+599) 542 97 82
e-mail: info@rapidexplorer.com
Web: www.rapidexplorer.com

High-speed craft operated

Type	Name	Seats	Delivered
Hydrocruiser 22.5 m	*Sea Shuttle I*	117	1994
Hydrocruiser 27 m	*HC Katia*	149	2004
Hydrocruiser 27 m	*HC Milancia*	149	2004

Operations
St Maarten to St Barths
Antigua to Montserrat.

New Caledonia

Compagnie Maritime des Iles

2 avenue Henri Lafleur, Quai des Caboteurs, BP-12241, 98802 Nouméa Cedex, New Caledonia

Tel: (+687) 27 04 05
Fax: (+687) 25 90 93
e-mail: cmisa@lagoon.nc

High-speed craft operated

Type	Name	Seats	Delivered
CAT Austal 52 m	*Betico*	366	1999

Operations
Noumea to the Loyalty Islands.

Mary-D Enterprises

Galerie Palm Beach, Promenade Roger Laroque, Anse Vata, BP 233-98845, Noumea, New Caledonia

Tel: (+687) 26 31 31
Fax: (+687) 26 39 79
e-mail: amedee@mary-d.nc
Web: www.amedee.ws

High-speed craft operated

Type	Name	Seats	Delivered
MH Image Marine 30 m	*Mary D Dolphin*	184	1998

Operations
Noumea to Amedee Island.

New Zealand

Dolphin Discoveries

PO Box 400, Paihia 0242, Bay of Islands, New Zealand

Tel : (+64 9) 402 82 34
Fax: (+64 9) 402 60 58
e-mail: info@dolphinz.co.nz
Web: www.dolphinz.co.nz

High-speed craft operated

Type	Name	Seats	Delivered
F-CAT Teknicraft 18	*Discovery IV*	105	1998

Operations
Dolphin and whale watching excursions.

Fullers Bay of Islands Ltd

PO Box 145, Maritime Building, Paihia, Bay of Islands, New Zealand

Tel: (+64 9) 402 74 21
Fax: (+64 9) 402 78 31
e-mail: info@fboi.co.nz
Web: www.fboi.co.nz

Kit Nixon, *Chief Executive Officer*
Tony Molloy, *Operating Manager*
Allan Field, *Sales Manager – Bay of Islands*
Dominic Strobel, *International Sales Manager*

The Fullers Company is owned by Tourism Holdings Ltd. Fullers has plied the Bay of Islands since 1886 and the company presently operates a fleet of eight vessels, including an 18 m 40-passenger underwater viewing craft.

High-speed craft operated

Type	Name	Seats	Delivered
CAT Terry Bailey 11 m	*Tutunui*	35	1993
CAT Circa Eng/ Technicraft 19 m	*Tangaroa*	165	1999
CAT Sabre 22.3 m	*Tiger V* (ex-*Stewart Islander*)	218	2004

Operations
Cape Brett 'Hole in the Rock' cruise, 4 hours, fare NZD65
Dolphin watching, swimming with dolphins, 4 hours NZD95
Cream Trip Supercruise, 6 hours NZD95.

Fullers Group Ltd

PO Box 1346, Auckland, New Zealand

Tel: (+64 9) 367 91 11
Fax: (+64 9) 379 91 48
e-mail: enquires@fullers.co.nz
Web: www.fullers.co.nz

George R Hudson, *Chairman*
Douglas G Hudson, *Chief Executive Officer*
Michael Fitchett, *General Manager, Support Services*

A subsidiary of the Stagecoach company.

High-speed craft operated

Type	Name	Seats	Delivered
CAT SBF Engineering 35 m	*Quickcat*	600+	March 1987
MH WaveMaster Int 37 m	*Jet Raider*	400	December 1990
CAT Fast Craft	*Seaflyte*	170	December 1993
CAT Sabre	*Quickcat II*	200	December 1993
CAT Wavemaster Int 41 m	*SuperFlyte Auckland*	600	November 1996
CAT Wavemaster Int 33 m	*StarFlyte*	300	1999

Operations
Inner and Outer Hauraki Gulf destinations
Half Moon Bay to Auckland CBD
Auckland to Waiheke Island.

Superflyte *operated by Fullers Group Ltd* 0007302

Pine Harbour Ferries

PO Box 57, Beachlands, Auckland, New Zealand

Tel: (+64 9) 536 47 25
Fax: (+64 9) 536 56 10
e-mail: ferry@pineharbour.co.nz
Web: www.pineharbour.co.nz

High-speed craft operated

Type	Name	Seats	Delivered
CAT Teknicraft/Q-West 15 m	*Clipper II*	49	2006

Operations
Pine Harbour to Auckland, 11 return trips per day, 35 minutes.
Pine Harbour to Waiheke and Rakino Island, weekends and public holidays.

Real Journeys

PO Box 1, Lakefront Drive, Te Anau, New Zealand

Tel: (+64 3) 249 74 16
Fax: (+64 3) 249 70 22
e-mail: dhawkey@fiordlandtravel.co.nz
Web: www.realjourneys.co.nz

Dave Hawkey, *Chief Executive*

Commander Peak *operated by Real Journeys* 0507012

23 m Luminosa *operated by Real Journeys* 1198278

High-speed craft operated

Type	Name	Seats	Delivered
CAT InCat 18 m	*Fiordland Flyer*	90	1984
CAT NQEA InCat 22 m	*Commander Peak*	140	1985
CAT InCat Crowther 32 m	*Patea Explorer*	195	2005
CAT RDM 23 m	*Luminosa*	95	2006

Operations
Lake Manapouri and Doubtful Sound.

Stewart Island Experience

PO Box 89, Stewart Island, New Zealand

Tel: (+64 3) 212 76 60
Fax: (+64 3) 212 83 77
e-mail: info@sie.co.nz
Web: www.sie.co.nz

High-speed craft operated

Type	Name	Seats	Delivered
CAT Ellis/Gough	*Southern Express*	62	1997
CAT Sabre 20 m	*Foveaux Express*	100	1999

Operations
Stewart Island to Bluff (Foveaux Strait).

Whale Watch Kaikoura

PO Box 89, Kaikoura, South Island, New Zealand

Tel: (+64 3) 319 67 67
Fax: (+64 3) 319 65 45
e-mail: res@whalewatch.co.nz
Web: www.whalewatch.co.nz

Marcus Solomon, *General Manager, Operations*
Lisa Thomas, *Administration*

Whale Watch Kaikoura operate a fleet of high-speed ferries on whale watch and nature tours from Kaikoura. The company is owned and operated by the indigenous Kati Kari people of Kaikoura.

High-speed craft operated

Type	Name	Seats	Delivered
CA Teknicraft F-Cat 18 m	*Aoraki*	54	2002
CA Teknicraft F-Cat 18 m	*Wheketere*	54	1999
CA Teknicraft F-Cat 18 m	*Te Ao Marama*	54	2000
CA Teknicraft F-Cat 18 m	*Paikea*	54	2004
CA Teknicraft F-Cat 18 m	*Tohora*	54	2005

Operations
Whale watch and nature tours from Kaikoura.

Northern Mariana Islands

Tinian Shipping & Transportation

One Broadway, PO Box 468 Tinian, MP96952 Saipan, Northern Mariana Islands

e-mail: tinian.shipping@vzpacifica.net
Web: www.expage.com/page/tinianshipping

High-speed craft operated

Type	Name	Seats	Delivered
CAT 40 m Flying Cat	*Saipan Express*	333	1996
CAT 40 m Flying Cat	*Tinian Express**	333	1996

*Currently undergoing repairs in dry dock.

Operations
Departing Tinian and Saipan Islands daily.

Norway

62° Nord Cruises

Postboks 836 Sentrum, N-6001 Ålesund, Norway

Tel: (+47 70) 11 44 30
Fax: (+47 70) 11 64 46
Web: www.62nord.net

Kari Mette, *General Manager*

High-speed craft operated

Type	Name	Seats	Delivered
CAT 22 m	*Fjord Aalesund*	90	2006

Operations
Fjord cruises.

Arctic Voyager

Grønnegt 114, PB 1282, N-9263 Tromsø, Norway

Tel: (+47 77) 62 4440
Fax: (+47 77) 62 4441
e-mail: info@arcticvoyager.com
Web: www.arcticvoyager.com

High-speed craft operated

Type	Name	Seats	Delivered
CAT Brodrene Aa 23 m	*Cetacea*	97	2006

Operations
Whale watching trips from Tromsø.

BNR Fjordcharter

Kaikanter, N-5286 Haus, Norway

Tel: (+47 56) 39 10 20
Fax: (+47 56) 39 10 21
e-mail: post@fjordcharter
Web: www.fjordcharter.no

High-speed craft operated

Type	Name	Seats	Delivered
CAT Lindstol 22.5 m	*Hardanger Prins*	97	1990
CAT Westamarin 3000	*Kyst Express*	187	1987
CAT Oma 25.75 m	*Snarveien*	180	2005

Operations
Tours of Fjords around Bergen.

FFR

Finnmark Fylkesrederi OG Ruteselskap
PO Box 308, N-9615 Hammerfest, Norway

Tel: (+47) 78 40 70 00
Fax: (+47) 78 40 70 74
e-mail: info@ffr.no
Web: www.ffr.no

Stig Solheim, *Managing Director*

High-speed craft operated

Type	Name	Seats	Additional payload	Delivered
CAT Brødrene Aa Båtbyggeri 25.5 m	*Ingøy*	48	4 cars	1987
CAT OMA 21 m	*Rypøy*	90	3.6 t cargo	1997
CAT OMA 24 m	*Marøy*	142	4.8 t cargo	1997
CAT OMA 24 m	*Rødøyløren*	92	–	1997
CAT OMA 24 m	*Kobbøy*	85	9 t cargo	1998
CAT OMA 30 M	*Renøy*	185	3 cars	1999
CAT OMA 29 m	Jernøy	145	3 cars	2007

Operations
Masry to Hammerfest to Sørøysundbass
Masry to Havrøysund to Hammerfest
Kirkenes to Murmansk, summer service.
Alta to Hammerfest.

Fjord 1 Fylkesbaatane AS

PO Box 354, Strandavegen, N-6900 Florø, Norway

Tel: (+47) 57 75 70 00
Fax: (+47) 57 75 70 01
e-mail: fylkesbaatane@fjord1.no
Web: www.fjord1.no

Stig Kristoffersen, *Director*

High-speed craft operated

Type	Name	Seats	Additional payload	Delivered
CAT Fjellstrand 38.8 m	*Fjordprins*	193	–	October 1987
CAT Båtrustning 19.2 m	*Delfinjr*	82	–	1988
CAT Fjellstrand 40 m Flying Cat	*Kommandøren*	282	–	June 1990
CAT Fjellstrand 38.8 m	*Solundir*	275	–	1989
CAT Lindstol 25.4 m	*Sylvarnes*	50	8 cars	2000
CAT Lindstøl 32 m	*Fjordtroll*	193	–	2002
CAT Lindstøl 25 m	*Fjordglytt*	81	8 cars	2000
CAT OMA 24.4	*Vetlefjord*	60	8 cars	2000
CAT Båtservice 29 m	*Tansoy*	96	10 cars	2007

Operations
Bergen to Nordfjord
Bergen to Sogn.

Fjordprins *operated by Fylkesbaatane* 0507093

Fjord 1 MRF A/S

Part of the Fjord1 Group
N-6405 Molde, Norway

Tel: (+47) 71 21 95 00
Fax: (+47) 71 21 95 01
email: mrf@fjord1.no
Web: www.fjord1.no/mrf

Anker Grøvdal, *Managing Director*
Frode Ohr, *Assistant Director*
Jøran Nyheim, *Economics Manager*
John Aåge Rasmussen, *Technical Director*
Ståle Tangen, *Head of Quality Department*
Tor Kristoffersen, *Nautical Personnel Manager*
Arild Reistadbakk, *Marketing Manager*

Hjørungavåg *(background) and* Lauparen 0506747

Fjørtoft 0506748

High-speed craft operated

Type	Name	Seats	Delivered
CAT Fjellstrand 31.5 m	*Lauparen*	232	1983
CAT Brødrene Aa Båtbyggeri CIRR 265P	*Fjørtoft*	120	1988
MH Brødrene Aa Båtbyggeri	*Romsdalsfjord*	92	1987
CAT K.Fjellstrand 35 m	*Hjørungavåg*	300	1998
CAT K.Fjellstrand 30	*Grip*	150	2000

Operations
Ålesund to Valderøy to Hareid.
Molde to Helland to Vikebukt.
Ålesund to Valderøy to Nordøyane.
Ålesund to Langevåg.

Fosen Trafikklag A/S

Pirterminalen, N-7486 Trondheim, Norway

Tel: (+47) 73 89 07 00
Fax: (+47) 73 89 07 50
e-mail: firmapost.fosen.no
Web: www.fosen.no

High-speed craft operated

Type	Name	Seats	Delivered
CAT Lindstøl	*Fosningen*	114	1991
CAT Mjosundet	*Frøyfart*	98 + cargo	1991
CAT Harding 29 m	*Agdenes*	210	1991
CAT Brodrene 24.5 m	–	130	(2007)
CAT Brodrene 24.5 m	–	130	(2007)

Operations
Trondheim to Brekstad to Storfosna to Fjellvoer to Fillan to Sistranda to Mausundvøer to Vadsøysund to Bogoyvøer to Sula
Sistranda to Sula to Froan
Trondheim to Kristansund, 3 hours 15 minutes
Trondheim to Vanvikan, 25 minutes.

Gjendebatene

N-2680 Vaga Norway

Tel: (+47 61) 23 85 09
Fax: (+47 61) 23 85 09
e-mail: post@gjende.no
Web: www.gjende.no

High-speed craft operated

Type	Name	Seats	Delivered
MH Båtutrustning 19 m	*Gjendine* (ex-*Clipper Skyss*)	88	1990

Operations
Gjendesheim to Memruba to Gjendebu.

Tide ASA

Postboks 6300, N-5893 Bergen, Norway

Tel: (+47) 552 38 79
Fax: (+47) 55 23 87 01
e-mail: post@tide.no
Web: www.tide.no

Per Atle Tellnes, *Managing Director*
Gier AGA, *Manager*
Terje Fivelstad, *Technical Manager*
Alexandra Angell, *Marketing Manager*

Tide ASA was formed in 2006 from the merger of Hardanger Sunnhordlandske Dampskipsselskap (HSD) with Gaia Trafikk.

High-speed craft operated

Type	Name	Seats	Delivered
CAT Harding 29 m	*Tedno*	173	1992
CAT Båtservice Sea Lord 32 m	*Beinveien*	177	1992
CAT Oma 19.8 m	*Vøringen*	80	1995
CAT K Fjellstrand 40 m	*Vingtor* (ex-*Jet Cat*, ex-*Pattaya Express*)	–	1996
CAT Austal P.Nautica 42 m	*Draupner**	358	1999
CAT Fjellstrand 38.8 m	*Tjelden*	195	1986
CAT Brodrene Aa 18.5 m	*Nordic Queen***	97	2005
CAT Fjellstrand 33 m	*Teisten*	180	2006
CAT Fjellstrand 33 m	*Tranen*	120	2006
CAT Oma 22.5 m	*Tedno*	85	2007

*The 42 m *Sleipner* (sister ship to *Draupner*) sank after hitting rocks in November 1999.
**Leased from Cruise Service.

Operations
Bergen to Austevoll to Sunnhordland approximately 80 n miles
Bergen to Kleppeestø, 2.8 n miles, 7 minutes
Os to Sunnhordland approximately 50 n miles
Flaggruten service, Bergen to Stavanger approximately 104 n miles.

Vingtor 0103292

Draupner 0097462

Helgeland SKA A/S

N-8805 Sandnessjøen, Norway

Tel: (+47) 75 06 41 00
Fax: (+47) 75 04 02 56
e-mail: epost@helgelandske.no
Web: www.helgelandske.no

Frode Neillson, *Managing Director*
Roger Hansen, *Marketing Manager*

High-speed craft operated

Type	Name	Seats	Additional payload	Delivered
CAT Rosendal 29 m	*Thorolf Kveldulfsøn*	100	–	1995
CAT Båtservice 36 m	*Helgeland*	200	14 t	1997
CAT Oma 26 m	*Donna*	60	8 cars	2003

Operations
Sandnessjøen to Nesna to Lurøy to Træna
Sandnessjøen to Engan to Austbø to Søndre to Heroysteder to Vega
Solfjellsjøen to Vandve to Herøy.

Hurtigruten Group AS

PO Box 43, N-8501, Narvik, Norway

Tel: (+47 76) 96 76 00
e-mail: firmapost@hurtigruten.com
Web: www.hurtgruten.com

Formed by the merger of Troms Fyles Dampskibsselskop and Ofoten og Vestertalens Dampskibsselskop in March 2006, the company operates a large fleet of ferries in Nordland and Troms.

High-speed craft operated

Type	Name	Seats	Additional payload	Delivered
MH Fjellstrand 25.5 m	*Øykongen*		94	1982
MH Fjellstrand 25.5 m	*Øydronningen*			1982
CAT Westamarin 3000	*Børtind*	130	30 m³	1989
CAT Harding Verft 29 m	*Skogoy*	130	30 m³	1991
CAT Fjellstrand 35 m	*Ofoten*	199	Cargo Pallets	1997
CAT Båtservice Sea Lord 28	*Fjordsol*	150	–	1991
CAT Austal 41.5 m	*Salten*	216	12 tonnes	2003
CAT Austal 41.5 m	*Steigtind*	216	12 tonnes	2003
CAT Batservice 38 m	*Fjorddronningen II*	336	–	1995
CAT Oma 24 M	*Rødøyløven*	92		1997
CAT OMA 24 m	*Vågsfjord*	146	–	1999
CAT OMA 30 m	*Fjordprinsessen*	198	–	1999
CAT Fjellstrand Flying Cat 46	*Fjorddronningen*	346	–	2004
CAT Brødrene Aa 23 m	*Kvaenangen*	50	1 vehicle	2005
CAT Fjellstrand Flying Cat 46	*Fjordkongen*	340	–	2005

Operations
Tromsø to Harstad
Bodø to Sandnessjøen, Svolvaer and Helnessund.

Jon Arne Helgøy A/S

4167 Helgøy i Ryfylke, Norway

Tel: (+47 51) 75 22 06
Fax: (+47 51) 75 22 67
Web: www.helgoy.no

High-speed craft operated

Type	Name	Seats	Delivered
MH GS Marine Scirocco 50	*Helgøy Prinsess*	50	2005

Operations
Fjord trips.

L Rødne and Sønner A/S

Skagenkaien 18, N-4006 Stavanger, Norway

Tel: (+47) 51 89 52 70
Fax: (+47) 51 89 52 02
e-mail: mail@rodne.no
Web: www.rodne.no

High-speed craft operated

Type	Name	Seats	Delivered
Brødrene Aa 6829 Hyen	*Clippercruise*	115	1981
MH Brødrene Aa 24.4 m	*Clipper*	120	1982
Brødrene Aa 6829 Hyen	*RygerAa*	30	1984
MH Brødrene Aa TMS 16.0 m	*Rygerspeed*	52	1985
Brødrene Aa Båtbyggeri AS	*Rygerskyss*	52	1986
MH Brødrene Aa 19.0 m	*Rygerdoktoren*	12	2002
CAT Brødrene Aa 18.5 m	*Rygerkatt*	62	2003
Brødrene Aa 6829 Hyen	*Rygerstril*	25	2004
CAT Brødrene Aa 20.0 m	*Rygerfjord*	97	2004
CAT Brødrene Aa 23.0 m	*Rygerkongen*	130	2006
CAT Brødrene 23 m	*Rygerfonn (Hardangerfjord Express)*	97	2007

Operations
Stavanger to Pulpit Rock and Lysebotn
Rosendal to Bergen (1 hour 55 minutes).
Sightseeing Lysøen and Rosendal.

Master Ferries

PB 593, N-4665 Kristiansand, Norway

Tel: (+47 815) 265 00
Fax: (+47 38) 09 20 62
e-mail: info@masterferries.com
Web: www.masterferries.com

Morten Larsen, *Operation Manager*

High-speed craft operated

Type	Name	Seats	Additional payload	Delivered
CAT InCat Tasmania 91 m	*Master Cat* (ex *Mad Mols*, ex *Cat-Link V*)	676	220 cars	1998

Operations
Kristiansand to Hanstholm.

Namsos Trafikkselskap ASA

PO Box 128, N-7801 Namsos, Norway

Tel: (+47) 74 21 63 00
Fax: (+47) 74 21 63 01
e-mail: firmapost@ntsasa.no
Web: www.ntsasa.no

High-speed craft operated

Type	Name	Seats	Delivered
CAT Lindstøl Skip 22.5 m	*Namdalingen*	107	1990

Operations
Leka to Rørvik to Namsos.

Namdalingen 0506749

Nesodden-Bundefjord Dampskipsselskap A/S

Stranden 1, N-0250 Oslo, Norway

Tel: (+47) 22 87 64 20
Fax: (+47) 22 83 09 48

Prinsessen 0507009

e-mail: nbds@online.no
Web: www.nbds.no

High-speed craft operated

Type	Name	Seats	Delivered
CAT Harding	*Prinsessen*	175	1990
CAT Fjellstrand/Admiral 36	*Prinsen*	290	1998

Operations
Nesodden to Lysaker
Oslo to Drøbak and Son (May to August)
Oslo to Vollen to Slemmestad
Oslo to Nesodden.

North Cape Minerals A/S

PO Box 45, N-1309 Rud, Norway

Tel: (+47) 67 15 22 00
Fax: (+47) 67 15 22 01
e-mail: sales@ncm.no
Web: www.ncm.no

Sven Fagerli, *Plant Manager*
Sigbjorn Solli, *Shipping Manager*

High-speed craft operated

Type	Name	Seats	Additional payload	Delivered
MH Fjellstrand 26 m	*Nefelin IV*	84	32 m^3	1980

Operations
This craft operates a 47 km route from Alta to Sjernoey three times a day with a crossing time of 1 hour 10 minutes.

Fjellstrand Nefelin IV *operated by North Cape Nefelin* 0506746

SeaAction AS

Postboks 6, 3159 Melsomvik, Tønsberg, Norway

Tel: (+47) 33 33 69 63
Fax: (+47) 33 33 62 77
email: post@seaaction.no
Web: www.seaaction.no

High-speed craft operated

Type	Name	Seats	Delivered
MH Cirrus M27 P/LC	*Klipperfjord*	49	1985

Stavangerske AS

A subsidiary of DSD

PO Box 839, Sentrum, N-4004 Stavanger, Norway

Tel: (+47) 51 86 87 00
Fax: (+47) 51 89 54 25
e-mail: post@stavangerske.no
Web: www.stavangerske.no

Egil Nylund, *Director*

High-speed craft operated

Type	Name	Seats	Delivered
MH Batutrustning 18 m	*Lyse Ekspress-*	58	1988
CAT Brodrene 27 m	*Fjordkatt*	180	2006
CAT Brodrene 27 m	*Fjordfart*	180	2007
CAT Brodrene 27 m	*Fjorddrott*	180	2007

Operations
Stavanger to Kalvøy.
Stavanger to Hommersåk.
Stavanger to Ryfylke and Hjelmelland.

Torghatten Trafikkselskap ASA

Havnegt 40, N-8905 Brønnøysund, Norway

Tel: (+47) 75 01 81 00
Fax: (+47) 75 01 81 01
e-mail: post@tts.no
Web: www.tts.no

High-speed craft operated

Type	Name	Seats	Additional payload	Delivered
CAT Westcraft 24 m	*Ortind*	30	8 cars	1998
CAT Rosendal Verft 29 m	*Vegtind*	100	–	1995
CAT OMA 28 m	*Strandafjord*	180	–	2002
CAT Båtservice 28 m	*Vindafjord*	180	–	2002

Operations
Brønnøysund to Vega to Sandnessjøen.

Pakistan

Pakistan Water and Power Development Authority

WAPDA House, Sharah-e-Quaid-e-Azam, Lahore, Pakistan

Fax: (+92) 024 54

High-speed craft operated

Type	Name	Seats	Delivered
HOV Griffon Hovercraft Ltd 1000TD (008) (GH 9456)	–	–	1987
HOV Griffon Hovercraft Ltd 1000TD	–	–	1993

Griffon Hovercraft 1000 TD (008) operated by the WAPDA of Pakistan 0506752

Philippines

Aquajet Maritime

Seaborne Bldg, 4203 Magsaysay Blvd, Metro Manila, Philippines

High-speed craft operated

Type	Name	Seats	Delivered
CAT MES CP 15	*Aquajet II*	196	1989
CAT MES CP 15	*Aquajet III*	190	1990
CAT Miho Shipyard Co Ltd	*Aquajet Super III*	370	1994

Aquajet III 0507005

Delta Fast Ferries

Water Jet terminal, Pier 4, Port Area, Cebu City, Philippines

Tel: (+63 32) 232 62 37
Fax: (+63 32) 232 62 95

High-speed craft operated

Type	Name	Seats	Delivered
MH 20 m	*Delta I*	78	1997
CAT Sabre 32 m	*Delta II*	300	1997

Operations
Dumaguete, Siquijor, Cebu and Larena.

El Greco Jet Ferries Inc

70 Sgt EA Esguerra Avenue, Quezon City 1107, Metro Manila, Philippines

Tel: (+63 2) 929 70 52
Fax: (+63 2) 921 33 54
e-mail: egif@elgrecomanila.com
Web: www.elgrecomanila.com

High-speed craft operated

Type	Name	Seats	Delivered
Wavemaster 30.5 m	*Kristen*	176	1998

Operations
Manila to Bataan.

Mt Samat Ferry Express Inc

2/F Skyfreight Bldg, Ninoy Aquino Ave, Parañaque City, Philippines

Tel: (+63 551) 52 90
Fax: (+63 551) 52 91

High-speed craft operated

Type	Name	Seats	Delivered
Cat Sabre 22 m	*Ok Ka Ferry I*	200	1997
CAT Crowther 28 m	*Mt Samat Ferry 3*	180	2000
CAT Crowther 28 m	*Mt Samat Ferry 5*	180	2000

Operations
Manila to Orion, 50 minutes.

Oceanjet

A division of Socor Shipping Lines
General McArthur Boulevard, Corner General Maxilom Street, Cebu City 6000, Philippines

Tel: (+63 32) 255 75 60
Fax: (+63 32) 255 01 15
email: customerservice@oceanjet.net
Web: www.oceanjet.net

High-speed craft operated

Type	Name	Seats	Delivered
MH Miho 28 m	*Oceanjet 1*	160	1995
MH Miho 30 m	*Oceanjet 2*	200	1997
MH Cheoy Lee 32 m	*Oceanjet 3*	285	2001
MH Cheoy Lee 32 m	*Oceanjet 5*	287	2002
MH Cheoy Lee 32 m	*Oceanjet 6*	285	2003

Operations
Cebu to neighbouring islands.

Supercat Fast Ferry Corporation

Pier 4, North Reclamation Area, Cebu City, Cebu 6000, Philippines

Tel: (+63 22) 232 45 11
Fax: (+63 22) 234 96 67
email: scat-customercare@supercat.com.ph
Web: www.supercat.com.ph

Supercat Fast Ferry Corporation (formerly Philippine Fast Ferry) is a company formed under a joint venture agreement between Aboitiz Transport Systems Inc, Parkview Group of Hong Kong.

High-speed craft operated

Type	Name	Seats	Delivered
CAT Marinteknik 41 m	*SuperCat 2* (ex-*Camoes*)	306	1995
CAT Marinteknik 41 m	*SuperCat 9* (ex-*St Cruz*)	306	1996
CAT FBM Tricat 50 m	*Supercat 2002* (ex-*Tricat 3*)	403	2002

Operations
Cebu to Dapitan, Dumagute, Larena, Ormoc and Tagbilaran
Calapan on Mindoro Island to Batangas City
Bacolod to Iloilo City.

Weesam Express

Amils Tower, Pilar Street, Zamboanga City, Philippines

Tel: (+63) 992 37 56
Fax: (+63) 992 07 30
e-mail: zam@weesamexpress.com
Web: www.weesamexpress.com

High-speed craft operated

Type	Name	Seats	Delivered
MH Wong 39.5 m	*Weesam Express*	278	1997
MH Wong 42 m	*Weesam Express 2*	165	1998
MH Wong 42 m	*Weesam Express 3*	197	1999
MH Wong 43 m	*Weesam Express 5*	288	1999
MH Wong 41 m	*Weesam Express 7*	300	2006

Operations
Cagayan De Oro to Camiguin
Bacolod to Iloilo
Ceby, Dumaguete and Siquijor.

Poland

Zegluga Gdanska

80-830 Gdansk ul, Ponczoszikow 2, Skrytka Pocztowa 91, Gdansk 1, Poland

Tel: (+48 58) 301 74 26
Fax: (+48 58) 301 74 20
e-mail: zegluga@zegluga.pl
Web: www.zegluga.gda.pl

High-speed craft operated

Type	Name	Seats	Delivered
HYD Kolkhida	*Delfin IV*	155	1986
HYD Raketa	*Raketa 02*	66	–
HYD Raketa	*Raketa 03*	66	–
HYD Raketa	*Raketa 05*	66	–
HYD Polesie	*Polesie 09*	46	–
HYD Polesie	*Polesie 11*	46	–
HYD Polesie	*Polesie 12*	46	–

Operations
Gdynia to Hel
Gdansk to Hel to Kaliningrad
Gdynia to Klajpeda
Gdynia to Bornholm (Sweden).

Portugal

Sociedade Fluvial De Transportes

SOFLUS
Estação Sul e Sueste, Avenue Infant D Henrique, P-1100-282 Lisbon, Portugal

Tel: (+351 218) 86 71 50
Fax: (+351 218) 88 37 04
e-mail: geral@soflusa.pt
Web: www.soflusa.pt

High-speed craft operated

Type	Name	Seats	Delivered
CAT Damen 4912	*Damião de Goes*	600	2003
CAT Damen 4912	*Augusto Gil*	600	2003
CAT Damen 4912	*Gil Vicente*	600	2003
CAT Damen 4912	*Fernando Namora*	600	2003

Type	Name	Seats	Delivered
CAT Damen 4912	*Miguel Torga*	600	2003
CAT Damen 4912	*Almeida Garrett*	600	2003
CAT Damen 4912	*Jorge de Sena*	600	2004
CAT Damen 4912	*Fernando Pessoa-*	600	2004
CAT Damen 4912	*Antero de Quental-*	600	2004

Operations
Lisbon to Barrerio.

Transtejo

Rua da Cintura do Porto de Lisbon, Terminal Fluvial do Cais do Sodré, 1249-249 Lisbon, Portugal

Tel: (+351 210) 42 24 00
Fax: (+351 210) 42 24 99
e-mail: geral@transtejo.pt
Web: www.transtejo.pt

Dr Pedro Rolo, *Financial Director*
Dr Teresa Gato, *Commercial Director*

High-speed craft operated

Type	Name	Seats	Delivered
CAT FBM Transcat 45 m	*Algés*	496	1995
CAT FBM Transcat 45 m	*Castelo*	496	1995
CAT FBM/ENM Transcat 45 m	*Chiado*	496	1995
CAT FBM/ENM Transcat 45 m	*Bica*	496	1996
CAT FBM Transcat 46 m	*St Julião*	496	1997
CAT FBM Transcat 46 m	*Sé*	496	1998
CAT FBM/ENM Transcat 46 m	*Aroeira*	496	1998
CAT FBM/ENM Transcat 46 m	*Carnide*	496	1998
CAT Lindstøl 28 m	*Bairro Alto**	202	1996
CAT Lindstøl 28 m	*Parque Das Nacöes**	202	1996
CAT BSC 25 m	*Fantasia*	146	1999
CAT Austal 37 m	*Pedro Nunes*	292	2002
CAT Austal 37 m	*Cesario Verde*	292	2002
CAT BSC 26.4	*Fantasia* (ex-*Suzanne*)	152	2002

*Leased to Cape Verde Islands.

Operations
Terreiro do Paço to Montijo, Seixal and Barreiro
Cais do Sodre to Cacilhas
Belém to Trafaria and Porto Brandão
Ribeira Das Naus to Cacilhas
Parque Das Nações to Seixal, Barreiro and Cacilhas.

Puerto Rico

Puerto Rico Ports Authority

PO Box 362829, San Juan, Puerto Rico 00936-2829

Tel: (+1 787) 863 07 05

Rafael Carrión, *Manager, Acuaexpreso Project*

High-speed craft operated

Type	Name	Seats	Delivered
CAT Nichols Bros InCat 22 m	*Martin Peña*	167	August 1989
CAT Nichols Bros InCat 22 m	*Amelia*	167	November 1989
CAT Nichols Bros InCat 22 m	*Covadonga*	167	December 1989
CAT Nichols Bros InCat 22 m	*San Geronimo*	167	April 1990
CAT Nichols Bros InCat 22 m	*Viejo San Juan*	167	April 1990
CAT Nichols Bros InCat 22 m	*Cristobal Colón*	167	June 1990

Operations
Culebra to Fajardo
Fajardo to Vieques.

Luxury Resorts

El Conquistador Resort, Fajardo, Puerto Rico

Tel: (+1 787) 863 1000
Web: www.luxuryresorts.com/puertorico

High-speed craft operated

Type	Name	Seats	Delivered
CAT inCat Crowther 22 m	*Taino Dancer*	118	2006
CAT inCat Crowther 22 m	*Yunque Princess*	118	2006

Operations
Trips from El Conquistador Resort to Palomino Island.

Russian Federation

Alien Shipping

127 Movkovsky Avenue, St Petersburg, Russian Federation

Tel: (+812) 387 66 67
e-mail: shipping@alientelecom.ru
Web: www.alienshipping.ru

High-speed craft operated

Type	Name	Seats	Delivered
HYD Meteor	*The Andromeda*	98	1982
HYD Voskhod	*The Icarus*	75	1983
HYD Voskhod	*The Penguin*	75	1984
HYD Meteor	*The Perseus*	122	1989
HYD Meteor	*The Hermes*	120	–
HYD Meteor	*The Medeya*	110	–
CAT Sokol 31.5 m	*Kapitan Korsak*	108	1995
HYD Meteor	*The Danaya*	125	–
HYD Meteor	*The Yason*	125	–
HYD Meteor	*The Triada*	125	–

Operations
St Petersburg to Kotka, 4 h 30 min.

Saudi Arabia

Saudi Aramco

PO Box 5000, Dhahran 31311, Saudi Arabia

Tel: (+966 3) 872 01 15
Fax: (+966 3) 873 81 90
Web: www.saudiaramco.com

High-speed craft operated

Type	Name	Delivered
CAT Båtservice Sea Lord 36	*Ain Dar 7*	1992
CAT Båtservice Sea Lord 36	*Ain Dar 8*	1992

Operations
Pollution control vessels.

United Company For Marine Lines

PO Box 3009, Braidah 81999, Saudi Arabia

Web: www.unitedmarinelines.com

Capt Mambouh Rashad, *Manager*

High-speed craft operated

Type	Name	Seats	Additional payload	Delivered
SES Cirrus 120PCAT Incat/Afai	*Red Sea Jet* (ex-*Alnafatha* ex-*Fjordkon Gen*)	–	–	1990
CAT Austal 82 m	*Al Mottahedah1* (ex-*Delphin*)	600	175 cars	1996

Operations
Hurghada to Dubai.

Seychelles

Inter Island Boats

Maison La Rosiere, PO Box 738, Victoria, Seychelles

Tel: (+248) 32 48 43
Fax: (+248) 32 48 45

e-mail: catcacos@seychelles.net
Web: www.catcacos.com

High-speed craft operated

Type	Name	Seats	Delivered
CAT Incat Crowther 35 m	*Cat Cacos*	352	2006

Operations
Mahé to Prashin, 45 mins.

Sierra Leone

Diamønd Airlines Hovercraft Services

Family Kingdom, Aberdeen, Freetown, Sierra Leone

Tel: (+232 22) 29 22 07
Fax: (+232 22) 29 22 08
e-mail: datt@sierratel.sl

Diamond Airlines purchased the AP1-88 hovercraft *Double-O-Seven* from Hovertravel UK to run between Freetown and Lungi, Sierra Leone.

High-speed craft operated

Type	Name	Seats	Delivered
HOV AP1-88	*Double-O-Seven*	98	1989

Operations
Freetown to Lungi Airport.

Singapore

Batam Fast Ferry Pte Ltd

1 Maritime Square, Harbourfront Centre, 09-57 Singapore 099253

Tel: (+65) 62 70 03 11
Fax: (+65) 62 70 03 22
e-mail: enquiries@batamfast.com
Web: www.batamfast.com

High-speed craft operated

Type	Name	Seats	Delivered
MH SBF 31 m	*Sea Flyte*	240	1985
MH WaveMaster 32 m	*Sea Raider II*	250	1986
MH WaveMaster 32 m	*Golden Raider* (ex-*Sea Raider III*, ex-*Sea Spirit*)	250	1989
MH WaveMaster 37 m	*Jet Raider*	330	1991
MH WaveMaster 39 m	*Ocean Raider*	339	1992
MH WaveMaster 31 m	*Ocean Flyte*	200	1995
MH WaveMaster 28.5 m	*Jet Flyte I*	162	1996
MH WaveMaster 28.5 m	*Jet Flyte II*	162	1996
MH WaveMaster 30 m	*Asean Raider I*	200	1997
MH WaveMaster 30 m	*Asean Raider II*	200	1997

Operations
Singapore Harbour Centre to Sekupang
Tanah Merah to Nongsa
Singapore to Teluk Senimba.

Berlian Ferries

02-08, SPI Building, 2 Maritime Square, Singapore 099255

Tel: (+65) 546 88 30
Fax: (+65) 546 88 31

High-speed craft operated

Type	Name	Seats	Delivered
MH Wavemaster 38 m	*Wavemaster 3*	170	2001
MH Wavemaster 38 m	*Wavemaster 5*	170	2001
MH Wavemaster 38 m	*Wavemaster 6*	170	2002
MH Wavemaster 38 m	*Wavemaster 7*	170	2002
MH Wavemaster 38 m	*Wavemaster 8*	170	2002
MH Wavemaster 38 m	*Wavemaster 9*	170	2002
MH Penguin 34 m	*Berlian 2*	230	1997
CAT InCat 30 m	*Berlian 6*	275	1987

Operations
Singapore to Batu Ampar and Tanjung Pinang.

Bintan Resort Ferries Pte Ltd

371 Beach Road, #25-02 Key Point, Singapore 199597

Tel: (+65) 62 94 60 08
Fax: (+65) 62 94 62 08
e-mail: enquiry@brf.com.sg
Web: www.brf.com.sg

High-speed craft operated

Type	Name	Seats	Delivered
CAT Fjellstrand 40 m	*Indera Bupala*	306	1994
CAT Fjellstrand 40 m	*Aria Bupala*	270	1995
CAT Damen DFF 3810	*Arung Mendora*	280	2002

Operations
Singapore to Bintan Island.

Lewek Shipping Pte Ltd

111 Defu Lane 10 #02-01, Victory Building, Singapore, 539226

Tel: (+65) 6285 9788
Fax: (+65) 6285 8755

High-speed craft operated

Type	Name	Seats	Additional payload	Delivered
MH Asia-Pac Geraldston	*Sarah Jade*	80	40 t	2002
MH Strategic Marine	*Sarah Pearl*	120	35 t	2003

Operations
Crewboat in Gulf of Thailand.

Pelican Offshore Services

18 Tuas Basin Link, 638784 Singapore

Tel: (+65) 68 68 06 74
Fax: (+65) 68 98 52 98
e-mail: enquiries@pelican-offshore.com
Web: www.penguin.com.sg

High-speed craft operated

Type	Name	Seats	Additional payload	Delivered
MH Penguin 41 m	*Green Ocean*	100	130 t	1993
MH Penguin 46 m	*Pelican Challenge*	100	150 t	1998
MH Penguin 46 m	*Pelican Glory*	100	150 t	1998
MH Penguin 46 m	*Pelican Venture*	80	150 t	1999
MH Penguin 46 m	*Pelican Pride*	80	150 t	1999
MH Penguin 46 m	*Pelican Vision*	80	150 t	1999
MH Penguin 50 m	*Pelican Champion*	200	–	2000
MH Penguin 40 m	–	70	70 t	2004
MH Penguin 36 m	–	80	75 t	2004

Operations
Crewboat operation in Indonesian waters.

Penguin Ferry Services Pte Ltd

1 Maritime Square, 09-59 Harbour Front Centre, Singapore 099253

Tel: (+65) 637 763 35
Fax: (+65) 627 164 96
e-mail: pfs@penguin.com.sg
Web: www.penguin.com.sg

High-speed craft operated

Type	Name	Seats	Delivered
MH SBF Engineering 25 m	*Penguin 5* (ex-*Auto Batam 5*, ex-*Sundowner 5*)	180	1986
MH SBF Engineering 30 m	*Penguin 6* (ex-*Auto Batam 6*)	200	1990
MH SBF Engineering 30 m	*Penguin 7* (ex-*Auto Batam 7*)	200	1991
MH Al Craft 32 m	*Penguin Success*	267	1992
MH SBF Engineering 30 m	*Penguin 1* (ex-*Auto Batam 1*)	233	1993
CAT NQEA InCat 30 m	*Island Jade*	310	1994
MH Greenbay 32 m	*Penguin 10* (ex-*Auto Batam 10*)	200	1995

Type	Name	Seats	Delivered
MH Greenbay 32 m	*Penguin 12* (ex-*Auto Batam 12*)	200	1995
MH Greenbay 32 m	*Penguin 31* (ex-*Auto Batam 13*)	200	1995
MH Al Craft 37 m	*Merbau ERA*	300	1995
MH Al Craft 35 m	*Penguin Teklong*	200	1995
MH Al Craft 35 m	*Penguin Ever*	250	1995
MH Al Craft 23 m	*Penguin Seletar*	100	1995
MH Greenbay 35 m	*Penguin Tioman*	220	1996
MH Greenbay 35.5 m	*Penguin 16* (ex-*Auto Batam 16*)	220	1996
MH Al Craft 35 m	*Penguin Sakra*	300	1996
MH Marinteknik 35 m	*Penguin 16* (ex-*Auto Batam 16*)	200	1997
CAT Al Craft 23 m	*Penguin Ubin*	100	1999

Operations
Singapore to Batam
Singapore to Bintan
Singapore to Karimun
Singapore to Malaysia.

Tian San Shipping (Pte) Ltd

No 6 Jalan Samulan, Singapore 629123

Tel: (+65) 62 61 62 40
Fax: (+65) 62 64 23 26
e-mail: tsshg@singnet.com.sg

High-speed craft operated

Type	Name	Seats	Delivered
MH Cheoy Lee 28 m	*Sea Falcon*	200	2001
MH Cheoy Lee 28 m	*Sea Osprey*	200	2001

Operations
Singapore to Palau Bukom for Shell Eastern Petroleum.

Waterfront City Resort

Singapore

High-speed craft operated

Type	Name	Seats	Delivered
MH WaveMaster 28.5 m	*Waterfront I*	–	1998
MH WaveMaster 28.5 m	*Waterfront II*	–	1998

Operations
Singapore to Batam Island (Waterfront City).

Slovakia

Slovak Shipping And Ports, Joint Stock Co

Division of Passenger Shipping
Fajnorovo nábr 2, 811 02, Bratislava, Slovakia

Tel: (+421 2) 529 322 26
Fax: (+421 2) 529 322 31
e-mail: travel@lod.sk
Web: www.lod.sk

Miroslav Gerhát, *Director*

High-speed craft operated

Type	Name	Seats	Delivered
HYD Voskhod	*Piestany*	65	1981
HYD Meteor	*Myjava*	112	1988
HYD Meteor	*Modra*	112	1988
HYD Meteor	*Bratislava*	112	1988

Piestany 0506930

Bratislava 1178200

Operations
Vienna to Bratislava, 1 hour and 30 minutes, (downstream)
Bratislava to Vienna, 1 hour 45 minutes (upstream)
Bratislava to Budapest, 4 hours and 15 minutes (downstream)
Budapest to Bratislava, 4 hours and 40 minutes (upstream).

Slovenia

Kompas d.d.

Prazakova 4, 1515 Ljubljana, Slovenia

Tel: (+386 1) 200 61 00
Fax: (+386 1) 200 64 34
e-mail: info@kompas-online.net
Web: www.kompas-online.net

Janez Pergar, *President and CEO*

High-speed craft operated

Type	Name	Seats	Delivered
WPC NQEA InCat	*Prince of Venice*	303	1989

Operations
Prince of Venice operates from Isola to Venice.
Service in operation from April to October each year.

Prince of Venice (van Ginderen Collection) 0045227

South Africa

Robben Island Museum

Private Bag, Robben Island, Cape Town 7400, South Africa

Tel: (+27 21) 409 51 00
Fax: (+27 21) 411 10 59
e-mail: shonik@robben-island.org.za
Web: www.robben-island.org.za

High-speed craft operated

Type	Name	Seats	Delivered
CAT Farocean/Damen 3209	–	300	(2007)

Operations
Cape Town to Robben Island.

Spain

Acciona Trasmediterranea SA

Avda de Europa 10, Parque Empresarial La Muraleja, CP 28108 Alcobendas, Madrid

Tel: (+34 91) 423 85 00
Fax: (+39 91) 423 85 55
e-mail: correom@trasmediterranea.es
Web: www.trasmediterranea.es

Alcantara *at speed* 0507095

Almundaina (Adolpho Ortigueira) 0114497

High-speed craft operated

Type	Name	Seats	Additional payload	Delivered
HYD Kawasaki Jetfoil 929-117	*Princesa Teguise*	286	–	1991
WPC InCat 74 m	*Patricia Olivia*	300	–	1991
MH EN Bazán Mestrel 96 m	*Alcantara*	450	76 cars	1995
CAT Austal 74 m	*Tallink Autoexpress*	550	150 cars	1995
CAT Austal 79 m	*Alcantara Dos* (ex-*Tallink Autoexpress*)	550	150 cars	1995
MH EN Bazán Mestrel 96 m	*Almudaina*	450	76 cars	1996
MH Fincantieri MDV 1200	*Almudaina Dos* (ex-*SuperSea Cat One*)	674	148 cars	1997
CAT InCat	*InCat K3*	398	89 cars	1998
WPC InCat 96 m	*Alborán* (ex-*Avemar*)	900	260 cars	1999
WPC InCat 96 m	*Milenium*	900	260 cars	2000
WPC InCat 98 m	*Milenium Dos*	900	260 cars	2003
WPC InCat 98 m	*Milenium Tres*	900	260 cars	(2007)

Operations

Las Palmas, Grand Canaria to Santa Cruz de Tenerife, 52 n miles, 1 hour 20 minutes
Las Palmas to Morro Jable, 1 hour 30 minutes
Algeciras to Ceuta, 15 n miles, 30 minutes
Cristianos to Gomera, Canary Islands, 21 n miles, 45 minutes
Valencia to Ibiza (July to September)
Barcelona to Palma De Mallorca
Las Palmas to Valencia.

Balearia Eurolineas Maritimes

Maritima s/n (Port Denia), E-3700 Alicante, Valencia, Spain

Tel: (+34 96) 642 86 00
Fax: (+34 96) 578 76 06
e-mail: info@balearia.com
Web: www.balearia.com

High-speed craft operated

Type	Name	Seats	Additional payload	Delivered
CAT Westamaran W95	*Rapido de Formentera* (ex-*Sunnhordland*)	180	–	1991
WPC InCat Tasmania 77 m	*Jaume I*	600	145 cars	1994
MH Rodriquez TMV 115	*Frederiko Garcia Lorca*	884	200 cars	2001
MH Rodriquez TMV 84	*Ramon Llull*	462	58 cars	2003
CAT Marinteknik 63 m	*Nixe*	600	70 cars	2004
CAT Marinteknik 63 m	*Nixe II*	536	60 cars	2004

Operations

Barcelona to Menorca and Mallorca
Denia to Ibiza to Formentera and to Palma
Ceuta (Morocco) to Algeciras.

Buquebus Espana

Avenida Virgen del Carmen 29, E-11380 Tarifa (Cadiz), Spain

Tel: (+34 956) 65 68 48
Fax: (+34 956) 66 83 32
e-mail: info@buquebus.es
Web: www.buquebus.es

High-speed craft operated

Type	Name	Seats	Additional payload	Delivered
WPC InCat Australia 74 m	*Patricia Olivia*	515	90 cars	1992
WPC InCat Australia 91 m	*Express**	900	225 cars	1998
CAT 91 m	*Avemar*	900	240	1988
CAT InCat 77 m	*Ronda Marina*	600	130	1994
CAT Marinteknik 33 m	*Flecha de Buenos Aires*	234	92	1966
CAT FBM TriCat 45 m	*Thomas Edison*	279	–	2000
CAT Austal Auto Express 82 m	*Avemar Dos* (ex *Superstar Express*)	900	175	1997

*On charter to another operator.

Avemar (Adolfo Ortigueira Gil) 0114498

Cape Balear

C/, Pizarro, E-49 07590 Cala Ratjada, Majorca, Spain

Tel: (+34 971) 81 86 68
Fax: (+34 971) 56 33 36
e-mail: info@calaratjadatours.es
Web: www.interilles.es

High-speed craft operated

Type	Name	Seats	Delivered
CAT Westamaran 95 30 m	*Rapido del Puerto*	200	1981

Operations

Cala Ratjada to Ciutadella.

Europa Ferrys SA

A subsidiary of Acciona Trasmediterranea
Avenida Virgen del Carmen 1-5, E-11201 Algeciras, Spain

Tel: (+34 956) 65 23 24
Fax: (+34 956) 66 69 05
e-mail: marketing@euroferrys.com
Web: www.euroferrys.com

EuroFerrys Pacifica 0102270

High-speed craft operated

Type	Name	Seats	Additional payload	Delivered
CAT Austal 101 m	*EuroFerrys Pacifica*	950	250 cars	2001

Operations
Algeciras to Ceuta
Algeciras to Tangiers
Almeria to Nador.

Fred Olsen SA

Edificio Fred Olsen, Poligono Industrial Añaza, Santa Cruz de Tenerife, Canary Islands.

Tel: (+34 922) 62 82 00
Fax: (+34 922) 62 82 32
e-mail: lineas@fredolsen.es
Web: www.fredolsen.es

Fred Olsen, *Director*
Guillermo Van De Waal, *Managing Director*
Juan Ramsden, *Operations Director*

High-speed craft operated

Type	Name	Seats	Additional payload	Delivered
WPC InCat 96 m	*Bonanza Express*	735	220 cars	1999
WPC InCat 98 m	*Bentago Express* (ex *Bentayga Express*)	717	227 cars	2000
WPC InCat 98 m	*Bencomo Express* (ex *Benchijigua*) *Express*	871	227 cars	2000
CAT Austal 66 m	*Bocayna Express*	450	69 cars	2003
TRI Austal 126 m	*Benchijigua Express*	1,350	341 cars	2005

Bentayga Express (Adolpho Ortigueira) 0114494

Bonanza Express (Adolpho Ortigueira) 0114495

Operations
Los Cristanos (Tenerife Island) to La Gomera Island (40 minutes)
Los Cristanos (Tenerife Island) to El Hierro Island (120 minutes)
Los Cristanos (Tenerife Island) to La Palma Island (120 minutes)
Santa Cruz de Tenerife to Gran Canaria (60 minutes)
Lanzarote Island to Fuerteventura Island (12 minutes).

FRS Iberia

(Ferrys Rapidos Del Sur)
Apartado de Correo 13, Avenida de la Constitucion 2-Blq 1C, E-11380 Tarifa (Cadiz), Spain

Tel: (+34 956) 62 74 40
Fax: (+34 956) 64 74 44
e-mail: info@frs.es
Web: www.frs.es

Götz Becker, *Managing Director*

A subsidiary of Förde Reederi Seetouristik of Germany.

High-speed craft operated

Type	Name	Seats	Vehicles	Delivered
CAT Fjellstrand Jumbo 60	*Tanger Jet* (ex-*Solidor 3*)	558	58	1996
CAT Fjellstrand Jumbo 60	*Nordic Jet*	450	49	1999
InCat 78 m	*Thundercat I* (ex-*Highspeed I*)	600	140 cars	1995
InCat 86 m	*Tarifa Jet* (ex-*Pescara Jet*, ex-*Sardinia Jet*, ex-*Sicilia Jet*)	800	175 cars	1997
CAT Autoexpress 86	*Tangier Jet II* (ex-*Spirit of Ontario*)	774	238	2004

Operations
Tangiers in Morocco to Tarifa in Spain
Algeciras to Tangiers
Gibraltar to Tangiers.

Garajonay Expres SA

Muelle de Vueltas s/n, E-38870 Valle Gran Rey, Spain

Tel: (+34) 922 80 70 24
Fax: (+34) 922 80 70 31
e-mail: info@garajonayexpres.com
Web: www.garajonayexpres.com

The group purchased the fast catamarans *Vittoria M* and *Garagonay* from their sister company Ustica Lines.

High-speed craft operated

Type	Name	Seats	Delivered
CAT Fjellstrand Flying Cat 40 m	*Garajonay*	268	1997
CAT Fjellstrand Flying Cat 40 m	*Orone* (ex-*Vittoria M*)	268	1997

Operations
Valle Gran Rey to Los Cristianos
Valle Gran Rey to S.S. de La Gomera
Valle Gran Rey to Playa Santiago
Playa Santiago to S.S. de La Gomera
Playa Santiago to Los Cristianos
S.S. de La Gomera to Los Cristianos.

Iscomar Ferrys

Islena Maritima de Contenedores, SA, Con domicilio social en la Tinglado Oeste, Prolongacion Muelle, Adosado s/n 07012, Palma de Mallorca, Spain

Tel: (+ 34 902) 11 91 28
Fax: (+ 34 902) 43 75 02
e-mail: iscomar.ferrys@ral.es
Web: www.iscomar.ferrys.com

High-speed craft operated

Type	Name	Seats	Additional payload	Delivered
CAT Westamaran 49.5 m	*Pitiusa Nova* (ex-*Solidor 4*)	302	36 cars	1987

Operations
Ibiza to Formentera.

Mediterranea Pitiusa

Paseo de la Marina S/n, Edificio Vincent Lluquinet, 07870 Formentera – La Savina, Islas Baleares, Spain

Tel: (+34 971) 32 24 43
Fax: (+34 971) 32 22 24
e-mail: medpitiusa@terra.es
Web: www.medpitiusa.net

High-speed craft operated

Type	Name	Seats	Delivered
CAT Westamaran 95 29 m	*Aigues de Formentera* (ex-*Trident 5*)	180	1974
CAT Rodriquez CityCat 40 m	*Blau de Formentera*	260	2006

Operations
Formentera to Ibiza.

Nautas Al Maghreb

38 C/O Donnell, E-28009 Madrid, Spain

Tel: (+34 902) 16 11 81
e-mail: info@balearia.com
Web: www.nautas-almaghreb.com

A subsidiary of Eurolineas Maritimes Balearia.

High-speed craft operated

Type	Name	Seats	Cars	Delivered
CAT Marinteknik 63 m	*Nixe II*	536	60	2004
CAT InCat 77 m	*Jaume I*	600	145	1994

Operations
Algeciras to Tangier.

Trasmapi

C/Aragon 71, Apd C1465, 07800 Ibiza, Spain

Tel: (+34 610) 421 431
e-mail: info@trasmapi.com
Web: www.trasmapi.com

High-speed craft operated

Type	Name	Seats	Delivered
CAT Westamaran 95	*Tagomago Jet*	180	1977
CAT Westamaran 95	*Ibiza Jet*	180	1978
CAT Fjellstrand 38.8 m	*Elvissa Jet*	312	1986
CAT Fjellstrand 38.8 m	*Formentera Jet*	312	1987
CAT Austal 42 m	*Espalmador Jet* (ex-*Draupner*)	358	1999

Operations
Ibiza to Formentera.

Sri Lanka

Ceylon Shipping Corporation

No 27, Bristol Street, Colombo 01, Sri Lanka

Tel: (+941) 32 87 72
Fax: (+941) 44 75 47
e-mail: cscl@cscl.lk
Web: www.cscl.lk

High-speed craft operated

Type	Name	Seats	Additional payload	Delivered
CAT K Fjellstrand (S)	*Lanka Bani*	300	10 tonnes	2000
CAT K Fjellstrand (S)	*Lanka Devi*	300	10 tonnes	2000

Operations
North Coast of Sri Lanka, Jaffna to Tricomalee.

Sweden

Destination Gotland

Box 1234, SE-62123 Visby, Sweden

Tel: (+46 498) 20 18 00
e-mail: info@destinationgotland.se
Web: www.destinationgotland.se

High-speed craft operated

Type	Name	Seats	Additional payload	Delivered
MH Alstom Corsaire 11500	*Gotlandia*	700	145 cars	1999
MH Fincantieri 1000500	*Gotlandia II*	780	160 cars	2006

Operations
Gotlandia II: Nynäshamn to Visby, 2 hours 50 minutes.
Gotlandia: Grankullavik on Öland to Visby, 2 hours.

Hoverline AB

PO Box 27049, SE-10251, Stockholm, Sweden

Tel: (+46 8) 54 58 96 35
Fax: (+45 8) 54 58 96 49
e-mail: info@hoverline.se
Web: www.hoverline.se

Peder Silfverhjelm, *President and CEO*
Philip von Mecklenburg, *Vice-President*
Eric Carlén, *Fleet Manager*
Paul Snow, *Engineer*
Jim Murphey, *Skirt Engineer*
Melvin Beale, *Engineering Coordinator*

Established in 2000, Hoverline is an international operator of hovercraft, principally in the offshore industry.

The company operates through Hoverline International AB, Hoverline Offshore IOM Ltd and Hoverline Offshore Brazil Ltda.

High-speed craft operated

Type	Name	Seats	Delivered
ACV AP1-88/100	*Hover Griffin*	100	1989
ACV AP1-88	*Hover Falcon*	70	1991
ACV AP1-88/100	*Hover Eagle*	60	1987

Operations
Brazil and West Africa.

Stena Line Scandinavia AB

SE-405 19 Göteborg, Sweden

Tel: (+46 31) 85 80 00
Fax: (+46 31) 24 10 38
e-mail: info@stenaline.com
Web: www.stenaline.com

Stena Carisma 0007951

High-speed craft operated

Type	Name	Seats	Additional payload	Delivered
CAT HSS 900	*Stena Carisma*	900	200 cars	1997

Operations
Gothenburg to Frederikshavn.

Stromma

Skeppsbron 22, SE-111 30, Stockholm, Sweden

Tel: (+46 8) 58 71 40 00
Fax: (+46 8) 58 71 40 44
e-mail: cinderellabatama@stromma.se
Web: www.stromma.se

Jan Larson, *Director*
Alf Norgrem, *Technical Officer*

High-speed craft operated

Type	Name	Seats	Delivered
MH Marinteknik 41 m	*Cinderella I*	450	1990
MH Marinteknik 41 m	*Cinderella II*	450	1987

Operations
Stockholm to Archipelago.

Switzerland

CGN

Compagnie Générale de Navigation sur le Lac Léman
Avenue de Rhodanie 17, Case Postale 116, 1000 Lausanne 6, Switzerland

Tel: (+41 848) 811848
e-mail: info@cgn.ch
Web: www.cgn.ch

High-speed craft operated

Type	Name	Seats	Delivered
MH Navibus 25 m	*Coppet*	120	2007

Operations
Nyon to Thonon, Yvoire, Chens-sur-Léman and Geneva.

Taiwan

Tung Hsin Steamship Company Ltd

Ping Tung, Taiwan

High-speed craft operated

Type	Name	Seats	Delivered
MH SBF 26.55 m	*Tung Hsin*	193	February 1993

Yeuam Shian Shipping

28F-1 No 3 Tzuchiang 3rd Road, Lane 3 Ling Ya District, Kaohsiung 802, Taiwan

Tel: (+88 67) 566 50 17
Fax: (+88 67) 566 50 11

High-speed craft operated

Type	Name	Seats	Delivered
CAT Fjellstrand 40 m	*Yeuam Shian Princess*	346	1996

Operations
Kaohsiung to Makung, 2½ hours.

Tanzania

Azam Marine

PO Box 744, Zanzibar, Tanzania

Tel: (+255) 222 18 01 94
Fax: (+255) 222 18 01 67
e-mail: azam@cats-net.com

High-speed craft operated

Type	Name	Seats	Delivered
CAT Westamaran 86	*Kilimanjaro*	150	1994
HYD Rodnquez RHS160F	*Kondor 5*	200	1994
MH WaveMaster 32 m	*Kondor 7*	200	1995
CAT Image Marine 24 m	*Seabus I*	–	1997
CAT Image Marine 24 m	*Seabus II*	–	1997
CAT Crowther/South Pacific Marine 31 m	*Super Seabus III*	282	1998

Operations
Dar es Salaam to Zanzibar.

Thailand

Andaman Wave Master

23/4 Luang Poh Wat-Chalong Road, Tambol Talad-yai, Ampur Mulang, Phuket 83000, Thailand

Tel: (+ 66 76) 232 561
Fax: (+66 76) 232 096
e-mail: info@andamanwavemaster.com
Web: www.andamanwavemaster.com

High-speed craft operated

Type	Name	Seats	Delivered
CAT Westamarin W100	*Jet Cruise 1* (ex- *Gibline1*)	250	1991

Operations
Phuket to Phi Phi Island.

Kon-tiki Diving & Snorkelling Co Ltd

1/21 Moo 9, Muang, Phukett 83130, Thailand.

Tel: (+66 76) 280 36 66
Fax: (+66 76) 28 03 57
e-mail: kontiki@loxinfo.co.th
Web: www.kon-tiki-diving.com

High-speed craft operated

Type	Name	Seats	Delivered
HYD RHS 170	*Flyer 1* (ex-*Colona 7*, ex-*Shearwater 5*)	50	1980

Operations
Similian, Kaoh Kag, Patong.

Lomprayah High Speed Ferry Company Ltd

100/1 Moo 1, Maenam Samui Island, Suratthani 84330, Thailand

Tel: (+66) 77 42 77 65
Fax: (+66) 77 42 77 66
e-mail: info@lomprayah.com
Web: www.lomprayah.com

High-speed craft operated

Type	Name	Seats	Delivered
CAT Crowther 20 m	*Pralarn*	120	2002
CAT Crowther 25 m	*Namuang*	228	2004
CAT Crowther 25 m	*Maenam Samui*	255	2006

Operations
Chumpom to Koh Nang Yuan to Koh Phangan to Koh Samui.

Seatran Ferry Company

A subsidiary of Seatran Travel Company
599/1 Rimtangrotfai-Chongnonsri Road, Klontoey, Bangkok 10110, Thailand

Tel: (+66 2) 240 25 82
Fax: (+66 2) 249 59 51
email: info@seatranferry.com
Web: www.seatranferry.com

High-speed craft operated

Type	Name	Seats	Delivered
MH Mitsubishi 48 m	*Seatran Express 4*	401	1977

Operations
Suratthani to Koh Samui to Pa Ngan.

Trinidad and Tobago

Port Authority Of Trinidad and Tobago

Administrative Building, 1 Dock Road, Port of Spain, Trinidad

Tel: (+868) 625 26 44
e-mail: kelvinh@patnt.com
Web: www.patnt.com

Kelvin Harris, *Divisional Manager Operations*

High-speed craft operated

Type	Name	Seats	Additional payload	Delivered
WPC InCat Australia 91 m	*T&T Express* (ex-*The Lynx*)	900	240 cars	1997
WPC InCat Australia 98 m	*T&T Spirit* ex- (*TSV-IX Spearhead*)	900	180 cars	2002

Operations
Trinidad to Tobago.

Turkey

Akgunler Shipping Ltd

Yeni Liman Yolu, Candemin 5 Apt No 3, Girne KKTC, Turkey

Tel: (+90 392) 815 79 29
Fax: (+90 392) 815 86 47
email: denizcilik@akgunler.com.tr
Web: www.akgunler.com.tr

High-speed craft operated

Type	Name	Seats	Delivered
HYD Kolkida	*Marude*	140	1988
HYD Kometa	*Ayse*	116	1977
CAT Marinteknik 34 m	*Akgunler 1*	297	1992
CAT Westamarin 27 m	*Akgunler 2*	180	1981
CAT Fjellstarnd 38.8 m	*Akgunler 3*	308	1988
HYD Kometa	*Kadiye*	116	1982
HYD Kometa	*Nazli*	116	1983

Operations
Girne and Tasucu.
Girne and Alanya.

Bodrum Express Lines

Kale Laddesi, No 18 Bodrum, Turkey

Tel: (+90 252) 316 10 87
Fax: (+90 252) 313 00 77
e-mail: info@bodrumexpresslines.com
Web: www.bodrumexpresslines.com

High-speed craft operated

Type	Name	Seats	Delivered
HYD Kometa	*Bodrum Princess*	116	1978
HYD Kometa	*Ege Princess*	116	1979
HYD Voskhod-2	*Didim Princess*	79	–
HYD Voskhod-2	*Datça Princess*	79	–
HYD Voskhod-2	*Knidos Princess*	79	–

Operations
Bodrum to Kos, Kalymnos, Dalyan, Gelibolu and Rhodes.
Turgutreis to Leros.

Istanbul Deniz Otobusleri

Kennedy Cad, Hizli Feribot Iskelesi, Yenikapi TR-34480, Istanbul, Turkey

Tel: (+90 212) 444 44 36
Fax: (+90 212) 517 38 00
e-mail: info@ido.com.tr
Web: www.ido.com.tr

Dr Ahmet Paksoy, *General Manager*

In 1987 there were 10 Fjellstrand 38.8 m, 449-passenger catamarans ordered to provide a ferry service for commuters across the Bosphorus. Five have a 24 kt cruise speed capability and five, 32 kt. Further monohull and catamaran craft, including car carrying craft have subsequently been delivered.

High-speed craft operated

Type	Name	Seats	Additional payload	Delivered
CAT Fjellstrand 38.8 m	*Umer Bey*	449	–	1987
CAT Fjellstrand 38.8 m	*Sarica Bey*	449	–	1987
CAT Fjellstrand 38.8 m	*Uluç Ali Reis*	449	–	1988
CAT Fjellstrand 38.8 m	*Nusret Bey*	449	–	1988
CAT Fjellstrand 38.8 m	*Hezarfen Çelebi*	449	–	1988
CAT Fjellstrand 38.8 m	*Çaka Bey*	449	–	1987
CAT Fjellstrand 38.8 m	*Yeditepe I*	449	–	1987
CAT Fjellstrand 38.8 m	*Ulubatli Hasan*	449	–	1987
CAT Fjellstrand 38.8 m	*Karamürsel Bey*	449	–	1988
CAT Fjellstrand 38.8 m	*Cavli Bey*	449	–	1988
MH Austal 30 m	*Aksemseddin*	155	–	1994
MH Austal 30 m	*Ertugrul Gazi*	155	–	1994
CAT Austal 40 m	*Sinan Pasa*	450	–	1996
CAT Austal 40 m	*Piyale Pasa*	450	–	1996
CAT Fjellstrand 35 m	*Piri Reis II*	400	–	1997
CAT Fjellstrand 35 m	*Kaptan Pasa*	400	–	1997
CAT Fjellstrand 35 m	*Sedyi Ali Reis I*	400	–	1997
CAT Fjellstrand 35 m	*Oruc Reis V*	400	–	1997
CAT Fjellstrand 35 m	*Hizir Reis III*	400	–	1997
CAT Fjellstrand 35 m	*Temel Reis II*	400	–	1998
CAT Fjellstrand 35 m	*Barbaros Hayreddin Pasa*	400	–	2000
CAT Fjellstrand 35 m	*Sokullu Mehmed Pasa*	400	–	2000
CAT Austal 60 m	*Turgut Reis 1*	450	94 cars	1997
CAT Austal 60 m	*Cezayirili Hasan Pasa 1*	450	94 cars	1997
CAT Austal 86 m	*Adnan Menderes*	800	200 cars	1998
CAT Austal 86 m	*Turgut Ozal*	800	200 cars	1999
CAT Austal Auto Express 88	*Osman Gazi I*	1,200	225 cars	2007
CAT Austal Auto Express 88	*Orhan Gazi I*	1,200	225 cars	2007
CAT Damen DFF 4212	To be named	449	–	(2007)
CAT Damen DFF 4212	To be named	449	–	(2007)
CAT Damen DFF 4212	To be named	449	–	(2007)
CAT Damen DFF 4212	To be named	449	–	(2007)
CAT Damen DFF 4212	To be named	449	–	(2007)

Oruc Reis *operated by Istanbul Deniz Otobüsleri* 0010484

Adnan Menderes *on trials prior to delivery to IDO* 0045228

Operations
The company operates within the Sea of Marmara area covering:
Bostanci to Kabatas, 8 n miles, 22 minutes
Bostanci to Karaköy, 8 n miles, 22 minutes
Bostanci to Bakirköy, 11 n miles, 27 minutes
Bostanci to Yenikapi, 7 n miles, 20 minutes
Bostanci to Yalova, 20 n miles, 45 minutes
Yalova to Kartal, 14 n miles, 30 minutes
Yalova to Kabatos, 28 n miles, 55 minutes
Kadiköy to Bakirköy, 6 n miles, 18 minutes
Kadiköy to Emenönii, 3 n miles, 10 minutes
Istanbul to Mudanya to Armuthi, 52 n miles, 1 hour 55 minutes
Yenikapi to Bursa, 75 minutes.

Yesil Marmaris Shipping

PO Box 8, Marmaris, TR-48700 Mugla, Turkey

Tel: (+90 252) 412 64 86
Fax: (+90 252) 412 50 77
e-mail: yesilmarmaris@superonline.com
Web: www.yesilmarmaris.com

High-speed craft operated

Type	Name	Seats	Delivered
Crowther 38 m	*Marmaris Express*	360	1995
Marinteknik 38 m	*Aegean Jet*	185	1980

Operations
Marmaris to Rhodes.

Ukraine

Black Sea Shipping Company

1 Lastochkina Str, 270026 Odessa, Ukraine

Tel: (+380 48) 25 21 60
Fax: (+380 48) 274 00 18
Telex: 232711; 412677

High-speed craft operated

Type	Name	Seats	Delivered
HYD Kolkhida	*Kolkhida 5*	155	1984
SES Sosnovka Rassvet	–	80	–
HYD-Feodosia Tsiklon	–	–	November 1987
CAT Westamarin W 4100S	*Krymskaya Strela*	298	April 1990
CAT Westamarin W 4100S	*Golubaya Strela*	300	April 1990

Operations
Odessa to Yevpatoria, Sevastopol, Yalta, Feodosia, Novorossitsk, Tuapse and Sochi.

Kolkhida craft at Odessa 0506754

Ukrainian Danube Shipping Company

28 Krasnoflotskaya St, Izmail, 272630 Odessa, Ukraine

Tel: (+380 48) 41 25 55
Fax: (+380 48) 41 25 35
Telex: 412699

High-speed craft operated

Type	Name	Delivered
HYD Kometa	*Kometa 35*	1978
HYD Kometa	*Kometa 36*	1978

United Arab Emirates

Danat Dubai Cruises

PO Box 12940 Dubai, United Arab Emirates

Tel: (+971 4) 351 11 17
Fax: (+971 4) 351 11 16
e-mail: danatdxb@emirates.net.ae
Web: www.danatdubaicruises.com

Chris Anns, *General Manager*

High-speed craft operated

Type	Name	Seats	Delivered
CAT Austal 34 m	*Danat Dubai* (ex-*Bali Hai I*)	196	1989

Operations
Day and evening cruises in Dubai.

United Kingdom

F H Bertling Ltd

5th Floor, York House, Empire Way, Wembley, Middlesex, HA9 0PA, United Kingdom

Tel: (+44 20) 89 00 20 60
Fax: (+44 20) 89 00 12 48
e-mail: sales@bertling.com
Web: www.bertling.com

These vessels are operated by Bertling Kazakh Logistics Hovercraft.

High-speed craft operated

Type	Name	Seats	Delivered
HOV BHC/AP1-88 G112024	*Idun Viking*	98	2003
HOV BHC/AP1-88 variant	*Manta*	–	–

Operations
Caspian Sea.

Condor Ferries Ltd

Condor House, New Harbour Road South, Hamworthy, Poole, Dorset BH15 4AJ, United Kingdom

Tel: (+44 1202) 20 72 07
Fax: (+44 1202) 68 51 84
e-mail: reservations@condorferries.co.uk
Web: www.condorferries.com

R A Provan, *Managing Director*
D Harbord, *Financial Director*
D Norman, *General Manager, Marine Services*
N R Dobbs, *General Manager, Sales and Marketing*
S Williams, *Operations Manager*
P Gardner, *Technical Manager*

Condor Ltd operates a fast car and passenger network between the UK and the Channel Islands as well as the UK and St Malo in France. The passenger/car ferry service employs two Incat Australia 86 m and one 74 m WPC operating on routes between the UK, the Channel Islands and France.

High-speed craft operated

Type	Name	Seats	Additional payload	Delivered
WPC InCat Australia 74 m	*Condor 10*	550	85 cars	1993
WPC InCat Australia 86 m	*Condor Express*	732	185 cars	1997
WPC InCat Australia 86 m	*Condor Vitesse*	732	185 cars	1998

Condor Ferries Ltd (Condor Express) 1048709

Operations
Weymouth to St Malo (via Guernsey), from 4 hours 30 minutes
Weymouth to Guernsey, 2 hours
Weymouth to Jersey, 3 hours 15 minutes
Poole to St Malo, 4 hours, 30 minutes
Poole to Guernsey, 2 hours 30 minutes
Poole to Jersey, 3 hours
Guernsey to St Malo, 1 hour 45 minutes
Jersey to St Malo, 1 hour 10 minutes
Guernsey to Jersey, 55 minutes
Poole to Cherbourg (for Brittany Ferries), 2 hours 15 minutes.

HD Ferries

Elizabeth Terminal, La Route du Port, Elizabeth, St Helier, Jersey, JE2 3NW, United Kingdom

Tel: (+44 870) 460 02 31
e-mail: mail@hdferries.com
Web: www.hdferries.co.uk

HD ferries started a service between Jersey, Guernsey and Saint Malo in March 2007.

High-speed craft operated

Type	Name	Seats	Additional payload	Delivered
CAT InCat K50	*HD1* (ex- *InCat K3*)	400	100 cars	1998

Operations
Jersey to Saint Malo, 1 hour.
Jersey to Guernsey, 1 hour.

HM Customs and Excise Marine Branch

Hamilton Road, Cosham, Portsmouth, PO6 4PX, United Kingdom

Web: www.hmrc.gov.uk

High-speed craft operated

Type	Name	Seats	Delivered
MH FBM Babcock Protector	*HMCC Vincent*	–	1993
MH FBM Babcock Protector	*HMCC Valiant*	–	1993
MH FBM Babcock Protector	*HMCC Venturous*	–	1993
MH FBM Babcock Protector	*HMCC Vigilant*	–	1993
MH VT 36 m	*HMCC Sentinel*	–	1993
MH Damen 4207	*HMCC Seeker*	–	2001
MH Damen 4207	*HMCC Searcher*	–	2002

HMCC Vincent 0103295

HMCC Sentinel 0103296

HMCC Seeker 0121767

Hoverspeed Ltd

Hoverspeed was the leading fast ferry operator on the English Channel until 2005 when it ceased operations due to financial losses. A subsidiary of Sea Containers Ltd, the company pioneered operations of the world's first car-carrying catamaran SeaCat, and the largest monohull fast car ferry operating in UK waters, SuperSeaCat.

Hovertravel Ltd

Quay Road, Ryde, Isle of Wight PO33 2HB, United Kingdom

Tel: (+44 1983) 81 10 00
Fax: (+44 1983) 56 22 16
e-mail: info@hovertravel.co.uk
Web: www.hovertravel.co.uk

C D J Bland, *Chairman*
R K Box, *Chief Executive*
J Gaggero, *Director (Non Executive)*
P White, *Finance Director and Company Secretary*
R H Barton, *Engineering Director*
B A Jehan, *Operations Manager*

Hovertravel Ltd was founded in 1965 and is the world's longest established commercial hovercraft operator. Operations began using two SR.N6 Winchester Class hovercraft and today two 95-seat diesel powered AP1-88 hovercraft are employed between Ryde, Isle of Wight and Portsmouth. The typical crossing time is under 10 minutes and,

Freedom 90 *in service* 0134521

Hovertravel terminal at Ryde, Isle of Wight 0554528

on an annual basis, over 20,000 hovercraft departures are made to carry more than 750,000 passengers. By July 2006, the accumulated passenger count had reached 22 million. During 1999 *Freedom 90* was modified with larger propulsion engines, and in 2001 *Freja Viking* (now *Island Express*) was uprated to the same specifications as *Freedom 90*. Hovertravel has ordered a new BHT 130 hovercraft from its wholly owned subsidiary Hoverwork, and this new craft is began service in June 2007.

High-speed craft operated

Type	Name	Seats	Delivered
HOV BHC/HW AP1-88 GH 2114	*Freedom 90*	95	1990
HOV BHC/HW AP1-88 GH 2132	*Island Express* (ex-*Freja Viking*)	95	1984
HOV HW 130	*Solent Express*	130	2007

Operations
Ryde, Isle of Wight to Southsea, Portsmouth.

Isle of Man Steam Packet Company

Imperial Buildings, Douglas, Isle of Man IM1 2BY, United Kingdom

Tel: (+44 1624) 64 56 45
Fax: (+44 1624) 64 56 09
e-mail: spc@steam-packet.com
Web: www.steam-packet.com

Juan Kelly, *Chairman*
Mark Woodward, *Chief Executive*
D Grant, *Financial Director*

In 2005 the company was acquired by Macquaries Bank Ltd, Australia. The company also operates one conventional freight ferry on the Isle of Man route.

High-speed craft operated

Type	Name	Seats	Additional payload	Delivered
MH Fincantieri 100 m	*Super SeaCat Two*	694	140 cars	1998

Operations
Douglas to Heysham, Liverpool, Dublin and Belfast.

P&O Irish Sea

Larne Harbour, Larne, Antrim, BT40 1AQ, United Kingdom

Tel: (+44 870) 242 47 77
Web: www.poirishsea.com

High-speed craft operated

Type	Name	Seats	Additional payload	Delivered
CAT InCat 91 m	*Express* (ex-*Cherbourg Express*, ex-*Portsmouth Express*)	920	225 cars	1998

Operations
P&O Irish Sea operate the following services between 15th March and 2nd October.
Cairnryan, Scotland to Larne, Northern Ireland, 60 minutes
Troon, Scotland to Larne, Northern Ireland, 109 minutes.

Express *(ex-*Cherbourg Express*, ex-*Portsmouth Express*)* 0093266

Red Funnel Ferries

12 Bugle Street, Southampton, Hampshire SO14 2JY, United Kingdom

Tel: (+44 2380) 33 30 42
Fax: (+44 2380) 23 05 92
e-mail: post@redfunnel.co.uk
Web: www.redfunnel.co.uk

T Docherty, *Managing Director*
C Hetherington, *Commercial Director*
R Scott, *Financial Director*

High-speed craft operated

Type	Name	Seats	Delivered
CAT FBM Marine 31.5 m	*Red Jet 1*	138	1991
CAT FBM Marine 31.5 m	*Red Jet 2*	138	1991
CAT FBM Marine 34.7 m	*Red Jet 3*	190	1998
CAT NWBS 39 m	*Red Jet 4*	277	2003

Operations
Southampton to West Cowes, Isle of Wight, 10.8 n miles, 22 minutes.

Passengers carried
In 2005 1,280,234 passengers were carried.

Red Jet 4 0550058

Sea Containers Ltd

Sea Containers House, 20 Upper Ground, London SE1 9PF, United Kingdom

Tel: (+44 20) 78 05 50 00
Fax: (+44 20) 78 05 59 00
Web: www.seacontainers.com

Robert MacKenzie, *President and CEO*
Daniel O'Sullivan, *Senior Vice-President and Chief Financial Officer*
Robert Ward, *Senior Vice-President, Containers*
Paul Benson, *Commercial Director of Ferries*
James Beveridge, *Vice-President, Administration and Property*

Sea Containers Ltd's subsidiary companies include: Aegean Speed Ferries, SuperSeaCat and Seastreak America Inc. The company used to own and run Hoverspeed and Sea Containers Ferries Scotland before they ceased operation in October 2005.

Speed Ferries

Hoverport, Western Docks, Dover, Kent CT17 9TG, United Kingdom

Tel: (+44 1304) 20 30 00
Fax: (+44 1304) 20 80 00
e-mail: mail@speedferries.com
Web: www.speedferries.com

This service commenced operations in May 2004.

High-speed craft operated

Type	Name	Seats	Additional payload	Delivered
CAT InCat Australia 86 m	*SpeedOne* (ex-*InCat 045*)	800	200 cars	1997

Operations
Dover (UK) to Boulogne (France), 50 minutes.

Stena Line

1 Suffolk Way, Sevenoaks, Kent TN13 1YL, United Kingdom

Tel: (+44 8705) 70 70 70
Fax: (+44 1233) 20 23 91
e-mail: info.uk@stenaline.com
Web: www.stenaline.com

Gunnar Blondahl, *Managing Director*
Alan Gordon, *Route Director*
Pim de Lange, *Route Director*
Vic Goodwin, *Route Director*

Stena Line operates a variety of large high-speed passenger/car ferries on its UK routes.

In January 2007 the *Stena Discovery* was withdrawn from its Harwich to Hook of Holland service due to competition from low cost airlines and high fuel prices.

High-speed craft operated

Type	Name	Seats	Additional payload	Delivered
WPC InCat 81 m	Stena Express (ex-*Stena Lynx III*, ex-*Elite*)	600	155 cars	1996
CAT Stena HSS 1500	*Stena Explorer*	1,500	375 cars	1996
CAT Stena HSS 1500	*Stena Voyager*	1,500	375 cars	1996
CAT Stena HSS 1500	*Stena Discovery*	1,500	375 cars	1997

Operations

In 2007 Stena Line will operate on the following routes:
Stena Express: Fishguard to Rosslare
Stena Explorer: Holyhead to Dun Laoghaire
Stena Voyager: Stranraer to Belfast.

Stena Express *(ex-*Stena Lynx III*)* 0507096

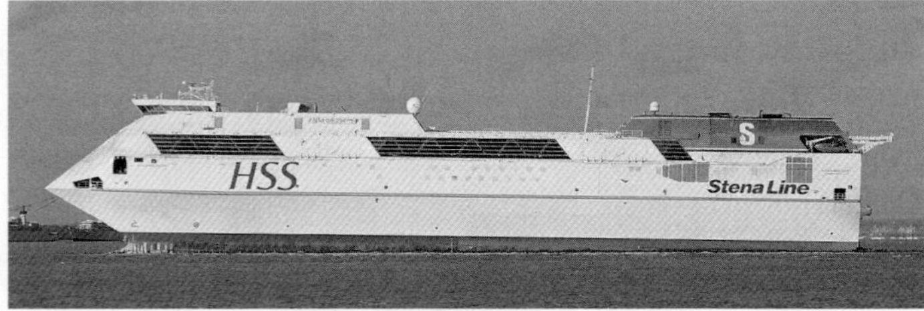

Stena Discovery (Maritime Photographic) 0109540

Thames Clippers

Nelson House, 265 Rotherhithe Street, London SE16 5HW, United Kingdom

Tel: (+44 870) 781 50 49
Fax: (+44 20) 723 740 19
Web: www.thamesclippers.com

Alan Woods, *Chairman*
Sean Collins, *Managing Director*
Jane Woods, *Company Secretary*

Started in 1999 as Collins River Enterprises, Thames Clippers now operates three vessels on the Thames. In 2003 the company gained the licence to operate the route from Tate Britain to Tate Modern via the BA London Eye.

High-speed craft operated

Type	Name	Seats	Delivered
CAT FBM Thames Class	*Star Clipper*	62	1992
CAT FBM Thames Class	*Storm Clipper*	62	1993
CAT FBM Thames Class	*Sky Clipper*	62	1993
CAT NQEA River Runner 200 Mk II	*Hurricane Clipper*	206	2002
CAT NQEA River Runner 150 MkIII	*Sun Clipper*	138	2001
CAT NQEA River Runner 150 MkIII	*Moon Clipper*	138	2001
CAT Aimtek	*02 Express*	–	2007

Storm Clipper, Sky Clipper *and* Hurricane Clipper 0132892

Operations

Savoy, Blackfriars, Bankside, London Bridge, St Katherines, Canary Wharf, Greenlands Passage, Mast House Terrace (also Hilton Hotel to Canary Wharf) and Greenwich
Tate Britain to BA London Eye and Tate Modern
Waterloo, London Bridge and O2.

Wightlink Ltd

A subsidiary of Macquarie Bank Ltd
Gunwharf Road, Portsmouth, Hampshire PO1 2LA, United Kingdom

Tel: (+44 23) 92 85 52 30
Fax: (+44 23) 92 85 52 57
e-mail: bookings@wightlink.co.uk
Web: www.wightlink.co.uk

Andrew Willson, *CEO*
Jonathan Pascoe, *Finance Director*

Wightlink Ltd operates ferry services on three routes across the Solent to the Isle of Wight. The three vessels on the Portsmouth-Ryde route are high-speed passenger catamarans. The eight conventional ferries on the other two routes, Portsmouth-Fishbourne and Lymington-Yarmouth, carry all sizes of vehicles as well as passengers.

High-speed craft operated

Type	Name	Seats	Delivered
CAT InCat 30 m	*Our Lady Pamela*	400	1986
CAT Fjellstrand 40 m	*FastCat Shanklin* (ex-*Water-jet 3*)	361	2000
CAT Fjellstrand 40 m	*FastCat Ryde* (ex-*Water-jet 5*)	361	2000

Operations

Portsmouth Harbour-Ryde Pier, Isle of Wight, in under 18 minutes.

Our Lady Pamela 0069466

FastCat Shanklin 0100070

United States

Alameda Harbor Bay Ferry

Tel: (+1 510) 769 5172
e-mail: office@alamedaharborbayferry.com
Web: www.alamedaharborbayferry.com

High-speed craft operated

Type	Name	Seats	Delivered
CAT InCat 29 m	*Bay Breeze*	250	1994
CAT	*Express II*	149	–

Operations

Commuter service: Alameda to San Francisco.

Alameda Oakland Ferry

950 West Mall Square, Alameda, California 94501, US

Tel: (+1 510) 749 59 72
e-mail: epsanche@ci.alameda.ca.us
Web: www.eastbayferry.com

Ernest Sanchez, *Manager*

High-speed craft operated

Type	Name	Seats	Delivered
CAT Nichols/Incat 35 m	*Peralta*	324	2001

Operations
San Francisco harbour Pier 39 to SF Ferry Building to Alameda, Angel Island and Oakland.

Alaska Fjordlines

PO Box 246, Haines, Alaska 99827, US

Tel: (+1 800) 320 01 46
e-mail: info@alaskafjordlines.com
Web: www.alaskafjordlines.com

High-speed craft operated

Type	Name	Seats	Delivered
CAT K&C 20 m	*Fjordland*	54	2001

Operations
Skagway to Haines to Juneau.

Alaska Hovercraft Ventures

6441 South Airpark Place, Anchorage, AK 99502-1809, US

Tel: (+1 907) 245 1931
Fax: (+1 907) 245 1744
e-mail: genmgr@ahv.lynden.com
Web: www.lynden.com/ahv

High-speed craft operated

Type	Name	Seats	Payload	Delivered
Hov AP 1-88	–	80	–	1983
Hov AP 1-88	–	80	–	1985
HOV LACV-30	–	–	30 tonnes	–

Operations
Hovercraft leasing.

Alaska Marine Highway System

755 North Tongass Avenue, Ketchikan, Alaska AK-99901, United States

Tel: (+1 907) 465 3955
Web: www.dot.state.ak.us/amhs

High-speed craft operated

Type	Name	Seats	Additional payload	Delivered
CAT Derecktor 73 m	*Fairweather*	250	35 cars	2004
CAT Derecktor 73 m	*Chenega*	250	35 cars	2005

Operations
Juneau to Sitka
Juneau to Haines to Skagway
Cordova to Valdez and Whittier.

Aleutians East Borough

3380 C Street, Suite 205, Anchorage, AK 99503, United States

Tel: (+1 907) 274 7555
e-mail: publicinfo@aleutianseast.org
Web: www.aleutianseast.org

High-speed craft operated

Type	Name	Seats	Additional payload	Delivered
HOV BHT 130	–	50	1 car	2006

Operations
King Cove to Cold Bay Airport.

Allen Marine Tours

PO Box 1049, Sitka, Alaska 99835-1049, United States

Tel: (+1 907) 747 81 00
Fax: (+1 907) 747 48 19
e-mail: infoa@allenmarinetours.com
Web: www.allenmarinetours.com

High-speed craft operated

Type	Name	Seats	Delivered
MH Allen Marine Otter 64 m	*Sea Otter Express*	49	1992
MH Allen Marine Otter 64 m	*Sea Lion Express*	49	1992
MH Allen Marine Otter 64	*Katlian Express*	49	1993
MH Allen Marine Otter 64 m	*Kalinin Express*	49	1993
CAT Allen 20 m	*St Michael*	110	1995
CAT Allen 20 m	*St Eugene*	110	1995
CAT Allen 20 m	*St Tatiana*	110	1995
CAT Allen 24 m	*Sit' Ku*	149	1996
CAT Allen 24 m	*St Gregory*	149	1996
CAT Allen 24 m	*St Herman*	149	1997
CAT Allen 24 m	*St Phillip*	149	1997
CAT Allen 20 m	*Majestic Fjord*	110	1998
CAT Allen 24 m	*St Nicholas*	149	1999
CAT Allen 24 m	*Cape Aialik*	149	1999
CAT Allen 24 m	*St Maria*	149	2003
CAT Allen 24 m	*St Aquilina*	162	2004
CAT Allen 24 m	*St John*	149	2005
CAT Allen 24 m	*St Nona*	149	2005

Operations
Seasonal. Excursions from Everett, Juneau, Seward, Sitka and Whittier.

Arnold Transit Company

PO Box 220, Mackinac Island, Michigan 49757, United States

Tel: (+1 850) 542 85 28
Fax: (+1 906) 847 38 92
e-mail: info@arnoldline.com
Web: www.arnoldline.com

Robert Brown, *General Manager*

High-speed craft operated

Type	Name	Seats	Delivered
CAT Gladding-Hearn InCat 25 m	*Mackinac Express*	350	1987
CAT Gladding-Hearn InCat 25 m	*Island Express*	300	1988
CAT Marinette 30 m	*Straits Express*	300	1995

Operations
Mackinaw City to Mackinac Island, May to October
St Ignace to Mackinac Island, 14 minutes, May to November.

Arnold Transit's Island Express 0506755

Auk Nu Tours

76 Egan Drive, Juneau, AK 99801, United States

Tel: (+1 800) 820 26 28
Fax: (+1 907) 586 13 37
Web: www.auknutours.com

High-speed craft operated

Type	Name	Seats	Delivered
CAT Allen Marine 78	*Keet*	148	1998

Operations
Tours in Misty Fjords National Park.

Bar Harbor Whale Watch Co

1 West Street, Bar Harbor, Maine 04609, United States

Tel: (+1 207) 288 23 86
Fax: (+1 207) 288 43 93
e-mail: info@barharborwhales.com
Web: www.barharborwhales.com

High-speed craft operated

Type	Name	Seats	Delivered
CAT Gladding-Hearn InCat 34 m	*Friendship V*	325	1995
CAT 27 m	*Bay King III*	149	2001

Operations
Whale watching in the gulf of Maine, nature tours, tender to cruise ships.

Bar Harbor Whale Watch Friendship V 0007308

Bay State Cruise Company

200 Seaport Boulevard, Suite 75, Boston, Massachusetts 02210, United States

Tel: (+1 617) 748 14 28
Fax: (+1 617) 748 14 25
e-mail: info@baystatecruises.com
Web: www.baystatecruises.com

Julie Doherty, *General Manager*

High-speed craft operated

Type	Name	Seats	Delivered
CAT Incat/Gladding Hearn 30 m	*Provincetown III*	149	2004

Operations
Boston to Providencetown, 90 minutes.

Billybey Ferry Company LLC

Pershing Road, Weehawken, New Jersey 07086, US

Tel: (+1 201) 902 87 00
Fax: (+1 201) 348 93 84
e-mail: customerservice@nywaterway.com
Web: www.nywaterway.com

In 2005, NY Waterway sold a number of vessels to the Billybey Ferry Company. The company has retained NY Waterway to provide personnel and to manage the ferry operations on the original routes.

High-speed craft operated

Type	Name	Seats	Delivered
Allen Marine 78	*Brooklyn*	148	2002
Allen Marine 78	*Yogi Berra*	148	2000
Allen Marine 78	*Fiorella La Guardia*	148	2000
Allen Marine 78	*Frank Sinatra*	148	2000
Allen Marine 78	*US Senator Frank R Lautenberg*	148	2002
Allen Marine 78	*Hoboken*	148	2002
Allen Marine 78	*Christopher Colombus*	148	2000
Allen Marine 78	*Peter R Weiss*	148	2001
Allen MarineJet Otter 65	*Father Mychal Judge*	87	2001
Allen MarineJet Otter 65	*Douglas B Gurian*	87	2002
Allen MarineJet Otter 65	*Fred V Morrone*	87	2002
Allen MarineJet Otter 65	*Enduring Freedom*	87	2002

Operations
Manhattan: to Hoboken
Manhattan: to Port Liberte.

Blue and Gold Fleet, L.P.

Pier 39 MarineTerminal, San Francisco, California 94133, United States

Tel: (+1 415) 705 82 00
Fax: (+1 415) 421 54 29
e-mail: info@blueandgoldfleet.com
Web: www.blueandgoldfleet.com

Key personnel
Carolyn Horgan, *Manager*

High-speed craft operated

Type	Name	Seats	Delivered
CAT Gladding-Hearn InCat 31 m	*Vallejo**	–	1994
CAT Dakota/AMD 360	*Intintoli**	300	1997
CAT Dakota/AMD 360	*Mare Island**	300	1997
CAT Nichols Brothers InCat 26 m	*Encinal*	400	1985
CAT Nichols Brothers InCat 26 m	*Zelinsky*	400	1986

**Vallejo, Intintoli* and *Mare Island* are in operation with Vallejo Baylink Ferry.

Operations
San Francisco to Alameda
San Francisco to Vallejo
San Francisco to Tiburon
San Francisco to Sausalito
San Francisco to Angel Island.

Boston Harbor Cruises

One Long Wharf, Boston, Massachusetts 02110, United States

Tel: (+1 617) 227 43 21
Fax: (+1 617) 723 20 11
e-mail: contact@bostonharborcruises.com
Web: www.bostonharborcruises.com

High-speed craft operated

Type	Name	Seats	Delivered
CAT InCat Gladding Hearn 37 m	*Nora Vittoria*	350	1998
CAT InCat Gladding Hearn 37 m	*Aurora*	350	1999
CAT InCat Gladding Hearn 44 m	*Salacia*	500	2000

Operations
Boston top Provincetown
Boston Harbor – whale watching and commuter services between Hingham and Downtown Boston.

Catalina Channel Express

Berth 95, San Pedro, California 90731, United States

Tel: (+1 310) 519 12 12; 79 71
Fax: (+1 310) 548 73 89
e-mail: mail@catalinaexpress.com
Web: www.catalinaexpress.com

Doug Bombard, *Chief Executive Officer*
Greg Bombard, *President*
Audrey Bombard, *Secretary/Treasurer*
Tom Rutter, *Vice-President of Operations*
Elaine Vaughan, *Vice-President of Marketing*
R McElroy, *Vice-President of Engineering*
P Sawasaki, *Controller*
R Lyman, *Vice-President Vessel Operations*

Jet Cat Express 0113366

Islander Express 0007309

Cat Express 0007310

Starship Express 0069464

High-speed craft operated

Type	Name	Seats	Delivered
MH Westport 95	*Super Express*	150	1989
MH Westport 95	*Avalon Express*	150	June 1990
MH Westport 100	*Catalina Express*	150	1994
MH Westport 100	*Islander Express*	150	1994
CAT InCat/Nichols 30 m	*Cat Express* (ex-*Glacier Explorer*)	360	1986
CAT Pequot/FBM 41 m	*Starship Express*	302	1999
CAT InCat/Nicholas 44 m	*Jet Cat Express*	381	2001
CAT Nichols 44 m	*Catalina Jet**	450	1999

*Chartered from Nichols Brothers Boatbuilders, US.

Operations
San Pedro to Avalon and Two Harbors, Santa Catalina Island, 1 hour 15 minutes
Long Beach to Avalon, Catalina Island, approximately 1 hour
Dana Point to Avalon, Catalina Island, 1 hour 30 minutes.

Catalina Ferries

13763 Fiji Way, Suite C2, Marina del Rey, California 90292, United States

Tel: (+1 310) 305 72 50
Fax: (+1 310) 305 72 00
e-mail: info@catalinaferries.com
Web: www.catalinaferries.com

Catalina Ferries operates two routes from Marina del Rey to Avalon and to Two Harbours, Catalina Island.

High-speed craft operated

Type	Name	Seats	Delivered
CAT Kitsap T-85	*Marina Flyer*	149	2006

Operations
Marina del Rey to Avalon and Two Harbours, Catalina Island.

Catalina Passenger Services

400 Main Street, Balboa, California 92661, United States

Tel: (+1 949) 673 52 45
Fax: (+1 949) 673 83 40
e-mail: bob@catalinainfo.com
Web: www.catalinainfo.com

High-speed craft operated

Type	Name	Seats	Delivered
CAT InCat/Nichols 36 m	*Catalina Flyer*	500	1988

Operations
Newport Beach to Catalina Island, 26 n miles, 1 hour 15 minutes.

Catalina Flyer 1120268

Chilkat Cruises And Tours

142 Beach Road, PO Box 509, Haines, Alaska 99827, US

Tel: (+1 907) 766 21 00
Fax: (+1 907) 766 21 01
email: fastferry@chilkatcruises.org
Web: www.chilkatcruises.com

High-speed craft operated

Type	Name	Seats	Delivered
CAT Allen Marine 24 m	*Fairweather Express*	150	1996
CAT Allen Marine 24 m	*Fairweather Express II*	150	2001
F-CAT Teknicraft 19 m	*Chilkat Express*	63	2001

Operations
Haines to Skagway, 35 minutes.

Circle Line Downtown

17 Battery Place, Suite 715, New York 10004, United States

Tel: (+1 212) 563 32 00
e-mail: info@circlelineferry.com
Web: www.circlelinedowntown.com

High-speed craft operated

Type	Name	Seats	Delivered
CAT Austal USA 43.5 m	*Zephyr*	460	2003
CAT Austal USA 25 m	*Patriot**	203	2002

*On lease to Titan Cruise Lines for 6 months

Operations
Manhattan to Statue of Liberty and Ellis Island.

Clipper Navigation Inc

2701 Alaskan Way, Pier 69, Seattle, Washington 98121, United States

Tel: (+1 206) 443 25 60
Fax: (+1 206) 443 25 83
e-mail: reservations@victoriaclipper.com
Web: www.victoriaclipper.com

254 Belleville Street, Victoria,
British Columbia V8V 1W9, Canada

Tel: (+1 250) 382 81 00
Fax: (+1 250) 382 21 52

Victoria Clipper IV 1150231

Merideth Tall, *President*
Darrell E Bryan, *Executive Vice-President and General Manager*

High-speed craft operated

Type	Name	Seats	Delivered
CAT Fjellstrand 38.8 m	*Victoria Clipper*	293	1986
CAT Gladding-Hearn InCat 31 m	*Victoria Clipper III* (ex-*Audubon Express*)	239	1991
CAT Fjellstrand Flying Cat 40 m	*Victoria Clipper IV*	330	1993

Operations
Seattle to Victoria, British Columbia, Canada, 71 n miles, 2 to 3 hours
Fares vary by season: single USD65-79, return USD79-130
Seattle to San Juan Islands.

Condor Cruises

Sealanding 301 West Cabrillo Boulevard, Santa Barbara 93109, California, United States

Tel: (+1 805) 882 00 88
Fax: (+1 805) 965 09 42
e-mail: info@condorcruises.com
Web: www.condorcruises.com

High-speed craft operated

Type	Name	Seats	Delivered
CAT All American Marine 23 m	*Condor Express*	149	2002

Operations
Whale watching excursions.

Condor Express 1107852

Cross Sound Ferry Services Inc

2 Ferry Street, PO Box 33, New London, Connecticut 06320, United States

Tel: (+1 860) 443 52 81
Fax: (+1 860) 440 34 92
e-mail: info@longislandferry.com
Web: www.longislandferry.com

High-speed craft operated

Type	Name	Seats	Delivered
WPC Nichols/InCat 37 m	*Sea Jet I*	400	1995
WPC Incat/Aluminium Shipbuilders 49 m	*Jessica W* (ex-*Condor 9*)	550	1990

Operations
New London (Connecticut) to Block Island (Rhode Island). Up to six daily return services
New London (Connecticut) to Long Island, New York.

Four Seasons Tours

13401 Glacier Highway, PO Box 211267, Auke Bay, Alaska 99821, United States

Tel: (+1 907) 790 66 71
Fax: (+1 907) 790 66 72
e-mail: info@4seasonsmarine.com
Web: www.4seasonsmarine.com

High-speed craft operated

Type	Name	Seats	Delivered
MH Wesport 100	*Admiralty Wind*	150	1998
CAT Incat/Nichols 22 m	*Baranof Wind*	212	1984
F-CAT Teknicraft 22 m	*Spirit*	149	2005

Operations
Transport and tours in southeast Alaska
Transport and tours in Puget Sound, State of Washington
Tours in Seward, Alaska.

Golden Gate Ferry

1011 Andersen Drive, San Rafael, California 94901, United States

Tel: (+1 415) 455 2000
Fax: (+1 415) 257 4411
e-mail: events@goldengate.org
Web: www.goldengate.org

John J Moylan, *President*
Albert J Boro, *First Vice President*
Celia G Kupersmith, *General Manager*
James Swindler, *Ferry Division Manager*
Charles (Rex) McCardell, *Marine Engineer*
Rebecca Wessling, *Terminal Operations Superintendent*
Greg Hansard, *Marine Superintendent*

The Ferry Division was formed in 1970 to operate waterborne mass transit on San Francisco Bay, operating between Marin County and San Francisco. In 1976, service was expanded with three semi-planing, triple gas-turbine, 25 kt, water-jet-propelled vessels. For economic reasons these three vessels were re-powered in 1985 with twin diesel engines using conventional propellers and rudders, with a resulting speed of 20.5 kt. The 51 m monohull Marin was refurbished in 2007.

High-speed craft operated

Type	Name	Seats	Delivered
MH Campbell S-165 51.2 m	*Marin*	715	December 1976
MH Campbell S-165 51.2 m	*Sonoma*	715	March 1977
MH Campbell S-165 51.2 m	*San Francisco*	715	September 1977
CAT Dakota/AMD 360B	*Del Norte*	390	1998
CAT InCat/Nichols 42 m	*Mendocino*	450	2001

The Golden Gate Ferry vessel Sonoma 0506756

Del Norte 0038856

Mendocino 0102868

Operations
San Francisco to Larkspur, Marin County, 30 minutes
San Francisco to Sausalito, Marin County, 30 minutes.
Approximately 1.8 million passengers are carried annually.

Harbor Bay Maritime

1141 Harbor Bay Parkway, Alameda, California 94501, US

Tel: (+1 510) 769 51 72
e-mail: office@alamedaharborbayferry.com
Web: www.harborbayferry.com

P Bishop, *General Manager*

High-speed craft operated

Type	Name	Seats	Delivered
CAT InCat/NRBB	*Bay Breeze*	250	1994
F-CAT USA Catamarans 20 m	*Harbor Bay Express II*	126	1993

Operations
Alameda to San Francisco.

Harbor Express

703 Washington Street, Quincy, Massachusetts 02169, United States

Tel: (+1 617) 222 69 99
e-mail: info@harborexpress.com
Web: www.harborexpress.com

Harbor Express is a water shuttle service implemented by Water Transportation Alternatives Inc (WTA) in conjunction with the Massachusetts Bay Transportation Authority. WTA also operates a whale watching charter and island transportation services.

High-speed craft operated

Type	Name	Seats	Delivered
CAT Gladding Hearn/ Incat 27 m	*Flying Cloud*	149	1997
CAT Gladding Hearn/ Incat 27 m	*Lightning*	149	1997
CAT Island Boats/Morelli Melvin 19.8 m	*Island Expedition*	200	2005
CAT Island Boats/Morelli Melvin 19.8 m	*Island Discovery*	200	2005

Operations
Quincy to Boston; Quincy to Logan Airport; Hull to Boston; Hull to Logan
Whale watching tours and island transportation.

Hawaii Superferry

One Waterfront Plaza, 500 Ala Moana Blvd, Suite 300, Honolul, Hawaii 96813, United States

Tel: (+1 808) 531 7400
Fax: (+1 808) 531 7410
Web: www.hawaiisuperferry.com

John L Garibaldi, *President and CEO*
Timothy W Dick, *Vice Chairman*
Terry White, *Vice-President Operations*

High-speed craft operated

Type	Name	Seats	Additional payload	Delivered
CAT Austal 105.3 m	*Alakai*	866	282 cars	2007
CAT Austal 105.3 m	–	866	282 cars	(2009)

Operations
Honolulu and Kahului, Maui
Honolulu and Kuua'I.
Honolulu and Lihue.

Holland America Westours

300 Elliot Avenue West, Seattle, WA 98119, US

Tel: (+1 206) 281 3535
Fax: (+1 206) 281 7110
Web: www.hollandamerica.com

High-speed craft operated

Type	Name	Seats	Delivered
CAT SBF 32 m	*Yukon Queen II*	110	1999

Hy-Line Cruises

22 Channel Point Road, Hyannis, Massachusetts MA-02601, United States

Tel: (+1 508) 778 26 00
Fax: (+1 508) 775 26 62
e-mail: betsyk@hylinecruises.com
Web: www.hylinecruises.com

Richard M Scudder, *President*

High-speed craft operated

Type	Name	Seats	Delivered
CAT InCat/Gladding Hearn 32 m	*Lady Martha*	149	1997
CAT InCat/Gladding Hearn 44 m	*Grey Lady*	312	2003

Operations
Hyannis to Nantucket Island and Martha's Vineyard.

Lady Martha 0038863

Grey Lady 1327492

Interstate Navigation

State Pier, Galilee, Narragansett, Rhode Island 02882, United States

Tel: (+1 401) 783 79 96
e-mail: info@blockislandferry.com
Web: www.blockislandferry.com

Interstate Navigation runs fast ferry services from Port Judith to Block Island from May to October.

High-speed craft operated

Type	Name	Seats	Delivered
CAT InCat Designs 29 m	*Athena*	256	2001

Operations
Port Judith to Block Island, 30 minutes, six crossings per day.

Island Hi-speed Ferry

PO Box 5447, Wakefield, Rhode Island 02880, United States

Tel: (+1 877) 733 94 25
e-mail: contact@islandhispeedferry.com
Web: www.islandhispeedferry.com

Island Hi-Speed ferry operates a service to Block Island from May and October and between Old San Juan and Culebra in Puerto Rico from November to April.

High-speed craft operated

Type	Name	Seats	Delivered
CAT Incat 30.4	*Athena*	256	2001

Operations
Galillee to Block Island.
Old San Juan to Culebra to Vieques.

Island Packers Cruises Inc

Ventura Harbor, 1691 Spinnaker Drive Suite 105B, Ventura, California 93001, United States

Tel: (+1 805) 642 13 93
Fax: (+1 805) 642 65 73
e-mail: info@islandpackers.com
Web: www.islandpackers.com

High-speed craft operated

Type	Name	Seats	Delivered
CAT AAM 20 m	*MV Islander*	149	2001
CAT 20 m	*MV Island Adventure*	149	–

Operations
Ventura to the California Channel Islands.

Jet Express

#3 North Monroe Street, Port Clinton, Ohio 43452, United States

Tel: (+1 800) 245 15 38
e-mail: fast@jet-express.com
Web: www.jet-express.com

High-speed craft operated

Type	Name	Seats	Delivered
CAT Gladding-Hearn InCat 28 m	*Jet Express I*	380	May 1989
CAT Gladding-Hearn InCat 29 m	*Jet Express II*	395	May 1992
CAT Gladding-Hearn InCat 27 m	*Jet Express III*	149	2002

Operations
Port Clinton to Put-In-Bay, Kelley's Island and Sandusky.

Kenai Fjords Tours

Division of Alaska Heritage Tours
PO Box 1889, Seward, Alaska 99664, United States

Tel: (+1 907) 224 80 68
Fax: (+1 907) 224 89 34
e-mail: info@kenaifjords.com
Web: www.kenaifjordtours.com

T Tougas, *President*

Since 1974, this operator has been serving the Kenai Fjords National Park, an area known for abundant wildlife and spectacular glaciers.

Kenai Fjords Tours' Alaskan Explorer *in front of Holgate Glacier located in Kenai Fjords National Park* 0084064

High-speed craft operated

Type	Name	Seats	Delivered
MH Westport 90	*Kenai Explorer*	149	1993
MH Westport 95	*Alaskan Explorer*	149	1995
MH Westport 95	*Glacier Explorer*	149	1996
MH Westport 95	*Coastal Explorer*	149	1998
CAT All American 25 m	*Aialik Voyager*	150	2007

Operations
Tours of Kenai Fjords National Park (March to October).

Key West Express

Marco River Marina, 951 Bald Eagle Drive, Marco Island, Florida 34145, United States

Tel: (+1 239) 394 97 00
e-mail: marcoi@seakeywestexpress.com
Web: www.keywestshuttle.com

Key West Express runs ferries from Key West to Marco Island and Ft Myers Beach.

High-speed craft operated

Type	Name	Seats	Delivered
CAT Crowther 38 m	*Atlanticat*	469	2003
CAT Crowther 44 m	*Big Cat Express*	378	2004
MH Gulf Craft 36.5 m	*Whale Watcher*	318	1999
CAT Crowther 52 m	*Key West Express*	500	2006

Operations
Fort Myers to Key West and Marco Island to Key West.
Miami to Key West.

Kitsap Ferry Co

385 Ericksen Ave NE, Suite 123, Bainbridge Island, Washington 98110, United States

Tel: (+1360) 377 03 37
e-mail: info@kitsapferryco.com
Web: www.kitsapferryco.com

High-speed craft operated

Type	Name	Seats	Delivered
CAT International Catamarans 26 m	*Spirit of Adventure*	250	1985

Operations
Seattle to Bremerton.*

*This route suspended until April 2008.

Lake Express LLC

2330 South Lincoln Memorial Drive, Milwaukee, Wisconsin 53207, United States.

Tel: (+1 866) 914 10 10
Web: www.lake-express.com

High-speed craft operated

Type	Name	Seats	Additional payload	Delivered
CAT Austal US 58.5 m	*Lake Express*	250	46 cars	2004

Operations
Milwaukee to Muskegon, Lake Michigan. Journey time two hours 30 minutes.

Miss Barnegat Light

PO Box 609, Barnegat Light, New Jersey 08006 United States

Tel: (+1 609) 494 20 94
e-mail: jlarson742@aol.com

High-speed craft operated

Type	Name	Seats	Delivered
CAT Breaux's Bay 27 m	*Miss Barnegat Light*	100	1973

Operations
Deep sea fishing.

New England Aquarium

Central Wharf, Boston, Massachusettes 02110, US

Tel: (+1 617) 973 52 81
Web: www.neaq.org

High-speed craft operated

Type	Name	Seats	Delivered
CAT InCat Gladding Hearn 34 m	*Voyager III*	325	1999

Operations
Whale watching trips from Boston.

New England Fast Ferry Company

200 Seaport Boulevard, Suite 75, Boston, Massachusetts 02210, United States

Tel: (+1 401) 453 68 00
Fax: (+1 617) 748 14 25
e-mail: info@nefastferry.com
Web: www.nefastferry.com

High-speed craft operated

Type	Name	Seats	Delivered
CAT Derecktors 28 m	*Whaling City Express*	149	2004
CAT Derecktors 28 m	*Martha's Vineyard Express*	149	(2004)

Operations
Providence to Newport
New Bedford to Martha's Vineyard.

New York Water Taxi

499 Van Brunt Street, Sec 8B, Brooklyn, New York, New York 11231, United States

Tel: (+1 212) 742 19 69
Fax: (+1 212) 742 19 70
e-mail: info@nywatertaxi.com
Web: www.nywatertaxi.com

High-speed craft operated

Type	Name	Seats	Delivered
CAT Derektor/N Gee 16 m	*Mickey Murphy*	74	2002
CAT Derektor/N Gee 16 m	*Michael Mann*	74	2002
CAT Derektor/N Gee 16 m	*Curt Berger*	74	2002
CAT Derektor/N Gee 16 m	*Schyler Meyer Jr*	74	2003
CAT Derektor/N Gee 16 m	*John Keith*	74	2003
CAT Derektor/N Gee 16 m	*Ed Rogowsky*	74	2003
CAT InCat Designs 30 m	*Provincetown III*	224	2004
CAT InCat Designs 22 m	*Seymour B Durst*	147	2005
CAT InCat Designs 22 m	*Sam Holmes*	147	2005
CAT InCat Crowther 22 m	*Marian S Heiskell*	147	2006
CAT InCat Crowther 22 m	*Ed Rogowsky*	147	2006
CAT InCat Crowther 22 m	–	147	(2008)
CAT InCat Crowther 22 m	–	147	(2008)

Operations
New York Harbour between Fulton Ferry Landing, South Street Seaport, Wall Street, Battery Park, North Cove, Chelsea Piers and Circle Line.
Hunters Point in Queens, New York, to East 34th Street, Manhattan, every 15 minutes during rush hours.
Yonkers to Manhattan, commenced in Spring 2007.

NY Waterway Tours

4800 Ave at Port Imperial, Weehawken, New Jersey 07086, United States

Tel: (+1 201) 902 87 00
Fax: (+1 201) 348 93 84
e-mail: customerservice@nywaterway.com
Web: www.nywaterway.com

Arthur E Imperatore, *President*

High-speed craft operated

Type	Name	Seats	Delivered
MH Blount Marine Corporation	*Port Imperial*	149	1984
CAT InCat/NBBB 22 m	*American Eagle* (ex-*Native Sun*, ex-*Klondike*)	149	1984

The 397-passenger 24 kt Empire State 0507097

Type	Name	Seats	Delivered
MH Gulf Craft	*Port Imperial Manhattan*	350	1986
MH Gulf Craft	*New Jersey*	350	1988
MH Gulf Craft	*George Washington*	399	1989
MH Gulf Craft	*Thomas Jefferson*	399	1989
MH Gulf Craft	*Alexander Hamilton*	399	1989
MH Gulf Craft	*Abraham Lincoln*	399	1989
MH Gulf Craft	*West New York*	149	1990
MH Gladding-Hearn	*Henry Hudson*	397	1992
MH Gladding-Hearn	*Robert Fulton*	397	1993
MH Gladding-Hearn	*Empire State*	397	1994
MH Gladding-Hearn	*Garden State*	397	1994
CAT Derecktor NGA 38 m	*Bravest*[1]	350	1996
CAT Derecktor	*Finest*[2]	350	1996
MH Gladding-Hearn	*John Stevens*	397	1997
CAT InCat/Gladding Hearn 28 m	*Monmouth* (ex-*Theodore Roosevelt*)	150	1998
CAT Allen 24 m	*Yogi Berra*	150	1999
CAT Allen 24 m	*Fiorello La Guardia*	150	1999
CAT Allen 24 m	*Frank Sinatra*	150	2000
CAT Allen 24 m	*Christopher Columbus* (ex-*Vince Lombardi*)	150	2000
CAT Yank Marine 31 m	*Middletown* (ex-*Royal Miss Belmar*)[3]	227	2000
CAT Allen 24 m	*Giovanni de Verrazono*	150	2001
MH Allen 20 m	*Austin Tobin*	97	2001
MH Allen 20 m	*Fr. Mychal Judge*	97	2001
MH Allen 20 m	*Moira Smith*	97	2001
MH Allen 20 m	*Douglas Gurian*	97	2002
CAT InCat 34 m	*Voyager III*[4]	325	2002
CAT InCat 29 m	*Athena*[5]	256	2002
CAT Allen 24 m	*Congressman Robert A Roe*	150	2003
CAT Allen 24 m	Jersey City	150	2003
CAT Allen 24 m	Bayonne	150	2003
CAT Allen 24 m	Queens	150	2003
CAT Allen 24 m	Admiral Richard E Bennis	150	2003

[1]Leased from Bar Harbour Whale Watch
[2]Leased from NY Fast Ferry
[3]Leased to Bay Shore Ferry
[4]Leased from Miss Belmar Fishing
[5]Leased from New England Aquarium

Operations
Weehawken (New Jersey), to 38th Street and 12th Avenue Manhattan, every 15 minutes from 06.45 to midnight
Weehawken (New Jersey), to slip 5 in Lower Manhattan during the rush hours
Hoboken (New Jersey), to the World Financial Centre, Manhattan, every 5 minutes during peak and every 20 minutes off-peak from 06.30 to 22.00.
Colgate pier in Jersey City (New Jersey), to the World Financial Centre, Manhattan every 15 minutes from 06.45 to 22.00.

Otto Candies LLC

PO Box 25, Des Allemands, Louisiana, 70030-0025, United States

Tel: (+1 504) 469 77 00
Fax: (+1 504) 469 77 40
e-mail: info@ottocandies.com
Web: www.ottocandies.com

High-speed craft operated

Type	Name	Seats	Additional payload	Delivered
CAT Austal 38 m	*Seba'an* (ex-*Speeder*)	330	–	1995
MH Swiftships 41 m	*Fernanda*	–	–	2001
MH Austal 46 m	*Veronica*	80	152 t	2002
MH Austal 46 m	*Ana Paula*	80	152 t	2002
MH 41 m	*Victoria*	–	–	2002

Operations
Serving the offshore oil industry.

Pacific Whale Foundation

300 Maalaea Road, Suite 211, Wailuku, Hawaii 96793, United States

Tel: (+1 808) 249 88 11
Fax: (+1 808) 243 90 21
e-mail: info@pacificwhale.org
Web: www.pacificwhale.org

Greg Kaufman, *Founder and President*

Established in 1980, the company runs tourist whale watching trips around Maui Island.

High-speed craft operated

Type	Name	Seats	Delivered
CAT Crowther 20 m	*Ocean Quest*	147	2000
CAT Crowther 20 m	*Ocean Odyssey*	149	2001
CAT Crowther 19 m	*Ocean Voyager*	149	2004
CAT One2Three 22 m	*Ocean Discovery*	149	2007

Operations
Maui Island, Hawaii.

Phillips Cruises & Tours LLC

519 West Fourth Avenue, Anchorage, Alaska 99501, United States

Tel: (+1 907) 274 27 23
Fax: (+1 907) 265 58 90
e-mail: info@26glaciers.com
Web: www.26glaciers.com

Robert G Neumann, *Owner*
Barrie Swanberg, *General Manager*

High-speed craft operated

Type	Name	Seats	Delivered
CAT Nichols Brothers InCat 40 m	*Klondike Express*	342	1999

Operations
Whittier, Alaska to College Fjord to Barry Arm to Harriman Fjord, 135 miles, 4½ hours. Fare (2007) USD129 + tax.

Prince William Sound Cruises and Tours

Division of Alaska Heritage Tours
PO Box 1297, Valdez, Alaska 99686, United States

Tel: (+1 907) 835 47 31
Fax: (+1 907) 835 37 65
e-mail: info@princewilliamsound.com
Web: www.princewilliamsound.com

Colleen Stephens, *Sales and Operations Manager*

MV Tanaina 0100064

High-speed craft operated

Type	Name	Seats	Delivered
MH Westport 95	*Tanaina*	149	1999
MH Westport 95	*Nunatak*	149	1996
MH Westport 80	*Chugach*	149	1992

Operations
Cruises and tours in Prince William Sound.

Royal Caribbean Cruise Lines

1050 Caribbean Way, Miami, Fl-33132-2096, United States

Web: www.royalcaribbean.com

The company operates four fast 21 m catamarans to ferry passengers from their cruise ships to the shore.

High-speed craft operated

Type	Name	Seats	Delivered
CAT Island Boats	–	230	2004
CAT Island Boats	–	230	2004
CAT Island Boats	–	230	2004
CAT Island Boats	–	230	2004

Operations
Ferry from cruise ship to shore.

Salem Ferry

Blaney St, Salem, Massachusetts 01970, US

Tel: (+1 978) 741 0220
Web: www.salemferry.com

High-speed craft operated

Type	Name	Seats	Delivered
CAT InCat/Gladding Hearn 27 m	Nathaniel Bowditch (ex-Friendship IV)	149	1996

Operations
Salem to Boston, 45 minutes.

Sayville Ferry Service Inc

41 River Road, PO Box 626, Sayville, Long Island, New York 11782-0626, United States

Tel: (+1 631) 589 08 10
Fax: (+1 631) 589 08 43
e-mail: sayvilleferry@erols.com
Web: www.sayvilleferry.com

Captain Kenneth Stein, *President*

Sayville Ferry Service has a total fleet of eight vessels with passenger capacities from 100 to 413. Most of them are operated between 15 and 20 kt but the *Fire Island Clipper, South Bay Clipper* and *Fire Island Empress* can cruise at higher speeds.

High-speed craft operated

Type	Name	Seats	Delivered
MH Derecktor Shipyard 22 m	*Fire Island Clipper*	350	1979
MH Gladding Hearn 25 m	*South Bay Clipper*	413	1996
MH Derecktor Shipyard 21.6 m	*Fire Island Empress*	270	2004

Operations
Sayville, Long Island (South Shore) to Fire Island, 5 n miles, serving three communities, stops at Cherry Grove, Water Island and Fire Island Pines. The ferries also serve Sunken Forest and Sailor Haven Federal Parks.

Sayville Ferry Service South Bay Clipper 0507098

Seastreak

A subsidiary of Sea Containers
Two First Avenue, Atlantic Highlands, New Jersey 07716, United States

Tel: (+1 800) 262 87 43
Fax: (+1 732) 872 96 91
e-mail: contact@seastreak.com
Web: www.seastreak.com

John Burrows, *General Manager*

High-speed craft operated

Type	Name	Seats	Delivered
CAT Incat Gladding Hearn 43 m	*Seastreak New York*	400	2001
CAT Incat Gladding Hearn 43 m	*Seastreak New Jersey*	400	2001
CAT Incat Gladding Hearn 43 m	*Seastreak Wall Street*	413	2003
CAT Incat Gladding Hearn 43 m	*Seastreak Highlands*	430	2004

Operations
Monmouth County, New Jersey and Middlesex County, New Jersey to Wall Street and East 34th Street, Manhatten, 45 minutes.

Passengers carried
Over 800,000 commuters were carried in 2005.

Seastreak New York 0129587

Shepler's Mackinac Island Ferry

556 E.Central, Mackinaw City, Michigan 49701, United States

Tel: (+1 231) 436 50 23
Fax: (+1 231) 436 75 21
e-mail: information@sheplersferry.com
Web: www.sheplersferry.com

High-speed craft operated

Type	Name	Seats	Delivered
MH Camcraft Boats Inc (Hargrave) 18.3 m	*The Welcome*	120	1969
MH Camcraft Boats Inc (Hargrave) 17.1 m	*Felicity*	150	1972
MH Bergeron (Hargrave) 19.9 m	*Hope*	150	1975
MH Bergeron (Hargrave) 25.3 m	*Wyandot*	265	1979
MH Aluminium Boats Inc (Hargrave) 25.6 m	*Captain Shepler*	265	1986

Operations
St Ignace (Upper Peninsula) and Mackinaw City (Lower Peninsula) to Mackinac Island, Michigan on Lake Huron.

Star Line

587 North State Street, St Ignace, Michigan 49781, United States

Tel: (+1 906) 643 76 35
Fax: (+1 906) 643 98 56
Web: www.mackinacferry.com

High-speed craft operated

Type	Name	Seats	Delivered
MH Gulf Craft 19.8 m	*Nicolet*	150	1985
MH Gulf Craft 25.9 m	*Radisson*	350	1988
MH Gulf Craft 19.8 m	*Cadillac*	150	1990
MH Gulf Craft 25 m	*Marquette II*	330	2005

Operations
St Ignace and Mackinaw City, Michigan to Mackinac Island.

Monohull Gulf Craft Radisson *in service with Star Line* 0506854

Trico Marine Services Inc

250 North American Court, PO Box 4097, Houma, Louisiana 70363, United States

Tel: (+1 985) 851 38 33
Fax: (+1 985) 851 43 21
Web: www.tricomarine.com

High-speed craft operated

Type	Name	Seats	Delivered
SWATH Eastern Shipbuilding Group	*Stillwater River*	250	1998

Operations
Crewboat operations in the Campos Basin, Brazil.

Stillwater River 0045232

Vallejo Baylink Ferries

PO Box 2287, Vallejo, California 94592 2287, United States

Tel: (+1 707) 562 31 40
Fax: (+1 707) 562 31 41
e-mail: mrobbins@baylinkferry.com
Web: www.baylinkferry.com

Martin Robbins, *General Manager*

High-speed craft operated

Type	Name	Seats	Delivered
CAT Dakota/AMD 360	*Intintoli*	300	1997
CAT Dakota/AMD 360	*Mare Island*	300	1997
CAT InCat/G Hearn 34 m	*Vallejo* (ex-*Jet Cat Express*)	300	1994
CAT Dakota/AMD 360	*Solano*	300	2004

Note: *Jet Cat Express* was lengthened from 31 to 34 m and renamed *Vallejo* in 2001.

Operations
Vallejo to San Francisco Ferry Building and Fisherman's Wharf, 1 hour; 14 round trips per day.

Vineyard Fast Ferry

Quonset Point, North Kingstown, Rhode Island 02852, United States

Tel: (+1 401) 295 40 40
Fax: (+1 401) 295 49 30
e-mail: info@vineyardfastferry.com
Web: www.vineyardfastferry.com

Aaron D Houls, *Controller*

High-speed craft operated

Type	Name	Seats	Delivered
CAT InCat/Gladding Hearn 37 m	*Millennium*	350	1998

Operations
Rhode Island to Martha's Vineyard.

Yankee Fleet

75 Essex Avenue, Gloucester, MA 01930, US

Tel: +1 978 283 0313
Fax: +1 978 283 6089
e-mail: office@yankeefleet.com
Web: www.yankeefleet.com

High-speed craft operated

Type	Name	Seats	Delivered
CAT Incat/Gladding Hearn 28 m	*Yankee Freedom II*	250	1989

Operations
Whale watching trips from Gloucester Harbour.

Yankee Freedom II 0081011

Washington State Ferries

2901 Third Avenue, Suite 500, Seattle, Washington 98121-3014, United States.

Tel: (+1 206) 515 34 00
e-mail: wsf@wsdot.wa.gov
Web: www.wsdot.wa.gov/ferries

Mike Anderson, *Executive Director*

Skagit *operated by Washington State Ferries* 0506757

Chinook *was delivered to WSF in 1998* 0045233

High-speed craft operated

Type	Name	Seats	Delivered
MH Equitable 34 m	*Kalama*	253	1990
MH Equitable 34 m	*Skagit*	253	1990
CAT Dakota Creek/AMD 385	*Chinook*	350	1998
CAT Dakota Creek/AMD 385	*Snohomish*	350	1999

Operations
Seattle to Vashon 9 n miles.

Woods Hole and Martha's Vineyard Steamship Authority

PO Box 284, Woods Hole, Massachusetts 02543, United States

Tel: (+1 508) 548 50 11
Fax: (+1 508) 548 09 20
Web: www.steamshipauthority.com

High-speed craft operated

Type	Name	Seats	Delivered
CAT NGA 41 m	*Flying Cloud*	336	2000
CAT InCat Crowther 47 m	*Iyanough*	389	2006

Operations
Wood Hole to Martha's Vineyard and Hyannis to Nantucket.

Uruguay

Ferrylineas

Terminal 3 XXX L27, Rio Negro 1400, Montevideo, Uruguay

Tel: (+598 11) 43 14 45 80
e-mail: ventas@ferryturismo.com.uy
Web: www.ferryturismo.com.uy

High-speed craft operated

Type	Name	Seats	Additional payload	Delivered
WPC InCat Australia 74 m	*Atlantic III* (ex-*Juan L*)	600	110 cars	1993

Vessel leased from Buquebus.

Operations
Four times daily service from Colonia and Montevideo to Buenos Aires.

Venezuela

Conferry

Avenida Casanova con Av Las Acacias, Torre Banhorient, Planta Baja, Sabana Grande, Caracas 1050 Venezuela

Tel: (+58 212) 793 20 30
Fax: (+58 212) 793 59 61
e-mail: webmasterconferry@dbaccess.com
Web: www.conferry.com

High-speed craft operated

Type	Name	Seats	Additional payload	Delivered
CAT Austal 86 m	Tallink Auto Express II	700	175 cars	1997
CAT Austal 86 m	*Carmen Ernestina*	800	200 cars	1999
CAT Austal 86 m	*Lilia Concepcion*	868	243 cars	2002

Operations
Puerto La Cruz to Punta De Piedras, La Guaira to Punta De Piedras.

Ferryven

Hotel Flamingo local 3, Calle El Cristo, Sector La Caranta, Pampatar, Venezuela

Tel: (+58 295) 262 26 72
Fax: (+58 295) 262 97 53
e-mail: ferryvenmargarita@yahoo.com
Web: www.ferryven.com

High-speed craft operated

Type	Name	Seats	Delivered
CAT Austal 40 m	*Opale Express*	304	1998
CAT Marinteknik 40 m	*Turquoise Express*	395	1995

Operations
Puerto La Cruz to Pampatar.

Gran Cacique

Tel: (+58 295) 239 8339
Web: www.grancacique.com.ve

High-speed craft operated

Type	Name	Seats	Additional payload	Delivered
CAT Marinteknik 55 m	*Josefa Camejo*	400	48 cars	1999
MH Swiftships 35 m	*Gran Cacique III*	347	–	1980

Vietnam

Duong Dong Express

Borough 1, Tran Hung Dao Street, Duong Dong Town, Phu Quoc District, Kien Giang Province, Vietnam

Tel: (+84 77) 98 16 50
Fax: (+84 77) 98 16 49
e-mail: sge@duongdongexpress.com.vn
Web: www.duongdongexpress.com.vn

Established in 2004, the company purchased a vessel from Indonesian operator PT Mabua Intan Express.

High-speed craft operated

Type	Name	Seats	Delivered
CAT Fjellstrand Flying Cat 40 m	*Duong Dong Express* (ex *Mabua Express*)	284	1992

Operations
From Rach Gia in Kien Giang Province to Phu Quoc Island, SW Vietnam, 65 miles, 2 hours 30 minutes.

Virgin Islands (UK)

M & M Transportation Services

British Virgin Islands

Tel: (+1 869) 466 67 34
email: mmts@hotmail.com

High-speed craft operated

Type	Name	Seats	Delivered
CAT CNB TMCAT 27	*Carib Surf*	243	2003

Operations
St Kitts to Nevis.

Speedy's

PO Box 35, The Valley, Virgin Gorda, British Virgin Islands

Tel: (+1 284) 495 52 40
Fax: (+1 284) 495 57 55

Peacemaker 1111648

e-mail: speedysbvi@surfbvi.com
Web: www.speedysbvi.com

High-speed craft operated

Type	Name	Seats	Delivered
CAT Sabre 22 m	*Peacemaker*	215	2002

Operations
Virgin Gorda, Tortola and St Thomas
Charters available throughout the British Virgin Islands and United States Virgin Islands.

Road Town Fast Ferry

Ferry Dock, Road Town, Tortola, British Virgin Islands

Tel: (+1 284) 494 23 23
e-mail: roadtownfastferry@hotmail.com
Web: www.tortolafastferry.com

High-speed craft operated

Type	Name	Seats	Delivered
CAT Marinteknik 30 m	*Tortola Fast Ferry*	230	1984

Operations
Daily return sailings between central Charlotte Amalie, St Thomas and Road Town, Tortola, 50 minutes.

Virgin Islands (US)

Island Lynx Ferry Company

Gallows Bay, PO Box 25320, Christiansted, St Croix, VI 00824, Virgin Islands, United States

e-mail: info@southerlandtours.com
Web: www.southerlandtours.com/islandlynx

High-speed craft operated

Type	Name	Seats	Delivered
CAT Yank Marine 31 m*	*Royal Miss Belmar*	227	2000

*Leased from Miss Belmar Fishing

Operations
St Thomas to St Croix.

Smith's Ferry Services Ltd

3400 Veterans Drive, Ste, 2 St Thomas, 00802 US Virgin Islands

Tel: (+340) 775 7292
Fax: (+340) 774 5532
email: info@smithsferry.com
Web: www.smithsferry.com

High-speed craft operated

Type	Name	Seats	Delivered
CAT Westamaran 26.2 m	*Mayflower*	149	1972
CAT Allen Marine 23.7 m	*Island Rocket III*	149	1997

Operations
St Thomas, Tortola, Virgin Gorda, Anegada.

Transportation Services of St John

PO Box 28, Crue Bay, St John, US Virgin Islands

Tel: (+1 340) 776 62 82

High-speed craft operated

Type	Name	Seats	Delivered
MH Aluminium Boats 29 m	*Caribe Time*	142	1979
MH Aluminium Boats 29 m	*Caribe Tide*	232	1995
MH Aluminium Boats 29 m	*Caribe Cay*	232	1995

Operations
St John to St Thomas
Fajardo, Puerto Rico (weekly).

VI Sea Trans

A subsidiary of Smith's Ferry Services Ltd.
3400 Veterans Drive, Ste, 2 St Thomas, 00802 US Virgin Islands

Tel: (+340) 775 7292
Fax: (+340) 774 5532
email: info@smithsferry.com

High-speed craft operated

Type	Name	Seats	Delivered
CAT Yank Marine 31 m	*Royal Miss Belmar*	277	2000

Operations
St Thomas to St Croix.

PRINCIPLE EQUIPMENT AND COMPONENTS FOR HIGH-SPEED CRAFT

ENGINES

Company listing by country

Canada
Pratt & Whitney Canada Corp

France
Baudouin Moteurs
MAN Diesel SAS
Turbomeca
Wärtsilä SACM Diesel

Germany
Deutz AG
MAN Diesel SE
MAN Nutzfahrzeuge AG
MTU

Italy
CRM Motori Marini SpA
Isotta Fraschini Motori SpA
Iveco Aifo SpA
Seatek SpA

Japan
Mitsubishi Heavy Industries Ltd

Norway
GE Energy (Norway) A/S

Russian Federation and Associated States (CIS)
Barnaultransmash Joint-Stock Company
Ivchenko Progress Design Bureau
NK Engines
Zvezda Joint Stock Company

Sweden
Scania
Volvo Penta AB

United Kingdom
MAN B&W Diesel Ltd
Rolls-Royce Marine Systems

United States
Caterpillar Inc
Cummins Marine Inc
General Electric Company
Pratt & Whitney
Turbo Power and Marine Systems Inc
Vericor Power Systems

Canada

Pratt & Whitney Canada Corp

A United Technologies company
1,000 Marie-Victorin Boulevard, Longueuil, Quebec J4G 1A1, Canada

Tel: (+1 450) 677 94 11
Fax: (+1 450) 647 36 20
e-mail: customerhelpdesk@pwc.ca
Web: www.pwc.ca

A M Bellemare, *President*
John Saabas, *Senior Vice President, Operations and Engineering*
Claude Paquette, *Vice President*
Monique Chaput, *Information Co-ordinator*

ST40 gas turbine 0077002

ST6L-90 gas turbine 0077003

In addition to its wide range of small gas-turbine aero-engines, Pratt & Whitney Canada (P&WC) also manufactures industrial and marine derivatives supplied by its Industrial Engine Division. Engines are rated from 410 kW upwards (see table) and are of the simple-cycle, free turbine type. They run on Nos 1 and 2 diesel or aviation turbine fuels.

Typical marine applications are the ST6K-77 driving the auxiliary power unit on the 'Flagstaff 2' Class hydrofoils built for the Israeli Navy, and the ST6T-76 Twin Pac powering LACV-30 hovercraft built by Bell Aerospace Textron (now Textron Marine and Land Systems) for the US Army. Including aero-engine installations, more than 43,800 of P&WC's turbines have been delivered. They have accumulated in excess of 290 million hours of operation.

The ST6L-90 engine is based on the largest and most efficient PT6 turboprop in current production at Pratt & Whitney Canada Inc.

Engine features: All ST6 engines have a single-spool gas generator and a multistage compressor (three axial plus one centrifugal stage) driven by a single-stage axial turbine. The larger models have cooled vanes. The radial diffuser incorporates P&WC patented diffuser pipes. The burner section has a reverse flow annular combustor. A single- or two-stage free turbine provides a direct high-speed output or drives through a reduction gearbox.

Air intake is located at the rear of the engine and is through a screened annular inlet.

The ST6T-76 Twin Pac is a dual version of the ST6 with two power sections mounted side by side coupled to a twinning reduction gearbox. The gearbox incorporates automatic clutches which permit emergency operation of one side independent of the other.

ST18

The ST18 is an industrial and marine version of the PW124 aero turboprop. This engine benefits from the experience gained by its aero counterparts with 12,000,000 hours of operation.

Engine features: The ST18 is a three-shaft engine. The low- and high-pressure centrifugal compressors (combined pressure ratio 15:1) are each driven independently by single-stage compressor turbines. This allows both compressors to operate at their optimum efficiencies without the need for complex variable geometry. Other features comprise P&WC patented pipe diffusers, reverse annular combustor, low-pressure and high-pressure stator cooling and turbine blade cooling. The two-stage power turbine retains the free turbine concept of the ST6 engine.

ST30 and ST40

The ST30 and its upgraded version the ST40 are free turbine turboshaft engines derived from the PW 150A aviation turboprop powerplant. The turbomachinery comprises three independent rotating assemblies mounted on concentric shafts.

Engine features: The compressor is made up of a three-stage axial unit followed by a single centrifugal impeller, each driven by a single-stage turbine. Power is extracted by a two-stage turbine and delivered to the inlet end of the engine. The combustion system comprises a reverse flow annular combustion chamber, multiple fuel nozzles and a spark igniter.

The engine incorporates a radial air intake which also provides the interface to the reduction gearbox or other driven load.

MFT8

The marine FT8 engine is a derivative of aviation's popular JT8D engine, which has more than half a billion hours of service operation. The engine has been specifically designed to operate in a corrosive environment with extensive use of protective coatings. The engine can be supplied for a variable speed direct mechanical application or as a generator for electric drive propulsion.

ENGINE DATA SUMMARY

Sea level standard pressure at 15°C inlet temperature

Model	Max kW	Max power SFC (kg/kWs-hr)	Normal kW	Normal power SFC (kg/kWs-hr)	RPM	Length (mm)	Width (mm)	Height (mm)	Engine dry weight (kg)
ST18M	2,128	0.28	1,803	0.291	18,900	1,530	660	810	350
ST40M	4,591	0.257	3,781	0.262	14,875	1,700	660	960	525
MFT8	28,101	0.221	26,210	0.223	3,600/5,000	7,620	2,590	3,048	7,258
ST6M	709	0.349	646	0.352	30,000	1335	483	559	132

France

Baudouin Moteurs

165 Boulevard de Pont-de-Vivaux, F-13010 Marseille, France

Tel: (+33 4) 91 83 85 00
Fax: (+33 4) 91 83 85 73
e-mail: hello@moteurs-baudouin.fr
Web: www.moteurs-baudouin.fr

Thomas Ghiani, *Export Manager*

Moteurs Baudouin produce a range of diesel engines for high-performance craft as well as for fishing vessels and barges. The high-performance models are described here.

Model	Power (kW)	Rpm	Weight (kg)
6R102 SR	250	2000	1085
6R123 SR	294	2000	1100
12M26 S	486	1800	3120
12M26 SRP	736	1900	3180
6 R102 S	250	2,000	680
6 R124 SR	375	1,900	1,400
6 M26 SR	478	2,000	1,990
12 M26 SR	957	2,000	3,180

Baudouin 12 M26 SR
0121770

MAN Diesel SAS

22 avenue des Nations, Le Ronsard Paris Nord 2, BP 84013 Villepinte, F-95931 Roissy CDG Cedex, France

Tel: (+33 1) 48 17 63 00
Fax: (+33 1) 48 17 63 49
e-mail: service_marketing@pielstick.com
Web: www.pielstick.com

MAN Diesel designs and develops four-stroke high- and medium-speed diesel engines in the output range of 1,000 to 26,500 kW per unit at speeds from 360 to 1,500 rpm.

These engines are manufactured in France and under license in 13 other countries.

The PA4-200VGA covers power requirements from 1,990 to 2,650 kW at 1,500 rpm and has been installed in numerous fast ferries.

The PA6 B STC is rated from 4,860 to 8,100 kW at 1,050 rpm. This is the basic engine for the propulsion of any ship requiring high power and light installed weight.

MAN Diesel has many years experience with the PA6 engines on board merchant ships, including high-speed ferries, naval ships, locomotives and for generating sets.

The PA6 B STC has particularly high torque characteristics at low rpm. With its single-stage sequential turbocharging, the PA6 B STC combines performance flexibility and high power.

MAN Diesel 20 PA6 B STC engine rated at 8,100 kW at 1,050 rpm 1033514

Engine model	Cyl	Rating kW	rpm	Length mm	Width mm	Height mm	Weight t
12 PA4 200 VGA	12	1,990	1,500	2,700	1,790	1,920	8.5
16 PA4 V 200 VGA	16	2,650	1,500	3,300	1,790	1,920	10.0
12 PA6 B STC	12	4,860	1,050	5,830	2,340	3,124	31
16 PA6 B STC	16	6,480	1,050	6,780	2,340	3,124	37
18 PA6 B STC	18	7,290	1,050	7,540	2,640	3,166	41
20 PA6 B STC	20	8,100	1,050	7,960	2,640	3,166	43

Turbomeca

Part of the Safran Group
F-64511 Bordes Cedex, France

Tel: (+33 5) 59 12 50 00
Fax: (+33 5) 59 53 15 12
Web: www.turbomeca.com

Emeric d'Arcimoles, *Chairman, President & CEO*
Pierre Fabre, *Executive Vice President*
Jacques Brochet, *Vice President Engineering*
Didier Blauwart, *Vice President Land & Marine Turbines*
Frederic Matha, *Marketing & Sales*
Valery Krol, *Business Development Manager, TMI 800*

TM 1800

The TM 1800 is a new genset for Navy applications, as defined by the UK MoD and French DGA, with a power range of 1,800 kW in S/L ISO conditions.

Compact, with a high reliability, the TM 1800 includes one single shaft gas turbine. It offers a thermodynamic design based on a recuperated cycle, with moderate life cycle costs.

The TM has a high-speed alternator, a dedicated monitoring and control system and requires no reduction gearbox. It is designed with a single stage centrifugal compressor in titanium, cast iron for most castings, a single can external combustor and a two stage axial turbine. Its modular design allows for a major overhaul to be completed within 24 hours.

There are two operating modes to meet mission requirements: Operating mode with fast response and no limitation in load impact with a constant speed datum, and Economy mode for optimum fuel efficiency with a variable speed.

The TM 1800 can also be adapted to commercial ship generators, cogeneration, train propulsion, portable gensets, oil and gas.

Specifications

Main characteristics

Air mass flow: 9 kg/sec

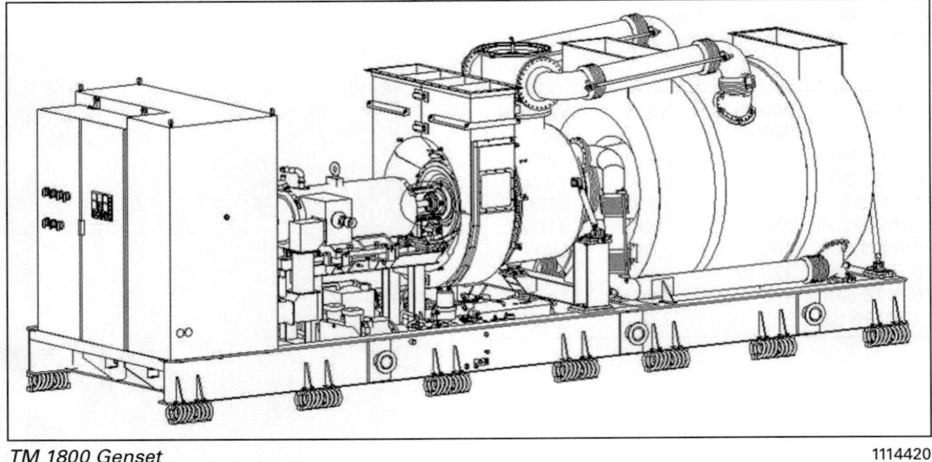

TM 1800 Genset 1114420

Pressure ration: 8.10
Speed: 22,500 rpm
TBO: 20,000 h
NO_x: <3.5 g/kWh
CO: <2.5 g/kWh
UHC: <1.0 g/kWh
Genset weight: 13 t

Wärtsilä SACM Diesel

Headquarters
1 rue de la Fonderie, BP 1210,
F-68054 Mulhouse Cedex, France

Tel: (+33 3) 89 66 68 68
Fax: (+33 3) 89 66 68 60
Web: www.wartsila.com

Class-Eink Strand, *President and Director General*
Patrice Flot, *Technical Director*
Peter Zoeteman, *Vice President, Marine*

Marine Department
La Combe, BP 1212, F-17700 Surgeres, France

Tel: (+33 5) 46 30 31 32
Fax: (+33 5) 46 07 35 37

Wärtsilä SACM Diesel designs, manufactures and markets a range of high-speed diesel engines from 300 to 4,000 kW. The company belongs to one of the world's largest manufacturers of medium- and high-speed engines: the Wärtsilä Diesel Group. Wärtsilä Diesel engines and related systems are used for marine propulsion and onboard generator sets. The Group has facilities in six European countries, India and the US, supported by a worldwide network of sales and service companies. The Group employs more than 6,000 people in 60 different countries.

Wärtsilä SACM Diesel is certified ISO 9001 by the international classification society, Det Norske Veritas.

In 2001 the company was contracted to supply four Wärtsilä 12V200 engines for the new Finnish Frontier Guard patrol vessels due for completion in 2002–04.

Type of engine	Length (mm)	Width (mm)	Height (mm)	Weight (kg)
Wärtsilä UD23				
UD23 12V M5D	1,975	1,380	1,450	2,650
UD23 12V M4	2,300	1,436	1,550	3,150
UD23 12V M5	2,300	1,436	1,550	3,150
Wärtsilä UD25				
UD25 6L M4	2,070	1,260	1,845	2,850
UD25 6L M5	2,070	1,260	1,845	2,850
UD25 12V M4	2,711	1,624	1,970	4,900
UD25 12V M5	2,711	1,624	1,970	4,900
Wärtsilä 200				
12V 200	3,640	1,636	2,345	12,600
16V 200	4,350	1,636	2,360	16,000
18V 200	4,700	1,636	2,360	17,500

Wärtsilä 12V UD23 M5, 970 kW engine 0507014

Wärtsilä 12V 200, 2,400 kW engine 0507015

Germany

Deutz AG

Ottostr. 1, 51149, Köln-Porz (EiL), Germany

Tel: (+49 221) 82 20
Fax: (+49 221) 822 35 25
e-mail: info@deutz.de
Web: www.deutz.de

Gordon Riske, *Chief Executive Officer*

Group companies:

Motoren-Werke Mannheim AG and Deutz Energy GmbH
Carl-Benz-Strasse 5, D-68167 Mannheim, Germany

Tel: (+49 621) 384 0
Fax: (+49 621) 384 86 12

UK subsidiary:

DEUTZ UK Ltd, 2 St Martin's Way, London SW17 0UT, United Kingdom

Tel: (+44 20) 87 81 72 00
Tx: 8954136 KHDLON G
Fax: (+44 20) 89 47 63 80
Web: www.deutz.com

Founded in 1898, Deutz AG produce a wide range of marine and industrial diesel engines, from 4 to 7,400 kW. The company has four divisions, diesel engines, service, energy supply and industrial plant construction.

Deutz water-cooled engines

Since 1985, when Motoren-Werke Mannheim AG was taken over by Klöckner-Humboldt-Deutz AG, the company has offered a combined range of medium-sized and large water-cooled engines from 300 to 7,400 kW. The principal engine ranges suitable for high-speed surface craft are in the accompanying table.

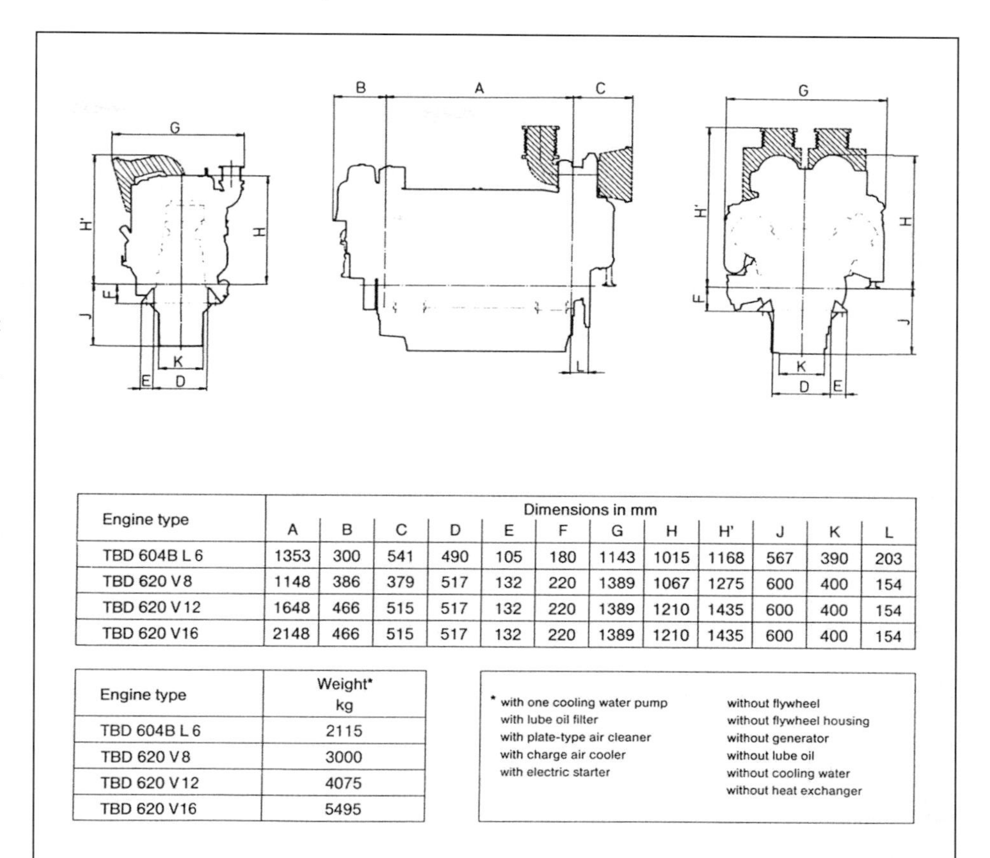

Engine type	Dimensions in mm											
	A	B	C	D	E	F	G	H	H'	J	K	L
TBD 604B L 6	1353	300	541	490	105	180	1143	1015	1168	567	390	203
TBD 620 V 8	1148	386	379	517	132	220	1389	1067	1275	600	400	154
TBD 620 V 12	1648	466	515	517	132	220	1389	1210	1435	600	400	154
TBD 620 V16	2148	466	515	517	132	220	1389	1210	1435	600	400	154

Engine type	Weight* kg
TBD 604B L 6	2115
TBD 620 V 8	3000
TBD 620 V 12	4075
TBD 620 V16	5495

* with one cooling water pump, with lube oil filter, with plate-type air cleaner, with charge air cooler, with electric starter; without flywheel, without flywheel housing, without generator, without lube oil, without cooling water, without heat exchanger

Deutz TBD 604B and 620 series weights and dimensions 0506855

Deutz-powered Griffon hovercraft, left to right: two 3000 TD (twin Deutz BF8L 513 diesels), two 2000 TDX (single Deutz BF8L 513 diesel) and one 2000 TD (single Deutz BF6L 913C diesel) 0506765

Deutz 616 series for propulsion output of up to 1,360 kW 0063229

Deutz 628 series for propulsion output of up to 3,600 kW 0063230

Deutz BF12L 513C air-cooled diesel as fitted to BHC and NQEA AP1-88 hovercraft 0506764

Model summary

Type	Number of cylinders and configuration	Power according to DIN 6271 and ISO 3046/1 Continuous net brake fuel stop power kW	Marine Propulsion Rpm	Length mm	Weight kg
BFM1013ECP	6	195	2,300	1,420	700
BF6M1015	6V	240	2,100	1,205	880
BF6M1015C	6V	330	2,100	1,480	950
BF8M1015C	8V	440	2,100	1,673	1,250
TBD 616 V8	V8	480	2,100	1,800	1,900
TBD 616 V12	V12	720	2,100	2,180	2,450
TBD 616 V16	V16	960	2,100	2,550	3,000
TBD 620 V8	V8	1,016	1,800	1,950	3,300
TBD 620 V12	V12	1,524	1,800	2,690	5,000
TBD 620 V16	V16	2,032	1,800	3,150	6,200
BV 6 M 628	L6	1,350	1,000	3,494	9,500
BV 8 M 628	L8	1,800	1,000	4,246	11,500
BV 9 M 628	L9	2,025	1,000	4,556	13,400
BV 12 M 626	V12	2,700	1,000	4,360	16,300
BV 16 M 628	V16	3,600	1,000	5,135	21,200
TBD 645 L6	L6	2,760	650	5,660	26,500
TBD 645 L8	L8	3,680	650	6,640	34,000
TBD 645 L9	L9	4,140	650	7,130	37,600
TBD 632 L9	L9	2,565	1,000	4,450	14,300
TBD 632 V12	V12	3,420	1,000	4,000	16,600
TBD 632 V16	V16	4,650	1,000	4,700	21,700
TBD 632 V18	V18	5,130	1,000	5,100	23,900
BV 12 M 640	V12	5,290	650	6,640	48,000
BV 16 M 640	V16	7,060	650	7,983	60,000

Recent applications
Deutz engines are installed by the following hovercraft manufacturers:

Manufacturer	Craft type	Engine type
Griffon Hovercraft Ltd	1000TD/1500TD/2000TD	BF6L 913C
	2500TD	BF6L 913C × 2
	2000TDX/M	BF8L 513/LC
	3000TD	BF8L 513 × 2
	4000TD	BF10L 513 × 2
Slingsby Aviation Ltd	SAH 2200	BF6L 913C
ABS Hovercraft Ltd	M10	BF8L 513
		BF12L 513C × 2
Westland Aerospace	WAP1-88	BF12L 513C × 4 or
		BF12L 513CP × 4

MARIC types 7210 and 716II use the BF6L 913C
Korea Tacoma Marine Industries Ltd and Mitsui-Deutz also use 913 and 513 engines for hovercraft.

DEUTZ air-cooled engines

Deutz air-cooled engines of the type 913C and 513 series have been in large-scale production for a number of years. The BF6L 913C engine is a 141 kW engine with a high power-to-weight ratio suitable for smaller hovercraft. For larger craft the 513 series is available as 8, 10 and 12-cylinder V-types in naturally aspirated, turbocharged and turbocharged/charge-cooled versions. The 513 engines encompass the power range 114 to 405 kW. The latest variation, the BF8L 513LC, a long-stroke cylinder version of the 513 series, meets EURO 1 exhaust emission requirements and has also recently been introduced into hovercraft applications.

BF6/8/10/12l 513C

8, 10 and 12-cylinder versions of this engine are installed in a variety of hovercraft.

Type: Air-cooled four-stroke diesel with direct fuel injection, naturally aspirated or turbocharged; BF12L 513C with air-charge cooling system.

Cylinders: Individually removable cylinders made in grey cast-iron alloy, each with one inlet and one exhaust valve and overhead type. The valves are controlled via tappets and pushrods by a camshaft running in three metal bearings in the upper part of the crankcase. The camshaft is crankshaft driven via helical spur gears arranged at the flywheel end of the engine.

Pistons: Each is equipped with two compression rings and one oil control ring and is force oil-cooled.

Cooling system: Air-cooled, mechanically driven axial type cooling air blower, with optional load-dependent electronic control.

Lubrication: Force-fed by gear type pump. Oil is cleaned by full-flow paper filters. A centrifugal filter is installed in the fan hub as a high-efficiency secondary flow filter.

Deutz 620 series for propulsion output of up to 2,240 kW 0063231

BF12L 513C
Power, max (intermittent duty): 386 kW at 2,300 rpm
Number of cylinders: 12
Bore/stroke: 125/130 mm
Capacity: 19.144 litres
Compression ratio: 15.8:1
Rotational speed: 2,300 rpm
Mean piston speed: 9.96 m/s
Specific fuel consumption (automotive rating flywheel net at max torque): 205 g/kW h
Shipping volume: 3.16 m^3

Dimensions
Length: 1,582 mm
Height: 1,243 mm
Width: 1,196 mm
User: British Hovercraft Corporation and NQEA Australia AP1-88

BF8/10L 513
Turbocharged eight and 10-cylinder engines with similar basic engine data to the BF12L 513C, except that the outputs vary as follows:
BF8L 513: 235 kW (intermittent) at 2,300 rpm 211 kW (cruise) at 2,300 rpm
BF10L 513: 294 kW (intermittent) at 2,300 rpm 263 kW (cruise) at 2,300 rpm

BF8L 513LC
Turbocharged, charge-cooled eight-cylinder long stroke engine derived from the 513 engine series but giving increased power with emissions below EURO 1 levels.
Max power: 265 kW at 2,100 rpm
Bore/stroke: 125/140 mm
Capacity: 13.744 litres

BF6L 913C
Built to the same specification as the 513, this six-cylinder in-line engine is also available for hovercraft applications. It is fitted with an exhaust turbocharger and charge air-cooler.
Max (intermittent duty) power: 141 kW at 2,500 rpm
Number of cylinders: 6
Bore/stroke: 102/125 mm
Capacity: 6.128 litres
Compression ratio: 15.5:1
Rotational speed: 2,500 rpm
Mean piston speed: 10.4 m/s
Specific fuel consumption (automotive rating at max torque): 214 g/kW h
Shipping volume: 0.8 m^3

Dimensions
Length: 1,245 mm
Height: 991 mm
Width: 711 mm
Dry weight: 510 kg

MAN Diesel SE

Stadtbachstrasse 1, D 86224, Augsburg, Germany

Tel: (+49 821) 3220
Fax: (+49 821) 322 3382
e-mail: mandiesel-de@mandiesel.com
Web: www.manbw.com

C B Foulkes, *Senior Vice President*
D Boughey, *Commercial Director Marine*

Founded in 1840, MAN Diesel has established itself as a leading supplier of propulsion engines for fast craft, primarily by the acquisition, in 2000, of the Ruston Diesel and Paxman Diesel engine ranges from Alstom Engines.

The company has experience in applications engineering enabling it to work closely with its customers to provide integrated marine packages throughout the vessel design and construction stages. Worldwide customer support is provided through the resources of MAN Diesel SE and authorised representatives and distributors. This support includes operator training, supply of spares and field service support. Service packages to meet the specific needs of the customer, including full maintenance contracts, are also available.

Since its involvement with Incat Tasmania with the first 74 m wavepiercing vessel, the company has developed considerable experience of the special requirements of fast ferry installations with respect to engine mounting systems, drive trains and ancillary equipment, as well as the engines and their components. Being at the forefront of this rapidly developing market the requirements for higher powers, reduced operating costs, reduced emissions, and the provision of higher levels of reliability in this unique operating environment poses particular demands. In 2001 the RK280 engine series was introduced for propulsion and electrical generation purposes.

MAN Diesel Ltd 20-cylinder RK270 engine 0109537

RK270 ENGINE
The installation of the 16- and 20-cylinder RK270 engines in the larger fast vessels has proved to be particularly demanding and considerable efforts have been made to satisfy the required standards of reliability. The lessons the company has learned since 1990 have already contributed to improving the reliability of its products in fast ferry installations. The company has incorporated developments in the range to further improve performance and the 20RK270 provides an SFC of 202 g/kWb at its continuous fast ferry rating of 7,080 kWb at 1,030 rpm.

The RK270 also offers the benefit of potential savings by utilisation of lower cost blended fuels. *Bonanza Express*, Incat 051, operated by Fred Olsen Lineas SA, in the Canary Islands, is equipped to operate on blended fuel by the incorporation of fuel heating and visco m to the engines and on shore treatment.

MAN Diesel RK270 Medium Speed Diesels
Dimensions, weights and ratings

Engine size		**6RK270**	**8RK270**	**12RK270**	**16RK270**	**20RK270**
Number of cylinders		6	8	12	16	20
Cylinder configuration		in-line	in-line	vee	vee	vee
Cycle		4-stroke	4-stroke	4-stroke	4-stroke	4-stroke
Bore × stroke	mm	270 × 305	270 × 305	270 × 305	270 × 305	270 × 305
Sump capacity	litres	340	410	650	690	775
Dry weight less flywheel	Kg	13,050	17,500	22,000	27,000	33,500
Length	mm	4,020	4,585	4,285	5,075	5,965
Width	mm	1,325	1,300	1,825	1,830	1,940
Height (overall)	mm	2,490	2,480	2,645	2,820	2,820
Speed	rpm	1,000	1,000	1,000	1,000	1,030
Rating	kW	2,065	2,750	4,125	5,500	7,080

MAN Diesel V28/33D Medium Speed Diesels
Dimensions, weights and ratings

Engine size		**12V28/33D**	**16V28/33D**	**20V28/33D**
Number of cylinders		12	16	20
Cylinder configuration		vee	vee	vee
Cycle		4-stroke	4-stroke	4-stroke
Bore × stroke	mm	280 × 330	280 × 330	280 × 330
Dry weight less flywheel	kg	30,000	37,000	46,000
Length	mm	5,490	6,410	7,330
Width	mm	2,100	2,100	2,100
Height (overall)	mm	3,180	3,180	3,180
Speed	rpm	1,000	1,000	1,000
Rating	kW	5,400	7,200	9,000
Fuel consumption	g/kWh	190	190	190

ALSTOM Leroux Naval's 110 m monohull vessel, built for Gotland Rederi, is also fitted with 4 × 20RK270 engines. The installation includes a Siemans SCR system to reduce NOx emissions below 2 g/kW/h. This low level of emission, less than 20 per cent of the IMO standard, was imposed by the Swedish authorities as the vessel will operate in a designated environmentally sensitive area. The company undertook responsibility for the supply of the system and proving it at works test, as well as on sea trials.

MAN engines powered the Incat 91 m vessel that enabled Cat-Link V to regain the 'Blue Riband' for the fastest Atlantic crossing. In doing so, four new records were established. These were for the fastest Atlantic crossing at 41.2 kt average speed; the first time 40 kt average speed was achieved; the first time the crossing was completed in under three days, 2 days 20 hours 9 minutes; and the longest distanced travelled in 24 hours (1,018.2 n miles). During the voyage the engines operated at an average of 93 per cent MCR.

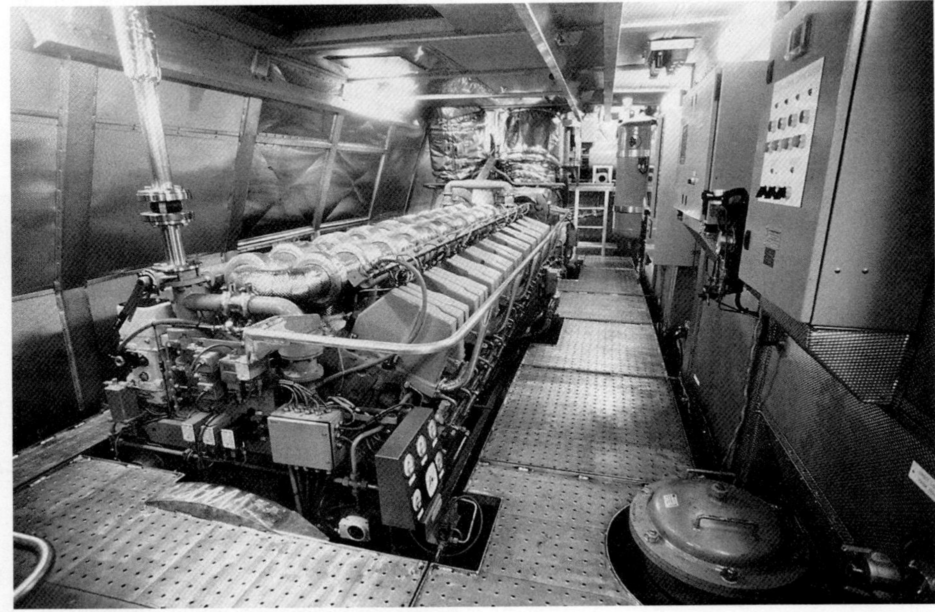

MAN Diesel Ltd 20RK270 in Condor Express engine room 0010059

Fuel specifications

BSMA 100 (ISO8217) Class DMA, DMB or equivalent
Installations (as at June 1999)

Name	**Type**	**No of engines**	**Engine power**
Hoverspeed Great Britain	InCat 74 m	4 × 16RK270	3,650 kW at 750 rpm
Atlantic II	InCat 74 m	4 × 16RK270	3,650 kW at 750 rpm
SeaCat Isle of Man	InCat 74 m	4 × 16RK270	3,650 kW at 750 rpm
SeaCat Danmark	InCat 74 m	4 × 16RK270	3,650 kW at 750 rpm
SeaCat Scotland	InCat 74 m	4 × 16RK270	3,650 kW at 750 rpm
Condor 10	InCat 76 m	4 × 16RK270	4,050 kW at 760 rpm
Stena Lynx I	InCat 76 m	4 × 16RK270	4,050 kW at 760 rpm
Catlink I	InCat 78 m	4 × 16RK270	4,400 kW at 805 rpm
Catlink 2	InCat 78 m	4 × 16RK270	4,400 kW at 805 rpm
Catlink 3	Austal 79 m	4 × 16RK270	5,500 kW at 1,000 rpm
Holyman Rapide	InCat 81 m	4 × 16RK270	5,500 kW at 1,000 rpm
Holyman Diamante	InCat 81 m	4 × 16RK270	5,500 kW at 1,000 rpm
Stena Lynx III	InCat 81 m	4 × 16RK270	5,500 kW at 1,000 rpm
Condor Express	InCat 86 m	4 × 20RK270	7,080 kW at 1,030 rpm
Condor Vitesse	InCat 86 m	4 × 20RK270	7,080 kW at 1,030 rpm
Sicilia Jet	InCat 86 m	4 × 20RK270	7,080 kW at 1,030 rpm
HMAS Jervis Bay	InCat 86 m	4 × 20RK270	7,080 kW at 1,030 rpm
SuperSeaCat one	Fincantieri 100 m	4 × 20RK270	6,875 kW at 1,015 rpm
SuperSeaCat two	Fincantieri 100 m	4 × 20RK270	6,875 kW at 1,015 rpm
SuperSeaCat three	Fincantieri 100 m	4 × 20RK270	6,875 kW at 1,015 rpm
SuperSeaCat four	Fincantieri 100 m	4 × 20RK270	6,875 kW at 1,015 rpm
NF 08	Afai Ships 50 m	4 × 16RK270	5,500 kW at 996 rpm
The Cat	InCat 91 m	4 × 20RK270	7,080 kW at 1,030 rpm
Max Mols	InCat 91 m	4 × 20RK270	7,080 kW at 1,030 rpm
Mads Mols	InCat 91 m	4 × 20RK270	7,080 kW at 1,030 rpm
HSC Gotland	ALSTOM Leroux Naval 115 m	4 × 20RK270	7,080 kW at 1,030 rpm
Bonanza Express	InCat 95 m	4 × 20RK270	7,080 kW at 1,030 rpm
Millennium	InCat 97 m	4 × 20RK270	7,080 kW at 1,030 rpm
Normandie Express (Hull No 057)	InCat 97 m	4 × 20RK270	7,080 kW at 1,030 rpm
Millennium Dos (Hull No 058)	InCat 97 m	4 × 20RK270	7,080 kW at 1,030 rpm
The Cat (Hull No 059)	InCat 97 m	4 × 20RK270	7,080 kW at 1,030 rpm
–	Fincantieri 122 m	4 × 20RK280	9,000 kW at 1,000 rpm
T&T Spirit (Hull No 060)	InCat 98 m	4 × 20RK270	7,080 kW at 1,030 rpm
Millennium Tres	InCat 98 m	4 × 16VRK280	7,200 kW at 1,000 rpm
Natchan Rera (Hull No 064)	InCat 112 m	4 × 20V28/33D	9,000 kW at 1,000 rpm

By 2002 there were over 130 × 20RK270 engines in service in 31 vessels which had accumulated over 1.5 million hours operation in varying conditions throughout the world, providing MAN B&W with significant operating experience for engines producing over 6,500 kW. These engines are now built under licence by Dalian Engines in China.

MAN V28/33D ENGINE

Developed from extensive experience with the RK270 engines, this new engine series was introduced in 2001 as the RK280. Now re-designated the V28/33D, it has fewer components than the RK270, extended service intervals and improved specific fuel consumption and power output. Four of these engines were delivered to Fincantieri in 2005 to power a new 122 m monohull fast ferry for Swedish operator Rederi AB Gotland and the engines were also supplied to Incat for the 112 m wave piercing catamaran.

VP185

This high-speed diesel engine is available in both 12 and 18 cylinder formats. Its high power to weight ratio means it has been extensively used in the marine market.

MAN Nutzfahrzeuge AG

Moterenwerk Nürnberg, Vogelweiherstr 33, D-90441 Nürnberg, Germany

Tel: (+49 911) 420 62 39
Fax: (+49 911) 420 19 15
e-mail: marinemotor@de.man-mn.com
Web: www.man-mn.com/engines

Kurt Heuser, *Director of Sales*
Marko Dekonia, *General Manager, Industrial & Vehicle Engines*
Arnd Loettgen, *Director, Business Unit*

Other offices:
Europe
MAN ERF UK Ltd
Frankland Road
Blagrove
Swindon
Wiltshire, SN5 8YU, United Kingdom

Tel: (+44 1793) 44 80 00
Fax: (+44 1793) 44 82 62
e-mail: engines@man.co.uk

Richard Noy, *General Manager*

United States
MAN Engines Components Inc
595 SW 13th Terrace
Pompano Beach, Florida 33069-3520

Tel: (+1 954) 946 90 92
Fax: (+1 954) 946 90 98
e-mail: man@manengines.us
Web: www.man-mec.com

Sven von Saalfeld, *President*

Asia
MAN B&W Diesel (Singapore) Pte Ltd
29 Tuas Avenue, Singapore 639460

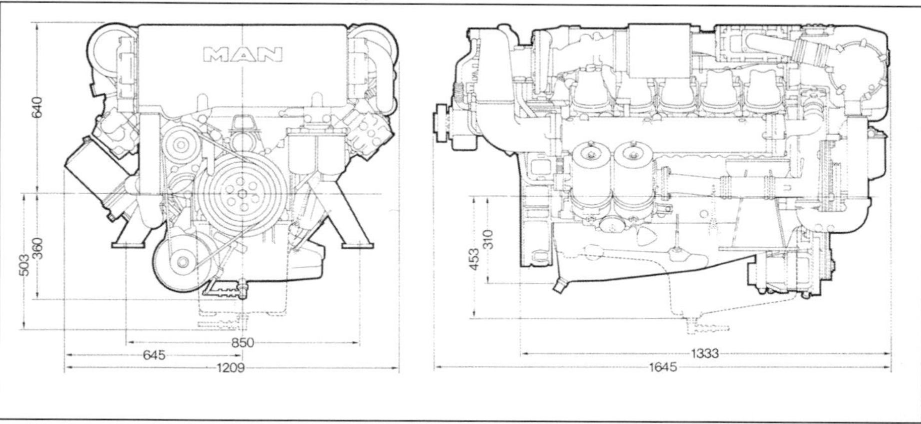

MAN Model D 2840 LE401, dimensions (mm) 0506766

MAN D2842 LE404 engine rated at 950 kW for light-duty applications 0069456

MAN D2842 LE403 engine rated at 529 kW for heavy-duty application 0016115

MAN Marine diesels are in widespread use and have been supplied to the following shipyards: Anne Wever; Azimut; BAIA; Camuffo; Canados; Eder/Knoche; Ferretti; Ladenstein; San Lorenzo; Lowland; Marchi; Mochi Craft; Neptunus; Posillipo; Riva; Sunseeker; Tecnomarine; Marine Projects; Gestione; Best Yachts; Rizzardi; Storeboo; Oskarshamns; Lux-Shipyard; Deggendorf Shipyard; Viking; Vosper Thornycroft; Hatteras; Bertram; Dalla Pieta; Golfo; Piantoni; Cantieri della Pasquaz; Engitalia; Fairline; Henriques and Guy Couach.

MAN marine diesel engines are blocked for different marine applications, rating 1, rating 2 and rating 3. Please see the accompanying table for rating definitions.

D 0826L

Type: Four-stroke diesel engine, six-cylinder vertical in-line water-cooled with turbocharger and intercooler. Direct injection system. High unit output, low-noise level and quiet running with low fuel consumption. Long service life and low upkeep and all parts to be serviced are readily accessible.

Model	No of Cylinders	Heavy duty rating kW	Medium duty rating kW	Light duty rating kW	Speed rpm	Length mm	Width mm	Height mm	Weight kg
D08636 LE 402	6 R	–	265	–	2,400	1,127	740	925	730
D0836 LE 401	6 R	–	–	331	2,500	1,127	740	925	730
D2866 LXE 40	6 R	190	–	–	1,800	1,448	897	1148	1,020
D2866 LXE 47	6 R	221	–	–	1,800	1,448	897	1,148	1,020
D2866 LE 403	6 R	–	368	–	2,100	1,320	869	998	1,160
D2866 LXE 40	6 R	250	279	–	1,800	1,448	897	1,148	1,020
D2066 LE 405	6 R	–	–	449	2,200	1,320	869	998	1,160
D287 LE 403	6 R	331	–	–	1,800	1,320	882	966	1,290
D2876 LE 402	6 R	–	412	–	2,100	1,320	882	966	1,290
D2876 LE 404	6 R	–	–	463	2,200	1,320	882	966	1,290
D2876 LE 401	6 R	–	–	515	2,200	1,320	882	966	1,290
D2876 LE 405	6 R	–	–	537	2,200	1,320	882	966	1,290
R 8-800	6 R	–	–	588	2,300	1,320	882	977	1,300
D2848 LE 405	8 V	–	478	–	2,100	1,176	1,230	1,078	1,390
D2848 LE 405	8 V	–	478	–	2,100	1,176	1,230	1,073	1,390
D2848 LE 403	8 V	–	–	588	2,300	1,176	1,230	1,063	1,390
V 8-900	8 V	–	–	662	2,300	1,178	1,230	1,120	1,500
D2840 LE 401	10 V	–	478	–	2,100	1,334	1,217	1,071	1,580
D2480 LE 403	10 V	–	–	772	2,300	1,334	1,228	1,117	1,630
V 10-1100	10 V	–	–	809	2,300	1,334	1,230	1,183	1,750
D2842 LE	12 V	441	–	–	1,800	1,638	1203	1171	1,720
D2842 LE 403	12 V	529	–	–	1,800	1,491	1,230	1,105	1,790
D2842 LE 401	12 V	558	–	–	1,800	1,491	1,230	1,105	1,790
D2842 LE 401	12 V	–	588	–	2,100	1,491	1,217	1,082	1,790
D2842 LE 405	12 V	662	–	–	2,100	1,491	1,230	1,105	1,790
D2842 LE 413	12 V	–	735	–	2,100	1,491	1,230	1,105	1,790
D2842 LE 410	12 V	–	809	–	2,100	1,491	1,230	1,105	1,860
D2842 LE 407	12 V	–	–	882	2,300	1,491	1,230	1,105	1,860
V 12-1224	12 V	–	–	900	2,300	1,491	1,230	1,185	2,000
D 2842 LE 404	12 V	–	–	956	2,300	1,491	1,230	1,105	1,860
V 12-1360	12 V	–	–	1,000	2,300	1,491	1,230	1,185	2,000
D 2842 LE 409	12 V	–	–	1,103	2,300	1,491	1,385	1,270	2,160
V 12-1550	12 V	–	–	1,140	2,300	1,491	1,380	1,270	2,160
R6-730	6R	–	–	537	2,300	1,605	882	977	1,305
R6-800	6R	–	–	588	2,300	1,605	882	977	1,305
R6-550	6R	–	_	404	2,600	1,622	867	939	900

Crankcase: One-piece crankcase and cylinder block. Replaceable dry cylinder liners. Seven bearing crankshaft with forged-on balance weights. Induction-hardened main bearing and connecting rod bearing journals, additional hardening of journal radii. Main and connecting rod bearing journals in ready-to-install three-component bearing. Torsional vibration damper fitted at front of six-cylinder engines. Die-forged, straight split connecting rods, which are removable through top of cylinder. Three-ring pistons of special aluminium alloy with ring carrier for the top piston ring. Piston crown cooled by lubricating oil jet.
Cylinder heads and valve train: Cross-flow cylinder head for each pair of cylinders with cast swirl inlet and exhaust channels on opposite sides. Shrunk-fit inlet and outlet valve seat inserts and replaceable pressed-in valve guides. One inlet and outlet valve per cylinder arranged in overhead position. Forged camshaft with induction-hardened cams and bearing supports.
Lubrication: Force-fed lubrication with gear oil pump for crankshaft, connecting rod and camshaft bearings as well as valve train and turbocharger. Oil filter/oil cooler combination with coolant-covered, flat-tube oil cooler. Lubricant cleaned in full flow by easy maintenance screw-on filters with fine disposable filter cartridge of paper.
Fuel system: Bosch injectors. Bosch-MW in-line injection pump. Mechanical engine speed governor and timing device. Switch-off via separate stop lever and solenoid. Fuel pre-supply pump, fuel filter with easy maintenance screw-on disposable cartridges.
Intake and exhaust system: Wet air filter, water-cooled manifold cooled by engine water.
Supercharging: Exhaust gas turbocharger, water-cooled by engine water, seawater-cooled intercooler.
Electrical system: Two-pole starter, 4 kW, 24 V and two-pole alternator 28 V, 35 A.
Applications: Workboats, customs and police patrol boats, yachts.

Technical data for 108 mm bore
Bore/stroke: 108/120 mm
Swept volume: 6.6 litres
Compression ratio: 17:1
Rotation looking on flywheel: anti-clockwise
Flywheel housing: SAE 2
Weight of engine dry, with cooling system: 640 kg
Max rating 3: 199 kW at 2,600 rpm
Mean effective pressure: 13.9 bar at 2,600 rpm
Torque: 860 Nm at 1,700 rpm
Starter motor: Bosch solenoid-operated starter Type KB, 24 V, 5.4 kW

Technical data for 128 mm bore
Bore/stroke: 128/155 mm
Swept volume: 11.97 litres
Compression ratio: 15.5:1
Max rating 3: 449 kW at 2,200 rpm
Mean effective pressure: 20.5 bar at 2,200 rpm
Torque: 1,949 Nm at 2,200 rpm
Fuel consumption (+5% tolerance) at max rating 3: 215 g/kW h at 2,200 rpm

D 2848 L
Type: Four-stroke, direct injection.
Cylinders: Eight-cylinder, V-form, wet replaceable cylinder liners.
Aspiration: Turbocharged, intercooled.
Cooling: Water circulation by centrifugal pump fitted on engine.
Lubrication: Force-fed lubrication by gear pump, lubrication oil cooler in cooling water circuit of engine.
Injection: Bosch in-line pump with mechanical Bosch speed governor fitted.
Generator: Bosch three-phase generator with rectifier and transistorised governor Type K1, 28 V, 35 A.
Starter motor: Bosch solenoid-operated starter, Type KB, 24 V, 6.5 kW.

Technical data
Bore/stroke: 128/142 mm
Volume: 14.62 litres
Compression ratio: 13.5:1
Max rating 3: 588 kW at 2,300 rpm
Mean effective pressure: 21 bar at 2,300 rpm
Torque: 2,441 Nm at 1,700 rpm
Fuel consumption (+5% tolerance): 230 g/kW h at 2,300 rpm.

D 2840
Type: Four-stroke marine diesel engine, 10-cylinder, V-form, water-cooled with turbocharger and intercooler. Direct injection system.

Crankcase: Grey cast-iron cylinder block, six-bearing crankshaft with screwed-on balance weights, three-layer type bearings, die-forged connecting rods. Replaceable wet-type cylinder liners.
Cylinder heads and valve train: Individual cylinder heads of grey cast-iron, overhead valves, one intake and one exhaust valve per cylinder, valve actuation via tappets, pushrods and rocker arms. Six-bearing camshaft, shrunk-fit valve seat inserts.
Lubrication: Force-fed lubrication by gear pump, oil-to-water oil cooler, full-flow oil filter, changeover type optional.
Fuel system: Bosch in-line injection pump with mechanical speed governor, fuel supply pump, fuel filter, changeover type optional.
Intake and exhaust system: Viscous air filter, water-cooled exhaust manifold connected in engine cooling circuit.
Supercharging: Turbochargers, water-cooled in freshwater circuit, seawater-cooled intercooler Waste Gate System.
Electrical system: Two-pole starter, 6.5 kW, 24 V, two-pole alternator 28 V, 120 A, additional alternator 28 V available with 35 A, 55 A or 120 A on request.
Applications: Yachts, customs and police patrol boats.

Technical data
Bore/stroke: 128/142 mm
Swept volume: 18.27 litres

Compression ratio: 13.5:1
Rotation looking on flywheel: anti-clockwise
Weight of engine, dry with cooling system:
1,570 kg
Speed: 2,300 rpm
Max rating 3: 772 kW/820 hp
Mean effective pressure: 22 bar
Fuel consumption: 229 g/kW h

D 2842 HLE402

Type: Four-stroke marine diesel engine, 12-cylinder, V-form, water-cooled with turbocharger and intercooler. Direct injection system.
Crankcase: Cylinder block of grey cast-iron. Replaceable wet-type cylinder liners. Seven-bearing crankshaft with screwed-on balance weights, three-layer type bearings, die-forged connecting rods.
Cylinder heads and valve train: Individual cylinder heads of grey cast-iron, overhead valves, one intake and one exhaust valve per cylinder, valve actuation via tappets, pushrods and rocker arms. Seven-bearing camshaft, shrunk-fit valve seat inserts.
Lubrication: Force-fed lubrication by gear pump, oil-to-water oil cooler, full flow oil filter, changeover type optional.
Fuel system: Bosch in-line injection pump with mechanical speed governor, fuel supply pump, fuel filter, changeover type optional.
Intake and exhaust system: Viscous air filter, water-cooled exhaust manifold connected in engine cooling circuit.
Supercharging: Turbochargers, water-cooled in freshwater circuit, seawater-cooled intercooler Waste Gate System.
Electrical system: Two-pole starter, 6.5 kW, 24V, two-pole alternator 28 V, 120 A, additional alternator 28 V with 35 A, 55 A or 120 A on request.
Applications: Yachts, customs and police boats.

Technical data

Bore/stroke: 128/142 mm
Volume: 21.93 litres
Compression ratio: 13.5:1
Rotation looking on flywheel: anti-clockwise
Weight of engine, dry with cooling system:
approx: 1,850 kg
Speed: 2,300 rpm
Max rating 3: 956 kW at 2,300 rpm
Mean effective pressure: 22.75 bar
Torque: 4,330 Nm at 1,700 rpm
Mean specific fuel consumption (+5%)
225 g/kW h

MTU – Motoren- und Turbinen-Union Friedrichshafen GmbH

A Tognum Group company
D-88040 Friedrichshafen, Germany

Tel: (+49 7541) 90 4741
Fax: (+49 7541) 903085
Web: www.mtu-online.com

Volker Heuer, *President and CEO*

MTU Friedrichshafen is a supplier of drive systems for land, water and railroad applications, as well as plants for electric power generation.

Since 1950, MTU and its predecessor companies have delivered more than 85,000 engines. The company's range of products includes compact diesel engines with a power ranging from 40 to 9,000 kW, gas engines with an output between 130 and 1,320 kW and gas turbines between 3 and 30 MW. Additionally MTU supplies electronic control and monitoring systems.

In the field of hydrofoils, surface-effect ships, catamarans and other high-speed craft, MTU can draw from decades of experience with nearly 2,800 engines having been supplied for the propulsion of hydrofoils and catamarans. As early as 1955, an MB 820 engine was delivered for the first

MTU 20V 8000 M70 0077579

MTU Diesel engines for catamaran, hydrofoil and Surface Efffect Ships (SES) propulsion and similar

Engine Model	Rated Power (kW)	Engine Speed (rpm)	Length (mm)	Width (mm)	Height (mm)	Engine dry weight (kg)
S 60	354	2,100	2,040	995	1,155	1,840
S 60	399	2,100	2,040	995	1,155	1,840
S 60	447	2,100	2,040	995	1,155	1,840
8V 2000 M72	720	2,250	2,080	1,130	1,225	2,400
10V 2000 M72	900	2,250	2,365	1,130	1,305	2,820
12V 2000 M70	788	2,100	2,305	1,400	1,230	3,110
12V 2000 M72	1,080	2,250	2,380	1,295	1,350	3,280
16V 2000 M70	1,050	2,100	2,895	1,400	1,290	3,870
16V 2000 M72	1,440	2,250	2,890	1,295	1,425	4,010
8V 396 TE74 L	1,000	1,900	2,435	1,530	1,540	4,145
12V 396 TE74 L	1,500	1,900	3,040	1,530	1,785	5,640
16V 396 TE74 L	2,000	1,900	3,905	1,540	1,755	7,070
8V 4000 M70	1,160	2,000	3,015	1,380	1,990	5,770
12V 4000 M70	1,740	2,000	3,620	1,520	1,830	7,550
12V 4000 M71	1,850	2,000	3,995	1,520	1,890	7,720
16V 4000 M70	2,320	2,000	4,525	1,520	1,890	8,905
16V 4000 M71	2,465	2,000	4,525	1,520	1,890	8,905
16V 595 TE70	3,600	1,750	3,980	1,660	2,870	13,000
16V 595 TE70 L	3,925	1,750	3,980	1,660	2,870	13,000
16V 1163 TB73 L	5,200	1,230	4,660	1,895	3,520	19,700
20V 1163 TB73	6,000	1,200	5,350	1,895	3,615	22,300
20V 1163 TB73 L	6,500	1,230	5,350	1,895	3,615	22,800
20V 8000 M71 R	8,200	1,150	6,620	1,885	3,330	446,000
20V 8000 M71	9,100	1,150	6,620	1,885	3,290	44,600

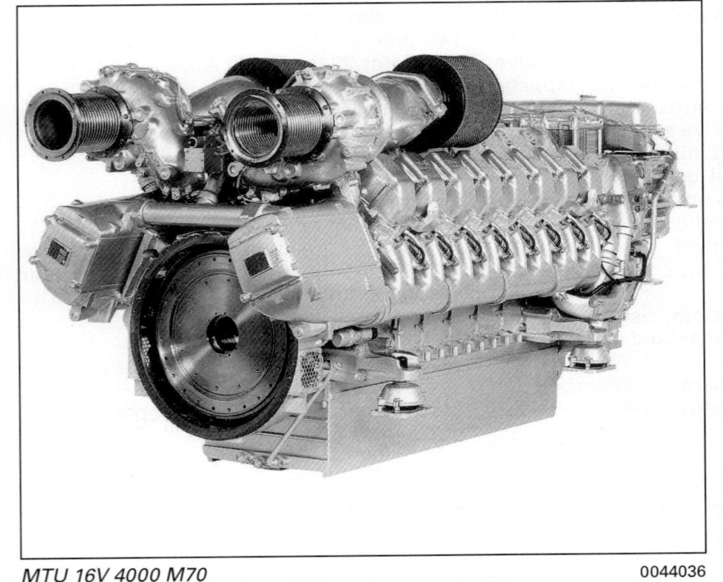

MTU 16V 4000 M70 0044036

MTU 12V 2000 M70 0536806

PT 20 hydrofoil built by Cantiere Navale Rodriquez (now Rodriquez Cantieri Navali SpA) in Messina.

Currently the 2000, 183, 396, 4000, 595, 1163 and 8000 engine series are offered for propulsion of these special types of craft. These engines cover a wide range of power. The standard power outputs are listed in the accompanying table. However, these may have to be adjusted, depending on the application, the power demand and the operating profile. The outputs are based on ISO 3046.

In addition to the delivery of the propulsion engine, MTU can lay out, design and deliver complete propulsion packages, including the interface engineering and the technical assistance during installation and the start up phase.

In 2002 MTU signed an agreement with Vericor of the US whereby MTU will package and distribute Vericor's TF40 and TF50 aeroderivative marine gas turbines.

At present over 50 per cent of all high-speed vessels are powered by MTU engines.

In 2004 MTU launched its new range of Series 2000 CR engines. Using a common rail system, the engines are capable of producing more power than the original 2000 series, with a much lighter weight (1.99 kg/kW).

Engine model	A	B	C
6R 099 TE 72 (OM 447LA)*	1,100	845	900
8V 183 TE 72 (OM 442LA)*	1,185	1,295	1,240
12V 183 TE 72 (OM 444LA)*	1,550	1,305	1,195

*original Mercedes-Benz designation on which types the corresponding MTU engines are based.

Main dimensions (mm) for 099/183 engine series 0109532

Engine model	A	B	C
8V 2000 M70	1,480	1,220	1,100
12V 2000 M70	1,890	1,400	1,280
16V 2000 M70	2,250	1,400	1,280

Main dimensions (mm) for 2000 M70 engine series 0109531

Engine model	A	B	C
8V 396TE 74L	2,010	1,525	1,540
12V 396TE 74L	2,535	1,540	1,600
16V 396TE 74L	2,735	1,540	1,750

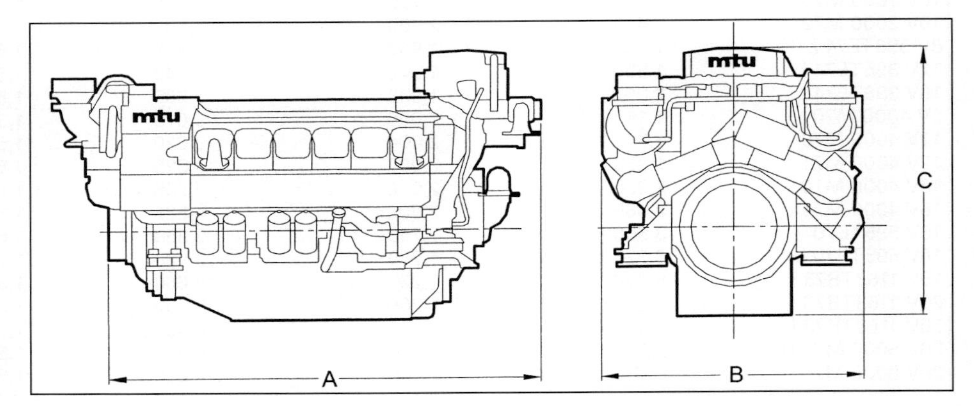

Main dimensions (mm) for 396TE74L engine series 0109530

Engine model	A	B	C
8V 4000 M70	2,280	1,380	1,990
12V 4000 M70	2,835	1,520	1,870
16V 4000 M70	3,310	1,520	1,870

Main dimensions (mm) for 4000 M70 engine series
0109533

Engine model	A	B	C
16V 595TE 70L	3,980	1,670	2,870

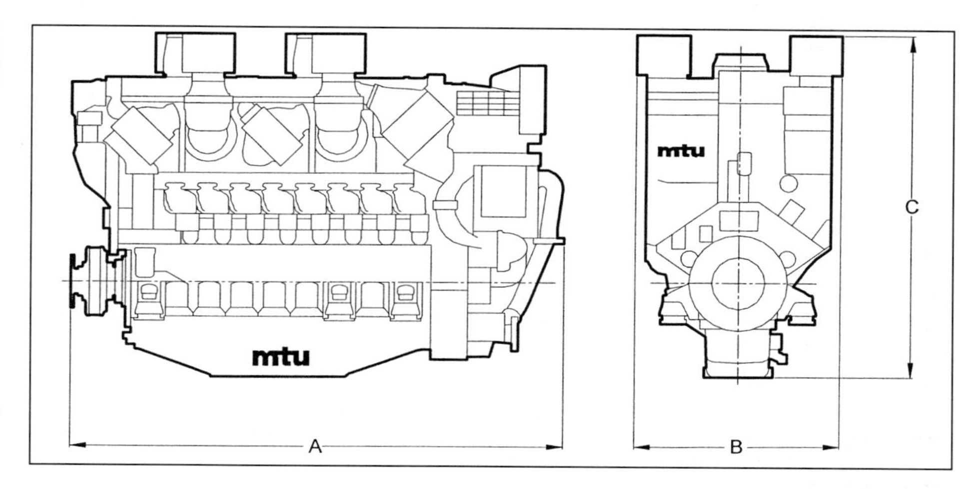

Main dimensions (mm) for 595TE 70L engine series
0109534

Engine model	A	B	C
16V 1163TB 73L	4,565	1,790	2,830
20V 1163TB73	5,140	1,790	2,830
20V 1163TB 73L	5,140	1,790	2,830

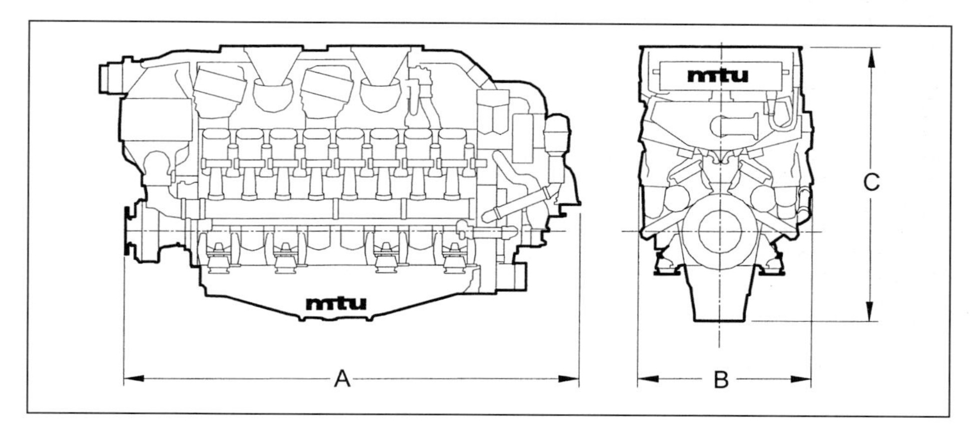

Main dimensions (mm) for 1163TB 73L engine series
0109536

Engine model	A	B	C
20V 8000 M70	6,650	1,925	3,290

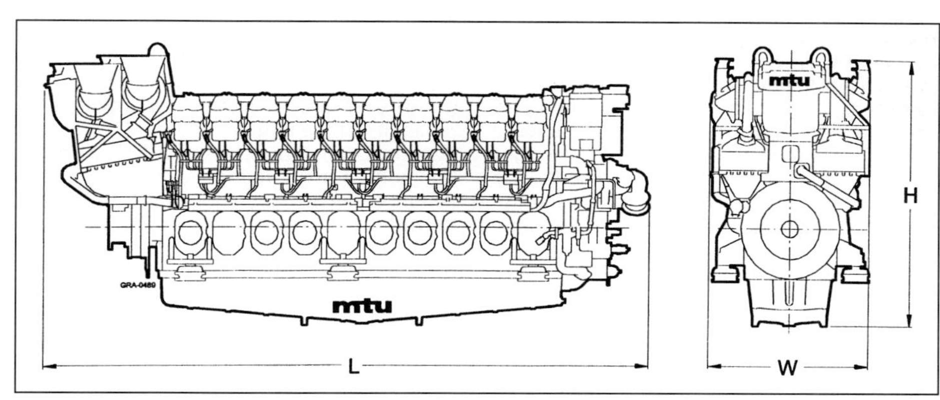

MTU 20V 8000 M70
0100073

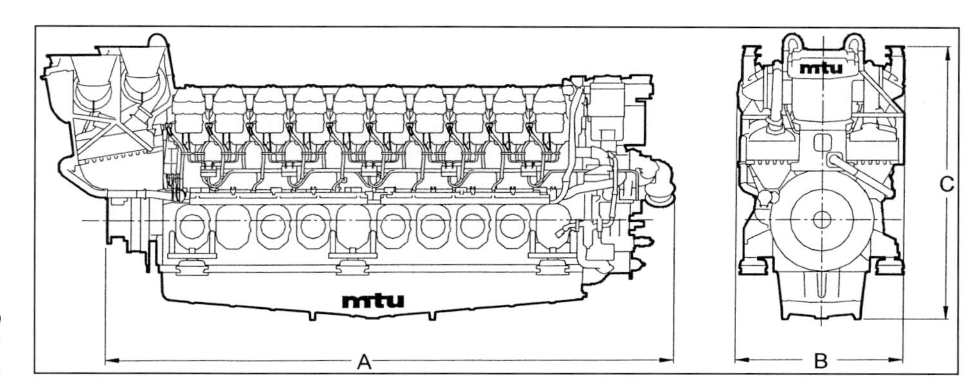

Main dimensions for 8000 M70 engine series 1
0109535

Italy

CRM Motori Marini SpA

Via Marnate 41, I-21053 Castellanza, Italy

Tel: (+39 033) 157 26 00
Fax: (+39 033) 150 55 01
e-mail: info@crmmotori.it
Web: www.crmmotori.it

Dott. Germano Mariani, *Managing Director, CEO and President*

CRM has specialised in building lightweight diesel engines for more than 50 years. The company's engines are used in large numbers of motor torpedo boats, coastal patrol craft and privately owned motor yachts. The engines have also been installed in hydrofoils.

The range of engines comprises the 12-cylinder 12 D/SS, 12 D/SS 1500 and the 18-cylinder 18 D/SS, BR-1 and BR-2. All are turbocharged with different supercharging ratios. The 12 cylinders are arranged in two banks of six (V-type) and the 18 cylinders are set out in a 'W' arrangement of three banks of six. Due to the extensive use of aluminium alloy, magnetic signature in standard engines is very low. For special purposes all engines are available in non-magnetic versions with the perturbation field reduced to an insignificant amount.

The most stringent tests have shown that gaseous emissions from all CRM engines are at least 30 per cent lower than limits laid down in IMO and 94/25 EC standards.

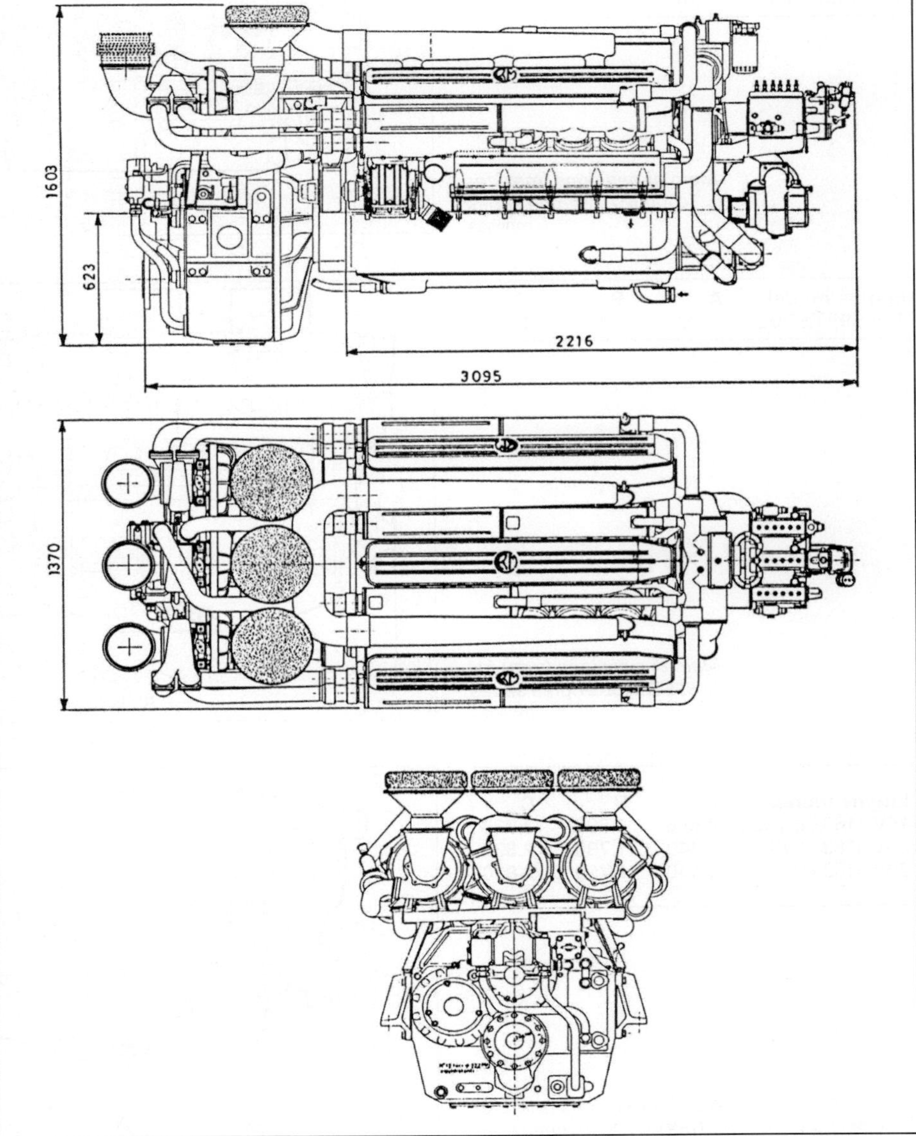

CRM 18 D/SS, BR-2 0507016

CRM 18-cylinder

First in CRM's series of low-weight, high-speed diesel engines, the CRM 18-cylinder is arranged in a W form with different ratings according to the type as shown on the table.

The following description relates to the 18D/SS, BR-1 and BR-2:

Type: 18-cylinder in-line W40°-type, four-stroke, water-cooled, turbocharged with different supercharging ratios: 2.4:1 (18 D/SS); 2.6:1 (BR-1) and 3:1 (BR-2).

Cylinders: Bore 150 mm. Stroke 180 mm. Swept volume 3.18 litres per cylinder. Total swept volume 57.3 litres. Geometric Compression ratio 14:1. Separate pressed-steel cylinder, blind at the top end, with gas-welded jacket for coolant circulation and internally chromium plated to minimise wear. The lower half of the cylinder is ringed by a drilled flange for bolting to the crankcase. Cylinder top also houses a spherical shaped pre-combustion chamber as well as inlet and exhaust valve seats. The pre-combustion chamber is of high-strength, corrosion and creep-resisting alloy. A single-cast light-alloy head, carrying valve guides, pre-combustion chambers and camshaft supports, bridges each bank of six cylinders. The cylinder head is attached to the cylinder bank by multiple studs. As a result of the unique design, the combustion process happens inside the blind steel cylinder and the light-alloy head is not submitted to the firing pressure therefore improving the reliability of the engine.

Pistons: Light-alloy forging with three rings, the bottom ring acting as an oil scraper. Piston crowns (hard anodised top) shaped to withstand high temperatures especially in the proximity of pre-combustion chamber outlet ports.

Connecting rods: Main and secondary articulated type, all being completely machined I-section steel forged. Big end of each main rod is bolted to the ribbed cap by six studs. Big end bearings are white metal-lined steel shells. Each secondary rod anchored at its lower end to a pivot pin inserted in two lugs protruding from big end of main connecting rod. Both ends of all articulated rods and small ends of main rods have bronze bushes.

Crankshafts: One-piece hollow shaft in forged and nitrided alloy steel, with six throws equally spaced at 120°. Seven main bearings with white metal lined steel shells. There are 12 balancing counterweights.

	18 D/SS	**BR-1**	**BR-2**	**BR-32**	**12 D/SS**	**12 D/SS 1500**
Dimensions:						
Length	2,700 mm	2,700 mm	2,700 mm	2,700 mm	2,400 mm	2,400 mm
Width	1,352 mm	1,352 mm	1,352 mm	1,352 mm	1,210 mm	1,210 mm
Height	1,290 mm	1,290 mm	1,290 mm	1,290 mm	1,206 mm	1,210 mm
Gearbox	655 mm	655 mm	767 mm	767 mm	575 mm	575 mm
Weights, dry:						
Engine	1,950 kg	1,950 kg	1,950 kg	2,100 kg	1,600 kg	1,600 kg
Gearbox	460 kg	460 kg	750 kg	750 kg	340 kg	340 kg
Ratings:						
Max power	1,213 kW at 2,075 rpm	1,335 kW at 2,120 rpm	1,544 kW at 2,120 rpm	1,764 kW at 2,350 rpm	1,010 kW at 2,075 rpm	1,103 kW 2,120 rpm
Continuous rating	1,103 kW at 2,020 rpm	1,213 kW at 2,050 rpm	1,403 kW at 2,050 rpm	1,544 kW at 2,050 rpm	918 kW at 2,050 rpm	1,003 kW at 2,050 rpm
Specific fuel consumption	0.220 ±5% kg/kW h	0.220 ±5% kg/kW h	0.220 ±5% kg/kW h	0.224 ±5% kg/kW h	0.226 ±5% kg/kW h	0.224 ±5% kg/kW h
Oil consumption	0.5±0.9% of fuel	0.5±0.9% of fuel	0.5±0.9% of fuel	0.5±0.9% of fuel	0.5±0.9% of fuel	0.5±0.9% of fuel

Crankcase: Cast light-alloy crankcase bolted by studs and tie bolts. Multiple integral reinforced ribs provide a rigid and strong structure. Both sides of each casting braced by seven cross ribs incorporating crankshaft bearing supports. Protruding sides of crankcase ribbed throughout length.
Valve gear: Hollow sodium-cooled valves of each bank of cylinders actuated by twin camshafts. Two inlet and two outlet valves per cylinder and one rocker for each valve. End of stem and facing of exhaust valves fitted with Stellite inserts. Valve cooling water forced through passage formed by specially shaped plate welded to top of cylinder.
Fuel injection: Three pumps fitted with common variable speed governor and pilot injection nozzle.
Pressure charger: Three turbochargers.
Accessories: Standard accessories include oil and freshwater heat exchangers with thermostats and oil filters mounted on the engine; air filters and freshwater tank with preheating system; engine room instruments panel with warning system; wheel room instrument panel; electronic monitoring system; prelubricating electric pump; saltwater and fuel hand pumps and expansion joints for exhaust piping separate from the engine.
Cooling system: Freshwater.
Fuel: Fuel oil with specific gravity of 0.83 to 0.84.
Lubrication: Dry sump type with circulating and scavenge oil pumps.
Oil: Mineral oil to MIL-L-2104E.
Oil cooling: Saltwater circulating through heat exchanger.
Starting: 24 V 15 kW electric motor and 85 A, 24 V generator for battery charge. Optional compressed air.
Mounting: At any transverse or longitudinal angle tilt to 20°. Due to low vibration levels, the engine can be rigidly mounted to the hull.

CRM 12-cylinder

The second in the CRM series of low-weight diesels, the CRM 12-cylinder is a unit with two blocks of six cylinders set at 60° to form a V assembly with different rating according to the type as indicated on the table. The bore and stroke are the same as in the CRM 18 series and many of the components are interchangeable including the crankshaft, lower crankcase, heads, valves, camshafts and several auxiliary components, cylinders and pistons. The upper crankcase and connecting rod assemblies are of a modified design; the secondary rod is anchored at its lower end to a pivot pin inserted on two lugs protruding from the big end of the main connecting rod. The fuel injection pump is single block type, with 12 pumping elements located between the cylinder banks.
Type: 12-cylinder 60° V-type, four-stroke, water-cooled, turbo-supercharged with different supercharging ratios (2.85:1 for 12 D/SS and 3.1 for 12 D/SS-1500).

CRM 18 D/SS, BR-2 0506856

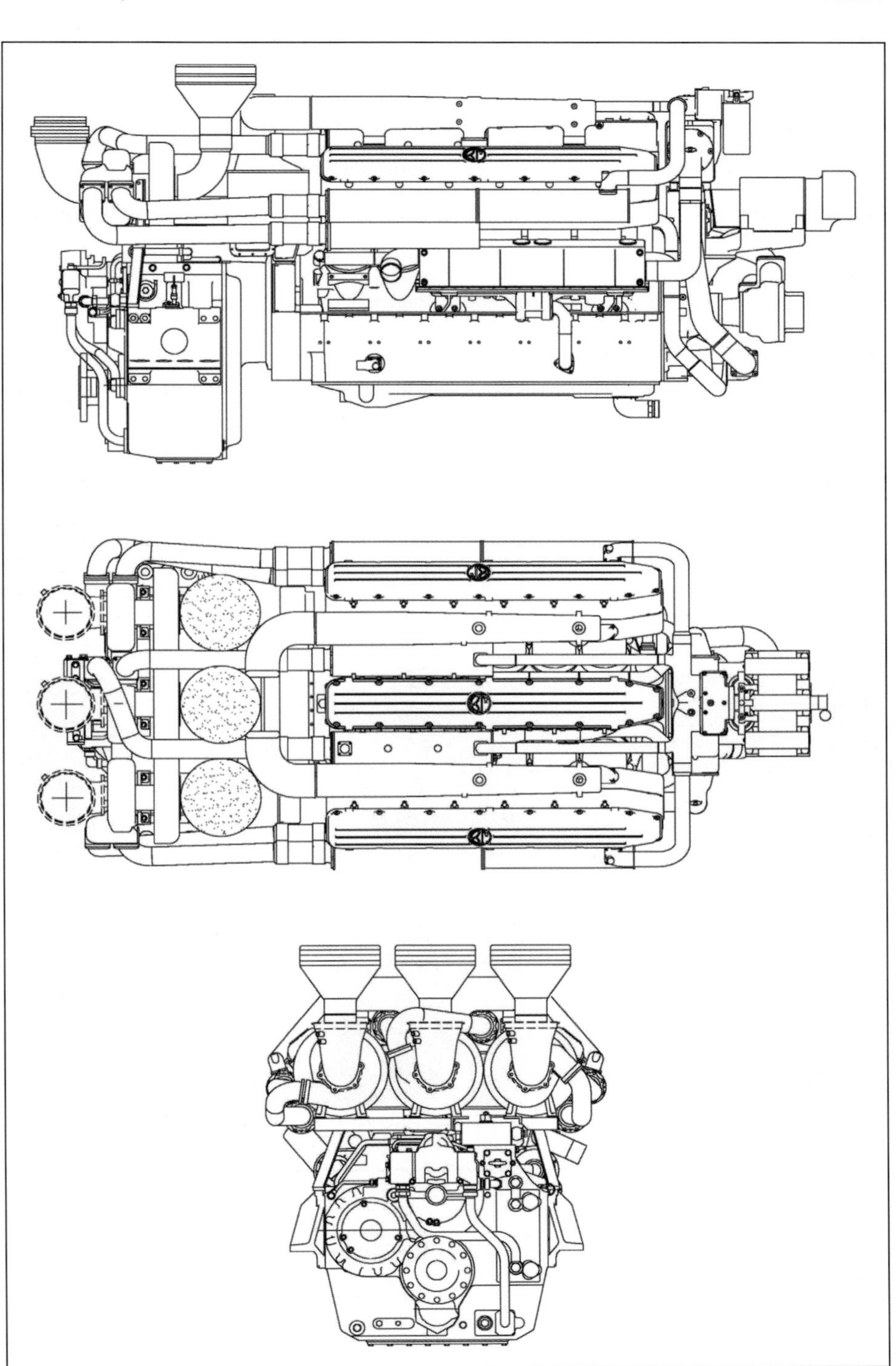

CRM 18 D/SS, BR-2 1120280

Isotta Fraschini Motori SpA

A subsidiary of Fincantieri-Cantieri Navali Italiani SpA
Via Francesco de Blasio 21, I-70123, Bari, Italy

Tel: (+39 080) 534 50 00
Fax: (+39 080) 531 10 09
e-mail: isottafrashini@isottafrashini.it
Web: www.isottafraschini.it

Gianpiero Riganti, *President*
Edgardo Bertoli, *Managing Director*

Following the merging of Isotta Fraschini Motori SpA, the trademark IF is now a range of engines covering power outputs from 220 to 2,350 kW per unit is now available.

The company has a production plant in Bari which has a total area of 200,000 m² of which 24,000 m² are covered.

IF 1716T2 16V marine engine 0506769

Isotta Fraschini 1300

The 1300 series comprises a six-cylinder in-line unit and 8- and 12-cylinder 90°V models. A 130 mm bore and a stroke of 142 mm on the in-line unit are principal features of the 1300 series. The stroke is reduced to 126 mm for these models. The in-line engine has a cylinder displacement of 1,885 cm³, while the V units have a cylinder of 1,672 cm³. The 1300 engines have an impressive power-to-weight ratio, being highly standardised and extremely reliable engines of modular type. Introduced as marine engines, these diesels will also serve the power generation and industrial market segments.

The 1300 series specification may be summarised as follows:

Four-stroke diesel engines, direct injection, exhaust gas turbo-charging, supercharging air-cooling through water/air exchangers, engine double-circuit water-cooling, complete with heat exchanger, cooling system complete with centrifugal pumps, thermostats for internal circuit water and oil control, engine water-cooled exhaust manifolds, force-feed lubrication through gear pump, injection pump complete with mechanical governor, fuel feed pump, oil exchanger with replaceable/cartridge oil filters, fuel oil filter, dry air filter, crankshaft vibration damper, flywheel and flywheel housing SAE standard, battery charger generator, 24 V electric starter.

Isotta Fraschini 1700

The 1700 series comprises 8 to 16 cylinder models arranged with a 90° V on a high-tensile alloy iron crankcase.

These engines feature a direct injection system with four valves per cylinder. They are built using a modular concept, with most components common to all units. A special magnetic version is available for mine warfare vessels.

Engine	Cylinder arrangement	Engine ratings (brake power) High-performance special craft kW	Fast performance special craft rpm	Weight (kg) kW	rpm	kg
L 1306T2	6 in line	441	2,400	400	2,400	940
V 1308T2	8 vee	551	2,700	500	2,700	980
V 1312T2	12 vee	882	2,700	800	2,700	1,450
V 1708T2	8 vee	955	2,000	900	2,000	2,950
V 1712T2	12 vee	1,680	2,000	1,540	2,000	4,330
V 1716T2	16 vee	2,350	2,100	2,140	2,030	7,300

IF L 1306T2 marine engine 0506767

IF V 1308T2 marine engine 0506768

Iveco Motors

Via Vanzago, 18/20, I-20010 Pregnana Milanese, Milan, Italy

Tel: (+39 02) 93 51 01
Fax: (+39 02) 935 19 08 81
e-mail: luciano.caprotti@iveco.com
Web: www.ivecomotors.com

Giorgio Bandera, *Commercial Operations Manager*
Luciano Caprotti, *Sales and Network Support*

Iveco Motors manufactures a range of marine diesel engines from 42 to 883 kW. The accompanying table details the more powerful professional range of engines.

Type	No of cylinders and arrangement	Output Light duty kW @ rpm	Output Heavy Duty kW @ rpm	Weight (kg)	Length × Width × Height (mm)
NEF 100	4L – NA	74 @ 2,800	63 @ 2,800	450	830 × 683 × 830
8140 SRM15	4L –TAA	110 @ 3,500	75 @ 3,600	–	705 × 640 × 860
NEF 150	6L – NA	110 @ 2,800	92 @ 2,800	530	1071 × 780 × 869
8040 SRM16	4L –TAA	118 @ 2,700	85 @ 2,500	–	875 × 860 × 830
8060 SM21	6L –TC	154 @ 2,700	118 @ 2,500	–	1044 × 660 × 830
SOFIM HPI 230	4L –TAA	169 @ 4,000	85 @ 3,500	330	941 × 775 × 753
NEF 220	6L –TC	162 @ 2,800	132 @ 2,800	605	1236 × 780 × 793
NEF 250	4L –TAA	184 @ 2,800	125 @ 2,800	490	983 × 780 × 785
NEF 280	6L –TAA	206 @ 2,800	169 @ 2,800	605	1236 × 780 × 793
NEF 370	6L –TAA	272 @ 2,800	199 @ 2,800	595	1333 × 805 × 774
NEF 400	6L –TAA	294 @ 3,000	199 @ 3,000	595	1349 × 843 × 788
NEF 450	6L –TAA	331 @ 3,000	258 @ 3,000	600	1333 × 805 × 774
CURSOR 500	6L –TAA	368 @ 2,600	257 @ 2,600	910	1771 × 1000 × 1077
CURSOR 550	6L –TAA	405 @ 2,600	294 @ 2,600	910	1771 × 1000 × 1077
8281 SRM70	V8 –TAA	515 @ 2,200	397 @ 2,200	–	1895 × 1136 × 1165
CURSOR 770	6L –TAA	567 @ 2,300	397 @ 2,300	1380	2015 × 1064 × 1039
CURSOR 825	6L –TAA	607 @ 2,400	442 @ 2,400	1400	2015 × 1064 × 1039
VECTOR 1100	V8 –TAA	–	635 @ 2,100	–	2166 × 1263 × 1160
VECTOR 1800	V12 –TAA	1325 @ 2,300	1001 @ 2,100	–	– × – × –

NA=Naturally aspirated
TC=Turbocharged
TAA=Turbocharged aftercooled

Seatek SpA

Via Provinciale 71, I-23841 Annone Brianza, Lecco, Italy

Tel: (+39 0341) 57 93 35
Fax: (+39 0341) 57 93 17
e-mail: sales@seatek-spa.com
Web: www.seatek-spa.com

P Fumagalli, *Managing Director*

Model	No of cylinders	Max Power Output kW	Speed (rpm)	Weight (kg)
620 plus	6R	456	2600	900
660 plus	6R	485	3100	900
725 plus	6R	533	3100	900
820 plus	8R	625	3100	925
950 plus	8R	700	3200	–

Seatek SpA build high-speed marine diesel engines which are noted for their high power-to-weight and power-to-size ratios as well as fuel economy and low emissions.

In 1988, a Seatek-powered boat won the Italian, European and World offshore Class One Championships and since this time Seatek powered craft have continually won Class One and Two offshore races worldwide.

The new Navy configuration of the 10 litre engines gives a power output of 485 kW and has been used in a number of applications including the Italian and Spanish Customs boats, the UK Royal Marines special craft and the Coastal Assault Craft of the US Navy. The newly released 600 Plus engine is aimed at the work boat and patrol boat markets where the additional reliability offered is of significance. The company's new 800 Plus engines offer a very high power to weight ratio (0.73 kW/kg) and a 12 cylinder 1,100 kW engine, weighing only 1,550 kg, is also under development.

Japan

Mitsubishi Heavy Industries Ltd

16-5 Konan 2-chome, Minato-ku, Tokyo 108-8215, Japan

Tel: (+81 3) 67 16 31 11
Fax: (+81 3) 67 16 58 00
e-mail: jjk100@mail.smw.mhi.co.jp
Web: www.mhi.co.jp

S16R-S

In 1992 Mitsubishi developed the S16R-S, a high-speed diesel engine aimed at the high-performance marine vessels market.
The S16R-S has been developed from the S16R series and has the following basic characteristics:

Specifications
Type: direct injection, four-cycle, water-cooled
Turbocharger: Mitsubishi TD15
Cylinders: 16 in 60° V
Base: 170 mm
Stroke: 180 mm
Displacement: 65.4 litres output, MCR: 2,100 kW at 2,000 rpm and bmep 19.3 bar
Specific fuel consumption: 218 g/kW h
Weight: 5,500 kg
Weight per kW: 2.54 kg/kW

Four of these engines power the Mitsubishi Super Shuttle 400 *Rainbow*, a 350-passenger, 40 kt catamaran.

Norway

GE Energy (Norway) A/S

Gasevikveien 6 (PO Box 125), N-2027 Kjeller, Norway

Tel: (+47) 63 82 31 00
Fax: (+47) 63 84 40 01
e-mail: jan.halle@ps.ge.com
Web: www.ge.com/no/

Jan Halle, *Manager, Marine Propulsion*

GE Energy supplies gas turbine packages for propulsion of high-speed vessels, built under Original Equipment Manufacturers Agreements with General Electric (GE).

GE/General Electric's marine and industrial gas-turbine modules are compact, high-performance power units. Each, from the LM 500 to the LM 6000, is a simple-cycle gas turbine derived from highly reliable aircraft engines.

GE Energy's workshop, which specialises in the overhaul, repair and testing of gas turbines, is located at Ågotnes, on the outskirts of Bergen. This workshop is one of the few of its kind worldwide with authorisation from GE to carry out maintenance and servicing on the LM 2500 and LM500 gas turbines at all levels.

LM 500

The LM 500 is a simple-cycle, two-shaft gas turbine, derived from GE's TF34 aircraft engine. This engine powers a Danish Navy patrol boat, a Japanese PG Class hydrofoil patrol craft and the Kværner Fjellstrand Flying Foilcat class passenger ferry.

LM 1600

The LM 1600 is a simple-cycle three-shaft gas-turbine engine, derived from GE's F404 engine. It is designed with 50 to 60 per cent fewer parts than other gas-turbines in its class. This engine has been selected as the power unit for a number of large high-speed vessels: the *Destriero* powered by three LM 1600 engines, *Eco* powered by one LM 1600 engine and the Danyard/NQEA Seajet 250 class of fast ferry. These engines are also used in the 'father and son' arrangement in the Stena HSS 1500 fast ferry.

The LM 1600 is a fuel efficient simple-cycle engine. Lightweight, compact and modular, it combines the latest in blade-loading design, cooling technology and corrosion-resistant materials and coatings.

LM 2500

The LM 2500 is a simple-cycle two-shaft gas-turbine engine, derived from GE's military TF39 and the commercial bypass turbofan engines. Two LM 2500s in 'father and son' configuration with two LM 1600s, power the HSS 1500 class, 124 m Stena high-speed catamarans. Its current experience base comprises over 250 ships and 19 navies.

Compact, lightweight and powerful, the LM 2500 is adaptable to a broad range of ships and is available in both the standard 2500 version and an LM2500+ model. The LM 2500+ has been used in the SNCM-Corsaire 1300 and NEL-Corsaire 1400 fast ferries. The engines are also due to be fitted to the US Navy X-Craft.

GE Energy LM 1600 Marine propulsion package 0507100

Model	Speed (rpm)	Thermal Efficiency%	Power Output (MW)	Weight (kg)	Length (mm)
LM1600	7000	37.7	14.92	3720	424
LM2500	3600	38	25	4670	852
LM2500+	3600	39	30.2	5236	670
LM6000	3600	42	42.75	8170	730
LM500	7000	31	4.47	900	296

LM 6000

The LM 6000 is a simple-cycle, two-shaft high-performance gas-turbine engine, derived from GE's most powerful and reliable aircraft engine, the CF6-80C2. Delivering more than 40 MW at over 40 per cent thermal efficiency, the powerful LM 6000 is very fuel efficient.

All components of the LM 6000 incorporate corrosion-resistant materials and coatings to provide maximum parts life and time between overhaul, regardless of the unit's operational environment.

Russian Federation

Barnaultransmash Joint-stock Company

28 Kalinin Avenue, 656037 Barnaul, Russian Federation

Tel: (+7 3852) 77 04 01
Fax: (+7 3852) 77 95 11
Web: www.barnaultransmash.ru

Victor Padas, *General Director*
Sergei Kurkin, *Technical Director*
Andre Kalyunov, *Chief Designer*

Barnaultransmash is a Russian Joint-stock company specialising in designing and building high-speed diesel engines with a power range up to 500 kW. The engines are used in river and sea-going vessels, heavy trucks, railways, oil and gas drilling rigs, road building machinery, excavational cranes and electric power stations.

3KD12H-520R DIESEL ENGINE

The 3KD12H-520R diesel engine is designed to operate as the main ship propulsion on vessels of different applications.

The engine is a four-stroke, 12-cylinder, V-type engine, with direct fuel injection and supercharging, a power take-off shaft and an outboard water pump.

The cooling system is of the double-loop type with a water circulating pump and an outboard water pump, water and water-oil coolers and temperature regulators.

The lubrication system is of a forced-circulation type with a 'dry' crankcase, with an electric motor for prestarting system priming.

The diesel engine is equipped with a reversing reduction gear which contains a reducer and a multi-plate clutch aimed at engaging and disengaging of the propeller and cranking as well as reversing of the propeller rotation direction.

Specifications
Cylinder diameter: 150 mm
Piston stroke: 180 mm
Full power: 368 kW
Crankshaft rotational speed: 1,500 rpm
Specific fuel consumption: 228 g/kW h
Starting system: Electric starter or compressed air

Dimensions
Length: 2,695 mm
Width: 1,108 mm
Height: 1,230 mm
Dry weight: 2,600 kg

3D12A DIESEL ENGINE

The 3D12A diesel engine is designed to be employed by sea and river ships of various applications as their main propulsion engine. The engine is a four-stroke, 12 cylinder, V-type diesel with direct fuel injection. The cooling system is of the double loop type with a water circulating pump, ware and water-oil coolers and a temperature regulator, and the engine has an integral reverse reduction gearbox. Lubrication is of a forced circulation type with a "dry" crankcase and an electric pump for pre-starting engine priming.

Specifications
Cylinder diameter: 150 mm
Piston stroke: 180 mm
Full power: 220 kW
Crankshaft rotation speed at full power: 1,500 rpm
Specific fuel consumption: 235 g/kWh
Starting system: Electric starter or compressed air

Dimensions
Length, with power take-off shaft: 2,464 mm
Length, without power take-off shaft: 2,390 mm
Width: 1,052 mm
Height: 1,225 mm
Dry weight: 1,890 kg

3D23 DIESEL ENGINE

The 3D23 diesel engine is designed for small high speed vessels, patrol and search and rescue service and ecological operations and is a high-speed, four stroke, supercharged 6-cylinder V-type (at an angle of 120°).

The crankcase unit of the engine is of tunnel type, made integral with the cylinder jackets. The crankshaft is mounted on roller bearings. The connecting rods are of central type. The cooling system is a water double loop with or without an outboard water pump. Lubrication is of a forced-circulation type with a feed oil tank and a dry crankcase.

Partial power take-off at the side opposite to the flywheel is possible, with the corresponding decrease of power taken at the flywheel side.

Specifications
Cylinder diameter: 150 mm
Piston stroke: 150 mm
Power: rated, 220 kW @ 2200 rpm
Specific Fuel Consumption: 232 g/kWh

Dimensions
Length: 1,010mm
Width: 1,150 mm
Height: 1,015 mm

Dry weight: 850 kg
Starting system: Electric starter or compressed air

3D6 DIESEL ENGINE
Designed for sea and river vessels the 3D6 diesel engine is a four stroke, in-line 6-cylinder with diesel injection.

The engine is liquid cooled and is equipped with an integral reverse reduction gearbox and clutch.

Specifications
Cylinder diameter: 150 mm
Piston stroke: 180 mm
Rated power: 110 kW @ 1,500 rpm
Specific fuel consumption: 224 g/kW h

Dimensions
Length: 2,462 mm
Width: 882 mm
Height: 1,179 mm
Dry weight: 1750 kg

3D29 DIESEL ENGINE
The 3D29 diesel engine is designed to be employed by sea and river vessels and is a four stroke, high-speed, 10-cylinder V-type diesel with a V-angle of 144°.

The engine is liquid cooled and can be fitted with an optional power take-off. Power can be increased by up to 550 kW by turbo-supercharging.

Specifications
Cylinder diameter: 150 mm
Piston stroke: 150 mm
Power: rated, 257 kW (370 kW with supercharging)
Specific fuel consumption: 231 g/kW h
Starting system: Electric starter or compressed air

Dimensions
Length: 1,090 mm
Width: 1,228 mm
Height: 598 mm
Dry weight: 950 kg

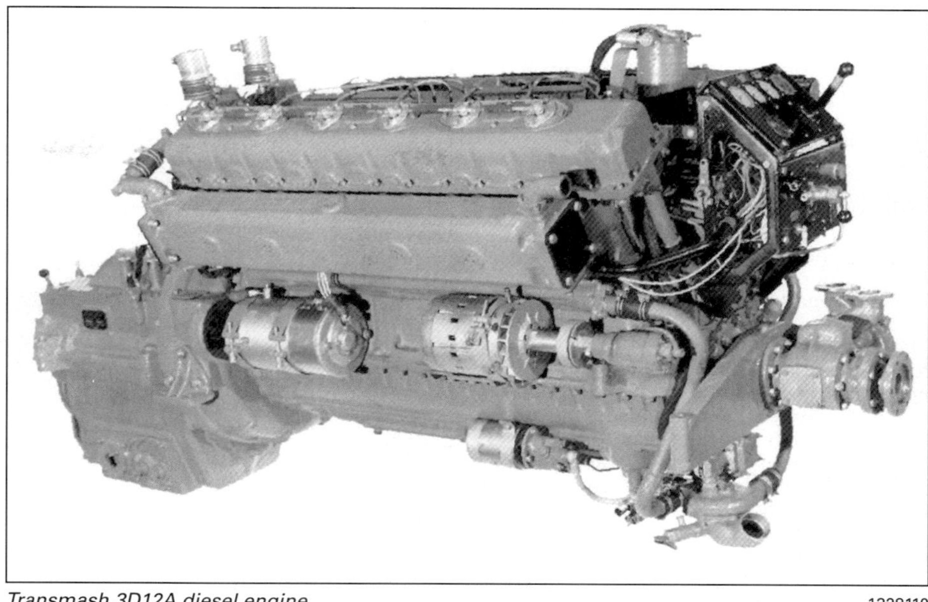

Transmash 3D12A diesel engine 1328110

Transmash 3D23 diesel engine 1328111

NK Engines

JSC N.D. Kuznetsov
2a S. Lazo St, Samara 443026, Russia

Tel: (+7 8462) 50 02 28
Fax: (+7 8462) 50 12 11
email: sntk@saminfo.ru

Yevgeniy A Gritsenko, *General Director and General Designer*

Description

NK-12/14
In its original form the NK-12M engine developed 8,948 kW. The later NK-12MV is rated at 11,033 kW and powers the Tupolev Tu-114 transport, driving four-blade, contrarotating propellers of 5.6 m diameter. The NK-12MA, rated at 11,185 kW, powers the Antonov An-22 military transport, with propellers of 6.2 m diameter.

The NK-12M has a single 14-stage axial flow compressor. Compression ratio varies from 9:1 to 13:1 and variable inlet guide vanes and blow off valves are necessary. A cannular-type combustion system is used. Each flame tube is mounted centrally on a downstream injector, but all tubes merge at their maximum diameter to form an annular secondary region. The single turbine is a five-stage axial. Mass flow is 56 kg/s.

The casing is made in four portions from precision-welded sheet steel. An electric control for variation of propeller pitch is incorporated to maintain constant engine speed.

The NK-14 is the modern derivative of the same engine, providing similar power outputs. The Kuznetsov Design Bureau is currently working on low NOx combustor designs for this engine.

Specifications
Dimensions
Length: 6,000 mm
Diameter: 1,150 mm
Dry weight: 2,350 kg
Performance ratings
Max power: 11,033 kW
Nominal power: 8,826 kW at 8,300 rpm
Idling speed: 6,600 rpm

Kuznetsov NK-14E (Ken Fulton) 0506932

Zvezda Joint Stock Company

123 Babushkina Str, 192012 St Petersburg,
Russian Federation

Tel: (+7 812) 3675745
Fax: (+7 812) 367 3776
e-mail: office@zvezda.spb.ru
Web: www.zvezda.spb.ru

Pavel Plavnik, *Director General*
Anton Goumenny, *Chief of International Sales Department*

The Zvezda Joint Stock Company was founded in 1945 as a large production engineering company.

Zvezda designs, produces and maintains high-speed lightweight diesel engines for marine, locomotive and industrial applications; marine gearboxes; gensets and power plants for various facilities.

The Zvezda high-speed four-stroke ChN16/17 series of radial configuration diesel engines offers low specific weights and high specific volumetric output powers. Engines are produced with 42 and 56 cylinders in addition to two 112 cylinder models made from back-to-back 56 cylinder engines.

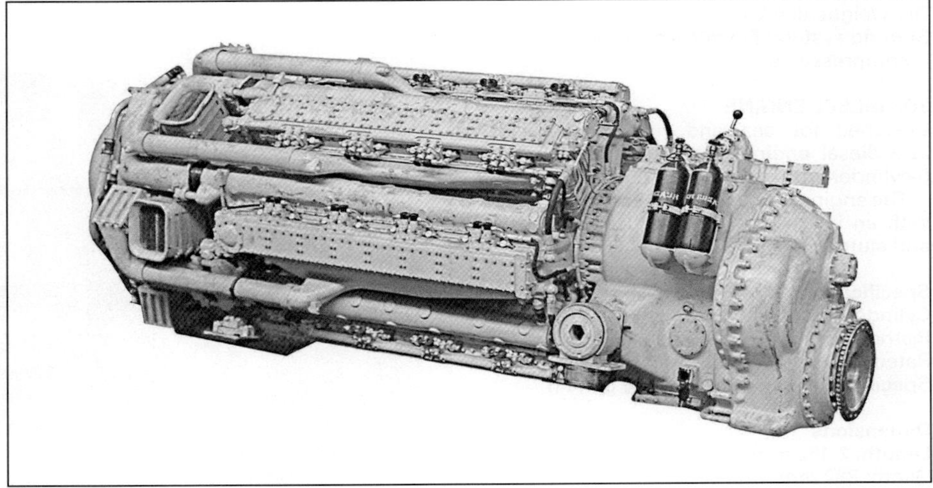

Zvezda M534 1111657

ChN16/17

Zvezda ChN16/17 engines are used as main propulsion engines in high-speed vessels where compact and powerful propulsion units are specified.

ChN18/20

The Zvezda high-speed four-stroke series of V12 configuration diesel engines (bore 180 mm, stroke 200mm) offers low specific weights and high specific volumetric output powers. The M480 diesel engine is produced with 6 cylinders in-line.

ChN18/20 engines are widely used for high-speed vessels, including custom vessels, patrol boats, rescue boats, ecological vessels, passenger hydrofoils, high speed transport vessels.

Zvezda M470M 1111656

Chn16/17 Series Engine Data Summary

Type	Model	Continuous (rated) power, kW	Rotation at continuous (rated) power, rpm	Fuel-stop power kW/rpm	Length mm	Width mm	Height mm	Weight kg
42 cylinders	M532	2,426	2,000	2,940/2,200	3,700	1,555	1,560	5,450
	M532A	1,765	1,700	1,985/1,800	3,700	1,555	1,560	5,450
	DRA-532A	2,074	1,800	2,074/1,800	–	1,555	1,560	5,800
	M503A	2,425	2,000	–	3,700	1,555	1,560	5,450
	M503B	1,840	1,780	–	3,900	1,555	1,560	5,600
56 cylinders	M534	3,675	2,000	–	4,400	1,675	1,655	7,250
	M504A	3,495	1,950	–	4,650	1,675	1,655	7,500
	M504B	3,675	2,000	–	4,400	1,675	1,655	7,250
	M510A 1st speed	3,530	2,000	4,965	4,965	1,740	1,925	9,050
	M510A 2nd speed	2,980	2,000					
	M517	4,750	2,000	–	4,400	1,675	1,655	7,250
	M520	3,600	1,900	–	4,400	1,675	1,655	7,250
	M533	3,676	2,000	–	4,400	1,685	1,655	7,300
2 × 56 cylinders	M507A	7,355	2,000	–	7,000	1,820	2,490	17,100
	M507D	5,882	1,750	–				16,900
	M511A 1st speed	7,355	2,000	–	7,000	1,805	2,480	18,200
	M511A 2nd speed	7,130	2,000					
	M521	6,323	2,000	–	7,000	1,820	2,490	17,100

Chn18/20 Series Engine Data Summary

Type	Model	Continuous (rated) power, kW	Rotation at continuous (rated) power, rpm	Fuel-stop power kW/rpm	Length mm	Width mm	Height mm	Weight kg
12 cylinders, V-engine	M401A*	736	1,550	810/1,600	2,680	1,260	1,250	2,200
	M417A*	736	1,550	810/1,600	2,825	1,260	1,250	2,200
	M419A	810	1,550	890/1,600	2,825	1,260	1,250	2,200
	M419B*	810	1,550	890/1,600	2,825	1,260	1,250	2,250
	DRA-210B	736	1,550	810/1,600	3,250	1,260	1,270	2,250
	DRA-210V	810	1,550	885/1,600	2,715	1,260	1,270	2,250
	M470*	990	1,550	1,100/1,600	2,730	1,150	1,285	2,700
	M470M*	990	1,550	1,100/1,600	2,950	1,250	1,455	3,000
	M470MK*	990	1,550	1,100/1,600	2,950	1,250	1,455	3,000
	DRA-470	990	1,550	1,100/1,600	3,545	1,150	1,285	3,200
	DRA-470M	990	1,550	1,100/1,600	–	1,250	1,455	3,450
	DRA-470MR	990	1,550	–	3,960	1,150	1,500	3,500
	DRA-472	990	1,550	1,100		1,150	1,285	3,150
	DRA-473	990	1,550	1,100/1,600	4,118	1,250	1,455	3,500
6 cylinders, inline emgine	M480	500	1,550	550/1,600	2,400	8,54	1,520	1,850

* depending on the intermediate shaft length.

Sweden

Scania

Industrial & Marine Engines
SE-151 87 Södertälje, Sweden

Tel: (+46 8) 55 38 10 00
Fax: (+46 8) 55 38 29 93
e-mail: industrial.marine@scania.com
Web: www.scania.com

Scania marine diesels cover a power output range of 155 to 588 kW with weights of 890 to 1,500 kg. The engine family consists of two six-cylinder in-line engines with displacements of 9 and 12 litres, and a 16 litre V8 engine. The DI12M EMS and the DI16M is equipped with both unit injectors and an electronic engine management system (EMS). This means that combustion and vital engine data can be optimised for all possible load conditions. Another important development goal was to meet the emission standards that are likely to be set for the future.

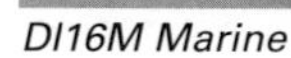

DI16M Marine 0538232

All Scania engines feature separate cylinder heads to allow easier and quicker maintenance and service repairs on board.

Scania's engine production is based on a modular approach where important components are found in several models. The component standardisation results in higher quality service and less risk for downtime.

Marine diesel engines
RATINGS

Engine type with heat exchanger	Turbo	Inter-cooled	Displacement dm³	Configuration	Propulsion High-speed workboats, patrol, and so on kW	rpm	Specific fuel consumption at 1,500 rpm g/kW h
D 9M	T	–	9.0	6 L	164	2,200	204
D 9M	T	–	9.0	6 L	201	2,200	205
DI 9M	T	I	9.0	6 L	259	2,200	201
DI 9M	T	I	9.0	6 L	309	2,200	202
DI 9M	T	I	9.0	6 L	331	2,200	205
DI 12M EMS	T	I	11.7	6 L	368	2,100	197
DI 12M EMS	T	I	11.7	6 L	478	2,200	203
DI 16M EMS	T	I	15.6	V 8	441	1,800	205@2,100 rpm
DI 16M EMS	T	I	15.6	V 8	578	2,100	207@2,100 rpm
DI 16M EMS	T	I	15.6	V 8	550	2,100	210@2,100 rpm
DI 16M EMS	T	I	15.6	V 8	588	2,200	216@2,200 rpm
With keel cooling							
D 9M	T	–	9.0	6 L	201	2,200	204
D I9M	T	I	9.0	6 L	232	2,200	205
DI 12M	T	I	11.7	6 L	313	2,100	196
DI 12M	T	I	11.7	6 L	368	2,100	197
DI 16M EM	T	I	15.6	V 8	478	2,100	205@2,100 rpm

WEIGHTS

Engine type	Max dimensions (mm)			Weight dry (kg)
With heat exchanger	Length	Width	Height	
D9M	1285	811	1,080	890
DI9M	1285	846	1,087	905
DI12M	1341	853	1128	1150
DI12M EMS	1358	870	1038	1150
DI16M EMS	1358	1172	1198	1550
With keel cooling				
D9M	1285	810	1080	855
DI9M	1285	815	1132	865
DI12M	1341	798	1128	1130
DI12M EMS	1358	820	1038	1130
DI16M EMS	1236	1172	1198	1550

Weights and dimensions. Quoted values are only a guide; there are variations for each engine type. Weights exclude oil and water.

Volvo Penta AB

SE-405 08 Gotenburg, Sweden

Tel: (+46 31) 23 54 60
Fax: (+46 31) 50 81 97
Web: www.volvopenta.com

Stefan Jufors, *President*
Huig van Barneveld, *Marine Commercial Sales*
Ovar Lundberg, *Marine Commercial Sales*
Anders Bevreus, *Marine Commercial Sales*

Part of the Volvo Group, AB Volvo Penta designs, manufactures and markets engines, transmissions, accessories and equipment for marine and industrial use.

The sales of Volvo Penta engines account for around 4 per cent of sales within the transport equipment sector. The engines are used in ferries, pilot vessels, fishing boats and all types of leisure craft, as well as for industrial propulsion or power generation applications.

Production facilities are in the US and Brazil as well as four locations in Sweden. Volvo Penta products are sold in 100 countries worldwide. Approximately 2,600 people are employed either directly or indirectly by Volvo Penta.

Volvo Penta diesels for commercial craft can provide up to 544 kW and up to 1,400 kW for combined Volvo and Mitsubishi Heavy Industry engines.

Diesel Aquamatic and Diesel Inboard, rating 1, 2, 3 and 4

Engine	Rating 1 Prop shaft power kW/hp	rpm max	Rating 2 Prop shaft power kW/hp	rpm max	Rating 3 Prop shaft power kW/hp	rpm max	Rating 4 Prop shaft power kW/hp	rpm max	No. of cyl.	Displ. litres	Weight kg
D3-110	–	–	–	–	–	–	81/110	3,000	5	2.4	292*
D3-130	–	–	–	–	–	–	95/130	4,000	5	2.4	292*
D3-160	–	–	–	–	–	–	120/163	4,000	5	2.4	292*
D4-210	–	–	–	–	–	–	154/210	3,500	4	3.6	546*
D4-260	–	–	–	–	–	–	191/260	3,500	4	3.6	546*
D6-280	–	–	–	–	–	–	206/280	3,500	6	5.5	656*
D6-310	–	–	–	–	–	–	228/310	3,500	6	5.5	656*
D6-370	–	–	–	–	–	–	272/370	3,500	6	5.5	656*
D5A T	7/98	1,900	83/113	1,900	–	–	–	–	4	4.8	510**
D5A TA	80/121	1,900	103/140	1,900	–	–	–	–	4	4.8	525**
D7A TA	–	–	174/237	2300	–	–	–	–	–	–	–
D7C TA	–	–	195/265	2300	–	–	–	–	–	–	–
D9 MH	269/365	1,800/2200	–	–	–	–	–	–	6	9.4	1,150**
D9 MH	221/300	–	313/425	2,200	–	–	–	–	6	9.4	1,150**
D9 425	261/355	–	–	–	313/425	2,200	–	–	6	9.4	1,150**
D9 500	–	1,800	–	–	–	–	368/500	2,600	6	9.4	1,150**
D12 300	220/300	1,800	–	–	–	–	–	–	6	12.1	1,400**
D12 350	267/350	1,800	–	–	–	–	–	–	6	12.1	1,400**
D12-400	294/400	1,800	–	–	–	–	–	–	6	12.1	1,400**
D12-450	331/450	–	–	–	–	–	–	–	6	12.1	1,400**
D12-500	–	–	368/500	1,800	–	–	–	–	6	12.1	1,400**
D12-550	–	–	405/551	1,900	–	–	478/550	2,900	6	12.1	1,400**
D12-615	–	–	–	–	452/615	2,100	–	–	6	12.1	1,400**
D12-675	–	–	–	–	–	–	496/675	2,300	6	12.1	1,400**
D16 MH	479/651	1800	–	–	–	–	–	–	–	–	–
D25A MS	440/598	1,600	495/660	1,650	–	–	–	–	6	24.5	2,890**
D25A MT	470/639	1,600	520/707	1,650	625/923	1,800	–	–	6	24.5	2,930**
D30A MS	445/605	1,350	490/696	1,400	–	–	–	–	6	30.0	2,950**
D30A MT	490/853	1,350	530/721	1,400	610/930	1,500	–	–	6	30.0	2,990**
D34A MS	634/962	1,940	701/953	2,000	–	–	–	–	V-12	33.9	3,400**
D34A MT	701/953	1,940	776/1055	2,000	858/1167	2,100	–	–	V-12	33.9	3,820**
D49A MS	890/1197	1,600	970/1319	1,650	–	–	–	–	V-12	49.0	5,250**
D49A MT	940/1278	1,600	1040/1414	1,650	1210/1646	1,800	–	–	V-12	49.0	5,420**
D65A MS	1170/1591	1,600	1290/1754	1,650	–	–	–	–	V-12	65.4	6,600**
D65A MT	1250/1700	1,600	1380/1877	1,650	1610/2190	1,800	–	–	V-12	65.4	6,800**

*Weight: dry including transmission
**Weight: dry excluding tramsission

Marine Commercial Ratings			
Rating 1 (Heavy Duty Commercial) For commercial vessels with displacement hulls in heavy operation. Typical boats: Larger trawlers, ferries, freighters, tugboats, passenger vessels with longer journeys. Load and speed could be constant, and full power can be used without interruption.	**Rating 2** (Medium Duty Commercial) For commercial vessels with semiplaning or displacement hulls in cyclical operation. Typical boats: Most patrol and pilot boats, coastal fishing boats in cyclical operation (gillnetters, purse seiners, light trawlers), passenger boats and coastal freighters with shorter trips. Full power could be utilised for a maximum of 4 hours per 12 hour operation period. Between full load operation periods, engine speed should be reduced at least 10% from the obtained full load engine speed.	**Rating 3** (Light Duty Commercial) For commercial vessels or craft with high demands on speed and acceleration, planing or semi-planing hulls in cyclical operation. Typical boats: Fast patrol, rescue, police, light fishing, fast passenger and taxi boats and so on. Full power could be utilised for a maximum of 2 hours per 12 hour operation period. Between full load periods, engine speed should be reduced at least 10% from the obtained full load engine speed.	**Rating 4** (Special Light Duty Commercial) For light planing craft in commercial operation. Typical boats: High speed patrol, rescue, navy, and special high-speed fishing boats. Recommended speed at cruising = 25 kt. Full power could be utilised for a maximum of 1 hour per 12 hour operation period. Between full load operation periods, engine speed should be reduced at least 10% from the obtained full load engine speed.

Ukraine

Ivchenko Progress Design Bureau

2 Ivanova Street, Zaporozhye, 69068, Ukraine

Tel: (+380 612) 65 03 27
Fax: (+380 612) 65 46 97
e-mail: progress@ivchenko-progress.com
Web: www.ivchenko-progress.com

This design team, which was headed by the late A Ivchenko, is based in a factory at Zaporozhye in Ukraine, where all prototypes and preproduction engines bearing the 'AI' prefix are developed and built.

The first engine with which Ivchenko was officially associated was the 40 kW AI-4G piston engine, used in the Kamov Ka-10 ultra-light helicopter. He later progressed via the widely used AI-14 and AI-26 piston engines, to become one of the former Soviet Union's leading designers of gas-turbine engines.

The *Burevestnik*, the first RFAS gas-turbine hydrofoil was powered by two AI-20s in derated, marinised form and driving two three-stage water-jets. The integrated lift and propulsion system of the Sormovich ACV was derived from the 1,894 kW gas-turbine AI-20 DK. The two AI-20s, each rated at about 2,700 kW continuous, are also thought to power the Lebed amphibious assault landing craft.

IVCHENKO AI-20

The Ivchenko design bureau is responsible for the AI-20 turboprop engine which powers the Antonovspring An-10, An-12 and Ilyushin Il-18 airliners and the Beriev M-12 Tchaika amphibian.

AI-20K Rated at 3,000 kW. Used in Il-18V, An-10A and An-12.

AI-20M Uprated version with T-O rating of 3,100 kW. Used in Il-18D/E, An-10A and An-12.

Conversion of the turboprop as a marine power unit for hydrofoil water-jet propulsion (as on the *Burevestnik*) involved a number of changes to the engine. In particular, it was necessary to hold engine rpm at a constant level during conditions of varying load from the water-jet pump. It was also necessary to be able to vary the thrust from the water-jet unit from zero to forward or rearwards thrust to facilitate engine starting and vessel manoeuvring.

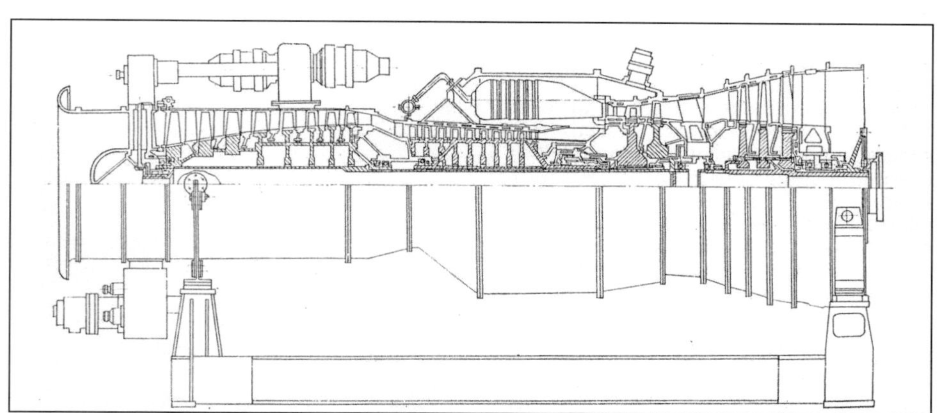

1,300 kW Ivchenko AI-23-CI marine gas turbine 0506760

The AI-20 is a single-spool turboprop, with a 10-stage axial flow compressor, cannular combustion chamber with 10 flame tubes and a three-stage turbine, of which the first two stages are cooled. Planetary reduction gearing, with a ratio of 0.08732:1, is mounted forward of the annular air intake. The fixed nozzle contains a central bullet fairing. All engine-driven accessories are mounted on the forward part of the compressor casing, which is of magnesium alloy.

The AI-20 was designed to operate reliably in all temperatures from –60 to +55°C at heights up to 10,000 m. It is a constant speed engine, the rotor speed being maintained at 21,300 rpm by automatic variation of propeller pitch. Gas temperature after turbine is 560°C in both current versions. TBO of the AI-20K was 4,000 hours in early-1966.

Specifications
Weights
Dry
AI-20K: 1,080 kg
AI-20M: 1,039 kg

Performance ratings
Max T-O
AI-20K: 3,000 kW
AI-20M: 3,100 kW
AI-20K: 1,700 kW
AI-20M: 2,000 kW

Specific fuel consumption
At cruise rating
AI-20K: 288 g/kW h
AI-20M: 264 g/kW h

Oil consumption
Normal: 1 litre/h

IVCHENKO AI-24

In general configuration this single-spool turboprop engine, which powers the An-24 transport aircraft, is very similar to the earlier and larger AI-20.

Specifications
Dimensions
Length, overall: 2,435 mm
Weight
Dry: 499 kg

Performance rating
Max power with water injection: 2,102 kW

United Kingdom

Rolls-Royce Marine Systems

Neptune House, PO Box 3, Filton, Bristol, Avon BS34 7QE, UK

Tel: (+44 117) 979 7242
Fax: (+44 117) 979 6722
Web: www.rolls-royce.com

Gordon Waddington, *Managing Director*
Roger Took, *Chief Applications Engineer*
Stuart Wilmshurst, *Sales Support*

Rolls-Royce brings together acknowledged leaders in the field of marine technology to offer a wide range of marine products from a single supplier. In addition to Rolls-Royce marine gas turbines, it includes Bergen diesels; Kamewa and Bird-Johnson propellers and thrusters; Kamewa waterjets; Mermaid™ podded propulsors; Intering and Brown Brothers stabilisers; Tenfjord Frydenbö steering gear systems; Michell propeller shaft bearings; and marine electrical power, distribution and automation systems.

A full range of deck machinery completes the range.

A complete service is offered from initial concept design, equipment selection and supply, to commissioning and through life support. Extensive experience in both ship and propulsion systems design enables Rolls-Royce to develop an optimum solution for each customer.

The market requirement for more power from a compact and flexible system was a key driver for the development of the latest addition to the Rolls-Royce marine gas turbine range, the 36 MW MT30, the world's most powerful marine gas turbine. It has been designed to provide lightweight, compact power for larger vessels where operational speed is important and/or space saving is required to release valuable on board revenue earning potential. The engine package, just 8.64m long, opens up new propulsion systems options to the vessel designer.

Over 2,700 Rolls-Royce marine and industrial gas turbines have been sold or are on order for customers around the world and have accumulated over 50 million hours of operating experience.

MT30 MARINE GAS TURBINE

The Rolls-Royce MT30 uses the latest aero-engine materials and technology to reliably deliver a broad band of powers in an exceptionally lightweight package. It has a maximum continuous rating of 36 MW flat rated to 26°C (commercial), and 38°C/100°F (Naval) with a specific fuel consumption of 210 g/kWhr. The engines' high power rating provides flexibility for ship growth or increased speed, and has a 25 per cent power growth potential. Use at lower powers results in extended engine life.

The MT30 is the marine version of the successful Trent 800 aero engine, which has evolved into a family of seven since it's introduction in 1995. Designed with 50 per cent to 60 per cent fewer parts than other aero-derivative gas turbines it maintains its competitive efficiency down to 25 MW; it is also designed to burn commercially available distillate fuels, giving a high degree of operational flexibility and associated through-life cost benefits. Det Norske Veritas (DNV) type certification was achieved in early 2004 with ABS certification following in early 2005.

The MT30 benefits from the advanced technology features common to the Trent aero engine family. It is a twin-spool, high-pressure ratio gas generator with an eight stage variable geometry Intermediate Pressure Compressor (IPC) and a six stage high-pressure compressor (HP). The four-stage free power turbine is derived from the Trent 800, which is supported on a robust bearing structure for optimum reliability. Proven components, incorporating the latest blade cooling technologies are used throughout. Key rotating parts are protectively coated for service in the marine environment to reduce maintenance and deliver long service life. The annular combustor is similar to the aero parent and ensures the MT30 meets all current and anticipated legislation on emissions and smoke.

A self contained, fully packaged module, the MT30 can be supplied for direct drive or power generation – complete with an alternator and its own acoustic enclosure. The design installs all items on the engine base plate with ancillary connections sited at convenient locations. The unit can be installed in a single lift for ease and speed of installation at the shipyard. The only additional items are the engine controller, the lube oil module, water wash skid and hydraulic start package, which can be situated at any convenient location in the

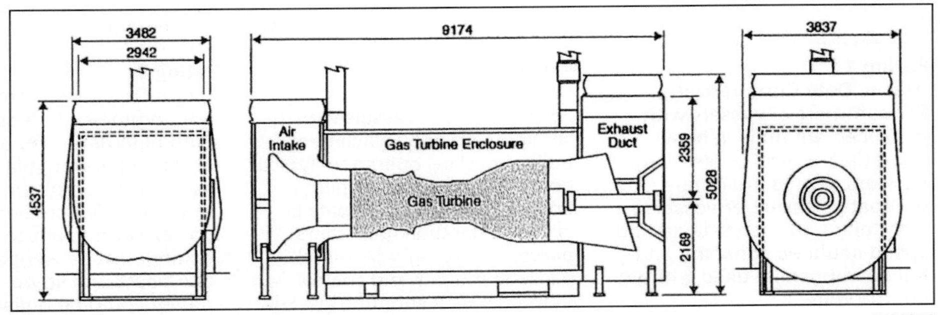

MT30 Propulsion Drive Module 0526356

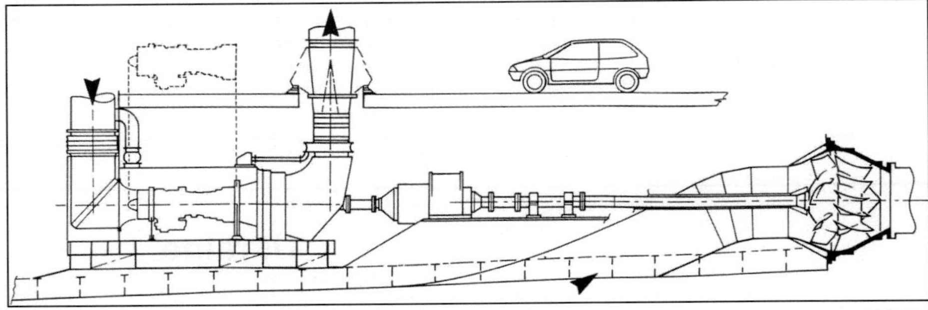
Typical installation showing engine removal route 0507017

Spey SM1C for HMS Brave *conversion* 0506775

engine room. The enclosure is also the fire boundary and is fitted with an automated fire suppression system.

MT30 Module dimensions
Length: 8,600 mm
Height: 4,021 mm
Width: 3,953 mm
With the gas generator change unit weighing 6,500 kg and a total module weight, including enclosure of 26,000 kg (dependent on application), the MT30 offers the best power-to-weight ratio in its class.

The MT30 delivers high availability and is designed for unmanned engine rooms. Condition Based Maintenance (CBM) is a feature of the engine design and routine maintenance is limited to checking fluid levels and visual examinations. Internal condition sensors enable the unit to be serviced 'on condition'. Modular construction speeds major maintenance when it becomes necessary, resulting in quicker turnaround times for repair with reduced maintenance costs and spares holding.

Performance (at ISO conditions – no loss)
Power (MCR): 36MW@26°C (Commercial) 36MW@38°C (Naval)
Specific fuel consumption: 0.21kg/kWhr
Power shaft speed: 3,300 rpm
Exhaust mass flow: 113 kg/sec
Exhaust temperature: 474°C

SPEY MARINE GAS TURBINE
The marine Spey is a complete propulsion package suitable for a wide range of naval surface ships that require a compact and powerful prime mover. Delivering up to 19.5 MW, the marine Spey is derived from the successful TF41 military aero engine, which has accumulated over 4.5 million hours in service. Of twin spool, modular design, the Spey incorporates the latest technological developments from the commercial Tay and RB211 engine programmes, combining improved reliability and operational flexibility with outstanding economy and performance.

Originally designed and developed for the UK Royal Navy, the Spey is suitable for both main and cruise propulsion.

The Spey brings substantial benefits to operators and shipbuilders alike. The module includes the gas generator, power turbine, ancillary systems and fire protection systems all mounted on a fabricated steel base. This allows simple installation with the principle interfaces limited to the intake and uptake flanges, the hull mountings and the output shaft flange. The fully insulated enclosure significantly reduces airborne sound/heat levels and weighs 25,700 kg. The unit has its own local digital controller as well as interfaces for remote control and surveillance from the ship control centre.

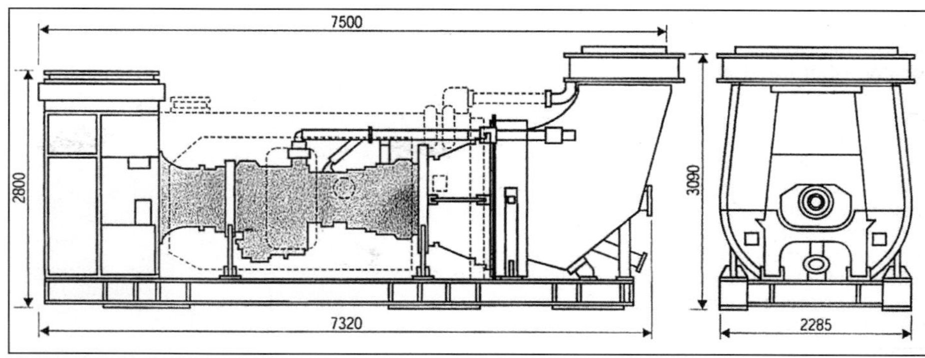

Marine Spey Module 0526357

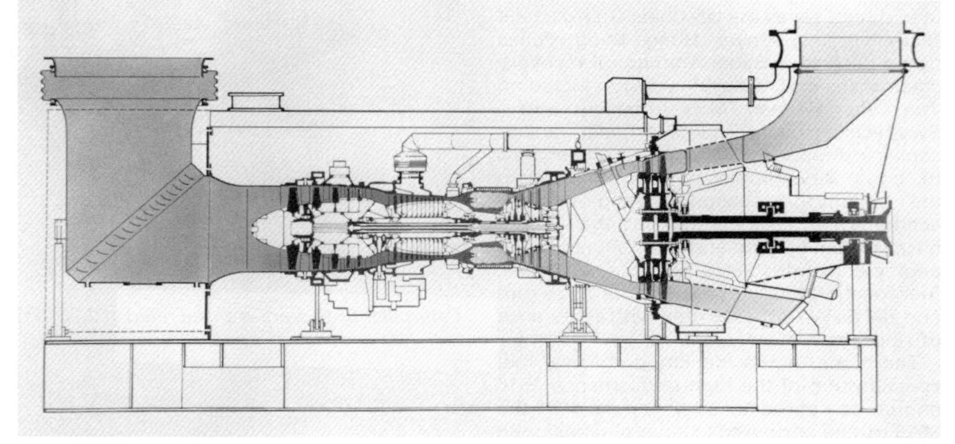
Spey SM1C module 0506773

Spey Marine Module dimensions
Length: 7,500 mm
Height: 3,100 mm
Width: 2,285 mm
In service around the world in nine ship classes with over 500,000 hours and operational experience in three navies, the Spey is a rugged and reliable marine propulsion package. The modular design inherited from its aero parent ensures ease of maintenance and reduced through-life costs. The power turbine is a long-life efficient two-stage design with short rugged shrouded blades capable of withstanding significant foreign object damage. It is available in either clockwise or anti-clockwise output shaft rotation.

The Spey is designed for optimum maintainability. Scheduled weekly maintenance is limited to just 40 minutes, with an additional 40 minutes each month. This reflects a change in prime mover maintenance philosophy, from planned to condition-based maintenance, as maximum running hours and cyclic lives are progressively extended. The Spey gas turbine change unit weighs 1,800 kg.

Performance (at ISO conditions – no loss)
Power: 18 MW
Max shaft power: 19.5 MW
Specific fuel consumption: 0.226 g/kWhr
Power shaft speed: 5,000 rpm
Exhaust mass flow: 66.7 kg/sec
Exhaust temperature: 458°C

United States

Caterpillar Inc

Marine Power Systems N4-AC6110, PO Box 610, Mossville, Illinois 61552-0610, US

Tel: (+1 309) 578 62 98
Fax: (+1 309) 578 25 59
e-mail: cat_power@cat.com
Web: www.cat-engines.com

CATERPILLAR SUBSIDIARIES
EUROPE
Caterpillar Overseas SA
PO Box 6000, CH-1211 Geneva 6, Switzerland

Tel: (+41 22) 849 44 44
Fax: (+41 22) 849 49 84

ASIA
Caterpillar Asia Pte Ltd
7 Tractor Road, Jurong, Singapore 627968

Tel: (+65) 66 62 84 00
Fax: (+65) 66 62 84 14

JAPAN
Caterpillar Power Systems Inc
Sanno Grand Building, 8 Floor, 2-14-2 Nagatacho Chiyoda-ku, Tokyo 100, Japan

Tel: (+81 3) 35 93 32 33
Fax: (+81 3) 35 93 32 38

AUSTRALIA
Caterpillar Australia Ltd
Private Mail Bag 4, Tullamarine, Victoria 3043, Australia

Tel: (+61 3) 99 53 93 01
Fax: (+61 3) 93 38 90 21

MEXICO, THE CARIBBEAN, SOUTH AMERICA
Caterpillar Americas Company
100 N E Adams, Peoria, Illinois 61629-9340, US

Tel: (+1 309) 675 17 62
Fax: (+1 309) 675 17 64

Cat China Limited
37/F The Lee Gardens, 33 Hysan Avenue, Causeway Bay, Hong Kong

Tel: (+852) 28 48 03 20
Fax: (+852) 28 48 02 23

Caterpillar is one of the world's largest manufacturers of medium- and high-speed diesel engines, with ratings from 40 to 10,000 kW. As well as main propulsion engines, Caterpillar also supplies marine gensets ranging from 11 kW to 15,750 kW.

3500 Diesel Product Group
Over 20,000 of the 3500 series engines have been produced, a large percentage of which are for marine applications.

The 3500 series family consists of vee 8-, 12- and 16-cylinder versions, all with a common bore and stroke of 170 × 190 mm. For fast craft applications the 1,800 and 1,925 rpm ratings yield the highest outputs with the lowest weight-to-power ratios. The 3516 high-performance version rated at 2,088 kW at 1,925 rpm for patrol craft, yachts and short-trip ferries, has a weight-to-power ratio of 3.56 kg/kW. This is a complete factory package, consisting of installed fuel, air and oil filters, flywheel and flywheel housing, fuel, oil, jacket water and seawater pumps and instrument panel.

Over 50 of the 3516 high-performance propulsion engines are currently in service in such vessels as the US Coast Guard Island Class vessels, Royal Hong Kong Police patrol vessels, a Saudi Aramco oil recovery catamaran and several yachts, including *The Other Woman*, *Miss Turnberry*, and a Swath Ocean yacht. This 16-cylinder version contains the newest 3500 Series refinements to boost horsepower without penalising the weight-to-horsepower ratio or fuel economy. At its rated power the engine burns 512.4 litres/h and at 1,600 rpm the engine consumes 286.8 litres/h. The specific fuel consumption is 206 g/kW h at 1,925 rpm and 201 g/kW h at 1,600 rpm, with a tolerance of +3 per cent.

The higher power density and lower fuel consumption of the high-performance 3516 engine compared to other versions of the 3516 can be attributed to several redesigned components, including an improved air intake and exhaust system, increased efficiency turbocharger, high-pressure unit injector fuel system and a two-piece piston design. While several components have been changed to increase the power and reduce the weight, the high-performance 3516 shares many parts with the proven standard versions of the 3500 Series.

Like all Caterpillar Marine Engines, the 3500 Series engines are supported by the Caterpillar worldwide network of dealers and parts warehouses.

Caterpillar 3516 marine diesel rated at 2,237 kW at 1,925 rpm 0044039

3600 Diesel Product Group

Production of the Caterpillar 3600 Series in-line engine commenced in 1985, with vee engine production following in 1986. By mid-1998, over 1,700 had been produced, approximately 500 of which were marine engines. The longest-running marine engine had 74,000 hours service. To meet the needs of fast, weight-sensitive vessels, Caterpillar offers a high-performance distillate version. The high-performance design incorporates numerous changes, (including reduced mass, special alloy components and lube and air systems optimised for use with distillate fuels), while not affecting the fundamental reliability of the basic engine.

September 1992 saw the first installation of 3600s in fast-ferry operation, with four 3616s being installed in the *Patricia Olivia,* a 74 m wave piercer, built by International Catamarans Tasmania Pty Ltd and operated by Buquebus Lines between Buenos Aires and Montevideo. Rated at 4,020 kW at 765 rpm, these engines propel the 780 tonne deadweight vessel up to a light ship top speed of 42 kt and a loaded cruising speed of 38 kt.

Since that time 42 Caterpillar 3616 engines have been specified for other fast ferries. The new 3616 was introduced in 1997 at 7,200 kW and, by the end of 1998, 24 of these engines had been delivered.

The 3,600 engine family includes in-line 6- and 8-cylinders and vee 12, 16 and 18 cylinders. All versions are turbocharged and aftercooled. Displacement is 18.5 litres/cylinder, with 280 × 300 mm bore and stroke. At rated speeds of 720 to 1,050 rpm, piston speeds average 7.2 to 10.5 m/s. Peak cylinder pressure is 162 to 185 bar and BMEP ranges from 22.0 to 24.7 bar at maximum continuous ratings. Specific fuel consumption with all pumps is 188 to 199 g/kW h, based on ISO 3046/1 with +5 per cent tolerance for fuel having an LHV of 42, 780 kJ/kg and weighing 838.9 g/litre.

Caterpillar 3616 marine diesel rated at 6,000 kW at 1,000 rpm 0044037

The advantage of the Caterpillar 3600 series is high power-to-weight ratio combined with long TBO and low maintenance costs.

3600 Fast Vessel Ratings

Specific Fast Vessel Ratings have been established for 3612, 3616 and 3618 engines, corresponding to the typical operating conditions of this type of vessel. The rating is intended for the typical fast-ferry load profile of 85 per cent of operating hours at 100 per cent load, 15 per cent of hours at manoeuvring load (<50 per cent). Under this rating condition, a TBO of 20,000 to 25,000 hours is expected, coinciding with typical vessel survey period. Ratings are valid for marine society unrestricted conditions: 45°C ambient, 32°C seawater, and operation at sealevel.

Weights include dry engine with all pumps, filters, flywheel, damper, air starter, manifold shields and resilient mounts.

Caterpillar 3600 engines include integral lube oil and cooling pumps, providing ease of installation and a clean engine room.

Four of the 3618 engines were recently specified for the Austal-built *Highspeed 5* catamaran. Driving four Kamewa 112 SII waterjets, the engines, each rated at 7200 kW propelled the 85 m craft to a maximum speed of 41 knots carrying 400 t of deadweight.

Fast Vessel Rating (kW)

Model	rpm	Rating (kW)	Weight kg
3618	1,050	7,200	35,000
3616	1,020	6,000	31,000
3616	1,000	5,650	29,900
3612	1,000	4,250	24,700
3516B	1,785	1,790	7,800
	1,835	1,939	7,800
	1,880	2,088	7,800
	1,925	2,237	7,800
3512B	1,785	1,342	6,550
	1,835	1,454	6,550
	1,885	1,566	6,550
	1,925	1,678	6,550
3508B	1,785	895	4,625
	1,835	970	4,625
	1,880	1,044	4,625
	1,925	1,119	4,625
3412C	2,100	690	2,313
	2,300	858	2,313

Higher ratings available for military sprint applications

Cummins Inc

4,500 Leeds Avenue, Suite 301, Charleston, South Carolina 29405, United States

Tel: (+1 843) 745 16 20
Fax: (+1 843) 745 15 49
e-mail: wavemaster@cummins.com
Web: www.marine.cummins.com

Rachel Bridges, *Communications Manager – Marine*

Formed in 1919 in Columbus, Indiana, Cummins Inc produces a wide range of marine diesel engines which are now manufactured and distributed internationally. In addition to manufacturing plants in the US, the company also produces diesel engines in Brazil, China, India, Japan, Mexico and the UK. All these plants build engines to the same specifications, thus ensuring interchangeability of parts and the same quality standards. These standards meet design approvals for worldwide agency certification.

Cummins QSK60-M diesel engine 0594231

CUMMINS COMMERCIAL PRODUCT LINE

Model	Cylinders	Intermittent kW at rpm	Medium Duty kW at rpm	Heavy duty kW at rpm	Continuous kW at rpm	Weight (kg)	Length (mm)	Width (mm)	Height (mm)
KTA19-M3	6	–	–	477 at 1,800	3,737 at 1,800	2,073	1,539	1,003	1,905
KTA19-M4	6	–	522 at 2,100	–	–	2,073	1,539	1,003	1,928
QSK19-M	6	–	597 at 2,100	567 at 2,100	497 at 1,800	2,373	1,692	1,011	1,649
QSK19-M Tier	6	–	–	522 at 2,100	373 at 1,800	2,463	1,792	1,168	–
KTA38-M0	12	–	–	–	597 at 1,600	4,218	2,388	1,462	1,661
		–	–	–	634 at 1,800	–	–	–	–
K38-M Tier 2	–	–	–	–	634 at 1,800	4,218	2,388	1,462	2,083
KTA38-M1	12	–	–	821 at 1,800	671 at 1,600	4,218	2,388	1,462	2,083
		–	–	–	745 at 1,800	–	–	–	–
KTA38-M2	12	1,119 at 2,050	1,044 at 1,950	969 at 1,800	783 at 1,600	4,218	2,388	1,462	2,083
		–	–	1,007 at 1,900	895 at 1,800	–	–	–	–
				1,007 at 1,950					
QSK38-M Tier 2	12	–	–	1,007 at 1,900	895 at 1,800	N/A	2,637	1,642	1,999
KTA50-M2	16	–	1,398 at 1,950	1,193 at 1,900	1,044 at 1,600	5,431	2,603	1,564	1,661
		–	–	1,268 at 1,800	1,193 at 1,800	–	–	–	–
				1,342 at 1,900					
QSK50-M Tier 2	16	–	–	1,342 at 1,900	1,193 at 1,800	N/A	3,206	1,642	2,105
QSK60-M	16	–	1,864 at 1,900	1,715 at 1,900	1,492 at 1,600	8,754	2,800	1,760	2,096
					1,492 at 1,800				
					1,641 at 1,800				

General Electric Company

GE Transportation – Marine Engines
1 Neumann Way, Mail Drop S156, Cincinnati (Evendale), Ohio 45215-6301, United States

Tel: (+1 814) 875 57 10
Fax: (+1 866) 420 18 05
Web: www.geae.com

Scott Donnely, *President and Chief Executive Officer, GE Aviation*
John Lynch, *Marketing, Diesels*
David Nolson, *Marketing, Marine Gas Turbines*

General Electric Company's Dr Sanford A Moss operated the first gas turbine in the US in 1903 and produced the aircraft turbo-supercharger, first flown in 1919 and mass produced in the Second World War for US fighters and bombers.

The company built its first aircraft gas turbine in 1942, when it began development of a Whittle-type turbojet, under an arrangement between the United Kingdom and United States governments.

Today, General Electric produces turbofan, turboprop and turboshaft engines for civil and military aircraft and helicopters, as well as aero-derivative gas turbines for marine and industrial applications.

General Electric offers six gas turbines for marine service: the LM500, LM1600, LM2500, LM2500+, LM2500+G4 and LM6000.

LM 6000

The LM 6000 aero-derivative gas turbine is one of the world's most efficient simple-cycle gas turbines and is available with a standard combustor (designated PD) or with a dry, low-emission combustor (designated DLE). Both offer over 40 mW of power with a 42 per cent ISO thermal efficiency. With a weight of 21 tonnes and an overall length of under 12 m, this marine gas turbine is being evaluated for a number of large fast freight and military vessel concepts.

LM 2500

The LM 2500 marine gas turbine is a two-shaft, simple-cycle, high-efficiency engine, derived from the General Electric military TF39 and the commercial CF6 high-bypass turbofan engines for the US Air Force C-5 Galaxy transport, and DC-10, 747 and A300 commercial jets. The compressor, combustor and turbine are designed to give maximum progression in reliability, parts life and time between overhauls. The engine has a simple-cycle efficiency of more than 37 per cent, which is due to advanced cycle pressures, temperatures and component efficiencies.

Fast ferry installations include two GE LM 2500s for the Rodriquez 101 m high-speed monohull passenger/car ferries designated Aquastrada, two GE LM 2500s and two GE LM 1600s for the Stena High-Speed Service (HSS) ferries and two further GE LM 2500s for the Bornsholms Trafikken Austal 86 m fast ferry. The MDV 3000 vessels built by Fincantieri are powered by two LM 2500 turbines in combination with MTU diesels, as is the SNCM Corsaire 13000. The Austal AutoExpress 86 m for Bornholms Trafikkan is also powered by two LM 2500 turbines.

Several other high-speed ferries are powered by GE aeroderivative gas turbines. Naval ship applications include the Perry Class FFG-7 frigates, Spruance' Class DD-963 destroyers, Kidd Class DDG-993 destroyers, Ticonderoga Class Aegis CG-47 cruisers, Burke Class DDG-51 Aegis destroyers, the US Navy Sealift ships and supply class AOE-6 auxiliary ships, the F124 German frigates and the Mursame lass destroyers of the Japanese Maritime Self Defence Force.

In combination with diesels the LM 2500 powers several classes of high-speed patrol boats with top speeds of more than 40 kt. LM 2500 gas turbines and diesel combinations also provide propulsive power for a broad cross-section of more than 45 ship programmes for 24 navies throughout the world, including aircraft carriers, cruisers, destroyers, frigates and corvettes.

It is understood that over 1,000 LM 2500 gas turbines have been installed in the world fleet, accounting for more than 380 ship installations.

Type: Two-shaft, axial flow, simple-cycle.
Air intake: Axial, inlet bell mouth or duct can be customised to installation.
Combustion chamber: Annular.
Fuel grade: Kerosene, JP4, JP5, diesel, distillate fuels and natural gas.
Turbine: Two-stage gas generator, six-stage power turbine.
Jet pipe: Vertical or customised to fit installation.
Oil specification: Synthetic turbine oil (MIL-L-23699) or equal.
Mounting: At power turbine and compressor front frame.
Starting: Pneumatic, hydraulic.

LM 2500+ 0007641

LM 6000 0007642

Dimensions
Length: 6,630 mm
Width: 2,100 mm
Performance rating: 25,060 kW at 15°C at sea level
Specific fuel consumption: 0.226 g/kW h

LM 2500+

The LM 2500+ is an upsized LM 2500, designed to offer the reliability of the LM 2500 as well as superior performance in the 30 MW range. Applications include the Vantage and Millenium Class cruise ships, the Corsaire 13,000 fast ferry and the US Navy's LH08 amphibious landing craft.

LM 2500+G4

GE's LM 2500+G4 aeroderivative gas turbine is the 4th generation of the LM 2500 marine gas turbine. Its main feature is increased power (17 per cent) compared to the third generation LM 2500+. The LM 2500+G4 system's modular design with its split compressor casing, in-place blade and vane replacement, in-place hot-section maintenance and external fuel nozzles provide for easy repair and refurbishment. The LM 2500+G4 was selected to power the French and Italian FREMM frigates.

LM 1600

The LM 1600 is derived from GE's F404 engine, which powers the F/A-18 Hornet and F-117 Stealth Fighter. A simple-cycle, three-shaft gas-turbine engine, the LM 1600 has been selected as the power-plant for a number of large, high-speed vessels. The *Destriero*, which set the speed record for a transatlantic crossing without refuelling, is powered by three LM 1600 engines developing up to a total 44,740 kW. Another vessel, designed by Martin Francis and built by the Blohm + Voss shipyard, named *Eco*, is powered by one LM 1600 engine in a CODOG configuration. Additionally, the two Danyard/NQEA Seajet 250 high-speed ferries are powered by two LM 1600 gas turbines each.

The LM 1600 is packaged by MTU (Motoren- und Turbinen-Union) of Germany and Kvaerner Energy of Norway, ACB of France, S&S Energy Products of the US and IHI (Ishikawajima-Harima Heavy Industries) of Japan.

Type: Three-shaft, axial flow, simple-cycle.
Air intake: Axial, inlet bellmouth or duct can be customised to installation.
Combustion chamber: Annular.
Fuel grade: Kerosene, JP4, JP5, diesel, distillate fuels and natural gas.
Turbine: Two-stage gas generator, two-stage power turbine.
Jet pipe: Vertical or customised to fit installation.
Oil specification: Synthetic turbine oil (MIL-L-23699) or equivalent.
Mounting: At turbine frame and front frame.
Starting: Pneumatic, hydraulic.

Dimensions
Length: 4,890 mm
Width: 2,032 mm
Performance rating: 14,913 kW (20,000 shp) at 15°C at sea level
Specific fuel consumption: 226 g/kW h

LM 500

The LM 500, derived from GE's TF34 high-bypass aircraft engine, is a simple-cycle, two-shaft gas-turbine engine that powers the Danish Navy's Flyvefisken (Stanflex 300) patrol boat, a Japanese PG Class hydrofoil patrol craft, and the Korean PKX patrol boats. Throughout its operating range, the LM 500 is characterised by outstanding efficiency, which is attributable to the high-pressure ratio of the compressor, high turbine inlet temperature, improved component efficiency and conservation of cooling airflow.

Type: Two-shaft, axial flow, simple-cycle.
Air intake: Axial; vertical inlet duct can be customised to installation.
Combustion chamber: Annular.

Fuel grade: Kerosene, JP4, JP5, diesel, distillate fuels, and natural gas.
Turbine: Two-stage gas generator, four-stage power turbine.
Jet pipe: Vertical or axial or customised to fit installation.
Oil specification: Synthetic turbine oil (MIL-L-23699) or equivalent.
Mounting: At turbine frame and front frame.

Dimensions
Length: 2,187 mm
Width: 864 mm
Performance rating: 4,474 kW at 15°C at sea level.
Specific fuel consumption: 270 g/kW h.

LM 1600 0007643

GE 7FDM
A medium-speed, four-stroke marine propulsion engine, produced in V8, V12 and V16 cylinder configurations. In 2003, EFI versions of the V12 and V16 models were introduced with reduced fuel consumption and exhaust emissions. During 2004, the V8 and V8 EFI models were introduced.

Cylinder block
The cylinder block and integral crank case carries two banks of cylinders at a V angle of 45°. The crankcase has large inspection doors offering access to the crankshaft, bearings and connecting rods.
Cylinders: Welded construction.
Pistons: Two-piece articulated assembly with hardened piston crown.
Fuel systems: Mechanical fuel injection system or EFI using GE's Powerstar controller.
Turbocharging: Exhaust gas turbocharger with twin water-cooled intercoolers.
Lubrication: Gear driven lubrication oil pump.

Dimensions
Bore: 229 mm
Stroke: 267 mm
Cylinder volume: 11 litres
Cycles: 4
Injection: Direct
Cooling: Fresh water

Pratt & Whitney

A subsidiary of United Technologies
Corporate Headquarters, 400 Main Street, East Hartford, Connecticut 06108, United States

Tel: (+1 860) 565 4321
e-mail: info@pratt-whitney.com
Web: www.pw.utc.com

Peter Christian, *Vice President and General Manager, Power Systems*

In 2004 it was announced that Pratt & Whitney Marine Systems of Connecticut had been contracted by Norwegian yard Umoe Mandal to supply propulsion systems for the six Skjold class surface effect ship fast patrol vessels due for delivery by 2007.

The gas turbine engines will be marinised versions of Pratt & Whitney Canada's PW100 and PW150A aviation power plants. Each vessel will be fitted with two of the gas turbines which, in their marine form, are free turbine turboshaft engines designated the ST18M and ST40M respectively.

Model	Power	rpm	SFC (kg/kW-hr)	Weight (kg)	Length (mm)	Width (mm)	Height (mm)
MFT8	26210	3600/5000	0.223	7258	7620	2590	3048
ST40M	3781	14,875	0.262	525	1700	660	965
ST18M	1803	18,900	0.291	350	1532	657	807
ST6M	646	30,000	0.352	132	1335	483	559

Turbo Power and Marine Systems Inc

A subsidiary of United Technologies International Inc
400 Aircraft Road, PO Box 611, Middletown, Connecticut 06457, US

Tel: (+1 860) 343 20 00
Fax: (+1 860) 343 22 66
Web: www.cirmcorp.com/cirmturbine/pwft2805.htm

M A Cvercko, *Director, Marketing and Sales*

Turbo Power and Marine Systems Inc (TPM) designs, builds and markets industrial and marine gas-turbine power plants and related systems throughout the world. It also provides systems support for its installations. The current gas-turbine product line includes the 25 MW class FT8, and a 50 MW class FT8 Twin Pac (electric power generation only). Support of the existing fleet of FT4 gas turbines is also continued.

The FT8 is also marketed by Turbo Power's licensee/engine packaging partners GHH Borsig, Ebara and Ormat Industries. The unit is marketed in the People's Republic of China by Turbo Power's licensee China National Aero-Technology Import and Export Corporation (CATIC).

MFT8 gas turbine 0506936

A modified version of the FT8 called the MFT8, which consists of the GG8 gas generator and a light-weight power turbine designed and manufactured by Mitsubishi Heavy Industries, is marketed for marine and other applications by MHI and TPM.

MARINE GAS TURBINES
TPM's FT4 marine gas turbines were first used for boost power in military vessels, including two Royal Danish Navy frigates, 12 US Coast Guard Hamilton class high-endurance cutters and four Canadian Forces DDH-280 'Iroquois' class destroyers. Another boost power application of the FT4 is in the Fast Escort and ASW vessel *Bras d'Or*, also built for the Canadian Armed Forces. Another application is in two 12,000 tonne Arctic icebreakers for the US Coast Guard. With three FT4 marine gas turbines, these vessels are capable of maintaining a continuous speed of 3 kt through ice 1.8 m thick, and are able to ram through ice 6.4 m thick.

TPM's marine gas turbines were used for both the main and boost propulsion in the four Canadian DDH-280 destroyers. These are the first military combatant vessels to be designed for complete reliance on gas-turbine power.

Twin FT4 gas turbines were also used in the *Finnjet,* a high-speed Finnlines passenger liner which cut the Baltic crossing time in half, routinely maintaining 30 kt, with an engine availability of over 99 per cent. The record time for an engine replacement is 1 hour 20 minutes (2 hours being the normal time).

FT8 MARINE GAS TURBINE

TPM introduced the FT8 marine gas turbine based on the Pratt & Whitney JT8D aircraft engine, which incorporates advanced technology from Pratt & Whitney PW4000 and PW2000 series commercial aircraft gas-turbine engines. There are over 40 FT8 units already in commercial operation in nine countries. The FT8's continuous rating at 25,204 kW or maximum rating of 27,500 kW is relevant for many marine applications.

Three MFT8 gas-turbine engines have been supplied to Mitsubishi Heavy Industries for the Techno-Superliner demonstration programme in Japan. This vessel achieved 55 kt on sea trials in 1994. The full-scale vessel is to be propelled by four MFT8s (plus one spare) at speeds of over 50 kt.

The FT8 aero-derivative marine gas turbine is of a modular design with major components designed to be easily removed for servicing. With a spare gas generator to replace the one removed, the ship's power plant can be changed in a matter of hours. Borescope inspection features allow internal inspection without disassembly.

FT8 marine gas turbine 0506782

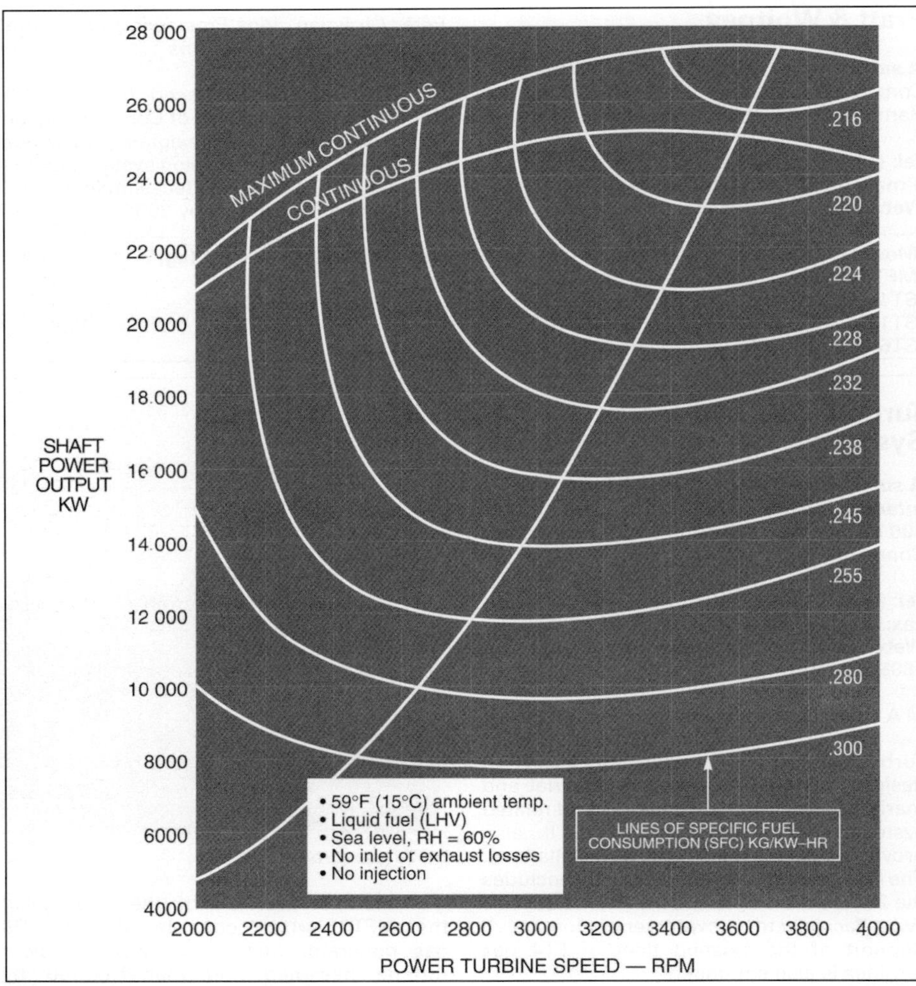

FT8 marine gas-turbine performance 0506783

GAS GENERATOR

Type: Simple-cycle two-spool gas generator. A low-pressure compressor is driven by a two-stage turbine and a high-pressure compressor is driven by a single-stage turbine. The burner section has nine burner cans each equipped with one fuel nozzle.
Air intake: Steel casing with radial struts supporting the front compressor bearing.
Low-pressure compressor: Eight-stage axial flow on inner of two concentric shafts driven by two-stage turbine and supported on ball and roller bearings. Variable geometry inlet guide vanes and the first two stages of vanes improve part-power efficiencies and reduce the start-up power requirements.
High-pressure compressor: Seven-stage axial flow on outer hollow shaft driven by single-stage turbne and running on ball and roller bearings.
Combustion chamber: Nine burner cans located in an annular arrangement and enclosed in a one-piece steel casing. Each burner has one fuel nozzle.
Turbines: Steel casing with hollow guide vanes. Turbine wheels are bolted to the compressor shafts and are supported on ball and roller bearings. A single-stage turbine drives the high-pressure compressor and a two-stage turbine drives the low-pressure compressor.

POWER TURBINE

Power turbine (MFT8): The MFT8 power turbine was developed by Mitsubishi Takasago Works as a lightweight substitute for those FT8 power turbines directed to land-based applications. The weight of the three-stage 5,000 rpm MFT8 power turbine is 2,600 kg. The MFT8 employs a cantilevered rotor design as opposed to the straddle arrangement used on the FT8.

Control system

The control system is a digital system based on the Woodward Governor Company's Netcon 5000 controller. Software can be readily integrated with ship systems.
Accessory drive: Starter, fluid power pump, tachometer drives for the low-pressure compressor shaft and the high-pressure compressor shaft.
Lubrication: Scavenge pump return system and supply system pumps.
Oil specifications: Synthetic lubricating oil, PWA-521.
Starting: Pneumatic, hydraulic or electric.

Dimensions

Length: 8,800 mm
Width: 2,650 mm
Height: 2,600 mm

Fuel specifications

Marine diesel, Aviation-grade kerosene, light distillate (naptha) TPM-FR-1.

Treated crude and residual oil, refer to manufacturer.

MFT8 DESIGN PARAMETERS

59°F, Sea Level, No Losses, 43 MJ/kg
Shaft Output (kW): 25,200
Heat rate (kg/kW h): 0.218
Airflow: 83.9 kg/s
Pressure ratio: 19.8:1
Combustor exit temperature: 1,190°C
Power turbine exhaust temperature: 459°C.

Vericor Power Systems

3625 Brookside Parkway, Suite 500, Alpharetta, Georgia, 30022 United States

Tel: (+1 770) 569 88 00
Fax: (+1 770) 569 03 99
e-mail: info@vericor.com
Web: www.vericor.com

Description

Vericor Power Systems is one of the leading producers of marine and industrial gas turbines in the 0.5 to 15 MW class. Vericor was launched in 1999 by AlliedSignal which had acquired the Textron Lycoming engines business in 1995 and with it, the TF40 marine turbine engine. Vericor became 100 per cent owned by MTU Aero Engines in 2002.

There have been over 600 TF marine engines delivered for a variety of commercial and military programmes, the largest is the US Navy LCAC programme with 400 engines. The TF50A engine delivers 3,804 kW MCP, began deliveries in 1999 and offers improved fuel efficiency as well as 20 per cent more power in the same frame size.

Recently the 43 m Derecktor fast ferry designed by Nigel Gee and built for Buquebus, as well as several high-speed mega-yachts, notably *Detroit Eagle* and the *Wallypower 118*, have selected TF40 and TF50A marine propulsion. The Royal Swedish Navy selected Twin TF50A propulsion packages for the composite 'Visby' class corvettes which are now being delivered. The Finnish Navy used TF40 engines in their T2000 hovercraft prototype. Vericor is also delivering ETF40B engines to the US Navy as part of the overall mid-life upgrade of the LCAC air cushion vehicle.

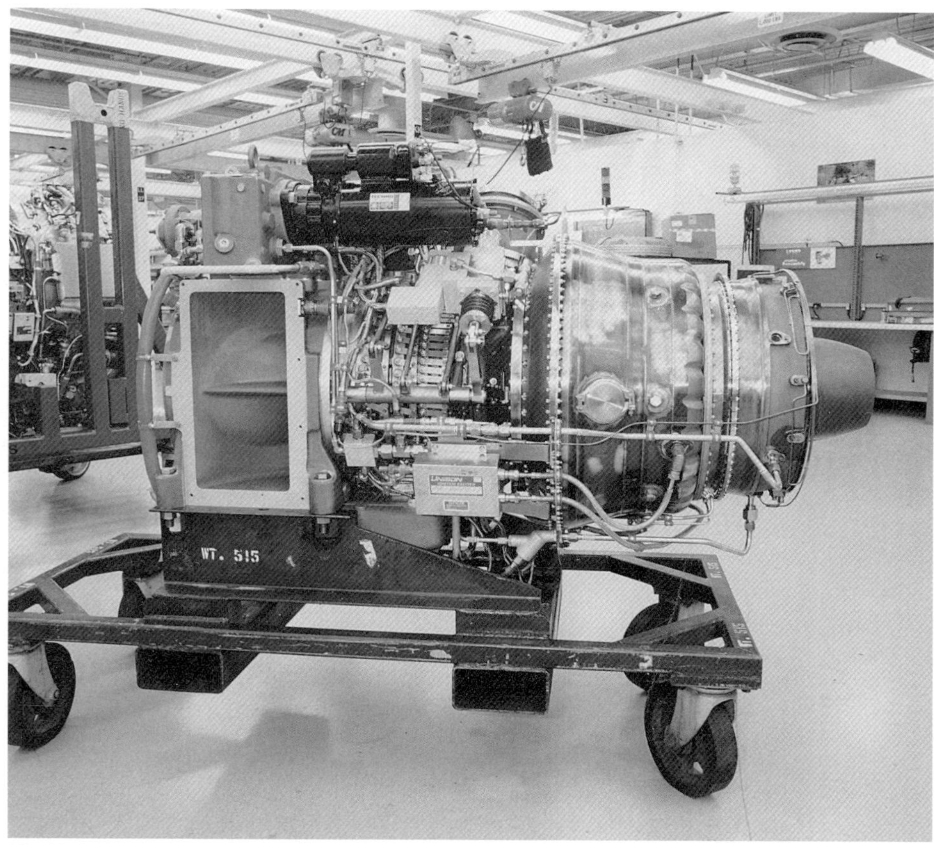

The Vericor TF50 marine gas-turbine engine 0109538

ETF40B

The ETF40B was specifically designed for the LCAC Upgrade programme and incorporates a higher pressure ratio compressor with variable inlet geometry and a digital control. The turbine section of the TF40 was retained. The engine was designed to fit into the same installation space of the TF40 on the LCAC and a compressor bleed function was added.

Air intake: Same as TF40.
Compressor: Longer diameter than TF40 with first three stages at higher pressure ratio and using variable geometry stators and stainless steel compressor housing.
Control system: Full authority digital electronic control.
Turbine: As TF40.
Accessories: As TF40.
Lubrication: Recirculating type with integral oil tank and cooler.
Performance ratings (uninstalled)
Max continuous(at 15°C, sea level): 3,700 kW
Max continuous (at 15°C, sea level): 4,073 kW

TF40

The TF40 engine is an aero-derivative gas turbine sharing power section experience with the T55 turboshaft and LF502 turbofan engines. The configuration is direct drive by a two-stage free power turbine with a combination seven-stage axial/centrifugal compressor and two-stage axial flow turbine gas generator. It also incorporates a reverse-flow annular combustor chamber assembly, oil cooler assembly, top-mounted accessory gearbox module, and an integral lubrication system. The engine features a design allowing removal and installation of engine modules without having visit a repair facility. The individual modules include gas producer, turbine accessory gearbox, inlet housing and sump. Each module is structurally interdependent and designed to interface with the other modules to allow interchangeability. The complete engine may be cantilever-mounted from a main reduction gearbox. Two TF engines are sometimes mounted to a single combining reduction gearbox.

The compact size of the TF engines makes them particularly suited to the Twin concept, where two gas turbines drive a single shaft line through a combining gearbox. A high percentage of the projects either under construction or under consideration use the twin concept.

Air intake: Intakes on both sides of the air inlet housing of aluminium alloy. This also serves to house internal gearing and support the power producer section and output drive shaft. Integral water-wash nozzles are provided.
Compressor: The compressor section consists of seven stainless steel axial stages, and a single titanium centrifugal stage. The compressor housing is two-piece aluminium alloy bolted to a steel alloy diffuser, to which the combustion chamber housing is attached.
Combustion chamber: The combuster is an annular reverse-flow type with a steel outer shell and inner liner. There are 28 fuel nozzles with downstream injection.
Control system: Electronic fuel-control system with gas producer and power shaft governors. Gear-type fuel pump, flow control and shut-off valve.
Turbine: The turbine section consists of a two-stage axial-flow gas generator with two-stage axial-flow power turbine. The gas generator features cooled first-stage segmented nozzle and cooled blades.
Accessories: Electric, air or hydraulic starter. High-energy ignition unit; two igniter plugs.
Lubrication: Recirculating type with integral oil tank and cooler.

Dimensions
Length: 1.321 m
Width: 0.889 m
Height: 1.046 m

Performance ratings
Boost Power (at 15°C, sea level): 3,430 kW
Max continuous (at 15°C, sea level): 2,983 kW
Fuel consumption
At max continuous rating: 299 g/kW-hr
Fuel grade: MIL-T-5624, JP-4, JP-5; MIL-F-1688 marine diesel and kerosene.

TF50A

The TF50 series builds on the ETF40B engine, adding a redesigned power turbine section for increased efficiency. The first operational marine TF50 was fitted to a high-speed superyacht *Detroit Eagle* in 2000. These engines have been selected for the new Swedish 'Visby' Class patrol boats currently under construction. The latest application to attract worldwide attention is the *Wallypower 118* ultra innovative high-speed luxury yacht which is driven by three FT50 gas turbines. The general description of the TF40 applies to the TF50 series with the following exceptions:

Compressor: First three stages of larger diameter with variable geometry stators. Larger diameter, leaned-back centrifugal compressor. Stainless steel compressor housing.
Control system: Full Authority Digital Electronic Control (FADEC).
Turbine: Redesigned low-pressure turbine including higher efficiency fourth stage.

Specifications
Performance ratings
Max intermittent (at 15°C, sea level): 4,170 kW
Max continuous (at 15°C, sea level): 3,804 kW
Fuel consumption
At max continuous rating: 276 g/kW-hr.

Dimensions
Length: 1.397 m
Width: 0.889 m
Height: 1.046 m

TRANSMISSIONS

Company listing by country

Australia
DBD Marine Pty Ltd

Austria
Geislinger & Co

France
France Helices

Germany
BHS-Cincinnati GmbH
Centa
Reintjes GmbH
Renk AG
Vulkan
ZF Friedrichshafen Ag

Italy
FB Design Srl
Flexitab Srl
Jolly Drive Srl
LA.ME Srl

Japan
Hitachi Nico Transmission Co Ltd

New Zealand
Q-SPD International Ltd
Seafury International Ltd

Norway
Servogear A/S

Sweden
Applied Composites AB

United Kingdom
Air Vehicles-Design and Engineering Ltd
Allen Power Engineering Ltd
Yellowfin Ltd

United States
Philadelphia Gear Corporation
Powervent
Pulse-Drive Systems International LLP
T-Torque Drive System Inc
Twin Disc Inc

Australia

DBD Marine Pty Ltd

PO Box 994, Edgecliff, New South Wales 2027, Australia

Tel: (+61 4) 159 634 64
Fax: (+61 3) 978 963 49
e-mail: sales@dbdmarine.com
Web: www.dbdmarine.com

Darren Finch, *Owner*

DBD Marine produces a range of three stern drive systems. Made from stainless steel and having just six internal moving parts, the products require little maintenance.

The GP300 is designed for petrol engines up to 265 kW or diesel engines up to 178 kW. Maximum torque is 700 nm and maximum speed is 5,000 rpm.

The CD300 Outdrive is designed for situations where underwater clearance and strong rugged design are critical. It is for use with diesel engines up to 260 kW and petrol engines up to 400 kW.

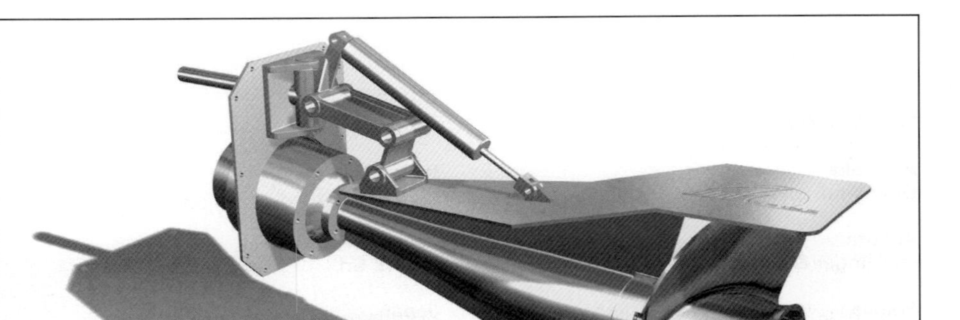

DBD Marine CD500 sterndrive 0580684

Austria

Geislinger & Co

Hallwanger Landesstrasse 3, 5300 Hallwang, Salzburg, Austria

Tel: (+43 662) 66 99 90
Fax: (+43 662) 669 99 40
e-mail: info@geislinger.com
Web: www.geislinger.com

Geislinger specialises in minimising torsional vibrations in transmission systems. Its product range includes torsional vibration dampers, elastic damping couplings, flexible link couplings and composite shafting. Geislco couplings are made of advanced composite material with low mass and reaction forces. With a damping factor of between 0.2 and 0.7, the torque range of the Geislinger couplings covers 1 to 6,520 kNm.

Geislinger has supplied advanced composite shafting to Austal Ships for main propulsion power transmission in the Auto Express 82 vessels and the Auto Express 86 *Spirit of Ontario.*

These shafts reduce the weight over conventional shafting and also reduce the number of support bearings required.

Geislinger torsional coupling
0081036

France

France Helices

ZI La Frayere, F-06150 Cannes La Bocca, France

Tel: (+33 4) 93 47 69 38
Fax: (+33 4) 93 47 08 59
e-mail: info@francehelices.fr
Web: www.francehelices.fr

France Helices produce a range of propellers and transmission equipment for high performance craft as well as for fishing vessels and barges.

Surface Drive System (SDS)
This is a transom mounted surface drive system and is produced in five versions for vessels with speeds above 25 knots and with a power to weight ratio of at least 40 kW/tonne.

Model	Max torque kg/m	Total weight (inc propeller) (kg)
SDS2	182	228
SDS3	275	345
SDS4	575	565
SDS5	1,190	775
SDS6	2,250	1,438

Germany

BHS-Cincinnati GmbH

Hans-Boeckler-Strasse 7, D-87527 Sonthofen, Germany

Tel: (+49 8321) 80 20
Fax: (+49 8321) 80 26 89
e-mail: info@bhs-cincinnati.de
Web: www.bhs-cincinnati.de

Dieter Groher, *Managing Director, Commercial*
Lother Hennauer, *Managing Director, Technical*
Alfred Siedl, *Vice President, Sales*
Jean-Claude Debout, *Group Manager Projects and Sales*

It is understood that BHS-Cincinnati GmbH continue to manufacture the range of gears formerly produced by The Cincinnati Gear Company in the US. Marine gears are produced for powers up to 50,000 kW.

A CINTI MA-107 reduction gear with close coupled TF40 gas turbine. This 3,700 kW reduction gear is one of the MA series gearboxes being produced for vessels such as high-speed ferries and megayachts
0043240

Centa

Centa Antriebe Kirschey GmbH
PO Box 1125, Bergische Strasse 7, D-42781 Haan, Germany

Tel: (+49 2) 129 91 20
Fax: (+49 2) 129 27 90
e-mail: info@centa.de
Web: www.centa.de

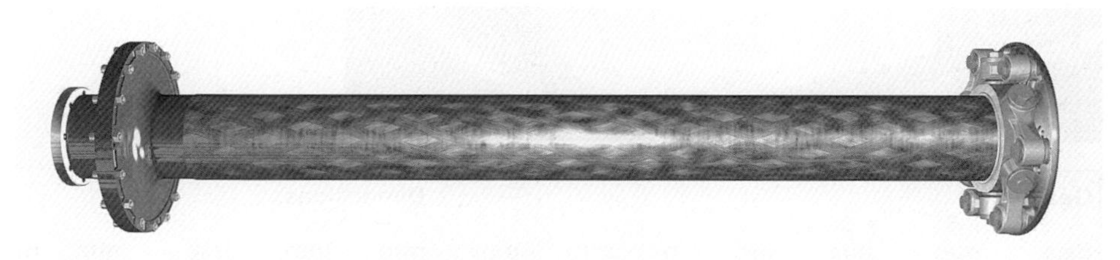

Centa's lightweight carbon fibre drive shaft
1048821

Centa was established in 1970 and is a designer of flexible couplings and shafts for advanced torsional vibration applications. Over 5 million Centa couplings are installed worldwide, serviced by subsidiary companies in 10 countries, licensees in two and national distributors in 30 countries.

Flexible couplings and shafts are manufactured for main propulsion and auxiliary drives including couplings for generator sets.

The main product lines are:
Centaflex-DS: dual-stage couplings for torques up to 12 kNm or about 2,000 kW power.
Centaflex-RS: progressive roller-type coupling for flange-mounted applications with torques up to 12 kNm or about 2,500 kW power.
Centaflex-A: high elastic series with flexibility for diesel engines, pumps and power take-offs for torques up to 12.5 kNm.
Centaflex-A-G: universal joint shafts dampen noise, torsional vibration and shock while compensating for misalignment. Torque up to 5 kNm.
Centax-SEC: this is a range of super-elastic couplings for torques up to 750 kNm.
Centax-N: economical coupling series for substantial misalignments and up to 18 kNm torque.
Centax-TT: for high torque range up to 500 kNm for ship propulsion.
Centax-FH: bearing housings for torques up to 20 kNm.
Centalink: torsionally stiff drive shaft for torques up to 360 kNm.
Centa-CP: integrated torsional coupling and e-clutch in an SAE flange housing. Torque up to 45 kNm.
Centaflex-D: flexible couplings for generator sets, centrifugal pumpsets and similar drives. The couplings are available in a range covering nominal torques of 0.25 kNm to 40 kNm.
Centamax: torsionally highly elastic couplings for diesel engines available in 15 sizes to cover a torque range from 0.1 kNm to 20 kNm.
Centa Carbon: carbon fibre driveshafts widely used on high-speed craft. Type P shafts have a torque rating up to 800 kNm and type S up to 200 kNm.
CentaDisc-C: couplings and driveshafts made from filter-reinforced plastics, available for torques up to 700,000 Nm.

Reintjes GmbH

Eugen-Reintjes Strasse 7, D-31785 Hameln, Germany

Tel: (+49 5151) 10 40
Fax: (+49 5151) 10 43 00
e-mail: info@reintjes-gears.de
Web: www.reintjes-gears.net

Christian Schliephack, *President*
Hartmut Hansmann, *Vice-President, Sales and Marketing*

Hartmut Hansmann

Established in 1929, Reintjes is one of the leading companies in marine gearbox manufacturing worldwide. Reintjes produces gearboxes for a wide range of vehicles in the 250 to 20,000 kW power range. The company, established in 1929, has wholly owned subsidiaries in Antwerp, Dubai, Madrid and Singapore, as well as sales and service centres all over the world. The company produces custom made gearboxes as well as serial production units.

Renk AG

Augsburg Plant
Gögginger Strasse 73, D-86159 Augsburg, Germany

Tel: (+49 821) 570 00
Fax: (+49 821) 570 04 60
e-mail: info.augsburg@renk.biz
Web: www.renk.de

Manfred Hirt, *CEO*
Ulrich Sauter, *Managing Director*
Dr Franz Hoppe, *Sales Director, Marine Gears*

Rheine Plant
Rodder Damm 170, D-48432 Rheine, Germany

Tel: (+49 5971) 79 00
Fax: (+49 5971) 79 02 08
e-mail: info.rheine@renk.biz

PLS and PWS gearboxes

Renk marine planetary gear units (PLS and PWS series) have been specifically developed for use in fast ships such as corvettes, speedboats, minesweepers, minelayers, OPVs and IPVs. Designed for a performance range between 800 and 10,000 kW, they cover a wide range of applications. The specific characteristics and advantages of disconnectable planetary gear units (PLS) and planetary reversing gear units (PWS) are:
Ratios 1.5 to 7.1;
PWS series efficiencies 97.5 to 98 per cent;
PLS series efficiencies 98.5 to 99 per cent;
Compact design, permitting favourable engine room concepts;
Low power-to-weight ratio of less than 0.5 kg/kW;
Coaxial input and output shafts for optimum power plant layouts;
High shock resistance due to planetary design;
Insensitivity of transmission elements to hull distortion, providing high-operational reliability even under extreme service conditions;
High efficiency due to epicyclic concept;
Starboard and port gear units offer identical connection interfaces and with PWS units this means completely identical starboard and port gear units, even for multipropeller ships;
Forward and reverse gears of PWS units are both designed for 100 per cent loads, engines can therefore operate with the same direction of rotation without affecting the gear unit;
On request, these gear units can be built under the supervision of any classification society.
Anti-magnetic versions are available;
High reliability;
Suitable for use with gas turbines;
If required, low-speed gears can be provided.
Renk gears have now been supplied for over 100 vessels (approximately 200 gear units). These have included a special ASL 2 × 80 gearbox model for Austal's 126 m Trimaran. This gearbox combines the drive from two 9,100 kW diesels to drive a single booster water-jet.

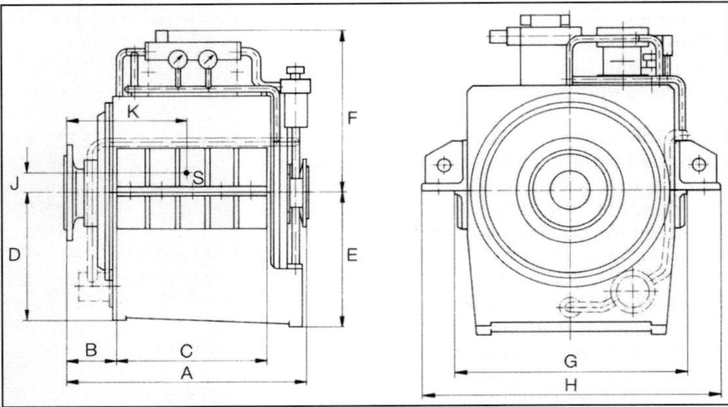

Renk PLS and PWS series gearboxes (see table for dimensions) 0506937

PLS gearbox
0007645

Gear unit			Dimensions										Weight, dry		Thrust	Oil content
	i>5	i<5											i>5	i<5		
Size	mm	mm	mm	mm	mm	mm	mm	mm	mm	mm	mm		kg	kg	kN	litres
PLS																
18	900	800	200	480	530	560	710	850	1,120	80	400		1,350	1,200	130	90
25	1,000	900	225	530	600	630	800	950	1,250	90	450		1,750	1,550	160	110
35.5	1,120	1,000	250	600	670	710	900	1,060	1,400	100	500		2,350	2,050	200	130
50	1,250	1,120	280	670	750	800	1,000	1,180	1,600	115	560		3,000	2,650	250	150
71	1,400	1,250	315	750	850	900	1,120	1,320	1,800	125	630		3,800	3,450	320	185
100	1,600	1,400	355	850	950	1,000	1,250	1,500	2,000	140	710		4,800	4,300	400	230
PWS																
18	1,060	950	200	600	530	560	710	850	1,120	80	480		1,600	1,450	130	120
25	1,180	1,060	225	670	600	630	800	950	1,250	90	530		2,150	1,950	160	145
35.5	1,320	1,180	250	750	670	710	900	1,060	1,400	100	600		3,000	2,800	200	170
50	1,500	1,320	280	850	750	800	1,000	1,180	1,600	112	670		4,200	3,900	250	200
71	1,700	1,500	315	950	850	900	1,120	1,320	1,800	125	750		5,750	5,300	320	235
100	1,900	1,700	355	1,060	950	1,000	1,250	1,500	2,000	140	850		8,100	7,300	400	280

Vulkan

Vulkan Kupplungs-und Getriebebau
Heerstrasse 66, D-44653 Herne, Germany

Tel: (+49 2325) 92 20
Fax: (+49 2325) 711 10
e-mail: info@vulkan-vkg.de
Web: www.vulkan24.com

Andreas Böhme, *Managing Director*

Vulkan is an international company with a worldwide reputation for flexible coupling.

Manufacturing takes place in Germany, Brazil, US, China and India and it includes such applications as: Ship's main and auxiliary drives, PTO's and PTI's, boat drives, generator drives and industrial drives. The performance spectrum offered ranges from 0.1 to 2,000 kNm.

The main Vulkan range of couplings/shafts comprises:

Vulastick-I

The Vulastik-L coupling (from 0.4 kNmn to 25 kNm) is a torsionally flexible rubber coupling that compensates axial, angular and, to a certain degree, radial displacement of the connected machines.

The RATO family comprises couplings from 2.5 kNm to 800 kNm. The most flexible coupling type in this family is the RATO-S. This torsionally, highly flexible rubber coupling compensates radial, axial and angular shaft displacements of the connected machinery.

Vulkardan

The Vulkardan-E coupling (from 0.16 kNm to 20 kNm) is a highly flexible rubber coupling with a linear stiffness characteristic.

Torflex

The Torflex coupling (from 0.25 kNm to 1.6 kNm) is a multistage coupling with progressive stiffness characteristics. The Torflex coupling is exclusively intended for installation in bell housings and can be used where the gearbox input shaft has an external spline or oil press fit.

MEGIFLEX-B

The highly flexible MEGIFLEX-B (from 10 Nm to 3,125 Nm) is torsionally flexible, offering multidirectional misalignment compensation capacity.

RATO-R

Designed for high speeds and/or high power density, it offers high levels of torsional flexibility and misalignment capability. It is available in the torque range of 2.5 to 270 kNm.

Composite shaft

Vulkan composite shafts are especially suited for drive systems requiring low mass and/or where there is a great distance between shaft bearings.

ZF Friedrichshafen AG

D-88038 Friedrichshafen, Germany

Tel: (+49 7541) 77 2207
Fax: (+49 7541) 77 4222
e-mail: info.fastcraft@zf.com
Web: www.zf-marine.com

R Heil, *Chief Executive Officer, ZF Marine Group*
P Herring, *Marketing Manager, ZF Marine Group*
W Bareth, *Head of Fast Craft Department, ZF Marine Group*

ZF produces marine transmission systems in its four plants in Friedrichshafen (Germany), Padova and Arco (Italy) and Sorocaba (Brazil) in a power range from 10 to 10,000 hp. Applications include sailing boats, ski-boats, luxury yachts, patrol craft, fast ferries, fishing boats, tugs and inland waterways vessels.

If required, transmissions can be supplied with classification approval and are offered with a wide variety of optional equipment including mechanical or electrically actuated trolling valves, ZF Autoroll automatic control system, PTOs, PTIs flexible couplings and comprehensive monitoring equipment.

Around half a million transmissions are currently in service in naval, commercial and pleasure craft throughout the world, and they are supported by a sales and service network with over 500 offices.

ZF is an international group of companies with a turnover of more than DM8 billion and 33,000 employees specialising in driveline, steering and chassis technology. ZF units are installed in all types of vehicles including cars, commercial vehicles, agricultural and construction machinery, marine craft and aircraft as well as precision machinery.

A recent addition to the ZF range is the SeaRex surface drive system. This has no exposed hydraulic hoses nor connectors and is available in four different sizes catering for torque ratings from 2,520 Nm up to 15,070 Nm.

The addresses of the three other ZF plants are:

ZF Padova SpA
Via Penghe 48, I-35030 Caselle di Selvazzano, Padova, Italy

P Herring, *Marine Sales Manager*

Tel: (+39 049) 829 93 11
Fax: (+39 049) 829 95 50

ZF Hurth Marine SpA
Via S.Andrea 16, I-38062 Arco, Trento, Italy

D Gianese, *Product Manager*

Tel: (+39 0464) 58 05 55
Fax: (+39 0464) 58 05 44

ZF do Brasil Lemforder Ltda
Avenida Conde Zeppelin, 1935, CEP 18103-000, Sorocaba – SP, Brazil

A S Italiano, *Manager, Off-Road Division*

Tel: (+55 15) 235 70 03
Fax: (+55 15) 235 70 70

ZF Marine Gears with parallel offset input and output shafts

Type	Ratio range	Max input power kW* Cont duty	Light duty	Max input speed rpm Cont Duty	Dry weight kg	SAE bell hsg
ZF 220	1.0–3.0	127	248	3,200	63	2,3
ZF 280-1	0.814–3.0	178	325	3,300	73	1,2,3
ZF 305	0.884–2.917	272	466	3,000	98	1,2,3
ZF 325-1	1.0–2.957	327	585	3,000	130	1
ZF 350	1.025–2.864	324	611	2,400	230	1
ZF 500	0.925–2.917	559	778	3,000	–	1,0
ZF 550	0.936–3.042	449	795	3,000	242	0,1
ZF 665	1.111–3.042	522	915	3,000	248	0,1
ZF 2000	1.086–2.958	507	995	2,600	342	0,1
ZF 2050	2.76–2.958	560	1,074	2,600	344	0,1
ZF 2200	3.0–3.519	507	995	2,600	404	1,0
ZF 2250	3.0–3.519	560	1,073	2,600	406	1,0
ZF 2350	3.028–4.760	560	1074	2,600	441	1,0
ZF 2500	1.306–2.75	716	1,161	2,500	459	0,1
ZF 3000	1.093–2.952	716	1,240	2,600	482	1,0
ZF 3050	1.093–2.952	839	1,375	2,600	482	1,0
ZF 3310	3.519–5.0	716	1,238	2,600	665	1,0
ZF 3350	3.519–5.00	839	1,375	2,600	665	1,0
ZF 4540	1.18–3.04	990	1,727	2,100	737	–
ZF 4600	1.18–3.040	1,137	2,106	2,100	741	–
ZF 4650	1.18–3.040	1,320	2,240	2,100	745	–
ZF 7500	1.33–3.231	1,489	2,265	2,300	2,410	–
ZF 23560 D	1.721–3.577	1,780	2,140	1,600	2,140	–

*The stated max power ratings and speeds do not apply to all gear ratios. Actual ratings depend on duty.
Note: Power capacity (kW) = (torque (Nm) × rated speed (rpm))/9,550.

ZF Marine Gears with coaxial input and output shafts

Type	Ratio range Cont/Med duty	Max input power kW* Light/pleasure duty		Max input speed rpm	Dry weight kg	SAE bell hsg
ZF 301 C	1.106–2.694	169	405	3,000	87	1,2,3
ZF 311	0.804–3.455	272	540	2,600	171	1,2
ZF 7500 C	3.641–4.615	1,042	–	2,000	1,135	–
7850 C	0.952–2.643	1,618	–	1,640	1,140	–
ZF 23560 C	1.722–5.122	1,780	–	1,600	2,600	–
ZF 23860 C	0.697–1.76	2,532	–	750	2,600	–

*The stated max power ratings and speeds do not apply to all gear ratios. Actual ratings depend on duty.

ZF Marine Gears with down-angled output shafts					
Type	**Ratio range Cont duty**	**Max input power kW* Light duty**	**Max input speed rpm**	**Dry weight kg**	**SAE bell hsg**
ZF 220 A	1.235–2.455	139	3,200	50	2,3
ZF 280-1 A	1.118–2.476	178	3,300	73	1,2,3
ZF 285 A	1.2–2.5	206	3,300	77	1,2,3
ZF 286 A	1.2–2.5	241	3,300	77	1,2,3
ZF 305 A	1.2–2.423	272	3,000	97	1,2,3
ZF 325-1 A	1.216–2.417	327	3,000	130	1
ZF 350 A	1.222–2.636	324	2,400	230	1
ZF 500-1 A	1.216–2.609	410	3,000	198	0,1
ZF 510-1 A	1.125–2.480	410	3,000	215	0,1
ZF 550 A	1.525–2.960	449	3,000	256	0,1
ZF 665 A	1.525–2.960	522	3,000	253	0,1
ZF 2000 A	1.267–3.250	507	2,600	344	1,0
ZF 2050 A	1.267–2.467	560	2,600	346	1,0
ZF 3000 A	1.351–3.280	716	2,600	485	1,0

*The stated max power ratings and speeds do not apply to all gear ratios. Actual ratings depend on duty.

Italy

FB Design Srl

Via Provinciale 73, I-23841 Annone Brianza (LC), Italy

Tel: (+39 341) 26 01 05
Fax: (+39 341) 26 01 08
e-mail: mail@fbdesign.it
Web: www.fbdesign.it

Ing Fabio Buzzi, *Managing Director*
Dr Alberto Nencha, *Sales Director*

FB Design is a design and construction company covering fast craft and their propulsion systems. The company employs a staff of over 40 within a 20,000 m² site.

Trimax surface drive

The company produces a range of surface piercing propeller drives which are marketed by ZF Friedrichshafen AG. The smallest of the range is the Trimax G-Drive with an integrated gearbox for powers up to approximately 300 kW. The largest of the range is the Trimax 5000I with a torque rating of 2,000 kgm (or a power range of 1,000 to 3,000 kW). The weight of this unit is 440 kg.

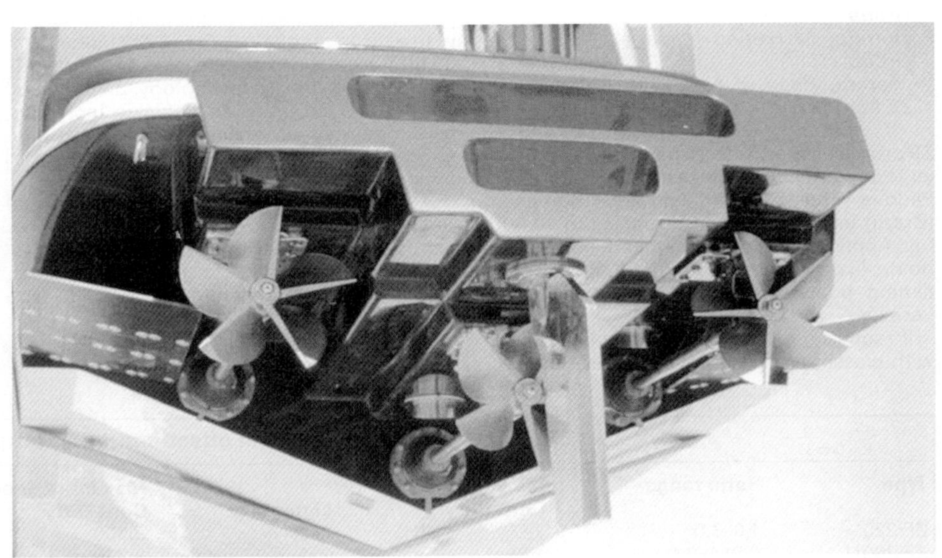

Triple Trimax installation 0109529

In the last 15 years, craft with Trimax drives have won over 25 world speed titles and set 51 world speed records.

Flexitab Srl

Via Melisurgo 15, 80133 Naples, Italy

Tel: +39 081 790 19 09
Fax: +39 081 420 69 03
e-mail: info@flexitab.com
Web: www.flexitab.com

Flexitab was created in 1989 by Italian naval architect, Brunello Acampara of Victory Design. Flexitab designs and produces the Flexitab® trim tab range and the Flexidrive® range of surface drives.

Flexidrive® Energy

The Flexidrive® Energy bolt-on surface drive range for 200 to 2,000 kW applications is a new concept for use on vessels with a surface speed of up to 30 kts. The drive features a fixed shaft and integrated rudder.

Flexidrive® Power

The Flexidrive® Power range is aimed at medium-speed vessels with a power range of 350 to 2,200 kW. The drive is trimmable and features a built-in rudder and a highly skewed surface piercing propeller that can be optimised for performance at various speeds.

Flexidrive® Speed

The Flexidrive® Speed range is designed for very high-speed craft with a power range of 350–1,200 kW. The drive is trimmable and uses a conventional surface piercing propeller.

Jolly Drive Srl

Via Nettuno 117/A, 04012 Cisterna di Latina, Italy

Tel: (+39 06) 96 91 10 30 **Fax:** (+39 06) 969 24 19
email: jdm@jollydrivemarine.com
Web: www.jollydrive.com

Model	Max Input Torque (Nm)	Weight (kg)	Length (mm)
JDM 19T	1600	102	1060
JDM 25T	2500	111	1170
JDM 25TL	2500	111	1370
JDM 31T	3500	130	1195
JDM 31TL	3500	130	1395
JDM 36T	5400	220	1555
JDM 33T	10,000	440	1810
JDM 34T	16,500	580	1980
JDM 32T	26,500	880	2200

The Jolly Drive uses a double universal joint arrangement in the shaft to reduce cost and complexity.

Fabricated from stainless steel, the transmissions can be installed on craft with an inclined transom between 6° and 15°.

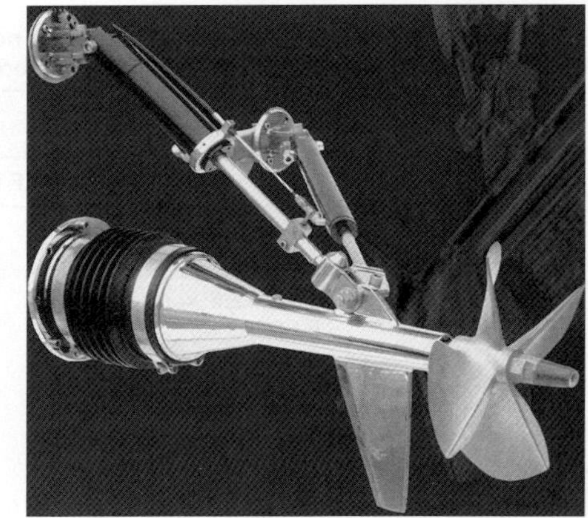

Stainless steel SSS Drive 0121765

LA. ME Srl

Via Zanica 17/0, I-214050 Grassobbio (BG), Italy

Tel: (+39 03) 53 84 46 04
Fax: (+39 03) 53 84 45 90
e-mail: info@lamemarine.com
Web: www.lamemarine.com

Giovanni Patrone Raggi, *Managing Director*

Sea Rider

Sea Rider is an evolution of the 'Step Drive' offering the advantages of the conventional 'Z' drive with the increased performance of surface propulsion; it is claimed to maintain all the advantages of the surface propeller and eliminates completely the negative aspects which prevented the acceptance of this system of propulsion for many years.

During its short life the Sea Rider has been successfully fitted on hundreds of craft in pleasure, workboat and military categories.

The engine may be fitted amidships or right aft without a V-drive and transmission shaft. The tunnel rudders over the top half of the propellers have been employed to overcome some of the disadvantages of conventional rudders or propeller power steering. The two vertical rudder blades, which are a continuation of the shroud, act as sidewalls and, operating below the hull, give positive control as required as well as protecting the propeller in shallow water.

To enhance the reverse thrust capabilities of a fixed-pitch propeller a new blade section profile has been devised giving, on the back of the blade, a concave area towards the trailing edge, producing improved section lift coefficient and hence thrust, when in reverse rotation.

The propeller incorporating this profile is called the Diamond Back surface propeller and overcomes the poor astern thrust associated with surface propellers.

The present range of Sea Rider units extends from 37 to 2,980 kW (indicative power limits).

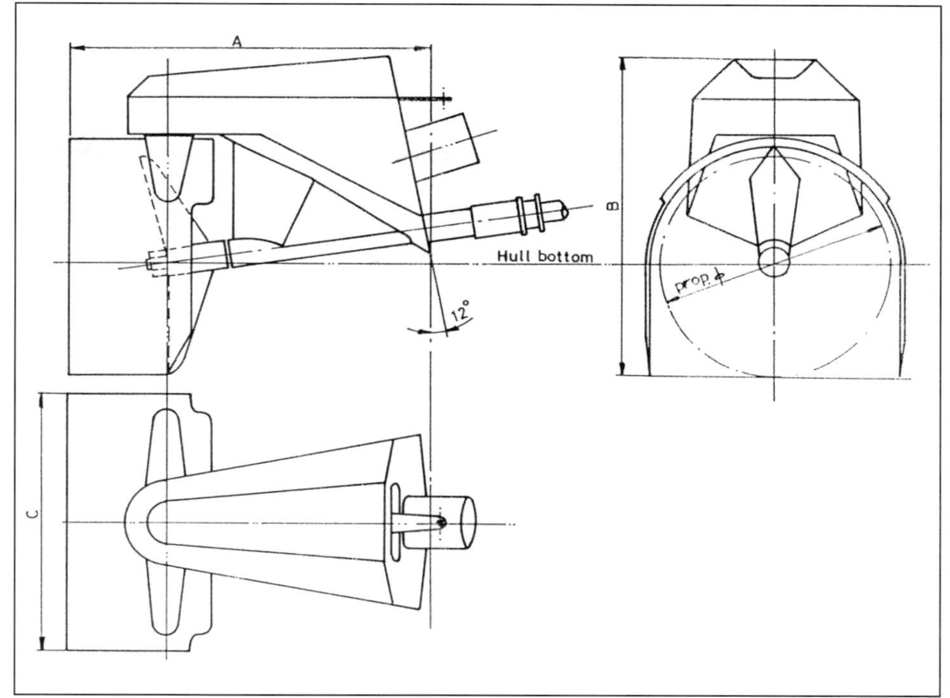

Sea Rider dimensions, see table 0506804

Principal features

Body of Sea Rider: monocoque structure in 316L stainless steel, welded and heat-treated
Rudder: of semi-circular design together with reinforcing plate and stock of 316L stainless steel plate and bar respectively
Rudder linkage: consisting of yoke, dummy tiller and tie rod terminals in cast 316L stainless steel, or nickel-aluminium-bronze
Propeller shaft: Armco Aquamet 17-18-22
Surface propeller: four-blade with Diamond Back sections cast in nickel-aluminium-bronze. For the Model TI 60 and above, propellers are five- and seven-blade
Shaft bearings: Water-lubricated in rubber

Recent applications

LA. ME Srl propellers have recently been fitted to a number of larger high-speed marine craft: the Brooke Marine 33 m, 50 kt pleasure boat *G-Whiz*; the Brooke Marine 50 kt *Virgin Atlantic Challenger II*, 20 Italian custom patrol boats 16.5 m, 52 kt, 14 other custom boats, three patrol craft for the United Arab Emirates, 20 patrol craft for Croatia, 27 craft for the Chinese Coast Guard fitted with model Ti25 and three passenger boats for Italian Navigazione Laghi fitted with modelTi90.

Model data

Model	Sea Rider Ti3	Sea Rider Ti6	Sea Rider Ti10	Sea Rider Ti15	Sea Rider Ti16	Sea Rider Ti25	Sea Rider Ti30	Sea Rider Ti60	Sea Rider Ti90	Sea Rider Ti120
Max torque, shaft, kg/m	45	105	175	205	270	400	470	850	850	1,500
rpm/kW indicative limits	30 kW at 650 rpm to 220 kW at 4,760 rpm	210 kW at 1,920 rpm to 285 kW at 2,600 rpm	283 kW at 1,554 rpm to 450 kW at 2,460 rpm	326 kW at 1,400 rpm to 525 kW at 2,270 rpm	373 kW at 1,460 rpm to 600 kW at 2,130 rpm	560 kW at 2,130 rpm to 746 kW at 1,800 rpm	522 kW at 1,070 rpm to 900 kW at 1,850 rpm	900 kW at 1,010 rpm to 1,650 kW at 1,650 rpm	895 kW at 1,010 rpm to 1,640 kW at 1,850 rpm	1,650 kW at 1,050 rpm to 2,600 kW at 1,650 rpm
Max torque, rudder stock, kg/m	30	75	125	150	150	150	250	500	500	800
Exhaust, outside diameter mm	120	130	168	168	168	200	200	300	300	460
Fitting to transom (studs)	22	22	22	22	22	22	22	22	24	24
Lubrication, linkage	molybdenum grease	molybdenum grease	molybdenum grease	molybdenum grease	molybdenum grease	molybdenum grease	molybdenum grease	molybdenum grease	molybdenum grease	molybdenum grease
Lubrication, shaft	water	water	water	water	water	water	water	water	water	water
Dimensions: A, mm	806	878	1,030	1,117	1,117	1,117	1,356	1,596	1,588	1,995
Dimensions: B, mm	670	805	961	1,019	1,019	1,120	1,231	1,450	1,663	1,845
Dimensions: C, mm	526	635	735	798	798	926	952	1,120	1,400	1,400
Weight, excluding propeller, kg	85	140	240	265	285	325	600	1,100	1,700	2,200

Japan

Hitachi Nico Transmission Co Ltd

405-3,Yoshino-cho, 1-chome, Kita-ku, Saitama 330-8646, Japan

Tel: (+81 48) 652 69 69
Fax: (+81 48) 663 49 48
Web: www.hitachi-nico.jp

Masaaki Morimoto, *President*

Hitachi Nico Transmission Company Ltd (HNT), known for the manufacture and marketing of diversified lines of marine products, has a series of marine reverse and reduction gears in lightweight and compact design utilising aluminium alloy housings. There are 34 models in both of the standard and V-drive versions available in the 334 to 2,858 kW range. These marine gears are for such vessels as pleasure craft, passenger ferries, patrol boats and crew boats for which the essential requirement is high speed.

Built-in hydraulic clutches, cooled by the same type of oil used in the engines, give the marine gears smooth and instantaneous shifting from ahead to astern and from astern to ahead.

Standard equipment on all models includes filters, pumps, temperature gauges, heat exchangers for sea or freshwater cooling, output companion flanges and manually actuated range selectors and control valves.

Optional equipment such as X-control or trailing pump is available on request.

MARINE GEARS FOR HIGH-SPEED VESSELS

SPECIFICATIONS

Model	Max capacity (kW at rpm)	Standard gear ratios	Max input speed (rpm)	In/out flange distance (mm)	Width (mm)	Weight (kg)
MGN 123G	334 at 2,500	1.52, 1.97, 2.57, 3.08, 3.46	3,300	350	426	154
MGN 133A	555 at 2,500	1.65, 2.00, 2.48, 2.92, 3.25, 3.43	2,800	415	500	220
MGN 153A	509 at 2,500	1.65, 2.00, 2.48, 2.92, 3.25, 3.43	2,800	415	500	220
MGX 293D	1,452 at 2,300	1.07, 1.66, 1.74, 2.45, 2.80	2,500	592	680	460
MGX 294B	1,462 at 2,300	2.46, 3.03, 3.48, 3.93, 4.43	2,500	592	720	580
MGN 335E	1,446 at 2,300	3.30, 4.11, 4.68, 4.72, 5.22, 6.05	2,300	784	750	794
MGX 353B	1,566 at 2,300	1.15, 1.33, 1.53, 1.73, 2.02, 2.44, 2.72	2,500	698	680	560
MGN 453A	2,222 at 2,100	1.18, 1.54, 2.06, 2.52, 2.92, 3.25, 3.43	2,100	754	890	1,030
MGN 493A	2,858 at 2,100	1.16, 1.52, 2.08, 2.47, 2.96	2,100	936	1,000	1,240
MGN 532	3,012 at 2,100	1.45, 1.57, 1.70, 1.84, 2.00, 2.36, 2.50	2,100	1,082	960	1,150
MGN 533	3,012 at 2,100	2.60, 2.79, 3.00, 3.24, 3.50	2,100	900	1,130	1,450
MGN 632	3,720 at 2,000	1.49, 1.68, 1.83, 2.00, 2.29. 2.52	2,000	960	1,120	1,670
MGN 633	3,720 at 2,000	2.64, 2.85, 3.09, 3.37, 3.52	2,000	970	1,280	2,150
MGN 732	4,604 at 2,000	1.46, 1.68, 1.84, 2.03, 2.25, 2.50	2,000	1,000	1,240	1,800
MGN 733	4,604 at 2,000	2.66, 2.90, 3.03, 3.33, 3.50	2,000	1,010	1,400	2,600
MGN 832	5,281 at 1,800	1.45, 1.72, 1.88, 2.06, 2.27, 2.50	1,800	1,100	1,320	2,630
MGNV 123	411 at 2,500	1.53, 2.03, 2.46, 2.96	3,300	356.1	490	180
MGNV 133	515 at 2,500	1.57, 1.97, 2.48, 2.96	2,800	409	500	225
MGNV 172E	732 at 2,500	1.56, 1.78, 2.03, 2.37	2,600	485.5	570	300
MGNV 172EC	732 at 2,500	1.56, 1.78, 2.03, 2.37	2,600	114	570	290
MGXV 293D	1,418 at 2,300	1.34, 1.51, 1.74, 2.03, 2.48, 2.80	2,500	594	680	460
MGXV 293DC	1,418 at 2,300	1.34, 1.51, 1.74, 2.03, 2.48, 2.80	2,500	202	680	450
MGXV 353B	1,566 at 2,300	1.55, 1.72, 2.09, 2.42, 2.73	2,500	653	680	565
MGXV 353BC	1,566 at 2,300	1.55, 1.72, 2.09, 2.42, 2.73	2,500	167	680	545
MGNV 453A	2,222 at 2,100	1.48, 1.97, 2.50, 2.79, 2.93	2,100	746.8	890	1,080
MGNV 453AC	2,222 at 2,100	1.48, 1.97, 2.50, 2.79, 2.93	2,100	300.2	890	910
MGNV 493A	2,858 at 2,100	1.49, 2.08, 2.44, 2.93	2,100	949.9	1,000	1,295
MGNV 493AC	2,858 at 2,100	1.49, 2.08, 2.44, 2.93	2,100	288.1	1,000	1,125

New Zealand

Q-SPD International Ltd

3/H Henry Rose Place, Albany, PO Box 34-881, Auckland, New Zealand

Tel: (+64 9) 448 5801
Fax: (+64 9) 448 5803
e-mail: sales@q-spd.com
Web: www.q-spd.com

Q-SPD offers a range of advanced marine propulsion systems designed to produce optimum performance and reliability in an innovative and low maintenance package. Capable of absorbing from 150 to 1,860 kW, the surface-piercing units are manufactured from composites.

Model	Max input power kW	Length mm	Weight, kg
SD 100	328	840	80
SD 150	522	885	100
SD 300	895	1140	150
FD 250	522	–	–
FD 300	745	–	–
FD 350	1120	–	–
FD 400	1865	–	–

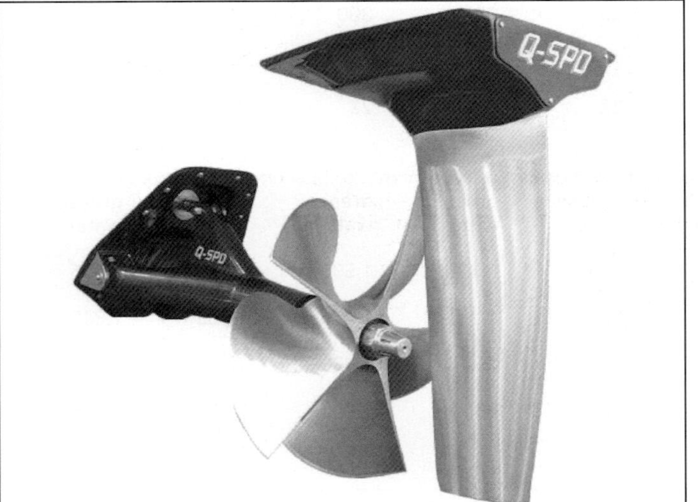

Q-SPD – FD 1120283

Q-SPD – SD 1120282

Seafury International Ltd

33 Forge Road, PO Box 336, Silverdale, Hibiscus Coast, New Zealand

Tel: (+64 9) 426 78 47
Fax: (+64 9) 426 78 07
e-mail: info@seafury.com
Web: www.seafury.com

Seafury International Ltd designs and manufactures drive systems, based on the use of surface piercing propellers, which require less maintenance, allow for a shallower draught and provide faster speeds than conventional drive systems.

Manufactured in four standard sizes, the Seafury units can be used in single, twin, multiple or mixed drive configurations.

Each system is supplied complete with propeller, rudder and steering tiller link, propeller shaft and coupling flange. Over 500 units have been sold since its introduction.

Model	Propeller diameter	Maximum input power @ 1,400 rpm	Displacement 1 drive	Displacement 2 drive	Displacement 3 drive	Displacement 4 drive
SF22	560 mm	225 kW	2.5 tonnes	5 tonnes	–	–
SF26	660 mm	410 kW	5.5 tonnes	13 tonnes	20 tonnes	–
SF30	760 mm	820 kW	12 tonnes	26 tonnes	36 tonnes	–
SF36	960 mm	1,640 kW	25 tonnes	65 tonnes	90 tonnes	125 tonnes

Larger vessels require multiple SF36 units or custom built units depending on application.

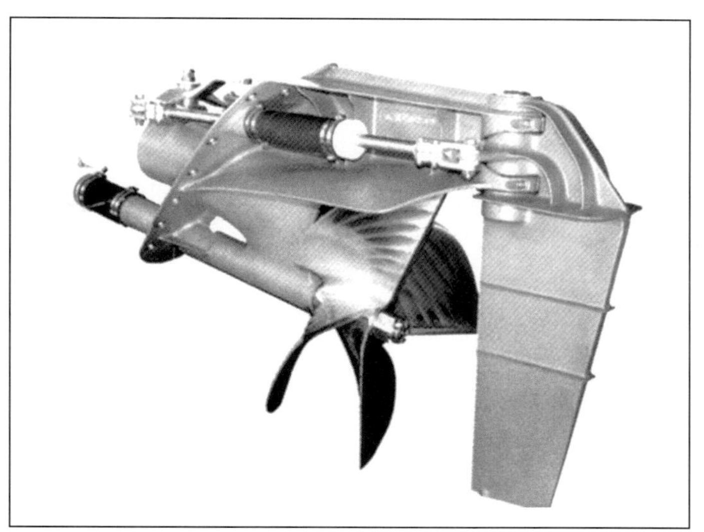

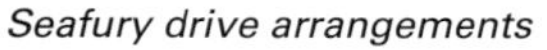

Seafury drive arrangements 0069471

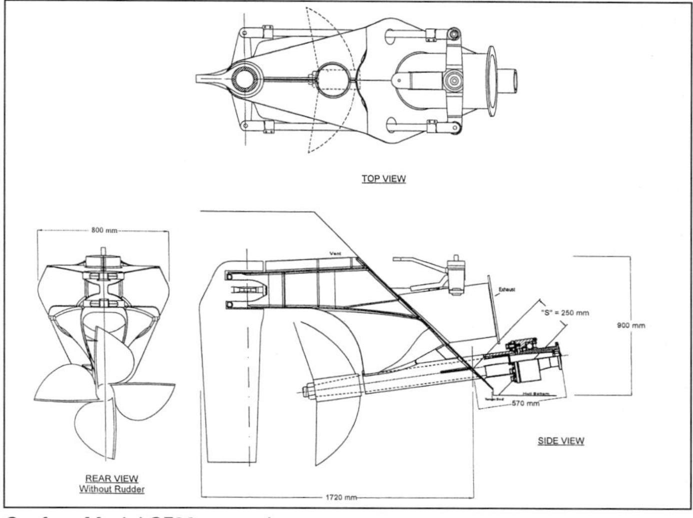

Seafury Model SF36 general arrangement 0100074

Norway

Servogear A/S

N-5420 Rubbestadneset, Norway

Tel: (+47) 53 42 39 50
Fax: (+47) 53 42 39 99
e-mail: main.office@servogear.no
Web: www.servogear.no

Magne Møklebust, *Managing Director*
Arvid Kvernøy, *Operations Director*

Servogear AS is a leading manufacturer of controllable-pitch propeller systems for fast vessels. The company has delivered over 1,100 propulsion systems worldwide for most types of fast vessels, including pilot boats, police/customs boats, lifeboats, ambulance boats and motor yachts, as well as mono-hulled and catamaran passenger vessels. Recent vessels to be equipped with Servogear propulsion components have included the Brodrene AA Za 5 m fast passenger catamaran and the 22.5 m Fosen class search and rescue boat built by Batservice Mandal AS.

Servogear propulsion system

Designed as a complete unit, the Servogear Propulsion System comprises:
CP propellers;
Reduction gearboxes;
Shaft brackets;
Effect rudders;
Stern tubes.

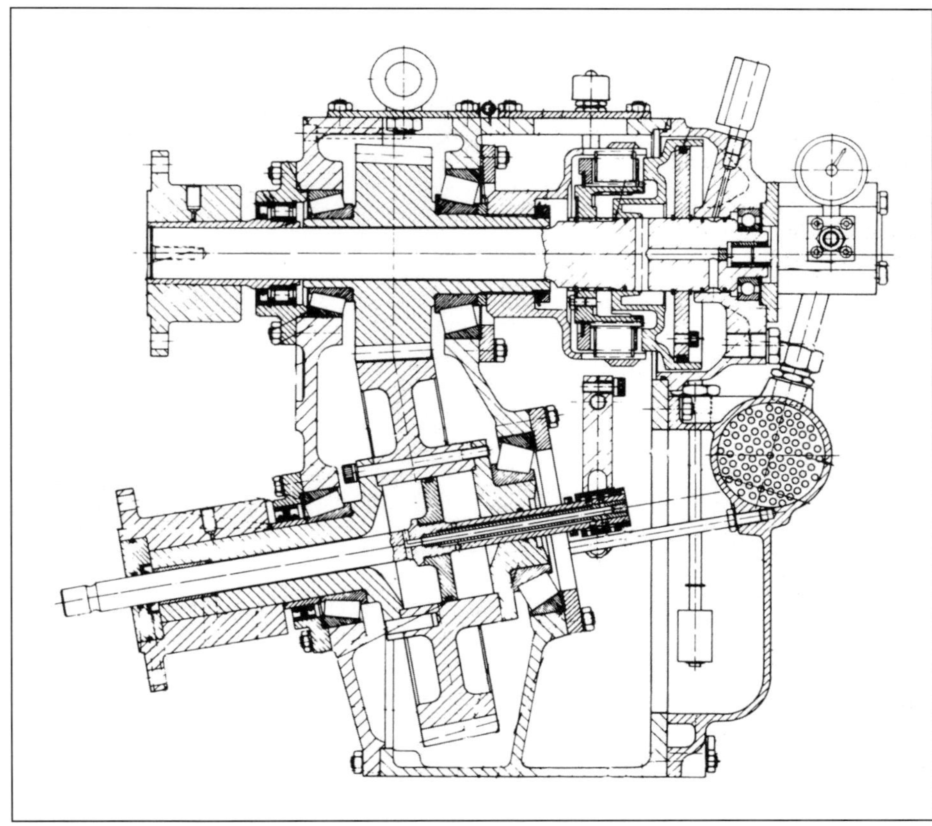

Section through a Servogear gearbox 0506787

They have a power absorption range of between 300 and 4,000 kW with speeds of up to 50 knots.

Reduction gearboxes
Servogear gearboxes are designed for high-speed craft where high torque, low weight and limited space are of importance. Integrated in each gearbox is an hydraulically operated clutch and a servo system to actuate the propeller pitch. The gearboxes, type approved by DnV, are available in several configurations, including a twin-input/single-output version (one engine aft).

Propeller tunnels
Unique to the Servogear propulsion system is the propeller tunnel design, which gives hydrodynamic advantages to the vessel as well as reduced hull resistance and draft. The tunnel is integrated into each hull above the shaft and CP propeller.

Russian Federation

Zvezda

123 Babushkina Str., 192012 St Petersburg, Russian Federation

Tel: (+7 812) 367 57 45
Fax: (+7 812) 367 37 76
e-mail: gumennyy@zvezda.spb.ru
Web: www.zvezda.spb.ru

Pavel Plavnik, *Director General*
Anton Gumenny, *Chief of International Sales Department*

JSC Zvezda produces reverse reduction gearboxes, reduction gearboxes and reverse clutches.

Compact multi-disc autonomous reverse reduction gearboxes can be coupled directly with a diesel engine or be coupled with it by an intermediate shaft, allowing angularity of shaft-line.

The gearboxes are completely autonomous, equipped with oil pumps, filter and oil-cooler with automated temperature control system. The gearboxes are equipped with a manual control desk and are adapted for attachment of an automated remote control. Reverse-reduction gearboxes have a built-in bearing for withstanding propeller thrust.

A new direction is the development and production of heavy-duty reverse-reduction gearboxes with transmitted power up to 8,830 kW.

Type	Model	Max power, kW	Max rotation speed (rpm)	Transmission ratio	Length (mm)	Width (mm	Height (mm))	Dry weight (kg)
Planetary gearbox	470.34	1,100	1,600	2,85	630	685	675	400
Reverse-reduction gearbox	473.15	1,100	1,600	1–2	830	1,150	845	800
Reverse-reduction gearbox	471.43	2,000	1,800	1–2	1,075	1,290	1,150	1,100
Two-speed reverse-reduction gearbox	RRD12000	8,830	1,000	6.1/4.6	2,890	2,600	2,900	24,500

Sweden

Applied Composites AB

Box 13070, SE-580 13 Linköping, Sweden

Tel: (+46 13) 20 97 00
Fax: (+46 13) 20 97 09
e-mail: info@acab.se
Web: www.acab.se

Bjorn Thundal, *President*
Kenneth Karlsson, *Marketing Manager*

Applied Composites produces a range of composite products including composite propulsion drive shafts for powers up to 30,000 kW. The scope of supply from this company includes the complete shafting system including bearings, couplings and seals. The company has supplied Bazan Shipbuilders with such a system for installation in the *Silvia Ana*, a large, fast ferry powered by six Caterpillar 3616 diesel engines each rated at 5,650 kW at 1,000 rpm.

Applied Composites AB Composite shafting and couplings
0010061

United Kingdom

Air Vehicles – Design and Engineering Ltd

Three Gates Road, Cowes, Isle of Wight, PO31 7UT, United Kingdom

Tel: (+44 1983) 29 31 94
Fax: (+44 1983) 29 19 87
e-mail: info@airvehicles.co.uk
Web: www.airvehicles.co.uk

Charles Eden, *Director*

Air Vehicles has been in the high speed, air supported marine market for over 30 years. Originally a builder of small/medium sized ACVs, the company now specialises in the design and manufacture of systems and components for fully amphibious ACVs and SESs. All Air Vehicles' systems and components are designed specifically to suit the particular project with emphasis on light weight and performance. All components are designed and manufactured to British

Hovercraft safety requirements which usually ensures compliance with all other classification societies.

Transmission systems
Air Vehicles has used toothed-belt drives for lift and propulsion systems for over 25 years and it designs and manufactures complete transmission systems for ACVs and SESs. The systems are designed for each specific project, are efficient and compact and can usually be designed and manufactured at lighter weights and lower costs than equivalent gearboxes.

Besides resolving the drive-system loads the structural element of the system often supports the driven components, absorbing the system thrust and inertia loads.

Currently, typical belt life is 7,500 to 10,000 hours and, with careful design, the belt is easily changed at main overhaul. Power levels are up to 1,000 kW.

The belt drive system is integrated into the structure of the lift fan casing. The complete assembly is resiliently mounted (350 kW)
0045323

Allen Power Engineering Ltd

Allen Gears
Atlas Works, Station Road, Pershore, Worcestershire WR10 2BZ, UK

Tel: (+44 1386) 55 22 11
Fax: (+44 1386) 55 44 91
e-mail: sales@allengears.com
Web: www.allengears.com

C F W Brimmell, *Managing Director*
G Horton, *Sales Manager*
P Davies, *Commercial Manager*
D Haycock, *Business Support Team Manager*
K Whittle, *Engineering Team Leader*
D Wilson, *Order Delivery Team Leader*

Allen Gears supply high-quality precision epicyclic, parallel shaft or combination gearing to worldwide industrial and marine markets. The company forms part of the Industrial Businesses Group within Rolls-Royce Plc.

To date the company has supplied many thousands of transmission systems from 2 to 45 MW for all types of combination drives using gas turbines and for diesel engines.

Significant applications include fast patrol vessels for various navies (including the German, Danish, Swedish, Malaysian, Libyan and Norwegian Navy), fast luxury yachts, merchant vessels such as oil tankers and container ships and a range of specialist

Allen Gears' gearbox for the Swedish Navy's Visby class patrol craft 0594825

vessels such as coastal survey craft, minesweeper craft and multirole vessels.

The company designs and manufactures precision gears and gearboxes for all types of high-speed rotating equipment to the highest international standards and classification society requirements.

C Form gearbox

In 1978 Allen Gears fitted its C form gearboxes in a Don Shead-designed 29 m, 45 kt luxury yacht. The yacht cruises on two wing, diesel engine-driven jets, and a gas turbine provides power for maximum speed. The turbine is a Textron Lycoming Super TF40 which produces 3,430 kW at 15,400 rpm and drives a Rocketdyne jet pump running at 1,664 rpm. The gearbox has a C drive configuration, both input and output shafts are at the aft end, and consist of a primary epicyclic train with a single helical parallel-shaft secondary train. An idler is required to cover the necessary centre distance from the gas turbine to the Rocketdyne jet pump and also matches their standard rotations. The jet pump houses the main thrust bearing, therefore secondary gearbox bearings need only accommodate the thrust imposed by single helical gearing. A caliper disc brake is fitted to the free end of the secondary pinion enabling the main jet pump to be held stationary when using wing engine propulsion.

Five-Shaft C Form drive gearbox

As a follow-on from uprated gears for the Royal Navy, Allen Gears constructed a combined epicyclic and parallel-shaft C drive gearbox for a 46 m, 45 kt luxury yacht built by Fr Lürssen Werft of Germany. The yacht was powered by two Allison 570 gas turbines driving KaMeWa water-jets. In addition there were two wing engines driving small water-jets of 1,119 kW. Total available power was 13,050 kW. The gas turbines were positioned aft, driving forward into the gearbox. Each of the turbines has a maximum input of 5,406 kW at 12,000 rpm giving an output speed, to the water-jet unit from gearbox, of 681 rpm. A five-shaft configuration gearbox was adopted to accommodate the centre distance between the gas turbines and the

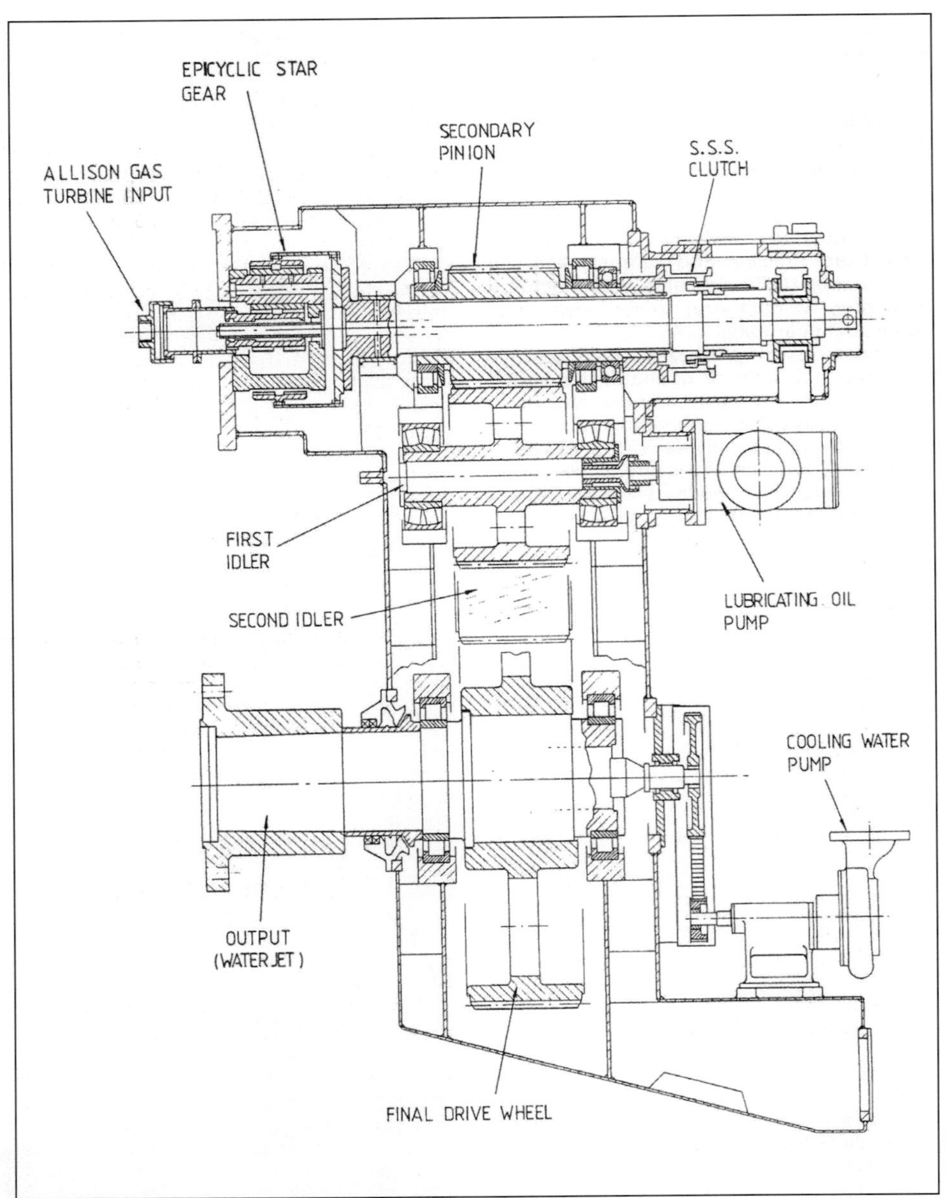

Sectioned view of one drive line of the five-shaft C form reduction gearbox 0506793

Port and starboard multistage hovercraft propulsion gearboxes 0506791

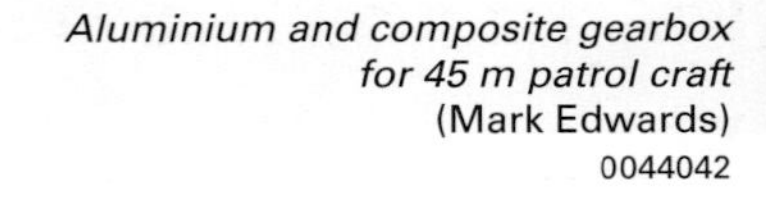

Aluminium and composite gearbox for 45 m patrol craft (Mark Edwards) 0044042

output shaft. The input is taken from the primary epicyclic train through quill-shafts, within the secondary pinion, to SSS self-synchronising clutches at the forward end of the gear case. This arrangement permits the second gas turbine to be introduced to the drive line or to be disconnected without interruption of power.

Hovercraft transmissions

Allen Gears' association with the Spanish company CHACONSA and its VCA-36 craft moved on to future developments with an overseas navy for a similar sized craft. This prototype vessel utilises Pratt & Whitney gas turbines for the lift system and AlliedSignal gas turbines for the propulsion system. In both cases the reduction gears are of Allen Gears' lightweight construction. In order to achieve the best machinery layout and to meet the necessary weight targets Allen provided multistage parallel shaft gears in aluminium gear cases, using gear components constructed according to proven aircraft methods.

Codag propulsion system

The first seven ships of the Royal Danish Navy 'Standard Flex 300' multirole vessel programme use a CODAG propulsion plant powered by MTU 16V 396 TB 94 diesels for cruise and a GE LM 500 gas turbine for boost. The engines drive through Allen Gears vertically offset custom-engineered lightweight parallel shaft gearboxes with integral lubricating system. Each gearbox is equipped with a hydraulic motor which forms part of a hydraulic system (powered by a General Motors diesel) to meet the requirements of auxiliary propulsion during silent minehunting and economic loitering on nearly stationary patrols.

FPB propulsion

In 1996, Allen Gears received an order from Volvo Aero Support to supply 2 × mirror image port and starboard codag gears for a 45 m Fast Patrol Boat (FPB).

Powered by Allison 571 and MTU diesel engines, the fully automatic codag gearboxes are lightweight featuring aluminium fabricated gearcases and a composite construction electron beam-welded bull gear. The gearboxes were custom designed to suit the very tight spatial constraints normal in FPB engine rooms and include integral oil systems and self-synchronising clutches on both the diesel and gas turbine inputs to achieve automatic 'hot shift' from diesel to gas turbine mode.

The gearbox output drives a single-element membrane coupling integral with a GRP shaft to a KaMeWa water-jet.

In 2003 the company supplied eight gearbox units for the Swedish Navy Stealth Vessel programme which uses a CODAG propulsion system.

Yellowfin Ltd

2 Saxon House, Saxon Wharf, Lower York Street, Southampton, SO14 5QF, United Kingdom

Tel: (+44 2380) 235 299
Fax: (+44 2380) 381 719
Web: www.yellowfin.com

In 2006 Yellowfin Ltd launched the Variable Surface drive (VSD) for high-speed craft. VSD is a smart, integrated propulsion package which delivers propulsion, ride, altitude, position and motion control. The system uses a surface piercing drive that has pitch/rake control allowing the drive to thrust in multiple directions. The system eliminates the need for rudders, flaps, stern thrusters and traditional gearboxes (gearbox is incorporated in the drive). Models are available to absorb up to 1,100 kW.

United States

Philadelphia Gear Corporation

901 East 8th Avenue, King of Prussia, Pennsylvania 19406, US

Tel: (+1 610) 265 30 00
Fax: (+1 610) 337 56 37
e-mail: jphinney@philagear.com
Web: www.philagear.com

Carl D Rapp, *President & CEO*
Jules DeBaecke, *Vice President Engineering*
John m Phinney, *General Manager, Marine*

Marine gear products ranging from 400 to 27,000 kW have been designed and manufactured by Philadelphia Gear Corporation. Applications include commercial and military main propulsion drives for propellers and water-jets, dredge pump, jet pump, bow thruster and power generation drives. Configurations include concentric, horizontal and vertical shaft offset gears, single and multiple reductions with either steel or aluminium housings. They are supplied in standardised or custom configurations; with or without clutches, compound inputs, articulated locked train gearing, epicyclic and reverse reduction gearing. Single- or double-helical gears with either through hardened or case hardened and ground teeth used depending on the application. Philadelphia marine drives can be supplied with either wet or dry clutches and brakes. Hydroviscous drive technology supplied by the Synchrotorque division allows the efficient integration of wet clutches and brakes into designs requiring disconnect, reversing or speed varying functions that demand energy dissipation capabilities beyond that of dry clutches and brakes.

In addition to Philadelphia Gear brand name products, the company also supplies sub-assemblies for main drive propulsion drives provided by other manufacturers: including complete gear sets, housings, gear tooth couplings, turning gears, power take-offs and other accessory drives in both parallel shaft and right angle (bevel and worm) configurations.

Philadelphia Gear is the exclusive source of most naval and commercial marine gear product lines formerly manufactured by Western Gear, WesTech Gear and the Delaval Turbine Division of IMO Delaval Inc. These lines include main propulsion reduction drives and ship service turbogenerator drives for US Navy and commercial applications up to 37,285 kW.

Philadelphia Gear has recently been selected to be the supplier of the reduction gearing for the Fast Ship Atlantic project. These drives will be used to transmit power from the 50 MW aeroderivative gas turbines to the propulsion water-jets for these high-speed container ships.

Powervent

2615 NE 5th Avenue, Pompano Beach, Florida 33064, US

Tel: (+1 954) 545 32 24
Fax: (+1 954) 943 78 77
e-mail: info@powervent.com
Web:www.powervent.com

Peter Caspiri, *Sales Manager*

Powervent has developed a surface propeller drive system which allows automatically controlled venting of the surface propeller. Demonstrated in a number of powerboat designs, the system has been evaluated by the Naval Surface Warfare Center.

In 2001 an advanced propulsion technology demonstrator vessel was ordered by the US Navy in order to evaluate the Powervent concept.

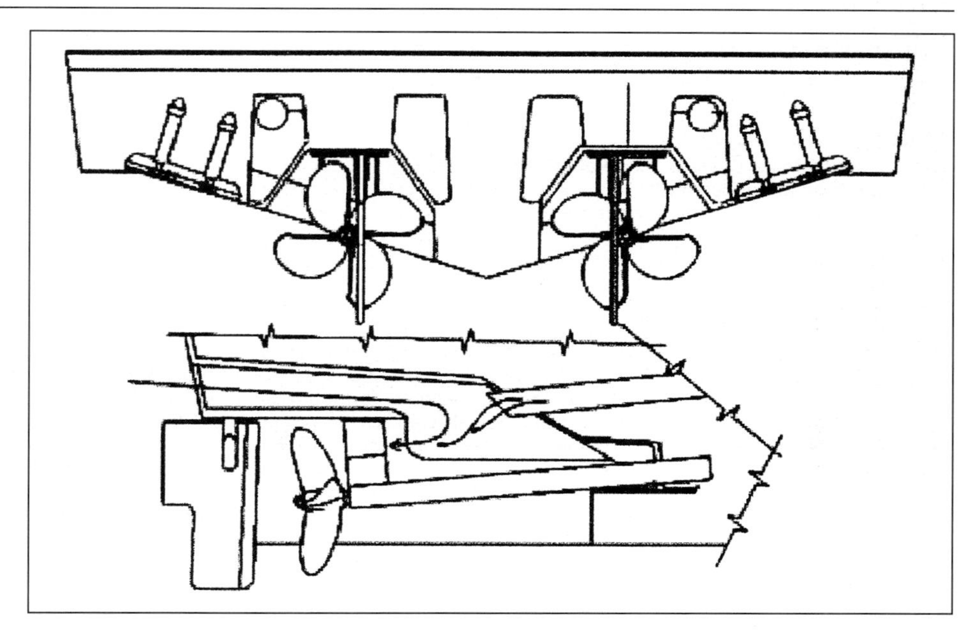

General arrangement of the Powervent system
0100065

Pulse-Drive Systems International LLP

601 NE 26th Court, Pompano Beach, Florida 33064, United States

Tel: (+1 954) 788 6363
Fax: (+1 954) 968 75 85
e-mail: info@pulsedrive.net
Web:www.pulsedrive.com

Paul LaCarubba, *President*
Tony Mills, *General Manager*

Pulse-Drive Systems International provides a range of trimmable surface propeller drives for small to medium-sized craft. The Pulse Drive range, as they are known, are made in six main sizes and are suitable for single or multiple unit installation from shaft powers of 75 kW to 1,100 kW.

The largest in the range, the Pulse Drive 3000 can take a maximum pleasure craft rating of 1,200 kW, having a weight of 450 kg.

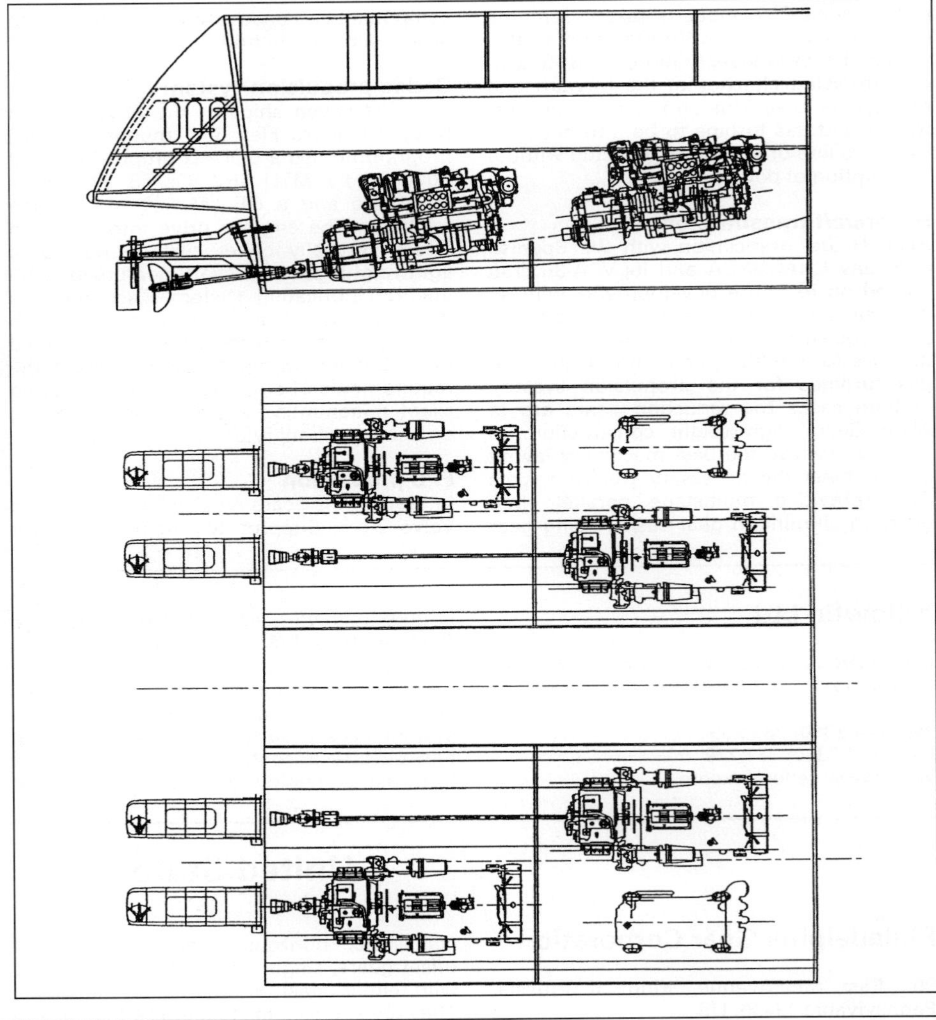

Typical pulse drive layout in a catamaran hull
0121766

Pulse Drive Model	1000	1250	1500	1750	2000	2500	3000
Power range (kW)	90	225	500	600	750	950	1,200
Propeller dia (mm)	300	460	460	460	600	700	850
Weight (kg)	20	78	100	115	160	270	450
Rudder angle	±40°	±40°	±40°	±40°	±40°	±40°	±40°
Propeller trim	6°	6°	6°	6°	6°	6°	6°

T-Torque Drive System Inc

Tempest Engineering
21173 NE 18 Place, North Miami Beach, Florida 33179, US

Tel: (+1 305) 527 90 60
Fax: (+1 305) 935 08 07
e-mail: tempestyachts@aol.com
Web: www.tempestyachts.net

Adam Erdberg, *President*

T-Torque Drive System

The T-Torque Drive System (US Patent No 4 919 630) is an advanced heavy-duty marine propulsion system based on surface-piercing operation of super-cavitating propellers. The system provides the recreational, commercial or military diesel-powered craft operator with high manoeuvrability, high-speed and shallow water operation capability. The system is highly reliable due to its construction in the highest marine grade polished stainless steel (316L) and simplistic concept. The shafts connect directly to the engine transmissions and pass through the transom supported by heavy-duty struts which are mounted above the bottom of the boat. Since the propeller shafts directly penetrate the transom, the shaft angle is far less than conventional inboard systems, thereby providing greatly reduced appendage drag and increased efficiency. Shaft centres are at normal separations depending upon the particular vessel and engines. This configuration allows the vessel to have the low-speed dockside precision manoeuvrability of a conventional inboard-powered vessel. The T-Torque rudder system is supported by a polished stainless steel T-Strut securing the rudders behind the propellers. The rudders turn via a hydraulic power steering system with hydraulic lines built into the T-Strut (see illustration).

The system has been in service since 1984 on various recreational, commercial and military vessels up to 32 m in length.

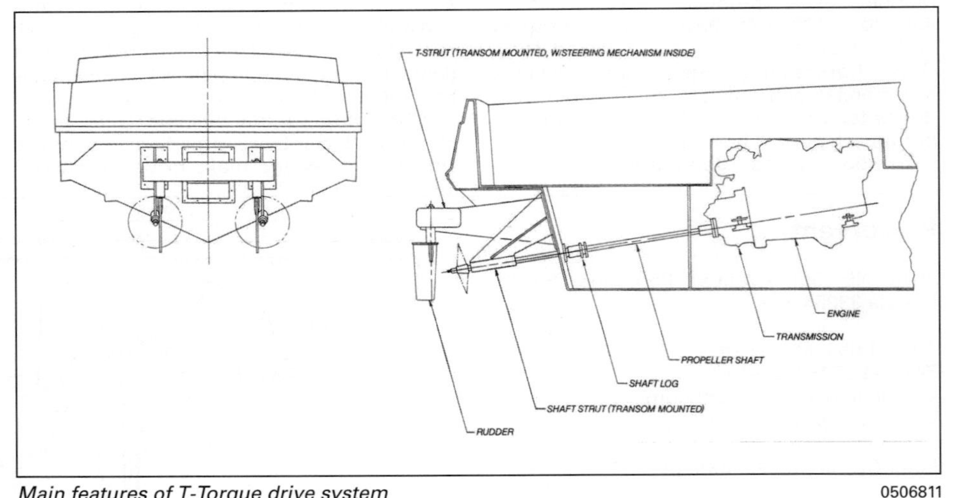

Main features of T-Torque drive system
0506811

Twin Disc Inc

1328 Racine Street, Racine, Wisconsin 53403, United States

Tel: (+1 262) 638 40 00
Fax: (+1 262) 638 44 81
Web: www.twindisc.com

M E Batten, *Chairman, Chief Executive Officer*
J E Feiertag, *Executive Vice President*
J H Batten, *Executive Vice President*
D Bratel, *Vice President, Engineering*

Twin Disc manufactures complete vessel management systems including marine transmissions, electronic propulsion controls, Arneson Surface Drives, propellers and marine hydraulics including bow thrusters, trim tabs and steering systems.

The QuickShift® marine transmission instantly applies power to the driveline at low torque when shifting from neutral to full ahead or full reverse. The system is designed to eliminate driveline shock while optimising power to the driveshaft, resulting in a steep but smooth power curve. QuickShift has the ability to regulate engine torque at extremely low speeds to slow propeller speed, enabling manoeuvring at less than 5 kt.

Twin Disc marine transmissions feature helical gearing in most models for quieter operation; hydraulic-controlled and oil-cooled clutches for smooth, fast shifting; identical ratios in forward and reverse with full power in forward and reverse in most cases and minimal external plumbing. Down angle output configurations provide for near level engine installation and space saving remote- and direct-mounted V-drive models are also available. Power ratings range from 50 kW in the MG-5010 models to 1,760 kW in the MG-5600 series with input rpm between 1,800 and 2,300.

The Twin Disc Marine Control Drive is a variable speed drive available in a range of sizes with power capacities from 160 to 4,500 kW at 1,800 rpm. The system provides operational advantages for any vessel requiring accurate positioning or a high degree of slow speed manoeuvrability while the main engine powers other action. There are three sizes with five models for each size to cover the range up to 4,500 kW.

Twin Disc Arneson Surface Drives combine surface-piercing propeller technology with

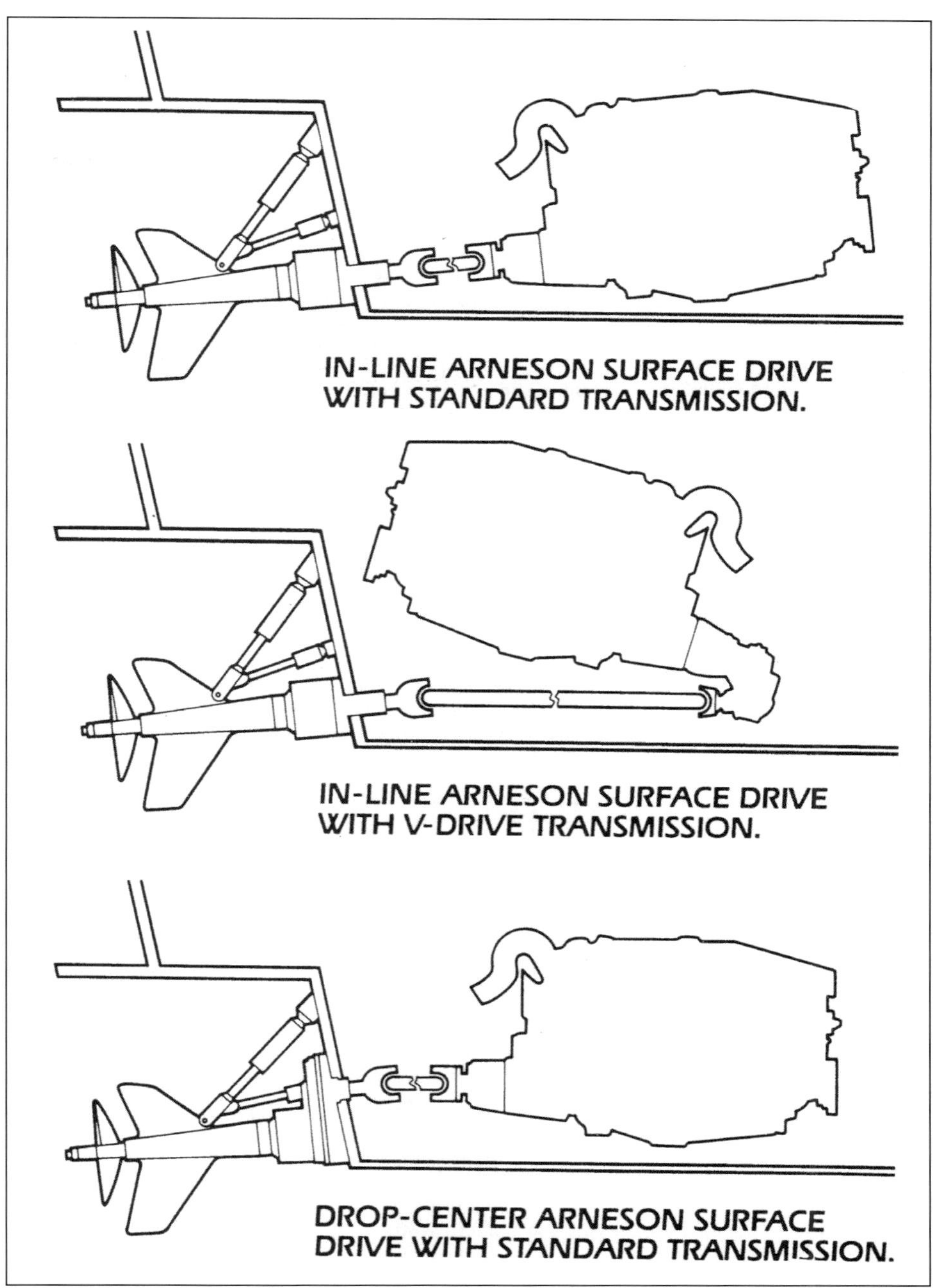

Three possible layouts for Twin Disc Arneson Surface Drives 0506794

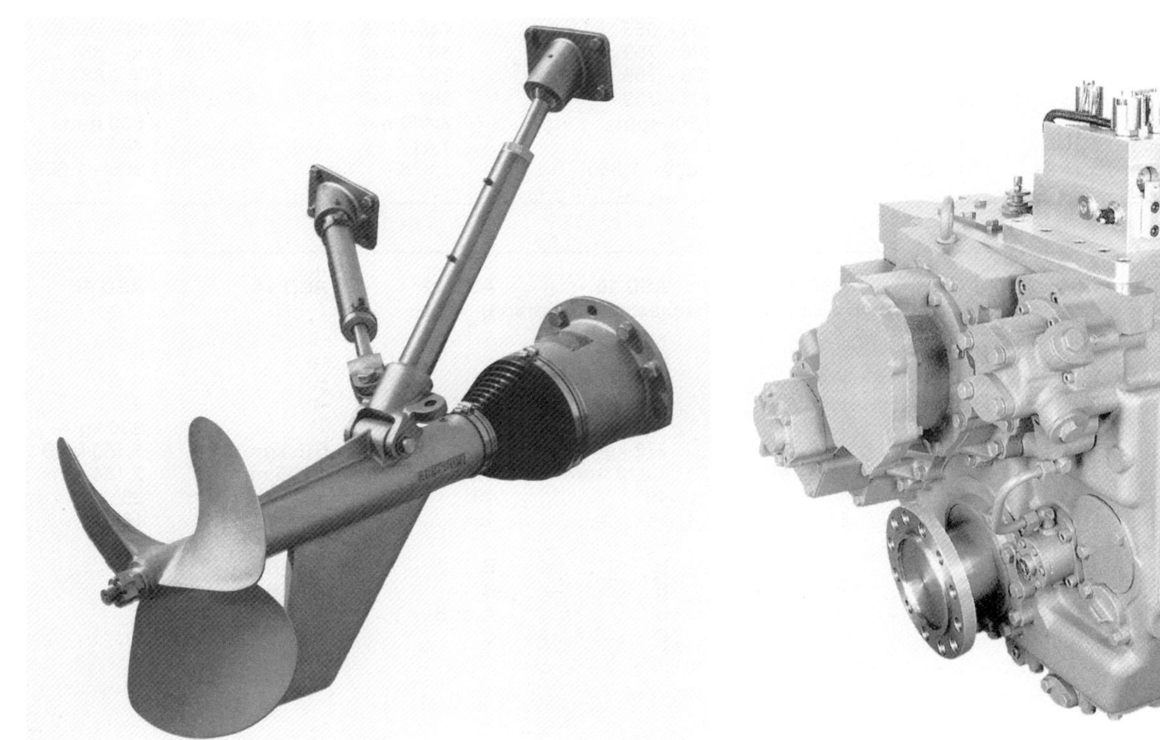

Twin Disc Arneson ASD 10 in-line unit 0506795

Twin Disc MG-6599A marine transmission 0044044

Model	Reduction ratio (:1)	Medium duty 1,800 rpm	Power – kW Intermediate 2,100 rpm	Pleasure craft 2,300 rpm
MG-5005 A′	1.54 – 2.47	–	47–35	101–60
MG-5015 A′	1.51 – 2.40	–	92 – 65	160 – 114
MG-5055 A′	1.53 – 2.08	–	147	258
MG-5011 A′	1.44 – 2.39	74 – 66	112 – 103	186 – 168
MG-5050 SC′	1.23 – 3.00	115 – 102	157 – 142	272 – 231
MG-5050 A′	1.12 – 2.50	115	157	272 – 239
MG-506-1	1.09 – 2.96	–	121	236 – 209
MG-5061 SC′	1.15 – 3.00	135 – 126	201	345 – 298
MG-5061 A′	1.13 – 2.47	135 – 126	201	345 – 321
MG-5062 V′	1.19 – 2.51	–	201	354 – 321
MG 5075 SC & MG-5075 A	1.06 – 2.88	205	268 – 261	453 – 434
MG-5085 SC	1.05 – 2.33	250 – 225	332 – 302	433 – 365
MG-5085 A	1.05 – 2.43	250 – 217	332 – 302	433 – 365
MG-5082 SC & MG-5082 A	1.06 – 2.88	273	396 – 384	492 – 447
MG-5091 SC	1.17 – 3.38	274 – 219	403 – 309	522 – 447
MG-5091 DC	3.82 – 5.10	262 – 225	377 – 298	–
MG-5114 SC	0.93 – 3.00	394	503 – 451	673 – 578
MG-5114 DC	3.28 – 4.86	361	503 – 429	673 – 604
MG-5114 A	1.03 – 2.50	357 – 338	466 – 400	673
MG-5114 RV	1.03 – 2.50	–	466 – 400	673
MG-5114 IV	1.05 – 2.57	405	466 – 400	673 – 649
MG-514C SC	1.51 – 3.50	328 – 283	380 – 336	–
MG-514C DC	4.13 – 6.00	298	380 – 330	–
MG-5145 SC & MG 5145 A	1.20 – 2.50	596 – 522	746 – 671	1,007 – 916
MG-5145 RV	1.26 – 2.50	–	746 – 671	1,007 – 916
MG-516 DC	3.06 – 6.00	447 – 406	573 – 470	–
MG-5170 DC	4.06 – 6.95	537 – 457	552 – 493	–
MG-5203 SC	1.17 – 3.48	600	660	969 – 800
MG-5222 DC	4.03 – 6.96	600	660	–
MG-5204 SC	1.17 – 3.48	–	771 – 660	1,104 – 1,081
MG-5225 DC	4.03 – 6.10	727	771	–
MG-5301 DC	3.35 – 6.93	982 – 802	1,101 – 892	–
MG-540	1.93 – 7.47	999 – 740	1,141 – 1,047	–
MG-5600	2.53 – 6.04	1,760 – 1,451	1,910 – 1,575	–
MG-5600 DR	6.02 – 7.01	1,760 – 1,566	1,858 – 1,653	–
MG-6449 A	1.51 – 2.95	741 – 489	822 – 530	1,104 – 782
MG-6557 SC	1.07 – 2.93	774 – 649	867 – 685	1,342 – 1,045
MG-6557 DC	2.46 – 4.48	779 – 635	881 – 718	1,342 – 1,104
MG-6557 A	1.29 – 2.92	779 – 641	882 – 676	1,342 – 1,104
MG-6600 DC	3.30 – 6.05	765 – 571	864 – 651	1,445 – 1,081
MG-6619 SC	1.06 – 2.93	919 – 746	1,007 – 812	1,417 – 1,141
MG-6619 A	1.55 – 2.95	862 – 768	976 – 836	1,491 – 1,230

LIGHT DUTY RATINGS FOR QUICKSHIFT® MARINE TRANSMISSIONS

Model	Reduction ratio (:1)		Power – kW	
		2,100 rpm	**2,300 rpm**	**2,500 rpm**
MGX-5114 SC	0.93 – 3.00	579 – 494	629 – 540	669 – 588
MGX-5114 A	1.03 – 2.50	580	629	655
MGX-5114 RV	1.03 – 2.50	580	629	655
MGX-5114 IV	1.05 – 2.57	580 – 560	629 – 607	655 – 632
MGX-5135 SC	1.00 – 2.90	704 – 622	746 – 671	788 – 720
MGX-5135 A	1.10 – 2.52	704 – 663	746 – 716	788 – 768
MGX-5135 RV	1.10 – 2.52	704 – 663	746 – 716	788 – 768
MGX-5145 SC	1.20 – 2.50	808 – 759	857 – 820	906 – 881
MGX-5145 A	1.26 – 2.50	808 – 759	857 – 820	906 – 881
MGX-5145 RV	1.26 – 2.50	808 – 759	857 – 820	906 – 881
		1,900 rpm	**2,100 rpm**	**2,500 rpm**
MGX-6650 SC & MGX-6690 SC	1.51 – 3.21	1,289	1,417	1,464
MGX-6848 SC	1.51 – 3.21	1,670 – 1,360	1,828 – 1,485	1,985 – 1,609

Model	ASD 6	ASD 8	ASD 10	ASD 12	ASD 14	ASD 16
	contact twin disc to 3,750 (subject to application)					
Power acceptance (kW)						
petrol at 5,200 rpm	to 410	to 740	to 1,100	–	–	–
diesel at 2,400 rpm	to 216	–	–	–	–	–
Unit weight, dry, with	–	–	–	–	–	–
hydraulic cylinders (in-line)	61 kg	129 kg	189 kg	–	–	900 kg A
(drop centre)	72 kg	220 kg	272 kg	352 kg	515 kg	1,150 kg B
Overall external length	914–991 mm	1,067 mm	1,270 mm	1,638 mm	1,805 mm	2,184 mm
Steering angle	40°	40°	40°	40°	40°	36°
Trim angle (max travel inclusive)	15°	15°	15°	15°	15°	15°
Materials	–	–	–	–	–	–
socket A or B	B	B	B	B	A or B	
thrust tube A or B	B	B	B	B	A or B	
ball B	B	B	B	A or B	A or B	
propeller shaft C	C	C	C	C	C	

A – aluminium alloy.
B – manganese bronze.
C – 17.4 PH stainless steel.

hydraulically-actuated steering and trim control through angular displacement of the propeller shaft, providing greater propulsion, manoeuvring effectiveness and shallow water capabilities. In most applications, elimination of underwater shafts, struts and rudders results in marked improvement in vessel performance and efficiency.

This concept allows complete flexibility of engine location, weight placement and effective reduction of noise and vibration. Hydraulic steering and propeller depth control provide manoeuvrability and shallow draught capability limited only by the draft of the vessel itself.

Twin Disc Arneson Surface Drives serve the commercial, military and pleasure craft markets. Differentiated by torque capacity, these drives are available for use with petrol, diesel and gas-turbine engines up to approximately 7,500 kW.

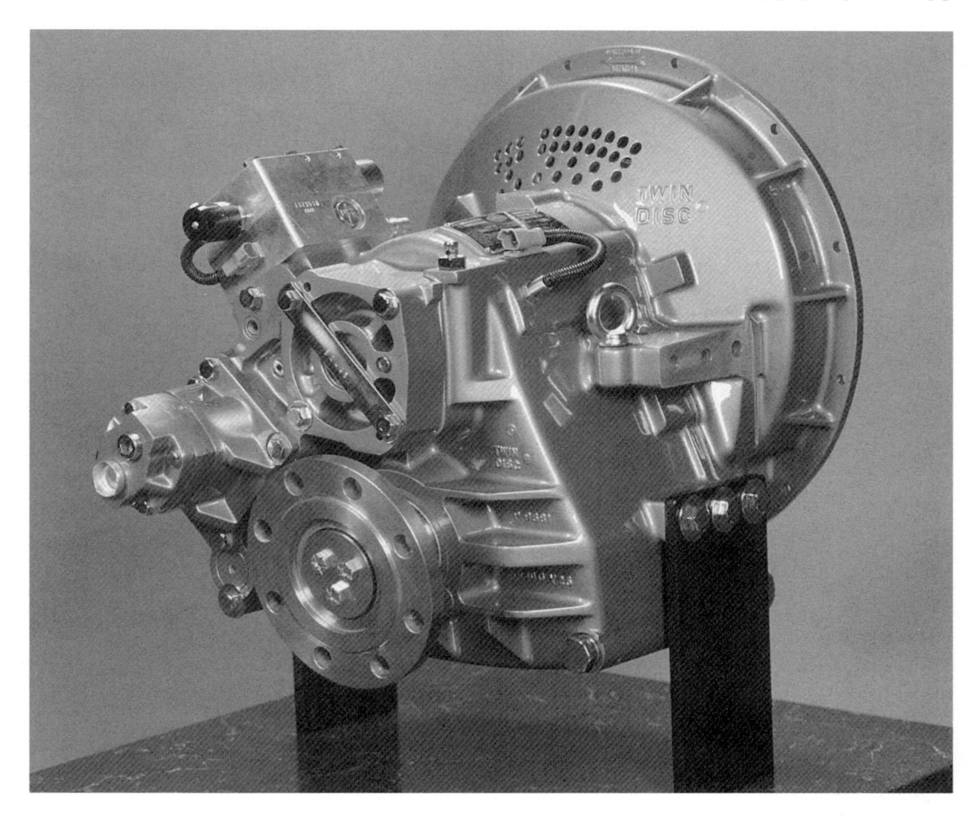

Twin Disc MG-5145 A QuickShift® marine transmission
0526567

AIR PROPELLERS

Company listing by country

Germany
Hoffmann Propeller GmbH & Co KG
MT-Propeller Entwicklung GmbH

United Kingdom
Air Vehicles Design and Engineering Ltd
Dowty Propellers

United States
Pacific Propeller Inc

Germany

Hoffmann Propeller GmbH & Co KG

PO Box 100339, Küpferlingstrasse 9, D-83022 Rosenheim, Germany

Tel: (+49 8031) 187 80
Fax: (+49 8031) 18 78 78
e-mail: info@hoffmann-prop.com
Web: www.hoffmann-prop.com

Richard Wurm, *Proprietor*
Johann Sterr, *Managing Director*
Peter Ihrenberger, *Sales Manager*

Hoffmann Propeller GmbH and Co KG was founded in 1955 by Ing Richard Wurm, starting with six employees. Now Hoffmann has 70 employees and the company is not only involved in the field of general aviation propellers but also in that of hovercraft. Propellers absorbing an input power up to 4,500 kW are designed and manufactured for these craft. Besides this, blades for wind energy converters, blowers and large fan blades for wind tunnel application in the automotive industry are in current production. Propeller overhaul and service is provided for all types of Hoffmann propellers as well as other manufacturers' propellers.

The company covers the following certifications: EASA 21J 083 for design and development of aircraft propellers; DE.21G.0014 for production of aircraft propellers and equipment; PAR 145 DE.145.0063 for repair of aircraft propellers and governors of all types; FAA BV5Y767M for repair and overhaul of aircraft propellers and accessories; CAA Hovercraft Approval for design and production of hovercraft propellers.

Hovercraft propellers manufactured by Hoffmann incorporate various special features. The erosion resistance of the composite materials used is greater than that of aluminium alloy. The wood laminations are reinforced with carbon or glass fibre which adds torsional strength and offers resistance to impact from particles entering the propeller disc.

For improved erosion protection the blades optionally have a special Irathane coating. The leading edge protection is made of stainless steel totally integrated into the airfoil profile. Additional protection against erosion is guaranteed by different types of erosion strip.

Since 1960, fixed-pitch, ground-adjustable and either hydraulically or mechanically controlled variable and reverse-pitch propellers, have been continuously built for air propeller driven boats, snow sledges and hovercraft.

ABS M-10 0007622

Hoffmann 3.6 m diameter five-blade propeller 0506798

Marine applications
Chaconsa, VCA-2/3: Two, two-blade ground-adjustable propellers, free, driven by about 150 kW, 2 m diameter.
Wärtsilä Larus: Four, four-blade hydraulically controlled propellers, forward and reverse pitch, driven by about 550 kW, 3 m diameter.
BHC AP1-88: Two four-blade ground-adjustable propellers per craft, ducted, driven by approximately 370 kW, 2.75 m diameter.
Chaconsa VCA-36: Two five-blade hydraulically controlled, forward and reverse pitch propellers (one left-hand, one right-hand rotation), free, driven by 1,000 kW, 4 m diameter.
Slingsby, Tropimere 6: Two three-blade ground-adjustable propellers, 1.10 m diameter.
Slingsby SAH 2200 (ex-1500): One three-blade reversible-pitch propeller, ducted, 1.40 m diameter.
Marineswift Thunderbolt 30: One five-blade ground-adjustable propeller, driven by approximately 200 kW, 1.12 m diameter.
CIS: Two hydraulically controlled propellers, ducted, driven by 257 kW, 3 m diameter and for a 25 m Hovercraft a five-blade hydraulically controlled propeller, ducted, driven by 1,800 kW, 3.6 m diameter.
Griffon 2000 TDXs for the Swedish Coast Guard: Two four-blade variable-pitch propellers, mechanically controlled, 1.80 m diameter, driven by 239 kW Deutz V8 diesel engines.
ABS M-10: Two, four-blade, low noise, ducted propellers, hydraulically controlled and with a diameter of 2.5 m. They are driven by Deutz 400 kW engines, type BF122513C.
Textron LACV-30: Two five-bladed, low-noise, hydraulically controlled propellers driven by Allison gas turbines at 970 kW each. Propeller diameter 3.35 m.
Griffon 8000 TD: Two four-bladed variable pitch propellers, hydraulically controlled with a diameter of 2.60 m absorbing 470 kW each at a maximum speed of 1,200 rpm.
BHT-130: Two 5-bladed, 3.5 m hydraulically variable pitch propellers are installed.
LSF-II: Two 5-bladed variable pitch propellers, hydraulically controlled with a diameter of 3.58 m, absorbing 3728 kW each at 1,300 rpm.

MT-Propeller Entwicklung GmbH

Flugplatz 1, D-94348 Atting, Germany

Tel: (+49 9429) 940 90
Fax: (+49 9429) 84 32
e-mail: sales@mt-propeller.com
Web: www.mt-propeller.com

Gerd Mühlbauer, *President*
Eric Greindl, *Sales*

This company was formed in 1982 and manufactures fixed, variable and ground adjustable propellers for aircraft, airships, flying boats and hovercraft. It has developed a range of 26 different models of electric hydraulic, constant-speed propellers. Most of the propellers are EASA and FAA approved. The largest diameter propeller built is understood to be 2.9 m for 650 kW, but designs can be undertaken up to 3.5 m for 2,500 kW and above. Company approvals include DE.21G.0008 for production and EASA.21J.020 for design and development.

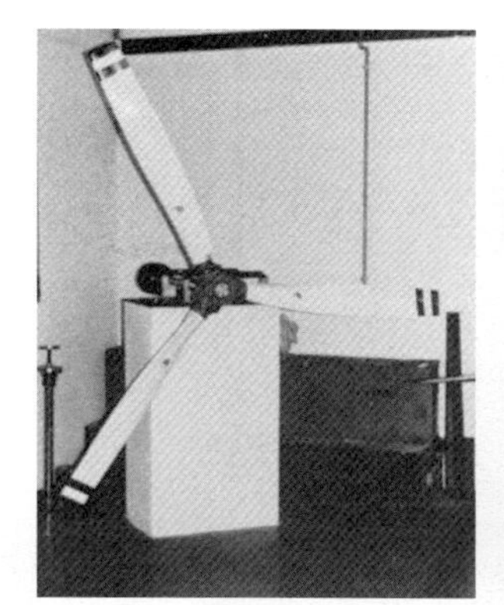

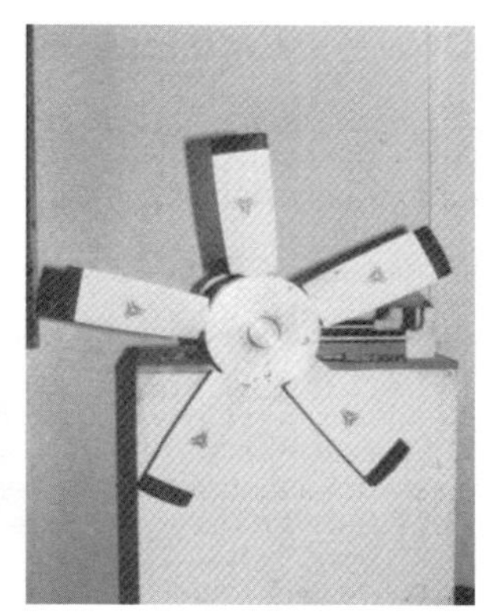

Examples of recent MT-Propeller Entwicklung propellers 0506799

United Kingdom

Air Vehicles Design and Engineering Ltd

Head office and factory:
Three Gates Road, Cowes, Isle of Wight, United Kingdom

Tel: (+44 1983) 29 31 94
Fax: (+44 1983) 29 19 87
e-mail: sales@airvehicles.co.uk
Web: www.airvehicles.co.uk

C B Eden, *Director*

Air Vehicles has been in the high-speed, air-supported marine market for over 30 years. Originally a builder of small/medium sized ACVs, the company now specialises in the design and manufacture of systems and components for fully amphibious ACVs and SESs. All Air Vehicles' systems and components are designed specifically to suit the particular project with emphasis on lightweight and performance. All components are designed and manufactured to comply with the British Hovercraft safety requirements which will usually ensure full compliance with all other classification societies.

Air propellers: Propellers are custom designed to suit the application and are usually the ducted type. Emphasis is placed on low tip speeds to give extremely low noise signatures but at high thrust levels. Current production includes large chord, five-blade types of 2.2 m diameter. Fixed and variable pitch types have been produced and larger diameters are available.

Propeller ducts: Associated propeller ducts are designed and manufactured currently up to a propeller diameter of 3.6 m and absorbing 2,163 kW. Ducts can be designed in aluminium or composite or a combination of materials to suit the project.

Lift fan systems: Air Vehicles is a leading designer and manufacturer of lift fans for ACVs and SESs. Custom designed for each project, axial, mixed flow and centrifugal types have been produced. Rotors are usually manufactured in marine aluminium alloy giving a robust fan with good fatigue life. Complete lift systems are also supplied with lightweight volutes and transmission systems designed to fully integrate with the craft design.

Interior systems: Air Vehicles designs and manufacturers lightweight bulkhead and partitioning systems for military and paramilitary ACV interiors. These include sleeping accommodation, galley, WC and dining areas.

Air Vehicles 2.2 m, five-bladed large chord propellers 0045324

Air Vehicles propeller duct for 3.6 m propeller 0045325

Dowty Propellers

A part of Smiths Aerospace
Anson Business Park, Cheltenham Road East, Gloucester GL2 9QN, UK

Tel: (+44 1452) 71 60 00
Fax: (+44 1452) 71 60 01
Web: www.smiths-aerospace.com

Jonathan Chestney, *Vice President and General Manager*
S Powers, *Operations Director*
M H Burden, *New Projects and Technical Director*
A R Cooper, *New Business and Spares Manager*

Dowty Propellers, an operating business of Smiths Aerospace Gloucester Limited, which is itself now part of the GE Aviation Group, has been designing and manufacturing propellers for aircraft since 1937 when the company was originally formed as Rotol Airscrews Ltd. For over 30 years the company has been actively engaged in propulsion systems for air cushion vehicles.

Applications include:

British Hovercraft Corporation SR. N5, SR. N6, 2.744 m diameter, four-blade hydraulic pitch control with reversing, single-propeller installation, produced in aluminium alloy and composite material construction
British Hovercraft Corporation SR. N6 Mk 6, 3.049 m diameter, four-blade, hydraulic pitch control with reversing, two-propeller installation, aluminium alloy blades
Mitsui PP15 3.201 m diameter, four-blade, hydraulic pitch control with reversing, single-propeller installation, aluminium alloy blades
Vosper Thornycroft VT2 4.116 m diameter, seven-blade, two-propeller installation (ducted), composite material blades
Textron Marine LCAC 3.582 m diameter, four-blade, two-propeller installation (ducted) composite material blades
Textron Marine C-7 2.438 m diameter, four-blade, two-propeller installation (ducted), composite material blades

The extensive corrosion and erosion problems associated with air cushion vehicles led Dowty to develop composite blades with all-over erosion protection. These blades, the latest of which incorporate advanced technology aerofoil sections unique to Dowty, offer the following advantages: low weight combined with high strength, internal carbon fibre spars for high integrity, freedom from corrosion, and easily repairable.

Dowty propellers for air cushion vehicles are designed to combine simple construction with safe operation. Techniques proven on ACVs have in turn been applied to and certificated on new generation general aviation, executive and commuter aircraft.

Dowty ducted propeller installation on the Textron Marine LCAC 0506800

In support of Dowty Propellers products worldwide, a network of repair and overhaul facilities are available. These facilities are strategically placed in the US, Europe and South East Asia.

Dowty Propellers Repair and Overhaul (UK)

Meteor Business Park, Cheltenham Road East, Gloucester GL2 9QL, UK

Tel: (+44 1452) 71 62 01
Fax: (+44 1452) 71 61 99

Dowty Propellers – Americas, Sterling (US)

114 Powers Court, Sterling, Virginia 20166, US

Tel: (+1 703) 421 44 30
Fax: (+1 703) 450 00 87

United States

Pacific Propeller Inc (PPI)

5802 South 228th Street, Kent, Washington 98032-1810, United States

Tel: (+1 253) 872 77 67
Fax: (+1 253) 872 72 21
e-mail: sales@pacprop.com
Web: www.pacificpropeller.com

Jeff Heikke, *President*

PPI manufactures, overhauls and sells propeller systems for aircraft and hovercraft use, having been in the business since 1946. It has manufactured over 10,000 metal propeller blades for single- and multi-engine installations and is the current supplier of propellers for the Canadian Coast Guard SR. N5, SR. N6 and Voyageur hovercraft. In addition, PPI-designed and -manufactured propeller blades are the only current production units approved for installation on the US Army LACV-30 hovercraft. By late 1991, over 500 AG200-1S blades had been supplied to the US Army for use on the LACV-30.

HC200-1S Propeller Blade

This is a hard-alloy derivative of the AG-series blades manufactured for aircraft use. It is designed to be tougher and more erosion-resistant to the effects of salt and sand spray. Typical hovercraft installations use the three-blade HSP 43D50 hub coupled to the Pratt & Whitney ST6 TwinPac or the Rolls-Royce Gnome turbine engines. Propeller diameter is 2.72 m. This configuration has been tested in excess of 1,193 kW and is safe for operation up to 2,300 rpm. Each blade weighs 22.2 kg and can be overhauled using standard propeller overhaul facilities.

MARINE PROPELLERS

Company listing by country

Australia
Veem Engineering Group

Denmark
Hundested Propeller A/S

Germany
Schottel-Werft
VA Tech Escher Wyss

Italy
Eliche Radice SpA

Japan
Kamome Propeller Company Ltd

Netherlands
DK Group Netherlands BV
Wärtsilä Marine Division

Norway
Servogear A/S

Sweden
Berg Propulsion AB
Rolls-Royce AB

Switzerland
Rolla SP Propellers SA

United Kingdom
Brunton's Propellers Ltd
Stone Manganese Marine Ltd
Teignbridge Propellers Ltd

United States
Michigan Wheel Corporation
Rolls-Royce Naval Marine Inc

Australia

Veem Engineering Group

22 Baile Road, Canning Vale, Western Australia 6155, Australia

Tel: (+61 8) 945 593 55
Fax: (+61 8) 945 593 33
e-mail: sales@veem.com.au
Web: www.veem.com.au

Brad Miocevich, *Managing Director – Marine Propulsion*
Greg Temby, *Marine Business Development Manager*

VEEM specialise in the design and CNC manufacture of fixed-pitch marine propellers for open water, nozzle and surface piercing configurations from 500 to 4,000 mm in diameter and up to 9 tonnes (cast weight).

Propellers are generally manufactured in nickel aluminium bronze, however, they can be manufactured in other materials if requested, for instance stainless steel.

VEEM also specialises in the design and manufacture of complete driveline packages from the gearbox coupling through to the rudder.

Propellers are manufactured in accordance with ISO-484 class 1 or class S as required.

VEEM is accredited to ISO-9001 and is also the agent for MJP Water-jets for the Australia and New Zealand region.

Veem fiveblade propeller 0044049

Denmark

Hundested Propeller A/S

Stadionvej 4, DK-3390 Hundested, Denmark

Tel: (+45 47 93) 71 17
Fax: (+45 47 93) 99 02
e-mail: hundested@hundestedpropeller.dk
Web: www.hundestedpropeller.dk

Mogens Christensen, *Managing Director*

Designer and manufacturer of controllable-pitch propellers since 1929. The current range of propellers suits engines from 120 to 2,500 kW.

Germany

Schottel GmbH & Co KG

Mainzer Strasse 99, D-56322 Spay/Rhein, Germany

Tel: (+49 2628) 610
Fax: (+49 2628) 613 00
e-mail: info@schottel.de
Web: www.schottel.de

Christophe Mourat, *Director, Marketing and Sales*

Schottel offers 360° azimuthing propulsion and manoeuvring systems as well as conventional propulsion packages up to 30 MW. The main area of application is vessels requiring the utmost in manoeuvrability. The company has a world wide sales and service network and a quality management system to DIN EN ISO 9001.

Seven passenger catamarans for Sydney Harbour, Australia, are successfully in operation. Each vessel is equipped with two Schottel Rudder-propellers (2 × 373 kW). The maximum speed is 24 kt. The ferries were built by NQEA and are operated by the State Transit Authority of Sydney on a regular service on the Paramatta River.

Dawn Frazer *and* Betty Cuthbert *fitted with Schottel Rudder-propellers* 0506803

VA Tech Escher Wyss

A subsidiary of the Andritz Group
Escher-Wyss-Strasse 25, D-88212 Ravensburg, Germany

Tel: (+49 751) 83 00
Fax: (+49 751) 83 23 96
e-mail: thomas.stehle@vatew.de
Web: www.escherwysspropellers.com
Telex: 732901

T Stehle, *Sales Manager*

The company has considerable experience in the design and production of controllable-pitch propellers, known as Escher Wyss CPPs. Many are supplied for frigates, corvettes, patrol boats, mine countermeasure vessels and special purpose vessels. Over 140 propellers have been supplied with an air ejection system to reduce noise from cavitation. Designs with seven-blade propellers developed in 1986/87 are in service successfully. Particular interest is paid to the propeller blade design with regard to hydroacoustic and hydrodynamic performance whereby five-axis CNC-machining of the complete blade is a standard procedure. Experience from the many designs made and employment of modern computer programs and techniques, secure a continuous development in this important field of high-performance propulsion systems. In total more than 1,900 CP propellers have been delivered with a maximum power per propeller of 34,000 kW and a maximum propeller diameter of 11 m.

A seven-blade controllable-pitch propeller design was supplied for the Blohm+Voss SES Corsair. The semi-submerged propellers for this vessel had the following characteristics:
Specifications
Diameter: 1,200 mm
Hub ratio: 0.32
Design pitch ratio: 1.75
Max shaft power: 2,560 kW
Rotational speed: 940 rpm

Two Escher Wyss 4.1 m diameter seven-bladed controllable-pitch frigate propellers 0069470

Italy

Eliche Radice SpA

Via Valtellina 45, I-20092 Cinisello Balsamo, Milan, Italy

Tel: (+39 02) 66 04 93 48
Fax: (+39 02) 612 76 88
e-mail: lucaradice@elicheradice.com
Web: www.elicheradice.com

Alfredo Radice, *Proprietor*
Carlo Radice, *Proprietor*

Manufacturer of fixed-pitch propellers for high-speed craft. Early Hovermarine HM2 craft were fitted with Radice propellers in stainless steel. Bronze propellers with two to five blades and diameters up to 3.5 m can be supplied.

Japan

Kamome Propeller Company Ltd

690 Kamiyabe-cho, Totsuka-ku, Yokohama 245-8542, Japan

Tel: (+81 45) 811 24 61
Fax: (+81 45) 811 94 44
e-mail: info@kamome-propeller.co.jp
Web: www.kamome-propeller.co.jp

Hiroshi Itazawa, *President*

Kamome Propeller Company designs and manufactures a wide range of controllable-pitch and fixed pitch propellers from 224 to 25,000 kW. Installations have included 5,600 mm diameter, controllable pitch propellers on a 170 m Ro-Ro vessel.

Netherlands

DK Group Netherlands BV

Weena 340, NL 3012 Rotterdam, The Netherlands

Tel: (+31) 10 213 18 45
e-mail: info@dkgroup.eu
Web: www.dkgroup.eu

Jorn Winkler, *Chief Executive Officer*

DK Group Netherlands BV was formed to commercialise the technical developments made by DKG in the use of surface piercing propellers. The Ventilated Wing Jet is a patented concept in marine propulsion and consists of an enclosed rotor and inlet much like a water-jet installation except that the rotor operates in the partially submerged condition.

Currently under development with MARIN, the device has considerable potential for use in air-lubricated vessels since it is an air-ventilated system. The concept has been tested on a Rigid Inflatable Boat (RIB) and development is continuing with use of Computational Fluid Dynamics (CFD). Engineering for production began at the end of 2005 and the first product versions were available in 2006.

Wärtsilä Marine Division

Lipsstraat 52, PO Box 6, NL-5150 BB Drunen, Netherlands

Tel: (+31 4163) 881 15
Fax: (+31 4163) 731 62
e-mail: jacqueline.peinenburg@wartsila.com
Web: www.wartsila.com

Established in 1934, the company produces fixed- and controllable-pitch propellers, side-thrusters and systems to its own design. There are 500 employees in the Dutch company. Wärtsilä also own 55 per cent of Wärtsilä CME Zhenjiang Propeller Co Ltd (W-CZP) which was established in 2004 in China, employing approximately 260 people.

The company has provided propellers for a number of high-speed craft including Fjellstrand catamarans for Turkey and Norway and the FBM Marine FDC 400 fast displacement catamaran.

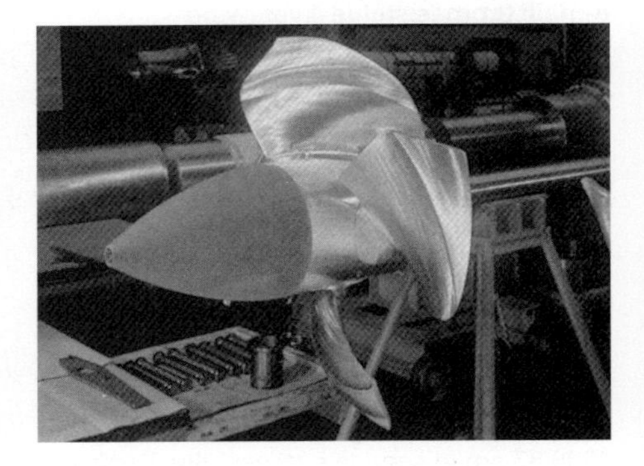

Lips transcavitating controllable-pitch propeller designed for the US Navy SES Sea Viking
0506805

Norway

Servogear A/S

N-5420 Rubbestadneset, Norway

Tel: (+47) 53 42 39 50
Fax: (+47) 53 42 39 99
e-mail: main.office@servogear.no
Web: www.servogear.no

Magne Møklebust, *Managing Director*
Arvld Kvernøy, *Operations Director*

Established in 1973, Servogear produces lightweight propulsion systems for high-speed craft. The systems are supplied as complete units comprising controllable-pitch propellers, shafting, gearboxes and rudders and also specifications for full shape in way of the underwater part of this system.

Since Servogear started production in 1975, over 900 gearboxes and propeller systems have been delivered of which more than 500 have been for high-speed craft. This range has covered engine power from 300 to 3,000 kW and ship speeds up to 50 kt.

Applications have been for seven triple-engine 45 kt patrol craft and six twin-engine 40 kt patrol craft, all for the Taiwanese Police. Fast ferry applications include the Holen and Lindstøl Skips catamarans and more recently on five 42 m catamarans being built by Damen Shipyards for Turkey. Brødrene Aa have also built fast 27 m and 23 m catamarans with Servogear Ecoflow Propulsors™.

Servogear propulsion system with 'Power Rudder' 0506806

Sweden

Berg Propulsion AB

Box 1005, SE-43090 Öckerö, Sweden

Tel: (+46 31) 97 65 00
Fax: (+46 31) 97 65 38
e-mail: info-eu@bergpropulsion.se
Web: www.bergpropulsion.se

Björn Svensson, *Chairman*
Håkan Svensson, *Managing Director*

Berg Propulsion AB is one of the world's leading companies in the design, development and manufacture of controllable-pitch propellers. Since 1928, Berg has manufactured and delivered more than 6,000 propulsion systems and has a worldwide service network.

Rolls-Royce AB

Kamewa – Controllable Pitch Propellers (CPP)
Kamewa Ulstein – Fixed Pitch Propellers (FPP)
PO Box 1010, SE-681 29 Kristinehamn, Sweden

Tel: (+46 550) 840 00
Fax: (+46 550) 181 90
Web: www.rolls-royce.com

Anders Johansson, *Commercial Marine*
Fredrik Bagge, *Naval Marine*

Rolls-Royce, through its Kamewa and Kamewa Ulstein product ranges, has accumulated experience from over 50 years of activity in the marine field. The company also has a cavitation laboratory with comprehensive facilities for advanced testing of various forms of marine propulsion devices.

Kamewa propellers are available for high-speed craft cover super-cavitating designs, which are modified to meet quiet and high-efficiency cruising conditions, wide-blade design for these conditions, skewed wide-blade propellers for extremely quiet cruising and the same design for ventilated blades. These propellers are manufactured in either stainless steel or nickel-aluminium-bronze. Propellers for patrol boats have been supplied over many years. Some of the most recent installations are for patrol boats and corvettes for Finland, Greece, Italy, South Korea, Malaysia, Morocco, Spain and Thailand.

Water-jet units are becoming an ever increasing alternative for high-speed craft propulsion. The Kamewa product range from Rolls-Royce has become a world leader (see

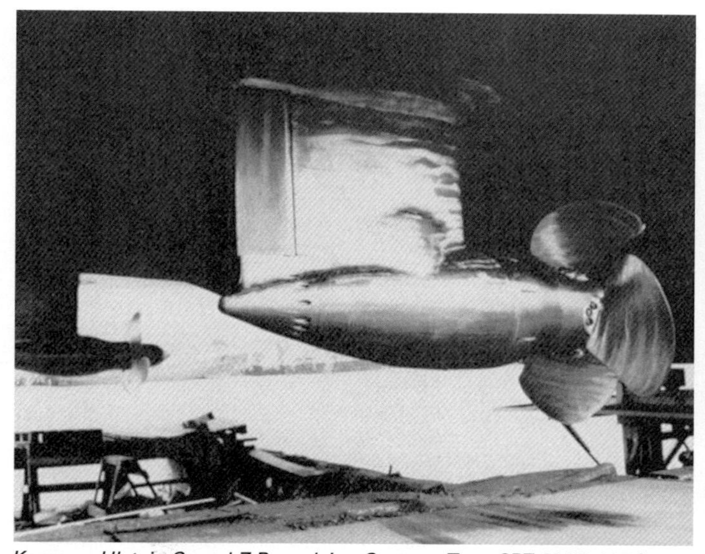

Kamewa Ulstein Speed-Z Propulsion Systems Type CPZ 60/40-125 fitted to the Westamaran 5000 Anne Lise, 28 kt thermo-cargo catamaran 0506808

Rolls-Royce Kamewa entry in the Water-jet units section).

Kamewa Ulstein designs and manufactures controllable-pitch propellers, tunnel thrusters, compass thrusters and the Speed-Z propulsion system. The Speed-Z system is marketed in the output range of 700 to 2,800 kW with controllable- and fixed-pitch propellers. Orders for the Speed-Z system have been received from the following shipyards: Westamarin A/S, Fjellstrand A/S, Broward Marine Inc, Baglietto Shipyard and CRN Shipyard. A total of 28 units had been delivered by 1998, with some having spent over 25,000 hours in operation. In comparison with conventional propeller systems an efficiency gain of 10 per cent is claimed. The system combines propulsion and steering functions. Other advantages of the Speed-Z propulsion system are reduced vibration and noise onboard, a cost-efficient compact installation, effective load control for protection of the drive motors and good manoeuvring capability.

Since Speed-Z is a traction (pulling) propeller there are no appendages in front of it and therefore the propeller acts in a homogeneous velocity field. These are ideal conditions for efficiency and avoidance of damaging cavitation. This has been substantiated by cavitation tests and prototypes in operation which produce no noticeable noise or vibration in the hull.

Another major advantage is the right-angle drive which allows the propeller to be installed in line with the water flow. A conventional propeller installation is always a compromise between keeping the shaft angle low to avoid harmful root cavitation and ensuring sufficient clearance between the propeller and the hull to avoid noise and vibration from the high-pressure pulses created by the propeller. The Speed-Z unit has, as mentioned, the best possible conditions of flow to the propeller giving stable cavitation conditions and minimal fluctuating forces.

A comparison has been made, based on model test tank results with the Speed-Z unit, with a conventional installation with sloping shaft, brackets and rudder. The propeller on the conventional installation had a diameter of 1.6 m with a maximum speed of 525 rpm, while the Speed-Z unit has a diameter of 1.25 m at 769 rpm. Even though the Speed-Z unit is higher loaded, that is, the power per unit area is greater, the propeller efficiency is 76 per cent at an engine output of 2,040 kW and ship speed of 28 kt. The corresponding efficiency of a conventional installation is about 72 per cent. When the resistance of the appendages was taken into account the improvement in efficiency was even more significant. The result of the model tests showed that the overall propulsive efficiency of the model fitted with the Speed-Z was 66 per cent at 28 kt which compared with 58 per cent for the conventional system as described previously. In this case the new Speed-Z gave an improvement in efficiency of eight per cent, that is to say a saving in installed power of 13.8 per cent to achieve the same speed.

The latest controllable pitch propeller, the Type-A, builds on the success of the well proven XF-5. It is available in a range of sizes from 0.5–75 MW. The CP-A/H hub has been specifically designed for vessels with a speed of over 30 knots.

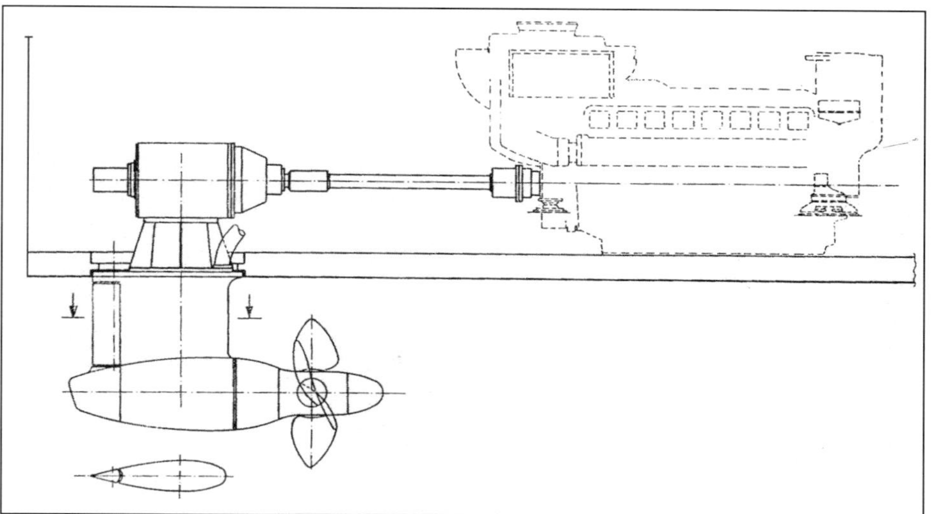

Layout of the Speed-Z Propulsion System Type CPZ 0506807

Finnish Navy's Turku *equipped with three Kamewa featherable controllable-pitch propellers, 3,000 kW per shaft, from Kamewa* 0506809

Switzerland

Rolla SP Propellers SA

Via Roncaglia 6, PO Box 109, CH-6883 Novazzano, Switzerland

Tel: (+41 91) 695 20 00
Fax: (+41 91) 695 20 01
e-mail: info@rolla-propellers.ch
Web: www.rolla-propellers.ch

Bruno Orlando, *President*
Otello Sattin, *Managing Director*
Franco Bertoni, *Finance Consultant*
Marzio Porro, *Design*
Franco Gotta, *Design*
Luca Libanori, *Sales*
Claudio Ghinlanda, *Research*
Mario Caponnetto, *Research, Consultant*
Philip Rolla, *Consultant*

Rolla SP Propellers SA designs and manufactures one-off and series surface piercing and fully submerged marine propellers for military, work and pleasure vessels.

The company was founded in 1963 by Philip Rolla. In 2004 Rolla SP Propellers SA was acquired by Twin Disc Inc., and is today a part of the Twin Disc Marine Group.

Surface piercing propellers are manufactured in NiBrAl up to 1.65 m diameter, in Stainless Steel up to 0.83 m diameter and fully submerged propellers are manufactured in NiBrAl up to 1.9 m diameter and cover a power range from 180 kW up to 4,200 kW. For each specific application, the propeller has a dedicated design and pattern. All the design codes are proprietary and have been developed in house. The propellers can be designed to meet any Classification Register rules.

Surface propellers are designed from systematic series based on tunnel testing. They permit Rolla to calculate torque-thrust co-efficients, efficiency and force figures for different shaft inclinations and propeller immersions. Fully submerged propellers are designed with a computer panel code method allowing the entire 3-D study and optimisation of the blades and the hub area. To properly interface a propeller with its intended application, the hull lines of any design of particular importance can be preliminarily analysed to predict resistance, trim and wake field at the propeller disc. Once the velocity field on the propeller has been computed, it is used as "inflow" for the design of the propeller with the panel method and CFD analysis.

All Rolla propellers are dynamically balanced and geometrically comply with the "S" Class of ISO 484/2 standards. A full CMM measurement report for Register purposes can be supplied. Other services provided by Rolla include the CFD performance and power prediction for any hull geometry including calm water and sea keeping study.

The CFD code used for hull analysis solves the Navier-Stokes equations for turbulent flow with finite volume methods. Both water and air are solved simultaneously. Waves and spray are computed with Volume Of Fluid (VOF) method called High Resolution Interface Capturing (HRIC). This method allows prediction of the pressure drag also at very high speed. Any geometry can be computed: hulls with tunnels, catamarans and so on. Complete CFD sea keeping analysis is also available.

The penetration of the large and fast motor yacht market has increased in recent years and includes customers such as Bertram, Ferretti Group, Cantieri Di Baia, Lazzara, Overmarine, Magnum, Sunseeker Azimut, Cantieri Navali di San Lorenzo and Cantieri di Pisa.

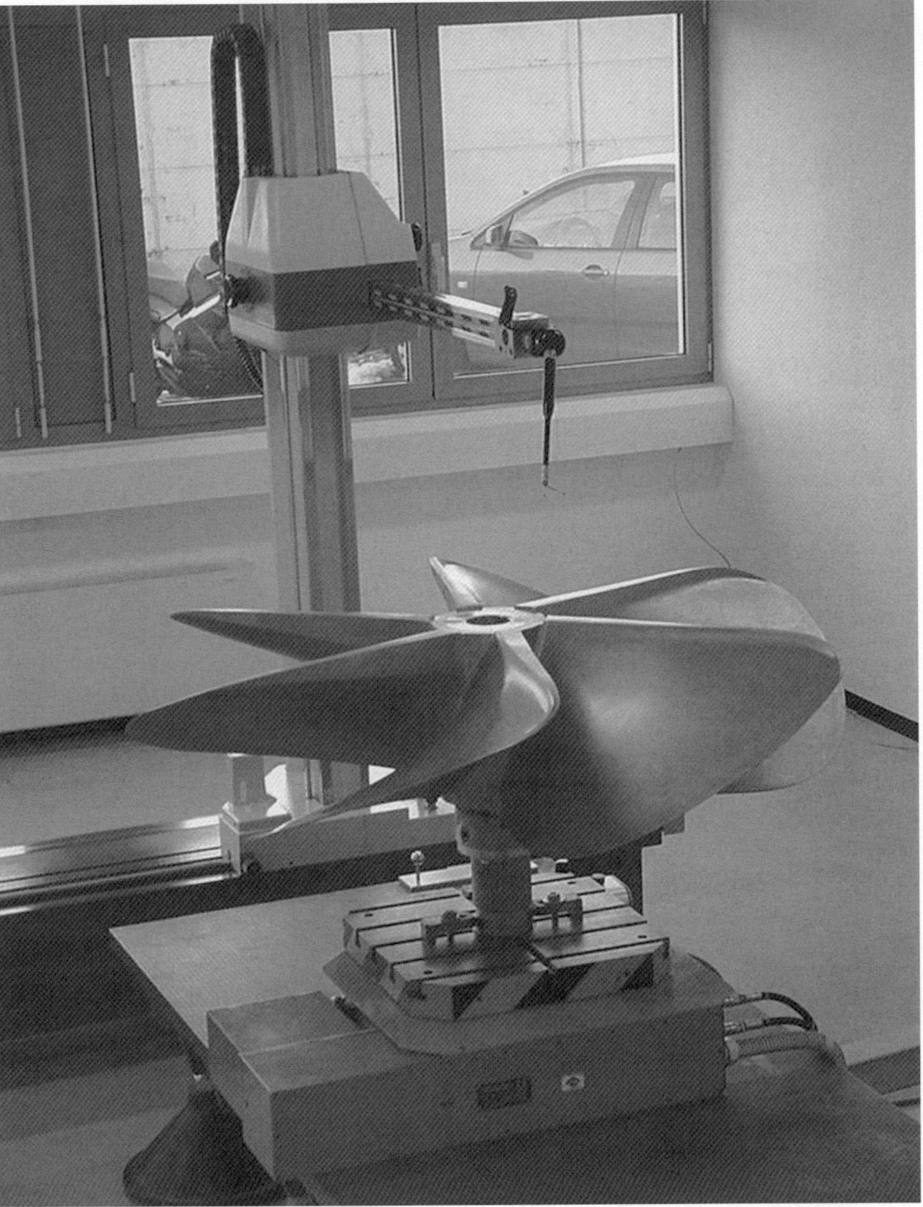

Propeller quality control at Rolla SP Propellers SA 1311257

Recent applications of Rolla propellers to military vessels

IAI Ramta, Dvora Mk1, Israel
21 m naval craft, two 1,400 kW, conventional shaft, 36 kt.
IAI Ramta, Super Dvora Mk2, Israel
23.4 m naval craft, two 1,700 kW/Arneson ASD 16 at 50 kt.
IAI Ramta, Super Dvora Mk3, Israel
25.5 m naval craft, two 2,040 kW/Arneson ASD-L 16 at 46 kt.
Intermarine-Biglianl V-GdF (Customs), Italy
35 m coast guard vessel, two 2,530 kW/ conventional shaft line at 35 kt.
Intermarine-FB Marine MIL 50-GdF (Customs), Italy
15 m coast guard vessel, four 530 kW/Trimax 2200 at 75 kt.
Intermarine-Bigliani VII – Gdf (Customs), Italy
36 m coast guard vessel, two 2,720 kW, conventional shaft line at 36 kt.
Yonka, Onuk MRPT 15, Turkey
15.4 m coast guard vessel, two 860 kW/ Arneson ASD 12-L at 62 kt.
Yonka, Onuk MRPT 16, Turkey
15.4 m Turkish Health Authority – Ambulance Vessel, two 790 kW Arneson ASD 12-L at 54 kt.

Rolla 1,168 mm diameter six-bladed surface piercing propeller 1311258

Yonka, Onuk MRTP 20, Turkey
20 m coast guard vessel, two 1200 kW/ Arneson ASD 14-at 60 kt
FB Marine RIB 33-Gdf (Customs), Italy
10 m coast guard vessel, two 162 kW/Trimax G-drive.
Halmatic, VSV, UK
16 m naval craft, two 530 kW/Arneson ASD 10 at 48 kt.
FB Marine/Halmatic, MIL 50, Italy/England
15 m naval vessel, three 470 kW/Trimax 2200 at 60 kt.
Lürssen, VSV, Germany
16 m naval craft, two 450 kW/Arneson ASD 10 at 50 kt.

United Kingdom

Brunton's Propellers Ltd

Oakwood Business Park, Stephenson Road West, Clacton-on-sea, Essex CO15 4TL, United Kingdom

Tel: (+44 1255) 42 00 05
Fax: (+44 1255) 42 77 75
e-mail: info@bruntons-propellers.com
Web: www.bruntons-propellers.com

Adrian Miles, *Managing Director*
Toby Ramsay, *Designer*

Owned by parent company Langham Industries Ltd., Brunton's Propellers Ltd is a company specialising in the design and manufacture of high performance propellers and sterngear for patrol boats, surface effect ships, hydrofoils, catamarans and luxury motor yachts. Propellers are manufactured to ISO Class S tolerances with fully CNC machined blade surfaces.

As well as meeting commercial requirements, Brunton's Propellers supplies the UK MoD and other navies around the world.

Propellers can be manufactured in high-tensile manganese bronze, nickel-aluminium bronze, Superston (a manganese aluminium bronze) and several grades of stainless steel.

Sterngear, tailshafts, bearings, seals, shaft brackets, sterntubes, thrustblocks and rudders are also produced enabling the shipyards to purchase the complete system from one source. Additionally Bruntons produce the Autoprop, an automatic pitch propeller and the Varifold folding propeller for sailing yachts and motor-sailers. They are also UK agents for the Q-SPD Surface Drive system for high-speed craft.

Brunton's Propellers is accredited to BS EN ISO 9001.2000 and manufactures to all major Classification Society approvals.

Other services offered include noise, vibration and propulsion analysis, repairs, spare parts and consultancy work.

Patrol boat propeller (class S tolerance) undergoing final inspection 0043254

A six blade Brunton CNC Machined Class S Propeller for a high-speed patrol craft 0142344

Stone Manganese Marine Ltd

A member of the Langham Industries Group
Dock Road, Birkenhead, Merseyside CH41 1DT, United Kingdom

Tel: (+44 151) 652 23 72
Fax: (+44 151) 652 23 77
e-mail: sales@stonemanganese.co.uk
Web: www.stonemanganese.co.uk

G Patience, *Managing Director*
L Bodger, *Technical Director*

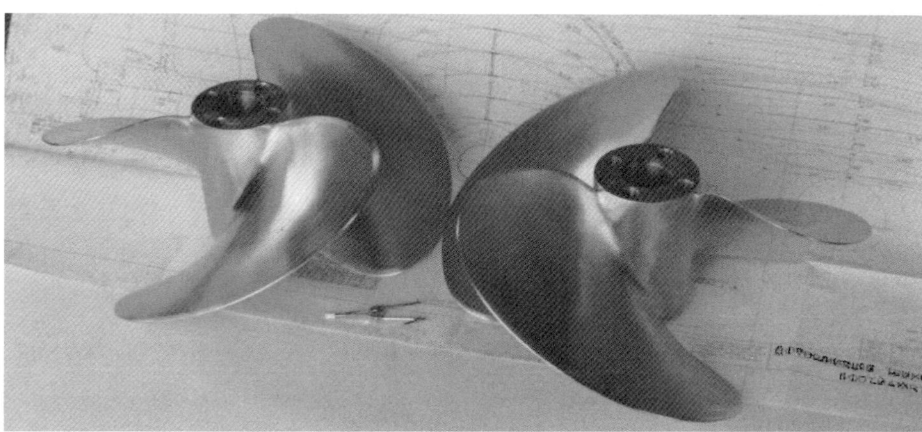

Stone Manganese Marine Ltd design propellers for high-speed craft 0506810

Stone Manganese Marine has been a leader in propeller design and manufacture for more than 100 years. The company and its associates operate a number of propeller manufacturing and repair facilities throughout the world. The product range covers all sizes of fixed-pitch marine propellers for all types of vessels, including military and high-speed craft.

The company's headquarters is in Birkenhead, UK, where a staff of qualified naval architects, engineers and metallurgists offer a comprehensive service to customers for the design and manufacture of fixed-pitch propellers. In addition to the servicing of fixed-pitch propellers, Stone Manganese Marine Shipcare is the sole European service agent for Nakashima Propeller Co of Japan and regularly services controllable pitch propellers and thrusters.

Stone Manganese Marine holds ISO 9001 accreditation and is approved by the major Classification Societies for the design, manufacture and servicing of propellers.

The company has a long association with the University of Newcastle upon Tyne, and has collaborated closely in the operation of the cavitation tunnel which has been responsible for significant research and development of propeller design technology for high-speed craft.

Teignbridge Propellers Ltd

Great Western Way, Ford Road, Brunel Industrial Estate, Newton Abbot, Devon TQ12 4AW, UK
Tel: (+44 1626) 33 33 77
Fax: (+44 1626) 36 07 83
e-mail: sales@teignbridge.co.uk
Web: www.teignbridge.co.uk

Sean Cummins, *Managing Director*
Jonathan Shaw, *Production Director*
T Hughes, *Sales and Marketing Director*
Endicott Fay, *Engineering Director*
John Orthaber, *Operations Strategy Director*
K Adams, *Finance Director*

Operating out of two separate manufacturing facilities in the UK, Teignbridge offers a

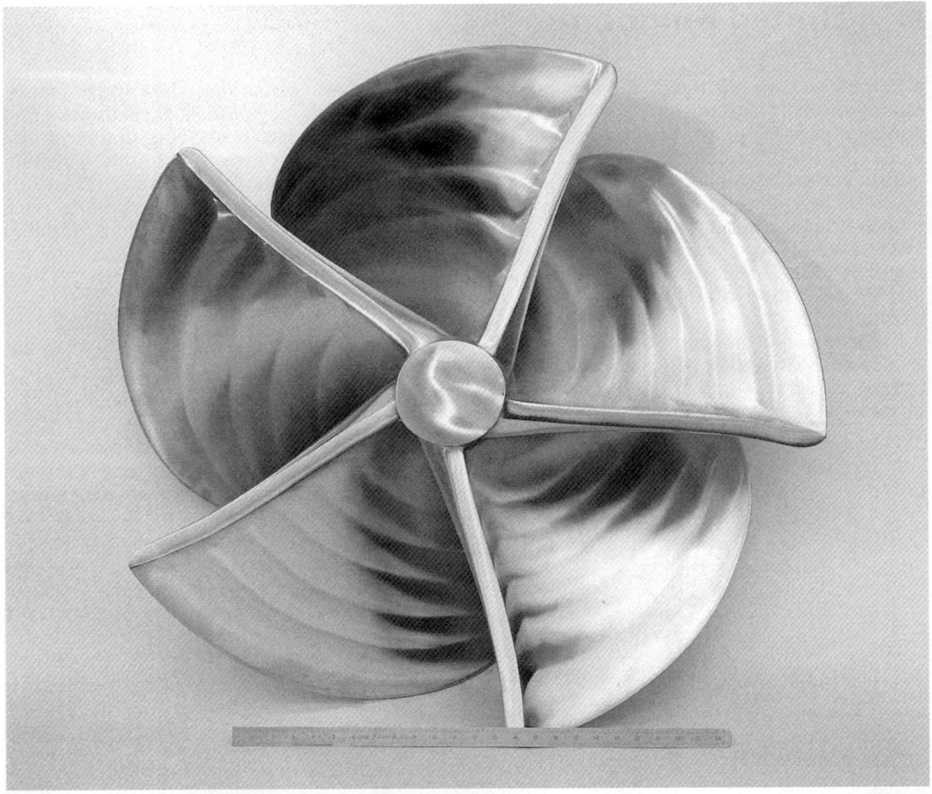
Aquaquin five-bladed surface-piercing high-speed propeller 0010063

complete design and manufacturing service to naval architects, operators and builders of vessels up to 25 m, focusing principally on production motor yachts.

For the manufacture of high-definition propellers to ISO Class I and Class S tolerances, the most up to date CAD/CAM technology is employed.

Propeller designs offered by Teignbridge include the Aquafoil, Aquaquad, Surface Piercing Propellers, Aquapoise 55, Aquapoise 65, Aquaqin, Highly Skewed propellers, Aquastar and Kaplon.

Teignbridge also produces a range of standard and custom-designed sterngear, rudders, shaft brackets and associated equipment. All products can be designed and manufactured to comply with the rules of any classification society.

Complete sterngear systems are produced up to 300 mm shaft diameter. Shaft materials used are Temet 25 duplex stainless steel, conventional stainless steels AISI316 and 304, mild steel and Temet 17, a high-strength stainless steel. Sterntubes are manufactured to suit GRP, aluminium, steel and wooden vessels and supplied with bearings and shaft seals for water or oil lubrication.

Shaft brackets and rudder systems are supplied either cast in nickel-aluminium bronze or fabricated in carbon or stainless steels. Sterngear components are produced in purpose-built machine shops on modern CNC machinery.

Standard range propellers

For fast vessels Teignbridge produces a standard series of 'sub-cavitating' propellers manufactured in nickel-aluminium bronze and manganese bronze. This range of propellers is manufactured as standard from

Aquaquin five-blade standard high-speed propeller 0507021

305 mm diameter to 1,020 mm diameter covering all common P/D ratios.

For non-planing applications propellers are generally manufactured to ISO Class 2 tolerances. For faster vessels Class 1 tolerances are applied.

Custom-designed propellers

These are specified for applications where standard series propellers cannot be used. Teignbridge custom-designed propellers fall into three categories: highly skewed; propellers for optimum performance in terms of speed, noise and vibration; and propellers for surface-piercing applications.

High-speed propellers are manufactured to Class S tolerances and are designed using the 'lifting surface' theory to optimise propeller geometry to suit each vessel's particular application.

Teignbridge's surface-piercing propellers are the result of many years of experience in designing partially submerged super-cavitating propellers. This has culminated in the design of a series of four- and five-blade propellers using the revolutionary 'cascade theory' for section design. This has allowed the design of very efficient, thin sections using nickel-aluminium bronze and Temet 25 duplex stainless steel. This propeller series has been optimised for use on high-speed, diesel powered ferries and military vessels between 30 and 55 kt. The four-blade series is used on highly loaded, high-speed applications.

United States

Michigan Wheel Corporation

1501 Buchanan Avenue South West, Grand Rapids, Michigan 49507, US

Tel: (+1 616) 452 69 41
Fax: (+1 616) 247 02 27
e-mail: info@miwheel.com
Web: www.miwheel.com

Stan Heide, *President and Chief Executive Officer*
Martin Ronis, *Vice-President of Marketing and Sales*
Ken Creech, *Chief Financial Officer*
William Herrick, *Director Outboard/Sterndrive Products*

Established in 1903, Michigan Wheel manufacturers marine fixed-pitch propellers from 0.075 to 2.5 m in diameter. Propellers are available in manganese bronze nibral (nickel bronze and aluminium) and stainless steel alloys. The company also manufactures a complete range of replacement propellers for inboard and stern-drive applications.

Michigan Wheel has now formed an association with the Federal Propellers and Commercial Propellers brands.

Rolls-Royce Naval Marine Inc

14850 Conference Centre Drive, Suite 100, Chantilly, VA 20151, US

Tel: (+1 703) 834 17 00
Fax: (+1 703) 709 60 86
e-mail: navalmarine.usa@rolls-royce.com
Web: www.rolls-royce.com

Patrick Marolda, *President*
Jorge Morales, *Vice President, Finance*
Christopher Cikanovich, *Vice President, Sales and Marketing*

Foundry location:
3719 Industrial Road, Pascagoula, Mississippi 39567, US

Tel: (+1 601) 762 07 28
Fax: (+1 601) 769 70 48

Peter J Lapp, *Director of Operations*

System engineering location:
190 Admiral Cochrane Drive, Annapolis, Maryland 21401, US

Tel: (+1 410) 224 21 30
Fax: (+1 410) 266 67 21

Thurman Harper, *VP Engineering and Technology*

Rolls-Royce is a global leader in naval and commercial marine integrated power, propulsion and motion control solutions. Over 2,000 commercial customers and 70 navies now use Rolls-Royce propulsion systems and products.

Products manufactured by Rolls-Royce in the USA include Bird-Johnson Naval Propulsors, New Generation Workwheels and 4–5.5 kW Gas Turbine alternators, KaMeWa propellers and water-jets, Ulstein and Aquamaster thrusters, Rauma Brattvaag deck machinery. Other products available include Bergen Diesels, Michell bearings and an extensive range of other equipment including underway replenishment equipment, stabilisers, steering gear, reduction gears, automation and control systems, modelling and simulation, and through-life support.

WATER-JET UNITS

Company listing by country

Canada
Alpha Power Jets

Finland
Rolls-Royce Oy AB

Italy
Castoldi SpA

Japan
Kawasaki Heavy Industries Ltd (GasTurbine & Machinery Company)
Mitsubishi Heavy Industries Ltd

Netherlands
Wärtsila Propulsion Jets BV

New Zealand
HamiltonJet

Sweden
MJP water-jets
Rolls-Royce AB

United Kingdom
Ultra Dynamics Ltd

United States
North American Marine Jet Inc
Rolls-Royce Commercial Marine
Rolls-Royce Naval Marine Inc

Canada

Alpha Power Jets

47 rue Dalhousie, Quebec G1K 8S3, Canada

Tel: (+1 418) 694 09 09
Fax: (+1 418) 694 07 61
e-mail: rtm@r-t-m.com
Web: www.r-t-m.com

Pierre Tremblay, *Sales and Marketing Director*

Alpha Power Jets introduced a stainless steel water-jet system in 1998. There are currently two models in the range, the Model 356 for fast planing vessels and the Model 420 for workboats, both capable of absorbing input powers of up to 400 kW. Both models are designed for commercial operation with the stainless steel construction offering low maintenance requirements.

The company is linked to Recherges et Travaux Maritimes Constructions which designs and builds small workboats and passenger vessels using Alpha Power water-jets.

Twin Alpha Power water-jet installation 0125136

Finland

Rolls-Royce OY AB

PO Box 579, FIN-67701 Kokkola, Finland

Tel: (+358 68) 32 45 00
Fax: (+358 68) 32 45 11
Web: www.rolls-royce.com

In 2002 Rolls-Royce aquired Kamewa/FF Jet AB and consolidated its position as a leading supplier of water-jets to this industry. This water-jet product range, manufactured in Kokkola, Finland is marketed through the Kamewa sales network (see Rolls-Royce Kamewa entry under Sweden).

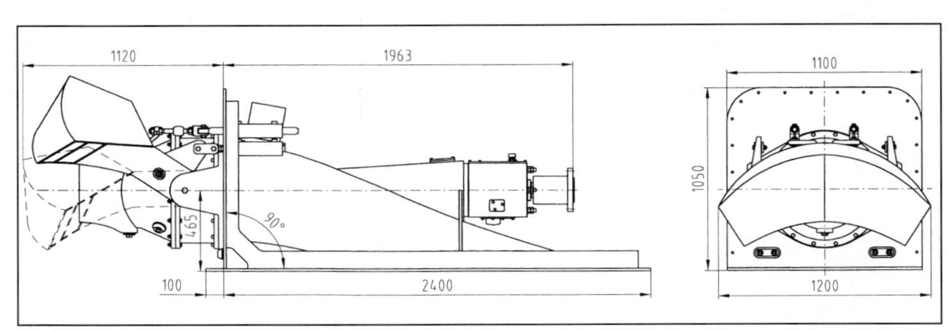

Arrangement of Kamewa FF 550 unit 0103310

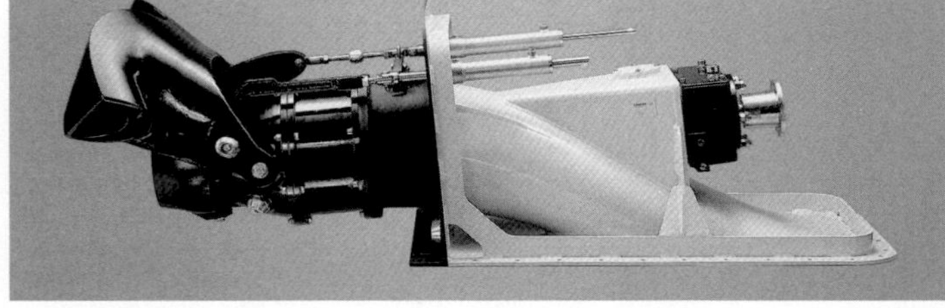

Kamewa FF375 unit 0536823

Application of Kamewa FF410 units 0536821

Model	Power (kW)	Impeller Diameter mm	Max rpm Impeller Shaft	Dry Weight (kg)
FF Series				
FF670	2,300	–	1,360	1,970
FF600	1,800	–	1,800	1,420
FF550	1,500	550	1,650	960
FF500	1,400	500	1,900	840
FF450S	–1,100	450	2,100	510
FF450	600	450	2,100	460
FF410S	1,000	410	2,300	470
FF410	600	410	2,300	420
FF375S	700	375	2,500	335
FF375	350	375	2,500	309
FF340	530	340	2,800	270
FF310	–500	310	3,000	242
FF270	370	270	3,500	155
FF240	260	240	4,000	127
A Series				
A40	1,400	Conical impeller	–	770
A45	1,650	Conical impeller	–	990
A50	2,100	Conical impeller	–	1,380
A56	2,800	Conical impeller	–	1,860

Italy

Castoldi SpA

Strada Provinciale 114 n.10, I-20080 Albairate, Milan, Italy

Tel: (+39 02) 940 18 81
Fax: (+39 029) 401 88 50
e-mail: info@castoldijet.it
Web: www.castoldijet.it

Development of Castoldi Water-jet units began in 1958 with over 30,000 sold to date. They are now available for fast craft in the 4 to 30 m range.

All the Castoldi units feature a single-stage axial flow impeller; the casings are built in lightweight aluminium alloy which is very durable being hard anodised up to 60 microns. The impeller, the impeller shaft and many other parts are made of high quality stainless steel.

The Castoldi drives have several features that make them stand out from other Water-jet units: a built-in gearbox with several gear wheel ratios for adapting the power and rpm characteristics of the engine to the jet drive; an integral mechanical disconnecting clutch (hydraulic multi-disc clutch for Turbodrive 238, 340, 400 and 490 H.C.) allowing a true neutral condition; a remotely operated movable debris screen rake for cleaning the water intake; all-oil lubricated bearings and duplex stainless steel micro-casted impeller rotating in an impeller housing, protected by a replacable titanium liner. The Castoldi Water-jet units are equipped with specially designed mechanical/electronic/hydraulic controls which make them easy to operate.

MC Series

Model	Dry weight, kg	Impeller diameter, mm	Number of gear wheel ratios in integral gear box	Maximum power input, kW
Jet 03	25	163	–	44
Jet 05	75	200	–	175
Turbodrive 238	123	238	25	250
Turbodrive 238 H.C.	130	238	18	309
Turbodrive 340 H.C.	285	340	25	625
Turbodrive 400 H.C.	469	400	21	882
Turbodrive 490 H.C.	890	490	20	1,324

Castoldi Turbodrive 238 H.C. water-jet drive 1195986

Castoldi Turbodrive 490 H.C. water-jet drive 1195987

Castoldi Turbodrive 400 H.C. water-jet drive 1195987

Castoldi Jet 05 water-jet drive 1195985

Japan

Kawasaki Heavy Industries, Ltd (Gas Turbine & Machinery Company)

Tokyo Head Office
World Trade Centre Building, 4-1 Hamamatsu-cho 2-chome, Minato-ku, Tokyo 105-6116, Japan

Tel: (+81 3) 34 35 21 11
Fax: (+81 3) 34 36 30 37
Web: www.khi.co.jp

Kobe Works
1-1 Higashi Kawasaki-cho 1-chome, Chuo-ku, Kobe Hyogo 650-8680, Japan

Tel: (+81 78) 371 95 30
Fax: (+81 78) 371 95 68

S Hasgawa, *Vice President*
T Yoshino, *Associate Director*
H Mizukawa, *Senior Manager of Marine Machinery Department*

Kawasaki KPJ-54A water-jet unit 0507104

Kawasaki has been manufacturing PJ-20 water-jet propulsors for Jetfoil since 1989 and has developed a KPJ-A series of water-jet propulsors for high-speed displacement vessels. The first KPJ-A unit was completed in 1993, four KPJ-169A units were delivered in 1994, four KPJ- 54A units were delivered in 1996 and an additional four in 1997.

PJ-20
This unit was designed to match the Rockwell International Corporation units supplied for hydrofoils such as the Jetfoil, and this business was transferred to Kawasaki in 1987.

Maximum continuous input power for these units is 2,795 kW at 2,060 rpm.

KPJ-A Series
Kawasaki completed the first unit (KPJ-1A) and made a load test in Kawasaki's dry dock to prove its efficiency and strength in 1993.

Model: KPJ-1A (43A equivalent)
Type: Single-stage axial flow, flush inlet
Input power: 515 kW

Four much larger units (KPJ-169A) were installed on the AMD1500 Mk II Kawasaki Jet Piercer in 1994.

Model: KPJ-169A
Type: Single-stage axial flow, flush inlet
Input power: 5,420 kW.

The KPJ-A series accepts power inputs from 200 to 20,000 kW.

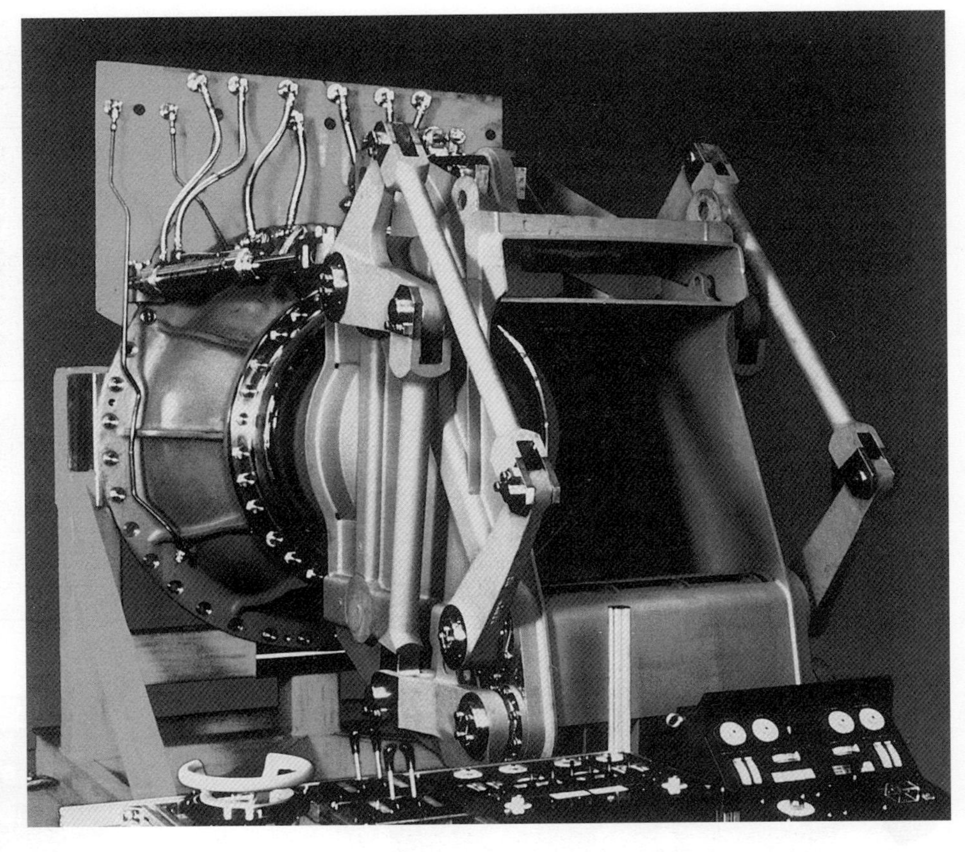

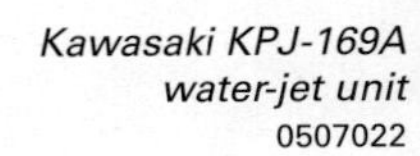

Kawasaki KPJ-169A water-jet unit
0507022

Mitsubishi Heavy Industries Ltd

16-5 Konan 2-chome, Minato-ku, Tokyo 108-8215, Japan

Tel: (+81 3) 67 16 31 11
Fax: (+81 3) 67 16 58 00
Web: www.mhi.co.jp

The Mitsubishi company has produced water-jet units for Japanese monohull fast ferries and for the Mitsubishi Super Shuttle 400 diesel-driven hydrofoil catamaran. Currently two models are offered: the MWJ-900A with a maximum input of 4042 kW and a rotating speed of 850 rpm., and the MWJ-800A with a maximum input of 2278 kW and a rotating speed of 900 rpm. A novel feature of the units is the use of a double-cascade type of impeller and a vane cascade arrangement for reverse thrust.

Netherlands

Wärtsila Propulsion Jets B.V.

(Formerly LIPS Jets BV)
Lipstraat 52, NL-5150 BB Drunen, Netherlands

Tel: (+31 416) 38 81 15
Fax: (+31 416) 37 31 62
e-mail: jacqueline.peinenburg@wartsila.com
Web:www.wartsila.com

Hanno Schoohman, *General Manager*
Johan Huber, *Sales and Products Manager*

LIPS's involvement in water-jets started in 1985 in co-operation with Milan-based Riva Calzoni. In 1998, the company merged with John Crane to form John Crane – LIPS Marine Propulsion Systems. In 2000 it was taken over by the Smiths Group and in 2002 by Wärtsilä. Wärtsila Propulsion Jets is an internationally recognised company with the expertise, technical staff and automated facilities to design and build water-jet propulsions systems that suit the market requirements.

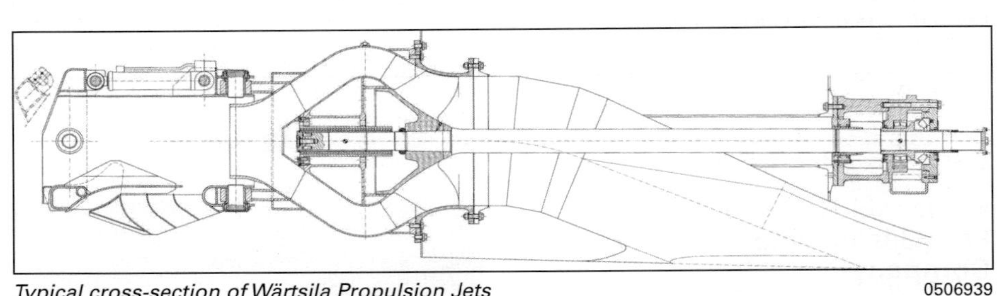

Typical cross-section of Wärtsila Propulsion Jets 0506939

Main features of Wärtsila water-jets
At present, Wärtsila Propulsion Jets designs and markets water-jet propulsion systems starting at a jet type suitable for approximately 500 kW up to the size demanded by the market. Specialist installations successfully operating included a 20 MW jet built according to NATO shock requirements with a specialised high-speed full power crash-stop facility, and a waterjet that supplies thrust in all directions through a 360° revolving nozzle.

Pump design
To be able to select and optimise a water-jet propulsion system, taking into consideration the special requirements of each specific application, it was decided to offer two different mixed-flow pump designs. The newly developed six-bladed water-jet (type E) combines high efficiency with an excellent cavitation margin. This cavitation margin allows selection of a very compact and light installation, making this water-jet ideal for relatively high ship speeds. As an alternative, the three-bladed water-jet (type D) is also offered. This pump offers the highest available efficiency at relatively low ship speeds. The cavitation characteristics of this more axial flow type of pump result in a slightly larger water-jet. The impellers are cast in high resistant stainless steel and dynamically balanced. The impeller is hydraulically fitted to the shaft.

Water-jet assembly
Each unit is fabricated in welded stainless steel. This allows selection of the optimum water-jet size for each specific project. This is illustrated by the fact that water-jet sizes 114E, 120E, 145D and 150D have been supplied for different vessel designs, but are all driven by the same engines rated at 7,200 kW.

Intake duct
For each specific application the optimum inlet is designed to match the required flow at the craft design speed and power with the optimum inlet velocity ratio to ensure high efficiency in a cavitation free condition. Full scale measurement, wind tunnel experiments and cavitation tunnel tests are used to validate the CFD models.

Thrust bearing
The thrust bearing is located inside the craft, which makes it easily accessible for inspection or maintenance without the need

for dry docking and water-jet dismantling. This position eliminates the risk of water penetration into the lubrication oil and eliminates the risk of water pollution by lubrication oil.

Water lubricated radial bearing
Inside the hub the shaft/impeller is supported by a water lubricated radial bearing. A very high reliability is achieved by avoiding complex mechanical components outside the craft, which require critical sealing in order to safeguard oil lubrication. The rubber cutlass bearing can be inspected and replaced without dry docking, is not subject to sudden failure and has proven to be a reliable long-life component.

Hydraulic system
The hydraulic-operated steering and reversing jetavator has been designed such that the equipment is supplied with inboard hydraulic cylinders to avoid damage to hydraulic cables and feedback cables. Each water-jet is delivered with individual hydraulic power pack, pump and lubrication system.

Four steerable water-jets type LJ-115-DL for Stena Sea Lynx II 0506938

Lipstronic/j control systems
An important and integrated part of the water-jet propulsion system is the Lipstronic/j control system. Safety and good manoeuverability are the key parameters.

The Lipstronic/j intelligent manoeuvring aid system allows the master to operate the craft by a single three-axis joystick. The water-jet steering-angles bucket positions and diesel-engine speed are automatically set such that the maximum thrust and turning moment becomes available in the sense as commanded by the master, with the minimum possible environmental burden.

The Lipstronic/j (DnV type approved) auto-steering control system is adaptive, freeing the master from setting autopilot parameters. Trials have showed a significant course keeping improvement compared with traditional autopilots.

There are also weight and installation cost savings by application of field-bus technologies. On conventional craft every single signal needs a wire. Wärtsila Propulsion Jets applies single wire (serial) communication for a multitude of signals.

Integrity is achieved by proper redundant design. Single (component) failures should not cause dangerous situations that is the Lipstronic/j system must be fault redundant. Wärtsila Propulsion Jets applies Failure Mode Effect Analysis (FMEA) techniques to achieve the highest possible level of system integrity.

New Zealand

HamiltonJet

C W F Hamilton & Company Ltd
Head office and factory
PO Box 709, Christchurch, New Zealand

Tel: (+64 3) 962 05 30
Fax: (+64 3) 962 05 34
e-mail: marketing@hamjet.co.nz
Web: www.hamjet.co.nz

European office
HamiltonJet (UK) Ltd
Unit 26, The Birches Industrial Estate, East Grinstead, West Sussex RH19 1XZ, United Kingdom

Tel: (+44 1342) 31 34 37
Fax: (+44 1342) 31 34 38
e-mail: info@hamjetuk.com

US office
HamiltonJet Inc
1111 North West Ballard Way, Seattle, WA, 98107, United States

Tel: (+1 206) 784 84 00
Fax: (+1 206) 783 73 23
e-mail: marketing@hamiltonjet.com
Web: www.hamiltonjet.com

Designer and manufacturer of water-jets for over 50 years, HamiltonJet now offers models capable of absorbing power inputs up to 3,500 kW. With in excess of 35,000 units installed worldwide, HamiltonJet water-jets are widely used in fast ferries, patrol craft, naval vessels, fishing boats, crew boats, rescue craft, pilot tenders, fire boats and pleasure cruisers. HamiltonJet liaises with designers, builders and operators from concept to commissioning, providing computer speed predictions, detailed installation advice, commissioning assistance and training programmes. Logistic support for all installations is provided by a worldwide network of factory trained sales and service distributors, who are supported by a factory-based field service team.

The current HamiltonJet product range includes the HJ Series and the larger HM Series.

Common design features
All HamiltonJet models are designed and manufactured to the standards of the world's leading certifying authorities and are manufactured from corrosion-resistant materials and supplied, assembled and tested as a complete propulsion package.

Each model incorporates:
matched intake and hull transition duct with integral protection screen
high-efficiency mixed-flow-type pump
optimised tailpipe and discharge nozzle
high resistance to cavitation
integral steering and split duct-type astern thrust deflectors
integral control systems
inboard thrust bearing and hydraulic control components for easy servicing.

Intake transition/Intake
Configured to mount inboard at the stern, Hamilton water-jets draw water through a factory-supplied intake transition which is manufactured in materials to suit the hull. A protection screen is integral with the intake.

HJ Series Specifications

Model	A (mm)	B (mm)	C (mm)	D (mm)	E (mm)	F (mm)	G (mm)
HJ212	450	221	762	732	440	386	450
HJ213	413	249	762	720	420	386	450
HJ241	424	284	829	824	491	502	431
HJ274	570	302	1,100	840	548	470	608
HJ292	681	330	1,180	862	550	495	608
HJ322	866	371	1,380	980	637	550	680
HJ364	937	420	1,634	901	708	621	747
HJ403	1,053	474	1,723	1,230	752	690	803
HJ422	1,380	484	2,082	1,090	940	960	964

The transition duct transmits the force of the generated thrust to the hull bottom and engine stringers so fore and aft propulsor loads are not borne by the transom or engine, eliminating the need for additional strengthening in these areas.

The main thrust bearing is incorporated inboard in the intake, a one-piece casting providing a rigid housing to ensure lifelong alignment.

Pump

Mounted on a precision stainless steel mainshaft, the mixed-flow-type impeller in each jet is rated to absorb full design engine power at full shaft rpm.

Power is therefore constantly absorbed regardless of boat speed, eliminating the possibility of overloading the engine. Impellers are selected to give good cavitation resistance throughout a wide speed range, ensuring full engine power can be applied at low boat speeds for quick acceleration.

The flexibility of the HamiltonJet design is evident in its ability to operate effectively in off-design conditions, such as that experienced below the planing threshold, in adverse sea conditions or with an overloaded or fouled hull.

Stator and nozzle

Water flow exiting the pump unit passes through the stator vanes where the rotation velocity components are removed to ensure a straight uniform flow is presented to the discharge nozzle. HamiltonJet selects the optimum nozzle size for each application.

Control functions

On Hamilton water-jets, the steering and astern deflectors are independent but, when used in conjunction with each other, can effect complex vessel manoeuvres without the need for complicated control systems.

Steering is affected by the deflection of the jet stream to port or starboard by an integral balanced JT steering deflector. This deflector is designed to maximise lateral thrust with minimum loss of forward thrust while maintaining lightest operating loads.

Ahead/astern function is achieved with an integral split duct astern deflector. Designed to provide maximum astern thrust under all conditions of boat speed, water depth and throttle opening, the split duct angles the astern jetstream down to clear the transom and to the sides to retain a steering thrust component. This arrangement vectors the jetstream away from the intake, avoiding recycling and resulting in astern thrust generation up to 60 per cent of ahead thrust.

With the astern deflector fully raised, full-ahead thrust is available. With the deflector in the lowered position, full-astern thrust is available. In both positions, full-independent steering effect is available for rotating the craft. By positioning the astern deflector at the intermediate 'zero-speed' setting; ahead and astern thrusts are equalised for holding the craft on station, but with full steering function still available for rotational control.

Control systems

A number of HamiltonJet-packaged control systems is available for interface between the control station(s) and the jet(s). Depending on the jet model, options include simple manual cable, hydraulic, electric or electronic. Where applicable, the hydraulic systems are an integral part of the jet including jet-driven hydraulic pump and oil cooler cast into the jet intake.

All systems are proportional so deflector movement follows control lever movement and the electronic options can be configured to interface with engine, gearbox and autopilot circuits.

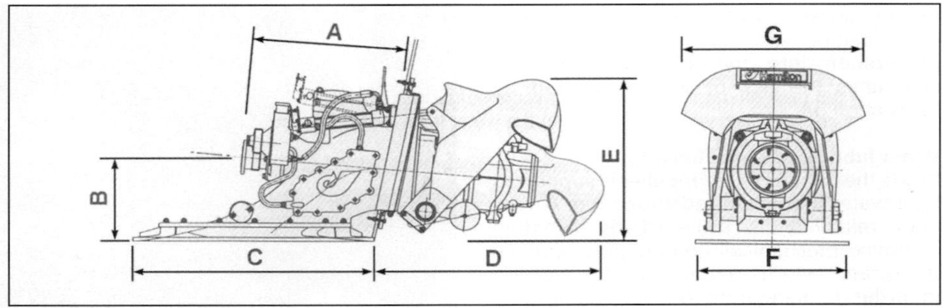

Hamilton Jet HJ Series 0125050

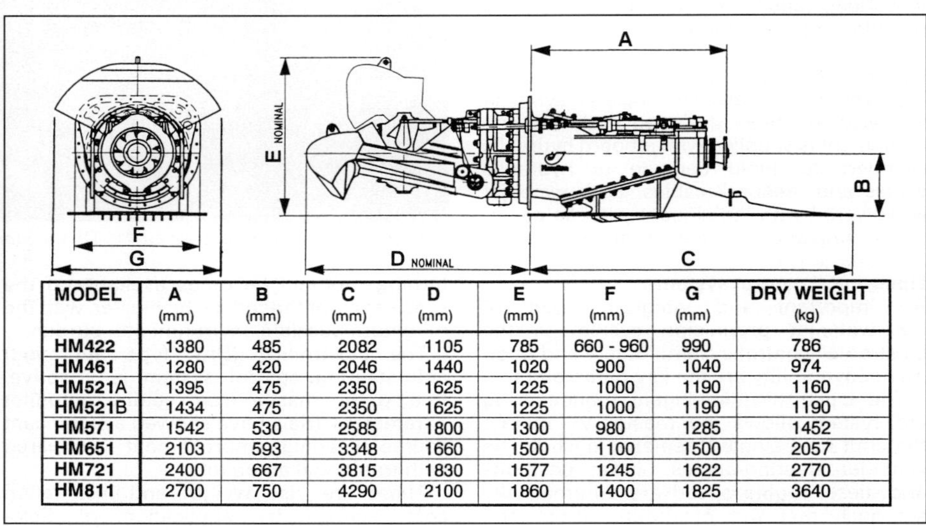

MODEL	A (mm)	B (mm)	C (mm)	D (mm)	E (mm)	F (mm)	G (mm)	DRY WEIGHT (kg)
HM422	1380	485	2082	1105	785	660 - 960	990	786
HM461	1280	420	2046	1440	1020	900	1040	974
HM521A	1395	475	2350	1625	1225	1000	1190	1160
HM521B	1434	475	2350	1625	1225	1000	1190	1190
HM571	1542	530	2585	1800	1300	980	1285	1452
HM651	2103	593	3348	1660	1500	1100	1500	2057
HM721	2400	667	3815	1830	1577	1245	1622	2770
HM811	2700	750	4290	2100	1860	1400	1825	3640

Hamilton Jet HM Series 0084069

HJ Series

These are high-efficiency single-stage units, primarily for high-speed work and patrol craft, fast ferries and pleasure cruisers typically up to 20 m. A large number of models and impeller rating combinations mean these jets can generally be direct driven by a wide range of engine and gearbox combinations, or in some cases, direct driven by petrol or diesel marine engines.

Model HJ212

(for trailerable craft only)
Max power input: 260 kW at 3,950 – 4,500 rpm
Impeller diameter: 215 mm (four rating options plus "Turbo" impeller options for use with high-power engines or in white water conditions).

Typical 212 application (twin)

Vessel: 7.5 m 28 passenger tour boat
Speed: 35 kt
Engines: Chevrolet 350 cld petrol
Drive: direct
Operator: Kawarau Jet, Queenstown, New Zealand

Model HJ213

Max power input: 260 kW at 3,950–4,500 rpm
Impeller diameter: 215 mm (eight rating options)

Typical 213 application (single)

Vessel: 7.6 m Recreational Fishing Boat
Speed: 37 kt
Engine: Redline 460 cld petrol
Drive: direct
Operator: Private, Washington, US

Model HJ241

Max power input: 260 kW at 3,250–4,000 rpm
Impeller diameter: 240 mm (12 rating options)

Typical 241 application (twin)

Vessel: 11.6 m Crew/Supply Boat
Speed: 37 kt
Engines: Mercruiser D4.2L diesels
Drive: direct
Builder: Motomarine SA, Greece

Model HJ274

Max power input: 330 kW at 2,930–3,300 rpm
Impeller diameter: 270 mm (20 rating options)

Typical 274 application (twin)

Vessel: Combat support boat
Speed: 30 kt
Engine: Yanmar 6LP-DTE diesels
Drive: Gearbox
Builder: RTK Marine, Poole, UK

Model HJ292

Max power input: 400 kW at 2,650–3,000 rpm
Impeller diameter: 290 mm (15 rating options)

Typical 292 application (single)

Vessel: 11.1 m Pleasure Cruiser
Speed: 29 kt
Engine: Yanmar 6LY-STE diesel
Drive: direct
Builder: The Hinckley Co, Massachusetts, US

Model HJ322

Max power input: 500 kW
Impeller diameter: 320 mm (15 rating options)

Typical 322 application (triple)

Vessel: Rescue boat
Speed: 33 kt
Engines: Volvo diesels
Drive: gearbox
Operator: Dutch rescue authority

Model HJ364

Max power input: 670 kW at 2,300–2,500 rpm
Impeller diameter: 360 mm (17 rating options)

Introduced in 2005 to replace the HJ362, the HJ364 features the HamiltonJet "Blue Arrow" control system option in addition to the standard hydraulic control option.

Model HJ 403

Max power input: 900 kW at 2,240–2,400 rpm
Impeller diameter: 400 mm

Typical 403 application (twin)

Vessel: 18.8 m Lobster fishing boat
Speed: 38 kt (max), 31 kt (cruise)

Engines: MAN diesels
Drive: gearbox
Builder: Mark Millman Marine, WA

This model was introduced in 2002 to replace the HJ 391 water-jet.

HM Series

The HM series of water-jets are larger units typically for craft in the 20 to 60 m range and are normally driven via a reduction gearbox.

As with all the models in the Hamilton range, vital control components are mounted inboard where they are serviceable without the need to dock the vessel and are protected from the elements and impact damage. All HM Series jets feature integral hydraulic power packs driven off the jet mainshaft. Transducers for feedback of steering and astern deflector positions use reliable inboard mechanical linkages, eliminating the need for chains or cables to pass through the hull.

The thrust bearing is located inboard, minimising the chance of water entering the oil and thus improving reliability. The bearing can be easily inspected and serviced from inside the vessel. Bearing and seals do not run directly on the mainshaft but on low-cost, easily replaced sleeves, ensuring the jet can be quickly reconditioned with minimal cost and downtime.

Model HJ 422

Max power input: 1,000 kW at 2,020–2,300 rpm
Impeller diameter: 420 mm

Typical 422 application (twin)

Vessel: 17.4 m Fast Patrol Boat
Speed: 48 kt
Engines: MTU diesels
Drive: gearbox
Builder: Motomarine SA, Greece

Model HM461

Continuous max power input: 900 kW at 1,680–1,800 rpm
Sprint max power input: 1,100 kW at 1,795–1,900 rpm
Impeller diameter: 460 mm

Model HM521

Continuous max power input: 1,200 kW at 1,508–1,624 rpm
Sprint max power input: 1,400 kW at 1,587 – 1,710 rpm
Impeller diameter: 520 mm

Model HM571

Continuous max power input: 1,400 kW at 1,357–1,470 rpm
Sprint max power input: 1,700 kW at 1,448–1,569 rpm
Impeller diameter: 570 mm

Model HM651

Continuous max power input: 1,800 kW at 1,220–1,316 rpm
Sprint max power input: 2,200 kW at 1,305–1,407 rpm
Impeller diameter: 650 mm

Model HM721

Continuous max power input: 2,200 kW at 1,073–1,154 rpm
Sprint max power input: 2,700 kW at 1,149–1,236 rpm
Impeller diameter: 720 mm

Model HM811

Continuous max power input: 2,800 kW at 955–1,025 rpm
Sprint max power input: 3,500 kW at 1,030–1,104 rpm
Impeller diameter: 810 mm

Typical single HM Series application

Vessel: 12.6 m Pilot/Rescue craft
Displacement: 13.8 t
Speed: 24 kt
Jet: Single HamiltonJet Model HM422
Engine: Volvo TAMD 163P diesel
Operator: Swedish National Administration of Shipping and Navigation

Typical twin HM Series application

Vessel: 49.0 m Catamaran Passenger Ferry
Speed: 30 kt
Jets: Twin HamiltonJet Model HM811
Engines: Twin MTU diesels
Operator: Soflusa, Portugal

Typical triple HM Series application

Vessel: 50 m Survey Boat
Displacement: 265 t
Speed: 32 kt
Jets: Triple HamiltonJet Model HM811
Engines: Triple Cummins KTA 38-M1
Operator: PG Charters, Freeport, US

Typical quadruple HM Series application

Vessel: 53.5 m Fast Support and Intervention vessel
Displacement: 478 t (laden)
Speed: 30 kt (lightship)
Jets: Quadruple HamiltonJet Model HM811
Engines: Quadruple Cummins KTA 50-M2
Operator: Sonasurf, France

Typical quintuple HM Series application

Vessel: 41.0 m Oil Rig Crew Boat
Displacement: 205 t (laden)
Speed: 27 kt
Jets: Quintuple HamiltonJet Model HM521
Engines: Quintuple, Cummins KTA-19 diesels
Operator: SeaMac Inc, Los Angeles, US

Typical HM Series loiter/boost application

Vessel: 44 m Crew/Supply Vessel
Displacement: 265 t
Water-jet: HM 571 boost jet
Speed: 20-24 kt loaded
Engine: Cummins KTA38-M1
Operator: Candy Fleet Corp, Morgan City, US

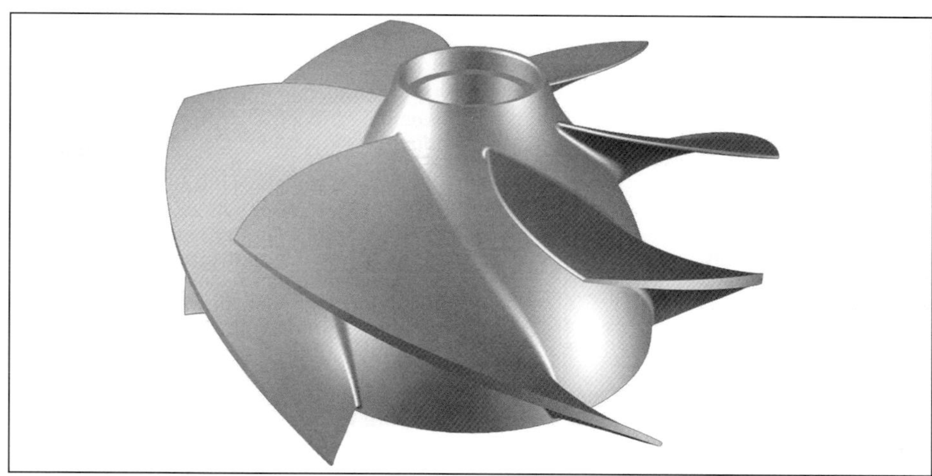

HamiltonJet mixed flow impeller 1197776

HamiltonJet HJ322 1197769

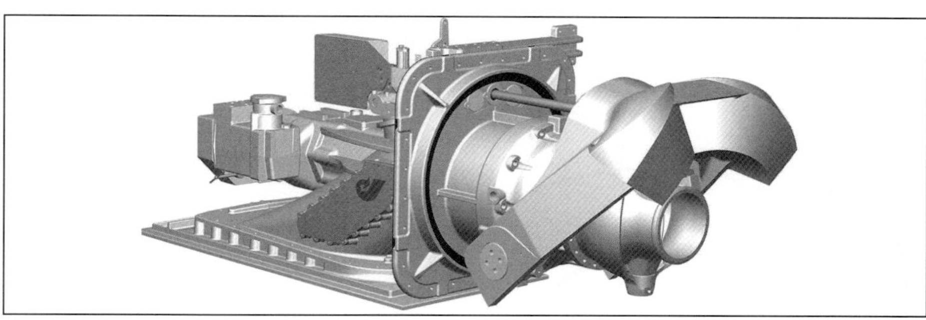

HamiltonJet HJ403 1197773

HM Series Specifications

Model	A (mm)	B (mm)	C (mm)	D (mm)	E (mm)	F (mm)	G (mm)
HM422	1,380	485	2,082	1,105	785	660-960	990
HM461	1,280	420	2,048	1,440	1,018	900	1,040
HM521	1,424	475	2,350	1,630	1,200	1,000	1,200
HM571	1,561	530	2,585	1,800	1,300	1,000	1,285
HM651	2,105	593	3,360	1,650	1,470	1,100	1,500
HM721	2,381	667	3,779	1,860	1,612	1,250	1,660
HM811	2,672	750	4,252	2,100	1,800	1,400	2,000

Sweden

MJP Water-Jets

Marine Jet Power AB
A division of Österbybruk Industrier AB
SE-748-01 Österbybruk, Sweden

Tel: (+46 295) 24 42 50
Fax: (+46 295) 24 42 60
e-mail: sales@mjp.se
Web: www.mjp.se

Eric Uhlander, *Product Manager*
Nils Morén, *Sales Manager*
Michael Näsström, *Service Manager*
Hans Lindgren, *After Sales Manager*
Niklas Widemark, *Project Engineer*

MJP Marine Jet Power AB is part of the Swedish industrial group Osterbybruk Industrier including Osterby Marine and Alcopropeller. Osterby Gjuteri is a well known foundry, specialising in castings for marine applications. Alcopropeller has 80 years experience in manufacturing propeller blades. The company has a reputation for developing and designing advanced propulsion systems for high-performance vessels in commercial, coast guard and navy operations. MJP is developing water-jets to absorb up to 10,000 kW.

The MJP design is an all stainless steel structure with an oil lubricated hub unit. The range of water-jets covers intake diameters from 450 mm to 1,350 mm which can absorb powers between 500 kW and 10 mW. The units are delivered in various configurations, namely DD (Double Drive), TD (Triple Drive), QD (Quadruple Drive, DD+SB (Single Booster) and DD+DB (Double Booster). The largest jet produced to date is the MJP J950R-DD installed in the 45 m Buquebus catamaran ferry which has a speed of over 52 knots in full load condition.

The company has representatives in 23 countries for water-jet supply and engineering support. Recent deliveries have included eight MJP 850 water-jets to Marinteknik for use on the fast passenger catamarans *Nixe* and *Nixe II*.

Size	kW*	A	B	C	D	E	F	G	H	** Steerable	** Booster	*** Intake V litres
450	1,100	428	1,247	708	450	1,850	3,100	570	470	540	350	300
500	1,200	510	1,290	855	500	2,050	3,400	575	460	815	575	350
550	2,000	560	1,412	940	550	2,250	3,790	630	505	1,090	710	400
650	2,750	665	1,680	1,120	650	2,675	4,510	750	600	1,600	1,035	750
750	3,600	765	1,900	1,255	750	3,075	5,180	860	690	2,100	1,460	1,200
850	7,100	880	2,100	1,440	870	3,540	5,890	990	765	3,400	2,320	1,900
950	7,500	1,010	2,460	1,610	970	4,070	6,590	1,100	915	4,300	3,000	2,700
1,100	8,700	1,200	2,845	1,890	1,180	4,820	7,630	1,350	1,090	7,000	5,400	4,700
1,350	9,700	1,355	3,490	2,110	1,360	5,560	9,370	1,555	1,250	12,000	8,300	7,000

*Maximum power for continuous operation
**Weight per unit (kg) including hydraulics, excluding shafting and intake
***Approximate volume (litres) depending on adaption to hull lines

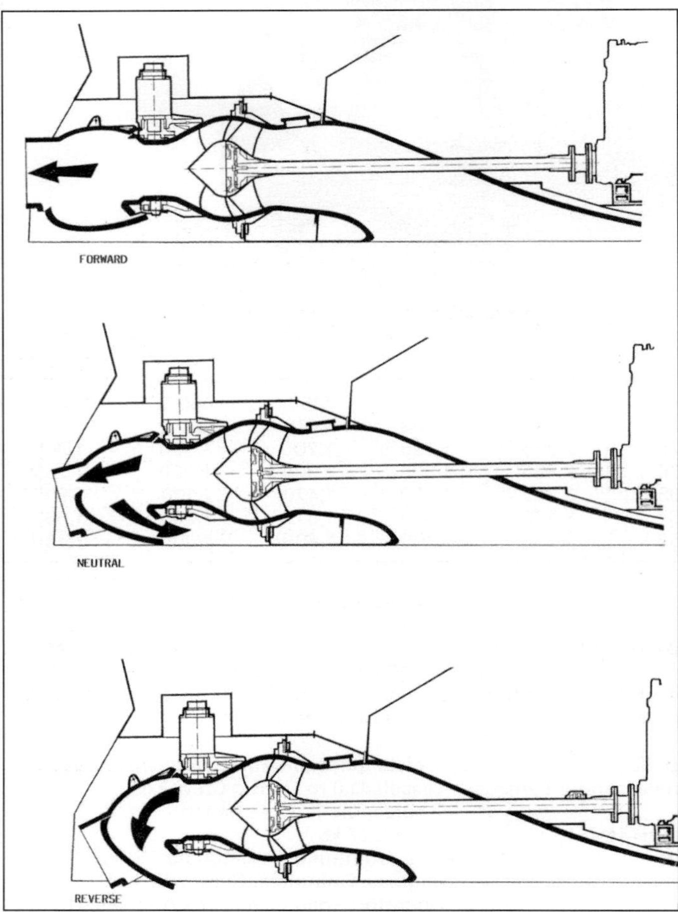

Mechanism of flow redirection on MJP water-jet units 0506813

Dimensions of MJP water-jet units (see table) 0044031

MJP J500R-DD water-jets for the Swedish Coast Guard 0506941

Rolls-Royce AB

Kamewa water-jets
PO Box 1010, SE-681 29 Kristinehamn, Sweden

Tel: (+46 550) 840 00
Fax: (+46 550) 181 90
Web: www.rolls-royce.com

Anders Johansson, *Commercial Marine*
Fredrik Bagge, *Naval Marine*

The Kamewa S-Series range is designed for the largest and most demanding applications and is manufactured entirely in stainless steel for maximum wear and corrosion resistance. They are equipped with a mixed-flow pump for maximum efficiency. To facilitate hydrodynamic performance of the pump the built-in bearings keep the impellor in the correct position and provide control of the important tip clearance. Five blade angles, eight nozzle sizes and at least 10 variations of inlet duct design are utilised to optimise the unit for each application. Each water-jet can be supplied as a booster, without steering and reversing gear.

The Kamewa VLWJ range is available for powers from 24 to 50 MW and features innovative thrust block technology. A pair of 2.35 m diameter water-jets (Kamewa 235) can each handle 27 MW, and are the largest units built to date. They will power the Japanese Techno-Superliner (TSL) and give a service speed of 37 kt.

A third range, the Kamewa A-series features aluminium water-jets with mixed-flow pump technology. Typical applications are rescue boats and smaller high speed naval craft.

In 2006 the Kamewa S3 range was launched, featuring a 10 per cent reduction in weight, reduced flange size (allowing two jets to be 5% closer together in twin installations), improved bearings and a new control system. The S3 range will be phased in over a period of 2 years, gradually replacing the SII range.

Model	Power (kW)	Dry Weight (kg)
63SII	1,000–4,000	1,790
71SII	2,500–7,000	2,420
80SII	3,000–8,000	3,230
90SII	4,000–10,000	4,530
110SII	5,000–12,000	6,380
112SII	6,000–16,000	7,650
125SII	8,000–15,000	11,360
140SII	9,000–18,000	15,210
160SII	12,000–20,000	22,870
180SII	14,000–22,000	30,880
200SII	16,000–24,000	41,220
VLWJ range	20,000–50,000	TBA

United Kingdom

Ultra Dynamics Ltd

Upperfield Road, Kingsditch Trading Estate, Cheltenham, Gloucestershire GL51 9NY, UK

Tel: (+44 1242) 70 79 00
Fax: (+44 1242) 70 79 01
e-mail: sales@ultrajet.co.uk
Web: www.ultradynamics.com

D Burton, *Chairman*
M J Lane, *Managing Director*
R Godfrey, *Finance Director*
S Middleton, *Marketing Manager*

US Office
Ultra Dynamics Marine, LLC
1110A Claycroft Road, Columbus, Ohio 43230, US

Tel: (+1 614) 759 90 00
Fax: (+1 614) 759 90 46
e-mail: sales@ultradynamics.com

G Scott, *President*

Ultra Dynamics market the UltraJet waterjet propulsion systems for high-speed craft, used for patrol, customs, port and harbour security, enforcement, pilot and search and rescue applications. The systems provide sprint capability plus high thrust at low boat speed (for towing or recovery).

Ultrajet propulsion systems are used by military and government bodies in Europe, North America, Asia and South America.

Ultra Dynamics has been designing and building waterjet propulsion equipment from the mid-1950's. Since then over 15,000 waterjets have been fitted worldwide.

The UltraJet range is capable of absorbing powers up to a maximum of 1,268 kW in high-speed craft. In many cases this can be achieved without the need for a gearbox.

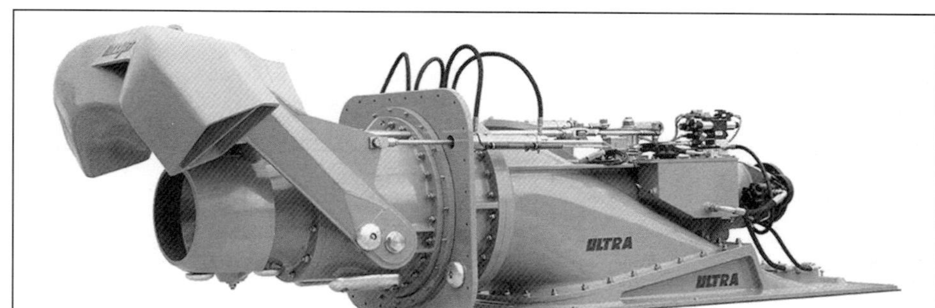
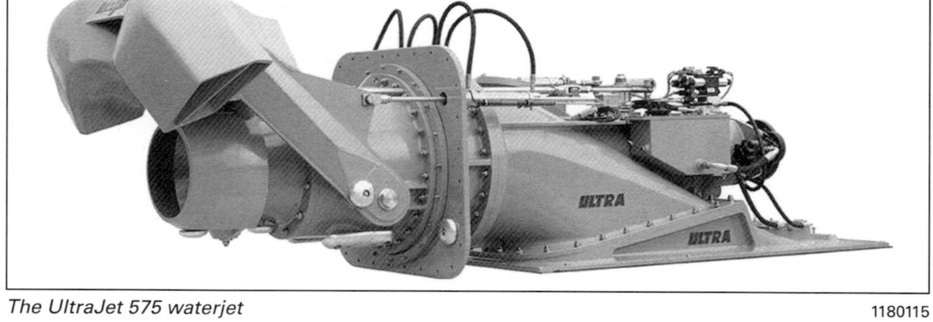

The UltraJet 575 waterjet 1180115

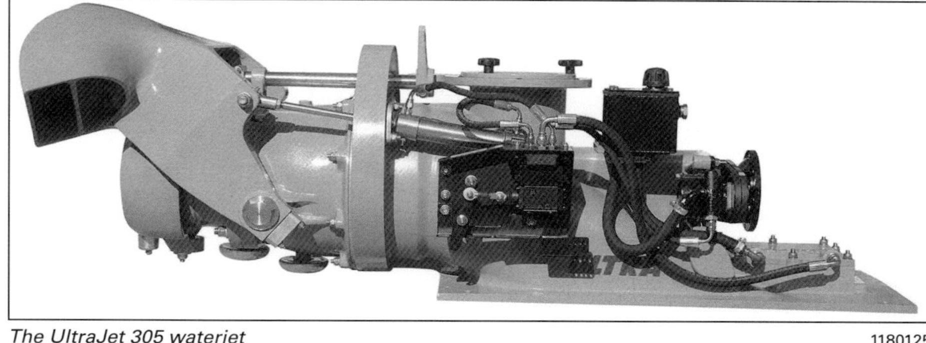

The UltraJet 305 waterjet 1180125

The basic components of the waterjet is an axial flow impeller (single or two stage), made of stainless steel to ensure maximum durability in contaminated water conditions. It is matched to the power curve of the selected engine/gearbox. The impeller is supplied in four- or five-blade options to suit the specified power and craft duty. Constructed from marine grade aluminium alloy, the impeller casing incorporates a stainless steel liner to provide maximum durability and to allow for service replacement, if required.

Steering and reverse control system options are available to best suit the type of craft and the operating environment. Hydro-mechanical reverse deflector controls operated by conventional cable systems can be supplied where electro-hydraulic systems would be considered undesirable for use.

Full electro-hydraulic controls can be supplied for the steering and reverse deflector operation. Options are available for either engine-driven or electrically driven pumps providing hydraulic power for the control systems.

United States

North American Marine Jet Inc

216 Sevier Street, Benton, Arkansas 72015, US

Tel: (+1 501) 778 41 51
Fax: (+1 501) 778 63 81
e-mail: jason@marinejet.com
Web:www.marinejet.com

Established in 1980, this company manufactures water-jets for high-speed craft, for the power range 110 to 560 kW. The company also supplies high-thrust, low-speed units not covered in this entry. Water-jets have been supplied to a variety of craft including US Navy patrol boats, fishing boats, workboats, dive boats, crew boats, fire boats and excursion boats.

The company has in-house facilities for the fabrication of steel, stainless steel and aluminium, backed up by analytical design software and a CAD system.

The Nomera range is offered to the high-speed craft market for speeds up to 40 kt. The Amerajet and Mach-1 jet models were introduced in 1997 to extend the speed range to 60 kt.

Nomera 12 Y

Designed to operate with diesel or petrol engines in the 3,200 to 4,800 rpm range, this unit provides high thrust and manoeuvrability. Straightforward construction allows for easier maintenance and increased reliability.

Specifications

Dry weight: 55 kg
Length: 1.1 m
Power: 75 to 260 kW
Shaft speed to: 4,800 rpm
Speed range: 30 to 50 kt
Bollard pull: 1,600 kg

Nomera 300
Specifications
Dry weight: 180 kg
Wet weight: 240 kg
Length: 1.6 m
Power: 75 to 260 kW
Shaft speed: to 3,000 rpm
Speed range: 20 to 50 kt

Nomera 800
Specifications
Dry weight: 340 kg
Wet weight: 450 kg
Length: 2.2 m
Power: 100 to 550 kW
Shaft speed: 1,800 to 2,400 rpm
Speed range: 20 to 50 kt

Rolls-Royce Commercial Marine

10255 Richmond Avenue, Suite 101, Houston, Texas, US

Tel: (+1 713) 273 77 00
Fax: (+1 713) 273 77 77
e-mail: commercialmarine.usa@rolls-royce.com
Web: www.rolls-royce.com

Richard Allinson, *Director*

The company offers manufacturing and sales support for the full range of Kamewa water-jets in the US.

Rolls-Royce Naval Marine Inc

110 Norfolk Street, Walpole, Massachusetts 02081, US

Tel: (+1 508) 668 96 10
Fax: (+1 508) 660 6152
e-mail: navalmarine.usa@rolls-royce.com
Web: www.rolls-royce.com

Patrick Marolda, *President*
Jorge Morales, *Vice-President, Finance*
Christopher Cikanovich, *Vice-President, Sales and Marketing*

The company engineers and builds ship propulsion systems.

RIDE CONTROL SYSTEMS

Company listing by country

Australia
Seastate Pty Ltd

Italy
La Me Marine Division

Japan
Universal Shipbuilding Corporation

Netherlands
Quantum Controls

Norway
Fjellstrand A/S

United Kingdom
Rolls-Royce Brown Brothers
VT Marine Products Ltd

United States
Naiad Marine Systems
VT Maritime Dynamics Inc

Australia

Seastate Pty Ltd

An Austal Group Company
100 Clarence Beach Road, Henderson, Perth, Werstern Australia 6166, Australia

Tel: (+61 8) 94 10 11 11
Fax: (+61 8) 94 94 10 25 64
e-mail: marketing@austal.com
Web: www.austal.com

Tony Elms, *General Manager*

Seastate Pty Ltd is part of the Austal Group and supplies complete motion control systems. The Seastate system is based on a family of devices: activators, standard hydraulic power packs and a range of control interfaces, all managed through a common control network. The main motion control devices include transom flaps, transom interceptors, rudders, T-foils and stabilised fins as well as any hydraulic deck equipment.

The company offers a motion simulation service for high-speed craft to enable the potential benefits of their system to be defined. Seastate motion control systems are fitted to a number of large fast catamaran and monohull craft.

Italy

LA ME srl

Marine Division, via Zanica 17/0, I-24050 Grassobbio (BG), Italy

Tel: (+39 03) 53 84 46 04
Fax: (+39 03) 53 84 45 90
e-mail: info@lamemarine.com
Web: www.lamemarine.com

La Me has used the interceptor concept for both steering and ride control as well as resistance reduction for high-speed craft.

The Stena HSS vessels have a La Me steering system and the Rodriquez TMV 103 Aquastrada has a La Me transom mounted ride control system.

Japan

Universal Shipbuilding Corporation

Head Office:
Muza-Kawasaki Central Tower, 1310, Omiya-cho, Saiwai-ku, Kawasaki, Kanagawa 212-8554, Japan

Tel: (+81 44) 543 27 00
Fax: (+81 44) 543 27 10
Web: www.u-zosen.co.jp

Universal Shipbuilding Corporation, formerly Hitachi Zosen developed a ride control system with Supramar for the PTS 50 type hydrofoil, *Housho*, a PTS 50 Mk II which was delivered to Hankyu Kisen KK on 19 January 1983.

The underside of the bow foil is fitted with two flapped fins to improve ride comfort. Operated by automatic sensors, the fins augment stability and provide side forces to dampen rolling and transverse motions.

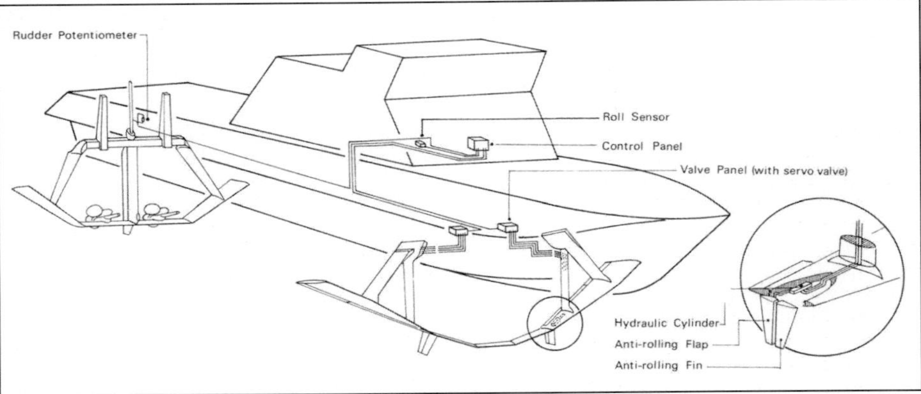

PTS 50 Mk II roll-stabilising system as fitted to Housho 0506816

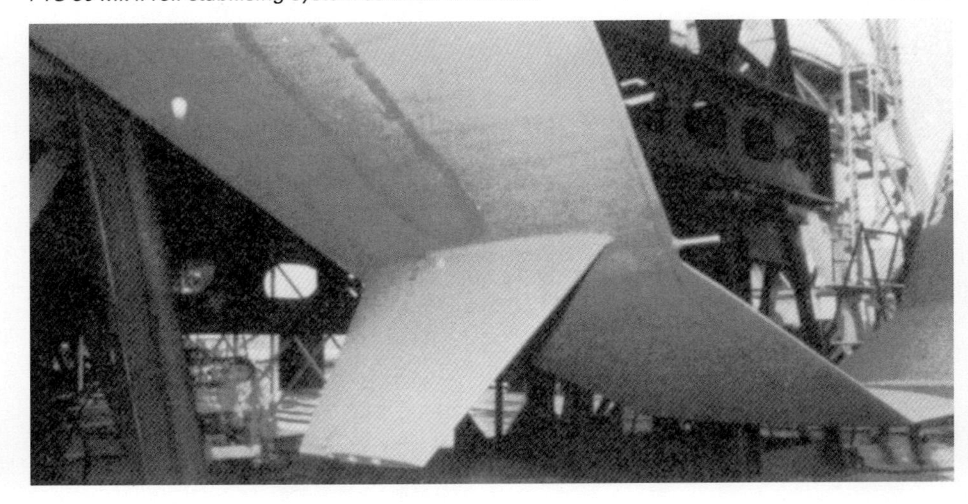

Flapped roll-stabilisation fin on Housho 0506817

Netherlands

Quantum Controls BV

Industriestraat 5, NL_6361 HD Nuth, Netherlands

Tel: (+31 045) 577 84 30
Fax: (+31 045) 577 84 31
e-mail: info@quantumcontrols.nl
Web: www.quantumcontrols.nl

Lambert Dinissen, *Chief Executive Officer*

Quantum Controls produce a range of ride control systems. Primarily designed for the luxury yacht market, they also provide systems for fast ferries and SWATH vessels in conjunction with Island Engineering, a company that specialises in commercial and military ride control systems.

The IMCS ride control system is designed for vessels that are capable of exceeding 24 kt. The system provides fully digital adaptive electronic control for stabilisers and trim tabs also developed by the company. These include the Archer™ active trim tab ride control system that incorporates retractable foils. This system is designed to optimise the performance of ride control over a greater range of speeds than was possible with fins or trim tabs alone. The system combines the roll damping benefits of fin stabilisers with the pitch and trim control attributes of trim tabs. At high vessel speeds the foils are retracted to reduce drag and at lower speeds are extended to increase motion control.

Quantum Controls also provide the Arrow™ fixed size active trim tabs system which increases the range of vessels that Quantum can provide ride control for. Quantum Controls have developed two upgrades for their control systems called ZeroSpeed™ and OnAnchor™ that provide roll control for vessels when at very low speeds and at anchor.

Norway

Fjellstrand A/S

N-5632 Omastrand, Norway

Tel: (+47 56) 55 76 00
Fax: (+47 56) 55 76 20
e-mail: fjellstrand@fjellstrand.no
Web: www.fjellstrand.no

Asbjørn Tolo, *Managing Director*
Bjarne Boerven, *Sales Director*
Stig Oma, *Technical Manager*

Fjellstrand Motion Dampening System (MDS)

Fjellstrand developed the MDS to improve the sea-keeping characteristics of its high-speed catamarans in rough seas and currently 50 systems are in operation worldwide. By counteracting the impact of the waves on the hull, Fjellstrand MDS minimises slamming, pitching and rolling.

The Fjellstrand MDS comprises a strut carrying the controlling fins, a hydraulic transmission system, a control system and sensors.

Struts on which the controlling fins are mounted are installed at the forward end of each hull at the point where they can provide maximum dampening effect. An adjustable fin with a surface area corresponding to 2 m^2 is mounted on each strut. Vessel motion data are acquired by sensors and then passed to a computer which, in turn, continuously adjusts the angle of the fins to counteract the motion of the vessel.

The system can be installed on Fjellstrand's 38.8 m Advanced Slender Catamarans and 40 m Flying Cats. Retrofitting is also possible on existing models.

One of the hull installations of Fjellstrand's first Motion Dampening System (MDS) mounted on Victoria Clipper 0506655

The first Fjellstrand MDS was mounted on *Victoria Clipper*, operated by Clipper Navigation Inc of Seattle, US.

Experience with the 38.8 m *Victoria Clipper* and other craft has shown that bow accelerations in 1.5 to 2.0 m significant wave heights are reduced by 36 per cent and with the Fjellstrand MDS in use in lower wave heights of 0.5 to 1.0 m the reduction is as much as 46 per cent. The operator has found that the Fjellstrand MDS dramatically reduces motion discomfort in 2.4 m seas (6.2 per cent of craft length) and that in following seas it is possible for the first time to continue on autopilot.

United Kingdom

Rolls-Royce Brown Brothers

A Rolls Royce company
Hillend Industrial Park, Dunfermline, Fife KY11 9JT, United Kingdom

Tel: (+44 1383) 82 31 88
Fax: (+44 1313) 82 40 38
e-mail: navalinfo@rolls-royce.com
Web: www.rolls-royce.com

John Nicholson, *Director*
Bob McFarlane, *Sales and Marketing Manager*

Rolls-Royce Brown Brothers' product range offers ride control systems for high-speed vessels, including: non-retractable fin stabilisers and trim tabs; controlling pitch and roll for monohulls and Swath motion control equipment; comprising non-retractable fin stabilisers with a multivariable gain control for independent or simultaneous control of pitch, roll and heave; Neptune and Aquarius retractable stabiliser fins for luxury yachts.

Both systems permit roll stabilisation of up to 90 per cent.

The company has recently introduced touch-screen control units to simplify operation.

VT Marine Products Ltd

A subsidiary of the VT Group
Hamilton Road, Cosham, Portsmouth, Hampshire PO6 4PX, United Kingdom

Tel: (+44 23) 92 53 97 50
Fax: (+44 23) 92 53 97 64
e-mail: scolliss@vtmp.co.uk
Web: www.vtplc.com/marine

Steve Colliss, *General Manager*

VT Marine Products Ltd is a world leader in the design and manufacture of fin stabilisers for fast marine craft. With over 3,500 installations worldwide, the company's product range covers fins for luxury yachts, commercial craft and military vessels.

The product range is characterised by fin area ranges as follows:

Product code	Fin area range (metres²)
100	0.42–1.00
150	0.75–1.50
175	0.75–1.75
200	1.00–2.00
300	1.50–3.00
400	3.0–4.50
500	4.5–6.00
600	5.5–7.00
700	6.5–8.00
800	8.0–12.00
Custom	Any size

United States

Naiad Marine Systems

A member of The VT Group, plc
50 Parrott Drive, Shelton, Connecticut 06484, United States

Tel: (+1 203) 929 63 55
Fax: (+1 203) 929 35 94
e-mail: sales@naiad.com
Web: www.naiad.com

Scott Anthony, *Global Sales Manager*

Naiad Marine Systems produce a range of fin stabiliser systems for the roll stabilisation of motor yachts from 10 to 100 m in length.

Model	Fin Size (m²)	Ship Length (m)	Ship Displacement (tonnes)
160	0.28–0.42	9–13	25
174	0.28–0.56	10–15	35
252	0.28–0.70	13–23	50
254	0.28–0.84	16–26	80
302	0.42–1.12	24–31	140
353	0.70–1.76	27–37	200
420	0.84–2.04	33–43	350
520	1.12–3.53	39–49	600
620	1.76–4.18	39–49	750
710	2.04–4.18	45–61	1,000
810	4.18–6.00	54–85	1,500

The hydraulic activation of the fins is controlled DATUM™, by a digital controller available for fin and other control surface use.

Naiad also offers bow and stern thrusters and integrated hydraulic systems as well as ride control systems.

VT Maritime Dynamics Inc

A subsidiary of VT Group
21001 Great Mills Road, Lexington Park, Maryland 20653, United States

Tel: (+1 301) 863 54 99
Fax: (+1 301) 863 02 54
e-mail: sales@maritimedynamics.com
Web: www.maritimedynamics.com

Joseph Kubinec, *General Manager*
Chris Swanton, *Naval Architect/Applications Engineer & Sales*
Dave Allen, *Service Manager*
Ruth Schumacher, *Marketing Coordinator*

Maritime Dynamics Inc (VTMD) brings 22 years of technical leadership to the development and manufacture of stabilisation and ride control systems for conventional and advanced marine vehicles. VTMD's capability includes the analysis and simulation of ship motions, control system design, software development, hardware fabrication, equipment installation and crew training. VTMD stabilisation and ride control systems are in operation on Surface Effect Ships (SES), Air Cushion Vehicles, Catamarans, Small Waterplane Area Twin Hull (SWATH) ships, foil supported vessels, and monohulls. The VTMD range of products includes fixed fin stabilisers, lightweight retractable fin stabilisers, active trim tabs, active interceptors, fixed T-foils, pivoting T-foils, retractable T-foils, lifting foils, lift fans, vent valves, and monitoring and control electronics.

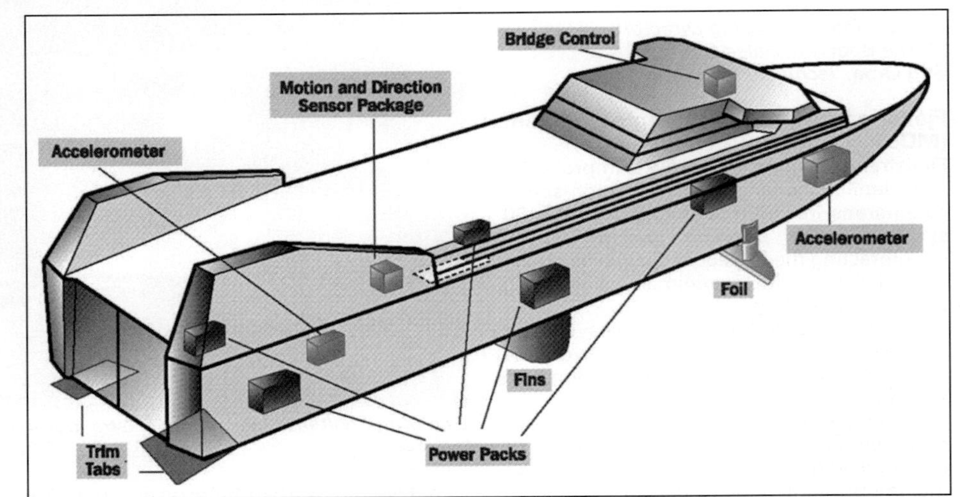

Typical components of an VTMD ride control system 0538547

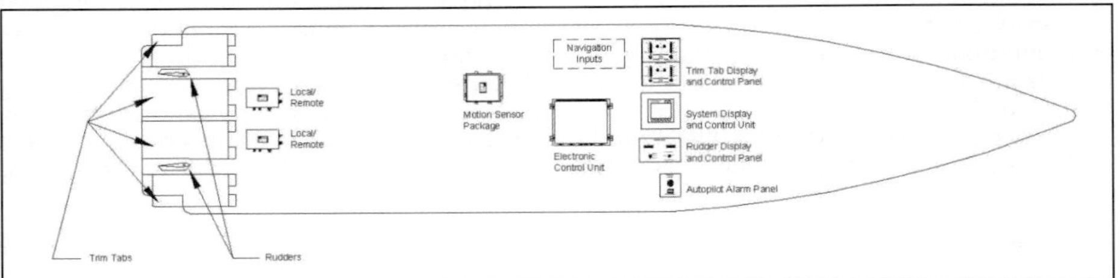

Major components of a typical Maritime Dynamics ride control system for a monohull 0538550

Surface Effect Ships (SES)

VTMD lift and ride control systems for SES provide motion control and lift system control for optimisation of overall craft performance. In 1980, VTMD designed and supplied a ride control system for the USN XR-1D SES. This was the first active ride control system and it demonstrated significant improvements in SES ride quality under rough sea conditions. VTMD has continued the development of the SES ride control systems and currently manufactures microprocessor-based systems for both military and civilian applications.

The SES ride control system minimises wave-induced pressure changes in the air cushion to reduce craft motions and vertical accelerations caused by 'wave pumping' of the cushion volume. The resulting attenuation of the vertical accelerations can significantly reduce fatigue and discomfort during moderate and high-speed cushionborne operations. The system Electronic Control Unit (ECU) controls hydraulically driven cushion vent valves and/or variable-flow fans to dynamically regulate the net cushion air flow and maintain a constant pressure.

Components of a typical SES ride control system consist of pressure, altitude, and acceleration sensors; a microprocessor-based ECU with embedded control algorithms; a hydraulic system; and multiple vent valve assemblies that are ducted to the air cushion. Each vent valve assembly consists of aerodynamically shaped louvers driven by a servo-controlled hydraulic cylinder. Components of a typical SES lift system consist of the main lift fans that are sometimes fitted with variable geometry inlets and stern boost fans that maintain the geometry of the stern seal.

The ECU uses microprocessor based electronics to implement sampled data control algorithms using cushion pressure and ship motion feedback signals, and to output servo control signals to each vent valve or fan inlet guide vane selected for active control. Fault monitoring of the system's electronic, hydraulic and mechanical components is performed between each control algorithm computation. VTMD ride control systems installed on SESs operating at 35 to 60 knots consistently achieve a 50 per cent reduction in heave accelerations and have demonstrated reductions as high as 70 per cent.

The ride control display also provides a menu-driven real-time display of measured craft parameters relating to air cushion and craft operating conditions (for example, means and standard deviations of trim, roll, cushion pressure, vent valve position and vertical acceleration). The data has proven to be extremely valuable to the vessel operator for optimising overall performance in different sea conditions.

The ECU can be used on any SES or ACV equipped with either vent valves or variable flow fans, by programming it with appropriate control algorithms. VTMD has developed a systematic technique for deriving these control algorithms, which is based on both classical and optimal control theory as well as extensive experimental testing.

Ride control systems have been installed on over 40 SESs, including the following vessels: USN SES-200; USCG WSES *Sea Hawk*; the CIRR 120P class of passenger ferries; Royal Schelde's 23 m SES; the Blohm + Voss SES *Corsair*; SEMO's 37 m *Democracy*; the Swedish Navy's *Smyge*; Beliard Polyship's 30 m *Manto*; the Norwegian Navy's Mine Countermeasures Vessels, and the Norwegian Navy's *Skjold* Class Fast Patrol Boat.

Future systems are expected to utilise actively controlled bow T-foils, interceptors, and fin units (for lower speed applications) to further improve motion reduction and enhance ride quality.

Catamarans

Ride control systems for catamarans consist of a microcomputer-based controller that measures the vessel's motions and commands hydraulically actuated effectors to reduce wave-induced pitch, roll, and heave motion. Effectors on catamarans can be trim tabs, interceptors, fixed fin units, retractable fin units, fixed T-foils, pivoting T-foils, and retractable T-foils. Fixed lifting foils can also be used as effectors to provide passive heave damping. When made movable or fitted with flaps, lifting foils will provide ride control. Controlling electronics can be integrated with the steering system, whether rudders or water-jets, for yaw damping and heading control. The VTMD system has been type approved by several regulatory bodies for this purpose.

In 1991, VTMD introduced the first ride control system to improve passenger comfort on catamarans. VTMD's first installation was on *Condor 9*, and Incat Design 49 m wave-piercing catamaran built by Aluminium Shipbuilders Ltd. This 450-passenger vessel was operated on the western end of the English Channel by Condor Ltd between Weymouth, UK and St. Malo, France. The vessel was fitted with bow fins mounted inboard and outboard on each hull and a stern fins mounted inboard on each hull. During tests in February 1992, in measured seas of 2.4 m significant wave height, the ride control system installed on *Condor 9* consistently reduced pitch and roll motions by 50 per cent relative to the uncontrolled case. In head seas, the vertical accelerations at the forward, mid, and aft passenger seats were reduced by 45 per cent, 35 per cent, and 20 per cent respectively. In co-operation with Vosper Thornycroft (UK) Ltd, VTMD provided a similar system for *Seajet 1*, a 37 m wave-piercing catamaran built by Nichols Brothers.

VTMD ride control systems have been installed on over 100 catamarans, including all of the International Catamarans Tasmania (Incat) wave-piercing catamarans in the 74 m to 98 m size range. Configurations include trim tab only systems, trim tabs plus T-foils, interceptors plus T-foils, trim tabs plus fixed fin units, and trim tabs plus a retractable T-foil.

The simplest systems for high-speed catamarans employ transom mounted effectors such as trim tabs or interceptors. For trim tab systems, the controller commands two very large hydraulically actuated trim tabs mounted on the transom of each hull. Without increasing overall resistance, the trim tab systems substantially reduce the pitch and roll accelerations that cause passenger discomfort and motion sickness. Where weight and available hydraulic power are limited, interceptors can be fitted to the transom in-lieu-of trim tabs with some loss of motion reduction capability.

Higher performance systems for high speed catamarans employ transom mounted effectors in conjunction with a single retractable T-foil on the centre body forward, or in conjunction with fixed or pivoting T-foils located forward on each hull. Trade-offs of motion reduction performance, acquisition cost, operational cost savings, and operational issues are conducted to validate a candidate configuration.

There are significant advantages gained with a retractable T-foil such as operational flexibility, reliability, and lower fuel consumption. For example, in calm sea conditions the VTMD retractable T-foil is stowed above the waterline, which increases vessel speed, allows operation in shallow water areas, reduces the risk of hitting submerged or floating objects and reduces wear on the system. In addition, the retractable T-foil system is fitted with hydraulic dampers that allow the foil to pivot upwards in the event of hitting submerged or floating objects. A further advantage of the retractable T-foil results from easier maintenance because service is performed with the vessel afloat as all mechanical and hydraulic components are above the waterline. Drydocking is not required. A disadvantage of the retractable T-foil is that it should be integrated with the vessel structural arrangement early in the vessel design process; retrofits are more difficult. Further acquisition cost is higher for a retractable, but they remain an attractive alternative because of improved shallow draught operability and drag reduction. Current generation VTMD retractable T-foils allow cavitation free operation well into the 45 knot speed range. Designs are being produced for higher speed operation.

VTMD has also fielded integrated lift and ride control systems to catamarans that employ lifting foils. Careful integration of the hull and lifting foils is required to avoid drag penalties, and careful tradeoffs between section thickness (strength) and cavitation are required.

SWATH

VTMD ride control systems for SWATH provide trim and list stabilisation (such as mean attitude control) and pitch, roll and relative bow motion control. Components for a typical SWATH installation consist of: ship motion sensors; a microprocessor-based ECU with applicable control algorithms; a fin control panel; a hydraulic system; a pair of forward fins or canards; and a pair of aft fins or stabilisers. Control of rudders for manoeuvring is provided if desired. The SWATH ECU and control panels provide automatic and manual control of the fins; real-time display of vessel pitch, heave, roll and fin motions; and display of mean values for trim and list.

VTMD ride control systems are currently installed on over 15 SWATH vessels worldwide, which range in size from 15 m to 73 m, and are in service as pilot, crewboat, yacht, military patrol, and military test applications. All of the vessels are in four fin installations, and approximately half have integrated steering and heading control provided by the ride control system. The VTMD controller has type approval from several regulatory bodies for this purpose.

Monohulls

Since early 1993, VTMD has offered integrated steering and ride control systems for high-speed monohulls. The monohull system integrates steering, trim and list stabilisation, and pitch, roll and yaw motion damping to provide superior control of vessel motions during high-speed operation

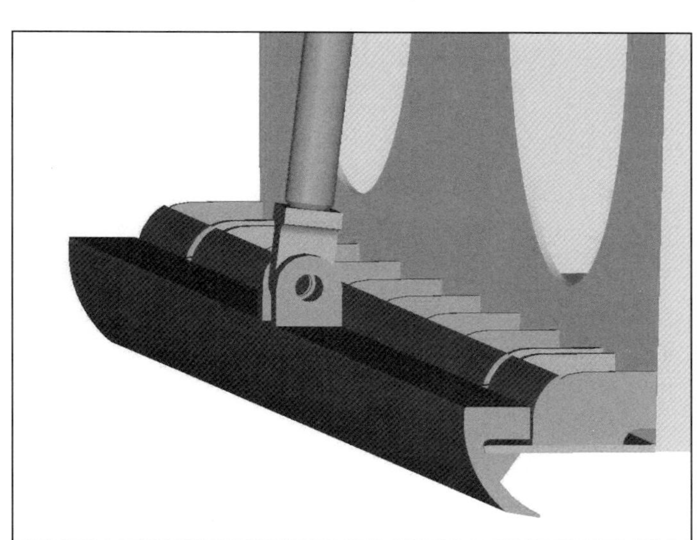

VTMD Hinged Interceptor 0538548

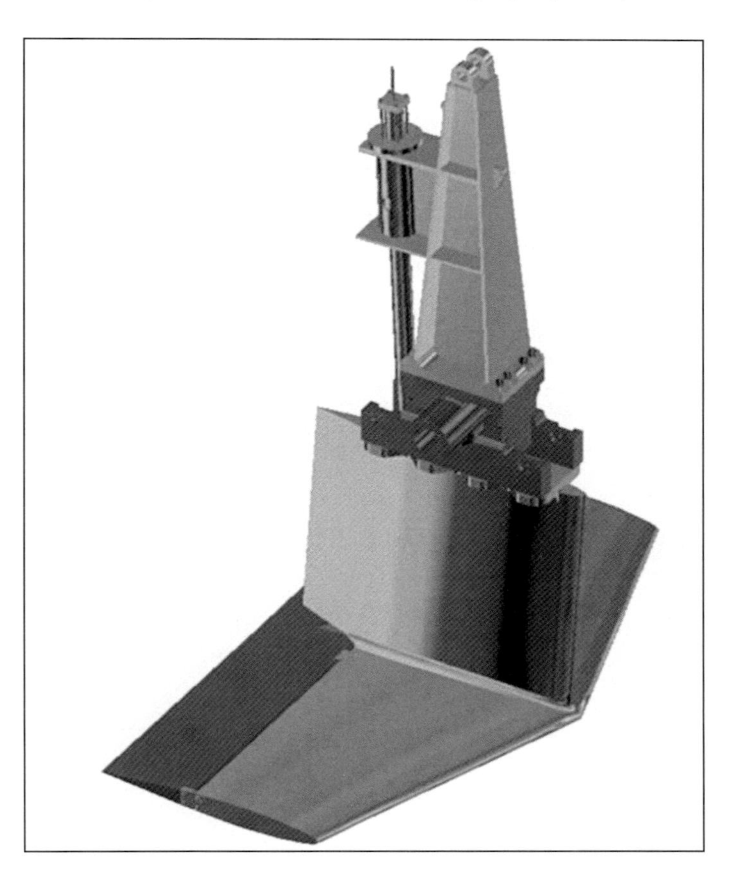

VTMD Pivoting 13 square m T-Foil 0538551

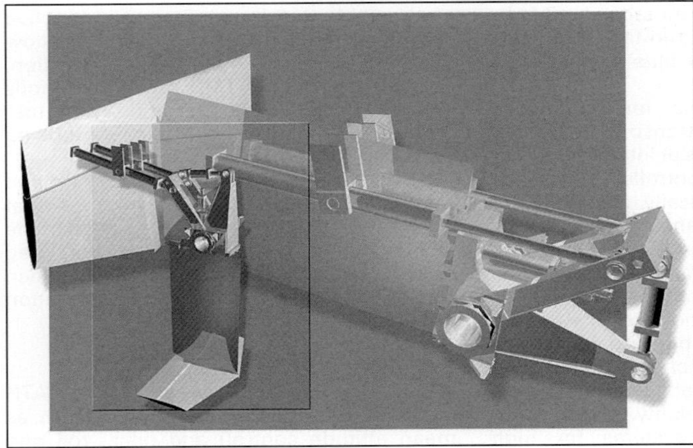

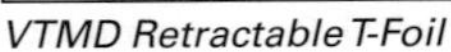

VTMD Retractable T-Foil 0538549

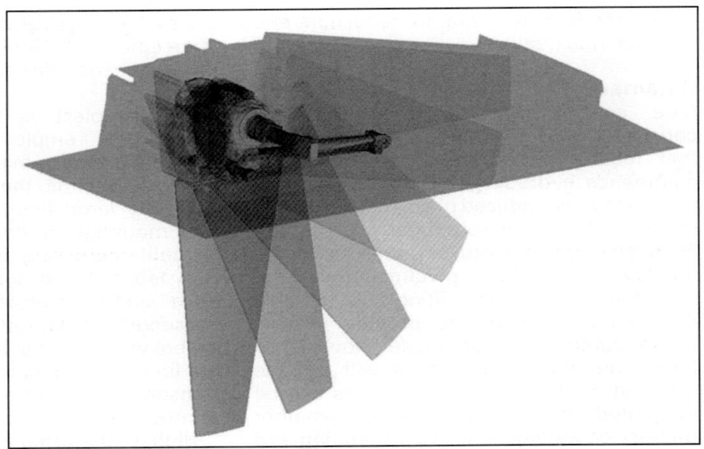

VTMD Lightweight Retractable Fin Unit 0538552

in calm and rough water. This is possible through integrated control of the vessel's rudders or water-jets, and control of stern mounted trim tabs in response to ship motions and helm commands. VTMD's first monohull installation was on WaveMaster International's 40 m monohull *Super Flyte*. In 1 to 2 m bow seas, the system has consistently demonstrated 60 per cent roll reduction and 40 per cent pitch motion reduction. VTMD has provided such systems on over 100 monohull vessels, including the largest fast ferry from Fincantieri having a length of 145 m.

Other effectors such as interceptors, fixed fin stabilisers, lightweight retractable fin stabilisers, and T-foils have been fitted to monohulls. The most complex system fielded by VTMD to-date, consisting of two trim tabs, two aft fixed fin stabiliser units, two forward fixed stabiliser units, and a pivoting T-foil on centreline forward, is the Alstom Leroux Naval Corsaire 14000 which is operated by Nels Line in Greece as *Aeolos Kenteris*. This 140 m vessel has an operating speed in excess of 40 knots and provides excellent passenger comfort. The ride control system is highly integrated with the steering system to optimise overall vessel propulsive efficiency. For example, the aft fins in the ride control system are employed to execute small heading changes instead of deflecting the water-jet buckets.

Retractable T-foils are difficult to integrate on monohulls due to structural considerations. As an alternative, VTMD fielded the first lightweight retractable fin units in 2002 for use in applications to approximately 70 knots.

Trimarans

Due to the relatively short distances between hulls, trimarans are excellent candidates for integrated lift and ride control systems. For example, the North West Bay Ships (Australia) built *Triumphant*, now operating between Japan and Korea as *Ulsan Dolphin*, is fitted with a VTMD lifting foil system and a trim tab on the centre hull. The foil system is located near midship, and is segmented into outboard and inboard halves for optimum angle of attack control and roll reduction. The foil system also provides heave damping.

Conventional stabilisation systems can also be fitted to trimarans. VTMD furnished a standard two-fin roll stabilisation system to UK research trimaran *Triton*. The stabilisers were fitted to the centre hull and provided reasonable motion reduction. Greater motion reduction could have been achieved with other combinations of effectors such as T-foils or trim tabs mounted to the outboard hulls, or merely relocating the fin units to the outboard hulls.

AIR CUSHION SKIRT SYSTEMS

Company listing by country

France
Zodiac

Spain
Neumar SA

United Kingdom
Avon Fabrications
Hovercraft Consultants Ltd
Icon Polymer Ltd

United States
Avon Engineered Fabrications Inc

France

Agnes 200 at high speed, fitted with Aerazur skirt systems (French Navy) 0506815

Zodiac

Formerly Aerazur SA
Division Applications des Elastomères, 2 rue Maurice Mallet, F-92130 Issy-Les-Moulineaux Cedex, France

Tel: (+33 1) 41 23 24 25
Fax: (+33 1) 46 48 74 79
e-mail: jmserein@zodiac.com
Web: www.elastomeres.zodiac.com

J M Serein, *Sales Manager*

In January 1990, Aerazur (the aerospace subsidiary of the Zodiac group, active in flexible material technologies and a world leader in inflatable boats) purchased the Coated Fabric Division of Kléber Industrie (part of the Michelin Group). The activities of this division and the corresponding division of Aerazur have been merged into a new 'Division Applications des Elastomères'.

The synergy between Aerazur (in charge of air cushion seal design on various French SES programmes since 1980, including Molenes and Agnes 200), and the extensive coated fabrics development and manufacturing capabilities of the former Coated Fabric Division of Kléber, resulted in the creation of a significant industrial base in the design and manufacturing of SES seals.

Design expertise is available within the company for various SES and hovercraft seal designs (loop and segments and full height segments or innovative designs such as a proprietary self-adjusting front-seal design used on Agnes 200). Various simulation software and computer-assisted design systems have been developed and are used for preliminary design, full-scale development or test data analysis. Installation services and full-scale test and maintenance support are available.

The division is one of the world's major producers of coated fabrics, specialising in high-quality products, conforming to the rigorous standards and specifications prevailing in the aerospace and marine (notably liferafts and military inflatables) fields.

For hovercraft and SES seals, a range of fabrics has been specially developed, using advanced base fabrics and rubber compounds (based on Hypalon, neoprene or natural rubber, according to the projected use). These products are tested and qualified on specially designed test rigs, reproducing the conditions encountered in real use. Various materials are available in surface weights ranging from 170 to 4,000 g/m^2.

These materials are used on the seals designed and manufactured by Aerazur or are sold to other skirt manufacturers like Griffon Hovercraft Ltd and Air Vehicles Ltd in the UK. Recently, Aerazur has designed, manufactured and installed the seal system for the French SES Agnes 200 and a new aft seal design for the Blohm+Voss Corsair SES.

During 1992, Agnes 200 was used commercially for a scheduled trans-Channel service between Brighton and Dieppe, a 77 n mile crossing often in difficult sea conditions (diagonal or transverse seas, waves over 2 m height encountered 17 per cent of the time). Normal service and full operational speed (35 kt) were maintained up to Sea State 5, with several crossings in Sea State 6. More than 500 hours of cushionborne service were logged, with negligible wear on the aft seal and very little on the front seal. The side fingers were replaced for inspection, as a precautionary measure, every 200 hours and returned to service within a week. The central fingers remained on the ship and showed no wear after 500 hours. All the inspection and replacement was conducted without dry-docking; due to the modular design the side fingers were replaced in less than 4 hours, between two trips. Of particular interest is the fact that the good performance of the seal system resulted in a stable and comfortable ride. This confirms the design option that no ride control system is needed with these seals (Agnes 200 has no RCS installed) and good pressure retention characteristics allowed frequent operation on a single supply fan. Subsequent design activity has resulted in an improved design with more than 500 hours of service life possible without maintenance.

The Division is also one of the world leaders in flexible fuel cells and fuel systems for aerospace and armoured vehicles and this technology has been applied to various high-speed marine vehicles. An example of this is that Aerazur has designed, manufactured and installed the complete fuel system of the French Blue Ribbon challenger *Jet Ruban Bleu* and supplied the fuel cells for the Spanish Chaconsa hovercraft.

Spain

Neumar SA

La Rinconada, Blq 6, E-28023 Madrid, Spain

Tel: (+34 91) 633 56 93
Fax: (+34 91) 633 52 32
e-mail: contact@neumar.com
Web: www.neumar.com

M de la Cruz, *Technical Director*
J A Barbeta, *Manufacturing Manager*

Neumar SA specialises in the research, development and design of air cushion lift systems, and in the manufacture of flexible structures for hovercraft. The company was formed to bring together a group of engineers and technicians all of whom had previous experience in hovercraft technology. This experience included the research, development, design and manufacture of the hovercraft lift system and skirt for the company Chaconsa under a contract for the Spanish Ministry of Defence.

Neumar has developed a hovercraft lift system offering very high stability with low-power requirements and reduced manufacturing and maintenance costs. It has been called an Automatic Transversal Air Distribution or ATAD lift system because of the main function it performs. Several two-dimensional models and two prototypes have been built and tested, with which the viability of this new lift system has already been demonstrated.

The company offers a wide range of services including design, feasibility studies, technical assistance and training for maintenance personnel and crew.

United Kingdom

Avon Fabrications

A division of the Checkmate Group
Unit 6, Pegasus Way, Bowerhill, Melksham, Wiltshire SN12 6TR, United Kingdom

Tel: (+44 1225) 70 54 65
Fax: (+44 1225) 70 74 97
e-mail: sales@avonfab.co.uk
Web: www.avonfabrications.com

T Roach, *Divisional Director*
Captain D Gosden, *Marine and Special Projects Director*
M E Prentice, *Technical Manager*

Avon Fabrications is a large air cushion vehicle skirt designer and maker which has facilities for comprehensive design, development, manufacture and material supply services to craft manufacturers and operators. Present development programmes include investigations relating to the design of seals for SES vehicle ferries and fast cargo craft as well as the Royal Norwegian Navy Fast Patrol Boat Building programme. The company acquired Air Cushion Ltd of the UK in 1999 as part of its development programme. Avon Fabrications is the sole supplier to civil operators of SES worldwide and the Royal Norwegian Navy's fleet of SES MCMVs and FPBs.

The majority of Avon's air cushion skirt materials are coated with natural synthetic rubber blends, or neoprene rubber calendered on to specially selected base fabrics. Avon can provide all the materials required for cost effective skirts of all weights.

Avon Fabrications also offers special experience in the design of the attachments of skirts to craft and the attachments between flexible components. Mechanical fastenings for these purposes are stocked and can be offered when these form part of a skirt system. In the design of skirt components, Avon provides for the rapid removal and attachment of all parts during maintenance periods. The company also offers design and construction services specialising in the use of large and small scale rubber fabrication.

Applications of Avon skirt materials include the following:

British Hovercraft Corporation SRN4 Mk II
Avon was the supplier of fingers (segments) for the four SRN4 Mk II craft, operated by Hoverspeed, having first supplied to Hoverspeed's predecessors (Hoverlloyd and Seaspeed) in 1972. Avon also supplied and developed fingers for Hoverlloyd and Seaspeed SRN4 Mk 1 craft dating back to 1969. Avon's involvement over the years has led to major increases in component life resulting from the development of improved coated fabric materials and bonding systems.

British Hovercraft Corporation SRN4 Mk III
Avon Fabrications has been a supplier of fingers/materials to Hoverspeed for the SRN4 Mk III since the two craft were stretched from the original Mk II craft in 1978. Avon Fabrications was involved in the specification of these components and since then has continued the development of materials for these craft. The company also manufactures the upper skirt components (segments) for the MK III.

LCAC (Textron Marine Systems)
Involvement with this programme started in 1980. Avon is an approved supplier of materials for the skirt system on this craft and has been heavily involved in the development and manufacture of components.

HM5/HM2 (Hovermarine International)
Avon developed the new bow and stern seals for the four HM5 SES craft, operating on the route between Hong Kong and Macau, and has supplied materials and components.

BH 110 SES-Bell Halter (now Textron Marine Systems)
Avon has been the sole supplier of bow and stern seals for Bell Halter's BH 110 craft since development craft were produced. It has also been manufacturing and developing components for the prototype since 1979, until the establishment of Bell Avon in the US.

CIRR 105 (Brødrene Aa A/S)
The detailed design of seals for the *Norcat* fast ferry air cushion catamaran (SES), was undertaken in 1984 and included the geometric and structural design of the flexible components and their attachments to the craft. Since then, Avon has manufactured seals for all successive craft of this type and supplied seals to its customers for operation in various parts of the world.

UT904 (Ulstein International A/S)
Sole supplier of bow and stern seals for the UT904 vessels worldwide.

Karlskronavarvet AB
Avon was awarded a contract in 1987 for the design and supply of the seal system for the two Jet Rider 3400 craft, followed by contracts for the SES 4000 vessels built by Westmarin A/S.

AP1-88 (Westland and licenses)
Avon Fabrications has supplied all types of material for this craft since its inception.

SES SEASWIFT 23 (Royal Schelde)
Design and supply of seals in 1988–89.

CORSAIR (Blohm+Voss AG)
Design and supply of seals for the Corsair SES test craft.

SMYGE (Karlskronavarvet/ Kockoms AB)
Development of materials with special underwater signature properties and the manufacture of seals for the SMYGE Test Craft 1990–91, plus spares and replacements.

MCMV (Royal Norwegian Navy)
Supply of bow and stern seals plus spares and replacements. Checkmate Avon now has a 20-year agreement with the Norwegian MoD to supply these vessels.

Samsung Heavy Industries
Supply of bow and stern seals plus spares for 37 m SES.

International Shipyards UT 928
Design and supply of bow fingers and three loop stern seals and subsequent spares supply.

Hovercraft Consultants Ltd

Unit 43, South Hampshire Industrial Park, Totton, Southampton, Hampshire SO40 3SA, United Kingdom

Tel: (+44 23) 80 87 11 88
Fax: (+44 23) 80 87 17 99
e-mail: sales@hovercraftconsultants.co.uk
Web: www.hovercraftconsultants.co.uk

M J Cox, *Director*
J F Cox, *Secretary*

Griffon 8000TD skirt system 1109911

In 2001 HCL expanded its operations to manufacture skirt systems and components to offer a design and build service. The company can provide complete skirt systems for new air cushion craft designs (amphibious or SES) or components for existing craft.

Skirt systems and components have been supplied for craft from less than one tonne to over 70 tonnes in weight. Components are built in a variety of materials including natural and synthetic rubbers, polyurethanes and PVCs. Fabrication is by hot bonding, cold adhesives, welding or wire stitching. The company holds stocks of materials with proven characteristics in hovercraft use.

HCL also manufactures other flexible components in its Duratank range of marine products. These include flexible tanks for water, fuel and effluent as well as seals, inflatables and flexible covers for hovercraft and other marine craft.

Icon Polymer Ltd

Retford, Nottinghamshire DN22 6HH, United Kingdom

Tel: (+44 1777) 71 43 00
Fax: (+44 1777) 70 97 39
e-mail: info@iconpolymer.com
Web: www.iconpolymer.com

T Pryce, *Managing Director*
M Tolliday, *Operations Director*
J Greaves, *Sales Director*
D Simpson, *Key Accounts Manager*

Icon Polymer Ltd works in co-operation with many major constructors of hovercraft and SES around the world, supplying skirt materials, components and complete fabrications for vehicles from the lightest hovercraft to some of the largest craft currently in service.

Experience gained during initial development of skirt fabrics in the UK, together with continuing development closely matched to the requirements of constructors and operators, has led to an established range of materials used for the complete requirements of skirt structures. This includes fingers (segments), cones, loops, doublers, spray suppressors and anti-bounce webs. Development continues with new fabrics and Icon Polymer has the technical resources, including computer-aided factorial experimental design (FED), to work closely with their customers on new products.

Recent applications for Icon Northern Polymer's materials include:

ABS M-10 Hovercraft

ABS Hovercraft selected Icon Polymer neoprene/nylon for the severe wear areas of the skirt for its M10, 20 m patrol vessels, first introduced in 1994. Designed for military and workboat operations, the craft is built from FRP and is able to exceed 50 kt. It can withstand heavy seas and clear obstacles up to 1 m high. M10 craft currently operate in Asia, Scandinavia and in Europe and the craft has been awarded a Millennium Product mark by the British government for advancing hovercraft technology into the 21st century. Icon Polymer continues to co-operate with ABS for the development of other materials on this, one of the most advanced hovercraft vehicles available today.

SR N6 Series (BHC)

Both civil and military variants of this series have incorporated skirt materials from the Icon Polymer range. The abilities of the materials, to withstand the most rigorous operating conditions, has contributed to the development of the SR. N6 Mk 6. It features significantly enhanced all-weather performance and improved manoeuverability, partly due to skirt construction, in which Icon Northern Rubber's materials feature. The craft are still in service but are no longer being built.

AP1-88 (BHC)

Icon Polymer supply skirt component materials to the UK and overseas constructors and operators of the AP1-88 and associated variants.

HM5/HM2 Series (Hovermarine International)

Many of these craft are in service in Asia where Icon Polymer skirt fabrics perform under conditions of high utilisation.

Griffon

Icon Polymer supply a wide range of materials for use on Griffon hovercraft in service around the world.

US Navy

Icon Polymer were awarded a development contract to provide a more durable skirt material and to provide assistance in the design of new skirt systems for future craft.

M10 Hovercraft fitted with Icon Polymer neoprene/nylon skirt 0507042

Icon Polymer hovercraft fabrics range

	Breaking strength N/25 mm (min)	Tear strength double tongue N (min)	Peel adhesion N/25 mm (min)	Total weight g/m²	Width (mm) (nominal)
ZTC11323 Neoprene					
Use: Segment fabric for hover trailers and inflatable craft					
Warp	1,785	313	67	950	1,500
Weft	1,785	313	–	–	–
ZTE11569 Neoprene					
Use: Skirt segments for hovercraft, water skates and heavy load trailers					
Warp	2,230	625	112	1,360	1,500
Weft	2,008	580	–	–	–
ZTE10863 Neoprene					
Use: Skirt and finger fabric for commercial passenger carrying vehicles					
Warp	2,450	890	134	2,515	1,500
Weft	2,250	890	–	–	–
ZTE11184 Neoprene					
Use: Skirt and finger fabric for passenger carrying vehicles					
Warp	4,018	2,230	223	2,890	1,500
Weft	3,795	2,008	–	–	–
ZTE11828 Natural Rubber					
Use: Skirt and segment fabric for heavy load transporters and passenger carrying vehicles					
Warp	3,795	2,230	200	3000	1,500
Weft	3,570	2,230	–	–	–
ZTE11748 Neoprene					
Segments for heavy load transporters and passenger carrying vehicles					
Warp	4,018	2,232	223	3,220	1,500
Weft	3,795	2,008	–	–	–

Materials

The range of composite flexible materials is manufactured in combinations of natural rubber or neoprene polymer and nylon substrates. All materials are tested in accordance with the highest standards covered by BS 4F100. To properly demonstrate the adhesive properties of the hovercraft materials, Icon Polymer has developed a system of testing the materials which gives a representative view of how the materials will perform in use.

Neoprene composite materials have outstanding oil and ozone resistance and good low temperature flexibility down to –30°C. Natural rubber composite materials combine excellent abrasion resistance and lower temperature flexibility to –50°C.

United States

Avon Engineered Fabrications Inc

1200 Martin Luther King Jr Boulevard, Picayune, Mississippi 39466-5427, United States

Tel: (+1 601) 799 12 17
Fax: (+1 601) 799 01 48
e-mail: enquiries@avon-rubber.com
Web: www.avonrubber.com

Mike Hammer, *General Manager*
Danny Rogers, *Technical Manager*
David Poole, *Business Development Manager*
Donald LeFebre, *Senior Contract Administrator*

As a subsidiary of Avon, Avon Engineered Fabrications has full access to the resources of the world's largest and most experienced manufacturer of hovercraft skirt materials.

In April 1985, Avon Engineered Fabrications opened its facility in Picayune, Mississippi, as the only specialised manufacturer of hovercraft skirt systems in North America.

Avon Engineered Fabrications's facility in South Mississippi currently occupies over 50,000 sq ft of manufacturing space. Specific manufacturing equipment includes:

High-precision hydraulic presses
Large beam presses
A range of 'C' frame presses
Thermoplastic welding machines
Dialectic welding equipment
Surface preparation machines
Die cutting presses
Specialised mixing equipment.

Manufacture and assembly is carried out at this purpose-built plant. Avon Engineered Fabrications utilises state-of-the-art bonding processes to produce an extremely well integrated composite structure. The proprietary thin adhesive layer provides considerably better flexural fatigue properties than bonding tapes.

Avon Engineered Fabrications and its parent company, Avon, offer full skirt design, including stress analysis, design appraisal, finite element analysis and reduced or full-scale model testing.

Avon Engineered Fabrications is able to offer customers many refurbishment options for prolonging the life of skirt components, often at a fraction of the cost of a new component.

LCAC: The LCAC skirt comprises more than 80 different components and it is supplied to craft manufacturers as assembled segments for ease of installation. The skirt is developed from the conventional bag and finger design, utilising lock bolts to facilitate the changing of worn or damaged components. The skirt design embraces many differing weights of coated fabric and calendered sheet to arrive at the optimum combination of weight, flexibility and operational life. Coated fabric weights vary between 1,390 g/m^2 and 3,050 g/m^2, and both natural and synthetic rubbers are used.

Avon Engineered Fabrications has rapidly become a major supplier of original equipment and spares for this programme. Since its inception, Avon Engineered Fabrications has produced, or has orders in hand to produce, over 72 of the original skirt sets used on the 91 LCACs currently operated or scheduled for construction.

Surface Effect Ships (SES): Bow fingers and stern seal components manufactured and assembled by Avon Engineered Fabrications have been installed on Bell Halter commercial craft operated in Egypt and South America, on US Coast Guard craft formerly based in Key West, Florida, on the US Navy's SES 200 and on the US Army corps of Engineers survey boat, *Rodolf*.

In 1990, the US Navy's SES 200 was upgraded by Textron Marine and Land Systems to include water-jet propulsion and a new three-lobe stern seal. The seal was designed utilising the vast experience gained from supplying the European SES market and in conjunction with Avon a new seal was fabricated. This seal has consistently exceeded all specification and craft requirements.

Commercial hovercraft: Avon Engineered Fabrications regularly works with customers on skirt design for both military and commercial ACVs. Skirt systems, utilising the vast range of coated fabrics available, have been produced for the Utility Air Cushion Vehicle (UACV) and the C7/FR7 craft designed and produced by Textron Marine and Land Systems. The C7 hovercraft is a commercial craft operating in Asia as a passenger craft for Freeport McMoran and the FR7 is used as a fire/rescue craft by the Singapore Aviation Authority.

Textron Marine Systems Landing Craft, Air Cushion (LCAC) for which Avon Engineered Fabrications has now become a major supplier of skirt systems and components 0506943

MARINE ESCAPE SYSTEMS

Company listing by country

Australia
Liferaft Systems Australia Pty Ltd (LSA)

Canada
DBC Marine Safety Systems

Denmark
Viking Lifesaving Equipment A/S

France
Zodiac International

United Kingdom
RFD Beaufort Ltd

Australia

Liferaft Systems Australia Pty Ltd (LSA)

5 Sunmont Street, Derwent Park, Tasmania 7009, Australia

Tel: (+61 3) 62 73 92 77
Fax: (+61 3) 62 73 92 81
e-mail: info@lsames.com
Web: www.liferaftsystems.com.au

European Office:
PO Box 306, Bangor, County Down, BT20 9BF, United Kingdom

Tel: (+44 28) 91 274 424
Fax: (+44 28) 91 240 138
e-mail: p.rea@lsames.com

North American Office:
Unit 102, 2965 Horley Street, Vancouver, BC V5R 6B9, Canada

Tel: (+1 604) 780 0016
Fax: (+1 604) 431 2924
email: v.prato@lsames.com

Mike Grainger, *Managing Director*

Liferaft Systems Australia's inflatable marine escape system 1175709

Liferaft Systems Australia (LSA) was formed specifically for the design and manufacture of inflatable Marine Evacuation Systems (MES) and large capacity liferafts for the passenger ferry and military markets.

LSA equipment is now installed on many fast car, passenger and conventional ferries and military vessels. Type approval has been received for LSA equipment from DNV (EU Marine Directive), US Coast Guard and Transport Canada. In recent years LSA has supplied slide-type MES for the InCat built US Navy high-speed vessels HSV-XI Joint Venture, TSV-IX Spearhead and HSV2 Swift, as well as the Nichols Brothers' X-craft for the US military.

Canada

DBC Marine Safety Systems Ltd

A division of Zodiac International
Unit 101, 3760 Jacombs Road, Richmond, British Columbia V6V 1Y6, Canada

Tel: (+1 604) 278 32 21
Fax: (+1 604) 278 78 12
e-mail: dbcsales@zodiac.com
Web: www.dbcmarine.com

Thierry Schwab *General Manager*

DBC Marine Safety Systems manufactures ship evacuation systems including an inflatable slide system for low freeboard craft, a vertical descent chute for higher freeboard craft and a range of davits and liferaft racking systems.

DBC was the first North American liferaft manufacturer to successfully complete testing and obtain CCG and USCG approvals for both the 50-person liferaft and 100-person platform to SOLAS specifications.

DBC Marine evacuation SOLAS chute system 0093265

Denmark

Viking Lifesaving Equipment A/S

Saedding Ringvej 13, DK-6710 Esbjerg V, Denmark

Tel: (+45) 76 11 81 00
Fax: (+45) 76 11 81 01
e-mail: viking@viking-life.com
Web: www.viking-life.com

Kjeld Amann, *Managing Director*
Henrik Uhd Christensen, *Global Sales and Marketing Director*

Viking supplies three main inflatable marine escape systems: the VES, which is a twin slide for drop heights of between 6 and 15 m; the VEC is a vertical chute available in single chute and side by side configurations and the VEMC which is a single-track chute for use with drop heights from 5 to 12 m, developed for small to medium-size vessels.

Slide

The dual-track VES slide is fabricated from twelve main longitudinal tubes within eight separate main compartments, each individually inflated. The slide, supplied in lengths from 12 to 25.5 m, is designed for embarkation heights of 6 to 15 m and is connected to a 100 person self draining evacuation platform which may serve as an open liferaft when evacuation has been completed and the connections to the slide cut. The capacity of the slide is tailored to the specific requirements but this slide has been tested with 500 passengers being evacuated in under 30 minutes. For ships built according to the high-speed craft code, the slide system evacuates 350 people in 17 minutes 40 seconds. The 30° angle between the slide and ship ensures the system can compensate and absorb excessive ship and sea movement. The slide is supplied in a stowage box designed for simple access, installation and maintenance.

Chute

The company produces a range of vertical chute MES's which provide for a completely dry evacuation of passengers directly from the ship into the liferafts via a vertical chute. This can be arranged in single or side-by-side chute configurations. The single chute system can evacuate 367 passengers in 30 minutes and the double chute system 734 passengers in 30 minutes. In 2003 Viking launched the VEMC, a minichute developed for small to medium-size vessels. The minichute has an evacuation capacity of 354 persons in 30 minutes and the evacuation height can range from 5 m to 20 m.

Liferafts

Viking offers a range of automatically self-righting liferafts for 25,51,101 and 150 persons, designed to be applied with slides or chutes.

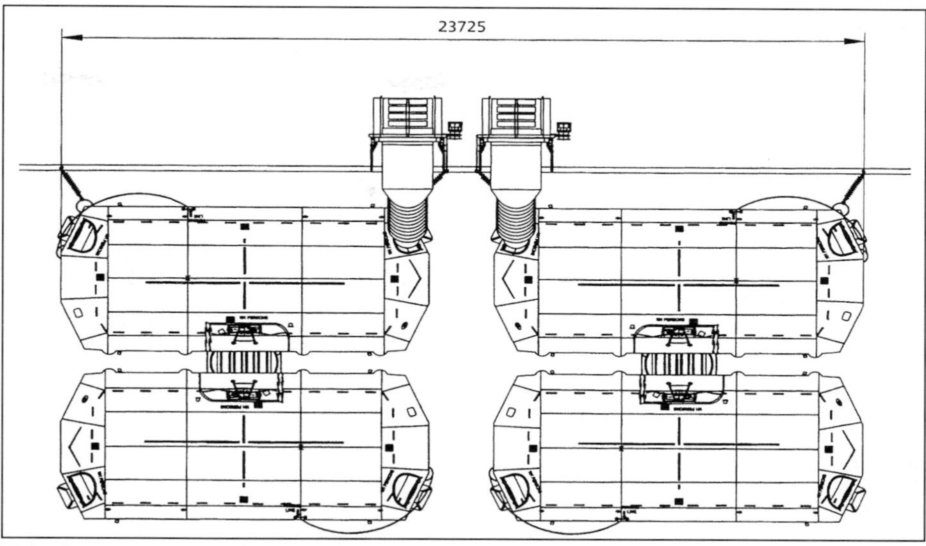

Twin chute arrangement for evacuation of over 700 passengers in less than 30 minutes 0103309

Viking vertical descent MES 0507110

France

Zodiac International

2 rue Maurice Mallet, F-92130, Issy-Les-Moulineaux Cedex, France

Tel: (+33 1) 41 23 23 23
Fax: (+33 1) 41 23 23 98
e-mail: solassales@zodiac.com
Web: www.zodiacsolas.com

Marc Lavorata, *Marketing Manager*

Zodiac International produce a range of throw over self righting liferafts from 25 to 150 man capacities. The larger liferafts have been designed to integrate with the Canadian DBC marine escape system. The company also manufactures evacuation chutes, and a small inflatable slide for situations where the height of the evacuation deck is less than 3.2 m.

United Kingdom

RFD Beaufort Ltd

A division of the Survitec Group
Kingsway, Dunmurry, Belfast, Northern Ireland BT17 9AF, United Kingdom

Tel: (+44 28) 90 30 15 31
Fax: (+44 28) 90 62 17 65
e-mail: info@rfdbeaufort.com
Web: www.rfdbeaufort.com

W McChesney, *Chief Executive Officer*
Moya Johnston, *Finance Director*
C Stocker, *Operations Director*
Alasdair MacIntyre, *Commercial Marine Director*
Richard McCormick, *MES Sales Manager*
Bethany Sinclair, *Marketing Officer*

RFD Beaufort Ltd manufactures and supplies a range of marine and aerospace survival systems and was the first company worldwide to supply Marine Evacuation Systems (MES).

RFD Marin-Ark MES

The RFD Marin-Ark MES comprises up to four 100+ person capacity *reversible* life rafts per station which are attached to two fully telescopic descent paths, which transfer evacuees dry-shod from the ship directly into the life raft. The complete MES is all contained within a single compact and lightweight stowage unit which is simple to install.

The RFD Marin-Ark MES can be configured to suit the needs of individual vessels and is available in combinations of two, three or four rafts per station. With its full complement of four rafts and two descent paths, the RFD Marin-Ark MES has a capacity for 430 people. It is designed to be the only required solution for a whole range of passenger vessels, making it the most cost effective fleet system.

Simple to deploy and operate, this MES requires less trained crew per station than old MES systems, which substantially reduces crew manning and crew training costs. With fewer inflatable components, the servicing costs and the time the ship needs to be alongside during the annual servicing period are also significantly reduced.

An RFD Marin-Ark 430 station with four life rafts and two descent paths weighs approximately 5,000 kg with stowage dimensions of 2.9 × 2.4 × 2.3 m. The descent paths are fully approved to a stowage height of 17 m.

An RFD Marin-Ark 109 person capacity liferaft (stand alone) weighs 840 kg and is stowed in a GRP container measuring 2.3 m in length and 1.0 m in diameter. To accommodate vessels with a freeboard of 5 m or less the RFD Marin-Ark liferaft is available with a simple, single-point browsing arrangement. This provides the advantage of fully complying with the new IMO HSC Code without the need to install a full marine evacuation system. There is also the option to install a mini-slide if the freeboard height is above 4.0 m.

RFD Marin-Ark system 0044050

CONSTRUCTION MATERIALS

Company listing by country

Australia
Ayres Composite Panels

Canada
Alcan Inc

France
Alcan Marine

Germany
Aleris Aluminium Koblenz

Norway
Devold AMT A/S

Sweden
SAPA Mass Transportation

Switzerland
Alcan Aluminium Valais Ltd

United Kingdom
Gurit (UK) Ltd

Australia

Ayres Composite Panels

25-27 Clune Street, Bayswater, Western Australia 6053

Tel: (+61 8) 92 79 34 26
Fax: (+61 8) 92 79 61 92
e-mail: asia-pacific@ayrescom.com
Web:www.ayrescom.com

European Office:
Rue de Dison 100, B-4840Welkenraedt, Belgium

Tel: (+32 87) 89 12 45
Fax: (+32 87) 89 12 46
e-mail: europe@ayrescom.com

United States Office and Manufacturing Unit:
PO Box 616, Berwick, Los Angeles 70342

Tel: (+1 985) 395 90 37
Fax: (+1 985) 395 42 50
e-mail: usa@ayrescom.com

Ayres Composite Panels supply much of the world's fast vessel market with Ayrlite™ lightweight composite panels systems for interior outfit. The range consists of a variety of aluminium honeycomb aore panels and assembly systems. Key products are SOLAS type approved and United States Coast Guard-approved as well as being wheel marked for European use. Vessels include fast ferry, naval and paramilitary.

The panels have applications as wall linings, partitions, bench tops, tables, shelving, toilet and kiosk modules.The panels can be supplied with a range of surface finishes meeting the IMO fire regulations for high-speed craft as well as a complete extrusion system.

Canada

Alcan Inc

1188 Sherbrooke Street West, Montreal, Quebec H3A 3G2, Canada

Tel: (+1 514) 848 80 00
Fax: (+1 514) 848 81 15
e-mail: maison.infocentre@alcan.com
Web: www.alcan.com

Alcan, formed in 1902, is one of the world's largest producers of aluminium and a leading manufacturer of aluminium-based products. Alcan operates in more than 55 countries and has a network of smelters in Australia, France, Canada, UK, and the US and with related companies in Japan and India. The company currently employs 68,000 people. As well as supplying the marine industry, Alcan also supplies materials for electrical, aerospace and other industries.

In December 2003 Alcan purchased the Pechiney Group of France.

Denmark

Sapa Mass Transportation

Rolshøyvej, 8500 Grenå, Denmark

Tel: (+45 86) 326 100
Fax: (+45 87) 585 402
e-mail: masstransportation@sapagroup.com
Web: www.sapagroup.com

Sven Lundin, *Sales Manager*

Sapa is an industrial group and specialises in value added products based on lightweight aluminium. The products include profiles and heat-transfer strip in aluminium. The company has a turnover of around EUR1.35 billion with 8,200 employees in Europe, United States and China. Sapa is one of Europe's leading companies within its field and customers are located in the transport, building, engineering and telecom industries.

Sapa Mass Transportation is the driving force within the group in development and exploitation of the market opportunities within the mass transportation industry, working with customers in the rolling stock and marine sectors.

With Friction Stir Welding (FSW) the company can join profiles into 14,300 × 3,000 mm panels. The prefabricated panels are used for decks and superstructures for cruise and fast ships and in rolling stock vehicles.

France

Alcan Marine

Formerly Pechiney Marine
2 rue Grande Bretangne, 4430 Nantes, France

Tel: (+33 2) 4030 4400
Fax: (+33 4) 73 55 50 60
e-mail: benoit.lancrenon@alcan.com
Web: www.alcan-marine.com

Jean-Luc Ruax, *Marketing Manager*

Alcan Marine is involved in innovation and development of marine shipbuilding. Pechiney Marine 5383 is extensively used for the hulls and superstructure of many high-speed vessels and the company has recently successfully introduced Sealium marine grade alloy.

Germany

Aleris Aluminium Koblenz

Part of Aleris International Inc
Carl-Spaeter Strasse 10, D-56070 Koblenz, Germany

Tel: (+49 261) 89 10
Fax: (+49 261) 891 73 42
e-mail: info.koblenz@aleris.com
Web:www.aleris-koblenz.com

Alfred Haszler, *President*
Manfred Rust, *Sales Director, Commercial Plate & Distribution*
Ursula Berndsen, *Marketing Manager*
Dieter Rätz, *Sales Manager*

Aleris Rolled Products – Europe is a leading manufacturer of high quality, precision aluminium plate, sheet and coil. To meet future shipbuilding needs, a major research programme was started to develop advanced material systems yielding substantial reductions in structural weight. A wrought Al-Mg alloy plate called Alustar® (AA5059) has been developed which is up to 26 per cent stronger than AA5083 while exhibiting a similar corrosion resistance and weldability. The high strength of Alustar® provides designers with the possibility of significantly reducing structural weights of all kinds of vessels.

The modern hot and cold rolling mills at Koblenz and Duffel (Belgium) produce Alustar® in extra wide/long plates and sheets and various tempers. Alustar® has been approved by the certification bodies ABS, DNV, GL, LRS and Class NK, and its global network of offices, subsidiaries, agencies, representatives and qualified trade companies enables it to support customers throughout the shipbuilding industry.

Norway

Devold AMT A/S

N-6030 Langevåg, Norway

Tel: (+47) 70 19 85 00
Fax: (+47) 70 19 85 01
e-mail: office@amt.no
Web: www.amt.no

Johan Fausa, *Managing Director*
Baard Røsvik, *Sales and Marketing Manager*

Devold AMT (Advanced Multiaxial Technology) supplies a wide range of reinforcing fibres for engineering structures including those for use in the marine industry. References in the high-speed craft industry include a range of monohull craft up to 74 m in length, SES craft up to 54 m and luxury yachts up to 50 m. Of particular interest is the multiaxial knitting technology employed by the company for its aramid, carbon, glass fibres, providing improved structural and handling performance. These have recently been used by Brødrene Aa for its *Rygerkatt* passenger ferry.

Devold AMT products are also approved by a wide range of classification societies.

Switzerland

Alcan Aluminium Valais Ltd

CH-3960 Sierre, Switzerland

Tel: (+41 27) 457 51 11
Fax: (+41 27) 457 68 75
e-mail: max.siegrist@alcan.com

Max Siegrist, *Product Manager*

Alcan has been producing aluminium for more than 100 years and the manufacture of large aluminium extrusions for the transport industry is one of its principal activities, as evidenced by three large presses of 7,500, 7,200 and 5,500 tons respectively.

Its wide range of products and specialised expertise have made Alcan a leading manufacturer of large extrusions for use in industry, transport and shipbuilding. The company is ISO/EN9001, 14001 and OHSAS 18001 accredited.

United Kingdom

Gurit (UK) Ltd

St Cross Business Park, Newport, Isle of Wight PO30 5WU, United Kingdom

Tel: (+44 1983) 82 80 00
Fax: (+44 1983) 82 81 00
e-mail: info@gurit.com
Web:www.gurit.com

Gurit is a developer and manufacturer of composite technology, enabling high-performance structures to be lighter, stronger and more cost-effective than ever before. The company combines skill in structural design, process engineering and materials science to offer complete composite solutions. Gurit has a history of innovation and excellence in the field of composites and continues to focus on research and development activities aimed at making high performance composites accessible for new applications. Gurit, formerly SP Systems in the United Kingdom, has contributed significantly to the structural design and materials supply of America's Cup racing yachts, the Royal National Lifeboat Institution (UK) Trent, Mersey and Severn class vessels as well as various fast patrol and leisure craft and the VSV military troop insertion craft, in service with the special forces in Germany, United Kingdom and the United States.

In 2004, the company was contracted to undertake structural engineering and supply all of the composite materials for the US Department of Defense's M80 (M-Hull) demonstrator craft to be built at the Knight and Carver Yacht centre in San Diego, United States.

The VSV uses Gurit materials 0097641

LIGHTWEIGHT SEATING

Company listing by country

Australia
Beurteaux Australia
Maxton Fox Commerical Furniture Pty Ltd
McConnell Seats Australia
Tecnoseat Australia Pty Ltd
Transport Seating Technology

Italy
Geven SRL

Norway
Georg Eknes Industrier AS
Modell-Møbler AS

United Kingdom
Merok Marine International

United States
Arduous Inc

Australia

Beurteaux Australia

8 SuccessWay, Henderson, Western Australia 6166, Australia

Tel: (+61 8) 94 94 68 88
Fax: (+61 8) 94 94 68 89
e-mail: enquiries@beurteaux.com
Web: www.beurteaux.com

Neil Howe, *Marketing Manager*

Established in 1954, Beurteaux has supplied seats for over 500 vessels worldwide. Offering a wide range of seat classes from tourist to first class seating systems with weights as low as 7 kg per seat. Beurteaux has offices in Australia, China, Europe, Hong Kong, Singapore and the United States, and is represented in Japan by Marubeni.

Beurteaux Australia's lightweight fast ferry seating
0068899

Maxton Fox Commercial Furniture Pty Ltd

87 Bay Street, Glebe, New South Wales 2037, Australia

Tel: (+61 2) 96 92 00 33
Fax: (+61 2) 96 60 11 91
e-mail: sales@maxtonfox.com.au
Web: www.maxtonfox.com.au

Maxton Fox designs and manufactures a range of slim, lightweight passenger seating suitable for fast-ferry applications ranging from a super light design at 4.75 kg per seat to a luxury version incorporating a range of accessories.

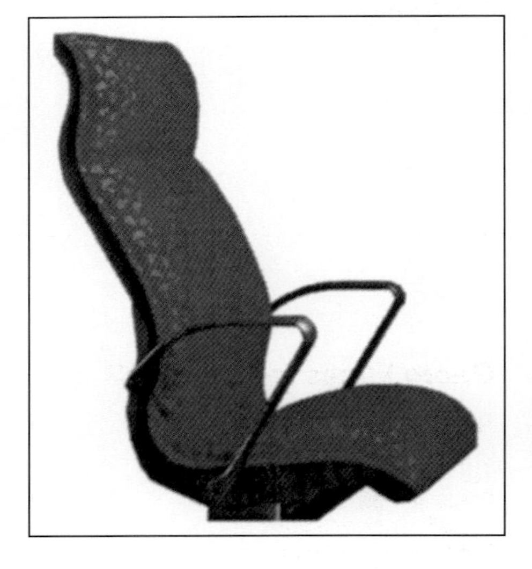

Maxton Fox slim fast-ferry seating
0079994

McConnell Seats Australia

100 Bakers Road, North Coburg, Victoria 3058, Australia

Tel: (+61 3) 93 50 72 77
Fax: (+61 3) 93 50 73 44
e-mail: info@mcconnellseats.com.au
Web: www.mcconnellseats.com.au

Denis McConnell, *Managing Director*
Craig Coulter, *Sales Manager*

Founded in 1952, the company manufactures lightweight seating for a range of transport vehicles including coaches, buses, railways and ferries. The Espace range of seating has been designed specifically for the fast ferry industry.

Tecnoseat Australia Pty Ltd

A division of UES (International) Pty Ltd
Locked Bag 2003, Lidcombe, New South Wales 2141, Australia

Tel: (+61 2) 87 45 90 60
Fax: (+61 2) 89 15 15 57
e-mail: uesnsw@uesint.com
Web: www.uesint.com

Johann Naess, *Design and Development Manager*

The company was acquired in 2004 by UES International of Australia and continues to supply a range of lightweight seating systems for fast ferry applications.

The super-lightweight Link range of marine seating features seat weights as low as 3.454 kg.

Transport Seating Technology

14 Angel Road, Stapylton, Queensland 4207, Australia

Tel: (+61 7) 38 07 80 08
Fax: (+61 7) 38 07 88 81
e-mail: info@transportseating.com
Web: www.transportseating.com

Rod Ferguson, *Managing Director*

Founded in 2000, the company staff has considerable experience in the design and construction of lightweight seating for fast ferries. The company produces a range of seats for marine use, including the alloy Waverider series which is manufactured in alloy with no welding.

Italy

Geven Srl

Via Boscofangone, Zona Industriale Nola Marigliano, 80035 Nola, Naples, Italy

Tel: (+39 081) 31 21 311
Fax: (+39 081) 31 21 321
e-mail: geven@geven.com
Web: www.geven.com

Tullio Veneruso, *Chief Executive Officer*
Alberto Veneruso, *Managing Director*
Angelo Romano, *Senior Vice-President, Commercial*
Giovanni Verrazzo, *Engineering Director*
Salvatore Cicolecchia, *Operations Manager*
Antonio Celoro, *Quality Manager*
Roberta Macciocca, *Customer Care Manager*
Doria De Chiara, *Business Development Manager*

Geven Srl has developed, manufactured, tested and certified a range of commercial marine seats and interiors based on its extensive experience as an aircraft seat manufacturer. The current range consists of several seat designs developed for economy and first class seating. The company also operates an overhaul and repair service worldwide.

Geven ferry seating 1030873

Norway

Georg Eknes Industrier AS

N-5913 Eikangervaag, Norway

Tel: (+47) 56 35 75 00
Fax: (+47) 56 35 75 01
e-mail: transit@eknes.no
Web: www.eknes.no

Jean Gerkens, *Sales Manager*

The company has supplied lightweight passenger seating since 1980 and offers an extensive range of fast ferry seating systems with individual seat weights as low as 4.5 kg. The range includes a low weight-to-strength ratio, all-anodised aluminium construction, no welded connections (which eliminates material cracking) and fire-retardant materials.

Georg Eknes transit seating system
(Georg Eknes Industrier AS)
0077029

Modell-Møbler AS

Skibåsen 7, N-4636 Kristiansand S, Norway

Tel: (+47) 38 12 79 90
Fax: (+47) 38 12 79 99
e-mail: office@modell-mobler.com
Web: www.modell-mobler.com

Frank Robertsen, *Marketing Manager*

The company has been supplying passenger seating for over 20 years and provides a wide range of lightweight fast-ferry seating systems.

Modell-Møbler AS fast-ferry seating
0581577

United Kingdom

Merok Marine International

PO Box 244, Hythe, Kent, CT21 5WR, United Kingdom

Tel: (+44 1303) 261 296
Fax: (+44 560) 115 57 79
e-mail: merok@btconnect.com
Web: www.merokmarine.com

Clive Sumner, *Managing Director*

Merok Marine International specialises in the design and manufacture of passenger seating for both fast ferry and Ro-Ro ferry applications, however some of the range is suitable for river boats and surface vessels such as hovercraft. The company has recently redesigned the whole range with emphasis on weight, interchangability and the ability to supply the seats in kit form.

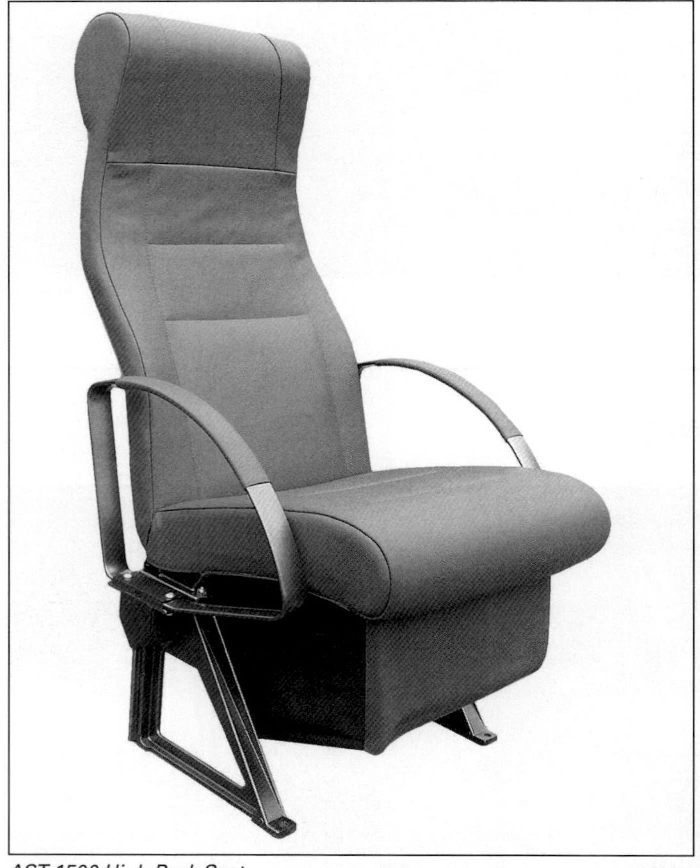

AST 1500 High Back Seat 1176218

Merok Marine Fast Ferry Seat 0093274

United States

Arduous Inc

22185 Highway 97, Tonasket, Washington 98855, US

Tel: (+1 509) 486 2010
Fax: (+1 509) 486 2010
e-mail: sales@arduousinc.com
Web: www.arduousinc.com

Dave Berry, *President*
Cliff Berry, *General Manager*

The company was formed in 1997 to produce a range of high-quality marine interior products including lightweight seats for fast ferries.

SERVICES

Marine craft regulatory authorities
Consultants and designers
Societies involved with high-speed craft

MARINE CRAFT REGULATORY AUTHORITIES

Argentina

Prefectura Naval Argentina
Avenida Eduardo Madero 235 piso 1
1106 Buenos Aires CF
Argentina

Tel: (+54 1) 4318 74 00
Fax: (+54 1 318) 76 43
e-mail: info@prefecturanaval.gov.ar
Web: www.prefecturanaval.gov.ar

Lieutenant Commander José Ramón Balmaceda, *Chief of the National Navigation Regulation Division*
Prefecto General Carlos Eduardo Fernandez, *Argentine Coast Guard Commandant*
Prefecto General Ricardo Rodriquez, *Deputy Commandant*

Australia

High-speed craft
Covers interstate and international voyages. Smaller craft come under jurisdiction of state or local authorities.

Australian Maritime Safety Authority
Ship and Personnel Safety Services
GPO Box 2181
Canberra ACT 2601
Australia

Tel: (+61 2) 62 79 50 00
Fax: (+61 2) 62 79 59 50
e-mail: shipsafetyreqs@amsa.gov.au
Web: www.amsa.gov.au

Greg Martin, *Chief Executive Officer*

New South Wales
Sydney Ports Corporation
Level 8
207 Kent Street
Sydney
New South Wales 2000
Australia
Postal address
PO Box 25
Millers Point
Sydney
New South Wales 2000
Australia

Tel: (+61 2) 92 96 49 99
Fax: (+61 2) 92 96 47 42
e-mail: spoh@sydneyports.com.au
Web: www.sydneyports.com.au

Captain R Heath, *Harbour Master and Navigation Services Manager*
Shane Hobday, *General Manager Marine Operations*

Northern Territory
Department of Infrastructure, Planning and Environment
Marine Safety Branch
PO Box 2520
Darwin
Northern Territory 0801
Australia

Tel: (+61 8) 89 99 52 85
Fax: (+61 8) 89 99 53 00
e-mail: marinesafety@nt.gov.au
Web: www.nt.gov.au

Sri Srinivas, *Principal Marine Surveyor*

South Australia
Flinders Ports Pty Ltd
PO Box 19
Port Adelaide
South Australia 5015
Australia

Tel: (+61 8) 84 47 06 11
Fax: (+61 8) 84 47 06 06
e-mail: flindersports@flindersports.com.au
Web: www.flindersports.com.au

Vincent Tremaine, *Chief Executive Officer*

Tasmania
Marine and Safety Tasmania (MAST)
GPO Box 607
Hobart
Tasmania 7001
Australia

Tel: (+61 3) 62 33 51 55
Fax: (+61 3) 62 33 56 62
e-mail: admin@mast.tas.gov.au
Web: www.mast.tas.gov.au

Colin Finch, *Chief Executive*

Victoria
The Port of Melbourne Corporation
GPO Box 261
Melbourne
Victoria 3001
Australia

Tel: (+61 3) 13 00 857 662
Fax: (+61 3) 96 83 15 70
e-mail: information@portofmelbourne.com
Web: www.portofmelbourne.com.au

Neil Edwards, *Executive Chairman*
Brendan Power, *Chief Operating Officer*

Belgium

Federal Public Service Mobility and Transport
Maritime Transport
Rue du Progres 56
B-1210 Brussels
Belgium

Tel: (+32 2) 277 35 02
Fax: (+32 2) 277 40 51
e-mail: info.mar@mobilit.fgov.be
Web: www.mobilit.fgov.be

Martine Stoquart, *Information Officer*

Canada

Transport Canada
Marine Safety
Tower C
Place de Ville
11th Floor
330 Sparks Street
Ottawa
Ontario K1A 0N8
Canada

Tel: (+1 613) 998 06 12
Fax: (+1 613) 954 10 32
Web: www.tc.gc.ca

William Nash, *Director General*

The administration of national and international laws designed to ensure the safe operation, navigation, design and maintenance of ships, protection of life and property and prevention of ship-source pollution.

Côte d'Ivoire

Ministère des Travaux Publics et des Transports
BP V6
Abidjan
Ivory Coast
Tel: (+225) 34 75 13
Fax: (+225) 34 73 29

Denmark

Danish Maritime Authority
Vermundsgade 38C
DK-2100 Copenhagen Ø
Denmark

Tel: (+45 3917) 44 00
Fax: (+45 3917) 44 01
e-mail: dma@dma.dk
Web: www.dma.dk

Egypt

Ministry of Transport
Egyptian Authority for Maritime Safety
Bab Gomrok-1
Ras el Tin
Alexandria
Egypt

Tel: (+20 3) 48 63 50/480 20 31/483 36 98
Fax: (+20 3) 483 76 27/483 20 41/487 04 01

Fiji

Marine Department
Department of Maritime Safety
Fiji Islands Maritime Safety Administration (FIMSA)
FIMSA House
Walu Bay
P.O. Box 326
Suva
Fiji

Tel: (+679) 331 52 66
Fax: (+679) 30 32 51
e-mail: fimsa@connect.com.fj
Web: www.fimsa.gov.fj j

J E Rounds, *Director of Maritime Safety*

Finland

Finnish Maritime Administration
Porkkalankatu 5
PO Box 171
FI-00181 Helsinki
Finland

Tel: (+358) 20 44 81
Fax: (+358) 204 48 43 55
e-mail: Kirjaamo@fma.fi
Web: www.fma.fi

France

Ministère de L'Ecologie, du Développement et de L'Arménagement durables
Arche Sud
F-92005 La Défense Cedex
France

Tel: (+33 1) 44 49 86 41
Fax: (+33 1) 44 49 86 40
e-mail: pierre.mitton@equipement.gouv.fr
Web: www.transportsequipement.gouv.fr

Gambia

Gambia Ports Authority
34 Liberation Avenue
PO Box 617
Banjul
Gambia
West Africa

Tel: (+220) 22 99 40
Fax: (+220) 22 72 68
e-mail: info@gamport.gm
Web: www.gamport.gm

M L Gibba, *Managing Director*

Germany
See-Berufsgenossenschaft
Ship Safety Division
PO Box 11 04 89
Reimerstwiete 2
D-20457 Hamburg
Germany

Tel: (+49 40) 36 13 70
Fax: (+49 40) 36 13 72 04
Web: www.see-bg.de/schiffssicherheit/division

Nikolai Woelki, *Managing Director*

Ghana
Shipping Commissioners
Division of Shipping and Navigation
Ministry of Transportation
PO Box M38
Accra
Ghana

Tel: (+23 3) 21 66 35 06
Fax: (+23 3) 21 66 71 14
Web: www.mrt.gov.gh

Greece
Ministry of Mercantile Marine
Merchant Ships Inspectorate (MSI)
K Paleologou 1
GR-18535 Piraeus
Greece

Tel: (+30 210) 419 17 00
e-mail: gov@yen.gr
Web: www.egov.yen.gr

S Bourliaskos, *Commodore*

Hong Kong
Director of Marine
Marine Department
Harbour Building
38 Pier Road
Central
Hong Kong

Tel: (+852) 25 42 37 11
Fax: (+852) 25 41 71 94
e-mail: mdenquiry@mardep.gov.hk
Web: www.mardep.gov.hk

Hungary
Ministry of Transport, Communications and Waterways Superintendence for Supply
PO Box 102
H-1389 Budapest 62
Hungary

Tel: (+36 1) 111 22 61
Fax: (+36 1) 111 14 12

Iceland
The Icelandic Maritime Administration
Vesturvör 2
PO Box 120
200 Kopavogur
Iceland

Tel: (+354) 560 00 00
Fax: (+354) 560 00 60
e-mail: upplysingdeild@sigling.is
Web: www.sigling.is

India
Directorate General of Shipping
Jahaz Bhavan
Walchand Hirachand Marg
Mumbai I 400 001
India

Tel: (+91 22) 22 61 36 51
Fax: (+91 22) 22 61 36 55
e-mail: dgship@dgshipping.com
Web: www.dgshipping.com

Indonesia
Directorate General of Sea Communications
Jalan Medan Merdelka Barat No 8
Jakarta-Pusat 10110
Indonesia

Tel: (+62 21) 345 6703
Fax: (+62 21) 350 65 34
e-mail: ditnav@indomarinav.com
Web: www.indomarinav.com

Ireland
Department of Transport
Marine Survey Office, Leeson Lane
Dublin 2
Ireland

Tel: (+353 1) 678 34 00
Fax: (+353 1) 678 34 09
e-mail: mso@transport.ie
Web: www.transport.ie

Brain Hogan, *Principal Officer*

Israel
Ministry of Transport
Administration of Shipping and Ports
Division of Engineering & Operations
PO Box 806
Haifa 31999
Israel

Tel: (+972 4) 863 2121
Fax: (+972 4) 863 21 19
e-mail: techni@mot.gov.il
Web: www.mot.gov.il

Italy
Ministero delle Infrastrutture e dei Trasporti
Sicurezza della Navigazione
Viale dell'Arte 16
I-00144 Rome
Italy

Tel: (+39 06) 59 08 42 07
Fax: (+39 06) 59 08 45 08
email: sicnavi@libero.it
Web: www.infrastrutturetrasporti.it

Jamaica
Mariitime Authority of Jamaica
40 Knutsford Boulevard
Kingston 5
Jamaica

Tel: (+1 876) 754 7260
Fax: (+1 876) 754 72 56
e-mail: registrar@jamaicaships.com
Web: www.jamaicaships.com

Eric Deans, *Registrar General*

Japan
Ministry of Land Infrastructure and Transport
Maritime Bureau
2-1-3 Kasumigaseki
Chiyoda-ku
Tokyo 100
Japan

Tel: (+81 3) 52 53 86 39
email: psc@mlit.go.jp
Web: www.mlit.go.jp/english/index/html

Korea, South
Ministry of Maritime Affairs and Fisheries
Maritime Technology Division
140-2 Gye-dong
Jongno-Gu
Seoul 110-793
Republic of Korea

Tel: (+82 2) 36 74 63 21
Fax: (+82 2) 36 74 63 27
e-mail: psc@momaf.go.kr
Web: www.momaf.go.kr

Kuwait
Department of Customs and Ports
PO Box 318
Safat 11111
Kuwait
Tel: (+965) 484 4222
Fax: (+965) 484 48 31
e-mail: office@customs.gov.kw
Web: www.kwtlogistics.com/customs.html

Lebanon
Ministry of Transport
Directorate General of Land and Maritime Transport
Sturco Building 3rd Floor
Beirut
Lebanon

Luxembourg
Ministère de l'Economie et du Commerce Extérieur
Immeuble Forum Royal
Boulevard Royal 19-21

Tel: (+352) 24 78 41 37
Fax: (+352) 46 04 48
e-mail: info@eco.public.lu
Web: www.eco.public.lu

Commissariat aux Affaires Maritimes
Immeuble Forum Royal
Boulevard Royal 19-21

Tel: (+352) 24 78 44 53
Fax: (+352) 29 91 40
e-mail: cam@cam.etat.lu
Web: www.cam.etat.lu

Madagascar
Ministère des Transports, de la Météorologie et du Tourisme
BP 4139
Anosy Antananarivo
Madagascar

Tel: (+261 2) 246 04
Fax: (+261 2) 240 01

Malawi
Ministry of Transport and Public Works
Department of Marine Services
Private Bag 322
Lilongwe Capital City
Malawi

Tel: (+265) 178 93 77
Fax: (+265) 178 93 28
e-mail: transport@malawi.gov.mw
Web: www.malawi.gov.mw/

Malaysia
Marine Department Headquarters
PO Box 12
Foreshore Road 42007
Port Klang
Selangor Darus Ehsan
Malaysia

Tel: (+60 3) 368 66 16
Fax: (+60 3) 255 70 41
Web: www.marine.gov.my

Malta
Malta Maritime Authority
Maritime Trade Centre
Xatt 1-Ghassara ta' l-Gheneb
Marsa MRS 1912
Malta

Tel: (+356) 21 222 203
Fax: (+356) 21 250 365
e-mail: info@mma.gov.mt
Web: www.mma.gov.mt

Dr M Bonello, *Chairman*

Mexico
Direccion de Marina Mercante
Avenida Nuevo Leon 210
3er Piso
Colonia Hipodromo CP 06100
Delegacion Cueuhtemoc
Mexico DF

Tel: (+52 55) 52 65 32 20
Fax: (+52 55) 52 65 32 33
e-mail: jtlozano@sct.gob.mx

Morocco
Ministère de l'Equipement et du Transport
Direction de la Marine Marchande
Blvd Felix Houphouet
Boigny
20 000 Casablanca
Morocco

Netherlands
Netherlands Shipping Inspectorate
PO Box 8634
3009 AP
Rotterdam
Netherlands

Tel: (+31 10) 266 8500
Fax: (+31 70) 202 3424

New Zealand
Maritime New Zealand
Level 10, Optimation House
1 Grey Street
PO Box 27006
Wellington 6141
New Zealand

Tel: (+64 4) 473 01 11
Fax: (+64 4) 494 12 63
e-mail: enquiries@maritimenz.govt.nz
Web: www.maritimenz.govt.nz

High-speed craft are subject to the NZ Maritime Rule Part 40A and B.

Norway
Norwegian Maritime Directorate
Smedasundet 50a
PO Box 2222
N-5509 Haugesund
Norway

Tel: (+47) 52 74 50 00
Fax: (+47) 52 74 50 01
e-mail: postmottak@sjofartdir.no
Web: www.sjofartsdir.no

South Africa
South African Maritime Safety Authority
Private Bag X 7025
Roggebaai 8012
South Africa

Tel: (+27 21) 402 89 86
Fax: (+27 21) 421 61 09
e-mail: samsaops@iafrica.com

Captain B Watt, *Chief Executive Officer*

Spain
Dirección General de la Marina Mercante
Inspección General Maritima
C/ Ruiz de Alacrón, num 1
E-28071 Madrid
Spain

Tel: (+34 91) 597 92 58
Fax: (+34 91) 597 90 03

Felipe Martinez, *Director General*

Sweden
The Swedish Maritime Administration
Sjofartsverket
S-601 78 Norrkoping
Sweden

Tel: (+46 11) 19 10 00
Fax: (+46 11) 10 19 49
e-mail: hk@sjofartsverket.se
Web: www.sjofartsverket.se

Switzerland
Seepolizei des Kantons Thurgau
Bleichestrasse 42
Postfach 2141
CH-8280 Kreuzlingen
Switzerland

Tel: (+41 71) 221 49 00
Fax: (+41 71) 221 49 02
e-mail: info@kapo.tg.ch
Web: www.kapo.tg.ch

Lake Constance
Strassenverkehrs und Schiffahrtsamt des Kantons St Gallen
Oberer Graben 32
9001 St Gallen
Switzerland

Tel: (+41 71) 229 3657
Fax: (+41 71) 846 60 71
e-mail: info@stva.sg.ch
Web: www.stva.sg.ch

Strassenverkehrs und Schiffahrtsamt des Kantons Schaffhausen
Rosengasse 8
CH-8200 Schaffhausen
Switzerland

Tel: (+41 52) 632 76 04
Fax: (+41 52) 632 78 11

Genève, Lake Geneva Service des Automobiles et de la Navigation
Case postale 1556
CH-1227 Carouge
Switzerland

Tel: (+41 22) 343 02 00
Fax: (+41 22) 343 81 11
Web: www.geneve.ch/san

Lake Geneva
Strassenverkehrsamt des Kantons Luzern
Schiffahrtsamt
Postfach 162
CH-6000 Luzern 4
Switzerland

Tel: (+41 41) 318 19 11
Fax: (+41 41) 318 18 45

Lake Lucerne
Sezione della Circolazione
Ufficio amministrativo
Capo Servizio Navigazione
CH-6528 Camorino
Switzerland

Tel: (+41 91) 814 93 02
Fax: (+41 91) 814 91 39
e-mail: di-sc.navigazione@ti.ch

E Regazzi, *Director*

Lake Lugano and Lake Locarno
Departement de Police
Service Cantonal des Automobiles et de la Navigation
Case postale 200
CH-2007 Neuchatel 7
Switzerland

Tel: (+41 32) 889 63 20
Fax: (+41 32) 889 60 77
Web: www.ne.ch/san www.ne.ch/san

Lake Neuchatel
Strassenverkehrs-und Schiffahrtsamt des Kantons Bern
Schermenweg 5
PO Box
CH-3001 Bern
Switzerland

Tel: (+41 31) 634 24 95
Fax: (+41 31) 634 22 67
e-mail: info.svsa@pom.be.ch
Web: www.pom.be.ch/svsa

H U Kuhn, *Director*

Lake Thoune, Lake Brienz and Lake Biel
Kantonspolizei Zürich
Seepolizei
Seestr. 87
CH-8942 Oberrieden
Switzerland

Tel: (+41 44) 722 58 30
Fax: (+41 44) 720 74 06

Turkey
General Directorate for Maritime Transport
Bulvari No 128
06570 Maltepe
Ankara
Turkey

Tel: (+90 312) 231 91 05
Fax: (+90 312) 232 08 23
e-mail: bbmdugm@isnet.net.tr
Web: www.denizcilik.gov.tr

T C Denizcilik Mustesarligi, *Undersecretariat for Maritime Affairs*

United Kingdom
Maritime and Coastguard Agency
(Hovercraft Operating Permits and Registration High-Speed Marine Craft)
Standards Directorate
Spring Place
105 Commercial Road
Southampton
Hampshire SO15 1EG
United Kingdom

Tel: (+44 23) 80 839 641
Fax: (+44 23) 80 32 92 51
e-mail: jenny.vines@mcga.gov.uk
Web: www.mcga.gov.uk

Jenny Vines, *Ship Surveyor, Shipping Safety Branch*

United States
Department of Transportation
Commandant (G-MSE-1)
US Coast Guard
Washington DC 20593-0001
United States

Tel: (+1 202) 372 1251
Fax: (+1 202) 372 1918
email: dot.comment@dot.gov
Web: www.dot.gov

Venezuela
Venezuelan Maritime Administration
Instituto Nacional de los Espacios, Acuaticos e Insulares
Avenida Rio Orinoco
Urbinizacion Las Mercedes
Zona Postale 1050
Caracas
Venezuela

Tel: (+58 212) 909 14 30
Fax: (+58 212) 909 15 27
e-mail: ineal@inea.gov.ve

CONSULTANTS AND DESIGNERS

Company listing by country

Australia
AMD Marine Consulting Pty Ltd
Australian Hovercraft Pty Ltd
Commercial Marine Consulting Services
Incat Crowther
Mark Ellis Marine Design
One2Three
Revolution Design

Canada
Aker Marine Inc
BMT Fleet Technology Ltd
The Mariport Group Ltd
Promaxis Systems Inc
Robert Allan Ltd

China
MARIC

Denmark
OSK ShipTech Marine Consultants

France
Bassin D'Essais
Bertin Technologies
Techni Carène

Greece
National Technical University of Athens

Italy
FB Design srl
Sciomachen Naval Architects
Victory Design

Korea, South
Korea Research and Development Institute

Netherlands
DK Group Netherlands BV
Nevesbu B.V.
Mulder Design

New Zealand
Malcolm Tennant Multihull Design
Teknicraft Design Ltd
Warwick Yacht Design Ltd

Norway
Amble & Stokke A/S
Marintek
Ola Lilloe-Olsen
Paradis Nautica
Rolls-Royce (NVC Design)
SES Europe AS

Russian Federation
AKS-Invest
Central Hydrofoil Design Bureau
FIG High Speed Ships
Krylov Shipbuilding Research Institute
Sudoexport

South Africa
Foil Assisted Ship Technologies cc

Spain
Neumar SA

Sweden
Marinteknik Design (E) AB
Pelmatic AB
SSPA Sweden AB

Switzerland
Supramar AG

United Kingdom
Air Vehicles Design and Engineering Ltd
Blyth Bridges Marine Consultants Ltd
BMT Group Ltd
BMT Nigel Gee
Francis Design Ltd
Hart Fenton and Co Ltd
Hovercraft Consultants Ltd
Hoverwork Ltd
Hypercraft Associates
Independent Maritime Assessment Associates Ltd
Inserve Ltd
International Hovercraft Ltd
JB Marine Consultancy
Lorne Campbell Design
Maritime Services International Ltd
Seaspeed Marine Consulting Ltd
White Young & Green
Wolfson Unit for Marine Technology and Industrial Aerodynamics
W S Atkins Consultants Ltd
Yg Consultants

United States
Air Ride Craft Inc
Apollonio Naval Architecture
Art Anderson Associates
CDI Marine Systems
Davidson Laboratory Stevens Institute of Technology
D F Dickins Associates Ltd
Donald L Blount & Associates Inc
Elliott Bay Design Group Ltd
FastShip Atlantic Inc
Fryco
Gibbs & Cox Inc
Guido Perla & Associates
J B Hargrave Yacht Design
Hornblower Marine Services
Morrelli & Melvin Design and Engineering Inc
Naval Surface Warfare Center
Quadtech Marine Inc
C Raymond Hunt Associates Inc
M Rosenblatt & Son Inc
Seacraft Design LLC
George G Sharp Inc
SSSCO
Swath International Ltd
Viking Fast Craft Solutions
VT Maritime Dynamics Inc

Australia

AMD Marine Consulting Pty Ltd

PO Box 878, Pymble, New South Wales 2073, Sydney, Australia

Tel: (+61 2) 9484 0288
Fax: (+61 2) 9481 9466
e-mail: amd@amd.com.au
Web: www.amd.com.au

John Szeto, *Director*
Allan Soars, *Director*

AMD Marine Consulting Pty Ltd continues to enhance its reputation as a designer of innovative and reliable high-speed vessels with the launch of a number of new vessels and the development of a number of new designs.

Luciano Federico L, built for Buquebus by Bazan in 1997, firmly established AMD as one of the leaders in ultra-high-speed ferry design. With a service speed of 57 kt and carrying 446 passengers and 52 cars, this vessel competes with airlines on the 110 n mile Buenos Aires to Montevideo route.

AMD is currently involved in a number of Middle East ferry projects, with one 52 m car/passenger wave-piercer design being built by Damen for Kuwait Gulf Link Transport, and three of the 52 m car/passenger wave-piercer design being built by Rodriquez for the Government of the Sultan of Oman. AMD is also acting as owner's technical representative for the Oman Government in the construction of two 50 kt vessels at Austal.

Following on from the three 50 m rescue vessels recently built in China for the Chinese Bureau of Salvage and Rescue, are two fast ocean rescue vessel wave-piercer designs being built by Rodriquez for the Royal Oman Police Coast Guard. These vessels have a loaded speed of 40 knots and can each carry 100 personnel or 200 survivors. Providing a unique level of versatility, the vessels are fitted with a hospital and special operations headquarters, have helicopter landing/refuelling facilities and a firefighting capability provided by three fire monitors.

Australian Hovercraft Pty Ltd

99 Glyde Road, Lesmurdie, Western Australia 6076, Australia

Tel: (+61 8) 92 91 08 77
Fax: (+61 8) 92 91 09 77
e-mail: hoverfane@yahoo.com
Web: www.australianhovercraft.com

Michael Fane, *Managing Director*

Australian Hovercraft Pty Ltd specialise in the design and manufacture of quiet, high-speed open-ocean hovercraft and surface effect craft built to IMO international standards, and wing in ground effect vessels. Designs are available to transport from 10 to 85 passengers or 1,000 to 8,500 kg of cargo.

Commercial Marine Consulting Services

PO Box 773, Aspley, Queensland 4034, Australia

Tel: (+61 7) 38 62 99 88
Fax: (+61 7) 38 62 94 88
e-mail: sales@seaspeeddesign.com
Web: www.seaspeeddesign.com

Paul Birgan, *Director*

This Australian – based company has been responsible for designing over 15 fast ferries including: *Whitsunday Experience, Whitsunday Connection II, Whitsunday Ambassador, ACG Express Liner, Seri Anna, Gren Gaden Sangiang I, Campo de Hielo Sur, Reef Quest, Daydream Express, Aqua Spirit, Crystal Spirit, First Ferry IX, First Ferry X, First ferry XI and Tasman Venture III.*

In October 2005 the vessel *Liang Yu*, a 33 m passenger catamaran, was launched by Wang Tak in Hong Kong and in August 2007 a 24 m dive catamaran with a top speed of 24 knots was delivered to Tusa Dive.

Incat Crowther

PO Box 204 Newport, New South Wales 2106, Australia

Tel: (+61 2) 94 50 04 47
Fax: (+61 2) 94 50 04 94
e-mail: design@incatcrowther.com
Web: www.incatcrowther.com

Brett Crowther, *Managing Director*
Ben Hercus, *Commercial Director*
Andrew Tuite, *Technical Manager*
Phil Hercus, *Technical Consultant*

52 m vessel Key West Express 1198277

Based in Sydney, Australia, Incat Crowther was created from the merging of International Catamaran Designs Pty Ltd (InCat) and Crowther Design Pty Ltd. These two companies are credited with many achievements and have a combined experience of over 50 years in marine design. Nearly 300 vessels have been built to InCat and Crowther designs.

InCat Designs was best known as the designer of the wave-piercing catamaran and their 30 m prototype was launched in 1985 as *Spirit of Victoria*. A turning point for InCat Designs, and also the fast-ferry industry as a whole, was the delivery of *Hoverspeed Great Britain* in 1990. This vessel was then by far the largest fast catamaran to have been designed and built, and it was ordered by Sea Containers for operation across the English Channel. To date over 30 wave-piercing catamarans and another 140 vessels have been built to InCat designs.

Crowther Design was founded by Lock Crowther, who was a pioneer in applying proven catamaran sailing design technology to high-speed catamaran ferries and work boats. Derivatives of his catamaran hullform are now used extensively in the industry. More recently, Crowther Design has become well known for its large range of successful commercial vessels which number in excess of 130.

Incat Crowther's expertise covers applications, including passenger ferries, work boats and yachts, in all hullforms, including monohull, catamaran and wave-piercing. Their experience extends to knowledge of construction materials (including aluminium, steel and GRP) and of propulsion (including sail, power and solar). Incat Crowther intends to advance technology and provide a professional service offering builders a design that is simple and efficient to construct.

Built by Gulf Craft in Louisiana, a 52 m, 40 kt catamaran with a capacity for 513 passengers was recently completed for US operator Key West Express, and in 2007 the company was awarded a contract for the design of a 32 m catamaran for Ishigaki Dream Tours of Japan. The 31 kt, 257 passenger vessel will operate in the Ryukyu islands on the southernmost tip of Japan.

Mark Ellis Marine Design

11 Mews Road, Fremantle, Western Australia 6160, Australia

Tel: (+61 8) 94 30 52 70
Fax: (+61 8) 94 30 56 01
e-mail: info@markellisdesign.com
Web: www.markellisdesign.com

Mark Ellis, *Naval Architect and Managing Director*

Established in 1987, MEMD are designers of high-speed catamaran and monohull craft. The range of vessels include passenger ferries, pilot craft, offshore supply craft, fishing and charter vessels constructed from aluminium, GRP and steel. Other services offered by the company include survey and stability assessment, project management and construction supervision, cost estimating and market research.

To complement MEMD designs, Inform Marine Technology was established in 1999 to supply lightweight aluminium and composite vessels as technology packages to builders. IMT packages can be tailored to suit individual builders from plasma cut aluminium to full equipment supply.

One2Three Pty Ltd

11 Phillips Road, Kogarah, NSW 2217, Australia

Tel: (+61 2) 95 53 03 66
Fax: (+61 2) 95 53 03 44
e-mail: design@one2three.com.au
Web: www.one2three.com.au

One2Three is a Sydney-based company offering vessel design and consultancy services, including equipment layout modelling, for high-speed vessels. Formed in 2005, the company employs 12 naval architects and draughtsmen.

Revolution Design

16 Bender Drive, Derwent Park, Tasmania 7009, Australia

Tel: (+61 3) 62 74 18 45
Fax: (+61 3) 62 74 18 46
e-mail: revolutiondesign@revolutiondesign.com.au
Web: www.revolutiondesign.com.au

Mark Dewey, *Managing Director*

Revolution Design was formed in 2002 and encompasses designing and building the Incat Australia range of wave piercing catamaran craft. In 2003, the company was retained to integrate the full conversion of *HSV 2 Swift* to US Navy and ABS standards. Built by InCat Tasmania and following on from *Joint Venture* and *Spearhead*, the vessel initially served as a Mine Warfare Command Support Ship (MCS).

The company's latest design is the Revolution 112 m built by InCat Tasmania. *Natchan Rera* was the first vessel delivered to Nigashi Nihon Ship Management in Japan for service between Honshuana and Hohaido. A sistership will follow in 2008.

Services offered by the company include:

Design of high speed vessels – complete design service for lightweight aluminium vessels to survey and classification society requirements for high speed craft.
Marine drafting services – structural and mechanical detailing, construction drawings and general arrangements.
Conceptual design and route feasibility – to provide information on suitable vessel types and operating costs for freight and passenger/car routes.
Structural design, monitoring and life assessment – finite element analysis, hull monitoring, strain gauging and fatigue assessment of structures.

Canada

Aker Marine Inc

Division of Aka Yards ASA, Oslo
1818 Cornwall Avenue, Vancouver, British Columbia V6J 1C7, Canada

Tel: (+1 604) 730 42 00
Fax: (+1 604) 730 42 97
e-mail: info@akermarine.com
Web: www.akermarine.com

Aker Marine Inc, formerly Kvaerner Masa Marine has been active in North America since 1983 and is owned by the Finnish shipbuilder AKA Yards ASA. The company has considerable knowledge of the fast ship industry, having had access to the Fjellstrand company in Norway and through its research department division has produced designs for a number of fast monohull and trimaran concepts.

BMT Fleet Technology Ltd

A subsidiary of BMT Ltd.
311 Legget Drive, Kanata, Ontario K2K 1Z8, Canada

Tel: (+1 613) 592 28 30
Fax: (+1 613) 592 49 50
e-mail: fleet@fleetech.com
Web: www.fleetech.com

Aaron Dinovitzer, *President*
Andrew Kendrick, *Vice President*

Established in 1973, BMT Fleet Technology Limited provides design and construction support for conventional and advanced marine vehicles. The company carries out performance prediction, concept development, vessel fitness for purpose, sea-keeping analysis, structural analysis, model testing and full-scale trials. In addition BMT Fleet Technology Ltd supports operator and business decisions through economic analysis, route analysis and risk analysis.

With a staff of about 60, the company operates laboratories and workshops for prototype/model development. This includes a materials and welding laboratory which develops production and repair procedures, carries out materials and fracture testing, fatigue analysis and undertakes materials and fabrication consulting in steels, aluminium and composites. An inventory of trials and data acquisition equipment is also maintained.

BMT Fleet Technology's advanced craft experience includes involvement in ACV operations in commercial and naval roles, concept studies for an SES Sovereignty vessel, catamaran design, manoeuvring and sea-keeping predictions, full-scale manoeuvring and sea-keeping trials on various small high-speed craft and SAR cutters, Finite Element analysis of multihulls, route and economic evaluation studies.

Mariport Group Ltd

61 Montague Row, PO Box 2295, Digby, Nova Scotia B0V 1AD, Canada

Tel: (+1 902) 245 4598
Fax: (+1 902) 245 4209
e-mail: info@mariport.com
Web: www.mariport.com

Christopher Wright, *President*

The Mariport Group Ltd offers a specialised advisory service for ports and the shipping industry. The company conducts marine systems analysis and work has included projects for Toronto Waterbus, Busan-Yeosu Ferry (South Korea), High-Speed Ferry (southeast Alaska), Mine Supply and Personnel Vessel (Alaska, Newfoundland and China); economic analysis of high-speed passenger vehicle ferries for Canadian and Great Lakes, Key West and Bahamas ferry routes.

Promaxis Systems Inc

2385 St Laurent Boulevard, Ottawa, Ontario K1G 6C3, Canada

Tel: (+1 613) 737 21 12
Fax: (+1 613) 737 02 29
e-mail: info@promaxis.com
Web: www.promaxis.com

Andrew Knight, *President*

Promaxis has developed a design for a Swath coastal patrol vessel for use in search and rescue, fisheries patrol, defence surveillance, MCM and other related missions in rough seas such as those that exist off the east and west coasts of Canada.

This 360 tonne all-aluminium vessel is arranged to facilitate handling and treatment of survivors by providing dedicated rescue zones, recessed into the port and starboard sides of the upper hulls. Within the accommodation, seating and bunks are provided for up to 25 survivors. A helicopter landing pad is designed to enable landing and take-off of a JetRanger helicopter in up to Sea State 5 conditions. For ship external firefighting, two monitors are provided.

The vessel will be driven at 15 kt in Sea State 5 head seas by two medium-speed diesels, developing a total of 3,600 hp. The main engines are located within the main deck structure with a mechanical transfer of drive to controllable-pitch propellers.

The design minimises development risks by employing standard, proven systems technology. The vessel will be built to Transport Canada regulations and classed with an international classification society.

Robert Allan Ltd

Suite 230, 1639 West Second Avenue, Vancouver, British Columbia V6J 1H3, Canada

Tel: (+1 604) 736 94 66
Fax: (+1 604) 736 94 83
e-mail: ral@ral.bc.ca
Web: www.ral.bc.ca

Robert Allan, *President*

Well known internationally for the company's tug designs, Robert Allan Ltd has also been closely involved in the development of high-speed craft and many other types of specialised workboats. These include the Raven Class fast patrol catamarans of which six vessels are now in service across Canada, and the new Rally series semi-displacement offshore crewboat/patrol craft.

The company has specialist experience in the use of CAD and CAM technology to assist shipyards in the planning and construction process.

Current projects and recent deliveries include a variety of twin-screw, ASD and VSP tugs and offshore vessels up to 6,500 kW, fast response fireboats for the cities of Philadelphia, Baltimore, Portland (Maine), and New York City; 10 m fast patrol craft for the New York City Police; several offshore crewboats; a number of RoRo ferry designs and numerous speciality product barges.

22 m fast patrol catamaran for New York City Police 1178280

China

MARIC

Marine Design and Research Institute of China
1688 Xizangnan Road, Shanghai 200011
People's Republic of China

Tel: (+86 21) 63 78 80 98
Fax: (+86 21) 63 77 97 44
Telex: 33029 MARIC CN
Web: www.maric.com.cn

Mrs Hu Ankang, *President*
Wang Zhongzheng, *Chief Engineer*

The Marine Design and Research Institute of China (MARIC) has been responsible for much of the ACV research, development and design programmes on both amphibious and sidewall (SES) hovercraft built in China. The table summarises the more recent developments in ACVs in China.

Since 1990, the fast marine transportation system in China has expanded rapidly with MARIC co-ordinating the domestic design and build of ACVs in conjunction with the Chinese Society of Naval Architects and Marine Engineers (CSNAME).

Summary of craft built

Builder and craft	Designer	Seats/payload	Launched
Bei Hai Shipyard			
Type 7217 (sidewall) two built	MARIC	260 seats	October 1995
Dagu Shipyard			
Type 722 (amphibious)	MARIC	15 t payload	August 1979
Type 7203 (sidewall) *Jinxiang*	MARIC	81 seats	September 1983
Type 7212	MARIC	33 seats	1992
Type 7224 II	MARIC	15 seats	1994
Dong Feng Shipyard			
Type 717 II (sidewall) *Ming Jiang*	MARIC	54–60 seats	October 1984
Type 717 III (sidewall) *Chongqing*	MARIC	70 seats	September 1984
Type 717 IIIC (sidewall) *Yu Xiang*	MARIC	–	September 1989
Type 7210 (amphibious) two built	MARIC	0.8 t	May 1985
Type 7218 (amphibious)	MARIC	70 seats	October 1994
Type 7224 (amphibious)	MARIC	15–28 seats	April 1993
Type 7224 (amphibious)	MARIC	15–28 seats	April 1993
Type 7224 (amphibious)	MARIC	15–28 seats	April 1993
Type 7215 (sidewall)	MARIC	126 seats	September 1993
Type 7226 (amphibious)	MARIC	40–50 seats	July 1995
Huangpu Shipyard			
Type 7211 (sidewall)	MARIC	162 seats	December 1992
Hudong Shipyard			
Type 716 II (amphibious)	MARIC	32 seats	1985
Jiang Hui Shipyard			
Type 7221 (sidewall)	MARIC	225 seats	October 1995
Ming Jian Shipyard			
Type 717 V	MARIC	400 seats	August 1993
Quog Hua Shipyard			
Type 719 III	MARIC	257 seats	August 1988
Shanghai Aircraft Factory			
Type 7212 *Zhengzhou*	MARIC	33 seats	1989
Zhong Hua Shipyard			
Type 719 II *Hong Xiang*	MARIC	257 seats	1988

Denmark

OSK-ShipTech A/S

Baldersgade 6, DK-2200 Copenhagen N, Denmark

Tel: (+45) 45 76 42 10
Fax: (+45) 45 76 42 20
e-mail: khm@shiptech.com
Web: www.shiptech.com

Kristian Holten Moeller, *Naval Architect, Dept. Manager*

Established in 1985 as independent marine consultants, the company specialises in the design of ships, advanced hydrodynamic and structural analysis, outfit arrangements and performance management. In particular, the company undertook the structural verification of the Fastship TG770 project's 260 m, 40 kt ferry.

In January 2007 Shiptech Marine Consultants was merged with Ole Steen Knudsen A/S under the new company name "OSK-ShipTech A/S Consulting Naval Architects & Marine Engineers.

France

Bassin D'essais Des Carenes (DGA)

Chaussée Du Vexin, BP-510, F-27100 Val De Reuil, France

Tel: (+33 2) 32 59 78 00
Fax: (+33 2) 32 59 77 77
e-mail: lars@bassin.fr
Web: www.bassin.fr

Philippe Lars, *Project Manager*

The Bassin D'Essais des Carenes is part of the French MOD research establishment, Délégation Géneral pour L'Armement (DGA).

The organisation offers expertise in hydrodynamics and hydroacoustics and in particular in the design of propellers, numerical simulations, tests on scale models and full scale trials.

The facilities available include three towing tanks, including the B600, which is 545 m long, 15 m wide and 7 m deep with a maximum tow speed of 12 m/s and a maximum wave height of 1 m, a rotating arm facility, a hydrodynamic tunnel and a hydrobalistic tank.

Bertin Technologies

Parc d'activities du Pas du lac, 10 Avenue Ampere, F-78180 Montagny-le-Bretonneux, France

Tel: (+33 1) 39 30 60 00
Fax: (+33 1) 39 30 61 45
e-mail: info@bertin.fr
Web: www.bertin.fr

Chantal Picoreau, *Industrial Services*

Bertin has been engaged in developing the Bertin Technologies principle of separately fed multiple plenum chambers, surrounded by flexible skirts since 1956. A research and design organisation, the company employs a staff of more than 350, mainly scientists and design engineers involved in many areas of industrial research, including air cushion techniques and applications.

The Bertin principle of multiple air cushions has led to the development of the Aérotrain high-speed tracked transport system and to numerous applications in the area of industrial handling and aeronautics.

Techni Carène

6 route de Bû, Les Christophes, F-28260 Sorel Moussel, France

Tel: (+33 2) 37 41 80 38
Fax: (+33 2) 37 41 73 81
e-mail: technicarene@wanadoo.fr

Philippe Nineuil, *Director*

Techni Carène is a naval architecture and engineering office established in 1979 by Philippe Nineuil, naval architect and a member of the Institut Français des Architectes Navals (IFAN).

The company is involved in the design of all types of ships from 15 to 60 m (in length) in all materials and adopted to a wide range of operational programmes including passenger transport, hydrographic, diving support, fishing, patrol, and offshore supply.

It has also developed a range of displacement and medium-speed catamarans named Catatrans, which includes fishing, hydrographic and passenger units.

Two examples of high-speed craft built to Techni Carène designs are: *Azenor*, a 25 m, 200-passenger aluminium catamaran operated in Brittany, France, delivered in 1992, and *Dravanteg*, which is a 27 m mixed cargo and passenger catamaran powered by two 895 kW diesels driving Kamewa water-jets, also operated in Brittany. Both of these vessels were built by the shipyard Navale Aluminium Atlantique, Saint Nazaire, France.

Greece

National Technical University of Athens

Ship Design Laboratory
9 Heroon Polytechniou Street, GR-15773 Athens, Greece

Tel: (+30 210) 772 14 16; 14 09
Fax: (+30 210) 772 14 08
e-mail: papa@deslab.ntua.gr
Telex: 221682 NTUA GR
Web: www.naval.ntua.gr/~sdl

Apostolos Papanikolaou, *Professor, Director*

In recent years, the Ship Design Laboratory (SDL) of the National Technical University of Athens (NTUA), has been established as an internationally known designer of advanced marine vehicles and a leading institution of Swath and hybrid catamaran technology, research and development. The range of activities of SDL includes research in various disciplines, contributing to the concurrent engineering approach of marine vehicles design and the system integration of engineering problems, including the consideration of technoeconomic aspects. Typical examples of the most recent research work of the SDL are as follows:

The fast marine transportation system SMUCC (Swath Multipurpose Container Carrier). The initial concept included the preliminary design of a 1,000 tonne displacement, 36 kt service speed, 60 TEU feeder container ship for short sea shipping operations in Europe and the outline of an innovative Intermodal Container Terminal facility. This concept received the first prize award at an international competition among university laboratories (SMM94, Hamburg). A second SMUCC prototype of 2000 tons displacement has now been completed including verification of performance by model tests.

The passenger/car ferry Swath design *Aegean Queen* for fast marine transport within the Aegean Sea: 1,000 tonne displacement, 30 kt service speed, 800 passengers and 88 cars for the route from Pireaus to Crete.

The development of the SIMICAT: a 450 tons displacement, 19 kt passenger ferry for the Dodecanese Islands. Following this design the ferry *Playtera* has since been built for operation in northern Greece.

The development of the hybrid catamaran passenger car ferry *Super Cat Haroula*, a 2,000 tons displacement, 22 kt service speed, 1,500 passenger and 240 car steel ferry for the route from Piraeus to the Cycladic Islands. The vessel is one of the largest steel catamarans ever built worldwide and the first one entirely designed and built in Greece. The preliminary design of a 30 kt, 950 tonne hybrid SWATH corvette with stealth technology features for military operations in the Aegean Sea is currently under development.

NTUA-SDL is a partner of an international European consortium (led by SES Europe) currently developing the innovative air lubricated hull SES concepts EFFISES E40 and E125, 70 kt passenger and 55 kt Ro-Ro ferries.

Italy

FB Design Srl

Via Provincial 73, Annone-Brianza (LC) I-23841, Italy

Tel: (+39 0341) 26 01 05
Fax: (+39 0341) 26 01 08
e-mail: fabiobuzzi@fbdesign.it
Web: www.fbdesign.it

Fabio Buzzi, *Managing Director*

FB Design was established in 1971 to design and build high-performance craft and their propulsion systems.

Currently, the core business of FB Design is to design very fast military vessels. The company has a range of customers including the UK's Special Boat Service (SBS), the Italian Guardia di Finanza, the Hellenic Coast Guard, the US, Rwandan and Ugandan forces.

The company has recently added a small towing tank and lake testing site to its facilities.

Sciomachen Naval Architects

Via Ferrarese 3, I-40128 Bologna, Italy

Tel: (+39) 051 419 84 60
Fax: (+39) 051 419 84 50
e-mail: info@sciomachen.com
Web: www.sciomachen.com

Ernesto Sciomachen, *Director*

The Sciomachen name has been associated with yachts and boats since 1951. After being mainly devoted to sailing craft in the early years, the company's projects today encompass a full range of marine designs from high-speed power yachts to fast ferries, excursion vessels, fishing vessels and displacement yachts, as well as racing and cruising sailboats. Construction materials include glass fibre, steel, aluminium and wood, with sizes from 7 to over 45 m.

The company's yacht design experience is reflected in the design of its commercial vessels, where functionality is always matched with aesthetically pleasing lines.

38 m fast ferry Eolian Queen 1279620

Eolian Queen is a new passenger vessel designed by Sciomachen currently under construction at Cantiere Navale Foschi of Cesenatico, Italy, for TarNav of Milazzo, Sicily. This is the fourth vessel for TarNav built by Foschi and designed by Sciomachen. At 38 m l.o.a., *Eolian Queen*, is the largest wooden vessel built by Foschi, using a combination of cold molding, multi-ply marine plywood and solid wood, according to the most advanced technology.

Victory Design

Via Melisurgo 15, I-80133 Naples, Italy

Tel: (+39 081) 252 82 43
Fax: (+39 081) 420 68 96
e-mail: victory@victory.it
Web: www.victory.it

Brunello Acampora, *Director*

Victory Design was founded in 1989 as a naval architecture and marine engineering company specialising in high-performance craft. Victory carries out full design work for complete boats, hull and superstructure lines, tank testing programmes, project management, surveying and development of all systems including hydraulics. The company has been successful in the offshore racing sector and with large fast monohull craft for Pershing and Ferretti, both market leaders in the field of fast luxury racing craft.

The company has recently been working closely with Flexitab in Italy to develop surface drive systems, flexible trim tab systems and advanced test boats.

Korea, South

Korea Research and Development Institute

Maritime and Ocean Engineering Research Institute
PO Box 23, Yusong, Daejon 305-600, South Korea

Tel: (+82 42) 868 77 01
Fax: (+82 42) 868 77 11
Web: www.kordi.re.kr

Seok Won Hong, *Director General*

This company was founded in 1976 and is supported by the South Korean government.

MOERI is a research institute specialised in high-performance ships, offshore structures, deep sea submersibles and underwater robots, maritime safety and conservation of the ocean environment.

A towing tank (216 × 16 × 7 m) is used for evaluation of hull and propulsor technology, sea-keeping and manoeuvring and for high-speed craft such as SESs, catamarans, hydrofoil catamarans and hybrid ships. Computational fluid dynamics are also widely used.

An ocean engineering basin (56 × 35 × 0.4–4.5 m) is used for simulating any harsh ocean environment for the development of design techniques and reliability tests of offshore structures and systems.

Other areas of interest being researched include advanced ship design technology, strength and vibration technology, ship product modelling and expert systems for shipbuilding, assessment of ship safety and salvage technology, ship handling simulator and intelligent navigation systems and key technologies for the prevention of oil spills.

Netherlands

DK Group Netherlands BV

Weena 340, NL-3012 NJ Rotterdam, Netherlands

Tel: (+31) 102 13 18 45
Fax: (+31) 104 12 61 15
e-mail: into@dkgroup.eu
Web: www.dkgroup.eu

Jorn Winkler, *Chief Executive Officer*

The DK Group was formed in 1999 to establish a consortium, supplying the marine industry with novel low-speed and high-speed craft technology based on the patented Ventilated Wing Jet (VWJ) and Air Cavity Ship designs. These vessel designs feature a novel air lubrication system in their hulls which is understood to substantially improve their fuel efficiency. The VWJ is based on a ventilated propeller operating within a controlled duct. Testing of both the vessel and VWJ have been undertaken during 2002 and 2003 at Marin in the Netherlands.

Nevesbu B.V.

Noordhoek 37, PO Box 1155, NL-3350 CD Papendrecht, Netherlands

Tel: (+31 78) 644 82 01
Fax: (+31 78) 644 86 67
e-mail: info@nevesbu.nl
Web: www.nevesbu.com

C.I.M. van Roosmalen, *Managing Director*

Nevesbu are naval architects and marine engineers specialising in naval, offshore and megayachts. Founded in 1935, the company was taken over by IV-Group, a Dutch private company, in 2004. Since 2006 shares are divided equally between IV-Group and Damen Netherlands Engineering.

The company provides a range of services including consultancy, conceptual and detailed design. Nevesbu is known for its work in the naval defence market, particularly in conventional submarines, however, destroyers, frigates, corvettes and aircraft carriers form only part of the product range. In addition, double hull fleet tankers and amphibious transport ships like HNIMSs *Rotterdam* were designed by Nevesbu.

Nevesbu's second market is the offshore market. Experience ranges from a series of FPSO and other floating unit projects as well as participation in research and development projects. Nevesbu usually concentrates on hull and marine systems together with the accommodation to provide a platform on which oil and gas processes take place.

Nevesbu was involved in the successful design of the ARKA, an oil recovery vessel recently used in the environmental disaster off the coast of Spain. The company also anticipated the development of double hull tankers which nowadays are a must on most transit routes.

Clients include: the Royal Netherlands Navy as well as foreign navies, ship owners, shipyards, offshore yards and contractors, petrochemical industries and governmental organisations.

Mulder Design BV

Benedeneind Zuidzijde, 289B, NL-3405 CK Benschop, Netherlands

Tel: (+31 348) 43 96 80
Fax: (+31 348) 43 96 81
email: info@mulderdesign.nl
Web: www.mulderdesign.nl

Founded in 1979, Mulder Design specialises in the design, styling, naval architecture and engineering of fast motor yachts and full-displacement motor yachts. Recent deliveries include a 43 m high-speed motor yacht and a 23 m Open Sport Boat.

Mulder Design has under construction a 127' fast motor yacht, as well as several other projects.

New Zealand

Craig Loomes Design Group Ltd

PO Box 147-027, Ponsonby, Auckland, New Zealand

Tel: (+64 9) 360 97 99
Fax: (+64 9) 360 97 96
e-mail: designer@cld.co.nz
Web: www.cld.co.nz

Craig Loomes, *Director*

Known for its sailing and powered craft from 8 m to 200 m, the company designed the novel bio-diesel-powered wavepiercing trimaran *Earthrace*, which in 2007 will attempt a record circumnavigation of the world via the Panama and Suez canals.

24 m Earthrace wavepiercing trimaran 1120270

Malcolm Tennant Multihull Design Ltd

PO Box 60513, Titirangi, Auckland 1007, New Zealand

Tel: (+64 9) 817 19 88
Fax: (+64 9) 817 60 80
e-mail: malcolm@tennantdesign.co.nz
Web: www.tennantdesign.co.nz

M Tennant, *Director*

Malcolm Tennant Multihull Design Ltd, known for its sailing and powered high-speed catamaran designs, has designed more than 250 vessels. The company has been involved in the design of multihulled craft for more than 30 years and much of the current workload is in the area of composite ferries, large motor sailors and ocean-going motor yachts; the largest to date is the 30 m yacht Spadaccino.

Teknicraft Design Ltd

PO Box 34-712, Birkenhead, Auckland, New Zealand

Tel: (+64 9) 482 33 31
Fax: (+64 9) 482 33 34
e-mail: info@teknicraft.com
Web: www.teknicraft.com

Nic De Waal, *Director*

Teknicraft Design Ltd, originally of South Africa, now provides a fast-craft design service in particular for foil assisted catamarans.

The company has had a number of fast ferries, workboats and pleasure craft built to its designs, including the 27 m *Hydrocruiser*, built by Vosper Thornycroft in the UK, and *Chilkat Express*, an 18.5 m, 45 kt passenger vessel built by All American Marine.

In 2007 the company saw delivery of a 25 m whale watching catamaran built by All American Marine, as well as a 16 m high-speed catamaran and a 17.6 m whale watch vessel which were built at Q-West in New Zealand.

Warwick Yacht Design Ltd

2B William Pickering Drive, PO Box 302 156, North Harbour, Auckland 1310, New Zealand

Tel: (+64 9) 410 96 20
Fax: (+64 9) 410 82 54
e-mail: wyd@wyd.co.nz
Web: www.warwickyachts.com

Warwick Yacht Design is a company of naval architects, engineers and stylists. A range of fast motor yachts and sports fishermen have been designed and built with lengths up to 40 m and speeds in excess of 30 kt.

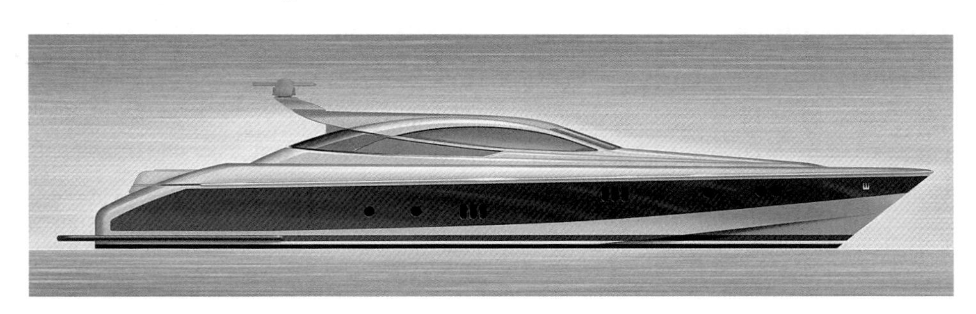

W90 Sport Boat 1148088

Norway

Amble & Stokke A/S

PO Box 1616, Lundsiden, N-4688 Kristiansand, Norway

Tel: (+47) 38 09 58 03
Fax: (+47) 38 09 58 71
e-mail: tor.stokke@amble-stokke.no
Web: www.amble-stokke.no

Eivind Amble, *Naval Architect*
Tor Stokke, *Naval Architect*

The company specialises in the design and development of high-speed marine craft and related research and development. With more than 36 years of experience as naval architects, the reference list of Amble and Stokke includes small dinghies, racing yachts, luxury motor yachts, Ro-Ro ships and military craft.

Since 1990 Amble & Stokke have been involved in most of the new ideas for the Norwegian Society for Sea Rescue.

NSSR 23 m advanced life boat Kaptein Skaugen 0580416

Marintek

Norwegian Marine Technology Research Institute
PO Box 4125, Valentinlyst, N-7450 Trondheim, Norway

Tel: (+47) 73 59 55 00
Fax: (+47) 73 59 57 76
e-mail: marintek@marintek.sintef.no
Web: www.marintek.sintef.no

Marintek has been involved in the development and testing of high-speed craft for many years and has a reference list of over 100 fast craft designs tested since 1985. The organisation has a range of hydrodynamic test facilities including three towing tanks, one ocean basin laboratory and a cavitation tunnel. The ocean basin facility (measuring 80 m × 50 m × 10 m) is provided with a unique carriage system that can follow free-running models which is of particular interest for the testing of high-speed craft.

OLA Lilloe-Olsen

N-6829 Hyen, Norway

Tel: (+47) 57 86 98 05
Fax: (+47) 57 86 99 25
e-mail: olal-o@online.no

Ola Lilloe-Olsen, *Director*

A well-established naval architect in the development and design of high-speed monohulls and catamarans in Norway. Designer of a large number of passenger and ambulance-vessels in operation, as well as the Norwegian Skomvaer-Class of rescue vessels, patrol vessels and passenger ferries.

Recent deliveries of Ola Lilloe-Olsen design include nine 20 m Pilot Vessels for Iran, 2 crew Supply Vessels for AISO Marine Services, a 32 m catamaran for operation on the Danube between Vienna and Bratislava, and the combined catamaran *Tansøy* for Norwegain operator Fjord 1. All the vessels were built by Båtservice Mandal AS.

Other new deliveries include a new catamaran for Talisman Energy (UK) Ltd, built by Maloy Werfty AS.

New projects include 2 pilot vessels for the Olympic Games in Beijing in 2008, and a second passenger catamaran for the Danube.

20 m SAR for Iran 1113591

24 m Crew Supply Vessel for ASO Marine Service 1340608

Paradis Nautica

PO Box 439, Nesttun, N-5853 Bergen, Norway

Tel: (+47) 55 92 74 70
Fax: (+47) 55 92 74 80
e-mail: post@paradis-nautica.com
Web: www.paradis-nautica.com

Eirik Neverdal, *Chairman*
Absjørn Tolo, *Managing Director*

Paradis Nautica is a ship design and technology company based in Bergen. The company specialises in the following areas: development and design of high-speed craft and ferries, engineering, project management, vessel sale and marketing. This range of service is offered to ship owners, operators, investors, yards and banks. The company also offers a range of safety products: FMEA, FMECA, Annex 9 and evacuation analysis.

Paradis Nautica has also developed 20, 24, 28, 30 and 42 m catamarans that are built and in operation.

Another product is the patented Nautica SMC – Ship Motion Control system. This system can be retrofitted and has reduced resistance compared with other systems. Unique control surfaces are fitted at the vessel's stern. The system can be combined with control surfaces in other locations.

In 2006, Paradis Nautica designed two 33 m catamarans for operator HSD. Carrying 180 and 120 passengers respectively, the vessels will be constructed by Fjellstrand in Norway.

Nautica 42 m Design
0097285

Rolls-Royce (NVC – Design)

(Formally Nordvestconsult) – A Rolls-Royce Subsidiary
PO Box 826, N-6001, Ålesund, Norway

Tel: (+47) 70 10 37 00
Fax: (+47) 70 10 37 01

Rolls-Royce NVC-Design (formally Nordvestconsult) is a ship technology consulting company and a subsidiary of Rolls-Royce Plc.

The company has developed a high-speed monohull ferry design capable of speeds up to 60 kts. The design is approximately 150 m in length with a slender beam of 22 m. The deadweight capacity will be approximately 1,250 tonnes. For speeds up to 45 kts an installed power of 80 MW is proposed and for 60 kts up to 150 MW would be required. With a passenger capacity of 1,500, the design also incorporates space for up to 320 cars or 180 cars with 22 trucks. Propulsion is envisaged at all Rolls-Royce gas turbines for 60 knots or a CODAG arrangement for slower speeds. In 2003 it was announced that this design has been developed and is now available as a cost-effective, deadweight maximised fast naval sealift vessel.

SES Europe AS

Thor Dahlsgt 1A, N-3210 Sandefjord, Norway

Tel: (+47) 33 46 56 50
Fax: (+47) 33 46 56 10
e-mail: ulf.tudem@brygga.no
Web: www.seseu.com

Ulf Tudem, *General Manager*

In 2002 SES Europe acquired the global development and exploitation rights to a new robust Air Assisted Vessel concept. The technology, owned by Effect Ships International AS is patented in more than 50 countries and may be applied to a wide range of fast ferries for commercial, naval and pleasure craft use.

Water-jet is the preferred propulsion system and the hull designs have been developed to deliver very good ride qualities at high speeds.

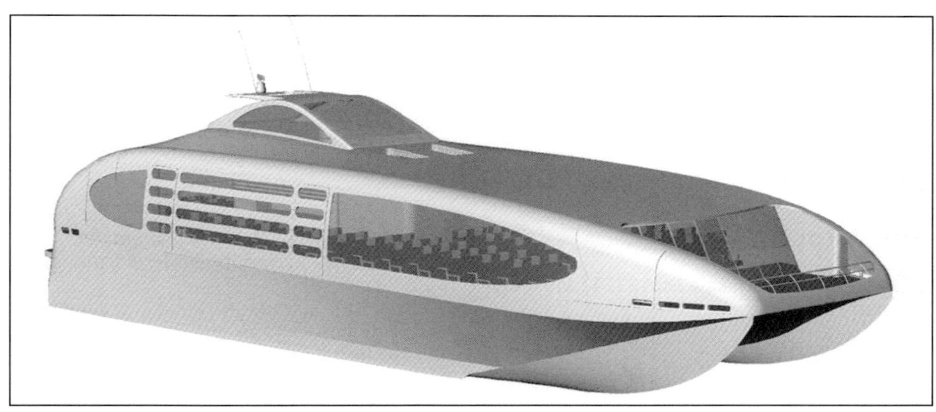

Artist's impression of the SES Air Assisted Vessel (Design JGE) 1020679

A 22 × 7.4 m demonstrator vessel is currently under construction at Batutrustning Bomlo AS in Norway. The vessel will undergo a comprehensive test and demonstration programme after the expected launch in 2006.

Russian Federation

Aks-Invest

Torfyanaya Street 34, 603037 Nizhny Novgorod, Russian Federation

Tel: (+7 8312) 25 02 46
Fax: (+7 8312) 25 21 51
e-mail: aks@r52.ru
Web: www.aks.r52.ru

Serguey Italyantser, *Director General*

Aks-Invest is a research, design and consultancy company specialising in high-performance marine and amphibious craft. Only limited information concerning the company's military and commercial designs is available, although it is clear that fast passenger transportation is one of the main markets being considered. One craft being discussed has a speed of over 300 kt with a passenger capacity of 50. The range is 250 n miles with a payload of 5 tonnes. Fuel consumption is predicted to be 0.054 kg/seat n miles.

AKS-Invest Mars-700 hovercraft 1036900

Central Hydrofoil Design Bureau

Part of the FIG High-Speed Ships Group
51 Svoboda Street, Nizhny Novgorod 603003, Russian Federation

Tel: (+7 8312) 73 19 09
Fax: (+7 8312) 73 02 48
e-mail: rostislav@sinn.ru
Web: www.hs-ships.ru

Sergey V Patonov, *General Director*
I M Vasilavsky, *Director*
M A Suslov, *Technical Director*

The Central Hydrofoil Design Bureau has extensive experience in the design and manufacture of hydrofoils, surface effect ships and Wing-In-Ground effect craft (WIG).

Some of the company's many designs are the successful Serna and Linda high speed air cavity vessels, and the Meteor, Comet and Kolkhida hydrofoils. The company is also known for the design of the Caspian Monster WIG.

More recently the company has designed the Lastochlea-M Riverina passenger hydrofoils, with the first vessel constructed in 2003 and the second due for completion in 2004. Based on their experience in the design of earlier hydrofoils, the Central Hydrofoil Design Bureau has also developed a new range of craft (29 to 39 m) called the Biriuza project. These vessels will use an advanced ride control system and either MTU or Deutz engines. A design for a larger (67.5 m), 45 knot hydrofoil is well advanced and known as project Izumrud.

Fig High Speed Ships

Financial Industrial Group High Speed Ships
6 Suvorovskaya Street, Moscow 107023, Russia

Tel: (+7 95) 964 00 88
Fax: (+7 95) 963 00 18
e-mail: hs-ships@aha.ru
Web: www.hs-ships.ru

Danilov Gennady, *Director General*

FIG High Speed Ships is a Russian financial group with specific interests in fast marine craft technology. It represents the following Russian joint stock companies: Alexeev Central Hydrofoil Design Bureau, Volga Shipbuilding Plant, Redan, Central Design Bureau Neptune and Zvezda.

Krylov Shipbuilding Research Institute

Moskovskoye Shosse, 44, 196158 St Petersburg, Russian Federation

Tel: +7 (812) 727 96 47
Fax: +7 (812) 727 95 94
e-mail: krylov@krylov.spb.ru
Web: www.krylov.com.ru

Valentin M Pashin, *Science Principal and Director*
Aleksander P Pylaev, *Marketing Director*

Established in 1894, the research institute has a wide range of test facilities and extensive experience. Its principal areas of expertise, services and R&D include:

Advanced ship concept design and ship construction progress.
Offshore and ocean engineering research and design.
R&D works, development of hardware and software in the fields of:

- Hydrodynamics of naval ships, commercial vessels, marine weapons and ocean engineering structures.
- Structural strength and reliability.
- Marine and industrial acoustics.
- Ocean physical fields and marine vehicle signatures.
- Marine power plants, nuclear, radiation and environmental safety.
- Corrosion protection.

Certification and standardisation.
Development of facilities for hydrodynamic, strength, acoustic and electromagnetic tests.

Sudoexport

11 Sadovaja-Kudrinskaja St, 123242 Moscow, Russian Federation

Tel: (+7 495) 252 44 91
Fax: (+7 495) 255 01 45
e-mail: info@sudoexport.ru
Web: www.sudoexport.ru/

Eugeny A Garrov, *General Director*
Anatoli Korolev, *Deputy General Director*

Sudoexport is a Russian organisation engaged in the export and import of all kinds of sea- going and riverine ships and ship's equipment, and the export and import of technological equipment for shipbuilding and associated engineering activities.

Sudoexport offers skilled services in ship repair including repair for the naval-defence fleet. The company is also an investor in various joint ventures and has a worldwide agent network.

South Africa

Foil Assisted Ship Technologies cc

FAST cc
45 Aster Street, Heldervue, Somerset West, 7130 Western Cape, South Africa

Tel: (+27 21) 855 21 16
Fax: (+27 21) 855 54 77
e-mail: fastcc@hydrofoildesign.com
Web: www.hydrofoildesign.com

The company was founded in 1998 by K G W Hoppe. The company offers naval architecture consultancy in the field of hydrofoil assisted ships. Professor Hoppe has developed foil assisted ships of different kinds and has been responsible for patenting the Hydrofoil-Supported-Catamaran (HYSUCAT) and more recently the Hydrofoil-Supported-Wavecraft (HYSUWAC). In the last five years several hundred catamarans have been equipped with foil systems to improve performance, and full ship designs have been completed for yachts and ferries.

At the end of 2006 a 16 m luxury vessel from Stealth Yachts Pty Ltd was launched in Cape Town. The vessel has fully asymmetrical demi-hulls with a standard HYUSCAT foil system inside the tunnel and 2 × 820 hp MANB diesel engines driving Q-SPD surface propellers. Fully laden to 21 t, the craft reached 47.3 kt in initial trials.

24m, 149 passenger ferry constructed at Geo Shipyard which uses HYSUCAT technology 1325694

The HYUSCAT technology is currently being applied to a 24 m, 149 passenger ferry built at Geo Shipyard Inc in Louisiana, US.

Spain

Neumar SA

La Rinconada B-6, E-28023 Madrid, Spain

Tel: (+34 91) 633 56 93
Fax: (+34 91) 633 52 32
e-mail: contact@neumar.com
Web: www.neumar.com

J Peire, *General Manager*
M de la Cruz, *Technical Director*
J A Barbeta, *Manufacturing Manager*

Neumar SA was founded in 1983 as a consulting engineers and designers company for air cushion technology. The services it offers include: research and development with an emphasis on lift systems; hovercraft design; project management, comprising construction subcontracting with collaborating shipyards; hovercraft skirt design; and manufacturing, technical and economic feasibility studies.

The company offers a range of hovercraft designs based on a new lift system recently developed and patented. This has been called an Automatic Transversal Air Distribution (ATAD) lift system because of the main function it performs, in providing very high stability with low-power requirements and reduced manufacturing and maintenance costs. To prove the viability of this lift system, several two-dimensional models and two prototypes were built and tested.

The company currently markets the NM-6 hovercraft designs, a range of 6 m hovercraft with specialist designs for surveillance and search and rescue as well as pleasure and scientific surveys.

Sweden

Marinteknik Design (E) AB

Vastergatan 61, SE-740 71 Öregrund, Sweden

Tel: (+46 173) 313 15
Fax: (+46 173) 313 65
e-mail: mde@swipnet.se
Web: www.marinteknikdesign.se

Hans Erikson, *Executive Manager*

Marinteknik Design undertakes research and development projects for Marinteknik Shipbuilders and other shipyards worldwide. The company offers a comprehensive marketing service for new and old fast ferries and is developing new designs based on vessels designed and built during the last 35 years.

Pelmatic AB

Drottninggatan 65, SE-371 33 Karlskrona, Sweden

Tel: (+46 455) 30 04 50
Fax: (+46 455) 30 04 55
email: info@pelmatic.se
Web: www.pelmatic.se

Stefan Johansson, *Managing Director*

Pelmatic AB is a firm of consulting engineers specialising in product development and design projects for the shipbuilding, offshore, aircraft and vehicle manufacturing industries. The company has offices in Sweden, Denmark and Singapore together employing 200 engineers.

Pelmatic has been involved in the design of high-speed military and commercial vessels, notably the structural design of the 95 m fast ferry 'Kattegat' and the 'Stockholm', 'Gothenburg' and 'Smyge' classes of fast patrol craft.

SSPA Sweden AB

PO Box 24001, SE-400 22 Gothenburg, Sweden

Tel: (+46 31) 772 90 00
Fax: (+46 31) 772 91 24
e-mail: postmaster@sspa.se
Web: www.sspa.se

Susanne Abrahamsson, *President*
Lars T Gustafsson, *Vice-President, Head of Ship Design*
Jim Sandkrist, *Vice-President, Head of Maritime Operations*
Claes Källström, *Vice-President, Head of Research*

SSPA has been involved in the testing of high-speed craft designs and their associated equipment for many years. Noted for its contribution to the success of the Royal Swedish Navy's high-speed vessels (the recently launched VISBY class), the company has also been closely involved in the testing and development of commercial high-speed craft, notably the Stena HSS catamaran, and the Fast Ship project for transatlantic freight shipping. The company has also been active in testing high-speed yacht designs for Italian shipyards and has carried out projects for Donald L Blount and Associates. Extensive studies have been undertaken as part of the European Research project (EFFISES) for a 70 kt air-supported vessel.

SSPA runs three scale-model test facilities and a simulator laboratory. The main areas of consulting expertise offered by SSPA are: hydrodynamics; hydroacoustics; marine environment; risk and safety analysis; and port and coastal development. Using its broad experience of marine operations, SSPA also develops customised simulators. Mathematical models are based on both full-scale experience as well as laboratory tests in both deep and shallow waters, with bank effects included.

An important tool for high-speed craft developed by SSPA is the Hydrodynamic Predictor (HdP). This is a navigation aid to be used for continuous dynamic position prediction within 30 to 600 seconds, with the ship's contour visualised on a chart display with up to 10 different predicted positions.

Switzerland

Supramar AG

Plattenstrasse 9, CH-8152, Glattbrugg, Switzerland

Tel: (+41 79) 220 21 69
Fax: (+41 1) 523 439 005
e-mail: info@supramar.ch
Web: www.supramar.ch

Volker Jost, *President*
Harry Trevisani, *General Manager*
Ulrich La Roche, *Research and Development*
Herrmann de Witt, *Hydrodynamics*
Otto Münch, *Stabilisation and Control*

(Patent Holder and Designer)

Since its foundation in 1952, Supramar has provided a worldwide consultancy service, covering not only its hydrofoil vessels but also other aspects of fast marine transport.

The company has been under contract to many governments and military services.

Supramar developed, on a commercial basis, the hydrofoil system introduced by the Schertel-Sachsenberg Hydrofoil syndicate and its licensee, the Gebrüder Sachsenberg Shipyard. The development started in the 1930s and led to the realisation of a number of military hydrofoils up to 80 tonne displacement and 41 kt in speed.

The inherently stable, rigid surface-piercing V-foil system which is typical for the Supramar type craft was developed by the late Baron Hanns von Schertel.

In May 1953, the world's first passenger hydrofoil service started on Lake Maggiore in Italy, with a Supramar Type PT 10 craft *Freccia d'Oro*. She was later transferred to Lake Lucerne. A larger craft, the PT 20, was built by Lürssen Shipyard in 1953 and named *Bremen Pioneer*. Since then many Supramar-type hydrofoils have been built under licence from Supramar, mainly by Rodriquez, Hitachi and Westermoen. Many Supramar PT 20 and PT 50 hydrofoil vessels are still in service.

Supramar Ltd is also engaged in new designs of high-speed craft such as fast monohulls and catamarans as well as general engineering services.

Due to the trend to higher speeds, Supramar has concentrated its recent investigations on the development of new profiles and research on solutions to eliminate cavitation. A patent has been registered to this effect and tests are taking place in a cavitation tunnel to confirm the company's claims. Parallel to this development, Supramar decided to extend its activity in the field of wing in ground effect flying ferries, adding a new team of specialists in surface effect to its long established hydrofoil experience. Wind tunnel and model testing of the Supramar WISE 2000 (Wing In Surface Effect) project is under way.

In December 1997, Supramar started collaboration with the SEABUS-HYDAER Consortium, a group of 13 specialised European companies. Funding was received from the European Commission in Brussels under Brite Euram. Supramar's task is the layout of the hydrofoils for this hydrofoil-assisted wing in ground effect fast ferry with a total displacement of 500 tonnes, 800 passengers and 100 cars and a range of 850 km with a speed of 120 kt.

Based on this new development and related patents, Supramar's activities today cover the field of high-speed hydraulic applications in general, such as superfast hydrofoils for speeds well above 60 kt, or propeller blades having tip speeds which would normally cause cavitation or supercavitation.

United Kingdom

Air Vehicles Design And Engineering Ltd

Head office and factory:
Three Gates Road, Cowes, Isle of Wight, PO31 7UT, UK

Tel: (+44 1983) 29 31 94
Fax: (+44 1983) 29 19 87
e-mail: info@airvehicles.co.uk
Web: www.airvehicles.co.uk

C B Eden, *Director*

Air Vehicles has been in the high speed, air supported, marine market for over 30 years. Originally the builder of small- to medium-sized air cushion vehicles, the company now specialises in the design and manufacture of systems and components for fully amphibious ACV and SES. These items include interior systems, lift fans and propeller ducts, air propeller and transmission systems.

Blyth Bridges Marine Consultants Ltd

Sarisbury Building, 180 Bridge Road, Sarisbury Green, Southampton, SO31 7EH, United Kingdom

Tel: (+44 1489) 57 44 32
Fax: (+44 1489) 57 44 32
e-mail: info@blythbridges.co.uk
Web: www.blythbridges.co.uk

Andrew G Blyth, *Director*
David C Bridges, *Director*

Independent naval architecture and marine engineering services for high-speed conventional and unconventional surface craft. Formed in 1990 by the joining of two established independent consultants, the directors have over 70 years experience in the research, design and construction of advanced monohulls and multihulls, including Small-Waterplane-Area Twin-Hull (SWATH) ships, Surface Effect Ships (SES) and Air Cushion Vehicles (ACV).

Andrew Blyth is a naval architect with a background of practical design and applied research, including model testing and project management. He has participated actively in the development of the IMO Code of Safety for High-Speed Craft, through membership of MCA, and RINA committees. He has recently been co-ordinator of the IMO correspondence group on stability matters for high-speed craft.

David Bridges is a marine engineer, with particular experience and expertise in the conceptual and detailed design and commissioning of gas-turbine and high-speed diesel machinery installations and associated systems in high-speed craft.

The company benefits from the experience and knowledge of a wide range of specialist associates, including experts in structural design and analysis, electrical design and ship operation; it has a full computer draughting capability.

The services offered by the company include:
Definition of ship operator requirements
Management of tendering, including requests for proposals
Objective comparative evaluation of alternative designs and tenders
Owner's representative during design, construction and trials
Technical and economic feasibility studies
Concept selection and design of ferries and patrol and work boats
Complete ship design service from concept selection, through contract specification to guidance or production drawings, and conducting of tests and trials.

Recent projects include:
Development work towards a code of safety for small craft hired out to the public
Assisting a client with re-entering three hovercraft into class
Expert witness in fast ferry and high-speed craft litigation
Machinery system design for four patrol boats
Managing a series of five research projects into the stability of high-speed craft for the UK Maritime and Coastguard Agency
Assisting the Isle of Sark government in procuring a new passenger/cargo ferry
Assisting the government of a Middle Eastern nation in assessing tenders for two 18 m fast-patrol boats
Developing the UK MCA Guidance for Instructions for Surveyors for the 2000 IMO Code of Safety for High-Speed Craft, and updating them in respect of the latest IMO amendments
Developing a new stability guidance booklet for small commercial craft (MCA)
High-speed water-jet trans-Atlantic container ship
Military and passenger amphibious hovercraft – various design aspects for a variety of clients
Review of pilot launch design for a UK port authority.

Specialist studies for:
Stability hazards and criteria for unconventional marine vehicles for the UK Ministry of Defence
Technical services for the purchase of a passenger-vehicle ferry including oversight of construction and trials
Concept design and costing of two sizes of offshore protection vessel and a light frigate
Long-term procurement and replacement strategy for a fleet of patrol craft
An ISO standard for the stability of craft up to 24 m length.

HM Customs and Excise (UK Marine Branch) has selected the company several times to assist in the procurement of new high-speed monohull surveillance cutters. These projects have included technical and commercial evaluation of competing design and build proposals for 34 and 26 m cutters, 15 and 12 m regional boats and oversight of the construction of the prototype regional boat.

The company has worked in close collaboration with SSPA Maritime Consulting AB on a number of major projects, including self-propulsion, sea-keeping and manoeuvring tests on water-jet propelled high-speed SES and monohull ships.

BMT Maritime Technology Ltd

Goodrich House, 1 Waldegrave Road, Teddington, Middlesex TW11 8LZ, United Kingdom

Tel: (+44 20) 89 43 55 44
Fax: (+44 20) 89 77 93 04
e-mail: enquiries@bmtmail.com
Web: www.bmt.org

Peter French, *Chief Executive*

British Maritime Technology (BMT) is a multi-disciplinary engineering and technology consultancy specialising in design, design support, risk and contract management and salvage, principally in the defence, environment, insurance, oil and gas and transportation sectors. BMT, with its headquarters in London, consists of 17 specialist subsidiary companies employing 950 staff based in over 50 locations across 18 countries.

Within the defence sector BMT provides a confidential and objective engineering design management consultancy that is independent from manufacturing interests, offering probably the greatest concentration of independent naval design and programme management expertise in Europe. Professional expertise encompasses naval architecture, mechanical and electrical engineering, combat systems engineering, formal safety assessments, programme and risk management, configuration and data management and technical documentation. The company is assessed to ISO 9001.

BMT holds multimillion pound (GBPmm) defence contracts from the UK MoD for in-service support for all classes of the Royal Navy's front line surface warships and the entire submarine fleet.

BMT is the platform design authority for the Thales team, developing a range of platform options for the UK's Future Aircraft Carrier (CVF) programme. Also, the company recently won a GBP12 million contract to supply and operate the Sea Trials Platform for the Principal Anti-Air Missile System (PAAMS) in the UK.

The MoD has asked BMT to identify design options for a Primary Casualty Receiving Ship and has also authorised BMT to provide safety certification services for surface warships and auxiliaries. BMT, in partnership with Frazer Nash Consultancy, has recently won a contract to provide specialist technical support to the MoD's Sea Technology Group (STG) for submarines and surface ships.

BMT forms a key part of the evaluation team for the US Coast Guard's Deepwater Program to replace all its coastal and ocean going vessels. The company has been providing support to the Canadian Navy in their development of concepts for the next generation logistics support vessels. Also, the Royal Navy has decided to use BMT's maritime search and rescue tool, SARIS, to help track floating mines.

The company continues to provide high-level hydrodynamic and aerodynamic model testing, for example for the T45 Destroyer. In aerospace, BMT is leading a 6-month risk and safety study into the life extension of the Tornado aircraft hydraulic system.

Combining its technical and management skills, the company offers customers a turnkey approach to meeting project needs. BMT is committed to partnering and to the transfer of marine technology to countries with established and growing naval needs.

BMT Nigel Gee Ltd

Floors 1-3, Building 14, Shamrock Quay, William Street, Southampton, Hampshire SO14 5QL, United Kingdom

Tel: (+44 23) 80 22 66 55
Fax: (+44 23) 80 22 88 55
e-mail: admin@ngal.co.uk
Web: www.ngal.co.uk

Peter French, *Chief Executive*
John Bonafoux, *Managing Director*
Ed Dudson, *Technical Director*
Ian Glen, *Director*

BMT Nigel Gee Ltd offers a comprehensive service for all aspects of specialised high-speed marine craft technology. A staff of 40 engineers and naval architects specialise in the complete range of high-speed craft design and consultancy activities. Over the last 20 years, the company has designed over 150 vessels that have either been built or are currently being built. The company joined the BMT group in 2003.

The company has experience across a broad spectrum of craft types and has design and build experience for monohulls, catamarans, Surface Effect Ships (SES), hovercraft, foil-assisted craft, SWATHs, trimarans, pentamarans and other combinations of hulls with air and foil assistance.

Recent design and consultancy activities carried out by the company include:

- Transport analysis and technical consultancy for new high-speed ferry services.
- Sea Fighter – complete design of an 80 m fast naval vessel.
- 38 m commuter ferry – complete design of a 350 passenger high-speed ferry for Bermuda.
- 40 m ferry – 350 passenger ferry for River Sado, Portugal.

The services offered to the marine industry include: market research; initial concept design and production of specifications for quotation purposes; preliminary design and naval architecture; design of model test programmes and model test supervision; structural design from first principles or to Classification Society rules: engineering design – machinery, systems, electrics and detailed fit-out design; presentation drawings and artist's impressions and interior artwork.

Fairweather *Alaska Marine Highway 72 m catamaran* 0589532

Sea Fighter *for the US Navy Office of Naval Research* 1194802

38 m commuter ferry for Bermuda 1324315

Francis Design Ltd

12 Regents Wharf, All Saints Street, London N1 9RL

Tel: (+44 20) 79 23 53 60
Fax: (+44 20) 79 23 53 61
e-mail: francis@francisdesign.com
Web: www.francisdesign.com

Martin Francis, *Principal*

Francis Design is a multidisciplinary practice, specialising in yacht design. Current projects include high speed super yachts.

Eco *at speed* 0098021

Hart Fenton and Co Ltd

Norman House, Kettering Terrace, Portsmouth, Hampshire PO2 7AE, United Kingdom

Tel: (+44 2392) 87 52 77
Fax: (+44 2392) 87 52 80
e-mail: hf@hart-fenton.com
Web: www.hart-fenton.demon.com

Mike Simpson, *Managing Director*

Hart Fenton and Co Ltd provides consultancy services covering structural analysis, hydrodynamic analysis, new building supervision, plan approval, FMEA, ship surveys including condition, expert witness for arbitration and litigation, collision and sale and purchase feasibility studies, project technical assessments and CAD drafting services.

Hart Fenton has extensive experience of passenger ferries, ro-ro container vessels, cable laying and offshore vessels, both new construction and conversion. The company has particular knowledge of high-speed craft, having overseen the build of Seacats and Superseacats, and has experience in new construction, operational aspects, refits and improvements/ retrofits.

Hart Fenton's staff are drawn from a broad cross section of the industry and staff members sit on a number of industry committees in the UK and internationally.

In October 2006, Hart Fenton became a subsidiary of Marine Consultants Houlder Limited.

Hovercraft Consultants Ltd

Unit 43, South Hampshire Industrial Park, Southampton, SO40 3SA, United Kingdom

Tel: (+44 23) 80 87 11 88
Fax: (+44 23) 80 87 17 99
e-mail: sales@hovercraftconsultants.co.uk
Web: www.hovercraftconsultants.co.uk

M J Cox, *Director*
J F Cox, *Secretary*

HCL specialises in providing design expertise in those areas of air cushion technology which are not covered by conventional naval architecture. Assistance and advice is available, not only at the conceptual stage on matters of craft size, weight and power, but in more detail in the areas of skirt design, skirt shift systems, lift systems and air propulsors.

The design process is computer based which reduces the time required to design new craft and minimises the costs of manufacture, particularly those of the skirt system which can be assembled from CNC cut panels. The company has designed and manufactured many different types of skirt system, including bag-and-finger, loop-and-segment and bag-and-pericell. Most new projects no longer need a comprehensive model test programme to predict full-scale performance. HCL can provide data from computer-based performance prediction to reduce the cost of new concept development. Recent work on propeller and lift fan design at HCL has been concentrated on the need to reduce noise levels in propulsion systems and improve lift system efficiency characteristics. HCL has also developed a mixed-flow fan geometry that is particularly well-suited to ACV lift systems.

Griffon 8100TD skirt 1197864

Hovercraft Consultants has expanded its operation to include the manufacture of complete skirt systems and skirt components. Other products include flexible tanks for water or fuel as well as flexible seals and covers.

Past air cushion vehicle projects have ranged from small recreational craft up to industrial hoverplatforms designed to carry hundreds of tonnes. Work has been carried out not only on amphibious craft but also SES, particularly in the areas of lift system and cushion seal design. In addition to design projects, HCL has carried out numerous feasibility studies, not only on proposed designs but also on potential routes and operations.

Hoverwork Ltd

Quay Road, Ryde, Isle of Wight, PO33 2HB, United Kingdom

Tel: (+44 1983) 56 30 51
Fax: (+44 1983) 81 28 59
e-mail: info@hoverwork.co.uk
Web: www.hoverwork.co.uk

C D J Bland, *Chairman*
R K Box, *Chief Executive*
R H Barton, *Engineering Director*
P White, *Finance Director and Company Secretary*
B A Jehan, *Operations Manager*

Hoverwork Ltd is a wholly owned subsidiary of Hovertravel Ltd, who provide a high-speed, year-round, hovercraft service between the Isle of Wight and Portsmouth. Founded in 1966, Hoverwork was established to provide hovercraft sales, charter services and consultancy on a worldwide basis. The company has trained over 60 hovercraft captains and has engaged numerous hovercraft charter operations in many diverse locations and environments. Charter duties include: passenger services, mineral surveys, medical evacuation duties, feature film production and logistics support.

During 1998 the company operated in Canada in support of the Canadian Coast Guard AP1-88/400 hovercraft when it was engaged on manufacturer's and customer acceptance trials. During July 1999, Hoverwork Ltd provided a hovercraft service between Carnoustie and St Andrews in support of the British Open Golf Championships.

Hoverwork Ltd has access to all hovercraft owned and operated by Hovertravel Ltd.

In 2002 the company released details of their 160 passenger hovercraft, the BHT 150 design, based on a development of the AP1-88 craft.

Hypercraft Associates Ltd

23 Wyndham Avenue, High Wycombe, Buckinghamshire HP13 5ER, UK

Tel: (+44 1494) 46 16 89
e-mail: mail@hypercraft-associates.com
Web: www.hypercraft-associates.com

Graham K Taylor, *Managing Director*

The consultancy exists to promote the commercialisation of the ekranoplan/wing-in-ground-effect concept. It provides a commercial/business strategy interface between designers and potential users. Services include:

Working with technology providers to give: better understanding of operator-client requirements, improved focus on technical and design criteria, improved business propositions, sales effectiveness and product marketing, market opportunity assessment.

Working with transport operators to: assess business propositions of transport technology providers, improve capital expenditure, investment and strategic decisions, assess market opportunity.

Independent Maritime Assessment Associates Ltd

Portland House, 35 Knights Bank Road, Hill Head, Fareham, Hampshire PO14 3HX, United Kingdom

Tel: (+44 1329) 66 32 02
Fax: (+44 1329) 66 81 76
e-mail: imaa.co@ntlworld.com
Web: www.imaa.co.uk

John Lewthwaite, *Naval Architecture*
Anthony Marchant, *Structural Design Engineering*
David Bridges, *Marine Engineering*
Darrol Stinton, *Aero-marine Engineering*
Jonathan Lewthwaite, *CAD and Model Making*

IMAA draws together advanced design and commercial talents from the marine engineering industry. Its purpose is to offer a completely integrated independent consultancy service to designers, builders, operators and owners of marine craft.

The consultancy offers expertise in all aspects of naval architecture, marine engineering, advanced structural design, aero-marine engineering, risk analysis, operational analysis, project management and commercial marketing disciplines.

IMAA's collective experience covers all configurations from conventional monohulls and fast planing hulls, to catamarans, Surface Effect Ships (SES), SWATH vessels and Wing-In-Ground-effect craft. IMAA's offices are fully equipped to offer an integrated advisory service on all aspects of craft design and performance, to technical audits on route analysis, operational assessments, procurement and cost/risk evaluations.

IMAA has developed the PACSCAT (Partial Air Cushion Support Catamaran) concept. Such vessels are designed to carry heavy payloads at moderately high speeds. The UK Ministry of Defence has recently ordered a 30 m technical demonstrator of a PACSAT fast landing craft for evaluation by the Royal Marines.

Inserve Ltd

Renown House, 33/34 Bury Street, London EC3A 5AT

Tel: (+20 7) 929 23 79
Fax: (+20 7) 929 24 79
e-mail: info@inserve.org
Web: www.inserve.org

Specialising in fast ferries, Inserve provides professional surveying and consultant services to insurers, operators and legal firms. A Shanghai office was opened in May 2005.

Recent projects include: technical risk appraisal, collision investigations, damage surveys and loss adjusting. A technical bias also enables Inserve to compile robust Failure Mode and Effects Analysis to the requirements of the High Speed Craft Code.

International Hovercraft Ltd

Atec House, Peel Street, Southampton, Hampshire SO14 5QT, United Kingdom

Tel: (+44 2380) 33 77 43
Fax: (+44 2380) 33 77 43
e-mail: sales@atec-hants.co.uk

Peter Hill, *Director*
Graham Banks, *Director*

International Hovercraft Ltd acquired the assets of Hovermarine International Ltd in July 1997, which include the detailed designs of all the previous company's craft. This covers the highly successful HM 2 family, utilising composite superstructures and diesel power, of which over 100 have been constructed. The company undertakes surveys, as well as providing a worldwide spares and technical support service.

International Hovercraft Ltd is keen to identify partners to introduce these fast ferry designs into their own local areas of operation with manufacturing taking place under license.

HM221-P with a capacity for 123 to 150 passengers 0100071

JB Marine Consultancy

Vue de Lion, La Giffardiere, Castel, Guernsey G75 7HN, United Kingdom

Tel: (+44 1481) 253053
Fax: (+44 1481) 253053
e-mail: john@jbmarine.biz
Web: www.jbmarine.biz

John Bate, *Director*

The company specialises in offering naval architecture, marine survey and engineering services to the small ship and boat industry.

Lorne Campbell Design

13A Moor Road, Broadstone, Dorset BH18 8AZ, United Kingdom

Tel: (+44 1202) 65 74 66
Fax: (+44 1202) 65 74 66
e-mail: lorne@lornecampbelldesign.com
Web: www.lornecampbelldesign.com

Lorne F Campbell, *Principal*

Lorne Campbell has been working in the area of high-speed marine craft for over 30 years, starting the present company in 1981. His experience covers all configurations of planing craft including monohull, trimaran and catamaran and speeds have ranged from just on plane up to 130 kt.

Ongoing work includes: the design of steps and stepped hulls for a wide variety of craft encompassing racing, military and leisure applications; the design of RIB craft and general consultancy and trouble shooting of planing craft.

The hull for a series of water-jet driven Combat Support Boats for the UK forces is in production as is a 10 m high speed AEM (Air Entrapment Monohull) leisure craft (speeds from 52 to 75 kt, with over 90 kt for the race version). This craft is built by ICE Marine and known as the Bladerunner. A 15.2 m fast AEM Bladerunner for both military and leisure applications has been completed and the first two production craft are about to be launched. The prototype achieved 71 kt on trials and took the record for the fastest circumnavigation of Britain (27 hours 10 minutes) in 2005.

A 15 m fast catamaran workboat hull design has now entered production and a 60 kt, 20 m stepped monohull patrol craft hull design has been completed and the first two craft delivered. Design of a 22 m stepped hull for the same client is underway; the hull lines for a prototype aluminium alloy Unmanned Surface Vehicle for the US Navy have been designed and prototypes are undergoing trials; in addition, an aluminium 12 m Sea Sled workboat has been delivered; the design of steps for a 14 m production leisure craft is underway, and the design of a fast, air transportable 11 m RIB is also proceeding, as is a 12 m AEM combatant.

The unique AEM Bladerunner hull design, which consists of a narrow stepped main hull flanked by stabilizing sponsons and twin air entrapment tunnels, undergoes constant development and is proving to be a very fast and stable concept for high-speed offshore conditions.

The company can offer a blend of practical and theoretical experience in the design, performance and naval architecture of fast planing craft for both calm and rough conditions. Designs are developed from existing craft or from scratch. Both wind tunnel, water tank and free running models are used to evaluate and develop designs which have no 'parent' form to work from. Wind tunnel testing has been an important part of this work since the higher speed craft are well into the aerodynamic ground effect regime. Considerable experience has been gained in propulsion systems – water-jet, submerged propeller, stern drive and surface propeller.

A service is offered covering the design and consultancy of planing craft up to 30 m ranging from concept design and preliminary investigation through troubleshooting, general arrangement, layout and styling, resistance, propulsion and stability.

Maritime Services International Ltd

Stone Lane, Gosport, Hampshire PO12 1SS, United Kingdom

Tel: (+44 23) 92 52 44 90
Fax: (+44 23) 92 52 44 98
e-mail: ian@maritimeservices.demon.co.uk

Ian C Biles, *Managing Director*

Established in 1988, Maritime Services International provides technical advice and assistance to owners and operators of high speed craft. Services include vessel refit planning and site overseeing of implementation, new building inspection and quality control supervision, performance trials (both builders and owners), failure mode and effect analysis (FMEA) trials, feasibility studies including route evaluation, damage surveys and survey for sale and purchase, development of operator manuals together with crew training and examination for type rating.

The company has experience of catamarans, SWATH, hydrofoils, hovercraft and surface effect, diesel engines, gas turbines and composite, aluminium and composite craft.

Seaspeed Marine Consulting Ltd

2 City Business Centre, Basin Road, Chichester, West Sussex PO19 8DU, UK

Tel: (+44 1243) 78 42 22
Fax: (+44 1243) 78 43 33
e-mail: info@seaspeed.co.uk
Web: www.seaspeed.co.uk

S J Phillips, *Managing Director*
D Hook, *Technical Director*
Richard Daltry, *Naval Architect*

Seaspeed Marine Consulting Ltd is an independent marine consultancy company specialising in the performance and safety technology of high-speed marine craft. The company provides a high level of academic, professional and practical experience over the complete range of design, build and operational technologies in both the commercial and military industry sectors including survey and valuation services for the fast ferry industry. The company is accredited to the BS EN ISO 9001 quality assurance standard.

Seaspeed Marine Consulting Ltd is also a member of a number of industrial networks with specialist interests in unmanned vehicles, hydrodynamics and risk assessment.

The main activities of the company include:
High-speed craft preliminary design;
Speed, manoeuvring and sea-keeping predictions;
Structural design and materials selection;
Computer simulations of multihull manoeuvring and sea-keeping;
Model construction, testing and data analysis;
Ship sea trials and instrumentation;
FSA, risk assessments and FMEA analysis for ship systems and equipment;
Condition and valuation survey;
Design assessments;
Market analysis.

Recent contracts have included:
Fast ferry inspection, valuation and survey;
Safety case provision for large fast non-commercial craft;
Arbitration and expert witness services for legal cases involving high speed craft;
Development of fast semi-submersible craft;
Risk analysis for the Department of Transport and industrial companies;
Radio-controlled open water model construction and testing;
Fast craft manoeuvring and sea-keeping simulations;
Passage plan risk assessments and HSC wash training;
HSC wash predictions and measurements;
Compilation of HSC Incident Database.

White Young Green Consulting Ltd

The Mill Yard, Nursling Street, Southampton, Hampshire SO16 0AJ, UK

Tel: (+44 870) 609 10 84
Fax: (+44 23) 80 74 37 62
e-mail: southampton@wyg.com
Web: www.wyg.com

Rhys Bowen, *Engineering Consultant at Specialist Structures*

CETEC was acquired by White Young Green in 2004, forming a new specialist area within the company. WYG is a recognised authority on the structural design and analysis of all types of high-speed marine vessels and marine transportation. Services are provided to naval architects, shipbuilders and operators on a world wide basis.

The MRTP 15 high-speed monohull for the Turkish Coastguard. The craft is constructed in composite materials employing a unique hull laminate developed by WYG 0113524

WYG has been involved in the development and design of structures for hovercraft, Surface-Effect Ships (SES), catamarans, monohulls, SWATH's, foil-assisted vessels, WIG's, trimarans and pentamarans. WYG has experience in all construction materials – steel, aluminium alloy and composites.

Because of the need to provide fully optimised and minimum weight structures for fast craft, WYG have developed extensive analytical tools for mathematical modelling and Finite Element Analysis (FEA). Full and detailed models of complete craft or ships can be quickly produced, thereby providing the design tools around which structural optimisation can be undertaken against stated or agreed acceptance criteria. The analysis can be used for Classification acceptance. Such a direct calculation approach has been used from small fast craft to ocean going fast ships.

WYG undertake project management for the design and build supervision of high-speed craft.

Wolfson Unit For Marine Technology and Industrial Aerodynamics

University of Southampton, Southampton, Hampshire SO17 1BJ, UK

Tel: (+44 23) 80 58 50 44
Fax: (+44 23) 80 67 15 32
e-mail: wumtia@soton.ac.uk
Web: www.wumtia.com

Ian Campbell, *Senior Engineer*
John Robinson, *Software Sales Manager*

The Wolfson Unit is a commercial consultancy, within the School of Engineering Sciences at the University of Southampton. The unit was established in 1967 to provide a comprehensive advisory service in marine technology and industrial aerodynamics. It is staffed by full-time qualified consulting engineers, with a wide range of academic and industrial experience. It is also able to draw on the experience of academic staff and other consulting engineers throughout the university to broaden the scope of the services offered.

The unit has always specialised in small fast vessels, with towing tank testing of high-speed motor yachts, patrol craft and pilot boats forming a significant proportion of the work in its early years. Test techniques have developed to keep pace with the demands of industry, and specialised equipment has been designed to address the particular requirements of catamarans, Swaths and dynamically supported craft.

With the importance of sea-keeping in high-speed craft design, much of the unit's work now includes assessment of motions and accelerations. These can be predicted using the Wolfson Unit's own computer software or measured in the towing tank. The unit has considerable experience in the measurement of motions at full scale, and has developed an acceleration meter for permanent installation on high-speed craft to monitor motions with regard to passenger comfort, crew performance, or structural limitations. For trial measurements the Wolfson Unit has a range of portable PC-based trials data acquisition systems tailored for routine acceptance trials, and instrumentation packages to cater for specialised requirements.

In the field of model testing the Unit has access to high-speed towing tanks, with carriage speeds up to 15 m/s, for resistance, sea-keeping and related testing, and two large aeronautical wind tunnels for measurement of aerodynamic forces and flow studies for funnel and superstructure designs. Radio-controlled models may be used in the study of manoeuvring or sea-keeping, and a variety of sheltered inland and coastal sites are available locally, in addition to controlled sea-keeping basins. The same facilities and expertise are employed by sailing yacht designers, and the Wolfson Unit's capabilities were demonstrated in 1995 and 2000 when the New Zealand yachts, which were tested in an extensive towing tank and wind tunnel programme at the Wolfson Unit, won the America's Cup.

All computer software developed by the unit is offered for sale and by bureau service. Programs include a comprehensive suite of design software for hydrostatics, stability and damage, a power prediction and propeller design suite, lines generation and fairing, sea-keeping, manoeuvring, and trials data acquisition. A recent addition to the range is an onboard loading program approved for use on passenger vessels by the UK Marine Safety Agency. The wide experience of the engineers makes them invaluable in troubleshooting problems with propulsion, vibration, sea-keeping, structures, stability and aerodynamics.

WS Atkins Consultants Ltd

Woodcote Grove, Ashley Road, Epsom, Surrey, KT18 5BW, UK

Tel: (+44 1372) 72 61 40
Fax: (+44 1372) 74 00 55
e-mail: info@atkinsglobal.com
Web: www.atkinsglobal.com

WS Atkins is a large, international, multidisciplinary consultancy offering a wide range of services useful to designers, builders and operators of fast craft. Although a relatively recent entrant into the fast craft arena, the company has been established for many years in the offshore industry.

Structural and naval architectural work is carried out by the Marine and Structural Technology division which specialises in engineering analysis, using software developed in-house. The main areas of work are analysis of hydrodynamic loads and motions in regular and irregular waves; linear and non-linear structural analysis, including fracture mechanics; and fatigue analysis.

There have been two major software packages developed; AQWA deals with hydrodynamic analysis whilst ASAS is a finite element package. Both have been widely used by the company and others for analysis of fixed- and floating marine structures, including offshore platforms, VLCCs and fast catamarans.

The Marine and Structural Technology division is supported by the other divisions in the group which offer a range of relevant skills, including computational fluid dynamics; noise and vibration; design of port and harbour facilities; safety and reliability; planning of transport systems and cargo management systems.

Recent experience has included work on large wave-piercing catamarans, in which hydrodynamic analyses were performed with waves from various directions, comparing sea-keeping and loads on different vessels. This was followed by extensive structural analysis involving global stress distribution, design of fatigue-resistant details and investigation of superstructure mounting systems.

YG Consultants

Plas Elyn, Llangoed, Beaunmaris, Ynys Mon, LL58 8SB, UK

Tel: (+44 1248) 49 01 25
Fax: (+44 1248) 49 01 95
e-mail: trevor.bailey@ygyg.com
Web: www.ygyg.com

Trevor Bailey, *Managing Director*

YG Consultants was established in 1998 to provide a worldwide consulting service for Maritime Safety and Training aimed at owners and operators, P&I Clubs and professional bodies. Managing Director Trevor Bailey is an experienced fast ferry master having also had command of a Stena HSS 1500 vessel.

The company offers a range of services covering ISM and STCW training programmes as well as routine crew training and the production of vessel documentation.

United States

Air Ride Craft Inc

15840 SW 84th Avenue, Miami, Florida 33157, United States

Tel: (+1 305) 233 43 06
Fax: (+1 305) 233 13 39
e-mail: info@seacoaster.com
Web: www.seacoaster.com

Don Burg, *President*

(Patent holders, designers and developers)

The Air Ride concept, designed and patented by Don Burg, has been configured in a number of ways on monohull and multihull designs. The concept is based on supporting a substantial percentage of the craft weight on a cushion of air, thereby reducing frictional resistance. In the catamaran designs it is reported that 85 per cent of the weight of the craft would be supported by the air cushion.

A number of models, prototypes and operational craft have been built using Air Ride's designs, the first test craft having been built in 1978 with a 12.8 m demonstration craft built in 1980. A 19.8 m crew/supply craft was launched in 1983, by 1991 two other Air Ride craft for passenger transportation had been built by Avondale and, in 1999, a 20 m fast ferry was developed.

33 m SeaCoaster built by Austal USA 1324309

There are a number of associated patents held by Don Burg, including a locking docking system, which uses the variation in height of the vessel in the water produced by variation in air cushion pressure. Recently the SWEEP™ concept for a ship with wave energy engulfing propulsors has been patented. This is estimated to reduce powering requirements by 40 per cent on high-speed displacement vessels, by greatly reducing wave drag, in addition to reducing sonar signature and environmental impact.

Air Ride has delivered a 20 m SeaCoaster ferry to Island Express Boat Lines and in October 2004 a 33 m Seacoaster craft, funded by the ONR and built by Austal US, was launched.

Apollonio Naval Architecture

1225 E Sunset Drive, No 514, Bellingham, Washington 98226-3597, US

Tel: (+1 360) 733 68 59
Fax: (+1 360) 715 94 74
e-mail: hapollo@az.com

Howard Apollonio, *Principal*

Apollonio Naval Architecture designs advanced commercial vessels and yachts to specific requirements to clients worldwide. Designs include fast passenger multihulls, high and low-speed Swath craft, work boats for unique applications and super yachts. Synthesising attributes between these craft types has been very successful for this office. Apollonio provides the benefits of 30 years experience with these craft and others, including patrol craft along with shipyard and at-sea experience. Vessels to its designs include 31 m, 400 passenger, 35 kt catamaran *Straits Express* 26.5 m, 16 kt Swath *Alison* and 31 m, 30 kt superyacht *El Lobo*, and a 24 m catamaran *Excellence* delivered to Jolly Roger Cruises of Antigua. New projects include very advanced 10 and 14 m catamarans and 21 and 24 m monohull military craft.

Art Anderson Associates

202 Pacific Avenue, Bremerton, Washington, 98337-1932, US

Tel: (+1 360) 479 56 00
Fax: (+1 360) 479 56 05
Web: www.artanderson.com

Eric Anderson, *Chief Executive Officer*

Art Anderson Associates specialises in facilities design, water transportation studies and vessel design. The company has recently completed a concept design for a new type of air cushion/hydrofoil passenger vessel and is having two technology demonstrators built by Radix Marine.

The company also holds the patent for Stolkraft, the registered trademark for an innovative hull form, developed in Australia for high-speed applications by Stolkraft International. The hull form makes use of both hydrodynamic and aerodynamic lift at high speeds achieved solely by the forward speed of the craft.

Performance has been proven by a number of model and prototype tests worldwide, initially by the MARIN Institute and more recently by the US Navy's Naval Surface Warfare Centre in Washington DC. Commercial applications to date include pilot boats, patrol boats, small ferries and a range of workboats.

CDI Marine Systems Development Division

900 Ritchie Highway, Suite 102, Severna Park, Maryland 21146, United States

Tel: (+1 410) 544 28 00; 30 12 61 10 30
Fax: (+1 410) 647 34 11
e-mail: bla@cdicorp.com
Web: www.cdicorp.com

David R Lavis, *Senior Vice-President, CDIM*
Brian G Forstell, *Director of Research & Development*
Daniel L Wilkins, *Director of Engineering*
Daniel G Bagnell, *Director of Naval Architecture*
Robert Percival, *Manager of Systems Integration*
Manish Gupta, *Manager of Engineering*
Robert Johnson, *Manager of Engineering Development*

CDI Marine Systems Development Division (CDIM-SDD), a division of CDI Government Services, is an engineering firm that offers a range of conventional and advanced marine engineering and naval architectural services. The company provides technical support to the US Navy's conventional and advanced craft programmes. Similar work is performed for the US Coast Guard's small boat and cutter programmes and for commercial companies both foreign and domestic.

CDI Government Services with over 700 employees, is known for its ship detail design expertise, while CDIM-SDD has worked extensively in innovative early stage design and research and development of marine systems.

The company offers particular expertise in the hydrodynamics and design of high-speed ships and marine craft. The company has a particular expertise in the design of water-jet pumps and systems for both conventional and advanced ship and craft applications. The company has a unique capability in the design of air cushion vehicles and surface effect ships. CDIM provides support for the design of ACV components using tools developed over the last 30 years plus the latest in computational fluid dynamics.

CDIM-SDD has developed a comprehensive ship design synthesis model designated Commercial Parametric Analysis of Ship Systems (ComPASS™). ComPASS permits simulation-based design for Displacement Monohulls, Planing Monohulls, Semi-Planing Monohulls, Catamarans, Trimarans, SWATH and Semi-SWATH, ACV and SES. For each hull form type, the model permits either whole ship or major subsystem design trade-offs to be examined for acquisition or operating costs and ship performance, including seakeeping. With support from ONR and technical oversight by NSWCCD and NAVSEA, ComPASS has been developed with a strong focus on the use of physics-based algorithms. This approach helps to ensure that extrapolations to new out-of-the-box designs will be accomplished with a higher degree of confidence than could be expected with methods that use empirical trends. The algorithms used have been extensively validated, and since 1978, have been used to support over 61 separate design projects.

The software is coded in C++, has an object-orientated architecture and runs within a Windows environment on a standard PC. It simulates the traditional ship design spiral and, via numerous iterations, can converge on balanced solutions within seconds. The impact of varying over 400 inputs can be examined, usually one at a time. For each individual segment of the route, the ship's speed, wave height, sea spectra, seaway modal period, ambient air temperature and percent of the time spent in the segment can be specified to create a complete route profile. All can then be varied to determine the overall impact on cost.

The company has access to over 300 AutoCAD stations, a unique fully catalogued product library to support high-speed vessel detail design, a prototype fabrication shop, experience in conducting ship surveys and ship checks for the US Navy, US Army and US Coast Guard, and significant experience in detail design, manufacturing and testing for the US Navy, US Army and commercial clients.

Recent projects

Support to US Navy LCAC and JMAC Programme Office; design of high-speed semi-SWATH ferry for a US yard; design of ultra-large water-jet propulsors; supervision of numerous model and full-scale test programmes; support of AKERT 2000 patrol ACV for the Finnish Navy; LCAC in-service support including design of the LCAC Deep

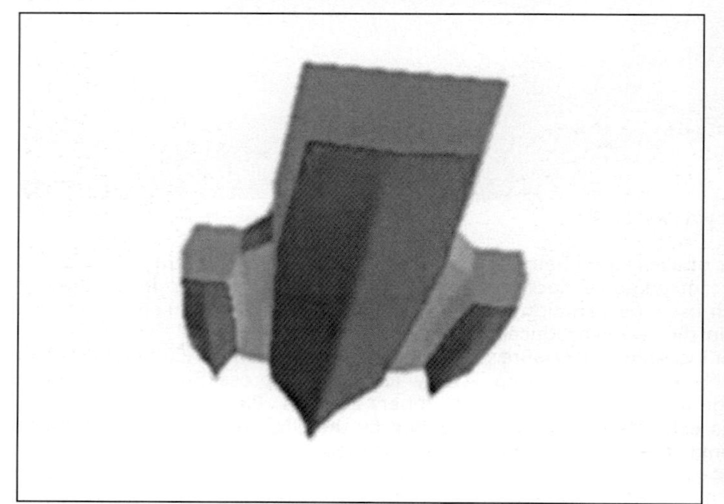

PASS-Generated Trimaran 1066430

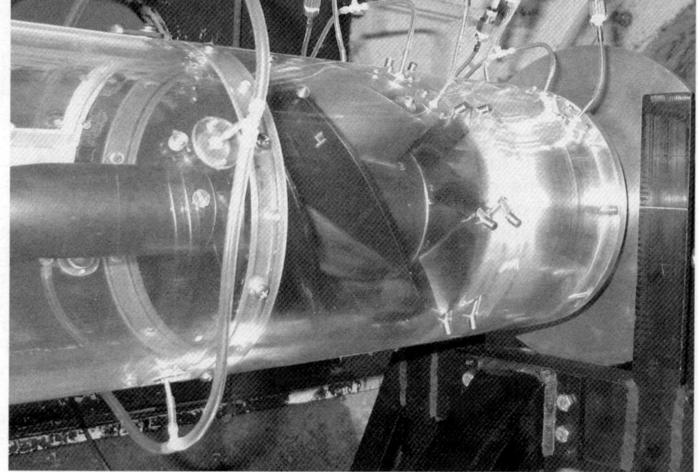

CDIM-SDD axial-flow water-jet pump 1172272

Skirt; development of MPF future ship concepts; USCG Deep Water programme; US Army Fast Sealift programme; Marine LHD 8 structural FEA; LHD 8 systems engineering and design of Littoral Combat Ship (LCS) for Northrop Grumman and NAVSEA; Heavy-Lift LCAC for NAVSEA; development of a Rapidly Deployable Intermodal Facility (RDIF); LCAC Integrated Landing Platform (ILP) and Mobile Landing Platform (MLP) for NSWCCD; and new inflatable boat concepts for NSWCCD.

Current projects

Supporting design studies for the JMAC ACV design alternatives for the US Navy; developing advanced ACV skirt concepts for ONR; multiple efforts involving the design and support of the US Coast Guard small boat fleet; supporting multiple prime contractors in the naval architecture and model testing for ships for the ONR High-Speed Sealift R&D programme; supporting multiple prime contractors to build a high-speed transport that can operate both at sea and in an amphibious mode for the ONR T-Craft Programme; providing CFD analysis support for several commercial yacht design firms; continuing with our design, CFD analysis and model testing for advanced axial-flow pump installations for CCDOTT; supporting the City of Baltimore for the requirements and acquisition of two Fireboats; support to NAVSEA on JHSV and JHSS programmes; and support of numerous contractors on the ONR HSSL Programme.

Davidson Laboratory

Stevens Institute of Technology
711 Hudson Street, Hoboken, New Jersey 07030, United States

Tel: (+1 201) 216 52 90
Fax: (+1 201) 216 82 14
e-mail: rdatla@stevens.edu
Web: www.stevens.edu/engineering/cms

Dr Raja V Datla, *Director*

Organised in 1935 as the Experimental Towing Tank, the laboratory is active in basic and applied hydrodynamic research, including smooth water performance and manoeuvrability, sea-keeping, propulsion and control of marine vehicles. This includes ACV, SES, hydrofoil craft, planing craft and so on. Special model test facilities are available to investigate the dynamic behaviour of all types of vessels and platforms in smooth water and waves.

Projects have included:

Towing tank studies of various seaplane model configurations for the analysis of bow spray, porpoising, high-speed resistance and impact loads.

Experimental studies on multi-hulls including trimarans for the optimization of side-hull location to minimise resistance.

An experimental investigation into the stability, course-keeping and manoeuvring characteristics of planing hulls and submerged oblects, performed in the laboratory's rotating arm facility and in the linear tank using planar motion mechanism.

D F Dickins Associates Ltd

9463 Poole Street, La Jolla, California 92037, United States

Tel: (+1 858) 453 86 88
Fax: (+1 858) 453 67 66
e-mail: info@dfdickins.com
Web: www.dfdickins.com

David Dickins, *President*

D F Dickins Associates Ltd is an engineering consulting company with a 27 year record of marine transportation and environmental projects for resource development companies in the US, Canada, Russia and Europe. Clients include international oil companies, large mining producers and government agencies.

Specialities include air cushion vehicle applications, marine transportation systems, oil pollution research and environmental assessment.

ACV projects include working with: Alaska Hovercraft on regulatory approvals for operations in a sensitive river environment; City of Anchorage to evaluate the potential use of an AP1-88 for emergency response; Cominco Metals to develop a successful four year AP1-88 hovercraft freight operation up the Stikine River in Alaska and British Columbia; East Aleutians Borough to assess technical and environmental issues surrounding a proposed new hovercraft ferry service (new craft on order); and a major oil company to evaluate using hovercraft for emergency evacuation over rough ice off Sakhalin Island, Russia. The Sakhalin project included leasing the Canadian Coast Guard AP.1-88/200 *Wabanaki* for tests over ice in the St Lawrence River in January 2002. In 2004, Dickins Associates completed a review of hovercraft applications in the North Caspian Sea for an international client, and in April 2007 the company completed a study for Defense Research and Development Canada, looking at the feasibility of using hovercraft to patrol the northwest passage through the Canadian Arctic.

Donald L Blount & Associates Inc

1316 Yacht Drive, Suite 305, Chesapeake, Virginia 23320, United States

Tel: (+1 757) 545 37 00
Fax: (+1 757) 545 82 27
e-mail: bblount@dlba-inc.com
Web: www.dlba-inc.com

Bill Blount, *Business Development Manager*

Donald L Blount & Associates Inc (DLBA) is an international marine design firm specialising in engineering applications of technologies for commercial and military craft. Many of DLBA's designs have speed requirements in excess of 40 kt with broad experience for engineering propulsion systems for gas turbines and lightweight diesel engines. High-performance, ocean-capable vessels with superior ride quality are a speciality.

DLBA was the design manager for *Destriero*, the 67.7 m vessel which holds the Blue Riband prize for the fastest Atlantic crossing, at an average speed of 53.1 kt.

DLBA was the design manager for *Fortuna*, the fastest private motor yacht in the world. *Fortuna* is a 41 m high-performance motor yacht capable of speeds in excess of 65 kt. DLBA has participated in the US Government's Littoral Combat Ship, Focused Mission High-Speed Ship, Landing Craft Utility (replacement) and Coastal Observation Project (OASIS) programmes. Additional projects include a concept design for a 46 m, 100 kt motor yacht, the design of a fast response fire/rescue boat, and a solar-powered unmanned research vessel.

Destriero 0009875

DLBA has recently completed a new hull design for the Mk 5 patrol boat intended to improve ride quality and reduce crew fatigue. Additionally, DLBA has completed the designs for three high performance motor yachts. These 30 m, 40 m and 45 m craft are currently under construction.

Elliott Bay Design Group Ltd

5305 Shilshole Avenue NW, Suite 100, Seattle, Washington 98107, United States

Tel: (+1 206) 782 30 82
Fax: (+1 206) 782 34 49
e-mail: info@ebdg.com
Web: www.ebdg.com

Gulf Office:
1100 Poydras Street, 20th Floor, New Orleans, Louisiana 70163, United States

Tel: (+1 504) 529 17 54
Fax: (+1 504) 529 19 74

John Waterhouse, *President*
Kenneth Lane, *Executive Vice-President*
Douglas Wolff, *Vice-President, Operations*
Brian King, *Vice-President, Engineering*
Joe Pritting, *Finance Manager*
Michael Walker, *Gulf Office, General Manager*
Christina Villiott, *Marketing Manager*

Elliott Bay Design Group is a firm dedicated to providing naval architecture, marine engineering analysis and shipyard support services. Using microcomputer technology and an extensive reference library, the company offers a wide range of services. The firm handles projects ranging from concept design of high-speed hulls to numerical lofting of structure. The full-time staff of 50 includes naval architects, eight marine engineers and three electrical engineers.

Elliott Bay Design Group's design experience is focused on commercial workboats and passenger boats. High-speed vessel designs up to 45 m and 50 kt have been created using both high-speed diesels and gas turbines. The firm has completed work on hydrofoils, catamarans, planing hulls and SWATHs. The staff are knowledgeable about the requirements of various regulatory bodies including USCG, ABS, IMO and DnV.

Fastship, Inc

1835 Market Street, Suite 2705, Philadelphia, Pennsylvania 19103, United States

Tel: (+1 215) 574 17 70
Fax: (+1 215) 574 17 75
e-mail: fastship@fastshipatlantic.com
Web: www.fastshipatlantic.com

Designed by Thornycroft Giles & Co Inc, the FastShip family of vessels has been developed to provide an enhanced logistics capability for sea freight for containerised, palletised and trailer traffic. The designs are based on an all-steel hull powered by gas turbines and water-jets capable of maintaining high service speeds at sea, with large payloads. The detailed design has been developed for two standard vessel configurations, the TG770 and TG560.

The TG770 has a trans ocean capability with a service speed of 38 kt and a range of 3,500 n miles. The vessel has an installed power of 225 MW and offers a payload of 10,500 tonnes with 1,448 TEU or 4,600 m of trailer lane.

The TG560 has short sea capabilities with a service speed of 32.5 kt and a range of 1,000 n miles. The vessel has an installed power of 65 MW and offers a payload of 5,250 tonnes with 736 TEU or 2,500 m of trailer lane.

The cargo stowage is designed for high terminal speed with the flexibilities to handle a wide range of cargo-type simultaneously.

Fryco

Fryco Associates Inc
5420 Waddell Hollow Road, Franklin, Tenessee, 37064-9422, US

Tel: (+1 615) 591 84 55
Fax: (+1 615) 591 84 54
e-mail: frycomar@aol.com
Web: www.members.aol.com/frycomar

Edward D Fry, *President*

Edward Fry established FRYCO in 1978 after 22 years of building experience. He has supervised the design and construction of over 700 commercial, military and pleasure craft including: US Navy high-speed combatants for Navy SEAL Teams; catamarans for commercial and US Army use; monohull yachts up to 50 kt and rig service vessels up to 30 kt. The company maintains a library of mode test results at various institutions and has data available for design study. Fryco designed equipment and wrote training manuals for oil pollution recovery equipment at five major ports and supervised a Middle East licensee shipyard for a US builder for four years.

FRYCO designs vessels up to 100 m and specialises in high-speed craft. The company is experienced in gas-turbine engine packaging and installation as well as diesel and petrol engines. Computer models are used for speed prediction and hydrostatics.

Gibbs & Cox Inc

2711 Jefferson Davis Highway, Arlington, Virginia 22202, United States

Tel: (+1 703) 416 36 00
Fax: (+1 703) 416 36 79
e-mail: info@gibbscox.com
Web: www.gibbscox.com

K E Moak, *Chairman and President*
William Post, *CEO*

New York Office:
3 West 35th Street, New York, New York 10001, United States

Tel: (+1 212) 366 39 00
Fax: (+1 212) 366 79 99

Hampton Office:
2101 Executive Drive, Suite 4c, Tower Box 86, Hampton, Virginia 23666, United States

Tel: (+1 757) 896 02 00

Washington Navy Office:
80M Street SE, Suite 410, Washington DC 20003-3538, United States

Tel: (+1 202) 863 36 00

Founded in 1929, Gibbs & Cox is one of the oldest naval architecture and marine engineering firms in the world. The company specialises in project management, co-ordination, consultation on conceptual and preliminary designs, contract drawings and specifications and construction drawings for commercial or naval ships of the SES/ACV or submerged hydrofoil systems, destroyers, escorts, frigates, corvettes and VTOL/Helicopter carriers.

The company has teamed with Lockheed Martin, Marinette Marine and Bollinger Shipyards to compete for the US Navy's Littoral Combat Ship Program.

Guido Perla & Associates

Colombia Centre, 701 5th Avenue, Suite 1200, Seattle, Washington 98104, US

Tel: (+1 206) 768 15 15
Fax: (+1 206) 768 97 00
e-mail: gpa@gpai.com
Web: www.gpai.com

Guido Perla, *Chairman*
Alex Loudon, *President*
Dan Koch, *Vice President of Engineering*
David Pasciuti, *Vice President of Naval Architecture*
Greg Hughes, *Ship Design Manager*

Well known in the United States and worldwide for the design of a wide variety of workboats and passenger vessels, the company currently has 51 platform supply vessels under construction in the US and Asia for Rigdon Marine, Bourbon Offshore and Trico Marine, while 21 of the company's platform supply vessels are already serving the offshore industry. The company is also developing designs for different workboats including tugs, pilot boats and the E-craft, a unique craft that incorporates lift technology, allowing the vessel to change from a SWATH mode to a barge mode by lowering or raising its central deck.

J B Hargrave Yacht Design

1887 West State Road 84, Fort Lauderdale, Florida 33315, US

Tel: (+1 954) 463 05 55
Fax: (+1 954) 463 86 21
Web: www.hargrave.org

Designer of a wide range of medium to high-speed monohull craft.

Hargrave launched five glass fibre yachts in 2000, ranging from 20.4 to 29 m in size. Sterling, Hargrave's Japanese yard is currently building a 25.8 m mega yacht with three further orders, all constructed of composite and metal, in the pipeline.

Hargrave currently has nine yachts, ranging from 10.4 to 34.5 m, under construction in Taiwan with delivery scheduled in 2003 and 2004.

Hornblower Marine Services Inc

115 E Market Street, New Albany, Indiana 47150, US

Tel: (+1 812) 941 99 90
Fax: (+1 812) 941 99 94
Web: www.hornblowermarine.com

John Waggoner, *President and CEO*

Hornblower provides ship management and consulting services to the high-speed craft industry in the US. The company was formed in 1994 to service the casino industry but soon expanded into other areas. Hornblower currently provides ship management and other consulting services to a number of high-speed craft owners and operators.

Morrelli & Melvin Design and Engineering Inc

4952 Warner Avenue, Suite 205, Huntingdon Beach, California 92649, United States

Tel: (+1 714) 861 13 20
Fax: (+1 714) 840 05 38
e-mail: info@morrellimelvin.com
Web: www.morrellimelvin.com

The company specialises in the design and engineering of high performance craft. The design fleet includes sailboats, high-speed ferries, dive boats, trawlers, sport-fishing boats, cruisers, race boats, power cats, trimarans and US Coast Guard certified passenger vessels. General work services include marine engineering, maritime work for production film companies, legal maritime work and project management. Engineering services include rig analysis, multihull velocity prediction (VPP), structures in composites, aluminium, wood epoxy, computer aided design, ship motion studies, 3-D renderings/animation, weight analysis and cost analysis.

Naval Surface Warfare Center (Carderock Division)

9500 MacArthur Boulevard, West Bethesda, Maryland 20817-5700, United States

Tel: (+1 301) 227 11 42
Web: www.dt.navy.mil

The Naval Surface Warfare Center (Carderock Division) was formed in 1992, bringing together the David Taylor Naval Ship Research and Development Center and the Naval Ship Systems Engineering Station. With its headquarters located in West Bethesda, Maryland, NSWC Carderock, one of eight divisions of the Naval Sea Systems Command (NAVSEA) Warfare Center Enterprise, is a full-spectrum organisation with its principal product area being ships and ship systems. It is the Navy's premier activity providing in-house technical capability in surface and under vehicle Hull, Mechanical and Electrical (HM&E) systems. Functions include research, development, test, evaluation, design, operational support, in-service engineering, maintenance and logistical support of naval platforms. Carderock Division employs approximately 3,400 scientists, engineers and support personnel working in more than 40 disciplines ranging from fundamental science to applied/in-service engineering. Carderock Division is considered the Navy's expert for maritime technology and is chartered by Congress to support America's maritime industry.

Carderock Division has laboratories, model and simulation facilities, at-sea assets and large scale, land-based engineering and test sites at ten locations across the country. The Division continues to solve challenging engineering problems to meet future fleet needs in the following areas: new hull forms, unmanned vehicles, all-electric warships, increased stealth, improved survivability, reduced manning, human systems integration and high-speed ships. Carderock Division addresses all aspects of ships and ship systems with a focus on:

Ship Integration and Design;
Hull Forms and Propulsors;
Machinery Systems and Components;
Structures and Materials;
Environmental Quality Systems;
Vulnerability and Survivability Systems;
Signatures and Silencing Systems.

QuadTech Marine Inc

15 Hunter Trace, Hampton, Virginia 23669, United States

Tel: (+1 757) 850 84 92
Fax: (+1 240) 214 44 97
e-mail: rheffley@cox.net
Web: www.quadtechmarine.com

Robert E Heffley, *President*

The company was formed in May 2004 to commercialise the quadrimaran design. Previous design work, model testing and prototype construction had already been conducted on the concept by other companies.

The patented quadrimaran design uses four wedge-shaped hulls under a main deck that can support a variety of structures. Each hull is divided into several watertight components. The proposed concept provides operating speeds of 45 to 60 knots while consuming less fuel than competing high-speed designs.

An 18 m, 14 t aluminium prototype quadrimaran has successfully completed a test and evaluation programme.

C Raymond Hunt Associates Inc

5 Dover Street, New Bedford, Massachusetts 02740, United States

Tel: (+1 508) 717 0600
Fax: (+1 508) 717 0620
e-mail: info@crhunt.com
Web: www.crhunt.com

John H Deknatel, *President*
Winn Willard, *Manager Commercial and Military Projects*

Raymond Hunt Associates created the deep-vee hull design in 1959 and has been refining the concept since. Specialising in the design of deep-vee fast yachts and commercial and military craft, the company's in-house capabilities include the full range of design and engineering services. The present staff of nine includes naval architects and engineers with expertise in all construction materials and methods of propulsion.

Other commercial applications of the Hunt deep-vee concept have included:

Sandy Hook, a 20 m, 24 kt, aluminium pilot boat for the Port of New York and New Jersey, built by Gladding-Hearn Shipbuilding Corporation
P-150, a 24 m, 27 kt FRP patrol vessel for Western Australia syndicate, now a fisheries patrol vessel in the Solomon Islands
TPB-86, a 26 m, 30 kt FRP water-jet patrol vessel built in Thailand by Technautic Company Ltd
1988, study for Massachusetts Port Authority concerning water transportation in Boston Harbor and Logan Airport focusing on high-speed ferries
1990, design of 22 and 30 kt 18.3 m FRP patrol boats for India
1990 design of 18 m very fast patrol and pursuit craft for Boston Whaler Commercial Products Division.

Recent projects include the 20 m, 25 kt *Golden Gate* pilot boat for San Francisco, the 29 m, 32 kt FRP sportsfishing yacht *Golden Odyssey II*, a 30 m, 25 kt ferry for SeaArk Marine, a 12 m, 40 kt RIB for the US Navy Seals, a 20 m motor cruiser and three 30 m fast motor yachts.

M Rosenblatt & Son Inc

An AMSEC LLC Group Company
Maritime Plaza, 1201 m Street, SE 20003, Washington, D.C., United States

Tel: (+1 202) 608 8400
Fax: (+1 202) 544 5645
Web: www.mrosenblatt.com

Carl Albero, *Chairman and Chief Executive Officer*
B Rosenblatt, *President and Group Manager*
A Baki, *Senior Vice-President Advanced Programs and Marketing*
J C Daidola, *Senior Vice-President of Engineering*
G Thompson, *Vice-President Western Division*

The M Rosenblatt & Son Group, recently merged with AMSEC LLC, is an established naval architectural and marine engineering firm with over 50 years of proven experience in all phases of ship and marine vehicle design.

With offices in ten US cities, the firm is close to the entire shipbuilding community and has a thorough understanding of its problems and needs. Its experience covers programme management, inspection of construction and integrated logistics support, as well as design.

A major portion of the company's design activities has been and is for the US Navy. Completed assignments are of the broadest possible variety, covering research and development, feasibility studies, preliminary, contract and detail design for all classes of major combatants, auxiliaries and high-performance craft. In addition, the company has provided extensive design services for the conversion, overhaul and repair of naval combatants, auxiliaries, submarines, amphibious warfare vessels, supply ships and landing craft.

The service to the maritime industry covers the new and modification design of oceanographic ships, container ships, tankers, general cargo ships, dredgers, bulk carriers, drilling platforms and ships, survey vessels, pipe-laying barges and a great variety of supporting craft.

Typical high-speed marine craft and ACV assignments have included:
Concept designs of a 20 t wheeled ACV and a 60 t wheeled hydrofoil for the US Army;
Preliminary and detail design of a 30 m combat hydrofoil for a foreign army;
Steering and manoeuvring system design for a tracked ACV for Bell Textron;
Surface Effect Ship Advanced Design and Technology handbook for US Navy;
PHM design, feasibility and cost reduction review for US Navy;
Development of Advanced Marine Vehicle (AMV) bibliography for US Navy;
Investigations for structure, stability and safety for a SWATH vessel. Design and analysis support for various fast marine craft.

ARPA Advanced Surface Effect Vehicles

Conceptual studies, parametric studies and propulsion machinery analysis for phase 'O' studies of Advanced Surface Effect Vehicles for the Advanced Research Project Agency. Work performed for the American Machine and Foundry Company.

JSESPO Surface Effect Ship Test Craft

Conceptual and feasibility design studies of candidate SES vehicles for the JSESPO sizing study for second-generation SES test craft in the 1,000 to 3,000 t range. The work included studies of various candidate versions of SES to identify and evaluate the unique operational and design capabilities; technological assessment of various structural materials and systems; and preparation of a proposed development programme with required supporting research and development. Work performed for the Joint Surface Effect Ship Program office.

AAV(P) Advanced Assault Amphibious Vehicle Personnel

Assistance to the FMC Corporation with development of alternative concepts for proposal to the US Navy.

Fast Response Fireboat for New York City

Feasibility design studies and preparation of performance specifications for a 40 m fast response fireboat for the Fire Department of the city of New York, US.

Seacraft Design LLC

61 Michigan Street, PO Box 234, Sturgeon Bay, Wisconsin 54235, United States

Tel: (+1 920) 746 8613
Fax: (+1 920) 746 8615
e-mail: mdp@seacraftdesign.com
Web: www.seacraftdesign.com

Mark D Pudlo, *President*

Seacraft Design started operating in 2007, having purchased the design business of Timothy Graul Marine Design. Building on 25 years of experience, the company specialises in passenger ferries, crew/supply boats, research vessels as well as yachts, barges and towboats.

Major design and consulting projects recently completed or in progress include several aluminium offshore crew/supply boats, ferries, passenger vessels and research craft. In addition the firm assists builders of many other craft with engineering services, regulatory body approvals and working plans.

George G Sharp Inc

22 Cortlandt St, 10FL New York, New York 10407-3518, United States

Tel: (+1 212) 732 28 00
Fax: (+1 212) 732 28 09
e-mail: ccyang@ggsharp.com
Web: www.ggsharp.com

I Hilary Rolih, *Chairman*
A Chin, *President*
Chi Cheng Yang, *Vice President, Naval Architecture*

George Sharp Inc is a leading ship designer in the US. The company developed the world's first container and ro-ro ship designs and the largest passenger ferry design in the US. Sharp also provides technical support services to the US Navy.

SSSCO – Semi-submerged Ship Corporation

417 Loma Larga Drive, Solana Beach, California 92075, United States

Tel: (+1 858) 481 64 17
Fax: (+1 858) 481 72 82
e-mail: tglang@adelphia.net

Thomas G Lang, *President*

SSSCO is an advanced ship design and patent licensing organisation founded in 1979 by Dr Thomas G Lang, the inventor of the Semi-Submerged Ship (S^3). The S^3 consists of two parallel torpedo-like hulls attached to which are two or more streamlined struts. These pierce the water surface and support an above-water platform. Stabilising fins are attached near the aft end of each hull and a pair of smaller fins are optionally located near their forward ends.

SSSCO six passenger recreational version of a planing SWATH 0576685

Semi-submerged ship technology (also known as SWATH) has been proven over the past 30 years by the 190 ton SSP *Kaimalino*, a US Navy-developed range-support vessel which has been operating in the rough seas off the Hawaiian islands since 1975. Following private development, Dr Lang introduced the concept into the US Navy. He led the navy's first research work and initiated and developed the hydrodynamic design for the SSP *Kaimalino*, the world's first high-performance, open-ocean SWATH ship. There are now over 50 SWATH vessels in the world.

The performance features that distinguish S³s from conventional vessels are greatly reduced motions with sustained speed even in heavy seas, lower hydrodynamic drag and reduced power requirements at moderate to high speeds, and superior course-keeping characteristics at all sea headings. A radio-controlled model of a four-strut US SWATH patent (No 4944238) issued in 1990 exhibits particularly low wavemaking drag and low motion in waves.

Planing SWATH Vessels
A high-speed version of SWATH, called a Planing SWATH, was developed and patented by Dr. Lang and son James T. Lang. This new concept is similar to a SWATH, but after reaching the maximum speed of a SWATH, the submerged hulls smoothly rise above the water surface and plane, permitting much higher speeds.

A Planing SWATH provides less motion in waves than a planing catamaran because its twin hulls punch through the higher waves instead of riding on top of them. Motion in waves at pre-planing speeds is extremely low, much like a typical SWATH vessel.

Kaimalino
The SSP *Kaimalino* has operated from near calm conditions to beyond Sea State 6 at speeds up to 25 kt. Its motion is small relative to a conventional monohull of similar payload capacity, either when at rest or under way. The SSP has made smooth transits in 4.57 m swells without any impacts; however, in short, steep 3.7 m waves occasional bow impacts have occurred. No structural damage has occurred, even during storm conditions when 7.6 to 9.2 m high waves were encountered.

In 1993 the two gas turbines used for propulsion were replaced by two 671 TA Detroit Diesel engines to provide greater economy but lower speeds as indicated.

SSP Kaimalino (US Navy) 0093273

Specifications

Length	27.0 m
Beam	14.0 m
Height	9.7 m
Displacement	217 t
Payload	50 t
Max speed	15 kt
Operational speed	13 kt
Range	400 n miles

Swath International Ltd

1661 Northrock Ct., Rockford, IL 61103, United States

Tel: (+1 815) 654 4800
Fax: (+1 815) 654 4810
e-mail: info@swath.com
Web: www.swath.com

Richard Holcomb, *Managing Director*

Swath International Ltd is a ship design company specialising in the technology of SWATH craft. The company markets a wide range of designs covering passenger and ro-ro ferries, cruise ships, hydrographic survey vessels, offshore crewboats and pilot/patrol craft ranging in size from 18 to 100 m.

Design classes include: the 80 m Super Regency Class that can carry 876 passengers and 258 cars at 37 kt or can be configured as a small cruise ship; the 25+ kt, 37 m 4000 Class that can be configured as a ro-ro, passenger-only ferry, offshore crewboat or as a mini-cruise ship; the 25 kt, 20 m Channel Class that can be configured as a 100 passenger-only ferry, pilot transfer vessel or VIP day boat.

A Super 4000 Class vessel, *Cloud X*, a 27 kt, 367 passenger ferry with onboard gaming, has recently been classed by Lloyd's Register of Shipping and issued a Certificate of Inspection by the US Coast Guard. It entered service with Party Line Cruise Company in October 2004 on a route between Florida and Grand Bahama island.

The company is also developing tri-hull small-waterplane-area designs for high-speed operation in the role of passenger ferries, ro-ro and cruise vessels.

The design of SWATH craft offers extremely low ship motions and minimal degradation in speed performance in rough weather as compared to all other multihull craft and monohulls.

Viking Fast Craft Solutions LLC

1069 Wolf Creek Ct, Staunton, IL 62088, United States

Tel: (+1 618) 635 2076
Fax: (+1 618) 635 2228
e-mail: g.smith@vikingfastcraft.com
Web: www.vikingfastcraft.com

Brennan J Smith, *President*

Viking Fast Craft Solutions was founded in 2000, specialising in the development, application and management of high-speed vessel technology and design. Brennan J Smith has over 20 year's experience in the design and construction of small and medium size vessels, including the Mk V Special Operations Craft and the Special Operations Craft Riverine (SOCR). The company owns the US rights to the HYSUCAT™ and HYSUWAC™ hull forms developed by Fast CC in South Africa. The company used this hull form on its recent design and demonstration programme for the USCG Response Medium (RBM). This 13.7 m boat has a top speed of 45 knots and an operational envelope of up to 2.5 m waves and up to 30 knot winds. other recent applications of this hull form include the *Catalina Adventure* fast ferry built by Geo shipyard.

VT Maritime Dynamics Inc

A subsidiary of Vosper Thornycroft (UK) Limited

21001 Great Mills Road, Lexington Park, Maryland 20653, US

Tel: (+1 301) 863 54 99
Fax: (+1 301) 863 02 54
e-mail: sales@maritimedynamics.com
Web: www.maritimedynamics.com

Joseph Kubinec, *General Manager*
Patricia Lane, *Chief Financial Operator*
Chris Swanton, *Naval Architect/Applications Engineer & Sales*
Dave Allen, *Service Manager*
Ruth Schumacher, *Marketing Coordinator*

VT Maritime Dynamics Inc (VTMD) provide motion control solutions to high performance vessels. In addition to VTMD's broad range of stabilisation and motion control products, the company provides specialised engineering consulting services to designers, shipyards and operators.

Founded in 1972 to provide engineering services to the US Navy for the design, development and testing of advanced marine vehicles and their subsystems, VTMD has, since the early 1990's, focused on motions and propulsion efficiency improvements for high performance monohulls, catamarans, Surface Effect Ships (SES), Air Cushion Vehicles (ACV) and foil supported craft. Typical vessel lengths are currently between 20 to 150 m, and displacements are between 50 to 4,000 tonnes. To date a total of 3700 stabiliser and ride control systems have been fitted.

During its 23 year history, VTMD has developed simulations to predict the motions of the following types of craft, operating in random seas, with and without active ride control, at rest and at speed:
SES
ACV
Catamarans
Monohulls
Foil Supported Craft.

VTMD offers engineering services in support of concept, preliminary, contract and detail design of high-speed craft and related systems in the following areas:

Sea-keeping predictions with and without ride control
Control surface design and hull integration
Lift system design for SES, ACV and foil-supported craft
Structural design and stress analysis
Electronic design of machinery, monitoring and craft control systems
Hydrodynamic analysis with state-of-art computational fluid dynamics tools
Vibration analysis
Instrumentation systems
Model fabrication, testing, analysis and full-scale prediction
Full-scale testing and analysis.

VTMD's engineering capabilities result from direct participation in military and commercial high-speed craft development programmes, design work on the company's stabilisation, lift and ride control products (see the Ride Control section of this publication), and research and development initiatives.

SOCIETIES INVOLVED WITH HIGH-SPEED CRAFT

Company listing by country

Belgium
International Association of Public Transport (UITP)

Canada
Canadian Air Cushion Technology Society
Interferry

United Kingdom
Association of Search and Rescue Hovercraft
The Hovercraft Museum Trust
High Speed Vessels Group

United States
The International Hydrofoil Society

Belgium

International Association of Public Transport (UITP)

Rue Sainte Marie 6, B-1080 Brussels, Belgium

Tel: (+32 2) 673 61 00
Fax: (+32 2) 660 10 72
e-mail: info@uitp.org
Web: www.uitp.org

Roberto Cavalieri, *President*
Hans Rat, *Secretary General*

UITP is an international association for public transport and represents over 2900 urban, regional and national transport professionals from more than 90 countries on all continents.

UITP brings together operating companies, local, regional and national authorities, the service and supply industry, and research and academic institutes. It acts as a point of reference and a centre of best practice and benchmarking for the public transport sector, including fast waterborne transport, gathering and monitoring public transport statistics and acting as a knowledge hub on current developments and future trends.

Canada

Canadian Air Cushion Technology Society

c/o Canadian Aeronautics and Space Institute, 350 Terry Fox Drive, Suite 104, Kanata, Ontario, Canada K2K 2W5

Tel: (+1 613) 591 8787
Fax: (+1 613) 591 7291
e-mail: casi@casi.ca
Web: www.casi.ca

Jacques Laframboise, *Chairman*
Geoffrey Languedoc, *Executive Director*

The Canadian Air Cushion Technology Society (CACTS) is a constituent society of the Canadian Aeronautics and Space Institute (CASI). It is devoted to the development and application of air cushion technology, principally with regard to the transportation domain, but also in industry and in other fields where this technology may be of benefit. The goal of the society is to keep its members abreast of developments in the field through information dissemination and exchange. This is done through periodic conferences to which participants from various countries involved in air cushion technology and hovercraft are invited.

Interferry

8-735 Moss Street, Victoria, British Columbia, Canada V8V 4N9

Tel: (+1 250) 592 96 12
Fax: (+1 250) 592 96 13
e-mail: len.roueche@interferry.com
Web: www.interferry.com

C Mitchell McLean, *Chairman*
Jan Eric Nilsson, *President*
Len Roueche, *Chief Executive Officer*

Interferry was originally known as IMTA (International Marine Transit Association). Formed in 1976, the organisation has grown rapidly over the past few years, now consisting of over 200 member companies from 25 countries. Members include owners/operators, ship builders and designers, equipment manufacturers and suppliers, consultants, ship brokers and others with an interest in the ferry industry. Annual conferences have been held recently in Australia, Bahamas, Greece, Netherlands, and the US. Interferry also represents the ferry industry at regulatory forums. It has consultative status at the International Maritime Organisation (IMO) and at the European Community Shipowners Association (ECSA).

United Kingdom

Association of Search and Rescue Hovercraft

Beacon House, Pyrford Road, West Byfleet, Surrey, KT14 6LD, UK

Fax: (+44 870) 705 14 49
e-mail: info@asrh.co.uk
Web: www.asrh.co.uk

Alan Orchard FCCA, *Trustee*
Bill Allen, *Trustee*
Gordon Rominson, *Trustee*
Dr K Davies PhD MSc Bsc, *Trustee*
Tony Drake, *Trustee*
Kip McCallum, *Trustee*

The Association of Search and Rescue Hovercraft (ASRH) is a registered charity formed in 1998 to provide hovercraft to assist in search and rescue missions.

Osprey *5 rescue hovercraft operated by Association of Search and Rescue Hovercraft* 0593249

High Speed Vessels Group

The Royal Institution of Naval Architects, 10 Upper Belgrave Street, London SW1X 8BQ, United Kingdom

Tel: (+44 20) 723 546 22
Fax: (+44 20) 724 569 59
e-mail: hq@rina.org.uk
Web: www.rina.org.uk

Lisa Staples, *Secretary*

The Royal Institution of Naval Architects (RINA) is an international professional institution whose members are involved in all aspects of the design, construction and maintenance of marine structures, including high-speed vessels. The Institution promotes and facilitates the exchange of technical information within the high-speed craft area through its conferences and publications covering all aspects of the sector. The Institution's online technical forum also povides an opportunity for individuals within the sector to raise and discuss matters of interest or concern. The Institution's High Speed Vessels Group brings together key individuals from the international marine industry to discuss current issues and to identify and encourage research, development or design work which is likely to contribute to improved and safer high-speed vessels.

The Hovercraft Society and Hovercraft Museum Trust

HMS Daedalus, Argus Gate, Chark Lane, Lee-on-Solent, Hampshire PO13 9NY, UK

Tel: (+44 23) 92 55 20 90
Fax: (+44 23) 92 55 20 90
e-mail: enquiries@hovercraft-museum.org
Web: www.hovercraft-museum.org

Chris Potter, *Bulletin Editor*
Warwick Jacobs, *Museum Manager*

Formed in 1971, The Hovercraft Society (THS) was the UK constituent of the 'International Air Cushion Engineering Society'. Membership is open to persons engaged in hovercraft related fields and to those having a genuine interest in hovercraft in the UK and overseas. Current membership is drawn from ACV manufacturers, ferry operators, design groups, government departments and agencies, financial and insurance organisations, consultants, journalists and universities in over 21 countries.

THS organises regular meetings at which talks are given on technical, commercial and military design and operation of amphibious craft, SES and other air cushion devices. It also produces a quarterly Hovercraft Bulletin issued to all members, containing information and news on air cushion related topics.

A large collection of references on the subject of hovercraft has been accumulated by the Society. Access to these documents is by prior arrangement with the Museum Manager.

Details of the Hovercraft Museum Trust, now housing a wide range of hovercraft and associated equipment, including the world's fastest ferry, the SRN4 Swift which reached 90 kt in service, can be obtained from THS.

United States

The International Hydrofoil Society

PO Box 51, Cabin John, Maryland 20818, United States

Tel: (+1 301) 519 90 43
e-mail: jr8meyer@comcast.net
Web: www.foils.org

John R Meyer, *President*
Mark R Bebar, *Vice-President*
George Jenkins, *Treasurer*
Kenneth B Spaulding, *Secretary*

A quarterly newsletter is published by the International Hydrofoil Society and is available from the above address.

The membership fee for this society is USD20 per annum for regular membership or USD250 per annum for sustaining membership.

IHS Board of Directors

2004–2007
Sumi Arima
John Meyer
Jeffrey Benson
William White

2005–2008
Jerry Gore
Joel Billingsley
John Monk
Kenneth B Spaulding Jr

2006–2009
Mark R Bebar
George Jenkins
William Hockberger
Denis Clark

BIBLIOGRAPHY

BIBLIOGRAPHY

Introduction

In general only papers specifically concerned with high-speed marine craft presented in the last five years are included in this bibliography. In a few instances some of the papers listed may not be available in published form.

Conference proceedings for five years from 2002 to 2007

Papers presented at the 18th Fast Ferry Conference, 26 to 28 February 2002, Nice, France

European Commission ferry research project – TOHPIC, Jacques Mezieres (The Alliance of Maritime Regional Interests in Europe)
Establishing an international high-speed ferry service; Political influences affecting implementation, William Price (Port Project Manager, City of Rochester, New York, US)
The development of the car carrying fast ferry industry since 1996, and its prospects for the future, Erik Ostergaard (Scandlines AG)
Villum Clausen – Two seasons of successful operating experience, Mads Kofod (Bornsholmstraffiken)
Fast freight ferry services: an alternative to long distance road transport, Dr Alf Baird (TRI Maritime Research Group, Napier University, Glasgow)
The future outlook for fast ferries in the Mediterranean, Peter Wild (GP Wild International Ltd)
Port adaptation for fast ferries, Laurent Monsaingeon (Riviera Ports Authority, France)
Fast ferry security – is it mandatory?, Brigadier Parritt, CBE (International Maritime Security)
Clandestine entry through the UK Channel Ports: partnerships and technology, Keith Rogers (UK Immigration Service)
Onboard gaming – an attractive option for increasing onboard spend, Jan Patrik Svahnstrom (PAF Casino)
Internet Booking engines; the ultimate solution for ferry ticketing, Anders Flensborg (Ocean 24)
Making ticketing pay for itself, Alistair Macleod (Clydefast, UK)
Safety aspects in the early design stage, Abraham Papakirillou (National Technical University of Athens, Greece)
The use of simulation to assess the evacuation of passengers from ships, Michael Starling (BMT Reliability Consultants, UK)
Performance Improvements in new monohull designs, Juan Clements (IZAR Construcciones Navales, Spain)
A new look at the fast monohull as a pax/cars/cargo carrier, Erik Andreassen (Rolls Royce Marine)
High-strength marine grade aluminium alloys; improving the properties of a new generation of 5883, Ronan Dif (Pechiney Marine)
The IZAR High-Deadweight Fast Ro-Pax Ship – An economic solution to the carriage of heavy trucks and cars at 40 knots, Nigel Gee (Nigel Gee & Associates, UK)
Are you missing the boat? The Ekranoplan in the 21st Century – its possibilities and limitations, Graham Taylor (Hypercraft Associates)
A commercial future for the air lifted catamaran, Ulf Tudem (SES Europe AS)
Fast marine vehicles in the Netherlands: past experiences, the status quo and future expectations, Jakob Pinkster (Delft University of Technology)
Hovercraft; a commercially viable technology for the 21st century, Brian Russell (Hovercraft Technical Services)
Better accessability through good design, Ragnar Domstad (Vasttrafik Goteborgsomradet AB).

Papers presented at Ro-Ro 2002, 28 to 30 May 2002, Lübeck, Germany

Short sea european transportation: now and in the future, Michael McGrath (Stena Line, Sweden)
The Anglo/Continental & Irish Sea ro-ro market: will growth continue? Graeme Dunlop (P&O Ferries, London)
Prerequisites for investment in new ro-ro terminals, Bo Lerenius (Associated British Ports, UK)
Creating an integrated European ro-ro service network, Emanuele Grimaldi (Grimaldi Group, Naples)
Ro-ro and short sea freight capacity in the Baltic: market size, market share and anticipated developments among ferry and port operators, Peter Baker (PRB Associates, UK)
Serving a discerning freight market in the Sweden/Germany trade, Hanns H Conzen (TT-Line, Hamburg)
The Via Baltica – options for the future, Martii Miettinen (Transys, Finland)
Danish/Lithuanian platform for the development of Baltic traffic, Peder Gallert Pedersen (DFDS Tor Line A/S Copenhagen)
The reality of the Autostrada Del Mare, Virgilio Cimaschi (Stradeblu, Genoa)
Integrating deep sea and short sea ro-ro with inland operations to achieve total logistics chains, Richard Lawson (Wallenius Wilhelmsen Logistics Services)
Freight tracking: an important new service tool, Stefan Jarlheim (Translogic Consulting & Service Co, London)
EU road haulage industry legislation and its impact on the European ro-ro business
Integrating road, rail and ro-ro, Klaus-Jurgen Uhl (Vienna Consult, Austria)
One port's response to the growth of intermodal traffic in the Baltic, Hans-Gerd Gielessen (Luebecker Hafen-Gesellschaft, Lübeck, Germany)
The merit of own control of ro-ro terminals, and ultimate benefits to shippers
The demands of operators and owners on port and terminal operators, Guilano Gallanti (Intermed)
A best practice study in shore terminal planning, Stephen Osborn (Posford Haskoning, UK)
Scope for the fast ship concept in European coastal trade? Enrique Llorens (IZAR, Spain)
Transforming the economics of fast freight, Richard Regan (Austal Ships, Australia)
Heavy freight catamaran with the accent on economy, Grant R Firth (Vosper Thornycroft, UK)
Taking the fast monohull Ro-Pax beyond 40 knots, Steinar Leganger (Rolls-Royce Nordvestconsult, Norway)
Future trends impacting on ro-ro ports and terminals, Bjorn Hanson (TTS) and Richard Marks (Posford Haskoning, UK)
Ship operating strategies and their implications on terminal operators
Procurement and the ro-ro terminal of the future. Getting what you want. Stephen Osborn (Posford Haskoning, UK)
Planning the ro-ro terminal of the future, Jan Van Beeman (Posford Haskoning, UK)
Tailor-made ro-ro: how far can yards go within the bounds of keen pricing? Eckart Knoth (SSW Bremerhaven, Germany)
The application of through life engineering in ro-ro construction, Markku Kanerva (Deltamarin, Finland)
Moby Wonder: a milestone in Korean cruise ferry design and construction? Anders Hansen (Knud E Hansen, Copenhagen)
Made in Germany: The route to real competitive advantage? Prof Dr-Ing Stefan Krueger (Flensburger Schiffbau, Germany)
Cost, competence and competition in ro-ro new buildings, Dr Werner Schoettelndreyer (German Shipbuilding and Ocean Industries Association)
Pure freight ro-ro systems for United States coastal trades, Asaf Ashar (National Ports and Waterways Institute, Washington DC, US)
Port friendly ro-ro ship designs, Richard Markes (Posford Haskoning, UK)
New design solutions for high-speed ro-ro with minimal wake wash, Claes Kallstrom (SSPA Gothenburg)
New ro-ro ship and barge designs promise great strides in handling efficiencies, Haakon Lindstad (Marintek, Norway)
Competing in the ro-ro market with high-tech solutions, Kai Levander (Kvaerner Masa-Yards Technology, Finland)
Latest developments in cargo access equipment for ro-ro vessels, Lars Brath (MacGregor RoRo, Sweden)
IPSI: simplifying the ship-shore interface, Goran Johansson and Per Fagerlund (Hamworthy KSE, Gothenburg)
Ferrycat: a new concept in short-haul double-enders, Ivar Myklebost (Fjellstrand Ferrycat AS, Norway)
Advances in medium-speed diesel engineering that can best benefit ro-ro ferry design and operations, Stefan Rother (Caterpillar Motoren, Germany)
Enhancing the manoeuvring performance of ro-ro vessels, Peter Tragardh (SSPA, Gothenburg)
An integrated European approach to safe European ro-ro passenger ship design, Pierre Sames (Germanisher Lloyd)
Stowaways: what practical measures can ro-ro operators take to improve security, Brigadier Brian Parritt CBE (International Maritime Security, UK)
The special requirements of ro-ro ship management and crewing, (V Ships)
International safety rules and their impact on the operation of ro-ro vessels.

Papers presented at the Seatrade Med Cruise & Ferry Convention, 11 to 13 September 2002, Genoa, Italy

European ferry market & technology evolution, Girgio Arena (Fincantieri Cantieri Navali, Italy)
High street lessons for ferry retail & catering, Michael Foley (Foley Cooke)
A view from the southern Tyrrhenian Sea, Hans Isler (Grimaldi Group)
New generation gas turbines for the fast ferry industry, Gerraint Lloyd (Rolls-Royce Plc)

Pentamaran – evolution or revolution? Juan Antonio Moret (IZAR, Spain)
A seismic change in maritime security, Major Doc Addison (International Maritime Security)
Anti-pollution systems of the latest generation, Vittorio Antonelli (ISIR)
Solid and liquid waste management – the complete green ship philosophy 21, Jochen Deerberg (Deerberg-Systems GmbH)
Operational aspects of cruise ship's safety and environmental protection: can classification societies help? Mario Dogliani (RINA SpA)
EnviroEngines: the cost-effective way to environmental friendliness, Daniel Paro (Wärtsilä Corporation)
Implementation of the Athens Convention, Ian Rose (Watson, Farley & Williams)
Security in Mediterranean ports: the MedCruise Project, Ana Karina Santini (MedCruise Association)
Maritime security lessons learned, Capt Ted Thompson (International Council of Cruise Lines)
Safety and Comfort on-board – a Modern Approach for an Old Problem, Allessandro Cappiello (Rodriquez Marine System Srl)
Passenger and crew accommodation comfort: Optimisation of the Shipboard environment, John Carlton (Lloyd's Register)
Designing marine interiors for the brand generation, Mark Hilferty (McNeece)
Cruise vessels and cruise ferries come closer together, Giuseppe De Jorio, (Studio De Jorio Srl)
The need for comfort onboard: the outfitter's answer to the architectural requests, Fabio Mosetti (Gerolamo Scorza SpA)
Podded drives in the cruise & ferry industry, Fritz Gaede (Siemens AG)
The cruise ship order book, Anne Kalosh (Seatrade Cruise Review)
European ship design and technology, Theodore C Kontes (Festival Cruises Inc)
Seven seas voyage: top luxury – large cruise ship innovation design, Raffaello Putti (V Ships Leisure)
Impact of a newbuilding project in the growth of a shipyard, Giangiacomo Zino (T Marriotti Spa).

Papers presented at the High-Speed Marine Vehicles Conference, 11 to 13 September 2002, Naples, Italy

Competition in short sea shipping between different ships, M Catalani (Università degli Studi Parthenope)
The special research program for the enhancement of Italian short sea shipping, L Mor B Della Loggia (CETENA)
European fast seas traffic (short sea shipping). Some suggestions for design of prototypes, A Gutierrez (ANIE)
Sea traffic control as enhancement of maritime navigation safety, P Fadda *et al* (Università di Cagliari, CETENA)
Are port terminal's facilities limiting potential benefits from fast ships? C Ferrari *et al* (Università di genova)
Review of technoeconomic characteristics of fast marine vehicles, G Zaraphonitis *et al* (National Technology University of Athens)
Application of Coanda effects to the hydrofoil ship, J K Oh *et al* (Seoul National University)
Improvement in resistance performance of a high-speed boat with air lubrication, J Jang *et al* (Seoul National University)
Analytical formulation for calculating thrust of a slant-inlet water-jet, V Quaggiotti and E Benini (Università di Padova)
Calm water resistance of semi-displacement high-speed catamarans through regression analysis, A Schwetz and P Sahoo (Wavemaster International, Australia)
Propeller wake velocity and propeller field, G Giordano *et al* (INSEAN)
Powering prediction on hybrid propulsion concepts for fast motor yachts, T Verwoest (MARIN)
Further investigations into the hydrodynamic performance eof the AXE-Bow concept, J A Keuning and J Pinkster (Technical University of Delft)
Experimental and CFD investigation of viscous flow around complex configurations of bulbous and fin keel, M Elefante *et al* (Marina Militaire Italiana)
One or two rudders? A dilemma in fast twin screw vessel design, G Avellino and A Coan (Fincantieri)
Multidisciplinary design optimization of hull form for naval surface ship, R Dattola *et al* (Marina Militaire Italiana)
Experimental and numerical study on the efficiency of trimaran configuration for high-speed very large ships, F Capizzano *et al* (Università degli Studi di Napoli Federico II)
An electric driven sub-jet for hydrofoil propulsion, E Benini *et al* (Università di Padova)
Information Technology for efficient ship's management, A Stefani (ABB)
The HSF-Arcos research project, PGM van der Klugt *et al* (Imtech Marine Offshore)
Analysis of the working conditions of a high-speed water-jet catamaran, F Balsamo *et al* (Università degli Studi di Napoli Federico II)
New diesel engine series 8000 for fast vessels, G Haussman (MTU)
The journey to the RK280, M Siberry (MAN B&W Diesel)
Design and testing of a gearbox vibration monitoring system on high-speed ferries, L Guarino and L Lecce (Università degli Studi di Napoli Federico II)
Multi-attribute concept design model of search and rescue vessels, I Grubisic *et al* (University of Zagreb)
Fast seaworthy express trimaran for coastal routes, I Mizine *et al* (SAIC)
An optimization method for high-speed craft design: technical problems and project financing, C Rizzo (Università di Genova)
Intact stability of SWATH ships, Y M Welaya *et al* (Alexandria University)
The Italian Navy projects for minor vessels, C Boccalatte (Marina Militaire Italiana)
Deadrise influence on rolling dynamic in planing and semi-planing range, F Balsamo *et al* (Università degli Studi di Napoli Federico II)
Seakeeping of motoryachts, R Dallinga and G Gaillarde (MARIN)
Comparison of different prediction tools for seakeeping performance of high-speed vessels, A Esteves *et al* (University of Newcastle)
Sea-keeping evaluation of fast monohull ferry, R Folso and J J Jensen (Registro Italiani Navale)
Global motion calculation for a high-speed catamaran in oblique seas and problems of full scale validation, G Guzsvany and P Sahoo (University of Tasmania)
Seakeeping optimisation of fast vessel by means of parametric modelling, F Valdenazzi *et al* (CETENA)
A system for the on-line evaluation of people comfort aboard, R Fazzeri *et al* (CETENA)
A preliminary study for trimaran ship structural design, T Coppola and M Mandarino (Università degli Studi di Napoli Federico II)
Transient dynamic analysis and slamming bending moment evaluation of a fast ferry, A Moggia Cappelletti *et al* (CETENA)
A rational approach to the structural design of large fast ships, L Sebastiani *et al* (CETENA)
Structural integrity of large HSC. Evaluation and criticity of main modes of collapses, E Thiberge (Bureau Veritas) Structural design of last generation ro-ro passenger displacing vessels, C A Zamburlini (Fincantieri)
Case study on the structural failure of a high-speed propeller, S Lee (Chungnam National University)
Local and global effects of wet deck slamming on a high-speed catamaran, O M Faltinsen and C Ge T Moan (Norway University of Science & Technology)
The economics of short sea shipping, E van de Voorde and H Meersman (University of Antwerp)
Slamming loads on a small high-speed catamaran, G Thomas *et al* (University of Tasmania)
Impact of imperfections on strength and reliability of marine structures, A Mansour (University of California).

Papers presented at HIPER 02 Third International Conference on High Performance Vehicles, 14 to 17 September 2002, Bergen, Norway

Aerodynamic flow computations for a superfast ferry, D Schmode *et al*
Aerodynamic flow simulations for an SES employing virtual reality post-processing techniques, O Lindenau *et al*
Automation control of the passenger eKranoplan craft, G V Alexandrov and V V Strelkov
Wind tunnel measurements of 6-component forces and moments acting on a model of new wing in surface effect craft, H Akimoto *et al*
The effect of length in a fast boat, F Perez
A scale effect check about resistance estimation of planing boats, Y Yoshida
High speed craft: investigation on the resistance towing test procedure, C Bertorello *et al*
Wave resistance of semi-displacement high-speed catamarans through CFD and regression analysis, A Schwetz and P Sahoo
Identification of hydrodynamic coefficient from standard manoeuvres, R Depascale *et al*
Manoeuvring aspects of fast ships with pods, S Toxopeus and G Loeff
An investigation into the course stability and control of a fast, pod drive Ro-Pax, D Clarke and M Woodward
Rudder loads for a fast ferry at unusually high rudder angles, O M El Moctar and V Bertram
Research and development for an advanced ride control system for a high-speed monohull, F v Walree and PGM Klugt
Hydrodynamic design of high efficiency fin stabilisers, S Brizzolara
On the effect of viscous forces on the motions of high-speed hulls, E Begovic *et al*
Numerical prediction of high-speed catamaran behaviour in waves, D Bruzzone and P Gauleni
Hydrodynamic loads based on non-linear methods, O Nakken
Seakeeping performance of a wave-piercing frigate, R Pereira and C Wagner
Slamming induced pressures on HSC, R Folso and F Torti
Impacts of composite cylinders, P B Gning *et al*
Full-scale measurements of global loads on SES vessels built in FRP sandwich using networks of fibre optic strain sensors, A E Jensen *et al*
Energy dissipation in sandwich structures during axial crushing, J Urban
Exploiting the non-linear response of sandwich panels in high-speed craft design, B Hayman *et al*
Finite element structural modelling of a composite-material multihull, M Salas *et al*
Design criteria for the use of high tensile steel in fast ships, T Gosch

High speed craft in service, K Fach
Fatigue of high-speed vessels including in-service experience, A O Lungstad
Evacuation analysis as a design tool, A Hellesoy
Safety assessment for equivalent design solutions, S Nusser *et al*
A risk based design methodology used for determining risk against capsize in damaged condition, J Tellkamp
Safety aspects of HSC – an overview, D Vassalos
Deisgn of military and commercial SES, G Rise
Resistance and structural modulus for multicriterial trimaran hull design, M Barone and C Bertorello
Power prediction and hydrodynamic improvement for fast conventional ships, M Fritsch and V Bertram
Advanced marine platforms – an overview, R Azcueta and V Bertram
Opportunities and limitations of using database/regression analysis of high performance craft and their hybrids, R MacGregor
A new tool for assessing the cost-benefit of emerging technologies for naval and commercial ships, B Lavis
Wake wash reduction on a new monohull design, F Clemente
Designing low wash monohulls, I Østvik
Wake wash minimisation by hull form optimisation using artificial intelligence, K Koushan *et al*
Experimental investigation on propeller wash using laser doppler anemometry in cavitation tunnel, D Wang *et al*
The excitation of ships by wave wash generated by high-speed craft, B Elsässer and T Whittaker
The environmental impacts of fast ferry wash on rocky shore communities in Loch Ryan, Scotland, S Hotchkiss *et al*
Environmental performance of land and marine transportation in inland shipping – impact of ship speed, K Hasegawa and K S Iqbal
Hi-speed revolution, A Osmundsvaag
Construction of the world's largest solar powered catamaran, B Bolinger and R Stadelmann
Paddle Steamer DS Gallia on Lake Lucerne: High-tech of 1913 or 2002? W Schäfer
An experience from commencing the Bergen-Stavanger HSC route, C Bovig.

Papers presented at the CACTS International Symposium on Air Cushion Technology, 16 to 18 September 2002, Indiana, United States

Flood rescue in Japan, Mitso Yamashita (Nihon Hovercraft Manufacture Corporation, Japan)
The 25 Heron: design and development of a competitive Formula 25 Hovercraft, Graham Spencer and Kevin Pratt (US)
Hydrodynamic lift effects with forward facing skirt systems, Ralph Dubose (US)
Evolution of the home-built entry level hovercraft, Bob Windt (Universal Hovercraft, US)
Manufacturing standards in Canada, Mike Dua (Transport Canada)
Design philosophy behind the development of the AVC-10, Dean Poliee (US)
An overview on curriculum development, exhibit design and youth activities related to hovercraft technologies
Aspects & development challenges pertaining to neoteric light hovercraft, Rob Wilson (Neoteric Affiliates Pty Ltd, Australia)
The hovercraft's future, Warwick Jacobs and Brian Russell (UK)
SAR hovercraft operations, Captain Brain Wotton (Canadian Coast Guard, Canada)
Recent developments in gimbal fan technology for air cushion vehicles. The Aeromobile-Aeroduct System of automated transportation and the AirTrack crawler tractor all-terrain vehicle, William R Bertelsen and John Grant (Aeromobile Inc, US)
Hovercraft – a proposal for the 21st Century, John Mitchell (Mitchell Research, US)
Stability, noise & efficiency study, 900 kg test vehicle, Bjorn-Ake Skold (Bjorn-Ake Skold Design AB, Sweden)
Griffon 8000TD, design and build, John Gifford (Gifford Hovercraft, UK)
Commercial certification of hovercraft in Canada, Ray Crick, (Transport Canada)
Ice breaking in the St Lawrence, David Dickins (DF Dickins, US) and Capt Daniel L'Heureuz (Canadian Coast Guard Hovercraft Unit)
Evolution of air cushion technology – from hovercraft to super high-speed, Amphibious Wing-in-Ground-Effect (AWIG) craft, Liang Yun (MARIC, China)
The Hovercraft Museum and its place in current hovercraft technology, Brian Russell (Hovercraft Museum, UK)
LCAC service life extension program, Leonard Maxwell (LCAC, US)
Current developments of Ekranoplans (WIG) and opportunities for introducing passenger WIG transportation in the 21st Century, D N Sinitsyn (Amphibious Transportation Technologies, Russian Federation)
First passenger Ekranoplan (WIG) Aquaglide: small wonder-enormous pleasure, D N Sinitsyn and R A Nagapetyan (Amphibious Transportation Technologies, Russian Federation)
Hovercraft ferry operations, Robert Wagner (Hover-United States).

Papers presented at the RINA High-Speed Craft Technology and Operation Conference, 11 to 12 November 2002, London

Ferries and fast craft; A North American Perspective, Bruce Hutchison (The Glosten Associates Inc, US)
The use of tactile navigation cues in high-speed craft operation, Trevor Dobbins & Shaun Samways (QinetiQ Centre for Human Sciences, UK)
Ocean transits in a 50 m, 45 kt catamaran – the minimisation of motions and speed loss, Ed Dudson & James Roy (Nigel Gee and Associates, UK)
Motion prediction and sea trial measurements for comfort rating of a catamaran, Axel Köhlmoos & Thomas Schellin (Germanischer Lloyd, Germany)
Prediction of motion sickness in high-speed craft, Christos Verveniotis & Osman Turan (Ship Stability Research Centre, Universities of Glasgow and Strathclyde, UK)
A conceptual model for a shipboard motion monitoring and operator guidance system, Paul Crossland & Michael Johnson (QinetiQ, UK)
High-speed ro-ro designs, Juan Antonio Moret (Izar Construcciones Navales, Spain)
Air crew training vessels: a study of the synergy between fast patrol and passenger craft, Andrew Bailey & Nigel Warren (FBM Babcock Marine Ltd, UK)
An investigation of the comparative costs of high-speed sealift versus airlift, Matthel Williamsen *et al* (Naval Surface Warfare Centre, Caderock Division, US)
Theoretical study of the tradeoff between stabilizer drag and hull motion, Lawrence Doctors (The University of New South Wales, Australia)
The IZAR pentamaran – tank testing, speed loss and parametric rolling, Nigel Gee & Ed Dudson (Nigel Gee & Associates, UK) and Jose Maria Gonzales, (IZAE Construcciones Navales, Spain)
Design building and trials of two high-speed landing craft prototypes LCM-1E for the Spanish Royal Navy, Jesús Alonzo Pérez *et al* (IZAR Construcciones Navales, Spain)
Speed, seakeeping and efficiency (Fast, Safe and Cheap – FSC): Three into one, Valentin Tugolukov (Feodosia Shipbuilding, Ukraine)
Rough seas and small passenger ferries: The Dames 3737 SWATH Solution, Rik Vrugt, Sovereign (Marine Services, NV, Belgium), Piet Hein Noordenbos (Damen Shipyards BV, The Netherlands), Ed Dudson (Nigel Gee and Associates, UK)
Apply Tanaka's nomogram to the PT thrust estimation, Yasushi Yoshida (Wessex Institute of Technology, UK)
Motion of wave-piercing trimaran: test data and a comparison with the motion of other vessels, Victor Dubrovsky (Independent Designer, Russian Federation).

Papers presented at FAST 2003, 7 to 10 October 2003, Naples, Italy

The fast ship fever...from Trondheim to Ischia, The Gulf of Naples, Kjell OHolden, Chairman of the FAST2003 International Committee (Marintek – Norway)
High-speed craft – regulations and efforts to minimise accidents, Eugenio Sicurezza, Head of Italian Coast Guards
Simulation based resistance and seakeeping performance of high-speed monohull and multihull vessels equipped with motion control lifting appendages, Paul D Sclavounos (speaker), Shiran Purvin, Talha Ulusoy, Sungeun Kim (MIT, Cambridge US)
Hydrodynamic developments of 154 m class ro-ro passenger ferry, Soon Ho Choi, Jung Joong Kim, Gun Ho Lee, Se Eun Kim, Sung Mok Ahn (Samsung Heavy Industries Co Ltd, Korea)
A comparative analysis of the resistance qualities of semi-displacement high-speed monohull forms, Jacques B Hadler and Richard Vanhooff (Webb Institute Glen Cove, New York-US)
A non-linear time domain simulation for predicting wet-deck slamming of catamaran Rong Zhao (Marintek Trondheim, Norway)
Series 64 parent hull displacement and static trim variation, Gabor Karafiath, Todd Carrico (Navy, Naval Surface Warfare Centre, Carderock Division, US)
A new generation of large fast ferry – from concept to contract reality, Tony Armstrong and Kjell Holden (Austal Ships-Australia, Marintek-Norway)
Design curves for selecting hat-type stiffeners in FRP boats, Nicholas Tsouvalis, George Spanopoulos (University of Athens, Greece)
Local and global finite element analysis of an adhesively bonded aluminium structure, Xinyu Wang[1], Alois Starlinger[1], Augusto Moggia Cappelletti[2], Giovanni Cusano[2] (1- Alcan Alesa Eng Ltd Switzerland, 2- CETENA Italian Ship Research Centre, Genoa, Italy)
Near and distant waves of fast ships in unlimited and limited depths, Dario Bruzzone (Invited Speaker), Stefano Brizzolara (DINAV University of Genoa, Italy)
The wash waves generated by high-speed ships prediction and directive influence, Anatoly Lyakhovitsky[1], Li Yun-bo[2], Huang De-bo[2], (1- St Petersburg State Marine Technical University, Russian Federation, 2- Harbin Engineering University, China)
Noise reduction for improved comfort onboard high-speed craft, Kai A Abrahamsen (DNV, Norway)
Hullform analysis and optimisation for a fast Ro-Pax ferry, Jochen Marzi, Scott Gatchell (Hamburger Schiffbau Versuchsanstalt, Germany)

A critical survey on performance assessment for classical and advanced ships, Luigi Iannone (INSEAN, Italian Model Basin, Rome, Italy)

The effect of demihull separation on the frictional resistance of catamarans, Tony Armstrong (Austal Ships, Australia)

Performance and cavitation analysis of a water-jet system on a cavitation tunnel, Klaas Kooiker1, Tom Van Terwisga[1], Rob Verbeek[2], Peter Van Terwisga[3] (1- MARIN, The Netherlands, 2- Wartsila Propulsion Netherlands, Royal Netherlands Navy, The Netherlands)

Evolution of high-speed ferries and Fincantieri experience: from the built dreams to the dreams to be built, Giorgio Arena (Invited Speaker- Fincantieri SpA Italy)

An investigation of the options for evaluating the compressive strength and reliability of aluminium stiffened panels, Mathew Collette, Atilla Incecik (University of Newcastle upon Tyne UK)

On the preliminary analysis for trimaran structural design, Tommaso Coppola, Masino Mandarino, Fabio Simeone (DIN University of Naples Federico II Italy)

Choosing affordable requirements and technology options for future surface ships, David R Lavis, Brian G Forstell (band, Lavis & A, US)

Technical and economic investigations of fast ferry operations (University of Athens, Greece and University of Southampton, UK)

Spin-Hsv: towards an improved high-speed maritime transport, Sverre Alvik, Richard Clements, Marilyne Guevel, Arnulf Hader, Stig Vaular (DNV-Norway, Marinetech South Ltd-UK, Mettle Groupe-France, Institut fur Seeverkehrswirtschaft und Logistik-Germany, Hardanger Sunnhordlandske Dampskipsselskap ASA-Norway)

Fast shuttle services for cruise vessels, Jari Sirvio[1], Olli Salmela[1], Egil Jullumstro[2] (1-Kvaerner Masa-Yards Technology-Finland, 2- Norwegian Marine Technology Research Institute (Marintek)-Norway)

The accuracy of wave pattern resistance determination from wave measurements in longitudinal cuts, Nastia Degiuli, Andreja Werner, Zdravko Doliner (University of Zagreb, Croatia)

Influence of hull shape on the resistance of a fast trimaran vessel, Antonio cardo[1], Marco Ferrando[2], Carlo Podenzana-Bonvino[2] (1- University of Trieste, Italy, 2- University of Genoa, Italy)

Hydrodynamic performance and exciting force of a surface piercing propeller, Kazuo Nozawa (Osaka University, Japan)

Drag reduction on a high-speed trimaran, Robert Latorre, Aaron Miller, Richard Phillips (University of New Orleans-US)

Comparison between the predicted and measured response of a planing vessel, GR Finch (Industrial Research Limited)

A crash-stop study for an azipod-propelled vessel, Vladimir Boushkovsky[1], Aexander Poustoshny[1], Andrey Vasiliev[1], Tomi Veikonheimo[2] (1- Krylov Shipbuilding Research Institute, Russian Federation, 2- Tomi Veikonheimo, ABB, Finland)

A new integrated methodology in course keeping and manoeuvring, Roberto Balestrieri, Antonio Scamardella, Mario Voltaggio (University of Naples Parthenope, Italy)

One or two rudders? A dilemma in fast twin screw vessel design, Gennaro Avellino, Alessia Coan (Fincantieri, Cantieri Navali Italiani SpA, Trieste Italy)

Analysis of the operations of a catamaran in service in the Neopolitan Gulf; comparison between full scale trials and towing tank tests. (Flacio Balsamo, Salvatore Miranda, Claudio Pensa, Franco Quaranta (DIN University of Naples Federico II Italy)

Hydrodynamic aspects of interest for the development of an integrated ride control system for high-speed monohulls, Albert Jurgens, Frans van Walree (MARIN, The Netherlands)

Application of risk based methodology to the preliminary design phase of high-speed craft, Severine Delautre[1], Richard Birmingham[1], Jean-Claude Astrugue[1], Jonathan McGregor[2] (1- Bureau Veritas, France 2- University of Newcastle upon Tyne, UK)

Optimising boarding and de-boarding processes with AENEAS, Ulf Petersen[1], Tim Meyer-Konig[2], Daniel Povel[1], (1- Germanischer Lloyd AG,Germany, 2- TraffGo, Germany)

Dynamic stall and cavitation on stabiliser fins and their influence on the ship behaviour, Guilheim Gaillarde (MARIN, The Netherlands)

On the application of risk analysis for short sea shipping at very high-speed, Massimo Figari, Paola Gualeni, Enrico Rizzato (University of Genoa, Italy)

Development of a crew seat system for high-speed rescue craft, Bob Cripps[1], Simon Rees[2], Holy Phillips[2], Colin Cain[2], David Richards[2], Jac Cross[2] (1- Royal National Lifeboat Institution UK, 2- Frazer Nash Consultancy Limited (FNC) UK)

Hydrodynamic aspects of high-speed vessels, Faltinsen OM[1] (Speaker), Claudio Lugni[2], Maurizio Landrini[2] (1- NTNU University of Trondheim, Norway, 2- INSEAN Italian Ship Model Basin, Rome, Italy)

Fast ships and boats-new concepts and designs, Nigel Gee (Nigel Gee & Associates Ltd UK)

The optimisation of trimaran sidehull position for minimum resistance, Lawrence J Doctors (invited speaker), Robert J Scrace (The University of New South Wales, Australia)

Transom hollow prediction for high-speed displacement vessels, Simon W Robards, Lawrence J Doctors (The University of New South Wales, Australia)

Fully non-linear wave-wave interactions between a high-speed vessel and incident waves (Ray-Qing Lin, Arthur M Reed, William Belknap David Taylor Model Basin, NSWC/CD, US)

Non-linear response of laterally loaded FRP sandwich panels for high-speed craft, Brian Hayman[1], Harold Osnes[2], Dag McGeorge[1] (1- DNV, Norway, 2- University of Oslo and DNV, Norway)

Service lifetime assurance for ship structures of aluminium alloys, Oleg M Paliy, Evgeny Ya Voronenok, Semion D Knoring (Krylov Shipbuilding Research Institute, Russian Federation)

Structural analysis of a trimaran fast ferry, Dario Boote, Tommaso Colaianni, Cesare Rizzo (DINAV University of Genoa, Italy)

Vibrational behaviour of a deck panel of a fast ferry, Alberto Ferrari, Enrico Rizzuto (DINAV University of Genoa, Italy)

Effect of sea, ride controls, hull form and spacing on motion and sickness incidence for high-speed catamarans, Michael Davis, Damien Holloway (University of Tasmania – Australia)

Evaluation method of ride control system for fast craft from the viewpoint of passenger comfort, Ritsuo Shigehiro[1], Takako Kuroda[2], Harald E Kagaruki[1] (1- University of Kagoshima, Japan, 2- Hokkaido University, Japan)

Automatic hull form optimisation towards lower resistance and wash using artificial intelligence, Kourosh Koushan (Marintek, Norway)

On the development of structural rules for the classification of naval vessels, Claudio Boccalatte[1], Dino Cervetto[2], Roberto Dattola[1], Stefano Ferraris[3], Rasmus Folso[2], Sergio Simone (1- Italian Navy General Staff, Rome, Italy, 2- RINA SpA, Italy, 3- Fincantieri, Genova, Italy)

Towards structural design of high-speed craft based on direct calculations, Torgeir Moan (NTNU, University Of Trondheim, Norway)

Developments on a probabilistic risk/cost model for large scale flooding consequence analysis of high-speed monohulls, Dracos Vassalos (speaker), Dimitris Konovessis, Daria Cabaj (University of Glasgow and Strathclyde)

A numerical method for prediction performance of hydrofoil-assisted catamarans, Michael Andrwartha[1], Lawrence Doctors[1], Kishore Kantimahanthi[2], Paul Bradner[2] (1- The University of New South Wales, Sydney, Australia, 2- Australian Maritime College, Australia)

Numerical prediction of steady flow around high-speed vessels with transom stern, S X Du[1], DA Hudson[2], WG Price[1], PTemarel[1], YS Wu[2] (1- University of Southampton, UK, 2- China Ship Scientific Research Centre, China)

A study on wave resistance of high-speed displacement hull forms in restricted depth, Prasanta K Sahoo, Lawrence J Doctors (Australian Maritime College-Australia, The University of New South Wales, Sydney-Australia)

A numerical model for hydrodynamics of planing surfaces, David Battistin, Allesandro Iafrati (INSEAN Rome, Italy)

Prediction of global wave induced loads for structural analysis of IZAR high-speed pentamaran, Axel Kohlmoos[1], Francisco Viejo de Francisco[2], Antonio Perez de Lucas[2] (1- Germanischer Lloyd, Germany, 2- IZAR Technical Direction, Spain)

Determination of design slamming loads on bow doors for ro-ro ships, Uh Wang, Xuekang Gu, Jinwei Shen (China Ship Scientific Centre, China)

Design load of bow flare impact pressures on container ship, Ryoju Matsunami[1], Yoshitaka Ogawa[2], Makato Arai[3], Atsushi Kumano[1] (1- Nippon Kaiji Kyokai, Japan, 2- National Maritime Research Institute, Japan, 3-Yokohama national University, Japan)

Estimation method for probability density function of the water impact pressure of post Panamax container ship due to bow flare slamming, Yoshitaka Ogawa[1], Ryoju Matsunami[2], Makiko Minami[1], Katsuji Tanizawa[1], Makoto Arai[3], Atsushi Kumano[2], Ryuji Miyake[2], (1- National Maritime Research Institute, Japan, 2- Nippon Kaiji Kyokai (Class Nk), Japan, 3- Yokohama National University, Japan)

Stern flap: an economical fuel-saving, go-faster, go-farther device for the commercial vessel market, Dominic S Cusanelli (NSWCCD, US)

The hovercraft starts the new millennium with increased utilisation, Brian J Russell (Hovercraft Technical Service, UK)

Prospects of unconventional hydrodynamic configurations for large high-speed marine ships, Victor N Anosov, Margarita V Galoushina, Alexander V Poustoshny, Sergey D Prokhorov, sergey O Rozhdestvensky, Andrey V Sverchkov (Krylov Shipbuilding Research Institute, Russian Federation)

Study on the compound effects of interceptor with stern flap for two fast monohulls with transom stern, Jing-Fa Tsai[1], Jeng-Lih Hwang[2], Shean-Kwang Chou[2] (1- National Taiwan University, Taiwan, 2- United Ship Design And Development Centre, Taiwan)

Ship design optimisation with variable fidelity models, Daniele Peri, Lanfranco Benedetti (INSEAN The Italian Ship Model Basin, Rome, Italy)

Turbulent steady flow around Kvlcc2 hull form using multiblock domains, Francesco

Capizzano, Agostino De Marco (University of Naples Federico II, Italy)

Simulation of floating body motions in viscous flow, Matthias Klemt[1], Milovan Peric[2], Gerhard Jensen[1] (1- Technical University of Hamburg, Germany, 2- CD Adapco Group, Nuremberg, Germany)

Model trials for studying hydrodynamic force components in following seas, Olov Lundback[1], Jianbo Hua[2] (1- Chalmers Lindholmen, Sweden, 2- SSPA, Sweden)

Wetdeck slamming of high-speed catamarans with a centre bow, James Whelan[1], Damien Holloway[1], Tim Roberts[2], Michael Davis[1] (1- University of Tasmania, Australia, 2- Revolution Design Pty Ltd, Australia)

Rational determination of non-linear design loads for advanced vessels, Wouter Pastoor, Trym Tveitnes (DNV, Norway)

Asymmetric water entry of a wedge (Yuri A Semenov, Alessandro Iafrati – Institute of Technical Mechanics – Ukraine, INSEAN Italian Ship Model Basin, Rome- Italy)

Rigid body dynamic hull bending moments, shear forces and Pcm in fast catamarans, Damien Holloway, Michael Davis, Giles Thomas (University of Tasmania, Australia)

The effects of length on the powering of large slender hull forms, Nigel Gee, James Roy (Nigel Gee & Associates Ltd, UK)

Optimisation of the wave-making characteristics of fast ferries, Justus Heimann, Stefan Harries (Technical University of Berlin, Germany)

Self-propulsion model tests of a wing-in-surface-effect-ship with canard configuration, Hiromichi Akimoto[1], Takahiro Taketsume[2], Keiichiro Iidal, Syozo Kubo[1], (1- Tottori University, Japan, 2- Tsuneishi Shipbuilding Co Ltd, Japan)

Containership evaluation and mega-container carriers development, Branislav Dragovic[1], Zoran Radmilovic[2], Vladislav Maras[1] (1- University of Montenegro, Maritime Faculty, Serbia and Montenegro, 2- University of Belgrade, Faculty of Traffic and Transport Engineering, Serbia and Montenegro)

Prediction of offshore and inshore wave spectra: a case study, Guido Benassai (University of Naples Parthenope)

Prediction of non-linear motions and wave loads of high-speed monohulls in oblique waves, Forng-Chen Chiu[1], Wen-Chuan Tiao[1], Yen-Hwa Lin[2], Chih-Chung Fang[3] (1- National Taiwan University, Taiwan, 2- Naval Shipbuilding and Development Centre, Taiwan, 3- United Ship Design and Development Centre, Taiwan)

Time domain prediction of steady and unsteady motions of high-speed crafts, Fuat Kara, Dracos Vassalos (University of Glasgow and Strathclyde, UK)

Experience on the application and validation of theoretical models for the seakeeping prediction of fast ships, Augusto Moggia Cappelletti[1], Enrico Pino[1], Luca Sebastiani[1], Claudio Boccalatte[2], Michel Viviani[3], Dario Buzzone[4], Stefano Brizzolara[4] (1- CETENA Italian Ship Research Centre, Genoa, Italy, 2- Ministero Difesa Marina, Rome, Italy, 3- Fincantieri Naval Vessels Business Unit, Genoa, Italy, 4- DINAV University of Genoa, Italy)

A parametric study for the structural behaviour of a lightweight car-carrier deck, Junbo Jia, Anders Ulfvarson (Chalmers University of Technology, Sweden)

Experimental analysis of the wave-induced response of a fast monohull via a segmented hull model, Daniele Dessi, Riccardo Mariani, Francesco La Gala, Lanfrancoi Benedetti (INSEAN, Rome, Italy)

Local dynamic strains in fast ship structures under slamming, Gennady Kryzhevich (Krylov Shipbuilding Research Institute, St Petersburg, Russian Federation)

Numerical assessment of segmented test model approach for measurement of whipping responses, Ole David Okland[1], Rong Zhao[1], Torgeir Moan[2] (1- Marintek Trondheim, Norway, 2- NTNU, University of Trondheim, Norway)

Concept development and model testing – new generation Air Assisted Vessels (AAV) with water-jet propulsion, Bjorn Allenstrom[1], Hans Liljenberg[1], Ulf Tudem[2] (1- SSPA, Goteborg, Sweden, 2- SES Europe AS, Norway)

Development of the four passenger class WIG ship, Myung-Soo Shin[1], Il-Sung Moon[1], Seung-Il Yang[1], Jae-Kuk Lee[2], Ju-Ho Park[2], Jae-Wan park[2] (1- Korea Research Institute of Ship & Ocean Engineering, Kordi, Korea, 2- Infinity Technologies Co Ltd, Korea)

High-speed monohull – concept optimisation and verification, Bjorn Ola Berge[1], Ingebjorn Aasheim[2] (1- Marintek, Trondheim, Norway, 2- Brodrene Aa AS, Norway)

Evaluation of a landing flying boat on the sea – development and applications of a high-speed experimental guiding system, Tsugukiyo Hirayama, Koji Miyaji, Takehiko Takayama, Yoshiaki Hirakawa, Tomonari Hayashi (Yokohama National University, Japan)

From model scale to full size investigation on turbulence stimulation in resistance model tests of high-speed craft, Carlo Bertorello[1] (invited speaker), Dario Bruzzone[2], Sebastiano Caldarella[1], Pasquale Cassella[1], Igor Zotti[3] (1- DIN, University of Naples Federico II, Italy, 2- DINAV, University of Genoa, Italy, 3- DINMA, University of Trieste, Italy)

Sea-keeping and structural behaviour of a fast patrol vessel in closed and open sea, R Folso[1], S Ferraris[2], R Grillo[2], F Albertoni[2], L Folso[3] (1- RINA SpA, Italy, 2- Fincantieri SpA, Italy, 3- DINAV, University of Genoa, Italy)

Human comfort and motion sickness onboard high-speed crafts, Osman Turan, A Ganguly, Seref Aksu and C Verveniotis (University of Glasgow And Strathclyde, Glasgow, UK)

Optimisation of the catamaran hull to minimise motions and maximise operability, Edward Dudson, Han Jorgen Rambech (Nigel Gee and Associates Ltd, Southampton, UK)

Hydrodynamic Improvements of catamaran hulls when using streamlined bodies of revolution, Igor Zotti (Invited speaker) (DINMA University of Trieste, Italy)

Applicability of a rim drive pod for high-speed ship propulsion, John Richards[1], Jon Eaton[2], Juergen Friesch[1], Michelle Lea[3], Donald Thompson[3], Bill Vanblarcom[3] (1- HSVZ, Hamburg, Germany, 2- Pennsylvania State University, US, 3- General Dynamics Electric Boats, Groton, CT)

Design of a water-jet propulsion system for an amphibious tracked vehicle, MC Kim, HH Chun W, G Park (Pusan National University, Korea)

Enhanced performance scaling methodology, Jon E Eaton, Eckhard Praefke, Friedrich Mewis (Pennsylvania State University, US)

High-speed craft in the Gulf of Naples, Antonio Scamardella[1] (invited speaker), Dario Coccoli[2], (1- University of Naples Parthenope , Italy, 2- University Of Naples Federico II, Italy)

Multi-attribute concept design model of patrol, rescue, and anti-terrorist craft, Izvor Grubisic[1], Ermina Begovic[2] (1- University of Zagreb, Croatia, 2- University of Naples Federico II, Italy)

The Saettia class patrol ships: a highly flexible Fincantieri design for the Italian Coast Guard, Alfonso Barbato[1], Andrea Castagno[1], Mauro Elefante[2], Vincenzo Farinetti[1] (1- Fincantieri-Naval Business Unit, Italy, 2- Maricogecap, Italy)

Study on the space motion of Wing-In-Ground-Effect craft, Chang-Hua Yuan, Yong-Lin Ye (China Ship Scientific Centre Wuxi, China)

Seakeeping assessment of trimaran hulls, Ermina Begovic, Carlo Bertorello, Guido Boccadamo (DIN University of Naples Federico II, Italy)

Effects of seakeeping quality on travel demand of a fast ferry, Yoshiho Ikeda, Takako Kuroda and Yuji Takeuchi (Osaka Prefecture University, Japan) transient dynamic slam response of large high-speed catamarans, Giles Thomas, Michael Davis, Damien Holloway, Tim Roberts (University of Tasmania, Australia)

Numerical estimation of ship wash waves in deep and shallow water, AF Molland, PA Wilson, DJ Taunton, S Chandraprabha, PA Ghani (University of Southampton, UK)

Erosion problems on fast high powered ships, Jurgen Friesch (HSMB, Hamburg, Germany)

Design of optimal inlet duct geometry based on vessel operational profile, NWH Bulten, R Verbeek (Wartsila Propulsion, The Netherlands)

Investigations about the use of podded drives for fast ships, Cornelia Heinke, Hans-Jurgen Heinke (Potsdam Model Basin- Germany)

Dynamic performance analysis of a gas turbine/water-jet propulsion system for a fast trimaran ferry, Giovanni Benvenuto, Ugo Campora (DINAV University of Genoa, Italy)

Large bow and stern bulbs installed on centre plane of catamaran and their wave making characteristics, Kazuo Suzuki, Hisashi Kai, Hisamitsu Tatsunami (National University of Yokohama, Japan)

Effect of hulls form variations on the hydrodynamic performances in waves of a trimaran ship, Stefano Brizzolara[1], Marco Capasso[1], Alberto Francescutto[2] (1- DINAV University of Genoa, Italy, 2- DINMA University of Trieste, Italy)

Determination of optimum position of outriggers of trimaran regarding minimum wave pattern resistance, Nastia Degiuli, Andreja Werner, Zdravko Doliner (University of Zagreb, Croatia)

Experimental and numerical investigation of 'Arrow' trimarans, Sandy Day, David Clelland, Edd Nixon (Universities of Glasgow and Strathclyde, Glasgow, UK)

A new ship motion control system for high-speed craft, Toru Katayama, Koji Suzuki, Yoshiho Ikeda (Osaka Prefecture University, Japan)

Test of a GPS RTK system for ship motion measurements, Giovanni Carrera (DINAV University of Genoa, Italy)

Ride control systems- reduced motions on the cost of increased sectional forces? Rasmus Folso[1], Ulrik Dam Nielsen[2], Francesco Torti[3] (1- RINA SpA Genoa, Italy, 2- DTU Lingby, Denmark, 3- Rodriquez Engineering Genoa, Italy)

Steady and unsteady RANSE simulations for planing crafts, Rodrigo Azcueta (MTG Marinetechnik GmbH, Germany)

Minimising the effects of transom geometry on water-jet propelled craft operating in the displacement and pre-planing regime, James Roy, John Bonafoux (Nigel Gee and Associates Ltd, UK)

Latest developments in fast ship gear technology, Franz Hoppe (Marine Gears RENK, Germany)

On the hydrodynamic performance of high-speed craft, C Bertorello[1], S Brizzolara[2], D Bruzzone[2], P Cassella[1], I Zotti[3] (1- DIN University of Naples Federico II, Italy, 2- DINAV University of Genoa, Italy, 3- DINMA University of Trieste, Italy)

Hydrodynamic analysis of interceptors with CFD methods, Stefano Brizzolara (DINAV University of Genoa, Italy)

Aerodynamic properties of a high-speed offshore racing catamaran, Hans Jorgen B Morch (Agder University College, Norway)
Design project of a trimaran multi-purpose frigate- study of the hydrodynamic aspects, Luca Sebastiani[1], Roberta Depascale[2], Michele Viviani, Roberto Dattola[3], (1- CETENA SpA, Italy, 2- Fincantieri Naval Vessel Business Unit, Italy, 3- Italian Navy General Staff, Italy)
Are fast ships competitive on short sea passenger multimodal routes? A Model Calibration On The Directrix Civitavecchia/Golfo Aranci
Structural design of the 100 knot yacht, Dean M Schleicher PE, Christopher J Swanhart (Donald L Blount and Associates Inc, US).

Papers presented at the RINA Surveillance, Pilot and Rescue Craft Conference, 17 to 18 March 2004, London

The philosophy for the design, procurement, operation and maintenance of a fleet of modern pilot boats, S Thomas and R Littler (VT Halmatic, UK)
Development of integrated design procedures for lifeboats, R Cripps, H Phillips and C Cain (RNLI, UK)
Revisiting a SWATH design 7 years on, A Crawford (SERCO Denholm Ltd UK) and N Warren and J Kecsmar (FBM Babcock, UK)
Design of SWATH vessels, K Spethmann (Abeking and Rasmussen, Germany)
Development of an integrated electronic system and the human machine interface in a new class of lifeboat, J Nusser and N Chaplin (RNLI, UK)
The development of lifting structures for daughter craft, A Humphries, S Lee and J Clapham (VT Halmatic, UK) and M Pollard (White Young Green Consulting, UK)
Evaluation of a water-jet propulsion for a boat operating directly from a beach, R J Moss and R H Cantrill (RNLI, UK)
Development of a new crew seat for all weather lifeboats, R Cripps, C Cain and H Phillips (RNLI, UK) and S Rees and D Richards (Frazer-Nash Consultancy Ltd, UK)
Health, safety & training for fast patrol & rescue craft, D Pike (Dag Pike Associates UK)
Human factors in high-speed vessel design, J Ullman (Ullman Human Design AB, Sweden)
High-speed craft design from a human centred perspective, T Dobbins (Human Sciences and Engineering Ltd, UK)
Vessels for civilian authority support in the EEZ, M Courts *et al* (VT Shipbuilding, UK)
Contemporary design status of german surveillance craft and future development project, K Spethmann, (Abeking & Rasmussen, Germany)
Fast patrol boats for HM Customs, C Dyas (Delta Power Services, UK)
Fast rescue boats, R Steen (Landsort Maritime AB, Sweden)
Design, build & trial of a 19 m self-righting SAR boat, E Yeh (Lung The Shipbuilding Co Ltd, Taiwan)
Design and development of a new inflatable lifeboat, S Austen & H F J Fogarty (RNLI, UK).

Papers presented at the 4th International Conference on High Performance Marine Vessels, 25 to 26 March 2004, Shanghai

The research on the grounding mode and grounding strength of SWATH, Cheng ZJ *et al* (China Ship Science Research Centre, China)
Investigation on increasing speed of SWATH, Zheng M *et al* (CSNAME, China)
Prediction of motion characteristics of SWATH, Guo ZX *et al* (China Ship Science Research Centre, China)
Propects for SWATH in chinese passenger water transportation, Wu ZL (Marine Design & Research Institute, China)
Research on intelligent control over a simulation model of WPC's water-jet propulsion system via double variables, Yi W (East China Shipbuilding Institute, China)
On the experimental equipment for ACV skirt vibration test
On the motion of high-speed trimaran in waves, Lu XP (Naval University of Engineering, China)
Integrate optimisation study of design method on intelligence propulsion system of adjustable screwpitch trimaran based on GA, Cheng SL (East China Shipbuilding Institute, China)
On the development and performance of high-speed air cavity craft, Pavlov GA (Shipbuilding Company Morye) and Yun L (CSNAME, China)
On the performance of high-speed craft with shallow draught, He SL (China Ship Science Research Centre, China)
Handling the compliance of ever more powerful craft, Ceproof (Southampton, UK)
Aerodynamic characteristic test on a model by high-speed aviation tank, Cu LT *et al* (Research Institute No 605 of Aviation Industrial, China)
An analytic approach to safety of ships, Zhong CH (Naval University of Engineering, China)
Shipform development for a cable delivering boat on shallow of three gorges, Fan XZ *et al* (Mid-China University of Technology, China)
Application and detail design of the structural transition joints in a ship with steel hull and aluminium alloy superstructure, Wu Han (Nanhua High-Speed Ship)
The application of superconducting materials in the high performance marine vessel, Zhang WY *et al* (Lu Yang Ship Material Institute, China)
Vibration reducing analysis of high-damping unreinforced U-shape bellows, Zhong YP (Lu Yang Ship Material Institute, China)
New foamed aluminium material and noise reducing of high-performance ship cabins, Wong Y (Lu Yang Ship Material Institute, China)
Duplex stainless steels for high-speed boats, Cheng GY (Shanghai Marine Steel and Structure Institute, China)
From the investigation of the biggest GRP ship of China to study control of the deformation, Cheng GY (Shanghai Marine Steel and Structure Institute, China).

Papers presented at the 56th UITP World Congress, 5 to 9 June 2004, Rome

Charting new horizons: air-sea interchange services at Hong Kong Airport, Jenning Wang (Shun Tak Ship Travel, Hong Kong)
Heritage as a trendsetter: transforming Abra ferries into modern services, Abdul Aziz Malik (Director Public Transport Department, Dubai)
Powered by the sun: a new cruising experience, Ulrich Sinzig (Aare Seeland Mobil AG, Switzerland) and Jens Wrage (HADAG AG, Germany)
Unlimited sailing pleasure: accessibility of waterborne transport, Paul Bunnell (Mersey Ferries, UK)

Papers presented at HIPER, the 4th International Conference on High-Performance Marine Vehicles, 27 to 29 September 2004, Rome

Simple design formulae for fast ships derived by artificial neural nets, V Bertram and E Mesbahi
Ecologic study of fast ships, K Hasegawa
An optimum design of monohull high-speed craft in still water, H Tanaka and Y Yoshida
A statistical analysis to the towing tank test data of monohull high-speed craft models and the applicable range of the empirical equations, H Tanaka and Y Yoshida
WIG continued research, H Akimoto *et al*
Manoeuvring prediction of HPMC including multihulls, H Yasukawa
Quantifying external cost to the environment due to shipping, J Isensee and V Bertram
Optimisation of fast ships using the DELPHI Optimisation Shell, H Gudenschwager and J Isensee
Hydrofoil-supported catamaran design support system, G Migeotte and G Thiart
Hydro-elastic analyses for fast ships, M Salas
Sensitivities on the ultimate strength of aluminium stiffened panel's parameters: comparative analysis, T Richir *et al*
Hydro-elastic simulation of simple geometrics in water entry, A Neme *et al*
HSC reglementation for fast yachts, L Laubenstein
Fire and ventilation simulations for the Olympic Voyager fast cruise vessels, S Nusser *et al*
Parametric design studies for SWATH ships, J Michalski
Fast feeder concept study, M Elsholz
CFD simulation and virtual reality visualisation for a fast monohull in waves, M Klemt *et al*
Advances in simulation of ditching of airplanes on water, O Lindenau and H Soeding
Development for new demonstrator fast air-cushion vehicle, U Tudem
Wash simulations for fast ferries, M Leeer-Andersen
Classification of HSC multihull vessels including trimarans, K Fach
Case studies for damages for aluminium alloys in fast ships, K Fach
Interceptors for fast monohulls, S Brizzolara
Design study for a trimaran using optimisation, S Brizzolara
Implemented multi-criteria design overview for fast ships, M Barone *et al*
Preliminary studies for hydrofoil boats, M Seif
New technologies for higher speed at sea, M Seif *et al*
Commercial viability of fast pod-driven ships, M Woodward
Rolls-Royce Norwegian Fast Navy Ship Project FNSLV, P Vedlog.

Papers presented at the 1st Fast Ferry Information Conference, 29 to 30 September 2004, Southampton

Operating fast vehicle ferries across Europe, David Benson (Sea Containers, UK)
Operating fast ferries throughout North America, John Waggoner (Hornblower Marine Services, United States)
The Greek fast ferry market, Christos Economou (Econ Shipping, Greece)
Operating fast ferries on Lisbon's commuter routes, João Franco (Transtejo/Soflusa, Portugal)
None sister vessels, advantages for a builder, Joris Neven (Damen Shipyards, The Netherlands)
Ups and downs of Scandinavian fast ferry services, Erik Østergaard (Scandlines Danmark, Denmark)
Establishing new ferry services in the Mediterranean, Sam Crockford (Rodriquez Cantieri Navali, Italy)
A new concept in fast ferry operations, Raymond Kalley (Hydrocruiser, UK)
The development of a fast freight ferry for European and American operations, Nigel Gee (BMT Nigel Gee & Associates, UK)
In search of passenger comfort, the Austal Ships trimaran range, James Bennett (Austal Ships, Australia)

The development of Incat's 112 m SeaFrame wave-piercing catamaran, Robert Clifford (Incat, Australia)
The commercial utility of modified commercial fast ferries, Larry S Ryder (United States Marines).

Papers presented at the 29th Interferry Conference, 1 to 4 November 2004, Bahamas

The future of ferry transportation in Cuba, Carlos Manuel Pazo Torrado (Minister of Transport, Cuba)
The evolution of inter-island ferries in Cuba, Khaalis Rolle (Bahama Ferries)
Transportation overview of the Caribbean, Gilbert Morris (The Landfall Centre, Bahamas)
Ferry developments in Brazil, Marco Santarelli (Rodriquez, Brazil)
New opportunities for private operators in British Columbia, David Hahn (BC Ferries, Canada)
Developing small-scale urban waterborne transportation markets, Tom Fox (New York Water Taxi, US)
Fast ferries on the Great Lakes, John Waggoner (Hornblower Marine, US)
Developing new markets using fast ferry technology – what is required, Richard Regan (Austal Ships, Australia)
EU policy and its application, Alfred Baird (Napier University, UK)
Government policy in North America, Robert Kunkel (SCOOP and Apex Shipping, US)
Developments in the Baltic, Erik Ostergaard (Scandlines, Denmark)
Opportunities on the Great Lakes, Christopher Wright (Mariport Group, Canada)
How revenue management can improve productivity, Ted Esse (Revenue Technology Services, US)
Airlines – pricing strategies, Keith Sherwood (Anite Systems, UK)
An overview on security, Peter Varnish (International Political Solutions Ltd, UK)
Dealing with security in the Philippines: an operator's perspective, Bob Gothong (Aboitiz Transport System, Philippines)
Security issues in the United States, Beth Gedney (Passenger Vessel Association, US)
Work of the DNV Ferry Committee on Standards, Håkan Enlund (Aker Finnyards, Finland)
Current issues at IMO, Len Roueche (Interferry, Canada)
Defending irrational regulations in shipping, Dracos Vassolos (Strathclyde University, UK).

Papers presented at the RINA High-Speed Craft Conference, 17 to 18 November 2004, London

The influences of new technology on the design and manufacture of high-speed craft with special reference to recent monohulls, multihulls, Air Cushion Vehicles and Surface Effect Ships, John L Allison *et al* (CDI Marine, US), Tormod Salveson (Umoe Mandal, Norway) and John G Stricker (Marine Propulsors Company, US)
High-speed multihull craft for medium distance marine transportation, Ermina Begovic *et al* (University of Naples, Italy)
Defect and damage assessment for ships built in FRP sandwich, Brian Hayman (Det Norske Veritas, Norway)
An inverse design procedure for hydrodynamic optimisation of high-speed hull form using commercial software, Nigel Koh *et al* (University of Newcastle, UK) and David Cooper (AVEVA Group, UK)
Unsteady influences on the damping from appendages of a trimaran during rolling motion, Lawrence Doctors (University of New South Wales, Australia) and Robert Scrace (QinetiQ Haslar, UK)
The development and validation of a hydrodynamic and structural design methodology for high-speed rescue craft, Holly Phillips and Bob Cripps (Royal National Lifeboat Institution, UK) and Simon Rees and Laurence Vaughan, (Frazer-Nash Consultancy Ltd, UK)
The influence of central bow bulbs on the resistance of catamarans, Carlo Bertorello and Pasquale Cassella (University of Naples, Italy), Dario Bruzzone (University of Genoa, Italy) and Igor Zotti (University of Trieste, Italy)
No stone unturned: a comprehensive approach to the development of a new class of high-speed ferry for Alaska, John Bonafoux and Jago Lawless (BMT Nigel Gee and Associates, UK), Gary Smith (Alaska Marine Highway Systems, US) and Gavin Higgins and David Rusnak (Derecktor Shipyards, US)
Seakeeping predictions for a 100 knot yacht, Jeffrey B Bowles and Dean M Scheicher (Donald L Blount and Associates, Inc, US)
The evolution of the 112 m wave-piercing catamaran design, Gary Davidson and Tim Roberts (Revolution Design, Australia)
Design of high-speed low wake hydrofoil passenger ferry, Endicott M Fay (Teignbridge Propellers Ltd, UK)
The X-Craft, Nigel Gee and Mike Machell (BMT Nigel Gee and Associates Ltd, UK)
The design of a low wash fast ferry for inland waters use: the TRIDELTA, J A Keuning, H Boonstra and M H van den Hoven (Delft University of Technology, The Netherlands) and P H Noordenbos, (DAMEN Shipyard, The Netherlands)
Design development of a 24 m Air-Supported Vessel (ASV) catamaran demonstrator, suitable for fast passenger ferries and various naval/paramilitary applications, Ulf Tudem and Andre Ellertsen (SES Europe AS, Norway)
An overview of current UK research projects into the stability of high-speed craft, Andrew G Blyth (Blyth Bridges Marine Consultants, UK) and Ronald Allen (Maritime and Coastguard Agency, UK)
HSC incident statistics, Stephen Phillips and Dan Hook (Seaspeed Technology Ltd, UK)
High-speed safety: the impact of human and organisational factors, Torkel Soma and Simen Heum Listerud (Det Norske Veritas, Norway)
Raking damage to high-speed craft: a proposal for the high-speed code, Bo Cerup Simonsen and Rikard Tömqvist (Det Norske Veritas, Norway) and Maria Lützen (MARTEC Maritime Training and Education Centre, Denmark).

Papers presented at FAST 2005 International Conference on Fast Sea Transportation, 27 to 30 June 2005, St Petersburg, Russian Federation

The world market for fast vessels, Nigel Gee (BMT Nigel Gee and Associates Ltd, UK)
Russian experience of creation of fast naval ships, Alexander Shlyakhtenko (Almaz Design Bureau, Russian Federation)
Innovative fast naval ship design, Jay Cohen (Office of Naval Research, US Navy)
Naval and commercial fast ship development experience and further prospects, A V Pustoshny, A A Rusetsky (Krylov Shipbuilding Research Institute, Russian Federation)
Developing high-speed aluminium ships to meet emerging commercial and defence needs, John Rothwell (Austal Ships, Australia)
Considerations in the selection of propulsion systems for fast naval ships, Frank Mungo, David J Bricknell, Andrew Tate (Rolls-Royce, UK)
Safety of ice-navigation in the Baltic region, Paavo Wihuri (Finnish Maritime Administration)
Numerical modeling of cavitating flows, Alexey G Terentiev (Cheboksary Institute of Moscow State Open University, Cheboksary, Russian Federation)
Wing-in-Ground-Effect vehicles, Kirill Rozhdestvensky (St Petersburg State Marine Technical University, Russian Federation)
Dynamic performance of a 140 m SES in simulations and model tests – an introduction of recent development of Techno Super Liner, Taketsune Matsumura, Susumu Yamashita, Naoki Ohba (Mitsui Engineering & Shipbuilding Co, Tokyo, Japan)
A statistical analysis of fast ferry incidents worldwide, Stephen Phillips and Daniel Hook (Seaspeed Technology Limited, UK)
Development of a family of fast monohulls for rapid deployment of logistics and troops, David J Bricknell, Per-Egil Vedlog (Rolls-Royce Marine, UK and Rolls-Royce Marine AS, Ship Technology, Norway)
Global optimisation of fast vessels, D Peri, A Pinto, EF Campana (INSEAN – The Italian Ship Model Basin, Roma, Italy)
The potential for the use of a novel craft, pacscat (partial air cushion supported catamaran), in inland European waterways, RJ Clements, JC Lewthwaite, P Ivanov, PA Wilson, AF Molland, (Marinetech South Ltd, UK, Independent Maritime Assessment Associates Ltd, Fareham, UK, Sovtransavto Deutschland GmbH, Germany, School of Engineering Sciences, University of Southampton, UK)
A study of the potential replacements of a vessel for the Port of London Authority, P A Wilson, A F Molland, R A Cartwright (University of Southampton and Port of London Authority, UK)
Designing the high-speed supercavitating vehicles, Yu N Savchenko, V N Semenenko, S I Putilin, G Yu Savchenko, and Ye I Naumova (Ukrainian National Academy of Sciences – Institute of Hydromechanics, Ukraine)
A hydrogen fuelled gas turbine powered high-speed container ship: a technical and economic investigation of the ship & associated port infrastructure, IJS Veldhuis, RN Richardson, HBJ Stone (Southampton University, UK)
On the development of a hull form with minimum wetted surface for high-speed catamarans and trimarans, Jacques B Hadler (Webb Institute New York, US)
Optimised design of high-speed catamarans with limiting the wash waves, Anatoly G Lyakhovitsky, Eduard B Sakhnovsky (St Petersburg State Marine Technical University, St Petersburg, Russian Federation)
Hydrodynamic fine-tuning of a pentamaran for high-speed sea transportation services, Ed Dudson, Stefan Harries (BMT Nigel Gee and Associates Ltd, UK and Friendship-Systems GmbH, Potsdam, Germany)
Springing investigation for a 205 m fast-ferry built of HTS 690, Hubertus von Selle (Germanischer Lloyd AG, Hamburg, Germany)
Automatic optimisation of a trimaran hull form configuration, Stefano Brizzolara, Dario Bruzzone, Emilio Tincani (University of Genoa, Genoa, Italy)
Effect of hull form variation on hydrodynamic performance of a trimaran ship for fast transportation, Stefano Brizzolara, Marco Capasso, Marco Ferrando, Carlo Podenzana-Bonvino (University of Genoa, Genoa, Italy)
URANS simulations for a high-speed transom-stern ship with breaking waves, Robert Wilson and Fred Stern (The University of Iowa, Iowa City, US)
An investigation into the hydrodynamic characteristics of a high-speed Partial Air Cushion Supported CATamaran (PACSCAT),

A F Molland, JC Lewthwaite, P A Wilson, D J Taunton (University of Southampton, Southampton, UK, Independent Maritime Assessment Associates Ltd, UK, and BMT SeaTech Ltd, Southampton, UK)
Wash effects of high-speed monohulls, Dimitris S Chalkias and Gregory J Grigoropoulos (National Technical University of Athens, Zografou, Greece)
Hull-form optimisation of high-speed monohulls, Gregory J Grigoropoulos and Dimitris S Chalkias (National Technical University of Athens, Zografou, Greece)
Current status and future trends of transport ground-effect machines (ekranoplanes), R A Nagapetjan, D N Sinitsyn (Amphibian TransportTechnologies, Nizhniy Novgorod, Russian Federation)
High-speed amphibian ACV of new generation – the vessels with an optimal aero-hydrodynamical lifting complex, V V Klichko (Krylov Shipbuilding Research Institute, St Petersburg, Russian Federation)
Air Cushion Vehicle (ACV) developments in the United States, David R Lavis and Brian G Forstell (CDI Marine Systems Development Division, Severna Park, US)
Dynamic stability of Wing-in-Ground-Effect vehicles: a general model, Caterina Grillo, Cinzia Gatto (University of Palermo, Italy)
Numerical simulation of a ship ventilation to collect landing craft air cushion, François Grosjean, Steven Kerampran (Ecole Nationale des Ingénieurs des Etudes et Techniques d'Armement, Brest, France)
Problematic questions of aerohydrodynamics and dynamics of movement of transport ekranoplanes, Al Maskalik (Amphibian Transport Technologies, Nizhniy Novgorod, Russian Federation)
A new field of wing-in-surface-effect craft, Syozo Kubo, Hiromichi Akimoto, Kei Ohtsubo, Gombodorjiin Batkhurel, and Kenichi Manabe Tottori University, Tottori, Japan and Mongolian University of Science and Technology, Ulaanbaatar, Mongolia)
Self-propulsion model test of a wing-in-surface-effect-ship with canard configuration, Part 2, Hiromichi Akimoto, Syozo Kubo, Motoki Tanaka and Manami Sakumasu (Tottori University, Tottori, Japan)
Experimental identification of wet bending modes with segmented model tests, Daniele Dessi, Riccardo Mariani, Giuliano Coppottelli, Marcello Rimondi (Italian Ship Model Basin and University of Rome La Sapienza, Rome, Italy)
RAO calculation of a hydrofoil boat in platforming and contouring modes of operation, Sunghoon Kim, Key-Pyo Rhee, Dong Jin Yeo, Seong Mo Yeon, Juhyuck Choi, Se Won Kim, Chang Seop Kwon (Seoul National University, Seoul, Republic of Korea)
Perspectives of application in Arctic's regions of the newest amphibian vehicles utilizing ground effect, Alexander I Bogdanov (Central Marine Research & Design Institute Ltd. St Petersburg, Russian Federation)
WIG – what are you waiting for? Graham Taylor (Hypercraft Associates, High Wycombe, Buckinghamshire, England)
Specific features of global vibration analysis for planing boats and ships operating in transient conditions, (Krylov Shipbuilding Research Institute and St Petersburg State Marine Technical University, St Petersburg, Russian Federation)
Experience and prospect of using aluminum alloys for construction of fast ships, Gennadij B Kryzhevich, Semion D Knoring, Valerij M Shaposhnikov (Krylov Shipbuilding Research Institute, St Petersburg, Russian Federation)
A study of sound transmission loss across orthotropic composite laminate, Chao-Nan Wang, Huei-Jeng Lin, Ya-Jung Lee, Yan-Min Guo (National Taiwan University, Tiawan, ROC)
Modelling the vacuum infusion process for a small composite powerboat, Nicholas G Tsouvalis and Ioannis S Ioannou (National Technical University of Athens, Athens, Greece)
Hydrodynamics of fast boats in transient speed region, Vladimir K Dyachenko (Krylov Shipbuilding Research Institute, St Petersburg, Russian Federation)
Some features of water resistance determination for planing and semi-planing ships, Victor N Anosov, Vladimir K Dyachenko, Tatiana A Djakova (Krylov Shipbuilding Research Institute, St Petersburg, Russian Federation)
The problem of final thickness jet, Shamil G Aliyev, Yuri N Kormilitzyn, Yuri M Khaliullin (St Petersburg State Marine Technical University, St Petersburg, Russian Federation)
On the added resistance of catamarans in waves, Tony Armstrong and Anton Schmieman (Austal Ships, Henderson, Australia)
The waves generated by a trimaran, Prasanta K Sahoo and Lawrence J Doctors (Australian Maritime College and The University of New South Wales, Sydney, Australia)
Transom-stern flow for high-speed craft, Kevin J Maki, Lawrence J Doctors, Robert F Beck, Armin W Troesch (University of Michigan, Ann Arbor, US and The University of New South Wales, Sydney, Australia)
Co-operative investigation into resistance of different trimaran hull forms and configurations, Ermina Begovic, Andrea Bove, Dario Bruzzone, Sebastiano Caldarella, Pasquale Cassella, Marco Ferrando, Emilio Tincani, Igor Zotti (The University of Naples, The University of Genoa and The University of Trieste, Italy)
An experimental investigation into the resistance components of trimaran configurations, Nastia Degiuli, Andreja Werner, Igor Zotti (The University of Zagreb, Croatia and Università degli studi di Trieste, Italy)
Turbulent boundary layer drag reduction at high Reynolds numbers with wall-injected polymer solution, Eric S Winkel, David R Dowling, Marc Perlin, and Steven L Ceccio (University of Michigan, Ann Arbor, US)
Influence of bubble size on micro-bubble drag reduction, Xiaochun Shen, Eric S Winkel, Steven L Ceccio, and Marc Perlin (University of Michigan, Ann Arbor, US)
Prospects of artificial cavities in resistance reduction for planing catamarans with asymmetric demihulls, Andrey V Sverchkov (Krylov Shipbuilding Research Institute, St Petersburg, Russian Federation)
Investigation of the hydrodynamic characteristics of a fast-ferry advancing in regular waves, Nuno Fonseca and Carlos Guedes Soares, Adolfo Marón (Technical University of Lisbon, Portugal and Canal de Experiências Hidrodinamicas de El Pardo, Madrid, Spain)
Classification of engineering systems for Wing-in-Ground-Effect craft, Norman Rattenbury (Research and Development Lloyd's Register, London, UK)
Experience and operation safety of the first small civil ekranoplan of type A Aquaglide. Value of its creation for the further development of ekranoplan building, A G Butlitscki, A I Maskalik, V V Tomilin, A I Lukjanov (Amphibian Transport Technologies, Nizhniy Novgorod, Russian Federation)
Wing-In-Ground (WIG) craft (ekranoplan) general safety aspects as contained in the international Instruments and RS rules. Perspectives Mikhail A Kuteynikov, Mikhail A Gappoev, Vladimir V Gadalov (Russian Maritime Register of Shipping, St Petersburg, Russian Federation)
A calculation method of marine ship's RCS, A M Dehziri, A Mallahzadeh, K Barkeshli (Malek-Ashtar University of Technology, Shiraz University, Sharif University of Technology, Iran)
Wind heeling moments on multihull vessels, Barry Deakin and Alexander M Wright (University of Southampton, UK)
Simulations and full-scale trials for a HSC linked by wave-height measurements, Karl Garme and Jakob Kuttenkeuler (Royal Institute of Technology, Stockholm, Sweden)
Study on auto-navigation system for berthing and de-berthing of ship using thrusters, Nam-sun Son, Sun-young Kim, Chang-min Lee, Seok-ho Van (Korea Research Institute of Ships and Ocean Engineering, Daejeon, Korea)
Comprehensive comparative studies on 3-D seakeeping methods with experimental results regarding a late post-Panamax container ship, Tingyao Zhu, Li Xu, Sanjay Pratap Singh, and TaeBum Ha (Nippon Kaiji Kyokai (ClassNK), Chiba, Japan, China Classification Society (CCS), Beijing, China, Indian Register of Shipping (IRS), Mumbai, India, and Korean Register of Shipping (KR), Taejon, Republic of Korea)
The integration of lifting foils into ride control systems for fast ferries, Alan J Haywood and Benton H Schaub (VT Maritime Dynamics, Inc, US)
Effects of actuator dynamics on stabilization of high-speed planing vessels with controllable transom flaps, Handa Xi and Jing Sun (University of Michigan, Ann Arbor, US)
Water-jet research and optimisation methods and new propulsor developments, MA Mavluidov, AA Rousetsky, OV Yakovleva (Krylov Shipbuilding Research Institute, St Petersburg, Russian Federation)
Supercavitating, highly cavitating and surface-piercing propellers design, Alexander S Achkinadze (St Petersburg State Marine Technical University, St Petersburg, Russian Federation)
Investigation of the phenomenon of domination of certain orders of the water-jet blade-passing frequency harmonics, L Berghult (Rolls-Royce Hydrodynamic Research Centre, Sweden)
Heat transfer in a rib-roughened rotating twin-pass trapezoidal-sectioned passage with cooling application to gas turbine rotor blade, Shyy Woei Chang, Tsun Lirng Yang, Yu, Ker-Wei, Shun-Hsyung Chang, and Wei Jen Wang (National Kaohsiung Marine University and Fortune Institute of Technology, Taiwan, China)
Podded propulsors for fast and large commercial vessels (FASTPOD project), Roberta Depascale, Mehmet Atlar, and Michael D Woodward (CETENA, Genoa, Italy and University of Newcastle, Newcastle upon Tyne, UK)
Experimental study on the hydrodynamic performance of a two-phase nozzle for water-jet propulsion system, Jing-Fa Tsai, Chuan-Cheung Tse, Shing-Yaur Wang, Da-Chen Yeh (National Taiwan University, Chung-Shan Institute of Science and Technology, and Lung Teh Shipbuilding Co Ltd, Taiwan, ROC)
Experience with the analysis of controllable-pitch propellers at off-design conditions using a boundary element method, Vladimir I Krasilnikov, Aage Berg and Fredrik A Bolstad (Marintek – Norwegian Marine Technology Research Institute and Norwegian Defense Materiel Agency, Norway)
Improvement of high-speed craft propulsion by using of propellers with shifted blade connection on the hub, Leonid I Vishnevsky, Anatoy R Togunjac, and

Andrew V Pechenuk (Krylov Shipbuilding Research Institute, St Petersburg, Russian Republic, Giprorybflot, St Petersburg, Russian Republic, and Digital Marine Technology Company, Odessa, Ukraine)
Computation of propeller wake on podded propulsors of fast ships, Vladimir A Boushkovsky, Liudmila A Moukhina, Aleksey Yu Yakovlev (Krylov Shipbuilding Research Institute, St Petersburg, Russian Federation)
Mathematical expressions of thrust and torque of Newton-Rader propeller series for high-speed crafts using artificial neural networks, Kourosh Koushan (Marintek (Norwegian Marine Technology Research Institute), Trondheim, Norway)
The Gowind 170 corvette, Letty Patrick, Bouvier Christophe (DCN, France)
70 kt with a high-speed motor yacht, not a concept but reality, Norbert W H Bulten (Wärtsilä Propulsion Netherlands BV/ Technical University Drunen, Netherlands)
Application of vortex methods to the design of vessels operating at transitional and planing Froude numbers, Nikolai Kornev, Günther Migeotte, Anna Nesterova (University of Rostock, Rostock, Germany, University of Stellenbosh, South Africa, and St Petersburg State Marine Technical University, St Petersburg, Russian Federation)
Development of hullform of a ship with high-performance in waves, Masashi Kashiwagi and Kazunari Sumi (Research Institute for Applied Mechanics, Kyushu University Fukuoka, Japan)
ULCC development: new trends in liner shipping networks, Branislav Dragović, Vladislav Maraš (University of Montenegro, Kotor, Serbia & Montenegro)
MEGA trends in container shipping, Zoran Radmilović and Branislav Dragović (University of Beograd, Beograde and University of Montenegro, Kotor, Serbia and Montenegro)
A high inertia, slowed overhead rotor as take off aid ensures an unequalled level of safety in the operation of Wing-In-Ground effect (WIG) craft, Luc J De Keyser (Denderleeuw, Belgium)
On a study of a WIG with propeller-deflected slipstream (PDS) par by using a radio-controlled model, Rinichi Murao, Shingo Seki, Nobuyuki Tomita (Air Cushion Vehicle Association of Japan and Musashi Institute of Technology, Tokyo, Japan)
Project of ekranoplan application for spaceplane assist at horizontal launch and landing, Alexander Nebylov, Nobuyuki Tomita (State University of Aerospace Instrumentation, St Petersburg, Russian Federation and Musashi Institute of Technology, Tokyo, Japan)
Optimisation of a split-cushion surface-effect ship, Lawrence J Doctors, Vidar Tregde, Changben Jiang, Chris B McKesson (The University of New South Wales, Sydney, Australia, Umoe Mandal, Mandal, Norway, John J McMullen Associates, Inc, US)
Interactive model of sea waves for pre-flight preparation of small class A ekranoplan, Sergey V Tarasov, Michael A Mikhaelov (St Petersburg State Marine Technical University, St Petersburg, Russian Federation)
Advantage and adaptation of Iranian fast ferry (hovercraft) in Persian Gulf island (Kish-Charak port), Ali Dehghanian, Kambiz Alempour, Hamid Javadi (Malek Ashtar University of Technology, Shiraz, Iran)
Computation of ship motions in waves and slamming loads for fast ships using RANSE, Ould A El Moctar, Jörg Brunswig, Andreas Brehm, Thomas E Schellin (Germanischer Lloyd, Hamburg, Germany)
Global and slam loads for a large wave-piercing catamaran design, G Davidson, T Roberts, G Thomas (Revolution Design, Australia and Australian Maritime College, Australia)
The influence of slamming and whipping on the fatigue life of a high-speed catamaran, Giles Thomas, Michael Davis, Damien Holloway, Timothy Roberts (Australian Maritime College, University of Tasmania, and Revolution Design Pty Ltd, Tasmania, Australia)
Explicit FE analysis of hull-water impacts, I Stenius and A Rosén (Royal Institute of Technology, Stockholm, Sweden)
A proposal for a simple fatigue design check procedure, Marco Biot, Alberto Marinò, Fabio Susmel (University of Trieste and Fincantieri SpA – Cruise Ship Business Unit passeggio, Trieste, Italy)
Manoeuvring induced loads on fast POD drives, Michael D Woodward, Mehmet Atlar, David Clarke (University of Newcastle, Newcastle-upon-Tyne, UK)
Validation of non-linear wave loads predicted by time domain method in sea trials of an 86 m catamaran, Michael R Davis, Damien S Holloway, and Nigel L Watson (University of Tasmania, Hobart and Ship Dynamics Pty Ltd, Henderson, Australia)
Experimental identification of wet bending modes with segmented model tests, Daniele Dessi, Riccardo Mariani, Giuliano Coppottelli, Marcello Rimondi (Italian Ship Model Basin and University of Rome La Sapienza, Rome, Italy)
Structural response operator for the fatigue assessment of a fast patrol vessel operating in the Mediterranean Sea, C Cartalemi, S Ferraris, A Moggia Cappelletti, L Sebastiani, D Boote (Fincantieri CNI Spa Naval Vessel Business Unit, CETENA, and DINAV, Università di Genova, Genoa, Italy)
An integrated reliability, risk analysis, and cost model for preliminary structural design: a module of the Safety@Speed design methodology, Matthew Collette, Martyn Cooper, Ana Mesbahi, Atilla Incecik (University of Newcastle upon Tyne, UK and American Bureau of Shipping Europe Ltd, London, UK)
On the assessment of motions and structural loads of a high-speed catamaran, Suqin Wang, Balji Menon, Derek Novak (American Bureau of Shipping, Houston, US)
Analysis of hydrodynamic longitudinal bending moments of air cushion vehicle, Kambiz Alempour, Ali Dehghanian, Hamid Javadi (Malek Ashtar University of Technology, Shiraz, Iran)
Water surface disturbance near the bow of high-speed, hard-chine hullforms, Jeffrey B Bowles and Stephen B Denny (Donald L Blount and Associates, Inc, Chesapeake, US)
Wave-wave interactions between a high-speed vessel and surface waves in a non-linear time-dependent, 6 DoF ship motion model, Ray-Qing Lin and Allen Engle (David Taylor Model Basin, West Bethesda, US)
Modeling flow around an ACV using volume of fluid interface tracking with Lagrangian propagation, AH Nikseresht, MM Alishahi, H Emdad (Yasuj University, Yasuj, Iran and Shiraz University, Shiraz, Iran)
Time domain computation of the wavemaking resistance of ships, F Kara and D Vassalos (The Universities of Glasgow and Strathclyde, Scotland, UK)
Improvement of seakeeping prediction methods for high-speed vessels with a transom stern, TM Ahmed, SX Du, DA Hudson, PW Kingsland and P Temarel (University of Southampton, Southampton, UK)
Tests with a 6 DoF forced oscillator to investigate the non-linear effects in ship motions, Pepijn de Jong, Jan Alexander Keuning (Delft University of Technology, Delft, Netherlands)
Planing craft in waves – full scale measurements, Hans Jørgen B Mørch, Ole A Hermundstad (Agder University College, Grimstad, Norway and Centre for Ships and Ocean Structures, Norwegian University of Science and Technology, Trondheim, Norway)
Simulation of a fast catamaran's manoeuvring motion based on a 6 DoF regression model, Serge Sutulo and Carlos Guedes Soares (Technical University of Lisbon, Portugal)
A practical approach to a fast displacement ship's stabilisation in head seas, Victor Sokolov and Serge Sutulo (Balttechnoprom Co, St Petersburg, Russian Federation)
Improvement of seakeeping prediction methods for high-speed vessels with a transom stern, TM Ahmed, SX Du, DA Hudson, PW Kingsland and P Temarel (University of Southampton, Southampton, UK)
Development of a simple system to measure 5 DoF ship motions in a seaway, Forng-Chen Chiu, Sao-Wei Liu, Wen-Chuan Tiao and Jenhwa Guo (National Taiwan University, Taipei, Taiwan)
Effect of the rudder and the bilge keels in the roll damping, M Felli, F Di Felice, R Dattola (INSEAN, Italian Ship Model Basin and Italian Navy Maristat SPMM, Roma, Italy)
The high-speed craft steering at off design motion regime, Marina P Lebedeva (NAVIC, St Petersburg, Russian Federation)
Assurance of digital control accuracy at non-centralised stochastic actions, J Ladisch, BP Lampe, EN Rosenwasser, VO Rybinskii, AV Smolnikov (St Petersburg State Marine Technical University, St Petersburg, Russian Federation)
Sea trials of high-speed crafts data processing and model identification, Victor MAmbrosovsky, Elena B Ambrosovskaya (St Petersburg State Electrical Engineering University and NAVIS, St Petersburg, Russian Federation)
Effects of running attitudes on manoeuvring hydrodynamic forces for planing hulls, Toru Katayama, Ryo Kimoto and Yoshiho Ikeda (Osaka Prefecture University, Osaka, Japan)
An experimental evaluation of the stability criteria of the HSC code, Barry Deakin (MTIA University of Southampton, UK)
Redundancy aspects for mono and multiple hull HSC, Jan Peter Securius (Germanischer Lloyd, Hamburg, Germany)
International regulations for high-speed craft – an overview, Heike Hoppe (International Maritime Organization (IMO), London, UK)
Safety management of high-speed craft, Chengi Kuo (University of Strathclyde, Glasgow, Scotland)
Static and dynamic effects of rudder-hull-propeller interaction on fast monohulls, Albert J Jurgens (Marin, Wageningen, Netherlands)
Optimisation of podded propulsor for fast Ro-Pax using RANSE solver with cavitation model, Antonio Sánchez-Caja & Jaakko V Pylkkänen (VTT Industrial Systems, Espoo, Finland)
Speed naval combatant – propulsion enablers, P Mountford, T Roberts, P Hopkins, D A Hofmann, J Haigh (Marine Propulsion Systems IPT, Warship Support Agency, UK MoD Defence Logistics Organisation, MIMechE, and University of Newcastle, UK)
Performance improvements of high-speed propellers at inclined shaft conditions by root unloading, Young-Zehr Kehr, Wen-Chi Lee, Ching-Yeh Hsin (National Taiwan Ocean University, Keelung, Taiwan)
A research on propulsive performance of high-speed craft equipped with surface piercing propellers, Kazuo Nozawa and Naohisa Takayama (Osaka University, Osaka, Japan)

Perfection of the monoswimfin, being the mover for the sportsman-submariner as a factor of increasing the speed of swimming for the biotechnical system Sportsman-monoswimfin, G N Orlov 9 St Petersburg State Academy of Physical Training named after P F Lesgaft, St Petersburg, Russian Federation)

Hydrodynamic modelling of unsteady interactions among podded propulsor components, Francesco Salvatore, Luca Greco, Danilo Calcagni (INSEAN – Italian Ship Model Basin, Rome, Italy)

Modern high-speed propulsion systems and their impact on marine transmission design Eric Luetjens (ZF Marine GmbH, Friedrichshafen, Germany)

Experimental and numerical investigation of a trimaran in calm water, Giuseppina Colicchio, Andrea Colagrossi, Claudio Lugni, Odd Magnus Faltinsen (Italian Ship Model Basin (INSEAN), Rome, Italy and Center for Ships and Ocean Structures (CeSOS), Trondheim, Norway)

Unsteady approach for curved and swept lifting-lines computation, T Gallois, Ph Devinant (Laboratoire de Mecanique et Energetique, Orleans, France)

The effects of bulbous bows, stern flaps and a wave-piercing bow on the resistance of a series 64 hull form, Siu C Fung, Garbor Karafiath (Naval Surface Warfare Center, West Bethesda and JJMA, US)

Numerical estimation of wavemaking characteristics of high-speed ships taking sinkage and trim effects into account, Kazuo Suzuki, Hisashi Kai (Yokohama National University, Yokohama, Japan)

RANSE simulations for high-speed ships in deep and shallow water, Nobuaki Sakamoto, Robert Wilson, and Fred Stern (IIHR – Hydroscience and Engineering, The University of Iowa, Iowa City, US)

Bulbous bow resistance reduction of semi-planing vessels, Richard A Royce (Webb Institute, New York, US)

Boundary layer propulsion and wake control by means of flapping wings, Max F Platzer and Claus Dohring (AeroHydro Research & Technology Associates, California, US and German Navy Bonn, Germany)

Experimental verification of the resistance of a split-cushion Surface-Effect Ship, Sverre Steen and Christian Adriaenssens (Norwegian University of Science and Technology, Trondheim, Norway)

Papers presented at the 7th Symposium on High-Speed Marine Vehicles, 21 to 23 September 2005, Naples

Induced Drag of an America's Cup Yacht's fin-bulb-winglet:numerical and experimental investigation, D P Coiro *et al* (University of Naples Federico II, Italy)

Hydrodynamic improvements by interceptors application: Fincantieri fast ferry experience, A Barbato *et al* (Fincantieri Naval Business Unit, Genoa, Italy)

Safety and dependability of machinery systems for high-speed marine vehicles, N Rattenbury (Lloyd's Registry)

Structural response of a fast patrol vessel when operating in several sea areas: comparison by means of a fatigue assessment, C Cartalemi *et al* (Fincantieri Naval Business Unit, Genoa, Italy)

Damage identification and health monitoring of composite structures using vibrations measurements and piezo-electric devices, I Bovio and L Lecce (University of Naples Federico II, Italy)

Thermografic method for fatigue prediction of friction stir welded light alloy panels in shipbuilding, M Biot *et al* (Trieste University, Italy)

Optimal manoeuvring of vessel in deterministic seas, F S Hover (MIT Cambridge Massachusetts, US)

A gas turbine modular model for ship propulsion studies, G Benvenuto and U Campora (University of Genoa, Italy)

Micro-emulsion technology for emission reduction of existing ship engines, M Civati *et al* (Mec System Como, Italy)

Performance prediction of a planing craft by dynamic numerical simulation, M Altosole *et al* (University of Genoa, Italy)

The X-Craft, a potential solution to littoral warfare requirements, N Gee and E Dudson (BMT Nigel Gee & Associates, UK)

Displacement Catamarans: experimental clearance data on stagger and clearance systematic variations, F Caprio *et al* (University of Naples Federico II, Italy)

A proposal for a fast passenger ferry operating in Italian coastal islands, A Migali and C Pensa (University of Naples Federico II, Italy)

Water-jet propulsion for a 3,500 t corvette from Blohm & Voss, J Wessel

CFD simulations of the flow through a water-jet installation, N W H Bulten and R Verbeek

Simulation research of turning course of water-jet propelled catamarans, J M Ding *et al*

Modelling of water-jets in a propulsion system, J E Buckingham

2-D analytical formulation to calculate water-jet performance under dynamic similitude, E Benini *et al*

Performance mapping of bubbly water ram-jet, M Mor and A Gany

A new proposal for a passenger fast ferry operating in the Italian archipelago: structural analysis by finite element method, T Coppola and F Simeone (University of Naples Federico II, Italy)

On the structural analysis of the staggered multihull, T Coppola (University of Naples Federico II, Italy)

CRP pod propulsion for fast Ro-Pax ship, R Depescale *et al* (CETENA, Italy)

Evaluation of wave loads on a SWATH in short crest sea, X B Chen *et al* (Bureau Veritas, France)

Experimental theoretic researches and optimal design of local strengthening for thin-walled structures of marine vehicles, V S Hudramovich *et al* (National Academy of Science, Ukraine)

Platform Management systems for high-speed naval vessels: designing a system for minimal manning, A Stefani (CAE Italy Sr, Genoa, Italy)

Experimental study of non-stationary flow around a profile: application to the hydrofoil, L Hammadi *et al* (Faculty of Architect and Civil Engineering, Oran, Algeria)

Hydrofoil drag reduction by partial cavitation and cross-flow effects, E L Amromin *et al* (Mechmath LLC, Prior Lake, US)

An alternative approach to the calculation of frictional resistance of the hull, M Simeone (University of Naples Federico II, Italy)

The Vedette V2000. Latest naval units of the Italian Revenue Guard Corp, V Giusti and G Zacchardi (Guardia di Finanza, Italy)

Redundance aspects of mono and multi-hull HSC, P Securius Germanischer Lloyd)

Hydrodynamic analysis (CFD) of a hull rigged with new stabilising systems, V Filardi *et al*

Development of a methodology for the environmental management of the maritime transport companies: the EMAS-SHIP project, F Beltrano *et al*

Performance and safety aspects of multihull design, A Nazarov (Sevastopol National University, Ukraine).

Papers presented at the 2nd Fast Ferry Information Conference, 28–29 September 2005, London

A new approach to waterborne transportation in New York harbor, Tom Fox (New York Water Taxi)

Developing the fast ferry market in the French West Indies, Roland Bellemare (L'Express des Îles)

The Halifax/Bedford fast ferry business case – a first step in risk management, Rob Gair (TDV Global) & Brian Taylor (Halifax Regional Municipality)

Travellers' preferences and hierarchy of needs, Arne Osmundsvaag (FjordExpressen)

Ferries in a changing world – a broker's perspective, John Aitkenhead (International Broking Services)

High-speed ships and the collision regulations, David Norman (Condor Ferries Marine Services)

The safety record of fast ferries, Stephen J Phillips (Seaspeed Technology)

Latest revisions to the IMO High-Speed Craft Code, Andrew Blyth (Blyth Bridges Marine Consultants)

Fast ferry expansion on the River Thames, Sean Collins (Thames Clippers)

Development of fast vehicle ferry business in North America and the Caribbean, Mark MacDonald (Northumberland Bay Ferries)

Operating fast passenger/car ferries in the Dodecanese Islands, Christos G Spanos (Dodekanisos Naftiliaki)

The development of a new high-speed Ro-Pax ferry for Alaska, John Bonafoux (BMT Nigel Gee & Associates)

The evolution of the FlyingCat 46 and FerryCat ranges, Edmund Tolo (Fjellstrand)

Austal Auto Express 127 trimaran – experiences to date, Michael Wake (Austal Ships)

Papers presented at the 30th Interferry Conference, 4–7 October 2005, Athens

Is there a future for fast ferries? David Benson (Sea Containers, UK)

Ferry safety in developing countries, Anders Hansen (Ole Steen Knudsen, Denmark)

Standardized rules for domestic ferries, Hilde Thunes (DNV, Norway)

Consolidation amongst ferry operators in the Mediterranean ferry markets – what does the future hold? Antonis Maniadakis (Minoan Line, Greece)

Conventional vs high-speed ferries in the Greek market, Gerasimos Strintzis (Hellenic Seaways, Greece)

Developments in Turkey, Dr Ahmet Paksoy (IDO, Turkey)

Opportunities in the Greek coastal market – are there any? George Xiradakis (XRTC, Greece)

Destination marketing in the high-speed ferry business, Mark McDonald (NFL-Bay Ferries, Canada)

Ferry development in Hawaii, Terry White (Hawaii Superferry, US)

Shipping policy developments in the EU, Alfons Guinier (ECSA, Belgium)

EU policy on short sea shipping from an operator's perspective, Paul Kyprianou (Grimaldi Napoli Ferries, Italy)

Ferry business development and regulatory obstacles, Mihalis Sakellis (Blue Star Ferries, Greece)

Sulphur scrubbers John Garner and Don Gregory (P&O Ferries, UK and BP Marine, UK)

LNG cruise ferry – a truly environmentally sound ship, Oskar Levander (Wärtsilä, Finland)

Gas as ship fuel for motors today; for fuel cells in future? Gerd Würsig (Germanischer Lloyd, Germany)

Emission credit trading, Mel Davies (British Maritime Technology, UK)

Papers presented at the 31st Interferry Conference, 3–6 October 2006, Long Beach, California

Packaging the product, Darrell Bryan (Clipper Navigation, US)

The value game, Soren Jespersen (DFDS, Denmark)

Branding to increase sales, Geoff Dickson (BC Ferries, Canada)
The role of PR in promoting the product, Paul Ellis (Direct Public Relations, UK)
Operating fast ferries in Alaska, Captain John Falvey (Alaska Marine Highway, US)
Trimaran operating experience, James Bennett (Austal, Australia)
Design priorities from an operator's perspective, Jen-ning Wang (Shun Tak Turbo Jet, Hong Kong)
Naval applications, Matt Nichols (Nichols Bros, US)
Aluminium corrosion, Zayna Connor (Alcan Rolled Products LLC, US)
Night vision, Doug Houghton (Current Corporation, Canada)
Learning from the low-cost airlines, (Revenue Technology Systems, US)
An australian experience, Jeff Ellison (Kangaroo Island Sealink, Australia)
Tourism and ferry operations: challenges and solutions, Ted Bell (Moffat & Nichol, US)
A new system in Hawaii, Terry White (Hawaii Superferry, US)
Great lakes: a european perspective, Ger van Langen (Wagenborg, Netherlands)
Developing Tasmanian tourism, Peter Simmons (TT Line, Australia)
The changing market in New York City, Tom Fox (New York Water Taxi, US)
Public/private partnerships in Puget Sound, Greg Dronkert (Hornblower Marine, US)
Security at ferry terminals, Captain Jim Bamberger (US Transportation Security Administration, US)
Balancing security with commerce, Beth Gedney (Passenger Vessel Association, US)
Meeting air quality standards: products, technology and trading, Kevin James (BP Marine, UK)
Emission trading in California, Mahesh Talwar (Ocean Air Environmental, US)
Roads vs ferries, Steve Castleberry (San Francisco Water Transit Authority, US)
Ship recycling – upcoming challenges for all players, Henning Gramann (Germanischer Lloyd, Germany)
Propellers and fuel efficiency, Robert Otto and Teus van Beek (Wärtsilä, Netherlands)
Energy management and emissions control, Zabi Bazari (Lloyd's Register, UK)

Papers presented at the 3rd Fast Ferry Information Conference, 26–27 September 2007, London

Developing the Austal Auto Express 107m catamaran, James Bennett (Austal Ships, Australia)
The Return of a World Class Bay Area Ferry System, Steve Castleberry (San Francisco Bay Area Water Transit Authority, USA)
The Design and Construction of Incat's 112m Wave Piercers, Robert Clifford (Incat, Australia)
A Change of Pace in Fast Ferry Services on the River Thames, Sean Collins (Thames Clippers, UK)
Operating a Fast Ferry Service in Queensland, Terry Dodd (Sun Ferries, Australia)
A New Medium Size Fast Ferry for Rough Water Applications, Ed Dudson (BMT Nigel Gee Fast Ferries, UK)
The Franchise Approach, Chris Howe-Davies (HD Ferries, UK)
Challenges Faced by Fast Ferry Operators in Hong Kong, John C. Y. Hui (New World First Ferry, Hong Kong)
The Introduction of New Ferry Routes in the Middle East, Steve Hunt (Thompson Clarke Shipping, Australia)
A Survey of Ride Control Effectors, Joe Kubinec (VT Maritime Dynamics, USA)
A Potential Role for Hovercraft on the Forth Estuary in Scotland, Alistair Macleod (Ferry Consultant to the Stagecoach Group, UK)
Two New Design Concepts from the Rodriquez Group, Marco Pavoncelli (Rodriquez, Italy)
River Runner Fleet for the Thames: Design Advances Marc Richards (AIMTEK, Australia)
Fast Ferries on the Norwegian coast, Magnar Svendsen (Fjord1 Fylkesbaatane International, Norway)
Coastal Fast Ferry Services in Croatia, Luka Tomic (UTO Kapetan Luka, Croatia)

Papers presented at the 9th International Conference on Fast Sea Transportation, 23–27 September 2007, Shanghai

An Overview of Yellow Sea Transportation System, Jae Wook Lee, Seung-Hee Lee (Inha University, Korea)
Advances in Technology of High Performance Ships in China, You-Sheng Wu, Qi-Jun Ni and Wei-Zhen Ge (China Ship Scientific Research Center, China)
Container Ship and Port Development: A Review of State-of-the-Art, Branislav Dragović and Dong-Keun Ryoo (Korea Maritime University, Korea)
JHSS (Joint High-Speed Sealift Ship) Hull Form Development, Test and Evaluation Siu C. Fung *et al* (Carderock Division, Naval Surface Warfare Center (NSWCCD), USA)
Hard Chine Design with Developable Surfaces, F. Péres-Arribaz (Naval Architecture School of Madrid, Universidad Politécnica de Madrid, Spain)
Fast Ship Motions in Coastal Regions, Ray-Qing Lin and John G. Hoyt (Naval Surface Warfare Center, Carderock Division, USA)
Seakeeping Analysis of the Lifting Body Technology Demonstrator Sea Flyer Using Advanced Time-Domain Hydrodynamics, Christopher J. Hart, Todd J. Peltzer (Navatek Ltd, USA), Kenneth M. Weems (Science Applications International Corporation, USA)
Predicting Motions of High-Speed Rigid Inflatable Boats: Improved Wedge Impact Prediction, D.A. Hudson, Stephen R. Turnock and Simon G. Lewis (University of Southampton, UK)
Erosion Damages on Propellers and Rudders Caused by Cavitation, Juergen Friesch (Hamburgische Schiffbau-Versuchsanstalt GmbH (HSVA), Germany)
Development of New Waterjet Installations for Applications with Reduced Transom Width, Norbert Bulten and Robert Verbeek (Wärtsilä Propulsion Netherlands, The Netherlands)
Very Large Waterjet with Adjustable Tip Clearance, Mats Heder (Kamewa Waterjets, Rolls-Royce AB, Sweden)
Development of an Integrated Monitoring System and Monitoring of Global Hull Loadings on High Speed Mono-Hull, Seppo Kivimaa and Antti Rantanen (VTT Vehicle Engineering, Finland)
Numerical Simulation of Whipping Responses induced by Stern Slamming Loads in Following Waves, Han-Bing Luo *et al* (China Ship Scientific Research Center, China)
Full-Scale Design Evaluation of the Visby Class Corvette, Anders Rosén, Karl Garme and Jakob Kuttenkeuler (KTH Centre for Naval Architecture, Sweden)
Design Development and Evaluation Of Affordable High Speed Naval Vessels for Offshore Service, Rubin Sheinberg *et al* (U.S. Coast Guard, USA), Lex Keuning (Delft Technical University, Netherlands)
The Development of ACV Technology in China, Tao Ma *et al* (Marine Design & Research Institute of China (MARIC), China)
Improvement of Taking-off and Alighting Performances of a Flying Boat Utilizing Hydrofoil, Yoshiaki Hirakawa *et al* (Yokohama National University, Japan)
Porpoising and Dynamic Behavior of Planing Vessels in Calm Water, Hui Sun and Odd M. Faltinsen (Norwegian University of Science and Technology, Norway)
Numerical Analysis of Seakeeping Performances for High Speed Catamarans in Waves, Yoshiyuki Inoue (Yokohama National University, Japan), Md. Kamruzzaman (Nippon Kaiji Kyokai, ClassNK, Japan)
Trimaran Motions and Hydrodynamic Interaction of Side Hulls, Yuefeng Wei, Wenyang Duan and Shan Ma (Harbin Engineering University, China)
Propeller Wake Evolution, Instability and Breakdown by Flow Measurements and High Speed Visualizations, Mario Felli (INSEAN, Italy), G. Guj and R. Camusi (University of "Roma Tre", Italy)
Prediction of Open Water Characteristics of Podded Propulsors Using a Coupled Viscous/Potential Solver, Vladimir I. Krasilnikov and Jia Ying Sun (MARINTEK, Norway), Alexander S. Achkinadze and Dmitry V. Ponkratov (State Marine Technical University, Russia)
Steady Analysis of Viscous Flow around Ducted Propellers: Validation and Study on Scale Effects, Vladimir Krasilnikov (MARINTEK, Norway), Zhi-Rong Zhang and Fang-Wen Hong (CSSRC, China), Dmitry V. Ponkratov (State Marine Technical University, Russia)
The Method for Evaluating the Design Wave Loads on SWATH Ships, Ji-ru Lin *et al* (China Ship Scientific Research Center, China)
Analysis of Bending Moments in Surface Effect Ship Structure by Russian Regulation, Ali Dehghanian, Kambiz Alempour (Hydro Aerostatic Dept, MT University, Iran), Hamid Amini (Sharif Technical University, Iran)
The Whipping Vibratory Response of a Hydroelastic Segmented Catamaran Model, Jason Lavroff *et al* (University of Tasmania, Australia), Giles Thomas (Australian Maritime College, Australia)
Practical Aspects of the Classification and Survey According to RS Instruments, Vladimir V. Gadalov, Mikhail A. Gappoev and Mikhail A. Kuteynikov (Russian Maritime Register of Shipping, Russia)
Development of a Wing-In-Surface-Effect Ship for Research Purposes in Cooperation Between Vietnam and Japan, Nguyen Tien Khiem *et al* (Institute of Mechanics, Vietnam Academy of Science and Technology, Vietnam) Hiromichi Akimoto (University of Tokyo, Japan)
Preliminary Conceptual Design of 20-Passenger Class WIG Craft, Myung-Soo Shin *et al* (Maritime and Ocean Engineering Research Institute, Korea)
Prediction of Hydrodynamics Performance of Catamarans Accounting for Viscous Effects, Xue-Liang Wang, Xue-Kang Gu and Quan-Ming Miao (China Ship Scientific Research Center (CSSRC), China)
A Comparison of Roll Prediction Algorithms with Model Test Data of a High Speed Trimaran, Allen Engle and Ray-Qing Lin (David Taylor Model Basin(NSWCCD), USA)
Catamaran Motions in Beam and Oblique Seas, Giles Thomas and Mani hackett (Australian Maritime College, Australia), Lawrence J. Doctors (The University of New South Wales, Australia), Patrick Couse (Sunnypowers Ltd, France)
Analysis and Design of a Hydrofoil for the Motion Control, Ching-Yeh Hsin (National Taiwan Ocean University, Taiwan), Hua-Tung Wu and Chun-Hsien Wu (United Ship Design and Development Center, Taiwan)
Research on Plane Maneuverability Stability of ACV by Phase Plane Method, Chunguang Liu, Pingping Tao and Tao Ma (Marine Design & Research Institute of China, China)

Validation of a 4DOF Manoeuvring Model of a High-speed Vehicle-Passenger Trimaran, Tristan Perez *et al* (Norwegian University of Science and Technology, Norway), Tim Mak and Tony Armstrong (Austal Ships, Australia)

A Practical Method for Evaluating Steady Flow about a Ship, Chi Yang and Hyun Yul Kim (George Mason University, USA), Francis Noblesse (NSWCCD, USA)

Simulations of Ship Flows at High Froude Numbers Using Smoothed Particles Hydrodynamics, Guillaume Oger *et al* (Ecole Centrale de Nantes, France)

Numerical Investigation of the Wave Pattern and Resistance of the Naval Combatant INSEAN 2340 Model, Andreja Werner *et al*, University of Zagreb, Croatia

Trajectory Tracking for an Ultralight WIG, Caterina Grillo *et al* (Flight Mechanics Division, Dept. of Transportation Engineering, University of Palermo, Italy)

Design Features of an Unconventional Passenger Vessel with Low Environmental Impact, Dario Boote and D.Mascia (University of Genova, Italy)

A New Paradigm for High-Speed Monohulls: the Bow Lifting Body Ship, Todd J. Peltzer *et al* (Navatek, Ltd., USA)

On the Parametric Rolling of Ships in Regular Seas Using a Numerical Simulation Method, Bor-Chau Chang (National Kaohsiung Marine University, Taiwan)

Experimental and Theoretical Study of the Roll Stability of Hovercraft Moving at Yaw, Zong-Ke Zhang, Ping-Ping Tao and Tao Ma (Marine Design & Research Institute of China (MARIC), China)

Active Motion Control of High-Speed Vessels in Waves by Hydrofoils, Jang-Whan Bai and Yonghwan Kim (Seoul National University, Korea)

Development of a Nonlinear Simulation for Testing of Control Systems in a General Class of Lifting Body Vessels, SWATHs, and Hydrofoils, Benjamin Rosenthal (Navatek Ltd., USA)

Analysis of Asymmetrical Shaft Power Increase during Tight Manoeuvres, Michele Viviani and Carlo Podenzana Bonvino (Genoa University, Italy), Salvatore Mauro (INSEAN, Rome, Italy) *et al*

Towards Numerical Dynamic Stability Predictions of Semi-Displacement Vessels, Wei Zhu and Odd M. Faltinsen (Norwegian University of Science and Technology, Norway)

Resistance and Research on Multi-hull's Configuration Based on New Slender-Ship Wave Resistance Theory, Duanfeng Han, Haipeng Zhang and Hongde Qin (College of Shipbuilding, Harbin Engineering University, China)

Experimental Investigations of the Waves Generated by High-speed Ferries, Dimitris S.Chalkias and Gregory J. Grigoropoulos (National Technical University of Athens, Greece)

Theory and Experimental Study on the Pentamaran Wave Making Resistance Characteristics, Junsong He, Zhen Chen and Xi Xiao (Shanghai Jiaotong University, China)

The Effect of Air Cushion on the Slamming Pressure Peak Value of Trimaran Cross Structure, Zhenglin Cao and Weiguo Wu (Wuhan University of Technology, China)

The Effect of Speed and Sea State for Probability of Ships Slamming, Zhen Chen and Xi Xiao (Shanghai Jiaotong University, China)

Computational Modelling of Wet Deck Slam Loads with Reference to Sea Trials, Michael R. Davis (University of Tasmania, Australia, James R. Whelan (INTEC Engineering Pty. Ltd., Australia), Giles A.Thomas (Australian Maritime College, Tasmania, Australia)

Passenger Comfort Assessment Method for High Speed Craft Design, Antti Rantanen and Seppo Kivimaa (VTT Vehicle Engineering, Finland)

Numerical and Experimental Study of Green Water on a Moving FPSO, Xiufeng Liang and Jianmin Yang (Shanghai Jiao Tong University, China), Chi Yang, Haidong Lu and Rainald Löhner, (George Mason University, USA)

Numerical Studies on the Hydrodynamic Performance and the Start-up Stability of High Speed Ship Hulls with Air Plenums and Air Tunnels, Jin-Keun Choi, Chao-Tsung Hsiao and Georges L. Chahine (Dynaflow, Inc., USA)

Concepts & Principles for Creating an Autonomous and Intelligent WIG Vehicle for Coastal Patrolling and Search & Rescue Operations, Alexander Nebylov and Sukrit Sharan (International Institute for Advanced Aerospace Technologies of State Univ. of Aerospace Instrumentation) Russia

Research on the Relationship between the Required Power for Level Flying and Flight Height Stability of WIG Craft, Chang-Hua Yuan and Ya-Jun Shi (China Ship Scientific Research Center, China)

Investigation on Numerical Prediction of WIG's Aerodynamics and Longitudinal Stability, Fu Xing, Chang-Hua Yuan and Bao-Shan Wu (China Ship Scientific Research Center, China)

The Effect of Draft on Bulbous Bow Performance, Richard A. Royce and Patrick J.Doherty (Webb Institute, USA)

Performance of a Stern Flap with Waterjet Propulsion, Michael B. Wilson, Scott Gowing and Cheng-Wen Lin (Naval Surface Warfare Center, Carderock Division, USA)

On the Effect of Transom Area on the Resistance of Hi-Speed Mono-Hulls, Jacques B. Hadler (Webb Institute, USA), Jessica L. Kleist (NSWCCD – Ship Systems Engineering Station, USA), Matthew L. Unger (Seaworthy Systems Inc., USA)

Research on FEM Generation Techniques in Ship CAE Analysis, Jian-hai Jin *et al* (China Ship Scientific Research Center, China), Hai Pu (Southern Yangtze University, China)

Influence of Wave-induced Ship Hull Vibrations on Fatigue Damage, Jong-Jin Jung *et al* (Maritime Research Institute, Hyundai Heavy Industries Co. Ltd. Korea)

Structural Design of Ramp in Aluminum Alloy for ACV, Ping Zhang *et al* (Marine Design & Research Institute of China (MARIC), China)

Development of IMO Requirements to Qualification of Officers on WIG Craft, Alexander I. Bogdanov (Central Marine Research & Design Institute, Russia)

The Generic Management System Approach for Addressing Maritime Emergency Scenario Situations, Chengi Kuo (University of Strathclyde, UK), Andy Humphreys and Stuart Wallace (Stena Line, U.K).

Robust Real-Time Microcontroller-based Control Hardware for a 21.3 m Bow Lifting Body Technology Demonstrator Craft, Robert Knapp, John Elm, and Brian Kays (Navatek, Ltd., USA)

Development of 5-blades SPP Series for Fast Speed Boats, A.V. Pustoshny *et al* (Krylov Shipbuilding Research Institute, Russia)

A Series of Surface Piercing Propellers and Its Application, Enbao Ding (China Ship Scientific Research Center (CSSRC), China)

Mathematical Expressions of Thrust and Torque of Gawn-Burrill Propeller Series for High Speed Crafts Using Artificial Neural Networks, Kourosh Koushan (MARINTEK, Norway)

The Decay of Catamaran Wave Wake in Shallow Water, Alex Robbins *et al* (Australian Maritime College, Australia), Ian Dand (BMT SeaTech Ltd, Southampton, England)

Combined Numerical and Experimental Evaluation of the Flow Field around a Racing Yacht, Stelios G. Perissakis, Gregory J. Grigoropoulos and Dimitris E. Liarokapis (National Technical University of Athens (NTUA), Greece)

Investigation of Planing Craft in Shallow Water, Benjamin Friedhoff (Institute of Ship Technology and Transport Systems (IST), Germany) Rupert Henn, Tao Jiang and Norbert Stuntz (Development Center for Ship Technology and Transport Systems (DST), Germany)

The Dynaplane Design for Planing Motorboats, Eugene P. Clement and John G Hoyt (Naval Surface Warfare Center, USA), Lawrence J. Doctors (The University of New South Wales, Australia

Optimization of Planing Hull Structure Design, Santini Julien *et al* (Alltair Engineering, Michigan, USA)

Experimental Investigation of a Composite Patch Reinforced Cracked Steel Plate in Static Loading, Lazaros S.Mirisiotis and Nicholas G. Tsouvalis (National Technical University of Athens, Greece)

The Right Level of Composite Technology, Richard Downs-Honey (High Modulus, Auckland, New Zealand)

Theory and Practice of Application of the Interceptors on High-speed Ships, Gregory Fridman and Kirill Rozhdestvensky (St.Petersburg, State Marine Technical University (SMTU), Russia) Alexander Shlyakhtenko (Marine Design Bureau "Almaz", Russia)

Experimental Investigation of Interceptor Performance, Sverre Steen (Norwegian University of Science and Technology (NTNU), Norway)

Study on the Gas Turbine Inlet System of a Hovercraft, Dejuan Chen, Weizhong Qian and Jun Sun (Marine Design & Research Institute of China (MARIC), China)

Influence of Increased Weight on SES-performance in a Seaway, Christian Wines and Hans Olav Midtun (Norwegian Defence Systems Management Division, Norway), Sverre Steen (Norwegian University of Science and Technology (NTNU), Norway), Magnus Tvete (Norwegian Marine Technology Research Institute (MARINTEK), Norway)

Research on Modeling and Simulation for WIG Craft Space Motion, Qian Zhou *et al* (China Ship Scientific Research Center (CSSRC), China)

Self-propulsion Model Test of a Wing-In-Surface-Effect-Ship with Canard Configuration, Hiromichi Akimoto (The University of Tokyo, Japan), Syozo Kubo and Masahide Kawakami (Tottori University, Tottori, Japan)

Experimental Study on the Hull Form of High-speed Air Cavity Craft, Wencai Dong *et al* (Naval University of Engineering, China)

Potential of the Artificial Air Cavity Technology for Raising the Economic Efficiency of China's Inland Waterway Shipping, Andrey V. Sverchkov (Krylov Shipbuilding Research Institute, Russia)

Experimental Method for Calculation Drags Reduction in Air Cavity Boat, Ahmad Fakhraee, Manucher Rad and Hamid Amini (Mechanical School, Sharif University of Technology, Iran)

Bibliographies and Glossaries

Bibliography and Proposed Symbols on Hydrodynamic Technology as Related to Model Tests of High-Speed Marine Vehicles, SSPA, Public Research Report No 101, 1984, SSPA, PO Box 24001, S-400 22, Gothenburg, Sweden

Bibliography on Hovercraft (ACV), Compiled by H G Russell, TIL Reports Centre, UK, September 1968, TIL/BIB/101

Glossary for High-Speed Surface Craft, The Society of Naval Architects and Marine Engineers, 1 World Trade Center, Suite 1369, New York, New York 10048, US

High-Speed Waterborne Passenger Operations and Craft: Bibliography, US Department of Transportation, Urban Mass Transportation Administration, Washington DC, US, UMTA-IT-32-0001- 84-2, August 1984

ITTC Dictionary of Ship Hydrodynamics, Maritime Technology Monograph No 6, 1978. The Royal Institution of Naval Architects, London, August 1978.

General interest books

Amazon Task Force, Peter Dixon. Hodder and Stoughton, London, UK, 1981. ISBN 0 340 32713 8 and 0 340 34578 0 Pbk

An Introduction to Hovercraft and Hoverports, Cross & O'Flaherty. Pitman Publishing/Juanita Kalerghi

A Quest for Speed at Sea, Christopher Dawson. Hutchinson & Company Ltd. ISBN 0 09 109720 7

Ekranoplanes – Controlled Flight Close to the Sea, A V Nebylov and P Wilson, WIT Press. ISBN 1 85312 831 7

Fast Ferries, Paul Hynds, Lloyds business intelligence centre. ISBN 1859780571, 1996

Fast Ferries – Shaping the Ferry Market for the 21st Century, Drewry Shipping Consultants, 1997

From Sea to Air – the Heritage of Sam Saunders, A E Tagg and R L Wheeler, Crossprint, Daish Way, Dodnor Industrial Estate, Newport, Isle of Wight, UK, pp 316, 1989. ISBN 0 9509739 3 9

High-Speed Car Ferries – Economic and Technical Performance, Paul Hynds, Lloyd's shipping economist management report. ISBN 1859788904, 1998

Hover Craft, Angela Croome. 4th edition, 1984 Hodder and Stoughton Ltd. ISBN 0-340- 33201-8, ISBN 0-340-33054-6 Pbk

Hovercraft and Hydrofoils, Roy McLeavy. Blandford Press Ltd. ISBN 0 7137 07674, pp 215

Hurtigbåten (The Fast Craft), Bjørn Foss, Published by: Nordvest Forlag, Norway, 1989, pp 160. In colour. ISBN 82 90330 464

Hydrofoils and Hovercraft, Bill Gunston. Aldous Books, London, UK, 1969. ISBN 490 00135 1 and 490 00136 X

INCAT – The first 40 years, Baird Publications, Melborne, Australia, 1998

Light Hovercraft Handbook, (ed) Neil MacDonald. Hoverclub of Great Britain Ltd (available from 45 St Andrews Road, Lower Bemerton, Salisbury, Wiltshire), 1976

Power Boat Speed, Racing and Record-Breaking: 1897 to the Present, Kevin Desmond. Conway Maritime Press Ltd, London, UK, 1988. ISBN 0 85177 427 X, pp 256

Ships and Shipping of Tomorrow, Rolf Schonknecht, Jurgen Lusch, Manfred Schelzel, Hans Obenaus, Faculty of Maritime Transport Economics, Wilhelm-Pieck University, Rostock, Germany. MacGregor Publications Ltd, Hounslow, UK, pp 240, 1987

The Great Himalayan Passage, Adventure Extraordinary by Hovercraft, Michel Peissel. William Collins Sons & Company Ltd, London, UK, 1974. ISBN 0 00 211841 6

The Hoverspeed Story, Miles Cowsill & John Hendy, Ferry Publications, 12 Millields Close, Pentlepoir, Kilgetty, Pembrokeshire SA68 0SA, UK

The Interservice Hovercraft (Trials) Unit, B J Russell. Hover Publications, 1979. ISBN 0 9506 4700 4

The Law of Hovercraft, L J Kovats. Lloyd's of London Press Ltd, 1989

The Law of Shipbuilding Contracts, Simon Curtis (Partner, Watson, Farley & Williams, London) Lloyds of London Press Ltd, Sheepen Place, Colchester, Essex C03 3LP, UK, pp330, 1991. ISBN 10 85044 334 3

Worldwide High-Speed Ferries, Paul Hynds (Hoverspeed), Conway Maritime Press Ltd, 101 Fleet Street, London EC4Y IDE, UK, pp 160, 1992. ISBN 0 0 85177 587 X

Tourism and the future role of ferries and fast ferries, G P Wild (International) Ltd, 15 Gander Hill, Haywards Heath, West Sussex, UK, 1998

Twin Deliveries, J Fogagnolo. International Catamarans Pty Ltd, Hobart, Tasmania. Paperback, 1986

Waterborne Transport – Potential and Future Prospects, International waterborne transport group of UITP, 1997

The World Fast Ferry Market, Baird Publications. ISBN 09581013OX (2 editions)

Market Forecasts for High-Speed Marine Craft, their Materials and Equipment, S Phillips, Jane's Information Group, 2000. ISBN 0 7106 2145 0

The World High Speed CargoVessel Market, Baird Publications, 2002

Fast Ferry Industry Outlook and Future Prospects to 2010, G P Wild (International) Ltd, 15 Gander Hill, Haywards Heath, West Sussex, UK, 2002.

Technical books

Air Cushion Craft Development (First Revision), P J Mantle (Mantle Engineering Company, Inc, Alexandria, Virginia, US). David W Taylor, Naval Ship Research and Development Center DTNSRDC-80/012, January 1980, pp 593

Design of High-Speed Boats. Vol 1 Planing, P R Payne. Fishergate Inc, Annapolis, Maryland, US, pp 244, 1988. ISBN 0 942 720 06 07

Dynamics of Marine Vehicles, Rameswar Bhattacharyya (US Naval Academy, Annapolis, Maryland, US). John Wiley & Sons, 1978, ISBN 0 471 07206 0, pp 498

Encyclopaedia of Composite Materials and Components, Editor: M Grayson, 1983. pp 1100, available from: AIAA, order number 57-8, ISBN 0 471 87357 8

Fibre Reinforced Composites 1986, Institution of Mechanical Engineers publication 1986, ISBN 0 8529 8589 4/297, pp 262

Global Wave Statistics, N Hogben (British Maritime Technology Ltd). Unwin Brothers Ltd, Woking, UK, 1987, pp 656, £295

High-Speed Small Craft by P du Cane. Temple Press Books Ltd, London, 3rd edition 1964, pp 470

Hovercraft Design and Construction, Elsley & Devereaux. David & Charles, Newton Abbot, UK, 1968, pp 262

Hovercraft Technology, Economics and Applications. Editor: J R Amyot, Elsevier Science Publishers BV, 1989, ISBN 0-444-88152-2 & 0-444-41872-5, pp 770, DFI.395.00

Hydrodynamics of High-speed Marine Vehicles, Odd M. Faltinsen (Norwegian University of Science and Technology), Cambridge University Press, USA, 0 521 84568 8

Industrial Fans-Aerodynamic Design, Papers presented at a seminar organised by the Fluid Machinery Committee of the Power Industries Division of the IMechE, held London, 9 April 1987. Published: MEP Ltd

Jane's High-Speed Marine Craft (annual 1986–1995) compiler and editor: Robert Trillo and Stephen J Phillips, published by Jane's Information Group

Jane's High-Speed Marine Transportation (annual 1996–2008) compiler and editor: Stephen J Phillips, published by Jane's Information Group

Jane's Surface Skimmers (annual 1967–1985) compiler and editor: Roy McLeavy, published by Jane's Publishing Company

Marine Gas Turbines, J B Woodward. Wiley-Interscience (1-95962-6), 1975, pp 390.

Marine Hovercraft Technology, Robert L Trillo, pp 245, 1971. ISBN 0 249 44036 9. Available from Robert L Trillo, Broadlands, Brockenhurst, Hampshire SO42 7SX.

Marine Rudders and Control Surfaces Tony Molland and Stephen Turnock, published by Butterworth-Heinemann, UK, 2007, ISBN 0 750 66944 6

Mechanics of Marine Vehicles, B R Clayton and R E D Bishop, Gulf Publishing Company, Houston, US (not available in Bangladesh, Europe, Ireland or the UK)

Multi-Hull Ships , V Dubrovsky and A Lyakhovitsky, ISBN 0-9644311-2-2

Resistance and Propulsion of Ships, Svend Aage Harvald (The Technical University of Denmark). John Wiley and Sons, 1983, ISSN 0275 8741 and ISBN 0-471-06353-3, pp 353

Ships with Outriggers, Victor Dubrovsky, ISBN 0-9742019-0-1

The Leading Star of Future Overwater Commuter Transport – WIG, Shigenori Ando, Professor of Aeronautical Engineering, Nagoya University, Japan (In Japanese)

The Marine Encyclopaedia Dictionary (2nd Edition), E Sullivan, May 1988, ISBN 1 85044 180 4, pp 468

Transport Ships on Hydrofoils, Blumin, Massejef and Ivanof Isdatelstvo Transport, Basmannij Tupik, D 6a, Moskowsaja Tipografija Nr 33, Glawpoligrafproma, Moscow, Russian Federation, 1964

Theory and Design of Hovercraft, Professor Yun Liang, Marine Design & Research Institute of China (MARIC), Shanghai, China, pp 344, ISBN 7-118-00745-5/U.62. National Defence Industry Press, China, June 1990 (in Chinese). Industry Press, China, June 1990.

Periodicals

Cruise Ferry Info, ISSN 1102-934X. Marine Trading AB, Brogantan 7, S-302 43 Halmstad, Sweden

Fast Ferry International, (10 issues annually) ISSN 0954 3988, Fast Ferry Information Ltd, 14 Marsden Gate, Winchester, SO23 7DS, UK

Ferry Technology, ISSN 1746-1111. Riviera Maritime Media Ltd, 1D Queen Ann's Place, Bush Hill Park, Enfield EN1 2QB, UK

Hovercraft Bulletin, The Hovercraft Society, 24 Jellicoe Avenue, Alverstoke, Gosport, Hampshire PO12 2PE, UK

Marine Log, (monthly) ISSN 08970491, USPS 576-910, Simmons – Boardman Publications, 345 Hudson Street, New York, New York 10014, US

Maritime Reporter and Engineering News, (monthly) ISSN 00253448, USPS 016 750, Maritime Activity Reports, 118 east 25th street, New York, New York 10010, US

Ship and Boat International, (10 issues annually) 10 Upper Belgrave Street, London SW1X 8BQ, UK

Small Ships, (bi-monthly) ISSN 0262 480X, International Trade Publications Ltd, Queensway House, 2 Queensway, Redhill, Surrey RH1 1QS, UK

Work Boat World, (monthly) ISSN 1037-3748, Baird Publications Ltd 1993, 4A Carmelite Street, London EC4Y 0BN, UK and 10 Oxford Street, South Yarra, Melbourne 3141, Australia.

INDEXES

Index of Organisations

C

D

Insel – und Halligreederei (Adler-Schiffe) *civil operators* (Germany) ... 380
Inserve Ltd *consultants/designers* (United Kingdom) ... 553
Inter Island Boats *civil operators* (Seychelles) ... 409
Inter Island Ferries *civil operators* (Australia) ... 367
Interferry *societies* (Canada) ... 564
Intermar Clarion Shipping Co Ltd *civil operators* (Cuba) ... 376
Intermarine SPA *fast interceptors* (Italy) ... 300
Intermarine SpA *high-speed patrol craft* (Italy) ... 348
International Association of Public Transport (UITP) *societies* (Belgium) ... 564
International Fast Ferries *civil operators* (Egypt) ... 377
International Hovercraft Ltd *consultants/designers* (United Kingdom) ... 553
Interstate Navigation *civil operators* (United States) ... 425
Introduction *bibliography* ... 569
Ionian Cruises *civil operators* (Greece) ... 382
Ireland *regulatory authorities* (Ireland) ... 536
Iris Catamarans *high-speed multihull* (France) ... 123
Irish Ferries *civil operators* (Ireland) ... 389
Iscomar Ferrys *civil operators* (Spain) ... 413
Ishizaki Kisen Company Ltd *civil operators* (Japan) ... 395
Island Boats Inc *high-speed multihull* (United States) ... 193
Island Ferries Teo *civil operators* (Ireland) ... 389
Island HI-speed Ferry *civil operators* (United States) ... 426
Island Lynx Ferry Company *civil operators* (Virgin Islands (US)) ... 431
Island Packers Cruises Inc *civil operators* (United States) ... 426
Isle of Man Steam Packet Company *civil operators* (United Kingdom) ... 419
Isotta Fraschini Motori SpA *engines* (Italy) ... 450
Israel *regulatory authorities* (Israel) ... 536
Israel Aerospace Industries *fast interceptors* (Israel) ... 298
Israel Shipyards *fast interceptors* (Israel) ... 298
Israel Shipyards *high-speed patrol craft* (Israel) ... 347
Istanbul Fast Ferries Inc *civil operators* (Turkey) ... 416
Italthai Marine Ltd *high-speed multihull* (Thailand) ... 170
Italy *regulatory authorities* (Italy) ... 536
Ivante *civil operators* (Croatia) ... 375
Ivchenko Progress Design Bureau *engines* (Ukraine) ... 457
Iveco Motors *engines* (Italy) ... 450

J

J B Hargrave Yacht Design *consultants/designers* (United States) ... 558
Jadrolinija *civil operators* (Croatia) ... 375
Jamaica *regulatory authorities* (Jamaica) ... 536
Japan *regulatory authorities* (Japan) ... 536
Japan Coast Guard *civil operators* (Japan) ... 395
JB Marine Consultancy *consultants/designers* (United Kingdom) ... 553
Jet Express *civil operators* (United States) ... 426
Jiang Men Passenger Shipping *civil operators* (China) ... 374
Jianghui Shipyard *air cushion vehicles* (China) ... 7
Jindo Transportation *civil operators* (Korea, South) ... 399
Jolly Drive Srl *transmissions* (Italy) ... 472
Jon Arne Helgøy A/S *civil operators* (Norway) ... 406
Jong Shyn Shipbuilding Co Ltd *High-Speed Patrol Craft* (Taiwan) ... 355
JR Kyushu Railway Company *civil operators* (Japan) ... 396

K

Kagoshima Shosen Company Ltd *civil operators* (Japan) ... 396
Kallisti Ferries *civil operators* (Greece) ... 382
Kamome Propeller Company Ltd *marine propellers* (Japan) ... 491
Kashima Futo Co Ltd *civil operators* (Japan) ... 396
Kawasaki Heavy Industries Ltd *high-speed multihull* (Japan) ... 132
Kawasaki Heavy Industries Ltd *hydrofoils* (Japan) ... 61
Kawasaki Heavy Industries, Ltd (Gas Turbine & Machinery Company) *water-jet units* (Japan) ... 501
Kay Marine SDN BHD *fast interceptors* (Malaysia) ... 303
Kay Marine SDN BHD *high-speed multihull* (Malaysia) ... 139
Kenai Fjords Tours *civil operators* (United States) ... 426
Keppel Bay Marina *civil operators* (Australia) ... 366
Key West Express *civil operators* (United States) ... 426
Kiriacoulis Maritime *civil operators* (Greece) ... 382
Kiso Shipyard *high-speed monohull* (Japan) ... 236
Kitsap Catamarans *high-speed multihull* (United States) ... 193
Kitsap Ferry Co *civil operators* (United States) ... 426
Knight and Carver Yachtcenter *high-speed multihull* (United States) ... 193
Kockums AB *fast interceptors* (Sweden) ... 315
Kompas International DD *civil operators* (Slovenia) ... 411
Kon-tiki Diving & Snorkelling Co Ltd *civil operators* (Thailand) ... 415
Korea Research and Development Institute *consultants/designers* (Korea, South) ... 544
Korea South *regulatory authorities* (Korea, South) ... 536
Koshikijima Shosen Ltd *civil operators* (Japan) ... 396
Krasnoye Sormovo Shipyard *air cushion vehicles* (Russian Federation) ... 24
Krylov Shipbuilding Research Institute *consultants/designers* (Russian Federation) ... 548
Kumamoto Ferry Company *civil operators* (Japan) ... 396
Kuwait *regulatory authorities* (Kuwait) ... 536
Kvichak Marine Industries *air cushion vehicles* (United States) ... 43
Kvichak Marine Industries *high-speed multihull* (United States) ... 194
Kyushu Shosen Company Ltd *civil operators* (Japan) ... 396
Kyushu Yusen Company Ltd *civil operators* (Japan) ... 396

L

L' Express Des Iles *civil operators* (Guadeloupe) ... 383
L Rødne and Sønner A/S *civil operators* (Norway) ... 406
LA ME Srl *ride control* (Italy) ... 510
LA. ME Srl *transmissions* (Italy) ... 473
Lürssen Werft GmbH & Co *high-speed monohull* (Germany) ... 223
Lada Langkawi Holdings *civil operators* (Malaysia) ... 400
Lady Musgrave Barrier Reef Cruises *civil operators* (Australia) ... 367
Lake Express LLC *civil operators* (United States) ... 426
Lancer Industries Ltd *fast interceptors* (New Zealand) ... 305
Langkawi Ferry Service SDN BHD *civil operators* (Malaysia) ... 400
Larivera Lines *civil operators* (Italy) ... 391
Laumzis Sun Cruises *civil operators* (Greece) ... 382
Lebanon *regulatory authorities* (Lebanon) ... 536
Lenco Marine S.A.L. *fast interceptors* (Lebanon) ... 302
Lewek Shipping Pte Ltd *civil operators* (Singapore) ... 410
Liferaft Systems Australia Pty Ltd (LSA) *marine escape* (Australia) ... 522
Lightning Boats *high-speed multihull* (Australia) ... 101
Linda Line Express *civil operators* (Finland) ... 378
Lindstøls Skips- & Båtbyggeri A/S *high-speed multihull* (Norway) ... 153
Lita Ocean PTE Ltd *fast interceptors* (Singapore) ... 310
Lomac Nautica SRL *fast interceptors* (Italy) ... 301
Lomprayah High Speed Ferry Company Ltd *civil operators* (Thailand) ... 415
Lorne Campbell Design *consultants/designers* (United Kingdom) ... 553
Lung Teh Shipbuilding Co Ltd *high-speed monohull* (Taiwan) ... 253
Lung Teh Shipbuilding Co Ltd *fast interceptors* (Taiwan) ... 316
Lung Teh Shipbuilding Co Ltd *high-speed patrol craft* (Taiwan) ... 355
Lurssen Werft *high-speed patrol craft* (Germany) ... 341
Luxury Resorts *civil operators* (Puerto Rico) ... 409
Luyt BV *high-speed multihull* (Netherlands) ... 141

M

M & M Transportation Services *civil operators* (Virgin Islands (UK)) ... 431
M Rosenblatt & Son Inc *consultants/designers* (United States) ... 559
Måløy Verft *high-speed monohull* (Norway) ... 245
Måløy Verft *high-speed multihull* (Norway) ... 156
Mackay Island and Reef Cruises *civil operators* (Australia) ... 367
Mackenzies Island Cruises *civil operators* (Australia) ... 367
Madagascar *regulatory authorities* (Madagascar) ... 536
Madera Ribs B.V. *fast interceptors* (Netherlands) ... 304
Magnum Marine Corporation *high-speed monohull* (United States) ... 263
MAHART Passnave Passenger Shipping Ltd *civil operators* (Hungary) ... 387
Malawi *regulatory authorities* (Malawi) ... 536
Malaysia *regulatory authorities* (Malaysia) ... 536
Malaysia Marine and Heavy Engineering *high-speed monohull* (Malaysia) ... 240
Malcolm Tennant Multihull Design *consultants/designers* (New Zealand) ... 545
Malta Maritime Authority *regulatory authorities* (Malta) ... 536
MAN *engines* (Germany) ... 443
MAN Diesel SAS *engines* (France) ... 437
MAN Diesel SE *engines* (Germany) ... 441
Mapso Marine *high-speed multihull* (Egypt) ... 122
Maric *wing-in-ground-effect* (China) ... 277
MARIC *consultants/designers* (China) ... 542
Marine Alutech Oy AB *fast interceptors* (Finland) ... 293
Marine Specialised Technology Ltd *fast interceptors* (United Kingdom) ... 320
Marinette Marine Corporation *high-speed monohull* (United States) ... 264
Marinette Marine Corporation *high-speed multihull* (United States) ... 195
Marintek *consultants/designers* (Norway) ... 546
Marinteknik Design (E) AB *consultants/designers* (Sweden) ... 549
Marinteknik Shipbuilders (S) Pte Ltd *high-speed multihull* (Singapore) ... 163
Marinteknik Shipbuilders (S) PTE Ltd *high-speed monohull* (Singapore) ... 248
Maritime Partner A/S *fast interceptors* (Norway) ... 307
Maritime Services International Ltd *consultants/designers* (United Kingdom) ... 554
Mark Ellis Marine Design *consultants/designers* (Australia) ... 540
Mary-d Enterprises *civil operators* (New Caledonia) ... 402
Master Ferries *civil operators* (Norway) ... 406
Maxton Fox Commercial Furniture Pty Ltd *lightweight seating* (Australia) ... 530
McConnell Seats Australia *lightweight seating* (Australia) ... 530
McTAY Marine *high-speed monohull* (United Kingdom) ... 256
McTay Marine *High-Speed Patrol Craft* (United Kingdom) ... 356
Mediterranea Pitiusa *civil operators* (Spain) ... 414
Merok Marine International *lightweight seating* (United Kingdom) ... 532
Metal Craft Marine Inc *fast interceptors* (Canada) ... 290
Mexico *regulatory authorities* (Mexico) ... 536
Miatours *civil operators* (Croatia) ... 375
Michigan Wheel Corporation *marine propellers* (United States) ... 496
Midhip Marine Inc *high-speed monohull* (United States) ... 265
Midship Marine Inc *high-speed multihull* (United States) ... 195
Ming Zhu Passenger Transport Co-op Co Ltd *civil operators* (China) ... 374
Ministry of Mercantile Marine *regulatory authorities* (Greece) ... 536
Mirejet *civil operators* (Korea, South) ... 399
Miss Barnegat Light *civil operators* (United States) ... 426
Mitsubishi Heavy Industries (MHI) Ltd *high-speed monohull* (Japan) ... 237
Mitsubishi Heavy Industries (MHI) Ltd *high-speed multihull* (Japan) ... 133
Mitsubishi Heavy Industries Ltd *air cushion vehicles* (Japan) ... 10
Mitsubishi Heavy Industries Ltd *hydrofoils* (Japan) ... 62
Mitsubishi Heavy Industries Ltd *High-Speed Patrol Craft* (Japan) ... 349
Mitsubishi Heavy Industries Ltd *engines* (Japan) ... 451
Mitsubishi Heavy Industries Ltd *water-jet units* (Japan) ... 502
Mitsui Engineering & Shipbuilding Company Ltd *High-Speed Patrol Craft* (Japan) ... 350
Mitsui Engineering & Shipbuilding Company Ltd *air cushion vehicles* (Japan) ... 11
Mitsui Engineering & Shipbuilding Company Ltd *high-speed multihull* (Japan) ... 134

Mitsui Engineering & Shipbuilding Company Ltd *swath vessels* (Japan) ... 202
MJP Water-Jets *water-jet units* (Sweden) ... 506
Modell-Møbler AS *lightweight seating* (Norway) ... 532
Mokubei Shipbuilding Company *high-speed monohull* (Japan) ... 238
Mols-linien A/S *civil operators* (Denmark) ... 376
Moose Boats *fast interceptors* (United States) ... 325
Moreton Bay Whalewatching *civil operators* (Australia) ... 367
Morocco *regulatory authorities* (Morocco) ... 537
Morrelli & Melvin Design and Engineering Inc *consultants/designers* (United States) ... 559
MOSCHINI – Cantieri Ing Moschina SpA *high-speed monohull* (Italy) ... 232
Motomarine SA *fast interceptors* (Greece) ... 296
Moura Company *civil operators* (Cape Verde) ... 373
Mt Samat Ferry Express Inc *civil operators* (Philippines) ... 408
MTE GmbH *wing-in-ground-effect* (Germany) ... 281
MT-Propeller Entwicklung GmbH *air propellers* (Germany) ... 487
MTU – Motoren- und Turbinen-Union Friedrichshafen GmbH *engines* (Germany) ... 445
Mulder Design BV *consultants/designers* (Netherlands) ... 545

N

Naiad Inflatables *fast interceptors* (New Zealand) ... 306
Naiad Marine Systems *ride control* (United States) ... 512
Nam Hae Express Company *civil operators* (Korea, South) ... 399
Namsos Trafikkselskap ASA *civil operators* (Norway) ... 406
Nansha Ferry Co Ltd *civil operators* (Hong Kong) ... 386
Nantong High-speed Passenger Ship Company *civil operators* (China) ... 374
National Technical University of Athens *consultants/designers* (Greece) ... 543
Nautas Al Maghreb *civil operators* (Spain) ... 414
Nautica International Inc *fast interceptors* (United States) ... 326
Naval Surface Warfare Center (Carderock Division) *consultants/designers* (United States) ... 559
Navantia *high-speed monohull* (Spain) ... 251
Navantia *high-speed multihull* (Spain) ... 168
Navantia *High-Speed Patrol Craft* (Spain) ... 354
Navatek Ships Ltd *hydrofoils* (United States) ... 75
Navigators *civil operators* (Australia) ... 367
Navatek Ships Ltd *high-speed monohull* (United States) ... 265
Navatek Ships Ltd *swath vessels* (United States) ... 208
Navigazione Golfo dei Poeti *civil operators* (Italy) ... 391
Navigazione Lago Di Como *civil operators* (Italy) ... 391
Navigazione Lago Maggiore *civil operators* (Italy) ... 392
Navigazione Libera Del Golfo SpA *civil operators* (Italy) ... 392
Navigazione Navigargano Srl *civil operators* (Italy) ... 392
Navigazione Sul Lago Di Garda *civil operators* (Italy) ... 392
Nel Lines *civil operators* (Greece) ... 382
Neptune CDB Corporation *air cushion vehicles* (Russian Federation) ... 25
Nesodden-Bundefjord Dampskipsselskap A/S *civil operators* (Norway) ... 406
Netherlands *regulatory authorities* (Netherlands) ... 537
Neumar SA *air cushion skirt* (Spain) ... 516
Neumar SA *consultants/designers* (Spain) ... 549
Nevesbu B.V. *consultants/designers* (Netherlands) ... 545
New England Aquarium *civil operators* (United States) ... 427
New England Fast Ferry Company *civil operators* (United States) ... 427
New Wave Catamarans Pty Ltd *high-speed multihull* (Australia) ... 102
New World First Ferry Services Ltd *civil operators* (Hong Kong) ... 386
New York Water Taxi *civil operators* (United States) ... 427
New Zealand *regulatory authorities* (New Zealand) ... 537
NGV Tech SDN BHD *high-speed multihull* (Malaysia) ... 139
NGV Tech SDN BHD *fast interceptors* (Malaysia) ... 303
Nichols Brothers Boat Builders Inc *high-speed multihull* (United States) ... 196
Nichols Brothers Boat Builders Inc *swath vessels* (United States) ... 210
Ningbo Fast Ferries Ltd *civil operators* (China) ... 374
Nordic Jet Line AS *civil operators* (Estonia) ... 377
Nordlund Boat Company *high-speed monohull* (United States) ... 265
Norsafe *fast interceptors* (Norway) ... 308
North American Marine Jet Inc *water-jet units* (United States) ... 507
North Cape Minerals A/S *civil operators* (Norway) ... 407
North Sea Boats *fast interceptors* (Indonesia) ... 297
North West Bay Ships Pty Ltd *high-speed multihull* (Australia) ... 103
Northrop Grumman Ship Systems *air cushion vehicles* (United States) ... 44
Northwind Marine Inc *fast interceptors* (United States) ... 326
Norway *regulatory authorities* (Norway) ... 537
Novamarine 2 SpA *fast interceptors* (Italy) ... 301
Novurania *fast interceptors* (Italy) ... 301
NQEA Australia Pty Ltd *high-speed multihull* (Australia) ... 106
NY Waterway Tours *civil operators* (United States) ... 427

O

OCEA *high-speed monohull* (France) ... 222
OCEA SA *High-Speed Patrol Craft* (France) ... 340
Ocean Dynamics *fast interceptors* (United Kingdom) ... 321
Oceanjet *civil operators* (Philippines) ... 408
Oita Hover Ferry Company Ltd *civil operators* (Japan) ... 396
OKI Kisen KK *civil operators* (Japan) ... 397
OLA Lilloe-Olsen *consultants/designers* (Norway) ... 546
Oma Baatbyggeri *high-speed multihull* (Norway) ... 156
Onbada Co Ltd *civil operators* (Korea, South) ... 399
One2Three Pty Ltd *consultants/designers* (Australia) ... 541
Oregon Iron Works *fast interceptors* (United States) ... 327
OSK-ShipTech A/S *consultants/designers* (Denmark) ... 543
Otech *fast interceptors* (United States) ... 327
Otto Candies LLC *civil operators* (United States) ... 427

P

P&O Irish Sea *civil operators* (United Kingdom) ... 419
Pacific Propeller Inc (PPI) *air propellers* (United States) ... 488
Pacific Whale Foundation *civil operators* (United States) ... 428
Pakistan Water and Power Development Authority *civil operators* (Pakistan) ... 407
Palm Beach Ferry Service *civil operators* (Australia) ... 368
Paradis Nautica *consultants/designers* (Norway) ... 546
Park Island Transportation Company *civil operators* (Hong Kong) ... 386
Parker Ribs *fast interceptors* (United Kingdom) ... 321
Patagonia Connection *civil operators* (Chile) ... 373
Paxos Hydrofoil Shipping *civil operators* (Greece) ... 382
Peel's Tourist and Ferry Services Pty Ltd *civil operators* (Australia) ... 368
Peene Werft *fast interceptors* (Germany) ... 295
Peene Werft *High-Speed Patrol Craft* (Germany) ... 342
Pelican Offshore Services *civil operators* (Singapore) ... 410
Pelmatic AB *consultants/designers* (Sweden) ... 549
Pelni *civil operators* (Indonesia) ... 388
Pendik Shipyards *high-speed multihull* (Turkey) ... 170
Penguin Ferry Services Pte Ltd *civil operators* (Singapore) ... 410
Penguin Shipyard International Ltd *high-speed monohull* (Singapore) ... 249
Penguin Shipyard International Ltd *high-speed multihull* (Singapore) ... 166
Philadelphia Gear Corporation *transmissions* (United States) ... 479
Phillips Cruises & Tours LLC *civil operators* (United States) ... 428
Pine Harbour Ferries *civil operators* (New Zealand) ... 403
Planet Seaways *civil operators* (Greece) ... 383
Plascoa *fast interceptors* (France) ... 294
Port Authority Of Trinidad and Tobago *civil operators* (Trinidad and Tobago) ... 416
Poseidon Cruises *civil operators* (Australia) ... 368
Powervent *transmissions* (United States) ... 479
Pratt & Whitney Canada Corp *engines* (Canada) ... 436
Pratt & Whitney *engines* (United States) ... 463
Precision Craft (88) PTE Ltd *high-speed monohull* (Singapore) ... 250
Precision Craft(88) Pte Ltd *high-speed multihull* (Singapore) ... 167
Prefectura Naval Argentina *regulatory authorities* (Argentina) ... 535
Prince William Sound Cruises and Tours *civil operators* (United States) ... 428
Promaxis Systems Inc *consultants/designers* (Canada) ... 541
PSC Naval Dockyard *High-Speed Patrol Craft* (Malaysia) ... 351
PT Pal Indonesia *High-Speed Patrol Craft* (Indonesia) ... 345
PT Palindo *High-Speed Patrol Craft* (Indonesia) ... 346
Puerto Rico Ports Authority *civil operators* (Puerto Rico) ... 409
Pulse-Drive Systems International LLP *transmissions* (United States) ... 480

Q

Qiuxin Shipyard *air cushion vehicles* (China) ... 8
Q-SPD International Ltd *transmissions* (New Zealand) ... 474
Quadtech Marine Inc *consultants/designers* (United States) ... 559
Quantum Controls *ride control* (Netherlands) ... 510
Queensland Government *civil operators* (Australia) ... 368
Quick Cat Cruises *civil operators* (Australia) ... 368
Quicksilver Beluga *civil operators* (Indonesia) ... 389
Quicksilver Connections Ltd *civil operators* (Australia) ... 368
Q-West *high-speed multihull* (New Zealand) ... 142

R

Raging Thunder Pty Ltd *civil operators* (Australia) ... 369
Rapid Explorer *civil operators* (Netherlands Antilles) ... 402
Rayglass Boats *fast interceptors* (New Zealand) ... 306
Real Journeys *civil operators* (New Zealand) ... 403
Red Funnel Ferries *civil operators* (United Kingdom) ... 419
Redbay Boats Inc *fast interceptors* (United Kingdom) ... 322
Rederij G Doeksen En Zonen Bv *civil operators* (Netherlands) ... 402
Reef Magic Cruises *civil operators* (Australia) ... 369
Reefjet *civil operators* (Australia) ... 369
Reflex Advanced Marine Corporation *fast interceptors* (Canada) ... 291
Reintjes GmbH *transmissions* (Germany) ... 469
Renk AG *transmissions* (Germany) ... 470
Revolution Design *consultants/designers* (Australia) ... 541
RFD Beaufort Ltd *marine escape* (United Kingdom) ... 524
RFI Bluvia *civil operators* (Italy) ... 392
Ribeye Ltd *fast interceptors* (United Kingdom) ... 322
Richardson Devine Marine Constructions Pty Ltd *high-speed multihull* (Australia) ... 113
Righetti Navi Srl *civil operators* (Italy) ... 392
Ring Powercraft *fast interceptors* (Sweden) ... 315
Riva *high-speed monohull* (Italy) ... 232
Robben Island Museum *civil operators* (South Africa) ... 411
Robert Allan Ltd *consultants/designers* (Canada) ... 542
Rodman Polyships SA *fast interceptors* (Spain) ... 312
Rodriquez Cantieri Navali Do Brazil *high-speed multihull* (Brazil) ... 119
Rodriquez Cantieri Navali SpA *high-speed multihull* (Italy) ... 130
Rodriquez Cantieri Navali SpA *hydrofoils* (Italy) ... 55
Rodriquez Cantieri Navali SpA *high-speed monohull* (Italy) ... 232
Rolla SP Propellers SA *marine propellers* (Switzerland) ... 494
Rolls-royce (NVC – Design) *consultants/designers* (Norway) ... 547
Rolls-Royce AB *marine propellers* (Sweden) ... 492
Rolls-Royce AB *water-jet units* (Sweden) ... 506
Rolls-Royce Brown Brothers *ride control* (United Kingdom) ... 511
Rolls-Royce Commercial Marine *water-jet units* (United States) ... 508
Rolls-royce Marine Systems *engines* (United Kingdom) ... 457

Rolls-Royce Naval Marine Inc *marine propellers* (United States) ... 496
Rolls-Royce Naval Marine Inc *water-jet units* (United States) ... 508
Rolls-Royce OY AB *water-jet units* (Finland) ... 500
Rottnest Express Pty Ltd *civil operators* (Australia) ... 369
Royal Caribbean Cruise Lines *civil operators* (Belize) ... 428
Royal Crown Yachts *high-speed multihull* (United States) ... 200
Ruta Nautica Del Caribe *civil operators* (Mexico) ... 401

S

Sabre Catamarans Pty Ltd *high-speed multihull* (Australia) ... 114
Sado Kisen Kaisha *civil operators* (Japan) ... 397
Safehaven Marine *fast interceptors* (Ireland) ... 297
Safeway Maritime Transportation Company *civil operators* (Honduras) ... 384
Salem Ferry *civil operators* (United States) ... 428
Samara Science and Technical Complex *engines* (Russian Federation) ... 453
Samos Hydrofoils *civil operators* (Greece) ... 383
Samsung Heavy Industries Co Ltd *air cushion vehicles* (Korea, South) ... 17
Samudra Link Ferry Shipping *civil operators* (India) ... 388
Sanyo Shosen *civil operators* (Japan) ... 397
Sapa Mass Transportation *construction materials* (Denmark) ... 526
Saronic Dolphins *civil operators* (Greece) ... 383
Saudi Aramco *civil operators* (Saudi Arabia) ... 409
Sayville Ferry Service Inc *civil operators* (United States) ... 428
SBF Shipbuilders *high-speed monohull* (Australia) ... 215
SBF Shipbuilders *high-speed multihull* (Australia) ... 117
Scania *engines* (Sweden) ... 455
Scat Technologies Pte Ltd *air cushion vehicles* (Singapore) ... 30
Schelde Naval Shipbuilding (Royal Schelde) *air cushion vehicles* (Netherlands) ... 19
Schelde Shipbuilding (Royal Schelde) *high-speed multihull* (Netherlands) ... 142
Schottel GmbH & Co KG *marine propellers* (Germany) ... 490
Sciomachen Naval Architects *consultants/designers* (Italy) ... 544
Sea Chrome Marine Eritrea Share Co *high-speed patrol craft* (Australia) ... 335
Sea Containers Ltd *civil operators* (United Kingdom) ... 419
SeaAction AS *civil operators* (Norway) ... 407
SeaArk Marine *fast interceptors* (United States) ... 328
Seaark Marine Inc *high-speed monohull* (United States) ... 266
Sea-Cat Ferries and Charters P/L *civil operators* (Australia) ... 369
Seacraft Design LLC *consultants/designers* (United States) ... 560
Seacraft Shipyard Corporation *high-speed monohull* (United States) ... 266
Seafury Propulsion Systems Ltd *transmissions* (New Zealand) ... 475
Seaspeed Marine Consulting Ltd *consultants/designers* (United Kingdom) ... 554
Seaspray Marine Services and Engineering *fast interceptors* (United Arab Emirates) ... 318
Seaspray Marine Services and Engineering FZC *high-speed monohull* (United Arab Emirates) ... 254
Seastate Pty Ltd *ride control* (Australia) ... 510
Seastreak *civil operators* (United States) ... 429
Seatek SpA *engines* (Italy) ... 451
Seatran Ferry Company *civil operators* (Thailand) ... 415
Semo Company Ltd *air cushion vehicles* (Korea, South) ... 18
Semo Company Ltd *high-speed monohull* (Korea, South) ... 239
Semo Company Ltd *high-speed multihull* (Korea, South) ... 138
Sensation Yachts *fast interceptors* (New Zealand) ... 307
Seogyeong Marine Co *civil operators* (Korea, South) ... 399
Servogear A/S *marine propellers* (Norway) ... 492
Servogear A/S *transmissions* (Norway) ... 475
SES Europe AS *consultants/designers* (Norway) ... 547
Setonaikai Kisen Company Ltd (Seto Inland Sea Lines) *civil operators* (Japan) ... 397
Shanghai Free Flying Transport *civil operators* (China) ... 374
Shanghai Huandoo Passenger Ship *civil operators* (China) ... 374
Shanghai Ya Tong Co Ltd *civil operators* (China) ... 374
Shenzhen Hengtong Shipping *civil operators* (China) . 374
Shenzhen Jianghua Marine *fast interceptors* (China) ... 292
Shenzhen Shipping *civil operators* (China) ... 374
Shenzhen Xun Long Transportation *civil operators* (China) ... 375
Shepler's Mackinac Island Ferry *civil operators* (United States) ... 429
Shipping Corporation of India *civil operators* (India) ... 388
Sillinger *fast interceptors* (France) ... 294
Silverships *fast interceptors* (United States) ... 329
Singapore Technologies Marine Ltd *air cushion vehicles* (Singapore) ... 31
Singapore Technologies Marine Ltd *high-speed multihull* (Singapore) ... 167
Singapore Technologies Marine Ltd *wing-in-ground-effect* (Germany) ... 280
Singapore Technologies Marine Ltd *fast interceptors* (Singapore) ... 310
Singapore Technologies Marine Ltd *high-speed patrol craft* (Singapore) ... 353
Siremar *civil operators* (Italy) ... 393
Slingsby Advanced Composites Ltd *air cushion vehicles* (United Kingdom) ... 42
Slovak Shipping And Ports, Joint Stock Co *civil operators* (Slovakia) ... 411
Smith's Ferry Services Ltd *civil operators* (Virgin Islands (US)) ... 431
SNAV Croazia Jet *civil operators* (Italy) ... 393
SNCM (Société National Corse Mediterranée) *civil operators* (France) ... 379
Société de Développement de Moorea *civil operators* (French Polynesia) ... 379
Société De Navigation De Normandie *civil operators* (France) ... 379
Société de Transports Maritimes Brudey Frères *civil operators* (Guadeloupe) ... 383
Société Morbihannaise de Navigation *civil operators* (France) ... 379
Sociedade Fluvial De Transportes *civil operators* (Portugal) ... 408
Sosnovka Shipyard *air cushion vehicles* (Russian Federation) ... 27
South Africa *regulatory authorities* (South Africa) ... 537
South Pacific Marine *high-speed multihull* (Australia) ... 118
South Sea Cruises Ltd *civil operators* (Fiji) ... 378
Spain *regulatory authorities* (Spain) ... 537
Speed Ferries *civil operators* (United Kingdom) ... 419
Speedy's *civil operators* (Virgin Islands (UK)) ... 431
Spirit of the Bay Cruises *civil operators* (Australia) ... 369
Split Tours dd *civil operators* (Croatia) ... 376
Sriwani Tours and Travel *civil operators* (Malaysia) ... 400
SSPA Sweden AB *consultants/designers* (Sweden) ... 549
SSSCO – Semi-submerged Ship Corporation *consultants/designers* (United States) ... 560
Star Line *civil operators* (United States) ... 429
Stavangerske AS *civil operators* (Norway) ... 407
Stena Line *civil operators* (United Kingdom) ... 419
Stena Line AB *civil operators* (Sweden) ... 414
Stewart Island Experience *civil operators* (New Zealand) ... 403
Stingray Marine *fast interceptors* (South Africa) ... 311
Stone Manganese Marine Ltd *marine propellers* (United Kingdom) ... 495
Storebro Bruks AB *fast interceptors* (Sweden) ... 316
Strategic Marine Pte Ltd *high-speed monohull* (Singapore) ... 250
Strategic Marine Pty Ltd *high-speed multihull* (Australia) ... 119
Strategic Marine PTY Ltd *high-speed monohull* (Australia) ... 218
Strategic Marine PTY Ltd *fast interceptors* (Australia) ... 288
Strategic Marine Pty Ltd *high-speed patrol craft* (Australia) ... 335
Stromma *civil operators* (Sweden) ... 415
Sudoexport *consultants/designers* (Russian Federation) ... 548
Sumidagawa Shipyard *high-speed patrol craft* (Japan) ... 350
Sumidagawa Shipyard Co Ltd *high-speed monohull* (Japan) ... 239
Sumitomo Heavy Industries Ltd *hydrofoils* (Japan) ... 62
Sunferries *civil operators* (Australia) ... 369
Sung Woo Ferry *civil operators* (Korea, South) ... 399
Sunlover Cruises *civil operators* (Australia) ... 370
Supercat Fast Ferry Corporation *civil operators* (Philippines) ... 408
SuperSeaCat *civil operators* (Estonia) ... 377
Supramar AG *consultants/designers* (Switzerland) ... 549
Swath International Ltd *consultants/designers* (United States) ... 561
Swath Ocean Systems LLC *swath vessels* (United States) ... 210
Swede Ship Marine AB *fast interceptors* (Sweden) ... 316
Sweden *regulatory authorities* (Sweden) ... 537
Swift Marine *fast interceptors* (Australia) ... 289
Swiftships Shipbuilders, LLC *high-speed monohull* (United States) ... 267
Swiftships Shipbuilders, LLC *high-speed patrol craft* (United States) ... 359
Swiftships Shipbuilders, LLC *fast interceptors* (United States) ... 329
Switzerland *regulatory authorities* (Switzerland) ... 537
Sydney Ferries *civil operators* (Australia) ... 370
Sylte Shipyards *high-speed multihull* (Canada) ... 120

T

Tahaa Transport Services *civil operators* (French Polynesia) ... 380
Tam Bac *high-speed multihull* (Vietnam) ... 200
Tangalooma WIld Dolphin Resort Ltd *civil operators* (Australia) ... 370
Taranto Navigazione *civil operators* (Italy) ... 393
Tasman Venture *civil operators* (Australia) ... 370
Techni Carène *consultants/designers* (France) ... 543
Technical Books *bibliography* 581
Tecnoseat Australia Pty Ltd *lightweight seating* (Australia) ... 530
Teignbridge Propellers Ltd *marine propellers* (United Kingdom) ... 495
Teknicraft Design Ltd *consultants/designers* (New Zealand) ... 545
Tenix Defence Pty Ltd *high-speed patrol craft* (Australia) ... 335
Textron Marine & Land Systems *air cushion vehicles* (United States) ... 45
Thames Clippers *civil operators* (United Kingdom) ... 420
The Hovercraft Society and Hovercraft Museum Trust *societies* (United Kingdom) ... 565
The International Hydrofoil Society *societies* (United States) ... 565
The Mariport Group Ltd *consultants/designers* (Canada) ... 541
Tian San Shipping (Pte) Ltd *civil operators* (Singapore) ... 411
Tide ASA *civil operators* (Norway) ... 405
Tiger Marine Ltd *fast interceptors* (United Kingdom) ... 322
Tilos 21st Century Shipping *civil operators* (Greece) ... 383
Tinian Shipping and Transportation *civil operators* (Northern Mariana Islands) ... 403
Tirrenia Navigazione *civil operators* (Italy) ... 393
Tokai Kisen *civil operators* (Japan) ... 397
Tokushima Buri University *wing-in-ground-effect* (Japan) ... 281
Tokushima Shuttle Line Company Ltd *civil operators* (Japan) ... 398
Toremar *civil operators* (Italy) ... 394
Torghatten Trafikkselskap ASA *civil operators* (Norway) ... 407
Tornado Boats International Ltd *fast interceptors* (United Kingdom) ... 323
Tortola Fast Ferry *civil operators* (Virgin Islands (UK)) ... 431
Tottori University *wing-in-ground-effect* (Japan) ... 281
Transport Seating Technology *lightweight seating* (Australia) ... 531
Transportation Services of St John *civil operators* (Virgin Islands (US)) ... 432
Transtejo *civil operators* (Portugal) ... 409
Transtur *civil operators* (Brazil) ... 372
Trasmapi *civil operators* (Spain) ... 414
Trico Marine Services Inc *civil operators* (United States) ... 429
Trinity Yachts Inc *high-speed monohull* (United States) ... 267
T-Torque Drive System Inc *transmissions* (United States) ... 480
Tung Hsin Steamship Company Ltd *civil operators* (Taiwan) ... 415

Turbo Power and Marine Systems Inc *engines* (United States) ... 463
Turbojet *civil operators* (Hong Kong) ... 386
Turbomeca *engines* (France) ... 437
Twin Disc Inc *transmissions* (United States) ... 481

U

Ukrainian Danube Shipping Company *civil operators* (Ukraine) ... 417
Ulstein Technology *air cushion vehicles* (Norway) ... 20
Ultra Dynamics Ltd *water-jet units* (United Kingdom) ... 507
Umoe Mandal AS *air cushion vehicles* (Norway) ... 21
Umoe Mandal AS *High-Speed Patrol Craft* (Norway) ... 352
United company for Marine Lines *civil operators* (Egypt) ... 377
United Company For Maritime Lines *civil operators* (Saudi Arabia) ... 409
United Kingdom *regulatory authorities* (United Kingdom) ... 537
United States *regulatory authorities* (United States) ... 537
United States Marine Inc *fast interceptors* (United States) ... 330
Universal Shipbuilding Corporation *hydrofoils* (Japan) ... 63
Universal Shipbuilding Corporation *high-speed multihull* (Japan) ... 135
Universal Shipbuilding Corporation *High-Speed Patrol Craft* (Japan) ... 351
Universal Shipbuilding Corporation *ride control* (Japan) ... 510
Uri Express Ferry Co Ltd *civil operators* (Korea, South) ... 399
USA Avenger Inc *fast interceptors* (United States) ... 330
Ustica Lines *civil operators* (Italy) ... 394
Uto Kapetan Luka *civil operators* (Croatia) ... 376

V

VA Tech Escher Wyss *marine propellers* (Germany) ... 491
Valfajr Shipping Company *civil operators* (Iran) ... 389
Valiant *fast interceptors* (Portugal) ... 308
Vallejo Baylink Ferries *civil operators* (United States) ... 429
Vasilopoulos Hydrofoils *civil operators* (Greece) ... 383
Veem Engineering Group *marine propellers* (Australia) ... 490
Venezia Lines SPA *civil operators* (Italy) ... 394
Venezuela *regulatory authorities* (Venezuela) ... 537
Veolia Transport *civil operators* (Netherlands) ... 401
Vericor Power Systems *engines* (United States) ... 465
Vetor Aliscafi Srl *civil operators* (Italy) ... 394
VI Sea Trans *civil operators* (Virgin Islands (US)) ... 432
Victory Design *consultants/designers* (Italy) ... 544
Viking Fast Craft Solutions LLC *consultants/designers* (United States) ... 561
Viking Lifesaving Equipment *fast interceptors* (Denmark) ... 292
Viking Lifesaving Equipment A/S *marine escape* (Denmark) ... 523
Vineyard Fast Ferry *civil operators* (United States) ... 430
Virtu Ferries Ltd *civil operators* (Malta) ... 400
Volga Shipyard *hydrofoils* (Russian Federation) ... 64
Volvo Penta AB *engines* (Sweden) ... 456
Voyager Marine *fast interceptors* (Bahrain) ... 289
Voyages *civil operators* (Australia) ... 370
VT Halmatic *fast interceptors* (United Kingdom) ... 323
VT Halter Marine *fast interceptors* (United States) ... 330
VT Halter Marine Inc *high-speed monohull* (United States) ... 268
VT Halter Marine Inc *high-speed multihull* (United States) ... 200
VT Halter Marine Inc *High-Speed Patrol Craft* (United States) ... 359
VT Maritime Dynamics Inc *consultants/designers* (United States) ... 561
VT Marine Products Ltd *ride control* (United Kingdom) ... 511
VT Maritime Dynamics Inc *ride control* (United States) ... 512
VT Shipbuilding *high-speed multihull* (United Kingdom) ... 177
VT Shipbuilding *High-Speed Patrol Craft* (United Kingdom) ... 357
Vulkan *transmissions* (Germany) ... 470
Vympel Central Design Bureau *air cushion vehicles* (Russian Federation) ... 29
Vympel Shipyard *fast interceptors* (Russian Federation) ... 310
Vympel Shipyard *High-Speed Patrol Craft* (Russian Federation) ... 353

W

Wärtsila Propulsion Jets B.V. *water-jet units* (Netherlands) ... 502
Wärtsilä Marine Division *marine propellers* (Netherlands) ... 492
Wärtsilä SACM Diesel *engines* (France) ... 438
Wally Yachts *high-speed monohull* (Monaco) ... 242
Wang Tak Engineering *High-Speed Patrol Craft* (Hong Kong) ... 344
Wang Tak Engineering and Shipbuilding Co Ltd *high-speed monohull* (Hong Kong) ... 225
Wang Tak Engineering and Shipbuilding Co Ltd *high-speed multihull* (Hong Kong) ... 129
Warwick Yacht Design Ltd *consultants/designers* (New Zealand) ... 545
Washington State Ferries *civil operators* (United States) ... 430
Waterbus *civil operators* (Netherlands) ... 402
Watercraft Hellas *fast interceptors* (Greece) ... 296
Waterfront City Resort *civil operators* (Singapore) ... 411
Weesam Express *civil operators* (Philippines) ... 408
West Coast Launch Ltd *civil operators* (Canada) ... 373
Westport Shipyard Inc *high-speed monohull* (United States) ... 269
Whale Watch Kaikoura *civil operators* (New Zealand) ... 403
White Young Green Consulting Ltd *consultants/designers* (United Kingdom) ... 554
Wightlink Ltd *civil operators* (United Kingdom) ... 420
Wilhelmshaven Helgoland Linie GmbH & Co KG *civil operators* (Germany) ... 380
Willard Marine Inc *fast interceptors* (United States) ... 331
Wolfson Unit For Marine Technology And Industrial Aerodynamics *consultants/designers* (United Kingdom) ... 554
Woods Hole and Martha's Vineyard Steamship Authority *civil operators* (United States) ... 430
Woody Marine Fabrication Pty Ltd *fast interceptors* (Australia) ... 289
World Heritage Cruises *civil operators* (Australia) ... 371
WS Atkins Consultants Ltd *consultants/designers* (United Kingdom) ... 555

X

XS-Ribs *fast interceptors* (United Kingdom) ... 324

Y

Yaeyama Kanko Ferry Company Ltd *civil operators* (Japan) ... 398
Yamaha Motor Company Ltd *high-speed monohull* (Japan) ... 239
Yamaha Motor Company Ltd *high-speed multihull* (Japan) ... 136
Yank Marine *high-speed multihull* (United States) ... 200
Yankee Fleet *civil operators* (United States) ... 430
Yasuda Sangyo Kisen Company Ltd *civil operators* (Japan) ... 398
Yellowfin Ltd *transmissions* (United Kingdom) ... 479
Yesil Marmaris Shipping *civil operators* (Turkey) ... 417
Yeuam Shian Shipping *civil operators* (Taiwan) ... 415
YG Consultants *consultants/designers* (United Kingdom) ... 555
Yonca Onuk *fast interceptors* (Turkey) ... 317
Yong Choo Kui Shipyard *high-speed monohull* (Malaysia) ... 240

Z

Zegluga Gdanska *civil operators* (Poland) ... 408
ZF Friedrichshafen AG *transmissions* (Germany) ... 471
Zodiac *air cushion skirt* (France) ... 516
Zodiac Hurricane Technologies *fast interceptors* (Canada) ... 291
Zodiac International *marine escape* (France) ... 523
Zvezda *transmissions* (Russian Federation) ... 476
Zvezda Joint Stock Company *engines* (Russian Federation) ... 454

Index of craft types

A

B

C

D

E

F

G

H

I

Index of craft names

NOTES